$\hat{q}$	$1 - \hat{p}$ where $\hat{p}$ is the sample proportion
$\overline{q}$	$1 - \overline{p}$ where $\overline{p}$ is the pooled sample proportion for two samples
Q_1, Q_2, Q_3	First, second, and third quartiles, respectively
R	(1) Number of rows in a contingency table; (2) number of runs for the runs test for randomness
r	Linear correlation coefficient for sample data
r^2	Coefficient of determination
r_s	Spearman *rho* rank correlation coefficient for sample data
ρ	(Greek letter *rho*) Linear correlation coefficient for population data
ρ_s	Spearman *rho* rank correlation coefficient for population data
S	Sample space
s	Sample standard deviation
s^2	Sample variance
s_b	Estimator of σ_b
s_d	Standard deviation of the paired differences for a sample
$s_{\overline{d}}$	Estimator of $\sigma_{\overline{d}}$
s_e	Standard deviation of errors for the sample regression model
s_p	Pooled standard deviation
$s_{\hat{p}}$	Estimator of $\sigma_{\hat{p}}$
$s_{\hat{p}_1 - \hat{p}_2}$	Estimator of $\sigma_{\hat{p}_1 - \hat{p}_2}$
$s_{\overline{x}}$	Estimator of $\sigma_{\overline{x}}$
$s_{\overline{x}_1 - \overline{x}_2}$	Estimator of $\sigma_{\overline{x}_1 - \overline{x}_2}$
$s_{\hat{y}_m}$	Estimator of $\sigma_{\hat{y}_m}$
$s_{\hat{y}_p}$	Estimator of $\sigma_{\hat{y}_p}$
Σ	(Greek letter capital *sigma*) summation notation
σ	(Greek letter lowercase *sigma*) Population standard deviation
σ^2	Population variance
σ_b	Standard deviation of the sampling distribution of b
σ_d	Standard deviation of the paired differences for the population
$\sigma_{\overline{d}}$	Standard deviation of the sampling distribution of $\overline{d}$
σ_ϵ	Standard deviation of errors for the population regression model
$\sigma_{\hat{p}}$	Standard deviation of the sampling distribution of $\hat{p}$
$\sigma_{\hat{p}_1 - \hat{p}_2}$	Standard deviation of the sampling distribution of $\hat{p}_1 - \hat{p}_2$
σ_R	Standard deviation of the sampling distribution of R for the runs test
σ_T	Standard deviation of the sampling distribution of T in the Wilcoxon signed-rank test and the Wilcoxon rank sum test for large samples
$\sigma_{\overline{x}}$	Standard deviation of the sampling distribution of $\overline{x}$
$\sigma_{\overline{x}_1 - \overline{x}_2}$	Standard deviation of the sampling distribution of $\overline{x}_1 - \overline{x}_2$
$\sigma_{\hat{y}_m}$	Standard deviation of $\hat{y}$ when estimating $\mu_{y|x}$
$\sigma_{\hat{y}_p}$	Standard deviation of $\hat{y}$ when predicting y_p
T	Test statistic for the Wilcoxon signed-rank test and the Wilcoxon rank sum test for a small sample
t	The t distribution
T_i	Sum of the values included in sample i in one-way ANOVA
T_U	The upper critical value for the Wilcoxon rank sum test
T_L	The lower critical value for the Wilcoxon rank sum test
X	Test statistic for the sign test for a small sample
x	(1) Variable; (2) random variable; (3) independent variable in a regression model
$\overline{x}$	Sample mean
y	(1) Variable; (2) dependent variable in a regression model
$\hat{y}$	Estimated or predicted value of y using a regression model
z	Units of the standard normal distribution

- Between-samples sum of squares:

$$SSB = \left(\frac{T_1^2}{n_1} + \frac{T_2^2}{n_2} + \frac{T_3^2}{n_3} + \cdots\right) - \frac{(\Sigma x)^2}{n}$$

- Within-samples sum of squares:

$$SSW = \Sigma x^2 - \left(\frac{T_1^2}{n_1} + \frac{T_2^2}{n_2} + \frac{T_3^2}{n_3} + \cdots\right)$$

- Total sum of squares: $SST = SSB + SSW = \Sigma x^2 - \frac{(\Sigma x)^2}{n}$

- Variance between samples: $MSB = \dfrac{SSB}{k-1}$

- Variance within samples: $MSW = \dfrac{SSW}{n-k}$

- Test statistic for a one-way ANOVA test: $F = \dfrac{MSB}{MSW}$

CHAPTER 13 • SIMPLE LINEAR REGRESSION

- Simple linear regression model: $y = A + Bx + \epsilon$
- Estimated simple linear regression model: $\hat{y} = a + bx$
- Sum of squares of xy, xx, and yy:

$$SS_{xy} = \Sigma xy - \frac{(\Sigma x)(\Sigma y)}{n}$$

$$SS_{xx} = \Sigma x^2 - \frac{(\Sigma x)^2}{n} \text{ and } SS_{yy} = \Sigma y^2 - \frac{(\Sigma y)^2}{n}$$

- Least squares estimates of A and B:

$$b = \frac{SS_{xy}}{SS_{xx}} \quad \text{and} \quad a = \bar{y} - b\bar{x}$$

- Standard deviation of the sample errors:

$$s_e = \sqrt{\frac{SS_{yy} - b\,SS_{xy}}{n-2}}$$

- Error sum of squares: $SSE = \Sigma e^2 = \Sigma(y - \hat{y})^2$

- Total sum of squares: $SST = \Sigma y^2 - \dfrac{(\Sigma y)^2}{n}$

- Regression sum of squares: $SSR = SST - SSE$

- Coefficient of determination: $r^2 = b\dfrac{SS_{xy}}{SS_{yy}}$

- Confidence interval for B:

$$b \pm t\,s_b \quad \text{where} \quad s_b = \frac{s_e}{\sqrt{SS_{xx}}}$$

- Test statistic for a test of hypothesis about B: $t = \dfrac{b - B}{s_b}$

- Linear correlation coefficient: $r = \dfrac{SS_{xy}}{\sqrt{SS_{xx}\,SS_{yy}}}$

- Test statistic for a test of hypothesis about ρ: $t = r\sqrt{\dfrac{n-2}{1-r^2}}$

- Confidence interval for $\mu_{y|x}$:

$$\hat{y} \pm ts_{\hat{y}_m} \quad \text{where} \quad s_{\hat{y}_m} = s_e\sqrt{\frac{1}{n} + \frac{(x_0 - \bar{x})^2}{SS_{xx}}}$$

- Prediction interval for y_p:

$$\hat{y} \pm t\,s_{\hat{y}_p} \quad \text{where} \quad s_{\hat{y}_p} = s_e\sqrt{1 + \frac{1}{n} + \frac{(x_0 - \bar{x})^2}{SS_{xx}}}$$

CHAPTER 14 • NONPARAMETRIC METHODS

- Test statistic for a sign test about the population proportion for a large sample:

$$z = \frac{(X \pm .5) - \mu}{\sigma} \quad \text{where} \quad \mu = np \text{ and } \sigma = \sqrt{npq}$$

Here, use $(X + .5)$ if $X \le n/2$ and $(X - .5)$ if $X > n/2$

- Test statistic for a sign test about the median for a large sample:

$$z = \frac{(X + .5) - \mu}{\sigma}$$

where $\mu = np$, $p = .5$, and $\sigma = \sqrt{npq}$

Let X_1 be the number of plus signs and X_2 the number of minus signs in a test about the median. Then, if the test is two-tailed, either of the two values can be assigned to X; if the test is left-tailed, $X = $ smaller of the values of X_1 and X_2; if the test is right-tailed, $X = $ larger of the values of X_1 and X_2.

- Test statistic for the Wilcoxon signed-rank test for a large sample:

$$z = \frac{T - \mu_T}{\sigma_T}$$

where $\mu_T = \dfrac{n(n+1)}{4}$ and $\sigma_T = \sqrt{\dfrac{n(n+1)(2n+1)}{24}}$

- Test statistic for the Wilcoxon rank sum test for large and independent samples:

$$z = \frac{T - \mu_T}{\sigma_T}$$

where $\mu_T = \dfrac{n_1(n_1 + n_2 + 1)}{2}$ and $\sigma_T = \sqrt{\dfrac{n_1 n_2(n_1 + n_2 + 1)}{12}}$

- Test statistic for the Kruskal–Wallis test:

$$H = \frac{12}{n(n+1)}\left[\frac{R_1^2}{n_1} + \frac{R_2^2}{n_2} + \cdots + \frac{R_k^2}{n_k}\right] - 3(n+1)$$

where $n_i = $ size of ith sample, $n = n_1 + n_2 + \cdots + n_k$, $k = $ number of samples, and $R_i = $ sum of ranks for ith sample

- Test statistic for the Spearman rho rank correlation coefficient test:

$$r_s = 1 - \frac{6\Sigma d^2}{n(n^2 - 1)}$$

where $d_i = $ difference in ranks of x_i and y_i

- Test statistic for the runs test for randomness for a large sample:

$$z = \frac{R - \mu_R}{\sigma_R}$$

where R is the number of runs in the sequence

$$\mu_R = \frac{2n_1 n_2}{n_1 + n_2} + 1 \quad \text{and} \quad \sigma_R = \sqrt{\frac{2n_1 n_2(2n_1 n_2 - n_1 - n_2)}{(n_1 + n_2)^2(n_1 + n_2 - 1)}}$$

KEY FORMULAS
Prem S. Mann • Introductory Statistics, Fourth Edition

CHAPTER 2 • ORGANIZING DATA

- Relative frequency of a class $= f/\Sigma f$

- Percentage of a class $=$ (Relative frequency) $\times$ 100

- Class midpoint or mark $=$ (Upper limit + Lower limit)/2

- Class width $=$ Upper boundary $-$ Lower boundary

- Cumulative relative frequency

$$= \frac{\text{Cumulative frequency}}{\text{Total observations in the data set}}$$

- Cumulative percentage

$$= \text{(Cumulative relative frequency)} \times 100$$

CHAPTER 3 • NUMERICAL DESCRIPTIVE MEASURES

- Mean for ungrouped data: $\mu = \Sigma x/N$ and $\bar{x} = \Sigma x/n$

- Mean for grouped data: $\mu = \Sigma mf/N$ and $\bar{x} = \Sigma mf/n$ where m is the midpoint and f is the frequency of a class

- Median for ungrouped data

$$= \text{Value of the } \left(\frac{n+1}{2}\right)\text{th term in a ranked data set}$$

- Range $=$ Largest value $-$ Smallest value

- Variance for ungrouped data:

$$\sigma^2 = \frac{\Sigma x^2 - \frac{(\Sigma x)^2}{N}}{N} \quad \text{and} \quad s^2 = \frac{\Sigma x^2 - \frac{(\Sigma x)^2}{n}}{n-1}$$

where σ^2 is the population variance and s^2 is the sample variance

- Standard deviation for ungrouped data:

$$\sigma = \sqrt{\frac{\Sigma x^2 - \frac{(\Sigma x)^2}{N}}{N}} \quad \text{and} \quad s = \sqrt{\frac{\Sigma x^2 - \frac{(\Sigma x)^2}{n}}{n-1}}$$

where σ and s are the population and sample standard deviations, respectively

- Variance for grouped data:

$$\sigma^2 = \frac{\Sigma m^2 f - \frac{(\Sigma mf)^2}{N}}{N} \quad \text{and} \quad s^2 = \frac{\Sigma m^2 f - \frac{(\Sigma mf)^2}{n}}{n-1}$$

- Standard deviation for grouped data:

$$\sigma = \sqrt{\frac{\Sigma m^2 f - \frac{(\Sigma mf)^2}{N}}{N}} \quad \text{and} \quad s = \sqrt{\frac{\Sigma m^2 f - \frac{(\Sigma mf)^2}{n}}{n-1}}$$

- Chebyshev's theorem:
  For any number k greater than 1, at least $(1 - 1/k^2)$ of the values for any distribution lie within k standard deviations of the mean.

- Empirical rule:
  For a specific bell-shaped distribution, about 68% of the observations fall in the interval $(\mu - \sigma)$ to $(\mu + \sigma)$, about 95% fall in the interval $(\mu - 2\sigma)$ to $(\mu + 2\sigma)$, and about 99.7% fall in the interval $(\mu - 3\sigma)$ to $(\mu + 3\sigma)$.

- Interquartile range: $\text{IQR} = Q_3 - Q_1$
  where Q_3 is the third quartile and Q_1 is the first quartile

- The kth percentile:

$$P_k = \text{Value of the } \left(\frac{kn}{100}\right)\text{th term in a ranked data set}$$

- Percentile rank of x_i

$$= \frac{\text{Number of values less than } x_i}{\text{Total number of values in the data set}} \times 100$$

CHAPTER 4 • PROBABILITY

- Classical probability rule for a simple event:

$$P(E_i) = \frac{1}{\text{Total number of outcomes}}$$

- Classical probability rule for a compound event:

$$P(A) = \frac{\text{Number of outcomes in } A}{\text{Total number of outcomes}}$$

- Relative frequency as an approximation of probability:

$$P(A) = \frac{f}{n}$$

- Conditional probability of an event:

$$P(A|B) = \frac{P(A \text{ and } B)}{P(B)} \quad \text{and} \quad P(B|A) = \frac{P(A \text{ and } B)}{P(A)}$$

- Condition for independence of events:

$$P(A) = P(A|B) \quad \text{and/or} \quad P(B) = P(B|A)$$

- For complementary events: $P(A) + P(\bar{A}) = 1$

- Multiplication rule for dependent events:

$$P(A \text{ and } B) = P(A)\,P(B|A)$$

- Multiplication rule for independent events:

$$P(A \text{ and } B) = P(A)\,P(B)$$

- Joint probability of two mutually exclusive events:

$$P(A \text{ and } B) = 0$$

- Addition rule for mutually nonexclusive events:

$$P(A \text{ or } B) = P(A) + P(B) - P(A \text{ and } B)$$

- Addition rule for mutually exclusive events:

$$P(A \text{ or } B) = P(A) + P(B)$$

CHAPTER 5 • DISCRETE RANDOM VARIABLES AND THEIR PROBABILITY DISTRIBUTIONS

- Mean of a discrete random variable x: $\mu = \Sigma x P(x)$
- Standard deviation of a discrete random variable x:

$$\sigma = \sqrt{\Sigma x^2 P(x) - \mu^2}$$

- n factorial: $n! = n(n-1)(n-2)\ldots 3 \cdot 2 \cdot 1$
- Number of combinations of n items selected x at a time:

$$\binom{n}{x} = \frac{n!}{x!(n-x)!}$$

- Binomial probability formula: $P(x) = \binom{n}{x} p^x q^{n-x}$

- Mean and standard deviation of the binomial distribution:

$$\mu = np \quad \text{and} \quad \sigma = \sqrt{npq}$$

- Poisson probability formula: $P(x) = \dfrac{\lambda^x e^{-\lambda}}{x!}$

- Mean, variance, and standard deviation of the Poisson probability distribution:

$$\mu = \lambda, \quad \sigma^2 = \lambda, \quad \text{and} \quad \sigma = \sqrt{\lambda}$$

CHAPTER 6 • CONTINUOUS RANDOM VARIABLES AND THE NORMAL DISTRIBUTION

- z value for an x value: $z = \dfrac{x - \mu}{\sigma}$

- Value of x when μ, σ, and z are known: $x = \mu + z\sigma$

CHAPTER 7 • SAMPLING DISTRIBUTIONS

- Mean of $\bar{x}$: $\mu_{\bar{x}} = \mu$
- Standard deviation of $\bar{x}$ when $n/N \le .05$: $\sigma_{\bar{x}} = \sigma/\sqrt{n}$

- z value for $\bar{x}$: $z = \dfrac{\bar{x} - \mu}{\sigma_{\bar{x}}}$

- Population proportion: $p = X/N$
- Sample proportion: $\hat{p} = x/n$
- Mean of $\hat{p}$: $\mu_{\hat{p}} = p$
- Standard deviation of $\hat{p}$ when $n/N \le .05$: $\sigma_{\hat{p}} = \sqrt{pq/n}$

- z value for $\hat{p}$: $z = \dfrac{\hat{p} - p}{\sigma_{\hat{p}}}$

CHAPTER 8 • ESTIMATION OF THE MEAN AND PROPORTION

- Margin of error for the point estimation of μ:

$$\pm 1.96\, \sigma_{\bar{x}} \quad \text{or} \quad \pm 1.96\, s_{\bar{x}}$$

where $\sigma_{\bar{x}} = \sigma/\sqrt{n}$ and $s_{\bar{x}} = s/\sqrt{n}$

Confidence interval for μ for a large sample:

$$\bar{x} \pm z\, \sigma_{\bar{x}} \quad \text{if } \sigma \text{ is known}$$
$$\bar{x} \pm z\, s_{\bar{x}} \quad \text{if } \sigma \text{ is not known}$$

where $\sigma_{\bar{x}} = \sigma/\sqrt{n}$ and $s_{\bar{x}} = s/\sqrt{n}$

- Confidence interval for μ for a small sample:

$$\bar{x} \pm ts_{\bar{x}} \quad \text{where} \quad s_{\bar{x}} = s/\sqrt{n}$$

- Margin of error for the point estimation of p:

$$\pm 1.96 s_{\hat{p}} \quad \text{where} \quad s_{\hat{p}} = \sqrt{\hat{p}\hat{q}/n}$$

- Confidence interval for p for a large sample:

$$\hat{p} \pm zs_{\hat{p}} \quad \text{where} \quad s_{\hat{p}} = \sqrt{\hat{p}\hat{q}/n}$$

- Maximum error of the estimate for μ:

$$E = z\sigma_{\bar{x}} \quad \text{or} \quad zs_{\bar{x}}$$

- Determining sample size for estimating μ: $n = z^2\sigma^2/E^2$
- Maximum error of the estimate for p:

$$E = zs_{\hat{p}} \quad \text{where} \quad s_{\hat{p}} = \sqrt{\hat{p}\hat{q}/n}$$

- Determining sample size for estimating p: $n = z^2 pq/E^2$

CHAPTER 9 • HYPOTHESIS TESTS ABOUT THE MEAN AND PROPORTION

- Test statistic z for a test of hypothesis about μ for a large sample:

$$z = \frac{\bar{x} - \mu}{\sigma_{\bar{x}}} \quad \text{if } \sigma \text{ is known, where } \sigma_{\bar{x}} = \frac{\sigma}{\sqrt{n}}$$

or $z = \dfrac{\bar{x} - \mu}{s_{\bar{x}}}$ if σ is not known, where $s_{\bar{x}} = \dfrac{s}{\sqrt{n}}$

- Test statistic for a test of hypothesis about μ for a small sample:

$$t = \frac{\bar{x} - \mu}{s_{\bar{x}}} \quad \text{where} \quad s_{\bar{x}} = \frac{s}{\sqrt{n}}$$

- Test statistic for a test of hypothesis about p for a large sample:

$$z = \frac{\hat{p} - p}{\sigma_{\hat{p}}} \quad \text{where} \quad \sigma_{\hat{p}} = \sqrt{\frac{pq}{n}}$$

CHAPTER 10 • ESTIMATION AND HYPOTHESIS TESTING: TWO POPULATIONS

- Mean of the sampling distribution of $\bar{x}_1 - \bar{x}_2$:

$$\mu_{\bar{x}_1 - \bar{x}_2} = \mu_1 - \mu_2$$

- Confidence interval for $\mu_1 - \mu_2$ for two large and independent samples:

$$(\bar{x}_1 - \bar{x}_2) \pm z\sigma_{\bar{x}_1 - \bar{x}_2} \quad \text{if } \sigma_1 \text{ and } \sigma_2 \text{ are known}$$

or $(\bar{x}_1 - \bar{x}_2) \pm zs_{\bar{x}_1 - \bar{x}_2}$ if σ_1 and σ_2 are not known

where $\sigma_{\bar{x}_1 - \bar{x}_2} = \sqrt{\dfrac{\sigma_1^2}{n_1} + \dfrac{\sigma_2^2}{n_2}}$ and $s_{\bar{x}_1 - \bar{x}_2} = \sqrt{\dfrac{s_1^2}{n_1} + \dfrac{s_2^2}{n_2}}$

- Test statistic for a test of hypothesis about $\mu_1 - \mu_2$ for two large and independent samples:

$$z = \frac{(\bar{x}_1 - \bar{x}_2) - (\mu_1 - \mu_2)}{\sigma_{\bar{x}_1 - \bar{x}_2}}$$

If σ_1 and σ_2 are not known, then replace $\sigma_{\bar{x}_1 - \bar{x}_2}$ by its point estimator $s_{\bar{x}_1 - \bar{x}_2}$.

- For two small and independent samples taken from two populations with equal standard deviations:

  Pooled standard deviation:
  $$s_p = \sqrt{\frac{(n_1 - 1)s_1^2 + (n_2 - 1)s_2^2}{n_1 + n_2 - 2}}$$

  Estimate of the standard deviation of $\bar{x}_1 - \bar{x}_2$:
  $$s_{\bar{x}_1 - \bar{x}_2} = s_p \sqrt{\frac{1}{n_1} + \frac{1}{n_2}}$$

  Confidence interval for $\mu_1 - \mu_2$: $(\bar{x}_1 - \bar{x}_2) \pm t s_{\bar{x}_1 - \bar{x}_2}$

  Test statistic: $t = \dfrac{(\bar{x}_1 - \bar{x}_2) - (\mu_1 - \mu_2)}{s_{\bar{x}_1 - \bar{x}_2}}$

- For two small and independent samples selected from two populations with unequal standard deviations:

  Degrees of freedom: $df = \dfrac{\left(\dfrac{s_1^2}{n_1} + \dfrac{s_2^2}{n_2}\right)^2}{\dfrac{\left(\dfrac{s_1^2}{n_1}\right)^2}{n_1 - 1} + \dfrac{\left(\dfrac{s_2^2}{n_2}\right)^2}{n_2 - 1}}$

  Estimate of the standard deviation of $\bar{x}_1 - \bar{x}_2$:
  $$s_{\bar{x}_1 - \bar{x}_2} = \sqrt{\frac{s_1^2}{n_1} + \frac{s_2^2}{n_2}}$$

  Confidence interval for $\mu_1 - \mu_2$: $(\bar{x}_1 - \bar{x}_2) \pm t s_{\bar{x}_1 - \bar{x}_2}$

  Test statistic: $t = \dfrac{(\bar{x}_1 - \bar{x}_2) - (\mu_1 - \mu_2)}{s_{\bar{x}_1 - \bar{x}_2}}$

- For two paired or matched samples:

  Sample mean for paired differences: $\bar{d} = \dfrac{\Sigma d}{n}$

  Sample standard deviation for paired differences:
  $$s_d = \sqrt{\frac{\Sigma d^2 - \frac{(\Sigma d)^2}{n}}{n - 1}}$$

  Mean and standard deviation of the sampling distribution of $\bar{d}$:
  $$\mu_{\bar{d}} = \mu_d \quad \text{and} \quad s_{\bar{d}} = \frac{s_d}{\sqrt{n}}$$

  Confidence interval for μ_d:
  $$\bar{d} \pm t s_{\bar{d}} \quad \text{where} \quad s_{\bar{d}} = \frac{s_d}{\sqrt{n}}$$

  Test statistic for a test of hypothesis about μ_d:
  $$t = \frac{\bar{d} - \mu_d}{s_{\bar{d}}}$$

- For two large and independent samples, confidence interval for $p_1 - p_2$:
  $$(\hat{p}_1 - \hat{p}_2) \pm z s_{\hat{p}_1 - \hat{p}_2}$$

  where $s_{\hat{p}_1 - \hat{p}_2} = \sqrt{\dfrac{\hat{p}_1 \hat{q}_1}{n_1} + \dfrac{\hat{p}_2 \hat{q}_2}{n_2}}$

- For two large and independent samples, for a test of hypothesis about $p_1 - p_2$ with $H_0: p_1 - p_2 = 0$:

Pooled sample proportion:
$$\bar{p} = \frac{x_1 + x_2}{n_1 + n_2} \quad \text{or} \quad \frac{n_1 \hat{p}_1 + n_2 \hat{p}_2}{n_1 + n_2}$$

Estimate of the standard deviation of $\hat{p}_1 - \hat{p}_2$:
$$s_{\hat{p}_1 - \hat{p}_2} = \sqrt{\bar{p}\,\bar{q}\left(\frac{1}{n_1} + \frac{1}{n_2}\right)}$$

Test statistic: $z = \dfrac{(\hat{p}_1 - \hat{p}_2) - (p_1 - p_2)}{s_{\hat{p}_1 - \hat{p}_2}}$

CHAPTER 11 • CHI-SQUARE TESTS

- Expected frequency for a category for a goodness-of-fit test:
  $$E = np$$

- Degrees of freedom for a goodness-of-fit test:
  $$df = k - 1 \quad \text{where } k \text{ is the number of categories}$$

- Expected frequency for a cell for an independence or homogeneity test:
  $$E = \frac{(\text{Row total})(\text{Column total})}{n}$$

- Degrees of freedom for a test of independence or homogeneity:
  $$df = (R - 1)(C - 1)$$

  where R and C are the total number of rows and columns, respectively, in the contingency table

- Test statistic for a goodness-of-fit test and a test of independence or homogeneity:
  $$\chi^2 = \Sigma \frac{(O - E)^2}{E}$$

- Confidence interval for the population variance σ^2:
  $$\frac{(n - 1)s^2}{\chi^2_{\alpha/2}} \quad \text{to} \quad \frac{(n - 1)s^2}{\chi^2_{1-\alpha/2}}$$

- Test statistic for a test of hypothesis about σ^2:
  $$\chi^2 = \frac{(n - 1)s^2}{\sigma^2}$$

CHAPTER 12 • ANALYSIS OF VARIANCE

Let:

k = the number of different samples (or treatments)

n_i = the size of sample i

T_i = the sum of the values in sample i

n = the number of values in all samples

$\quad = n_1 + n_2 + n_3 + \cdots$

Σx = the sum of the values in all samples

$\quad = T_1 + T_2 + T_3 + \cdots$

Σx^2 = the sum of the squares of values in all samples

- For the F distribution:

  Degrees of freedom for the numerator $= k - 1$

  Degrees of freedom for the denominator $= n - k$

FOURTH EDITION
INTRODUCTORY STATISTICS

PREM S. MANN
EASTERN CONNECTICUT STATE UNIVERSITY

Test No. 1 =

JOHN WILEY & SONS, INC.
NEW YORK • CHICHESTER • WEINHEIM • BRISBANE • SINGAPORE • TORONTO

To my parents

ACQUISITIONS EDITOR   Debbie Berridge
MARKETING MANAGER   Julie Z. Lindstrom
PRODUCTION SERVICES MANAGER   Jeanine Furino
SENIOR DESIGNER   Kevin Murphy
ILLUSTRATION EDITOR   Anna Melhorn
DEVELOPMENT EDITOR   Carolyn Smith
PRODUCTION SERVICES   Susan L. Reiland
This book was set in Times Roman by York Graphic Services and printed and bound by Von Hoffmann Press.
The cover was printed by Lehigh Press.
Cover photo by: © Lois Ellen Frank/CORBIS.
This book is printed on acid-free paper.

Library of Congress Cataloging in Publication Data:
Mann, Prem S.
  Introductory statistics / Prem S. Mann—4th ed.
     p.   cm.
  Includes index.
  ISBN 0-471-37353-2 (cloth: alk. paper)
  1. Statistics.   I. Title.

  QA276.12.M29 2001
  519.5—dc21

Printed in the United States of America

10 9 8 7 6 5 4 3 2 1

PREFACE

Introductory Statistics is written for a first course in applied statistics. The book is intended for students who do not have a strong background in mathematics. The only prerequisite for this text is a knowledge of elementary algebra.

Today, college students from almost all fields of study are required to take at least one course in statistics. Consequently, the study of statistical methods has taken on a prominent role in the education of students majoring in all fields of study. The goal of this text is to make the subject of statistics both interesting and accessible to such a wide and varied audience. Three major characteristics of this text support this goal: the realistic content of its examples and exercises that draw on a comprehensive range of applications from all facets of life, the clarity and brevity of its presentation, and the soundness of its pedagogical approach. These characteristics are exhibited through the interplay of a variety of significant text features.

The feedback received from the users of the third edition of *Introductory Statistics* has been very encouraging. This feedback serves as evidence of the success of the text in making statistics interesting and accessible—a goal of the author from the very first edition. The author has pursued the same goal through the refinements and updates in this fourth edition so that *Introductory Statistics* will continue to provide a successful experience in statistics to growing numbers of students and professors.

NEW TO THE FOURTH EDITION . . .

The following are some of the many significant changes made in the fourth edition of the text without changing its main features.

- A large number of exercises of greater difficulty have been added to all chapters except Chapter 1. These exercises appear under the heading **Advanced Exercises** as part of the supplementary exercises.
- All problems containing real data have been updated or replaced by new problems with real data.
- A large number of examples and case studies have been changed and updated to capture the interest of students and maintain relevancy.
- New chapter-ending **Mini-Projects** are data oriented, and can be given as group assignments during class or as homework.
- The MINITAB sections have been updated to the 13th Version. These sections now appear in **Appendix B** and include both the MINITAB PULL-DOWN MENUS and the MINITAB COMMAND LANGUAGE instructions.
- New **Excel Adventures** have been added to the text and they appear in **Appendix C**. These adventures show how to use Excel to solve statistics problems.

- For those who wish to have the option of using a graphing calculator in solving statistical exercises, a few exercises from each chapter that are suitable for this purpose have been identified with an icon. Graphing Calculator solutions for the odd-numbered problems are given as part of the answer section at the back of the book. However, these exercises can also be solved without using the graphing calculator.
- All data sets that appeared in Appendix B in the third edition have been updated. These data sets now appear in the Instructor's Solution Manual, Student's Solution Manual, on the Instructor's Resource CD that accompanies the text, and on the Web site for the text, www.wiley.com/college/mann.
- A new section on the test for correlation coefficient has been added to Chapter 13.
- A new Chapter 14 on Nonparametric Methods has been added to the text.
- Presentation of concepts has been refined at many places to enhance clarity.

HALLMARK FEATURES OF THIS TEXT

Style and Pedagogy

Clear and Concise Exposition   The explanation of statistical methods and concepts is clear and concise. Moreover, the style is user-friendly and easy to understand. In chapter introductions and in transitions from section to section, new ideas are related to those discussed earlier.

Abundant Examples

Examples   The text contains a wealth of examples, a total of 211 in 14 chapters and Appendix A. The examples are usually given in a format showing a problem and its solution. They are well-sequenced and thorough, displaying all facets of concepts. Furthermore, the examples capture students' interest because they cover a wide variety of relevant topics. They are based on situations practicing statisticians encounter every day. Finally, a large number of examples are based on real data that are taken from sources such as books, government and private data sources and reports, magazines, newspapers, and professional journals.

Realistic Settings

Solutions   A clear, concise solution follows each problem presented in an example. When the solution to an example involves many steps, it is presented in a step-by-step format. For instance, examples related to tests of hypotheses contain five steps that are consistently used to solve such examples in all chapters. Thus, procedures are presented in the concrete settings of applications rather than as isolated abstractions. Frequently, solutions contain highlighted remarks that recall and reinforce ideas critical to the solution of the problem. Such remarks add to the clarity of presentation.

Guideposts

Margin Notes for Examples   A margin note appears beside each example that briefly describes what is being done in that example. Students can use these margin notes to assist them as they read through sections and to quickly locate appropriate model problems as they work through exercises.

Frequent Use of Diagrams   Concepts can often be made more understandable by describing them visually, with the help of diagrams. This text uses diagrams frequently to help students understand concepts and solve problems. For example, tree diagrams are used extensively in Chapters 4 and 5 to assist in explaining probability concepts and in computing probabilities. Similarly, solutions to all examples about tests of hypotheses contain diagrams showing rejection regions, nonrejection regions, and critical values.

Highlighting Definitions of important terms, formulas, and key concepts are enclosed in color boxes so that students can easily locate them. A similar use of color is found in the *Using MINITAB* sections where a color tint highlights MINITAB solutions. Important terms appear in the text either in boldface or italic type.

☞ **Cautions** Certain items need special attention. These may deal with potential trouble spots that commonly cause errors. Or they may deal with ideas that students often overlook. Special emphasis is placed on such items through the headings: *Remember, An Observation,* or *Warning*. An icon is used to identify such items.

Realistic Applications

Case Studies Case studies, which appear in many chapters, provide additional illustrations of the applications of statistics in research and statistical analysis. Most of these case studies are based on articles/snapshots published in journals, magazines, or newspapers. All case studies are based on real data.

Abundant Exercises

Exercises and Supplementary Exercises The text contains an abundance of exercises, a total of 1486 in 14 chapters and Appendix A (excluding Computer Assignments). Moreover, a large number of these exercises contain several parts. Exercise sets appearing at the end of each section (or sometimes at the end of two or three sections) include problems on the topics of that section. These exercises are divided into two parts: **Concepts and Procedures** that emphasize key ideas and techniques, and **Applications** that use these ideas and techniques in concrete settings. Supplementary exercises appear at the end of each chapter and contain exercises on all sections and topics discussed in that chapter. A large number of these exercises are based on real data taken from varied data sources such as books, government and private data sources and reports, magazines, newspapers, and professional journals. Exercises given in the text do not merely provide practice for students, but the real data contained in exercises provide interesting information and insight into economic, political, social, psychological, and other aspects of life. The exercise sets also contain many problems that demand critical thinking skills. The answers to selected odd-numbered exercises appear in the *Answers Section* at the back of the book. **Optional exercises** are indicated by an asterisk (*).

Advanced Exercises All chapters (except Chapter 1) now have a set of exercises that are of greater difficulty. Such exercises appear under the heading *Advanced Exercises* as part of the *Supplementary Exercises*.

Summary and Review

Glossary Each chapter has a glossary that lists the key terms introduced in that chapter, along with a brief explanation of each term. Almost all the terms that appear in boldface type in the text are in the glossary.

Key Formulas Each chapter contains a list of key formulas used in that chapter. This list appears before the *Supplementary Exercises* for each chapter.

Self-Review Tests Each chapter contains a *Self-Review Test*, which appears immediately after the *Supplementary Exercises*. These problems can help students to test their grasp of the concepts and skills presented in respective chapters and to monitor their understanding of statistical methods. The problems marked by an asterisk (*) in the *Self-Review Tests* are **optional**. The answers to almost all problems of the *Self-Review Tests* appear in the *Answer Section*.

Formula Card A formula card that contains key formulas from all chapters is included at the beginning of the book.

Technology

Computer Usage   Another feature of this text is the detailed instructions on the use of **MINITAB**.[1] *Using MINITAB* sections for 13 of the 14 chapters and Appendix A appear in **Appendix B**. Each of these *Using MINITAB* sections contains a detailed description of the MINITAB commands that are used to perform the statistical analysis presented in the text. These sections now include both the MINITAB PULL-DOWN MENUS step-by-step instructions and the MINITAB COMMAND LANGUAGE instructions to perform statistical data analysis. In addition, each of these sections contains several illustrations that demonstrate how MINITAB can be used to solve statistical problems. These MINITAB instructions and illustrations are so complete that students do not need to purchase any other MINITAB supplement. Computer assignments are also given at the end of each chapter so that students can further practice MINITAB. A total of 80 computer assignments are contained in these sections.

Excel Adventures   Written by Jeff Mock of Diablo Valley College, **Appendix C** contains worked examples, called *Excel Adventures*, which show the various uses of Excel in performing statistical analysis. Three different methods are used depending on the content of the chapter—the student may be required to enter a formula or to utilize Excel's Data Analysis Command or Statistical Analysis ToolPak. Basic manipulative skills such as entering data, copying, cut-paste, opening files, and saving files as well as more extensive statistical applications can be obtained in the supplemental *Excel Manual* accompanying this text.

Graphing Calculator   More and more instructors are now using graphing calculators, such as the TI-83 or the TI-83+, to teach statistics. A few exercises from each chapter in the text have been identified and marked with a **calculator icon** as suitable exercises for graphing calculators. The complete solutions to these exercises along with the TI-83 screen reproductions are provided in the *Instructor's Solutions Manual*. Graphing Calculator answers are provided for such odd-numbered exercises as part of the answer section at the back of the book. Note that these exercises can also be solved without using a graphing calculator.

Calculator Usage   The text contains many footnotes that explain how a calculator can be used to evaluate complex mathematical expressions.

Data Sets   Four data sets appear in the Instructor's Solutions Manual, Student Solutions Manual, on CD, and on the Web site for the text. These data sets, collected from different sources, contain information on many variables. They can be used to perform statistical analysis with statistical computer software such as MINITAB or Excel. **These data sets are available from the publisher on a diskette/CD in MINITAB format, Excel format, and in ASCII format**. These data sets can also be downloaded from the Web site for the text.

WEB SITE

http://www.wiley.com/college/mann
The Mann Statistics Web site provides resources for instructors and students. The graphing calculator programs for running complex statistical tests on the TI-83 can be downloaded

---

[1]MINITAB is a registered trademark of Minitab, Inc., 3081 Enterprise Drive, State College, PA 16801. Phone: 814-238-3280; fax: 814-238-4383; telex: 881612. The author would like to take this opportunity to thank Minitab, Inc., for their help.

from this site. Other resources include four large data sets that accompany the text and links to interesting statistical sites on the Web. Instructors can find information on changes in the new edition and contact Wiley for examination copies and supplements.

OPTIONAL SECTIONS

Because each instructor has different preferences, the text does not indicate optional sections. This decision has been left to the instructor. Instructors may cover the sections or chapters that they think are important. However, a few alternative one-semester syllabi are available on the Web site for this text.

*Complete Learning
System*

SUPPLEMENTS

The following supplements are available to accompany this text.

Instructor's Solutions Manual   This manual contains complete solutions to all exercises and the *Self-Review Test* problems.

Instructor's Resource CD   Contains the data sets, full solutions manual, as well as:

- **Instructor's Resource Guide**   Sample syllabi, discussion questions, sample exams and suggested homework assignments, and transparency masters are included in this resource. Written for both the experienced and inexperienced instructor, the manual offers a clear overview of the course while outlining major topics and key points of each chapter.
- **Test Bank**   The printed copy of the test bank contains a large number of multiple choice questions, essay questions, and quantitative problems for each chapter.
- **Computerized Test Bank**   All questions that are in the printed *Test Bank* are available on a diskette. This diskette can be obtained from the publisher.

Students' Solutions Manual   This manual contains complete solutions to all of the odd-numbered exercises, a few even-numbered exercises, and all the *Self-Review Test* problems.

Student Study Guide   This guide contains review material about studying and learning patterns for a first course in statistics. Special attention is given to the critical material of each chapter. Reviews of mathematical notation, formulas, and table reading are also included.

Wiley E Grade for Statistics   Practice quizzes and tests for students and instructors appear on the Web site.

Graphing Calculator Manual   This manual is a basic guide for beginners on the TI-83 and TI-83+. The authors guide students through the important facets of using a graphing calculator in statistics by presenting clear examples, pictures of actual calculator screens, and programs specific to each calculator.

Excel Manual   This manual is an extension of the Excel Adventures given in Appendix C. In contains worked examples using current real data and Excel for the data analysis. The manual gives step-by-step instructions, screen captures to guide the students, as well as numerous applications of how to use Excel in statistics. The version of Excel used is Microsoft Excel 2000. Earlier versions of Excel are similar, and exact keystrokes may differ slightly.

ACKNOWLEDGMENTS

I thank the following reviewers of this, and/or previous editions of this book, whose comments and suggestions were invaluable in improving the text.

K. S. Asal
Broward Community College
Louise Audette
Manchester Community College
Nicole Betsinger
Arapahoe Community College
Joan Bookbinder
Johnson & Wales University
Dean Burbank
Gulf Coast Community College
Jayanta Chandra
University of Notre Dame
James Curl
Modesto Community College
Fred H. Dorner
Trinity University, San Antonio
William D. Ergle
Roanoke College, Salem, Virginia
Ruby Evans
Santa Fe Community College
Ronald Ferguson
San Antonio College
Frank Goulard
Portland Community College
Robert Graham
Jacksonville State University, Jacksonville, Alabama

Larry Griffey
Florida Community College, Jacksonville
A. Eugene Hileman
Northeastern State University, Tahlequah, Oklahoma
John G. Horner
Cabrillo College
Virginia Horner
Diablo Valley College
Ina Parks S. Howell
Florida International University
Gary S. Itzkowitz
Rowan State College
Jean Johnson
Governors State University
Michael Karelius
American River College, Sacramento
Dix J. Kelly
Central Connecticut State University
Linda Kohl
University of Michigan, Ann Arbor
Martin Kotler
Pace University, Pleasantville, New York
Marlene Kovaly
Florida Community College, Jacksonville

Carlos de la Lama
San Diego City College
Richard McGowan
University of Scranton
Daniel S. Miller
Central Connecticut State University
Satya N. Mishra
University of South Alabama
Jeffrey Mock
Diablo Valley College
Luis Moreno
Broome Community College, Binghamton
Robert A. Nagy
University of Wisconsin, Green Bay
Paul T. Nkansah
Florida Agricultural and Mechanical University
Joyce Oster
Johnson and Wales University
Lindsay Packer
College of Charleston
Mary Parker
Austin Community College
Roger Peck
University of Rhode Island, Kingston

Chester Piascik
Bryant College, Smithfield
Joseph Pigeon
Villanova University
Gerald Rogers
New Mexico State University, Las Cruces
Phillis Schumacher
Bryant College, Smithfield
Kathryn Schwartz
Scottsdale Community College
Ronald Schwartz
Wilkes University, Wilkes-Barre
David Stark
University of Akron
Larry Stephens
University of Nebraska, Omaha
Bruce Trumbo
California State University, Hayward
Jean Weber
University of Arizona, Tucson
Terry Wilson
San Jacinto College, Pasadena
K. Paul Yoon
Fairleigh Dickinson University, Madison

I express my special thanks to Professor Gerald Geissert of Eastern Connecticut State University and Professor Daniel S. Miller of Central Connecticut State University. Professor Geissert checked solutions to all examples for mathematical accuracy, prepared answers to selected odd-numbered exercises for the answer section, and read proofs. Professor Miller and Professor Geissert both helped in writing the problems with greater difficulty that appear in the *Advanced Exercises*. In addition, I thank Eastern Connecticut State University for all the support I received. I thank many of my students who were of immense help during this revision. I also offer my thanks to my colleagues, friends, and family whose support was a source of encouragement during the period when I spent long hours working on this revision.

It is of utmost importance that a textbook is accompanied by complete and accurate supplements. I take pride in mentioning that the supplements prepared for this text possess these qualities and much more. I appreciatively thank the following authors of these supplements: Professor Jim Curl for writing the Study Guide, Professor John Horner and Professor Virginia Horner for preparing the Graphing Calculator Manual for TI-83 and TI-83+, Professor Jeff Mock for writing the Excel Manual, Professor Gerald Geissert for preparing the Solutions Manuals, and Professor Edwin C. Hackleman for writing the Test Bank.

It is my pleasure to thank all the professionals at John Wiley with whom I enjoyed working. Their names appear earlier in this text.

Any suggestions from readers for future revisions would be greatly appreciated. Such suggestions can be sent to the author through Internet address: MANN@ECSU.CTSTATEU.EDU.

Prem S. Mann
Willimantic, CT 06226
June 2000

CONTENTS

Chapter 1 INTRODUCTION **1**

1.1 What Is Statistics? 2

1.2 Types of Statistics 3

    1.2.1 Descriptive Statistics 3

Case Study 1-1 Plastic in Your Pocket 3

    1.2.2 Inferential Statistics 4

Case Study 1-2 Public Schools Making Grade? 5

Exercises 6

1.3 Population Versus Sample 6

Case Study 1-3 Americans Expect Good, Bad, and Catastrophic 7

Exercises 9

1.4 Basic Terms 10

Exercises 11

1.5 Types of Variables 12

    1.5.1 Quantitative Variables 12

    1.5.2 Qualitative or Categorical Variables 13

Exercises 14

1.6 Cross-Section Versus Time-Series Data 15

    1.6.1 Cross-Section Data 15

    1.6.2 Time-Series Data 15

1.7 Sources of Data 16

Exercises 17

1.8 Summation Notation 17

Exercises 19

Glossary / Supplementary Exercises / Self-Review Test / Mini-Project / Computer Assignments

Chapter 2 ORGANIZING DATA **25**

2.1 Raw Data 26

2.2 Organizing and Graphing Qualitative Data 26

    2.2.1 Frequency Distributions 26

    2.2.2 Relative Frequency and Percentage Distributions 28

    2.2.3 Graphical Presentation of Qualitative Data 29

Case Study 2-1 Job Stress Can Be Satisfying 30

Case Study 2-2 Why Eat Out? 32

Exercises 32

2.3 Organizing and Graphing Quantitative Data 34

    2.3.1 Frequency Distributions 34

    2.3.2 Constructing Frequency Distribution Tables 36

    2.3.3 Relative Frequency and Percentage Distributions 39

    2.3.4 Graphing Group Data 40

Case Study 2-3 Boomers Go Fish 41

    2.3.5 More on Classes and Frequency Distributions 43

2.4 Shapes of Histograms 45

Case Study 2-4 Using Truncated Axes 47

Exercises 47

2.5 Cumulative Frequency Distributions 52

Exercises 54

2.6 Stem-and-Leaf Displays 55

Exercises **58**

*Glossary / Key Formulas / Supplementary
Exercises / Advanced Exercises / Self-Review
Test / Mini-Projects / Computer Assignments*

Chapter 3 NUMERICAL DESCRIPTIVE MEASURES **69**

3.1 Measures of Central Tendency for
Ungrouped Data **70**

3.1.1 Mean **70**

Case Study 3–1 Earnings by Degree **73**

3.1.2 Median **74**

Case Study 3–2 Cars, Trucks Older than
Ever **75**

3.1.3 Mode **76**

3.1.4 Relationships Among the Mean,
Median, and Mode **78**

Exercises **79**

3.2 Measures of Dispersion for Ungrouped
Data **82**

3.2.1 Range **83**

3.2.2 Variance and Standard
Deviation **84**

3.2.3 Population Parameters and Sample
Statistics **88**

Exercises **88**

3.3 Mean, Variance, and Standard Deviation
for Grouped Data **91**

3.3.1 Mean for Grouped Data **91**

3.3.2 Variance and Standard Deviation
for Grouped Data **93**

Exercises **96**

3.4 Use of Standard Deviation **98**

3.4.1 Chebyshev's Theorem **98**

3.4.2 Empirical Rule **100**

Case Study 3–3 Here Comes the SD **101**

Exercises **102**

3.5 Measures of Position **104**

3.5.1 Quartiles and Interquartile
Range **104**

3.5.2 Percentiles and Percentile
Rank **106**

Exercises **108**

3.6 Box-and-Whisker Plot **109**

Exercises **111**

*Glossary / Key Formulas / Supplementary
Exercises / Advanced Exercises / Appendix
3.1 / Self-Review Test / Mini-Project /
Computer Assignments*

Chapter 4 PROBABILITY **127**

4.1 Experiment, Outcomes, and Sample
Space **128**

4.1.1 Simple and Compound Events **130**

Exercises **132**

4.2 Calculating Probability **133**

4.2.1 Three Conceptual Approaches to
Probability **134**

Exercises **139**

4.3 Counting Rule **141**

4.4 Marginal and Conditional
Probabilities **142**

Case Study 4–1 Stress Busters **145**

4.5 Mutually Exclusive Events **146**

4.6 Independent versus Dependent
Events **147**

4.7 Complementary Events **149**

Exercises **151**

4.8 Intersection of Events and the
Multiplication Rule **154**

4.8.1 Intersection of Events **154**

4.8.2 Multiplication Rule **155**

Case Study 4–2 Baseball Players Have
"Slumps" and "Streaks" **161**

Exercises **162**

4.9 Union of Events and the Addition Rule **165**

4.9.1 Union of Events **165**

4.9.2 Addition Rule **166**

Exercises **171**

*Glossary / Key Formulas / Supplementary
Exercises / Advanced Exercises / Self-Review
Test / Mini-Projects*

Chapter 5 DISCRETE RANDOM VARIABLES AND THEIR PROBABILITY DISTRIBUTIONS **184**

5.1 Random Variables 185

    5.1.1 Discrete Random Variable 185

    5.1.2 Continuous Random Variable 186

Exercises 187

5.2 Probability Distribution of a Discrete Random Variable 187

Exercises 192

5.3 Mean of a Discrete Random Variable 194

Case Study 5-1 Aces High Instant Lottery 196

5.4 Standard Deviation of a Discrete Random Variable 198

Exercises 201

5.5 Factorials and Combinations 202

    5.5.1 Factorials 203

    5.5.2 Combinations 204

Case Study 5-2 Playing Lotto 206

    5.5.3 Using the Table of Combinations 207

Exercises 208

5.6 The Binomial Probability Distribution 209

    5.6.1 The Binomial Experiment 209

    5.6.2 The Binomial Probability Distribution and Binomial Formula 210

    5.6.3 Using the Table of Binomial Probabilities 215

    5.6.4 Probability of Success and the Shape of the Binomial Distribution 218

    5.6.5 Mean and Standard Deviation of the Binomial Distribution 219

Exercises 219

5.7 The Poisson Probability Distribution 222

Case Study 5-3 Ask Mr. Statistics 226

    5.7.1 Using the Table of Poisson Probabilities 227

    5.7.2 Mean and Standard Deviation of the Poisson Probability Distribution 228

Exercises 229

Glossary / Key Formulas / Supplementary Exercises / Advanced Exercises / Self-Review Test / Mini-Projects / Computer Assignments

Chapter 6 CONTINUOUS RANDOM VARIABLES AND THE NORMAL DISTRIBUTION **240**

6.1 Continuous Probability Distribution 241

Case Study 6-1 Distribution of Time Taken to Run a Road Race 244

6.2 The Normal Distribution 247

6.3 The Standard Normal Distribution 249

Exercises 256

6.4 Standardizing a Normal Distribution 257

Exercises 263

6.5 Applications of the Normal Distribution 264

Exercises 268

6.6 Determining the z and x Values When an Area Under the Normal Distribution Curve Is Known 270

Exercises 273

6.7 The Normal Approximation to the Binomial Distribution 275

Exercises 280

Glossary / Key Formulas / Supplementary Exercises / Advanced Exercises / Self-Review Test / Mini-Projects / Computer Assignments

Chapter 7 SAMPLING DISTRIBUTIONS **290**

7.1 Population and Sampling Distributions 291

    7.1.1 Population Distribution 291

    7.1.2 Sampling Distribution 292

7.2 Sampling and Nonsampling Errors 294

Exercises 296

7.3 Mean and Standard Deviation of $\bar{x}$ 297

Exercises 300

7.4 Shape of the Sampling Distribution of $\bar{x}$ 301

    7.4.1 Sampling from a Normally Distributed Population 302

    7.4.2 Sampling from a Population That Is Not Normally Distributed 304

Exercises 307

7.5 Applications of the Sampling Distribution of $\bar{x}$ 308

Exercises 312

7.6 Population and Sample Proportions 314

7.7 Mean, Standard Deviation, and Shape of the Sampling Distribution of $\hat{p}$ 315

    7.7.1 Sampling Distribution of $\hat{p}$ 315

    7.7.2 Mean and Standard Deviation of $\hat{p}$ 317

    7.7.3 Shape of the Sampling Distribution of $\hat{p}$ 318

Case Study 7–1 Clinton Poll Numbers Are Not Pulled from a Hat 319

Exercises 320

7.8 Applications of the Sampling Distribution of $\hat{p}$ 322

Exercises 324

Glossary / Key Formulas / Supplementary Exercises / Advanced Exercises / Self-Review Test / Mini-Projects / Computer Assignments

Chapter 8 ESTIMATION OF THE MEAN AND PROPORTION 333

8.1 Estimation: An Introduction 334

8.2 Point and Interval Estimates 335

    8.2.1 A Point Estimate 335

    8.2.2 An Interval Estimate 336

8.3 Interval Estimation of a Population Mean: Large Samples 337

Case Study 8–1 Cost of Running a Household 342

Exercises 343

8.4 Interval Estimation of a Population Mean: Small Samples 347

    8.4.1 The t Distribution 348

    8.4.2 Confidence Interval for μ Using the t Distribution 350

Case Study 8–2 Cardiac Demands of Heavy Snow Shoveling 352

Exercises 353

8.5 Interval Estimation of a Population Proportion: Large Samples 356

Exercises 358

8.6 Determining the Sample Size for the Estimation of Mean 361

Exercises 362

8.7 Determining the Sample Size for the Estimation of Proportion 363

Exercises 365

Glossary / Key Formulas / Supplementary Exercises / Advanced Exercises / Self-Review Test / Mini-Projects / Computer Assignments

Chapter 9 HYPOTHESIS TESTS ABOUT THE MEAN AND PROPORTION 376

9.1 Hypothesis Tests: An Introduction 377

    9.1.1 Two Hypotheses 377

    9.1.2 Rejection and Nonrejection Regions 378

    9.1.3 Two Types of Errors 379

    9.1.4 Tails of a Test 381

Exercises 385

9.2 Hypothesis Tests About a Population Mean: Large Samples 386

Exercises 391

9.3 Hypothesis Tests Using the p-Value Approach 395

Exercises 398

9.4 Hypothesis Tests About a Population Mean: Small Samples 400

Exercises 405

9.5   Hypothesis Tests About a Population
        Proportion: Large Samples   **408**

Case Study 9-1   Older Workers Most
        Content   **413**

Exercises   **415**

*Glossary / Key Formulas / Supplementary
Exercises / Advanced Exercises / Self-Review
Test / Mini-Projects / Computer Assignments*

Chapter 10   ESTIMATION AND
HYPOTHESIS TESTING: TWO
POPULATIONS   **429**

10.1   Inferences About the Difference
        Between Two Population Means for
        Large and Independent Samples   **430**

    10.1.1   Independent Versus Dependent
                Samples   **430**

    10.1.2   Mean, Standard Deviation,
                and Sampling Distribution of
                $\bar{x}_1 - \bar{x}_2$ **431**

    10.1.3   Interval Estimation of $\mu_1 - \mu_2$ **432**

    10.1.4   Hypothesis Testing About
                $\mu_1 - \mu_2$ **434**

Exercises   **437**

10.2   Inferences About the Difference
        Between Two Population Means for
        Small and Independent Samples:
        Equal Standard Deviations   **440**

    10.2.1   Interval Estimation of $\mu_1 - \mu_2$ **441**

    10.2.2   Hypothesis Testing About
                $\mu_1 - \mu_2$ **443**

Exercises   **446**

10.3   Inferences About the Difference
        Between Two Population Means for
        Small and Independent Samples:
        Unequal Standard Deviations   **449**

    10.3.1   Interval Estimation of $\mu_1 - \mu_2$ **450**

    10.3.2   Hypothesis Testing About
                $\mu_1 - \mu_2$ **451**

Exercises   **453**

10.4   Inferences About the Difference
        Between Two Population Means for
        Paired Samples   **455**

    10.4.1   Interval Estimation of μ_d **457**

    10.4.2   Hypothesis Testing About μ_d **459**

Exercises   **463**

10.5   Inferences About the Difference
        Between Two Population Proportions for
        Large and Independent Samples   **465**

    10.5.1   Mean, Standard Deviation,
                and Sampling Distribution of
                $\hat{p}_1 - \hat{p}_2$ **465**

    10.5.2   Interval Estimation of $p_1 - p_2$ **466**

    10.5.3   Hypothesis About $p_1 - p_2$ **467**

Case Study 10-1   Worried About Paying the
        Bills   **471**

Exercises   **473**

*Glossary / Key Formulas / Supplementary
Exercises / Advanced Exercises / Self-Review
Test / Mini-Projects / Computer Assignments*

Chapter 11   CHI-SQUARE
TESTS   **487**

11.1   The Chi-Square Distribution   **488**

Exercises   **490**

11.2   A Goodness-of-Fit Test   **490**

Case Study 11-1   Office Order   **496**

Exercises   **498**

11.3   Contingency Tables   **500**

11.4   A Test of Independence or
        Homogeneity   **501**

    11.4.1   A Test of Independence   **501**

    11.4.2   A Test of Homogeneity   **507**

Exercises   **510**

11.5   Inferences About the Population
        Variance   **513**

    11.5.1   Estimation of the Population
                Variance   **514**

    11.5.2   Hypothesis Tests About the
                Population Variance   **516**

Exercises   519

Glossary / Key Formulas / Supplementary Exercises / Advanced Exercises / Self-Review Test / Mini-Projects / Computer Assignments

Chapter 12   ANALYSIS OF VARIANCE   530

12.1   The *F* Distribution   531

Exercises   532

12.2   One-Way Analysis of Variance   533

    12.2.1   Calculating the Value of the Test Statistic   535

    12.2.2   One-Way Anova Test   538

Exercises   542

Glossary / Key Formulas / Supplementary Exercises / Advanced Exercises / Self-Review Test / Mini-Projects / Computer Assignments

Chapter 13   SIMPLE LINEAR REGRESSION   554

13.1   Simple Linear Regression Model   555

    13.1.1   Simple Regression   555

    13.1.2   Linear Regression   555

13.2   Simple Linear Regression Analysis   557

    13.2.1   Scatter Diagram   559

    13.2.2   Least Squares Line   560

    13.2.3   Interpretation of *a* and *b*   563

Case Study 13–1   Regression of Heights and Weights of NBA Players   565

    13.2.4   Assumptions of the Regression Model   566

    13.2.5   A Note on the Use of Simple Linear Regression   568

Exercises   568

13.3   Standard Deviation of Random Errors   573

13.4   Coefficient of Determination   575

Exercises   578

13.5   Inferences About *B*   580

    13.5.1   Sampling Distribution of *b*   580

    13.5.2   Estimation of *B*   581

    13.5.3   Hypothesis Testing About *B*   582

Exercises   583

13.6   Linear Correlation   586

    13.6.1   Linear Correlation Coefficient   586

    13.6.2   Hypothesis Testing About the Linear Correlation Coefficient   588

Exercises   590

13.7   Regression Analysis: A Complete Example   593

Exercises   598

13.8   Using the Regression Model   600

    13.8.1   Using the Regression Model for Estimating the Mean Value of *y*   600

    13.8.2   Using the Regression Model for Predicting a Particular Value of *y*   602

13.9   Cautions in Using Regression   604

Exercises   605

Glossary / Key Formulas / Supplementary Exercises / Advanced Exercises / Self-Review Test / Mini-Projects / Computer Assignments

Chapter 14   NONPARAMETRIC METHODS   618

14.1   The Sign Test   619

    14.1.1   Tests About Categorical Data   619

    14.1.2   Tests About the Median of a Single Population   624

    14.1.3   Tests About the Median Difference Between Paired Data   627

Exercises   631

14.2   The Wilcoxon Signed-Rank Test for Two Dependent Samples   635

Exercises   640

14.3   The Wilcoxon Rank Sum Test for Two Independent Samples   642

Exercises   647

14.4   The Kruskal–Wallis Test   649

Exercises   652

14.5   The Spearman Rho Rank Correlation
        Coefficient Test   654

Exercises   657

14.6   The Runs Test for Randomness   659

Exercises   663

*Glossary / Key Formulas / Supplementary
Exercises / Advanced Exercises / Self-Review
Test / Mini-Projects / Computer Assignments*

Appendix A   SAMPLE SURVEYS, SAMPLING TECHNIQUES, AND DESIGN OF EXPERIMENTS   **A0**

*Available for download on the Web site at
www.wiley.com/college/mann, and printed in
the Student Solutions Manual and the
Instructor Solutions Manual.*

A.1   Sources of Data

   1.   Internal Sources

   2.   External Sources

   3.   Surveys and Experiments

        I.   Surveys

Case Study A-1   Is It a Simple Question?

        II.   Experiments

A.2   Sample Surveys and Sampling Techniques

   A.2.1   Why Sample?

   A.2.2   Random and Nonrandom Samples

   A.2.3   Sampling and Nonsampling Errors

        1.   Sampling or Chance Error

        2.   Nonsampling or Systematic
             Errors

   A.2.4   Random Sampling Techniques

        1.   Simple Random Sampling

        2.   Systematic Random Sampling

        3.   Stratified Random Sampling

        4.   Cluster Sampling

A.3   Design of Experiments

Case Study A-2   Scientists Probe Mystery of
        the Placebo Effect

Exercises

Glossary / Computer Assignments

Appendix B   USING MINITAB   **B0**

B.1   An Introduction   B1

B.2   Organizing Data   B7

B.3   Numerical Descriptive Measures   B11

B.4   The Binomial and Poisson Probability
        Distributions   B13

B.5   The Normal Probability Distribution   B18

B.6   Sampling Distributions   B20

B.7   Estimation of the Mean and
        Proportion   B21

B.8   Hypothesis Tests About the Mean and
        Proportion   B25

B.9   Estimation and Hypothesis Testing: Two
        Populations   B30

B.10   Chi-Square Tests   B39

B.11   Analysis of Variance   B42

B.12   Simple Linear Regression   B44

B.13   Nonparametric Methods   B49

B.14   Generating Random Numbers and
        Selecting a Sample   B57

Appendix C   EXCEL ADVENTURES   **C0**

C.1   An Introduction   C1

C.2   Organizing Data   C2

C.3   Numerical Descriptive Measures   C7

C.4   Probability   C9

C.5   Discrete Probabilty Distributions   C14

C.6   The Normal Distribution   C20

C.7   Sampling from a Population   C22

C.8   Confidence Interval   C25

C.9   Hypothesis Testing   C27

C.10 Inferences on Two Population Means C29

C.11 Chi-Square Applications C34

C.12 Analysis of Variance C40

C.13 Regression Analysis C43

C.14 The Spearman Rank Correlation Coefficient C49

XII Critical Values of T for the Wilcoxon Signed-Rank Test D31

XIII Critical Values of T for the Wilcoxon Rank Sum Test D32

XIV Critical Values for the Spearman Rho Rank Correlation Coefficient Test D33

XV Critical Values for a Two-Tailed Runs Test with $\alpha = .05$ D34

Appendix D STATISTICAL TABLES D0

I Random Numbers D1

II Factorials D5

III Values of $\binom{n}{x}$ (Combination) D6

IV Table of Binomial Probabilities D7

V Values of $e^{-\lambda}$ D15

VI Table of Poisson Probabilities D16

VII Standard Normal Distribution Table D22

VIII The t Distribution Table D23

IX Chi-Square Distribution Table D25

X The F Distribution Table D26

XI Critical Values of X for the Sign Test D30

ANSWERS TO SELECTED ODD-NUMBERED EXERCISES AND SELF-REVIEW TESTS AN1

ANSWERS TO ODD-NUMBERED GRAPHING CALCULATOR EXERCISES AN20

INDEX I1

1 INTRODUCTION

1.1 WHAT IS STATISTICS? ✓

1.2 TYPES OF STATISTICS

CASE STUDY 1-1 PLASTIC IN YOUR POCKET

CASE STUDY 1-2 PUBLIC SCHOOLS MAKING GRADE?

1.3 POPULATION VERSUS SAMPLE

CASE STUDY 1-3 AMERICANS EXPECT GOOD, BAD, AND CATASTROPHIC

1.4 BASIC TERMS

1.5 TYPES OF VARIABLES

1.6 CROSS-SECTION VERSUS TIME-SERIES DATA

1.7 SOURCES OF DATA

1.8 SUMMATION NOTATION

GLOSSARY

SUPPLEMENTARY EXERCISES

SELF-REVIEW TEST

MINI-PROJECT

COMPUTER ASSIGNMENTS

The study of statistics has become more popular than ever during the past three decades or so. The increasing availability of computers and statistical software packages has enlarged the role of statistics as a tool for empirical research. As a result, statistics is used for research in almost all professions, from medicine to sports. Today, college students in almost all disciplines are required to take at least one statistics course.

Every field of study has its own terminology. Statistics is no exception. This introductory chapter explains the basic terms of statistics. These terms will bridge our understanding of the concepts and techniques presented in subsequent chapters.

1.1 WHAT IS STATISTICS?

The word **statistics** has two meanings. In the more common usage, *statistics* refers to numerical facts. The numbers that represent the income of a family, the age of a student, the percentage of passes completed by the quarterback of a football team, and the starting salary of a typical college graduate are examples of statistics in this sense of the word. A 1988 article in *U.S. News & World Report* declared "Statistics are an American obsession."[1] During the 1988 baseball World Series between the Los Angeles Dodgers and the Oakland A's, the then NBC commentator Joe Garagiola reported to the viewers numerical facts about the players' performances. In response, fellow commentator Vin Scully said, "I love it when you talk statistics." In these examples, the word *statistics* refers to numbers. The following examples present some statistics.

1. In 1999, the average salary of major league baseball players was $1,567,873.
2. In 1999, the *Fortune* 500 companies had a total of 11,681 corporate officers and 11.9% of them were women.
3. Nearly 17 million U.S. adults live in households with annual incomes of $100,000 or more.
4. According to a study by Professors Jon Hanson of Harvard Law School and Kyle Logue of the University of Michigan Law School, the cost of smoking to society and to smokers themselves comes to at least $7 a pack, once you include illness, premature death, lost workdays, and so on.
5. About $383 million of taxpayer money is spent every year on rescuing boaters, hikers, and campers in the United States.
6. In 1998, 41,480 people were killed in highway traffic accidents in the United States, and alcohol was involved in 15,936 of those deaths.
7. In 1998, comedian/producer Jerry Seinfeld's earnings totaled $267 million.
8. An employed woman in the United States spends an average of 13.2 hours per week on housework.
9. An average 6-year-old laughs 300 times each day (*USA TODAY*, July 29, 1998).

The second meaning of *statistics* refers to the field or discipline of study. In this sense of the word, *statistics* is defined as follows.

> **STATISTICS** *Statistics* is a group of methods used to collect, analyze, present, and interpret data and to make decisions.

Every day we make decisions that may be personal, business related, or of some other kind. Usually these decisions are made under conditions of uncertainty. Many times, the situations or problems we face in the real world have no precise or definite solution. Statistical methods help us to make scientific and intelligent decisions in such situations. Decisions made by using statistical methods are called *educated guesses*. Decisions made without using statistical (or scientific) methods are *pure guesses* and, hence, may prove to be unreliable.

Like almost all fields of study, statistics has two aspects: theoretical and applied. *Theoretical* or *mathematical statistics* deals with the development, derivation, and proof of sta-

---

[1]"The Numbers Racket: How Polls and Statistics Lie," *U.S. News & World Report*, July 11, 1988, pp. 44–47.

tistical theorems, formulas, rules, and laws. *Applied statistics* involves the applications of those theorems, formulas, rules, and laws to solve real-world problems. This text is concerned with applied statistics and not with theoretical statistics. By the time you finish studying this book, you will learn how to think statistically and how to make educated guesses.

1.2 TYPES OF STATISTICS

Broadly speaking, applied statistics can be divided into two areas: *descriptive statistics* and *inferential statistics*.

1.2.1 DESCRIPTIVE STATISTICS

Suppose we have information on the test scores of students enrolled in a statistics class. In statistical terminology, the whole set of numbers that represents the scores of students is called a **data set**, the name of each student is called an **element**, and the score of each student is called an **observation**. (These terms are defined in more detail in Section 1.4.)

A data set in its original form is usually very large. Consequently, such a data set is not very helpful in drawing conclusions or making decisions. It is easier to draw conclusions from summary tables and diagrams than from the original version of a data set. So, we reduce data to a manageable size by constructing tables, drawing graphs, or calculating summary measures such as averages. The portion of statistics that helps us to do this type of statistical analysis is called **descriptive statistics**.

> **DESCRIPTIVE STATISTICS** *Descriptive statistics* consists of methods for organizing, displaying, and describing data by using tables, graphs, and summary measures.

Case Study 1–1 presents an example of descriptive statistics. It shows a chart and a table that are constructed to organize a data set.

Both Chapters 2 and 3 discuss descriptive statistical methods. In Chapter 2, we learn how to construct tables like the one presented in Case Study 1–1 and how to graph data. In Chapter 3, we learn to calculate numerical summary measures such as averages.

CASE STUDY 1-1 PLASTIC IN YOUR POCKET

The chart, reproduced from *USA TODAY*, shows the percentage of adults who carry different numbers of plastic cards. Using the information in this chart, we can write the table shown on the next page. The first column in this table lists the number of cards that adults carry, and the second column shows the percentage of adults who belong to each group. Thus, for example, 50% of adults carry 1 to 3 cards, 30% carry 4 to 6 cards, and so forth.

Source: *USA TODAY*, August 30, 1999. Copyright © 1999, *USA TODAY*. Chart reproduced with permission.

Number of Cards	Percentage of Adults
1 to 3	50
4 to 6	30
7 to 9	7
10 or more cards	13

1.2.2   INFERENTIAL STATISTICS

In statistics, the collection of all elements of interest is called a **population**. The selection of a few elements from this population is called a **sample**. (Population and sample are discussed in more detail in Section 1.3.)

A major portion of statistics deals with making decisions, inferences, predictions, and forecasts about populations based on results obtained from samples. For example, we may make some decisions about the political views of all college and university students based on the political views of 1000 students selected from a few colleges and universities. As another example, we may want to find the starting salary of a typical college graduate. To do so, we may select 2000 recent college graduates, find their starting salaries, and make a decision based on this information. The area of statistics that deals with such decision-making procedures is referred to as **inferential statistics**. This branch of statistics is also called *inductive reasoning* or *inductive statistics*.

> **INFERENTIAL STATISTICS**   *Inferential statistics* consists of methods that use sample results to help make decisions or predictions about a population.

Case Study 1–2 presents an example of inferential statistics. It shows the views of people who were surveyed about the job done by public schools.

CASE STUDY 1-2 PUBLIC SCHOOLS MAKING GRADE?

The chart, reproduced from *USA TODAY*, shows the grading of public schools by adults. Thus, 31% of adults think that public schools are doing an excellent or good overall job, 24% think that public schools are doing an excellent or good job preparing students for work, and 33% think that public schools are doing an excellent or good job preparing students for college.

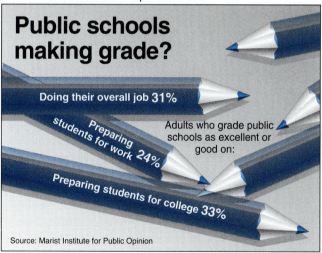

USA SNAPSHOTS®

A look at statistics that shape the nation

Public schools making grade?

Doing their overall job **31%**

Preparing students for work **24%**

Adults who grade public schools as excellent or good on:

Preparing students for college **33%**

Source: Marist Institute for Public Opinion

By Cindy Hall and Quin Tian, USA TODAY

Source: *USA TODAY*, September 10, 1998. Copyright © 1998, *USA TODAY*. Chart reproduced with permission.

Chapters 8 through 13 and parts of Chapter 7 deal with inferential statistics.

Probability, which gives a measurement of the likelihood that a certain outcome will occur, acts as a link between descriptive and inferential statistics. Probability is used to make statements about the occurrence or nonoccurrence of an event under uncertain conditions. Probability and probability distributions are discussed in Chapters 4 through 6 and parts of Chapter 7.

EXERCISES

■ **Concepts and Procedures**

1.1   Briefly describe the two meanings of the word *statistics*.

1.2   Briefly explain the types of statistics.

1.3   POPULATION VERSUS SAMPLE

We will encounter the terms *population* and *sample* on almost every page of this text.[2] Consequently, understanding the meaning of each of these two terms and the difference between them is crucial.

Suppose a statistician is interested in knowing

1.   The percentage of all voters in a city who will vote for a particular candidate in an election

2.   The 1999 gross sales of all companies in New York City

3.   The prices of all houses in California

In these examples, the statistician is interested in *all* voters, *all* companies, and *all* houses. Each of these groups is called the population for the respective example. In statistics, a population does not necessarily mean a collection of people. It can, in fact, be a collection of people or of any kind of item such as houses, books, television sets, or cars. The population of interest is usually called the **target population**.

> **POPULATION OR TARGET POPULATION**   A *population* consists of all elements—individuals, items, or objects—whose characteristics are being studied. The population that is being studied is also called the *target population*.

Most of the time, decisions are made based on portions of populations. For example, the election polls conducted in the United States to estimate the percentage of voters who favor various candidates in any presidential election are based on only a few hundred or a few thousand voters selected from across the country. In this case, the population consists of all registered voters in the United States. The sample is made up of the few hundred or few thousand voters who are included in an opinion poll. Thus, the collection of a few elements selected from a population is called a **sample**.

> **SAMPLE**   A portion of the population selected for study is referred to as a *sample*.

Figure 1.1 illustrates the selection of a sample from a population.

The collection of information from the elements of a population or a sample is called a **survey**. A survey that includes every element of the target population is called a **census**. Often the target population is very large. Hence, in practice, a census is rarely taken because

---

[2]To learn more about sampling and sampling techniques, refer to the Web site, Instructor Solutions Manual, or the Student Solutions Manual.

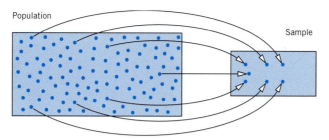

Figure 1.1 Population and sample.

it is expensive and time-consuming. In many cases, it is even impossible to identify each element of the target population. Usually, to conduct a survey, we select a sample and collect the required information from the elements included in that sample. We then make decisions based on this sample information. Such a survey conducted on a sample is called a **sample survey**. As an example, if we collect information on the 1999 incomes of all families in Connecticut, it will be referred to as a census. On the other hand, if we collect information on the 1999 incomes of 50 families from Connecticut, it will be called a sample survey.

> **CENSUS AND SAMPLE SURVEY** A survey that includes every member of the population is called a *census*. The technique of collecting information from a portion of the population is called a *sample survey*.

Case Study 1–3, which is based on a poll conducted by *USA TODAY*, presents an example of a sample survey.

CASE STUDY 1-3 AMERICANS EXPECT GOOD, BAD, AND CATASTROPHIC

The United States will elect a black president and people routinely will live past 100 in the 21st century, Americans predict. But they also say a series of major problems, from crime and terrorism to violations of personal privacy and freedom, will get worse.

A *USA TODAY* poll focused on the future finds that a 53% majority say they're optimistic that the quality of life for average Americans will be better by 2025.

But the survey shows a mixture of high hopes and deep pessimism in predictions about personal lives, social trends, international relations and the future of the earth.

The survey, a *USA TODAY* Forum and two Kansas City focus groups, were conducted as part of a look at American attitudes toward the new century and the millennium.

Among Americans' fears:

- Significant majorities say it will be harder to raise children to be good people, to afford medical care and to find and keep a good job. The gap between rich and poor will widen: 69% say life will be better for the rich; 56% say it will be worse for the poor.
- Three of four say a deadly new disease will emerge by 2025, and two out of three predict a global environmental catastrophe.

■ A 51% majority say it is very or somewhat likely that civilization will be destroyed by a nuclear or other man-made disaster during the 21st century. What's more, 38% say it's likely the earth will be destroyed by an asteroid or other environmental disaster; 39% say the world will come to an end because of judgment day or another religious cataclysm.

But there's also optimism. Solid majorities expect AIDS and cancer to be cured, the average life span to stretch, and personal health to improve. About two out of three Americans say race relations will get better and that a black and a woman will be elected president by 2025.

The survey of 1,055 adults, conducted Sep. 29–Oct. 1, has a margin of error of plus or minus 3 percentage points.

On the bright side: Of the one in four Americans who say humans will have contact with alien life by 2025, about 78% predict the aliens will be friendly, not hostile.

The chart gives summaries of results for certain questions asked in this poll.

Do you expect conditions in the following areas to be better or worse in 2025 than they are today?	Better	Worse	Same	No opinion
Race relations	66%	26%	6%	2%
Moral values in society	31%	62%	5%	2%
Availability of good medical care	66%	29%	3%	2%
Threat of terrorism	23%	70%	4%	3%
The crime rate	35%	57%	6%	2%
The quality of the environment	40%	54%	4%	2%

Source: USA TODAY/CNN/Gallup telephone poll of 1,055 adults conducted Sept. 29 to Oct. 1. Margin of error ±3 percentage points.

USA TODAY

Source: Susan Page, *USA TODAY*, October 13, 1998. Copyright © 1998, *USA TODAY*. Reproduced with permission.

The purpose of conducting a sample survey is to make decisions about the corresponding population. It is important that the results obtained from a sample survey closely match the results that we would obtain by conducting a census. Otherwise, any decision based on a sample survey will not apply to the corresponding population. As an example, to find the average income of families living in New York City by conducting a sample survey, the sample must contain families who belong to different income groups in almost the same proportion as they exist in the population. Such a sample is called a **representative sample**. Inferences derived from a representative sample will be more reliable.

> **REPRESENTATIVE SAMPLE** A sample that represents the characteristics of the population as closely as possible is called a *representative sample*.

A sample may be random or nonrandom. In a **random sample**, each element of the population has some chance of being included in the sample. However, in a nonrandom sample this may not be the case.

> **RANDOM SAMPLE** A sample drawn in such a way that each element of the population has some chance of being selected is called a *random sample*. If the chance of being selected is the same for each element of the population, it is called a **simple random sample**.

One way to select a random sample is by lottery or draw. For example, if we are to select five students from a class of 50, we write each of the 50 names on a separate piece of paper. Then we place all 50 slips in a box and mix them thoroughly. Finally, we randomly draw five slips from the box. The five names drawn give a random sample. On the other hand, if we arrange all 50 names alphabetically and then select the first five names on the list, it is a nonrandom sample because the students listed sixth to fiftieth have no chance of being included in the sample.

A sample may be selected with or without replacement. In sampling **with replacement**, each time we select an element from the population, we put it back in the population before we select the next element. Thus, in sampling with replacement, the population contains the same number of items each time a selection is made. As a result, we may select the same item more than once in such a sample. Consider a box that contains 25 balls of different colors. Suppose we draw a ball, record its color, and put it back in the box before drawing the next ball. Every time we draw a ball from this box, the box contains 25 balls. This is an example of sampling with replacement.

Sampling **without replacement** occurs when the selected element is not replaced in the population. In this case, each time we select an item, the size of the population is reduced by one element. Thus, we cannot select the same item more than once in this type of sampling. Most of the time, samples taken in statistics are without replacement. Consider an opinion poll based on a certain number of voters selected from the population of all eligible voters. In this case, the same voter is not selected more than once. Therefore, this is an example of sampling without replacement.

EXERCISES

■ **Concepts and Procedures**

1.3 Briefly explain the terms *population, sample, representative sample, random sample, sampling with replacement*, and *sampling without replacement*.

1.4 Give one example each of sampling with and sampling without replacement.

1.5 Briefly explain the difference between a census and a sample survey. Why is conducting a sample survey preferable to conducting a census?

■ **Applications**

1.6   Explain whether each of the following constitutes a population or a sample.

a.   Ages of all members of a family
b.   Number of days missed by all employees of a company during the past month
c.   Marital status of 50 persons selected from a large city
d.   Number of VCRs owned by all families in Chicago
e.   Weights of 100 packages mailed from a post office

1.7   Explain whether each of the following constitutes a population or a sample

a.   Scores of all students in a statistics class
b.   Yield of potatoes per acre for 10 pieces of land
c.   Weekly salaries of all employees of a company
d.   Cattle owned by 100 farmers in Iowa
e.   Number of computers sold during the past week at all computer stores in Los Angeles

1.4 BASIC TERMS

It is very important to understand the meaning of some basic terms that will be used frequently in this text. This section explains the meaning of an element (or member), a variable, an observation, and a data set. An element and a data set were briefly defined in Section 1.2. This section defines these terms formally and illustrates them with the help of an example.

Table 1.1 gives information on the 1998 profits (in millions of U.S. dollars) of six U.S. companies. We can call this group of companies a sample of six companies. Each company listed in this table is called an **element** or a **member** of the sample. Table 1.1 contains information on six elements. Note that elements are also called *observational units*.

Table 1.1   1998 Profits of Six U.S. Companies

Company	1998 Profits (millions of dollars) ←——— Variable
Exxon	6370
General Electric	9296
Philip Morris	5372 ←——— An observation or measurement
Kmart	568
Intel	6068
Coca-Cola	3533

An element or a member ——→ (points to Philip Morris / 5372)

> **ELEMENT OR MEMBER**   An *element* or *member* of a sample or population is a specific subject or object (for example, a person, firm, item, state, or country) about which the information is collected.

The *1998 profits of companies* in our example is called a **variable**. The *1998 profits* is a characteristic of companies that we are investigating.

> **VARIABLE**   A *variable* is a characteristic under study that assumes different values for different elements. In contrast to a variable, the value of a *constant* is fixed.

A few other examples of variables are the incomes of households, the number of houses built in a city per month during the past year, the makes of cars owned by people, the gross sales of companies, and the number of insurance policies sold by a salesperson per day during the past month.

In general, a variable assumes different values for different elements, as does the 1998 profits of the six companies in Table 1.1. For some elements in a data set, however, the value of the variable may be the same. For example, if we collect information on incomes of households, these households are expected to have different incomes, although some of them may have the same income.

A variable is often denoted by x, y, or z. For instance, in Table 1.1, the 1998 profits of companies may be denoted by any one of these letters. Starting with Section 1.8, we will begin to use these letters to denote variables.

Each of the values representing the 1998 profits of the six companies in Table 1.1 is called an **observation** or **measurement**.

> **OBSERVATION OR MEASUREMENT** The value of a variable for an element is called an *observation* or *measurement*.

From Table 1.1, the 1998 profits of Coca-Cola were $3533 million. The value $3533 million is an observation or measurement. Table 1.1 contains six observations, one for each of the six companies.

The information given in Table 1.1 on 1998 profits of companies is called the **data** or a **data set**.

> **DATA SET** A *data set* is a collection of observations on one or more variables.

Other examples of data sets are a list of the prices of 25 recently sold homes, scores of 15 students, opinions of 100 voters, and ages of all employees of a company.

EXERCISES

■ Concepts and Procedures

1.8 Explain the meaning of an element, a variable, an observation, and a data set.

■ Applications

1.9 The following table gives the scores of five students on a statistics test.

Student	Score
Bill	83
Susan	91
Allison	80
Jeff	69
Neil	87

Briefly explain the meaning of a member, a variable, a measurement, and a data set with reference to this table.

1.10   The following table lists the amounts of garbage generated in 1998 by the eight states that produced the most garbage.

State	Garbage (millions of tons)
California	56.0
Texas	33.8
New York	30.2
Florida	23.8
Michigan	19.5
Illinois	13.3
North Carolina	12.6
Ohio	12.3

Source: *Bio Cycle* magazine, cited in *PARADE* magazine, June 23, 1999.

Briefly explain the meaning of a member, a variable, a measurement, and a data set with reference to this table.

1.11   Refer to the data set in Exercise 1.9.

a.   What is the variable for this data set?
b.   How many observations are in this data set?
c.   How many elements does this data set contain?

1.12   Refer to the data set in Exercise 1.10.

a.   What is the variable for this data set?
b.   How many observations are in this data set?
c.   How many elements does this data set contain?

1.5   TYPES OF VARIABLES

In Section 1.4, we learned that a variable is a characteristic under investigation that assumes different values for different elements. The incomes of families, heights of persons, gross sales of companies, prices of college textbooks, makes of cars owned by families, number of accidents, and status (freshman, sophomore, junior, or senior) of students enrolled at a university are a few examples of variables.

A variable may be classified as quantitative or qualitative. These two types of variables are explained next.

1.5.1   QUANTITATIVE VARIABLES

Some variables can be measured numerically, whereas others cannot. A variable that can assume numerical values is called a **quantitative variable**.

> **QUANTITATIVE VARIABLE**   A variable that can be measured numerically is called a *quantitative variable*. The data collected on a quantitative variable are called *quantitative data*.

Incomes, heights, gross sales, prices of homes, number of cars owned, and accidents are examples of quantitative variables because each of them can be expressed numerically. For instance, the income of a family may be $41,520.75 per year, the gross sales for a company may be $567 million for the past year, and so forth. Such quantitative variables may be classified as either *discrete variables* or *continuous variables*.

Discrete Variables

The values that a certain quantitative variable can assume may be countable or noncountable. For example, we can count the number of cars owned by a family but we cannot count the height of a family member. The variable whose values are countable is called a **discrete variable**. Note that there are no possible intermediate values between consecutive values of a discrete variable.

> **DISCRETE VARIABLE** A variable whose values are countable is called a *discrete variable*. In other words, a discrete variable can assume only certain values with no intermediate values.

For example, the number of cars sold on any day at a car dealership is a discrete variable because the number of cars sold must be 0, 1, 2, 3, ... The number of cars sold cannot be between 0 and 1, or between 1 and 2. A few other examples of discrete variables are the number of people visiting a bank on any day, the number of cars in a parking lot, the number of cattle owned by a farmer, and the number of students in a class.

Continuous Variables

A **continuous variable** can assume any value over a certain range, and we cannot count these values.

> **CONTINUOUS VARIABLE** A variable that can assume any numerical value over a certain interval or intervals is called a *continuous variable*.

The time taken to complete an examination is an example of a continuous variable because it can assume any value, let us say, between 30 and 60 minutes. The time taken may be 42.6 minutes, 42.67 minutes, or 42.674 minutes. (Theoretically, we can measure time as precisely as we want.) Similarly, the height of a person can be measured to the tenth of an inch or to the hundredth of an inch. However, neither time nor height can be counted in a discrete fashion. A few other examples of continuous variables are weights of people, amount of soda in a 12-ounce can (note that a can does not contain exactly 12 ounces of soda), and yield of potatoes per acre. Note that any variable that involves money is considered a continuous variable.

1.5.2 QUALITATIVE OR CATEGORICAL VARIABLES

Variables that cannot be measured numerically but can be divided into different categories are called **qualitative** or **categorical variables**.

> **QUALITATIVE OR CATEGORICAL VARIABLE** A variable that cannot assume a numerical value but can be classified into two or more nonnumeric categories is called a *qualitative* or *categorical variable*. The data collected on such a variable are called *qualitative data*.

For example, the status of an undergraduate college student is a qualitative variable because a student can fall into any one of four categories: freshman, sophomore, junior, or senior. Other examples of qualitative variables are the gender of a person, hair color, and the make of a car.

Figure 1.2 illustrates the types of variables.

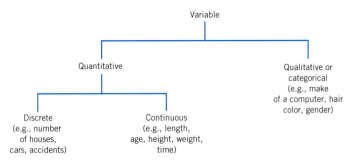

Figure 1.2 Types of variables.

EXERCISES

■ **Concepts and Procedures**

1.13 Explain the meaning of the following terms.

a. Quantitative variable
b. Qualitative variable
c. Discrete variable
d. Continuous variable
e. Quantitative data
f. Qualitative data

■ **Applications**

1.14 Indicate which of the following variables are quantitative and which are qualitative.

a. Number of persons in a family
b. Color of cars
c. Marital status of people
d. Length of a frog's jump
e. Number of students in a class

1.15 Indicate which of the following variables are quantitative and which are qualitative.

a. Number of homes owned
b. Rent paid by tenants
c. Types of cars owned by families
d. Monthly phone bills
e. Color of eyes

1.16 Classify the quantitative variables in Exercise 1.14 as discrete or continuous.

1.17 Classify the quantitative variables in Exercise 1.15 as discrete or continuous.

1.6 CROSS-SECTION VERSUS TIME-SERIES DATA

Based on the time over which they are collected, data can be classified as either cross-section or time-series data.

1.6.1 CROSS-SECTION DATA

Cross-section data contain information on different elements of a population or sample for the *same* period of time. The information on incomes of 100 families for the year 1999 is an example of cross-section data. All examples of data already presented in this chapter have been cross-section data.

> **CROSS-SECTION DATA** Data collected on different elements at the same point in time or for the same period of time are called *cross-section data*.

Table 1.2 shows the 1998 earnings of six celebrities. Because this table presents data on the earnings of six different celebrities for the same period (1998), it is an example of cross-section data.

Table 1.2 1998 Earnings of Six Celebrities

Celebrity	1998 Earnings (millions of dollars)
Jerry Seinfeld	267
Steven Spielberg	175
Oprah Winfrey	125
Michael Jordan	69
Master P.	56.5
Eddie Murphy	47.5

Source: *Forbes* magazine.

1.6.2 TIME-SERIES DATA

Time-series data contain information on the same element for *different* periods of time. Information on U.S. exports for the years 1980 to 1999 is an example of time-series data.

> **TIME-SERIES DATA** Data collected on the same element for the same variable at different points in time or for different periods of time are called *time-series data*.

The data in Table 1.3 on the next page are an example of time-series data. This table lists the average salaries of all major league baseball players for the years 1991 through 1999.

Table 1.3   Average Salaries of Major League Baseball Players for the Years 1991–1999

Year	Average Salary
1991	$ 845,383
1992	1,012,424
1993	1,062,780
1994	1,154,486
1995	1,094,440
1996	1,101,455
1997	1,314,420
1998	1,384,530
1999	1,567,873

1.7   SOURCES OF DATA

The availability of accurate and appropriate data is essential for deriving reliable results.[3] Data may be obtained from internal sources, external sources, or surveys and experiments.

Many times data come from *internal sources*, such as a company's own personnel files or accounting records. For example, a company that wants to forecast the future sales of its product may use the data of past periods from its own records. For most studies, however, all the data that are needed are not usually available from internal sources. In such cases, one may have to depend on outside sources to obtain data. These sources are called *external sources*. For instance, the *Statistical Abstract of the United States* (published annually), which contains various kinds of data on the United States, is an external source of data.

A large number of government and private publications can be used as external sources of data. The following is a list of some of the government publications.[4]

1. *Statistical Abstract of the United States*
2. *Employment and Earnings*
3. *Handbook of Labor Statistics*
4. *Source Book of Criminal Justice Statistics*
5. *Economic Report of the President*
6. *County & City Data Book*
7. *State & Metropolitan Area Data Book*
8. *Digest of Education Statistics*
9. *Health United States*
10. *Agricultural Statistics*

Besides these government publications, a large number of private publications (e.g., *Standard & Poors' Security Owner's Stock Guide* and *World Almanac and Book of Facts*) and periodicals (e.g., *The Wall Street Journal*, *USA TODAY*, *Fortune*, *Forbes*, and *Business Week*) can be used as external data sources.

---

[3]Sources of data are discussed in more detail in Appendix A.

[4]All of these books, and the catalog of U.S. government publications, can be obtained from the Superintendent of Documents, U.S. Government Printing Office, Washington, D.C. 20402.

Sometimes the needed data may not be available from either internal or external sources. In such cases, the investigator may have to conduct a survey or experiment to obtain the required data.

EXERCISES

■ Concepts and Procedures

1.18 Explain the difference between cross-section and time-series data. Give an example of each of these two types of data.

1.19 Briefly describe internal and external sources of data.

■ Applications

1.20 Classify the following as cross-section or time-series data.

a. IBM's gross sales for the period 1985 to 1999
b. Weights of 20 chickens
c. Poverty rates in the Unites States for 1980 to 1999
d. Auto insurance premiums paid by 100 persons

1.21 Classify the following as cross-section or time-series data.

a. Average prices of houses in 100 cities
b. Salaries of 50 employees
c. Number of cars sold each year by General Motors from 1980 to 1999
d. Number of employees employed by a company each year from 1985 to 1999

1.8 SUMMATION NOTATION

Sometimes mathematical notation helps to express a mathematical relationship concisely. This section describes the **summation notation** that is used to denote the sum of values.

Suppose a sample consists of five books and the prices of these five books are $25, $60, $37, $53, and $16. The variable *price of a book* can be denoted by x. The prices of the five books can be written as follows:

$$\text{Price of the first book} = x_1 = \$25$$

$$\uparrow$$

Subscript of x denotes the number of the book

Similarly,

$$\text{Price of the second book} = x_2 = \$60$$

$$\text{Price of the third book} = x_3 = \$37$$

$$\text{Price of the fourth book} = x_4 = \$53$$

$$\text{Price of the fifth book} = x_5 = \$16$$

In this notation, x represents the price and the subscript denotes a particular book.

Now, suppose we want to add the prices of all five books. We have

$$x_1 + x_2 + x_3 + x_4 + x_5 = 25 + 60 + 37 + 53 + 16 = \$191$$

The uppercase Greek letter Σ (pronounced *sigma*) is used to denote the sum of all values. Using Σ notation, we can write the foregoing sum as follows:

$$\Sigma x = x_1 + x_2 + x_3 + x_4 + x_5 = \$191$$

The notation Σx in this expression represents the sum of all the values of x and is read as "sigma x" or "sum of all values of x."

Using summation notation: one variable.

EXAMPLE 1–1   Suppose the ages of four persons are 35, 47, 28, and 60 years. Find

(a) Σx (b) $(\Sigma x)^2$ (c) Σx^2

Solution   Let x_1, x_2, x_3, and x_4 be the ages (in years) of the first, second, third, and fourth person, respectively. Then

$$x_1 = 35, \quad x_2 = 47, \quad x_3 = 28, \quad \text{and} \quad x_4 = 60$$

(a) $\Sigma x = x_1 + x_2 + x_3 + x_4 = 35 + 47 + 28 + 60 = \mathbf{170}$

(b) Note that $(\Sigma x)^2$ is the square of the sum of all x values. Thus,

$$(\Sigma x)^2 = (170)^2 = \mathbf{28{,}900}$$

(c) The expression Σx^2 is the sum of the squares of x values. To calculate Σx^2, we first square each of the x values and then add these squared values. Thus,

$$\Sigma x^2 = (35)^2 + (47)^2 + (28)^2 + (60)^2 = 1225 + 2209 + 784 + 3600 = \mathbf{7818}$$ ■

Using summation notation: two variables.

EXAMPLE 1–2   The following table lists four pairs of m and f values:

m	12	15	20	30
f	5	9	10	16

Compute the following:

(a) Σm (b) Σf^2 (c) Σmf (d) $\Sigma m^2 f$

Solution   We can write

$$m_1 = 12 \quad m_2 = 15 \quad m_3 = 20 \quad m_4 = 30$$
$$f_1 = 5 \quad\quad f_2 = 9 \quad\quad f_3 = 10 \quad\quad f_4 = 16$$

(a) $\Sigma m = 12 + 15 + 20 + 30 = \mathbf{77}$

(b) $\Sigma f^2 = (5)^2 + (9)^2 + (10)^2 + (16)^2 = 25 + 81 + 100 + 256 = \mathbf{462}$

(c) To compute Σmf, we multiply the corresponding values of m and f and then add the products as follows:

$$\Sigma mf = m_1 f_1 + m_2 f_2 + m_3 f_3 + m_4 f_4$$
$$= 12(5) + 15(9) + 20(10) + 30(16) = \mathbf{875}$$

(d) To calculate $\Sigma m^2 f$, we square each m value, then multiply the corresponding m^2 and f values, and add the products. Thus,

$$\Sigma m^2 f = (m_1)^2 f_1 + (m_2)^2 f_2 + (m_3)^2 f_3 + (m_4)^2 f_4$$
$$= (12)^2(5) + (15)^2(9) + (20)^2(10) + (30)^2(16) = \mathbf{21{,}145}$$

The calculations done in parts (a) through (d) to find the values of Σm, Σf^2, Σmf, and $\Sigma m^2 f$ can be performed in tabular form, as shown in Table 1.4.

Table 1.4

m	f	f^2	mf	m^2f
12	5	$5 \times 5 = 25$	$12 \times 5 = 60$	$12 \times 12 \times 5 = 720$
15	9	$9 \times 9 = 81$	$15 \times 9 = 135$	$15 \times 15 \times 9 = 2{,}025$
20	10	$10 \times 10 = 100$	$20 \times 10 = 200$	$20 \times 20 \times 10 = 4{,}000$
30	16	$16 \times 16 = 256$	$30 \times 16 = 480$	$30 \times 30 \times 16 = 14{,}400$
$\Sigma m = 77$	$\Sigma f = 40$	$\Sigma f^2 = 462$	$\Sigma mf = 875$	$\Sigma m^2f = 21{,}145$

The columns of Table 1.4 can be explained as follows:

1. The first column lists the values of m. The sum of these values gives $\Sigma m = 77$.
2. The second column lists the values of f. The sum of this column gives $\Sigma f = 40$.
3. The third column lists the squares of the f values. For example, the first value, 25, is the square of 5. The sum of the values in this column gives $\Sigma f^2 = 462$.
4. The fourth column records products of the corresponding m and f values. For example, the first value, 60, in this column is obtained by multiplying 12 by 5. The sum of the values in this column gives $\Sigma mf = 875$.
5. Next, the m values are squared and multiplied by the corresponding f values. The resulting products, denoted by m^2f, are recorded in the fifth column. For example, the first value, 720, is obtained by squaring 12 and multiplying this result by 5. The sum of the values in this column gives $\Sigma m^2f = 21{,}145$. ∎

EXERCISES

■ Concepts and Procedures

1.22 The following table lists five pairs of m and f values.

m	5	10	17	20	25
f	12	8	6	16	4

Compute the value of each of the following:

a. Σm b. Σf^2 c. Σmf d. Σm^2f

1.23 The following table lists six pairs of m and f values.

m	3	6	25	12	15	18
f	16	11	16	8	4	14

Calculate the value of each of the following:

a. Σf b. Σm^2 c. Σmf d. Σm^2f

1.24 The following table lists five pairs of x and y values.

x	15	22	11	8	5
y	10	12	14	9	18

Compute   a. Σx b. Σy c. Σxy d. Σx^2 e. Σy^2

1.25   The following table lists six pairs of x and y values.

x	4	18	25	9	12	20
y	12	5	14	7	12	8

Compute   **a.** Σx **b.** Σy **c.** Σxy **d.** Σx^2 **e.** Σy^2

■ Applications

1.26   The scores of five students in a statistics class are 75, 80, 97, 91, and 63. Find
a. Σx **b.** $(\Sigma x)^2$ **c.** Σx^2

1.27   The numbers of cars owned by six families are 3, 2, 1, 4, 1, and 2. Find
a. Σx **b.** $(\Sigma x)^2$ **c.** Σx^2

1.28   The electric heating bills for November 1999 for four families were $122, $72, $96, and $110. Find
a. Σx **b.** $(\Sigma x)^2$ **c.** Σx^2

1.29   The weights of seven newborn babies are 7, 9, 6, 12, 10, 9, and 8 pounds. Find
a. Σx **b.** $(\Sigma x)^2$ **c.** Σx^2

GLOSSARY

Census   A survey that includes all members of the population.

Continuous variable   A (quantitative) variable that can assume any numerical value over a certain interval or intervals.

Cross-section data   Data collected on different elements at the same point in time or for the same period of time.

Data or **data set**   Collection of observations or measurements on a variable.

Descriptive statistics   Collection of methods for organizing, displaying, and describing data using tables, graphs, and summary measures.

Discrete variable   A (quantitative) variable whose values are countable.

Element or **member**   A specific subject or object included in a sample or population.

Inferential statistics   Collection of methods that help make decisions about a population based on sample results.

Observation or **measurement**   The value of a variable for an element.

Population or **target population**   The collection of all elements whose characteristics are being studied.

Qualitative or **categorical data**   Data generated by a qualitative variable.

Qualitative or **categorical variable**   A variable that cannot assume numerical values but is classified into two or more categories.

Quantitative data   Data generated by a quantitative variable.

Quantitative variable   A variable that can be measured numerically.

Random sample   A sample drawn in such a way that each element of the population has some chance of being included in the sample.

Representative sample   A sample that contains the same characteristics as the corresponding population.

Sample   A portion of the population of interest.

Sample survey   A survey that includes elements of a sample.

Simple random sample   A sample drawn in such a way that each element of the population has the same chance of being selected for the sample.

Statistics   Group of methods used to collect, analyze, present, and interpret data and to make decisions.

Survey   Collection of data on the elements of a population or sample.

Time-series data   Data that give the values of the same variable for the same element at different points in time or for different periods of time.

Variable   A characteristic under study or investigation that assumes different values for different elements.

SUPPLEMENTARY EXERCISES

1.30   The following table gives the U.S. average SAT scores in mathematics for the years 1994 to 1999.

Year	Average SAT Score in Mathematics
1994	504
1995	506
1996	508
1997	511
1998	512
1999	511

Describe the meaning of a variable, a measurement, and the data set with reference to this table.

1.31   The following table lists the annual salaries as of October 1998 of the governors of all six New England states.

State	Governor's Salary
Connecticut	$ 78,000
Maine	70,000
Massachusetts	90,000
New Hampshire	90,547
Rhode Island	69,900
Vermont	105,402

Source: *The World Almanac and Book of Facts*, 1999.

Describe the meaning of a variable, a measurement, and a data set with reference to this table.

1.32   Refer to Exercises 1.30 and 1.31. Classify these data sets as either cross-section or time-series.

1.33   Indicate whether each of the following examples refers to a population or to a sample.

a.   A group of 25 patients selected to test a new drug
b.   Total items produced on a machine in a year
c.   Yearly expenditures on clothes for 50 persons
d.   Number of houses sold by each of the 10 employees of a real estate agency during 1999

1.34   Indicate whether each of the following examples refers to a population or to a sample.

a.   Ages of CEOs of all companies in New York City
b.   Prices of 100 houses selected from Boston
c.   Number of subscribers to five magazines
d.   Salaries of all employees of a bank

1.35 State which of the following is an example of sampling with replacement and which is an example of sampling without replacement.

a. Selecting 10 patients out of 100 to test a new drug

b. Selecting one professor to be a member of the university senate and then selecting one professor from the same group to be a member of the curriculum committee

1.36 State which of the following is an example of sampling with replacement and which is an example of sampling without replacement.

a. Selecting five students to form a committee

b. Selecting one student for a leadership award and then selecting one student from the same group for a scholarship award

1.37 The number of ties owned by six persons are 10, 9, 14, 12, 7, and 4. Find

a. Σx **b.** $(\Sigma x)^2$ **c.** Σx^2

1.38 The number of hours worked during the past week by five employees of a company are 43, 39, 44, 31, and 40. Find

a. Σx **b.** $(\Sigma x)^2$ **c.** Σx^2

1.39 The following table lists five pairs of m and f values.

m	3	16	11	9	20
f	7	32	17	12	34

Compute the value of each of the following:

a. Σm **b.** Σf^2 **c.** Σmf **d.** $\Sigma m^2 f$ **e.** Σm^2

1.40 The following table lists six pairs of x and y values.

x	7	11	8	4	14	28
y	5	15	7	10	9	19

Compute the value of each of the following:

a. Σy **b.** Σx^2 **c.** Σxy **d.** $\Sigma x^2 y$ **e.** Σy^2

SELF-REVIEW TEST

1. A population in statistics means

    a. a collection of all men and women

    b. a collection of all subjects or objects of interest

    c. a collection of all people living in a country

2. A sample in statistics means

    a. a portion of the people selected from the population of a country

    b. a portion of the people selected from the population of an area

    c. a portion of the population of interest

3. Indicate which of the following is an example of a sample with replacement and which is a sample without replacement.

    a. Ten students are selected from a statistics class in such a way that as soon as a student is selected, his or her name is deleted from the list before the next student is selected.

b. A box contains five balls of different colors. A ball is drawn from this box, its color is recorded, and it is put back into the box before the next ball is drawn. This experiment is repeated 12 times.

4. Indicate which of the following variables are quantitative and which are qualitative. Classify the quantitative variables as discrete or continuous.

    a. Brand of coffee   **b.** Number of TV sets owned by families

    c. Weekly earnings of employees

5. The following table gives the starting salaries of five recent college graduates.

Name	Salary
Matt	$29,200
Lucia	42,450
Kristen	49,500
Warren	32,350
Brooke	62,800

Explain the meaning of a member, a variable, a measurement, and a data set with reference to this table.

6. The values (in thousands of dollars) of cars owned by six persons are 13, 9, 3, 28, 7, and 46. Calculate

    a. Σx **b.** Σx^2 **c.** $(\Sigma x)^2$

7. The following table lists five pairs of m and f values.

m	3	6	9	12	15
f	15	25	40	20	12

Calculate   **a.** Σm **b.** Σf **c.** Σm^2 **d.** Σmf **e.** $\Sigma m^2 f$ **f.** Σf^2

MINI-PROJECT

Mini-Project 1–1

Collect information on any variable for 30 members. The following are some possible ways to obtain this information (you may use other sources if you wish).

1. Take a sample of 30 people and ask them to keep track of the number of telemarketing calls they receive during the next week. At the end of the week, collect the data from all 30 individuals.
2. Examine the real estate ads in the classified section of your local newspaper and record the rent for 30 one-bedroom apartments.
3. Check the ads for cars for sale in the classified section of your local newspaper and collect information on the asking prices for 30 different cars.
4. Examine the real estate ads in the classified section of your local newspaper and find the asking prices for 30 different homes.

Once you have collected the information, write a brief report that includes answers to these questions.

a. List the data.

b. What is a reasonable target population for the sample you used?

c. Is your sample a random sample from this target population?

d. Do you think your sample is representative of this population?

e. Is this an example of sampling with replacement or without replacement?

f. Are the sample data discrete or continuous?

g. Describe the meaning of an element, a variable, and a measurement for this data set.

h. Describe any problems you had in collecting the data.
i. Were any of the data values unusable? If yes, explain why.

Save this data set for projects in Chapters 2 and 3.

For instructions on using MINITAB for this chapter, please see Section B.1 in Appendix B.

See Appendix C for *Excel Adventure 1.*

COMPUTER ASSIGNMENTS

CA1.1 Refer to Data Set III on the heights and weights of all NBA players. (Note that the data sets that accompany this text appear in the Solutions Manuals and on the Web site for this text. These data sets can be downloaded from this Web site. Please see the Preface for details.) Enter the data on names, heights, and weights of all players in the MINITAB data window. If you are using the data disk that contains these data, you do not have to enter these data again. Take a sample of 15 players using the pull-down menus and the command language. Then print the sample data.

CA1.2 The following table gives the names, hours worked, and salary for the past week for five workers.

Name	Hours Worked	Salary
John	42	$725
Shannon	33	483
Kathy	28	355
David	47	790
Steve	40	820

a. Enter these data into MINITAB. Save the data file as WORKER.MTW. Exit MINITAB. Then restart MINITAB and retrieve the file WORKER.MTW.

b. Try the following commands for this assignment using command language and analyze the output for each command. Then print a hard copy of the session window.

```
MTB > PRINT C1-C3
MTB > NAME C1 'NAME' C2 'HOURS' C3 'SALARY'
MTB > PRINT 'NAME' 'HOURS' 'SALARY'
MTB > LET K2 = SUM(C2)
MTB > PRINT K2
MTB > LET K3 = SUM(C3)
MTB > PRINT K3
MTB > LET C4 = C3*C3
MTB > LET K4 = SUM(C4)
MTB > PRINT C2-C4
MTB > PRINT K2-K4
MTB > SAMPLE 2 C1-C3 C5-C7
MTB > PRINT C5-C7
```

2 ORGANIZING DATA

2.1 RAW DATA

2.2 ORGANIZING AND GRAPHING QUALITATIVE DATA

CASE STUDY 2–1 JOB STRESS CAN BE SATISFYING

CASE STUDY 2–2 WHY EAT OUT?

2.3 ORGANIZING AND GRAPHING QUANTITATIVE DATA

CASE STUDY 2–3 BOOMERS GO FISH

2.4 SHAPES OF HISTOGRAMS

CASE STUDY 2–4 USING TRUNCATED AXES

2.5 CUMULATIVE FREQUENCY DISTRIBUTIONS

2.6 STEM-AND-LEAF DISPLAYS

GLOSSARY

KEY FORMULAS

SUPPLEMENTARY EXERCISES

SELF-REVIEW TEST

MINI-PROJECTS

COMPUTER ASSIGNMENTS

In addition to thousands of private organizations and individuals, a large number of U.S. government agencies such as the Bureau of the Census, the Bureau of Labor Statistics, the National Agricultural Statistics Service, the National Center for Education Statistics, the National Center for Health Statistics, and the Bureau of Justice Statistics conduct hundreds of surveys every year. The data collected from each of these surveys fill hundreds of thousands of pages. In their original form, these data sets may be so large that they do not make sense to most of us. Descriptive statistics, however, supplies the techniques that help to condense large data sets by using tables, graphs, and summary measures. We see such tables, graphs, and summary measures in newspapers and magazines every day. At a glance, these tabular and graphical displays present information on every aspect of life. Consequently, descriptive statistics is of immense importance because it provides efficient and effective methods for summarizing and analyzing information.

This chapter explains how to organize and display data using tables and graphs. We will learn how to prepare frequency distribution tables for qualitative and quantitative data; how to construct bar graphs, pie charts, histograms, and polygons for such data; and how to prepare stem-and-leaf displays.

2.1   RAW DATA

When data are collected, the information obtained from each member of a population or sample is recorded in the sequence in which it becomes available. This sequence of data recording is random and unranked. Such data, before they are grouped or ranked, are called **raw data.**

> **RAW DATA**   Data recorded in the sequence in which they are collected and before they are processed or ranked are called *raw data.*

Suppose we collect information on the ages (in years) of 50 students selected from a university. The data values, in the order they are collected, are recorded in Table 2.1. For instance, the first student's age is 21, the second student's age is 19 (second number in the first row), and so forth. The data in Table 2.1 are quantitative raw data.

Table 2.1   Ages of 50 Students

21	19	24	25	29	34	26	27	37	33
18	20	19	22	19	19	25	22	25	23
25	19	31	19	23	18	23	19	23	26
22	28	21	20	22	22	21	20	19	21
25	23	18	37	27	23	21	25	21	24

Suppose we ask the same 50 students about their student status. The responses of the students are recorded in Table 2.2. In this table, F, SO, J, and SE are the abbreviations for freshman, sophomore, junior, and senior, respectively. This is an example of qualitative (or categorical) raw data.

Table 2.2   Status of 50 Students

J	F	SO	SE	J	J	SE	J	J	J
F	F	J	F	F	F	SE	SO	SE	J
J	F	SE	SO	SO	F	J	F	SE	SE
SO	SE	J	SO	SO	J	J	SO	F	SO
SE	SE	F	SE	J	SO	F	J	SO	SO

The data presented in Tables 2.1 and 2.2 are also called **ungrouped data**. An ungrouped data set contains information on each member of a sample or population individually.

2.2   ORGANIZING AND GRAPHING QUALITATIVE DATA

This section discusses how to organize and display qualitative (or categorical) data. Data sets are organized into tables, and data are displayed in graphs.

2.2.1   FREQUENCY DISTRIBUTIONS

A sample of 100 students enrolled at a university were asked what they intended to do after graduation. Forty-four said they wanted to work for private companies/businesses, 16 said they wanted to work for the federal government, 23 wanted to work for state or local governments, and 17 intended to start their own businesses. Table 2.3 lists the types of employment and the number of students who intend to engage in each type of employment. In this table, the variable is the *type of employment*, which is a qualitative variable. The categories (representing the type of employment) listed in the first column are mutually exclusive. In other words, each of the 100 students belongs to one and only one of these categories. The number of students who belong to a certain category is called the *frequency* of that category. A **frequency distribution** exhibits how the frequencies are distributed over various categories. Table 2.3 is called a *frequency distribution table* or simply a *frequency table*.

Table 2.3 Type of Employment Students Intend to Engage In

	Type of Employment	Number of Students	
Variable ⟶	**Type of Employment**	**Number of Students**	⟵ Frequency column
	Private companies/businesses	44	
Category ⟶	Federal government	16	⟵ Frequency
	State/local government	23	
	Own business	17	
		Sum = 100	

> **FREQUENCY DISTRIBUTION FOR QUALITATIVE DATA** A *frequency distribution* for qualitative data lists all categories and the number of elements that belong to each of the categories.

Example 2–1 illustrates how a frequency distribution table is constructed for qualitative data.

Constructing a frequency distribution table for qualitative data.

EXAMPLE 2-1 A sample was taken of 25 high school seniors who were planning to go to college. Each of the students was asked which of the following majors he or she intended to choose: business, economics, management information systems (MIS), behavioral sciences (BS), other. The responses of these students are as follows:

Economics	MIS	Economics	Business	Business
Business	Business	Other	Other	Other
BS	BS	MIS	Other	MIS
Other	Business	MIS	Business	Other
Economics	MIS	Other	Other	MIS

Construct a frequency distribution table for these data.

Solution Note that the *major a student intends to choose* is the variable in this example. This variable is classified into five categories: business, economics, MIS, BS, and other. We record these categories in the first column of Table 2.4 on the next page. Then we read each student's response from the given data and mark a *tally*, denoted by the symbol |, in the second column of Table 2.4 next to the corresponding category. For example, the first student intends to major in economics. We show this in the frequency distribution table by marking a tally in the second column next to the category *Economics*. Note that the tallies are marked in blocks of five for counting convenience. Finally, we record the total of the tallies for each category in the third column of the table. This column is called the *column of frequencies* and is usually denoted by *f*. The sum of the entries in the frequency column gives the sample size or total frequency. In Table 2.4, this total is 25, which is the sample size.

Table 2.4 Frequency Distribution of Majors

Major	Tally	Frequency (f)
Business	‖‖	6
Economics	‖‖‖	3
MIS	‖‖	6
BS	‖‖	2
Other	‖‖ ‖‖‖	8
		Sum = 25

2.2.2 RELATIVE FREQUENCY AND PERCENTAGE DISTRIBUTIONS

The **relative frequency** of a category is obtained by dividing the frequency of that category by the sum of all the frequencies. Thus, the relative frequency shows what fractional part or proportion of the total frequency belongs to the corresponding category. A *relative frequency distribution* lists the relative frequencies for all categories.

> **RELATIVE FREQUENCY OF A CATEGORY**
>
> $$\text{Relative frequency of a category} = \frac{\text{Frequency of that category}}{\text{Sum of all frequencies}}$$

The **percentage** for a category is obtained by multiplying the relative frequency of that category by 100. A *percentage distribution* lists the percentages for all categories.

> **PERCENTAGE**
>
> $$\text{Percentage} = (\text{Relative frequency}) \cdot 100$$

Constructing relative frequency and percentage distributions.

EXAMPLE 2-2 Determine the relative frequency and percentage distributions for the data in Table 2.4.

Solution The relative frequencies and percentages from Table 2.4 are calculated and listed in Table 2.5. Based on this table, we can state that .24 or 24% of the students in the sample said that they intend to major in business. By adding the percentages for the first two categories, we can state that 36% of the students said they intend to major in business or economics. The other numbers in Table 2.5 can be interpreted in the same way.

Notice that the sum of the relative frequencies is always 1.00 (or approximately 1.00 if the relative frequencies are rounded), and the sum of the percentages is always 100 (or approximately 100 if the percentages are rounded).

Table 2.5 Relative Frequency and Percentage
Distributions for Table 2.4

Major	Relative Frequency	Percentage
Business	6/25 = .24	.24 (100) = 24
Economics	3/25 = .12	.12 (100) = 12
MIS	6/25 = .24	.24 (100) = 24
BS	2/25 = .08	.08 (100) = 8
Other	8/25 = .32	.32 (100) = 32
	Sum = 1.00	Sum = 100

2.2.3 GRAPHICAL PRESENTATION OF QUALITATIVE DATA

All of us have heard the saying "a picture is worth a thousand words." A graphic display can reveal at a glance the main characteristics of a data set. The *bar graph* and the *pie chart* are two types of graphs used to display qualitative data.

Bar Graphs

To construct a **bar graph** (also called a *bar chart*), we mark the various categories on the horizontal axis as in Figure 2.1. Note that all categories are represented by intervals of the same width. We mark the frequencies on the vertical axis. Then we draw one bar for each category such that the height of the bar represents the frequency of the corresponding category. We leave a small gap between adjacent bars. Figure 2.1 gives the bar graph for the frequency distribution of Table 2.4.

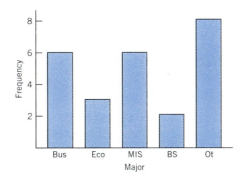

Figure 2.1 Bar graph for the frequency distribution of Table 2.4.

BAR GRAPH A graph made of bars whose heights represent the frequencies of respective categories is called a *bar graph*.

The bar graphs for relative frequency and percentage distributions can be drawn simply by marking the relative frequencies or percentages, instead of the class frequencies, on the vertical axis.

Sometimes a bar graph is constructed by marking the categories on the vertical axis and the frequencies on the horizontal axis.

CASE STUDY 2-1   JOB STRESS CAN BE SATISFYING

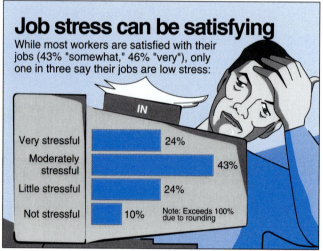

USA SNAPSHOTS®

A look at statistics that shape your finances

Job stress can be satisfying

While most workers are satisfied with their jobs (43% "somewhat," 46% "very"), only one in three say their jobs are low stress:

IN

Very stressful   24%

Moderately stressful   43%

Little stressful   24%

Not stressful   10%   Note: Exceeds 100% due to rounding

Source: *The Wirthlin Report*   By Cindy Hall and Genevieve Lynn, USA TODAY

The chart, reproduced from *USA TODAY*, is a bar graph for the data obtained in a survey about job stress. Using the information, we can create the accompanying table, which lists the percentage of workers in the sample who belong to different categories based on the amount of stress they experience at their jobs. Note that the percentages add up to 101% because of rounding.

Category	Percentage of Workers
Very stressful	24
Moderately stressful	43
Little stressful	24
Not stressful	10

Source: *USA TODAY*, April 19, 1999. Copyright © 1999, *USA TODAY*. Chart reproduced with permission.

Pie Charts

A **pie chart** is more commonly used to display percentages, although it can be used to display frequencies or relative frequencies. The whole pie (or circle) represents the total sample or population. The pie is divided into different portions that represent the percentages of the population or sample belonging to different categories.

> **PIE CHART** A circle divided into portions that represent the relative frequencies or percentages of a population or a sample belonging to different categories is called a *pie chart.*

As we know, a circle contains 360 degrees. To construct a pie chart, we multiply 360 by the relative frequency for each category to obtain the degree measure or size of the angle for the corresponding category. Table 2.6 shows the calculation of angle sizes for the various categories of Table 2.5.

Table 2.6 Calculating Angle Sizes for the Pie Chart

Major	Relative Frequency	Angle Size
Business	.24	360 (.24) = 86.4
Economics	.12	360 (.12) = 43.2
MIS	.24	360 (.24) = 86.4
BS	.08	360 (.08) = 28.8
Other	.32	360 (.32) = 115.2
	Sum = 1.00	Sum = 360

Figure 2.2 shows the pie chart for the percentage distribution of Table 2.5, which uses the angle sizes calculated in Table 2.6.

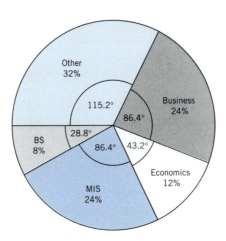

Figure 2.2 Pie chart for the percentage distribution of Table 2.5.

CASE STUDY 2–2   WHY EAT OUT?

USA SNAPSHOTS®

A look at statistics that shape our lives

Why eat out?

The average American ate 131 restaurant meals last year. Their reasons for eating dinners at restaurants:

Wanted food couldn't cook at home **29%**

Atmosphere **20%**

Quicker **12%**

Less work/ no cleanup **17%**

Enjoy eating out **19%**

3%

Don't know

Source: Market Facts for Tyson Foods

By Cindy Hall and Kevin Rechin, USA TODAY

The pie chart shows the reasons that people give for eating dinner at restaurants and the percentage of people in each category. As can be observed from this pie chart, 29% said that they eat out because they could not cook at home, 20% eat out because of the atmosphere, and so forth.

Source: USA TODAY, January 26, 1998. Copyright © 1998, *USA TODAY*. Chart reproduced with permission.

EXERCISES

■ **Concepts and Procedures**

2.1   Why do we need to group data in the form of a frequency table? Explain briefly.

2.2   How are the relative frequencies and percentages of categories obtained from the frequencies of categories? Illustrate with the help of an example.

2.3   The following data give the results of a sample survey. The letters A, B, and C represent the three categories.

A	B	B	A	C	B	C	C	C	A
C	B	C	A	C	C	B	C	C	A
A	B	C	C	B	C	B	A	C	A

a. Prepare a frequency distribution table.
b. Calculate the relative frequencies and percentages for all categories.
c. What percentage of the elements in this sample belong to category B?
d. What percentage of the elements in this sample belong to category A or C?
e. Draw a bar graph for the frequency distribution.

2.4 The following data give the results of a sample survey. The letters Y, N, and D represent the three categories.

```
D   N   N   Y   Y   Y   N   Y   D   Y
Y   Y   Y   Y   N   Y   Y   N   N   Y
N   Y   Y   N   D   N   Y   Y   Y   Y
Y   Y   N   N   Y   Y   N   N   D   Y
```

a. Prepare a frequency distribution table.
b. Calculate the relative frequencies and percentages for all categories.
c. What percentage of the elements in this sample belong to category Y?
d. What percentage of the elements in this sample belong to category N or D?
e. Draw a pie chart for the percentage distribution.

■ Applications

2.5 The data on the status of 50 students given in Table 2.2 of Section 2.1 are reproduced here.

```
J    F    SO   SE   J    J    SE   J    J    J
F    F    J    F    F    F    SE   SO   SE   J
J    F    SE   SO   SO   F    J    F    SE   SE
SO   SE   J    SO   SO   J    J    SO   F    SO
SE   SE   F    SE   J    SO   F    J    SO   SO
```

a. Prepare a frequency distribution table.
b. Calculate the relative frequencies and percentages for all categories.
c. What percentage of these students are juniors or seniors?
d. Draw a bar graph for the frequency distribution.

2.6 The following are the responses of 20 students from a statistics class who were asked to evaluate their instructor. The students were asked to choose one of five answers: Excellent (E), Above average (AA), Average (A), Below average (B), and Poor (P).

```
AA   B    A    A    E    AA   P    E    AA   B
E    AA   E    B    E    A    B    P    AA   E
```

a. Construct a frequency distribution table.
b. Calculate the relative frequencies and percentages for all categories.
c. What percentage of these students ranked this instructor as excellent or above average?
d. Draw a bar graph for the relative frequency distribution.

2.7 Employees from a wide range of occupations were surveyed by the University of Michigan, the University of Pennsylvania, Bryn Mawr College, and Swarthmore College (*Glamour*, May 1998). The employees were asked whether they viewed their work as a job (J) (necessary rather than fulfilling), a career (C) (focused on advancement, not pleasure), or a calling (F) (fulfilling and socially useful). Suppose a sample of 15 such responses are listed below.

```
F   J   J   F   C   C   F   C
F   J   J   C   F   F   J
```

a. Prepare a frequency distribution table.
b. Compute the relative frequencies and percentages for all categories.

c. What percentage of the employees in this sample regard their work as a calling (F)?

d. Draw a pie chart for the percentage distribution.

2.8 Twelve persons were asked to taste two types of soft drinks, *A* and *B*, and indicate if the taste of *A* was superior to (S), the same as (M), or inferior to (I) that of *B*. Their responses are listed next.

S　I　I　M　M　S　M　S　S　I　S　S

a. Construct a frequency distribution table.

b. Calculate the relative frequencies and percentages for all categories.

c. Draw a pie chart for the percentage distribution.

2.9 In a poll by Bruskin-Goldring for Direct Marketing Association, American women were asked why they shop from catalogs. The table below shows the percentage of women who gave each reason for shopping from catalogs.

Reason	Percentage
Convenience	50
Variety	20
Price	9
Service	3
Don't buy	17
Don't know	1

Source: USA TODAY, December 17, 1997.

Draw a pie chart for this percentage distribution.

2.10 In an April 1999 telephone poll conducted by Yankelovich Partners Inc. for *Time*/CNN, 13–17-year-old American teenagers were asked the question: "Do your parents have rules about your Internet use, and do you follow them?" The table below presents a summary of the responses obtained.

Response	Percentage
Yes, and always follow	31
Yes, but don't always follow	26
No, parents don't have rules	43

Source: Time, May 10, 1999.

Draw a bar graph to display these data.

2.3　ORGANIZING AND GRAPHING QUANTITATIVE DATA

In the previous section we learned how to group and display qualitative data. This section explains how to group and display quantitative data.

2.3.1　FREQUENCY DISTRIBUTIONS

Table 2.7 gives the weekly earnings of 100 employees of a large company. The first column lists the *classes*, which represent the (quantitative) variable *weekly earnings*. For quantitative data, an interval that includes all the values that fall within two numbers, the lower and upper limits, is called a **class**. Note that the classes always represent a variable. As we can observe, the classes are nonoverlapping; that is, each value on earnings belongs to one and only one class. The second column in the table lists the number of employees who have earnings within each class. For example, nine employees of this company earn $301 to $400 per week. The numbers listed in the second column are called the **frequencies**, which give the number of values that belong to different classes. The frequencies are denoted by f.

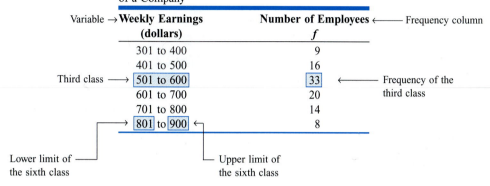

Table 2.7 Weekly Earnings of 100 Employees of a Company

Weekly Earnings (dollars)	Number of Employees f
301 to 400	9
401 to 500	16
501 to 600	33
601 to 700	20
701 to 800	14
801 to 900	8

Variable → Weekly Earnings (dollars)

Frequency column

Third class ⟶ 501 to 600 — Frequency of the third class

801 to 900

Lower limit of the sixth class

Upper limit of the sixth class

For quantitative data, the frequency of a class represents the number of values in the data set that fall in that class. Table 2.7 contains six classes. Each class has a *lower limit* and an *upper limit*. The values 301, 401, 501, 601, 701, and 801 give the lower limits, and the values 400, 500, 600, 700, 800, and 900 are the upper limits of the six classes, respectively. The data presented in Table 2.7 are an illustration of a **frequency distribution table** for quantitative data. Data presented in the form of a frequency distribution table are called **grouped data.**

> **FREQUENCY DISTRIBUTION FOR QUANTITATIVE DATA** A *frequency distribution* for quantitative data lists all the classes and the number of values that belong to each class. Data presented in the form of a frequency distribution are called *grouped data.*

To find the midpoint of the upper limit of the first class and the lower limit of the second class in Table 2.7, we divide the sum of these two limits by 2. Thus, this midpoint is

$$\frac{400 + 401}{2} = 400.5$$

The value 400.5 is called the *upper boundary* of the first class and the *lower boundary* of the second class. By using this technique, we can convert the class limits of Table 2.7 to **class boundaries**, which are also called *real class limits*. The second column of Table 2.8 lists the boundaries for Table 2.7.

> **CLASS BOUNDARY** The *class boundary* is given by the midpoint of the upper limit of one class and the lower limit of the next class.

Table 2.8 Class Boundaries, Class Widths, and Class Midpoints for Table 2.7

Class Limits	Class Boundaries	Class Width	Class Midpoint
301 to 400	300.5 to less than 400.5	100	350.5
401 to 500	400.5 to less than 500.5	100	450.5
501 to 600	500.5 to less than 600.5	100	550.5
601 to 700	600.5 to less than 700.5	100	650.5
701 to 800	700.5 to less than 800.5	100	750.5
801 to 900	800.5 to less than 900.5	100	850.5

The difference between the two boundaries of a class gives the **class width**. The class width is also called the **class size**.

> **CLASS WIDTH**
>
> Class width = Upper boundary − Lower boundary

Thus, in Table 2.8,

$$\text{Width of the first class} = 400.5 - 300.5 = 100$$

The class widths for the frequency distribution of Table 2.7 are listed in the third column of Table 2.8. Each class in Table 2.8 (and Table 2.7) has the same width of 100.

The **class midpoint** or **mark** is obtained by dividing the sum of the two limits (or the two boundaries) of a class by 2.

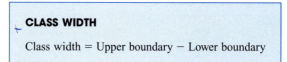

> **CLASS MIDPOINT OR MARK**
>
> $$\text{Class midpoint or mark} = \frac{\text{Lower limit} + \text{Upper limit}}{2}$$

Thus, the midpoint of the first class in Table 2.7 or Table 2.8 is calculated as follows:

$$\text{Midpoint of the first class} = \frac{301 + 400}{2} = 350.5$$

The class midpoints for the frequency distribution of Table 2.7 are listed in the fourth column of Table 2.8.

Note that in Table 2.8, when we write classes using class boundaries, we write *to less than* in order to ensure that each value belongs to one and only one class. As we can see, the upper boundary of the preceding class and the lower boundary of the succeeding class are the same.

2.3.2 CONSTRUCTING FREQUENCY DISTRIBUTION TABLES

While constructing a frequency distribution table, we need to make the following three major decisions.

Number of Classes

Usually the number of classes for a frequency distribution table varies from 5 to 20, depending mainly on the number of observations in the data set.[1] It is preferable to have more classes as the size of a data set increases. The decision about the number of classes is arbitrarily made by the data organizer.

Class Width

Although it is not uncommon to have classes of different sizes, most of the time it is preferable to have the same width for all classes. To determine the class width when all classes are the same size, first find the difference between the largest and the smallest values in the data. Then, the approximate width of a class is obtained by dividing this difference by the number of desired classes.

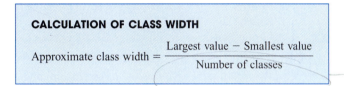

CALCULATION OF CLASS WIDTH

$$\text{Approximate class width} = \frac{\text{Largest value} - \text{Smallest value}}{\text{Number of classes}}$$

Usually this approximate class width is rounded to a convenient number, which is then used as the class width. Note that rounding this number may slightly change the number of classes initially intended.

Lower Limit of the First Class or the Starting Point

Any convenient number that is equal to or less than the smallest value in the data set can be used as the lower limit of the first class.

Example 2–3 illustrates the procedure for constructing a frequency distribution table for quantitative data.

Constructing a frequency distribution table for quantitative data.

EXAMPLE 2–3   Table 2.9 on the next page gives the 1999 total payrolls (rounded to millions) for all 30 major league baseball teams. Construct a frequency distribution table.

Solution   In these data, the minimum value is 15 and the maximum value is 92. Suppose we decide to group these data using five classes of equal width. Then,

$$\text{Approximate width of a class} = \frac{92 - 15}{5} = 15.4$$

Now we round this approximate width to a convenient number—say, 16. The lower limit of the first class can be taken as 15 or any number less than 15. Suppose we take 15 as the lower limit of the first class. Then our classes will be

$$15\text{–}30, \quad 31\text{–}46, \quad 47\text{–}62, \quad 63\text{–}78, \quad \text{and} \quad 79\text{–}94$$

---

[1]One rule to help decide on the number of classes is Sturge's formula:

$$c = 1 + 3.3 \log n$$

where c is the number of classes and n is the number of observations in the data set. The value of "$\log n$" can be obtained by entering the value of n on the calculator and pressing the *log* key.

Table 2.9 Total Payrolls of Major League Baseball Teams for 1999

Team	Total Payroll (million dollars)	Team	Total Payroll (million dollars)
Anaheim	51	Milwaukee	43
Arizona	70	Minnesota	16
Atlanta	79	Montreal	15
Baltimore	75	New York Mets	72
Boston	72	New York Yankees	92
Chicago Cubs	55	Oakland	25
Chicago White Sox	25	Philadelphia	30
Cincinnati	38	Pittsburgh	24
Cleveland	74	St. Louis	46
Colorado	54	San Diego	47
Detroit	37	San Francisco	46
Florida	15	Seattle	45
Houston	56	Tampa Bay	38
Kansas City	17	Texas	81
Los Angeles	77	Toronto	49

We record these five classes in the first column of Table 2.10.

Now we read each value from the given data and make a tally mark in the second column of Table 2.10 next to the corresponding class. The first value in our original data is 51, which belongs to the 47–62 class. To record it, we make a tally mark in the second column next to the 47–62 class. We continue this process until all the data values have been read and entered in the tally column. Note that tallies are marked in blocks of fives for counting convenience. After the tally column is completed, we count the tally marks for each class and write those numbers in the third column. This gives the column of frequencies. These frequencies represent the number of teams that belong to each of the five different classes representing the total payroll. For example, 8 of the 30 major league baseball teams had a total payroll of $15–$30 million in 1999.

Table 2.10 Frequency Distribution for Table 2.9

Total Payroll (million dollars)	Tally	f
15–30	∭ ∣∣∣	8
31–46	∭ ∣∣	7
47–62	∭ ∣	6
63–78	∭ ∣	6
79–94	∣∣∣	3
		$\Sigma f = 30$

In Table 2.10, we can denote the frequencies of the five classes by f_1, f_2, f_3, f_4, and f_5, respectively. Therefore,

$$f_1 = \text{Frequency of the first class} = 8$$

Similarly,

$$f_2 = 7, \quad f_3 = 6, \quad f_4 = 6, \quad \text{and} \quad f_5 = 3$$

Using the Σ notation (see Section 1.8 of Chapter 1), we can denote the sum of the frequencies of all classes by Σf. Hence,

$$\Sigma f = f_1 + f_2 + f_3 + f_4 + f_5 = 8 + 7 + 6 + 6 + 3 = 30$$

The number of observations in a sample is usually denoted by n. Thus, for the sample data, Σf is equal to n. The number of observations in a population is denoted by N. Consequently, Σf is equal to N for population data. Because the data set on the total payrolls of major league baseball teams in Table 2.10 is for all 30 teams, it represents the population. Therefore, in Table 2.10 we can denote the sum of frequencies by N instead of Σf. ■

Note that when we present the data in the form of a frequency distribution table, as in Table 2.10, we lose the information on individual observations. We cannot know the exact payroll of any particular major league baseball team from Table 2.10. All we know is that eight of these teams had a total payroll of $15–$30 million in 1999, and so forth.

2.3.3 RELATIVE FREQUENCY AND PERCENTAGE DISTRIBUTIONS

Using Table 2.10, we can compute the relative frequency and percentage distributions the same way we did for qualitative data in Section 2.2.2. The relative frequencies and percentages for a quantitative data set are obtained as follows.

RELATIVE FREQUENCY AND PERCENTAGE

$$\text{Relative frequency of a class} = \frac{\text{Frequency of that class}}{\text{Sum of all frequencies}} = \frac{f}{\Sigma f}$$

$$\text{Percentage} = (\text{Relative frequency}) \cdot 100$$

Example 2–4 illustrates how to construct relative frequency and percentage distributions.

Constructing relative frequency and percentage distributions.

EXAMPLE 2–4   Calculate the relative frequencies and percentages for Table 2.10.

Solution   The relative frequencies and percentages for the data in Table 2.10 are calculated and listed in the third and fourth columns, respectively, of Table 2.11. Note that the class boundaries are listed in the second column of Table 2.11.

Table 2.11   Relative Frequency and Percentage Distributions for Table 2.10

Total Payroll (million dollars)	Class Boundaries	Relative Frequency	Percentage
15–30	14.5 to less than 30.5	.267	26.7
31–46	30.5 to less than 46.5	.233	23.3
47–62	46.5 to less than 62.5	.200	20.0
63–78	62.5 to less than 78.5	.200	20.0
79–94	78.5 to less than 94.5	.100	10.0
		Sum = 1.0	Sum = 100%

Using Table 2.11, we can make statements about the percentage of teams with payrolls within a certain interval. For example, about 26.7% of the major league baseball teams in this population had a total payroll of $15–$30 million in 1999. By adding the percentages for the first two classes, we can state that 50% of these teams had a total payroll of $15–$46 million in 1999. Similarly, by adding the percentages of the last two classes, we can state that 30% of these teams had a total payroll of $63–$94 million in 1999.   ■

2.3.4   GRAPHING GROUPED DATA

Grouped (quantitative) data can be displayed in a *histogram* or a *polygon*. This section describes how to construct such graphs. We can also draw a pie chart to display the percentage distribution for a quantitative data set. The procedure to construct a pie chart is similar to the one for qualitative data explained in Section 2.2.3; it will not be repeated in this section.

Histograms

A **histogram** is a certain kind of graph that can be drawn for a frequency distribution, a relative frequency distribution, or a percentage distribution. To draw a histogram, we first mark classes on the horizontal axis and frequencies (or relative frequencies or percentages) on the vertical axis. Next, we draw a bar for each class so that its height represents the frequency of that class. The bars in a histogram are drawn adjacent to each other with no gap between them. A histogram is called a **frequency histogram**, a **relative frequency histogram**, or a **percentage histogram** depending on whether frequencies, relative frequencies, or percentages are marked on the vertical axis.

> **HISTOGRAM**   A *histogram* is a graph in which classes are marked on the horizontal axis and the frequencies, relative frequencies, or percentages are marked on the vertical axis. The frequencies, relative frequencies, or percentages are represented by the heights of the bars. In a histogram, the bars are drawn adjacent to each other.

Figures 2.3 and 2.4 show the frequency and the relative frequency histograms, respectively, for the data of Tables 2.10 and 2.11 of Sections 2.3.2 and 2.3.3. The two histograms look alike because they represent the same data. A percentage histogram can be drawn for the percentage distribution of Table 2.11 by marking the percentages on the vertical axis.

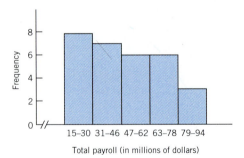

Figure 2.3   Frequency histogram for Table 2.10.

The symbol –//– used in the horizontal axes of Figures 2.3 and 2.4 represents a break, called the **truncation**, in the horizontal axis. It indicates that the entire horizontal axis is not shown in these figures. Notice that the 0 to 14 portion of the horizontal axis has been omitted in each figure.

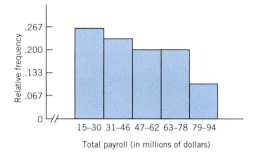

Figure 2.4 Relative frequency histogram for Table 2.11.

In Figures 2.3 and 2.4, we have used class limits to mark classes on the horizontal axis. However, we can show the intervals on the horizontal axis by using the class boundaries instead of the class limits.

CASE STUDY 2-3   BOOMERS GO FISH

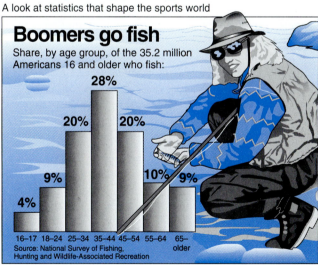

The chart, reproduced from *USA TODAY*, gives the percentage distribution of Americans age 16 and older who fish. As can be seen from the chart, 4% of those who go fishing are 16–17-years-old, 9% are 18–24-years-old, and so forth. Note that the classes in the chart have different widths. For example, class 16–17 has a width of two years, 18–24 has a width equal to seven years, and so on. Also, notice that the last class is *65 and older*. This class does not have an upper limit. Such classes are called **open-ended classes**.

Source: *USA TODAY*, July 29, 1998. Copyright © 1998, *USA TODAY*. Chart reproduced with permission.

Polygons

A **polygon** is another device that can be used to present quantitative data in graphic form. To draw a **frequency polygon**, we first mark a dot above the midpoint of each class at a height equal to the frequency of that class. This is the same as marking the midpoint at the top of each bar in a histogram. Next, we mark two more classes, one at each end, and mark their midpoints. Note that these two classes have zero frequencies. In the last step, we join the adjacent dots with straight lines. The resulting line graph is called a frequency polygon or simply a polygon.

A polygon with relative frequencies marked on the vertical axis is called a *relative frequency polygon*. Similarly, a polygon with percentages marked on the vertical axis is called a *percentage polygon*.

> **POLYGON**   A graph formed by joining the midpoints of the tops of successive bars in a histogram with straight lines is called a *polygon*.

Figure 2.5 shows the frequency polygon for the frequency distribution of Table 2.10.

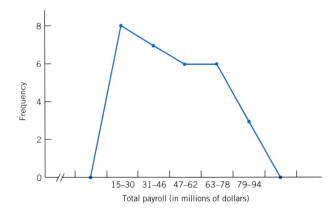

Figure 2.5   Frequency polygon for Table 2.10.

For a very large data set, as the number of classes is increased (and the width of classes is decreased), the frequency polygon eventually becomes a smooth curve. Such a curve is called a *frequency distribution curve* or simply a *frequency curve*. Figure 2.6 shows the frequency curve for a large data set with a large number of classes.

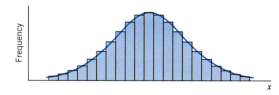

Figure 2.6 Frequency distribution curve.

2.3.5 MORE ON CLASSES AND FREQUENCY DISTRIBUTIONS

This section presents two alternative methods for writing classes to construct a frequency distribution for quantitative data.

Less Than Method for Writing Classes

The classes in the frequency distribution given in Table 2.10 for the data on payrolls of base-ball teams were written as 15–30, 31–46, and so on. Alternatively, we can write the classes in a frequency distribution table using the *less than* method. The technique for writing classes shown in Table 2.10 is more commonly used for data sets that do not contain fractional values. The *less than* method is more appropriate when a data set contains fractional values. Example 2–5 illustrates the *less than* method.

Constructing a frequency distribution using the less than method.

EXAMPLE 2–5 The following data are the gasoline tax rates (in cents per gallon) as of January 1, 1999, for 50 states[2] (U.S. Federal Highway Administration).

18.00	8.00	18.00	18.70	18.00	22.00	32.00	23.00	13.10	7.50
16.00	26.00	19.30	15.00	20.00	18.00	16.40	20.00	19.00	23.50
21.00	19.00	20.00	18.40	17.05	27.00	24.40	24.00	18.70	10.50
18.00	8.00	21.60	20.00	22.00	17.00	24.00	30.77	29.00	16.00
21.00	21.00	20.00	24.75	20.00	17.50	23.00	25.35	25.40	14.00

Construct a frequency distribution table. Calculate the relative frequencies and percentages for all classes.

Solution   The minimum value in this data set is 7.50 and the maximum value is 32.00. Suppose we decide to group these data using six classes of equal width. Then,

$$\text{Approximate width of a class} = \frac{32.00 - 7.50}{6} = 4.08$$

We round this number to a more convenient number—say, 5. Then we take 5 as the width of each class. If we start the first class at 5, the classes are written as 5 *to less than* 10, 10 *to less than* 15, and so on. The six classes, which cover all the data, are recorded in the first column of Table 2.12 on the next page. The second column lists the frequencies of the classes. A value in the data set that is 5 or larger but less than 10 belongs to the first class, a value that is 10 or larger but less than 15 falls in the second class, and so on. The relative fre-

[2]The data for the 50 states are entered (by row) in the following order: Alabama, Alaska, Arizona, Arkansas, California, Colorado, Connecticut, Delaware, Florida, Georgia, Hawaii, Idaho, Illinois, Indiana, Iowa, Kansas, Kentucky, Louisiana, Maine, Maryland, Massachusetts, Michigan, Minnesota, Mississippi, Missouri, Montana, Nebraska, Nevada, New Hampshire, New Jersey, New Mexico, New York, North Carolina, North Dakota, Ohio, Oklahoma, Oregon, Pennsylvania, Rhode Island, South Carolina, South Dakota, Tennessee, Texas, Utah, Vermont, Virginia, Washington, West Virginia, Wisconsin, Wyoming.

quencies and percentages for classes are recorded in the third and fourth columns, respectively, of Table 2.12. Note that the table does not contain a column of tallies.

Table 2.12 Gasoline Tax Rates for 50 States

Gasoline Tax Rate (cents per gallon)	f	Relative Frequency	Percentage
5 to less than 10	3	.06	6
10 to less than 15	3	.06	6
15 to less than 20	18	.36	36
20 to less than 25	19	.38	38
25 to less than 30	5	.10	10
30 to less than 35	2	.04	4
	$\Sigma f = 50$	Sum = 1.00	Sum = 100%

A histogram and a polygon for the data of Table 2.12 can be drawn in the same way as for the data of Tables 2.10 and 2.11.

Single-Valued Classes

If the observations in a data set assume only a few distinct values, it may be appropriate to prepare a frequency distribution table using *single-valued classes*—that is, classes that are made of single values and not of intervals. This technique is especially useful in cases of discrete data with only a few possible values. Example 2–6 exhibits such a situation.

Constructing a frequency distribution using single-valued classes.

EXAMPLE 2–6   The administration in a large city wanted to know the distribution of vehicles owned by households in that city. A sample of 40 randomly selected households from this city produced the following data on the number of vehicles owned:

$$
\begin{array}{cccccccccc}
5 & 1 & 1 & 2 & 0 & 1 & 1 & 2 & 1 & 1 \\
1 & 3 & 3 & 0 & 2 & 5 & 1 & 2 & 3 & 4 \\
2 & 1 & 2 & 2 & 1 & 2 & 2 & 1 & 1 & 1 \\
4 & 2 & 1 & 1 & 2 & 1 & 1 & 4 & 1 & 3
\end{array}
$$

Construct a frequency distribution table for these data using single-valued classes.

Solution   The observations in this data set assume only six distinct values: 0, 1, 2, 3, 4, and 5. Each of these six values is used as a class in the frequency distribution in Table 2.13, and these six classes are listed in the first column of that table. To obtain the frequencies of these classes, the observations in the data that belong to each class are counted, and the results are recorded in the second column of Table 2.13. Thus, in these data, 2 households own no vehicle, 18 own one vehicle each, 11 own two vehicles each, and so on.

Table 2.13 Frequency Distribution of Vehicles Owned

Vehicles Owned	Number of Households (f)
0	2
1	18
2	11
3	4
4	3
5	2
	$\Sigma f = 40$

The data of Table 2.13 can also be displayed in a bar graph, as shown in Figure 2.7. To construct a bar graph, we mark the classes, as intervals, on the horizontal axis with a little gap between consecutive intervals. The bars represent the frequencies of respective classes.

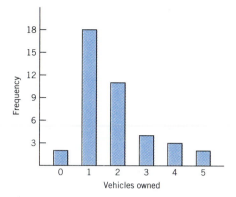

Figure 2.7 Bar graph for Table 2.13.

The frequencies of Table 2.13 can be converted to relative frequencies and percentages the same way as in Table 2.11. Then, a bar graph can be constructed to display the relative frequency or percentage distribution by marking the relative frequencies or percentages, respectively, on the vertical axis.

2.4 SHAPES OF HISTOGRAMS

A histogram can assume any one of a large number of shapes. The most common of these shapes are

1. Symmetric
2. Skewed
3. Uniform or rectangular

A **symmetric histogram** is identical on both sides of its central point. The histograms shown in Figure 2.8 are symmetric around the dashed lines that represent their central points.

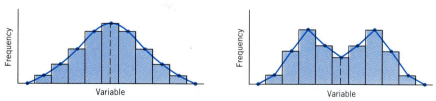

Figure 2.8 Symmetric histograms.

A **skewed histogram** is nonsymmetric. For a skewed histogram, the tail on one side is longer than the tail on the other side. A **skewed-to-the-right histogram** has a longer tail on the right side (see Figure 2.9a on the next page). A **skewed-to-the-left histogram** has a longer tail on the left side (see Figure 2.9b).

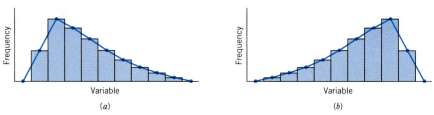

Figure 2.9 (*a*) A histogram skewed to the right. (*b*) A histogram skewed to the left.

A **uniform** or **rectangular histogram** has the same frequency for each class. Figure 2.10 is an illustration of such a case.

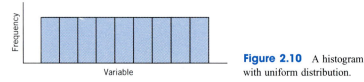

Figure 2.10 A histogram with uniform distribution.

Figures 2.11*a* and 2.11*b* display symmetric frequency curves. Figures 2.11*c* and 2.11*d* show frequency curves skewed to the right and to the left, respectively.

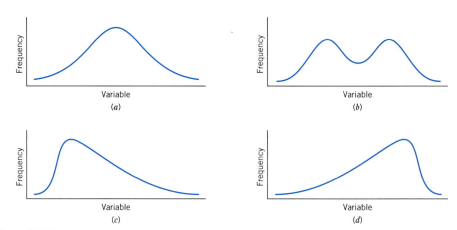

Figure 2.11 (*a*) and (*b*) Symmetric frequency curves. (*c*) Frequency curve skewed to the right. (*d*) Frequency curve skewed to the left.

☞ *Warning* Describing data using graphs helps to give us insights into the main characteristics of the data. But graphs, unfortunately, can also be used, intentionally or unintentionally, to distort the facts and to deceive the reader. The following are two ways to manipulate graphs to convey a particular opinion or impression.

1. *Changing the scale* either on one or on both axes—that is, shortening or stretching one or both of the axes.

2. *Truncating the frequency axis*—that is, starting the frequency axis at a number greater than zero

When interpreting a graph, we should be very cautious. We should observe carefully whether the frequency axis has been truncated or whether any axis has been unnecessarily shortened or stretched.

CASE STUDY 2-4   USING TRUNCATED AXES

In April 1999, Yankelovich Partners Inc. conducted a telephone poll of 13–17-year-old American teenagers for *Time*/CNN. One of the questions these teenagers were asked was: "What do your parents know about the websites you visit?" The responses of the teenagers who used the Internet are summarized in the following table.

Response	Percentage
A lot	38
A little	45
Nothing	17

Source: Time, May 10, 1999.

We constructed two bar graphs for the percentage distribution given in the table. The figure on the left side shows the complete vertical axis, and the one on the right side shows a truncated vertical axis. As is obvious, the two graphs give completely different impressions of the results of the survey and of the differences in the percentages for the various categories.

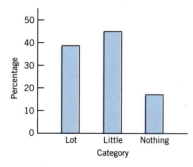 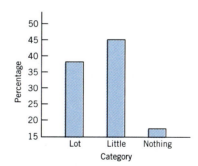

EXERCISES

■ Concepts and Procedures

2.11   Briefly explain the three decisions that have to be made to group a data set in the form of a frequency distribution table.

2.12   How are the relative frequencies and percentages of classes obtained from the frequencies of classes? Illustrate with the help of an example.

2.13   Three methods—writing classes using limits, using the *less than* method, and grouping data using single-valued classes—were discussed to group quantitative data into classes. Explain these three methods and give one example of each.

■ Applications

2.14   The following table gives the frequency distribution of weekly earnings for a sample of 100 employees selected from a large company.

Weekly Earnings (dollars)	Number of Employees
200 to 349	12
350 to 499	23
500 to 649	34
650 to 799	19
800 to 949	12

a. Find the class boundaries and class midpoints.
b. Do all classes have the same width? If yes, what is that width?
c. Prepare the relative frequency and percentage distribution columns.
d. What percentage of these employees earn $650 or more per week?

2.15 The following table gives the frequency distribution of ages for all 50 employees of a company.

Age	Number of Employees
18 to 30	12
31 to 43	19
44 to 56	14
57 to 69	5

a. Find the class boundaries and class midpoints.
b. Do all classes have the same width? If yes, what is that width?
c. Prepare the relative frequency and percentage distribution columns.
d. What percentage of the employees of this company are age 43 or younger?

2.16 A data set on money spent on lottery tickets during the past year by 200 households has a lowest value of $1 and a highest value of $1167. Suppose we want to group these data into six classes of equal widths.

a. Assuming we take the lower limit of the first class as $1 and the width of each class equal to $200, write the class limits for all six classes.
b. What are the class boundaries and class midpoints?

2.17 A data set on weekly expenditures (rounded to the nearest dollar) on bakery products for a sample of 500 households has a minimum value of $1 and a maximum value of $18. Suppose we want to group these data into five classes of equal widths.

a. Assuming we take the lower limit of the first class as $1 and the upper limit of the fifth class as $20, write the class limits for all five classes.
b. Determine the class boundaries and class widths.
c. Find the class midpoints.

2.18 The following table gives estimates of the average numbers of hours spent annually in gridlock on the road by motorists in 20 U.S. cities.

City	Hours	City	Hours
Los Angeles	82	Austin, TX	53
Washington	76	Portland–Vancouver	52
Seattle–Everett	69	St. Louis	52
Atlanta	68	Indianapolis	52
Boston	68	San Bernardino	47
Detroit	62	Baltimore	47
San Francisco	58	Nashville	46
Houston	58	San Jose, CA	45
Dallas	58	Denver	45
Miami–Hialeah	57	Chicago	44

Source: Texas Transportation Institute, *USA TODAY*, November 17, 1999.

a. Construct a frequency distribution table. Take the classes as 44–50, 51–57, 58–64, 65–71, 72–78, and 79–85.
b. Calculate the relative frequencies and percentages for all classes.
c. Based on the frequency distribution, can you say whether the data are symmetric or skewed?
d. In what percentage of these cities do the motorists spend an average of 65 hours or more annually in gridlock?

2.19 Nixon Corporation manufactures computer terminals. The following data are the numbers of computer terminals produced at the company for a sample of 30 days.

24	32	27	23	33	33	29	25	23	28
21	26	31	22	27	33	27	23	28	29
31	35	34	22	26	28	23	35	31	27

a. Construct a frequency distribution table using the classes 21–23, 24–26, 27–29, 30–32, and 33–35.
b. Calculate the relative frequencies and percentages for all classes.
c. Construct a histogram and a polygon for the percentage distribution.
d. For what percentage of the days is the number of computer terminals produced in the interval 27–29?

2.20 The following data give the numbers of computer keyboards assembled at the Twentieth Century Electronics Company for a sample of 25 days.

45	52	48	41	56	46	44	42	48	53	51	53	51
48	46	43	52	50	54	47	44	47	50	49	52	

Here is the frequency distribution table for these data.

Classes	Frequency
41–44	5
45–48	8
49–52	8
53–56	4

a. Calculate the relative frequencies for all classes.
b. Construct a histogram for the relative frequency distribution.
c. Construct a polygon for the relative frequency distribution.

2.21 The following data give the numbers of people who were unemployed (rounded to the nearest thousand) as of June 1998 in 20 northeastern and midwestern states (U.S. Bureau of Labor Statistics).

65	278	88	40	50	26	111	181	70	123
16	18	202	8	261	262	21	11	12	87

a. Construct a frequency distribution table. Take the classes as 1–60, 61–120, and so on.
b. Prepare the relative frequency and percentage columns for the table in part a.

Exercises 2.22 through 2.26 are based on the following data.

The table on the next page, based on the American Chamber of Commerce Researchers Association Survey for the second quarter of 1999, gives the prices of five items (in dollars) in 25 urban areas across the United States. (See Data Set I that accompanies the text.)

City	Apartment Rent	Price of House	Phone Bill	Cost of Hospital Room	Price of Wine
Huntsville (AL)	$ 502	$119,280	$22.05	$380.00	$6.04
Anchorage (AK)	788	187,337	14.58	766.50	6.28
Phoenix (AZ)	664	141,484	19.21	580.69	5.29
Little Rock (AR)	594	116,675	28.10	282.00	4.99
San Diego (CA)	936	244,000	16.89	741.60	4.85
Denver (CO)	740	179,740	21.05	643.71	5.16
Orlando (FL)	655	132,624	20.21	515.50	4.89
Bloomington (IN)	574	131,848	21.30	499.00	4.89
Wichita (KN)	515	131,008	22.80	574.75	6.27
Boston (MA)	1264	246,975	23.00	694.00	6.28
Minneapolis (MN)	663	141,080	22.99	830.17	5.43
Lincoln (NE)	477	169,687	22.64	398.33	4.73
Reno–Sparks (NV)	676	175,912	15.75	610.33	4.97
Manchester (NH)	882	167,450	21.13	479.00	4.99
Albuquerque (NM)	741	140,875	23.66	371.25	6.49
Buffalo (NY)	542	123,095	25.55	526.80	5.89
Charlotte (NC)	548	148,500	17.97	405.67	5.58
Cincinnati (OH)	665	137,783	22.85	418.20	6.49
Salem (OR)	561	169,823	19.92	415.00	4.95
Sioux Falls (SD)	602	128,040	23.29	454.50	5.84
Memphis (TN)	642	124,121	19.84	277.60	6.19
Houston (TX)	694	116,244	18.54	459.55	5.95
Cedar City (UT)	435	111,000	20.60	428.20	5.95
Charleston (WV)	602	142,660	28.08	267.25	6.32
Green Bay (WI)	545	142,500	17.77	367.33	4.81

Explanation of variables

Apartment rent   Monthly rent of an unfurnished two-bedroom apartment (excluding all utilities except water), 1½ or 2 baths, approximately 950 square feet

Price of house   Purchase price of a new house with 1800 square feet of living area, on an 8000-square-foot lot in an urban area with all utilities

Phone bill   Monthly telephone charges for a private residential line (customer owns instruments)

Cost of hospital room   Average cost per day of a semiprivate room in a hospital

Price of wine   Price of Paul Masson Chablis, 1.5-liter bottle

2.22   **a.**   Prepare a frequency distribution table for apartment rents using five classes of equal widths.
   b.   Construct the relative frequency and percentage distribution columns.
   c.   Write the class midpoints.

2.23   **a.**   Prepare a frequency distribution table for the purchase prices of houses using five classes of equal widths.
   b.   Calculate the relative frequencies and percentages for all classes.
   c.   Write the class midpoints.

2.24   **a.**   Prepare a frequency distribution table for phone bills.
   b.   Construct the relative frequency and percentage distribution columns.
   c.   Draw a histogram and a polygon for the relative frequency distribution.

2.25   **a.**   Prepare a frequency distribution table for the average costs per day of rooms in hospitals.
   b.   Calculate the relative frequencies and percentages for all classes.
   c.   Draw a histogram and a polygon for percentage distribution.

2.26   **a.**   Prepare a frequency distribution table for the prices of wine. Take $4.50 as the lower boundary of the first class and $.75 as the width of each class.
   b.   Construct the relative frequency and percentage distribution columns.

2.27 The following data set includes the pay (in thousands of dollars, not including benefits) for 20 presidents of private U.S. colleges and universities in 1996–1997 (*The Chronicle of Higher Education*, October 23, 1998).

397	430	352	335	360	499	342	385	472	350
358	428	338	340	401	373	350	353	357	351

a. Construct a frequency distribution table. Take $300 thousand as the lower limit of the first class and $50 thousand as the width of each class.

b. Prepare the relative frequency and percentage columns for the table in part a.

2.28 The following data give the number of women who held statewide elective executive offices in each of the 50 states in January 1997 (*Statistical Abstract of the United States*, 1998). The data are entered (by row) for the 50 states arranged in alphabetical order.

3	1	4	2	2	3	2	4	1	1
1	2	2	3	1	3	0	1	0	1
0	2	3	0	2	2	2	1	1	1
1	1	1	4	2	4	1	1	1	1
4	1	1	2	0	0	4	0	0	2

a. Prepare a frequency distribution table for these data using single-valued classes.

b. Calculate the relative frequencies and percentages for all classes.

c. How many states had three or four women in statewide elective executive offices in 1997?

d. Draw a bar graph for the frequency distribution in part a.

2.29 The following data give the numbers of children less than 18 years of age for 30 randomly selected families.

2	1	2	0	3	1	1	2	2	1
1	2	0	1	0	2	1	2	0	0
1	0	0	2	1	2	3	2	0	1

a. Prepare a frequency distribution table for these data using single-valued classes.

b. Calculate the relative frequencies and percentages for all classes.

c. How many families in this sample have two or three children under 18 years of age?

d. Draw a bar graph for the frequency distribution.

2.30 The following table gives the frequency distribution for the numbers of parking tickets received on the campus of a university during the past week for 200 students.

Number of Tickets	Number of Students
0	59
1	44
2	37
3	32
4	28

Draw two bar graphs for these data, the first without truncating the frequency axis and the second by truncating the frequency axis. In the second case, mark the frequencies on the vertical axis starting with 25. Briefly comment on the two bar graphs.

2.31 The following table lists the frequency distribution for the scores of students in a class.

Score	Frequency
61–70	12
71–80	13
81–90	22
91–100	15

Draw two histograms for these data, the first without truncating the frequency axis and the second by truncating the frequency axis. In the second case, mark the frequencies on the vertical axis starting with 10. Briefly comment on the two histograms.

2.5 CUMULATIVE FREQUENCY DISTRIBUTIONS

Consider again Example 2–3 of Section 2.3.2 about the payrolls of major league baseball teams. Suppose we want to know how many teams had payrolls of $62 million or less. Such a question can be answered using a **cumulative frequency distribution**. Each class in a cumulative frequency distribution table gives the total number of values that fall below a certain value. A cumulative frequency distribution is constructed for quantitative data only.

> **CUMULATIVE FREQUENCY DISTRIBUTION**   A *cumulative frequency distribution* gives the total number of values that fall below the upper boundary of each class.

In a *less than* cumulative frequency distribution table, each class has the same lower limit but a different upper limit. Example 2–7 illustrates the procedure to prepare a cumulative frequency distribution.

Constructing a cumulative frequency distribution table.

EXAMPLE 2–7   Using the frequency distribution of Table 2.10, reproduced here, prepare a cumulative frequency distribution for the 1999 payrolls of major league baseball teams.

Total Payroll (million dollars)	f
15–30	8
31–46	7
47–62	6
63–78	6
79–94	3

Solution   Table 2.14 gives the cumulative frequency distribution for the payrolls of major league baseball teams. As we can observe, 15 (which is the lower limit of the first class in Table 2.10) is taken as the lower limit of each class in Table 2.14. The upper limits of all classes in Table 2.14 are the same as those in Table 2.10. To obtain the cumulative frequency of a class, we add the frequency of that class in Table 2.10 to the frequencies of all preceding classes. The cumulative frequencies are recorded in the third column of Table 2.14. The second column of this table lists the class boundaries.

Table 2.14   Cumulative Frequency Distribution of Payrolls of Baseball Teams

Class Limits	Class Boundaries	Cumulative Frequency
15–30	14.5 to less than 30.5	8
15–46	14.5 to less than 46.5	8 + 7 = 15
15–62	14.5 to less than 62.5	8 + 7 + 6 = 21
15–78	14.5 to less than 78.5	8 + 7 + 6 + 6 = 27
15–94	14.5 to less than 94.5	8 + 7 + 6 + 6 + 3 = 30

From Table 2.14, we can determine the number of observations that fall below the upper boundary of each class. For example, 21 major league baseball teams had 1999 total payrolls of $62 million or less. ∎

The **cumulative relative frequencies** are obtained by dividing the cumulative frequencies by the total number of observations in the data set. The **cumulative percentages** are obtained by multiplying the cumulative relative frequencies by 100.

CUMULATIVE RELATIVE FREQUENCY AND CUMULATIVE PERCENTAGE

$$\text{Cumulative relative frequency} = \frac{\text{Cumulative frequency of a class}}{\text{Total observations in the data set}}$$

$$\text{Cumulative percentage} = (\text{Cumulative relative frequency}) \cdot 100$$

Table 2.15 contains both the cumulative relative frequencies and the cumulative percentages for Table 2.14. We can observe, for example, that 70% of the major league baseball teams had 1999 payrolls of $62 million or less.

Table 2.15 Cumulative Relative Frequency and Cumulative Percentage Distributions for Payrolls of Baseball Teams

Class Limits	Cumulative Relative Frequency	Cumulative Percentage
15–30	8/30 = .267	26.7
15–46	15/30 = .500	50.0
15–62	21/30 = .700	70.0
15–78	27/30 = .900	90.0
15–94	30/30 = 1.000	100.0

Ogives

When plotted on a diagram, the cumulative frequencies give a curve that is called an **ogive** (pronounced *o-jive*). Figure 2.12 gives an ogive for the cumulative frequency distribution of Table 2.14. To draw the ogive in Figure 2.12, the variable, total payroll, is marked on the horizontal axis and the cumulative frequencies on the vertical axis. Then, the dots are marked above the upper boundaries of various classes at the heights equal to the corresponding cumulative frequencies. The ogive is obtained by joining consecutive points with straight lines. Note that the ogive starts at the lower boundary of the first class and ends at the upper boundary of the last class.

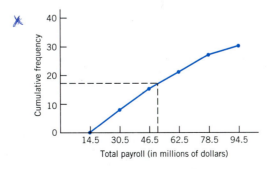

Figure 2.12 Ogive for the cumulative frequency distribution in Table 2.14.

> **OGIVE** An *ogive* is a curve drawn for the cumulative frequency distribution by joining with straight lines the dots marked above the upper boundaries of classes at heights equal to the cumulative frequencies of respective classes.

One advantage of an ogive is that it can be used to approximate the cumulative frequency for any interval. For example, we can use Figure 2.12 to find the number of major league baseball teams with payrolls of $50 million or less. First, draw a vertical line from 50 on the horizontal axis up to the ogive. Then draw a horizontal line from the point where this line intersects the ogive to the vertical axis. This point gives the cumulative frequency of the class 15–50. In Figure 2.12, this cumulative frequency is 17. Therefore, 17 baseball teams had 1999 payrolls of $50 million or less.

We can draw an ogive for cumulative relative frequency and cumulative percentage distributions in the same way we did for the cumulative frequency distribution.

EXERCISES

■ **Concepts and Procedures**

2.32 Briefly explain the concept of cumulative frequency distribution. How are the cumulative relative frequencies and cumulative percentages calculated?

2.33 Explain for what kind of frequency distribution an ogive is drawn. Can you think of any use for an ogive? Explain.

■ **Applications**

2.34 The following table, reproduced from Exercise 2.14, gives the frequency distribution of weekly earnings for a sample of 100 employees selected from a company.

Weekly Earnings (dollars)	Number of Employees
200 to 349	12
350 to 499	23
500 to 649	34
650 to 799	19
800 to 949	12

a. Prepare a cumulative frequency distribution table.
b. Calculate the cumulative relative frequencies and cumulative percentages for all classes.
c. What percentage of the workers earned $649 or less per week?
d. Draw an ogive for the cumulative percentage distribution.
e. Using the ogive, find the percentage of workers who earned $600 or less per week.

2.35 The following table, reproduced from Exercise 2.15, gives the frequency distribution of ages for all 50 employees of a company.

Age	Number of Employees
18 to 30	12
31 to 43	19
44 to 56	14
57 to 69	5

a. Prepare a cumulative frequency distribution table.
b. Calculate the cumulative relative frequencies and cumulative percentages for all classes.
c. What percentage of the employees of this company are 44 years of age or older?
d. Draw an ogive for the cumulative percentage distribution.
e. Using the ogive, find the percentage of employees who are age 40 or younger.

2.36 Using the frequency distribution table constructed in Exercise 2.18, prepare the cumulative frequency, cumulative relative frequency, and cumulative percentage distributions.

2.37 Using the frequency distribution table constructed in Exercise 2.19, prepare the cumulative frequency, cumulative relative frequency, and cumulative percentage distributions.

2.38 Using the frequency distribution table constructed in Exercise 2.20, prepare the cumulative frequency, cumulative relative frequency, and cumulative percentage distributions.

2.39 Prepare the cumulative frequency, cumulative relative frequency, and cumulative percentage distributions using the frequency distribution constructed in Exercise 2.23.

2.40 Using the frequency distribution table constructed for the data of Exercise 2.25, prepare the cumulative frequency, cumulative relative frequency, and cumulative percentage distributions.

2.41 Refer to the frequency distribution table constructed in Exercise 2.26. Prepare the cumulative frequency, cumulative relative frequency, and cumulative percentage distributions by using that table.

2.42 Using the frequency distribution table constructed for the data of Exercise 2.21, prepare the cumulative frequency, cumulative relative frequency, and cumulative percentage distributions. Draw an ogive for the cumulative frequency distribution. Using the ogive, find the (approximate) number of northeastern and midwestern states for which the number of unemployed people is 160 thousand or less.

2.43 Refer to the frequency distribution table constructed in Exercise 2.27. Prepare the cumulative frequency, cumulative relative frequency, and cumulative percentage distributions. Draw an ogive for the cumulative percentage distribution. Using the ogive, find the (approximate) percentage of presidents in this sample whose pay was $375 thousand or less.

✳ 2.6 STEM-AND-LEAF DISPLAYS

Another technique that is used to present quantitative data in condensed form is the **stem-and-leaf display**. An advantage of a stem-and-leaf display over a frequency distribution is that by preparing a stem-and-leaf display we do not lose information on individual observations. A stem-and-leaf display is constructed only for quantitative data.

> **STEM-AND-LEAF DISPLAY** In a *stem-and-leaf display* of quantitative data, each value is divided into two portions—a stem and a leaf. The leaves for each stem are shown separately in a display.

Example 2–8 describes the procedure for constructing a stem-and-leaf display.

Constructing a stem-and-leaf display for two-digit numbers.

EXAMPLE 2–8 The following are the scores of 30 college students on a statistics test:

75	52	80	96	65	79	71	87	93	95
69	72	81	61	76	86	79	68	50	92
83	84	77	64	71	87	72	92	57	98

Construct a stem-and-leaf display.

Solution   To construct a stem-and-leaf display for these scores, we split each score into two parts. The first part contains the first digit, which is called the *stem*. The second part contains the second digit, which is called the *leaf*. Thus, for the score of the first student, which is 75, 7 is the stem and 5 is the leaf. For the score of the second student, which is 52, the stem is 5 and the leaf is 2. We observe from the data that the stems for all scores are 5, 6, 7, 8, and 9 because all the scores lie in the range 50 to 98. To create a stem-and-leaf display, we draw a vertical line and write the stems on the left side of it, arranged in increasing order, as shown in Figure 2.13.

After we have listed the stems, we read the leaves for all scores and record them next to the corresponding stems on the right side of the vertical line. For example, for the first score we write the leaf 5 next to the stem 7; for the second score we write the leaf 2 next to the stem 5. The recording of these two scores in a stem-and-leaf display is shown in Figure 2.13.

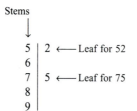

Figure 2.13 Stem-and-leaf display.

Now, we read all the scores and write the leaves on the right side of the vertical line in the rows of corresponding stems. The complete stem-and-leaf display for scores is shown in Figure 2.14.

```
5 | 2 0 7
6 | 5 9 1 8 4
7 | 5 9 1 2 6 9 7 1 2
8 | 0 7 1 6 3 4 7
9 | 6 3 5 2 2 8
```

Figure 2.14 Stem-and-leaf display of test scores.

By looking at the stem-and-leaf display of Figure 2.14, we can observe how the data values are distributed. For example, the stem 7 has the highest frequency, followed by stems 8, 9, 6, and 5.

The leaves of the stem-and-leaf display of Figure 2.14 are *ranked* (in increasing order) and presented in Figure 2.15.

```
5 | 0 2 7
6 | 1 4 5 8 9
7 | 1 1 2 2 5 6 7 9 9
8 | 0 1 3 4 6 7 7
9 | 2 2 3 5 6 8
```

Figure 2.15 Ranked stem-and-leaf display of test scores. ■

As mentioned earlier, one advantage of a stem-and-leaf display is that we do not lose information on individual observations. We can rewrite the individual scores of the 30 college students from the stem-and-leaf display of Figure 2.14 or 2.15. By contrast, the information on individual observations is lost when data are grouped into a frequency table.

Constructing a stem-and-leaf display for three- and four-digit numbers. **EXAMPLE 2-9** The following data are the monthly rents paid by a sample of 30 households selected from a city.

429	585	732	675	550	989	1020	620	750	660
540	578	956	1030	1070	930	871	765	880	975
650	1020	950	840	780	870	900	800	750	820

Construct a stem-and-leaf display for these data.

Solution Each of the values in the data set contains either three or four digits. We will take the first digit for three-digit numbers and the first two digits for four-digit numbers as stems. Then, we will use the last two digits of each number as a leaf. Thus, for the first value, which is 429, the stem is 4 and the leaf is 29. The stems for the entire data set are 4, 5, 6, 7, 8, 9, and 10. They are recorded on the left side of the vertical line in Figure 2.16. The leaves for the numbers are recorded on the right side.

```
 4 | 29
 5 | 85 50 40 78
 6 | 75 20 60 50
 7 | 32 50 65 80 50
 8 | 71 80 40 70 00 20
 9 | 89 56 30 75 50 00
10 | 20 30 70 20
```

Figure 2.16 Stem-and-leaf display of rents. ∎

Sometimes a data set may contain too many stems, with each stem containing only a few leaves. In such cases, we may want to condense the stem-and-leaf display by *grouping the stems*. Example 2–10 describes this procedure.

Preparing a grouped stem-and-leaf display. **EXAMPLE 2-10** The following stem-and-leaf display is prepared for the stock prices of 21 companies.

```
1 | 3 5
2 | 2 5 6
3 | 0 1
4 | 2 3 6
5 | 0
6 | 5 6
7 | 0 3 9
8 | 1 5 7
9 | 2 6
```

Prepare a new stem-and-leaf display by grouping the stems.

Solution To condense the given stem-and-leaf display, we can combine the first three rows, the middle three rows, and the last three rows, thus getting the stems 1–3, 4–6, and 7–9. The leaves for each stem of a group are separated by an asterisk (∗), as shown in Figure 2.17. Thus, the leaves 3 and 5 in the first row correspond to stem 1; the leaves 2, 5, and 6 correspond to stem 2; and leaves 0 and 1 belong to stem 3.

```
1–3 | 3 5 ∗ 2 5 6 ∗ 0 1
4–6 | 2 3 6 ∗ 0 ∗ 5 6
7–9 | 0 3 9 ∗ 1 5 7 ∗ 2 6
```

Figure 2.17 Grouped stem-and-leaf display. ∎

If a stem does not contain a leaf, this can be indicated in the grouped stem-and-leaf display by two consecutive asterisks. For example, in the following stem-and-leaf display there is no leaf for 3; that is, there is no number in the 30s. The numbers in this display are 21, 25, 43, 48, and 50.

$$2-5 \mid 1\ 5\ *\ *\ 3\ 8\ *\ 0$$

EXERCISES

Concepts and Procedures

2.44   Briefly explain how to prepare a stem-and-leaf display for a data set. You may use an example to illustrate.

2.45   What advantage does preparing a stem-and-leaf display have over grouping a data set using a frequency distribution? Give one example.

2.46   Consider this stem-and-leaf display.

$$
\begin{array}{c|l}
4 & 3\ 6 \\
5 & 0\ 1\ 4\ 5 \\
6 & 3\ 4\ 6\ 7\ 7\ 7\ 8\ 9 \\
7 & 2\ 2\ 3\ 5\ 6\ 6\ 9 \\
8 & 0\ 7\ 8\ 9 \\
\end{array}
$$

Write the data set that is represented by the display.

2.47   Consider this stem-and-leaf display.

$$
\begin{array}{c|l}
2-3 & 18\ 45\ 56\ *\ 29\ 67\ 83\ 97 \\
4-5 & 04\ 27\ 33\ 71\ *\ 23\ 37\ 51\ 63\ 81\ 92 \\
6-8 & 22\ 36\ 47\ 55\ 78\ 89\ *\ *\ 10\ 41 \\
\end{array}
$$

Write the data set that is represented by the display.

Applications

2.48   The following data give the time (in minutes) that each of 20 students took to complete a statistics test.

55	49	53	59	38	56	39	58	47	53
58	42	37	43	47	44	55	51	46	45

Construct a stem-and-leaf display for these data. Arrange the leaves for each stem in increasing order.

2.49   Following are the SAT scores (out of a maximum possible score of 1600) of 12 students who took this test recently:

785	890	996	1169	881	1042	995	1083	773	980	1128	1066

Prepare a stem-and-leaf display. Arrange the leaves for each stem in increasing order.

2.50   Reconsider the data on the numbers of computer terminals produced at the Nixon Corporation for a sample of 30 days given in Exercise 2.19. Prepare a stem-and-leaf display for those data. Arrange the leaves for each stem in increasing order.

2.51   Reconsider the data on the numbers of computer keyboards assembled at the Twentieth Century Electronics Company given in Exercise 2.20. Prepare a stem-and-leaf display for those data. Arrange the leaves for each stem in increasing order.

2.52   Refer to Exercise 2.25. Rewrite the data on the daily average costs of rooms in hospitals by rounding each observation to the nearest dollar. Prepare a stem-and-leaf display for those data. Arrange the leaves for each stem in increasing order.

2.53   These data give the times (in minutes) taken to commute from home to work for 20 workers:

10	50	65	33	48	5	11	23	39	26
26	32	17	7	15	19	29	43	21	22

Construct a stem-and-leaf display for these data. Arrange the leaves for each stem in increasing order. (*Note:* To prepare a stem-and-leaf display, each number in this data set can be written as a two-digit number. For example, 5 can be written as 05, for which the stem is 0 and the leaf is 5.)

2.54   The following data give the times served (in months) by 35 prison inmates who were released recently.

37	6	20	5	25	30	24	10	12	20
24	8	26	15	13	22	72	80	96	33
84	86	70	40	92	36	28	90	36	32
72	45	38	18	9					

a.   Prepare a stem-and-leaf display for these data.
b.   Condense the stem-and-leaf display by grouping the stems as 0–2, 3–5, and 6–9.

2.55   The following data give the money (in dollars) spent on textbooks by 35 college students during the 1998–99 academic year.

475	418	180	110	155	288	110	175	250
295	420	610	380	98	230	415	357	357
409	611	455	318	395	612	468	610	380
450	280	490	490	626	350	188	388	

a.   Prepare a stem-and-leaf display for these data using the last two digits as leaves.
b.   Condense the stem-and-leaf display by grouping the stems as 0–1, 2–3, and 4–6.

GLOSSARY

Bar graph   A graph made of bars whose heights represent the frequencies of respective categories.

Class   An interval that includes all the values in a (quantitative) data set that fall within two numbers, the lower and upper limits of the class.

Class boundary   The midpoint of the upper limit of one class and the lower limit of the next class.

Class frequency   The number of values in a data set that belong to a certain class.

Class midpoint or **mark**   The class midpoint or mark is obtained by dividing the sum of the lower and upper limits (or boundaries) of a class by 2.

Class width or **size**   The difference between the two boundaries of a class.

Cumulative frequency   The frequency of a class that includes all values in a data set that fall below the upper boundary of that class.

Cumulative frequency distribution   A table that lists the total number of values that fall below the upper boundary of each class.

Cumulative percentage   The cumulative relative frequency multiplied by 100.

Cumulative relative frequency   The cumulative frequency of a class divided by the total number of observations.

Frequency distribution   A table that lists all the categories or classes and the number of values that belong to each of these categories or classes.

Grouped data   A data set presented in the form of a frequency distribution.

Histogram   A graph in which classes are marked on the horizontal axis and frequencies, relative frequencies, or percentages are marked on the vertical axis. The frequencies, relative frequencies, or percentages of various classes are represented by bars that are drawn adjacent to each other.

Ogive   A curve drawn for a cumulative frequency distribution.

Percentage   The percentage for a class or category is obtained by multiplying the relative frequency of that class or category by 100.

Pie chart   A circle divided into portions that represent the relative frequencies or percentages of different categories or classes.

Polygon   A graph formed by joining the midpoints of the tops of successive bars in a histogram by straight lines.

Raw data   Data recorded in the sequence in which they are collected and before they are processed.

Relative frequency   The frequency of a class or category divided by the sum of all frequencies.

Skewed-to-the-left histogram   A histogram with a longer tail on the left side.

Skewed-to-the-right histogram   A histogram with a longer tail on the right side.

Stem-and-leaf display   A display of data in which each value is divided into two portions—a stem and a leaf.

Symmetric histogram   A histogram that is identical on both sides of its central point.

Uniform or **rectangular histogram**   A histogram with the same frequency for all classes.

KEY FORMULAS

1. **Relative frequency of a class**

$$\text{Relative frequency of a class} = \frac{\text{Frequency of that class}}{\text{Sum of all frequencies}} = \frac{f}{\Sigma f}$$

2. **Percentage of a class**

$$\text{Percentage} = (\text{Relative frequency}) \times 100$$

3. **Class midpoint or mark**

$$\text{Class midpoint} = \frac{\text{Lower limit} + \text{Upper limit}}{2}$$

4. **Class width or size**

$$\text{Class width} = \text{Upper boundary} - \text{Lower boundary}$$

5. **Cumulative relative frequency**

$$\text{Cumulative relative frequency} = \frac{\text{Cumulative frequency}}{\text{Total observations in the data set}}$$

6. **Cumulative percentage**

$$\text{Cumulative percentage} = (\text{Cumulative relative frequency}) \times 100$$

SUPPLEMENTARY EXERCISES

2.56   The following data give the political party of each of the first 30 U.S. presidents. In the data, D stands for Democrat, DR for Democratic Republican, F for Federalist, R for Republican, and W for Whig.

F	F	DR	DR	DR	DR	D	D	W	W
D	W	W	D	D	R	D	R	R	R
R	D	R	D	R	R	R	D	R	R

a.   Prepare a frequency distribution table for these data.
b.   Calculate the relative frequency and percentage distributions.
c.   Draw a bar graph for the relative frequency distribution and a pie chart for the percentage distribution.
d.   What percentage of these presidents were Whigs?

2.57   The following data indicate the geographic location of each of the top 40 emerging-market companies in *Business Week's* 1998 rankings (*Business Week*, July 13, 1998). The ranks are based on the market values of the companies.

Europe	S. America	Asia	N. America	S. America
S. America	Asia	Europe	Asia	S. America
Africa	S. America	Africa	Africa	S. America
S. America	S. America	Africa	S. America	S. America
Asia	Europe	N. America	Africa	Europe
Asia	N. America	Asia	Asia	Asia
Africa	Asia	Africa	S. America	N. America
Asia	Asia	Europe	Asia	N. America

a.   Prepare a frequency distribution table for these data.
b.   Calculate the relative frequency and percentage distributions.
c.   Draw a bar graph for the frequency distribution and a pie chart for the percentage distribution.
d.   What percentage of these companies are located in Africa?

2.58   The following data give the numbers of television sets owned by 40 randomly selected households.

1	1	2	3	2	4	1	3	2	1
3	0	2	1	2	3	2	3	2	2
1	2	1	1	1	3	1	1	1	2
2	4	2	3	1	3	1	2	2	4

a.   Prepare a frequency distribution table for these data using single-valued classes.
b.   Compute the relative frequency and percentage distributions.
c.   Draw a bar graph for the frequency distribution.
d.   What percentage of the households own two or more television sets?

2.59   The following data give the numbers of persons in each of 30 groups who made reservations at a restaurant on a recent Friday night.

2	2	2	2	1	2	4	4	2	1
5	1	3	2	2	3	1	4	2	4
4	2	5	4	1	2	2	4	3	1

a.   Prepare a frequency distribution table for these data using single-valued classes.
b.   Compute the relative frequency and percentage distributions.

c. What percentage of the groups in this sample have two or three persons?
d. Draw a bar graph for the relative frequency distribution.

2.60 The following data give the amounts spent on video rentals (in dollars) during 1999 by 30 households randomly selected from those households that rented videos in 1999.

595	24	6	100	100	40	622	405	90
55	155	760	405	90	205	70	180	88
808	100	240	127	83	310	350	160	22
111	70	15						

a. Construct a frequency distribution table. Take $1 as the lower limit of the first class and $200 as the width of each class.
b. Calculate the relative frequencies and percentages for all classes.
c. What percentage of the households in this sample spent more than $400 on video rentals in 1999?

2.61 The following data give the numbers of orders received for a sample of 30 hours at the Timesaver Mail Order Company.

34	44	31	52	41	47	38	35	32	39
28	24	46	41	49	53	57	33	27	37
30	27	45	38	34	46	36	30	47	50

a. Construct a frequency distribution table. Take 23 as the lower limit of the first class and 7 as the width of each class.
b. Calculate the relative frequencies and percentages for all classes.
c. For what percentage of the hours in this sample was the number of orders more than 36?

2.62 The following data give the weekly expenditures (in dollars) on fruit and vegetables for 30 households randomly selected from the households that incurred such expenses.

4.57	3.95	6.95	3.80	1.50	2.99	3.84	5.05
8.00	14.75	9.33	1.05	5.08	7.00	9.60	18.99
9.15	11.32	4.75	9.95	3.63	1.99	1.39	17.09
19.31	11.15	7.73	12.00	7.58	16.35		

a. Construct a frequency distribution table using the *less than* method to write classes. Take $0 as the lower boundary of the first class and $4 as the width of each class.
b. Calculate the relative frequencies and percentages for all classes.
c. Draw a histogram for the frequency distribution.

2.63 The following data give the repair costs (in dollars) for 30 cars randomly selected from a list of cars that were involved in collisions.

2300	750	2500	410	555	1576
2460	1795	2108	897	989	1866
2105	335	1344	1159	1236	1395
6108	4995	5891	2309	3950	3950
6655	4900	1320	2901	1925	6896

a. Construct a frequency distribution table. Take $1 as the lower limit of the first class and $1400 as the width of each class.
b. Compute the relative frequencies and percentages for all classes.
c. Draw a histogram and a polygon for the relative frequency distribution.
d. What are the class boundaries and the width of the fourth class?

2.64 Refer to Exercise 2.60. Prepare the cumulative frequency, cumulative relative frequency, and cumulative percentage distributions by using the frequency distribution table of that exercise.

2.65 Refer to Exercise 2.61. Prepare the cumulative frequency, cumulative relative frequency, and cumulative percentage distributions using the frequency distribution table constructed for the data of that exercise.

2.66 Refer to Exercise 2.62. Prepare the cumulative frequency, cumulative relative frequency, and cumulative percentage distributions using the frequency distribution table constructed for the data of that exercise.

2.67 Construct the cumulative frequency, cumulative relative frequency, and cumulative percentage distributions by using the frequency distribution table constructed for the data of Exercise 2.63.

2.68 Refer to Exercise 2.60. Prepare a stem-and-leaf display for the data of that exercise.

2.69 Construct a stem-and-leaf display for the data given in Exercise 2.61.

2.70 The following table gives the percentages of the population using wireless phones in six countries.

Country	Percentage of Population
Finland	58
Norway	49
Sweden	46
Japan	37
Italy	36
United States	26

Source: Dataquest, Inc., *Business Week*, May 3, 1999.

Draw two bar graphs for these data, the first without truncating the axis on which the percentage is marked and the second by truncating this axis. In the second case, mark the percentages on the vertical axis starting with 20%. Briefly comment on the two bar graphs.

2.71 The following table gives the 1999 average math SAT scores for the six New England states.

State	Average Math SAT Score
Connecticut	509
Maine	503
Massachusetts	511
New Hampshire	518
Rhode Island	499
Vermont	506

Draw two bar graphs for these data, the first without truncating the axis on which the SAT scores are marked and the second by truncating this axis. In the second case, mark the SAT scores on the vertical axis, starting with 490. Briefly comment on the two bar graphs.

■ Advanced Exercises

2.72 The following frequency distribution table gives the age distribution of drivers who were at fault in auto accidents that occurred during a one-week period in a city.

Age	f
18 to less than 20	7
20 to less than 25	12
25 to less than 30	18
30 to less than 40	14
40 to less than 50	15
50 to less than 60	16
60 and over	35

a.   Draw a relative frequency histogram for this table.

b.   In what way(s) is this histogram misleading?

c.   How can you change the frequency distribution so that the resulting histogram gives a clearer picture?

2.73   Refer to the data presented in Exercise 2.72. Note that there were 50% more accidents in the *25 to less than 30* age group than in the *20 to less than 25* age group. Does this suggest that the older group of drivers in this city is more accident-prone than the younger group? What other explanation might account for the difference in accident rates?

2.74   Suppose a data set contains the ages of 135 autoworkers ranging from 20 to 53.

a.   Using Sturge's formula given in footnote 1 at the beginning of Section 2.3.2, find an appropriate number of classes for a frequency distribution for this data set.

b.   Find an appropriate class width based on the number of classes in part a.

2.75   The following table shows the percentages of men and women who were assaulted at work in eight countries. These data were obtained from a study of on-the-job violence conducted by the International Labor Organization.

Country	Percentage Reporting Attacks at Work	
	Men	**Women**
France	11.2	8.6
Romania	8.7	4.1
Argentina	6.1	11.8
Switzerland	4.3	1.6
Canada	3.9	5.0
Netherlands	3.6	3.8
England	3.2	6.3
United States	1.0	4.2

Source: Business Week, August 17, 1998.

a.   Construct a bar graph for this data set using a pair of bars side by side for the men and women in each country.

b.   Are the pie charts (one for men and one for women) appropriate for this data set? Why or why not?

c.   Can you find the percentage of all workers (regardless of gender) who were attacked on the job in France? If so, find this percentage. If not, what additional information do you need to find this percentage?

2.76   In some applications, it may be desirable to use open-ended classes or classes of different widths in order to accommodate the data. Consider the following data on the ages of 35 students enrolled in an introductory statistics class:

17	18	18	18	18	19	19	19	19	19	19	19
19	19	20	20	20	20	20	20	20	21	21	21
22	22	23	26	28	30	38	43	51	62	64	

a.   Construct a stem-and-leaf display for these data.

b.   Do you think a stem-and-leaf display is appropriate for this data set? Explain.

c.   Construct a frequency distribution for these data using equal class widths. What problems do you encounter?

d.   Construct a frequency distribution using the first four classes with a width of 2 each, the next two classes having a width of 10 each, and the last class being an open-ended class.

e.   Draw a histogram based on the frequency distribution table in part d.

2.77 Statisticians often need to know the shape of a population in order to make inferences. Suppose that you are asked to specify the shape of the population of weights of all college students.

a. Sketch a graph of what you think the weights of all college students would look like.

b. The following data give the weights (in pounds) of a random sample of 44 college students. (Here, F and M indicate female or male.)

123 F	195 M	138 M	115 F	179 M	119 F
148 F	147 F	180 M	146 F	179 M	189 M
175 M	108 F	193 M	114 F	179 M	147 M
108 F	128 F	164 F	174 M	128 F	159 M
193 M	204 M	125 F	133 F	115 F	168 M
123 F	183 M	116 F	182 M	174 M	102 F
123 F	99 F	161 M	162 M	155 F	202 M
110 F	132 M				

i. Construct a stem-and-leaf display for these data.

ii. Can you explain why these data appear the way they do?

c. Now sketch a graph of what you think the weights of all college students look like. Is this similar to your sketch in part a?

SELF-REVIEW TEST

1. Briefly explain the difference between ungrouped and grouped data and give one example of each type.

2. The following table gives the frequency distribution of the durations (in minutes) of 100 long-distance phone calls made by persons using TVI long-distance service.

Duration (minutes)	Frequency
0 to 4	8
5 to 9	20
10 to 14	35
15 to 19	22
20 to 24	15

Circle the correct answer for each of the following questions, which are based on this table.

a. The number of classes in the table is   5, 100, 80
b. The class width is   4, 5, 10
c. The midpoint of the third class is   11.5, 12, 12.5
d. The lower boundary of the second class is   4.5, 5, 5.5
e. The upper limit of the second class is   8.5, 9, 9.5
f. The sample size is   5, 100, 50
g. The relative frequency of the first class is   .04, .08, .16

3. Briefly explain and illustrate with the help of graphs a symmetric histogram, a histogram skewed to the right, and a histogram skewed to the left.

4. Twenty elementary school children were asked if they live with both parents (B), father only (F), mother only (M), or someone else (S). The responses of the children follow:

M	B	B	M	F	S	B	M	F	M
B	F	B	M	M	B	B	F	B	M

  a. Construct a frequency distribution table.
  b. Write the relative frequencies and percentages for all categories.
  c. What percentage of the children in this sample live with their mothers only?
  d. Draw a bar graph for the frequency distribution and a pie chart for the percentages.

5. The following data set gives the numbers of years for which 24 workers have been with their current employers.

15	12	9	10	5	12	3	7	16	13	11	4
3	8	7	14	11	8	4	13	2	18	6	19

  a. Construct a frequency distribution table. Take 1 as the lower limit of the first class and 4 as the width of each class.
  b. Calculate the relative frequencies and percentages for all classes.
  c. What percentage of the employees have been with their current employers for 8 or fewer years?
  d. Draw the frequency histogram and polygon.

6. Refer to the frequency distribution prepared in Problem 5. Prepare the cumulative percentage distribution using that table. Draw an ogive for the cumulative percentage distribution. Using the ogive, find the percentage of employees in the sample who have been with their current employers for 11 or fewer years.

7. Construct a stem-and-leaf display for the following data, which give the times (in minutes) taken by an accountant to prepare 20 income tax returns.

34	21	67	78	18	38	45	56	62	74
75	58	43	69	56	38	71	50	42	36

8. Consider this stem-and-leaf display:

```
3 | 0 3 7
4 | 2 4 6 7 9
5 | 1 3 3 6
6 | 0 7 7
7 | 1 9
```

Write the data set that was used to construct the display.

MINI-PROJECTS

Mini-Project 2–1

Using the data you gathered for the mini-project in Chapter 1, prepare a summary of that data set which includes the following.

  a. Prepare an appropriate type of frequency distribution table and then compute relative frequencies and cumulative relative frequencies.
  b. Draw a histogram, a frequency polygon, and an ogive. Comment on any symmetry or skewness in your histogram.

c. Make a stem-and-leaf display, if possible, and comment on the skewness of the data based on this display.

Mini-Project 2–2

Watch 4 hours of television (not necessarily all on the same day). Record the duration of each commercial break to the nearest second. Time the total length of the break from the time the program is first interrupted until it resumes (or from the time a program ends until the next program begins). List all these durations, and then write a brief report that covers the following.

a. Prepare an appropriate type of frequency distribution table including relative frequencies and cumulative relative frequencies.
b. Draw a histogram. Comment on the shape of the histogram. Is there evidence of skewness?
c. Make a stem-and-leaf display and comment on the skewness of the data based on this display.

For instructions on using MINITAB for this chapter, please see Section B.2 in Appendix B.

See Appendix C for *Excel Adventure 2* on organizing data.

COMPUTER ASSIGNMENTS

CA2.1 Using MINITAB or any other statistical software, construct a bar graph and a pie chart for the frequency distribution prepared in Exercise 2.5.

CA2.2 Using MINITAB or any other statistical software, construct a bar graph and a pie chart for the frequency distribution prepared in Exercise 2.6.

CA2.3 Refer to Data Set IV that accompanies this text (see preface) on the times taken to run the Manchester Road Race for a sample of 500 participants. From that data set, select the 6th value and then select every 10th value after that (i.e., select the 6th, 16th, 26th, 36th, . . . values). This subsample will give you 50 measurements. (Such a sample selected from a population is called a *systematic random sample*.) Using MINITAB or any other statistical software, construct a histogram for these data. Let the software you use decide on classes and class limits.

CA2.4 Refer to Data Set I that accompanies this text on the prices of various products in different cities across the country. Using MINITAB or any other statistical software, select a subsample of 60 from column C7 (telephone charges) and then construct a histogram for these data.

CA2.5 Using MINITAB or any other statistical software, construct a histogram for the data from Exercise 2.20 on the numbers of computer keyboards assembled. Use the classes given in that exercise. Use the midpoints to mark the horizontal axis in the histogram.

CA2.6 Using MINITAB or any other statistical software, prepare a stem-and-leaf display for the data given in Exercise 2.48.

CA2.7 Using MINITAB or any other statistical software, prepare a stem-and-leaf display for the data of Exercise 2.53.

CA2.8 These data give the weights (in pounds) of 20 parcels mailed from a post office during the past week.

1.8	7.5	8.2	3.4	5.1	9.3	1.9	2.5	7.3	5.8
6.2	8.6	2.0	6.3	8.5	.7	3.8	7.3	5.2	3.7

Using MINITAB, prepare a stem-and-leaf display for these data. Use INCREMENT = 1. Observe the value of the leaf unit in the output.

CA2.9 Using MINITAB or any other statistical software, prepare a bar graph for the frequency distribution obtained in Exercise 2.28.

CA2.10 Using MINITAB or any other statistical software, prepare a bar graph for the frequency distribution obtained in Exercise 2.29.

CA2.11 Using MINITAB or any other statistical software, make a pie chart for the frequency distribution obtained in Exercise 2.19.

CA2.12 Using MINITAB or any other statistical software, make a pie chart for the frequency distribution obtained in Exercise 2.29.

3

NUMERICAL DESCRIPTIVE MEASURES

3.1 MEASURES OF CENTRAL TENDENCY FOR UNGROUPED DATA

CASE STUDY 3–1 EARNINGS BY DEGREE

CASE STUDY 3–2 CARS, TRUCKS OLDER THAN EVER

3.2 MEASURES OF DISPERSION FOR UNGROUPED DATA

3.3 MEAN, VARIANCE, AND STANDARD DEVIATION FOR GROUPED DATA

3.4 USE OF STANDARD DEVIATION

CASE STUDY 3–3 HERE COMES THE SD

3.5 MEASURES OF POSITION

3.6 BOX-AND-WHISKER PLOT

GLOSSARY

KEY FORMULAS

SUPPLEMENTARY EXERCISES

APPENDIX 3.1

SELF-REVIEW TEST

MINI-PROJECT

COMPUTER ASSIGNMENTS

In Chapter 2 we discussed how to organize and display large data sets. The techniques presented in that chapter, however, are not helpful when we need to describe verbally the main characteristics of a data set. The numerical summary measures, such as the ones that give the center and spread of a distribution, provide us with the main features of a data set. For example, the techniques learned in Chapter 2 can help us to graph data on family incomes. However, we may want to know the income of a "typical" family (given by the center of the distribution), the spread of the distribution of incomes, or the location of a family with a particular income. Figure 3.1 (page 70) shows these three concepts. Such questions can be answered using the numerical summary measures discussed in this chapter. Included among these are (1) measures of central tendency, (2) measures of dispersion, and (3) measures of position.

Exam #2 Start here

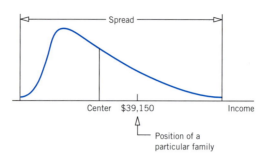

Figure 3.1

3.1 MEASURES OF CENTRAL TENDENCY FOR UNGROUPED DATA

We often represent a data set by numerical summary measures, usually called the *typical values*. A **measure of central tendency** gives the center of a histogram or a frequency distribution curve. This section discusses three different measures of central tendency: the mean, the median, and the mode. We will learn how to calculate each of these measures for ungrouped data. Recall from Chapter 2, the data that give information on each member of the population or sample individually are called *ungrouped data*, whereas *grouped data* are presented in the form of a frequency distribution table.

3.1.1 MEAN

The **mean**, also called the *arithmetic mean*, is the most frequently used measure of central tendency. This book will use the words *mean* and *average* synonymously. For ungrouped data, the mean is obtained by dividing the sum of all values by the number of values in the data set.

$$\text{Mean} = \frac{\text{Sum of all values}}{\text{Number of values}}$$

The mean calculated for sample data is denoted by $\bar{x}$ (read as "x bar"), and the mean calculated for population data is denoted by μ (Greek letter *mu*). We know from the discussion in Chapter 2 that the number of values in a data set is denoted by n for a sample and by N for a population. In Chapter 1, we learned that a variable is denoted by x and the sum of all values of x is denoted by Σx. Using these notations, we can write the following formulas for the mean.

MEAN FOR UNGROUPED DATA   The *mean for ungrouped data* is obtained by dividing the sum of all values by the number of values in the data set. Thus,

$$\text{Mean for population data:} \quad \mu = \frac{\Sigma x}{N}$$

$$\text{Mean for sample data:} \quad \bar{x} = \frac{\Sigma x}{n}$$

where Σx is the sum of all values, N is the population size, n is the sample size, μ is the population mean, and $\bar{x}$ is the sample mean.

Calculating the sample mean for ungrouped data.

EXAMPLE 3-1 Table 3.1 gives the 1998 earnings (in millions of dollars) of five celebrities.

Table 3.1

Celebrity	1998 Earnings (millions of dollars)
Mel Gibson	55
Jerry Seinfeld	267
Steven Spielberg	175
John Travolta	47
Oprah Winfrey	125

Source: Forbes magazine

Find the mean 1998 earnings for these celebrities.

Solution The variable in this example is *1998 earnings*. Let us denote it by x. Then the five values of x are

$$x_1 = 55, \quad x_2 = 267, \quad x_3 = 175, \quad x_4 = 47, \quad \text{and} \quad x_5 = 125$$

where x_1 represents the 1998 earnings of the first celebrity (Mel Gibson) listed in Table 3.1, x_2 represents the 1998 earnings of the second celebrity (Jerry Seinfeld), and so on. The sum of the 1998 earnings of these five celebrities is

$$\Sigma x = x_1 + x_2 + x_3 + x_4 + x_5$$

$$= 55 + 267 + 175 + 47 + 125 = \$669 \text{ million}$$

Note that the given information is on the 1998 earnings of only five celebrities. Hence, it represents a sample. Because the data set contains five values, $n = 5$. Substituting the values of Σx and n in the sample formula, we get the mean 1998 earnings of these five celebrities:

$$\bar{x} = \frac{\Sigma x}{n} = \frac{669}{5} = \textbf{\$133.8 million}$$

Thus, the mean 1998 earnings of these five celebrities are \$133,800,000. ∎

Calculating the population mean for ungrouped data.

EXAMPLE 3-2 The following are the ages of all eight employees of a small company:

$$53 \quad 32 \quad 61 \quad 27 \quad 39 \quad 44 \quad 49 \quad 57$$

Find the mean age of these employees.

Solution Since the given data set includes *all* eight employees of the company, it represents the population. Hence, $N = 8$.

$$\Sigma x = 53 + 32 + 61 + 27 + 39 + 44 + 49 + 57 = 362$$

The population mean is

$$\mu = \frac{\Sigma x}{N} = \frac{362}{8} = \textbf{45.25 years}$$

Thus, the mean age of all eight employees of this company is 45.25 years, or 45 years and 3 months. ∎

Reconsider Example 3–2. If we take a sample of three employees from this company and calculate the mean age of those three employees, this mean will be denoted by $\bar{x}$. Suppose the three values included in the sample are 32, 39, and 57. Then, the mean age for this sample is

$$\bar{x} = \frac{32 + 39 + 57}{3} = 42.67 \text{ years}$$

If we take a second sample of three employees of this company, the value of $\bar{x}$ will (most likely) be different. Suppose the second sample includes the values 53, 27, and 44. Then, the mean age for this sample is

$$\bar{x} = \frac{53 + 27 + 44}{3} = 41.33 \text{ years}$$

Consequently, we can state that the value of the population mean μ is constant. However, the value of the sample mean $\bar{x}$ varies from sample to sample. The value of $\bar{x}$ for a particular sample depends on what values of the population are included in that sample.

Sometimes a data set may contain a few very small or a few very large values. Such values are called **outliers** or **extreme values**.

OUTLIERS OR EXTREME VALUES   Values that are very small or very large relative to the majority of the values in a data set are called *outliers* or *extreme values*.

A major shortcoming of the mean as a measure of central tendency is that it is very sensitive to outliers. Example 3–3 illustrates this point.

Illustrating the effect of an outlier on the mean.

EXAMPLE 3–3   Table 3.2 lists the 1997 populations (in thousands) of the five Pacific states.

Table 3.2

State	Population (thousands)
Washington	5,610
Oregon	3,243
Alaska	609
Hawaii	1,187
California	32,268 ← An outlier

Notice that the population of California is very large compared to the populations of the other four states. Hence, it is an outlier. Show how the inclusion of this outlier affects the value of the mean.

Solution   If we do not include the population of California (the outlier), the mean population of the remaining four states (Washington, Oregon, Alaska, and Hawaii) is

$$\text{Mean} = \frac{5610 + 3243 + 609 + 1187}{4} = \mathbf{2662.25 \ thousand}$$

Now, to see the impact of the outlier on the value of the mean, we include the population of California and find the mean population of all five Pacific states. This mean is

$$\text{Mean} = \frac{5610 + 3243 + 609 + 1187 + 32{,}268}{5} = \textbf{8583.4 thousand}$$

Thus, including California causes more than a threefold increase in the value of the mean, as it changes from 2662.25 thousand to 8583.4 thousand. ∎

The preceding example should encourage us to be cautious. We should remember that the mean is not always the best measure of central tendency because it is heavily influenced by outliers. Sometimes other measures of central tendency give a more accurate impression of a data set.

CASE STUDY 3–1   EARNINGS BY DEGREE

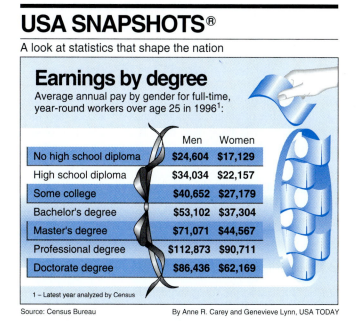

USA SNAPSHOTS®
A look at statistics that shape the nation

Earnings by degree
Average annual pay by gender for full-time, year-round workers over age 25 in 1996[1]:

	Men	Women
No high school diploma	$24,604	$17,129
High school diploma	$34,034	$22,157
Some college	$40,652	$27,179
Bachelor's degree	$53,102	$37,304
Master's degree	$71,071	$44,567
Professional degree	$112,873	$90,711
Doctorate degree	$86,436	$62,169

1 – Latest year analyzed by Census

Source: Census Bureau   By Anne R. Carey and Genevieve Lynn, USA TODAY

The chart, reproduced from *USA TODAY*, lists the mean annual pay for men and women over age 25 who were employed full time in 1996. As we can observe from the data, mean earnings increase tremendously with an increase in education level. However, not only do women earn less than men with the same educational qualifications, but also the increase in average pay for women is much less than the increase in average pay for men with an increase in education level.

Source: *USA TODAY*, September 11, 1998. Copyright © 1998, *USA TODAY*. Chart reproduced with permission.

3.1.2 MEDIAN

Another important measure of central tendency is the **median**. It is defined as follows.

MEDIAN   The *median* is the value of the middle term in a data set that has been ranked in increasing order.

As is obvious from the definition of the median, it divides a ranked data set into two equal parts. The calculation of the median consists of the following two steps:

1. Rank the data set in increasing order.
2. Find the middle term. The value of this term is the median.[1]

The position of the middle term in a data set with n values is obtained as follows:

$$\text{Position of the middle term} = \frac{n + 1}{2}$$

Thus, we can redefine the median as follows.

MEDIAN FOR UNGROUPED DATA

$$\text{Median} = \text{Value of the } \left(\frac{n + 1}{2}\right)\text{th term in a ranked data set}$$

If the given data set represents a population, replace n by N.

If the number of observations in a data set is *odd*, then the median is given by the value of the middle term in the ranked data. If the number of observations is *even*, then the median is given by the average of the values of the two middle terms.

Calculating the median for ungrouped data: odd number of data values.

EXAMPLE 3–4   The following data give the weight lost (in pounds) by a sample of five members of a health club at the end of two months of membership.

$$10 \quad 5 \quad 19 \quad 8 \quad 3$$

Find the median.

Solution   First, we rank the given data in increasing order as follows:

$$3 \quad 5 \quad 8 \quad 10 \quad 19$$

There are five observations in the data set. Consequently, $n = 5$ and

$$\text{Position of the middle term} = \frac{n + 1}{2} = \frac{5 + 1}{2} = 3$$

---

[1]The value of the middle term in a data set ranked in decreasing order will also give the value of the median.

Therefore, the median is the value of the third term in the ranked data.

$$3 \quad 5 \quad \boxed{8} \quad 10 \quad 19$$
$$\uparrow$$
$$\text{Median}$$

The median weight loss for this sample of five members of this health club is **8 pounds.**

■

CASE STUDY 3-2 CARS, TRUCKS OLDER THAN EVER

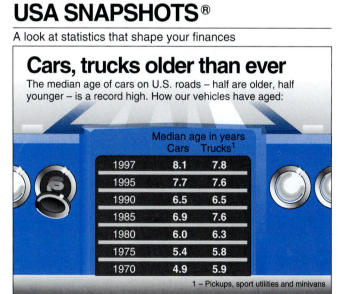

USA SNAPSHOTS®

A look at statistics that shape your finances

Cars, trucks older than ever

The median age of cars on U.S. roads – half are older, half younger – is a record high. How our vehicles have aged:

| | Median age in years | |
	Cars	Trucks[1]
1997	8.1	7.8
1995	7.7	7.6
1990	6.5	6.5
1985	6.9	7.6
1980	6.0	6.3
1975	5.4	5.8
1970	4.9	5.9

1 – Pickups, sport utilities and minivans

Source: Polk Co.   By Anne R. Carey and Jerry Mosemak, USA TODAY

The chart shows that the median age of cars in the United States was 8.1 years in 1997 and that of the trucks (pickups, sport utilities, and minivans) was 7.8 years. As the data indicate, the median age of cars and trucks in the United States has increased quite a bit since 1970. This suggests that people are holding on to their cars and other vehicles a lot longer than they used to. The median is a better measure than the mean here because the frequency distribution of ages of cars and other vehicles is skewed to the right, with many outliers; many people drive very old vehicles.

Source: *USA TODAY*, September 10, 1998. Copyright © 1998, *USA TODAY*. Chart reproduced with permission.

Calculating the median for ungrouped data: even number of data values.

EXAMPLE 3–5 Table 3.3 lists the 1998 total pay (which includes salary, bonus, and long-term compensation) of the then 10 top-paid chief executives in the United States. (Note that the pay has been rounded to the nearest million dollars. Some of these CEOs may have retired since 1998.)

Table 3.3

Chief Executive and Company Name	1998 Total Pay (millions of dollars)
Craig Barrett, Intel	116.5
Stephen Case, America Online	159.2
Michael Eisner, Walt Disney	575.6
M. Douglas Ivester, Coca-Cola	57.3
Mel Karmazin, CBS	201.9
L. Dennis Kozlowski, Tyco International	65.3
Henry Schacht, Lucent Technologies	67.0
Henry Silverman, Cendant	63.9
Sanford Weill, Citigroup	167.1
John Welch, General Electric	83.7

Find the median 1998 total pay of these 10 chief executives.

Solution First, we rank the data on the 1998 total pay of these 10 executives in increasing order, as follows:

57.3   63.9   65.3   67.0   83.7   116.5   159.2   167.1   201.9   575.6

There are 10 values in the data set. Hence, $n = 10$ and

$$\text{Position of the middle term} = \frac{n+1}{2} = \frac{10+1}{2} = 5.5$$

Therefore, the median is given by the mean of the fifth and the sixth values in the ranked data.

57.3   63.9   65.3   67.0   83.7   116.5   159.2   167.1   201.9   575.6

$$\text{Median} = \frac{83.7 + 116.5}{2} = \textbf{100.1}$$

Thus, the median 1998 total pay of these 10 CEOs was $100.1 million, or $100,100,000. ■

The median gives the center of a histogram, with half of the data values to the left of the median and half to the right of the median. The advantage of using the median as a measure of central tendency is that it is not influenced by outliers. Consequently, the median is preferred over the mean as a measure of central tendency for data sets that contain outliers.

3.1.3 MODE

Mode is a French word that means *fashion*—an item that is most popular or common. In statistics, the mode represents the most common value in a data set.

> **MODE** The *mode* is the value that occurs with the highest frequency in a data set.

Calculating the mode for ungrouped data.

EXAMPLE 3-6 The following data give the speeds (in miles per hour) of eight cars that were stopped on I-95 for speeding violations.

77   69   74   81   71   68   74   73

Find the mode.

Solution In this data set, 74 occurs twice and each of the remaining values occurs only once. Because 74 occurs with the highest frequency, it is the mode. Therefore,

Mode = 74 miles per hour ∎

A major shortcoming of the mode is that a data set may have none or may have more than one mode, whereas it will have only one mean and only one median. For instance, a data set with each value occurring only once has no mode. A data set with only one value occurring with the highest frequency has only one mode. The data set in this case is called **unimodal**. A data set with two values that occur with the same (highest) frequency has two modes. The distribution, in this case, is said to be **bimodal**. If more than two values in a data set occur with the same (highest) frequency, then the data set contains more than two modes and it is said to be **multimodal**.

Data set with no mode.

EXAMPLE 3-7 Last year's incomes of five randomly selected families were $26,150, $65,750, $34,985, $47,490, and $13,740. Find the mode.

Solution Because each value in this data set occurs only once, this data set contains **no mode**. ∎

Data set with two modes.

EXAMPLE 3-8 The prices of the same brand of television set at eight stores are found to be $495, $486, $503, $495, $470, $505, $470, and $499. Find the mode.

Solution In this data set, each of the two values $495 and $470 occurs twice and each of the remaining values occurs only once. Therefore, this data set has two modes: **$495** and **$470**. ∎

Data set with three modes.

EXAMPLE 3-9 The ages of 10 randomly selected students from a class are 21, 19, 27, 22, 29, 19, 25, 21, 22, and 30. Find the mode.

Solution This data set has three modes: **19, 21**, and **22**. Each of these three values occurs with a (highest) frequency of 2. ∎

One advantage of the mode is that it can be calculated for both kinds of data, quantitative and qualitative, whereas the mean and median can be calculated for only quantitative data.

Finding the mode for qualitative data.

EXAMPLE 3-10 The status of five students who are members of the student senate at a college are senior, sophomore, senior, junior, senior. Find the mode.

Solution Because **senior** occurs more frequently than the other categories, it is the mode for this data set. We cannot calculate the mean and median for this data set. ∎

To sum up, we cannot say for sure which of the three measures of central tendency is a better measure overall. Each of them may be better under different situations. Probably the mean is the most used measure of central tendency, followed by the median. The mean has the advantage that its calculation includes each value of the data set. The median is a better measure when a data set includes outliers. The mode is simple to locate, but it is not of much use in practical applications.

3.1.4 RELATIONSHIPS AMONG THE MEAN, MEDIAN, AND MODE

As discussed in Chapter 2, two of the many shapes that a histogram or a frequency distribution curve can assume are symmetric and skewed. This section describes the relationships among the mean, median, and mode for three such histograms and frequency curves. Knowing the values of the mean, median, and mode can give us some idea about the shape of a frequency curve.

1. For a symmetric histogram and frequency curve with one peak (see Figure 3.2), the values of the mean, median, and mode are identical, and they lie at the center of the distribution.

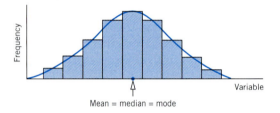

Figure 3.2 Mean, median, and mode for a symmetric histogram and frequency curve.

2. For a histogram and a frequency curve skewed to the right (see Figure 3.3), the value of the mean is the largest, that of the mode is the smallest, and the value of the median lies between these two. (Notice that the mode always occurs at the peak point.) The value of the mean is the largest in this case because it is sensitive to outliers that occur in the right tail. These outliers pull the mean to the right.

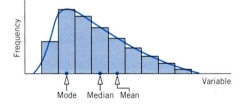

Figure 3.3 Mean, median, and mode for a histogram and frequency curve skewed to the right.

3. If a histogram and a distribution curve are skewed to the left (see Figure 3.4), the value of the mean is the smallest and that of the mode is the largest, with the value of the median lying between these two. In this case, the outliers in the left tail pull the mean to the left.

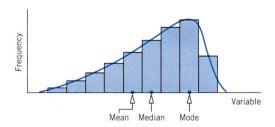

Figure 3.4 Mean, median, and mode for a histogram and frequency curve skewed to the left.

EXERCISES

■ **Concepts and Procedures**

3.1 Explain how the value of the median is determined for a data set that contains an odd number of observations and for a data set that contains an even number of observations.

3.2 Briefly explain the meaning of an outlier. Is the mean or the median a better measure of central tendency for a data set that contains an outlier? Illustrate with the help of an example.

3.3 Using an example, show how an outlier can affect the value of the mean.

3.4 Which of the three measures of central tendency (the mean, the median, and the mode) can be calculated for quantitative data only, and which can be calculated for both quantitative and qualitative data? Illustrate with examples.

3.5 Which of the three measures of central tendency (the mean, the median, and the mode) can assume more than one value for a data set? Give an example of a data set for which this summary measure assumes more than one value.

3.6 Is it possible for a (quantitative) data set to have no mean, no median, or no mode? Give an example of a data set for which this summary measure does not exist.

3.7 Explain the relationships among the mean, median, and mode for symmetric and skewed histograms. Illustrate these relationships with graphs.

3.8 Prices of cars have a distribution that is skewed to the right with outliers in the right tail. Which of the measures of central tendency is the best to summarize this data set? Explain.

3.9 The following data set belongs to a population:

$$5 \quad -7 \quad 2 \quad 0 \quad -9 \quad 16 \quad 10 \quad 7$$

Calculate the mean, median, and mode.

3.10 The following data set belongs to a sample:

$$14 \quad 18 \quad -10 \quad 8 \quad 8 \quad -16$$

Calculate the mean, median, and mode.

■ **Applications**

Exercises 3.11 through 3.15 are based on the following data.

The following table, based on the American Chamber of Commerce Researchers Association Survey, gives the prices of five items in 12 urban areas across the United States. (See Data Set I that accompanies the text.)

City	Apartment Rent	Price of House	Phone Bill	Cost of a Hospital Room	Price of Wine
Anchorage (AK)	$ 788	$187,337	$14.58	$766.50	$6.28
Phoenix (AZ)	664	141,484	19.21	580.69	5.29
San Diego (CA)	936	244,000	16.89	741.60	4.85
Denver (CO)	740	179,740	21.05	643.71	5.16
Bloomington (IN)	574	131,848	21.30	499.00	4.89
Boston (MA)	1264	246,975	23.00	694.00	6.28
Manchester (NH)	882	167,450	21.13	479.00	4.99
Albuquerque (NM)	741	140,875	23.66	371.25	6.49
Buffalo (NY)	542	123,095	25.55	526.80	5.89
Charlotte (NC)	548	148,500	17.97	405.67	5.58
Salem (OR)	561	169,823	19.92	415.00	4.95
Charleston (WV)	602	142,660	28.08	267.25	6.32

Explanation of variables

Apartment rent   Monthly rent of an unfurnished two-bedroom apartment (excluding all utilities except water), 1½ or 2 baths, approximately 950 square feet

Price of house   Purchase price of a new house with 1800 square feet of living area, on an 8000-square-foot lot in an urban area with all utilities

Phone bill   Monthly telephone charges for a private residential line (customer owns instruments)

Cost of a hospital room   Average cost per day of a semiprivate room in a hospital

Price of wine   Price of Paul Masson Chablis, 1.5-liter bottle

3.11   Calculate the mean and median for the data on apartment rents.

3.12   Find the mean and median for the data on prices of houses.

3.13   Calculate the mean and median for the data on phone bills. Do these data have a mode?

3.14   Calculate the mean and median for the data on costs of hospital rooms. Do these data have a mode?

3.15   Find the mean and median for the data on prices of wine.

3.16   The following data give the numbers of car thefts that occurred in a city during the past 12 days:

| 6 | 3 | 7 | 11 | 4 | 3 | 8 | 7 | 2 | 6 | 9 | 15 |

Find the mean, median, and mode.

3.17   The following data are the salaries (in millions of dollars) of the infielders of the New York Yankees baseball team for the 1999 season (*USA TODAY*, November 19, 1999). These salaries, entered in that order, are for the players Brosius, Degroote, Jeter, Knoblauch, Martinez, and Sojo.

| 5.25 | .20 | 5.00 | 6.00 | 4.30 | .80 |

Compute the mean and median. Do these data have a mode? Why or why not? Explain.

3.18   The following data give the 1998 profits (in millions of dollars) of the 10 airlines listed in *Fortune* magazine's top 1000 U.S. corporations (*Fortune*, April 26, 1999). A number with a negative sign represents a loss for that airline. The profits are for AMR, UAL, DELTA, NWA, US Airways, Continental, Southwest, TWA, America West Holdings, and Alaska Air Group, respectively.

| 1314 | 821 | 1001 | −286 | 538 | 383 | 433 | −120 | 109 | 124 |

Compute the mean and median. Do these data have a mode? Why or why not? Explain.

3.19   The following data give the numbers of hours spent partying by 10 randomly selected college students during the past week.

| 7 | 14 | 5 | 0 | 9 | 7 | 10 | 4 | 0 | 8 |

Compute the mean, median, and mode.

3.20   The following data give the salaries (in millions of dollars) of the Atlanta Braves baseball team pitchers for the 1999 season (*USA TODAY*, November 19, 1999). The salaries, entered in that order, are for the pitchers Cather, Chen, Glavine, Ligtenberg, Maddux, McGlinchy, Millwood, Mulholland, Perez, Remlinger, Rocker, Seanez, Smoltz, and Springer.

| .23 | .20 | 7.00 | .26 | 10.60 | .20 | .23 |
| 2.93 | .20 | 1.10 | .22 | .78 | 7.75 | .95 |

Calculate the mean and median. Do these data have any mode(s)? Why or why not? Explain.

3.21 Nixon Corporation manufactures computer terminals. The following data are the numbers of computer terminals produced at the company for a sample of 10 days.

24   32   27   23   35   33   29   40   23   28

Calculate the mean, median, and mode for these data.

3.22 The following data give the numbers of computer keyboards assembled at the Twentieth Century Electronics Company for a sample of 12 days.

45   52   48   41   56   46   44   42   48   53   43   50

Calculate the mean, median, and mode for these data.

3.23 The following data give the 1998 revenues (in millions of dollars) of the eight automotive retailing and service companies listed in *Fortune* magazine's top 1000 U.S. companies (*Fortune*, April 26, 1999). The data, entered in that order, are for Republic Industries, Ryder Systems, United Auto Group, Budget Group, Avis Rent a Car, Group 1 Automotive, Sonic Automotive, and AMERCO.

17,487   5189   3045   2616   2298   1630   1604   1410

a. Calculate the mean and median for these data.
b. Do these data contain an outlier? If so, drop the outlier and recalculate the mean and median. Which of these two summary measures changes by a larger amount when you drop the outlier?
c. Which is the better summary measure for these data, the mean or the median? Explain.

3.24 The following data give the 1998 revenues (in millions of dollars) for the nine soap and cosmetics companies in *Fortune* magazine's top 1000 U.S. companies (*Fortune*, April 26, 1999). The data, entered in that order, are for Procter & Gamble, Colgate-Palmolive, Avon Products, Estee Lauder, Clorox, Revlon, Alberto-Culver, Dial, and International Flavors & Fragrance.

37,154   8972   5213   3618   2741   2252   1835   1525   1407

a. Calculate the mean and median for these data.
b. Do these data contain an outlier? If yes, drop the outlier and recalculate the mean and median. Which of the two summary measures changes by a larger amount when you drop the outlier?
c. Is the mean or the median a better summary measure for these data? Explain.

*3.25 One property of the mean is that if we know the means and sample sizes of two (or more) data sets, we can calculate the **combined mean** of both (or all) data sets. The combined mean for two data sets is calculated by using the formula

$$\text{Combined mean} = \bar{x} = \frac{n_1\bar{x}_1 + n_2\bar{x}_2}{n_1 + n_2}$$

where n_1 and n_2 are the sample sizes of the two data sets and $\bar{x}_1$ and $\bar{x}_2$ are the means of the two data sets, respectively. Suppose a sample of 10 statistics books gave a mean price of $71 and a sample of 8 mathematics books gave a mean price of $75. Find the combined mean. (*Hint:* For this example $n_1 = 10$, $n_2 = 8$, $\bar{x}_1 = \$71$, $\bar{x}_2 = \$75$.)

*3.26 The mean score of 20 female students on a statistics test is 77 and the mean score of 15 male students on the same test is 74. Find the combined mean score.

*3.27 For any data, the sum of all values is equal to the product of the sample size and mean; that is, $\Sigma x = n\bar{x}$. Suppose the average amount of money spent on shopping by 10 persons during a given week is $85.50. Find the total amount of money spent on shopping by these 10 persons.

*3.28 The mean 1999 income for five families was $49,520. What was the total 1999 income of these five families?

*3.29 The mean age of six persons is 46 years. The ages of five of these six persons are 57, 39, 44, 51, and 37 years. Find the age of the sixth person.

*3.30 Seven airline passengers in economy class on the same flight paid an average of $361 per ticket. Because the tickets were purchased at different times and from different sources, the prices varied. The first five passengers paid $420, $210, $333, $695, and $485. The sixth and seventh tickets were purchased by a couple who paid identical fares. What price did each of them pay?

*3.31 Consider the following two data sets.

Data Set I:	12	25	37	8	41
Data Set II:	19	32	44	15	48

Notice that each value of the second data set is obtained by adding 7 to the corresponding value of the first data set. Calculate the mean for each of these two data sets. Comment on the relationship between the two means.

*3.32 Consider the following two data sets.

Data Set I:	4	8	15	9	11
Data Set II:	8	16	30	18	22

Notice that each value of the second data set is obtained by multiplying the corresponding value of the first data set by 2. Calculate the mean for each of these two data sets. Comment on the relationship between the two means.

*3.33 The **trimmed mean** is calculated by dropping a certain percentage of values from each end of a ranked data set. The trimmed mean is especially useful as a measure of central tendency when a data set contains a few outliers at each end. Suppose the following data give the ages of 10 employees of a company:

$$47 \quad 53 \quad 38 \quad 26 \quad 39 \quad 49 \quad 19 \quad 67 \quad 31 \quad 23$$

To calculate the 10% trimmed mean, first rank these data values in increasing order; then drop 10% of the smallest values and 10% of the largest values. The mean of the remaining 80% of the values will give the 10% trimmed mean. This data set contains 10 values, and 10% of 10 is 1. Hence, drop the smallest value and the largest value from this data set. The mean of the remaining 8 values will be the 10% trimmed mean. Calculate the 10% trimmed mean for this data set.

*3.34 The following data give the prices (in thousands of dollars) of 20 houses sold recently in a city.

184	197	145	209	245	187	169	238	161	290
223	278	310	179	107	271	357	195	259	199

Find the 20% trimmed mean for this data set.

3.2 MEASURES OF DISPERSION FOR UNGROUPED DATA

The measures of central tendency, such as the mean, median, and mode, do not reveal the whole picture of the distribution of a data set. Two data sets with the same mean may have completely different spreads. The variation among the values of observations for one data set may be much larger or smaller than for the other data set. (Note that the words *dispersion*, *spread*, and *variation* have the same meaning.) Consider the following two data sets on the ages of all workers in each of two small companies.

Company 1:	47	38	35	40	36	45	39
Company 2:		70	33	18	52	27	

The mean age of workers in both these companies is the same, 40 years. If we do not know the ages of individual workers in these two companies and are told only that the mean age of the workers in both companies is the same, we may deduce that the workers in these two companies have a similar age distribution. But, as we can observe, the variation in the workers' ages for each of these two companies is very different. As illustrated in the diagram, the ages of the workers in the second company have a much larger variation than the ages of the workers in the first company.

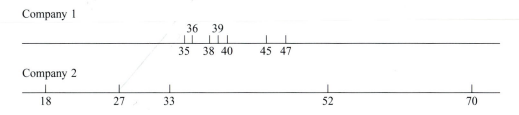

Thus, the mean, median, or mode is usually not by itself a sufficient measure to reveal the shape of the distribution of a data set. We also need a measure that can provide some information about the variation among data values. The measures that help us to know about the spread of a data set are called the **measures of dispersion**. The measures of central tendency and dispersion taken together give a better picture of a data set than the measures of central tendency alone. This section discusses three measures of dispersion: range, variance, and standard deviation.

3.2.1 RANGE

The **range** is the simplest measure of dispersion to calculate. It is obtained by taking the difference between the largest and the smallest values in a data set.

RANGE FOR UNGROUPED DATA

Range = Largest value − Smallest value

Calculating the range for ungrouped data.

EXAMPLE 3–11 Table 3.4 gives the total areas in square miles of the four western South-Central states of the United States.

Table 3.4

State	Total Area (square miles)
Arkansas	53,182
Louisiana	49,651
Oklahoma	69,903
Texas	267,277

Find the range for this data set.

Solution   The maximum total area for a state in this data set is 267,277 square miles, and the smallest area is 49,651 square miles. Therefore,

$$\text{Range} = \text{Largest value} - \text{Smallest value}$$

$$= 267{,}277 - 49{,}651 = \textbf{217{,}626 square miles}$$

Thus, the total areas of these four states are spread over a range of 217,626 square miles.

∎

The range, like the mean, has the disadvantage of being influenced by outliers. In Example 3–11, if the state of Texas with a total area of 267,277 square miles is dropped, the range decreases from 217,626 square miles to 20,252 square miles. Consequently, the range is not a good measure of dispersion to use for a data set that contains outliers.

Another disadvantage of using the range as a measure of dispersion is that its calculation is based on two values only: the largest and the smallest. All other values in a data set are ignored while calculating the range. Thus, the range is not a very satisfactory measure of dispersion.

3.2.2   VARIANCE AND STANDARD DEVIATION

The **standard deviation** is the most used measure of dispersion. The value of the standard deviation tells how closely the values of a data set are clustered around the mean. In general, a lower value of the standard deviation for a data set indicates that the values of that data set are spread over a relatively smaller range around the mean. In contrast, a larger value of the standard deviation for a data set indicates that the values of that data set are spread over a relatively larger range around the mean.

The *standard deviation is obtained by taking the positive square root of the* **variance**. The variance calculated for population data is denoted by σ^2 (read as *sigma squared*),[2] and the variance calculated for sample data is denoted by s^2. Consequently, the standard deviation calculated for population data is denoted by σ, and the standard deviation calculated for sample data is denoted by s. Following are the *basic formulas* that are used to calculate the variance:[3]

$$\sigma^2 = \frac{\Sigma(x - \mu)^2}{N} \quad \text{and} \quad s^2 = \frac{\Sigma(x - \bar{x})^2}{n - 1}$$

where σ^2 is the population variance and s^2 is the sample variance.

The quantity $x - \mu$ or $x - \bar{x}$ in the above formulas is called the *deviation* of the x value from the mean. The sum of the deviations of the x values from the mean is always zero; that is, $\Sigma(x - \mu) = 0$ and $\Sigma(x - \bar{x}) = 0$.

For example, suppose the midterm scores of a sample of four students are 82, 95, 67, and 92. Then the mean score for these four students is

$$\bar{x} = \frac{82 + 95 + 67 + 92}{4} = 84$$

The deviations of the four scores from the mean are calculated in Table 3.5. As we can observe from the table, the sum of the deviations of the x values from the mean is zero; that is, $\Sigma(x - \bar{x}) = 0$. For this reason we square the deviations to calculate the variance and standard deviation.

---

[2]Note that Σ is uppercase sigma and σ is lowercase sigma of the Greek alphabet.
[3]From the formula for σ^2, it can be stated that the population variance is the mean of the squared deviations of x values from the mean. However, this is not true for the variance calculated for a sample data set.

Table 3.5

x	$x - \bar{x}$
82	$82 - 84 = -2$
95	$95 - 84 = +11$
67	$67 - 84 = -17$
92	$92 - 84 = +8$
	$\Sigma(x - \bar{x}) = 0$

From the computational point of view, it is easier and more efficient to use *short-cut formulas* to calculate the variance and standard deviation. By using the short-cut formula, we reduce the computation time and round-off errors. Use of the basic formulas for ungrouped data is illustrated in Section A3.1.1 of Appendix 3.1 of this chapter. The short-cut formulas for calculating the variance and standard deviation are given next.

SHORT-CUT FORMULAS FOR THE VARIANCE AND STANDARD DEVIATION FOR UNGROUPED DATA

$$\sigma^2 = \frac{\Sigma x^2 - \dfrac{(\Sigma x)^2}{N}}{N} \quad \text{and} \quad s^2 = \frac{\Sigma x^2 - \dfrac{(\Sigma x)^2}{n}}{n - 1}$$

where σ^2 is the population variance and s^2 is the sample variance.

The standard deviation is obtained by taking the positive square root of the variance.

Population standard deviation: $\sigma = \sqrt{\sigma^2}$

Sample standard deviation: $s = \sqrt{s^2}$

Note that the denominator in the formula for the population variance is N but that in the formula for the sample variance it is $n - 1$.[4]

Calculating the variance and standard deviation for ungrouped data.

EXAMPLE 3–12 Refer to the data in Table 3.1 on the 1998 earnings (in millions of dollars) of five celebrities. Those data are reproduced here.

Celebrity	1998 Earnings (millions of dollars)
Mel Gibson	55
Jerry Seinfeld	267
Steven Spielberg	175
John Travolta	47
Oprah Winfrey	125

Find the variance and standard deviation for these data.

---

[4]The reason that the denominator in the sample formula is $n - 1$ and not n follows: The sample variance underestimates the population variance when the denominator in the sample formula for variance is n. However, the sample variance does not underestimate the population variance if the denominator in the sample formula for variance is $n - 1$. In Chapter 8 we will learn that $n - 1$ is called the degrees of freedom.

Solution   Let x denote the 1998 earnings (in millions of dollars) of these five celebrities. The values of Σx and Σx^2 are calculated in Table 3.6.

Table 3.6

x	x^2
55	3,025
267	71,289
175	30,625
47	2,209
125	15,625
$\Sigma x = 669$	$\Sigma x^2 = 122,773$

The calculation of the variance involves the following steps.

Step 1.   *Calculate Σx.*

The sum of the values in the first column of Table 3.6 gives the value of Σx, which is 669.

Step 2.   *Find Σx^2.*

The value of Σx^2 is obtained by squaring each value of x and then adding the squared values. The results of this step are shown in the second column of Table 3.6. Notice that $\Sigma x^2 = 122,773$.

Step 3.   *Determine the variance.*

Substitute all the values in the variance formula and simplify. Because the given data are on the earnings of a sample of five celebrities, we use the formula for the sample variance.

$$s^2 = \frac{\Sigma x^2 - \dfrac{(\Sigma x)^2}{n}}{n-1} = \frac{122,773 - \dfrac{(669)^2}{5}}{5-1} = \frac{122,773 - 89,512.2}{4} = \mathbf{8315.2}$$

Step 4.   *Obtain the standard deviation.*

The standard deviation is obtained by taking the positive square root of the variance:

$$s = \sqrt{8315.2} = \mathbf{91.187718} = \mathbf{\$91,187,718}$$

Thus, the standard deviation of the 1998 earnings of these five celebrities is $91,187,718.

☞   **TWO OBSERVATIONS**

1.   **The values of the variance and the standard deviation are never negative.** That is, the numerator in the formula for the variance should never produce a negative value. Usually the values of the variance and standard deviation are positive, but if a data set has no variation, then the variance and standard deviation are both zero. For example, if four persons in a group are the same age—say, 35 years—then the four values in the data set are

35   35   35   35

If we calculate the variance and standard deviation for these data, their values are zero. This is because there is no variation in the values of this data set.

2. **The measurement units of variance are always the square of the measurement units of the original data.** This is so because the original values are squared to calculate the variance. In Example 3–12, the measurement units of the original data are millions of dollars. However, the measurement units of the variance are squared millions of dollars, which, of course, does not make any sense. Thus, the variance of the 1998 earnings of the five celebrities in Example 3–12 is 8315.2 squared million dollars. But the measurement units of the standard deviation are the same as the measurement units of the original data because the standard deviation is obtained by taking the square root of the variance.

Calculating the variance and standard deviation for ungrouped data.

EXAMPLE 3–13 Following are the 1999 earnings (in thousands of dollars) before taxes for *all* six employees of a small company.

$$29.50 \quad 16.20 \quad 35.45 \quad 21.35 \quad 49.70 \quad 24.60$$

Calculate the variance and standard deviation for these data.

Solution Let x denote the 1999 earnings before taxes of employees of this company. The values of Σx and Σx^2 are calculated in Table 3.7.

Table 3.7

x	x^2
29.50	870.2500
16.20	262.4400
35.45	1256.7025
21.35	455.8225
49.70	2470.0900
24.60	605.1600
$\Sigma x = 176.80$	$\Sigma x^2 = 5920.4650$

Since the data on earnings are for *all* employees of this company, we use the population formula to compute the variance. Thus, the variance is

$$\sigma^2 = \frac{\Sigma x^2 - \dfrac{(\Sigma x)^2}{N}}{N} = \frac{5920.4650 - \dfrac{(176.80)^2}{6}}{6} = \mathbf{118.4597}$$

The standard deviation is obtained by taking the (positive) square root of the variance:

$$\sigma = \sqrt{118.4597} = \textbf{\$10.884 thousand} = \textbf{\$10,884}$$

Thus, the standard deviation of the 1999 earnings of all six employees of this company is $10,884. ∎

☞ *Warning* Note that Σx^2 is not the same as $(\Sigma x)^2$. The value of Σx^2 is obtained by squaring the x values and adding them. The value of $(\Sigma x)^2$ is obtained by squaring the value of Σx.

The uses of the standard deviation are discussed in Section 3.4. Later chapters explain how the mean and the standard deviation taken together can help in making inferences about the population.

3.2.3 POPULATION PARAMETERS AND SAMPLE STATISTICS

A numerical measure such as the mean, median, mode, range, variance, or standard deviation calculated for a population data set is called a *population parameter*, or simply a **parameter**. A summary measure calculated for a sample data set is called a *sample statistic*, or simply a **statistic**. Thus, μ and σ are population parameters, and $\bar{x}$ and s are sample statistics. As an illustration, $\bar{x} = \$133.8$ million in Example 3–1 is a sample statistic, and $\mu = 45.25$ years in Example 3–2 is a population parameter. Similarly, $s = \$91,187,718$ in Example 3–12 is a sample statistic, whereas $\sigma = \$10,884$ in Example 3–13 is a population parameter.

EXERCISES

■ **Concepts and Procedures**

3.35 The range, as a measure of spread, has the disadvantage of being influenced by outliers. Illustrate this with an example.

3.36 Can the standard deviation have a negative value? Explain.

3.37 When is the value of the standard deviation for a data set zero? Give one example. Calculate the standard deviation for the example and show that its value is zero.

3.38 Briefly explain the difference between a population parameter and a sample statistic. Give one example of each.

3.39 The following data set belongs to a population:

$$5 \quad -7 \quad 2 \quad 0 \quad -9 \quad 16 \quad 10 \quad 7$$

Calculate the range, variance, and standard deviation.

3.40 The following data set belongs to a sample:

$$14 \quad 18 \quad -10 \quad 8 \quad 8 \quad -16$$

Calculate the range, variance, and standard deviation.

■ **Applications**

3.41 The following data give the weekly food expenditures for a sample of five families.

$$\$82 \quad 116 \quad 65 \quad 170 \quad 92$$

a. Find the mean for these data. Calculate the deviations of the data values from the mean. Is the sum of these deviations zero?
b. Calculate the range, variance, and standard deviation.

3.42 A sample of seven statistics books produced the following prices:

$$\$56 \quad 75 \quad 68 \quad 81 \quad 71 \quad 96 \quad 78$$

a. Find the mean for these data. Calculate the deviations of the data values from the mean. Is the sum of these deviations zero?
b. Calculate the range, variance, and standard deviation.

3.43 The following data give the numbers of car thefts that occurred in a city in the past 12 days.

$$6 \quad 3 \quad 7 \quad 11 \quad 4 \quad 3 \quad 8 \quad 7 \quad 2 \quad 6 \quad 9 \quad 15$$

Calculate the range, variance, and standard deviation.

3.44 The following table gives the 1999–2000 average ticket prices for the 10 National Basketball Association teams with the highest prices.

Team	Average Price of a Ticket	Team	Average Price of a Ticket
New York	$86.82	New Jersey	$59.22
L. A. Lakers	81.89	Utah	54.60
Seattle	64.60	Chicago	52.84
Houston	62.63	Portland	52.28
Washington	59.65	Boston	49.79

Source: USA TODAY, November 12, 1999.

Find the range, variance, and standard deviation for these data.

3.45 The following data give the numbers of pieces of junk mail received by 10 families during the past month.

41   33   28   21   29   19   14   31   39   36

Find the range, variance, and standard deviation.

3.46 The following data give the numbers of new cars sold at a dealership during a 12-day period.

13   5   9   6   8   11   9   15   4   11   7   5

Find the range, variance, and standard deviation.

3.47 The following data give the weight (in pounds) lost by 15 new members of a health club at the end of their first two months of membership.

5   10   8   7   25   12   5   14
11   10   21   9   8   11   18

Compute the range, variance, and standard deviation.

3.48 The following data give the speeds (in miles per hour), as measured by radar, of 13 cars traveling on interstate highway I-84.

82   72   73   66   76   69   71
76   80   79   68   78   71

Calculate the range, variance, and standard deviation.

3.49 Following are the temperatures (in degrees Fahrenheit) observed during eight wintry days in a midwestern city:

23   14   6   −7   −2   11   16   19

Compute the range, variance, and standard deviation.

3.50 The following data give the numbers of hours spent partying by 10 randomly selected college students during the past week.

7   14   5   0   9   7   10   4   0   8

Compute the range, variance, and standard deviation.

3.51 The following data give the market values (in billions of U.S. dollars) of the 10 largest Canadian companies as of May 29, 1998 (*Business Week*, July 13, 1998). The data, entered in that order, are for Northern Telecom, BCE, Royal Bank of Canada, Thomson, Seagram, Bank of Montreal, Canadian Imperial Bank of Commerce, Toronto-Dominion Bank, Bank of Nova Scotia, and Canadian Pacific.

| 33.8 | 29.4 | 18.9 | 17.2 | 15.2 | 14.7 | 14.1 | 13.3 | 13.2 | 9.8 |

Find the range, variance, and standard deviation.

3.52   The following data from the U.S. Department of Commerce are the 1998 per-capita incomes for eight western states (*USA TODAY*, April 28, 1999). The per-capita incomes, entered in that order, are for Arizona, Colorado, Idaho, Montana, Nevada, New Mexico, Utah, and Wyoming.

| 23,060 | 28,657 | 21,801 | 20,172 | 27,200 | 19,936 | 21,019 | 23,167 |

Find the range, variance, and standard deviation.

3.53   The following data give the hourly wage rates of eight employees of a company.

| 12 | 12 | 12 | 12 | 12 | 12 | 12 | 12 |

Calculate the standard deviation. Is its value zero? If yes, why?

3.54   The following data are the ages (in years) of six students.

| 19 | 19 | 19 | 19 | 19 | 19 |

Calculate the standard deviation. Is its value zero? If yes, why?

*3.55   One disadvantage of the standard deviation as a measure of dispersion is that it is a measure of absolute variability and not of relative variability. Sometimes we may need to compare the variability of two different data sets that have different units of measurement. The **coefficient of variation** is one such measure. The coefficient of variation, denoted by CV, expresses standard deviation as a percentage of the mean and is computed as follows:

$$\text{For population data:}\quad CV = \frac{\sigma}{\mu} \times 100\%$$

$$\text{For sample data:}\quad CV = \frac{s}{\bar{x}} \times 100\%$$

The yearly salaries of all employees who work for a company have a mean of $42,350 and a standard deviation of $3,820. The years of schooling for the same employees have a mean of 15 years and a standard deviation of 2 years. Is the relative variation in the salaries greater or less than that in years of schooling for these employees?

*3.56   The SAT scores of 100 students have a mean of 915 and a standard deviation of 105. The GPAs of the same 100 students have a mean of 3.06 and a standard deviation of .22. Is the relative variation in SAT scores greater or less than that in GPAs?

*3.57   Consider the following two data sets.

| Data Set I: | 12 | 25 | 37 | 8 | 41 |
| Data Set II: | 19 | 32 | 44 | 15 | 48 |

Note that each value of the second data set is obtained by adding 7 to the corresponding value of the first data set. Calculate the standard deviation for each of these two data sets using the formula for sample data. Comment on the relationship between the two standard deviations.

*3.58   Consider the following two data sets.

| Data Set I: | 4 | 8 | 15 | 9 | 11 |
| Data Set II: | 8 | 16 | 30 | 18 | 22 |

Note that each value of the second data set is obtained by multiplying the corresponding value of the first data set by 2. Calculate the standard deviation for each of these two data sets using the formula for population data. Comment on the relationship between the two standard deviations.

3.3 MEAN, VARIANCE, AND STANDARD DEVIATION FOR GROUPED DATA

In Sections 3.1.1 and 3.2.2, we learned how to calculate the mean, variance, and standard deviation for ungrouped data. In this section, we will learn how to calculate the mean, variance, and standard deviation for grouped data.

3.3.1 MEAN FOR GROUPED DATA

We learned in Section 3.1.1 that the mean is obtained by dividing the sum of all values by the number of values in a data set. However, if the data are given in the form of a frequency table, we no longer know the values of individual observations. Consequently, in such cases, we cannot obtain the sum of individual values. We find an approximation for the sum of these values using the procedure explained in the next paragraph and example. The formulas used to calculate the mean for grouped data follow.

MEAN FOR GROUPED DATA

Mean for population data: $\mu = \dfrac{\Sigma mf}{N}$

Mean for sample data: $\bar{x} = \dfrac{\Sigma mf}{n}$

where m is the midpoint and f is the frequency of a class.

To calculate the mean for grouped data, first find the midpoint of each class and then multiply the midpoints by the frequencies of the corresponding classes. The sum of these products, denoted by Σmf, gives an approximation for the sum of all values. To find the value of the mean, divide this sum by the total number of observations in the data.

Calculating the population mean for grouped data.

EXAMPLE 3–14 Table 3.8 gives the frequency distribution of the daily commuting times (in minutes) from home to work for *all* 25 employees of a company.

Table 3.8

Daily Commuting Time (minutes)	Number of Employees
0 to less than 10	4
10 to less than 20	9
20 to less than 30	6
30 to less than 40	4
40 to less than 50	2

Calculate the mean of the daily commuting times.

Solution   Note that because the data set includes *all* 25 employees of the company, it represents the population. Table 3.9 shows the calculation of Σmf. Note that in Table 3.9, *m* denotes the midpoints of the classes.

Table 3.9

Daily Commuting Time (minutes)	*f*	*m*	*mf*
0 to less than 10	4	5	20
10 to less than 20	9	15	135
20 to less than 30	6	25	150
30 to less than 40	4	35	140
40 to less than 50	2	45	90
	$N = 25$		$\Sigma mf = 535$

To calculate the mean, we first find the midpoint of each class. The class midpoints are recorded in the third column of Table 3.9. The products of the midpoints and the corresponding frequencies are listed in the fourth column. The sum of the fourth column values, denoted by Σmf, gives the approximate total daily commuting time (in minutes) for all 25 employees. The mean is obtained by dividing this sum by the total frequency. Therefore,

$$\mu = \frac{\Sigma mf}{N} = \frac{535}{25} = \textbf{21.40 minutes}$$

Thus, the employees of this company spend an average of 21.40 minutes a day commuting from home to work. ∎

What do the numbers 20, 135, 150, 140, and 90 in the column labeled *mf* in Table 3.9 represent? We know from this table that 4 employees spend 0 to less than 10 minutes commuting per day. If we assume that the time spent commuting by these 4 employees is evenly spread in the interval 0 to less than 10, then the midpoint of this class (which is 5) gives the mean time spent commuting by these 4 employees. Hence, $4 \times 5 = 20$ is the approximate total time (in minutes) spent commuting per day by these 4 employees. Similarly, 9 employees spend 10 to less than 20 minutes commuting per day, and the total time spent commuting by these 9 employees is approximately 135 minutes a day. The other numbers in this column can be interpreted in the same way. Note that these numbers give the approximate commuting times for these employees based on the assumption of an even spread within classes. The total commuting time for all 25 employees is approximately 535 minutes. Consequently, 21.40 minutes is an approximate and not the exact value of the mean. We can find the exact value of the mean only if we know the exact commuting time for each of the 25 employees of the company.

Calculating the sample mean for grouped data.   **EXAMPLE 3–15**   Table 3.10 gives the frequency distribution of the number of orders received each day during the past 50 days at the office of a mail-order company.

Table 3.10

Number of Orders	Number of Days
10–12	4
13–15	12
16–18	20
19–21	14

Calculate the mean.

Solution Because the data set includes only 50 days, it represents a sample. The value of Σmf is calculated in Table 3.11.

Table 3.11

Number of Orders	f	m	mf
10–12	4	11	44
13–15	12	14	168
16–18	20	17	340
19–21	14	20	280
	$n = 50$		$\Sigma mf = 832$

The value of the sample mean is

$$\bar{x} = \frac{\Sigma mf}{n} = \frac{832}{50} = \textbf{16.64 orders}$$

Thus, this mail-order company received an average of 16.64 orders per day during these 50 days. ∎

3.3.2 VARIANCE AND STANDARD DEVIATION FOR GROUPED DATA

Following are the *basic formulas* used to calculate the population and sample variances for grouped data:

$$\sigma^2 = \frac{\Sigma f(m - \mu)^2}{N} \quad \text{and} \quad s^2 = \frac{\Sigma f(m - \bar{x})^2}{n - 1} ✓$$

where σ^2 is the population variance, s^2 is the sample variance, and m is the midpoint of a class.

In either case, the standard deviation is obtained by taking the positive square root of the variance.

Again, the *short-cut formulas* are more efficient for calculating the variance and standard deviation. Section A3.1.2 of Appendix 3.1 at the end of this chapter shows how to use the basic formulas to calculate the variance and standard deviation for grouped data.

SHORT-CUT FORMULAS FOR THE VARIANCE AND STANDARD DEVIATION FOR GROUPED DATA

$$\sigma^2 = \frac{\Sigma m^2 f - \dfrac{(\Sigma mf)^2}{N}}{N} \quad \text{and} \quad s^2 = \frac{\Sigma m^2 f - \dfrac{(\Sigma mf)^2}{n}}{n - 1}$$

where σ^2 is the population variance, s^2 is the sample variance, and m is the midpoint of a class.

The standard deviation is obtained by taking the positive square root of the variance.

Population standard deviation: $\sigma = \sqrt{\sigma^2}$

Sample standard deviation: $s = \sqrt{s^2}$

Examples 3–16 and 3–17 illustrate the use of these formulas to calculate the variance and standard deviation.

Calculating the population variance and standard deviation for grouped data.

EXAMPLE 3–16   The following data, reproduced from Table 3.8, give the frequency distribution of the daily commuting times (in minutes) from home to work for all 25 employees of a company.

Daily Commuting Time (minutes)	Number of Employees
0 to less than 10	4
10 to less than 20	9
20 to less than 30	6
30 to less than 40	4
40 to less than 50	2

Calculate the variance and standard deviation.

Solution   All four steps needed to calculate the variance and standard deviation for grouped data are shown after Table 3.12.

Table 3.12

Daily Commuting Time (minutes)	f	m	mf	m^2f
0 to less than 10	4	5	20	100
10 to less than 20	9	15	135	2,025
20 to less than 30	6	25	150	3,750
30 to less than 40	4	35	140	4,900
40 to less than 50	2	45	90	4,050
	$N = 25$		$\Sigma mf = 535$	$\Sigma m^2f = 14,825$

Step 1.   *Calculate the value of Σmf.*
To calculate the value of Σmf, first find the midpoint m of each class (see the third column in Table 3.12) and then multiply the corresponding class midpoints and class frequencies (see the fourth column). The value of Σmf is obtained by adding these products. Thus,

$$\Sigma mf = 535$$

Step 2.   *Find the value of Σm^2f.*
To find the value of Σm^2f, square each m value and multiply this squared value of m by the corresponding frequency (see the fifth column in Table 3.12). The sum of these products (that is, the sum of the fifth column) gives Σm^2f. Hence,

$$\Sigma m^2f = 14,825$$

Step 3.   *Calculate the variance.*
Because the data set includes all 25 employees of the company, it represents the population. Therefore, we use the formula for the population variance:

$$\sigma^2 = \frac{\Sigma m^2f - \dfrac{(\Sigma mf)^2}{N}}{N} = \frac{14,825 - \dfrac{(535)^2}{25}}{25} = \frac{3376}{25} = \mathbf{135.04}$$

Step 4. *Calculate the standard deviation.*

To obtain the standard deviation, take the (positive) square root of the variance.

$$\sigma = \sqrt{\sigma^2} = \sqrt{135.04} = \textbf{11.62 minutes}$$

Thus, the standard deviation of the daily commuting times for these employees is 11.62 minutes. ■

Note that the values of the variance and standard deviation calculated in Example 3–16 for grouped data are approximations. The exact values of the variance and standard deviation can be obtained only by using the ungrouped data on the daily commuting times of the 25 employees.

Calculating the sample variance and standard deviation for grouped data.

EXAMPLE 3-17   The following data, reproduced from Table 3.10, give the frequency distribution of the number of orders received each day during the past 50 days at the office of a mail-order company.

Number of Orders	f
10–12	4
13–15	12
16–18	20
19–21	14

Calculate the variance and standard deviation.

Solution   All the information required for the calculation of the variance and standard deviation appears in Table 3.13.

Table 3.13

Number of Orders	f	m	mf	m^2f
10–12	4	11	44	484
13–15	12	14	168	2,352
16–18	20	17	340	5,780
19–21	14	20	280	5,600
	$n = 50$		$\Sigma mf = 832$	$\Sigma m^2f = 14{,}216$

Because the data set includes only 50 days, it represents a sample. Hence, we use the sample formulas to calculate the variance and standard deviation. By substituting the values into the formula for the sample variance, we obtain

$$s^2 = \frac{\Sigma m^2 f - \dfrac{(\Sigma mf)^2}{n}}{n - 1} = \frac{14{,}216 - \dfrac{(832)^2}{50}}{50 - 1} = \textbf{7.5820}$$

Hence, the standard deviation is

$$s = \sqrt{s^2} = \sqrt{7.5820} = \textbf{2.75 orders}$$

Thus, the standard deviation of the number of orders received at the office of this mail-order company during the past 50 days is 2.75. ■

EXERCISES

■ Concepts and Procedures

3.59 Are the values of the mean and standard deviation that are calculated using grouped data exact or approximate values of the mean and standard deviation, respectively? Explain.

3.60 Using the population formulas, calculate the mean, variance, and standard deviation for the following grouped data.

x	2–4	5–7	8–10	11–13	14–16
f	5	9	14	7	5

3.61 Using the sample formulas, find the mean, variance, and standard deviation for the grouped data displayed in the following table.

x	f
0 to less than 4	17
4 to less than 8	23
8 to less than 12	15
12 to less than 16	11
16 to less than 20	8
20 to less than 24	6

■ Applications

3.62 The following table gives the frequency distribution of the amounts of telephone bills for October 1999 for a sample of 50 families.

Amount of Telephone Bill (dollars)	Number of Families
20 to less than 40	8
40 to less than 60	13
60 to less than 80	17
80 to less than 100	9
100 to less than 120	3

Calculate the mean, variance, and standard deviation.

3.63 The following table gives the frequency distribution of total hours spent studying statistics during the semester for all 40 students enrolled in an introductory statistics course during Spring 2000.

Hours of Study	Number of Students
24 to less than 40	3
40 to less than 56	5
56 to less than 72	10
72 to less than 88	12
88 to less than 104	5
104 to less than 120	5

Find the mean, variance, and standard deviation.

3.64 The following table gives the grouped data on the weights of all 100 babies born at a hospital in 1999.

Weight (pounds)	Number of Babies
3 to less than 5	5
5 to less than 7	30
7 to less than 9	40
9 to less than 11	20
11 to less than 13	5

Find the mean, variance, and standard deviation.

3.65   The following table gives the frequency distribution of the numbers of personal computers sold during the past month at 40 computer stores in New York City.

Computers Sold	Number of Stores
4 to 12	6
13 to 21	9
22 to 30	14
31 to 39	7
40 to 48	4

Calculate the mean, variance, and standard deviation. Give a brief interpretation of the values in the column labeled *mf* in your table of calculations. What does Σ*mf* represent?

3.66   The following table gives information on the amounts (in dollars) of electric bills for August 1999 for a sample of 50 families.

Amount of Electric Bill (dollars)	Number of Families
0 to less than 20	5
20 to less than 40	16
40 to less than 60	11
60 to less than 80	10
80 to less than 100	8

Find the mean, variance, and standard deviation. Give a brief interpretation of the values in the column labeled *mf* in your table of calculations. What does Σ*mf* represent?

3.67   For 50 airplanes that arrived late at an airport during a week, the time by which they were late was observed. In the following table, *x* denotes the time (in minutes) by which an airplane was late and *f* denotes the number of airplanes.

x	*f*
0 to less than 20	14
20 to less than 40	18
40 to less than 60	9
60 to less than 80	5
80 to less than 100	4

Find the mean, variance, and standard deviation.

3.68   The table at the top of the next page gives the frequency distribution of the numbers of cars owned by 100 households.

Number of Cars Owned	Number of Households
0	12
1	40
2	30
3	15
4	3

Calculate the mean, variance, and standard deviation. (*Hint:* The classes in this example are single-valued. These values of classes [the number of cars owned] will be used as values of m in the formula for the mean, variance, and standard deviation.)

3.69   The following table gives the numbers of television sets owned by 80 households.

Number of Television Sets Owned	Number of Households
0	4
1	33
2	28
3	10
4	5

Find the mean, variance, and standard deviation. (*Hint:* The classes in this example are single-valued. These values of classes [the number of television sets owned] will be used as values of m in the formula for the mean, variance, and standard deviation.)

3.4   USE OF STANDARD DEVIATION

By using the mean and standard deviation, we can find the proportion or percentage of the total observations that fall within a given interval about the mean. This section briefly discusses Chebyshev's theorem and the empirical rule, both of which demonstrate this use of the standard deviation.

3.4.1   CHEBYSHEV'S THEOREM

Chebyshev's theorem gives a lower bound for the area under a curve between two points that are on opposite sides of the mean and at the same distance from the mean.

> **CHEBYSHEV'S THEOREM**   For any number k greater than 1, at least $(1 - 1/k^2)$ of the data values lie within k standard deviations of the mean.

Figure 3.5 illustrates Chebyshev's theorem.

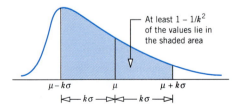

Figure 3.5   Chebyshev's theorem.

Thus, for example, if $k = 2$, then

$$1 - \frac{1}{k^2} = 1 - \frac{1}{(2)^2} = 1 - \frac{1}{4} = 1 - .25 = .75 \text{ or } 75\%$$

Therefore, according to Chebyshev's theorem, at least .75 or 75% of the values of a data set lie within two standard deviations of the mean. This is shown in Figure 3.6.

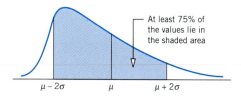

At least 75% of the values lie in the shaded area

Figure 3.6 Percentage of values within two standard deviations of the mean for Chebyshev's theorem.

If $k = 3$, then,

$$1 - \frac{1}{k^2} = 1 - \frac{1}{(3)^2} = 1 - \frac{1}{9} = 1 - .11 = .89 \text{ or } 89\% \text{ approximately}$$

According to Chebyshev's theorem, at least .89 or 89% of the values fall within three standard deviations of the mean. This is shown in Figure 3.7.

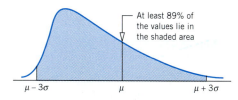

At least 89% of the values lie in the shaded area

Figure 3.7 Percentage of values within three standard deviations of the mean for Chebyshev's theorem.

Although in Figures 3.5 through 3.7 we have used the population notation for the mean and standard deviation, the theorem applies to sample and population data. Note that Chebyshev's theorem is applicable to a distribution of any shape. However, Chebyshev's theorem can be used only for $k > 1$. This is so because when $k = 1$, the value of $1 - 1/k^2$ is zero, and when $k < 1$, the value of $1 - 1/k^2$ is negative.

Applying Chebyshev's theorem.

EXAMPLE 3-18 For a statistics class, the mean for the midterm scores is 75 and the standard deviation is 8. Using Chebyshev's theorem, find the percentage of students who scored between 59 and 91.

Solution Let μ and σ be the mean and the standard deviation, respectively, of the midterm scores. Then from the given information,

$$\mu = 75 \quad \text{and} \quad \sigma = 8$$

To find the percentage of students who scored between 59 and 91, the first step is to determine k. As shown below, each of the two points, 59 and 91, is 16 units away from the mean.

$$59 - 75 = -16 \qquad 91 - 75 = 16$$

59	75	91

The value of k is obtained by dividing the distance between the mean and each point by the standard deviation. Thus,

$$k = 16/8 = 2$$

$$1 - \frac{1}{k^2} = 1 - \frac{1}{(2)^2} = 1 - \frac{1}{4} = 1 - .25 = .75 \text{ or } \mathbf{75\%}$$

Hence, according to Chebyshev's theorem, at least 75% of the students scored between 59 and 91. This percentage is shown in Figure 3.8.

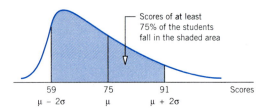

Scores of at least 75% of the students fall in the shaded area

| 59 | 75 | 91 | Scores |
| $\mu - 2\sigma$ | μ | $\mu + 2\sigma$ | |

Figure 3.8 Percentage of students who scored between 59 and 91.

3.4.2 EMPIRICAL RULE

Whereas Chebyshev's theorem is applicable to any kind of distribution, the **empirical rule** applies only to a specific type of distribution called a *bell-shaped distribution*, as shown in Figure 3.9. More will be said about such a distribution in Chapter 6, where it is called a *normal curve*. In this section, only the following three rules for the curve are given.

EMPIRICAL RULE For a bell-shaped distribution, approximately

1. 68% of the observations lie within one standard deviation of the mean
2. 95% of the observations lie within two standard deviations of the mean
3. 99.7% of the observations lie within three standard deviations of the mean

---

Figure 3.9 illustrates the empirical rule. Again, the empirical rule applies to population data as well as to sample data.

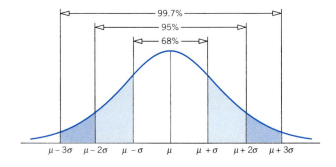

| $\mu - 3\sigma$ | $\mu - 2\sigma$ | $\mu - \sigma$ | μ | $\mu + \sigma$ | $\mu + 2\sigma$ | $\mu + 3\sigma$ |

Figure 3.9 Illustration of the empirical rule.

Applying the empirical rule.

EXAMPLE 3–19 The age distribution of a sample of 5000 persons is bell-shaped with a mean of 40 years and a standard deviation of 12 years. Determine the approximate percentage of people who are 16 to 64 years old.

Solution We use the empirical rule to find the required percentage because the distribution of ages follows a bell-shaped curve. From the given information, for this distribution,

$$\bar{x} = 40 \text{ years} \quad \text{and} \quad s = 12 \text{ years}$$

Each of the two points, 16 and 64, is 24 units away from the mean. Dividing 24 by 12, we convert the distance between each of the two points and the mean in terms of standard deviations. Thus, the distance between 16 and 40 and between 40 and 64 is each equal to $2s$. Consequently, as shown in Figure 3.10, the area from 16 to 64 is the area from $\bar{x} - 2s$ to $\bar{x} + 2s$.

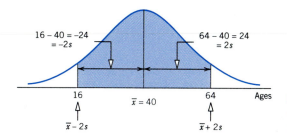

Figure 3.10 contents:

16 – 40 = –24 = –2s

64 – 40 = 24 = 2s

16

$\bar{x} = 40$

64

Ages

$\bar{x} - 2s$

$\bar{x} + 2s$

Figure 3.10 Percentage of people who are 16 to 64 years old.

Because the area within two standard deviations of the mean is approximately 95% for a bell-shaped curve, approximately **95%** of the people in the sample are 16 to 64 years old.

■

The following case study, reproduced from *Fortune* magazine, describes the application of the standard deviation to annual returns on mutual funds, assuming these returns are bell-shaped. Read this case study again after finishing Chapter 6.

CASE STUDY 3-3   HERE COMES THE SD

When your servant first became a *Fortune* writer several decades ago, it was hard doctrine that "several" meant three to eight, also that writers must not refer to "gross national product" without pausing to define this arcane term. GNP was in fact a relatively new concept at the time, having been introduced to the country only several years previously—in Roosevelt's 1944 budget message—so the presumption that readers had to be told repeatedly it was the "value of all goods and services produced by the economy" seemed entirely reasonable to this young writer, who personally had to look up the definition every time.

Numeracy lurches on. Nowadays the big question for editors is whether an average college-educated bloke needs a handhold when confronted with the term "standard deviation." The SD is suddenly onstage because the Securities and Exchange Commission is wondering aloud whether investment companies should be required to tell investors the standard deviation of their mutual funds' total returns over various past periods. Barry Barbash, SEC director of investment management, favors the requirement but confessed to the Washington

Post that he worries about investors who will think a standard deviation is the dividing line on a highway or something.

The view around here is that the SEC is performing a noble service, but only partly because the requirement would enhance folks' insights into mutual funds. The commission's underlying idea is to give investors a better and more objective measure than is now available of the risk associated with different kinds of portfolios. The SD is a measure of variability, and funds with unusually variable returns—sometimes very high, sometimes very low—are presumed to be more risky.

What one really likes about the proposal, however, is the prospect that it will incentivize millions of greedy Americans to learn a little elementary statistics. One already has a list of issues that could be discussed much more thrillingly if only your average liberal arts graduate had a glimmer about the SD and the normal curve. The bell-shaped normal curve, or rather, the area underneath the curve, shows you how Providence arranged for things to be distributed in our world—with people's heights, or incomes, or IQs, or investment returns bunched around middling outcomes, and fewer and fewer cases as you move down and out toward the extremes. A line down the center of the curve represents the mean outcome, and deviations from the mean are measured by the SD.

An amazing property of the SD is that exactly 68.26% of all normally distributed data are within one SD of the mean. We once asked a professor of statistics a question that seemed to us quite profound, to wit, why that particular figure? The Prof answered dismissively that God had decided on 68.26% for exactly the same reason He had landed on 3.14 as the ratio between circumferences and diameters—because He just felt like it. The Almighty has also proclaimed that 95.44% of all data are within two SDs of the mean, and 99.73% within three SDs. When you know the mean and SD of some outcome, you can instantly establish the percentage probability of its occurrence. White men's heights in the U.S. average 69.2 inches, with an SD of 2.8 inches (according to the National Center for Health Statistics), which means that a 6-foot-5 chap is in the 99th percentile. In 1994, scores on the verbal portion of the Scholastic Assessment Test had a mean of 423 and an SD of 113, so if you scored 649—two SDs above the mean—you were in the 95th percentile.

As the SEC is heavily hinting, average outcomes are interesting but for many purposes inadequate; one also yearns to know the variability around that average. From 1926 through 1994, the S&P 500 had an average annual return of just about 10%. The SD accompanying that figure was just about 20%. Since returns will be within 1 SD some 68% of the time, they will be more than 1 SD from the mean 32% of the time. And since half these swings will be on the downside, we expect fund owners to lose more than 10% of their money about one year out of six and to lose more than 30% (two SDs below the mean) about one year out of 20. If your time horizon is short and you can't take losses like that, you arguably don't belong in stocks. If you think SDs are highway dividers, you arguably don't belong in cars.

EXERCISES

■ **Concepts and Procedures**

3.70   Briefly explain Chebyshev's theorem and its applications.

3.71   Briefly explain the empirical rule. To what kind of distribution is it applied?

3.72 A sample of 2000 observations has a mean of 74 and a standard deviation of 12. Using Chebyshev's theorem, find at least what percentage of the observations fall in the intervals $\bar{x} \pm 2s$, $\bar{x} \pm 2.5s$, and $\bar{x} \pm 3s$. Note that here the interval $\bar{x} \pm 2s$ represents the interval $\bar{x} - 2s$ to $\bar{x} + 2s$, and so on.

3.73 A large population has a mean of 230 and a standard deviation of 41. Using Chebyshev's theorem, find at least what percentage of the observations fall in the intervals $\mu \pm 2\sigma$, $\mu \pm 2.5\sigma$, and $\mu \pm 3\sigma$.

3.74 A large population has a mean of 310 and a standard deviation of 37. Using the empirical rule, find what percentage of the observations fall in the intervals $\mu \pm 1\sigma$, $\mu \pm 2\sigma$, and $\mu \pm 3\sigma$.

3.75 A sample of 3000 observations has a mean of 82 and a standard deviation of 16. Using the empirical rule, find what percentage of the observations fall in the intervals $\bar{x} \pm 1s$, $\bar{x} \pm 2s$, and $\bar{x} \pm 3s$.

■ **Applications**

3.76 The mean time taken by all participants to run a road race was found to be 220 minutes with a standard deviation of 20 minutes. Using Chebyshev's theorem, find the percentage of runners who ran this road race in

a. 180 to 260 minutes   **b.** 160 to 280 minutes   **c.** 170 to 270 minutes

3.77 The 1999 gross sales of all firms in a large city have a mean of $2.3 million and a standard deviation of $.6 million. Using Chebyshev's theorem, find at least what percentage of firms in this city had 1999 gross sales of

a. $1.1 to $3.5 million   **b.** $.8 to $3.8 million   **c.** $.5 to $4.1 million

3.78 According to data from the College Board, the average cost to students for books and supplies was $634 at public 4-year colleges in the United States for the 1997–1998 school year (*The Chronicle of Higher Education Almanac*, August 28, 1998). Assume that these costs have a mean of $634 and a standard deviation of $140.

  a. Using Chebyshev's theorem, find at least what percentage of all such costs are between
     i. $284 and $984   ii. $214 and $1054
*b. Using Chebyshev's theorem, find the interval that contains at least 75% of all such costs.

3.79 The mean monthly mortgage paid by all home owners in a city is $1365 with a standard deviation of $240.

  a. Using Chebyshev's theorem, find at least what percentage of all home owners in the city pay a monthly mortgage of
     i. $885 to $1845   ii. $645 to $2085
*b. Using Chebyshev's theorem, find the interval that contains the monthly mortgage payments of at least 84% of all home owners.

3.80 The mean life of a certain brand of auto batteries is 44 months with a standard deviation of 3 months. Assume that the lives of all auto batteries of this brand have a bell-shaped distribution. Using the empirical rule, find the percentage of auto batteries of this brand that have a life of

a. 41 to 47 months   **b.** 38 to 50 months   **c.** 35 to 53 months

3.81 According to the American Federation of Teachers, the mean salary of teachers was $38,436 in 1996–1997 (*Education Week,* May 27, 1998). Assume that the 1996–1997 salaries of all teachers had a bell-shaped distribution with a mean of $38,436 and a standard deviation of $3500. Using the empirical rule, find the percentage of teachers whose 1996–1997 salaries were between

a. $31,436 and $45,436   **b.** $34,936 and $41,936   **c.** $27,936 and $48,936

3.82 According to CACI, a geodemographic firm, in 1997 the average household donation to United Way for 161 counties in the southwestern and northeastern United States was $64.73 (*American Demographics*, August 1998). Assume that these donations have a bell-shaped distribution with a mean of $64.73 and a standard deviation of $18.

  a. Using the empirical rule, find the percentage of such donations that were between
     i. $46.73 and $82.73   ii. $28.73 and $100.73
*b. Using the empirical rule, find the interval that contains 99.7% of all such donations.

3.83   The ages of cars owned by all employees of a large company have a bell-shaped distribution with a mean of 7 years and a standard deviation of 2 years.

   a.   Using the empirical rule, find the percentage of cars owned by these employees that are
      i.   5 to 9 years old   ii.   1 to 13 years old

*b.   Using the empirical rule, find the interval that contains the ages of the cars owned by 95% of the employees of this company.

3.5   MEASURES OF POSITION

A **measure of position** determines the position of a single value in relation to other values in a sample or a population data set. There are many measures of position; however, only quartiles, percentiles, and percentile rank are discussed in this section.

3.5.1   QUARTILES AND INTERQUARTILE RANGE

Quartiles are the summary measures that divide a ranked data set into four equal parts. Three measures will divide any data set into four equal parts. These three measures are the **first quartile** (denoted by Q_1), the **second quartile** (denoted by Q_2), and the **third quartile** (denoted by Q_3). The data should be ranked in increasing order before the quartiles are determined. The quartiles are defined as follows.

> **QUARTILES**   *Quartiles* are three summary measures that divide a ranked data set into four equal parts. The second quartile is the same as the median of a data set. The first quartile is the value of the middle term among the observations that are less than the median, and the third quartile is the value of the middle term among the observations that are greater than the median.

Figure 3.11 describes the positions of the three quartiles.

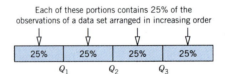

Each of these portions contains 25% of the observations of a data set arranged in increasing order

| 25% | 25% | 25% | 25% |

Q_1 Q_2 Q_3

Figure 3.11   Quartiles.

Approximately 25% of the values in a ranked data set are less than Q_1 and about 75% are greater than Q_1. The second quartile, Q_2, divides a ranked data set into two equal parts; hence, the second quartile and the median are the same. Approximately 75% of the data values are less than Q_3 and about 25% are greater than Q_3.

The difference between the third quartile and the first quartile for a data set is called the **interquartile range (IQR).**

> **INTERQUARTILE RANGE**   The difference between the third and the first quartiles gives the *interquartile range*; that is,
>
> $$\text{IQR} = \text{Interquartile range} = Q_3 - Q_1$$

Examples 3–20 and 3–21 show the calculation of the quartiles and the interquartile range.

Finding quartiles and the interquartile range.

EXAMPLE 3-20 The following are the scores of 12 students in a mathematics class:

75   80   68   53   99   58   76   73   85   88   91   79

(a) Find the values of the three quartiles. Where does the score of 88 lie in relation to these quartiles?
(b) Find the interquartile range.

Solution

Finding quartiles for an even number of data values.

(a) First, we rank the given scores in increasing order. Then we calculate the three quartiles as follows:

| Values less than the median | Values greater than the median |

53   58   68   73   75   76   79   80   85   88   91   99

$$Q_1 = \frac{68 + 73}{2}$$ $$Q_2 = \frac{76 + 79}{2}$$ $$Q_3 = \frac{85 + 88}{2}$$

$$= 70.5$$ $$= 77.5$$ $$= 86.5$$

Also the median

The value of Q_2, which is also the median, is given by the value of the middle term in the ranked data set. For the data of this example, this value is the average of the sixth and seventh terms. Consequently, Q_2 is 77.5. The value of Q_1 is given by the value of the middle term of the six values that fall below the median (or Q_2). Thus, it is obtained by taking the average of the third and fourth terms. So, Q_1 is 70.5. The value of Q_3 is given by the value of the middle term of the six values that fall above the median. For the data of this example, Q_3 is obtained by taking the average of the ninth and tenth terms, and it is 86.5.

The value of $Q_1 = 70.5$ indicates that the scores of (approximately) 25% of the students in this sample are less than 70.5 and those of (approximately) 75% of the students are greater than this value. Similarly, we can state that the scores of about half the students are less than 77.5 (which is Q_2) and those of the other half are greater than this value. The value of $Q_3 = 86.5$ indicates that the scores of (approximately) 75% of the students in this sample are less than 86.5 and those of (approximately) 25% of the students are greater than this value.

By looking at the position of 88, we can state that the score of 88 lies in the **top 25%** of the scores.

Finding the interquartile range.

(b) The interquartile range is given by the difference between the values of the third and the first quartiles. Thus,

IQR = Interquartile range = $Q_3 - Q_1 = 86.5 - 70.5 =$ **16**   ■

Finding quartiles and the interquartile range.

EXAMPLE 3-21 The following are the ages of nine employees of an insurance company:

47   28   39   51   33   37   59   24   33

(a) Find the values of the three quartiles. Where does the age of 28 fall in relation to the ages of these employees?
(b) Find the interquartile range.

Solution

Finding quartiles for an odd number of data values.

(a) First we rank the given data in increasing order. Then we calculate the three quartiles as follows:

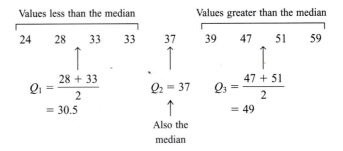

Thus, the values of the three quartiles are

$$Q_1 = \mathbf{30.5} \quad Q_2 = \mathbf{37}, \quad \text{and} \quad Q_3 = \mathbf{49}$$

The age of 28 falls in the **lowest 25%** of the ages.

Finding the interquartile range.

(b) The interquartile range is

$$\text{IQR} = \text{Interquartile range} = Q_3 - Q_1 = 49 - 30.5 = \mathbf{18.5} \qquad \blacksquare$$

3.5.2   PERCENTILES AND PERCENTILE RANK

Percentiles are the summary measures that divide a ranked data set into 100 equal parts. Each (ranked) data set has 99 percentiles that divide it into 100 equal parts. The data should be ranked in increasing order to compute percentiles. The kth percentile is denoted by P_k, where k is an integer in the range 1 to 99. For instance, the 25th percentile is denoted by P_{25}. Figure 3.12 shows the positions of the 99 percentiles.

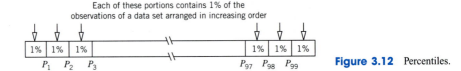

Figure 3.12   Percentiles.

Thus, the kth percentile, P_k, can be defined as a value in a data set such that about $k\%$ of the measurements are smaller than the value of P_k and about $(100 - k)\%$ of the measurements are greater than the value of P_k.

The approximate value of the kth percentile is determined as explained next.

PERCENTILES   The (approximate) value of the kth *percentile*, denoted by P_k, is

$$P_k = \text{Value of the } \left(\frac{kn}{100} \right) \text{th term in a ranked data set}$$

where k denotes the number of the percentile and n represents the sample size.

Example 3–22 describes the procedure to calculate the percentiles.

Finding the percentile for a data set. **EXAMPLE 3-22** The following are the test scores of 12 students in a mathematics class:

| 75 | 80 | 68 | 53 | 99 | 58 | 76 | 73 | 85 | 88 | 91 | 79 |

Find the value of the 62nd percentile. Give a brief interpretation of the 62nd percentile.

Solution First, we arrange the given scores in increasing order.

| 53 | 58 | 68 | 73 | 75 | 76 | 79 | 80 | 85 | 88 | 91 | 99 |

The position of the 62nd percentile is

$$\frac{kn}{100} = \frac{62(12)}{100} = 7.44\text{th term}$$

The value of the 7.44th term can be approximated by the average of the seventh and eighth terms in the ranked data. Therefore,

$$P_{62} = 62\text{nd percentile} = \frac{79 + 80}{2} = \mathbf{79.5}$$

Thus, approximately 62% of the scores are less than 79.5 and 38% are greater than 79.5 in the given data.

Note that if a data set contains only a few observations, then the number of values less than the 62nd percentile may not be exactly 62% and the number of values greater than the 62nd percentile may not be exactly 38%. For example, in our data on 12 scores, 7 scores (which is approximately 58% of 12) are less than 79.5 and 5 scores (which is approximately 42% of 12) are greater than 79.5. However, these percentages will be more accurate in a larger data set. ∎

We can also calculate the **percentile rank** for a particular value x_i of a data set by using the following formula. The percentile rank of x_i gives the percentage of values in the data set that are less than x_i.

PERCENTILE RANK OF A VALUE

$$\text{Percentile rank of } x_i = \frac{\text{Number of values less than } x_i}{\text{Total number of values in the data set}} \times 100$$

Example 3–23 shows how the percentile rank is calculated for a data value.

Finding the percentile rank for a data value. **EXAMPLE 3-23** Refer to the math scores of the 12 students given in Example 3–22 and reproduced here.

| 75 | 80 | 68 | 53 | 99 | 58 | 76 | 73 | 85 | 88 | 91 | 79 |

Find the percentile rank for the score 85. Give a brief interpretation of this percentile rank.

Solution First, we arrange the scores in increasing order.

| 53 | 58 | 68 | 73 | 75 | 76 | 79 | 80 | 85 | 88 | 91 | 99 |

In this data set, 8 of the 12 scores are less than 85. Hence,

$$\text{Percentile rank of } 85 = \frac{8}{12} \times 100 = \textbf{66.67\%}$$

Rounding this answer to the nearest integral value, we can state that about 67% of the scores in this sample are less than 85. In other words, about 67% of the students scored lower than 85. ■

EXERCISES

■ Concepts and Procedures

3.84 Briefly describe how the three quartiles are calculated for a data set. Illustrate by calculating the three quartiles for two examples, the first with an odd number of observations and the second with an even number of observations.

3.85 Explain how the interquartile range is calculated. Give one example.

3.86 Briefly describe how the percentiles are calculated for a data set.

3.87 Explain the concept of the percentile rank for an observation of a data set.

■ Applications

3.88 The following data give the weight lost by 15 members of a health club at the end of two months after joining the club.

5	10	8	7	25	12	5	14
11	10	21	9	8	11	18	

a. Compute the values of the three quartiles and the interquartile range.
b. Calculate the (approximate) value of the 82nd percentile.
c. Find the percentile rank of 10.

3.89 The following data give the speeds of 13 cars, measured by radar, traveling on interstate highway I-84.

73	75	69	68	78	69	74
76	72	79	68	77	71	

a. Find the values of the three quartiles and the interquartile range.
b. Calculate the (approximate) value of the 35th percentile.
c. Compute the percentile rank of 71.

3.90 The following data give the numbers of computer keyboards assembled at the Twentieth Century Electronics Company for a sample of 25 days.

45	52	48	41	56	46	44	42	48	53
51	53	51	48	46	43	52	50	54	47
44	47	50	49	52					

a. Calculate the values of the three quartiles and the interquartile range.
b. Determine the (approximate) value of the 53rd percentile.
c. Find the percentile rank of 50.

3.91 The following data give the hours worked last week by 30 employees of a company.

42	45	40	38	35	47	40	27	39	43
40	53	23	51	42	48	40	36	51	40
48	34	21	40	31	34	16	39	41	36

a. Calculate the values of the three quartiles and the interquartile range.
b. Find the (approximate) value of the 79th percentile.
c. Calculate the percentile rank of 39.

3.92 The following data are the numbers of car thefts that occurred in a city during the past 12 days.

6   3   7   11   4   3   8   7   2   6   9   15

a. Determine the values of the three quartiles and the interquartile range. Where does the value of 9 lie in relation to these quartiles?
b. Calculate the (approximate) value of the 55th percentile.
c. Find the percentile rank of 8.

3.93 Nixon Corporation manufactures computer terminals. The following data give the numbers of computer terminals produced at the company for a sample of 30 days.

24	32	27	23	33	33	29	25	23	36
26	26	31	20	27	33	27	23	28	29
31	35	34	22	37	28	23	35	31	43

a. Calculate the values of the three quartiles and the interquartile range. Where does the value of 31 lie in relation to these quartiles?
b. Find the (approximate) value of the 65th percentile. Give a brief interpretation of this percentile.
c. For what percentage of the days was the number of computer terminals produced 32 or higher? Answer by finding the percentile rank of 32.

3.94 The following data give the numbers of new cars sold at a dealership during a 20-day period.

8	5	12	3	9	10	6	12	8	8
4	16	10	11	7	7	3	5	9	11

a. Calculate the values of the three quartiles and the interquartile range. Where does the value of 4 lie in relation to these quartiles?
b. Find the (approximate) value of the 25th percentile. Give a brief interpretation of this percentile.
c. Find the percentile rank of 10. Give a brief interpretation of this percentile rank.

3.95 The following data are the scores of 19 students in a statistics class.

84	92	63	75	81	97	73	69	46	58
94	84	78	43	77	82	69	98	88	

a. Calculate the values of the three quartiles and the interquartile range. Where does the value of 81 lie in relation to these quartiles?
b. Compute the (approximate) value of the 93rd percentile. Give a brief interpretation of this percentile.
c. Find the percentile rank of 82. Give a brief interpretation of this percentile rank.

3.6 BOX-AND-WHISKER PLOT

A **box-and-whisker plot** gives a graphic presentation of data using five measures: the median, the first quartile, the third quartile, and the smallest and the largest values in the data set between the lower and the upper inner fences. A box-and-whisker plot can help us visualize the center, the spread, and the skewness of a data set. It also helps detect outliers. We can compare different distributions by making box-and-whisker plots for each of them.

> **BOX-AND-WHISKER PLOT**   A plot that shows the center, spread, and skewness of a data set. It is constructed by drawing a box and two whiskers that use the median, the first quartile, the third quartile, and the smallest and the largest values in the data set between the lower and the upper inner fences.

Example 3–24 explains all the steps needed to make a box-and-whisker plot.

Constructing a box-and-whisker plot.

EXAMPLE 3–24   The following data are the incomes (in thousands of dollars) for a sample of 12 households.

$$23 \quad 17 \quad 32 \quad 60 \quad 22 \quad 52 \quad 29 \quad 38 \quad 42 \quad 92 \quad 27 \quad 46$$

Construct a box-and-whisker plot for these data.

Solution   The following five steps are performed to construct a box-and-whisker plot.

Step 1.   First, rank the data in increasing order and calculate the values of the median, the first quartile, the third quartile, and the interquartile range. The ranked data are

$$17 \quad 22 \quad 23 \quad 27 \quad 29 \quad 32 \quad 38 \quad 42 \quad 46 \quad 52 \quad 60 \quad 92$$

For these data,

$$\text{Median} = (32 + 38)/2 = 35$$

$$Q_1 = (23 + 27)/2 = 25$$

$$Q_3 = (46 + 52)/2 = 49$$

$$\text{IQR} = Q_3 - Q_1 = 49 - 25 = 24$$

Step 2.   Find the points that are $1.5 \times \text{IQR}$ below Q_1 and $1.5 \times \text{IQR}$ above Q_3. These two points are called the **lower** and the **upper inner fences**, respectively.

$$1.5 \times \text{IQR} = 1.5 \times 24 = 36$$

$$\text{Lower inner fence} = Q_1 - 36 = 25 - 36 = -11$$

$$\text{Upper inner fence} = Q_3 + 36 = 49 + 36 = 85$$

Step 3.   Determine the smallest and the largest values in the given data set within the two inner fences. These two values for our example are as follows:

$$\text{Smallest value within the two inner fences} = 17$$

$$\text{Largest value within the two inner fences} = 60$$

Step 4.   Draw a horizontal line and mark the income levels on it such that all the values in the given data set are covered. Above the horizontal line, draw a box with its left side at the position of the first quartile and the right side at the position of the third quartile. Inside the box, draw a vertical line at the position of the median. The result of this step is shown in Figure 3.13.

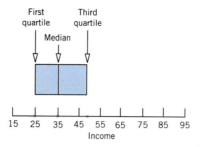

Figure 3.13

Step 5. By drawing two lines, join the points of the smallest and the largest values within the two inner fences to the box. These values are 17 and 60 in this example as listed in Step 3. The two lines that join the box to these two values are called **whiskers**. A value that falls outside the two inner fences is shown by marking an asterisk and is called an outlier. This completes the box-and-whisker plot, as shown in Figure 3.14.

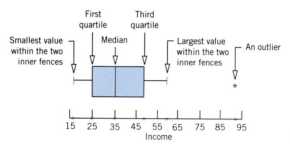

Figure 3.14

In Figure 3.14, about 50% of the data values fall within the box, about 25% of the values fall on the left side of the box, and about 25% fall on the right side of the box. Also, 50% of the values fall on the left side of the median and 50% lie on the right side of the median. The data of this example are skewed to the right because the lower 50% of the values are spread over a smaller range than the upper 50% of the values. ∎

The observations that fall outside the two inner fences are called outliers. These outliers can be classified into two kinds of outliers—mild and extreme outliers. To do so, we define two outer fences—a **lower outer fence** at $3.0 \times$ IQR below the first quartile and an **upper outer fence** at $3.0 \times$ IQR above the third quartile. If an observation is outside either of the two inner fences but within either of the two outer fences, it is called a *mild outlier.* An observation that is outside either of the two outer fences is called an *extreme outlier.* For the previous example, the outer fences are at -47 and 121. Since 92 is outside the upper inner fence but inside the upper outer fence, it is a mild outlier.

For a symmetric data set, the line representing the median will be in the middle of the box and the spread of the values will be over almost the same range on both sides of the box.

EXERCISES

■ **Concepts and Procedures**

3.96 Briefly explain what summary measures are used to construct a box-and-whisker plot.

3.97 Prepare a box-and-whisker plot for the following data:

36	43	28	52	41	59	47	61
24	55	63	73	32	25	35	49
31	22	61	42	58	65	98	34

Does this data set contain any outliers?

3.98 Prepare a box-and-whisker plot for the following data:

11	8	26	31	62	19	7	3	14	75
33	30	42	15	18	23	29	13	16	6

Does this data set contain any outliers?

■ **Applications**

3.99 The following data give the times (in minutes) taken to complete a statistics test by 14 students.

93	87	96	77	73	91	82
71	98	74	95	89	79	88

Prepare a box-and-whisker plot. Comment on the skewness of these data.

3.100 The following data give the scores of 19 students in a statistics class.

84	92	63	75	81	97	73	69	46	58
94	84	78	43	77	82	69	98	84	

Make a box-and-whisker plot. Are the data skewed in any direction?

3.101 The following data give the weight (in pounds) lost by 15 members of a health club at the end of two months after joining the club.

5	10	8	7	25	12	5	14
11	10	21	9	8	11	18	

Construct a box-and-whisker plot. Are the data symmetric or skewed?

3.102 The following data give the numbers of computer keyboards assembled at the Twentieth Century Electronics Company for a sample of 25 days.

45	52	48	41	56	46	44	42	48	53
51	53	51	48	46	43	52	50	54	47
44	47	50	49	52					

Prepare a box-and-whisker plot. Comment on the skewness of these data.

3.103 The following data give the hours worked last week by 30 employees of a company.

42	45	40	38	35	47	40	27	39	43
40	53	23	51	42	48	40	36	51	40
48	34	21	40	31	34	16	39	41	36

Make a box-and-whisker plot. Comment on the skewness of these data.

3.104 The following data are the numbers of car thefts that occurred in a city during the past 12 days.

6	3	7	11	5	3	8	7	2	6	9	13

Make a box-and-whisker plot. Comment on the skewness of these data.

3.105 Nixon Corporation manufactures computer terminals. The following are the numbers of computer terminals produced at the company for a sample of 30 days:

24	32	27	23	33	33	29	25	23	28
21	26	31	20	27	33	27	23	28	29
31	35	34	22	26	28	23	35	31	27

Prepare a box-and-whisker plot. Comment on the skewness of these data.

3.106 The following data give the numbers of new cars sold at a dealership during a 20-day period.

8	5	12	3	9	10	6	3	8	8
4	6	10	11	7	7	3	5	9	11

Make a box-and-whisker plot. Comment on the skewness of these data.

GLOSSARY

Bimodal distribution   A distribution that has two modes.

Box-and-whisker plot   A plot that shows the center, spread, and skewness of a data set with a box and two whiskers using the median, the first quartile, the third quartile, and the smallest and the largest values in the data set between the lower and the upper inner fences.

Chebyshev's theorem   For any number k greater than 1, at least $(1 - 1/k^2)$ of the values for any distribution lie within k standard deviations of the mean.

Coefficient of variation   A measure of relative variability that expresses standard deviation as a percentage of the mean.

Empirical rule   For a specific bell-shaped distribution, about 68% of the observations fall in the interval $(\mu - \sigma)$ to $(\mu + \sigma)$, about 95% fall in the interval $(\mu - 2\sigma)$ to $(\mu + 2\sigma)$, and about 99.7% fall in the interval $(\mu - 3\sigma)$ to $(\mu + 3\sigma)$.

First quartile   The value in a ranked data set such that about 25% of the measurements are smaller than this value and about 75% are larger. It is the median of the values that are smaller than the median of the whole data set.

Interquartile range   The difference between the third and the first quartiles.

Lower inner fence   The value in a data set that is $1.5 \times$ IQR below the first quartile.

Lower outer fence   The value in a data set that is $3.0 \times$ IQR below the first quartile.

Mean   A measure of central tendency calculated by dividing the sum of all values by the number of values in the data set.

Measures of central tendency   Measures that describe the center of a distribution. The mean, median, and mode are three of the measures of central tendency.

Measures of dispersion   Measures that give the spread of a distribution. The range, variance, and standard deviation are three such measures.

Measures of position   Measures that determine the position of a single value in relation to other values in a data set. Quartiles, percentiles, and percentile rank are examples of measures of position.

Median   The value of the middle term in a ranked data set. The median divides a ranked data set into two equal parts.

Mode   The value (or values) that occurs with highest frequency in a data set.

Multimodal distribution   A distribution that has more than two modes.

Outliers or **extreme values**   Values that are very small or very large relative to the majority of the values in a data set.

Parameter   A summary measure calculated for population data.

Percentile rank   The percentile rank of a value gives the percentage of values in the data set that are smaller than this value.

Percentiles   Ninety-nine values that divide a ranked data set into 100 equal parts.

Quartiles   Three summary measures that divide a ranked data set into four equal parts.

Range   A measure of spread obtained by taking the difference between the largest and the smallest values in a data set.

Second quartile   Middle or second of the three quartiles that divide a ranked data set into four equal parts. About 50% of the values in the data set are smaller and about 50% are larger than the second quartile. The second quartile is the same as the median.

Standard deviation   A measure of spread that is given by the positive square root of the variance.

Statistic   A summary measure calculated for sample data.

Third quartile   Third of the three quartiles that divide a ranked data set into four equal parts. About 75% of the values in a data set are smaller than the value of the third quartile and about 25% are larger. It is the median of the values that are greater than the median of the whole data set.

Trimmed mean   The k% trimmed mean is obtained by dropping k% of the smallest values and k% of the largest values from the given data and then calculating the mean of the remaining $(100 - 2k)$% of the values.

Unimodal distribution   A distribution that has only one mode.

Upper inner fence   The value in a data set that is $1.5 \times \text{IQR}$ above the third quartile.

Upper outer fence   The value in a data set that is $3.0 \times \text{IQR}$ above the third quartile.

Variance   A measure of spread.

KEY FORMULAS

1. **Mean for ungrouped data**

   For population data: $\mu = \dfrac{\Sigma x}{N}$

   For sample data: $\bar{x} = \dfrac{\Sigma x}{n}$

2. **Mean for grouped data**

   For population data: $\mu = \dfrac{\Sigma mf}{N}$

   For sample data: $\bar{x} = \dfrac{\Sigma mf}{n}$

   where m is the midpoint and f is the frequency of a class

3. **Median for ungrouped data**

   $$\text{Median} = \text{Value of the } \left(\frac{n + 1}{2}\right)\text{th term in a ranked data set}$$

4. **Range**

   $$\text{Range} = \text{Largest value} - \text{Smallest value}$$

5. **Variance for ungrouped data**

   For population data: $\sigma^2 = \dfrac{\Sigma x^2 - \dfrac{(\Sigma x)^2}{N}}{N}$

   For sample data: $s^2 = \dfrac{\Sigma x^2 - \dfrac{(\Sigma x)^2}{n}}{n - 1}$

6. **Variance for grouped data**

For population data: $\sigma^2 = \dfrac{\Sigma m^2 f - \dfrac{(\Sigma mf)^2}{N}}{N}$

For sample data: $s^2 = \dfrac{\Sigma m^2 f - \dfrac{(\Sigma mf)^2}{n}}{n - 1}$

where m is the midpoint and f is the frequency of a class

7. **Standard deviation**

For population data: $\sigma = \sqrt{\sigma^2}$

For sample data: $s = \sqrt{s^2}$

8. **Coefficient of variation**

For population data: $CV = \dfrac{\sigma}{\mu} \times 100\%$

For sample data: $CV = \dfrac{s}{\bar{x}} \times 100\%$

9. **Interquartile range**

$$IQR = \text{Interquartile range} = Q_3 - Q_1$$

10. **Percentiles**

The kth percentile is given by:

$$P_k = \text{Value of the } \left(\dfrac{kn}{100}\right)\text{th term in a ranked data set}$$

11. **Percentile rank of a value**

The percentile rank for a particular value x_i of a data set is calculated as follows:

$$\text{Percentile rank of } x_i = \dfrac{\text{Number of values less than } x_i}{\text{Total number of values in the data set}} \times 100$$

SUPPLEMENTARY EXERCISES

3.107 The following data give the 1998 profits (in millions of U.S. dollars) for the nine trucking companies listed in *Fortune* magazine's top 1000 U.S. companies (*Fortune*, April 26, 1999). The data, entered in that order, are for CNF Transportation, Yellow, Roadway Express, Consolidated Freightways, C. H. Robinson Worldwide, J. B. Hunt Transport Svcs, US Freightways, Arkansas Best, and Landstar System. A number with a negative sign indicates a net loss for that company.

139   −29   26   26   43   47   71   29   12

a. Calculate the mean and median.
b. Does this data set contain any outlier(s)? If yes, drop the outlier(s) and recalculate the mean and median. Which of these measures changes by a greater amount when you drop the outlier(s)?
c. Is the mean or the median a better summary measure for these data? Explain.

3.108   The following table gives the mayor's salary for each of 10 U.S. cities of comparable size.

City	Mayor's Salary	City	Mayor's Salary
Flint, MI	$101,640	Hayward, CA	$24,000
Lansing, MI	90,000	Savannah, GA	35,000
Hartford, CT	30,000	Pasadena, TX	77,130
Pasadena, CA	18,000	New Haven, CT	92,000
Hollywood, FL	16,000	Bridgeport, CT	98,500

Source: U.S. Census Bureau, National League of Cities, individual cities, *Hartford Courant*, November 15, 1999.

a.   Find the mean and median for these data. Do these data have a mode?
b.   Find the range, variance, and standard deviation.

3.109   The following data give the number of nuclear reactors as of May 1998 for each of the 11 countries that have 10 or more reactors (International Atomic Energy Agency). The data, entered in that order, are for Canada, France, Germany, India, Japan, South Korea, Russia, Sweden, Ukraine, United Kingdom, and the United States, respectively.

16   59   20   10   54   12   29   12   16   35   107

a.   Calculate the mean and median. Do these data have a mode?
b.   Find the range, variance, and standard deviation.

3.110   The following data give the numbers of driving citations received by 12 drivers.

4   8   0   3   11   7   4   14   8   13   7   9

a.   Find the mean, median, and mode for these data.
b.   Calculate the range, variance, and standard deviation.
c.   Are the values of the summary measures in parts a and b population parameters or sample statistics?

3.111   The following table gives the distribution of the amounts of rainfall (in inches) for July 1999 for 50 cities.

Rainfall	Number of Cities
0 to less than 2	6
2 to less than 4	10
4 to less than 6	20
6 to less than 8	7
8 to less than 10	4
10 to less than 12	3

Find the mean, variance, and standard deviation. Are the values of these summary measures population parameters or sample statistics?

3.112   The following table gives the distribution of the number of days for which *all* 40 employees of a company were absent during the last year.

Number of Days Absent	Number of Employees
0 to 2	10
3 to 5	14
6 to 8	9
9 to 11	4
12 to 14	3

Calculate the mean, variance, and standard deviation. Are the values of these summary measures population parameters or sample statistics?

3.113 The mean time taken to learn the basics of a word processor by all students is 200 minutes with a standard deviation of 20 minutes.

a. Using Chebyshev's theorem, find at least what percentage of students will learn the basics of this word processor in
  i. 160 to 240 minutes   ii. 140 to 260 minutes

*b. Using Chebyshev's theorem, find the interval that contains the time taken by at least 75% of all students to learn this word processor.

3.114 According to data from the NPD Group, the average amount spent by Americans for breakfast in restaurants was $3.12 per person in 1999 (*Newsweek*, May 17, 1999). Assume that the standard deviation of these breakfast expenditures was $.70.

a. Using Chebyshev's theorem, find at least what percentage of these expenditures were between
  i. $1.72 and $4.52   ii. $1.02 and $5.22

*b. Using Chebyshev's theorem, find the interval that contains at least 84% of all breakfast expenditures.

3.115 Refer to Exercise 3.113. Suppose the times taken to learn the basics of this word processor by all students have a bell-shaped distribution with a mean of 200 minutes and a standard deviation of 20 minutes.

a. Using the empirical rule, find the percentage of students who learn the basics of this word processor in
  i. 180 to 220 minutes   ii. 160 to 240 minutes

*b. Using the empirical rule, find the interval that contains the time taken by 99.7% of all students to learn this word processor.

3.116 Assume that the salaries of accounting professors at public 4-year colleges in the United States have a bell-shaped distribution with a mean of $63,947 and a standard deviation of $6600.

a. Using the empirical rule, find the percentage of such professors whose salaries are between
  i. $44,147 and $83,747   ii. $50,747 and $77,147

*b. Using the empirical rule, find the interval that contains the salaries of 68% of all such professors.

3.117 Refer to the data of Exercise 3.107 on the 1998 profits (in millions of U.S. dollars) of the nine trucking companies listed in *Fortune* magazine's top 1000 U.S. companies.

a. Determine the values of the three quartiles and the interquartile range. Where does the value of 12 lie in relation to these quartiles?

b. Calculate the (approximate) value of the 70th percentile. Give a brief interpretation of this percentile.

c. Find the percentile rank of 12. Give a brief interpretation of this percentile rank.

3.118 Refer to the data of Exercise 3.110 on the numbers of driving citations received by 12 drivers.

a. Determine the values of the three quartiles and the interquartile range. Where does the value of 4 lie in relation to these quartiles?

b. Calculate the (approximate) value of the 70th percentile. Give a brief interpretation of this percentile.

c. Find the percentile rank of 3. Give a brief interpretation of this percentile rank.

3.119 The following data give the ages of 15 employees of a company.

36	47	23	55	42	31	27	19
38	65	52	47	39	25	44	

Prepare a box-and-whisker plot. Is this data set skewed in any direction? If yes, is it skewed to the right or to the left? Does this data set contain any outliers?

3.120 The following data are the prices (in thousands of dollars) of 16 recently sold houses in an area.

141	163	127	104	197	203	113	179
256	228	183	119	133	199	871	191

Make a box-and-whisker plot. Comment on the skewness of this data set. Does this data set contain any outliers?

■ Advanced Exercises

3.121 Melissa's grade in her math class is determined by three 100-point tests and a 200-point final exam. To determine the grade for a student in this class, the instructor will add the four scores together and divide this sum by 5 to obtain a percentage. This percentage must be at least 80 for a grade of B. If Melissa's three test scores are 75, 69, and 87, what is the minimum score she needs on the final exam to obtain a B grade?

3.122 Jeffrey is serving on a six-person jury for a personal-injury lawsuit. All six jurors want to award damages to the plaintiff but cannot agree on the amount of the award. The jurors have decided that each of them will suggest an amount that he or she thinks should be awarded; then they will use the mean of these six numbers as the award to recommend to the plaintiff.

a. Jeffrey thinks the plaintiff should receive $20,000, but he thinks the mean of the other five jurors' recommendations will be about $12,000. He decides to suggest an inflated amount so that the mean for all six jurors is $20,000. What amount would Jeffrey have to suggest?

b. How might this jury revise its procedure to prevent a juror like Jeffrey from having an undue influence on the amount of damages to be awarded to the plaintiff?

3.123 The heights of five starting players on a basketball team have a mean of 76 inches, a median of 78 inches, and a range of 11 inches.

a. If the tallest of these five players is replaced by a substitute who is 2 inches taller, find the new mean, median, and range.

b. If the tallest player is replaced by a substitute who is 4 inches shorter, which of the new values (mean, median, range) could you determine, and what would their new values be?

3.124 The ZT500 car gets 12 miles per gallon in city driving and 23 miles per gallon on highways. Rick owns a ZT500 car. He does 70% of his driving in the city and 30% on highways. What mean mileage per gallon should he expect?

3.125 A small country bought oil from three different sources in one week, as shown in the following table.

Source	Barrels Purchased	Price Per Barrel
Mexico	1000	$20
Venezuela	200	25
Spot Market	100	40

Find the mean price per barrel for all 1300 barrels of oil purchased in that week.

3.126 An investor bought 47 ounces of gold bullion during 1999 as shown in the following table.

Month	Ounces Bought	Price Per Ounce
January	10	$400
March	5	410
July	12	390
October	20	380

The investor tells you that the mean price he paid for the gold is $(400 + 410 + 390 + 380)/4 = \$395$ per ounce. Do you agree? If not, explain to the investor why this method of calculating the mean is wrong in this case. Then find the correct value of the mean.

3.127 In the Olympic Games, when events require a subjective judgment of an athlete's performance, the highest and lowest of the judges' scores may be dropped. Consider a gymnast whose performance is judged by seven judges and the highest and lowest of the seven scores are dropped.

a. Gymnast A's scores in this event are 9.4, 9.7, 9.5, 9.5, 9.4, 9.6, and 9.5. Find this gymnast's mean score after dropping the highest and the lowest scores.

b. The answer to part a is an example of what percentage of trimmed mean?

c. Write another set of scores for a gymnast B so that gymnast A has a higher mean score than gymnast B based on the trimmed mean, but gymnast B would win if all seven scores were counted. Do not use any scores lower than 9.0.

3.128 A survey of young people's shopping habits in a small city during the summer months of 1998 showed the following: Shoppers aged 12–14 took an average of 8 shopping trips per month and spent an average of $14 per trip. Shoppers aged 15–17 took an average of 11 trips per month and spent an average of $18 per trip. Assume that this city has 1100 shoppers aged 12–14 and 900 shoppers aged 15–17.

a. Find the total amount spent per month by all these 2000 shoppers in both age groups.

b. Find the mean number of shopping trips per person per month for these 2000 shoppers.

c. Find the mean amount spent per person per month by shoppers aged 12–17 in this city.

3.129 Records of surgeries on 1000 patients were obtained to compare the safety of two anesthetics, A and B. These 1000 patients were classified as low- or high-risk based on their age, state of health, and the nature of the surgery. The following table gives the number of patients in each risk group who were administered each type of anesthetic and the number of deaths for each of the two anesthetics and risk groups.

	Low-Risk Patients		High-Risk Patients	
	A	B	A	B
Number of patients	100	500	300	100
Number of deaths	1	10	15	8

a. Compare the death rates for the two anesthetics for the low-risk surgeries.

b. Compare the death rates for the two anesthetics in the high-risk surgeries.

c. Compare the overall death rates for the two anesthetics.

d. Explain how the anesthetic with the higher death rates for both levels of risk can have a lower overall death rate. This phenomenon is called Simpson's Paradox.

3.130 In a study of distances traveled to a college by commuting students, data from 100 commuters yielded a mean of 8.73 miles. After the mean was calculated, data came in late from three students, with distances of 11.5, 7.6, and 10.0 miles. Calculate the mean distance for all 103 students.

3.131 The test scores for a large statistics class have an unknown distribution with a mean of 70 and a standard deviation of 10.

a. Find k so that at least 50% of the scores are within k standard deviations of the mean.

b. Find k so that at most 10% of the scores are more than k standard deviations above the mean.

3.132 The test scores for a very large statistics class have a bell-shaped distribution with a mean of 70 points.

a. If 16% of all students in the class scored above 85, what is the standard deviation of the scores?

b. If 95% of the scores are between 60 and 80, what is the standard deviation?

3.133 How much does the typical American family spend to go away on vacation each year? Twenty-five randomly selected households reported the following vacation expenditures during the past year (rounded to the nearest hundred dollars):

2500	500	800	0	100
0	200	2200	0	200
0	1000	900	321,500	400
500	100	0	8200	900
0	1700	1100	600	3400

a. Using both graphical and numerical methods, organize and interpret these data.

b. What measure of central tendency best answers the original question?

3.134 Actuaries at an insurance company must determine a premium for a new type of insurance. A random sample of 40 potential purchasers of this type of insurance were found to have suffered the following values of losses during the past year. These losses would have been covered by the insurance if it were available.

100	32	0	0	470	50	0	14,589	212	93
0	0	1127	421	0	87	135	420	0	250
12	0	309	0	177	295	501	0	143	0
167	398	54	0	141	0	3709	122	0	0

a. Use both graphical and numerical methods to organize and interpret these data. Write a report summarizing the loss characteristics for the insurance company.

b. Which measures of central tendency and variation should the actuaries use when determining the premium for this insurance?

3.135 A local golf club has men's and women's summer leagues. The following data give the scores for a round of 18 holes of golf for 17 men and 15 women randomly selected from their respective leagues.

Men	87	68	92	79	83	67	71	92	112
	75	77	102	79	78	85	75	72	
Women	101	100	87	95	98	81	117	107	103
	97	90	100	99	94	94			

a. Make a box-and-whisker plot for each of the data sets and use them to discuss the similarities and differences between the scores of the men and women golfers.

b. Compute the various descriptive measures you have learned for each sample. How do they compare?

3.136 In each situation, find the missing quantity.

a. The total weight of all pieces of luggage loaded onto an airplane is 12,372 pounds, which works out to be an average of 51.55 pounds per piece. How many pieces of luggage are on the plane?

b. A group of seven friends, having just gotten back a chemistry exam, discuss their scores. Six of the students reveal that they received grades of 81, 75, 93, 88, 82, and 85, but the seventh student is reluctant to say what grade she received. After some calculation she announces that the group averaged 81 on the exam. What is her score?

3.137 Suppose that there are 150 freshmen engineering majors at a college and each of them will take the same five courses next semester. Four of these courses will be taught in small sections of 25 students each, while the fifth course will be taught in one section containing all 150 freshmen. To accommodate all 150 students, there must be six sections of each of the four courses taught in 25-student sections. Thus, there are 24 classes of 25 students each and one class of 150 students.

a. Find the mean size of these 25 classes.

b. Find the mean class size from a student's point of view, noting that each student has five classes containing 25, 25, 25, 25, and 150 students.

Are the means in parts a and b equal? If not, why not?

3.138 The following data give the weights (in pounds) of a random sample of 44 college students. (Here F and M indicate female or male.)

123 F	195 M	138 M	115 F	179 M	119 F
148 F	147 F	180 M	146 F	179 M	189 M
175 M	108 F	193 M	114 F	179 M	147 M
108 F	128 F	164 F	174 M	128 F	159 M
193 M	204 M	125 F	133 F	115 F	168 M
123 F	183 M	116 F	182 M	174 M	102 F
123 F	99 F	161 M	162 M	155 F	202 M
110 F	132 M				

Compute the mean, median, and standard deviation for the weights of all students, of men only, and of women only. Of the mean and median, which is the more informative measure of central tendency? Write a brief note comparing the three measures for all students, men only, and women only.

3.139   The distribution of the lengths of fish in a certain lake is not known, but it is definitely not bell-shaped. It is estimated that the mean length is 6 inches with a standard deviation of 2 inches.

a.   At least what proportion of fish in the lake are between 3 inches and 9 inches long?
b.   What is the smallest interval that will contain the lengths of at least 84% of the fish?
c.   Find an interval so that fewer than 36% of the fish have lengths outside this interval.

3.140   The following stem-and-leaf diagram gives the distances (in thousands of miles) driven during the past year by a sample of drivers in a city.

```
0 | 3 6 9
1 | 2 8 5 1 0 5
2 | 5 1 6
3 | 8
4 | 1
5 |
6 | 2
```

a.   Compute the sample mean, median, and mode for the data on distances driven.
b.   Compute the range, variance, and standard deviation for these data.
c.   Compute the first and third quartiles.
d.   Compute the interquartile range. Describe what properties the interquartile range has. When would it be preferable to using the standard deviation when measuring variation?

APPENDIX 3.1

A3.1.1 BASIC FORMULAS FOR THE VARIANCE AND STANDARD DEVIATION FOR UNGROUPED DATA

Example 3–25 illustrates how to use the basic formulas to calculate the variance and standard deviation for ungrouped data. From Section 3.2.2, the basic formulas for variance for ungrouped data are

$$\sigma^2 = \frac{\Sigma(x - \mu)^2}{N} \quad \text{and} \quad s^2 = \frac{\Sigma(x - \bar{x})^2}{n - 1}$$

where σ^2 is the population variance and s^2 is the sample variance.

In either case, the standard deviation is obtained by taking the square root of the variance.

Calculating the variance and standard deviation for ungrouped data using basic formulas.

EXAMPLE 3-25   Refer to Example 3–12, where we used the short-cut formulas to compute the variance and standard deviation for the data on 1998 earnings (in millions of dollars) of five celebrities. Calculate the variance and standard deviation for those data using the basic formula.

Solution   Let x denote the 1998 earnings (in millions of dollars) of these five celebrities. Table 3.14 (page 122) shows all the required calculations to find the variance and standard deviation.

Table 3.14

x	$(x - \bar{x})$	$(x - \bar{x})^2$
55	$55 - 133.80 = -78.80$	6,209.44
267	$267 - 133.80 = 133.20$	17,742.24
175	$175 - 133.80 = 41.20$	1,697.44
47	$47 - 133.80 = -86.80$	7,534.24
125	$125 - 133.80 = -8.80$	77.44
$\Sigma x = 669$		$\Sigma(x - \bar{x})^2 = 33,260.80$

The following steps are performed to compute the variance and standard deviation.

Step 1.   Find the mean as follows:

$$\bar{x} = \frac{\Sigma x}{n} = \frac{669}{5} = 133.80$$

Step 2.   Calculate $x - \bar{x}$, the deviation of each value of x from the mean. The results are shown in the second column of Table 3.14.

Step 3.   Square each of the deviations of x from $\bar{x}$; that is, calculate each of the $(x - \bar{x})^2$ values. These values are called the *squared deviations*, and they are recorded in the third column.

Step 4.   Add all the squared deviations to obtain $\Sigma(x - \bar{x})^2$; that is, sum all the values given in the third column of Table 3.14. This gives

$$\Sigma(x - \bar{x})^2 = 33,260.80$$

Step 5.   Obtain the sample variance by dividing the sum of the squared deviations by $n - 1$. Thus,

$$s^2 = \frac{\Sigma(x - \bar{x})^2}{n - 1} = \frac{33,260.80}{5 - 1} = 8315.20$$

Step 6.   Obtain the sample standard deviation by taking the positive square root of the variance. Hence,

$$s = \sqrt{8315.20} = 91.187718 = \$91,187,718 \quad \blacksquare$$

A3.1.2 BASIC FORMULAS FOR THE VARIANCE AND STANDARD DEVIATION FOR GROUPED DATA

Example 3–26 demonstrates how to use the basic formulas to calculate the variance and standard deviation for grouped data. The basic formulas for these calculations are

$$\sigma^2 = \frac{\Sigma f(m - \mu)^2}{N} \quad \text{and} \quad s^2 = \frac{\Sigma f(m - \bar{x})^2}{n - 1}$$

where σ^2 is the population variance, s^2 is the sample variance, m is the midpoint of a class, and f is the frequency of a class.

In either case, the standard deviation is obtained by taking the square root of the variance.

Calculating the variance and standard deviation for grouped data using basic formulas.

EXAMPLE 3-26   In Example 3–17, we used the short-cut formula to compute the variance and standard deviation for the data on the numbers of orders received each day during the past 50 days at the office of a mail-order company. Calculate the variance and standard deviation for those data using the basic formula.

Solution All the required calculations to find the variance and standard deviation appear in Table 3.15.

Table 3.15

Number of Orders	f	m	mf	$m - \bar{x}$	$(m - \bar{x})^2$	$f(m - \bar{x})^2$
10–12	4	11	44	−5.64	31.8096	127.2384
13–15	12	14	168	−2.64	6.9696	83.6352
16–18	20	17	340	.36	.1296	2.5920
19–21	14	20	280	3.36	11.2896	158.0544
	$n = 50$		$\Sigma mf = 832$			$\Sigma f(m - \bar{x})^2 = 371.5200$

The following steps are performed to compute the variance and standard deviation using the basic formula.

Step 1. Find the midpoint of each class. Multiply the corresponding values of m and f. Find Σmf. From Table 3.15, $\Sigma mf = 832$.

Step 2. Find the mean as follows:

$$\bar{x} = \Sigma mf/n = 832/50 = 16.64$$

Step 3. Calculate $m - \bar{x}$, the deviation of each value of m from the mean. These calculations are done in the fifth column of Table 3.15.

Step 4. Square each of the deviations $m - \bar{x}$; that is, calculate each of the $(m - \bar{x})^2$ values. These are called *squared deviations*, and they are recorded in the sixth column.

Step 5. Multiply the squared deviations by the corresponding frequencies (see the seventh column of Table 3.15). Adding the values of the seventh column, we obtain

$$\Sigma f(m - \bar{x})^2 = 371.5200$$

Step 6. Obtain the sample variance by dividing $\Sigma f(m - \bar{x})^2$ by $n - 1$. Thus,

$$s^2 = \frac{\Sigma f(m - \bar{x})^2}{n - 1} = \frac{371.5200}{50 - 1} = \textbf{7.5820}$$

Step 7. Obtain the standard deviation by taking the positive square root of the variance.

$$s = \sqrt{s^2} = \sqrt{7.5820} = \textbf{2.75 orders} \qquad \blacksquare$$

SELF-REVIEW TEST

1. The value of the middle term in a ranked data set is called the

   a. mean   **b.** median   **c.** mode

2. Which of the following summary measures is/are influenced by extreme values?

   a. mean   **b.** median   **c.** mode   **d.** range

3. Which of the following summary measures can be calculated for qualitative data?

   a. mean   **b.** median   **c.** mode

4. Which of the following can have more than one value?

    a. mean   **b.** median   **c.** mode

5. Which of the following is obtained by taking the difference between the largest and the smallest values of a data set?

    a. variance   **b.** range   **c.** mean

6. Which of the following is the mean of the squared deviations of x values from the mean?

    a. standard deviation   **b.** population variance   **c.** sample variance

7. The values of the variance and standard deviation are

    a. never negative   **b.** always positive   **c.** never zero

8. A summary measure calculated for the population data is called

    a. a population parameter   **b.** a sample statistic   **c.** an outlier

9. A summary measure calculated for the sample data is called

    a. a population parameter   **b.** a sample statistic   **c.** a box-and-whisker plot

10. Chebyshev's theorem can be applied to

    a. any distribution   **b.** bell-shaped distributions only   **c.** skewed distributions only

11. The empirical rule can be applied to

    a. any distribution   **b.** bell-shaped distributions only   **c.** skewed distributions only

12. The first quartile is a value in a ranked data set such that

    a. about 75% of the values are smaller and about 25% are larger than this value
    b. about 50% of the values are smaller and about 50% are larger than this value
    c. about 25% of the values are smaller and about 75% are larger than this value

13. The third quartile is a value in a ranked data set such that

    a. about 75% of the values are smaller and about 25% are larger than this value
    b. about 50% of the values are smaller and about 50% are larger than this value
    c. about 25% of the values are smaller and about 75% are larger than this value

14. The 75th percentile is a value in a ranked data set such that

    a. about 75% of the values are smaller and about 25% are larger than this value
    b. about 25% of the values are smaller and about 75% are larger than this value

15. The following data give the numbers of times 10 persons used their credit cards during the past three months.

$$9 \quad 6 \quad 28 \quad 14 \quad 2 \quad 18 \quad 7 \quad 3 \quad 16 \quad 6$$

Calculate the mean, median, mode, range, variance, and standard deviation.

16. The mean, as a measure of central tendency, has the disadvantage of being influenced by extreme values. Illustrate this point with an example.

17. The range, as a measure of spread, has the disadvantage of being influenced by extreme values. Illustrate this point with an example.

18. When is the value of the standard deviation for a data set zero? Give one example of such a data set. Calculate the standard deviation for that data set to show that it is zero.

19. The following table gives the frequency distribution of the numbers of computers sold during the past 25 weeks at a computer store.

Computers Sold	Frequency
4 to 9	2
10 to 15	4
16 to 21	10
22 to 27	6
28 to 33	3

    a. What does the frequency column in the table represent?

    b. Calculate the mean, variance, and standard deviation.

20. The cars owned by all people living in a city are, on average, 7.3 years old with a standard deviation of 2.2 years.

    a. Using Chebyshev's theorem, find at least what percentage of the cars in this city are

        i. 1.8 to 12.8 years old   ii. .7 to 13.9 years old

    b. Using Chebyshev's theorem, find the interval that contains the ages of at least 75% of the cars owned by all people in this city.

21. The ages of cars owned by all people living in a city have a bell-shaped distribution with a mean of 7.3 years and a standard deviation of 2.2 years.

    a. Using the empirical rule, find the percentage of cars in this city that are

        i. 5.1 to 9.5 years old   ii. .7 to 13.9 years old

    b. Using the empirical rule, find the interval that contains the ages of 95% of the cars owned by all people in this city.

22. The following data give the numbers of books purchased by 16 adults during the past year.

8	12	28	16	0	19	18	4
10	6	17	24	15	9	2	16

    a. Calculate the three quartiles and the interquartile range. Where does the value of 4 lie in relation to these quartiles?

    b. Find the (approximate) value of the 68th percentile. Give a brief interpretation of this value.

    c. Calculate the percentile rank of 17. Give a brief interpretation of this value.

23. Make a box-and-whisker plot for the data on the numbers of books purchased by 16 adults during the past year given in Problem 22. Comment on the skewness of this data set.

*24. The mean weekly wages of a sample of 15 employees of a company are $435. The mean weekly wages of a sample of 20 employees of another company are $490. Find the combined mean for these 35 employees.

*25. The mean GPA of five students is 3.21. The GPAs of four of these five students are 3.85, 2.67, 3.45, and 2.91. Find the GPA of the fifth student.

*26. The following are the prices (in thousands of dollars) of 10 houses sold recently in a city:

179	166	58	207	287	149	193	2534	163	238

Calculate the 10% trimmed mean for this data set. Do you think the 10% trimmed mean is a better summary measure for these data than the (simple) mean (i.e., the mean of all 10 values)? Briefly explain why or why not.

*27. Consider the following two data sets.

Data Set I:	8	16	20	35
Data Set II:	5	13	17	32

Note that each value of the second data set is obtained by subtracting 3 from the corresponding value of the first data set.

    a. Calculate the mean for each of these two data sets. Comment on the relationship between the two means.

    b. Calculate the standard deviation for each of these two data sets. Comment on the relationship between the two standard deviations.

MINI-PROJECT

Mini-Project 3–1

Refer to the data you collected for Mini-Project 1–1 of Chapter 1. Write a report summarizing those data, including the following activities:

a. Calculate appropriate measures of location and dispersion.
b. Do these data have a mode?
c. Draw a box-and-whisker plot. Are there any outliers? Does the plot show any evidence of skewness?
d. Suppose that you collected the data on telemarketing. If this report were written by a consumer who feels that there is too much telemarketing activity, which measure of location (mean or median) might that person favor?
e. If you collected data on rents in Mini-Project 1–1, which measure might a person favor if he or she feels that rents are too high?

For instructions on using MINITAB for this chapter, please see Section B.3 in Appendix B.

See Appendix D for *Excel Adventure 3* on numerical descriptive measures.

COMPUTER ASSIGNMENTS

CA3.1 Refer to the subsample taken in the Computer Assignment CA2.3 of Chapter 2 from the sample data on the time taken to run the Manchester Road Race. Using MINITAB or any other statistical software, find the mean, median, range, and standard deviation for those data.

CA3.2 Refer to the data on phone charges given in column C7 of Data Set I. From that data set, select the 4th value and then select every 10th value after that (i.e., select the 4th, 14th, 24th, 34th, . . . values). This subsample will give you 20 measurements. (Such a sample taken from a population is called a *systematic random sample*.) Using MINITAB or any other statistical software, find the mean, median, standard deviation, first quartile, and third quartile for the phone charges for these 20 observations.

CA3.3 Refer to Data Set I on the prices of various products in different cities across the country. Using MINITAB or any other statistical software, select a subsample of the prices of regular unleaded gas (column C8) for 40 cities. Using MINITAB or any other statistical software, find the mean, median, and standard deviation for the data of this subsample.

CA3.4 Refer to the data on the 1999 gasoline tax rates (in cents per gallon) for 50 states given in Example 2–5 of Chapter 2. Using MINITAB or any other statistical software, compute the mean, median, and standard deviation for those data.

CA3.5 Refer to Data Set II, which contains data on different variables for 50 states. Using MINITAB or any other statistical software, make a box-and-whisker plot for the data on the 1998 per capita incomes for 50 states.

CA3.6 Refer to Data Set I on the prices of various products in different cities across the country. Using MINITAB or any other statistical software, make a box-and-whisker plot for the data on the monthly telephone charges given in column C7.

CA3.7 Refer to the data on the numbers of computer keyboards assembled at the Twentieth Century Electronics Company for a sample of 25 days given in Exercise 3.102. Using MINITAB or any other statistical software, prepare a box-and-whisker plot for those data.

4

PROBABILITY

4.1 EXPERIMENT, OUTCOMES, AND SAMPLE SPACE

4.2 CALCULATING PROBABILITY

4.3 COUNTING RULE

4.4 MARGINAL AND CONDITIONAL PROBABILITIES

CASE STUDY 4–1 STRESS BUSTERS

4.5 MUTUALLY EXCLUSIVE EVENTS

4.6 INDEPENDENT VERSUS DEPENDENT EVENTS

4.7 COMPLEMENTARY EVENTS

4.8 INTERSECTION OF EVENTS AND THE MULTIPLICATION RULE

CASE STUDY 4–2 BASEBALL PLAYERS HAVE "SLUMPS" AND "STREAKS"

4.9 UNION OF EVENTS AND THE ADDITION RULE

GLOSSARY

KEY FORMULAS

SUPPLEMENTARY EXERCISES

SELF-REVIEW TEST

MINI-PROJECTS

We often make statements about probability. For example, a weather forecaster may predict that there is an 80% chance of rain tomorrow. A health-news reporter may state that a smoker has a much greater chance of getting cancer than a nonsmoker does. A college student may ask an instructor about the chances of passing a course or getting an A if he or she did not do well on the midterm examination.

Probability, which measures the likelihood that an event will occur, is an important part of statistics. It is the basis of inferential statistics, which will be introduced in later chapters. In inferential statistics, we make decisions under conditions of uncertainty. Probability theory is used to evaluate the uncertainty involved in those decisions. For example, estimating next year's sales for a company is based on many assumptions, some of which may happen to be true and others may not. Probability theory will help us make decisions under such conditions of imperfect information and uncertainty. Combining probability and probability distributions (which are discussed in Chapters 5 through 7) with descriptive statistics will help us make decisions about populations based on information obtained from samples. This chapter presents the basic concepts of probability and the rules for computing probability.

4.1   EXPERIMENT, OUTCOMES, AND SAMPLE SPACE

Quality control inspector Jack Cook of Tennis Products Company picks up a tennis ball from the production line to check whether it is good or defective. Jack Cook's act of inspecting a tennis ball is an example of a statistical **experiment**. The result of his inspection will be that the ball is either "good" or "defective." Each of these two observations is called an **outcome** (also called a *basic* or *final outcome*) of the experiment, and these outcomes taken together constitute the **sample space** for this experiment.

> **EXPERIMENT, OUTCOMES, AND SAMPLE SPACE**   An *experiment* is a process that, when performed, results in one and only one of many observations. These observations are called the *outcomes* of the experiment. The collection of all outcomes for an experiment is called a *sample space*.

A sample space is denoted by **S**. The sample space for the example of inspecting a tennis ball is written as

$$S = \{\text{good, defective}\}$$

The elements of a sample space are called **sample points**.

Table 4.1 lists some examples of experiments, their outcomes, and their sample spaces.

Table 4.1   Examples of Experiments, Outcomes, and Sample Spaces

Experiment	Outcomes	Sample Space
Toss a coin once	Head, Tail	$S = \{\text{Head, Tail}\}$
Roll a die once	1, 2, 3, 4, 5, 6	$S = \{1, 2, 3, 4, 5, 6\}$
Toss a coin twice	HH, HT, TH, TT	$S = \{HH, HT, TH, TT\}$
Birth of a baby	Boy, Girl	$S = \{\text{Boy, Girl}\}$
Take a test	Pass, Fail	$S = \{\text{Pass, Fail}\}$
Select a student	Male, Female	$S = \{\text{Male, Female}\}$

The sample space for an experiment can also be illustrated by drawing either a Venn diagram or a tree diagram. A **Venn diagram** is a picture (a closed geometric shape such as a rectangle, a square, or a circle) that depicts all the possible outcomes for an experiment. In a **tree diagram**, each outcome is represented by a branch of the tree. Venn and tree diagrams help us understand probability concepts by presenting them visually. Examples 4–1 through 4–3 describe how to draw these diagrams for statistical experiments.

Drawing Venn and tree diagrams: one toss of a coin.

EXAMPLE 4–1   Draw the Venn and tree diagrams for the experiment of tossing a coin once.

Solution   This experiment has two possible outcomes: head and tail. Consequently, the sample space is given by

$$S = \{H, T\} \qquad \text{where } H = \text{Head} \quad \text{and} \quad T = \text{Tail}$$

To draw a Venn diagram for this example, we draw a rectangle and mark two points inside this rectangle that represent the two outcomes, head and tail. The rectangle is labeled *S* because it represents the sample space (see Figure 4.1a). To draw a tree diagram, we draw two branches starting at the same point, one representing the head and the second representing the tail. The two final outcomes are listed at the ends of the branches (see Figure 4.1b).

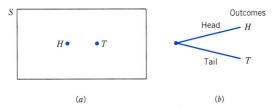

Figure 4.1 (a) Venn diagram and (b) tree diagram for one toss of a coin. ■

Drawing Venn and tree diagrams: two tosses of a coin.

EXAMPLE 4–2 Draw the Venn and tree diagrams for the experiment of tossing a coin twice.

Solution This experiment can be split into two parts: the first toss and the second toss. Suppose the first time the coin is tossed we obtain a head. Then, on the second toss, we can still obtain a head or a tail. This gives us two outcomes: *HH* (head on both tosses) and *HT* (head on the first toss and tail on the second toss). Now suppose we observe a tail on the first toss. Again, either a head or a tail can occur on the second toss, giving the remaining two outcomes: *TH* (tail on the first toss and head on the second toss) and *TT* (tail on both tosses). Thus, the sample space for two tosses of a coin is

$$S = \{HH, HT, TH, TT\}$$

The Venn and tree diagrams are given in Figure 4.2. Both these diagrams show the sample space for this experiment.

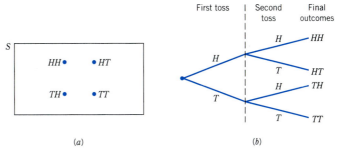

Figure 4.2 (a) Venn diagram and (b) tree diagram for two tosses of a coin. ■

Drawing Venn and tree diagrams: two selections.

EXAMPLE 4–3 Suppose we randomly select two persons from the members of a club and observe whether the person selected each time is a man or a woman. Write all the outcomes for this experiment. Draw the Venn and tree diagrams for this experiment.

Solution Let us denote the selection of a man by *M* and that of a woman by *W*. We can compare the selection of two persons to two tosses of a coin. Just as each toss of a coin can result in one of two outcomes, head or tail, each selection from the members of this club can result in one of two outcomes, man or woman. As we can see from the Venn and tree diagrams of Figure 4.3 (page 130), there are four final outcomes: *MM, MW, WM, WW*. Hence, the sample space is written as

$$S = \{MM, MW, WM, WW\}$$

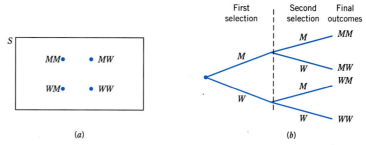

Figure 4.3   (*a*) Venn diagram and (*b*) tree diagram for selecting two persons.   ■

4.1.1   SIMPLE AND COMPOUND EVENTS

An **event** consists of one or more of the outcomes of an experiment.

> **EVENT**   An *event* is a collection of one or more of the outcomes of an experiment.

An event may be a *simple event* or a *compound event*. A simple event is also called an *elementary event*, and a compound event is also called a *composite event*.

Simple Event

Each of the final outcomes for an experiment is called a **simple event**. In other words, a simple event includes one and only one outcome. Usually simple events are denoted by E_1, E_2, E_3, and so forth. However, we can denote them by any of the other letters, too—that is, by A, B, C, and so forth.

> **SIMPLE EVENT**   An event that includes one and only one of the (final) outcomes for an experiment is called a *simple event* and is usually denoted by E_i.

Example 4–4 describes simple events.

Illustrating simple events.

EXAMPLE 4–4   Reconsider Example 4–3 on selecting two persons from the members of a club and observing whether the person selected each time is a man or a woman. Each of the final four outcomes (*MM*, *MW*, *WM*, and *WW*) for this experiment is a simple event. These four events can be denoted by E_1, E_2, E_3, and E_4, respectively. Thus,

$$E_1 = (MM), \quad E_2 = (MW), \quad E_3 = (WM), \quad \text{and} \quad E_4 = (WW)$$ ■

Compound Event

A **compound event** consists of more than one outcome.

> **COMPOUND EVENT**   A *compound event* is a collection of more than one outcome for an experiment.

Compound events are denoted by $A, B, C, D, \ldots$ or by $A_1, A_2, A_3, \ldots, B_1, B_2, B_3, \ldots,$ and so forth. Examples 4–5 and 4–6 describe compound events.

Illustrating a compound event: two selections.

EXAMPLE 4-5 Reconsider Example 4–3 on selecting two persons from the members of a club and observing whether the person selected each time is a man or a woman. Let A be the event that at most one man is selected. Event A will occur if either no man or one man is selected. Hence, the event A is given by

$$A = \{MW, WM, WW\}$$

Since event A contains more than one outcome, it is a compound event. The Venn diagram in Figure 4.4 gives a graphic presentation of compound event A.

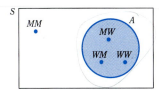

Figure 4.4 Venn diagram for event A.

Illustrating simple and compound events: two selections.

EXAMPLE 4-6 In a group of people, some are in favor of genetic engineering and others are against it. Two persons are selected at random from this group and asked whether they are in favor of or against genetic engineering. How many distinct outcomes are possible? Draw a Venn diagram and a tree diagram for this experiment. List all the outcomes included in each of the following events and mention whether they are simple or compound events.

(a)   Both persons are in favor of genetic engineering.

(b)   At most one person is against genetic engineering.

(c)   Exactly one person is in favor of genetic engineering.

Solution   Let

$$F = \text{a person is in favor of genetic engineering}$$

$$A = \text{a person is against genetic engineering}$$

This experiment has the following four outcomes:

$$FF = \text{both persons are in favor of genetic engineering}$$

$$FA = \text{the first person is in favor and the second is against}$$

$$AF = \text{the first person is against and the second is in favor}$$

$$AA = \text{both persons are against genetic engineering}$$

The Venn and tree diagrams in Figure 4.5 (page 132) show these four outcomes.

(a)   The event "both persons are in favor of genetic engineering" will occur if FF is obtained. Thus,

$$\text{Both persons are in favor of genetic engineering} = \{FF\}$$

Because this event includes only one of the final four outcomes, it is a **simple** event.

(b)   The event "at most one person is against genetic engineering" will occur if either none or one of the persons selected is against genetic engineering. Consequently,

$$\text{At most one person is against genetic engineering} = \{FF, FA, AF\}$$

Because this event includes more than one outcome, it is a **compound** event.

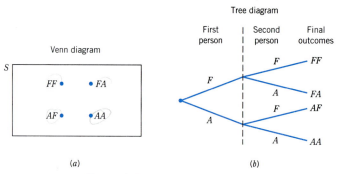

Figure 4.5 Venn and tree diagrams.

(c) The event "exactly one person is in favor of genetic engineering" will occur if one of the two persons selected is in favor and the other is against genetic engineering. Hence, it includes the following two outcomes:

Exactly one person is in favor of genetic engineering = {*FA, AF*}

Because this event includes more than one outcome, it is a **compound** event. ■

EXERCISES

■ Concepts and Procedures

4.1 Define the following terms: *experiment, outcome, sample space, simple event,* and *compound event.*

4.2 List the simple events for each of the following statistical experiments in a sample space *S.*

a. One roll of a die

b. Three tosses of a coin

c. One toss of a coin and one roll of a die

4.3 A box contains three items that are labeled A, B, and C. Two items are selected at random (without replacement) from this box. List all the possible outcomes for this experiment. Write the sample space *S.*

Both , one , no one

■ Applications

4.4 Two students are randomly selected from a statistics class, and it is observed whether or not they suffer from math anxiety. How many total outcomes are possible? Draw a tree diagram for this experiment. Draw a Venn diagram.

4.5 A hat contains a few red and a few green marbles. If two marbles are randomly drawn and the colors of these marbles are observed, how many total outcomes are possible? Draw a tree diagram for this experiment. Show all the outcomes in a Venn diagram.

4.6 A test contains two multiple-choice questions. If a student makes a random guess to answer each question, how many outcomes are possible? Depict all these outcomes in a Venn diagram. Also draw a tree diagram for this experiment. (*Hint:* Consider two outcomes for each question—either the answer is correct or it is wrong.)

4.7 A box contains a certain number of computer parts, a few of which are defective. Two parts are selected at random from this box and inspected to determine if they are good or defective. How many total outcomes are possible? Draw a tree diagram for this experiment.

4.8 In a group of people, some are in favor of a tax increase on rich people to reduce the federal deficit and others are against it. (Assume that there is no other outcome such as "no opinion" and "do not know.") Three persons are selected at random from this group and their opinions in favor or against raising such taxes are noted. How many total outcomes are possible? Write these outcomes in a sample space *S*. Draw a tree diagram for this experiment.

4.9 Draw a tree diagram for three tosses of a coin. List all outcomes for this experiment in a sample space *S*.

4.10 Refer to Exercise 4.4. List all the outcomes included in each of the following events. Indicate which are simple and which are compound events.

a. Both students suffer from math anxiety.
b. Exactly one student suffers from math anxiety.
c. The first student does not suffer and the second suffers from math anxiety.
d. None of the students suffers from math anxiety.

4.11 Refer to Exercise 4.5. List all the outcomes included in each of the following events. Indicate which are simple and which are compound events.

a. Both marbles are different colors.
b. At least one marble is red.
c. Not more than one marble is green.
d. The first marble is green and the second is red.

4.12 Refer to Exercise 4.6. List all the outcomes included in each of the following events and mention which are simple and which are compound events.

a. Both answers are correct.
b. At most one answer is wrong.
c. The first answer is correct and the second is wrong.
d. Exactly one answer is wrong.

4.13 Refer to Exercise 4.7. List all the outcomes included in each of the following events. Indicate which are simple and which are compound events.

a. At least one part is good.
b. Exactly one part is defective.
c. The first part is good and the second is defective.
d. At most one part is good.

4.14 Refer to Exercise 4.8. List all the outcomes included in each of the following events and mention which are simple and which are compound events.

a. At most one person is against a tax increase on rich people.
b. Exactly two persons are in favor of a tax increase on rich people.
c. At least one person is against a tax increase on rich people.
d. More than one person is against a tax increase on rich people.

4.2 CALCULATING PROBABILITY

Probability, which gives the likelihood of occurrence of an event, is denoted by *P*. The probability that a simple event E_i will occur is denoted by $P(E_i)$, and the probability that a compound event *A* will occur is denoted by $P(A)$.

> **PROBABILITY** *Probability* is a numerical measure of the likelihood that a specific event will occur.

☞ **TWO PROPERTIES OF PROBABILITY**

1. The probability of an event always lies in the range 0 to 1.

Whether it is a simple or a compound event, the probability of an event is never less than 0 or greater than 1. Using mathematical notation, we can write this property as follows.

$$0 \leq P(E_i) \leq 1$$

$$0 \leq P(A) \leq 1$$

An event that cannot occur has zero probability; such an event is called an **impossible event**. An event that is certain to occur has a probability equal to 1 and is called a **sure event**. That is,

For an impossible event M: $P(M) = 0$
For a sure event C: $P(C) = 1$

2. The sum of the probabilities of all simple events (or final outcomes) for an experiment, denoted by $\Sigma P(E_i)$, is always 1.

Thus, for an experiment:

$$\Sigma P(E_i) = P(E_1) + P(E_2) + P(E_3) + \cdots = 1$$

From this property, for the experiment of one toss of a coin,

$$P(H) + P(T) = 1$$

For the experiment of two tosses of a coin,

$$P(HH) + P(HT) + P(TH) + P(TT) = 1$$

For one game of football by a National Football League team,

$$P(\text{win}) + P(\text{loss}) + P(\text{tie}) = 1$$

4.2.1 **THREE CONCEPTUAL APPROACHES TO PROBABILITY**

The three conceptual approaches to probability are (1) classical probability, (2) the relative frequency concept of probability, and (3) the subjective probability concept. These three concepts are explained next.

Classical Probability

Outcomes that have the same probability of occurrence are called **equally likely outcomes**. The classical probability rule is applied to compute the probabilities of events for an experiment in which all outcomes are equally likely.

> **EQUALLY LIKELY OUTCOMES** Two or more outcomes (or events) that have the same probability of occurrence are said to be *equally likely outcomes* (or events).

According to the **classical probability rule**, the probability of a simple event is equal to 1 divided by the total number of outcomes for the experiment. This is obvious because the sum of the probabilities of all final outcomes for an experiment is 1, and all the final outcomes are equally likely. In contrast, the probability of a compound event A is equal to the number of outcomes favorable to event A divided by the total number of outcomes for the experiment.

CLASSICAL PROBABILITY RULE

$$P(E_i) = \frac{1}{\text{Total number of outcomes for the experiment}}$$

$$P(A) = \frac{\text{Number of outcomes favorable to } A}{\text{Total number of outcomes for the experiment}}$$

Examples 4–7 through 4–9 illustrate how probabilities of events are calculated using the classical probability rule.

Calculating the probability of a simple event.

EXAMPLE 4-7   Find the probability of obtaining a head and the probability of obtaining a tail for one toss of a coin.

Solution   The two outcomes, head and tail, are equally likely outcomes. Therefore,[1]

$$P(\text{head}) = \frac{1}{\text{Total number of outcomes}} = \frac{1}{2} = \mathbf{.50}$$

Similarly,

$$P(\text{tail}) = \frac{1}{2} = \mathbf{.50} \qquad \blacksquare$$

Calculating the probability of a compound event.

EXAMPLE 4-8   Find the probability of obtaining an even number in one roll of a die.

Solution   This experiment has a total of six outcomes: 1, 2, 3, 4, 5, and 6. All these outcomes are equally likely. Let A be an event that an even number is observed on the die. Event A includes three outcomes: 2, 4, and 6; that is,

$$A = \{2, 4, 6\}$$

If any one of these three numbers is obtained, event A is said to occur. Hence,

$$P(A) = \frac{\text{Number of outcomes included in } A}{\text{Total number of outcomes}} = \frac{3}{6} = \mathbf{.50} \qquad \blacksquare$$

---

[1]If the final answer for the probability of an event does not terminate within three decimal places, usually it is rounded to four decimal places.

*Calculating the
probability of a
compound event.*

EXAMPLE 4-9   A club has 100 members, 60 men and 40 women. Suppose one of these members is randomly selected to be the president of the club. What is the probability that a woman is selected?

Solution   Because the selection is to be made randomly, each of the 100 members of the club has the same probability of being selected. Consequently, this experiment has a total of 100 equally likely outcomes. Forty of these 100 outcomes are included in the event that "a woman is selected." Hence,

$$P(\text{a woman is selected}) = \frac{40}{100} = \mathbf{.40} \qquad \blacksquare$$

Relative Frequency Concept of Probability

Suppose we want to calculate the following probabilities:

1.   The probability that the next car that comes out of an auto factory is a "lemon"
2.   The probability that a randomly selected family owns two cars
3.   The probability that the next baby born at a hospital is a girl
4.   The probability that an 80-year-old person will live for at least one more year
5.   The probability that the tossing of an unbalanced coin will result in a head
6.   The probability that we will observe a 1-spot if we roll a loaded die

These probabilities cannot be computed using the classical probability rule because the various outcomes for the corresponding experiments are not equally likely. For example, the next car manufactured at an auto factory may or may not be a lemon. The two outcomes, "it is a lemon" and "it is not a lemon," are not equally likely. If they were, then (approximately) half the cars manufactured by this company would be lemons, and this might prove disastrous to the survival of the firm.

Although the various outcomes for each of these experiments are not equally likely, each of these experiments can be performed again and again to generate data. In such cases, to calculate probabilities, we either use past data or generate new data by performing the experiment a large number of times. The relative frequency of an event is used as an approximation for the probability of that event. This method of assigning a probability to an event is called the **relative frequency concept of probability**. Because relative frequencies are determined by performing an experiment, the probabilities calculated using relative frequencies may change almost each time an experiment is repeated. For example, every time a new sample of 500 cars is selected from the production line of an auto factory, the number of lemons in those 500 cars is expected to be different. However, the variation in the percentage of lemons will be small if the sample size is large. Note that if we are considering the population, the relative frequency will give an exact probability.

RELATIVE FREQUENCY AS AN APPROXIMATION OF PROBABILITY   If an experiment is repeated n times and an event A is observed f times, then, according to the relative frequency concept of probability:

$$P(A) = \frac{f}{n}$$

Examples 4–10 and 4–11 illustrate how the probabilities of events are approximated using the relative frequencies.

Approximating probability by relative frequency: sample data.

EXAMPLE 4-10   Ten of the 500 randomly selected cars manufactured at a certain auto factory are found to be lemons. Assuming that the lemons are manufactured randomly, what is the probability that the next car manufactured at this auto factory is a lemon?

Solution   Let n denote the total number of cars in the sample and f the number of lemons in n. Then,

$$n = 500 \quad \text{and} \quad f = 10$$

Using the relative frequency concept of probability, we obtain

$$P(\text{next car is a lemon}) = \frac{f}{n} = \frac{10}{500} = .02$$

This probability is actually the relative frequency of lemons in 500 cars. Table 4.2 lists the frequency and relative frequency distributions for this example.

Table 4.2   Frequency and Relative Frequency Distributions for the Sample of Cars

Car	f	Relative Frequency
Good	490	490/500 = .98
Lemon	10	10/500 = .02
	$n = 500$	Sum = 1.00

The column of relative frequencies in Table 4.2 is used as the column of approximate probabilities. Thus, from the relative frequency column,

$$P(\text{next car is a lemon}) = .02$$

$$P(\text{next car is a good car}) = .98$$

■

Note that relative frequencies are not probabilities but approximate probabilities. However, if the experiment is repeated again and again, this approximate probability of an outcome obtained from the relative frequency will approach the actual probability of that outcome. This is called the **Law of Large Numbers**.

> **LAW OF LARGE NUMBERS**   If an experiment is repeated again and again, the probability of an event obtained from the relative frequency approaches the actual or theoretical probability.

Approximating probability by relative frequency.

EXAMPLE 4-11   Allison wants to determine the probability that a randomly selected family from New York State owns a home. How can she determine this probability?

Solution   There are two outcomes for a randomly selected family from New York State: "This family owns a home" and "this family does not own a home." These two events are not equally likely. (Note that these two outcomes will be equally likely if exactly half of the families in New York State own homes and exactly half do not own homes.) Hence, the classical probability rule cannot be applied. However, we can repeat this experiment again and again. In other words, we can select a sample of families from New York State and observe whether or not each of them owns a home. Hence, we will use the relative frequency approach to probability.

Suppose Allison selects a random sample of 1000 families from New York State and observes that 670 of them own homes and 330 do not own homes. Then,

$$n = \text{sample size} = 1000$$

$$f = \text{number of families who own homes} = 670$$

Consequently,

$$P(\text{a randomly selected family owns a home}) = \frac{f}{n} = \frac{670}{1000} = \textbf{.670}$$

Again, note that .670 is just an approximation of the probability that a randomly selected family from New York State owns a home. Every time Allison repeats this experiment she may obtain a different probability for this event. However, because the sample size ($n = 1000$) in this example is large, the variation is expected to be very small. ■

Subjective Probability

Many times we face experiments that neither have equally likely outcomes nor can be repeated to generate data. In such cases, we cannot compute the probabilities of events using the classical probability rule or the relative frequency concept. For example, consider the following probabilities of events:

1. The probability that Carol, who is taking statistics, will earn an A in this course
2. The probability that the Dow Jones Industrial Average will be higher at the end of the next trading day
3. The probability that the New York Giants will win the Super Bowl next season
4. The probability that Joe will lose the lawsuit he has filed against his landlord

Neither the classical probability rule nor the relative frequency concept of probability can be applied to calculate probabilities for these examples. All these examples belong to experiments that have neither equally likely outcomes nor the potential of being repeated. For example, Carol, who is taking statistics, will take the test (or tests) only once, and based on that she will either earn an A or not. The two events "she will earn an A" and "she will not earn an A" are not equally likely. The probability assigned to an event in such cases is called **subjective probability**. It is based on the individual's own judgment, experience, information, and belief. Carol may assign a high probability to the event that she will earn an A in statistics, whereas her instructor may assign a low probability to the same event.

> **SUBJECTIVE PROBABILITY** *Subjective probability* is the probability assigned to an event based on subjective judgment, experience, information, and belief.

Subjective probability is assigned arbitrarily. It is usually influenced by the biases, preferences, and experience of the person assigning the probability.

EXERCISES

■ **Concepts and Procedures**

4.15 Briefly explain the two properties of probability.

4.16 Briefly describe an impossible event and a sure event. What is the probability of the occurrence of each of these two events?

4.17 Briefly explain the three approaches to probability. Give one example of each approach.

4.18 Briefly explain for what kind of experiments we use the classical approach to calculate probabilities of events and for what kind of experiments we use the relative frequency approach.

4.19 Which of the following values cannot be probabilities of events and why?

$$1/5 \quad .97 \quad -.55 \quad 1.56 \quad 5/3 \quad 0.0 \quad -2/7 \quad 1.0$$

4.20 Which of the following values cannot be probabilities of events and why?

$$.46 \quad 2/3 \quad -.09 \quad 1.42 \quad .96 \quad 9/4 \quad -1/4 \quad .02$$

■ **Applications**

4.21 Suppose a family is selected at random from New York City. Consider the following two outcomes: This family's yearly income is less than $50,000; this family's yearly income is $50,000 or higher. Are these two outcomes equally likely? Explain why or why not. If you are to find the probabilities of these two outcomes, would you use the classical approach or the relative frequency approach? Explain why.

4.22 A class has a total of 35 students. Of them, 15 are business majors and 20 are nonbusiness majors. Suppose one student is selected at random from this class. Consider the following two events: This student is a business major; this student is a nonbusiness major. If you are to find the probabilities of these two events, would you use the classical approach or the relative frequency approach? Explain why.

4.23 The president of a company has a hunch that there is a .80 probability that the company will be successful in marketing a new brand of ice cream. Is this a case of classical, relative frequency, or subjective probability? Explain why.

4.24 The coach of a college football team thinks there is a .75 probability that the team will win the national championship this year. Is this a case of classical, relative frequency, or subjective probability? Explain why.

4.25 A hat contains 40 marbles. Of them, 18 are red and 22 are green. If one marble is randomly selected out of this hat, what is the probability that this marble is

a. red   **b.** green?

4.26 A die is rolled once. What is the probability that

a. a number less than 5 is obtained
b. a number 3 to 6 is obtained?

4.27 A random sample of 2000 adults showed that 860 of them have shopped at least once on the Internet. What is the (approximate) probability that a randomly selected adult has shopped on the Internet?

4.28 In a statistics class of 45 students, 12 have a strong interest in statistics. Find the probability that a randomly selected student from this class has a strong interest in statistics.

4.29 In a group of 50 executives, 29 have a type A personality. If one executive is selected at random from this group, what is the probability that this executive has a type A personality?

4.30 Out of the 3000 families who live in an apartment complex in New York City, 600 paid no income tax last year. What is the probability that a randomly selected family from these 3000 families paid income tax last year?

4.31 A multiple-choice question on a test has five answers. If Dianne chooses one answer based on "pure guess," what is the probability that her answer is

a. correct   **b.** wrong?

Do these two probabilities add up to 1.0? If yes, why?

4.32 A university has a total of 320 professors and 105 of them are women. What is the probability that a randomly selected professor from this university is a

a. woman   **b.** man?

Do these two probabilities add up to 1.0? If yes, why?

4.33 A company that plans to hire one new employee has prepared a final list of six candidates, all of whom are equally qualified. Four of these six candidates are women. If the company decides to select at random one person out of these six candidates, what is the probability that this person will be a woman? What is the probability that this person will be a man? Do these two probabilities add up to 1.0? If yes, why?

4.34 A sample of 500 large companies showed that 120 of them offer free psychiatric help to their employees who suffer from psychological problems. If one company is selected at random from this sample, what is the probability that this company offers free psychiatric help to its employees who suffer from psychological problems? What is the probability that this company does not offer free psychiatric help to its employees who suffer from psychological problems? Do these two probabilities add up to 1.0? If yes, why?

4.35 A sample of 400 large companies showed that 130 of them offer free health fitness centers to their employees on the company premises. If one company is selected at random from this sample, what is the probability that this company offers a free health fitness center to its employees on the company premises? What is the probability that this company does not offer a free health fitness center to its employees on the company premises? Do these two probabilities add up to 1.0? If yes, why?

4.36 According to the U.S. Bureau of Labor Statistics, in May 1999, there were 34,104,000 Americans 16 to 24 years of age in the civilian noninstitutional population. Of these, 19,902,000 were employed, 11,985,000 were not in the labor force (i.e., they were not actively seeking employment), and the remaining were unemployed. If one of these 34,104,000 persons is selected at random, find the probability that he or she is

a. employed   **b.** unemployed   **c.** not in the labor force

Do these three probabilities add up to 1.0? If so, why?

4.37 A sample of 1000 families showed that 34 of them own no cars, 208 own one car each, 376 own two cars each, 265 own three cars each, and 117 own four or more cars each. Write the frequency distribution table for this problem. Calculate the relative frequencies for all categories. Suppose one family is randomly selected from these 1000 families. Find the probability that this family owns

a. two cars   **b.** four or more cars

4.38 In a sample of 500 families, 90 have a yearly income of less than $20,000, 270 have a yearly income of $20,000 to $50,000, and the remaining families have a yearly income of more than $50,000. Write the frequency distribution table for this problem. Calculate the relative frequencies for all classes. Suppose one family is randomly selected from these 500 families. Find the probability that this family has a yearly income of

a. less than $20,000   **b.** more than $50,000

4.39 Suppose you want to find the (approximate) probability that a randomly selected family from Los Angeles earns more than $75,000 a year. How would you find this probability? What procedure would you use? Explain briefly.

4.40 Suppose you have a loaded die and you want to find the (approximate) probabilities of different outcomes for this die. How would you find these probabilities? What procedure would you use? Explain briefly.

4.3 COUNTING RULE

The experiments dealt with so far in this chapter have had only a few outcomes, which were easy to list. However, for experiments with a large number of outcomes, it may not be easy to list all outcomes. In such cases, we may use the **counting rule** to find the total number of outcomes.

> ✳ **COUNTING RULE** If an experiment consists of three steps and if the first step can result in m outcomes, the second step in n outcomes, and the third step in k outcomes, then
>
> $$\text{Total outcomes for the experiment} = m \cdot n \cdot k$$

The counting rule can easily be extended to apply to an experiment that has less or more than three steps.

Applying the counting rule: 3 steps.

EXAMPLE 4–12 Suppose we toss a coin three times. This experiment has three steps: the first toss, the second toss, and the third toss. Each step has two outcomes: a head and a tail. Thus,

$$\text{Total outcomes for three tosses of a coin} = 2 \times 2 \times 2 = \mathbf{8}$$

The eight outcomes for this experiment are *HHH, HHT, HTH, HTT, THH, THT, TTH,* and *TTT*. ∎

Applying the counting rule: 2 steps.

EXAMPLE 4–13 A prospective car buyer can choose between a fixed and a variable interest rate and can also choose a payment period of 36 months, 48 months, or 60 months. How many total outcomes are possible?

Solution This experiment is made up of two steps: choosing an interest rate and selecting a loan payment period. There are two outcomes (a fixed or a variable interest rate) for the first step and three outcomes (a payment period of 36 months, 48 months, or 60 months) for the second step. Hence,

$$\text{Total outcomes} = 2 \times 3 = \mathbf{6}$$
∎

Applying the counting rule: 16 steps.

EXAMPLE 4–14 A National Football League team will play 16 games during a regular season. Each game can result in one of three outcomes: a win, a loss, or a tie. The total possible outcomes for 16 games are calculated as follows:

$$\text{Total outcomes} = 3 \cdot 3 \cdot 3 \cdot 3 \cdot 3 \cdot 3 \cdot 3 \cdot 3 \cdot 3 \cdot 3 \cdot 3 \cdot 3 \cdot 3 \cdot 3 \cdot 3 \cdot 3$$

$$= 3^{16} = \mathbf{43{,}046{,}721}$$

One of the 43,046,721 possible outcomes is all 16 wins.[2] ∎

---

[2]Using a calculator to evaluate 3^{16}: If your calculator contains a y^x or an x^y key, you can use that key to simplify 3^{16} as follows: First enter 3 on the calculator, then press the y^x key, next enter 16, and finally press the $=$ key. The screen of the calculator will display 43046721 as the answer.

4.4   MARGINAL AND CONDITIONAL PROBABILITIES

Suppose all 100 employees of a company were asked whether they are in favor of or against paying high salaries to CEOs of U.S. companies. Table 4.3 gives a two-way classification of the responses of these 100 employees.

Table 4.3   Two-Way Classification of Employee Responses

	In Favor	Against
Male	15	45
Female	4	36

Table 4.3 shows the distribution of 100 employees based on two variables or characteristics: gender (male or female) and opinion (in favor or against). Such a table is called a *contingency table*. In Table 4.3, each box that contains a number is called a *cell*. Notice that there are four cells. Each cell gives the frequency for two characteristics. For example, 15 employees in this group possess two characteristics: "male" and "in favor of paying high salaries to CEOs." We can interpret the numbers in other cells the same way.

By adding the row totals and the column totals to Table 4.3, we write Table 4.4.

Table 4.4   Two-Way Classification of Employee Responses with Totals

	In Favor	Against	Total
Male	15	45	60
Female	4	36	40
Total	19	81	100

Suppose one employee is selected at random from these 100 employees. This employee may be classified either on the basis of gender alone or on the basis of opinion. If only one characteristic is considered at a time, the employee selected can be a male, a female, in favor, or against. The probability of each of these four characteristics or events is called **marginal probability** or *simple probability*. These probabilities are called marginal probabilities because they are calculated by dividing the corresponding row margins (totals for the rows) or column margins (totals for the columns) by the grand total.

> **MARGINAL PROBABILITY**   *Marginal probability* is the probability of a single event without consideration of any other event. Marginal probability is also called *simple probability*.

For Table 4.4, the four marginal probabilities are calculated as follows:

$$P(\text{male}) = \frac{\text{Number of males}}{\text{Total number of employees}} = \frac{60}{100} = .60$$

As we can observe, the probability that a male will be selected is obtained by dividing the total of the row labeled "Male" (60) by the grand total (100). Similarly,

$$P(\text{female}) = 40/100 = .40$$

$$P(\text{in favor}) = 19/100 = .19$$

$$P(\text{against}) = 81/100 = .81$$

These four marginal probabilities are shown along the right side and along the bottom of Table 4.5.

Table 4.5 Listing the Marginal Probabilities

	In Favor (A)	Against (B)	Total	
Male (M)	15	45	60	$P(M) = 60/100 = .60$
Female (F)	4	36	40	$P(F) = 40/100 = .40$
Total	19	81	100	

$P(A) = 19/100$ $P(B) = 81/100$
= .19   = .81

Now suppose that one employee is selected at random from these 100 employees. Furthermore, assume it is known that this (selected) employee is a male. In other words, the event that the employee selected is a male has already occurred. What is the probability that the employee selected is in favor of paying high salaries to CEOs? This probability is written as follows:

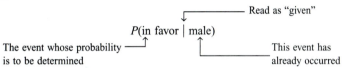

Read as "given"

$P(\text{in favor} \mid \text{male})$

The event whose probability is to be determined

This event has already occurred

This probability, $P(\text{in favor} \mid \text{male})$, is called the **conditional probability** of "in favor." It is read as "the probability that the employee selected is in favor given that this employee is a male."

> **CONDITIONAL PROBABILITY** *Conditional probability* is the probability that an event will occur given that another event has already occurred. If A and B are two events, then the conditional probability of A given B is written as
>
> $$P(A \mid B)$$
>
> and read as "the probability of A given that B has already occurred."

Calculating the conditional probability: two-way table.

EXAMPLE 4-15 Compute the conditional probability $P(\text{in favor} \mid \text{male})$ for the data on 100 employees given in Table 4.4.

Solution The probability $P(\text{in favor} \mid \text{male})$ is the conditional probability that a randomly selected employee is in favor given that this employee is a male. It is known that the event "male" has already occurred. Based on the information that the employee selected is a male, we can infer that the employee selected must be one of the 60 males and, hence, must belong to the first row of Table 4.4. Therefore, we are concerned only with the first row of that table.

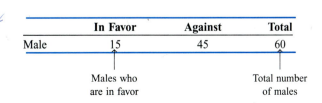

	In Favor	Against	Total
Male	15	45	60

Males who are in favor   Total number of males

✳ The required conditional probability is calculated as follows:

$$P(\text{in favor} \mid \text{male}) = \frac{\text{Number of males who are in favor}}{\text{Total number of males}} = \frac{15}{60} = .25$$

As we can observe from this computation of conditional probability, the total number of males (the event that has already occurred) is written in the denominator and the number of males who are in favor (the event whose probability we are to find) is written in the numerator. Note that we are considering the row of the event that has already occurred. The tree diagram in Figure 4.6 illustrates this example.

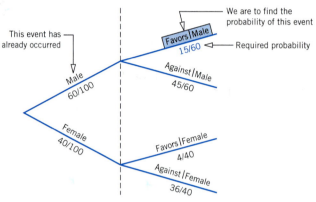

We are to find the probability of this event

This event has already occurred

Favors | Male
15/60 ← Required probability

Against | Male
45/60

Male
60/100

Female
40/100

Favors | Female
4/40

Against | Female
36/40

Figure 4.6 Tree diagram. ■

*Calculating the
conditional probability:
two-way table.*

EXAMPLE 4–16 For the data of Table 4.4, calculate the conditional probability that a randomly selected employee is a female given that this employee is in favor of paying high salaries to CEOs.

Solution We are to compute the probability, $P(\text{female} \mid \text{in favor})$. Since it is known that the employee selected is in favor of paying high salaries to CEOs, this employee must belong to the first column (the column labeled "in favor") and must be one of the 19 employees who are in favor.

In Favor

15

4 ⟵——— Females who are in favor

19 ⟵——— Total number of employees who are in favor

Hence, the required probability is

$$P(\text{female} \mid \text{in favor}) = \frac{\text{Number of females who are in favor}}{\text{Total number of employees who are in favor}}$$

$$= \frac{4}{19} = .2105$$

The tree diagram in Figure 4.7 illustrates this example.

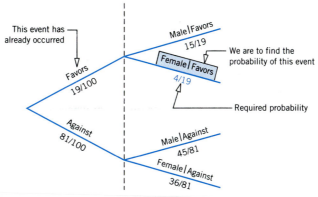

Figure 4.7 Tree diagram.

Case Study 4–1 illustrates the conditional probabilities of different activities that men and women participate in to relieve stress.

CASE STUDY 4-1 STRESS BUSTERS

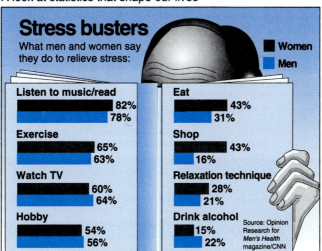

USA SNAPSHOTS®

A look at statistics that shape our lives

Stress busters
What men and women say they do to relieve stress:

- Women
- Men

Listen to music/read
82%
78%

Exercise
65%
63%

Watch TV
60%
64%

Hobby
54%
56%

Eat
43%
31%

Shop
43%
16%

Relaxation technique
28%
21%

Drink alcohol
15%
22%

Source: Opinion Research for *Men's Health* magazine/CNN

By Cindy Hall and Marcy E. Mullins, USA TODAY

From the chart, we can observe that 82% of women and 78% of men listen to music or read to relieve stress. These percentages can be written as conditional probabilities as follows. Suppose that one person is selected at random. Then, given that this person is a female, the probability is .82 that she listens to music or reads to relieve stress. On the other hand, if this selected person is a male, the same probability is .78. These probabilities can be written as follows:

$$P(\text{listens to music or reads to relieve stress} \mid \text{female}) = .82$$

$$P(\text{listens to music or reads to relieve stress} \mid \text{male}) = .78$$

Similarly, we can write many other conditional probabilities using the data given in this chart. Two more conditional probabilities are

$$P(\text{exercises to relieve stress} \mid \text{female}) = .65$$

$$P(\text{exercises to relieve stress} \mid \text{male}) = .63$$

Note that these are approximate probabilities because the data given in the chart are based on a sample survey. The reader should list all the conditional probabilities for men and women based on this chart.

Source: *USA TODAY*, December 21, 1998. Copyright © 1998, *USA TODAY*. Chart reproduced with permission.

4.5 MUTUALLY EXCLUSIVE EVENTS

Events that cannot occur together are called **mutually exclusive events**. Such events do not have any common outcomes. If two or more events are mutually exclusive, then at most one of them will occur every time we repeat the experiment. Thus, the occurrence of one event excludes the occurrence of the other event or events.

> **MUTUALLY EXCLUSIVE EVENTS** Events that cannot occur together are said to be *mutually exclusive events*.

For any experiment, the final outcomes are always mutually exclusive because one and only one of these outcomes is expected to occur in one repetition of the experiment. For example, consider tossing a coin twice. This experiment has four outcomes: *HH*, *HT*, *TH*, and *TT*. These outcomes are mutually exclusive because one and only one of them will occur when we toss this coin twice.

Illustrating mutually exclusive and mutually nonexclusive events.

EXAMPLE 4–17 Consider the following events for one roll of a die:

$$A = \text{an even number is observed} = \{2, 4, 6\}$$

$$B = \text{an odd number is observed} = \{1, 3, 5\}$$

$$C = \text{a number less than 5 is observed} = \{1, 2, 3, 4\}$$

Are events *A* and *B* mutually exclusive? Are events *A* and *C* mutually exclusive?

Solution Figures 4.8 and 4.9 show the diagrams of events A and B and events A and C, respectively.

As we can observe from the definitions of events A and B and from Figure 4.8, events A and B have no common element. For one roll of a die, only one of the two events, A and B, can happen. Hence, these are two mutually exclusive events. We can observe from the definitions of events A and C and from Figure 4.9 that events A and C have two common outcomes: 2-spot and 4-spot. Thus, if we roll a die and obtain either a 2-spot or a 4-spot, then A and C happen at the same time. Hence, events A and C are not mutually exclusive.

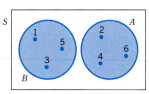

Figure 4.8 Mutually exclusive events A and B.

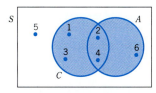

Figure 4.9 Mutually nonexclusive events A and C. ■

Illustrating mutually exclusive events.

EXAMPLE 4-18 Consider the following two events for a randomly selected adult:

Y = this adult has shopped at least once on the Internet

N = this adult has never shopped on the Internet

Are events Y and N mutually exclusive?

Solution Note that event Y consists of all adults who have shopped at least once on the Internet and event N includes all adults who have never shopped on the Internet. These two events are illustrated in the Venn diagram in Figure 4.10.

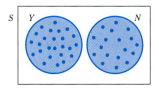

Figure 4.10 Mutually exclusive events Y and N.

As we can observe from the definitions of events Y and N and from Figure 4.10, events Y and N have no common outcome. They represent two distinct sets of adults: the ones who have shopped at least once on the Internet and the ones who have never shopped on the Internet. Hence, these two events are mutually exclusive. ■

4.6 INDEPENDENT VERSUS DEPENDENT EVENTS

In the case of two **independent events**, the occurrence of one event does not change the probability of the occurrence of the other event.

> **INDEPENDENT EVENTS** Two events are said to be *independent* if the occurrence of one does not affect the probability of the occurrence of the other. In other words, *A* and *B* are *independent events* if
>
> $$\text{either} \quad P(A \mid B) = P(A) \quad \text{or} \quad P(B \mid A) = P(B)$$

It can be shown that if one of these two conditions is true, then the second will also be true, and if one is not true, then the second will also not be true.

If the occurrence of one event affects the probability of the occurrence of the other event, then the two events are said to be **dependent events**. In probability notation, the two events are dependent if either $P(A \mid B) \neq P(A)$ or $P(B \mid A) \neq P(B)$.

Illustrating two dependent events: two-way table.

EXAMPLE 4–19 Refer to the information on 100 employees given in Table 4.4 in Section 4.4. Are events "female (*F*)" and "in favor (*A*)" independent?

Solution Events *F* and *A* will be independent if

$$P(F) = P(F \mid A)$$

Otherwise they will be dependent.

Using the information given in Table 4.4, we compute the following two probabilities:

$$P(F) = 40/100 = \textbf{.40} \quad \text{and} \quad P(F \mid A) = 4/19 = \textbf{.2105}$$

Because these two probabilities are not equal, the two events are dependent. Here, dependence of events means that the percentages of males who are in favor of and against paying high salaries to CEOs are different from the percentages of females who are in favor and against.

In this example, the dependence of *A* and *F* can also be proved by showing that the probabilities $P(A)$ and $P(A \mid F)$ are not equal. ■

Illustrating two independent events.

EXAMPLE 4–20 A box contains a total of 100 CDs that were manufactured on two machines. Of them, 60 were manufactured on Machine I. Of the total CDs, 15 are defective. Of the 60 CDs that were manufactured on Machine I, 9 are defective. Let *D* be the event that a randomly selected CD is defective, and let *A* be the event that a randomly selected CD was manufactured on Machine I. Are events *D* and *A* independent?

Solution From the given information,

$$P(D) = 15/100 = .15 \quad \text{and} \quad P(D \mid A) = 9/60 = .15$$

Hence,

$$P(D) = P(D \mid A)$$

Consequently, the two events, *D* and *A*, are independent.

Independence, in this example, means that the probability of any CD being defective is the same, .15, irrespective of the machine on which it is manufactured. In other words, the two machines are producing the same percentage of defective CDs. For example, 9 of the 60 CDs manufactured on Machine I are defective and 6 of the 40 CDs manufactured on Machine II are defective. Thus, for each of the two machines, 15% of the CDs produced are defective.

Actually, using the given information, we can prepare Table 4.6. The numbers in the shaded cells are given to us. The remaining numbers are calculated by doing some arithmetic manipulations.

Table 4.6   Two-Way Classification Table

	Defective (D)	Good (G)	Total
Machine I (A)	9	51	60
Machine II (B)	6	34	40
Total	15	85	100

Using this table, we can find the following probabilities:

$$P(D) = 15/100 = .15$$

$$P(D \mid A) = 9/60 = .15$$

Since these two probabilities are the same, the two events are independent. ■

☞ **TWO IMPORTANT OBSERVATIONS**

We can make two important observations about mutually exclusive, independent, and dependent events.

1. Two events are either mutually exclusive or independent.[3]
   a. Mutually exclusive events are always dependent.
   b. Independent events are never mutually exclusive.
2. Dependent events may or may not be mutually exclusive.

4.7 COMPLEMENTARY EVENTS

Two mutually exclusive events that taken together include all the outcomes for an experiment are called **complementary events**. Note that two complementary events are always mutually exclusive.

> **COMPLEMENTARY EVENTS**   The complement of event A, denoted by $\overline{A}$ and read as "A bar" or "A complement," is the event that includes all the outcomes for an experiment that are not in A.

Events A and $\overline{A}$ are complements of each other. The Venn diagram in Figure 4.11 (page 150) shows the complementary events A and $\overline{A}$.

Because two complementary events, taken together, include all the outcomes for an experiment and because the sum of the probabilities of all outcomes is 1, it is obvious that

$$P(A) + P(\overline{A}) = 1$$

[3]The exception to this rule occurs when at least one of the two events has a zero probability.

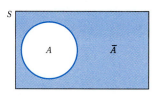

Figure 4.11   Venn diagram
of two complementary events.

From this equation we can deduce that

$$P(A) = 1 - P(\overline{A}) \quad \text{and} \quad P(\overline{A}) = 1 - P(A)$$

Thus, if we know the probability of an event, we can find the probability of its complementary event by subtracting the given probability from 1.

Calculating probabilities
of complementary events.

EXAMPLE 4–21   In a group of 2000 taxpayers, 400 have been audited by the IRS at least once. If one taxpayer is randomly selected from this group, what are the two complementary events for this experiment, and what are their probabilities?

Solution   The two complementary events for this experiment are

A = the selected taxpayer has been audited by the IRS at least once

$\overline{A}$ = the selected taxpayer has never been audited by the IRS

Note that here event A includes the 400 taxpayers who have been audited by the IRS at least once, and $\overline{A}$ includes the 1600 taxpayers who have never been audited by the IRS. Hence, the probabilities of events A and $\overline{A}$ are

$$P(A) = 400/2000 = \textbf{.20} \quad \text{and} \quad P(\overline{A}) = 1600/2000 = \textbf{.80}$$

As we can observe, the sum of these two probabilities is 1. Figure 4.12 shows a Venn diagram for this example.

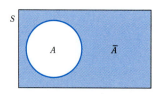

Figure 4.12   Venn diagram.   ■

Calculating probabilities
of complementary events.

EXAMPLE 4–22   In a group of 5000 adults, 3500 are in favor of stricter gun control laws, 1200 are against such laws, and 300 have no opinion. One adult is randomly selected from this group. Let A be the event that this adult is in favor of stricter gun control laws. What is the complementary event of A? What are the probabilities of the two events?

Solution   The two complementary events for this experiment are

A = the selected adult is in favor of stricter gun control laws

$\overline{A}$ = the selected adult either is against such laws or has no opinion

Note that here event $\overline{A}$ includes 1500 adults who either are against stricter gun control laws or have no opinion. Also notice that events A and $\overline{A}$ are complements of each other. Since 3500 adults in the group favor stricter gun control laws and 1500 either are against stricter gun control laws or have no opinion, the probabilities of events A and $\overline{A}$ are

$$P(A) = 3500/5000 = \textbf{.70} \quad \text{and} \quad P(\overline{A}) = 1500/5000 = \textbf{.30}$$

As we can observe, the sum of these two probabilities is 1. Also, once we find $P(A)$, we can find the probability of $P(\overline{A})$ as

$$P(\overline{A}) = 1 - P(A) = 1 - .70 = .30$$

Figure 4.13 shows a Venn diagram for this example.

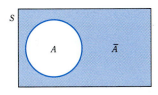

Figure 4.13   Venn diagram.   ■

EXERCISES

■ **Concepts and Procedures**

4.41   Briefly explain the difference between the marginal and conditional probabilities of events. Give one example of each.

4.42   What is meant by two mutually exclusive events? Give one example of two mutually exclusive events and another example of two mutually nonexclusive events.

4.43   Briefly explain the meaning of independent and dependent events. Suppose A and B are two events. What formula can you use to prove whether A and B are independent or dependent?

4.44   What is the complement of an event? What is the sum of the probabilities of two complementary events?

4.45   How many different outcomes are possible for four rolls of a die?

4.46   How many different outcomes are possible for 10 tosses of a coin?

4.47   A statistical experiment has eight equally likely outcomes that are denoted by 1, 2, 3, 4, 5, 6, 7, and 8. Let event $A = \{2, 5, 7\}$ and event $B = \{2, 4, 8\}$.

a.   Are events A and B mutually exclusive events?
b.   Are events A and B independent events?
c.   What are the complements of events A and B, respectively, and their probabilities?

4.48   A statistical experiment has 10 equally likely outcomes that are denoted by 1, 2, 3, 4, 5, 6, 7, 8, 9, and 10. Let event $A = \{3, 4, 6, 9\}$ and event $B = \{1, 2, 5\}$.

a.   Are events A and B mutually exclusive events?
b.   Are events A and B independent events?
c.   What are the complements of events A and B, respectively, and their probabilities?

■ **Applications**

4.49   A specific model of a car comes in five exterior colors and four interior colors. All exterior colors can be combined with any of the interior colors. How many different selections of one exterior and one interior color are possible?

4.50   A man just bought four suits, eight shirts, and 12 ties. All of these suits, shirts, and ties coordinate with each other. If he is to randomly select one suit, one shirt, and one tie to wear on a certain day, how many different outcomes (selections) are possible?

4.51   A restaurant menu has four kinds of soups, eight kinds of main courses, five kinds of desserts, and six kinds of drinks. If a customer randomly selects one item from each of these four categories, how many different outcomes are possible?

4.52   A student is to select three courses for next semester. If this student decides to randomly select one course from each of eight economics courses, six mathematics courses, and five computer courses, how many different outcomes are possible?

4.53   A sample of 2000 adults were asked whether or not they have ever shopped on the Internet. The following table gives a two-way classification of the responses.

	Have Shopped	**Have Never Shopped**
Male	300	900
Female	200	600

a.   If one adult is selected at random from these 2000 adults, find the probability that this adult
    i.   has never shopped on the Internet
    ii. is a male
    iii. has shopped on the Internet given that this adult is a female
    iv. is a male given that this adult has never shopped on the Internet
b.   Are the events "male" and "female" mutually exclusive? What about the events "have shopped" and "male"? Why or why not?
c.   Are the events "female" and "have shopped" independent? Why or why not?

4.54   The following table gives a two-way classification, based on gender and employment status, of the civilian labor force age 16 to 24 years as of July 1999. The numbers in the table are in thousands.

	Employed	**Unemployed**
Male	11,638	1337
Female	10,540	1157

Source: U.S. Bureau of Labor Statistics, August 26, 1999.

a.   If one person is selected at random from these young persons, find the probability that this person is
    i.   unemployed
    ii. a female
    iii. employed given the person is male
    iv. a female given the person is unemployed
b.   Are the events "employed" and "unemployed" mutually exclusive? What about the events "unemployed" and "male"?
c.   Are the events "female" and "employed" independent? Why or why not?

4.55   A group of 2000 randomly selected adults were asked if they are in favor of or against cloning. The following table gives the responses.

	In Favor	**Against**	**No Opinion**
Male	395	405	100
Female	300	680	120

a.   If one person is selected at random from these 2000 adults, find the probability that this person is
    i.   in favor of cloning
    ii. against cloning
    iii. in favor of cloning given the person is a female
    iv. a male given the person has no opinion
b.   Are the events "male" and "in favor" mutually exclusive? What about the events "in favor" and "against"? Why or why not?
c.   Are the events "female" and "no opinion" independent? Why or why not?

4.56   Five hundred employees were selected from a city's large private companies, and they were asked whether or not they have any retirement benefits provided by their companies. Based on this information, the following two-way classification table was prepared.

	Have Retirement Benefits	
	Yes	**No**
Men	225	75
Women	150	50

a. If one employee is selected at random from these 500 employees, find the probability that this employee
    i. is a woman
    ii. has retirement benefits
    iii. has retirement benefits given the employee is a man
    iv. is a woman given that she does not have retirement benefits

b. Are the events "man" and "yes" mutually exclusive? What about the events "yes" and "no"? Why or why not?

c. Are the events "woman" and "yes" independent? Why or why not?

4.57 The following table gives the two-way classification of 2000 randomly selected employees from a city based on gender and commuting time from home to work.

	Commuting Time from Home to Work		
	Less Than 30 Minutes	**30 Minutes to One Hour**	**More Than One Hour**
Men	524	455	221
Women	413	263	124

a. If one employee is selected at random from these 2000 employees, find the probability that this employee
    i. commutes for more than one hour
    ii. commutes for less than 30 minutes
    iii. is a man given that he commutes for 30 minutes to one hour
    iv. commutes for more than one hour given the employee is a woman

b. Are the events "man" and "more than one hour" mutually exclusive? What about the events "less than 30 minutes" and "more than one hour"? Why or why not?

c. Are the events "woman" and "30 minutes to one hour" independent? Why or why not?

4.58 Two thousand randomly selected adults were asked if they think they are financially better off than their parents. The following table gives the two-way classification of the responses based on the education levels of the persons included in the survey and whether they are financially better off, the same, or worse off than their parents.

	Less Than High School	**High School**	**More Than High School**
Better off	140	450	420
Same	60	250	110
Worse off	200	300	70

a. If one adult is selected at random from these 2000 adults, find the probability that this adult is
    i. financially better off than his/her parents
    ii. financially better off than his/her parents given he/she has less than high school education
    iii. financially worse off than his/her parents given he/she has high school education
    iv. financially the same as his/her parents given he/she has more than high school education

b. Are the events "better off" and "high school" mutually exclusive? What about the events "less than high school" and "more than high school"? Why or why not?

c. Are the events "worse off" and "more than high school" independent? Why or why not?

4.59   There are a total of 160 practicing physicians in a city. Of them, 75 are female and 25 are pediatricians. Of the 75 females, 20 are pediatricians. Are the events "female" and "pediatrician" independent? Are they mutually exclusive? Explain why or why not.

4.60   Of a total of 100 CDs manufactured on two machines, 20 are defective. Sixty of the total CDs were manufactured on Machine I and 10 of these 60 are defective. Are the events "machine type" and "defective CDs" independent? (*Note:* Compare this exercise with Example 4–20.)

4.61   A company hired 30 new graduates last week. Of these, 16 are female and 11 are business majors. Of the 16 females, 9 are business majors. Are the events "female" and "business major" independent? Are they mutually exclusive? Explain why or why not.

4.62   Define the following two events for two tosses of a coin:

$$A = \text{at least one head is obtained}$$

$$B = \text{both tails are obtained}$$

a.   Are A and B mutually exclusive events? Are they independent? Explain why or why not.
b.   Are A and B complementary events? If yes, first calculate the probability of B and then calculate the probability of A using the complementary event rule.

4.63   Let A be the event that a number less than 3 is obtained if we roll a die once. What is the probability of A? What is the complementary event of A, and what is its probability?

4.64   According to the U.S. Bureau of the Census, 44.3 million Americans had no health insurance in 1998 (*Newsweek*, November 8, 1999). Suppose that the U.S. population at that time was 275 million. If one person was selected at random from the 1998 U.S. population, what are the two complementary events and their probabilities?

4.65   The probability that an adult reads a newspaper every day is .65. What is its complementary event? What is the probability of this complementary event?

4.8   INTERSECTION OF EVENTS AND THE MULTIPLICATION RULE

This section discusses the intersection of two events and the application of the multiplication rule to compute the probability of the intersection of events.

4.8.1   INTERSECTION OF EVENTS

The **intersection of two events** is given by the outcomes that are common to both events.

> **INTERSECTION OF EVENTS**   Let A and B be two events defined in a sample space. The *intersection* of A and B represents the collection of all outcomes that are common to both A and B and is denoted by
>
> $$A \text{ and } B$$

The intersection of events A and B is also denoted by either $A \cap B$ or AB. Let

$$A = \text{event that a family owns a VCR}$$

$$B = \text{event that a family owns a telephone answering machine}$$

Figure 4.14 illustrates the intersection of events A and B. The shaded area in this figure gives the intersection of events A and B, and it includes all the families who own both a VCR and a telephone answering machine.

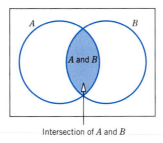

Intersection of A and B

Figure 4.14 Intersection of events A and B.

4.8.2 MULTIPLICATION RULE

The probability that events A and B happen together is called the **joint probability** of A and B and is written as $P(A$ and $B)$.

> **JOINT PROBABILITY**   The probability of the intersection of two events is called their *joint probability*. It is written as
>
> $$P(A \text{ and } B)$$

The probability of the intersection of two events is obtained by multiplying the marginal probability of one event by the conditional probability of the second event. This rule is called the **multiplication rule**.

> **MULTIPLICATION RULE**   The probability of the intersection of two events A and B is
>
> $$P(A \text{ and } B) = P(A)\, P(B \mid A)$$

The joint probability of events A and B can also be denoted by $P(A \cap B)$ or $P(AB)$.

Calculating the joint probability of two events: two-way table.

EXAMPLE 4–23   Table 4.7 gives the classification of all employees of a company by gender and college degree.

Table 4.7   Classification of Employees by Gender and Education

	College Graduate (G)	Not a College Graduate (N)	Total
Male (M)	7	20	27
Female (F)	4	9	13
Total	11	29	40

If one of these employees is selected at random for membership on the employee–management committee, what is the probability that this employee is a female and a college graduate?

Solution   We are to calculate the probability of the intersection of the events "female" (denoted by F) and "college graduate" (denoted by G). This probability may be computed using the formula

$$P(F \text{ and } G) = P(F)\, P(G \mid F)$$

The shaded area in Figure 4.15 shows the intersection of the events "female" and "college graduate."

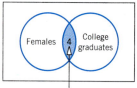

Females and college graduates

Figure 4.15   Intersection of events F and G.

Notice that there are 13 females among 40 employees. Hence, the probability that a female is selected is

$$P(F) = 13/40$$

To calculate the probability $P(G \mid F)$, we know that F has already occurred. Consequently, the employee selected is one of the 13 females. In the table, there are 4 college graduates among 13 female employees. Hence, the conditional probability of G given F is

$$P(G \mid F) = 4/13$$

The joint probability of F and G is

$$P(F \text{ and } G) = P(F)\, P(G \mid F) = (13/40)\,(4/13) = \mathbf{.100}$$

Thus, the probability is .100 that a randomly selected employee is a female and a college graduate.

The probability in this example can also be calculated without using the multiplication rule. As we can notice from Figure 4.15 and from the table, four employees out of a total of 40 are female and college graduates. Hence, if any of these four employees is selected, the events "female" and "college graduate" both happen. The required probability therefore is

$$P(F \text{ and } G) = 4/40 = \mathbf{.100}$$ ■

We can compute three other joint probabilities for the table in Example 4–23 as follows:

$$P(M \text{ and } G) = P(M)\, P(G \mid M) = (27/40)\,(7/27) = .175$$

$$P(M \text{ and } N) = P(M)\, P(N \mid M) = (27/40)\,(20/27) = .500$$

$$P(F \text{ and } N) = P(F)\, P(N \mid F) = (13/40)\,(9/13) = .225$$

The tree diagram in Figure 4.16 shows all four joint probabilities for this example. The joint probability of F and G is highlighted.

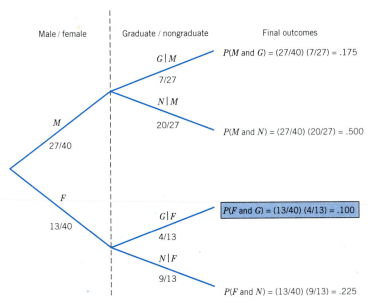

Figure 4.16 Tree diagram for joint probabilities.

Calculating the joint probability of two events.

EXAMPLE 4-24 A box contains 20 DVDs, 4 of which are defective. If 2 DVDs are selected at random (without replacement) from this box, what is the probability that both are defective?

Solution Let us define the following events for this experiment:

$$G_1 = \text{event that the first DVD selected is good}$$

$$D_1 = \text{event that the first DVD selected is defective}$$

$$G_2 = \text{event that the second DVD selected is good}$$

$$D_2 = \text{event that the second DVD selected is defective}$$

We are to calculate the joint probability of D_1 and D_2, which is given by

$$P(D_1 \text{ and } D_2) = P(D_1) P(D_2 \mid D_1)$$

As we know, there are 4 defective DVDs in 20. Consequently, the probability of selecting a defective DVD at the first selection is

$$P(D_1) = 4/20$$

To calculate the probability $P(D_2 \mid D_1)$, we know that the first DVD selected is defective because D_1 has already occurred. Because the selections are made without replacement, there are 19 total DVDs, and 3 of them are defective at the time of the second selection. Therefore,

$$P(D_2 \mid D_1) = 3/19$$

Hence, the required probability is

$$P(D_1 \text{ and } D_2) = P(D_1) P(D_2 \mid D_1) = (4/20) (3/19) = \textbf{.0316}$$

The tree diagram in Figure 4.17 (page 158) shows the selection procedure and the final four outcomes for this experiment along with their probabilities. The joint probability of D_1 and D_2 is highlighted in the tree diagram.

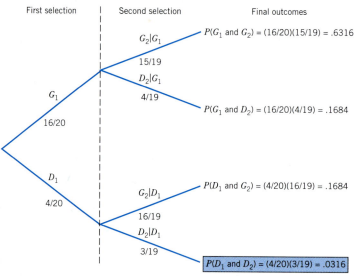

First selection | Second selection | Final outcomes

$G_2|G_1$

15/19

$P(G_1 \text{ and } G_2) = (16/20)(15/19) = .6316$

$D_2|G_1$

4/19

$P(G_1 \text{ and } D_2) = (16/20)(4/19) = .1684$

G_1

16/20

D_1

4/20

$G_2|D_1$

16/19

$P(D_1 \text{ and } G_2) = (4/20)(16/19) = .1684$

$D_2|D_1$

3/19

$P(D_1 \text{ and } D_2) = (4/20)(3/19) = .0316$

Figure 4.17 Selecting two DVDs.

Conditional probability was discussed in Section 4.4. It is obvious from the formula for joint probability that if we know the probability of an event A and the joint probability of events A and B, then we can calculate the conditional probability of B given A.

CONDITIONAL PROBABILITY If A and B are two events, then,

$$P(B \mid A) = \frac{P(A \text{ and } B)}{P(A)} \quad \text{and} \quad P(A \mid B) = \frac{P(A \text{ and } B)}{P(B)}$$

given that $P(A) \neq 0$ and $P(B) \neq 0$.

Calculating the conditional probability of an event.

EXAMPLE 4-25 The probability that a randomly selected student from a college is a senior is .20, and the joint probability that the student is a computer science major and a senior is .03. Find the conditional probability that a student selected at random is a computer science major given that he/she is a senior.

Solution Let us define the following two events:

$A =$ the student selected is a senior

$B =$ the student selected is a computer science major

From the given information,

$$P(A) = .20 \quad \text{and} \quad P(A \text{ and } B) = .03$$

Hence,

$$P(B \mid A) = \frac{P(A \text{ and } B)}{P(A)} = \frac{.03}{.20} = \mathbf{.15}$$

Thus, the (conditional) probability is .15 that a student selected at random is a computer science major given that he or she is a senior. ■

Multiplication Rule for Independent Events

The foregoing discussion of the multiplication rule was based on the assumption that the two events are dependent. Now suppose that events A and B are independent. Then,

$$P(A) = P(A \mid B) \quad \text{and} \quad P(B) = P(B \mid A)$$

By substituting $P(B)$ for $P(B \mid A)$ into the formula for the joint probability of A and B, we obtain

$$P(A \text{ and } B) = P(A) P(B)$$

> **MULTIPLICATION RULE FOR INDEPENDENT EVENTS** The probability of the intersection of two independent events A and B is
> $$P(A \text{ and } B) = P(A) P(B)$$

Calculating the joint probability of two independent events.

EXAMPLE 4–26 An office building has two fire detectors. The probability is .02 that any fire detector of this type will fail to go off during a fire. Find the probability that both of these fire detectors will fail to go off in case of a fire.

Solution In this example, the two fire detectors are independent because whether or not one fire detector goes off during a fire has no effect on the second fire detector. We define the following two events:

$$A = \text{the first fire detector fails to go off during a fire}$$

$$B = \text{the second fire detector fails to go off during a fire}$$

Then the joint probability of A and B is

$$P(A \text{ and } B) = P(A) P(B) = (.02) (.02) = \mathbf{.0004} \qquad \blacksquare$$

The multiplication rule can be extended to calculate the joint probability of more than two events. Example 4–27 illustrates such a case for independent events.

Calculating the joint probability of three events.

EXAMPLE 4–27 The probability that a patient is allergic to penicillin is .20. Suppose this drug is administered to three patients.

(a) Find the probability that all three of them are allergic to it.
(b) Find the probability that at least one of them is not allergic to it.

Solution

(a) Let A, B, and C denote the events that the first, second, and third patients, respectively, are allergic to penicillin. We are to find the joint probability of A, B, and C. All three events are independent because whether or not one patient is allergic does not depend on whether or not any of the other patients is allergic. Hence,

$$P(A \text{ and } B \text{ and } C) = P(A) P(B) P(C) = (.20) (.20) (.20) = \mathbf{.008}$$

The tree diagram in Figure 4.18 (page 160) shows all the outcomes for this experiment. Events $\overline{A}$, $\overline{B}$, and $\overline{C}$ are the complementary events of A, B, and C, respectively. They represent the events that the patients are not allergic to penicillin. Note that the intersection of events A, B, and C is written as ABC in the tree diagram.

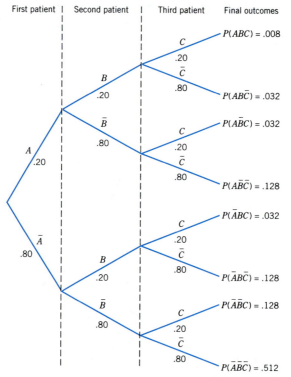

Figure 4.18 Tree diagram for joint probabilities.

(b) Let us define the following events:

$$G = \text{all three patients are allergic}$$

$$H = \text{at least one patient is not allergic}$$

Events G and H are two complementary events. Event G consists of the intersection of events A, B, and C. Hence, from part (a),

$$P(G) = P(A \text{ and } B \text{ and } C) = .008$$

Therefore, using the complementary event rule, we obtain

$$P(H) = 1 - P(G) = 1 - .008 = \textbf{.992} \qquad \blacksquare$$

Case Study 4–2 calculates the probability of a hitless streak in baseball by using the multiplication rule.

CASE STUDY 4–2 BASEBALL PLAYERS HAVE "SLUMPS" AND "STREAKS"

Going '0 for July,' as former infielder Bob Aspromonte once put it, is enough to make a baseball player toss out his lucky bat or start seriously searching for flaws in his hitting technique. But the culprit is usually just simple mathematics.

Statistician Harry Roberts of the University of Chicago's Graduate School of Business studied the records of major-league baseball players and found that a batter is no more likely to hit worse when he is in a slump than when he is in a hot streak. The occurrences of hits followed the same pattern as purely random events such as pulling marbles out of a hat. If there were one white marble and three black ones in the hat, for example, then a white marble would come out about one quarter of the time—a .250 average. In the same way, a player who hits .250 will in the long run get a hit every four times at bat.

But that doesn't mean the player will hit the ball exactly every fourth time he comes to the plate—just as it's unlikely that the white marble will come out exactly every fourth time.

Even a batter who goes hitless 10 times in a row might safely be able to pin the blame on statistical fluctuations. The odds of pulling a black marble out of a hat 10 times in a row are about 6 percent—not a frequent occurrence, but not impossible, either. Only in the long run do these statistical fluctuations even out. . . .

If we assume a player hits .250 in the long run, the probability that this player does not hit during a specific trip to the plate is .75. Hence, we can calculate the probability that he goes hitless 10 times in a row as follows.

$$P(\text{hitless 10 times in a row}) = (.75)(.75) \cdots (.75) \text{ ten times}$$

$$= (.75)^{10} = .0563$$

Note that each trip to the plate is independent, and the probability that a player goes hitless 10 times in a row is given by the intersection of 10 hitless trips. This probability has been rounded off to "about 6%" in this illustration.

Source: *U.S. News & World Report*, July 11, 1988, p. 46. Copyright © 1988, by U.S. News & World Report, Inc. Excerpts reprinted with permission.

Joint Probability of Mutually Exclusive Events

We know from an earlier discussion that two mutually exclusive events cannot happen together. Consequently, their joint probability is zero.

> **JOINT PROBABILITY OF MUTUALLY EXCLUSIVE EVENTS** The joint probability of two mutually exclusive events is always zero. If A and B are two mutually exclusive events, then
>
> $$P(A \text{ and } B) = 0$$

EXAMPLE 4–28   Consider the following two events for an application filed by a person to obtain a car loan:

$$A = \text{event that the loan application is approved}$$

$$R = \text{event that the loan application is rejected}$$

What is the joint probability of A and R?

Solution   The two events A and R are mutually exclusive. Either the loan application will be approved or it will be rejected. Hence,

$$P(A \text{ and } R) = 0 \qquad \blacksquare$$

EXERCISES

■ **Concepts and Procedures**

4.66   Explain the meaning of the intersection of two events. Give one example.

4.67   What is meant by the joint probability of two or more events? Give one example.

4.68   How is the multiplication rule of probability for two dependent events different from the rule for two independent events?

4.69   What is the joint probability of two mutually exclusive events? Give one example.

4.70   Find the joint probability of A and B for the following.
a. $P(A) = .40$ and $P(B \mid A) = .25$
b. $P(B) = .65$ and $P(A \mid B) = .36$

4.71   Find the joint probability of A and B for the following.
a. $P(B) = .59$ and $P(A \mid B) = .77$
b. $P(A) = .28$ and $P(B \mid A) = .35$

4.72   Given that A and B are two independent events, find their joint probability for the following.
a. $P(A) = .61$ and $P(B) = .27$
b. $P(A) = .39$ and $P(B) = .63$

4.73   Given that A and B are two independent events, find their joint probability for the following.
a. $P(A) = .20$ and $P(B) = .76$
b. $P(A) = .57$ and $P(B) = .32$

4.74   Given that A, B, and C are three independent events, find their joint probability for the following.
a. $P(A) = .20$, $P(B) = .46$,   and $P(C) = .25$
b. $P(A) = .44$, $P(B) = .27$,   and $P(C) = .43$

4.75   Given that A, B, and C are three independent events, find their joint probability for the following.
a. $P(A) = .49$, $P(B) = .67$,   and $P(C) = .75$
b. $P(A) = .71$, $P(B) = .34$,   and $P(C) = .45$

4.76   Given that $P(A) = .30$ and $P(A \text{ and } B) = .24$, find $P(B \mid A)$.

4.77   Given that $P(B) = .65$ and $P(A \text{ and } B) = .45$, find $P(A \mid B)$.

4.78   Given that $P(A \mid B) = .40$ and $P(A \text{ and } B) = .36$, find $P(B)$.

4.79   Given that $P(B \mid A) = .80$ and $P(A \text{ and } B) = .58$, find $P(A)$.

■ **Applications**

4.80   The following table gives a two-way classification, based on age and health insurance coverage, of the U.S. population aged 18 to 34 years based on U.S. Bureau of the Census estimates for the year 1998. The numbers in the table are in millions.

		Health Insurance	
		Covered (C)	Not Covered (N)
Age	18–24 (A)	17.8	7.2
	25–34 (B)	31.3	9.0

a. Suppose one person is randomly selected from this population. Find the following probabilities.
  i. $P(A$ and $C)$ ii. $P(N$ and $B)$
b. Find $P(C$ and $N)$. Is this probability zero? Explain why or why not.

4.81 The following table gives a two-way classification of all faculty members of a university based on gender and tenure.

	Tenured	Nontenured
Male	74	28
Female	29	12

a. If one of these faculty members is selected at random, find the following probabilities.
  i. P(male and nontenured)   ii. P(tenured and female)
b. Find P(tenured and nontenured). Is this probability zero? If yes, why?

4.82 Five hundred employees were selected from a city's large private companies and asked whether or not they have any retirement benefits provided by their companies. Based on this information, the following two-way classification table was prepared.

	Have Retirement Benefits	
	Yes	No
Men	225	75
Women	150	50

a. Suppose one employee is selected at random from these 500 employees. Find the following probabilities.
  i. Probability of the intersection of events "woman" and "yes"
  ii. Probability of the intersection of events "no" and "man"
b. Mention what other joint probabilities you can calculate for this table and then find them. You may draw a tree diagram to find these probabilities.

4.83 A sample of 2000 adults were asked whether or not they have ever shopped on the Internet. The following table gives a two-way classification of the responses obtained.

	Have Shopped	Have Never Shopped
Male	300	900
Female	200	600

a. Suppose one adult is selected at random from these 2000 adults. Find the following probabilities.
  i. P(has never shopped on the Internet and is a male)
  ii. P(has shopped on the Internet and is a female)
b. Mention what other joint probabilities you can calculate for this table and then find those. You may draw a tree diagram to find these probabilities.

4.84 The table at the top of the next page gives the two-way classification of 2000 randomly selected employees from a city based on gender and commuting time from home to work.

	Commuting Time		
	Less Than 30 Minutes	**30 Minutes to One Hour**	**More Than One Hour**
Men	524	455	221
Women	413	263	124

a.   Suppose one employee is selected at random from these 2000 employees. Find the following probabilities.
  i. P(commutes for more than one hour and is a man)
  ii. P(is a woman and commutes for less than 30 minutes)
b.   Find the joint probability of events "30 minutes to one hour" and "more than one hour." Is this probability zero? Explain why or why not.

4.85   Two thousand randomly selected adults were asked if they think they are financially better off than their parents. The following table gives the two-way classification of the responses based on the education levels of the persons included in the survey and whether they are financially better off, the same, or worse off than their parents.

	Less Than High School	**High School**	**More Than High School**
Better off	140	450	420
Same	60	250	110
Worse off	200	300	70

a.   Suppose one adult is selected at random from these 2000 adults. Find the following probabilities.
  i. P(better off and high school)
  ii. P(more than high school and worse off)
b.   Find the joint probability of the events "worse off" and "better off." Is this probability zero? Explain why or why not.

4.86   In a statistics class of 45 students, 10 have a strong interest in statistics. If two students are selected at random from this class, what is the probability that both of them have a strong interest in statistics? Draw a tree diagram for this problem.

4.87   In a group of 15 students, 6 have liberal views. If 2 students are randomly selected from this group, what is the probability that the first of them has liberal views and the second does not? Draw a tree diagram for this problem.

4.88   A company is to hire two new employees. They have prepared a final list of eight candidates, all of whom are equally qualified. Of these eight candidates, five are women. If the company decides to select two persons randomly from these eight candidates, what is the probability that both of them are women? Draw a tree diagram for this problem.

4.89   In a group of 10 persons, 4 have a type A personality and 6 have a type B personality. If 2 persons are selected at random from this group, what is the probability that the first of them has a type A personality and the second has a type B personality? Draw a tree diagram for this problem.

4.90   The probability is .25 that an adult has never flown in an airplane. If two adults are selected at random, what is the probability that the first of them has never flown in an airplane and the second has? Draw a tree diagram for this problem.

4.91   The probability is .76 that a family owns a house. If two families are randomly selected, what is the probability that neither of them owns a house?

4.92   A contractor has submitted bids for two state construction projects. The probability that he will win any contract is .25, and it is the same for each of the two contracts.

a.   What is the probability that he will win both contracts?
b.   What is the probability that he will win neither contract?
Draw a tree diagram for this problem.

4.93 Five percent of all items sold by a mail-order company are returned by customers for a refund. Find the probability that of two items sold during a given hour by this company

a. both will be returned for a refund

b. neither will be returned for a refund

Draw a tree diagram for this problem.

4.94 The probability that any given person is allergic to a certain drug is .03. What is the probability that none of three randomly selected persons is allergic to this drug? Assume that all three persons are independent.

4.95 The probability that a farmer is in debt is .80. What is the probability that three randomly selected farmers are all in debt? Assume independence of events.

4.96 The probability that a family owns a house is .76. The probability that a family owns a house and is a married couple is .69. Find the conditional probability that a randomly selected family is a married couple given that it owns a house.

4.97 The probability that an employee at a company is a female is .36. The probability that an employee is a female and married is .19. Find the conditional probability that a randomly selected employee from this company is married given that she is a female.

4.98 According to projections by the U.S. Department of Education, the total U.S. college enrollment in the year 2001 will be 14,992,000, of whom 8,505,000 will be women. Moreover, 3,721,000 of the women will be part-time students (*The Chronicle of Higher Education Almanac*, August 28, 1998). If these projections are accurate, find the probability that in the year 2001 a randomly selected U.S. college student will be part-time given that this student will be a female.

4.99 According to calculations made from the Centers for Disease Control and Prevention data (April 16, 1998), the probability that a randomly selected home health and hospice care patient is a female is .67, and the probability that this patient is a female and in hospice care is .02. What is the probability that this patient is in hospice care given that she is a female?

4.9 UNION OF EVENTS AND THE ADDITION RULE

This section discusses the union of events and the addition rule that is applied to compute the probability of the union of events.

4.9.1 UNION OF EVENTS

The **union of two events** A and B includes all outcomes that are either in A or in B or in both A and B.

> **UNION OF EVENTS** Let A and B be two events defined in a sample space. The *union of events* A and B is the collection of all outcomes that belong either to A or to B or to both A and B and is denoted by
>
> $$A \text{ or } B$$

The union of events A and B is also denoted by $A \cup B$. Example 4–29 illustrates the union of events A and B.

Illustrating the union of two events.

EXAMPLE 4–29 According to the U.S. Department of Education, 14,262 thousand students are enrolled at institutions of higher education in the United States. Of them, 7919 thousand are female, 8129 thousand are full-time students, and 4321 thousand are female and full-time students. Describe the union of the events "female" and "full-time."

Solution   The union of the events "female" and "full-time" for students enrolled at institutions of higher education includes all students who are either female or full-time or both. The number of such students is

$$7919 + 8129 - 4321 = 11,727 \text{ thousand}$$

Thus, there are a total of 11,727 thousand students enrolled at institutions of higher education who are either female or full-time students or both.

Why did we subtract 4321 thousand from the sum of 7919 thousand and 8129 thousand? The reason is that 4321 thousand students (which represent the intersection of events "female" and "full-time") are common to both events "female" and "full-time" and, hence, are counted twice. To avoid double counting, we subtracted 4321 thousand from the sum of the other two numbers. We can observe this double counting from Table 4.8, which is constructed using the given information. The sum of the numbers in the three shaded cells gives the students who are either female or full-time or both. However, if we add the totals of the row labeled "Female" and the column labeled "Full-time," we count 4321 twice. Note that the numbers in the table are in thousands.

Table 4.8

	Full-time	Part-time	Total
Male	3808	2535	6343
Female	4321 ←	3598	7919
Total	8129	6133	14,262

Counted twice

Figure 4.19 shows the diagram for the union of the events "female" and "full-time."

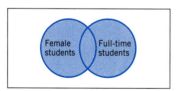

Shaded area gives the union of two events and includes 11,727 thousand students

Figure 4.19   Union of events "female" and "full-time."

4.9.2   ADDITION RULE

The method used to calculate the probability of the union of events is called the **addition rule**. It is defined as follows.

> **ADDITION RULE**   The probability of the union of two events A and B is
>
> $$P(A \text{ or } B) = P(A) + P(B) - P(A \text{ and } B)$$

Thus, to calculate the probability of the union of two events A and B, we add their marginal probabilities and subtract their joint probability from this sum. We must subtract the joint probability of A and B from the sum of their marginal probabilities to avoid double counting because of common outcomes in A and B.

Calculating the probability of the union of two events: two-way table.

EXAMPLE 4-30 A university president has proposed that all students must take a course in ethics as a requirement for graduation. Three hundred faculty members and students from this university were asked about their opinion on this issue. Table 4.9 gives a two-way classification of the responses of these faculty members and students.

Table 4.9 Two-Way Classification of Responses

	Favor	Oppose	Neutral	Total
Faculty	45	15	10	70
Student	90	110	30	230
Total	135	125	40	300

Find the probability that one person selected at random from these 300 persons is a faculty member or is in favor of this proposal.

Solution Let us define the following events:

$$A = \text{the person selected is a faculty member}$$

$$B = \text{the person selected is in favor of the proposal}$$

From the information given in Table 4.9,

$$P(A) = 70/300 = .2333$$

$$P(B) = 135/300 = .4500$$

$$P(A \text{ and } B) = P(A) P(B \mid A) = (70/300)(45/70) = .1500$$

Using the addition rule, we have

$$P(A \text{ or } B) = P(A) + P(B) - P(A \text{ and } B) = .2333 + .4500 - .1500 = \textbf{.5333}$$

Thus, the probability that a randomly selected person from these 300 persons is a faculty member or is in favor of this proposal is .5333.

The probability in this example can also be calculated without using the addition rule. The total number of persons in Table 4.9 who are either faculty members or in favor of this proposal is

$$45 + 15 + 10 + 90 = 160$$

Hence, the required probability is

$$P(A \text{ or } B) = 160/300 = \textbf{.5333}$$ ■

Calculating the probability of the union of two events.

EXAMPLE 4-31 A total of 7955 thousand persons have multiple jobs in the United States. Of them, 4237 thousand are male, 3522 thousand are single, and 1562 thousand are male and single (U.S. Bureau of Labor Statistics). What is the probability that a randomly selected person with multiple jobs is a male or single? Note that here single persons include never-married single, widowed, divorced, and separated.

Solution Let us define the following two events:

$$M = \text{the randomly selected person is a male}$$

$$A = \text{the randomly selected person is single}$$

From the given information,

$$P(M) = 4237/7955 = .5326$$

$$P(A) = 3522/7955 = .4427$$

$$P(M \text{ and } A) = 1562/7955 = .1964$$

Hence,

$$P(M \text{ or } A) = P(M) + P(A) - P(M \text{ and } A) = .5326 + .4427 - .1964 = \mathbf{.7789}$$

Actually, using the given information, we can prepare Table 4.10 for this example. The numbers in the shaded cells are given to us. The remaining numbers are calculated by doing some arithmetic manipulations.

Table 4.10   Two-Way Classification Table

	Single (A)	Married (B)	Total
Male (M)	1562	2675	4237
Female (F)	1960	1758	3718
Total	3522	4433	7955

Now, using Table 4.10, we can find the required probabilities.

$$P(M) = 4237/7955 = .5326$$

$$P(A) = 3522/7955 = .4427$$

$$P(M \text{ and } A) = 1562/7955 = .1964$$

$$P(M \text{ or } A) = P(M) + P(A) - P(M \text{ and } A) = .5326 + .4427 - .1964 = \mathbf{.7789}$$ ■

Addition Rule for Mutually Exclusive Events

We know from an earlier discussion that the joint probability of two mutually exclusive events is zero. When A and B are mutually exclusive events, the term $P(A \text{ and } B)$ in the addition rule becomes zero and is dropped from the formula. Thus, the probability of the union of two mutually exclusive events is given by the sum of their marginal probabilities.

> **ADDITION RULE FOR MUTUALLY EXCLUSIVE EVENTS**   The probability of the union of two mutually exclusive events A and B is
>
> $$P(A \text{ or } B) = P(A) + P(B)$$

Calculating the probability of the union of two mutually exclusive events: two-way table.

EXAMPLE 4-32   A university president has proposed that all students must take a course in ethics as a requirement for graduation. Three hundred faculty members and students from this university were asked about their opinion on this issue. The following table, reproduced from Table 4.9 in Example 4–30, gives a two-way classification of the responses of these faculty members and students.

	Favor	Oppose	Neutral	Total
Faculty	45	15	10	70
Student	90	110	30	230
Total	135	125	40	300

What is the probability that a randomly selected person from these 300 faculty members and students is in favor of the proposal or is neutral?

Solution   Let us define the following events:

$$F = \text{the person selected is in favor of the proposal}$$

$$N = \text{the person selected is neutral}$$

As shown in Figure 4.20, events F and N are mutually exclusive because a person selected can be either in favor or neutral but not both.

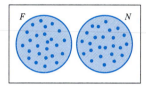

Figure 4.20   Venn diagram of mutually exclusive events.

From the given information,

$$P(F) = 135/300 = .4500$$

$$P(N) = 40/300 = .1333$$

Hence,

$$P(F \text{ or } N) = P(F) + P(N) = .4500 + .1333 = \mathbf{.5833}$$ ∎

The addition rule formula can easily be extended to apply to more than two events. The following example illustrates this.

Calculating the probability of the union of three mutually exclusive events.

EXAMPLE 4-33   Consider the experiment of rolling a die twice. Find the probability that the sum of the numbers obtained on two rolls is 5, 7, or 10.

Solution   The experiment of rolling a die twice has a total of 36 outcomes, which are listed in Table 4.11. Assuming that the die is balanced, these 36 outcomes are equally likely.

Table 4.11   Two Rolls of a Die

		Second Roll of the Die					
		1	**2**	**3**	**4**	**5**	**6**
	1	(1,1)	(1,2)	(1,3)	(1,4)	(1,5)	(1,6)
	2	(2,1)	(2,2)	(2,3)	(2,4)	(2,5)	(2,6)
First Roll of the Die	**3**	(3,1)	(3,2)	(3,3)	(3,4)	(3,5)	(3,6)
	4	(4,1)	(4,2)	(4,3)	(4,4)	(4,5)	(4,6)
	5	(5,1)	(5,2)	(5,3)	(5,4)	(5,5)	(5,6)
	6	(6,1)	(6,2)	(6,3)	(6,4)	(6,5)	(6,6)

The events that give the sum of two numbers equal to 5 or 7 or 10 are circled in the table. As we can observe, the three events "the sum is 5," "the sum is 7," and "the sum is

10" are mutually exclusive. Four outcomes give a sum of 5, six give a sum of 7, and three outcomes give a sum of 10. Thus,

$$P(\text{sum is 5 or 7 or 10}) = P(\text{sum is 5}) + P(\text{sum is 7}) + P(\text{sum is 10})$$

$$= 4/36 + 6/36 + 3/36 = 13/36 = \textbf{.3611} \qquad \blacksquare$$

Calculating the probability of the union of three mutually exclusive events.

EXAMPLE 4-34   The probability that a person is in favor of genetic engineering is .55 and that a person is against it is .45. Two persons are randomly selected, and it is observed whether they favor or oppose genetic engineering.

(a)   Draw a tree diagram for this experiment.

(b)   Find the probability that at least one of the two persons favors genetic engineering.

Solution

(a)   Let

$$F = \text{a person is in favor of genetic engineering}$$

$$A = \text{a person is against genetic engineering}$$

This experiment has four outcomes: both persons are in favor (*FF*), the first person is in favor and the second is against (*FA*), the first person is against and the second is in favor (*AF*), and both persons are against genetic engineering (*AA*). The tree diagram in Figure 4.21 shows these four outcomes and their probabilities.

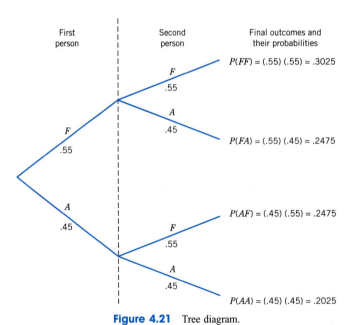

Figure 4.21   Tree diagram.

(b)   The probability that at least one person favors genetic engineering is given by the union of events *FF*, *FA*, and *AF*. These three outcomes are mutually exclusive. Hence,

$$P(\text{at least one person favors}) = P(FF \text{ or } FA \text{ or } AF)$$

$$= P(FF) + P(FA) + P(AF)$$

$$= .3025 + .2475 + .2475 = \textbf{.7975} \qquad \blacksquare$$

EXERCISES

■ **Concepts and Procedures**

4.100   Explain the meaning of the union of two events. Give one example.

4.101   How is the addition rule of probability for two mutually exclusive events different from the rule for two mutually nonexclusive events?

4.102   Consider the following addition rule to find the probability of the union of two events A and B:

$$P(A \text{ or } B) = P(A) + P(B) - P(A \text{ and } B)$$

When and why is the term $P(A \text{ and } B)$ subtracted from the sum of $P(A)$ and $P(B)$? Give one example where you might use this formula.

4.103   When is the following addition rule used to find the probability of the union of two events A and B?

$$P(A \text{ or } B) = P(A) + P(B)$$

Give one example where you might use this formula.

4.104   Find $P(A \text{ or } B)$ for the following.

a. $P(A) = .58,$ $P(B) = .66,$ and $P(A \text{ and } B) = .57$
b. $P(A) = .72,$ $P(B) = .42,$ and $P(A \text{ and } B) = .39$

4.105   Find $P(A \text{ or } B)$ for the following.

a. $P(A) = .18,$ $P(B) = .49,$ and $P(A \text{ and } B) = .11$
b. $P(A) = .73,$ $P(B) = .71,$ and $P(A \text{ and } B) = .68$

4.106   Given that A and B are two mutually exclusive events, find $P(A \text{ or } B)$ for the following.

a. $P(A) = .47$ and $P(B) = .32$
b. $P(A) = .16$ and $P(B) = .59$

4.107   Given that A and B are two mutually exclusive events, find $P(A \text{ or } B)$ for the following.

a. $P(A) = .25$ and $P(B) = .27$
b. $P(A) = .58$ and $P(B) = .09$

■ **Applications**

4.108   The following table gives a two-way classification, based on age and health insurance coverage, of the U.S. population aged 18 to 34 based on U.S. Bureau of the Census estimates for the year 1998. The numbers in the table are in millions.

		Health Insurance	
		Covered (C)	Not Covered (N)
Age	18–24 (A)	17.8	7.2
	25–34 (B)	31.3	9.0

Suppose one person is randomly selected from this population. Find the following probabilities.

a. $P(A \text{ or } N)$ **b.** $P(C \text{ or } B)$

4.109   The following table gives a two-way classification of all faculty members of a university based on gender and tenure.

	Tenured	**Nontenured**
Male	74	28
Female	29	12

If one of these faculty members is selected at random, find the following probabilities.

a. P(female or nontenured)   **b.** P(tenured or male)

4.110   Five hundred employees were selected from a city's large private companies and they were asked whether or not they have any retirement benefits provided by their companies. Based on this information, the following two-way classification table was prepared.

	Have Retirement Benefits	
	Yes	**No**
Men	225	75
Women	150	50

Suppose one employee is selected at random from these 500 employees. Find the following probabilities.

a.   The probability of the union of events "woman" and "yes"
b.   The probability of the union of events "no" and "man"

4.111   A sample of 2000 adults were asked whether or not they have ever shopped on the Internet. The following table gives a two-way classification of the responses.

	Have Shopped	Have Never Shopped
Male	300	900
Female	200	600

Suppose that one adult is selected at random from these 2000 adults. Find the following probabilities.

a. P(has never shopped on the Internet or is a female)
b. P(is a male or has shopped on the Internet)
c. P(has shopped on the Internet or has never shopped on the Internet)

4.112   The following table gives the two-way classification of 2000 randomly selected employees from a city based on gender and commuting time from home to work.

	Commuting Time		
	Less Than 30 Minutes	**30 Minutes to One Hour**	**More Than One Hour**
Men	524	455	221
Women	413	263	124

If one employee is selected at random from these 2000 employees, find the following probabilities.

a. P(commutes for more than one hour or is a man)
b. P(is a woman or commutes for less than 30 minutes)
c. P(man or woman)

4.113   Two thousand randomly selected adults were asked if they think they are financially better off than their parents. The following table gives the two-way classification of the responses based on the education levels of the persons included in the survey and whether they are financially better off, the same, or worse off than their parents.

	Less Than High School	High School	More Than High School
Better off	140	450	420
Same	60	250	110
Worse off	200	300	70

Suppose one adult is selected at random from these 2000 adults. Find the following probabilities.

a. P(better off or high school)
b. P(more than high school or worse off)
c. P(better off or worse off)

4.114 The probability that a football player weighs more than 230 pounds is .69, that he is at least 75 inches tall is .55, and that he weighs more than 230 pounds and is at least 75 inches tall is .43. Find the probability that a randomly selected football player weighs more than 230 pounds or is at least 75 inches tall.

4.115 The probability that a family owns a washing machine is .68, that it owns a VCR is .81, and that it owns both a washing machine and a VCR is .58. What is the probability that a randomly selected family owns a washing machine or a VCR?

4.116 The probability that a person has a checking account is .74, a savings account is .31, and both accounts is .22. Find the probability that a randomly selected person has a checking or a savings account.

4.117 The probability that a randomly selected elementary or secondary school teacher from a city is a female is .68, holds a second job is .38, and is a female and holds a second job is .29. Find the probability that an elementary or secondary school teacher selected at random from this city is a female or holds a second job.

4.118 According to the U.S. Bureau of Labor Statistics, there were 709 workplace homicides in 1998 (*Bureau of Labor Statistics News*, August 4, 1999). Of these 709 homicides, 286 occurred in retail trade and 50 occurred in police protection. If one of these 709 homicides is selected at random, find the probability that it occurred in retail trade or in police protection. Explain why this probability is not equal to 1.0.

4.119 According to preliminary data from the U.S. Centers for Disease Control, there were 484,975 live births to U.S. mothers aged 15 to 19 in 1998. Of these mothers, for 375,285 the child was her first and for 87,786 the child was her second. If one mother is randomly selected from these 484,975 mothers, what is the probability that her child born in 1998 was her first or second? Explain why this probability is not equal to 1.0.

4.120 The probability of a student getting an A grade in an economics class is .24 and that of getting a B grade is .28. What is the probability that a randomly selected student from this class will get an A or a B in this class? Explain why this probability is not equal to 1.0.

4.121 Seventy-two percent of a town's voters favor the recycling issue, 15% oppose it, and 13% are indifferent. What is the probability that a randomly selected voter from this town will either favor recycling or be indifferent? Explain why this probability is not equal to 1.0.

4.122 The probability that a corporation makes charitable contributions is .72. Two corporations are selected at random, and it is noted whether or not they make charitable contributions.

a. Draw a tree diagram for this experiment.
b. Find the probability that at most one corporation makes charitable contributions.

4.123 The probability that an open-heart operation is successful is .84. What is the probability that in two randomly selected open-heart operations at least one will be successful? Draw a tree diagram for this experiment.

GLOSSARY

Classical probability rule The method of assigning probabilities to outcomes or events of an experiment with equally likely outcomes.

Complementary events Two events that taken together include all the outcomes for an experiment but do not contain any common outcome.

Compound event An event that contains more than one outcome of an experiment. It is also called a *composite event*.

Conditional probability   The probability of an event subject to the condition that another event has already occurred.

Dependent events   Two events for which the occurrence of one changes the probability of the occurrence of the other.

Equally likely outcomes   Two (or more) outcomes or events that have the same probability of occurrence.

Event   A collection of one or more outcomes of an experiment.

Experiment   A process with well-defined outcomes that, when performed, results in one and only one of the outcomes per repetition.

Impossible event   An event that cannot occur.

Independent events   Two events for which the occurrence of one does not change the probability of the occurrence of the other.

Intersection of events   The intersection of events is given by the outcomes that are common to two (or more) events.

Joint probability   The probability that two (or more) events occur together.

Law of Large Numbers   If an experiment is repeated again and again, the probability of an event obtained from the relative frequency approaches the actual or theoretical probability.

Marginal probability   The probability of one event or characteristic without consideration of any other event.

Mutually exclusive events   Two or more events that do not contain any common outcome and, hence, cannot occur together.

Outcome   The result of the performance of an experiment.

Probability   A numerical measure of the likelihood that a specific event will occur.

Relative frequency as an approximation of probability   Probability assigned to an event based on the results of an experiment or based on historical data.

Sample point   An outcome of an experiment.

Sample space   The collection of all sample points or outcomes of an experiment.

Simple event   An event that contains one and only one outcome of an experiment. It is also called an *elementary event*.

Subjective probability   The probability assigned to an event based on the information and judgment of a person.

Sure event   An event that is certain to occur.

Tree diagram   A diagram in which each outcome of an experiment is represented by a branch of a tree.

Union of two events   All outcomes that belong either to one or to both events.

Venn diagram   A picture that represents a sample space or specific events.

KEY FORMULAS

1. Classical probability rule

For a simple event E_i:

$$P(E_i) = \frac{1}{\text{Total number of outcomes for the experiment}}$$

For a compound event A:

$$P(A) = \frac{\text{Number of outcomes favorable to } A}{\text{Total number of outcomes for the experiment}}$$

2. **Relative frequency as an approximation of probability**

$$P(\text{an event}) = \frac{\text{Frequency of that event}}{\text{Sample size}} = \frac{f}{n}$$

3. **Counting rule**

If an experiment consists of three steps and if the first step can result in m outcomes, the second step in n outcomes, and the third step in k outcomes, then

$$\text{Total outcomes for the experiment} = m \cdot n \cdot k$$

This rule can be extended to apply to an experiment with any number of steps.

4. **Conditional probability of an event**

$$P(B \mid A) = \frac{P(A \text{ and } B)}{P(A)} \quad \text{and} \quad P(A \mid B) = \frac{P(A \text{ and } B)}{P(B)}$$

5. **Independent events**

Two events A and B are independent if

$$P(A) = P(A \mid B) \quad \text{and/or} \quad P(B) = P(B \mid A)$$

6. **Complementary events**

For two complementary events A and $\overline{A}$:

$$P(A) + P(\overline{A}) = 1, \quad P(A) = 1 - P(\overline{A}), \quad \text{and} \quad P(\overline{A}) = 1 - P(A)$$

7. **Multiplication rule for the joint probability of events**

If A and B are dependent events, then

$$P(A \text{ and } B) = P(A) \, P(B \mid A)$$

If A and B are independent events, then

$$P(A \text{ and } B) = P(A) \, P(B)$$

8. **Joint probability of two mutually exclusive events**

For two mutually exclusive events A and B:

$$P(A \text{ and } B) = 0$$

9. **Addition rule for the probability of the union of events**

If A and B are mutually nonexclusive events, then

$$P(A \text{ or } B) = P(A) + P(B) - P(A \text{ and } B)$$

If A and B are mutually exclusive events, then

$$P(A \text{ or } B) = P(A) + P(B)$$

SUPPLEMENTARY EXERCISES

4.124   A lawyers' association has 90 members. Of them, 12 are corporate lawyers. One laywer is selected at random. Find the probability that this lawyer is

a.   a corporate lawyer   **b.**   not a corporate lawyer

4.125   In a class of 35 students, 13 are seniors, 9 are juniors, 8 are sophomores, and 5 are freshmen. If one student is selected at random from this class, what is the probability that this student is

a.   a junior   **b.**   a freshman?

4.126 A group of 150 randomly selected CEOs was tested for personality type. The table gives the results of this survey.

	Type A	Type B
Men	78	40
Women	19	13

a. If one CEO is selected at random from this group, find the probability that this CEO
   i. has a type A personality
   ii. is a woman
   iii. is a man given that he has a type A personality
   iv. has a type B personality given that she is a woman
   v. has a type A personality and is a woman
   vi. is a man or has a type B personality

b. Are the events "woman" and "type A" mutually exclusive? What about the events "type A" and "type B"? Why or why not?

c. Are the events "type A" and "man" independent? Why or why not?

4.127 A random sample of 250 adults was taken, and they were asked whether they prefer watching sports or opera on television. The following table gives the two-way classification of these adults.

	Prefers Watching Sports	Prefers Watching Opera
Male	96	24
Female	45	85

a. If one adult is selected at random from this group, find the probability that this adult
   i. prefers watching opera
   ii. is a male
   iii. prefers watching sports given that the adult is a female
   iv. is a male given that he prefers watching sports
   v. is a female and prefers watching opera
   vi. prefers watching sports or is a male

b. Are the events "female" and "prefers watching sports" independent? Are they mutually exclusive? Explain why or why not.

4.128 A random sample of 80 lawyers was taken, and they were asked if they are in favor of or against capital punishment. The following table gives the two-way classification of these lawyers.

	Favors Capital Punishment	Opposes Capital Punishment
Male	32	24
Female	13	11

a. If one lawyer is randomly selected from this group, find the probability that this lawyer
   i. favors capital punishment
   ii. is a female
   iii. opposes capital punishment given that the lawyer is a female
   iv. is a male given that he favors capital punishment
   v. is a female and favors capital punishment
   vi. opposes capital punishment or is a male

b. Are the events "female" and "opposes capital punishment" independent? Are they mutually exclusive? Explain why or why not.

4.129 A random sample of 400 college students was asked if college athletes should be paid. The following table gives a two-way classification of the responses.

	Should Be Paid	Should Not Be Paid
Student athlete	90	10
Student nonathlete	210	90

a. If one student is randomly selected from these 400 students, find the probability that this student
  i. is in favor of paying college athletes
  ii. favors paying college athletes given that the student selected is a nonathlete
  iii. is an athlete and favors paying student athletes
  iv. is a nonathlete or is against paying student athletes

b. Are the events "student athlete" and "should be paid" independent? Are they mutually exclusive? Explain why or why not.

4.130 A survey conducted about job satisfaction showed that 20% of workers are not happy with their current jobs. Assume that this result is true for the population of all workers. Two workers are selected at random, and it is observed whether or not they are happy with their current jobs. Draw a tree diagram. Find the probability that in this sample of two workers

a. both are not happy with their current jobs

b. at least one of them is happy with the current job

4.131 In July 1998, Princeton Survey Research Associates interviewed 752 adults by telephone for *Newsweek* magazine. Fifty-three percent of these adults felt that news organizations are "often inaccurate" in their reporting (*Newsweek*, July 20, 1998). Assume that this result is true for the current population of adults. Two adults are selected at random and asked whether news organizations report their facts correctly or are often inaccurate. Draw a tree diagram for this problem. Find the probability that in this sample of two adults

a. both feel that news organizations are often inaccurate

b. at most one feels that news organizations are often inaccurate

4.132 Refer to Exercise 4.124. Two lawyers are selected at random from this group of 90 lawyers. Find the probability that both of these lawyers are corporate lawyers.

4.133 Refer to Exercise 4.125. Two students are selected at random from this class of 35 students. Find the probability that the first student selected is a junior and the second is a sophomore.

4.134 A company has installed a generator to back up the power in case there is a power failure. The probability that there will be a power failure during a snowstorm is .30. The probability that the generator will stop working during a snowstorm is .09. What is the probability that during a snowstorm the company will lose both sources of power?

4.135 Terry & Sons Inc. makes bearings for autos. The production system involves two independent processing machines so that each bearing passes through these two processes. The probability that the first processing machine is not working properly at any time is .08, and the probability that the second machine is not working properly at any time is .06. Find the probability that both machines will not be working properly at any given time.

■ **Advanced Exercises**

4.136 A player plays a roulette game in a casino by betting on a single number each time. Since the wheel has 38 numbers, the probability that the player will win in a single play is 1/38. Note that each play of the game is independent of all previous plays.

a. Find the probability that the player will win for the first time on the 10th play.

b. Find the probability that it takes the player more than 50 plays to win for the first time.

c. The gambler claims that since he has one chance in 38 of winning each time he plays, he is certain to win at least once if he plays 38 times. Does this sound reasonable to you? Find the probability that he will win at least once in 38 plays.

4.137   A certain state's auto license plate has three letters of the alphabet followed by a three-digit number.

a.   How many different license plates are possible if all three-letter sequences are permitted and any number from 000 to 999 is allowed?

b.   Arnold witnessed a hit-and-run accident. He knows that the first two letters on the license plate were BX and that the last number was a 7. How many license plates fit this description?

4.138   The median life of Brand LT5 batteries is 100 hours. What is the probability that in a set of three such batteries, exactly two will last longer than 100 hours?

4.139   Powerball is a game of chance that has generated intense interest because of its large jackpots. For example, on July 29, 1998, the jackpot was estimated at $250 million. To play this game, a player selects five different numbers from 1 through 49, and then picks a powerball number from 1 through 42. The lottery organization randomly draws five different white balls from 49 balls numbered 1 through 49, and then randomly picks a powerball number from 1 through 42. Note that it is possible for the powerball number to be the same as one of the first five numbers.

a.   If the player's first five numbers match the numbers on the five white balls drawn by the lottery organization and the player's powerball number matches the powerball number drawn by the lottery organization, the player wins the jackpot. Find the probability that a player who buys one ticket will win the jackpot. (Note that the order in which the five white balls are drawn is unimportant.)

b.   If the player's first five numbers match the numbers on the five white balls drawn by the lottery organization but the powerball number does not match the one drawn by the lottery organization, the player wins about $100,000. Find the probability that a player who buys one ticket will win this prize.

4.140   A trimotor plane has three engines—a central engine and an engine on each wing. The plane will crash only if the central engine fails *and* at least one of the two wing engines fails. The probability of failure during any given flight is .005 for the central engine and .008 for each of the wing engines. Assuming that the three engines operate independently, what is the probability that the plane will crash during a flight?

4.141   A box contains 10 red marbles and 10 green marbles.

a.   Sampling at random from the box five times with replacement, you have drawn a red marble all five times. What is the probability of drawing a red marble the sixth time?

b.   Sampling at random from the box five times without replacement, you have drawn a red marble all five times. Without replacing any of the marbles, what is the probability of drawing a red marble the sixth time?

c.   You have tossed a fair coin five times and have obtained heads all five times. A friend argues that, according to the law of averages, a tail is due to occur and, hence, the probability of obtaining a head on the sixth toss is less than .50. Is he right? Is coin tossing mathematically equivalent to the procedure mentioned in part a or in part b? Explain.

4.142   A gambler has four cards—two diamonds and two clubs. The gambler proposes the following game to you: You will leave the room and the gambler will put the cards face down on a table. When you return to the room, you will pick two cards at random. You will win $10 if both cards are diamonds, you will win $10 if both are clubs, and for any other outcome you will lose $10. Assuming that there is no cheating, should you accept this proposition? Support your answer by calculating your probability of winning $10.

4.143   A thief has stolen Roger's automatic teller card. The card has a four-digit personal identification number (PIN). The thief knows that the first two digits are 3 and 5, but he does not know the last two digits. Thus, the PIN could be any number from 3500 to 3599. To protect the customer, the automatic teller machine will not allow more than three unsuccessful attempts to enter the PIN. After the third wrong PIN, the machine keeps the card and allows no further attempts.

a.   What is the probability that the thief will find the correct PIN within three tries? (Assume that the thief will not try the same wrong PIN twice.)

b.   If the thief knew that the first two digits were 3 and 5 and that the third digit was either 1 or 7, what is the probability of guessing the correct PIN in three attempts?

4.144 Consider the following games with two dice.

a. A gambler is going to roll a die four times. If he rolls at least one 6, you must pay him $5. If he fails to roll a 6 in four tries, he will pay you $5. Find the probability that you must pay the gambler. Assume that there is no cheating.

b. The same gambler offers to let you roll a pair of dice 24 times. If you roll at least one double 6, he will pay you $10. If you fail to roll a double 6 in 24 tries, you must pay him $10. The gambler says that you have a better chance of winning because your probability of success on each of the 24 rolls is 1/36 and you have 24 chances. Thus, he says, your probability of winning $10 is 24(1/36) = 2/3. Do you agree with this analysis? If so, indicate why. If not, point out the fallacy in his argument, and then find the correct probability that you will win.

4.145 Suppose that a history class has 25 students. Let A be the event that two or more of these students share the same birthday (the same day and month; the years need not match). Neglect the effect of February 29 and assume that all 365 possible birthdays are equally likely.

a. Would you expect $P(A)$ to be less than .50?
b. Find $P(A)$.
(*Hint:* It may be helpful to consider complementary events.)

4.146 A gambler has two coins—a standard penny and a penny with "heads" on both sides. She picks one of the two pennies at random and tosses it; it shows "heads." What is the probability that the other side of the coin is also "heads"? (*Warning:* The solution is not as obvious as it may seem. Analyze the problem carefully, using conditional probability. A tree diagram may be helpful.)

4.147 A club with 10 members consists of nine college professors and one business executive. If a committee of three is selected at random from the 10 members, find the probability that the business executive will be on the committee.

4.148 The company that produces Top Taste breakfast cereal is putting prize coupons in some of the Top Taste cereal boxes to increase sales. When it opens on a certain day, Ricardo's Supermarket has nine boxes of Top Taste cereal on its shelves, and two of these nine boxes contain prize coupons. Suppose that each customer who buys a cereal box picks it from the shelf at random. The store sells six of these nine boxes on that day.

a. Find the probability that none of the six boxes sold contains a prize coupon.
b. Find the probability that at least one of the six boxes sold contains a prize coupon.
c. Find the probability that both of the boxes with prizes are sold on that day. In other words, find the probability that two of the six boxes sold contain prize coupons.

4.149 A hotel owner has determined that 83% of the hotel's guests eat either dinner or breakfast in the hotel restaurant. Further investigation reveals that 30% of the guests eat dinner and 60% of the guests eat breakfast in the hotel restaurant.

a. What proportion of the hotel guests eat both dinner and breakfast in the hotel restaurant?
b. What proportion of the hotel guests eat neither dinner nor breakfast in the hotel restaurant?
c. What proportion of the hotel guests eat dinner but not breakfast in the hotel restaurant?

4.150 Eighty percent of all the apples picked on Andy's Apple Acres are satisfactory, but the other 20% are not suitable for market. Andy inspects each apple his pickers pick. Even though he has been in the apple business for 40 years, Andy is prone to make mistakes. He judges as suitable for market 5% of all unsatisfactory apples and judges as unsuitable for market 1% of all satisfactory apples.

a. What proportion of all apples picked on Andy's Apple Acres are both satisfactory and judged by Andy as being satisfactory for market?
b. What proportion of all apples picked are marketed?
c. What proportion of apples that are marketed are actually satisfactory?

4.151 The principal of a large high school would like to determine the proportion of students in her school who used drugs during the past week. Since the results of directly asking each student "Have you used drugs during the past week?" would be unreliable, the principal might use the following randomized response scheme. Each student rolls a fair die once, and only he or she knows the outcome. If a 1 or 2 is rolled, the student must answer the sensitive question truthfully. However, if a 3, 4, 5, or 6 is rolled, the student must answer the question with the opposite of the true answer. In this way, the principal would not know whether a *yes* response means the student used drugs and answered truthfully or the student did not use drugs and answered untruthfully. If 60% of the students in the school respond *yes*, what proportion of the students actually did use drugs during the past week? (*Note:* Splitting the outcomes of the roll of the die as described represents only one possibility. The outcomes could have been split differently but not in such a way that the probability of answering truthfully is .5. In this case, it would not be possible to estimate the proportion of all students who actually used drugs over the past week.)

SELF-REVIEW TEST

1. The collection of all outcomes for an experiment is called

   a. a sample space   **b.** the intersection of events   **c.** joint probability

2. A final outcome of an experiment is called

   a. a compound event   **b.** a simple event   **c.** a complementary event

3. A compound event includes

   a. all final outcomes   **b.** exactly two outcomes
   c. more than one outcome for an experiment

4. Two equally likely events

   a. have the same probability of occurrence
   b. cannot occur together
   c. have no effect on the occurrence of each other

5. Which of the following probability approaches can be applied only to experiments with equally likely outcomes?

   a. Classical probability   **b.** Empirical probability   **c.** Subjective probability

6. Two mutually exclusive events

   a. have the same probability   **b.** cannot occur together
   c. have no effect on the occurrence of each other

7. Two independent events

   a. have the same probability   **b.** cannot occur together
   c. have no effect on the occurrence of each other

8. The probability of an event is always

   a. less than 0   **b.** in the range 0 to 1.0   **c.** greater than 1.0

9. The sum of the probabilities of all final outcomes of an experiment is always

   a. 100   **b.** 1.0   **c.** 0

10. The joint probability of two mutually exclusive events is always

    a. 1.0   **b.** between 0 and 1   **c.** 0

11. Two independent events are

    a. always mutually exclusive   **b.** never mutually exclusive
    c. always complementary

12. A magazine editor is refurbishing her office. She can choose one of three kinds of paint, one of four kinds of furniture, and one of three kinds of telephone. If she randomly selects one paint, one kind of furniture, and one telephone, how many different outcomes are possible?

13. Lucia graduated this year with an accounting degree from Eastern Connecticut State University. She has received job offers from an accounting firm, an insurance company, and an airline. She cannot decide which of the three job offers she should accept. Suppose she decides to randomly select one of these three job offers. Find the probability that the job offer selected is

    a. from the insurance company

    b. not from the accounting firm

14. A company has 500 employees. Of them, 300 are male and 280 are married. Of the 300 males, 190 are married.

    a. Are the events "male" and "married" independent? Are they mutually exclusive? Explain why or why not.

    b. If one employee of this company is selected at random, what is the probability that this employee is

        i. a female   ii. a male given that he is married?

15. Reconsider Problem 14. If one employee is selected at random from this company, what is the probability that this employee is a male or married?

16. Reconsider Problem 14. If two employees are selected at random from this company, what is the probability that both of them are female?

17. The probability that an American adult has ever experienced a migraine headache is .35. If two American adults are randomly selected, what is the probability that neither of them has ever experienced a migraine headache?

18. A hat contains five green, eight red, and seven blue marbles. Let A be the event that a red marble is drawn if we randomly select one marble out of this hat. What is the probability of A? What is the complementary event of A, and what is its probability?

19. The probability that a student is a male is .48 and that a student likes statistics is .35. If these two events are independent, what is the probability that a student selected at random is

    a. a male and likes statistics

    b. a female or likes statistics?

20. Five hundred married men and women were asked whether or not they would marry their current spouses if they were given a chance to do it over again. Their responses are recorded in the table.

	Would Marry the Current Spouse?	
	Yes (Y)	No (N)
Male (M)	125	175
Female (F)	55	145

    a. If one person is selected at random from these 500 persons, find the following probabilities.

        i. Yes

        ii. Yes given female

        iii. Male and no

        iv. Yes or female

    b. Are the events "male" and "yes" independent? Are they mutually exclusive? Explain why or why not.

MINI-PROJECTS

Mini-Project 4–1

The following question and answer appeared in the "Ask Marilyn" column in *Parade Magazine* on December 27, 1998. (Reproduced with permission.)

Question: At a monthly "casino night," there is a game called Chuck-a-Luck: Three dice are rolled in a wire cage. You place a bet on any number from 1 to 6. If any one of the three dice comes up with your number, you win the amount of your bet. (You also get your original stake back.) If more than one die comes up with your number, you win the amount of your bet for each match. For example, if you had a $1 bet on number 5 and each of the three dice came up with 5, you would win $3. It appears that the odds of winning are 1 in 6 for each of the three dice, for a total of 3 out of 6—or 50%. Adding the possibility of having more than one die come up with your number, the odds would seem to be slightly in the gambler's favor. What are the odds of winning at this game? I can't believe that a casino game would favor the gambler.

—Daniel Reisman, Niverville, N.Y.

Answer: Chuck-a-Luck has a subtle trick that many people don't recognize unless they analyze it. If the three dice always showed different numbers, the game would favor no one. To illustrate, say that each of six people bets $1 on a different number. If the three dice showed different numbers, the operator would take in $3 from the losers and pay out that same $3 to the winners. But often, the three dice will show a doublet or triplet: two or three of the dice will show the same number. That's when the operator makes his money.

For example, say the dice show 3, 3, and 5. The operator collects $4 ($1 each from the people who bet on 1, 2, 4, and 6) but pays out only $3 ($2 to the person who bet on 3, and $1 to the person who bet on 5). Or say the dice show 4, 4, and 4. The operator collects $5 ($1 each from the people who bet on 1, 2, 3, 5, and 6) but still pays out only $3 (to the person who bet on 4). The collections and pay-offs change according to the bets (sometimes the house wins, sometimes the gamblers win), but with this game, you can still expect to lose about 8 cents with every $1 you bet over the long run.

Project for the class: Carry out a simulation of Chuck-a-Luck to see if it favors the gambler or the casino. The instructor may wish to divide the class into groups of four, with one student rolling the dice while the other three act as gamblers placing imaginary $1 bets on any of the six numbers. After each roll of the dice, a record is made of how much money would have been paid to the gamblers. If time permits, roll the dice 50 times; then determine how much money was bet and how much was paid to the gamblers. Combine these totals for all groups in the class and use this information to estimate the mean amount won (or lost) by the gamblers per $1 bet. Alternatively, students may simulate the game at home.

a. Do the results agree (approximately) with the columnist's assertion that the gambler should expect to lose about 8 cents per $1 bet?

b. The reader who asked the question conjectured that the chance of having one of the dice show the number the gambler bets is $1/6 + 1/6 + 1/6 = 3/6$. What is the fallacy in this calculation?

Mini-Project 4–2

Suppose that a small chest contains three drawers. The first drawer contains two $1 bills, the second drawer contains two $100 bills, and the third drawer contains one $1 bill and one $100 bill. Suppose that first a drawer is selected at random and then one of the two bills inside that drawer is selected at random. We can define these events:

$$A = \text{the first drawer is selected}$$

$$B = \text{the second drawer is selected}$$

$$C = \text{the third drawer is selected}$$

$$D = \text{a \$1 bill is selected}$$

a. Suppose when you randomly select one drawer and then one bill from that drawer, the bill you obtain is a $1 bill. What is the probability that the second bill in this drawer is a $100 bill? In other words, find the probability $P(C \mid D)$. Answer this question intuitively without making any calculations.

b. Use the relative frequency concept of probability to estimate $P(C \mid D)$ as follows. First select a drawer by rolling a die once. If either 1 or 2 occurs, the first drawer is selected; if either 3 or 4 occurs, the second drawer is selected; and if either 5 or 6 occurs, the third drawer is selected. Next select a bill by tossing a coin once. If you obtain a head, assume that you select a \$1 bill; if you obtain a tail, assume that you select a \$100 bill. Repeat this process 100 times. How many times in these 100 repetitions did the event D occur? What proportion of the time did C occur when D occurred? Use this proportion to estimate $P(C \mid D)$. Does this estimate support your guess of $P(C \mid D)$ in part a?

c. Calculate $P(C \mid D)$ using the procedures developed in this chapter (a tree diagram may be helpful). Was your estimate in part b close to this value? Explain.

 See Appendix D for *Excel Adventure 4* on probability.

5

DISCRETE RANDOM VARIABLES AND THEIR PROBABILITY DISTRIBUTIONS

5.1 RANDOM VARIABLES

5.2 PROBABILITY DISTRIBUTION OF A DISCRETE RANDOM VARIABLE

5.3 MEAN OF A DISCRETE RANDOM VARIABLE

CASE STUDY 5-1 ACES HIGH INSTANT LOTTERY

5.4 STANDARD DEVIATION OF A DISCRETE RANDOM VARIABLE

5.5 FACTORIALS AND COMBINATIONS

CASE STUDY 5-2 PLAYING LOTTO

5.6 THE BINOMIAL PROBABILITY DISTRIBUTION

5.7 THE POISSON PROBABILITY DISTRIBUTION

CASE STUDY 5-3 ASK MR. STATISTICS

GLOSSARY

KEY FORMULAS

SUPPLEMENTARY EXERCISES

SELF-REVIEW TEST

MINI-PROJECTS

COMPUTER ASSIGNMENTS

Chapter 4 discussed the concepts and rules of probability. This chapter extends the concept of probability to explain probability distributions. As we saw in Chapter 4, any given statistical experiment has more than one outcome. It is impossible to predict which of the many possible outcomes will occur if an experiment is performed. Consequently, decisions are made under uncertain conditions. For example, a lottery player does not know in advance whether or not he is going to win that lottery. If he knows that he is not going to win, he will definitely not play. It is the uncertainty about winning (some positive probability of winning) that makes him play. This chapter shows that if the outcomes and their probabilities for a statistical experiment are known, we can find out what will happen, on average, if that experiment is performed many times. For the lottery example, we can find out what a lottery player can expect to win (or lose), on average, if he continues playing this lottery again and again.

In this chapter, first, random variables and types of random variables are explained. Then, the concept of a probability distribution and its mean and standard deviation are discussed. Finally, two special probability distributions for a discrete random variable—the binomial probability distribution and the Poisson probability distribution—are developed.

5.1 RANDOM VARIABLES

Suppose Table 5.1 gives the frequency and relative frequency distributions of the number of vehicles owned by all 2000 families in a small town.

Table 5.1 Frequency and Relative Frequency Distributions of the Number of Vehicles Owned by Families

Number of Vehicles Owned	Frequency	Relative Frequency
0	30	30/2000 = .015
1	470	470/2000 = .235
2	850	850/2000 = .425
3	490	490/2000 = .245
4	160	160/2000 = .080
	$N = 2000$	Sum = 1.000

Suppose one family is randomly selected from this population. The act of randomly selecting a family is called a *random* or *chance experiment*. Let x denote the number of vehicles owned by the selected family. Then x can assume any of the five possible values (0, 1, 2, 3, and 4) listed in the first column of Table 5.1. The value assumed by x depends on which family is selected. Thus, this value depends on the outcome of a random experiment. Consequently, x is called a **random variable** or a **chance variable**. In general, a random variable is denoted by x or y.

> **RANDOM VARIABLE** A *random variable* is a variable whose value is determined by the outcome of a random experiment.

As we explain next, a random variable can be discrete or continuous.

5.1.1 DISCRETE RANDOM VARIABLE

A **discrete random variable** assumes values that can be counted. In other words, the consecutive values of a discrete random variable are separated by a certain gap.

> **DISCRETE RANDOM VARIABLE** A *random variable* that assumes countable values is called a *discrete random variable*.

In Table 5.1, *the number of vehicles owned by a family* is an example of a discrete random variable because the values of the random variable *x* are countable: 0, 1, 2, 3, and 4. Here are some other examples of discrete random variables:

1. The number of cars sold at a dealership during a given month
2. The number of houses in a certain block
3. The number of pairs of shoes a person owns
4. The number of complaints received at the office of an airline on a given day
5. The number of customers who visit a bank during any given hour
6. The number of heads obtained in three tosses of a coin

5.1.2 CONTINUOUS RANDOM VARIABLE

A random variable whose values are not countable is called a **continuous random variable**. A continuous random variable can assume any value over an interval or intervals.

> **CONTINUOUS RANDOM VARIABLE**   A random variable that can assume any value contained in one or more intervals is called a *continuous random variable.*

Because the number of values contained in any interval is infinite, the possible number of values that a continuous random variable can assume is also infinite. Moreover, we cannot count these values. Consider the life of a battery. We can measure it as precisely as we want. For instance, the life of this battery may be 40 hours, or 40.25 hours, or 40.247 hours. Assume that the maximum life of a battery is 200 hours. Let *x* denote the life of a randomly selected battery of this kind. Then, *x* can assume any value in the interval 0 to 200. Consequently, *x* is a continuous random variable. As shown in the diagram, every point on the line representing the interval 0 to 200 gives a possible value of *x*.

Every point on this line represents a possible value of *x* that denotes the life of a battery. There are an infinite number of points on this line. The values represented by points on this line are uncountable.

The following are some examples of continuous random variables:

1. The height of a person
2. The time taken to complete an examination
3. The amount of milk in a gallon (Note that we do not expect a gallon to contain exactly one gallon of milk but either slightly more or slightly less than a gallon.)
4. The weight of a baby
5. The price of a house

This chapter is limited to a discussion of discrete random variables and their probability distributions. Continuous random variables will be discussed in Chapter 6.

EXERCISES

■ **Concepts and Procedures**

5.1 Explain the meaning of a random variable, a discrete random variable, and a continuous random variable. Give one example each of a discrete random variable and a continuous random variable.

5.2 Classify each of the following random variables as discrete or continuous.

a. The number of students in a class   *discrete*

b. The amount of soda in a 12-ounce can   *continuous*

c. The number of cattle owned by a farmer   ? *discrete/contin.*

d. The age of a house   *discrete/conti*

e. The number of pages in a book that contain at least one error   *discrete*

f. The time spent by a physician examining a patient

5.3 Indicate which of the following random variables are discrete and which are continuous.

a. The number of new accounts opened at a bank during a certain month

b. The time taken to run a marathon

c. The price of a concert ticket   *discrete cont.*

d. The number of rotten eggs in a randomly selected box   *discret.*

e. The points scored in a football game   *discrete*

f. The weight of a randomly selected package   *conti.*

■ **Applications**

5.4 A household can watch news on any of the three networks—ABC, CBS, or NBC. On a certain day, five households randomly and independently decide which channel to watch. Let *x* be the number of households among these five that decide to watch news on ABC. Is *x* a discrete or a continuous random variable? Explain.

5.5 One of the four gas stations located at an intersection of two major roads is a Texaco station. Suppose the next six cars that stop at any of these four gas stations make the selections randomly and independently. Let *x* be the number of cars in these six that stop at the Texaco station. Is *x* a discrete or a continuous random variable? Explain.

5.2 PROBABILITY DISTRIBUTION OF A DISCRETE RANDOM VARIABLE

Let *x* be a discrete random variable. The **probability distribution** of *x* describes how the probabilities are distributed over all the possible values of *x*.

> **PROBABILITY DISTRIBUTION OF A DISCRETE RANDOM VARIABLE** The *probability distribution of a discrete random variable* lists all the possible values that the random variable can assume and their corresponding probabilities.

Example 5–1 illustrates the concept of the probability distribution of a discrete random variable.

Writing the probability distribution of a discrete random variable.

EXAMPLE 5–1 Recall the frequency and relative frequency distributions of the number of vehicles owned by families given in Table 5.1. That table is reproduced as Table 5.2 (page 188). Let *x* be the number of vehicles owned by a randomly selected family. Write the probability distribution of *x*.

Table 5.2 Frequency and Relative Frequency Distributions of the Number of Vehicles Owned by Families

Number of Vehicles Owned	Frequency	Relative Frequency
0	30	.015
1	470	.235
2	850	.425
3	490	.245
4	160	.080
	$N = 2000$	Sum = 1.000

Solution   In Chapter 4, we learned that the relative frequencies obtained from an experiment or a sample can be used as approximate probabilities. However, when the relative frequencies are known for the population, as in Table 5.2, they give the actual (theoretical) probabilities of outcomes. Using the relative frequencies of Table 5.2, we can write the *probability distribution* of the discrete random variable *x* in Table 5.3.

Table 5.3 Probability Distribution of the Number of Vehicles Owned by Families

Number of Vehicles Owned x	Probability $P(x)$
0	.015
1	.235
2	.425
3	.245
4	.080
	$\Sigma P(x) = 1.000$

The probability distribution of a discrete random variable possesses the following *two characteristics*.

1. The probability assigned to each value of a random variable *x* lies in the range 0 to 1; that is, $0 \leq P(x) \leq 1$ for each *x*.

2. The sum of the probabilities assigned to all possible values of *x* is equal to 1.0; that is, $\Sigma P(x) = 1$. (Remember, if the probabilities are rounded, the sum may not be exactly 1.0.)

TWO CHARACTERISTICS OF A PROBABILITY DISTRIBUTION   The probability distribution of a discrete random variable possesses the following two characteristics.

1. $0 \leq P(x) \leq 1$ for each value of *x*

2. $\Sigma P(x) = 1$

These two characteristics are also called the *two conditions* that a probability distribution must satisfy. Notice that in Table 5.3, each probability listed in the column labeled $P(x)$ is between 0 and 1. Also, $\Sigma P(x) = 1.0$. Because both conditions are satisfied, Table 5.3 represents the probability distribution of *x*.

From Table 5.3, we can read the probability for any value of *x*. For example, the probability that a randomly selected family from this town owns two vehicles is .425. This probability is written as

$$P(x = 2) = .425$$

The probability that the selected family owns more than two vehicles is given by the sum of the probabilities of owning three and four vehicles. This probability is $.245 + .080 = .325$, which can be written as

$$P(x > 2) = P(x = 3) + P(x = 4) = .245 + .080 = .325$$

The probability distribution of a discrete random variable can be presented in the form of a *mathematical formula*, a *table*, or a *graph*. Table 5.3 presented the probability distribution in tabular form. Figure 5.1 shows the graphical presentation of the probability distribution of Table 5.3. In this figure, each value of x is marked on the horizontal axis. The probability for each value of x is exhibited by the height of the corresponding line. Such a graph is called a **line graph**. This section does not discuss the presentation of a probability distribution in a mathematical formula.

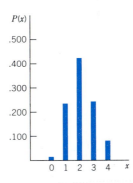

Figure 5.1 Graphical presentation of the probability distribution of Table 5.3.

Verifying the conditions of a probability distribution.

EXAMPLE 5-2 Each of the following tables lists certain values of x and their probabilities. Determine whether or not each table represents a valid probability distribution.

(a)

x	$P(x)$
0	.08
1	.11
2	.39
3	.27

(b)

x	$P(x)$
2	.25
3	.34
4	.28
5	.13

(c)

x	$P(x)$
7	.70
8	.50
9	−.20

Solution

(a) Because each probability listed in this table is in the range 0 to 1, it satisfies the first condition of a probability distribution. However, the sum of all probabilities is not equal to 1.0 because $\Sigma P(x) = .08 + .11 + .39 + .27 = .85$. Therefore, the second condition is not satisfied. Consequently, this table does not represent a valid probability distribution.

(b) Each probability listed in this table is in the range 0 to 1. Also, $\Sigma P(x) = .25 + .34 + .28 + .13 = 1.0$. Consequently, this table represents a valid probability distribution.

(c) Although the sum of all probabilities listed in this table is equal to 1.0, one of the probabilities is negative. This violates the first condition of a probability distribution. Therefore, this table does not represent a valid probability distribution. ■

EXAMPLE 5-3 The following table lists the probability distribution of the number of breakdowns per week for a machine based on past data.

Breakdowns per week	0	1	2	3
Probability	.15	.20	.35	.30

(a) Present this probability distribution graphically.

(b) Find the probability that the number of breakdowns for this machine during a given week is

    (i) exactly 2   (ii) 0 to 2

    (iii) more than 1   (iv) at most 1

Solution Let x denote the number of breakdowns for this machine during a given week. Table 5.4 lists the probability distribution of x.

Table 5.4 Probability Distribution of the Number of Breakdowns

x	$P(x)$
0	.15
1	.20
2	.35
3	.30
	$\Sigma P(x) = 1.00$

Graphing a probability distribution.

(a) Figure 5.2 shows the line graph of the probability distribution of Table 5.4.

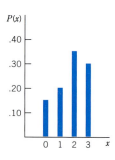

Figure 5.2 Graphical presentation of the probability distribution of Table 5.4.

Finding the probabilities of events for a discrete random variable.

(b) Using Table 5.4, we can calculate the required probabilities as follows.

    (i) The probability of exactly two breakdowns is

$$P(\text{exactly 2 breakdowns}) = P(x = 2) = \textbf{.35}$$

    (ii) The probability of zero to two breakdowns is given by the sum of the probabilities of 0, 1, and 2 breakdowns.

$$P(\text{0 to 2 breakdowns}) = P(0 \leq x \leq 2)$$
$$= P(x = 0) + P(x = 1) + P(x = 2)$$
$$= .15 + .20 + .35 = \textbf{.70}$$

(iii)   The probability of more than one breakdown is obtained by adding the probabilities of 2 and 3 breakdowns.

$$P(\text{more than 1 breakdown}) = P(x > 1)$$
$$= P(x = 2) + P(x = 3)$$
$$= .35 + .30 = \mathbf{.65}$$

(iv)   The probability of at most one breakdown is given by the sum of the probabilities of 0 and 1 breakdown.

$$P(\text{at most 1 breakdown}) = P(x \leq 1)$$
$$= P(x = 0) + P(x = 1)$$
$$= .15 + .20 = \mathbf{.35}$$

■

Constructing a
probability distribution.

EXAMPLE 5-4   According to a survey, 60% of all students at a large university suffer from math anxiety. Two students are randomly selected from this university. Let x denote the number of students in this sample who suffer from math anxiety. Develop the probability distribution of x.

Solution   Let us define the following two events:

N = the student selected does not suffer from math anxiety

M = the student selected suffers from math anxiety

As we can observe from the tree diagram of Figure 5.3, there are four possible outcomes for this experiment: NN (neither of the students suffers from math anxiety), NM (the first student does not suffer from math anxiety and the second does), MN (the first student suffers from math anxiety and the second does not), and MM (both students suffer from math anxiety). The probabilities of these four outcomes are listed in the tree diagram. Because 60% of the students suffer from math anxiety and 40% do not, the probability is .60 that any student selected suffers from math anxiety and .40 that he or she does not.

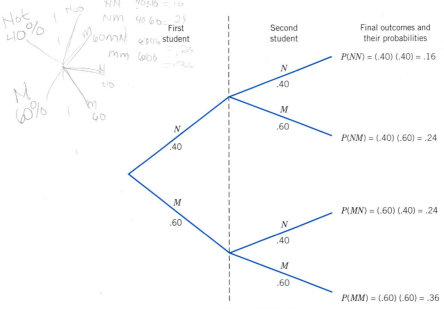

Figure 5.3   Tree diagram.

In a sample of two students, the number who suffer from math anxiety can be 0 (*NN*), 1 (*NM* or *MN*), or 2 (*MM*). Thus, *x* can assume any of three possible values: 0, 1, or 2. The probabilities of these three outcomes are calculated as follows:

$$P(x = 0) = P(NN) = .16$$

$$P(x = 1) = P(NM \text{ or } MN) = P(NM) + P(MN) = .24 + .24 = .48$$

$$P(x = 2) = P(MM) = .36$$

Using these probabilities, we can write the probability distribution of *x* as in Table 5.5.

Table 5.5 Probability Distribution of the Number of Students with Math Anxiety in a Sample of Two Students

x	*P(x)*
0	.16
1	.48
2	.36
	$\Sigma P(x) = 1.00$

EXERCISES

■ **Concepts and Procedures**

5.6 Explain the meaning of the probability distribution of a discrete random variable. Give one example of such a probability distribution. What are the three ways to present the probability distribution of a discrete random variable? Discrete, continuous, random

5.7 Briefly explain the two characteristics (conditions) of the probability distribution of a discrete random variable. $0 \le P(x) = 1.0$

5.8 Each of the following tables lists certain values of *x* and their probabilities. Verify whether or not each represents a valid probability distribution.

a.
x	*P(x)*
0	.10
1	.05
2	.45
3	.40

1.00

b.
x	*P(x)*
2	.35
3	.28
4	.20
5	.14

.97

c.
x	*P(x)*
7	−.25
8	.85
9	.40

1.25

5.9 Each of the following tables lists certain values of *x* and their probabilities. Determine whether or not each one satisfies the two conditions required for a valid probability distribution.

a.
x	*P(x)*
5	−.36
6	.48
7	.62
8	.26

1.30

b.
x	*P(x)*
1	.27
2	.24
3	.49

1.00

c.
x	*P(x)*
0	.15
1	.08
2	.20
3	.50

.93

5.10 The following table gives the probability distribution of a discrete random variable x.

x	0	1	2	3	4	5	6
$P(x)$	.11	.19	.28	.15	.12	.09	.06

Find the following probabilities.

a. $P(x = 3)$ **b.** $P(x \le 2)$ **c.** $P(x \ge 4)$ **d.** $P(1 \le x \le 4)$

e. Probability that x assumes a value less than 4

f. Probability that x assumes a value greater than 2

g. Probability that x assumes a value in the interval 2 to 5

5.11 The following table gives the probability distribution of a discrete random variable x.

x	0	1	2	3	4	5
$P(x)$	.03	.17	.22	.31	.15	.12

Find the following probabilities.

a. $P(x = 1)$ **b.** $P(x \le 1)$ **c.** $P(x \ge 3)$ **d.** $P(0 \le x \le 2)$

e. Probability that x assumes a value less than 3

f. Probability that x assumes a value greater than 3

g. Probability that x assumes a value in the interval 2 to 4

■ Applications

5.12 Elmo's Sporting Goods sells exercise machines as well as other sporting goods. On different days, it sells different numbers of these machines. The following table, constructed using past data, lists the probability distribution of the number of exercise machines sold per day at Elmo's.

Machines sold per day	4	5	6	7	8	9	10
Probability	.08	.11	.14	.19	.20	.16	.12

a. Graph the probability distribution.

b. Determine the probability that the number of exercise machines sold at Elmo's on a given day is
   i. exactly 6   **ii.** more than 8   **iii.** 5 to 8   **iv.** at most 6

5.13 Let x denote the number of auto accidents that occur in a city during a week. The following table lists the probability distribution of x.

x	0	1	2	3	4	5	6
$P(x)$	.12	.16	.22	.20	.14	.12	.04

a. Draw a line graph for this probability distribution.

b. Determine the probability that the number of auto accidents that will occur during a given week in this city is
   i. exactly 4   **ii.** at least 3   **iii.** less than 3   **iv.** 3 to 5

5.14 A consumer agency surveyed all 2500 families in a small town to collect data on the number of television sets owned by them. The following table lists the frequency distribution of the data collected by this agency.

Number of TV sets owned	0	1	2	3	4
Number of families	120	970	730	410	270

a. Construct a probability distribution table for the numbers of television sets owned by these families. Draw a line graph of the probability distribution.

b. Are the probabilities listed in the table of part a exact or approximate probabilities of various outcomes? Explain.

c. Let x denote the number of television sets owned by a randomly selected family from this town. Find the following probabilities.

i. $P(x = 1)$　　**ii.** $P(x > 2)$　　**iii.** $P(x \leq 1)$　　**iv.** $P(1 \leq x \leq 3)$

5.15 The Webster Mail Order Company sells expensive stereos by mail. The following table lists the frequency distribution of the number of orders received per day by this company during the past 100 days.

Number of orders received per day	2	3	4	5	6
Number of days	12	21	34	19	14

a. Construct a probability distribution table for the number of orders received per day. Draw a graph of the probability distribution.

b. Are the probabilities listed in the table of part a exact or approximate probabilities of various outcomes? Explain.

c. Let x denote the number of orders received on any given day. Find the following probabilities.

i. $P(x = 3)$　　**ii.** $P(x \geq 3)$　　**iii.** $P(2 \leq x \leq 4)$　　**iv.** $P(x < 4)$

5.16 Five percent of all cars manufactured at a large auto company are lemons. Suppose two cars are selected at random from the production line of this company. Let x denote the number of lemons in this sample. Write the probability distribution of x. Draw a tree diagram for this problem.

5.17 According to a CBS News study, 68% of adults think that DNA tests are "very reliable" for identifying an individual (*USA TODAY*, August 20, 1998). Assume that this result holds true for the current population of all adults. Suppose that two adults are selected at random. Let x denote the number of adults in this sample who believe that DNA tests are "very reliable" for identifying an individual. Construct the probability distribution table of x. Draw a tree diagram for this problem.

5.18 According to a survey, 30% of adults are against using animals for research. Assume that this result holds true for the current population of all adults. Let x be the number of adults who are against using animals for research in a random sample of two adults. Obtain the probability distribution of x. Draw a tree diagram for this problem.

5.19 In a 1999 poll conducted by the polling firm of Penn, Schoen and Berland for Discovery Health Media, 40% of the respondents strongly agreed that health care in the United States is no longer affordable (*Newsweek*, August 2, 1999). Assume that this result holds true for the current population of U.S. adults. Suppose that two adults are randomly selected from this population, and let x denote the number of adults in this sample who strongly agree with this view. Construct the probability distribution table of x. Draw a tree diagram for this problem.

*5.20 In a group of 12 persons, three are left-handed. Suppose that two persons are randomly selected from this group. Let x denote the number of left-handed persons in this sample. Write the probability distribution of x. You may draw a tree diagram and use it to write the probability distribution. (*Hint:* Note that the draws are made without replacement from a small population. Hence, the probabilities of outcomes do not remain constant for each draw.)

*5.21 A statistics class has 20 students; 12 of them are female. Suppose two students are randomly selected from this class. Let x denote the number of females in this sample. Find the probability distribution of x. You may draw a tree diagram and use that to write the probability distribution. (*Hint:* Note that the draws are made without replacement from a small population. Hence, the probabilities of outcomes do not remain constant for each draw.)

5.3　MEAN OF A DISCRETE RANDOM VARIABLE

The **mean of a discrete random variable**, denoted by μ, is actually the mean of its probability distribution. The mean of a discrete random variable x is also called its *expected value* and is denoted by $E(x)$. The mean (or expected value) of a discrete random variable is the value that we expect to observe per repetition, on average, if we perform an experiment a large number of times. For example, we may expect a car salesperson to sell, on average, 2.4 cars per week. This does not mean that every week this salesperson will sell exactly 2.4 cars. (Obviously she cannot sell exactly 2.4 cars.) This simply means that if we observe for many weeks, this salesperson will sell a different number of cars during different weeks; however, the average for all these weeks will be 2.4 cars.

To calculate the mean of a discrete random variable x, we multiply each value of x by the corresponding probability and sum the resulting products. This sum gives the mean (or expected value) of the discrete random variable x.

> **MEAN OF A DISCRETE RANDOM VARIABLE** The *mean of a discrete random variable x* is the value that is expected to occur per repetition, on average, if an experiment is repeated a large number of times. It is denoted by μ and calculated as
>
> $$\mu = \Sigma x P(x)$$
>
> The mean of a discrete random variable x is also called its expected value and is denoted by $E(x)$; that is,
>
> $$E(x) = \Sigma x P(x)$$

Example 5–5 illustrates the calculation of the mean of a discrete random variable.

Calculating and interpreting the mean of a discrete random variable.

EXAMPLE 5-5 Recall Example 5–3 of Section 5.2. The probability distribution Table 5.4 from that example is reproduced below. In this table, x represents the number of breakdowns for a machine during a given week, and $P(x)$ is the probability of the corresponding value of x.

x	$P(x)$
0	.15
1	.20
2	.35
3	.30

Find the mean number of breakdowns per week for this machine.

Solution   To find the mean number of breakdowns per week for this machine, we multiply each value of x by its probability and add these products. This sum gives the mean of the probability distribution of x. The products $xP(x)$ are listed in the third column of Table 5.6. The sum of these products gives $\Sigma x P(x)$, which is the mean of x.

Table 5.6   Calculating the Mean for the Probability Distribution of Breakdowns

x	$P(x)$	$xP(x)$
0	.15	$0(.15) = .00$
1	.20	$1(.20) = .20$
2	.35	$2(.35) = .70$
3	.30	$3(.30) = .90$
		$\Sigma x P(x) = 1.80$

The mean is

$$\mu = \Sigma xP(x) = \mathbf{1.80}$$

Thus, on average, this machine is expected to break down 1.80 times per week over a period of time. In other words, if this machine is used for many weeks, then for certain weeks we will observe no breakdowns; for some other weeks we will observe one breakdown per week; and for still other weeks we will observe two or three breakdowns per week. The mean number of breakdowns is expected to be 1.80 per week for the entire period.

Note that $\mu = 1.80$ is also the expected value of x. It can also be written as

$$E(x) = 1.80$$ ■

Case Study 5–1 illustrates the calculation of the mean amount that an instant lottery player is expected to win.

CASE STUDY 5-1   ACES HIGH INSTANT LOTTERY

EXAMPLE: Ticket as Printed

EXAMPLE: Scratched-off Ticket

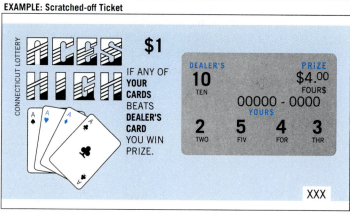

Currently (in 1999) the state of Connecticut has in circulation an instant lottery game called *Aces High*. The cost of each ticket for this lottery is $1. A player can instantly win $1000, $500, $25, $10, $4, $2, or a free ticket (which is equivalent to winning $1). Each ticket has six erasable spots, one of which contains the dealer's card, four spots contain the player's cards, and one spot shows the prize won by the player. A player will win the prize shown in the prize spot if any of the player's cards beats the dealer's card.

The following table lists the number of tickets with different prizes in a total of 9,600,000 tickets printed in the first batch for this lottery. Note that the lottery corporation could print more tickets but the proportions of tickets with these prizes remain the same in each batch. As is obvious from this table, out of a total of 9,600,000 tickets, 7,111,032 are nonwinning tickets (the ones with a prize of $0 in this table). Of the remaining tickets, 1,412,720 have a prize of a free ticket, 654,920 have a prize of $2 each, and so on.

Prize (dollars)	Number of Tickets
0	7,111,032
Free ticket	1,412,720
2	654,920
4	253,040
10	125,760
25	42,240
500	192
1000	96
	Total = 9,600,000

The net gain to a player for each of the instant winning tickets is equal to the amount of the prize minus $1, which is the cost of the ticket. Thus, the net gain for each of the non-winning tickets is $-$1, which is the cost of the ticket. Let

$x =$ the net amount a player wins by playing this lottery

The following table shows the probability distribution of x, and all the calculations required to compute the mean of x for that probability distribution. The probability of an outcome (net winnings) is calculated by dividing the number of tickets with that outcome by the total number of tickets.

x (dollars)	$P(x)$	$x P(x)$
-1	7,111,032/9,600,000 = .74073250	$-.7407325$
0	1,412,720/9,600,000 = .14715833	.0000000
1	654,920/9,600,000 = .06822083	.0682208
3	253,040/9,600,000 = .02635833	.0790750
9	125,760/9,600,000 = .01310000	.1179000
24	42,240/9,600,000 = .00440000	.1056000
499	192/9,600,000 = .00002000	.0099800
999	96/9,600,000 = .00001000	.0099900
		$\Sigma x P(x) = -.3499667$

Hence, the mean of x is

$$\mu = \Sigma x P(x) = -\$.3499667 \approx -\$.35$$

This mean gives the expected value of the random variable x, that is,

$$E(x) = \Sigma x P(x) = -\$.35$$

Thus, the mean of net winnings for this lottery is $-\$.35$. In other words, all players taken together will lose an average of $\$.35$ (or 35 cents) per ticket. This can also be interpreted as follows: Only 65% ($= 100 - 35$) of the total money spent by all players on buying lottery tickets for this lottery will be returned to them in the form of prizes and 35% will not be returned. (The money that will not be returned to players will cover the costs of operating the lottery, the commission paid to agents, and revenue to the state of Connecticut.)

Source: Connecticut Lottery Corporation Publications. Lottery ticket reproduced with permission.

5.4   STANDARD DEVIATION OF A DISCRETE RANDOM VARIABLE

The **standard deviation of a discrete random variable**, denoted by σ, measures the spread of its probability distribution. A higher value for the standard deviation of a discrete random variable indicates that x can assume values over a larger range about the mean. In contrast, a smaller value for the standard deviation indicates that most of the values that x can assume are clustered closely about the mean. The basic formula to compute the standard deviation of a discrete random variable is

$$\sigma = \sqrt{\Sigma[(x - \mu)^2 \cdot P(x)]}$$

However, it is more convenient to use the following short-cut formula to compute the standard deviation of a discrete random variable.

> **STANDARD DEVIATION OF A DISCRETE RANDOM VARIABLE**   The *standard deviation of a discrete random variable x* measures the spread of its probability distribution and is computed as
>
> $$\sigma = \sqrt{\Sigma x^2 P(x) - \mu^2}$$

Note that the variance σ^2 of a discrete random variable is obtained by squaring its standard deviation.

Example 5–6 illustrates how to use the short-cut formula to compute the standard deviation of a discrete random variable.

Calculating the standard deviation of a discrete random variable.

EXAMPLE 5–6   Baier's Electronics manufactures computer parts that are supplied to many computer companies. Despite the fact that two quality control inspectors at Baier's Electronics check every part for defects before it is shipped to another company, a few defective parts do pass through these inspections undetected. Let x denote the number of defective computer parts in a shipment of 400. The following table gives the probability distribution of x.

x	0	1	2	3	4	5
$P(x)$	.02	.20	.30	.30	.10	.08

Compute the standard deviation of x.

Solution　Table 5.7 shows all the calculations required for the computation of the standard deviation of x.

Table 5.7　Computations to Find the Standard Deviation

x	$P(x)$	$xP(x)$	x^2	$x^2P(x)$
0	.02	.00	0	.00
1	.20	.20	1	.20
2	.30	.60	4	1.20
3	.30	.90	9	2.70
4	.10	.40	16	1.60
5	.08	.40	25	2.00
		$\Sigma xP(x) = 2.50$		$\Sigma x^2P(x) = 7.70$

We perform the following steps to compute the standard deviation of x.

Step 1.　Compute the mean of the discrete random variable.
　　The sum of the products $xP(x)$, recorded in the third column of Table 5.7, gives the mean of x.

$$\mu = \Sigma xP(x) = 2.50 \text{ defective computer parts in 400}$$

Step 2.　Compute the value of $\Sigma x^2P(x)$.
　　First we square each value of x and record it in the fourth column of Table 5.7. Then we multiply these values of x^2 by the corresponding values of $P(x)$. The resulting values of $x^2P(x)$ are recorded in the fifth column of Table 5.7. The sum of this column is

$$\Sigma x^2P(x) = 7.70$$

Step 3.　Substitute the values of μ and $\Sigma x^2P(x)$ in the formula for the standard deviation of x and simplify.
　　By performing this step, we obtain

$$\sigma = \sqrt{\Sigma x^2P(x) - \mu^2} = \sqrt{7.70 - (2.50)^2} = \sqrt{1.45}$$

$$= \mathbf{1.204} \text{ defective computer parts}$$

　　Thus, a given shipment of 400 computer parts is expected to contain an average of 2.50 defective parts with a standard deviation of 1.204.　■

☞　*Remember*　Because the standard deviation of a discrete random variable is obtained by taking the positive square root, its value is never negative.

EXAMPLE 5–7　Loraine Corporation is planning to market a new makeup product. According to the analysis made by the financial department of the company, it will earn an annual profit of $4.5 million if this product has high sales and an annual profit of $1.2 million if the sales are mediocre, and it will lose $2.3 million a year if the sales are low. The probabilities of these three scenarios are .32, .51, and .17, respectively.

(a)　Let x be the profits (in millions of dollars) earned per annum by the company from this product. Write the probability distribution of x.

(b)　Calculate the mean and standard deviation of x.

Solution

Writing the probability distribution of a discrete random variable.

(a)   The following table lists the probability distribution of x. Note that since x denotes profits earned by the company, the loss is written as a *negative profit* in the table.

x	$P(x)$
4.5	.32
1.2	.51
-2.3	.17

Calculating the mean and standard deviation of a discrete random variable.

(b)   Table 5.8 shows all the calculations needed for the computation of the mean and standard deviation of x.

Table 5.8   Computations to Find the Mean and Standard Deviation

x	$P(x)$	$xP(x)$	x^2	$x^2P(x)$
4.5	.32	1.440	20.25	6.4800
1.2	.51	.612	1.44	.7344
-2.3	.17	$-.391$	5.29	.8993
		$\Sigma xP(x) = 1.661$		$\Sigma x^2P(x) = 8.1137$

The mean of x is

$$\mu = \Sigma xP(x) = \textbf{\$1.661 million}$$

The standard deviation of x is

$$\sigma = \sqrt{\Sigma x^2P(x) - \mu^2} = \sqrt{8.1137 - (1.661)^2} = \textbf{\$2.314 million}$$

Thus, it is expected that Loraine Corporation will earn an average of \$1.661 million in profits per year from the new makeup product with a standard deviation of \$2.314 million. ■

☞   **INTERPRETATION OF THE STANDARD DEVIATION**

The standard deviation of a discrete random variable can be interpreted or used the same way as the standard deviation of a data set in Section 3.4 of Chapter 3. In that section, we learned that according to Chebyshev's theorem, at least $[1 - (1/k^2)] \times 100\%$ of the total area under a curve lies within k standard deviations of the mean, where k is any number greater than 1. Thus, if $k = 2$, then 75% of the area under a curve lies between $\mu - 2\sigma$ and $\mu + 2\sigma$. In Example 5–6,

$$\mu = 2.50 \quad \text{and} \quad \sigma = 1.204$$

Hence,

$$\mu - 2\sigma = 2.50 - 2(1.204) = .092$$

$$\mu + 2\sigma = 2.50 + 2(1.204) = 4.908$$

Using Chebyshev's theorem, we can state that at least 75% of the shipments (each containing 400 computer parts) are expected to contain .092 to 4.908 defective computer parts each.

EXERCISES

■ **Concepts and Procedures**

of the probability distribution / expected value

5.22 Briefly explain the concept of the mean and standard deviation of a discrete random variable.

5.23 Find the mean and standard deviation for each of the following probability distributions.

a.

x	P(x)
0	.16
1	.27
2	.39
3	.18

b.

x	P(x)
6	.40
7	.26
8	.21
9	.13

5.24 Find the mean and standard deviation for each of the following probability distributions.

a.

x	P(x)
3	.09
4	.21
5	.34
6	.23
7	.13

b.

x	P(x)
0	.43
1	.31
2	.17
3	.09

■ **Applications**

5.25 Let x be the number of errors that appear on a randomly selected page of a book. The following table lists the probability distribution of x.

x	0	1	2	3	4
P(x)	.73	.16	.06	.04	.01

Find the mean and standard deviation of x.

5.26 Let x be the number of newborn pigs per month on a pig farm. The following table lists the probability distribution of x.

x	4	5	6	7	8	9	10
P(x)	.17	.26	.19	.13	.11	.10	.04

Find the mean and standard deviation of x.

5.27 The following table gives the probability distribution of the number of camcorders sold on a given day at an electronics store.

Camcorders sold	0	1	2	3	4	5	6
Probability	.05	.12	.19	.30	.20	.10	.04

Calculate the mean and standard deviation for this probability distribution. Give a brief interpretation of the value of the mean.

5.28 The following table, reproduced from Exercise 5.12, lists the probability distribution of the number of exercise machines sold per day at Elmo's Sporting Goods store.

Machines sold per day	4	5	6	7	8	9	10
Probability	.08	.11	.14	.19	.20	.16	.12

Calculate the mean and standard deviation for this probability distribution. Give a brief interpretation of the value of the mean.

5.29 Let x be the number of heads obtained in two tosses of a coin. The following table lists the probability distribution of x.

x	0	1	2
$P(x)$	.25	.50	.25

Calculate the mean and standard deviation of x. Give a brief interpretation of the value of the mean.

5.30 Let x be a random variable that represents the number of students who are absent on a given day from a class of 25. The following table lists the probability distribution of x.

x	0	1	2	3	4	5
$P(x)$	.08	.18	.32	.22	.14	.06

Calculate the mean and standard deviation for this probability distribution and give a brief interpretation of the value of the mean.

5.31 Refer to Exercise 5.14. Find the mean and standard deviation for the probability distribution you developed for the number of television sets owned by all 2500 families in a town. Give a brief interpretation of the values of the mean and standard deviation.

5.32 Refer to Exercise 5.15. Find the mean and standard deviation for the probability distribution you developed for the number of orders received per day for the past 100 days at the Webster Mail Order Company. Give a brief interpretation of the values of the mean and standard deviation.

5.33 Refer to the probability distribution you developed in Exercise 5.16 for the number of lemons in two selected cars. Calculate the mean and standard deviation of x for that probability distribution.

5.34 Refer to the probability distribution you developed in Exercise 5.17 for the number of adults in a sample of two who thought that DNA tests are "very reliable" for identifying an individual. Compute the mean and standard deviation of x for that probability distribution.

5.35 A farmer will earn a profit of $30 thousand next year in case of heavy rain, $60 thousand in case of moderate rain, and $15 thousand in case of little rain. A meteorologist forecasts that the probability is .35 for heavy rain, .40 for moderate rain, and .25 for little rain next year. Let x be the random variable that represents next year's profits in thousands of dollars for this farmer. Write the probability distribution of x. Find the mean and standard deviation of x. Give a brief interpretation of the values of the mean and standard deviation.

 5.36 An instant lottery ticket costs $2. Out of a total of 10,000 tickets printed for this lottery, 1000 tickets contain a prize of $5 each, 100 tickets have a prize of $10 each, 5 tickets have a prize of $1000 each, and 1 ticket has a prize of $5000. Let x be the random variable that denotes the net amount a player wins by playing this lottery. Write the probability distribution of x. Determine the mean and standard deviation of x. How will you interpret the values of the mean and standard deviation of x?

*5.37 Refer to the probability distribution you developed in Exercise 5.20 for the number of left-handed persons in a sample of two persons. Calculate the mean and standard deviation of x for that distribution.

*5.38 Refer to the probability distribution you developed in Exercise 5.21 for the number of females in a random sample of two students selected from a class. Calculate the mean and standard deviation of x for that distribution.

5.5 FACTORIALS AND COMBINATIONS

This section introduces factorials and combinations, which will be used in the binomial formula discussed in Section 5.6.

5.5.1 FACTORIALS

The symbol ! (read as *factorial*) is used to denote **factorials**. The value of the factorial of a number is obtained by multiplying all the integers from that number to 1. For example, 7! is read as "seven factorial" and is evaluated by multiplying all the integers from 7 to 1.

> **FACTORIALS** The symbol $n!$, read as "n factorial," represents the product of all the integers from n to 1. In other words,
>
> $$n! = n(n-1)(n-2)(n-3) \cdots 3 \cdot 2 \cdot 1$$
>
> By definition,
>
> $$0! = 1$$

Evaluating a factorial.

EXAMPLE 5–8   Evaluate 7!.

Solution   To evaluate 7!, we multiply all the integers from 7 to 1.

$$7! = 7 \cdot 6 \cdot 5 \cdot 4 \cdot 3 \cdot 2 \cdot 1 = \mathbf{5040}$$

Thus, the value of 7! is 5040.[1]

Evaluating a factorial.

EXAMPLE 5–9   Evaluate 10!.

Solution   The value of 10! is given by the product of all the integers from 10 to 1. Thus,

$$10! = 10 \cdot 9 \cdot 8 \cdot 7 \cdot 6 \cdot 5 \cdot 4 \cdot 3 \cdot 2 \cdot 1 = \mathbf{3{,}628{,}800}$$

Evaluating a factorial of the difference between two numbers.

EXAMPLE 5–10   Evaluate $(12 - 4)!$.

Solution   The value of $(12 - 4)!$ is

$$(12 - 4)! = 8! = 8 \cdot 7 \cdot 6 \cdot 5 \cdot 4 \cdot 3 \cdot 2 \cdot 1 = \mathbf{40{,}320}$$

Evaluating a factorial of zero.

EXAMPLE 5–11   Evaluate $(5 - 5)!$.

Solution   The value of $(5 - 5)!$ is 1.

$$(5 - 5)! = 0! = \mathbf{1}$$

Note that 0! is always equal to 1.

We can read the value of $n!$ for $n = 1$ through $n = 25$ from Table II of Appendix D. Example 5–12 illustrates how to read that table.

Using the table of factorials.

EXAMPLE 5–12   Find the value of 15! by using Table II of Appendix D.

Solution   To find the value of 15! from Table II, we locate 15 in the column labeled n. Then we read the value in the column for $n!$ entered next to 15. Thus,

$$15! = \mathbf{1{,}307{,}674{,}368{,}000}$$

[1]**Using a calculator to evaluate $n!$:** Most calculators have an $n!$ or $x!$ function key. If your calculator has such a key, you can use it to evaluate 7! as follows: First enter 7 and then press the $n!$ or $x!$ key. The calculator screen will display 5040 as the answer.

5.5.2   COMBINATIONS

Quite often we face the problem of selecting a few elements from a large number of distinct elements. For example, a student may be required to attempt any two questions out of four in an examination. As another example, the faculty in a department may need to select 3 professors from 20 to form a committee. Or a lottery player may have to pick 6 numbers from 49. The question arises: In how many ways can we make the selections in each of these examples? For instance, how many possible selections exist for the student who is to choose any two questions out of four? The answer is six. Let the four questions be denoted by the numbers 1, 2, 3, and 4. Then the six selections are

$$\text{(1 and 2)}\quad\text{(1 and 3)}\quad\text{(1 and 4)}\quad\text{(2 and 3)}\quad\text{(2 and 4)}\quad\text{(3 and 4)}$$

The student can choose questions 1 and 2, or 1 and 3, or 1 and 4, and so on.

Each of the possible selections in this list is called a **combination**. All six combinations are distinct; that is, each combination contains a different set of questions. It is important to remember that the order in which the selections are made is not significant in the case of combinations. Thus, whether we write (1 and 2) or (2 and 1), both these arrangements represent only one combination.

COMBINATIONS NOTATION   *Combinations* give the number of ways x elements can be selected from n elements. The notation used to denote the total number of combinations is

$$\binom{n}{x}$$

which is read as "the number of combinations of n elements selected x at a time."

Suppose there are a total of n elements from which we want to select x elements. Then,

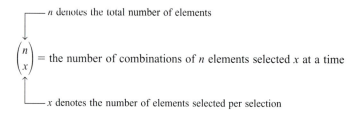

NUMBER OF COMBINATIONS   The *number of combinations* for selecting x from n distinct elements is given by the formula

$$\binom{n}{x} = \frac{n!}{x!(n-x)!}$$

where $n!$, $x!$, and $(n-x)!$ are read as "n factorial," "x factorial," and "n minus x factorial," respectively.

In the combinations formula,

$$n! = n(n-1)(n-2)(n-3) \cdots 3 \cdot 2 \cdot 1$$

$$x! = x(x-1)(x-2) \cdots 3 \cdot 2 \cdot 1$$

$$(n-x)! = (n-x)(n-x-1)(n-x-2) \cdots 3 \cdot 2 \cdot 1$$

Note that in combinations, n is always greater than or equal to x. If n is less than x, then we cannot select x distinct elements from n.

Finding the number of combinations using the formula.

EXAMPLE 5-13 An ice cream parlor has six flavors of ice cream. Kristen wants to buy two flavors of ice cream. If she randomly selects two flavors out of six, how many possible combinations are there?

Solution For this example,

$$n = \text{total number of ice cream flavors} = 6$$

$$x = \text{number of ice cream flavors to be selected} = 2$$

Therefore, the number of ways in which Kristen can select two flavors of ice cream out of six is

$$\binom{6}{2} = \frac{6!}{2!(6-2)!} = \frac{6!}{2!4!} = \frac{6 \cdot 5 \cdot 4 \cdot 3 \cdot 2 \cdot 1}{2 \cdot 1 \cdot 4 \cdot 3 \cdot 2 \cdot 1} = \mathbf{15}$$

Thus, there are 15 ways for Kristen to select two ice cream flavors out of six.[2] ■

Finding the number of combinations and listing them.

EXAMPLE 5-14 Three members of a jury will be randomly selected from five persons. How many different combinations are possible?

Solution There are a total of five persons, and we are to select three of them. Hence,

$$n = 5 \quad \text{and} \quad x = 3$$

Applying the combinations formula, we get

$$\binom{5}{3} = \frac{5!}{3!(5-3)!} = \frac{5!}{3! \, 2!} = \frac{120}{6 \cdot 2} = \mathbf{10}$$

If we assume that the five persons are A, B, C, D, and E, then the 10 possible combinations for the selection of three members of the jury are

ABC   ABD   ABE   ACD   ACE   ADE   BCD   BCE   BDE   CDE   ■

Case Study 5–2 describes the number of ways a lottery player can select 6 numbers in a lotto game.

---

[2]**Using a calculator to evaluate $\binom{n}{x}$:** Most calculators have a $C_{n,r}$ or $C_{n,x}$ or $\binom{n}{r}$ or $\binom{n}{x}$ function key. (Note that all these notations are used to denote combinations.) If your calculator has such a key, read the manual that accompanies the calculator to find out how this key functions and then evaluate $\binom{6}{2}$ using that key.

CASE STUDY 5-2 PLAYING LOTTO

During the past few years, many states have initiated the popular lottery game called lotto. To play lotto, a player picks any 6 numbers from a list of numbers usually starting with 1—for example, from 1 through 49. At the end of the lottery period, the state lottery commission randomly selects 6 numbers from the same list. If all 6 numbers picked by a player are the same as the ones randomly selected by the lottery commission, the player wins.

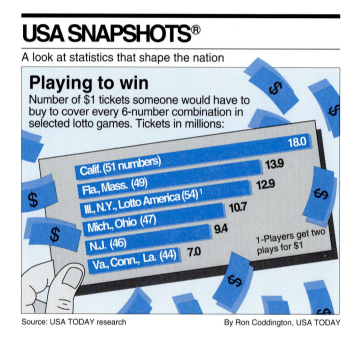

USA SNAPSHOTS®

A look at statistics that shape the nation

Playing to win
Number of $1 tickets someone would have to buy to cover every 6-number combination in selected lotto games. Tickets in millions:

	18.0
Calif. (51 numbers)	13.9
Fla., Mass. (49)	12.9
Ill., N.Y., Lotto America (54)[1]	10.7
Mich., Ohio (47)	9.4
N.J. (46)	
Va., Conn., La. (44)	7.0

1-Players get two plays for $1

Source: USA TODAY research   By Ron Coddington, USA TODAY

The chart shows the number of combinations (in millions) for picking 6 numbers for lotto games played in a few states. For example, in California a player has to pick 6 numbers from 1 through 51. As shown in the chart, there are approximately 18 million ways (combinations) to select 6 numbers from 1 through 51. In Florida and Massachusetts, a player has to pick 6 numbers from 1 through 49. For this lotto, there are approximately 13.9 million combinations.

Let us find the probability that a player who picks 6 numbers from 49 wins this game. The total combinations of selecting 6 numbers from 49 numbers are obtained as follows:

$$\binom{49}{6} = \frac{49!}{6!(49-6)!} = 13,983,816$$

Thus, there are a total of 13,983,816 different ways to select 6 numbers from 49 numbers. Hence, the probability that a player (who plays this lottery once) wins is

$$P(\text{player wins}) = 1/13,983,816 = .0000000715$$

5.5.3 USING THE TABLE OF COMBINATIONS

Table III in Appendix D lists the number of combinations of n elements selected x at a time. The following example illustrates how to read that table to find combinations.

Using the table of combinations.

EXAMPLE 5–15   Marv & Sons advertised to hire a financial analyst. The company has received applications from 10 candidates who seem to be equally qualified. The company manager has decided to call only 3 of these candidates for an interview. If she randomly selects 3 candidates from the 10, how many total selections are possible?

Solution   The total number of ways to select 3 applicants from 10 is given by $\binom{10}{3}$. To find the value of $\binom{10}{3}$ from Table III, we locate 10 in the column labeled n and 3 in the row labeled x. The relevant part of that table is reproduced here as Table 5.9.

Table 5.9   Determining the Value of $\binom{10}{3}$

n	x	0	1	2	3	$\cdots$	20
1		1	1				
2		1	2	1			
3		1	3	3	1		
.		.	.	.	.		
.		.	.	.	.		
.		.	.	.	.		
$n = 10 \longrightarrow$ 10		1	10	45	120 $\leftarrow$	$\cdots$	$\cdots$
.		.	.	.	.	$\cdots$	$\cdots$
.		.	.	.	.	$\cdots$	$\cdots$

The value of $\binom{10}{3}$

The number at the intersection of the row for $n = 10$ and the column for $x = 3$ gives the value of $\binom{10}{3}$, which is

$$\binom{10}{3} = 120$$

Thus, the company manager can select 3 applicants from 10 in 120 ways.   ∎

☞ *Remember*   If the total number of elements and the number of elements to be selected are the same, then there is only one combination. In other words,

$$\binom{n}{n} = 1$$

Also, the number of combinations for selecting zero items from n is 1; that is,

$$\binom{n}{0} = 1$$

For example,

$$\binom{5}{5} = \frac{5!}{5!(5-5)!} = \frac{5!}{5!0!} = \frac{120}{(120)(1)} = 1$$

$$\binom{8}{0} = \frac{8!}{0!(8-0)!} = \frac{8!}{0!8!} = \frac{40,320}{(1)(40,320)} = 1$$

EXERCISES

■ Concepts and Procedures

5.39 Determine the value of each of the following using the appropriate formula.

$$3! \qquad (9-3)! \qquad 9! \qquad (14-12)! \qquad \binom{5}{3} \qquad \binom{7}{4} \qquad \binom{9}{3} \qquad \binom{4}{0} \qquad \binom{3}{3}$$

Verify the calculated values by using Tables II and III of Appendix D.

5.40 Find the value of each of the following using the appropriate formula.

$$6! \qquad 11! \qquad (7-2)! \qquad (15-5)! \qquad \binom{8}{2} \qquad \binom{5}{0} \qquad \binom{5}{5} \qquad \binom{6}{4} \qquad \binom{11}{7}$$

Verify the calculated values by using Tables II and III of Appendix D.

■ Applications

5.41 An English department at a university has 16 faculty members. Two of the faculty members will be randomly selected to represent the department on a committee. In how many ways can the department select 2 faculty members from 16? Use the appropriate formula.

5.42 A person will randomly select 3 varieties of cookies from 10 varieties for a party. In how many ways can this person select 3 varieties from 10? Use the appropriate formula.

5.43 A superintendent of schools has 12 schools under her jurisdiction. She plans to visit 3 of these schools during the next week. If she randomly selects 3 schools from these 12, how many total selections are possible? Use the appropriate formula. Verify your answer by using Table III of Appendix D.

5.44 An environmental agency will randomly select 4 houses from a block containing 25 houses for a radon check. How many total selections are possible? Use the appropriate formula. Verify your answer by using Table III of Appendix D.

5.45 An investor will randomly select 6 stocks from 20 for an investment. How many total combinations are possible? Use the appropriate formula. Verify your answer by using Table III of Appendix D.

5.46 A company employs a total of 16 workers. The management has asked these employees to select two workers who will negotiate a new contract with management. The employees have decided to select the two workers randomly. How many total selections are possible? Use the appropriate formula. Verify your answer by using Table III of Appendix D.

5.47 In how many ways can a sample (without replacement) of 9 items be selected from a population of 20 items?

5.48 In how many ways can a sample (without replacement) of 5 items be selected from a population of 15 items?

5.6 THE BINOMIAL PROBABILITY DISTRIBUTION

The **binomial probability distribution** is one of the most widely used discrete probability distributions. It is applied to find the probability that an outcome will occur x times in n performances of an experiment. For example, given that the probability is .05 that a VCR manufactured at a firm is defective, we may be interested in finding the probability that in a random sample of three VCRs manufactured at this firm, exactly one will be defective. As a second example, we may be interested in finding the probability that a baseball player with a batting average of .250 will have no hits in 10 trips to the plate.

To apply the binomial probability distribution, the random variable x must be a discrete dichotomous random variable. In other words, the variable must be a discrete random variable and each repetition of the experiment must result in one of two possible outcomes. The binomial distribution is applied to experiments that satisfy the four conditions of a *binomial experiment*. (These conditions are described in Section 5.6.1.) Each repetition of a binomial experiment is called a **trial** or a **Bernoulli trial** (after Jacob Bernoulli). For example, if an experiment is defined as one toss of a coin and this experiment is repeated 10 times, then each repetition (toss) is called a trial. Consequently, there are 10 total trials for this experiment.

5.6.1 THE BINOMIAL EXPERIMENT

An experiment that satisfies the following four conditions is called a **binomial experiment**.

1. There are n identical trials. In other words, the given experiment is repeated n times. All these repetitions are performed under identical conditions.
2. Each trial has two and only two outcomes. These outcomes are usually called a *success* and a *failure*.
3. The probability of success is denoted by p and that of failure by q, and $p + q = 1$. The probabilities p and q remain constant for each trial.
4. The trials are independent. In other words, the outcome of one trial does not affect the outcome of another trial.

CONDITIONS OF A BINOMIAL EXPERIMENT A binomial experiment must satisfy the following four conditions.

1. There are n identical trials.
2. Each trial has only two possible outcomes.
3. The probabilities of the two outcomes remain constant.
4. The trials are independent.

Note that one of the two outcomes of a trial is called a *success* and the other a *failure*. Notice that a success does not mean that the corresponding outcome is considered favorable or desirable. Similarly, a failure does not necessarily refer to an unfavorable or undesirable outcome. Success and failure are simply the names used to denote the two possible outcomes of a trial. The outcome to which the question refers is usually called a success; the outcome to which it does not refer is called a failure.

Verifying the conditions of a binomial experiment.

EXAMPLE 5-16   Consider the experiment consisting of 10 tosses of a coin. Determine whether or not it is a binomial experiment.

Solution   The experiment consisting of 10 tosses of a coin satisfies all four conditions of a binomial experiment.

1.   There are a total of 10 trials (tosses), and they are all identical. All 10 tosses are performed under identical conditions.
2.   Each trial (toss) has only two possible outcomes: a head and a tail. Let a head be called a success and a tail be called a failure.
3.   The probability of obtaining a head (a success) is 1/2 and that of a tail (a failure) is 1/2 for any toss. That is,

$$p = P(H) = 1/2 \quad \text{and} \quad q = P(T) = 1/2$$

The sum of these two probabilities is 1.0. Also, these probabilities remain the same for each toss.
4.   The trials (tosses) are independent. The result of any preceding toss has no bearing on the result of any succeeding toss.
     Consequently, the experiment consisting of 10 tosses is a binomial experiment. ■

Verifying the conditions of a binomial experiment.

EXAMPLE 5-17   Five percent of all VCRs manufactured by a large electronics company are defective. Three VCRs are randomly selected from the production line of this company. The selected VCRs are inspected to determine whether each of them is defective or good. Is this experiment a binomial experiment?

Solution

1.   This example consists of three identical trials. A trial represents the selection of a VCR.
2.   Each trial has two outcomes: a VCR is defective or a VCR is good. Let a defective VCR be called a success and a good VCR be called a failure.
3.   Five percent of all VCRs are defective. So, the probability p that a VCR is defective is .05. As a result, the probability q that a VCR is good is .95. These two probabilities add up to 1.
4.   Each trial (VCR) is independent. In other words, if one VCR is defective, it does not affect the outcome of another VCR being defective or good. This is so because the size of the population is very large compared to the sample size.

Since all four conditions of a binomial experiment are satisfied, this is an example of a binomial experiment. ■

5.6.2   THE BINOMIAL PROBABILITY DISTRIBUTION AND BINOMIAL FORMULA

The random variable x that represents the number of successes in n trials for a binomial experiment is called a *binomial random variable*. The probability distribution of x in such experiments is called the **binomial probability distribution** or simply the *binomial distribution*. Thus, the binomial probability distribution is applied to find the probability of x successes in n trials for a binomial experiment. The number of successes x in such an experiment is a discrete random variable. Consider Example 5–17. Let x be the number of defective VCRs in a sample of three. Since we can obtain any number of defective VCRs from zero to three in a sample of three, x can assume any of the values 0, 1, 2, and 3. Since the values of x are countable, it is a discrete random variable.

BINOMIAL FORMULA   For a binomial experiment, the probability of exactly x successes in n trials is given by the binomial formula

$$P(x) = \binom{n}{x} p^x q^{n-x}$$

where

n = total number of trials

p = probability of success

$q = 1 - p$ = probability of failure

x = number of successes in n trials

$n - x$ = number of failures in n trials

In the binomial formula, n is the total number of trials and x is the total number of successes. The difference between the total number of trials and the total number of successes, $n - x$, gives the total number of failures in n trials. The value of $\binom{n}{x}$ gives the number of ways to obtain x successes in n trials. As mentioned earlier, p and q are the probabilities of success and failure, respectively. Again, although it does not matter which of the two outcomes is called a success and which one a failure, usually the outcome to which the question refers is called a success.

To solve a binomial problem, we determine the values of n, x, $n - x$, p, and q and then substitute these values in the binomial formula. To find the value of $\binom{n}{x}$, we can use either the combinations formula from Section 5.5.2 or the table of combinations (Table III of Appendix D).

To find the probability of x successes in n trials for a binomial experiment, the only values needed are those of n and p. These are called the *parameters of the binomial probability distribution* or simply the **binomial parameters**. The value of q is obtained by subtracting the value of p from 1.0. Thus, $q = 1 - p$.

Next we solve a binomial problem, first without using the binomial formula and then by using the binomial formula.

Calculating the probability using a tree diagram and the binomial formula.

EXAMPLE 5–18   Five percent of all VCRs manufactured by a large electronics company are defective. A quality control inspector randomly selects three VCRs from the production line. What is the probability that exactly one of these three VCRs is defective?

Solution   Let

D = a selected VCR is defective

G = a selected VCR is good

As the tree diagram in Figure 5.4 (page 212) shows, there are a total of eight outcomes, and three of them contain exactly one defective VCR. These three outcomes are

DGG, GDG,   and GGD

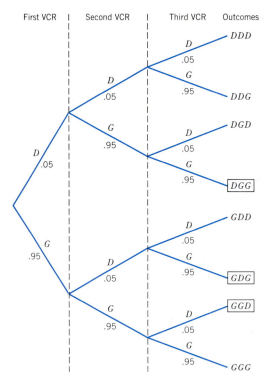

Figure 5.4 Tree diagram for selecting three VCRs.

We know that 5% of all VCRs manufactured at this company are defective. As a result, 95% of all VCRs are good. So, the probability that a randomly selected VCR is defective is .05 and the probability that it is good is .95.

$$P(D) = .05 \quad \text{and} \quad P(G) = .95$$

Because the size of the population is large (note that it is a large company), the selections can be considered to be independent. The probability of each of the three outcomes, which give exactly one defective VCR, is calculated as follows:

$$P(DGG) = P(D) \cdot P(G) \cdot P(G) = (.05)(.95)(.95) = .0451$$
$$P(GDG) = P(G) \cdot P(D) \cdot P(G) = (.95)(.05)(.95) = .0451$$
$$P(GGD) = P(G) \cdot P(G) \cdot P(D) = (.95)(.95)(.05) = .0451$$

Note that *DGG* is simply the intersection of the three events *D*, *G*, and *G*. In other words, *P(DGG)* is the joint probability of three events: the first VCR selected is defective, the second is good, and the third is good. To calculate this probability, we use the multiplication rule for independent events we learned in Chapter 4. The same is true about the probabilities of the other two outcomes: *GDG* and *GGD*.

Exactly one defective VCR will be selected if *DGG* or *GDG* or *GGD* occurs. These are three mutually exclusive outcomes. Therefore, from the addition rule of Chapter 4, the probability of the union of these three outcomes is simply the sum of their individual probabilities.

$$P(1 \text{ VCR is defective in } 3) = P(DGG \text{ or } GDG \text{ or } GGD)$$
$$= P(DGG) + P(GDG) + P(GGD)$$
$$= .0451 + .0451 + .0451 = \textbf{.1353}$$

Now let us use the binomial formula to compute this probability. Let us call the selection of a defective VCR a *success* and the selection of a good VCR a *failure*. The reason we have called a defective VCR a *success* is that the question refers to selecting exactly one defective VCR. Then,

$$n = \text{total number of trials} = 3 \text{ VCRs}$$

$$x = \text{number of successes} = \text{number of defective VCRs} = 1$$

$$n - x = \text{number of failures} = \text{number of good VCRs} = 3 - 1 = 2$$

$$p = P(\text{success}) = .05$$

$$q = P(\text{failure}) = 1 - p = .95$$

The probability of 1 success is denoted by $P(x = 1)$ or simply by $P(1)$. By substituting all the values in the binomial formula, we obtain

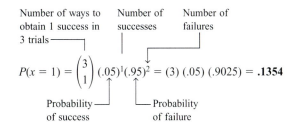

Note that the value of $\binom{3}{1}$ in the formula either can be read from Table III of Appendix D or can be computed as follows:

$$\binom{3}{1} = \frac{3!}{1!(3-1)!} = \frac{3 \cdot 2 \cdot 1}{1 \cdot 2 \cdot 1} = 3$$

In the above computation, $\binom{3}{1}$ gives the three ways to select one defective VCR in three selections. As listed earlier, these three ways to select one defective VCR are *DGG*, *GDG*, and *GGD*. The probability .1354 is slightly different from the earlier calculation (.1353) because of rounding. ∎

Calculating the probability using the binomial formula.

EXAMPLE 5–19 At the Express House Delivery Service, providing high-quality service to customers is the top priority of the management. The company guarantees a refund of all charges if a package it is delivering does not arrive at its destination by the specified time. It is known from past data that despite all efforts, 2% of the packages mailed through this company do not arrive at their destinations within the specified time. Suppose a corporation mails 10 packages through Express House Delivery Service on a certain day.

(a) Find the probability that exactly one of these 10 packages will not arrive at its destination within the specified time.

(b) Find the probability that at most one of these 10 packages will not arrive at its destination within the specified time.

Solution   Let us call it a success if a package does not arrive at its destination within the specified time and a failure if it does arrive within the specified time. Then,

$$n = \text{total number of packages mailed} = 10$$

$$p = P(\text{success}) = .02$$

$$q = P(\text{failure}) = 1 - .02 = .98$$

(a)   For this part,

$$x = \text{number of successes} = 1$$

$$n - x = \text{number of failures} = 10 - 1 = 9$$

Substituting all values in the binomial formula, we obtain

$$P(x = 1) = \binom{10}{1}(.02)^1(.98)^9 = \frac{10!}{1!\,(10-1)!}(.02)^1(.98)^9$$

$$= (10)(.02)(.83374776) = \mathbf{.1667}$$

Thus, there is a .1667 probability that exactly one of the 10 packages mailed will not arrive at its destination within the specified time.[3]

(b)   The probability that at most one of the 10 packages will not arrive at its destination within the specified time is given by the sum of the probabilities of $x = 0$ and $x = 1$. Thus,

$$P(x \le 1) = P(x = 0) + P(x = 1)$$

$$= \binom{10}{0}(.02)^0(.98)^{10} + \binom{10}{1}(.02)^1(.98)^9$$

$$= (1)(1)(.81707281) + (10)(.02)(.83374776)$$

$$= .8171 + .1667 = \mathbf{.9838}$$

Thus, the probability that at most one of the 10 packages will not arrive at its destination within the specified time is .9838.   ■

Constructing a binomial probability distribution and its graph.

EXAMPLE 5–20   According to a poll of women aged 18 and older conducted by Gallup for *USA TODAY*, 60% of the women said that they do not spend enough time on themselves (*USA TODAY*, February 17, 1999). Assume that this result holds true for the current population of all women 18 and older. Let x denote the number in a random sample of three women 18 and older who hold this view. Write the probability distribution of x and draw a line graph for this probability distribution.

Solution   Let x denote the number of women 18 and older in a sample of three who hold the given view. Then $n - x$ is the number of women 18 and older in the sample who do not hold this view. Note that $n - x$ is the number of women who either say that they spend enough time on themselves or have no opinion. From the given information,

$$n = \text{total number of women in the sample} = 3$$

$$p = P(\text{a woman holds the given view}) = .60$$

$$q = P(\text{a woman does not hold the given view}) = 1 - .60 = .40$$

---

[3]**Using a calculator to simplify the expression for $P(x = 1)$:** As explained in Chapter 4, you can evaluate $(.98)^9$ by using the y^x key on your calculator. To evaluate $(.98)^9$, first enter .98, then press the y^x key, then enter 9, and finally press the $=$ key. The calculator screen will display the answer.

The possible values that x can assume are 0, 1, 2, and 3. In other words, the number of women in a sample of three who hold the given view can be 0, 1, 2, or 3. The probability of each of these four outcomes is calculated as follows.

If $x = 0$, then $n - x = 3$. From the binomial formula, the probability of $x = 0$ is

$$P(x = 0) = \binom{3}{0} (.60)^0 (.40)^3 = (1)(1)(.064) = \mathbf{.0640}$$

Note that $\binom{3}{0}$ is equal to 1 by definition and $(.60)^0$ is equal to 1 because any number raised to the power zero is always 1.

If $x = 1$, then $n - x = 2$. The probability $P(x = 1)$ is

$$P(x = 1) = \binom{3}{1} (.60)^1 (.40)^2 = (3)(.60)(.16) = \mathbf{.2880}$$

Similarly, if $x = 2$, then $n - x = 1$, and if $x = 3$, then $n - x = 0$. Then the probabilities $P(x = 2)$ and $P(x = 3)$ are

$$P(x = 2) = \binom{3}{2} (.60)^2 (.40)^1 = (3)(.36)(.40) = \mathbf{.4320}$$

$$P(x = 3) = \binom{3}{3} (.60)^3 (.40)^0 = (1)(.216)(1) = \mathbf{.2160}$$

These probabilities are written in tabular form in Table 5.10. Figure 5.5 shows the line graph for the probability distribution of Table 5.10.

Table 5.10 Probability Distribution of x

x	$P(x)$
0	.0640
1	.2880
2	.4320
3	.2160

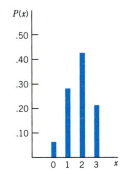

Figure 5.5 Line graph of the probability distribution of x. ■

5.6.3 USING THE TABLE OF BINOMIAL PROBABILITIES

The probabilities for a binomial experiment can also be read from Table IV, the table of binomial probabilities, in Appendix D. That table lists the probabilities of x for $n = 1$ to $n = 25$ and for selected values of p. Example 5–21 illustrates how to read Table IV.

Using the binomial table to find probabilities and construct the probability distribution and graph.

EXAMPLE 5–21 In a telephone poll of U.S. women aged 18 and older conducted by Princeton Survey Research Associates for *Newsweek*, 20% of women said that they exercise or work at being fit every day (*Newsweek* Special Issue, Spring/Summer 1999). Suppose that 20% of all current U.S. women 18 and older exercise or work at being fit every day. A sample of six such women is selected. Using Table IV of Appendix D, do the following.

(a)   Find the probability that exactly three women in this sample exercise or work at being fit every day.

(b)   Find the probability that at most two women in this sample exercise or work at being fit every day.

(c)   Find the probability that at least three women in this sample exercise or work at being fit every day.

(d)   Find the probability that one to three women in this sample exercise or work at being fit every day.

(e)   Let x be the number of women in this sample who exercise or work at being fit every day. Write the probability distribution of x and draw a line graph for this probability distribution.

Solution

(a)   To read the required probability from Table IV of Appendix D, we first determine the values of n, x, and p. For this example,

n = number of women in the sample = 6

x = number of women in six who exercise or work at being fit every day = 3

p = P(a woman exercises or works at being fit every day) = .20

Then we locate $n = 6$ in the column labeled n in Table IV. The relevant portion of Table IV with $n = 6$ is reproduced here as Table 5.11. Next, we locate 3 in the column for x in the portion of the table for $n = 6$ and locate $p = .20$ in the row for p at the top of the table. The entry at the intersection of the row for $x = 3$ and the column for $p = .20$ gives the probability of 3 successes in 6 trials when the probability of success is .20. From Table IV or Table 5.11,

$$P(x = 3) = \textbf{.0819}$$

Table 5.11   Determining $P(x = 3)$ for $n = 6$ and $p = .20$

$p = .20$

n	x	.05	.10	.20	$\cdots$	.95
$n = 6 \longrightarrow$ 6	0	.7351	.5314	.2621	$\cdots$	.0000
	1	.2321	.3543	.3932	$\cdots$	.0000
	2	.0305	.0984	.2458	$\cdots$	.0001
$x = 3 \longrightarrow$ 3	3	.0021	.0146	.0819 ←	$\cdots$	.0021
	4	.0001	.0012	.0154	$\cdots$	.0305
	5	.0000	.0001	.0015	$\cdots$	.2321
	6	.0000	.0000	.0001	$\cdots$	.7351

$P(x = 3) = .0819$

Using Table IV or Table 5.11, we write Table 5.12, which can be used to answer the remaining parts of this example.

Table 5.12   Portion of Table
IV for $n = 6$ and $p = .20$

			p
n	x		**.20**
6	0		.2621
	1		.3932
	2		.2458
	3		.0819
	4		.0154
	5		.0015
	6		.0001

(b)   The event that at most two women in this sample exercise or work at being fit every day will occur if x is equal to 0, 1, or 2. From Table IV of Appendix D or Table 5.12, the required probability is

$$P(\text{at most } 2) = P(0 \text{ or } 1 \text{ or } 2) = P(x = 0) + P(x = 1) + P(x = 2)$$
$$= .2621 + .3932 + .2458 = \mathbf{.9011}$$

(c)   The probability that at least three women in this sample exercise or work at being fit every day is given by the sum of the probabilities of 3, 4, 5, or 6 women. Using Table IV of Appendix D or Table 5.12,

$$P(\text{at least } 3) = P(3 \text{ or } 4 \text{ or } 5 \text{ or } 6)$$
$$= P(x = 3) + P(x = 4) + P(x = 5) + P(x = 6)$$
$$= .0819 + .0154 + .0015 + .0001 = \mathbf{.0989}$$

(d)   The probability that one to three women in this sample exercise or work at being fit every day is given by the sum of the probabilities $x = 1, 2,$ or 3. Using Table IV of Appendix D or Table 5.12,

$$P(1 \text{ to } 3) = P(x = 1) + P(x = 2) + P(x = 3)$$
$$= .3932 + .2458 + .0819 = \mathbf{.7209}$$

(e)   Using Table IV of Appendix D or Table 5.12, we list the probability distribution of x for $n = 6$ and $p = .20$ in Table 5.13. Figure 5.6 shows the line graph of the probability distribution of x.

Table 5.13   Probability
Distribution of x for
$n = 6$ and $p = .20$

x	$P(x)$
0	.2621
1	.3932
2	.2458
3	.0819
4	.0154
5	.0015
6	.0001

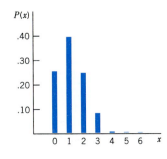

Figure 5.6   Line graph for the probability distribution of x.

5.6.4 PROBABILITY OF SUCCESS AND THE SHAPE OF THE BINOMIAL DISTRIBUTION

For any number of trials n:

1. The binomial probability distribution is symmetric if $p = .50$.
2. The binomial probability distribution is skewed to the right if p is less than .50.
3. The binomial probability distribution is skewed to the left if p is greater than .50.

These three cases are illustrated next with examples and graphs.

1. Let $n = 4$ and $p = .50$. Using Table IV of Appendix D, we have written the probability distribution of x in Table 5.14 and plotted it in Figure 5.7. As we can observe from Table 5.14 and Figure 5.7, the probability distribution of x is symmetric.

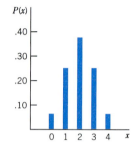

Table 5.14 Probability Distribution of x for $n = 4$ and $p = .50$

x	$P(x)$
0	.0625
1	.2500
2	.3750
3	.2500
4	.0625

Figure 5.7 Line graph for the probability distribution of Table 5.14.

2. Let $n = 4$ and $p = .30$ (which is less than .50). Table 5.15, which is written by using Table IV of Appendix D, and the graph of the probability distribution in Figure 5.8 show that the probability distribution of x for $n = 4$ and $p = .30$ is skewed to the right.

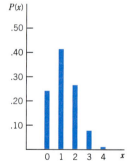

Table 5.15 Probability Distribution of x for $n = 4$ and $p = .30$

x	$P(x)$
0	.2401
1	.4116
2	.2646
3	.0756
4	.0081

Figure 5.8 Line graph for the probability distribution of Table 5.15.

3. Let $n = 4$ and $p = .80$ (which is greater than .50). Table 5.16, which is written by using Table IV of Appendix D, and the graph of the probability distribution in Figure 5.9 show that the probability distribution of x for $n = 4$ and $p = .80$ is skewed to the left.

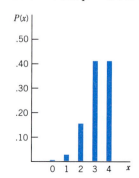

Table 5.16 Probability Distribution of x for $n = 4$ and $p = .80$

x	$P(x)$
0	.0016
1	.0256
2	.1536
3	.4096
4	.4096

Figure 5.9 Line graph for the probability distribution of Table 5.16.

5.6.5 MEAN AND STANDARD DEVIATION OF THE BINOMIAL DISTRIBUTION

Sections 5.3 and 5.4 explained how to compute the mean and standard deviation, respectively, for a probability distribution of a discrete random variable. When a discrete random variable has a binomial distribution, the formulas learned in Sections 5.3 and 5.4 could still be used to compute its mean and standard deviation. However, it is simpler and more convenient to use the following formulas to find the mean and standard deviation in such cases.

> **MEAN AND STANDARD DEVIATION OF A BINOMIAL DISTRIBUTION** The *mean and standard deviation of a binomial distribution* are
>
> $$\mu = np \quad \text{and} \quad \sigma = \sqrt{npq}$$
>
> where n is the total number of trials, p is the probability of success, and q is the probability of failure.

Example 5–22 describes the calculation of the mean and standard deviation of a binomial distribution.

Calculating the mean and standard deviation of a binomial random variable.

EXAMPLE 5-22 According to a study by the accounting consultant firm RHI Management Resources, 25% of chief financial officers said that their favorite luxury car is Lexus (*USA TODAY*, May 13, 1999). Assume that this result holds true for the current population of all U.S. chief financial officers. A sample of 40 chief financial officers is selected. Let x denote the number of chief financial officers in this sample who hold this view. Find the mean and standard deviation of the probability distribution of x.

Solution This is a binomial experiment with a total of 40 trials. Each trial has two outcomes: (1) the selected chief financial officer holds the view that his or her favorite luxury car is Lexus, or (2) the selected chief financial officer does not hold this view. (Note that the second outcome includes the possibility of a chief financial officer saying that his or her favorite luxury car is not a Lexus or having no opinion.) The probabilities p and q for these two outcomes are .25 and .75, respectively. Thus,

$$n = 40, \quad p = .25, \quad \text{and} \quad q = .75$$

Using the formulas for the mean and standard deviation of the binomial distribution, we obtain

$$\mu = np = 40(.25) = \mathbf{10}$$

$$\sigma = \sqrt{npq} = \sqrt{(40)(.25)(.75)} = \mathbf{2.739}$$

Thus, the mean of the probability distribution of x is 10 and the standard deviation is 2.739. The value of the mean is what we expect to obtain, on average, per repetition of the experiment. In this example, if we take many samples of 40 chief financial officers, we expect that each sample will contain an average of 10 chief financial officers, with a standard deviation of 2.739, who would say that their favorite luxury car is Lexus. ∎

EXERCISES

■ **Concepts and Procedures**

5.49 Briefly explain the following.

a. A binomial experiment   **b.** A trial   **c.** A binomial random variable

5.50   What are the parameters of the binomial probability distribution and what do they mean?

5.51   Which of the following are binomial experiments? Explain why.

a.   Rolling a die many times and observing the number of spots
b.   Rolling a die many times and observing whether the number obtained is even or odd
c.   Selecting a few voters from a very large population of voters and observing whether or not each of them favors a certain proposition in an election when 54% of all voters are known to be in favor of this proposition.

5.52   Which of the following are binomial experiments? Explain why.

a.   Drawing 3 balls with replacement from a box that contains 10 balls, 6 of which are red and 4 blue, and observing the colors of the drawn balls
b.   Drawing 3 balls without replacement from a box that contains 10 balls, 6 of which are red and 4 blue, and observing the colors of the drawn balls
c.   Selecting a few households from New York City and observing whether or not they own stocks when it is known that 28% of all households in New York City own stocks

5.53   Let x be a discrete random variable that possesses a binomial distribution. Using the binomial formula, find the following probabilities.

a. $P(x = 5)$ for $n = 8$ and $p = .70$
b. $P(x = 3)$ for $n = 4$ and $p = .40$
c. $P(x = 2)$ for $n = 6$ and $p = .30$

Verify your answers by using Table IV of Appendix D.

5.54   Let x be a discrete random variable that possesses a binomial distribution. Using the binomial formula, find the following probabilities.

a. $P(x = 0)$ for $n = 5$ and $p = .05$
b. $P(x = 4)$ for $n = 7$ and $p = .90$
c. $P(x = 7)$ for $n = 10$ and $p = .60$

Verify your answers by using Table IV of Appendix D.

5.55   Let x be a discrete random variable that possesses a binomial distribution.

a.   Using Table IV of Appendix D, write the probability distribution of x for $n = 7$ and $p = .30$ and graph it.
b.   What are the mean and standard deviation of the probability distribution developed in part a?

5.56   Let x be a discrete random variable that possesses a binomial distribution.

a.   Using Table IV of Appendix D, write the probability distribution of x for $n = 5$ and $p = .80$ and graph it.
b.   What are the mean and standard deviation of the probability distribution developed in part a?

5.57   The binomial probability distribution is symmetric for $p = .50$, skewed to the right for $p < .50$, and skewed to the left for $p > .50$. Illustrate each of these three cases by writing a probability distribution table and drawing a graph. Choose any values of n and p and use the table of binomial probabilities (Table IV of Appendix D) to write the probability distribution tables.

■ **Applications**

5.58   According to an August 1998 nationwide survey of U.S. workers, 59% of the respondents said that they were very concerned about job security for "those currently at work" (Rutgers University's Heldrich Center for Workforce Development and the University of Connecticut's Center for Survey Research & Analysis, *Business Week*, September 21, 1998). Suppose that this result is true for all current U.S. workers.

a.   Let x be a binomial random variable that denotes the number of U.S. workers in a random sample of 10 who hold this view. What are the possible values that x can assume?
b.   Find the probability that exactly six U.S. workers in a random sample of 10 will hold this view.

5.59   A survey of U.S. public school teachers in grades K–12 (ages 21–51) found that 66% of them felt "less valued" today than when they began teaching (Hughes Research for Horace Mann Educators, *USA TODAY*, February 10, 1998). Assume that this percentage is true for the current population of all U.S. teachers.

a. Let *x* be a binomial random variable that denotes the numbers of teachers in a sample of 12 who feel "less valued" today than when they began teaching. What are the possible values that *x* can assume?
b. Find the probability that in a random sample of 12 teachers, exactly 9 will say that they feel "less valued" today than when they began teaching.

5.60 In a 1998 poll, 20% of Americans were satisfied with their household incomes (Roper Starch for *Sunset*, *USA TODAY*, April 6, 1998). Assume that this result is true for the current population of U.S. adults. Using the binomial probabilities table (Table IV of Appendix D), find the probability that the number of Americans in a random sample of 15 who are satisfied with their household incomes is

a. at most 6   **b.** at least 8   **c.** 5 to 8

5.61 Eight hundred U.S. adults responded to a 1998 survey in which they were asked about the measures that should be taken to improve the quality of teachers. According to the results of the survey, 50% of these adults listed "requiring teachers to be tested would be most helpful" as their first or second choice (Association for Supervision and Curriculum Development, *Education Week*, September 23, 1998). Assume that this percentage is true for the current population of all U.S. adults. Using the binomial probabilities table (Table IV of Appendix D), find the probability that the number of adults in a random sample of 15 who would list required testing as their first or second choice is

a. at least 8   **b.** 7 to 10   **c.** at most 5

5.62 According to a 1998 American Management Association survey, 35% of companies use one or more surveillance tactics on employees, such as videotaping workers' performance, reviewing e-mail, taping phone calls, and so on (*USA TODAY*, July 20, 1998). Assume that this percentage is valid for the current population of American companies. Find the probability that in a random sample of eight American companies, the number that use such tactics is

a. exactly 6   **b.** none   **c.** exactly 8

5.63 In a 1998 telephone poll of adult Americans conducted by Yankelovich Partners, Inc., for *Time*/CNN, 37% of adults with health insurance were satisfied with their coverage (*Time*, July 13, 1998). Assuming that this result holds true for the current population of all adult Americans, find the probability that in a random sample of 10 adults who have health insurance, the number who are satisfied with their coverage is

a. exactly 3   **b.** none   **c.** exactly 7

5.64 According to Case Study 4–2 in Chapter 4, the probability that a baseball player will have no hits in 10 trips to the plate is .056, given that this player has a batting average of .250. Using the binomial formula, show that this probability is indeed .056.

5.65 A professional basketball player makes 85% of the free throws he tries. Assuming this percentage will hold true for future attempts, find the probability that in the next eight tries the number of free throws he will make is

a. exactly 8   **b.** exactly 5

5.66 According to Aragon Consulting Group, 20% of adults say that they most often use their vacation time for international travel (*USA TODAY*, July 22, 1997). Assume that this result is true for the current population of all American adults.

a. Using the binomial formula, find the probability that in a random sample of 12 American adults, the number who most often use their vacation time for international travel is
      i. exactly 6   **ii.** none

b. Using the binomial probabilities table, find the probability that in a random sample of 12 American adults, the number who most often use their vacation time for international travel is
      i. at least 6   **ii.** at most 3   **iii.** 2 to 5

5.67 The National Television Violence Study, released in 1998, was based on an analysis of 2750 shows on network, independent, public, and cable television stations. This study claims that 6 out of 10 shows contain violent acts (*Education Week*, April 22, 1998). Assume that this result is true for the current population of all such television shows.

a. Using the binomial formula, find the probability that in a random sample of 10 such shows, the number that contain violent acts is
      i. exactly 4   **ii.** exactly 10

b. Using the binomial probabilities table, find the probability that in a random sample of 10 such programs, the number that contain violent acts is

    i. less than 6   **ii.** more than 5   **iii.** 4 to 7

5.68 Johnson Electronics makes calculators. Consumer satisfaction is one of the top priorities of the company's management. The company guarantees a refund or a replacement for any calculator that malfunctions within 2 years from the date of purchase. It is known from past data that despite all efforts, 5% of the calculators manufactured by the company malfunction within a 2-year period. The company mailed a package of 10 randomly selected calculators to a store.

a. Let x denote the number of calculators in this package of 10 that will be returned for refund or replacement within a 2-year period. Using the binomial probabilities table, obtain the probability distribution of x and draw a graph of the probability distribution. Determine the mean and standard deviation of x.
b. Using the probability distribution of part a, find the probability that exactly 2 of the 10 calculators will be returned for refund or replacement within a 2-year period.

5.69 A fast food chain store conducted a taste survey before marketing a new hamburger. The results of the survey showed that 70% of the people who tried this hamburger liked it. Encouraged by this result, the company decided to market the new hamburger. Assume that 70% of all people like this hamburger. On a certain day, eight customers bought it.

a. Let x denote the number of customers in this sample of eight who will like this hamburger. Using the binomial probabilities table, obtain the probability distribution of x and draw a graph of the probability distribution. Determine the mean and standard deviation of x.
b. Using the probability distribution of part a, find the probability that exactly three of the eight customers will like this hamburger.

5.7 THE POISSON PROBABILITY DISTRIBUTION

The **Poisson probability distribution**, named after the French mathematician Simeon D. Poisson, is another important probability distribution of a discrete random variable that has a large number of applications. Suppose a washing machine in a laundromat breaks down an average of three times a month. We may want to find the probability of exactly two breakdowns during the next month. This is an example of a Poisson probability distribution problem. Each breakdown is called an *occurrence* in Poisson probability distribution terminology. The Poisson probability distribution is applied to experiments with random and independent occurrences. The occurrences are random in the sense that they do not follow any pattern and, hence, they are unpredictable. Independence of occurrences means that one occurrence (or nonoccurrence) of an event does not influence the successive occurrences or nonoccurrences of that event. The occurrences are always considered with respect to an interval. In the example of the washing machine, the interval is one month. The interval may be a time interval, a space interval, or a volume interval. The actual number of occurrences within an interval is random and independent. If the average number of occurrences for a given interval is known, then by using the Poisson probability distribution, we can compute the probability of a certain number of occurrences, x, in that interval. Note that the number of actual occurrences in an interval is denoted by x.

> **CONDITIONS TO APPLY THE POISSON PROBABILITY DISTRIBUTION**   The following three conditions must be satisfied to apply the Poisson probability distribution.
>
> **1.** x is a discrete random variable.
> **2.** The occurrences are random.
> **3.** The occurrences are independent.

The following are three examples of discrete random variables for which the occurrences are random and independent. Hence, these are examples to which the Poisson probability distribution can be applied.

1. Consider the number of telemarketing phone calls received by a household during a given day. In this example, the receiving of a telemarketing phone call by a household is called an occurrence, the interval is one day (an interval of time), and the occurrences are random (that is, there is no specified time for such a phone call to come in). The total number of telemarketing phone calls received by a household during a given day may be 0, 1, 2, 3, 4, ... The independence of occurrences in this example means that the telemarketing phone calls are received individually and none of two (or more) of these phone calls are related.

2. Consider the number of defective items in the next 100 items manufactured on a machine. In this case, the interval is a volume interval (100 items). The occurrences (number of defective items) are random because there may be 0, 1, 2, 3, . . . , 100 defective items in 100 items. We can assume the occurrence of defective items to be independent of one another.

3. Consider the number of defects in a 5-foot-long iron rod. The interval, in this example, is a space interval (5 feet). The occurrences (defects) are random because there may be any number of defects in a 5-foot iron rod. We can assume that these defects are independent of one another.

The following examples also qualify for the application of the Poisson probability distribution.

1. The number of accidents that occur on a given highway during a one-week period
2. The number of customers entering a grocery store during a one-hour interval
3. The number of television sets sold at a department store during a given week

In contrast, consider the arrival of patients at a physician's office. These arrivals are nonrandom if the patients have to make appointments to see the doctor. The arrival of commercial airplanes at an airport is nonrandom because all planes are scheduled to arrive at certain times, and airport authorities know the exact number of arrivals for any period (although this number may change slightly because of late or early arrivals and cancellations). The Poisson probability distribution cannot be applied to these examples.

In the Poisson probability distribution terminology, the average number of occurrences in an interval is denoted by λ (Greek letter *lambda*). The actual number of occurrences in that interval is denoted by x. Then, using the Poisson probability distribution, we find the probability of x occurrences during an interval given that the mean occurrences during that interval are λ.

POISSON PROBABILITY DISTRIBUTION FORMULA According to the *Poisson probability distribution*, the probability of x occurrences in an interval is

$$P(x) = \frac{\lambda^x e^{-\lambda}}{x!}$$

where λ (pronounced *lambda*) is the mean number of occurrences in that interval and the value of e is approximately 2.71828.

The mean number of occurrences in an interval, denoted by λ, is called the *parameter of the Poisson probability distribution* or the **Poisson parameter**. As is obvious from the Poisson probability distribution formula, we need to know only the value of λ to compute the probability of any given value of x. We can read the value of $e^{-\lambda}$ for a given λ from Table V of Appendix D. Examples 5–23 through 5–25 illustrate the use of the Poisson probability distribution formula.

Using the Poisson formula: x equals a specific value.

EXAMPLE 5-23   On average, a household receives 9.5 telemarketing phone calls per week. Using the Poisson distribution formula, find the probability that a randomly selected household receives exactly six telemarketing phone calls during a given week.

Solution   Let λ be the mean number of telemarketing phone calls received by a household per week. Then, $\lambda = 9.5$. Let x be the number of telemarketing phone calls received by a household during a given week. We are to find the probability of $x = 6$. Substituting all the values in the Poisson formula, we obtain

$$P(x = 6) = \frac{\lambda^x e^{-\lambda}}{x!} = \frac{(9.5)^6 e^{-9.5}}{6!} = \frac{(735,091.8906)(.00007485)}{720} = \textbf{.0764}$$

To do these calculations, we can find the value of 6! from Table II of Appendix D, and we can find the value of $e^{-9.5}$ by using the e^x key on a calculator.[4]   ∎

Calculating probabilities using the Poisson formula.

EXAMPLE 5-24   A washing machine in a laundromat breaks down an average of three times per month. Using the Poisson probability distribution formula, find the probability that during the next month this machine will have

(a)   exactly two breakdowns   (b)   at most one breakdown

Solution   Let λ be the mean number of breakdowns per month, and let x be the actual number of breakdowns observed during the next month for this machine. Then,

$$\lambda = 3$$

(a)   The probability that exactly two breakdowns will be observed during the next month is

$$P(x = 2) = \frac{\lambda^x e^{-\lambda}}{x!} = \frac{(3)^2 e^{-3}}{2!} = \frac{(9)(.04978707)}{2} = \textbf{.2240}$$

(b)   The probability that at most one breakdown will be observed during the next month is given by the sum of the probabilities of zero and one breakdown. Thus,

$$P(\text{at most 1 breakdown}) = P(0 \text{ or } 1 \text{ breakdown}) = P(x = 0) + P(x = 1)$$

$$= \frac{(3)^0 e^{-3}}{0!} + \frac{(3)^1 e^{-3}}{1!}$$

$$= \frac{(1)(.04978707)}{1} + \frac{(3)(.04978707)}{1}$$

$$= .0498 + .1494 = \textbf{.1992}$$ ∎

[4]**Using a calculator to simplify the expression for $P(x = 6)$:** In the expression for $P(x = 6)$, you can evaluate $(9.5)^6$ and 6! by using the y^x and $n!$ keys, respectively. If your calculator has the e^x key, you can evaluate $e^{-9.5}$ as follows. First enter 9.5, then press the $+/-$ key, and finally press the e^x key.

☞ *Remember* One important point about the Poisson probability distribution is that *the intervals for λ and x must be equal*. If they are not, the mean λ should be redefined to make them equal. Example 5–25 illustrates this point.

Calculating a probability using the Poisson formula.

EXAMPLE 5–25 Cynthia's Mail Order Company provides free examination of its products for seven days. If not completely satisfied, a customer can return the product within that period and get a full refund. According to past records of the company, an average of 2 of every 10 products sold by this company are returned for a refund. Using the Poisson probability distribution formula, find the probability that exactly 6 of the 40 products sold by this company on a given day will be returned for a refund.

Solution Let x denote the number of products in 40 that will be returned for a refund. We are to find $P(x = 6)$. The given mean is defined per 10 products, but x is defined for 40 products. As a result, we should first find the mean for 40 products. Because, on average, 2 out of 10 products are returned, the mean number of products returned out of 40 will be 8. Thus, $\lambda = 8$. Substituting $x = 6$ and $\lambda = 8$ in the Poisson probability distribution formula, we obtain

$$P(x = 6) = \frac{\lambda^x e^{-\lambda}}{x!} = \frac{(8)^6 e^{-8}}{6!} = \frac{(262{,}144)(.00033546)}{720} = \mathbf{.1221}$$

Thus, the probability is .1221 that exactly 6 products out of 40 sold on a given day will be returned. ∎

Note that Example 5–25 is actually a binomial problem with $p = 2/10 = .20$, $n = 40$, and $x = 6$. In other words, the probability of success (that is, the probability that a product is returned) is .20 and the number of trials (products sold) is 40. We are to find the probability of 6 successes (returns). However, we used the Poisson distribution to solve this problem. This is referred to as *using the Poisson distribution as an approximation to the binomial distribution*. We can also use the binomial distribution to find this probability as follows:

$$P(x = 6) = \binom{40}{6}(.20)^6(.80)^{34} = \frac{40!}{6!\,(40 - 6)!}(.20)^6(.80)^{34}$$

$$= (3{,}838{,}380)(.000064)(.00050706) = \mathbf{.1246}$$

Thus, the probability $P(x = 6)$ is .1246 when we use the binomial distribution.

As we can observe, simplifying the above calculations for the binomial formula is quite complicated when n is large. It is much easier to solve this problem using the Poisson distribution. As a general rule, if it is a binomial problem with $n > 25$ but $\mu \leq 25$, then we can use the Poisson distribution as an approximation to the binomial distribution. However, if $n > 25$ and $\mu > 25$, we prefer to use the normal distribution as an approximation to the binomial. The latter case will be discussed in Chapter 6.

Case Study 5–3 presents applications of the binomial and Poisson probability distributions.

CASE STUDY 5-3 ASK MR. STATISTICS

Fortune magazine frequently publishes a column titled *Ask Mr. Statistics*, which contains questions and answers to statistical problems. The following excerpts are reprinted from one such column.

Dear Oddgiver: I am in the seafood distribution business and find myself endlessly wrangling with supermarkets about appropriate order sizes, especially with high-end tidbit products like our matjes herring in superspiced wine, which we let them have for $4.25, and still they take only a half-dozen jars, thereby running the risk of getting sold out early in the week and causing the better class of customers to storm out empty-handed. How do I get them to realize that lowballing on inventories is usually bad business, also to at least try a few jars of our pickled crappie balls?

—HEADED FOR A BREAKDOWN

Dear Picklehead: The science of statistics has much to offer people puzzled by seafood inventory problems. Your salvation lies in the Poisson distribution, "poisson" being French for fish and, of arguably greater relevance, the surname of a 19th-century French probabilist.

Simeon Poisson's contribution was to develop a method for calculating the likelihood that a specified number of successes will occur given that (a) the probability of success on any one trial is very low but (b) the number of trials is very high. A real world example often mentioned in the literature concerns the distribution of Prussian cavalry deaths from getting kicked by horses in the period 1875–94.

As you would expect of Teutons, the Prussian military kept meticulous records on horse-kick deaths in each of its army corps, and the data are neatly summarized in a 1963 book called *Lady Luck*, by the late Warren Weaver. There were a total of 196 kicking deaths—these being the, er, "successes." The "trials" were each army corps' observations on the number of kicking deaths sustained in the year. So with 14 army corps and data for 20 years, there were 280 trials. We shall not detain you with the Poisson formula, but it predicts, for example, that there will be 34.1 instances of a corps' having exactly two deaths in a year. In fact, there were 32 such cases. Pretty good, eh?

Back to seafood. The Poisson calculation is appropriate to your case, since the likelihood of any one customer's buying your overspiced herring is extremely small, but the number of trials—i.e., customers in the store during a typical week—is very large. Let us say that one customer in 1,000 deigns to buy the herring, and 6,000 customers visit the store in a week. So six jars are sold in an average week.

But the store manager doesn't care about average weeks. What he's worried about is having too much or not enough. He needs to know the probabilities assigned to different sales levels. Our Poisson distribution shows the following morning line: The chance of fewer than three sales—only 6.2%. Of four to six sales: 45.5%. Chances of losing some sales if the store elects to start the week with six jars because that happens to be the average: 39.4%. If the store wants to be 90% sure of not losing sales, it needs to start with nine jars.

There is no known solution to the problem of pickled crappie balls.

Source: Daniel Seligman, "Ask Mr. Statistics," *Fortune*, March 7, 1994. Copyright © 1994, Time Inc. Reprinted with permission. All rights reserved.

Quiz: Using the Poisson probability distribution, calculate the probabilities mentioned at the end of this case study.

5.7.1 USING THE TABLE OF POISSON PROBABILITIES

The probabilities for a Poisson distribution can also be read from Table VI, the table of Poisson probabilities, in Appendix D. The following example describes how to read that table.

Using the table of Poisson probabilities.

EXAMPLE 5-26   On average, two new accounts are opened per day at an Imperial Savings Bank branch. Using Table VI of Appendix D, find the probability that on a given day the number of new accounts opened at this bank will be

(a)   exactly 6   (b)   at most 3   (c)   at least 7

Solution   Let

$$\lambda = \text{mean number of new accounts opened per day at this bank}$$

$$x = \text{number of new accounts opened at this bank on a given day}$$

(a)   The values of λ and x are

$$\lambda = 2 \quad \text{and} \quad x = 6$$

In Table VI of Appendix D, we first locate the column that corresponds to $\lambda = 2$. In this column, we then read the value for $x = 6$. The relevant portion of that table is shown here as Table 5.17. The probability that exactly 6 new accounts will be opened on a given day is .0120. Therefore,

$$P(x = 6) = \textbf{.0120}$$

Table 5.17   Portion of Table VI for $\lambda = 2.0$

			λ		
x	1.1	1.2	. . .	**2.0**	$\longleftarrow \lambda = 2.0$
0				.1353	
1				.2707	
2				.2707	
3				.1804	
4				.0902	
5				.0361	
$x = 6 \longrightarrow$ 6				.0120	$\longleftarrow P(x = 6)$
7				.0034	
8				.0009	
9				.0002	

Actually, Table 5.17 gives the probability distribution of x for $\lambda = 2.0$. Note that the sum of the 10 probabilities given in Table 5.17 is .9999 and not 1.0. This is so for two reasons. First, these probabilities are rounded to four decimal places. Second, on a given day more than 9 new accounts might be opened at this bank. However, the probabilities of 10, 11, 12, . . . new accounts are very small and they are not listed in the table.

(b)   The probability that at most three new accounts are opened on a given day is obtained by adding the probabilities of 0, 1, 2, and 3 new accounts. Thus, using Table VI of Appendix D or Table 5.17, we obtain

$$P(\text{at most 3}) = P(x = 0) + P(x = 1) + P(x = 2) + P(x = 3)$$

$$= .1353 + .2707 + .2707 + .1804 = \textbf{.8571}$$

(c) The probability that at least 7 new accounts are opened on a given day is obtained by adding the probabilities of 7, 8, and 9 new accounts. Note that 9 is the last value of x for $\lambda = 2.0$ in Table VI of Appendix D or Table 5.17. Hence, 9 is the last value of x whose probability is included in the sum. However, this does not mean that on a given day more than 9 new accounts cannot be opened. It simply means that the probability of 10 or more accounts is close to zero. Thus,

$$P(\text{at least } 7) = P(x = 7) + P(x = 8) + P(x = 9)$$

$$= .0034 + .0009 + .0002 = \mathbf{.0045}$$

Constructing a Poisson probability distribution and graphing it.

EXAMPLE 5-27   An auto salesperson sells an average of .9 car per day. Let x be the number of cars sold by this salesperson on any given day. Using the Poisson probability distribution table, write the probability distribution of x. Draw a graph of the probability distribution.

Solution   Let λ be the mean number of cars sold per day by this salesperson. Hence, $\lambda = .9$. Using the portion of Table VI corresponding to $\lambda = .9$, we write the probability distribution of x in Table 5.18. Figure 5.10 shows the line graph for the probability distribution of Table 5.18.

Table 5.18   Probability Distribution of x for $\lambda = .9$

x	$P(x)$
0	.4066
1	.3659
2	.1647
3	.0494
4	.0111
5	.0020
6	.0003

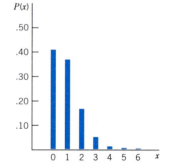

Figure 5.10   Line graph for the probability distribution of Table 5.18.

Note that 6 is the largest value of x for $\lambda = .9$ listed in Table VI for which the probability is greater than zero. However, this does not mean that this salesperson cannot sell more than six cars on a given day. What this means is that the probability of selling seven or more cars is very small. Actually, the probability of $x = 7$ for $\lambda = .9$ calculated by using the Poisson formula is .000039. When rounded to four decimal places, this probability is .0000, as listed in Table VI.

5.7.2   MEAN AND STANDARD DEVIATION OF THE POISSON PROBABILITY DISTRIBUTION

For the Poisson probability distribution, the mean and variance both are equal to λ, and the standard deviation is equal to $\sqrt{\lambda}$. That is, for the Poisson probability distribution,

$$\mu = \lambda$$

$$\sigma^2 = \lambda$$

$$\sigma = \sqrt{\lambda}$$

For Example 5–27, $\lambda = .9$. Therefore, for the probability distribution of x in Table 5.18, the mean, variance, and standard deviation are

$$\mu = \lambda = .9 \text{ car}$$

$$\sigma^2 = \lambda = .9$$

$$\sigma = \sqrt{\lambda} = \sqrt{.9} = .949 \text{ car}$$

EXERCISES

■ **Concepts and Procedures**

5.70 What are the conditions that must be satisfied to apply the Poisson probability distribution?

5.71 What is the parameter of the Poisson probability distribution, and what does it mean?

5.72 Using the Poisson formula, find the following probabilities.

a. $P(x \leq 1)$ for $\lambda = 5$ **b.** $P(x = 2)$ for $\lambda = 2.5$

Verify these probabilities using Table VI of Appendix D.

5.73 Using the Poisson formula, find the following probabilities.

a. $P(x < 2)$ for $\lambda = 3$ **b.** $P(x = 8)$ for $\lambda = 5.5$

Verify these probabilities using Table VI of Appendix D.

5.74 Let x be a Poisson random variable. Using the Poisson probabilities table, write the probability distribution of x for each of the following. Find the mean, variance, and standard deviation for each of these probability distributions. Draw a graph for each of these probability distributions.

a. $\lambda = 1.3$ **b.** $\lambda = 2.1$

5.75 Let x be a Poisson random variable. Using the Poisson probabilities table, write the probability distribution of x for each of the following. Find the mean, variance, and standard deviation for each of these probability distributions. Draw a graph for each of these probability distributions.

a. $\lambda = .6$ **b.** $\lambda = 1.8$

■ **Applications**

5.76 A mail-order company receives an average of 7.4 orders per day. Find the probability that it will receive exactly 11 orders on a certain day. Use the Poisson formula.

5.77 A commuter airline receives an average of 9.7 complaints per day from its passengers. Using the Poisson formula, find the probability that on a certain day this airline will receive exactly six complaints.

5.78 On average, 5.4 shoplifting incidents occur per week at an electronics store. Find the probability that exactly three such incidents will occur during a given week at this store.

5.79 On average, 12.5 rooms stay vacant per day at a large hotel in a city. Find the probability that on a given day exactly three rooms will be vacant. Use the Poisson formula.

5.80 An average of 3.2 employees of a telephone company are absent per day.

a. Find the probability that at most one employee will be absent on a given day at this company.

b. Find the probability that on a given day the number of employees who will be absent at this company is

　　i.　1 to 5　　ii.　at least 7　　iii.　at most 3

5.81 A large proportion of small businesses in the United States fail during the first few years of operation. On average, 1.6 businesses file for bankruptcy per day in a large city.

a. Using the Poisson formula, find the probability that exactly three businesses will file for bankruptcy on a given day in this city.

b. Using the Poisson probabilities table, find the probability that the number of businesses that will file for bankruptcy on a given day in this city is
    i.   2 to 3   ii.   more than 3   iii.   less than 3

5.82 Despite all efforts by the quality control department, the fabric made at Benton Corporation always contains a few defects. A certain type of fabric made at this corporation contains an average of .5 defect per 500 yards.

a. Using the Poisson formula, find the probability that a given piece of 500 yards of this fabric will contain exactly one defect.

b. Using the Poisson probabilities table, find the probability that the number of defects in a given 500-yard piece of this fabric will be
    i.   2 to 4   ii.   more than 3   iii.   less than 2

5.83 The reception office at Tom's Building Corporation receives an average of 4.7 phone calls per half hour.

a. Using the Poisson formula, find the probability that exactly six phone calls will be received at this office during a certain hour.

b. Using the Poisson probabilities table, find the probability that the number of phone calls received at this office during a certain hour will be
    i.   less than 8   ii.   more than 12   iii.   5 to 8

5.84 An average of 4.8 customers come to Columbia Savings and Loan every half hour.

a. Find the probability that exactly two customers will come to this savings and loan during a given hour.

b. Find the probability that during a given hour, the number of customers who will come to this savings and loan is
    i.   2 or fewer   ii.   10 or more

5.85 A certain newspaper contains an average of 1.1 typographical errors per page.

a. Using the Poisson formula, find the probability that a randomly selected page of this newspaper will contain exactly 4 typographical errors.

b. Using the Poisson probabilities table, find the probability that the number of typographical errors on a randomly selected page will be
    i.   more than 3   ii.   less than 4

5.86 An insurance salesperson sells an average of 1.4 insurance policies per day.

a. Using the Poisson formula, find the probability that this salesperson will sell no insurance policy on a certain day.

b. Let x denote the number of insurance policies that this salesperson will sell on a given day. Using the Poisson probabilities table, write the probability distribution of x.

c. Find the mean, variance, and standard deviation of the probability distribution developed in part b.

5.87 An average of .8 accident occurs per day in a large city.

a. Find the probability that no accident will occur in this city on a given day.

b. Let x denote the number of accidents that will occur in this city on a given day. Write the probability distribution of x.

c. Find the mean, variance, and standard deviation of the probability distribution developed in part b.

5.88 On average, 20 households in 50 own answering machines.

a. Using the Poisson formula, find the probability that in a random sample of 50 households, exactly 25 will own answering machines.

b. Using the Poisson probabilities table, find the probability that the number of households in 50 who own answering machines is
    i.   at most 12   ii.   13 to 17   iii.   at least 30

*5.89 Fifteen percent of the students who take a standardized test fail.

a. Using the Poisson formula, find the probability that in a random sample of 100 students who took this test, exactly 20 will fail.

b. Using the Poisson probabilities table, find the probability that the number of students who fail this test in a randomly selected 100 examinees is

    i.   at most 9   ii.   10 to 16   iii.   at least 20

GLOSSARY

Bernoulli trial   One repetition of a binomial experiment. Also called a *trial*.

Binomial experiment   An experiment that contains n identical trials such that each of these n trials has only two possible outcomes, the probabilities of these two outcomes remain constant for each trial, and the trials are independent.

Binomial parameters   The total trials n and the probability of success p for the binomial probability distribution.

Binomial probability distribution   The probability distribution that gives the probability of x successes in n trials when the probability of success is p for each trial of a binomial experiment.

Combinations   The number of ways x elements can be selected from n elements.

Continuous random variable   A random variable that can assume any value in one or more intervals.

Discrete random variable   A random variable whose values are countable.

Factorial   Denoted by the symbol !. The product of all the integers from a given number to 1. For example, $n!$ (read as "n factorial") represents the product of all the integers from n to 1.

Mean of a discrete random variable   The mean of a discrete random variable x is the value that is expected to occur per repetition, on average, if an experiment is performed a large number of times. The mean of a discrete random variable is also called its *expected value*.

Poisson parameter   The average occurrences, denoted by λ, during an interval for a Poisson probability distribution.

Poisson probability distribution   The probability distribution that gives the probability of x occurrences in an interval when the average occurrences in that interval are λ.

Probability distribution of a discrete random variable   A list of all the possible values that a discrete random variable can assume and their corresponding probabilities.

Random variable   A variable, denoted by x, whose value is determined by the outcome of a random experiment. Also called a *chance variable*.

Standard deviation of a discrete random variable   A measure of spread for the probability distribution of a discrete random variable.

KEY FORMULAS

1. **Mean of a discrete random variable x**

$$\mu = \Sigma x P(x)$$

The mean of a discrete random variable x is also called its expected value and is denoted by $E(x)$.

2. **Standard deviation of a discrete random variable x**

$$\sigma = \sqrt{\Sigma x^2 P(x) - \mu^2}$$

3. **Factorials**

$$n! = n(n-1)(n-2) \cdots 3 \cdot 2 \cdot 1$$

4. **Number of combinations of *n* items selected *x* at a time**

$$\binom{n}{x} = \frac{n!}{x!(n-x)!}$$

5. **Binomial probability formula**

$$P(x) = \binom{n}{x} p^x q^{n-x}$$

6. **Mean of the binomial probability distribution**

$$\mu = np$$

7. **Standard deviation of the binomial probability distribution**

$$\sigma = \sqrt{npq}$$

8. **Poisson probability distribution formula**

$$P(x) = \frac{\lambda^x e^{-\lambda}}{x!}$$

9. **Mean of the Poisson probability distribution**

$$\mu = \lambda$$

10. **Variance of the Poisson probability distribution**

$$\sigma^2 = \lambda$$

11. **Standard deviation of the Poisson probability distribution**

$$\sigma = \sqrt{\lambda}$$

SUPPLEMENTARY EXERCISES

5.90 Let *x* be the number of cars that a randomly selected auto mechanic repairs on a given day. The following table lists the probability distribution of *x*.

x	2	3	4	5	6
P(x)	.05	.22	.40	.23	.10

Find the mean and standard deviation of *x*. Give a brief interpretation of the value of the mean.

5.91 Let *x* be the number of shopping trips made during a given week by a randomly selected family from a city. The following table lists the probability distribution of *x*.

x	0	1	2	3	4	5
P(x)	.08	.24	.32	.18	.14	.04

Calculate the mean and standard deviation of *x*. Give a brief interpretation of the value of the mean.

5.92 Based on its analysis of the future demand for its products, the financial department at Tipper Corporation has determined that there is a .17 probability that the company will lose $1.2 million during the next year, a .21 probability that it will lose $.7 million, a .37 probability that it will make a profit of $.9 million, and a .25 probability that it will make a profit of $2.3 million.

a. Let *x* be a random variable that denotes the profit earned by this corporation during the next year. Write the probability distribution of *x*.

b. Find the mean and standard deviation of the probability distribution of part a. Give a brief interpretation of the value of the mean.

5.93 GESCO Insurance Company charges a $350 premium per annum for a $100,000 life insurance policy for a 40-year-old female. The probability that a 40-year-old female will die within one year is .002.

a. Let x be a random variable that denotes the gain of the company for next year from a $100,000 life insurance policy sold to a 40-year-old female. Write the probability distribution of x.

b. Find the mean and standard deviation of the probability distribution of part a. Give a brief interpretation of the value of the mean.

5.94 Spoke Weaving Corporation has eight weaving machines of the same kind and of the same age. The probability is .04 that any weaving machine will break down at any time. Find the probability that at any given time

a. all eight weaving machines will be broken down

b. exactly two weaving machines will be broken down

c. none of the weaving machines will be broken down

5.95 At the Bank of California, past data show that 8% of all credit card holders default at some time in their lives. On one recent day, this bank issued 12 credit cards to new customers. Find the probability that of these 12 customers, eventually

a. exactly 3 will default

b. exactly 1 will default

c. none will default

5.96 Maine Corporation buys motors for electric fans from another company that guarantees that at most 5% of its motors are defective and that it will replace all defective motors at no cost to Maine Corporation. The motors are received in large shipments. The quality control department at Maine Corporation randomly selects 20 motors from each shipment and inspects them for being good or defective. If this sample contains more than 2 defective motors, the entire shipment is rejected.

a. Using the appropriate probabilities table from Appendix D, find the probability that a given shipment of motors received by Maine Corporation will be accepted. Assume that 5% of all motors received by Maine Corporation are defective.

b. Using the appropriate probabilities table from Appendix D, find the probability that a given shipment of motors received by Maine Corporation will be rejected.

5.97 One of the toys made by Dillon Corporation is called Speaking Joe, which is sold only by mail. Consumer satisfaction is one of the top priorities of the company's management. The company guarantees a refund or a replacement for any Speaking Joe toy if the chip that is installed inside becomes defective within 1 year from the date of purchase. It is known from past data that 10% of these chips become defective within a 1-year period. The company sold 15 Speaking Joes on a given day.

a. Let x denote the number of Speaking Joes in these 15 that will be returned for a refund or a replacement within a 1-year period. Using the appropriate probabilities table from Appendix D, obtain the probability distribution of x and draw a graph of the probability distribution. Determine the mean and standard deviation of x.

b. Using the probability distribution constructed in part a, find the probability that exactly 5 of the 15 Speaking Joes will be returned for a refund or a replacement within a 1-year period.

5.98 An average of 10 videos are rented per hour at a video-rental store.

a. Using the appropriate formula, find the probability that during a given hour exactly five videos will be rented at this store.

b. Using the appropriate probabilities table from Appendix D, find the probability that during a given hour the number of videos that will be rented at this store is
    i. at least 10   ii. at most 4   iii. 6 to 9

5.99 An average of 6.3 robberies occur per day in a large city.

a. Using the Poisson formula, find the probability that on a given day exactly three robberies will occur in this city.

b. Using the appropriate probabilities table from Appendix D, find the probability that on a given day the number of robberies that will occur in this city is
    i. at least 12   ii. at most 3   iii. 2 to 6

5.100 An average of 1.4 private airplanes arrive per hour at an airport.

a. Find the probability that during a given hour no private airplane will arrive at this airport.

b. Let x denote the number of private airplanes that will arrive at this airport during a given hour. Write the probability distribution of x.

5.101 A machine produces an average of .8 item per minute.

a. Using the appropriate formula, find the probability that during a certain minute this machine will produce exactly three items.

b. Let x denote the number of items that this machine will produce during a given minute. Using the appropriate probabilities table from Appendix D, write the probability distribution of x.

■ Advanced Exercises

5.102 Scott offers you the following game: You will roll two fair dice. If the sum of the two numbers obtained is 2, 3, 4, 9, 10, 11, or 12, Scott will pay you $20. However, if the sum of the two numbers is 5, 6, 7, or 8, you will pay Scott $20. Scott points out that you have seven winning numbers and only four losing numbers. Is this game fair to you? Should you accept this offer? Support your conclusion with appropriate calculations.

5.103 Suppose the owner of a salvage company is considering raising a sunken ship. If successful, the venture will yield a net profit of $10 million. Otherwise, the owner will lose $4 million. Let p denote the probability of success for this venture. Assume the owner is willing to take the risk to go ahead with this project provided the expected net profit is at least $500,000. What is the smallest value of p for which the owner will risk undertaking this project?

5.104 Two teams, A and B, will play a best-of-seven series, which will end as soon as one of the teams wins four games. Thus, the series may end in four, five, six, or seven games. Assume that each team has an equal chance of winning each game and that all games are independent of one another. Find the following probabilities.

a. Team A wins the series in four games.

b. Team A wins the series in five games.

c. Seven games are required for a team to win the series.

5.105 York Steel Corporation produces a special bearing that must meet rigid specifications. When the production process is running properly, 10% of the bearings fail to meet the required specifications. Sometimes problems develop with the production process that cause the rejection rate to exceed 10%. To guard against this higher rejection rate, samples of 15 bearings are taken periodically and carefully inspected. If more than two bearings in a sample of 15 fail to meet the required specifications, the production is suspended for necessary adjustments.

a. If the true rate of rejection is 10% (that is, the production process is working properly), what is the probability that the production will be suspended based on a sample of 15 bearings?

b. What assumptions did you make in part a?

5.106 Residents in an inner-city area are concerned about drug dealers entering their neighborhood. Over the past 14 nights, they have taken turns watching the street from a darkened apartment. Drug deals seem to take place randomly at various times and locations on the street and average about three per night. The residents of this street contacted the local police, who informed them that they do not have sufficient resources to set up surveillance. The police suggested videotaping the activity on the street, and if the residents are able to capture five or more drug deals on tape, the police will take action. Unfortunately, none of the residents on this street owns a video camera and, hence, they would have to rent the equipment. Inquiries at the local dealers indicated that the best available rate for renting a video camera is $75 for the first night and $40 for each additional night. To obtain this rate, the residents must sign up in advance for a specified number of nights. The residents hold a neighborhood meeting and invite you to help them decide on the length of the rental period. Since it is difficult for them to pay the rental fees, they want to know the probability of taping at least five drug deals on a given number of nights of videotaping.

a. Which of the probability distributions you have studied might be helpful here?

b. What assumption(s) would you have to make?

c. If the residents tape for two nights, what is the probability they will film at least five drug deals?

d. For how many nights must the camera be rented so that there is at least .90 probability that five or more drug deals will be taped?

5.107 A high school history teacher gives a 50-question multiple-choice examination in which each question has four choices. The scoring includes a penalty for guessing. Each correct answer is worth 1 point, and each wrong answer costs 1/2 point. For example, if a student answers 35 questions correctly, 8 questions incorrectly, and does not answer 7 questions, the total score for this student will be $35 - (1/2)(8) = 31$.

a. What is the expected score of a student who answers 38 questions correctly and guesses on the other 12 questions? Assume the student randomly chooses one of the four answers for each of the 12 guessed questions.

b. Does a student increase his expected score by guessing on a question if he has no idea what the correct answer is? Explain.

c. Does a student increase her expected score by guessing on a question for which she can eliminate one of the wrong answers? Explain.

5.108 A baker who makes fresh cheesecakes daily sells an average of five such cakes per day. How many cheesecakes should he make each day so that the probability of running out and losing one or more sales is less than .10? Assume that the number of cheesecakes sold each day follows a Poisson probability distribution. You may use the Poisson probabilities table from Appendix D.

5.109 Suppose that a certain casino has the "money wheel" game. The money wheel is divided into 50 sections, and the wheel has an equal probability of stopping on each of the 50 sections when it is spun. Twenty-two of the sections on this wheel show a $1 bill, 14 show a $2 bill, seven show a $5 bill, three show a $10 bill, two show a $20 bill, one shows a flag, and one shows a joker. A gambler may place a bet on any of the seven possible outcomes. If the wheel stops on the outcome that the gambler bet on, he or she wins. The net payoffs for these outcomes for $1 bets are as follows.

Symbol bet on	$1	$2	$5	$10	$20	Flag	Joker
Payoff (dollars)	1	2	5	10	20	40	40

a. If the gambler bets on the $1 outcome, what is the expected net payoff?

b. Calculate the expected net payoffs for the other six outcomes.

c. Which bet(s) is best in terms of expected net payoff? Which is worst?

5.110 A history teacher has given her class a list of seven essay questions to study before the next test. The teacher announced that she will choose four of the seven questions to give on the test, and each student will have to answer three of those four questions.

a. In how many ways can the teacher choose four questions from the set of seven?

b. Suppose that a student has enough time to study only five questions. In how many ways can the teacher choose four questions from the set of seven so that the four selected questions include both questions that the student did not study?

c. What is the probability that the student in part b will have to answer a question that he or she did not study? That is, what is the probability that the four questions on the test will include both questions that the student did not study?

5.111 Refer to Mini-Project 4–1 in Chapter 4. Using the concepts you learned in Chapter 5, find the expected value of x, the amount won on a $1 bet. Note that you will first have to determine the probability distribution of x, for which the possible values are -1, 1, 2, and 3. Does your result agree with the columnist's assertion that the gambler should expect to lose about 8 cents for every $1 bet over the long run?

5.112 Consider the following three games. Which one would you most like to play? Which one would you least like to play? Explain your answer mathematically.

    Game I: You toss a fair coin once. If a head appears you receive $3, but if a tail appears you have to pay $1.

Game II: You receive a single ticket for a raffle that has a total of 500 tickets. Two tickets are chosen without replacement from the 500. The holder of the first ticket selected receives $300, and the holder of the second ticket selected receives $150.

Game III: You toss a fair coin once. If a head appears you receive $1,000,002, but if a tail appears you have to pay $1,000,000.

5.113   The random variable x represents the number of exams a randomly selected student has on any given day. The probability distribution function of x is given by the following formula:

$$P(x) = \frac{x^2 - x + 2}{8} \qquad \text{for } x = 0, 1, 2$$

a.   Find the probability that a student has at least one exam on a given day.
b.   On average, how many exams do you expect a randomly selected student will have on a given day?
c.   Find the standard deviation of x.

5.114   Brad Henry is a stone products salesman. Let x be the number of contacts he visits on a particular day. The following table gives the probability distribution of x.

x	1	2	3	4
$P(x)$	.12	.25	.56	.07

Let y be the total number of contacts Brad visits on two randomly selected days. Write the probability distribution for y.

5.115   The number of calls that come into a small mail-order company follows a Poisson distribution. Currently, these calls are serviced by a single operator. The manager knows from past experience that an additional operator will be needed if the rate of calls exceeds 20 per hour. The manager observes that 9 calls came into the mail-order company during a randomly selected 15-minute period.

a.   If the rate of calls is actually 20 per hour, what is the probability that 9 or more calls will come in during a given 15-minute period?
b.   If the rate of calls is really 30 per hour, what is the probability that 9 or more calls will come in during a given 15-minute period?
c.   Would you advise the manager to hire a second operator? Explain.

SELF-REVIEW TEST

1.   Briefly explain the meaning of a random variable, a discrete random variable, and a continuous random variable. Give one example each of a discrete and a continuous random variable.

2.   What name is given to a table that lists all the values that a discrete random variable x can assume and their corresponding probabilities?

3.   For the probability distribution of a discrete random variable, the probability of any single value of x is always
    a.   in the range 0 to 1   **b.**   1.0   **c.**   less than zero

4.   For the probability distribution of a discrete random variable, the sum of the probabilities of all possible values of x is always
    a.   greater than 1   **b.**   1.0   **c.**   less than 1.0

5.   The number of combinations of 10 items selected 7 at a time is
    a.   120   **b.**   200   **c.**   80

6.   State the four conditions of a binomial experiment. Give one example of such an experiment.

7. The parameters of the binomial probability distribution are

    a. n, p, and q **b.** n and p **c.** n, p, and x

8. The mean and standard deviation of a binomial probability distribution with $n = 25$ and $p = .20$ are

    a. 5 and 2   **b.** 8 and 4   **c.** 4 and 3

9. The binomial probability distribution is symmetric if

    a. $p < .5$ **b.** $p = .5$ **c.** $p > .5$

10. The binomial probability distribution is skewed to the right if

    a. $p < .5$ **b.** $p = .5$ **c.** $p > .5$

11. The binomial probability distribution is skewed to the left if

    a. $p < .5$ **b.** $p = .5$ **c.** $p > .5$

12. The parameter/parameters of the Poisson probability distribution is/are

    a. λ **b.** λ and x **c.** λ and e

13. Describe the three conditions that must be satisfied to apply the Poisson probability distribution.

14. Let x be the number of homes sold per week by all four real estate agents who work at a Century 21 office. The following table lists the probability distribution of x.

x	0	1	2	3	4	5
$P(x)$	.15	.24	.29	.14	.10	.08

Calculate the mean and standard deviation of x. Give a brief interpretation of the value of the mean.

15. According to a survey, 70% of adults believe that every college student should be required to take at least one course in ethics. Assume that this percentage is true for the current population of all adults.

    a. Find the probability that the number of adults in a random sample of 12 who hold this view is

        i. exactly 10 (use the appropriate formula)

        ii. at least 7 (use the appropriate table from Appendix D)

        iii. less than 4 (use the appropriate table from Appendix D)

    b. Let x be the number of adults in a random sample of 12 who believe that every college student should be required to take at least one course in ethics. Using the appropriate table from Appendix D, write the probability distribution of x. Find the mean and standard deviation of this probability distribution.

16. A department store sells an average of 2.5 electric appliances per day.

    a. Find the probability that on a given day this store will sell

        i. exactly five electric appliances (use the appropriate formula)

        ii. at most four electric appliances (use the appropriate table from Appendix D)

        iii. five to nine electric appliances (use the appropriate table from Appendix D)

    b. Let x be the number of electric appliances sold at this store on a given day. Write the probability distribution of x. Use the appropriate table from Appendix D.

17. The binomial probability distribution is symmetric when $p = .50$, it is skewed to the right when $p < .50$, and it is skewed to the left when $p > .50$. Illustrate these three cases by writing three probability distributions and graphing them. Choose any values of n and p and use the table of binomial probabilities (Table IV of Appendix D).

MINI-PROJECTS

■ **Mini-Project 5–1**

Consider the NBA data given in Data Set III that accompanies this text.

a. What proportion of these players are less than 74 inches tall?

b. Suppose a random sample of 25 of these players is taken and x is the number of players in the sample who are less than 74 inches tall. Find $P(x = 0)$, $P(x = 1)$, $P(x = 2)$, $P(x = 3)$, $P(x = 4)$, and $P(x = 5)$.

c. Note that x in part b has a binomial distribution with $\lambda = np$. Use the Poisson probabilities table of Appendix D to approximate $P(x = 0)$, $P(x = 1)$, $P(x = 2)$, $P(x = 3)$, $P(x = 4)$, and $P(x = 5)$.

d. Are the probability distributions of parts b and c consistent, or is the Poisson approximation inaccurate? Explain why.

■ Mini-Project 5–2

Refer to Mini-Project 4–1 in Chapter 4, which is based on the game Chuck-a-luck. Using the data obtained in the simulation there, calculate the net amount won by the gamblers (the amount paid to the gamblers minus the amount bet) for each of the 50 rolls of the dice. Note that some of these net amounts will be negative. Divide each of these net amounts by 3 to obtain the average net amount won per gambler.

a. Calculate the mean and standard deviation of these average net amounts.

b. Let the random variable x be the net amount won by a gambler on a \$1 bet. Find the probability distribution of x, noting that x has four possible values: 3, 2, 1, and -1.

c. Using the probability distribution of part b, find the mean and standard deviation of x.

d. Compare the theoretical values of the mean and standard deviation of x found in part c with the estimates of the mean and standard deviation in part a based on the simulation.

■ Mini-Project 5–3

Obtain information on the odds and payoffs of one of the instant lottery games in your state. Let the random variable x be the net amount won on one ticket (payoffs minus purchase price). Using the concepts presented in this chapter, find the probability distribution of x. Then calculate the mean and standard deviation of x. What is the player's average net gain (or loss) per ticket purchased?

For instructions on using MINITAB for this chapter, please see Section B.4 in Appendix B.

See Appendix C for *Excel Adventure 5* on discrete probability distributions.

COMPUTER ASSIGNMENTS

CA5.1   Forty-five percent of the adult population in a large city are women. A court is to randomly select a jury of 12 adults from the population of all adults of this city.

a. Using MINITAB or any other statistical software, find the probability that none of the five jurors is a woman.

b. Using the MINITAB CDF command, find the probability that at most 4 of the 12 jurors are women.

c. Let x denote the number of women in 12 adults selected for this jury. Using MINITAB, or any other statistical software, obtain the probability distribution of x.

d. Using the probability distribution obtained in part c, find the following probabilities.
     i. $P(x > 6)$ **ii.** $P(x \leq 3)$ **iii.** $P(2 \leq x \leq 7)$

CA5.2   A survey of U.S. public school teachers (ages 21–51) in grades K–12 found that 66% of them felt "less valued" today than when they began teaching (Hughes Research for Horace Mann Educators, *USA TODAY*, February 10, 1998). Assume that this percentage is true for the current population of all U.S. teachers.

a. Using MINITAB or any other statistical software, find the probability that the number of teachers in a sample of 12 who feel "less valued" today is 8.

b. Using the MINITAB CDF command, find the probability that at most 5 teachers in a sample of 12 feel "less valued."

c. Let x denote the number of teachers in a sample of 12 who feel "less valued." Using MINITAB or any other statistical software, obtain the probability distribution of x.

d. Using the probability distribution obtained in part c, find the following probabilities.

   i. $P(x \geq 7)$ **ii.** $P(x < 6)$ **iii.** $P(3 < x < 9)$

CA5.3 A mail-order company receives an average of 40 orders per day.

a. Using MINITAB or any other statistical software, find the probability that it will receive exactly 55 orders on a certain day.

b. Using the MINITAB CDF command, find the probability that it will receive at most 29 orders on a certain day.

c. Let x denote the number of orders received by this company on a given day. Using MINITAB or any other statistical software, obtain the probability distribution of x.

d. Using the probability distribution obtained in part c, find the following probabilities.

   i. $P(x \geq 45)$ **ii.** $P(x < 33)$ **iii.** $P(36 < x < 52)$

CA5.4 A commuter airline receives an average of 13 complaints per week from its passengers. Let x denote the number of complaints received by this airline during a given week.

a. Using MINITAB or any other statistical software, find $P(x = 0)$. If your answer is zero, does it mean that this cannot happen? Explain.

b. Using the MINITAB CDF command, find $P(x \leq 10)$.

c. Using MINITAB or any other statistical software, obtain the probability distribution of x.

d. Using the probability distribution obtained in part c, find the following probabilities.

   i. $P(x > 18)$ **ii.** $P(x \leq 9)$ **iii.** $P(10 \leq x \leq 17)$

6

CONTINUOUS RANDOM VARIABLES AND THE NORMAL DISTRIBUTION

6.1 **CONTINUOUS PROBABILITY DISTRIBUTION**

CASE STUDY 6–1 **DISTRIBUTION OF TIME TAKEN TO RUN A ROAD RACE**

6.2 **THE NORMAL DISTRIBUTION**

6.3 **THE STANDARD NORMAL DISTRIBUTION**

6.4 **STANDARDIZING A NORMAL DISTRIBUTION**

6.5 **APPLICATIONS OF THE NORMAL DISTRIBUTION**

6.6 **DETERMINING THE z AND x VALUES WHEN AN AREA UNDER THE NORMAL DISTRIBUTION CURVE IS KNOWN**

6.7 **THE NORMAL APPROXIMATION TO THE BINOMIAL DISTRIBUTION**

GLOSSARY

KEY FORMULAS

SUPPLEMENTARY EXERCISES

SELF-REVIEW TEST 286

MINI-PROJECTS

COMPUTER ASSIGNMENTS

Discrete random variables and their probability distributions were presented in Chapter 5. Section 5.1 defined a continuous random variable as a variable that can assume any value in one or more intervals.

The possible values that a continuous random variable can assume are infinite and uncountable. For example, the variable that represents the time it takes a worker to commute from home to work is a continuous random variable. Suppose 5 minutes is the minimum time and 130 minutes is the maximum time taken by all workers to commute from home to work. Let x be a continuous random variable that denotes the time it takes to commute from home to work for a randomly selected worker. Then x can assume any value in the interval 5 to 130 minutes. This interval contains an infinite number of values that are uncountable.

A continuous random variable can possess one of many probability distributions. In this chapter, we discuss the normal probability distribution and the normal distribution as an approximation to the binomial distribution.

6.1 CONTINUOUS PROBABILITY DISTRIBUTION

In Chapter 5, we defined a **continuous random variable** as a random variable whose values are not countable. A continuous random variable can assume any value over an interval or intervals. Because the number of values contained in any interval is infinite, the possible number of values that a continuous random variable can assume is also infinite. Moreover, we cannot count these values. In Chapter 5, it was stated that the life of a battery, heights of persons, time taken to complete an examination, amount of milk in a gallon, weights of babies, and prices of houses are all examples of continuous random variables. Note that although money can be counted, all variables involving money usually are considered to be continuous random variables. This is so because a variable involving money often has a very large number of outcomes.

Suppose 5000 female students are enrolled at a university, and x is the continuous random variable that represents the heights of these female students. Table 6.1 lists the frequency and relative frequency distributions of x.

Table 6.1 Frequency and Relative Frequency Distributions of Heights of Female Students

Height of a Female Student (inches) x	f	Relative Frequency
60 to less than 61	90	.018
61 to less than 62	170	.034
62 to less than 63	460	.092
63 to less than 64	750	.150
64 to less than 65	970	.194
65 to less than 66	760	.152
66 to less than 67	640	.128
67 to less than 68	440	.088
68 to less than 69	320	.064
69 to less than 70	220	.044
70 to less than 71	180	.036
	$N = 5000$	Sum $= 1.0$

The relative frequencies given in Table 6.1 can be used as the probabilities of the respective classes. Note that these are exact probabilities because we are considering the population of all female students.

Figure 6.1 (page 242) displays the histogram and polygon for the relative frequency distribution of Table 6.1. Figure 6.2 shows the smoothed polygon for the data of Table 6.1. The smoothed polygon is an approximation of the *probability distribution curve* of the continuous random variable x. Note that each class in Table 6.1 has a width equal to 1 inch. If the width of classes is more than 1 unit, we first obtain the *relative frequency densities* and then graph these relative frequency densities to obtain the distribution curve. The relative frequency density of a class is obtained by dividing the relative frequency of that class by the class width. The relative frequency densities are calculated to make the sum of the areas of all rectangles in the histogram equal to 1.0. Case Study 6–1, which appears later in this section, illustrates this procedure. The probability distribution curve of a continuous random variable is also called its *probability density function.*

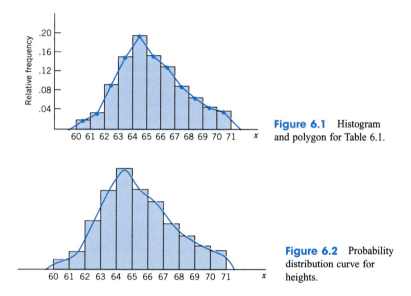

Figure 6.1   Histogram and polygon for Table 6.1.

Figure 6.2   Probability distribution curve for heights.

The probability distribution of a continuous random variable possesses the following *two characteristics.*

1. The probability that x assumes a value in any interval lies in the range 0 to 1.
2. The total probability of all the (mutually exclusive) intervals within which x can assume a value is 1.0.

The first characteristic states that the area under the probability distribution curve of a continuous random variable between any two points is between 0 and 1, as shown in Figure 6.3. The second characteristic indicates that the total area under the probability distribution curve of a continuous random variable is always 1.0 or 100%, as shown in Figure 6.4.

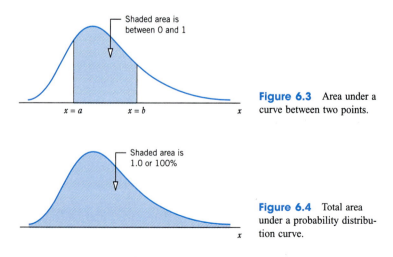

Figure 6.3   Area under a curve between two points.

Figure 6.4   Total area under a probability distribution curve.

The probability that a continuous random variable x assumes a value within a certain interval is given by the area under the curve between the two limits of the interval, as shown in Figure 6.5. The shaded area under the curve from a to b in this figure gives the probability that x falls in the interval a to b; that is,

$$P(a \leq x \leq b) = \text{Area under the curve from } a \text{ to } b$$

Note that the interval $a \leq x \leq b$ states that x is greater than or equal to a but less than or equal to b.

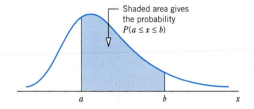

Shaded area gives the probability $P(a \leq x \leq b)$

Figure 6.5 Area under the curve as probability.

Reconsider the example on the heights of all female students at a university. The probability that the height of a randomly selected female student from this university lies in the interval 65 to 68 inches is given by the area under the distribution curve of the heights of all female students from $x = 65$ to $x = 68$, as shown in Figure 6.6. This probability is written as

$$P(65 \leq x \leq 68)$$

which states that x is greater than or equal to 65 but less than or equal to 68.

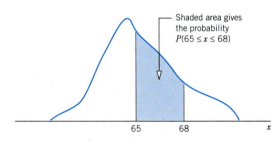

Shaded area gives the probability $P(65 \leq x \leq 68)$

Figure 6.6 Probability that x lies in the interval 65 to 68.

For a continuous probability distribution, the probability is always calculated for an interval. For example, in Figure 6.6, the interval representing the shaded area is from 65 to 68. Consequently, the shaded area in that figure gives the probability for the interval $65 \leq x \leq 68$.

The probability that a continuous random variable x assumes a single value is always zero. This is so because the area of a line, which represents a single point, is zero. For example, if x is the height of a randomly selected female student from that university, then the probability that this student is exactly 67 inches tall is zero; that is,

$$P(x = 67) = 0$$

This probability is shown in Figure 6.7 (page 244). Similarly, the probability for x to assume any other single value is zero.

In general, if a and b are two of the values that x can assume, then

$$P(a) = 0 \quad \text{and} \quad P(b) = 0$$

From this we can deduce that for a continuous random variable,

$$P(a \leq x \leq b) = P(a < x < b)$$

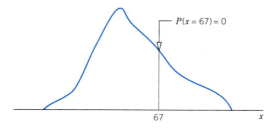

Figure 6.7   The probability of a single value of x is zero.

In other words, the probability that x assumes a value in the interval a to b is the same whether or not the values a and b are included in the interval. For the example on the heights of female students, the probability that a randomly selected female student is between 65 and 68 inches tall is the same as the probability that this female is 65 to 68 inches tall. This is shown in Figure 6.8.

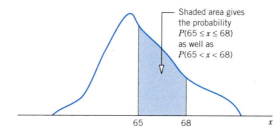

Figure 6.8   Probability "from 65 to 68" and "between 65 and 68."

Note that the interval "between 65 and 68" represents "$65 < x < 68$" and it does not include 65 and 68. On the other hand, the interval "from 65 to 68" represents "$65 \leq x \leq 68$" and it does include 65 and 68. However, as mentioned previously, in the case of a continuous random variable, both of these intervals contain the same probability or area under the curve.

Case Study 6–1 describes how we obtain the probability distribution curve of a continuous random variable.

CASE STUDY 6–1   DISTRIBUTION OF TIME TAKEN TO RUN A ROAD RACE

The accompanying table gives the frequency and relative frequency distributions for the time (in minutes) taken to complete the Manchester Road Race (held on November 26, 1998) by a total of 8821 participants who finished that race. This event is held every year on Thanksgiving Day in Manchester, Connecticut. The total distance of the course is 4.748 miles. The relative frequencies in the table are used to construct the histogram and polygon in Figure 6.9.

Class	Frequency	Relative Frequency
20 to less than 25	37	.0042
25 to less than 30	261	.0296
30 to less than 35	641	.0727
35 to less than 40	1011	.1146
40 to less than 45	1489	.1688
45 to less than 50	1841	.2087
50 to less than 55	1502	.1703
55 to less than 60	811	.0919
60 to less than 65	349	.0396
65 to less than 70	234	.0265
70 to less than 75	186	.0211
75 to less than 80	175	.0198
80 to less than 85	134	.0152
85 to less than 90	100	.0113
90 to less than 95	33	.0037
95 to less than 100	17	.0019
	$\Sigma f = 8821$	Sum $= .9999$

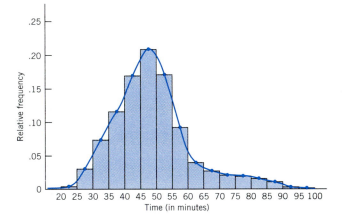

Figure 6.9 Histogram and polygon for the road race data.

To derive the probability distribution curve for these data, we calculate the relative frequency densities by dividing the relative frequencies by the class widths. The width of each class in the table is 5. By dividing the relative frequencies by 5, we obtain the relative frequency densities, which are recorded in the table at the top of page 246. Using the relative frequency densities, we draw a histogram and smoothed polygon, as shown in Figure 6.10. The curve in this figure is the probability distribution curve for the road race data.

Note that the areas of the rectangles in Figure 6.9 do not give probabilities (which are approximated by relative frequencies). Rather, it is the heights of these rectangles that give the probabilities. This is so because the base of each rectangle is 5 in this histogram. Consequently, the area of any rectangle is given by its height multiplied by 5. Thus, the total area of all the rectangles in Figure 6.9 is 5.0, not 1.0. However, in Figure 6.10, it is the areas, not the heights, of rectangles that give the probabilities of the respective classes. Thus, if we add the areas of all the rectangles in Figure 6.10, we obtain the sum of all probabilities

Class	Relative Frequency Density
20 to less than 25	.00084
25 to less than 30	.00592
30 to less than 35	.01453
35 to less than 40	.02292
40 to less than 45	.03376
45 to less than 50	.04174
50 to less than 55	.03406
55 to less than 60	.01839
60 to less than 65	.00791
65 to less than 70	.00531
70 to less than 75	.00422
75 to less than 80	.00397
80 to less than 85	.00304
85 to less than 90	.00227
90 to less than 95	.00075
95 to less than 100	.00039

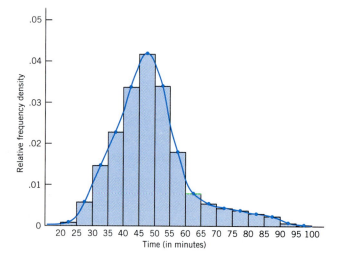

Figure 6.10   Probability distribution curve for the road race data.

equal to .999, which is approximately 1.0. Consequently, the total area under the curve is equal to 1.0.

The probability distribution of a continuous random variable has a mean and a standard deviation, denoted by μ and σ, respectively. The mean and standard deviation of the probability distribution curve of Figure 6.10 are 48.777 and 12.434 minutes, respectively. These values of μ and σ are calculated by using the raw data on 8821 participants.

Source: This case study is based on data published in *The Hartford Courant*, November 30, 1998.

6.2 THE NORMAL DISTRIBUTION

The normal distribution is one of the many probability distributions that a continuous random variable can possess. The normal distribution is the most important and most widely used of all probability distributions. A large number of phenomena in the real world are normally distributed either exactly or approximately. The continuous random variables representing heights and weights of people, scores on an examination, weights of packages (e.g., cereal boxes, boxes of cookies), amount of milk in a gallon, life of an item (such as a lightbulb or a television set), and time taken to complete a certain job have all been observed to have a (approximate) normal distribution.

The **normal probability distribution** or the *normal curve* is a bell-shaped (symmetric) curve. Such a curve is shown in Figure 6.11. Its mean is denoted by μ and its standard deviation by σ. A continuous random variable x that has a normal distribution is called a *normal random variable*. Note that not all bell-shaped curves represent a normal distribution curve. Only a specific kind of bell-shaped curve represents a normal curve.

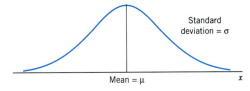

Standard
deviation = σ

Mean = μ x

Figure 6.11 Normal distribution with mean μ and standard deviation σ.

NORMAL PROBABILITY DISTRIBUTION A *normal probability distribution*, when plotted, gives a bell-shaped curve such that

1. The total area under the curve is 1.0.
2. The curve is symmetric about the mean.
3. The two tails of the curve extend indefinitely.

A normal distribution possesses the following three characteristics.

1. The total area under a normal distribution curve is 1.0 or 100%, as shown in Figure 6.12.

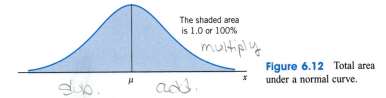

The shaded area
is 1.0 or 100%

multiply

μ x

sub.   add.

Figure 6.12 Total area under a normal curve.

2. A normal distribution curve is symmetric about the mean, as shown in Figure 6.13 (page 248). Consequently, 1/2 of the total area under a normal distribution curve lies on the left side of the mean, and 1/2 lies on the right side of the mean.

3. The tails of a normal distribution curve extend indefinitely in both directions without touching or crossing the horizontal axis. Although a normal distribution curve never meets the horizontal axis, beyond the points represented by $\mu - 3\sigma$ and $\mu + 3\sigma$ it be-

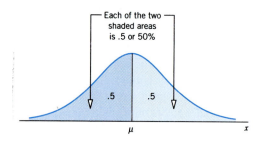

Figure 6.13   A normal curve is symmetric about the mean.

comes so close to this axis that the area under the curve beyond these points in both directions can be taken as virtually zero. These areas are shown in Figure 6.14.

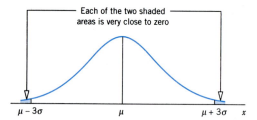

Figure 6.14   Areas of the normal curve beyond $\mu \pm 3\sigma$.

The mean, μ, and the standard deviation, σ, are the *parameters* of the normal distribution. Given the values of these two parameters, we can find the area under a normal distribution curve for any interval. Remember, there is not just one normal distribution curve but rather a *family* of normal distribution curves. Each different set of values of μ and σ gives a different normal distribution. The value of μ determines the center of a normal distribution curve on the horizontal axis, and the value of σ gives the spread of the normal distribution curve. The three normal distribution curves drawn in Figure 6.15 have the same mean but different standard deviations. By contrast, the three normal distribution curves in Figure 6.16 have different means but the same standard deviation.

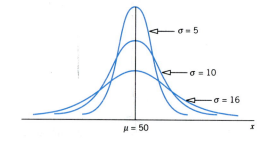

Figure 6.15   Three normal distribution curves with the same mean but different standard deviations.

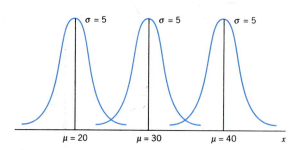

Figure 6.16   Three normal distribution curves with different means but the same standard deviation.

Like the binomial and Poisson probability distributions discussed in Chapter 5, the normal probability distribution can also be expressed by a mathematical equation.[1] However, we will not use this equation to find the area under a normal distribution curve. Instead, we will use Table VII of Appendix D.

6.3 THE STANDARD NORMAL DISTRIBUTION

The **standard normal distribution** is a special case of the normal distribution. For the standard normal distribution, the value of the mean is equal to zero, and the value of the standard deviation is equal to 1.

> **STANDARD NORMAL DISTRIBUTION**   The normal distribution with $\mu = 0$ and $\sigma = 1$ is called the *standard normal distribution.*

Figure 6.17 displays the standard normal distribution curve. The random variable that possesses the standard normal distribution is denoted by z. In other words, the units for the standard normal distribution curve are denoted by z and are called the z **values** or z **scores**. They are also called *standard units* or *standard scores*.

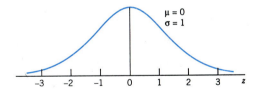

Figure 6.17   The standard normal distribution curve.

> **z VALUES OR z SCORES**   The units marked on the horizontal axis of the standard normal curve are denoted by z and are called the z *values* or z *scores*. A specific value of z gives the distance between the mean and the point represented by z in terms of the standard deviation.

In Figure 6.17, the horizontal axis is labeled z. The z values on the right side of the mean are positive and those on the left side are negative. *The z value for a point on the horizontal axis gives the distance between the mean and that point in terms of the standard deviation.* For example, a point with a value of z = 2 is two standard deviations to the right of

---

[1]The equation of the normal distribution is

$$f(x) = \frac{1}{\sigma\sqrt{2\pi}}\, e^{-(1/2)[(x - \mu)/\sigma]^2}$$

where $e = 2.71828$ and $\pi = 3.14159$ approximately; $f(x)$, called the probability density function, gives the vertical distance between the horizontal axis and the curve at point x. For the information of those who are familiar with integral calculus, the definite integral of this equation from a to b gives the probability that x assumes a value between a and b.

the mean. Similarly, a point with a value of $z = -2$ is two standard deviations to the left of the mean.

The standard normal distribution table, Table VII of Appendix D, lists the areas under the standard normal curve between $z = 0$ and the values of z from .00 to 3.09. To read the standard normal distribution table, we always start at $z = 0$, which represents the mean of the standard normal distribution. We learned earlier that the total area under a normal distribution curve is 1.0. We also learned that, because of symmetry, the area on each side of the mean is .5. This is shown in Figure 6.18.

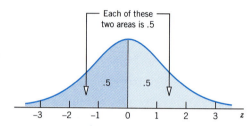

Figure 6.18   Area under the standard normal curve.

☞   *Remember*   Although the values of z on the left side of the mean are negative, the area under the curve is always positive.

The area under the standard normal curve between any two points can be interpreted as the probability that z assumes a value within that interval. Examples 6–1 through 6–4 describe how to read Table VII of Appendix D to find areas under the standard normal curve.

Finding the area between
$z = 0$ and a positive z.

EXAMPLE 6–1   Find the area under the standard normal curve between $z = 0$ and $z = 1.95$.

Solution   We divide the number 1.95 into two portions: 1.9 (the digit before the decimal and one digit after the decimal) and .05 (the second digit after the decimal). (Note that $1.95 = 1.9 + .05$.) To find the required area under the standard normal curve, we locate 1.9 in the column for z on the left side of Table VII and .05 in the row for z at the top of Table VII. The entry where the row for 1.9 and the column for .05 intersect gives the area under the standard normal curve between $z = 0$ and $z = 1.95$. The relevant portion of Table VII is reproduced as Table 6.2. From Table VII or Table 6.2, the entry where the row for 1.9 and the column for .05 cross is .4744. Consequently, the area under the standard normal curve between $z = 0$ and $z = 1.95$ is .4744. This area is shown in Figure 6.19. (It is always helpful to sketch the curve and mark the area we are determining.)

Table 6.2   Area Under the Standard Normal Curve from $z = 0$ to $z = 1.95$

z	.00	.01	...	.05	...	.09
0.0	.0000	.0040	...	.0199	...	.0359
0.1	.0398	.0438	...	.0596	...	.0753
0.2	.0793	.0832	...	.0987	...	.1141
.	.	.	...	.	...	.
.	.	.	...	.	...	.
.	.	.	...	.	...	.
1.9	.4713	.4719	...	.4744 ←	...	.4767
.	.	.	...	.	...	.
.	.	.	...	.	...	.
.	.	.	...	.	...	.
3.0	.4987	.4987	...	.4989	...	.4990

Required area

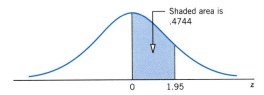

Figure 6.19 Area between $z = 0$ and $z = 1.95$.

The area between $z = 0$ and $z = 1.95$ can be interpreted as the probability that z assumes a value between 0 and 1.95; that is,

$$\text{Area between 0 and 1.95} = P(0 < z < 1.95) = .4744$$

As mentioned in Section 6.1, the probability that a continuous random variable assumes a single value is zero. Therefore,

$$P(z = 0) = 0 \quad \text{and} \quad P(z = 1.95) = 0$$

Hence,

$$P(0 < z < 1.95) = P(0 \le z \le 1.95) = .4744 \qquad \blacksquare$$

Finding the area between a negative z and $z = 0$.

EXAMPLE 6–2 Find the area under the standard normal curve from $z = -2.17$ to $z = 0$.

Solution Because the normal distribution is symmetric about the mean, the area from $z = -2.17$ to $z = 0$ is the same as the area from $z = 0$ to $z = 2.17$, as shown in Figure 6.20.

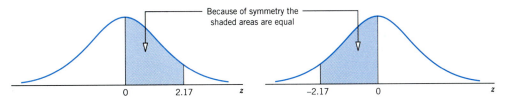

Figure 6.20 Area from $z = 0$ to $z = 2.17$ equals area from $z = -2.17$ to $z = 0$.

To find the area from $z = -2.17$ to $z = 0$, we look for the area from $z = 0$ to $z = 2.17$ in the standard normal distribution table (Table VII). To do so, first we locate 2.1 in the column for z and .07 in the row for z in that table. Then, we read the number at the intersection of the row for 2.1 and the column for .07. The relevant portion of Table VII is reproduced as Table 6.3. As shown in Table 6.3 and Figure 6.21 (page 252), this number is .4850.

Table 6.3 Area Under the Standard Normal Curve from $z = 0$ to $z = 2.17$

z	.00	.01	. . .	.07	. . .	.09
0.0	.0000	.0040	. . .	.0279	. . .	.0359
0.1	.0398	.0438	. . .	.0675	. . .	.0753
0.2	.0793	.0832	. . .	.1064	. . .	.1141
.	.	.	. . .	.	. . .	.
.	.	.	. . .	.	. . .	.
.	.	.	. . .	.	. . .	.
2.1	.4821	.4826	. . .	.4850 ←	. . .	.4857
.	.	.	. . .	.	. . .	.
.	.	.	. . .	.	. . .	.
3.0	.4987	.4987	. . .	.4989	. . .	.4990

Required area

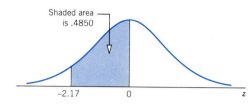

Figure 6.21 Area from $z = -2.17$ to $z = 0$.

The area from $z = -2.17$ to $z = 0$ gives the probability that z lies in the interval -2.17 to 0; that is,

$$\text{Area from } -2.17 \text{ to } 0 = P(-2.17 \leq z \leq 0) = \textbf{.4850} \qquad \blacksquare$$

Finding areas in the right and left tails.

EXAMPLE 6-3 Find the following areas under the standard normal curve.

(a)   Area to the right of $z = 2.32$

(b)   Area to the left of $z = -1.54$

Solution

(a)   As mentioned earlier, to read the normal distribution table, we must start with $z = 0$. To find the area to the right of $z = 2.32$, first we find the area between $z = 0$ and $z = 2.32$. Then we subtract this area from .5, which is the total area to the right of $z = 0$. From Table VII, the area between $z = 0$ and $z = 2.32$ is .4898. Consequently, the required area is $.5 - .4898 = .0102$, as shown in Figure 6.22.

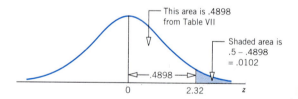

Figure 6.22 Area to the right of $z = 2.32$.

The area to the right of $z = 2.32$ gives the probability that z is greater than 2.32. Thus,

$$\text{Area to the right of } 2.32 = P(z > 2.32) = .5 - .4898 = \textbf{.0102}$$

(b)   To find the area under the standard normal curve to the left of $z = -1.54$, first we find the area between $z = -1.54$ and $z = 0$ and then we subtract this area from .5, which is the total area to the left of $z = 0$. From Table VII, the area between $z = -1.54$ and $z = 0$ is .4382. Hence, the required area is $.5 - .4382 = .0618$. This area is shown in Figure 6.23.

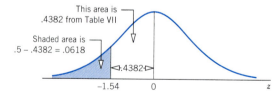

Figure 6.23 Area to the left of $z = -1.54$.

The area to the left of $z = -1.54$ gives the probability that z is less than -1.54. Thus,

$$\text{Area to the left of } -1.54 = P(z < -1.54) = .5 - .4382 = \textbf{.0618} \qquad \blacksquare$$

EXAMPLE 6-4   Find the following probabilities for the standard normal curve.

(a) $P(1.19 < z < 2.12)$ (b) $P(-1.56 < z < 2.31)$ (c) $P(z > -.75)$

Solution

Finding the area between two positive values of z.

(a)   The probability $P(1.19 < z < 2.12)$ is given by the area under the standard normal curve between $z = 1.19$ and $z = 2.12$, which is the shaded area in Figure 6.24.

Both of the points, $z = 1.19$ and $z = 2.12$, are on the same (right) side of $z = 0$. To find the area between $z = 1.19$ and $z = 2.12$, first we find the areas between $z = 0$ and $z = 1.19$ and between $z = 0$ and $z = 2.12$. Then we subtract the smaller area (the area between $z = 0$ and $z = 1.19$) from the larger area (the area between $z = 0$ and $z = 2.12$).

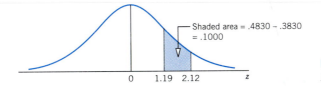

Shaded area = .4830 – .3830
= .1000

Figure 6.24   Finding $P(1.19 < z < 2.12)$.

From Table VII for the standard normal distribution, we find

Area between 0 and 1.19 = .3830

Area between 0 and 2.12 = .4830

Then, the required probability is

$$P(1.19 < z < 2.12) = \text{Area between 1.19 and 2.12}$$

$$= .4830 - .3830 = \mathbf{.1000}$$

☞   *Remember*   As a general rule, when the two points are on the same side of the mean, first find the areas between the mean and each of the two points. Then subtract the smaller area from the larger area.

Finding the area between a positive and a negative value of z.

(b)   The probability $P(-1.56 < z < 2.31)$ is given by the area under the standard normal curve between $z = -1.56$ and $z = 2.31$, which is the shaded area in Figure 6.25.

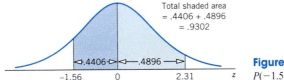

Total shaded area
= .4406 + .4896
= .9302

.4406   .4896

Figure 6.25   Finding $P(-1.56 < z < 2.31)$.

The two points, $z = -1.56$ and $z = 2.31$, are on different sides of $z = 0$. The area between $z = -1.56$ and $z = 2.31$ is obtained by adding the areas between $z = -1.56$ and $z = 0$ and between $z = 0$ and $z = 2.31$.

From Table VII for the standard normal distribution, we have

Area between -1.56 and 0 = .4406

Area between 0 and 2.31 = .4896

The required probability is

$$P(-1.56 < z < 2.31) = \text{Area between } -1.56 \text{ and } 2.31$$

$$= .4406 + .4896 = \textbf{.9302}$$

☞   *Remember*   As a general rule, when the two points are on different sides of the mean, first find the areas between the mean and each of the two points. Then add these two areas.

Finding the area to the right of a negative value of z.

(c)   The probability $P(z > -.75)$ is given by the area under the standard normal curve to the right of $z = -.75$, which is the shaded area in Figure 6.26.

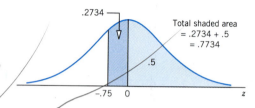

Figure 6.26   Finding $P(z > -.75)$.

The area to the right of $z = -.75$ is obtained by adding the area between $z = -.75$ and $z = 0$ and the area to the right of $z = 0$.

$$\text{Area to the right of } 0 = P(z > 0) = .5$$

From Table VII for the standard normal distribution,

$$\text{Area between } -.75 \text{ and } 0 = .2734$$

The required probability is

$$P(z > -.75) = \text{Area to the right of } -.75$$

$$= .2734 + .5 = \textbf{.7734}$$ ∎

In the discussion in Section 3.4 of Chapter 3 on the use of the standard deviation, we also discussed the empirical rule for a bell-shaped curve. That empirical rule is based on the standard normal distribution table. By using the normal distribution table, we can now verify the empirical rule as follows.

1.   The total area within one standard deviation of the mean is 68.26%. This area is given by the sum of the areas between $z = -1.0$ and $z = 0$ and between $z = 0$ and $z = 1.0$. As shown in Figure 6.27, each of these two areas is .3413 or 34.13%. Consequently, the total area between $z = -1.0$ and $z = 1.0$ is 68.26%.

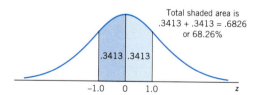

Figure 6.27   Area within one standard deviation of the mean.

2.   The total area within two standard deviations of the mean is 95.44%. This area is given by the sum of the areas between $z = -2.0$ and $z = 0$ and between $z = 0$ and $z = 2.0$. As shown in Figure 6.28, each of these two areas is .4772 or 47.72%. Hence, the total area between $z = -2.0$ and $z = 2.0$ is 95.44%.

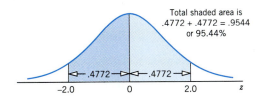

Total shaded area is
.4772 + .4772 = .9544
or 95.44%

Figure 6.28 Area within two standard deviations of the mean.

3. The total area within three standard deviations of the mean is 99.74%. This area is given by the sum of the areas between $z = -3.0$ and $z = 0$ and between $z = 0$ and $z = 3.0$. As shown in Figure 6.29, each of these two areas is .4987 or 49.87%. Therefore, the total area between $z = -3.0$ and $z = 3.0$ is 99.74%.

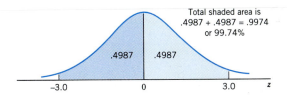

Total shaded area is
.4987 + .4987 = .9974
or 99.74%

Figure 6.29 Area within three standard deviations of the mean.

Again, note that only a specific bell-shaped curve represents the normal distribution. Now we can state that a bell-shaped curve that contains (about) 68.26% of the total area within one standard deviation of the mean, (about) 95.44% of the total area within two standard deviations of the mean, and (about) 99.74% of the total area within three standard deviations of the mean represents a normal distribution curve.

The standard normal distribution table, Table VII of Appendix D, goes up to only $z = 3.09$. In other words, that table can be read only for $z = 0$ to $z = 3.09$ (or to $z = -3.09$). Consequently, if we need to find the area between $z = 0$ and a z value greater than 3.09 (or between a z value less than -3.09 and $z = 0$) under the standard normal curve, we cannot obtain it from the normal distribution table because the table does not contain a z value greater than 3.09. In such cases, the area under the normal distribution curve between $z = 0$ and any z value greater than 3.09 (or less than -3.09) is approximated by .5. From the normal distribution table, the area between $z = 0$ and $z = 3.09$ is .4990. Hence, the area between $z = 0$ and any value of z greater than 3.09 is larger than .4990 and can be approximated by .5. Example 6–5 illustrates this procedure.

EXAMPLE 6–5 Find the following probabilities for the standard normal curve.

(a) $P(0 < z < 5.67)$ (b) $P(z < -5.35)$

Solution

Finding the area between $z = 0$ and a value of z greater than 3.09.

(a) The probability $P(0 < z < 5.67)$ is given by the area under the standard normal curve between $z = 0$ and $z = 5.67$. Because $z = 5.67$ is greater than 3.09 and is not in Table VII, the area under the standard normal curve between $z = 0$ and $z = 5.67$ can be approximated by .5. This area is shown in Figure 6.30.

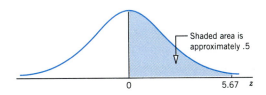

Shaded area is approximately .5

Figure 6.30 Area between $z = 0$ and $z = 5.67$.

The required probability is

$$P(0 < z < 5.67) = \text{Area between 0 and 5.67} = \mathbf{.5} \text{ approximately}$$

Note that the area between $z = 0$ and $z = 5.67$ is not exactly .5 but very close to .5.

Finding the area to the left of a z that is less than −3.09.

(b) The probability $P(z < -5.35)$ represents the area under the standard normal curve to the left of $z = -5.35$. The area between $z = -5.35$ and $z = 0$ is approximately .5. Consequently, the area under the standard normal curve to the left of $z = -5.35$ is approximately zero, as shown in Figure 6.31.

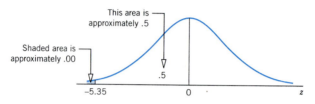

Figure 6.31 Area to the left of $z = -5.35$.

The required probability is

$$P(z < -5.35) = \text{Area to the left of } -5.35$$

$$= .5 - .5 = \mathbf{.00} \text{ approximately}$$

Again, note that the area to the left of $z = -5.35$ is not exactly .00 but very close to .00 ■

EXERCISES

■ **Concepts and Procedures**

6.1 What is the difference between the probability distribution of a discrete random variable and that of a continuous random variable? Explain.

6.2 Let x be a continuous random variable. What is the probability that x assumes a single value, such as a?

6.3 For a continuous probability distribution, why is $P(a < x < b)$ equal to $P(a \le x \le b)$?

6.4 Briefly explain the main characteristics of a normal distribution. Illustrate with the help of graphs.

6.5 Briefly describe the standard normal distribution curve.

6.6 What are the parameters of the normal distribution?

6.7 How do the width and height of a normal distribution change when its mean remains the same but its standard deviation decreases?

6.8 Do the width and/or height of a normal distribution change when its standard deviation remains the same but its mean increases?

6.9 For a standard normal distribution, what does z represent?

6.10 For a standard normal distribution, find the area within one standard deviation of the mean—that is, the area between $\mu - \sigma$ and $\mu + \sigma$.

6.11 For a standard normal distribution, find the area within 1.5 standard deviations of the mean—that is, the area between $\mu - 1.5\sigma$ and $\mu + 1.5\sigma$.

6.12 For a standard normal distribution, what is the area within two standard deviations of the mean?

6.13 For a standard normal distribution, what is the area within 2.5 standard deviations of the mean?

6.14 For a standard normal distribution, what is the area within three standard deviations of the mean?

6.15 Find the area under the standard normal curve

a. between $z = 0$ and $z = 1.95$ b. between $z = 0$ and $z = -1.85$
c. between $z = 1.15$ and $z = 2.37$ d. from $z = -1.53$ to $z = -2.88$
e. from $z = -1.67$ to $z = 2.44$

6.16 Find the area under the standard normal curve

a. from $z = 0$ to $z = 2.34$ b. between $z = 0$ and $z = -2.78$
c. from $z = .84$ to $z = 1.95$ d. between $z = -.57$ and $z = -2.39$
e. between $z = -2.15$ and $z = 1.67$

6.17 Find the area under the standard normal curve

a. to the right of $z = 1.56$ b. to the left of $z = -1.97$
c. to the right of $z = -2.05$ d. to the left of $z = 1.86$

6.18 Obtain the area under the standard normal curve

a. to the right of $z = 1.73$ b. to the left of $z = -1.55$
c. to the right of $z = -.65$ d. to the left of $z = .89$

6.19 Find the area under the standard normal curve

a. between $z = 0$ and $z = 4.28$ b. from $z = 0$ to $z = -3.75$
c. to the right of $z = 7.43$ d. to the left of $z = -4.49$

6.20 Find the area under the standard normal curve

a. from $z = 0$ to $z = 3.94$ b. between $z = 0$ and $z = -5.16$
c. to the right of $z = 5.42$ d. to the left of $z = -3.68$

6.21 Determine the following probabilities for the standard normal distribution.

a. $P(-1.83 \leq z \leq 2.57)$ b. $P(0 \leq z \leq 2.02)$
c. $P(-1.99 \leq z \leq 0)$ d. $P(z \geq 1.48)$

6.22 Determine the following probabilities for the standard normal distribution.

a. $P(-2.46 \leq z \leq 1.68)$ b. $P(0 \leq z \leq 1.86)$
c. $P(-2.58 \leq z \leq 0)$ d. $P(z \geq .83)$

6.23 Find the following probabilities for the standard normal distribution.

a. $P(z < -2.14)$ b. $P(.67 \leq z \leq 2.49)$
c. $P(-2.07 \leq z \leq -.93)$ d. $P(z < 1.78)$

6.24 Find the following probabilities for the standard normal distribution.

a. $P(z < -1.31)$ b. $P(1.03 \leq z \leq 2.89)$
c. $P(-2.24 \leq z \leq -1.09)$ d. $P(z < 2.12)$

6.25 Obtain the following probabilities for the standard normal distribution.

a. $P(z > -.98)$ b. $P(-2.47 \leq z \leq 1.19)$
c. $P(0 \leq z \leq 4.25)$ d. $P(-5.36 \leq z \leq 0)$
e. $P(z > 6.07)$ f. $P(z < -5.27)$

6.26 Obtain the following probabilities for the standard normal distribution.

a. $P(z > -1.06)$ b. $P(-.68 \leq z \leq 1.84)$
c. $P(0 \leq z \leq 3.85)$ d. $P(-4.34 \leq z \leq 0)$
e. $P(z > 4.82)$ f. $P(z < -6.12)$

6.4 STANDARDIZING A NORMAL DISTRIBUTION

As was shown in the previous section, Table VII of Appendix D can be used to find areas under the standard normal curve. However, in real-world applications, a (continuous) random variable may have a normal distribution with values of the mean and standard deviation that are different from 0 and 1, respectively. The first step in such a case is to convert the given normal distribution to the standard normal distribution. This procedure is called *standardizing a normal distribution*. The units of a normal distribution (which is not the standard normal distribution) are denoted by *x*. We know from Section 6.3 that units of the standard normal distribution are denoted by *z*.

CONVERTING AN *x* VALUE TO A *z* VALUE   For a normal random variable *x*, a particular value of *x* can be converted to its corresponding *z* value by using the formula

$$z = \frac{x - \mu}{\sigma}$$

where μ and σ are the mean and standard deviation of the normal distribution of *x*, respectively.

Thus, to find the *z* value for an *x* value, we calculate the difference between the given *x* value and the mean, μ, and divide this difference by the standard deviation, σ. If the value of *x* is equal to μ, then its *z* value is equal to zero. Note that we will always round *z* values to two decimal places.

☞ *Remember*   The *z* value for the mean of a normal distribution is always zero.

Examples 6–6 through 6–10 describe how to convert *x* values to the corresponding *z* values and how to find areas under a normal distribution curve.

Converting x values to the corresponding z values.

EXAMPLE 6–6   Let *x* be a continuous random variable that has a normal distribution with a mean of 50 and a standard deviation of 10. Convert the following *x* values to *z* values.

(a)   *x* = 55   (b)   *x* = 35

Solution   For the given normal distribution, $\mu = 50$ and $\sigma = 10$.

(a)   The *z* value for *x* = 55 is computed as follows:

$$z = \frac{x - \mu}{\sigma} = \frac{55 - 50}{10} = \textbf{.50}$$

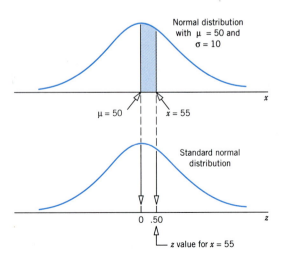

Figure 6.32   *z* value for *x* = 55

Thus, the z value for $x = 55$ is .50. The z values for $\mu = 50$ and $x = 55$ are shown in Figure 6.32. Note that the z value for $\mu = 50$ is zero. The value $z = .50$ for $x = 55$ indicates that the distance between the mean, $\mu = 50$, of the given normal distribution and the point given by $x = 55$ is 1/2 of the standard deviation, $\sigma = 10$. Consequently, we can state that the z value represents the distance between μ and x in terms of the standard deviation. Because $x = 55$ is greater than $\mu = 50$, its z value is positive.

From this point on, we will usually show only the z axis below the x axis and not the standard normal curve itself.

(b) The z value for $x = 35$ is computed as follows and is shown in Figure 6.33:

$$z = \frac{x - \mu}{\sigma} = \frac{35 - 50}{10} = -1.50$$

Because $x = 35$ is on the left side of the mean (i.e., 35 is less than $\mu = 50$), its z value is negative. As a general rule, whenever an x value is less than the value of μ, its z value is negative.

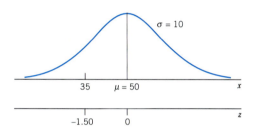

Figure 6.33 z value for $x = 35$.

☞ *Remember* The z value for an x value that is greater than μ is positive, the z value for an x value that is equal to μ is zero, and the z value for an x value that is less than μ is negative.

To find the area between two values of x for a normal distribution, we first convert both values of x to their respective z values. Then we find the area under the standard normal curve between those two z values. The area between the two z values gives the area between the corresponding x values.

EXAMPLE 6–7 Let x be a continuous random variable that is normally distributed with a mean of 25 and a standard deviation of 4. Find the area

(a) between $x = 25$ and $x = 32$ (b) between $x = 18$ and $x = 34$

Solution For the given normal distribution, $\mu = 25$ and $\sigma = 4$.

Finding the area between the mean and a point to its right.

(a) The first step in finding the required area is to standardize the given normal distribution by converting $x = 25$ and $x = 32$ to their respective z values using the formula

$$z = \frac{x - \mu}{\sigma}$$

The z value for $x = 25$ is zero because it is the mean of the normal distribution. The z value for $x = 32$ is

$$z = \frac{32 - 25}{4} = 1.75$$

As shown in Figure 6.34 (page 260), the area between $x = 25$ and $x = 32$ under the given normal distribution curve is equivalent to the area between $z = 0$ and $z = 1.75$

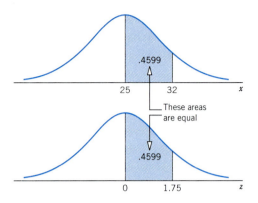

Figure 6.34   Area between $x = 25$ and $x = 32$.

under the standard normal curve. This area from Table VII is .4599. The area between $x = 25$ and $x = 32$ under the normal curve gives the probability that x assumes a value between 25 and 32. This probability can be written as

$$P(25 < x < 32) = P(0 < z < 1.75) = \mathbf{.4599}$$

Finding the area between two points on different sides of the mean.

(b)   First, we calculate the z values for $x = 18$ and $x = 34$ as follows:

$$\text{For } x = 18: \quad z = \frac{18 - 25}{4} = -1.75$$

$$\text{For } x = 34: \quad z = \frac{34 - 25}{4} = 2.25$$

The area under the given normal distribution curve between $x = 18$ and $x = 34$ is given by the area under the standard normal curve between $z = -1.75$ and $z = 2.25$. This area is shown in Figure 6.35. The two values of z are on different sides of $z = 0$. Consequently, the total area is obtained by adding the areas between $z = -1.75$ and $z = 0$ and between $z = 0$ and $z = 2.25$. From Table VII, the area between $z = -1.75$ and $z = 0$ is .4599 and the area between $z = 0$ and $z = 2.25$ is .4878. Hence,

$$P(18 < x < 34) = P(-1.75 < z < 2.25) = .4599 + .4878 = \mathbf{.9477}$$

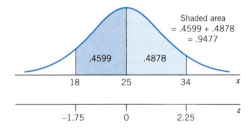

Figure 6.35   Area between $x = 18$ and $x = 34$.

EXAMPLE 6–8   Let x be a normal random variable with its mean equal to 40 and standard deviation equal to 5. Find the following probabilities for this normal distribution.

(a) $P(x > 55)$ (b) $P(x < 49)$

Solution   For the given normal distribution, $\mu = 40$ and $\sigma = 5$.

Calculating the probability of x falling in the right tail.

(a)   The probability that x assumes a value greater than 55 is given by the area under the normal distribution curve to the right of $x = 55$, as shown in Figure 6.36. This area is calculated by subtracting the area between $\mu = 40$ and $x = 55$ from .5, which is the total area to the right of the mean.

$$\text{For } x = 55: \quad z = \frac{55 - 40}{5} = 3.00$$

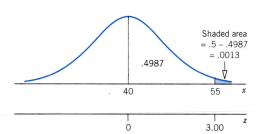

Figure 6.36 Finding $P(x > 55)$.

The required probability is given by the area to the right of $z = 3.00$. To find this area, first we find the area between $z = 0$ and $z = 3.00$, which is .4987. Then we subtract this area from .5. Thus,

$$P(x > 55) = P(z > 3.00) = .5 - .4987 = \textbf{.0013}$$

Calculating the probability that x is less than a value to the right of the mean.

(b) The probability that x will assume a value less than 49 is given by the area under the normal distribution curve to the left of 49, which is the shaded area in Figure 6.37. This area is given by the sum of the area to the left of $\mu = 40$ and the area between $\mu = 40$ and $x = 49$.

$$\text{For } x = 49: \quad z = \frac{49 - 40}{5} = 1.80$$

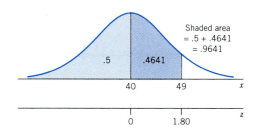

Figure 6.37 Finding $P(x < 49)$.

The required probability is given by the area to the left of $z = 1.80$. This area is obtained by adding the area to the left of $z = 0$ and the area between $z = 0$ and $z = 1.80$. The area to the left of $z = 0$ is .5. From Table VII, the area between $z = 0$ and $z = 1.80$ is .4641. Therefore, the required probability is

$$P(x < 49) = P(z < 1.80) = .5 + .4641 = \textbf{.9641} \qquad \blacksquare$$

Finding the area between two x values that are less than the mean.

EXAMPLE 6-9 Let x be a continuous random variable that has a normal distribution with $\mu = 50$ and $\sigma = 8$. Find the probability $P(30 \leq x \leq 39)$.

Solution For this normal distribution, $\mu = 50$ and $\sigma = 8$. The probability $P(30 \leq x \leq 39)$ is given by the area from $x = 30$ to $x = 39$ under the normal distribution curve. As shown in Figure 6.38 (page 262), this area is given by the difference between the area from $x = 30$ to $\mu = 50$ and the area from $x = 39$ to $\mu = 50$.

$$\text{For } x = 30: \quad z = \frac{30 - 50}{8} = -2.50$$

$$\text{For } x = 39: \quad z = \frac{39 - 50}{8} = -1.38$$

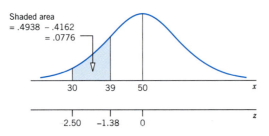

Shaded area
= .4938 − .4162
= .0776

Figure 6.38 Finding $P(30 \leq x \leq 39)$.

To find the required area, we first find the areas from $z = -2.50$ to $z = 0$ and from $z = -1.38$ to $z = 0$ and then take the difference between these two areas. From Table VII, the area from $z = -2.50$ to $z = 0$ is .4938 and the area from $z = -1.38$ to $z = 0$ is .4162. Thus, the required probability is

$$P(30 \leq x \leq 39) = P(-2.50 \leq z \leq -1.38) = .4938 - .4162 = \mathbf{.0776} \qquad \blacksquare$$

EXAMPLE 6–10 Let x be a continuous random variable that has a normal distribution with a mean of 80 and a standard deviation of 12. Find the area under the normal distribution curve

(a)   from $x = 70$ to $x = 135$ (b)   to the left of 27

Solution   For the given normal distribution, $\mu = 80$ and $\sigma = 12$.

Finding the area between two x values that are on different sides of the mean.

(a)   The area from $x = 70$ to $x = 135$ is obtained by adding the areas from $x = 70$ to $\mu = 80$ and from $\mu = 80$ to $x = 135$. This total area is given by the sum of the two shaded areas shown in Figure 6.39.

$$\text{For } x = 70: \quad z = \frac{70 - 80}{12} = -.83$$

$$\text{For } x = 135: \quad z = \frac{135 - 80}{12} = 4.58$$

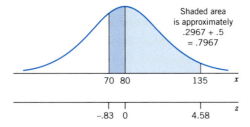

Shaded area
is approximately
.2967 + .5
= .7967

Figure 6.39 Area between $x = 70$ and $x = 135$.

Thus, the required area is obtained by adding the areas from $z = -.83$ to $z = 0$ and from $z = 0$ to $z = 4.58$ under the standard normal curve. From Table VII, the area from

$z = -.83$ to $z = 0$ is .2967 and the area from $z = 0$ to $z = 4.58$ is approximately .5. Hence,

$$P(70 \le x \le 135) = P(-.83 \le z \le 4.58)$$

$$= .2967 + .5 = \mathbf{.7967} \text{ approximately}$$

Finding an area in the left tail.

(b) The area to the left of $x = 27$ is obtained by subtracting the area from $x = 27$ to $\mu = 80$ from .5, which is the total area to the left of the mean. This area is calculated as follows:

$$\text{For } x = 27: \quad z = \frac{27 - 80}{12} = -4.42$$

As shown in Figure 6.40, the required area is given by the area under the standard normal distribution curve to the left of $z = -4.42$. This area is

$$P(x < 27) = P(z < -4.42) = .5 - P(-4.42 < z < 0)$$

$$= .5 - .5 = \mathbf{.00} \text{ approximately}$$

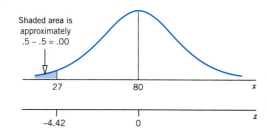

Shaded area is approximately
.5 − .5 = .00

Figure 6.40   Area to the left of $x = 27$.   ■

EXERCISES

■ Concepts and Procedures

6.27 Find the z value for each of the following x values for a normal distribution with $\mu = 30$ and $\sigma = 5$.

a. $x = 39$ **b.** $x = 17$ **c.** $x = 22$ **d.** $x = 42$

6.28 Determine the z value for each of the following x values for a normal distribution with $\mu = 16$ and $\sigma = 3$.

a. $x = 12$ **b.** $x = 21$ **c.** $x = 19$ **d.** $x = 13$

6.29 Find the following areas under a normal distribution curve with $\mu = 20$ and $\sigma = 4$.

a. Area between $x = 20$ and $x = 27$
b. Area from $x = 23$ to $x = 25$
c. Area between $x = 9.5$ and $x = 17$

6.30 Find the following areas under a normal distribution curve with $\mu = 12$ and $\sigma = 2$.

a. Area between $x = 7.76$ and $x = 12$
b. Area between $x = 14.48$ and $x = 16.54$
c. Area from $x = 8.22$ to $x = 10.06$

6.31 Determine the area under a normal distribution curve with $\mu = 55$ and $\sigma = 7$

a. to the right of $x = 58$
b. to the right of $x = 43$
c. to the left of $x = 67$
d. to the left of $x = 24$

6.32　Find the area under a normal distribution curve with $\mu = 37$ and $\sigma = 3$

a.　to the left of $x = 30$

b.　to the right of $x = 52$

c.　to the left of $x = 44$

d.　to the right of $x = 32$

6.33　Let x be a continuous random variable that is normally distributed with a mean of 25 and a standard deviation of 6. Find the probability that x assumes a value

a.　between 29 and 36　　b.　between 22 and 33

6.34　Let x be a continuous random variable that has a normal distribution with a mean of 40 and a standard deviation of 4. Find the probability that x assumes a value

a.　between 29 and 35　　b.　from 34 to 51

6.35　Let x be a continuous random variable that is normally distributed with a mean of 80 and a standard deviation of 12. Find the probability that x assumes a value

a.　greater than 69　　　b.　less than 74

c.　greater than 101　　　d.　less than 88

6.36　Let x be a continuous random variable that is normally distributed with a mean of 65 and a standard deviation of 15. Find the probability that x assumes a value

a.　less than 43　　　　b.　greater than 74

c.　greater than 56　　　d.　less than 71

6.5　APPLICATIONS OF THE NORMAL DISTRIBUTION

Sections 6.2 through 6.4 discussed the normal distribution, how to convert a normal distribution to the standard normal distribution, and how to find areas under a normal distribution curve. This section presents examples that illustrate the applications of the normal distribution.

Using the normal distribution: the area between two points on different sides of the mean.

EXAMPLE 6–11　According to an estimate, the average starting salary of college graduates with a computer science major is \$41,949 (*Newsweek*, February 1, 1999). Assume that the current starting salaries of all college graduates with a computer science major have a normal distribution with a mean of \$41,949 and a standard deviation of \$2500. Find the probability that the starting salary of a randomly selected college graduate with a computer science major is between \$38,624 and \$43,299.

Solution　Let x denote the starting salary of a randomly selected college graduate with a computer science major. Then, x is normally distributed with

$$\mu = \$41{,}949 \quad \text{and} \quad \sigma = \$2500$$

The probability that the starting salary of a randomly selected college graduate with a computer science major is between \$38,624 and \$43,299 is given by the area under the normal distribution curve of x that falls between $x = \$38{,}624$ and $x = \$43{,}299$. Because these two points are on different sides of the mean, the required probability is obtained by adding the two shaded areas shown in Figure 6.41.

$$\text{For } x = \$38{,}624: \quad z = \frac{38{,}624 - 41{,}949}{2500} = -1.33$$

$$\text{For } x = \$43{,}299: \quad z = \frac{43{,}299 - 41{,}949}{2500} = .54$$

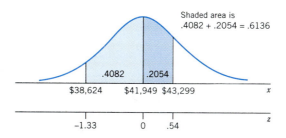

Shaded area is
.4082 + .2054 = .6136

.4082   .2054

$38,624 $41,949 $43,299   x

−1.33   0   .54   z

Figure 6.41   Area between
$x = \$38,624$ and $x = \$43,299$.

The required probability is given by the area under the standard normal curve between $z = -1.33$ and $z = .54$, which is obtained by adding the area between $z = -1.33$ and $z = 0$ and the area between $z = 0$ and $z = .54$. From Table VII in Appendix D, the area between $z = -1.33$ and $z = 0$ is .4082 and the area between $z = 0$ and $z = .54$ is .2054. Hence, the required probability is

$$P(\$38,624 < x < \$43,299) = P(-1.33 < z < .54) = .4082 + .2054 = \mathbf{.6136}$$

Thus, the probability is .6136 that the starting salary of a randomly selected college graduate with a computer science major is between $38,624 and $43,299. Converting this probability into a percentage, we can also state that the starting salaries of about 61.36% of college graduates with a computer science major would be between $38,624 and $43,299. ■

Using the normal distribution: probability that x is less than a value that is to the right of the mean.

EXAMPLE 6-12   A racing car is one of the many toys manufactured by Mack Corporation. The assembly times for this toy follow a normal distribution with a mean of 55 minutes and a standard deviation of 4 minutes. The company closes at 5 P.M. every day. If one worker starts to assemble a racing car at 4 P.M., what is the probability that she will finish this job before the company closes for the day?

Solution   Let x denote the time this worker takes to assemble a racing car. Then, x is normally distributed with

$$\mu = 55 \text{ minutes} \quad \text{and} \quad \sigma = 4 \text{ minutes}$$

We are to find the probability that this worker can assemble this car in 60 minutes or less (between 4 and 5 P.M.). This probability is given by the area under the normal curve to the left of $x = 60$. This area is obtained by adding the two shaded areas shown in Figure 6.42.

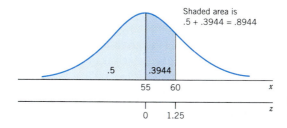

Shaded area is
.5 + .3944 = .8944

.5   .3944

55   60   x

0   1.25   z

Figure 6.42   Area to the
left of $x = 60$.

$$\text{For } x = 60: \quad z = \frac{60 - 55}{4} = 1.25$$

The required probability is given by the area under the standard normal curve to the left of $z = 1.25$, which is obtained by adding .5 and the area between $z = 0$ and $z = 1.25$. From

Table VII of Appendix D, the area between $z = 0$ and $z = 1.25$ is .3944. Therefore, the required probability is

$$P(x \le 60) = P(z \le 1.25) = .5 + .3944 = \mathbf{.8944}$$

Thus, the probability is .8944 that this worker will finish assembling this racing car before the company closes for the day. ∎

Using the normal distribution.

EXAMPLE 6–13   Hupper Corporation produces many types of soft drinks, including Orange Cola. The filling machines are adjusted to pour 12 ounces of soda into each 12-ounce can of Orange Cola. However, the actual amount of soda poured into each can is not exactly 12 ounces; it varies from can to can. It has been observed that the net amount of soda in such a can has a normal distribution with a mean of 12 ounces and a standard deviation of .015 ounce.

Calculating the probability between two points that are to the left of the mean.

(a) What is the probability that a randomly selected can of Orange Cola contains 11.97 to 11.99 ounces of soda?

(b) What percentage of the Orange Cola cans contain 12.02 to 12.07 ounces of soda?

Solution   Let x be the net amount of soda in a can of Orange Cola. Then, x has a normal distribution with $\mu = 12$ ounces and $\sigma = .015$ ounce.

(a) The probability that a randomly selected can contains 11.97 to 11.99 ounces of soda is given by the area under the normal distribution curve from $x = 11.97$ to $x = 11.99$. This area is shown in Figure 6.43.

$$\text{For } x = 11.97: \quad z = \frac{11.97 - 12}{.015} = -2.00$$

$$\text{For } x = 11.99: \quad z = \frac{11.99 - 12}{.015} = -.67$$

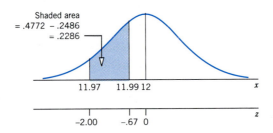

Figure 6.43   Area between $x = 11.97$ and $x = 11.99$.

The required probability is given by the area under the standard normal curve between $z = -2.00$ and $z = -.67$, which is obtained by subtracting the area from $z = -.67$ to $z = 0$ from the area from $z = -2.00$ to $z = 0$. From Table VII of Appendix D, the area from $z = -.67$ to $z = 0$ is .2486 and the area from $z = -2.00$ to $z = 0$ is .4772. Hence, the required probability is

$$P(11.97 \le x \le 11.99) = P(-2.00 \le z \le -.67) = .4772 - .2486 = \mathbf{.2286}$$

Thus, the probability is .2286 that any randomly selected can of Orange Cola will contain 11.97 to 11.99 ounces of soda. We can also state that about 22.86% of Orange Cola cans contain 11.97 to 11.99 ounces of soda.

(b) The percentage of Orange Cola cans that contain 12.02 to 12.07 ounces of soda is given by the area under the normal distribution curve from $x = 12.02$ to $x = 12.07$, as shown in Figure 6.44.

$$\text{For } x = 12.02: \quad z = \frac{12.02 - 12}{.015} = 1.33$$

$$\text{For } x = 12.07: \quad z = \frac{12.07 - 12}{.015} = 4.67$$

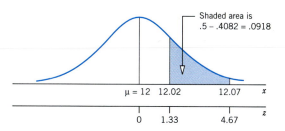

Figure 6.44 Area from $x = 12.02$ to $x = 12.07$.

The required probability is given by the area under the standard normal curve between $z = 1.33$ and $z = 4.67$, which is obtained by subtracting the area from $z = 0$ to $z = 1.33$ from the area from $z = 0$ to $z = 4.67$. From Table VII of Appendix D, the area from $z = 0$ to $z = 1.33$ is .4082. The area from $z = 0$ to $z = 4.67$ is approximately .5. Hence, the required probability is

$$P(12.02 \leq x \leq 12.07) = P(1.33 \leq z \leq 4.67) = .5 - .4082 = \textbf{.0918}$$

Converting this probability to a percentage, we can state that approximately 9.18% of all Orange Cola cans are expected to contain 12.02 to 12.07 ounces of soda. ■

EXAMPLE 6-14 The life span of a calculator manufactured by Intal Corporation has a normal distribution with a mean of 54 months and a standard deviation of 8 months. The company guarantees that any calculator that starts malfunctioning within 36 months of the purchase will be replaced by a new one. About what percentage of calculators made by this company are expected to be replaced?

Solution Let x be the life span of such a calculator. Then x has a normal distribution with $\mu = 54$ and $\sigma = 8$ months. The probability that a randomly selected calculator will start to malfunction within 36 months is given by the area under the normal distribution curve to the left of $x = 36$, as shown in Figure 6.45.

$$\text{For } x = 36: \quad z = \frac{36 - 54}{8} = -2.25$$

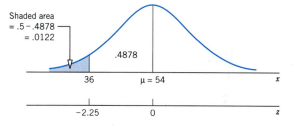

Figure 6.45 Area to the left of $x = 36$.

The required percentage is given by the area under the standard normal curve to the left of $z = -2.25$, which is obtained by subtracting the area between $z = -2.25$ and $z = 0$ from .5. From Table VII of Appendix D, the area between $z = -2.25$ and $z = 0$ is .4878. Hence, the required probability is

$$P(x < 36) = P(z < -2.25) = .5 - .4878 = \mathbf{.0122}$$

The probability that any randomly selected calculator manufactured by Intal Corporation will start to malfunction within 36 months is .0122. Converting this probability to a percentage, we can state that approximately 1.22% of all calculators manufactured by this company are expected to start malfunctioning within 36 months. Hence, 1.22% of the calculators are expected to be replaced. ∎

EXERCISES

■ **Applications**

6.37 Let x denote the time it takes to run a road race. Suppose x is approximately normally distributed with a mean of 190 minutes and a standard deviation of 21 minutes. If one runner is selected at random, what is the probability that this runner will complete this road race

a. in less than 150 minutes   **b.** in 205 to 245 minutes

6.38 The U.S. Bureau of Labor Statistics conducts periodic surveys to collect information on the labor market. According to an estimate by the bureau, workers in the private sector earned an average of $13.21 per hour in August 1999. Assume that the current hourly wages for all workers in the private sector have a normal distribution with a mean of $13.21 and a standard deviation of $2. Find the percentage of these workers whose current hourly wages are

a. between $15 and $17   **b.** between $11 and $14

6.39 According to the American Association of University Professors, the mean 1997–1998 salary of full-time instructional faculty at two-year colleges was $43,760. Suppose these salaries have a normal distribution with a mean of $43,760 and a standard deviation of $4200. Find the probability that the 1997–1998 salary of a randomly selected faculty member from such a college was

a. more than $45,000   **b.** between $40,000 and $45,000

6.40 The stress scores (on a scale of 1 to 10) of students before a statistics test are found to be approximately normally distributed with a mean of 7.08 and a standard deviation of .63. Find the probability that the stress score before a statistics test for a randomly selected student will be

a. greater than 6.25   **b.** between 7.40 and 8.90

6.41 The speeds of cars traveling on Interstate Highway I-15 are normally distributed with a mean of 73 miles per hour and a standard deviation of 3.5 miles per hour. Find what percentage of the cars traveling on this highway have a speed of

a. 64 to 69 miles per hour   **b.** 69 to 79 miles per hour

6.42 The Bank of Connecticut issues Visa and Mastercard credit cards. It is estimated that the balances on all Visa credit cards issued by the Bank of Connecticut have a mean of $845 and a standard deviation of $270. Assume that the balances on all these Visa cards follow a normal distribution.

a. What is the probability that a randomly selected Visa card issued by this bank has a balance between $1000 and $1400?

b. What percentage of the Visa cards issued by this bank have a balance of $750 or more?

6.43 According to data from Runzheimer International, the average cost of suburban day care (for a 3-year-old in a for-profit center, 8 hours per day for 5 days a week) in the Boston area was $622 per month (*USA TODAY*, August 18, 1998). Assume that current such day-care costs in the Boston area are normally distributed with a mean of $622 and a standard deviation of $120.

a. Find the probability that a randomly selected couple with a 3-year-old child in the Boston area pays more than $700 per month for suburban day care.
b. What percentage of such couples spend between $500 and $650 per month on day care?

6.44 The transmission on a model of a specific car has a warranty for 40,000 miles. It is known that the life of such a transmission has a normal distribution with a mean of 72,000 miles and a standard deviation of 12,000 miles.

a. What percentage of the transmissions will fail before the end of the warranty period?
b. What percentage of the transmissions will be good for more than 100,000 miles?

6.45 According to the records of an electric company serving the Boston area, the mean electric consumption for all households during winter is 1650 kilowatt-hours per month. Assume that the monthly electric consumptions during winter by all households in this area have a normal distribution with a mean of 1650 kilowatt-hours and a standard deviation of 320 kilowatt-hours.

a. Find the probability that the monthly electric consumption during winter by a randomly selected household from this area is less than 1850 kilowatt-hours.
b. What percentage of the households in this area have a monthly electric consumption of 900 to 1340 kilowatt-hours?

6.46 The management of a supermarket wants to adopt a new promotional policy of giving a free gift to every customer who spends more than a certain amount per visit at this supermarket. The expectation of the management is that after this promotional policy is advertised, the expenditures for all customers at this supermarket will be normally distributed with a mean of $95 and a standard deviation of $21. If the management decides to give free gifts to all those customers who spend more than $130 at this supermarket during a visit, what percentage of the customers are expected to get free gifts?

6.47 According to a College Board report, the average tuition at four-year private institutions in the United States was $15,380 for the year 1999–2000 (*USA TODAY*, October 5, 1999). Assume that the 1999–2000 tuitions at all four-year private institutions in the United States were normally distributed with a mean of $15,380 and a standard deviation of $1500.

a. Find the probability that the 1999–2000 tuition at a randomly selected four-year private institution was more than $16,400.
b. What percentage of all such institutions had 1999–2000 tuitions between $11,700 and $14,000?

6.48 The management at a large insurance company believes that workers are more productive if they are happy with their jobs. To keep track of workers' satisfaction, the company regularly conducts surveys. According to a recent survey, the mean job satisfaction score for all workers at this company was 13.10 (on a scale of 1 to 20) and the standard deviation was 1.95. Assume that the job satisfaction scores of workers are normally distributed.

a. Find the probability that the job satisfaction score for a randomly selected worker from this company is less than 11.50.
b. What percentage of the workers have a job satisfaction score between 14.50 and 18.50?
c. A worker with a score of 8 or less is considered to be very unhappy with the job. What percentage of the workers are very unhappy with their jobs?

6.49 According to data from The College Board, the average score on the verbal portion of the SAT nationwide in 1999 was 505 (*USA TODAY*, September 1, 1999). Suppose that the verbal SAT scores of all students who took the exam in 1999 were normally distributed with a mean of 505 and a standard deviation of 85.

a. Sue Hopern scored 550 on this test. What percentage of the examinees scored higher than Sue?
b. Joe Merck scored 490 on this test. What percentage of the examinees scored lower than Joe?
c. A particular liberal arts college requires that a student's verbal SAT score be 600 or higher for the student to be eligible to be considered for admission. What percentage of the examinees are eligible to be considered for admission to this college?

6.50 Fast Auto Service guarantees that the maximum waiting time for its customers is 20 minutes for oil and lube service on their cars. It also guarantees that any customer who has to wait longer than 20 minutes for this service will receive a 50% discount on the charges. It is estimated that the mean time taken for oil and lube service at this garage is 15 minutes per car and the standard deviation is 2.4 minutes. Suppose the time taken for oil and lube service on a car follows a normal distribution.

a. What percentage of the customers will receive the 50% discount on their charges?

b. Is it possible that a car may take longer than 25 minutes for oil and lube service? Explain.

6.51 The lengths of 3-inch nails manufactured on a machine are normally distributed with a mean of 3.0 inches and a standard deviation of .009 inch. The nails that are either shorter than 2.98 inches or longer than 3.02 inches are unusable. What percentage of all the nails produced by this machine are unusable?

6.52 Lewis Corporation manufactures bicycles. The wheels of one type of bicycle are supposed to have a diameter of 30 inches. The processing system that makes these wheels does not produce every wheel with exactly a 30-inch diameter. The diameters of wheels made by this system have a normal distribution with a mean of 30 inches and a standard deviation of .10 inch. The wheels that have a diameter of either less than 29.75 inches or greater than 30.25 inches are discarded. What percentage of all the wheels produced by this processing system are discarded?

6.6   DETERMINING THE z AND x VALUES WHEN AN AREA UNDER THE NORMAL DISTRIBUTION CURVE IS KNOWN

So far in this chapter we have discussed how to find the area under a normal distribution curve for an interval of z or x. Now we reverse this procedure and learn how to find the corresponding value of z or x when an area under a normal distribution curve is known. Examples 6–15 through 6–17 describe this procedure for finding the z value.

Finding z when the area between the mean and z is known.

EXAMPLE 6–15 Find a point z such that the area under the standard normal curve between 0 and z is .4251 and the value of z is positive.

Solution As shown in Figure 6.46, we are to find the z value such that the area between 0 and z is .4251.

Table 6.4 Finding the z Value When Area Is Known

z	.00	.01	. . .	.04	. . .	.09
0.0	.0000	.0040	. . .		. . .	.0359
0.1	.0398	.0438	. . .		. . .	.0753
0.2	.0793	.0832	. . .		. . .	.1141
.	.	.	. . .		. . .	.
.	.	.	. . .		. . .	.
.	.	.	. . .		. . .	.
1.4 ←				.4251 ←	. . .	. . .
.	.	.	. . .	.	. . .	.
.	.	.	. . .	.	. . .	.
3.0	.4987	.4987	. . .	.4988	. . .	.4990

We locate this value in Table VII of Appendix D

To find the required value of z, we locate .4251 in the body of the normal distribution table, Table VII of Appendix D. The relevant portion of that table is reproduced as Table 6.4. Next we read the numbers in the column and row for z that correspond to .4251. As shown in Table 6.4, these numbers are 1.4 and .04, respectively. Combining these two numbers, we obtain the required value of $z = $ **1.44**.

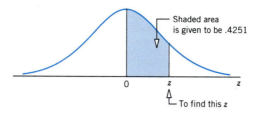

Figure 6.46 Finding the z value.

■

Finding z when the area in the right tail is known.

EXAMPLE 6-16   Find the value of z such that the area under the standard normal curve in the right tail is .0050.

Solution   To find the required value of z, we first find the area between 0 and z. The total area to the right of $z = 0$ is .5. Hence,

$$\text{Area between 0 and } z = .5 - .0050 = .4950$$

This area is shown in Figure 6.47.

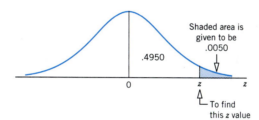

Figure 6.47 Finding the z value.

Now we look for .4950 in the body of the normal distribution table. Table VII does not contain .4950. So we find the value closest to .4950, which is either .4949 or .4951. We can use either of these two values. If we choose .4951, the corresponding z value is 2.58. Hence, the required value of z is **2.58,** and the area to the right of $z = 2.58$ is approximately .0050. Note that there is no apparent reason to choose .4951 and not to choose .4949. We can use either of the two values. If we choose .4949, the corresponding z value will be 2.57.   ■

Finding z when the area in the left tail is known.

EXAMPLE 6-17   Find the value of z such that the area under the standard normal curve in the left tail is .05.

Solution   Because .05 is less than .5 and it is the area in the left tail, the value of z is negative. To find the required value of z, first we find the area between 0 and z. The total area to the left of $z = 0$ is .5. Hence,

$$\text{Area between 0 and } z = .5 - .05 = .4500$$

This area is shown in Figure 6.48.

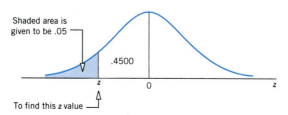

Figure 6.48 Finding the z value.

Next, we look for .4500 in the body of the normal distribution table. The value closest to .4500 in the normal distribution table is either .4495 or .4505. Suppose we use the value .4505. The corresponding z value is 1.65. Because the value of z is negative in our example (see Figure 6.48), the required value of z is **−1.65** and the area to the left of $z = -1.65$ is approximately .05. ■

To find an x value when an area under a normal distribution curve is given, first we find the z value corresponding to that x value from the normal distribution table. Then, to find the x value, we substitute the values of μ, σ, and z in the following formula, which is obtained from $z = (x - \mu)/\sigma$ by doing some algebraic manipulations. Also, if we know the values of x, z, and σ, we can find μ using this same formula. Exercises 6.63 and 6.64 present such cases.

> **FINDING AN x VALUE FOR A NORMAL DISTRIBUTION**   For a normal curve, with known values of μ and σ and for a given area under the curve between the mean and x, the x value is calculated as
>
> $$x = \mu + z\sigma$$

Examples 6–18 and 6–19 illustrate how to find an x value when an area under a normal distribution curve is known.

Finding x when the area in the left tail is known.

EXAMPLE 6–18   Recall Example 6–14. It is known that the life of a calculator manufactured by Intal Corporation has a normal distribution with a mean of 54 months and a standard deviation of 8 months. What should the warranty period be to replace a malfunctioning calculator if the company does not want to replace more than 1% of all the calculators sold?

Solution   Let x be the life of a calculator. Then, x follows a normal distribution with $\mu = 54$ months and $\sigma = 8$ months. The calculators that would be replaced are the ones that start malfunctioning during the warranty period. The company's objective is to replace at most 1% of all the calculators sold. The shaded area in Figure 6.49 gives the proportion of calculators that are replaced. We are to find the value of x so that the area to the left of x under the normal curve is 1% or .01.

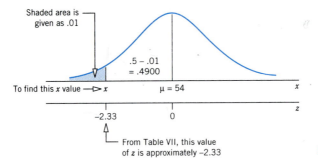

Figure 6.49   Finding an x value.

In the first step, we find the z value that corresponds to the required x value. To find this z value, we need to know the area between the mean and the x value. This area is calculated as follows:

Area between the mean and the x value $= .5 - .01 = .4900$

Next, we find the z value from the normal distribution table for .4900. Table VII of Appendix D does not contain a value that is exactly .4900. The value closest to .4900 in the table is .4901, and the z value for .4901 is 2.33. Since z is on the left side of the mean (see Figure 6.49), the value of z is negative. Hence,

$$z = -2.33$$

Substituting the values of μ, σ, and z in the formula $x = \mu + z\,\sigma$, we obtain

$$x = \mu + z\,\sigma = 54 + (-2.33)\,(8) = 54 - 18.64 = \mathbf{35.36}$$

Thus, the company should replace all the calculators that start to malfunction within 35.36 months (which can be rounded to 35 months) of the date of purchase so that they will not have to replace more than 1% of the calculators. ■

Finding x when the area in the right tail is known.

EXAMPLE 6-19 Almost all high school students who intend to go to college take the SAT test. In 1999, the mean SAT score (in verbal and mathematics) of all students was 1016. Debbie Sears is planning to take this test soon. Suppose the SAT scores of all students who take this test with Debbie will have a normal distribution with a mean of 1016 and a standard deviation of 153. What should her score be on this test so that only 10% of all examinees score higher than she does?

Solution Let x represent the SAT scores of examinees. Then, x follows a normal distribution with $\mu = 1016$ and $\sigma = 153$. We are to find the value of x such that the area under the normal distribution curve to the right of x is 10%, as shown in Figure 6.50.

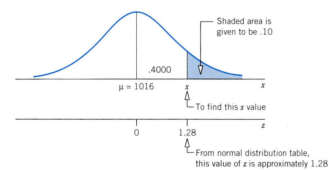

Figure 6.50 Finding an x value.

First, we find the area under the normal distribution curve between the mean and the x value.

$$\text{Area between } \mu \text{ and the } x \text{ value} = .5 - .10 = .4000$$

To find the z value that corresponds to the required x value, we look for .4000 in the body of the normal distribution table. The value closest to .4000 in Table VII is .3997, and the corresponding z value is 1.28. Hence, the value of x is computed as

$$x = \mu + z\,\sigma = 1016 + 1.28(153) = 1016 + 195.84 = 1211.84 \approx \mathbf{1212}$$

Thus, if Debbie Sears scores 1212 on the SAT, only about 10% of the examinees are expected to score higher than she does. ■

EXERCISES

■ **Concepts and Procedures**

6.53 Find the value of z so that the area under the standard normal curve

a. from 0 to z is .4772 and z is positive

b. between 0 and z is (approximately) .4785 and z is negative

  c. in the left tail is (approximately) .3565

  d. in the right tail is (approximately) .1530

6.54 Find the value of z so that the area under the standard normal curve

a. from 0 to z is (approximately) .1965 and z is positive

b. between 0 and z is (approximately) .2740 and z is negative

c. in the left tail is (approximately) .2050

d. in the right tail is (approximately) .1053

6.55 Determine the value of z so that the area under the standard normal curve

a. in the right tail is .0500

b. in the left tail is .0250

c. in the left tail is .0100

d. in the right tail is .0050

6.56 Determine the value of z so that the area under the standard normal curve

a. in the right tail is .0250

b. in the left tail is .0500

c. in the left tail is .0010

d. in the right tail is .0100

6.57 Let x be a continuous random variable that follows a normal distribution with a mean of 200 and a standard deviation of 25.

a. Find the value of x so that the area under the normal curve to the left of x is approximately .6330.

b. Find the value of x so that the area under the normal curve to the right of x is approximately .05.

c. Find the value of x so that the area under the normal curve to the right of x is .8051.

d. Find the value of x so that the area under the normal curve to the left of x is .0150.

e. Find the value of x so that the area under the normal curve between μ and x is .4525 and the value of x is less than μ.

f. Find the value of x so that the area under the normal curve between μ and x is approximately .4800 and the value of x is greater than μ.

6.58 Let x be a continuous random variable that follows a normal distribution with a mean of 550 and a standard deviation of 75.

a. Find the value of x so that the area under the normal curve to the left of x is .0250.

b. Find the value of x so that the area under the normal curve to the right of x is .9345.

c. Find the value of x so that the area under the normal curve to the right of x is approximately .0275.

d. Find the value of x so that the area under the normal curve to the left of x is approximately .9600.

e. Find the value of x so that the area under the normal curve between μ and x is approximately .4700 and the value of x is less than μ.

f. Find the value of x so that the area under the normal curve between μ and x is approximately .4100 and the value of x is greater than μ.

■ **Applications**

6.59 Fast Auto Service provides oil and lube service for cars. It is known that the mean time taken for oil and lube service at this garage is 15 minutes per car and the standard deviation is 2.4 minutes. The management wants to promote the business by guaranteeing a maximum waiting time for its customers. If a customer's car is not serviced within that period, the customer will receive a 50% discount on the charges. The company wants to limit this discount to at most 5% of the customers. What should the maximum guaranteed waiting time be? Assume that the times taken for oil and lube service for all cars have a normal distribution.

6.60 The management of a supermarket wants to adopt a new promotional policy of giving a free gift to every customer who spends more than a certain amount per visit at this supermarket. The expectation of the management is that after this promotional policy is advertised, the expenditures for all customers at this supermarket will be normally distributed with a mean of $95 and a standard deviation of $21. If the management wants to give free gifts to at most 10% of the customers, what should the amount be above which a customer would receive a free gift?

6.61   According to the records of an electric company serving the Boston area, the mean electric consumption during winter for all households is 1650 kilowatt-hours per month. Assume that the monthly electric consumptions during winter by all households in this area have a normal distribution with a mean of 1650 kilowatt-hours and a standard deviation of 320 kilowatt-hours. The company sent a notice to Bill Johnson informing him that about 90% of the households use less electricity per month than he does. What is Bill Johnson's monthly electric consumption?

6.62   Rockingham Corporation makes electric shavers. The life (period before which a shaver does not need a major repair) of Model J795 of an electric shaver manufactured by this corporation has a normal distribution with a mean of 70 months and a standard deviation of 8 months. The company is to determine the warranty period for this shaver. Any shaver that needs a major repair during this warranty period will be replaced free by the company.

a.   What should the warranty period be if the company does not want to replace more than 1% of the shavers?

b.   What should the warranty period be if the company does not want to replace more than 5% of the shavers?

*6.63   A study has shown that 20% of all college textbooks have a price of $80 or higher. It is known that the standard deviation of the prices of all college textbooks is $9.50. Suppose the prices of all college textbooks have a normal distribution. What is the mean price of all college textbooks?

*6.64   A machine at Keats Corporation fills 64-ounce detergent jugs. The machine can be adjusted to pour, on average, any amount of detergent into these jugs. However, the machine does not pour exactly the same amount of detergent into each jug; it varies from jug to jug. It is known that the net amount of detergent poured into each jug has a normal distribution with a standard deviation of .35 ounce. The quality control inspector wants to adjust the machine such that at least 95% of the jugs have more than 64 ounces of detergent. What should the mean amount of detergent poured by this machine into these jugs be?

6.7   THE NORMAL APPROXIMATION TO THE BINOMIAL DISTRIBUTION

Recall from Chapter 5 that

1.   The binomial distribution is applied to a discrete random variable.
2.   Each repetition, called a trial, of a binomial experiment results in one of two possible outcomes, either a success or a failure.
3.   The probabilities of the two (possible) outcomes remain the same for each repetition of the experiment.
4.   The trials are independent.

The binomial formula, which gives the probability of x successes in n trials, is

$$P(x) = \binom{n}{x} p^x q^{n-x}$$

The use of the binomial formula becomes very tedious when n is large. In such cases, the normal distribution can be used to approximate the binomial probability. Note that for a binomial problem, the exact probability is obtained by using the binomial formula. If we apply the normal distribution to solve a binomial problem, the probability that we obtain is an approximation to the exact probability. The approximation obtained by using the normal distribution is very close to the exact probability when n is large and p is very close to .50. However, this does not mean that we should not use the normal approximation when p is not close to .50. The reason the approximation is closer to the exact probability when p is close to .50 is that the binomial distribution is symmetric when $p = .50$. The normal distribution is always symmetric. Hence, the two distributions are very close to each other when n is large and p is close to .50. However, this does not mean that whenever $p = .50$, the binomial distribution is the same as the normal distribution because not every symmetric bell-shaped curve is a normal distribution curve.

> **NORMAL DISTRIBUTION AS AN APPROXIMATION TO BINOMIAL DISTRIBUTION**   Usually, the normal distribution is used as an approximation to the binomial distribution when np and nq are both greater than 5—that is, when
>
> $$np > 5 \quad \text{and} \quad nq > 5$$

Table 6.5 gives the binomial probability distribution of x for $n = 12$ and $p = .50$. This table is constructed using Table IV of Appendix D. Figure 6.51 shows the histogram and the smoothed polygon for the probability distribution of Table 6.5. As we can observe, the histogram in Figure 6.51 is symmetric, and the curve obtained by joining the upper midpoints of the rectangles is approximately bell-shaped.

Table 6.5   The Binomial Probability Distribution for $n = 12$ and $p = .50$

x	$P(x)$
0	.0002
1	.0029
2	.0161
3	.0537
4	.1208
5	.1934
6	.2256
7	.1934
8	.1208
9	.0537
10	.0161
11	.0029
12	.0002

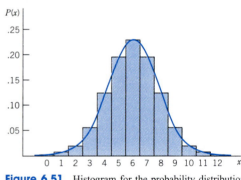

Figure 6.51   Histogram for the probability distribution of Table 6.5.

Examples 6–20 through 6–22 illustrate the application of the normal distribution as an approximation to the binomial distribution.

Using the normal approximation to the binomial: x equals a specific value.

EXAMPLE 6-20   According to an estimate, 50% of the people in America have at least one credit card. If a random sample of 30 persons is selected, what is the probability that 19 of them will have at least one credit card?

Solution   Let n be the total number of persons in the sample, x be the number of persons in the sample who have at least one credit card, and p be the probability that a person has at least one credit card. Then, this is a binomial problem with

$$n = 30, \quad p = .50, \quad q = 1 - p = .50,$$
$$x = 19, \quad n - x = 30 - 19 = 11$$

From the binomial formula, the exact probability that 19 persons in a sample of 30 have at least one credit card is

$$P(19) = \binom{30}{19} (.50)^{19}(.50)^{11} = \mathbf{.0509}$$

Now let us solve this problem using the normal distribution as an approximation to the binomial distribution. For this example,

$$np = 30(.50) = 15 \quad \text{and} \quad nq = 30(.50) = 15$$

Since np and nq are both greater than 5, we can use the normal distribution as an approximation to solve this binomial problem. We perform the following three steps.

Step 1. *Compute μ and σ for the binomial distribution.*

To use the normal distribution, we need to know the mean and standard deviation of the distribution. Hence, the first step in using the normal approximation to the binomial distribution is to compute the mean and standard deviation of the binomial distribution. As we know from Chapter 5, the mean and standard deviation of a binomial distribution are given by np and $\sqrt{npq}$, respectively. Using these formulas, we obtain

$$\mu = np = 30(.50) = 15$$

$$\sigma = \sqrt{npq} = \sqrt{30(.50)(.50)} = 2.73861279$$

Step 2. *Convert the discrete random variable to a continuous random variable.*

The normal distribution applies to a continuous random variable, whereas the binomial distribution applies to a discrete random variable. The second step in applying the normal approximation to the binomial distribution is to convert the discrete random variable to a continuous random variable by making the **correction for continuity.**

CONTINUITY CORRECTION FACTOR   The addition of .5 and/or subtraction of .5 from the value(s) of x when the normal distribution is used as an approximation to the binomial distribution, where x is the number of successes in n trials, is called the *continuity correction factor.*

As shown in Figure 6.52, the probability of 19 successes in 30 trials is given by the area of the rectangle for $x = 19$. To make the correction for continuity, we use the interval 18.5 to 19.5 for 19 persons. This interval is actually given by the two boundaries of the rectangle for $x = 19$, which are obtained by subtracting .5 from 19 and by adding .5 to 19. Thus,

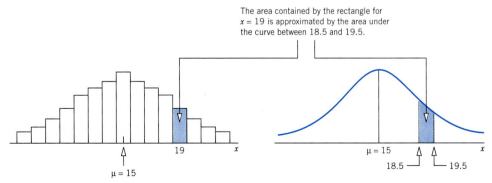

Figure 6.52

$P(x = 19)$ for the binomial problem will be approximately equal to $P(18.5 \leq x \leq 19.5)$ for the normal distribution.

Step 3.   *Compute the required probability using the normal distribution.*

As shown in Figure 6.53, the area under the normal distribution curve between $x = 18.5$ and $x = 19.5$ will give us the (approximate) probability that 19 persons have at least one credit card. We calculate this probability as follows:

$$\text{For } x = 18.5: \quad z = \frac{18.5 - 15}{2.73861279} = 1.28$$

$$\text{For } x = 19.5: \quad z = \frac{19.5 - 15}{2.73861279} = 1.64$$

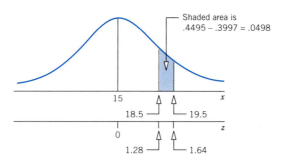

Figure 6.53 Area between $x = 18.5$ and $x = 19.5$.

The required probability is given by the area under the standard normal curve between $z = 1.28$ and $z = 1.64$. This area is obtained by subtracting the area between $z = 0$ and $z = 1.28$ from the area between $z = 0$ and $z = 1.64$. From Table VII of Appendix D, the area between $z = 0$ and $z = 1.28$ is .3997 and the area between $z = 0$ and $z = 1.64$ is .4495. Hence, the required probability is

$$P(18.5 \leq x \leq 19.5) = P(1.28 \leq z \leq 1.64) = .4495 - .3997 = \textbf{.0498}$$

Thus, based on the normal approximation, the probability that 19 persons in a sample of 30 will have at least one credit card is approximately .0498. Earlier, using the binomial formula, we obtained the exact probability .0509. The error due to using the normal approximation is .0509 − .0498 = .0011. Thus, the exact probability is underestimated by .0011 if the normal approximation is used. ∎

☞ *Remember*   When applying the normal distribution as an approximation to the binomial distribution, always make a *correction for continuity*. The continuity correction is made by subtracting .5 from the lower limit of the interval and/or by adding .5 to the upper limit of the interval. For example, the binomial probability $P(7 \leq x \leq 12)$ will be approximated by the probability $P(6.5 \leq x \leq 12.5)$ for the normal distribution; the binomial probability $P(x \geq 9)$ will be approximated by the probability $P(x \geq 8.5)$ for the normal distribution; and the binomial probability $P(x \leq 10)$ will be approximated by the probability $P(x \leq 10.5)$ for the normal distribution. Note that the probability $P(x \geq 9)$ has only the lower limit of 9 and no upper limit, and the probability $P(x \leq 10)$ has only the upper limit of 10 and no lower limit.

Using the normal approximation to the binomial: x assumes a value in an interval.

EXAMPLE 6–21   According to a *USA TODAY* survey of 20,000 American middle and high school students, 70% admitted to cheating on an exam (*Forbes*, January 25, 1999). Suppose that this result is true for the current population of all U.S. middle and high school students. What is the probability that in a random sample of 100 such students, 61 to 65 would admit to cheating on an exam?

Solution   Let n be the total number of middle and high school students in the sample, x be the number of middle and high school students in the sample who would admit to cheating on an exam, and p be the probability that a middle or high school student would admit to cheating on an exam. Then, this is a binomial problem with

$$n = 100, \quad p = .70, \quad \text{and} \quad q = 1 - .70 = .30$$

We are to find the probability of 61 to 65 successes in 100 trials. Because n is large, it is easier to apply the normal approximation than to use the binomial formula. We can check that np and nq are both greater than 5. The mean and standard deviation of the binomial distribution are

$$\mu = np = 100(.70) = 70$$

$$\sigma = \sqrt{npq} = \sqrt{100(.70)(.30)} = 4.58257570$$

To make the continuity correction, we subtract .5 from 61 and add .5 to 65 to obtain the interval 60.5 to 65.5. Thus, the probability that 61 to 65 middle and high school students in the sample would admit to cheating on an exam is approximated by the area under the normal distribution curve from $x = 60.5$ to $x = 65.5$. This area is shown in Figure 6.54.

$$\text{For } x = 60.5: \quad z = \frac{60.5 - 70}{4.58257570} = -2.07$$

$$\text{For } x = 65.5: \quad z = \frac{65.5 - 70}{4.58257570} = -.98$$

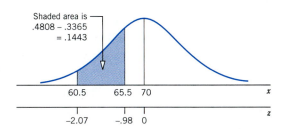

Shaded area is .4808 − .3365 = .1443

Figure 6.54   Area from $x = 60.5$ to $x = 65.5$.

The required probability is given by the area under the standard normal curve between $z = -2.07$ and $z = -.98$. This area is obtained by subtracting the area between $z = -.98$ and $z = 0$ from the area between $z = -2.07$ and $z = 0$. From Table VII in Appendix D, the area between $z = -.98$ and $z = 0$ is .3365 and the area between $z = -2.07$ and $z = 0$ is .4808. Hence,

$$P(60.5 < x < 65.5) = P(-2.07 < z < -.98) = .4808 - .3365 = \mathbf{.1443}$$

Thus, the probability that 61 to 65 American middle and high school students in a random sample of 100 would admit to cheating on an exam is approximately .1443.   ∎

Using the normal approximation to the binomial: x is greater than or equal to a value.

EXAMPLE 6-22   According to U.S. Census Bureau data, 41% of married couples have no children (*USA TODAY*, April 22, 1999). Assume that this result is true for the current population of all U.S. married couples. What is the probability that in a random sample of 200 married couples, 90 or more have no children?

Solution   Let n be the total number of married couples in the sample, x be the number of married couples in the sample who have no children, and p be the probability that a randomly selected married couple has no children. Then, this is a binomial problem with

$$n = 200, \quad p = .41, \quad \text{and} \quad q = 1 - .41 = .59$$

We are to find the probability of 90 or more successes in 200 trials. The mean and standard deviation of the binomial distribution are

$$\mu = np = 200(.41) = 82$$

$$\sigma = \sqrt{npq} = \sqrt{200(.41)(.59)} = 6.95557331$$

For the continuity correction, we subtract .5 from 90, which gives 89.5. Thus, the probability that 90 or more married couples in a random sample of 200 have no children is approximated by the area under the normal distribution curve to the right of $x = 89.5$, as shown in Figure 6.55.

$$\text{For } x = 89.5: \quad z = \frac{89.5 - 82}{6.95557331} = 1.08$$

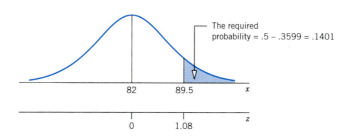

The required
probability = .5 − .3599 = .1401

Figure 6.55   Area to the right of $x = 89.5$.

To find the required probability, we find the area between $z = 0$ and $z = 1.08$ and subtract this area from .5, which is the total area to the right of $z = 0$. From Table VII in Appendix D, the area between $z = 0$ and $z = 1.08$ is .3599. Hence,

$$P(x \geq 89.5) = P(z \geq 1.08) = .5 - .3599 = \mathbf{.1401}$$

Thus, the probability that 90 or more married couples in a random sample of 200 have no children is approximately .1401.   ■

EXERCISES

■ **Concepts and Procedures**

6.65   Under what conditions is the normal distribution usually used as an approximation to the binomial distribution?

6.66   For a binomial probability distribution, $n = 20$ and $p = .60$.

a.   Find the probability $P(x = 12)$ by using the table of binomial probabilities (Table IV of Appendix D).
b.   Find the probability $P(x = 12)$ by using the normal distribution as an approximation to the binomial distribution. What is the difference between this approximation and the exact probability calculated in part a?

6.67   For a binomial probability distribution, $n = 25$ and $p = .40$.

a.   Find the probability $P(8 \leq x \leq 13)$ by using the table of binomial probabilities (Table IV of Appendix D).

b. Find the probability $P(8 \leq x \leq 13)$ by using the normal distribution as an approximation to the binomial distribution. What is the difference between this approximation and the exact probability calculated in part a?

6.68 For a binomial probability distribution, $n = 80$ and $p = .50$. Let x be the number of successes in 80 trials.

a. Find the mean and standard deviation of this binomial distribution.
b. Find $P(x \geq 40)$ using the normal approximation.
c. Find $P(41 \leq x \leq 48)$ using the normal approximation.

6.69 For a binomial probability distribution, $n = 120$ and $p = .60$. Let x be the number of successes in 120 trials.

a. Find the mean and standard deviation of this binomial distribution.
b. Find $P(x \leq 72)$ using the normal approximation.
c. Find $P(67 \leq x \leq 73)$ using the normal approximation.

6.70 Find the following binomial probabilities using the normal approximation.

a. $n = 140,$ $p = .45,$ $P(x = 67)$
b. $n = 100,$ $p = .55,$ $P(52 \leq x \leq 60)$
c. $n = 90,$ $p = .42,$ $P(x \geq 40)$
d. $n = 104,$ $p = .75,$ $P(x \leq 72)$

6.71 Find the following binomial probabilities using the normal approximation.

a. $n = 70,$ $p = .30,$ $P(x = 18)$
b. $n = 200,$ $p = .70,$ $P(133 \leq x \leq 145)$
c. $n = 85,$ $p = .40,$ $P(x \geq 30)$
d. $n = 150,$ $p = .38,$ $P(x \leq 62)$

■ Applications

6.72 According to CDB Research & Consulting, 68% of employees say that they are satisfied with their employers (*USA TODAY*, July 8, 1998). Suppose that this result holds true for the current population of employees in the United States. Find the probability that in a random sample of 800 such employees, 540 to 550 say that they are satisfied with their employers.

6.73 According to the Business Work-Life Study survey of 1000 American companies by the Families and Work Institute, 33% of these companies offered maternity leaves of over 13 weeks (*The Hartford Courant*, July 27, 1998). Assume that 33% of all American companies offer such leaves. Find the probability that in a random sample of 200 companies, 50 to 60 offer such leaves.

6.74 According to data from the NPD Group, 42% of Americans eat breakfast every day of the week (*Newsweek*, May 17, 1999). Assume that this percentage is true for the current population of all Americans. Find the probability that in a random sample of 300 Americans, the number who eat breakfast every day is

a. exactly 120   **b.** at most 100   **c.** 130 to 140

6.75 A 1998 study by Mathtech, Inc., for the U.S. Department of Education found that high school students from low-income families who scored high on a standard test were less likely to attend college than all students from the highest income group. Fifty-seven percent of students with high test scores from low-income families who were not planning to attend college said they were not going to because they could not afford it (*The Willimantic Chronicle*, August 11, 1998). Assume that this percentage holds true for all such students. Find the probability that in a random sample of 480 students from low-income families but with high test scores who do not plan to attend college, the number who state that they cannot afford college is

a. exactly 270   **b.** 300 or more   **c.** 260 to 280

6.76 The 1999 Catalyst Census of Women Corporate Officers and Top Earners found that about 11.9% of 11,681 corporate officers working in the top 500 U.S. companies were women (*The Willimantic Chronicle*, November 13, 1999). Suppose a random sample of 200 corporate officers is taken from these 11,681 corporate officers.

a. Find the probability that this sample contains exactly 25 women.
b. Find the probability that this sample contains at most 30 women.
c. What is the probability that this sample contains 25 to 35 women?

6.77 A fast food chain store conducted a taste survey before marketing a new hamburger. The results of the survey showed that 70% of the people who tried this hamburger liked it. Encouraged by this result, the company decided to market the new hamburger. Assume that 70% of all people like this hamburger. On a certain day, 100 customers bought this hamburger.

a. Find the probability that exactly 64 of the 100 customers will like this hamburger.
b. What is the probability that 61 or fewer of the 100 customers will like this hamburger?
c. What is the probability that 75 to 82 of the 100 customers will like this hamburger?

6.78 Johnson Electronics makes calculators. Consumer satisfaction is one of the top priorities of the company's management. The company guarantees the refund of money or a replacement for any calculator that malfunctions within 2 years from the date of purchase. It is known from past data that despite all efforts, 5% of the calculators manufactured by this company malfunction within a 2-year period. The company recently mailed 500 such calculators to its customers.

a. Find the probability that exactly 29 of the 500 calculators will be returned for refund or replacement within a 2-year period.
b. What is the probability that 27 or more of the 500 calculators will be returned for refund or replacement within a 2-year period?
c. What is the probability that 15 to 22 of the 500 calculators will be returned for refund or replacement within a 2-year period?

6.79 Hurbert Corporation makes font cartridges for laser printers that it sells to Alpha Electronics Inc. The cartridges are shipped to Alpha Electronics in large volumes. The quality control department at Alpha Electronics randomly selects 100 cartridges from each shipment and inspects them for being good or defective. If this sample contains seven or more defective cartridges, the entire shipment is rejected. Hurbert Corporation promises that of all the cartridges, only 5% are defective.

a. Find the probability that a given shipment of cartridges received by Alpha Electronics will be accepted.
b. Find the probability that a given shipment of cartridges received by Alpha Electronics will not be accepted.

GLOSSARY

Continuity correction factor   Addition of .5 and/or subtraction of .5 from the value(s) of x when the normal distribution is used as an approximation to the binomial distribution, where x is the number of successes in n trials.

Continuous random variable   A random variable that can assume any value in one or more intervals.

Normal probability distribution   The probability distribution of a continuous random variable that, when plotted, gives a specific bell-shaped curve. The parameters of the normal distribution are the mean μ and the standard deviation σ.

Standard normal distribution   The normal distribution with $\mu = 0$ and $\sigma = 1$. The units of the standard normal distribution are denoted by z.

z **value** or *z* **score**   The units of the standard normal distribution that are denoted by z.

KEY FORMULAS

1. **The z value for an x value to standardize a normal distribution**

$$z = \frac{x - \mu}{\sigma}$$

2. **The value of x for a normal distribution, when the values of μ, σ, and z are known**

$$x = \mu + z\,\sigma$$

3. **Normal approximation to the binomial distribution**

The normal distribution can be used as an approximation to the binomial distribution when

$$np > 5 \quad \text{and} \quad nq > 5$$

4. **Mean and standard deviation of the binomial distribution**

$$\mu = np \quad \text{and} \quad \sigma = \sqrt{npq}$$

SUPPLEMENTARY EXERCISES

6.80 The management at Ohio National Bank does not want its customers to wait in line for service for too long. The manager of a branch of this bank estimated that the customers currently have to wait an average of 8 minutes for service. Assume that the waiting times for all customers at this branch have a normal distribution with a mean of 8 minutes and a standard deviation of 2 minutes.

a. Find the probability that a randomly selected customer will have to wait for less than 3 minutes.
b. What percentage of the customers have to wait for 10 to 13 minutes?
c. What percentage of the customers have to wait for 6 to 12 minutes?
d. Is it possible that a customer may have to wait longer than 16 minutes for service? Explain.

6.81 The research department at Colorado Bank estimated that the mean daily balance of all checking accounts at this bank is $870. Suppose the balances of all checking accounts at this bank have a normal distribution with a mean of $870 and a standard deviation of $350.

a. Find the probability that on a certain day a randomly selected checking account will have a balance of more than $1450.
b. What percentage of the checking accounts at this bank will have a balance of $200 to $350 on a certain day?
c. What percentage of the checking accounts at this bank will have a balance of $1050 to $1400 on a certain day?
d. Is it possible that a checking account will have a balance of more than $2550 on a certain day? Explain.

6.82 At Jen and Perry Ice Cream Company, the machine that fills 1-pound cartons of Top Flavor ice cream is set to dispense 16 ounces of ice cream into every carton. However, some cartons contain slightly less than and some contain slightly more than 16 ounces of ice cream. The amounts of ice cream in all such cartons have a normal distribution with a mean of 16 ounces and a standard deviation of .18 ounce.

a. Find the probability that a randomly selected carton contains 16.20 to 16.50 ounces of ice cream.
b. What percentage of such cartons contain less than 15.70 ounces of ice cream?
c. Is it possible for a carton to contain less than 15.20 ounces of ice cream? Explain.

6.83 A machine at Kasem Steel Corporation makes iron rods that are supposed to be 50 inches long. However, the machine does not make all rods of exactly the same length. It is known that the probability distribution of the lengths of rods made on this machine is normal with a mean of 50 inches and a standard deviation of .06 inch. The rods that are either shorter than 49.85 inches or longer than 50.15 inches are discarded. What percentage of the rods made on this machine are discarded?

6.84 The time taken by new employees at Shia Corporation to learn a packaging procedure is normally distributed with a mean of 24 hours and a standard deviation of 2.5 hours.

a. Find the time beyond which only 1% of the new employees take to learn this procedure.
b. What is the time below which only 5% of the new employees take to learn this procedure?

6.85 The print on the package of 100-watt General Electric soft-white lightbulbs states that these bulbs have an average life of 750 hours. Assume that the lives of all such bulbs have a normal distribution with a mean of 750 hours and a standard deviation of 50 hours.

a.   Let x be the life of such a lightbulb. Find x so that only 2.5% of such lightbulbs have lives longer than this value.

b.   Let x be the life of such a lightbulb. Find x so that about 80% of such lightbulbs have lives shorter than this value.

6.86   Mong Corporation makes auto batteries. The company claims that 80% of its LL70 batteries are good for 70 months or longer.

a.   What is the probability that in a sample of 100 such batteries, exactly 85 will be good for 70 months or longer?

b.   Find the probability that in a sample of 100 such batteries, at most 74 will be good for 70 months or longer.

c.   What is the probability that in a sample of 100 such batteries, 75 to 87 will be good for 70 months or longer?

d.   Find the probability that in a sample of 100 such batteries, 72 to 77 will be good for 70 months or longer.

6.87   Stress on the job is a major concern of a large number of people who go into managerial positions. It is estimated that 80% of the managers of all companies suffer from job-related stress.

a.   What is the probability that in a sample of 200 managers of companies, exactly 150 suffer from job-related stress?

b.   Find the probability that in a sample of 200 managers of companies, at least 170 suffer from job-related stress.

c.   What is the probability that in a sample of 200 managers of companies, 165 or less suffer from job-related stress?

d.   Find the probability that in a sample of 200 managers of companies, 164 to 172 suffer from job-related stress.

■ Advanced Exercises

6.88   It is known that 15% of all homeowners pay a monthly mortgage of more than $2500 and that the standard deviation of the monthly mortgage payments of all homeowners is $350. Suppose that the monthly mortgage payments of all homeowners have a normal distribution. What is the mean monthly mortgage paid by all homeowners?

6.89   At Jen and Perry Ice Cream Company, a machine fills 1-pound cartons of Top Flavor ice cream. The machine can be set to dispense, on average, any amount of ice cream into these cartons. However, the machine does not put exactly the same amount of ice cream into each carton; it varies from carton to carton. It is known that the amount of ice cream put into each such carton has a normal distribution with a standard deviation of .18 ounce. The quality control inspector wants to set the machine such that at least 90% of the cartons have more than 16 ounces of ice cream. What should be the mean amount of ice cream put into these cartons by this machine?

6.90   Two companies, A and B, drill wells in a rural area. Company A charges a flat fee of $3500 to drill a well regardless of its depth. Company B charges $1000 plus $12 per foot to drill a well. The depths of wells drilled in this area have a normal distribution with a mean of 250 feet and a standard deviation of 40 feet.

a.   What is the probability that Company B would charge more than Company A to drill a well?

b.   Find the mean amount charged by Company B to drill a well.

6.91   Otto is trying out for the javelin throw to compete in the Olympics. The lengths of his javelin throws are normally distributed with a mean of 290 feet and a standard deviation of 10 feet. What is the probability that the longest of three of his throws is 320 feet or more?

6.92   Lori just bought a new set of four tires for her car. The life of each tire is normally distributed with a mean of 45,000 miles and a standard deviation of 2000 miles. Find the probability that all four tires will last at least 46,000 miles. Assume that the life of each of these tires is independent of the lives of other tires.

6.93   The Jen and Perry Ice Cream company makes a gourmet ice cream. Although the law allows ice cream to contain up to 50% air, this product is designed to contain only 20% air. Because of variability inherent in the manufacturing process, management is satisfied if each pint contains between 18% and 22% air. Currently two of Jen and Perry's plants are making gourmet ice cream. At Plant A, the mean amount of air per pint is 20% with a standard deviation of 2%. At Plant B, the mean amount of air per pint is 19% with a standard deviation of 1%. Assuming the amount of air is normally distributed at both plants, which plant is producing the greater proportion of pints that contain between 18% and 22% air?

6.94 The highway police in a certain state are using aerial surveillance to control speeding on a highway with a posted speed limit of 55 miles per hour. Police officers watch cars from helicopters above a straight segment of this highway that has large marks painted on the pavement at 1-mile intervals. After the police officers observe how long a car takes to cover 1 mile, a computer estimates that car's speed. Assume that the errors of these estimates are normally distributed with a mean of 0 and a standard deviation of 2 miles per hour.

a. The state police chief has directed his officers not to issue a speeding citation unless the aerial unit's estimate of speed is at least 65 miles per hour. What is the probability that a car traveling at 60 miles per hour or less will be cited for speeding?

b. Suppose the chief does not want his officers to cite a car for speeding unless they are 99% sure that it is traveling at 60 miles per hour or faster. What is the minimum estimate of speed at which a car should be cited for speeding?

6.95 Ashley knows that the time it takes her to commute to work is approximately normally distributed with a mean of 45 minutes and a standard deviation of 3 minutes. What time must she leave home in the morning so that she is 95% sure of arriving at work by 9 A.M.?

6.96 A soft-drink vending machine is supposed to pour 8 ounces of the drink into a paper cup. However, the actual amount poured into a cup varies. The amount poured into a cup follows a normal distribution with a mean that can be set to any desired amount by adjusting the machine. The standard deviation of the amount poured is always .07 ounce regardless of the mean amount. If the owner of the machine wants to be 99% sure that the amount in each cup is 8 ounces or more, to what level should she set the mean?

6.97 A newspaper article reported that the mean mathematics score on SAT tests for local high school students was 480 and that 20% of the students scored below 400. Assume that the SAT scores for students from this school follow a normal distribution.

a. Find the standard deviation of the mathematics SAT scores for students from this school.

b. Find the percentage of students at this school whose mathematics SAT scores were above 500.

6.98 Alpha Corporation is considering two suppliers to secure the large amounts of steel rods that it uses. Company A produces rods with a mean diameter of 8 mm and a standard deviation of .15 mm and sells 10,000 rods for $400. Company B produces rods with a mean diameter of 8 mm and a standard deviation of .12 mm and sells 10,000 rods for $460. A rod is usable only if its diameter is between 7.8 mm and 8.2 mm. Assume that the diameters of the rods produced by each company have a normal distribution. Which of the two companies should Alpha Corporation use as a supplier? Justify your answer with appropriate calculations.

6.99 A gambler is planning to make a sequence of bets on a roulette wheel. Note that a roulette wheel has 38 numbers of which 18 are red, 18 are black, and 2 are green. Each time the wheel is spun, each of the 38 numbers is equally likely to occur. The gambler will choose one of the following two sequences.

Single-number bet: The gambler will bet $5 on a particular number before each spin. He will win a net amount of $175 if that number comes up and lose $5 otherwise.

Color bet: The gambler will bet $5 on the red color before each spin. He will win a net amount of $5 if a red number comes up and lose $5 otherwise.

a. If the gambler makes a sequence of 25 bets, which of the two betting schemes do you think offers him a better chance of coming out ahead (winning more money than losing) after the 25 bets?

b. Now compute the probability of coming out ahead after 25 single-number bets of $5 each and after 25 color bets of $5 each. Do these results confirm your guess in part a? (Before using an approximation to find either probability, be sure to check whether it is appropriate.)

6.100 A charter bus company is advertising a singles outing on a bus that holds 60 passengers. The company has found that, on average, 10% of ticket holders do not show up for such trips; hence, the company routinely overbooks such trips. Assume that passengers act independently of one another.

a. If the company sells 65 tickets, what is the probability that the bus can hold all the ticket holders who actually show up? In other words, find the probability that 60 or fewer passengers show up.

b. What is the largest number of tickets the company can sell and still be at least 95% sure that the bus can hold all the ticket holders who actually show up?

6.101 The amount of time taken by a bank teller to serve a randomly selected customer has a normal distribution with a mean of 2 minutes and a standard deviation of .5 minute.

a. What is the probability that both of two randomly selected customers will take less than 1 minute each to be served?

b. What is the probability that at least one of four randomly selected customers will need more than 2.25 minutes to be served?

SELF-REVIEW TEST

1. The normal probability distribution is applied to

    a. a continuous random variable   **b.** a discrete random variable   **c.** any random variable

2. For a continuous random variable, the probability of a single value of x is always

    a. zero   **b.** 1.0   **c.** between 0 and 1

3. Which of the following is not a characteristic of the normal distribution?

    a. The total area under the curve is 1.0.
    b. The curve is symmetric about the mean.
    c. The two tails of the curve extend indefinitely.
    d. The value of the mean is always greater than the value of the standard deviation.

4. The parameters of a normal distribution are

    a. μ, z, and σ **b.** μ and σ **c.** μ, x, and σ

5. For the standard normal distribution,

    a. $\mu = 0$ and $\sigma = 1$ **b.** $\mu = 1$ and $\sigma = 0$ **c.** $\mu = 100$ and $\sigma = 10$

6. The z value for μ for a normal distribution curve is always

    a. positive   **b.** negative   **c.** 0

7. For a normal distribution curve, the z value for an x value that is less than μ is always

    a. positive   **b.** negative   **c.** 0

8. Usually the normal distribution is used as an approximation to the binomial distribution when

    a. $n \geq 30$ **b.** $np > 5$ and $nq > 5$ **c.** $n > 20$ and $p = .50$

9. Find the following probabilities for the standard normal distribution.

    a. $P(.85 \leq z \leq 2.33)$ **b.** $P(-2.97 \leq z \leq 1.49)$ **c.** $P(z \leq -1.29)$ **d.** $P(z > -.74)$

10. Find the value of z for the standard normal curve such that the area

    a. in the left tail is .1000
    b. between 0 and z is .2291 and z is positive
    c. in the right tail is .0500
    d. between 0 and z is .3571 and z is negative

11. A study of time spent on housework found that men who are employed spend an average of 8.2 hours per week doing housework (Americans' Use of Time Project, University of Maryland, *American Demographics*, November 1998). Assume that the amount of time spent on housework per week by all employed men in the United States is normally distributed with a mean of 8.2 hours and a standard deviation of 2.1 hours.

    a. Find the probability that a randomly selected employed man spends between 6 and 10 hours per week on housework.

    b. What is the probability that a randomly selected employed man spends 9 or more hours per week on housework?

    c. What is the probability that a randomly selected employed man spends 7 hours or less on housework per week?

    d. Find the probability that a randomly selected employed man spends between 9 and 12 hours per week on housework.

12. Refer to Problem 11.

    a. The 5% of employed men who spend the least time on housework spend fewer than how many hours per week on housework?

    b. The 10% of employed men who spend the most time on housework spend more than how many hours per week on housework?

13. In a 1999 poll of a sample of parents with children under 18 years of age, 53% of the parents used the TV ratings system to guide their children's selection of TV programs (Brushin/Goldring Research for Court TV, *USA TODAY*, September 2, 1999). Assume that this percentage is true for the current population of all such parents. A random sample of 420 such parents is selected.

    a. Find the probability that exactly 200 of them use these ratings.

    b. What is the probability that between 200 and 215 of them use these ratings?

    c. Find the probability that at least 225 of them use these ratings.

    d. What is the probability that at most 200 of them use these ratings?

    e. Find the probability that between 220 and 240 of them use these ratings.

    f. Find the probability that at most 220 of them do not use these ratings.

    g. Find the probability that 200 to 250 of them do not use these ratings.

MINI-PROJECTS

■ Mini-Project 6–1

Consider the data on heights of NBA players that accompany this text.

a. Use MINITAB or any other statistical software to obtain a histogram. Do these heights appear to be symmetrically distributed? If not, in which direction do they seem to be skewed?

b. Compute μ and σ.

c. What percentage of these heights lie in the interval $\mu - \sigma$ to $\mu + \sigma$? What about in the interval $\mu - 2\sigma$ to $\mu + 2\sigma$? In the interval $\mu - 3\sigma$ to $\mu + 3\sigma$?

d. How do the percentages in part c compare to the corresponding percentages for a normal distribution (68%, 95%, and 99.7%, respectively)?

■ Mini-Project 6–2

Consider the data on weights of NBA players.

a. Use MINITAB or any other statistical software to obtain a histogram. Do these weights appear to be symmetrically distributed? If not, in which direction do they seem to be skewed?

b. Compute μ and σ.

c. What percentage of these weights lie in the interval $\mu - \sigma$ to $\mu + \sigma$? What about in the interval $\mu - 2\sigma$ to $\mu + 2\sigma$? In the interval $\mu - 3\sigma$ to $\mu + 3\sigma$?

d. How do the percentages in part c compare to the corresponding percentages for a normal distribution (68%, 95%, and 99.7%, respectively)?

■ **Mini-Project 6–3**

Using weather records from back issues of a local newspaper or from some other source, record the maximum temperatures in your town for a period of 60 days.

a. Calculate $\bar{x}$ and s.
b. What percentage of the temperatures are in the interval $\bar{x} - s$ to $\bar{x} + s$?
c. What percentage are in the interval $\bar{x} - 2s$ to $\bar{x} + 2s$?
d. How do these percentages compare to the corresponding percentages for a normal distribution (68% and 95%, respectively)?
e. Now find the minimum temperatures in your town for 60 days by using the same source that you used to find the maximum temperatures or by using a different source. Then repeat parts a through d for this data set.

For instructions on using MINITAB for this chapter, please see Section B.5 in Appendix B.

See Appendix C for *Excel Adventure 6* on the normal distribution.

COMPUTER ASSIGNMENTS

CA6.1   Using MINITAB or any other statistical software, find the area under the standard normal curve

a.   to the left of $z = -1.94$ **b.**   to the left of $z = .83$
c.   to the right of $z = 1.45$ **d.**   to the right of $z = -1.65$
e.   between $z = .75$ and $z = 1.90$ **f.**   between $z = -1.20$ and $z = 1.55$

CA6.2   Using MINITAB or any other statistical software, find the following areas under a normal curve with $\mu = 86$ and $\sigma = 14$.

a.   Area to the left of $x = 71$ **b.**   Area to the left of $x = 96$
c.   Area to the right of $x = 90$ **d.**   Area to the right of $x = 75$
e.   Area between $x = 65$ and $x = 75$ **f.**   Area between $x = 72$ and $x = 95$

CA6.3   According to the American Association of University Professors, the mean 1997–1998 salary of full-time instructional faculty at two-year colleges was $43,760. Suppose that these salaries have a normal distribution with a mean of $43,760 and a standard deviation of $4200.

a.   Using MINITAB or any other statistical software, find the probability that the 1997–1998 salary of a randomly selected faculty member from such colleges was less than $40,000.

b.   Using MINITAB or any other statistical software, find the probability that the 1997–1998 salary of a randomly selected faculty member from such colleges was more than $37,000.

c.   Using MINITAB or any other statistical software, find the probability that the 1997–1998 salary of a randomly selected faculty member from such colleges was between $45,000 and $53,000.

CA6.4   According to data from Runzheimer International, the average cost of suburban day care (for a 3-year-old in a for-profit center, 8 hours per day for 5 days a week) in the Boston area was $622 per month (*USA TODAY*, August 18, 1998). Assume that current such day-care costs in the Boston area are normally distributed with a mean of $622 and a standard deviation of $120. Do the following using MINITAB or any other statistical software.

a.   Find the probability that a randomly selected couple with a 3-year-old child in the Boston area pays more than $500 per month for day care.

b.   What percentage of such couples with a 3-year-old child in the Boston area spend between $350 and $500 per month for day care?

c.   Find the probability that a randomly selected couple with a 3-year-old child in the Boston area spends less than $750 per month for day care.

CA6.5 The transmission on a particular model of car has a warranty for 40,000 miles. It is known that the life of such a transmission has a normal distribution with a mean of 72,000 miles and a standard deviation of 12,000 miles. Answer the following questions using MINITAB or any other statistical software.

a. What percentage of the transmissions will fail before the end of the warranty period?
b. What percentage of the transmissions will be good for more than 100,000 miles?
c. What percentage of the transmissions will be good for 80,000 to 100,000 miles?

7 | SAMPLING DISTRIBUTIONS

7.1 POPULATION AND SAMPLING DISTRIBUTIONS

7.2 SAMPLING AND NONSAMPLING ERRORS

7.3 MEAN AND STANDARD DEVIATION OF $\bar{x}$

7.4 SHAPE OF THE SAMPLING DISTRIBUTION OF $\bar{x}$

7.5 APPLICATIONS OF THE SAMPLING DISTRIBUTION OF $\bar{x}$

7.6 POPULATION AND SAMPLE PROPORTIONS

7.7 MEAN, STANDARD DEVIATION, AND SHAPE OF THE SAMPLING DISTRIBUTION OF $\hat{p}$

CASE STUDY 7–1 CLINTON POLL NUMBERS ARE NOT PULLED FROM A HAT

7.8 APPLICATIONS OF THE SAMPLING DISTRIBUTION OF $\hat{p}$

GLOSSARY

KEY FORMULAS

SUPPLEMENTARY EXERCISES

SELF-REVIEW TEST

MINI-PROJECTS

COMPUTER ASSIGNMENTS

Chapters 5 and 6 discussed probability distributions of discrete and continuous random variables. This chapter extends the concept of probability distribution to that of a sample statistic. As we discussed in Chapter 3, a sample statistic is a numerical summary measure calculated for sample data. The mean, median, mode, and standard deviation calculated for sample data are called *sample statistics*. On the other hand, the same numerical summary measures calculated for population data are called *population parameters*. A population parameter is always a constant, whereas a sample statistic is always a random variable. Since every random variable must possess a probability distribution, each sample statistic possesses a probability distribution. The probability distribution of a sample statistic is more commonly called its *sampling distribution*. This chapter discusses the sampling distributions of the sample mean and the sample proportion. The concepts covered in this chapter are the foundation of the inferential statistics discussed in succeeding chapters.

7.1 POPULATION AND SAMPLING DISTRIBUTIONS

This section introduces the concepts of population distribution and sampling distribution. Subsection 7.1.1 explains the population distribution, and Subsection 7.1.2 describes the sampling distribution of $\bar{x}$.

7.1.1 POPULATION DISTRIBUTION

The **population distribution** is the probability distribution derived from the information on all elements of a population.

> **POPULATION DISTRIBUTION** The *population distribution* is the probability distribution of the population data.

Suppose there are only five students in an advanced statistics class and the midterm scores of these five students are

$$70 \quad 78 \quad 80 \quad 80 \quad 95$$

Let x denote the score of a student. Using single-valued classes (because there are only five data values, there is no need to group them), we can write the frequency distribution of scores as in Table 7.1 along with the relative frequencies of classes, which are obtained by dividing the frequencies of classes by the population size. Table 7.2, which lists the probabilities of various x values, presents the probability distribution of the population. Note that these probabilities are the same as the relative frequencies.

Table 7.1 Population Frequency and Relative Frequency Distributions

x	f	Relative Frequency
70	1	1/5 = .20
78	1	1/5 = .20
80	2	2/5 = .40
95	1	1/5 = .20
	$N = 5$	Sum = 1.00

Table 7.2 Population Probability Distribution

x	$P(x)$
70	.20
78	.20
80	.40
95	.20
	$\Sigma P(x) = 1.00$

The values of the mean and standard deviation calculated for the probability distribution of Table 7.2 give the values of the population parameters μ and σ. These values are

$\mu = 80.60$ and $\sigma = 8.09$. The values of μ and σ for the probability distribution of Table 7.2 can be calculated using the formulas given in Sections 5.3 and 5.4 of Chapter 5 (see Exercise 7.6).

7.1.2   SAMPLING DISTRIBUTION

As mentioned at the beginning of this chapter, the value of a population parameter is always constant. For example, for any population data set, there is only one value of the population mean, μ. However, we cannot say the same about the sample mean, $\bar{x}$. We would expect different samples of the same size drawn from the same population to yield different values of the sample mean, $\bar{x}$. The value of the sample mean for any one sample will depend on the elements included in that sample. Consequently, *the sample mean, $\bar{x}$, is a random variable.* Therefore, like other random variables, the sample mean possesses a probability distribution, which is more commonly called the **sampling distribution of $\bar{x}$**. Other sample statistics such as the median, mode, and standard deviation also possess sampling distributions.

> **SAMPLING DISTRIBUTION OF $\bar{x}$**   The probability distribution of $\bar{x}$ is called its sampling distribution. It lists the various values that $\bar{x}$ can assume and the probability of each value of $\bar{x}$.
>
> In general, the probability distribution of a sample statistic is called its *sampling distribution.*

Reconsider the population of midterm scores of five students given in Table 7.1. Consider all possible samples of three scores each that can be selected, without replacement, from that population. The total number of possible samples, given by the combinations formula discussed in Chapter 5, is 10; that is,

$$\text{Total number of samples} = \binom{5}{3} = \frac{5!}{3! \, (5-3)!} = \frac{5 \cdot 4 \cdot 3 \cdot 2 \cdot 1}{3 \cdot 2 \cdot 1 \cdot 2 \cdot 1} = 10$$

Suppose we assign the letters A, B, C, D, and E to the scores of the five students so that

$$A = 70, \quad B = 78, \quad C = 80, \quad D = 80, \quad E = 95$$

Then the 10 possible samples of three scores each are

$$\text{ABC, \quad ABD, \quad ABE, \quad ACD, \quad ACE, \quad ADE, \quad BCD, \quad BCE, \quad BDE, \quad CDE}$$

These 10 samples and their respective means are listed in Table 7.3. Note that the first two samples have the same three scores. The reason for this is that two of the students (C and D) have the same score and, hence, the samples ABC and ABD contain the same values. The mean of each sample is obtained by dividing the sum of the three scores included in that sample by 3. For instance, the mean of the first sample is $(70 + 78 + 80)/3 = 76$. Note that the values of the means of samples in Table 7.3 are rounded to two decimal places.

By using the values of $\bar{x}$ given in Table 7.3, we record the frequency distribution of $\bar{x}$ in Table 7.4. By dividing the frequencies of the various values of $\bar{x}$ by the sum of all frequencies, we obtain the relative frequencies of classes, which are listed in the third column of Table 7.4. These relative frequencies are used as probabilities and listed in Table 7.5. This table gives the sampling distribution of $\bar{x}$.

Table 7.3 All Possible Samples and Their Means When the Sample Size Is 3

Sample	Scores in the Sample	$\bar{x}$
ABC	70, 78, 80	76.00
ABD	70, 78, 80	76.00
ABE	70, 78, 95	81.00
ACD	70, 80, 80	76.67
ACE	70, 80, 95	81.67
ADE	70, 80, 95	81.67
BCD	78, 80, 80	79.33
BCE	78, 80, 95	84.33
BDE	78, 80, 95	84.33
CDE	80, 80, 95	85.00

Table 7.4 Frequency and Relative Frequency Distributions of $\bar{x}$ When the Sample Size Is 3

$\bar{x}$	f	Relative Frequency
76.00	2	2/10 = .20
76.67	1	1/10 = .10
79.33	1	1/10 = .10
81.00	1	1/10 = .10
81.67	2	2/10 = .20
84.33	2	2/10 = .20
85.00	1	1/10 = .10
	$\Sigma f = 10$	Sum = 1.00

Table 7.5 Sampling Distribution of $\bar{x}$ When the Sample Size Is 3

$\bar{x}$	$P(\bar{x})$
76.00	.20
76.67	.10
79.33	.10
81.00	.10
81.67	.20
84.33	.20
85.00	.10
	$\Sigma P(\bar{x}) = 1.00$

If we select just one sample of three scores from the population of five scores, we may draw any of the 10 possible samples. Hence, the sample mean, $\bar{x}$, can assume any of the values listed in Table 7.5 with the corresponding probability. For instance, the probability that the mean of a randomly selected sample of three scores is 81.67 is .20. This probability can be written as

$$P(\bar{x} = 81.67) = .20$$

7.2 SAMPLING AND NONSAMPLING ERRORS

Usually, different samples selected from the same population will give different results because they contain different elements. This is obvious from Table 7.3, which shows that the mean of a sample of three scores depends on which three of the five scores are included in the sample. The result obtained from any one sample will generally be different from the result obtained from the corresponding population. The difference between the value of a sample statistic obtained from a sample and the value of the corresponding population parameter obtained from the population is called the **sampling error.** Note that this difference represents the sampling error only if the sample is random and no nonsampling error has been made. Otherwise, only a part of this difference will be due to the sampling error.

> **SAMPLING ERROR**   *Sampling error* is the difference between the value of a sample statistic and the value of the corresponding population parameter. In the case of the mean,
>
> $$\text{Sampling error} = \bar{x} - \mu$$
>
> assuming that the sample is random and no nonsampling error has been made.

It is important to remember that *a sampling error occurs because of chance.* The errors that occur for other reasons, such as errors made during collection, recording, and tabulation of data, are called **nonsampling errors**. These errors occur because of human mistakes, and not chance. Note that there is only one kind of sampling error—the error that occurs due to chance. However, there is not just one nonsampling error but many nonsampling errors that may occur for different reasons.

> **NONSAMPLING ERRORS**   The errors that occur in the collection, recording, and tabulation of data are called *nonsampling errors*.

The following paragraph, reproduced from the *Current Population Reports* of the U.S. Bureau of the Census, explains how nonsampling errors can occur.

Nonsampling errors can be attributed to many sources, e.g., inability to obtain information about all cases in the sample, definitional difficulties, differences in the interpretation of questions, inability or unwillingness on the part of the respondents to provide correct information, inability to recall information, errors made in collection such as in recording or coding the data, errors made in processing the data, errors made in estimating values for missing data, biases resulting from the differing recall periods caused by the interviewing pattern used, and failure of all units in the universe to have some probability of being selected for the sample (undercoverage).

The following are the main reasons for the occurrence of nonsampling errors.

1.   If a sample is nonrandom (and, hence, nonrepresentative), the sample results may be too different from the census results. The following quote from *U.S. News & World Report* describes how even a randomly selected sample can become nonrandom if some of the members included in the sample cannot be contacted.

A test poll conducted in the 1984 presidential election found that if the poll were halted after interviewing only those subjects who could be reached on the first try, Reagan showed a 3-percentage-point lead over Mondale. But when interviewers made a determined effort to reach everyone on their lists of randomly selected subjects—calling some as many as 30 times before finally reaching them—Reagan showed a 13 percent lead, much closer to the actual election result. As it turned out, people who were planning to vote Republican were simply less likely to be at home. ("The Numbers Racket: How Polls and Statistics Lie," *U.S. News & World Report*, July 11, 1988. Copyright © 1988 by U.S. News & World Report, Inc. Reprinted with permission.)

2. The questions may be phrased in such a way that they are not fully understood by the members of the sample or population. As a result, the answers obtained are not accurate.

3. The respondents may intentionally give false information in response to some sensitive questions. For example, people may not tell the truth about their drinking habits, incomes, or opinions about minorities. Sometimes the respondents may give wrong answers because of ignorance. For example, a person may not remember the exact amount he or she spent on clothes during the last year. If asked in a survey, he or she may give an inaccurate answer.

4. The poll taker may make a mistake and enter a wrong number in the records or make an error while entering the data on a computer.

Note that nonsampling errors can occur both in a sample survey and in a census, whereas sampling error occurs only when a sample survey is conducted. Nonsampling errors can be minimized by preparing the survey questionnaire carefully and handling the data cautiously. However, it is impossible to avoid sampling error.

Example 7–1 illustrates the sampling and nonsampling errors using the mean.

Illustrating sampling and nonsampling errors.

EXAMPLE 7–1 Reconsider the population of five scores given in Table 7.1. The scores of the five students are 70, 78, 80, 80, and 95. The population mean is

$$\mu = \frac{70 + 78 + 80 + 80 + 95}{5} = 80.60$$

Now suppose we take a random sample of three scores from this population. Assume that this sample includes the scores 70, 80, and 95. The mean for this sample is

$$\bar{x} = \frac{70 + 80 + 95}{3} = 81.67$$

Consequently,

$$\text{Sampling error} = \bar{x} - \mu = 81.67 - 80.60 = \mathbf{1.07}$$

That is, the mean score estimated from the sample is 1.07 higher than the mean score of the population. Note that this difference occurred due to chance—that is, because we used a sample instead of the population.

Now suppose, when we select the sample of three scores, we mistakenly record the second score as 82 instead of 80. As a result, we calculate the sample mean as

$$\bar{x} = \frac{70 + 82 + 95}{3} = 82.33$$

Consequently, the difference between this sample mean and the population mean is

$$\bar{x} - \mu = 82.33 - 80.60 = 1.73$$

However, this difference between the sample mean and the population mean does not represent the sampling error. As we calculated earlier, only 1.07 of this difference is due to the sampling error. The remaining portion, which is equal to $1.73 - 1.07 = .66$, represents the nonsampling error because it occurred due to the error we made in recording the second score in the sample. Thus, in this case,

$$\text{Sampling error} = \mathbf{1.07}$$

$$\text{Nonsampling error} = \mathbf{.66}$$

Figure 7.1 shows the sampling and nonsampling errors for these calculations.

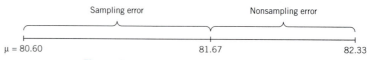

Figure 7.1 Sampling and nonsampling errors.

Thus, the sampling error is the difference between the correct value of $\bar{x}$ and μ, where the correct value of $\bar{x}$ is the value of $\bar{x}$ that does not contain any nonsampling errors. In contrast, the nonsampling error(s) is (are) obtained by subtracting the correct value of $\bar{x}$ from the incorrect value of $\bar{x}$, where the incorrect value of $\bar{x}$ is the value that contains the nonsampling error(s). For our example,

$$\text{Sampling error} = \bar{x} - \mu = 81.67 - 80.60 = 1.07$$

$$\text{Nonsampling error} = \text{Incorrect } \bar{x} - \text{Correct } \bar{x} = 82.33 - 81.67 = .66$$ ∎

Note that in the real world we do not know the mean of a population. Hence, we select a sample to use the sample mean as an estimate of the population mean. Consequently, we never know the size of the sampling error.

EXERCISES

■ **Concepts and Procedures**

7.1 Briefly explain the meaning of a population distribution and a sampling distribution. Give an example of each.

7.2 Explain briefly the meaning of sampling error. Give an example. Does such an error occur only in a sample survey or can it occur in both a sample survey and a census?

7.3 Explain briefly the meaning of nonsampling errors. Give an example. Do such errors occur only in a sample survey or can they occur in both a sample survey and a census?

7.4 Consider the following population of six numbers.

$$15 \quad 13 \quad 8 \quad 17 \quad 9 \quad 12$$

a. Find the population mean.
b. Liza selected one sample of four numbers from this population. The sample included the numbers 13, 8, 9, and 12. Calculate the sample mean and sampling error for this sample.
c. Refer to part b. When Liza calculated the sample mean, she mistakenly used the numbers 13, 8, 6, and 12 to calculate the sample mean. Find the sampling and nonsampling errors in this case.
d. List all samples of four numbers (without replacement) that can be selected from this population. Calculate the sample mean and sampling error for each of these samples.

7.5   Consider the following population of 10 numbers.

20   25   13   19   9   15   11   7   17   30

a.   Find the population mean.
b.   Rich selected one sample of nine numbers from this population. The sample included the numbers 20, 25, 13, 9, 15, 11, 7, 17, and 30. Calculate the sample mean and sampling error for this sample.
c.   Refer to part b. When Rich calculated the sample mean, he mistakenly used the numbers 20, 25, 13, 9, 15, 11, 17, 17, and 30 to calculate the sample mean. Find the sampling and nonsampling errors in this case.
d.   List all samples of nine numbers (without replacement) that can be selected from this population. Calculate the sample mean and sampling error for each of these samples.

■ Applications

7.6   Using the formulas of Sections 5.3 and 5.4 of Chapter 5 for the mean and standard deviation of a discrete random variable, verify that the mean and standard deviation for the population probability distribution of Table 7.2 are 80.60 and 8.09, respectively.

7.7   The following data give the ages of all six members of a family.

55   53   28   25   21   15

a.   Let x denote the age of a member of this family. Write the population distribution of x.
b.   List all the possible samples of size five (without replacement) that can be selected from this population. Calculate the mean for each of these samples. Write the sampling distribution of $\bar{x}$.
c.   Calculate the mean for the population data. Select one random sample of size five and calculate the sample mean $\bar{x}$. Compute the sampling error.

7.8   The following data give the years of teaching experience for all five faculty members of a department at a university.

7   8   14   7   20

a.   Let x denote the years of teaching experience for a faculty member of this department. Write the population distribution of x.
b.   List all the possible samples of size four (without replacement) that can be selected from this population. Calculate the mean for each of these samples. Write the sampling distribution of $\bar{x}$.
c.   Calculate the mean for the population data. Select one random sample of size four and calculate the sample mean $\bar{x}$. Compute the sampling error.

7.3   MEAN AND STANDARD DEVIATION OF $\bar{x}$

The mean and standard deviation calculated for the sampling distribution of $\bar{x}$ are called the **mean** and **standard deviation of $\bar{x}$.** Actually, the mean and standard deviation of $\bar{x}$ are, respectively, the mean and standard deviation of the means of all samples of the same size selected from a population. The standard deviation of $\bar{x}$ is also called the *standard error of $\bar{x}$.*

> **MEAN AND STANDARD DEVIATION OF $\bar{x}$**   The mean and standard deviation of the sampling distribution of $\bar{x}$ are called the *mean and standard deviation of $\bar{x}$* and are denoted by $\mu_{\bar{x}}$ and $\sigma_{\bar{x}}$, respectively.

If we calculate the mean and standard deviation of the 10 values of $\bar{x}$ listed in Table 7.3, we obtain the mean, $\mu_{\bar{x}}$, and the standard deviation, $\sigma_{\bar{x}}$, of $\bar{x}$. Alternatively, we can calculate the mean and standard deviation of the sampling distribution of $\bar{x}$ listed in Table 7.5. These will also be the values of $\mu_{\bar{x}}$ and $\sigma_{\bar{x}}$. From these calculations, we will obtain $\mu_{\bar{x}} = 80.60$ and $\sigma_{\bar{x}} = 3.30$ (see Exercise 7.25 at the end of this section).

The mean of the sampling distribution of $\bar{x}$ is always equal to the mean of the population.

MEAN OF THE SAMPLING DISTRIBUTION OF $\bar{x}$ The *mean of the sampling distribution of $\bar{x}$ is always equal to the mean of the population. Thus,*

$$\mu_{\bar{x}} = \mu$$

Hence, if we select all possible samples (of the same size) from a population and calculate their means, the mean ($\mu_{\bar{x}}$) of all these sample means will be the same as the mean (μ) of the population. If we calculate the mean for the population probability distribution of Table 7.2 and the mean for the sampling distribution of Table 7.5 by using the formula learned in Section 5.3 of Chapter 5, we get the same value of 80.60 for μ and $\mu_{\bar{x}}$ (see Exercise 7.25).

The sample mean, $\bar{x}$, is called an **estimator** of the population mean, μ. When the expected value (or mean) of a sample statistic is equal to the value of the corresponding population parameter, that sample statistic is said to be an **unbiased estimator**. For the sample mean $\bar{x}$, $\mu_{\bar{x}} = \mu$. Hence, $\bar{x}$ is an unbiased estimator of μ. This is a very important property that an estimator should possess.

However, the standard deviation, $\sigma_{\bar{x}}$, of $\bar{x}$ is not equal to the standard deviation, σ, of the population distribution (unless $n = 1$). The standard deviation of $\bar{x}$ is equal to the standard deviation of the population divided by the square root of the sample size; that is,

$$\sigma_{\bar{x}} = \frac{\sigma}{\sqrt{n}}$$

This formula for the standard deviation of $\bar{x}$ holds true only when the sampling is done either with replacement from a finite population or with or without replacement from an infinite population. These two conditions can be replaced by the condition that the above formula holds true if the sample size is small in comparison to the population size. The sample size is considered to be small compared to the population size if the sample size is equal to or less than 5% of the population size—that is, if

$$\frac{n}{N} \le .05$$

If this condition is not satisfied, we use the following formula to calculate $\sigma_{\bar{x}}$:

$$\sigma_{\bar{x}} = \frac{\sigma}{\sqrt{n}} \sqrt{\frac{N-n}{N-1}}$$

where the factor

$$\sqrt{\frac{N-n}{N-1}}$$

is called the finite population correction factor.

In most practical applications, the sample size is small compared to the population size. Consequently, in most cases, the formula used to calculate $\sigma_{\bar{x}}$ is $\sigma_{\bar{x}} = \sigma/\sqrt{n}$.

> **STANDARD DEVIATION OF THE SAMPLING DISTRIBUTION OF $\bar{x}$**   The *standard deviation of the sampling distribution of $\bar{x}$* is
>
> $$\sigma_{\bar{x}} = \frac{\sigma}{\sqrt{n}}$$
>
> where σ is the standard deviation of the population and n is the sample size. This formula is used when $n/N \leq .05$, where N is the population size.

Following are two important observations regarding the sampling distribution of $\bar{x}$.

1. *The spread of the sampling distribution of $\bar{x}$ is smaller than the spread of the corresponding population distribution.* In other words, $\sigma_{\bar{x}} < \sigma$. This is obvious from the formula for $\sigma_{\bar{x}}$. When n is greater than 1, which is usually true, the denominator in $\sigma/\sqrt{n}$ is greater than 1. Hence, $\sigma_{\bar{x}}$ is smaller than σ.

2. *The standard deviation of the sampling distribution of $\bar{x}$ decreases as the sample size increases.* This feature of the sampling distribution of $\bar{x}$ is also obvious from the formula

$$\sigma_{\bar{x}} = \frac{\sigma}{\sqrt{n}}$$

↳ consistant estimator.

If the standard deviation of a sample statistic decreases as the sample size is increased, that statistic is said to be a **consistent estimator**. This is another important property that an estimator should possess. It is obvious from the above formula for $\sigma_{\bar{x}}$ that as n increases, the value of $\sqrt{n}$ also increases and, consequently, the value of $\sigma/\sqrt{n}$ decreases. Thus, the sample mean $\bar{x}$ is a consistent estimator of the population mean μ. Example 7–2 illustrates this feature.

Finding the mean and standard deviation of $\bar{x}$.

EXAMPLE 7–2   The mean wage per hour for all 5000 employees who work at a large company is $13.50 and the standard deviation is $2.90. Let $\bar{x}$ be the mean wage per hour for a random sample of certain employees selected from this company. Find the mean and standard deviation of $\bar{x}$ for a sample size of

(a)   30   (b)   75   (c)   200

Solution   From the given information, for the population of all employees,

$$N = 5000, \quad \mu = \$13.50, \quad \text{and} \quad \sigma = \$2.90$$

(a)   The mean, $\mu_{\bar{x}}$, of the sampling distribution of $\bar{x}$ is

$$\mu_{\bar{x}} = \mu = \mathbf{\$13.50}$$

In this case, $n = 30$, $N = 5000$, and $n/N = 30/5000 = .006$. Because n/N is less than .05, the standard deviation of $\bar{x}$ is obtained by using the formula $\sigma/\sqrt{n}$. Hence,

$$\sigma_{\bar{x}} = \frac{\sigma}{\sqrt{n}} = \frac{2.90}{\sqrt{30}} = \mathbf{\$.529}$$

Thus, we can state that if we take all possible samples of size 30 from the population of all employees of this company and prepare the sampling distribution of $\bar{x}$, the mean and standard deviation of this sampling distribution of $\bar{x}$ will be $13.50 and $.529, respectively.

(b) In this case, $n = 75$ and $n/N = 75/5000 = .015$, which is less than .05. The mean and standard deviation of $\bar{x}$ are

$$\mu_{\bar{x}} = \mu = \$13.50 \quad \text{and} \quad \sigma_{\bar{x}} = \frac{\sigma}{\sqrt{n}} = \frac{2.90}{\sqrt{75}} = \$.335$$

(c) In this case, $n = 200$ and $n/N = 200/5000 = .04$, which is less than .05. Therefore, the mean and standard deviation of $\bar{x}$ are

$$\mu_{\bar{x}} = \mu = \$13.50 \quad \text{and} \quad \sigma_{\bar{x}} = \frac{\sigma}{\sqrt{n}} = \frac{2.90}{\sqrt{200}} = \$.205$$

From the preceding calculations we observe that the mean of the sampling distribution of $\bar{x}$ is always equal to the mean of the population whatever the size of the sample. However, the value of the standard deviation of $\bar{x}$ decreases from $.529 to $.335 and then to $.205 as the sample size increases from 30 to 75 and then to 200. ■

EXERCISES

■ **Concepts and Procedures**

7.9 Let $\bar{x}$ be the mean of a sample selected from a population.

a. What is the mean of the sampling distribution of $\bar{x}$ equal to?
b. What is the standard deviation of the sampling distribution of $\bar{x}$ equal to? Assume $n/N \leq .05$.

7.10 What is an estimator? When is an estimator unbiased? Is the sample mean, $\bar{x}$, an unbiased estimator of μ? Explain.

7.11 When is an estimator said to be consistent? Is the sample mean, $\bar{x}$, a consistent estimator of μ? Explain.

7.12 How does the value of $\sigma_{\bar{x}}$ change as the sample size increases? Explain.

7.13 Consider a large population with $\mu = 60$ and $\sigma = 10$. Assuming $n/N \leq .05$, find the mean and standard deviation of the sample mean, $\bar{x}$, for a sample size of

a. 18   **b.** 90

7.14 Consider a large population with $\mu = 90$ and $\sigma = 18$. Assuming $n/N \leq .05$, find the mean and standard deviation of the sample mean, $\bar{x}$, for a sample size of

a. 10   **b.** 35

7.15 A population of $N = 5000$ has $\sigma = 25$. In each of the following cases, which formula will you use to calculate $\sigma_{\bar{x}}$ and why? Using the appropriate formula, calculate $\sigma_{\bar{x}}$ for each of these cases.

a. $n = 300$ **b.** $n = 100$

7.16 A population of $N = 100,000$ has $\sigma = 40$. In each of the following cases, which formula will you use to calculate $\sigma_{\bar{x}}$ and why? Using the appropriate formula, calculate $\sigma_{\bar{x}}$ for each of these cases.

a. $n = 2500$ **b.** $n = 7000$

*7.17 For a population, $\mu = 125$ and $\sigma = 36$.

a. For a sample selected from this population, $\mu_{\bar{x}} = 125$ and $\sigma_{\bar{x}} = 3.6$. Find the sample size. Assume $n/N \leq .05$.
b. For a sample selected from this population, $\mu_{\bar{x}} = 125$ and $\sigma_{\bar{x}} = 2.25$. Find the sample size. Assume $n/N \leq .05$.

*7.18 For a population, $\mu = 46$ and $\sigma = 10$.

a. For a sample selected from this population, $\mu_{\bar{x}} = 46$ and $\sigma_{\bar{x}} = 2.0$. Find the sample size. Assume $n/N \leq .05$.

b. For a sample selected from this population, $\mu_{\bar{x}} = 46$ and $\sigma_{\bar{x}} = 1.6$. Find the sample size. Assume $n/N \leq .05$.

■ **Applications**

7.19 According to the NPD Group, the average amount Americans spend per person on breakfast in restaurants is $3.12 (*Newsweek,* May 17, 1999). Assume that the current mean of all such expenditures is $3.12 with a standard deviation of $.75. Let $\bar{x}$ be the mean amount spent per person on 400 randomly selected breakfasts in restaurants. Find the mean and standard deviation of the sampling distribution of $\bar{x}$.

7.20 According to the U.S. Bureau of Labor Statistics, nonsupervisory workers in the manufacturing sector worked an average of 41.8 hours per week in August 1999. Assume that the current mean working hours per week for all nonsupervisory workers in the manufacturing sector are 41.8 and the standard deviation is 4 hours per week. Let $\bar{x}$ be the mean hours worked per week by a random sample of 36 such workers. Find the mean and standard deviation of the sampling distribution of $\bar{x}$.

7.21 The mean annual salary for all 1050 professors at a university is $51,400 and the standard deviation of their annual salaries is $7400. Let $\bar{x}$ be the mean salary of a random sample of 16 professors selected from this university. Find the mean and standard deviation of the sampling distribution of $\bar{x}$.

7.22 According to the most recent data available from the U.S. Department of Labor, the mean retirement age for U.S. men is 62.2 years (*The Wall Street Journal,* February 2, 1999). Assume that the current mean retirement age for all U.S. men is 62.2 years and the standard deviation is 4 years. Let $\bar{x}$ be the mean retirement age of a random sample of 700 recently retired American men. Find the mean and standard deviation of $\bar{x}$.

∗**7.23** The standard deviation of the prices of all cars is known to be $3600. Let $\bar{x}$ be the mean price of a sample of cars. What sample size will produce the standard deviation of $\bar{x}$ equal to $180?

∗**7.24** The standard deviation of the 1999 gross sales of all corporations is known to be $139.50 million. Let $\bar{x}$ be the mean of the 1999 gross sales of a sample of corporations. What sample size will produce the standard deviation of $\bar{x}$ equal to $15.50 million?

∗**7.25** Consider the sampling distribution of $\bar{x}$ given in Table 7.5.

a. Calculate the value of $\mu_{\bar{x}}$ using the formula $\mu_{\bar{x}} = \Sigma \bar{x} P(\bar{x})$. Is the value of μ calculated in Exercise 7.6 the same as the value of $\mu_{\bar{x}}$ calculated here?

b. Calculate the value of $\sigma_{\bar{x}}$ by using the formula

$$\sigma_{\bar{x}} = \sqrt{\Sigma \bar{x}^2 P(\bar{x}) - (\mu_{\bar{x}})^2}$$

c. From Exercise 7.6, $\sigma = 8.09$. Also, our sample size is 3 so that $n = 3$. Therefore, $\sigma/\sqrt{n} = 8.09/\sqrt{3} = 4.67$. From part b, you should get $\sigma_{\bar{x}} = 3.30$. Why does $\sigma/\sqrt{n}$ not equal $\sigma_{\bar{x}}$ in this case?

d. In our example (given in the beginning of Section 7.1.1) on scores, $N = 5$ and $n = 3$. Hence, $n/N = 3/5 = .60$. Because n/N is greater than .05, the appropriate formula to find $\sigma_{\bar{x}}$ is

$$\sigma_{\bar{x}} = \frac{\sigma}{\sqrt{n}} \sqrt{\frac{N-n}{N-1}}$$

Show that the value of $\sigma_{\bar{x}}$ calculated by using this formula gives the same value as the one calculated in part b.

7.4   SHAPE OF THE SAMPLING DISTRIBUTION OF $\bar{x}$

The shape of the sampling distribution of $\bar{x}$ relates to the following two cases.

1. The population from which samples are drawn has a normal distribution.

2. The population from which samples are drawn does not have a normal distribution.

7.4.1 SAMPLING FROM A NORMALLY DISTRIBUTED POPULATION

When the population from which samples are drawn is normally distributed with its mean equal to μ and standard deviation equal to σ, then

1. The mean of $\bar{x}$, $\mu_{\bar{x}}$, is equal to the mean of the population, μ.
2. The standard deviation of $\bar{x}$, $\sigma_{\bar{x}}$, is equal to $\sigma/\sqrt{n}$, assuming $n/N \leq .05$.
3. The shape of the sampling distribution of $\bar{x}$ is normal, whatever the value of n.

SAMPLING DISTRIBUTION OF $\bar{x}$ WHEN THE POPULATION HAS A NORMAL DISTRIBUTION
If the population from which the samples are drawn is normally distributed with mean μ and standard deviation σ, then the sampling distribution of the sample mean, $\bar{x}$, will also be normally distributed with the following mean and standard deviation, irrespective of the sample size:

$$\mu_{\bar{x}} = \mu \quad \text{and} \quad \sigma_{\bar{x}} = \frac{\sigma}{\sqrt{n}}$$

☞ *Remember*   For $\sigma_{\bar{x}} = \sigma/\sqrt{n}$ to be true, n/N must be less than or equal to .05.

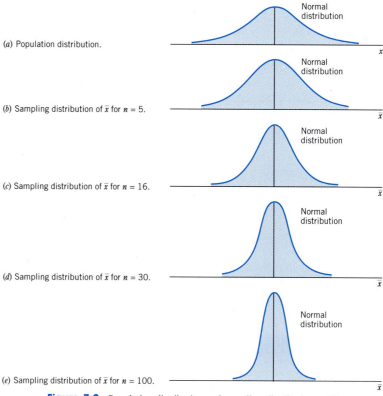

(*a*) Population distribution.

(*b*) Sampling distribution of $\bar{x}$ for $n = 5$.

(*c*) Sampling distribution of $\bar{x}$ for $n = 16$.

(*d*) Sampling distribution of $\bar{x}$ for $n = 30$.

(*e*) Sampling distribution of $\bar{x}$ for $n = 100$.

Figure 7.2 Population distribution and sampling distributions of $\bar{x}$.

Figure 7.2a shows the probability distribution curve for a population. The distribution curves in Figure 7.2b through Figure 7.2e show the sampling distributions of $\bar{x}$ for different sample sizes taken from the population of Figure 7.2a. As we can observe, the population has a normal distribution. Because of this, the sampling distribution of $\bar{x}$ is normal for each of the four cases illustrated in parts b through e of Figure 7.2. Also notice from Figure 7.2b through Figure 7.2e that the spread of the sampling distribution of $\bar{x}$ decreases as the sample size increases.

Example 7–3 illustrates the calculation of the mean and standard deviation of $\bar{x}$ and the description of the shape of its sampling distribution.

Finding the mean, standard deviation, and sampling distribution of $\bar{x}$: normal population.

EXAMPLE 7–3 In a recent SAT test, the mean score for all examinees was 1016. Assume that the distribution of SAT scores of all examinees is normal with a mean of 1016 and a standard deviation of 153. Let $\bar{x}$ be the mean SAT score of a random sample of certain examinees. Calculate the mean and standard deviation of $\bar{x}$ and describe the shape of its sampling distribution when the sample size is

(a) 16   (b) 50   (c) 1000

Solution Let μ and σ be the mean and standard deviation of SAT scores of all examinees, and let $\mu_{\bar{x}}$ and $\sigma_{\bar{x}}$ be the mean and standard deviation of the sampling distribution of $\bar{x}$. Then, from the given information,

$$\mu = 1016 \quad \text{and} \quad \sigma = 153$$

(a) The mean and standard deviation of $\bar{x}$ are

$$\mu_{\bar{x}} = \mu = \mathbf{1016} \quad \text{and} \quad \sigma_{\bar{x}} = \frac{\sigma}{\sqrt{n}} = \frac{153}{\sqrt{16}} = \mathbf{38.250}$$

Because the SAT scores of all examinees are assumed to be normally distributed, the sampling distribution of $\bar{x}$ for samples of 16 examinees is also normal. Figure 7.3 shows the population distribution and the sampling distribution of $\bar{x}$. Note that because σ is greater than $\sigma_{\bar{x}}$, the population distribution has a wider spread and less height than the sampling distribution of $\bar{x}$ in Figure 7.3.

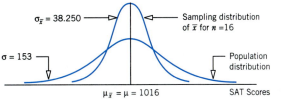

$\sigma_{\bar{x}} = 38.250$ Sampling distribution of $\bar{x}$ for $n = 16$

$\sigma = 153$ Population distribution

$\mu_{\bar{x}} = \mu = 1016$ SAT Scores   **Figure 7.3**

(b) The mean and standard deviation of $\bar{x}$ are

$$\mu_{\bar{x}} = \mu = \mathbf{1016} \quad \text{and} \quad \sigma_{\bar{x}} = \frac{\sigma}{\sqrt{n}} = \frac{153}{\sqrt{50}} = \mathbf{21.637}$$

Again, because the SAT scores of all examinees are assumed to be normally distributed, the sampling distribution of $\bar{x}$ for samples of 50 examinees is also normal. The population distribution and the sampling distribution of $\bar{x}$ are shown in Figure 7.4.

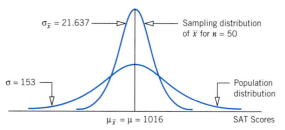

$\sigma_{\bar{x}} = 21.637$ ——→ ←—— Sampling distribution of $\bar{x}$ for $n = 50$

$\sigma = 153$ ——→ Population distribution

$\mu_{\bar{x}} = \mu = 1016$ SAT Scores   **Figure 7.4**

(c)   The mean and standard deviation of $\bar{x}$ are

$$\mu_{\bar{x}} = \mu = \mathbf{1016} \quad \text{and} \quad \sigma_{\bar{x}} = \frac{\sigma}{\sqrt{n}} = \frac{153}{\sqrt{1000}} = \mathbf{4.838}$$

Again, because the SAT scores of all examinees are assumed to be normally distributed, the sampling distribution of $\bar{x}$ for samples of 1000 examinees is also normal. The two distributions are shown in Figure 7.5.

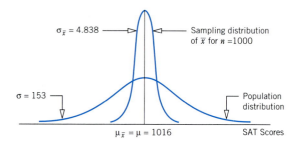

$\sigma_{\bar{x}} = 4.838$ ——→ ←—— Sampling distribution of $\bar{x}$ for $n = 1000$

$\sigma = 153$ ——→ Population distribution

$\mu_{\bar{x}} = \mu = 1016$ SAT Scores   **Figure 7.5**

Thus, whatever the sample size, the sampling distribution of $\bar{x}$ is normal when the population from which the samples are drawn is normally distributed.   ■

7.4.2   SAMPLING FROM A POPULATION THAT IS NOT NORMALLY DISTRIBUTED

Most of the time the population from which the samples are selected is not normally distributed. In such cases, the shape of the sampling distribution of $\bar{x}$ is inferred from a very important theorem called the **central limit theorem**.

CENTRAL LIMIT THEOREM   According to the *central limit theorem,* for a large sample size, the sampling distribution of $\bar{x}$ is approximately normal, irrespective of the shape of the population distribution. The mean and standard deviation of the sampling distribution of $\bar{x}$ are

$$\mu_{\bar{x}} = \mu \quad \text{and} \quad \sigma_{\bar{x}} = \frac{\sigma}{\sqrt{n}}$$

The sample size is usually considered to be large if $n \geq 30$.

Note that when the population does not have a normal distribution, the shape of the sampling distribution is not exactly normal but is approximately normal for a large sample size. The approximation becomes more accurate as the sample size increases. Another point to remember is that the central limit theorem applies to *large* samples only. Usually, if the sample size is 30 or more, it is considered sufficiently large to apply the central limit theorem to the sampling distribution of $\bar{x}$. Thus, according to the central limit theorem,

1. When $n \geq 30$, the shape of the sampling distribution of $\bar{x}$ is approximately normal irrespective of the shape of the population distribution.
2. The mean of $\bar{x}$, $\mu_{\bar{x}}$, is equal to the mean of the population, μ.
3. The standard deviation of $\bar{x}$, $\sigma_{\bar{x}}$, is equal to $\sigma/\sqrt{n}$.

Again, remember that for $\sigma_{\bar{x}} = \sigma/\sqrt{n}$ to apply, n/N must be less than or equal to .05.

Figure 7.6a shows the probability distribution curve for a population. The distribution curves in Figure 7.6b through Figure 7.6e show the sampling distributions of $\bar{x}$ for different sample sizes taken from the population of Figure 7.6a. As we can observe, the population is not normally distributed. The sampling distributions of $\bar{x}$ shown in parts b and c, when $n < 30$, are not normal. However, the sampling distributions of $\bar{x}$ shown in parts d and e, when $n \geq 30$, are (approximately) normal. Also notice that the spread of the sampling distribution of $\bar{x}$ decreases as the sample size increases.

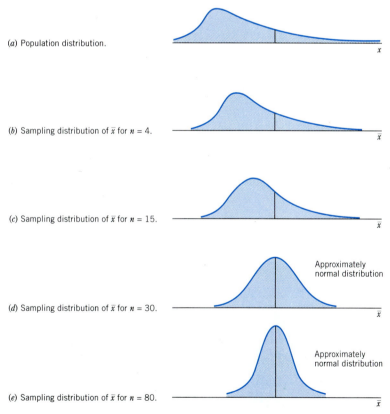

(a) Population distribution.

(b) Sampling distribution of $\bar{x}$ for $n = 4$.

(c) Sampling distribution of $\bar{x}$ for $n = 15$.

(d) Sampling distribution of $\bar{x}$ for $n = 30$.

Approximately normal distribution

(e) Sampling distribution of $\bar{x}$ for $n = 80$.

Approximately normal distribution

Figure 7.6 Population distribution and sampling distributions of $\bar{x}$.

Example 7–4 illustrates the calculation of the mean and standard deviation of $\bar{x}$ and describes the shape of the sampling distribution of $\bar{x}$ when the sample size is large.

*Finding the mean,
standard deviation, and
sampling distribution of
$\bar{x}$: nonnormal
population.*

EXAMPLE 7-4 The mean rent paid by all tenants in a large city is $1250 with a standard deviation of $225. However, the population distribution of rents for all tenants in this city is skewed to the right. Calculate the mean and standard deviation of $\bar{x}$ and describe the shape of its sampling distribution when the sample size is

(a)   30   (b)   100

Solution   Although the population distribution of rents paid by all tenants is not normal, in each case the sample size is large ($n \geq 30$). Hence, the central limit theorem can be applied to infer the shape of the sampling distribution of $\bar{x}$.

(a)   Let $\bar{x}$ be the mean rent paid by a sample of 30 tenants. Then, the sampling distribution of $\bar{x}$ is approximately normal with the values of the mean and standard deviation as

$$\mu_{\bar{x}} = \mu = \$1250 \quad \text{and} \quad \sigma_{\bar{x}} = \frac{\sigma}{\sqrt{n}} = \frac{225}{\sqrt{30}} = \$41.079$$

Figure 7.7 shows the population distribution and the sampling distribution of $\bar{x}$.

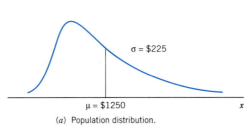
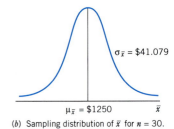

(a) Population distribution.

(b) Sampling distribution of $\bar{x}$ for $n = 30$.

Figure 7.7

(b)   Let $\bar{x}$ be the mean rent paid by a sample of 100 tenants. Then, the sampling distribution of $\bar{x}$ is approximately normal with the values of the mean and standard deviation as

$$\mu_{\bar{x}} = \mu = \$1250 \quad \text{and} \quad \sigma_{\bar{x}} = \frac{\sigma}{\sqrt{n}} = \frac{225}{\sqrt{100}} = \$22.500$$

Figure 7.8 shows the population distribution and the sampling distribution of $\bar{x}$.

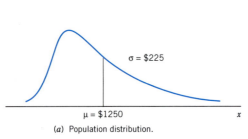
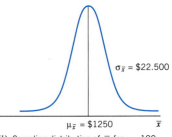

(a) Population distribution.

(b) Sampling distribution of $\bar{x}$ for $n = 100$.

Figure 7.8

EXERCISES

■ Concepts and Procedures

7.26   What condition or conditions must hold true for the sampling distribution of the sample mean to be normal when the sample size is less than 30?

7.27   Explain the central limit theorem.

7.28   A population has a distribution that is skewed to the left. Indicate in which of the following cases the central limit theorem will apply to describe the sampling distribution of the sample mean.

a. $n = 400$ **b.** $n = 25$ **c.** $n = 36$

7.29   A population has a distribution that is skewed to the right. A sample of size n is selected from this population. Describe the shape of the sampling distribution of the sample mean for each of the following cases.

a. $n = 25$ **b.** $n = 80$ **c.** $n = 29$

7.30   A population has a normal distribution. A sample of size n is selected from this population. Describe the shape of the sampling distribution of the sample mean for each of the following cases.

a. $n = 94$ **b.** $n = 11$

7.31   A population has a normal distribution. A sample of size n is selected from this population. Describe the shape of the sampling distribution of the sample mean for each of the following cases.

a. $n = 23$ **b.** $n = 450$

■ Applications

7.32   The time taken to complete a statistics test by all students is normally distributed with a mean of 120 minutes and a standard deviation of 12 minutes. Let $\bar{x}$ be the mean time taken to complete this test by a random sample of 16 students. Calculate the mean and standard deviation of $\bar{x}$ and describe the shape of its sampling distribution.

7.33   The speeds of all cars traveling on a stretch of Interstate Highway I-95 are normally distributed with a mean of 71 miles per hour and a standard deviation of 3.2 miles per hour. Let $\bar{x}$ be the mean speed of a random sample of 20 cars traveling on this highway. Calculate the mean and standard deviation of $\bar{x}$ and describe the shape of its sampling distribution.

7.34   The amounts of electric bills for all households in a city have an approximately normal distribution with a mean of \$45 and a standard deviation of \$8. Let $\bar{x}$ be the mean amount of electric bills for a random sample of 25 households selected from this city. Find the mean and standard deviation of $\bar{x}$ and comment on the shape of its sampling distribution.

7.35   The GPAs of all 5540 students enrolled at a university have an approximately normal distribution with a mean of 3.02 and a standard deviation of .29. Let $\bar{x}$ be the mean GPA of a random sample of 48 students selected from this university. Find the mean and standard deviation of $\bar{x}$ and comment on the shape of its sampling distribution.

7.36   The weights of all people living in a town have a distribution that is skewed to the right with a mean of 133 pounds and a standard deviation of 24 pounds. Let $\bar{x}$ be the mean weight of a random sample of 45 persons selected from this town. Find the mean and standard deviation of $\bar{x}$ and comment on the shape of its sampling distribution.

7.37   The amounts of telephone bills for all households in a large city have a distribution that is skewed to the right with a mean of \$75 and a standard deviation of \$27. Let $\bar{x}$ be the mean amount of telephone bills for a random sample of 90 households selected from this city. Calculate the mean and standard deviation of $\bar{x}$ and describe the shape of its sampling distribution.

7.38   The balances of checking accounts at a local bank have a distribution that is skewed to the right with its mean equal to \$410 and standard deviation equal to \$150. Let $\bar{x}$ be the mean balance of a random sample of 60 checking accounts selected from this bank. Calculate the mean and standard deviation of $\bar{x}$ and describe the shape of its sampling distribution.

7.39   According to the American Federation of Teachers, the average salary of 1998 college graduates with a degree in accounting was $33,702 (*USA TODAY*, June 18, 1999). Assume that the distribution of current salaries of recent college graduates with a degree in accounting is skewed to the right with a mean of $33,702 and a standard deviation of $3900. Let $\bar{x}$ be the mean current salary of a random sample of 700 recent college graduates with a degree in accounting. Calculate the mean and standard deviation of $\bar{x}$ and describe the shape of its sampling distribution.

7.5   APPLICATIONS OF THE SAMPLING DISTRIBUTION OF $\bar{x}$

From the central limit theorem, for large samples, the sampling distribution of $\bar{x}$ is approximately normal. Based on this result, we can make the following statements about $\bar{x}$ for large samples. The areas under the curve of $\bar{x}$ mentioned in these statements are found from the normal distribution table.

1. *If we take all possible samples of the same (large) size from a population and calculate the mean for each of these samples, then about 68.26% of the sample means will be within one standard deviation of the population mean.* Or we can state that if we take one sample (of $n \geq 30$) from a population and calculate the mean for this sample, the probability that this sample mean will be within one standard deviation of the population mean is .6826. That is,

$$P(\mu - 1\sigma_{\bar{x}} \leq \bar{x} \leq \mu + 1\sigma_{\bar{x}}) = .6826$$

This probability is shown in Figure 7.9.

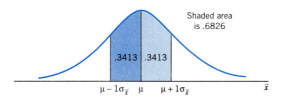

Figure 7.9
$P(\mu - 1\sigma_{\bar{x}} \leq \bar{x} \leq \mu + 1\sigma_{\bar{x}})$

2. *If we take all possible samples of the same (large) size from a population and calculate the mean for each of these samples, then about 95.44% of the sample means will be within two standard deviations of the population mean.* Or we can state that if we take one sample (of $n \geq 30$) from a population and calculate the mean for this sample, the probability that this sample mean will be within two standard deviations of the population mean is .9544. That is,

$$P(\mu - 2\sigma_{\bar{x}} \leq \bar{x} \leq \mu + 2\sigma_{\bar{x}}) = .9544$$

This probability is shown in Figure 7.10.

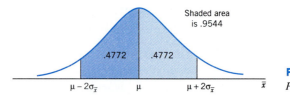

Figure 7.10
$P(\mu - 2\sigma_{\bar{x}} \leq \bar{x} \leq \mu + 2\sigma_{\bar{x}})$

3. *If we take all possible samples of the same (large) size from a population and calculate the mean for each of these samples, then about 99.74% of the sample means will be within three standard deviations of the population mean.* Or we can state that if we take

one sample (of $n \geq 30$) from a population and calculate the mean for this sample, the probability that this sample mean will be within three standard deviations of the population mean is .9974. That is,

$$P(\mu - 3\sigma_{\bar{x}} \leq \bar{x} \leq \mu + 3\sigma_{\bar{x}}) = .9974$$

This probability is shown in Figure 7.11.

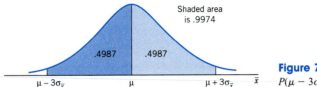

Shaded area
is .9974

.4987 .4987

$\mu - 3\sigma_{\bar{x}}$ μ $\mu + 3\sigma_{\bar{x}}$ $\bar{x}$

Figure 7.11
$P(\mu - 3\sigma_{\bar{x}} \leq \bar{x} \leq \mu + 3\sigma_{\bar{x}})$

When conducting a survey, we usually select one sample and compute the value of $\bar{x}$ based on that sample. We never select all possible samples of the same size and then prepare the sampling distribution of $\bar{x}$. Rather, we are more interested in finding the probability that the value of $\bar{x}$ computed from one sample falls within a given interval. Examples 7–5 and 7–6 illustrate this procedure.

Calculating the probability of $\bar{x}$ in an interval: normal population.

EXAMPLE 7–5   Assume that the weights of all packages of a certain brand of cookies are normally distributed with a mean of 32 ounces and a standard deviation of .3 ounce. Find the probability that the mean weight, $\bar{x}$, of a random sample of 20 packages of this brand of cookies will be between 31.8 and 31.9 ounces.

Solution   Although the sample size is small ($n < 30$), the shape of the sampling distribution of $\bar{x}$ is normal because the population is normally distributed. The mean and standard deviation of $\bar{x}$ are

$$\mu_{\bar{x}} = \mu = 32 \text{ ounces} \quad \text{and} \quad \sigma_{\bar{x}} = \frac{\sigma}{\sqrt{n}} = \frac{.3}{\sqrt{20}} = .06708204 \text{ ounce}$$

We are to compute the probability that the value of $\bar{x}$ calculated for one randomly drawn sample of 20 packages is between 31.8 and 31.9 ounces—that is,

$$P(31.8 < \bar{x} < 31.9)$$

This probability is given by the area under the normal distribution curve for $\bar{x}$ between the points $\bar{x} = 31.8$ and $\bar{x} = 31.9$. The first step in finding this area is to convert the two $\bar{x}$ values to their respective z values.

---

z VALUE FOR A VALUE OF $\bar{x}$ The *z value for a value of $\bar{x}$ is calculated as*

$$z = \frac{\bar{x} - \mu}{\sigma_{\bar{x}}}$$

The z values for $\bar{x} = 31.8$ and $\bar{x} = 31.9$ are computed next, and they are shown on the z scale below the normal distribution curve for $\bar{x}$ in Figure 7.12 on page 310.

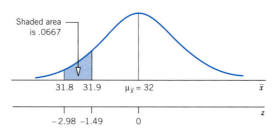

Figure 7.12
$P(31.8 < \bar{x} < 31.9)$

For $\bar{x} = 31.8$: $z = \dfrac{31.8 - 32}{.06708204} = -2.98$

For $\bar{x} = 31.9$: $z = \dfrac{31.9 - 32}{.06708204} = -1.49$

The probability that $\bar{x}$ is between 31.8 and 31.9 is given by the area under the standard normal curve between $z = -2.98$ and $z = -1.49$. Thus, the required probability is

$$P(31.8 < \bar{x} < 31.9) = P(-2.98 < z < -1.49)$$

$$= P(-2.98 < z < 0) - P(-1.49 < z < 0)$$

$$= .4986 - .4319 = \mathbf{.0667}$$

Therefore, the probability is .0667 that the mean weight of a sample of 20 packages will be between 31.8 and 31.9 ounces. ∎

Calculating the probability of $\bar{x}$ in an interval: $n > 30$.

EXAMPLE 7–6   According to a study by the Bureau of Labor Statistics, the mean annual earnings of physicians and surgeons in the United States is $100,920 (*The Wall Street Journal,* February 2, 1999). Suppose that the current annual earnings of all U.S. physicians and surgeons have a probability distribution that is skewed to the right with a mean of $100,920 and a standard deviation of $22,500. Let $\bar{x}$ be the mean annual earnings of a sample of 400 U.S. physicians and surgeons.

(a) What is the probability that the mean annual earnings of U.S. physicians and surgeons obtained from this sample will be within $2000 of the population mean?

(b) What is the probability that the mean annual earnings of U.S. physicians and surgeons obtained from this sample will be less than the population mean by $1500 or more?

Solution   From the given information, for the annual earnings of all U.S. physicians and surgeons,

$$\mu = \$100{,}920 \quad \text{and} \quad \sigma = \$22{,}500$$

Although the shape of the probability distribution of the population (annual earnings of all U.S. physicians and surgeons) is skewed to the right, the sampling distribution of $\bar{x}$ is approximately normal because the sample is large ($n > 30$). Remember that when the sample is large, the central limit theorem applies. The mean and standard deviation of the sampling distribution of $\bar{x}$ are

$$\mu_{\bar{x}} = \mu = 100{,}920 \quad \text{and} \quad \sigma_{\bar{x}} = \frac{\sigma}{\sqrt{n}} = \frac{22{,}500}{\sqrt{400}} = \$1125$$

(a) The probability that the mean annual earnings obtained from a sample of 400 U.S. physicians and surgeons will be within $2000 of the population mean is written as

$$P(98{,}920 \le \bar{x} \le 102{,}920)$$

This probability is given by the area under the normal curve for $\bar{x}$ between $\bar{x} = \$98,920$ and $\bar{x} = \$102,920$, as shown in Figure 7.13. We find this area as follows:

$$\text{For } \bar{x} = \$98,920: \quad z = \frac{98,920 - 100,920}{1125} = -1.78$$

$$\text{For } \bar{x} = \$102,920: \quad z = \frac{102,920 - 100,920}{1125} = 1.78$$

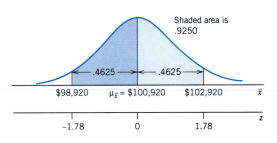

Figure 7.13
$P(\$98,920 \leq \bar{x} \leq \$102,920)$

Hence, the required probability is

$$P(98,920 \leq \bar{x} \leq 102,920) = P(-1.78 \leq z \leq 1.78)$$

$$= P(-1.78 \leq z \leq 0) + P(0 \leq z \leq 1.78)$$

$$= .4625 + .4625 = \mathbf{.9250}$$

Therefore, the probability that the mean annual earnings obtained from a sample of 400 U.S. physicians and surgeons will be within \$2000 of the population mean is .9250.

(b) The probability that the mean annual earnings obtained from a sample of 400 U.S. physicians and surgeons will be less than the population mean by \$1500 or more is written as

$$P(\bar{x} \leq 99,420)$$

This probability is given by the area under the normal curve for $\bar{x}$ to the left of $\bar{x} = \$99,420$, as shown in Figure 7.14. We find this area as follows:

$$\text{For } \bar{x} = 99,420: \quad z = \frac{99,420 - 100,920}{1125} = -1.33$$

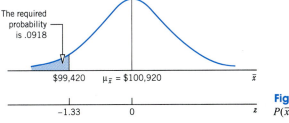

Figure 7.14
$P(\bar{x} \leq \$99,420)$

Hence, the required probability is

$$P(\bar{x} \leq 99,420) = P(z \leq -1.33)$$

$$= .5 - P(-1.33 \leq z \leq 0)$$

$$= .5 - .4082 = \mathbf{.0918}$$

Therefore, the probability that the mean annual earnings obtained from a sample of 400 U.S. physicians and surgeons will be less than the population mean by $1500 or more is .0918.   ■

EXERCISES

■ **Concepts and Procedures**

7.40   If all possible samples of the same (large) size are selected from a population, what percent of all the sample means will be within 2.5 standard deviations of the population mean?

7.41   If all possible samples of the same (large) size are selected from a population, what percent of all the sample means will be within 1.5 standard deviations of the population mean?

7.42   For a population, $N = 10,000$, $\mu = 124$, and $\sigma = 18$. Find the z value for each of the following for $n = 36$.

a. $\bar{x} = 128.60$ **b.** $\bar{x} = 119.30$ **c.** $\bar{x} = 116.88$ **d.** $\bar{x} = 132.05$

7.43   For a population, $N = 205,000$, $\mu = 66$, and $\sigma = 7$. Find the z value for each of the following for $n = 49$.

a. $\bar{x} = 68.44$ **b.** $\bar{x} = 58.75$ **c.** $\bar{x} = 62.35$ **d.** $\bar{x} = 71.82$

7.44   Let x be a continuous random variable that has a normal distribution with $\mu = 75$ and $\sigma = 14$. Assuming $n/N \leq .05$, find the probability that the sample mean, $\bar{x}$, for a random sample of 20 taken from this population will be

a.   between 68.5 and 77.3   **b.**   less than 72.4

7.45   Let x be a continuous random variable that has a normal distribution with $\mu = 48$ and $\sigma = 8$. Assuming $n/N \leq .05$, find the probability that the sample mean, $\bar{x}$, for a random sample of 16 taken from this population will be

a.   between 49.6 and 52.2   **b.**   more than 45.7

7.46   Let x be a continuous random variable that has a distribution skewed to the right with $\mu = 60$ and $\sigma = 10$. Assuming $n/N \leq .05$, find the probability that the sample mean, $\bar{x}$, for a random sample of 40 taken from this population will be

a.   less than 62.20   **b.**   between 61.4 and 64.2

7.47   Let x be a continuous random variable that follows a distribution skewed to the left with $\mu = 90$ and $\sigma = 18$. Assuming $n/N \leq .05$, find the probability that the sample mean, $\bar{x}$, for a random sample of 64 taken from this population will be

a.   less than 82.3   **b.**   greater than 86.7

■ **Applications**

7.48   The speeds of all cars traveling on a stretch of Interstate Highway I-95 are normally distributed with a mean of 71 miles per hour and a standard deviation of 3.2 miles per hour. Find the probability that the mean speed of a random sample of 16 cars traveling on this stretch of this interstate highway is

a.   less than 69 miles per hour
b.   more than 72 miles per hour
c.   70 to 72 miles per hour

7.49   The GPAs of all students enrolled at a large university have an approximately normal distribution with a mean of 3.02 and a standard deviation of .29. Find the probability that the mean GPA of a random sample of 20 students selected from this university is

a.   3.10 or higher   **b.**   2.90 or lower   **c.**   2.95 to 3.11

7.50   The times taken to complete a statistics test by all students are normally distributed with a mean of 120 minutes and a standard deviation of 12 minutes. Find the probability that the mean time taken to complete this test by a random sample of 16 students would be

a. between 122 and 125 minutes
b. within 4 minutes of the population mean
c. less than the population mean by 3 minutes or more

7.51 According to the U.S. Bureau of Labor Statistics, nonsupervisory workers in the manufacturing sector worked an average of 41.8 hours per week in June 1998. Assume that the weekly hours for all such workers are normally distributed with a mean of 41.8 and a standard deviation of 4 hours. Find the probability that the mean weekly working hours for a random sample of 25 such workers would be

a. between 42 and 44 hours
b. within 1.5 hours of the population mean
c. less than the population mean by 1 hour or more

7.52 The times that college students spend studying per week have a distribution that is skewed to the right with a mean of 8.4 hours and a standard deviation of 2.7 hours. Find the probability that the mean time spent studying per week for a random sample of 45 students would be

a. between 8 and 9 hours   b. less than 8 hours

7.53 The ages of all college students follow a distribution that is skewed to the right with a mean of 27 years and a standard deviation of 4.4 years. Find the probability that the mean age for a random sample of 36 students would be

a. between 25.5 and 28 years   b. less than 25.5 years

7.54 According to the American Federation of Teachers, the average salary of 1998 college graduates with a degree in accounting was \$33,702 (*USA TODAY*, June 18, 1999). Assume that the distribution of current salaries of recent college graduates with a degree in accounting is skewed to the right with a mean of \$33,702 and a standard deviation of \$3900. Find the probability that the mean current salary of a random sample of 200 recent college graduates with a degree in accounting would be

a. between \$34,000 and \$34,500
b. within \$400 of the population mean
c. less than the population mean by \$500 or more

7.55 The amounts of electric bills for all households in a city have a skewed probability distribution with a mean of \$65 and a standard deviation of \$25. Find the probability that the mean amount of electric bills for a random sample of 75 households selected from this city will be

a. between \$58 and \$63
b. within \$6 of the population mean
c. more than the population mean by at least \$5

7.56 The balances of all savings accounts at a local bank have a distribution that is skewed to the right with its mean equal to \$12,450 and standard deviation equal to \$4300. Find the probability that the mean balance of a sample of 50 savings accounts selected from this bank will be

a. more than \$11,500
b. between \$12,000 and \$13,800
c. within \$1500 of the population mean
d. more than the population mean by at least \$1000

7.57 The heights of all adults in a large city have a distribution that is skewed to the right with a mean of 68 inches and a standard deviation of 4 inches. Find the probability that the mean height of a random sample of 100 adults selected from this city would be

a. less than 67.8 inches
b. between 67.5 inches and 68.7 inches
c. within .6 inch of the population mean
d. less than the population mean by .5 inch or more

7.58 Johnson Electronics Corporation makes electric tubes. It is known that the standard deviation of the lives of these tubes is 150 hours. The company's research department takes a sample of 100 such tubes and finds that the mean life of these tubes is 2250 hours. What is the probability that this sample mean is within 25 hours of the mean life of all tubes produced by this company?

7.59 A machine at Katz Steel Corporation makes 3-inch-long nails. The probability distribution of the lengths of these nails is normal with a mean of 3 inches and a standard deviation of .1 inch. The quality control inspector takes a sample of 25 nails once a week and calculates the mean length of these nails. If the mean of this sample is either less than 2.95 inches or greater than 3.05 inches, the inspector concludes that the machine needs an adjustment. What is the probability that based on a sample of 25 nails the inspector will conclude that the machine needs an adjustment?

7.6 POPULATION AND SAMPLE PROPORTIONS

The concept of proportion is the same as the concept of relative frequency discussed in Chapter 2 and the concept of probability of success in a binomial experiment. The relative frequency of a category or class gives the proportion of the sample or population that belongs to that category or class. Similarly, the probability of success in a binomial experiment represents the proportion of the sample or population that possesses a given characteristic.

The **population proportion**, denoted by *p*, is obtained by taking the ratio of the number of elements in a population with a specific characteristic to the total number of elements in the population. The **sample proportion**, denoted by $\hat{p}$ (pronounced *p hat*), gives a similar ratio for a sample.

> **POPULATION AND SAMPLE PROPORTIONS** The *population* and *sample proportions*, denoted by p and $\hat{p}$, respectively, are calculated as
>
> $$p = \frac{X}{N} \quad \text{and} \quad \hat{p} = \frac{x}{n}$$
>
> where
>
> N = total number of elements in the population
>
> n = total number of elements in the sample
>
> X = number of elements in the population that possess a specific characteristic
>
> x = number of elements in the sample that possess a specific characteristic

Example 7–7 illustrates the calculation of the population and sample proportions.

Calculating population and sample proportions.

EXAMPLE 7–7 Suppose a total of 789,654 families live in a city and 563,282 of them own homes. Then,

$$N = \text{population size} = 789{,}654$$

$$X = \text{families in the population who own homes} = 563{,}282$$

The proportion of all families in this city who own homes is

$$p = \frac{X}{N} = \frac{563{,}282}{789{,}654} = \textbf{.71}$$

Now, suppose a sample of 240 families is taken from this city and 158 of them are homeowners. Then,

$$n = \text{sample size} = 240$$

$$x = \text{families in the sample who own homes} = 158$$

The sample proportion is

$$\hat{p} = \frac{x}{n} = \frac{158}{240} = .66$$

■

As in the case of the mean, the difference between the sample proportion and the corresponding population proportion gives the sampling error, assuming that the sample is random and no nonsampling error has been made. That is, in case of the proportion,

$$\text{Sampling error} = \hat{p} - p$$

For instance, for Example 7–7,

$$\text{Sampling error} = \hat{p} - p = .66 - .71 = -.05$$

7.7 MEAN, STANDARD DEVIATION, AND SHAPE OF THE SAMPLING DISTRIBUTION OF $\hat{p}$

This section discusses the sampling distribution of the sample proportion and the mean, standard deviation, and shape of this sampling distribution.

7.7.1 SAMPLING DISTRIBUTION OF $\hat{p}$

Just like the sample mean $\bar{x}$, the sample proportion $\hat{p}$ is a random variable. Hence, it possesses a probability distribution, which is called its **sampling distribution**.

> **SAMPLING DISTRIBUTION OF THE SAMPLE PROPORTION, $\hat{p}$** The probability distribution of the sample proportion, $\hat{p}$, is called its *sampling distribution*. It gives the various values that $\hat{p}$ can assume and their probabilities.

The value of $\hat{p}$ calculated for a particular sample depends on what elements of the population are included in that sample. Example 7–8 illustrates the concept of the sampling distribution of $\hat{p}$.

Illustrating the sampling distribution of $\hat{p}$.

EXAMPLE 7–8 Boe Consultant Associates has five employees. Table 7.6 gives the names of these five employees and information concerning their knowledge of statistics.

Table 7.6 Information on the Five Employees of Boe Consultant Associates

Name	Knows Statistics
Ally	yes
John	no
Susan	no
Lee	yes
Tom	yes

If we define the population proportion, p, as the proportion of employees who know statistics, then

$$p = 3/5 = .60$$

Now, suppose we draw all possible samples of three employees each and compute the proportion of employees, for each sample, who know statistics. The total number of samples of size three that can be drawn from the population of five employees is

$$\text{Total number of samples} = \binom{5}{3} = \frac{5!}{3! \, (5 - 3)!} = \frac{5 \cdot 4 \cdot 3 \cdot 2 \cdot 1}{3 \cdot 2 \cdot 1 \cdot 2 \cdot 1} = 10$$

Table 7.7 lists these 10 possible samples and the proportion of employees who know statistics for each of those samples. Note that we have rounded the values of $\hat{p}$ to two decimal places.

Table 7.7 All Possible Samples of Size 3 and the Value of $\hat{p}$ for Each Sample

Sample	Proportion Who Know Statistics $\hat{p}$
Ally, John, Susan	$1/3 = .33$
Ally, John, Lee	$2/3 = .67$
Ally, John, Tom	$2/3 = .67$
Ally, Susan, Lee	$2/3 = .67$
Ally, Susan, Tom	$2/3 = .67$
Ally, Lee, Tom	$3/3 = 1.00$
John, Susan, Lee	$1/3 = .33$
John, Susan, Tom	$1/3 = .33$
John, Lee, Tom	$2/3 = .67$
Susan, Lee, Tom	$2/3 = .67$

Using Table 7.7, we prepare the frequency distribution of $\hat{p}$ as recorded in Table 7.8, along with the relative frequencies of classes, which are obtained by dividing the frequencies of classes by the population size. The relative frequencies are used as probabilities and listed in Table 7.9. This table gives the sampling distribution of $\hat{p}$.

Table 7.8 Frequency and Relative Frequency Distributions of $\hat{p}$ When the Sample Size Is 3

$\hat{p}$	f	Relative Frequency
.33	3	$3/10 = .30$
.67	6	$6/10 = .60$
1.00	1	$1/10 = .10$
	$\Sigma f = 10$	Sum $= 1.00$

Table 7.9 Sampling Distribution of $\hat{p}$ When the Sample Size Is 3

$\hat{p}$	$P(\hat{p})$
.33	.30
.67	.60
1.00	.10
	$\Sigma P(\hat{p}) = 1.00$

7.7.2 MEAN AND STANDARD DEVIATION OF $\hat{p}$

The **mean of $\hat{p}$**, which is the same as the mean of the sampling distribution of $\hat{p}$, is always equal to the population proportion, p, just as the mean of the sampling distribution of $\bar{x}$ is always equal to the population mean, μ.

> **MEAN OF THE SAMPLE PROPORTION** The *mean of the sample proportion, $\hat{p}$,* is denoted by $\mu_{\hat{p}}$ and is equal to the population proportion, p. Thus,
>
> $$\mu_{\hat{p}} = p$$

The sample proportion, $\hat{p}$, is called an **estimator** of the population proportion, p. As mentioned earlier in this chapter, when the expected value (or mean) of a sample statistic is equal to the value of the corresponding population parameter, that sample statistic is said to be an **unbiased estimator**. Since for the sample proportion, $\mu_{\hat{p}} = p$, $\hat{p}$ is an unbiased estimator of p.

The **standard deviation of $\hat{p}$**, denoted by $\sigma_{\hat{p}}$, is given by the following formula. This formula is true only when the sample size is small compared to the population size. As we know from Section 7.3, the sample size is said to be small compared to the population size if $n/N \leq .05$.

> **STANDARD DEVIATION OF THE SAMPLE PROPORTION** The *standard deviation of the sample proportion, $\hat{p}$,* is denoted by $\sigma_{\hat{p}}$ and is given by the formula
>
> $$\sigma_{\hat{p}} = \sqrt{\frac{pq}{n}}$$
>
> where p is the population proportion, $q = 1 - p$, and n is the sample size. This formula is used when $n/N \leq .05$, where N is the population size.

However, if n/N is greater than .05, then $\sigma_{\hat{p}}$ is calculated as follows.

$$\sigma_{\hat{p}} = \sqrt{\frac{pq}{n}} \, \sqrt{\frac{N-n}{N-1}}$$

where the factor

$$\sqrt{\frac{N-n}{N-1}}$$

is called the finite population correction factor.

In almost all cases, the sample size is small compared to the population size and, consequently, the formula used to calculate $\sigma_{\hat{p}}$ is $\sqrt{pq/n}$.

As mentioned earlier in this chapter, if the standard deviation of a sample statistic decreases as the sample size is increased, that statistic is said to be a **consistent estimator**. It is obvious from the above formula for $\sigma_{\hat{p}}$ that as n increases, the value of $\sqrt{pq/n}$ decreases. Thus, the sample proportion, $\hat{p}$, is a consistent estimator of the population proportion, p.

7.7.3   SHAPE OF THE SAMPLING DISTRIBUTION OF $\hat{p}$

The shape of the sampling distribution of $\hat{p}$ is inferred from the central limit theorem.

> **CENTRAL LIMIT THEOREM FOR SAMPLE PROPORTION**   According to the central limit theorem, the *sampling distribution of* $\hat{p}$ is approximately normal for a sufficiently large sample size. In the case of proportion, the sample size is considered to be sufficiently large if np and nq are both greater than 5—that is, if
>
> $$np > 5 \quad \text{and} \quad nq > 5$$

Note that the sampling distribution of $\hat{p}$ will be approximately normal if $np > 5$ and $nq > 5$. This is the same condition that was required for the application of the normal approximation to the binomial probability distribution in Chapter 6.

Example 7–9 shows the calculation of the mean and standard deviation of $\hat{p}$ and describes the shape of its sampling distribution.

Finding the mean, standard deviation, and sampling distribution of $\hat{p}$.

EXAMPLE 7–9   Thirty-four percent of the adult U.S. women in a telephone poll conducted by Yankelovich Partners, Inc., for *Time*/CNN said that the main problem facing women (that men do not face) today is inequality in the workplace (*Time*, June 29, 1998). Assume that this result is true for the current population of all adult U.S. women. Let $\hat{p}$ be the proportion in a random sample of 500 adult U.S. women who hold this view. Find the mean and standard deviation of $\hat{p}$ and describe the shape of its sampling distribution.

Solution   Let p be the proportion of all adult U.S. women who think that the main problem facing women today is inequality in the workplace. Then,

$$p = .34 \quad \text{and} \quad q = 1 - p = 1 - .34 = .66$$

The mean of the sampling distribution of $\hat{p}$ is

$$\mu_{\hat{p}} = p = \mathbf{.34}$$

The standard deviation of $\hat{p}$ is

$$\sigma_{\hat{p}} = \sqrt{\frac{pq}{n}} = \sqrt{\frac{(.34)(.66)}{500}} = \mathbf{.021}$$

The values of np and nq are

$$np = 500(.34) = 170 \quad \text{and} \quad nq = 500(.66) = 330$$

Because np and nq are both greater than 5, we can apply the central limit theorem to make an inference about the shape of the sampling distribution of $\hat{p}$. Therefore, the sampling distribution of $\hat{p}$ is approximately normal with a mean of .34 and a standard deviation of .021, as shown in Figure 7.15.

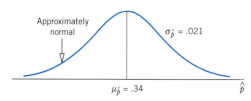

Figure 7.15

CASE STUDY 7-1   CLINTON POLL NUMBERS ARE NOT PULLED FROM A HAT

The phone keeps ringing here at *USA TODAY,* and the faxes keep rolling in. A lot of Americans want to know just whom the paper is polling, and how.

Polls on all kinds of subjects are notoriously controversial. What has people fired up these days are polls we've done in the past two weeks on President Clinton's approval ratings staying around 65%.

Readers have raised reasonable questions about whether the polls should be believed. Here's an attempt to answer:

Whom are you polling—just people who favor Clinton?

All our polls are national. The goal is to wind up with a group of respondents that closely resembles a cross-section of the population. The first key to accomplishing this is a good set of phone numbers.

The Gallup Organization, which conducts our polls, buys their phone numbers from a firm that keeps a constantly updated list of every possible number in the U.S. (listed and unlisted, and of course without anybody's name attached). For a typical three-day poll of 1,000 people, Gallup will buy several thousand numbers, which will be divided proportionately among the nation's 3,143 counties and purged of business phones and nonworking numbers.

When a poll is conducted, responses are closely monitored to make sure that:

- The proportions of responses in each of the four major regions of the U.S.—Northeast, South, Midwest, and West—closely match the proportions of people who live there.
- The number of men and women is close to the nation's adult population split of 52% women, 48% men.

Once the poll is completed, Gallup also makes sure the poll accurately mirrors the racial, age, and educational makeup of the United States. For instance, a recent poll underrepresented 18- to 29-year-olds and overrepresented 50- to 64-year-olds. To correct the imbalance, responses of the younger group were given more weight in the mathematical calculations, while those of the older ones were given less weight. The end result is a poll that isn't overloaded with liberals, minorities, women, or any other group with which Clinton normally scores higher in polls.

Why haven't I been polled?

The Roper Center, which archives most major national polls, estimates that about 400,000 people were interviewed by the major polling organizations last year. There are 200 million adults in the U.S. Therefore, your odds of not being polled in 1997 were 500 to 1.

How can a poll of 600 U.S. adults represent 200 million?

The assertion may seem farfetched, but it's based on a math formula for judging "sampling error" in scientific experiments.

Say you have 200 million U.S. adults' opinions. And say you can only sample 600 adults because of the cost or the time it takes. According to the formula, 19 out of 20 times, a poll of 600 will yield results within four percentage points of the results pollsters would get if they polled all U.S. adults. A poll of 1,000 would get the poll within three percentage points, and a poll of 10,000 could be expected to be within one percentage point.

A caveat: While the principle of "sampling error" is the basis for pollsters' confidence that samples of 600 or 1,000 are acceptable, there are many other potential sources for error in any poll.

Where's the proof polls work?

Comparing polls to actual votes in elections is the most obvious test of accuracy, and the results are excellent. In the 1992 and 1996 presidential races, almost all major polling organizations in the country were within the margin of sampling error.

Why do Internet polls and call-in radio polls show majorities opposing Clinton?

Call-ins and write-ins can be wildly inaccurate because there's no attempt to balance for age, gender, etc. Those who feel most intensely about an issue are the ones most likely to participate in such nonpolls, and right now the people who have the most intense feelings about Clinton apparently are those who are maddest at him . . . and at the polls.

Source: Jim Norman, *USA TODAY,* September 2, 1998. Copyright © 1998, *USA TODAY.* Reproduced with permission.

EXERCISES

■ Concepts and Procedures

7.60   In a population of 1000 subjects, 640 possess a certain characteristic. A sample of 40 subjects selected from this population has 24 subjects who possess the same characteristic. What are the values of the population and sample proportions?

7.61   In a population of 5000 subjects, 600 possess a certain characteristic. A sample of 120 subjects selected from this population contains 18 subjects who possess the same characteristic. What are the values of the population and sample proportions?

7.62   In a population of 18,700 subjects, 30% possess a certain characteristic. In a sample of 250 subjects selected from this population, 25% possess the same characteristic. How many subjects in the population and sample, respectively, possess this characteristic?

7.63   In a population of 9500 subjects, 75% possess a certain characteristic. In a sample of 400 subjects selected from this population, 78% possess the same characteristic. How many subjects in the population and sample, respectively, possess this characteristic?

7.64   Let $\hat{p}$ be the proportion of elements in a sample that possess a characteristic.

a.   What is the mean of $\hat{p}$?
b.   What is the standard deviation of $\hat{p}$? Assume $n/N \leq .05$.
c.   What condition(s) must hold true for the sampling distribution of $\hat{p}$ to be approximately normal?

7.65   For a population, $N = 12,000$ and $p = .71$. A random sample of 900 elements selected from this population gave $\hat{p} = .66$. Find the sampling error.

7.66   For a population, $N = 2800$ and $p = .29$. A random sample of 80 elements selected from this population gave $\hat{p} = .33$. Find the sampling error.

7.67   What is the estimator of the population proportion? Is this estimator an unbiased estimator of p? Explain why or why not.

7.68   Is the sample proportion a consistent estimator of the population proportion? Explain why or why not.

7.69   How does the value of $\sigma_{\hat{p}}$ change as the sample size increases? Explain. Assume $n/N \leq .05$.

7.70   Consider a large population with $p = .63$. Assuming $n/N \leq .05$, find the mean and standard deviation of the sample proportion $\hat{p}$ for a sample size of

a.   100   **b.**   900

7.71   Consider a large population with $p = .21$. Assuming $n/N \leq .05$, find the mean and standard deviation of the sample proportion $\hat{p}$ for a sample size of

a.   400   **b.**   750

7.72   A population of $N = 4000$ has a population proportion equal to .12. In each of the following cases, which formula will you use to calculate $\sigma_{\hat{p}}$ and why? Using the appropriate formula, calculate $\sigma_{\hat{p}}$ for each of these cases.

a. $n = 800$ **b.** $n = 30$

7.73 A population of $N = 1400$ has a population proportion equal to .47. In each of the following cases, which formula will you use to calculate $\sigma_{\hat{p}}$ and why? Using the appropriate formula, calculate $\sigma_{\hat{p}}$ for each of these cases.

a. $n = 90$ **b.** $n = 50$

7.74 According to the central limit theorem, the sampling distribution of $\hat{p}$ is approximately normal when the sample is large. What is considered a large sample in the case of the proportion? Briefly explain.

7.75 Indicate in which of the following cases the central limit theorem will apply to describe the sampling distribution of the sample proportion.

a. $n = 400$ and $p = .28$ **b.** $n = 80$ and $p = .05$
c. $n = 60$ and $p = .12$ **d.** $n = 100$ and $p = .035$

7.76 Indicate in which of the following cases the central limit theorem will apply to describe the sampling distribution of the sample proportion.

a. $n = 20$ and $p = .45$ **b.** $n = 75$ and $p = .22$
c. $n = 350$ and $p = .01$ **d.** $n = 200$ and $p = .022$

■ **Applications**

7.77 A company manufactured six television sets on a given day and these television sets were inspected for being good or defective. The results of the inspection follow.

Good   Good   Defective   Defective   Good   Good

a. What proportion of these television sets are good?
b. How many total samples (without replacement) of size five can be selected from this population?
c. List all the possible samples of size five that can be selected from this population and calculate the sample proportion, $\hat{p}$, of television sets that are good for each sample. Prepare the sampling distribution of $\hat{p}$.
d. For each sample listed in part c, calculate the sampling error.

7.78 The following data give the information on all five employees of a company.

Male   Female   Female   Male   Female

a. What proportion of employees of this company are female?
b. How many total samples (without replacement) of size three can be drawn from this population?
c. List all the possible samples of size three that can be selected from this population and calculate the sample proportion, $\hat{p}$, of the employees who are female for each sample. Prepare the sampling distribution of $\hat{p}$.
d. For each sample listed in part c, calculate the sampling error.

7.79 A poll of women aged 18 and older conducted by Princeton Survey Research Associates for *Newsweek* asked the question: "Suppose there were a genetic test that shows whether you are likely to develop a serious disease. Would you want to take the test to find out if you are at risk?" Sixty-six percent of the women polled answered yes (*Newsweek* Special Issue, Spring/Summer 1999). Assume that 66% of all women in the United States would answer yes to this question. Let $\hat{p}$ be the proportion of women in a random sample of 300 who would answer yes. Find the mean and standard deviation of $\hat{p}$ and describe the shape of its sampling distribution.

7.80 According to a poll of women aged 18 and older conducted by Gallup for *USA TODAY* in February 1999, 70% of the women polled felt that their financial situation was better than that of their mothers (*USA TODAY,* February 17, 1999). Assume that this percentage is true for the current population of all women 18 and older. Let $\hat{p}$ be the proportion of women in a random sample of 250 women who hold this view. Calculate the mean and standard deviation of $\hat{p}$ and comment on the shape of its sampling distribution.

7.81   According to a nationwide U.S. Department of Education Survey of kindergarten through 12th-grade teachers, 78% of teachers said they have training in technology but only 20% felt "very well prepared" to teach it (*USA TODAY,* January 29, 1999). Suppose that this result holds true for all kindergarten through 12th-grade teachers. Let $\hat{p}$ be the proportion of teachers in a random sample of 50 who feel "very well prepared" to teach technology. Calculate the mean and standard deviation of $\hat{p}$ and describe the shape of its sampling distribution.

7.82   Data from the National Highway Traffic Safety Administration show that of all injuries caused by aggressive driving during the past decade, 12% were life-threatening (*USA TODAY,* November 23, 1998). Assume that this percentage is true for all current injuries caused by aggressive driving. Let $\hat{p}$ be the proportion of life-threatening injuries in a random sample of 100 injuries caused by aggressive driving. Calculate the mean and standard deviation of $\hat{p}$ and describe the shape of its sampling distribution.

7.8   APPLICATIONS OF THE SAMPLING DISTRIBUTION OF $\hat{p}$

As mentioned in Section 7.5, when we conduct a study, we usually take only one sample and make all decisions or inferences on the basis of the results of that one sample. We use the concepts of the mean, standard deviation, and shape of the sampling distribution of $\hat{p}$ to determine the probability that the value of $\hat{p}$ computed from one sample falls within a given interval. Examples 7–10 and 7–11 illustrate this application.

Calculating the probability that $\hat{p}$ is in an interval.

EXAMPLE 7-10   According to the U.S. Transportation Department, drivers who run red lights are involved in about 89,000 crashes a year. Researchers at Old Dominion University surveyed licensed drivers and found that 56% of them admitted running red lights (*USA TODAY,* October 5, 1999). Assume that this result is true for the current population of all licensed drivers in the United States. Let $\hat{p}$ be the proportion in a random sample of 400 licensed drivers selected from the United States who will admit to running red lights. Find the probability that the value of $\hat{p}$ is between .58 and .61.

Solution   From the given information,

$$n = 400, \quad p = .56, \quad \text{and} \quad q = 1 - p = 1 - .56 = .44$$

where p is the proportion of all licensed drivers in the United States who will admit to running red lights. The mean of the sample proportion $\hat{p}$ is

$$\mu_{\hat{p}} = p = .56$$

The standard deviation of $\hat{p}$ is

$$\sigma_{\hat{p}} = \sqrt{\frac{pq}{n}} = \sqrt{\frac{(.56)(.44)}{400}} = .02481935$$

The values of np and nq are

$$np = 400(.56) = 224 \quad \text{and} \quad nq = 400(.44) = 176$$

Because np and nq are both greater than 5, we can infer from the central limit theorem that the sampling distribution of $\hat{p}$ is approximately normal. The probability that $\hat{p}$ is between .58 and .61 is given by the area under the normal curve for $\hat{p}$ between $\hat{p} = .58$ and $\hat{p} = .61$, as shown in Figure 7.16.

The first step in finding the area under the normal curve between $\hat{p} = .58$ and $\hat{p} = .61$ is to convert these two values to their respective z values. The z value for $\hat{p}$ is computed using the following formula.

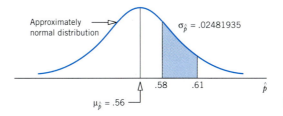

Figure 7.16

> **z VALUE FOR A VALUE OF $\hat{p}$** The *z value for a value of $\hat{p}$* is calculated as
>
> $$z = \frac{\hat{p} - p}{\sigma_{\hat{p}}}$$

The two values of $\hat{p}$ are converted to their respective z values and then the area under the normal curve between these two points is found using the normal distribution table.

$$\text{For } \hat{p} = .58: \quad z = \frac{.58 - .56}{.02481935} = .81$$

$$\text{For } \hat{p} = .61: \quad z = \frac{.61 - .56}{.02481935} = 2.01$$

Thus, the probability that $\hat{p}$ is between .58 and .61 is given by the area under the standard normal curve between $z = .81$ and $z = 2.01$. This area is shown in Figure 7.17. The required probability is

$$P(.58 < \hat{p} < .61) = P(.81 < z < 2.01)$$

$$= P(0 < z < 2.01) - P(0 < z < .81)$$

$$= .4778 - .2910 = \textbf{.1868}$$

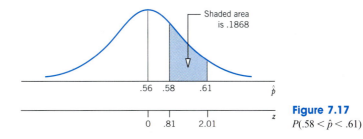

Figure 7.17

$P(.58 < \hat{p} < .61)$

Thus, the probability is .1868 that the proportion of drivers in a random sample of 400 who will admit to running red lights is between .58 and .61. ∎

Calculating the probability that $\hat{p}$ is less than a certain value.

EXAMPLE 7–11 Maureen Webster, who is running for mayor in a large city, claims that she is favored by 53% of all eligible voters of that city. Assume that this claim is true. What is the probability that in a random sample of 400 registered voters taken from this city, less than 49% will favor Maureen Webster?

Solution Let p be the proportion of all eligible voters who favor Maureen Webster. Then,

$$p = .53 \quad \text{and} \quad q = 1 - p = 1 - .53 = .47$$

The mean of the sampling distribution of the sample proportion $\hat{p}$ is

$$\mu_{\hat{p}} = p = .53$$

The population of all voters is large (because the city is large) and the sample size is small compared to the population. Consequently, we can assume that $n/N \leq .05$. Hence, the standard deviation of $\hat{p}$ is calculated as

$$\sigma_{\hat{p}} = \sqrt{\frac{pq}{n}} = \sqrt{\frac{(.53)\,(.47)}{400}} = .02495496$$

From the central limit theorem, the shape of the sampling distribution of $\hat{p}$ is approximately normal. The probability that $\hat{p}$ is less than .49 is given by the area under the normal distribution curve for $\hat{p}$ to the left of $\hat{p} = .49$, as shown in Figure 7.18. The z value for $\hat{p} = .49$ is

$$z = \frac{\hat{p} - p}{\sigma_{\hat{p}}} = \frac{.49 - .53}{.02495496} = -1.60$$

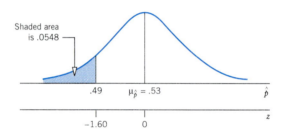

Figure 7.18 $P(\hat{p} < .49)$

Thus, the required probability is

$$P(\hat{p} < .49) = P(z < -1.60)$$

$$= .5 - P(-1.60 < z < 0)$$

$$= .5 - .4452 = \mathbf{.0548}$$

Hence, the probability that less than 49% of the voters in a random sample of 400 will favor Maureen Webster is .0548. ∎

EXERCISES

■ **Concepts and Procedures**

7.83 If all possible samples of the same (large) size are selected from a population, what percentage of all sample proportions will be within 2.0 standard deviations of the population proportion?

7.84 If all possible samples of the same (large) size are selected from a population, what percentage of all sample proportions will be within 3.0 standard deviations of the population proportion?

7.85 For a population, $N = 30,000$ and $p = .59$. Find the z value for each of the following for $n = 100$.

a. $\hat{p} = .56$ **b.** $\hat{p} = .68$ **c.** $\hat{p} = .53$ **d.** $\hat{p} = .65$

7.86 For a population, $N = 18,000$ and $p = .25$. Find the z value for each of the following for $n = 70$.

a. $\hat{p} = .26$ **b.** $\hat{p} = .32$ **c.** $\hat{p} = .17$ **d.** $\hat{p} = .20$

■ **Applications**

7.87 In a survey conducted by the American Medical Association, upper respiratory tract infections accounted for 21% of antibiotic prescriptions written by physicians in the United States (*Statistical Bulletin,* January–March 1998). Assume that this percentage is true for all current prescriptions for antibiotics. Let $\hat{p}$ be the proportion in a random sample of 400 antibiotic prescriptions that are for upper respiratory tract infections. Find the probability that the value of $\hat{p}$ will be

a. between .15 and .25   **b.** less than .20

7.88 A survey of all medium- and large-sized corporations showed that 64% of them offer retirement plans to their employees. Let $\hat{p}$ be the proportion in a random sample of 50 such corporations that offer retirement plans to their employees. Find the probability that the value of $\hat{p}$ will be

a. between .54 and .61   **b.** greater than .71

7.89 According to the American Association of University Professors, 33.8% of all faculty at all types of institutions of higher learning were female in 1997–1998 (*USA TODAY,* February 4, 1999). Assume that this percentage is true for all current faculty at such institutions and that $\hat{p}$ is the proportion of females in a random sample of 100 faculty members selected from such institutions. Find the probability that the value of $\hat{p}$ will be

a. between .30 and .35   **b.** greater than .32

7.90 Dartmouth Distribution Warehouse makes deliveries of a large number of products to its customers. It is known that 85% of all the orders it receives from its customers are delivered on time. Let $\hat{p}$ be the proportion of orders in a random sample of 100 that are delivered on time. Find the probability that the value of $\hat{p}$ will be

a. between .81 and .88   **b.** less than .87

7.91 Brooklyn Corporation manufactures computer diskettes. The machine that is used to make these diskettes is known to produce 6% defective diskettes. The quality control inspector selects a sample of 100 diskettes every week and inspects them for being good or defective. If 8% or more of the diskettes in the sample are defective, the process is stopped and the machine is readjusted. What is the probability that based on a sample of 100 diskettes the process will be stopped to readjust the machine?

7.92 Mong Corporation makes auto batteries. The company claims that 80% of its LL70 batteries are good for 70 months or longer. Assume that this claim is true. Let $\hat{p}$ be the proportion in a sample of 100 such batteries that are good for 70 months or longer.

a. What is the probability that this sample proportion is within .05 of the population proportion?

b. What is the probability that this sample proportion is less than the population proportion by .06 or more?

c. What is the probability that this sample proportion is greater than the population proportion by .07 or more?

GLOSSARY

Central limit theorem   The theorem from which it is inferred that for a large sample size ($n \geq 30$), the shape of the sampling distribution of $\bar{x}$ is approximately normal. Also, by the same theorem, the shape of the sampling distribution of $\hat{p}$ is approximately normal for a sample for which $np > 5$ and $nq > 5$.

Consistent estimator   A sample statistic with a standard deviation that decreases as the sample size increases.

Estimator   The sample statistic that is used to estimate a population parameter.

Mean of $\hat{p}$ The mean of the sampling distribution of $\hat{p}$, denoted by $\mu_{\hat{p}}$, is equal to the population proportion p.

Mean of $\bar{x}$ The mean of the sampling distribution of $\bar{x}$, denoted by $\mu_{\bar{x}}$, is equal to the population mean μ.

Nonsampling errors The errors that occur during the collection, recording, and tabulation of data.

Population distribution The probability distribution of the population data.

Population proportion p The ratio of the number of elements in a population with a specific characteristic to the total number of elements in the population.

Sample proportion $\hat{p}$ The ratio of the number of elements in a sample with a specific characteristic to the total number of elements in that sample.

Sampling distribution of $\hat{p}$ The probability distribution of all the values of $\hat{p}$ calculated from all possible samples of the same size selected from a population.

Sampling distribution of $\bar{x}$ The probability distribution of all the values of $\bar{x}$ calculated from all possible samples of the same size selected from a population.

Sampling error The difference between the value of a sample statistic calculated from a random sample and the value of the corresponding population parameter. This type of error occurs due to chance.

Standard deviation of $\hat{p}$ The standard deviation of the sampling distribution of $\hat{p}$, denoted by $\sigma_{\hat{p}}$, is equal to $\sqrt{pq/n}$ when $n/N \leq .05$.

Standard deviation of $\bar{x}$ The standard deviation of the sampling distribution of $\bar{x}$, denoted by $\sigma_{\bar{x}}$, is equal to $\sigma/\sqrt{n}$ when $n/N \leq .05$.

Unbiased estimator An estimator with an expected value (or mean) that is equal to the value of the corresponding population parameter.

KEY FORMULAS

1. **Mean of the sampling distribution of $\bar{x}$**

$$\mu_{\bar{x}} = \mu$$

2. **Standard deviation of the sampling distribution of $\bar{x}$ when $n/N \leq .05$**

$$\sigma_{\bar{x}} = \frac{\sigma}{\sqrt{n}}$$

3. **The z value for a value of $\bar{x}$**

$$z = \frac{\bar{x} - \mu}{\sigma_{\bar{x}}} \quad \text{where } \sigma_{\bar{x}} = \sqrt{\frac{\sigma}{\sqrt{n}}}$$

4. **Population proportion**

$$p = \frac{X}{N}$$

where

N = total number of elements in the population

X = number of elements in the population that possess a specific characteristic

5. **Sample proportion**

$$\hat{p} = \frac{x}{n}$$

where

n = total number of elements in the sample

x = number of elements in the sample that possess a specific characteristic

6. **Mean of the sampling distribution of $\hat{p}$**

$$\mu_{\hat{p}} = p$$

7. **Standard deviation of the sampling distribution of $\hat{p}$ when $n/N \le .05$**

$$\sigma_{\hat{p}} = \sqrt{\frac{pq}{n}}$$

8. **The z value for a value of $\hat{p}$**

$$z = \frac{\hat{p} - p}{\sigma_{\hat{p}}} \quad \text{where } \sigma_{\hat{p}} = \frac{pq}{n}$$

SUPPLEMENTARY EXERCISES

7.93 The print on the package of 100-watt General Electric soft-white lightbulbs claims that these bulbs have an average life of 750 hours. Assume that the lives of all such bulbs have a normal distribution with a mean of 750 hours and a standard deviation of 55 hours. Let $\bar{x}$ be the mean life of a random sample of 25 such bulbs. Find the mean and standard deviation of $\bar{x}$ and describe the shape of its sampling distribution.

7.94 The weekly earnings of all 2480 employees of a company have a distribution that is skewed to the right with its mean equal to $438 and standard deviation equal to $45. Let $\bar{x}$ be the mean weekly earnings of a random sample of 100 employees selected from this company. Calculate the mean and standard deviation of $\bar{x}$ and comment on the shape of its sampling distribution.

7.95 Refer to Exercise 7.93. The print on the package of 100-watt General Electric soft-white lightbulbs says that these bulbs have an average life of 750 hours. Assume that the lives of all such bulbs have a normal distribution with a mean of 750 hours and a standard deviation of 55 hours. Find the probability that the mean life of a random sample of 25 such bulbs will be

a. greater than 735 hours
b. between 725 and 740 hours
c. within 15 hours of the population mean
d. less than the population mean by 20 hours or more

7.96 Refer to Exercise 7.94. The weekly earnings of all 2480 employees of a company have a distribution that is skewed to the right with its mean equal to $438 and standard deviation equal to $45. Find the probability that the mean weekly earnings of a random sample of 100 employees selected from this company will be

a. less than $433
b. between $435 and $440
c. within $7 of the population mean
d. greater than the population mean by $2.50 or more

7.97 According to a Priority Management survey, adults spend an average of 10 hours a day at work and commuting. Let the daily work and commute times for all adults have a mean of 10 hours and a standard deviation of 2.1 hours. Find the probability that the mean of the daily work and commute times for a random sample of 80 adults will be

a. greater than 10.45 hours
b. between 9.75 and 10.50 hours

c. within .25 hour of the population mean

d. less than the population mean by .50 hour or more

7.98 A machine at Keats Corporation fills 64-ounce detergent jugs. The probability distribution of the amount of detergent in these jugs is normal with a mean of 64 ounces and a standard deviation of .4 ounce. The quality control inspector takes a sample of 16 jugs once a week and measures the amount of detergent in these jugs. If the mean of this sample is either less than 63.75 ounces or greater than 64.25 ounces, the inspector concludes that the machine needs an adjustment. What is the probability that based on a sample of 16 jugs the inspector will conclude that the machine needs an adjustment when actually it does not?

7.99 Ten percent of all items produced on a machine are defective. Let $\hat{p}$ be the proportion of defective items in a random sample of 80 items selected from the production line. Calculate the mean and standard deviation of $\hat{p}$ and describe the shape of its sampling distribution.

7.100 Seventy percent of adults favor some kind of government control on the prices of medicines. Assume that this percentage is true for the current population of all adults. Let $\hat{p}$ be the proportion of adults in a random sample of 400 who favor government control on the prices of medicines. Calculate the mean and standard deviation of $\hat{p}$ and describe the shape of its sampling distribution.

7.101 Refer to Exercise 7.100. Seventy percent of adults favor some kind of government control on the prices of medicines. Assume that this percentage is true for the current population of all adults.

a. Find the probability that the proportion of adults in a random sample of 400 who favor some kind of government control on the prices of medicines is
    i. less than .65   **ii.** between .73 and .76

b. What is the probability that the proportion of adults in a random sample of 400 who favor some kind of government control is within .06 of the population proportion?

c. What is the probability that the sample proportion is greater than the population proportion by .05 or more?

7.102 A poll conducted by Gallup for *USA TODAY*/CNN in February 1999 asked whether people were more concerned about the nation's economic problems or about its moral problems (*USA TODAY,* February 23, 1999). Of those polled, 38% were more concerned about economic problems, while 58% were more concerned about morality. Assume that 58% of all U.S. adults are currently more concerned about the nation's moral problems. Let $\hat{p}$ be the proportion who are more concerned about moral problems in a random sample of 100 U.S. adults.

a. What is the probability that this sample proportion is within .08 of the population proportion?

b. What is the probability that this sample proportion is not within .08 of the population proportion?

c. What is the probability that this sample proportion is less than the population proportion by .10 or more?

d. What is the probability that this sample proportion is greater than the population proportion by .09 or more?

■ Advanced Exercises

7.103 Let μ be the mean annual salary of major league baseball players for 1999. Assume that the standard deviation of the salaries of these players is $65,000. What is the probability that the 1999 mean salary of a random sample of 100 baseball players was within $10,000 of the population mean, μ? Assume that $n/N \leq .05$.

7.104 The test scores for 300 students were entered into a computer, analyzed, and stored in a file. Unfortunately, someone accidentally erased a major portion of this file from the computer. The only information that is available is that 30% of the scores were below 65 and 15% of the scores were above 90. Assuming the scores are normally distributed, find their mean and standard deviation.

7.105 A chemist has a 10-gallon sample of river water taken just downstream from the outflow of a chemical plant. He is concerned about the concentration, c (in parts per million), of a certain toxic substance in the water. He wants to take several measurements, find the mean concentration of the toxic substance for this sample, and have a 95% chance of being within .5 part per million of the true mean value of c. If the concentration of the toxic substance in all measurements is normally distributed with $\sigma = .8$ part per million, how many measurements are necessary to achieve this goal?

7.106 A television reporter is covering the election for mayor of a large city and will conduct an exit poll (interviews with voters immediately after they vote) to make an early prediction of the outcome. Assume that the eventual winner of the election will get 60% of the votes.

a. What is the probability that a prediction based on an exit poll of a random sample of 25 voters will be correct? In other words, what is the probability that 13 or more of the 25 voters in the sample will have voted for the eventual winner?
b. How large a sample would the reporter have to take so that the probability of correctly predicting the outcome would be .95 or more?

7.107 A city is planning to build a hydroelectric power plant. A local newspaper found that 53% of the voters in this city favor the construction of this plant. Assume that this result holds true for the population of all voters in this city.

a. What is the probability that more than 50% of the voters in a random sample of 200 voters selected from this city will favor the construction of this plant?
b. A politician would like to take a random sample of voters in which more than 50% would favor the plant construction. How large a sample should be selected so that the politician is 95% sure of this outcome?

7.108 Refer to Exercise 6.91. Otto is trying out for the javelin throw to compete in the Olympics. The lengths of his javelin throws are normally distributed with a mean of 290 feet and a standard deviation of 10 feet. What is the probability that the total length of three of his throws will exceed 885 feet?

7.109 A certain elevator has a maximum legal carrying capacity of 6000 pounds. Suppose that the population of all people who ride this elevator have a mean weight of 160 pounds with a standard deviation of 25 pounds. If 35 of these people board the elevator, what is the probability that their combined weight will exceed 6000 pounds? Assume that the 35 people constitute a random sample from the population.

SELF-REVIEW TEST

1. A sampling distribution is the probability distribution of
 a. a population parameter   **b.** a sample statistic   **c.** any random variable

2. Nonsampling errors are
 a. the errors that occur because the sample size is too large in relation to the population size
 b. the errors made while collecting, recording, and tabulating data
 c. the errors that occur because an untrained person conducts the survey

3. A sampling error is
 a. the difference between the value of a sample statistic based on a random sample and the value of the corresponding population parameter
 b. the error made while collecting, recording, and tabulating data
 c. the error that occurs because the sample is too small

4. The mean of the sampling distribution of $\bar{x}$ is always equal to
 a. μ **b.** $\mu - 5$ **c.** $\sigma/\sqrt{n}$

5. The condition for the standard deviation of the sample mean to be $\sigma/\sqrt{n}$ is that
 a. $np > 5$ **b.** $n/N \leq .05$ **c.** $n > 30$

6. The standard deviation of the sampling distribution of the sample mean decreases when
 a. x increases   **b.** n increases   **c.** n decreases

7. When samples are selected from a normally distributed population, the sampling distribution of the sample mean has a normal distribution

    a. if $n \geq 30$ **b.** if $n/N \leq .05$ **c.** all the time

8. When samples are selected from a nonnormally distributed population, the sampling distribution of the sample mean has an approximately normal distribution

    a. if $n \geq 30$ **b.** if $n/N \leq .05$ **c.** always

9. In a sample of 200 customers of a mail-order company, 174 are found to be satisfied with the service they receive from the company. The proportion of customers in this sample who are satisfied with the company's service is

    a. .87   **b.** .174   **c.** .148

10. The mean of the sampling distribution of $\hat{p}$ is always equal to

    a. p **b.** μ **c.** $\hat{p}$

11. The condition for the standard deviation of the sampling distribution of the sample proportion to be $\sqrt{pq/n}$ is

    a. $np > 5$ and $nq > 5$ **b.** $n > 30$ **c.** $n/N \leq .05$

12. The sampling distribution of $\hat{p}$ is (approximately) normal if

    a. $np > 5$ and $nq > 5$ **b.** $n > 30$ **c.** $n/N \leq .05$

13. Briefly state and explain the central limit theorem.

14. The weights of all students at a large university have an approximately normal distribution with a mean of 145 pounds and a standard deviation of 18 pounds. Let $\bar{x}$ be the mean weight of a random sample of certain students selected from this university. Calculate the mean and standard deviation of $\bar{x}$ and describe the shape of its sampling distribution for a sample size of

    a. 25   **b.** 100

15. The times taken to run a certain road race for all participants have an unknown distribution with a mean of 47 minutes and a standard deviation of 8.4 minutes. Let $\bar{x}$ be the mean time taken to run this road race for a random sample of certain participants. Find the mean and standard deviation of $\bar{x}$ and describe the shape of its sampling distribution for a sample size of

    a. 20   **b.** 70

16. According to information provided in the 1998 instructions for IRS Form 1040, it takes an average of 694 minutes to complete and file this form (including record keeping, learning about the form, and preparing, copying, assembling, and mailing the form to the IRS). Assume that the times taken by all taxpayers to complete and file this form have an unknown distribution with a mean of 694 minutes and a standard deviation of 140 minutes. Find the probability that the mean time taken to complete and file Form 1040 by a random sample of 60 taxpayers would be

    a. between 660 and 680 minutes   **b.** more than 730 minutes
    c. less than 715 minutes   **d.** between 675 and 725 minutes

17. At Jen and Perry Ice Cream Company, the machine that fills 1-pound cartons of Top Flavor ice cream is set to dispense 16 ounces of ice cream into every carton. However, some cartons contain slightly less than and some contain slightly more than 16 ounces of ice cream. The amounts of ice cream in all such cartons have a normal distribution with a mean of 16 ounces and a standard deviation of .18 ounce.

    a. Find the probability that the mean amount of ice cream in a random sample of 16 such cartons will be

        i. between 15.90 and 15.95 ounces
        ii. less than 15.95 ounces
        iii. more than 15.97 ounces

    b. What is the probability that the mean amount of ice cream in a random sample of 16 such cartons will be within .10 ounce of the population mean?

c. What is the probability that the mean amount of ice cream in a random sample of 16 such cartons will be less than the population mean by .135 ounce or more?

18. According to a MacArthur Foundation survey of people aged 35 to 64, 8% considered themselves "not at all religious" (*USA TODAY,* February 16, 1999). Suppose that this result holds true for the current population of all people in this age group. Let $\hat{p}$ be the proportion in a random sample of people in this age group who feel this way. Calculate the mean and standard deviation of $\hat{p}$ and describe the shape of its sampling distribution when the sample size is

    a. 50   **b.** 200   **c.** 1800

19. In a poll of 923 women aged 18 and older conducted by Gallup in February 1999 for *USA TODAY,* 70% of the women polled felt that their financial situation was better than that of their mothers (*USA TODAY,* February 17, 1999). Assume that this percentage is true for the current population of all women 18 and older.

    a. Find the probability that the proportion of women in a random sample of 400 who hold this view is

       i.   greater than .73   ii.   between .66 and .74

       iii. less than .73   iv. between .65 and .68

    b. What is the probability that the proportion of women in a random sample of 400 who hold this view is within .05 of the population proportion?

    c. What is the probability that the sample proportion for a random sample of 400 women is less than the population proportion by .04 or more?

    d. What is the probability that the sample proportion for a random sample of 400 women is greater than the population proportion by .03 or more?

MINI-PROJECTS

Mini-Project 7–1

Consider the data on heights of NBA players.

a. Use MINITAB or any other statistical software to compute μ and σ for this data set. (You may use your μ and σ from Mini-Project 6–1, part b, in Chapter 6 if they are available.)

b. Take 20 random samples of five players each and find $\bar{x}$ for each sample.

c. Compute the mean and standard deviation for the set of 20 sample means.

d. Using the formulas given in Section 7.3, find $\mu_{\bar{x}}$ and $\sigma_{\bar{x}}$ for $n = 5$.

e. How do your values of $\mu_{\bar{x}}$ and $\sigma_{\bar{x}}$ in part d compare with those in part c?

f. What percentage of the 20 sample means found in part b lie in the interval $\mu_{\bar{x}} - \sigma_{\bar{x}}$ to $\mu_{\bar{x}} + \sigma_{\bar{x}}$? In the interval $\mu_{\bar{x}} - 2\sigma_{\bar{x}}$ to $\mu_{\bar{x}} + 2\sigma_{\bar{x}}$? In the interval $\mu_{\bar{x}} - 3\sigma_{\bar{x}}$ to $\mu_{\bar{x}} + 3\sigma_{\bar{x}}$?

g. How do the percentages in part f compare to the corresponding percentages for a normal distribution (68%, 95%, and 99.7%, respectively)?

h. Repeat parts b through g using 20 samples of 10 players each.

Mini-Project 7–2

Consider Data Set II, Data on States, that accompanies this text. Let p denote the proportion of the 50 states that have a per capita income of less than $22,000.

a. Find p.

b. Select 20 random samples of five states each and find the sample proportion $\hat{p}$ for each sample.

c. Compute the mean and standard deviation for the set of 20 sample proportions.

d. Using the formulas given in Section 7.7.2, compute $\mu_{\hat{p}}$ and $\sigma_{\hat{p}}$. Is the finite population correction factor required here?

e. Compare your mean and standard deviation of $\hat{p}$ from part c with the values calculated in part d.

f. Repeat parts b through e using 20 samples of 10 states each.

Mini-Project 7–3

You are to conduct the experiment of sampling 10 times (with replacement) from the digits 0, 1, 2, 3, 4, 5, 6, 7, 8, and 9. You can do this in a variety of ways. One way is to write each digit on a separate piece of paper, place all the slips in a hat, and select 10 times from the hat, returning each selected slip before the next pick. As alternatives, you can use a random numbers table such as Table I of Appendix D (see Section A.2.4 of Appendix A for instructions on how to use this table), a 10-sided die, MINITAB, any other statistical software, or a calculator that generates random numbers. Perform the experiment using any of these methods and compute the sample mean $\bar{x}$ for the 10 numbers obtained. Now repeat the procedure 49 more times. When you are done, you will have 50 sample means.

a. Make a table of the population distribution for the 10 digits and display it using a graph.
b. Make a stem-and-leaf display of your 50 sample means. What shape does it have?
c. What does the central limit theorem say about the shape of the sampling distribution of $\bar{x}$? What mean and standard deviation does the sampling distribution of $\bar{x}$ have in this problem?

For instructions on using MINITAB for this chapter, please see Section B.6 of Appendix B.

See Appendix C for *Excel Adventure 7* on sampling from a population.

COMPUTER ASSIGNMENTS

CA7.1   Using MINITAB or any other statistical software, create 200 samples, each containing the results of 30 rolls of a die. Calculate the means of these 200 samples. Construct the histogram and calculate the mean and standard deviation of these 200 sample means.

CA7.2   Using the following MINITAB commands, create 150 samples each containing the results of selecting 35 numbers from 1 through 100.

```
MTB > RANDOM 150 C1-C35;
SUBC > INTEGER 1 100.
```

Calculate the means of these 150 samples and put those in column C36. Construct the histogram and calculate the mean and standard deviation of these 150 sample means.

8

ESTIMATION OF THE MEAN AND PROPORTION

8.1 ESTIMATION: AN INTRODUCTION

8.2 POINT AND INTERVAL ESTIMATES

8.3 INTERVAL ESTIMATION OF A POPULATION MEAN: LARGE SAMPLES

CASE STUDY 8-1 COST OF RUNNING A HOUSEHOLD

8.4 INTERVAL ESTIMATION OF A POPULATION MEAN: SMALL SAMPLES

CASE STUDY 8-2 CARDIAC DEMANDS OF HEAVY SNOW SHOVELING

8.5 INTERVAL ESTIMATION OF A POPULATION PROPORTION: LARGE SAMPLES

8.6 DETERMINING THE SAMPLE SIZE FOR THE ESTIMATION OF MEAN

8.7 DETERMINING THE SAMPLE SIZE FOR THE ESTIMATION OF PROPORTION

GLOSSARY

KEY FORMULAS

SUPPLEMENTARY EXERCISES

SELF-REVIEW TEST

MINI-PROJECTS

COMPUTER ASSIGNMENTS

Now we are entering that part of statistics called *inferential statistics.* In Chapter 1 inferential statistics was defined as the part of statistics that helps us to make decisions about some characteristics of a population based on sample information. In other words, inferential statistics uses the sample results to make decisions and draw conclusions about the population from which the sample is drawn. Estimation is the first topic to be considered in our discussion of inferential statistics. Estimation and hypothesis testing (discussed in Chapter 9) taken together are usually referred to as inference making. This chapter explains how to estimate the population mean and population proportion for a single population.

8.1 ESTIMATION: AN INTRODUCTION

Estimation is a procedure by which a numerical value or values are assigned to a population parameter based on the information collected from a sample.

> **ESTIMATION**   The assignment of value(s) to a population parameter based on a value of the corresponding sample statistic is called *estimation*.

In inferential statistics, μ is called the *true population mean* and p is called the *true population proportion*. There are many other population parameters, such as the median, mode, variance, and standard deviation.

The following are a few examples of estimation: an auto company may want to estimate the mean fuel consumption for a particular model of a car; a manager may want to estimate the average time taken by new employees to learn a job; the U.S. Census Bureau may want to find the mean housing expenditure per month incurred by households; and the AWAH (Association of Wives of Alcoholic Husbands) may want to find the proportion (or percentage) of all husbands who are alcoholic.

The examples about estimating the mean fuel consumption, estimating the average time taken to learn a job by new employees, and estimating the mean housing expenditure per month incurred by households are illustrations of estimating the *true population mean, μ*. The example about estimating the proportion (or percentage) of all husbands who are alcoholic is an illustration of estimating the *true population proportion, p*.

If we can conduct a *census* (a survey that includes the entire population) each time we want to find the value of a population parameter, then the estimation procedures explained in this and subsequent chapters are not needed. For example, if the U.S. Census Bureau can contact every household in the United States to find the mean housing expenditure incurred by households, the result of the survey (which will actually be a census) will give the value of μ, and the procedures learned in this chapter will not be needed. However, it is too expensive, very time consuming, or virtually impossible to contact every member of a population to collect information to find the true value of a population parameter. Therefore, we usually take a sample from the population and calculate the value of the appropriate sample statistic. Then we assign a value or values to the corresponding population parameter based on the value of the sample statistic. This chapter (and subsequent chapters) explains how to assign values to population parameters based on the values of sample statistics.

For example, to estimate the mean time taken to learn a certain job by new employees, the manager will take a sample of new employees and record the time taken by each of these employees to learn the job. Using this information, he or she will calculate the sample mean, $\bar{x}$. Then, based on the value of $\bar{x}$, he or she will assign certain values to μ. As another example, to estimate the mean housing expenditure per month incurred by all households in the United States, the U.S. Census Bureau will take a sample of certain households, collect the information on the housing expenditure that each of these households incurs per month, and compute the value of the sample mean, $\bar{x}$. Based on this value of $\bar{x}$, the bureau will then assign values to the population mean, μ. Similarly, the AWAH will take a sample of husbands and determine the value of the sample proportion, $\hat{p}$, which represents the proportion of husbands in the sample who are alcoholic. Then, using this value of the sample proportion, $\hat{p}$, AWAH will assign values to the population proportion, p.

The value(s) assigned to a population parameter based on the value of a sample statistic is called an **estimate** of the population parameter. For example, suppose the manager takes a sample of 40 new employees and finds that the mean time, $\bar{x}$, taken to learn this job for these employees is 5.5 hours. If he or she assigns this value to the population mean, then 5.5 hours is called an estimate of μ. The sample statistic used to estimate a population parameter is called an **estimator**. Thus, the sample mean, $\bar{x}$, is an estimator of the population mean, μ, and the sample proportion, $\hat{p}$, is an estimator of the population proportion, p.

> **ESTIMATE AND ESTIMATOR** The value(s) assigned to a population parameter based on the value of a sample statistic is called an *estimate*. The sample statistic used to estimate a population parameter is called an *estimator*.

The estimation procedure involves the following steps.

1. Select a sample.
2. Collect the required information from the members of the sample.
3. Calculate the value of the sample statistic.
4. Assign value(s) to the corresponding population parameter.

8.2 POINT AND INTERVAL ESTIMATES

An estimate may be a point estimate or an interval estimate. These two types of estimates are described in this section.

8.2.1 A POINT ESTIMATE

If we select a sample and compute the value of the sample statistic for this sample, then this value gives the **point estimate** of the corresponding population parameter.

> **POINT ESTIMATE** The value of a sample statistic that is used to estimate a population parameter is called a *point estimate*.

Thus, the value computed for the sample mean, $\bar{x}$, from a sample is a point estimate of the corresponding population mean, μ. For the example mentioned earlier, suppose the U.S. Census Bureau takes a sample of 10,000 households and determines that the mean housing expenditure per month, $\bar{x}$, for this sample is $874. Then, using $\bar{x}$ as a point estimate of μ, the bureau can state that the mean housing expenditure per month, μ, for all households is about $874. This procedure is called **point estimation**.

Usually, whenever we use point estimation, we calculate the **margin of error** associated with that point estimation. For the estimation of the population mean, the margin of error is calculated as follows:

$$\text{Margin of error} = \pm 1.96\sigma_{\bar{x}} \quad \text{or} \quad \pm 1.96 s_{\bar{x}}$$

That is, we find the standard deviation of the sample mean and multiply it by 1.96. Here $s_{\bar{x}}$ is a point estimator of $\sigma_{\bar{x}}$; it will be discussed later in this chapter.

Each sample selected from a population is expected to yield a different value of the sample statistic. Thus, the value assigned to a population mean, μ, based on a point estimate depends on which of the samples is drawn. Consequently, the point estimate assigns a value to μ that almost always differs from the true value of the population mean.

8.2.2   AN INTERVAL ESTIMATE

In the case of **interval estimation**, instead of assigning a single value to a population parameter, an interval is constructed around the point estimate and then a probabilistic statement that this interval contains the corresponding population parameter is made.

> **INTERVAL ESTIMATION**   In *interval estimation,* an interval is constructed around the point estimate, and it is stated that this interval is likely to contain the corresponding population parameter.

For the example about the mean housing expenditure (Section 8.2.1), instead of saying that the mean housing expenditure per month for all households is $874, we may obtain an interval by subtracting a number from $874 and adding the same number to $874. Then we state that this interval contains the population mean, μ. For purposes of illustration, suppose we subtract $110 from $874 and add $110 to $874. Consequently, we obtain the interval ($874 − $110) to ($874 + $110), or $764 to $984. Then we state that the interval $764 to $984 is likely to contain the population mean, μ, and that the mean housing expenditure per month for all households in the United States is between $764 and $984. This procedure is called *interval estimation.* The value $764 is called the *lower limit* of the interval and $984 is called the *upper limit* of the interval. Figure 8.1 illustrates the concept of interval estimation.

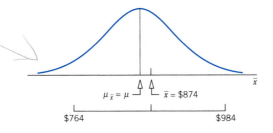

Figure 8.1   Interval estimation.

The question arises, What number should we subtract from and add to a point estimate to obtain an interval estimate? The answer to this question depends on two considerations:

1.   The standard deviation $\sigma_{\bar{x}}$ of the sample mean, $\bar{x}$

2.   The level of confidence to be attached to the interval

First, the larger the standard deviation of $\bar{x}$, the greater is the number subtracted from and added to the point estimate. Thus, it is obvious that if the range over which $\bar{x}$ can assume values is larger, then the interval constructed around $\bar{x}$ must be wider to include μ.

Second, the quantity subtracted and added must be larger if we want to have a higher confidence in our interval. We always attach a probabilistic statement to the interval estimation. This probabilistic statement is given by the **confidence level**. An interval that is constructed based on this confidence level is called a **confidence interval**.

> **CONFIDENCE LEVEL AND CONFIDENCE INTERVAL** Each interval is constructed with regard to a given *confidence level* and is called a *confidence interval.* The confidence level associated with a confidence interval states how much confidence we have that this interval contains the true population parameter. The confidence level is denoted by $(1 - \alpha)100\%$.

The confidence level is denoted by $\mathbf{(1 - \alpha)100\%}$ (α is the Greek letter *alpha*). When expressed as probability, it is called the *confidence coefficient* and is denoted by $1 - \alpha$. In passing, note that α is called the *significance level,* which will be explained in detail in Chapter 9.

Although any value of the confidence level can be chosen to construct a confidence interval, the more common values are 90%, 95%, and 99%. The corresponding confidence coefficients are .90, .95, and .99. The next section describes how to actually construct a confidence interval for the population mean for a large sample.

8.3 INTERVAL ESTIMATION OF A POPULATION MEAN: LARGE SAMPLES

This section explains how to construct a confidence interval for the population mean μ when the sample size is large.[1] Recall from the discussion in Chapter 7 that in the case of $\bar{x}$, the sample size is considered to be large when n is 30 or larger. According to the central limit theorem, for a large sample the sampling distribution of the sample mean, $\bar{x}$, is (approximately) normal irrespective of the shape of the population from which the sample is drawn. Therefore, *when the sample size is 30 or larger, we will use the normal distribution to construct a confidence interval for μ.* We also know from Chapter 7 that the standard deviation of $\bar{x}$ is $\sigma_{\bar{x}} = \sigma/\sqrt{n}$. However, if the population standard deviation, σ, is not known, then we use the sample standard deviation, s, for σ. Consequently, we use

$$s_{\bar{x}} = \frac{s}{\sqrt{n}}$$

for $\sigma_{\bar{x}} = \sigma/\sqrt{n}$. Note that the value of $s_{\bar{x}}$ is a point estimate of $\sigma_{\bar{x}}$.

> **CONFIDENCE INTERVAL FOR μ FOR LARGE SAMPLES** The $(1 - \alpha)100\%$ *confidence interval for μ* is
>
> $$\bar{x} \pm z\sigma_{\bar{x}} \quad \text{if } \sigma \text{ is known}$$
>
> $$\bar{x} \pm zs_{\bar{x}} \quad \text{if } \sigma \text{ is not known}$$
>
> where $\qquad \sigma_{\bar{x}} = \sigma/\sqrt{n} \quad \text{and} \quad s_{\bar{x}} = s/\sqrt{n}$
>
> The value of z used here is read from the standard normal distribution table for the given confidence level.

[1] Some statisticians prefer to discuss estimation and tests of hypotheses based on whether or not the population standard deviation σ is known. However, we prefer to use the large sample and small sample criteria. The reason is that σ is almost always unknown. Hence, discussing estimation and tests of hypotheses based on large and small samples makes more sense than whether or not σ is known.

The quantity $z\sigma_{\bar{x}}$ (or $zs_{\bar{x}}$ when σ is not known) in the confidence interval formula is called the **maximum error of estimate** and is denoted by E.

MAXIMUM ERROR OF ESTIMATE FOR μ   The *maximum error of estimate for μ,* denoted by E, is the quantity that is subtracted from and added to the value of $\bar{x}$ to obtain a confidence interval for μ. Thus,

$$E = z\sigma_{\bar{x}} \quad \text{or} \quad zs_{\bar{x}}$$

The value of z in the confidence interval formula is obtained from the standard normal distribution table (Table VII of Appendix D) for the given confidence level. To illustrate, suppose we want to construct a 95% confidence interval for μ. A 95% confidence level means that the total area under the normal curve for $\bar{x}$ between two points (at the same distance) on different sides of μ is 95% or .95, as shown in Figure 8.2. To find the value of z, we first divide the given confidence coefficient by 2. Then we look for this number in the body of the normal table. The corresponding value of z is the value we use in the confidence interval. Thus, to find the z value for a 95% confidence level, we perform the following two steps:

1. Divide .95 by 2, which gives .4750.
2. Locate .4750 in the body of the normal distribution table and record the corresponding value of z. This value of z is 1.96.

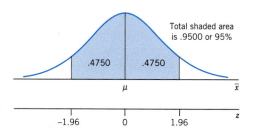

Figure 8.2   Finding z for a 95% confidence level.

For a $(1 - \alpha)100\%$ confidence level, the area between $-z$ and z is $1 - \alpha$. Because the total area under the normal curve is 1.0, the total area under the curve in the two tails is α. This, as mentioned earlier, is called the significance level. In the example of Figure 8.2, $\alpha = 1 - .95 = .05$. Therefore, as shown in Figure 8.3, the area under the curve in each of the two tails is $\alpha/2$. The value of z associated with a $(1 - \alpha)100\%$ confidence level is sometimes denoted by $z_{\alpha/2}$. However, this text will denote this value simply by z.

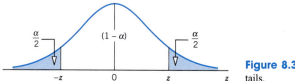

Figure 8.3   Area in the tails.

Example 8–1 describes the procedure used to construct a confidence interval for μ for a large sample.

Finding the point estimate and confidence interval for μ: σ known and $n > 30$.

EXAMPLE 8–1 A publishing company has just published a new college textbook. Before the company decides the price at which to sell this textbook, it wants to know the average price of all such textbooks in the market. The research department at the company took a sample of 36 such textbooks and collected information on their prices. This information produced a mean price of $70.50 for this sample. It is known that the standard deviation of the prices of all such textbooks is $4.50.

(a) What is the point estimate of the mean price of all such college textbooks? What is the margin of error for this estimate?

(b) Construct a 90% confidence interval for the mean price of all such college textbooks.

Solution From the given information,

$$n = 36, \quad \bar{x} = \$70.50, \quad \text{and} \quad \sigma = \$4.50$$

The standard deviation of $\bar{x}$ is

$$\sigma_{\bar{x}} = \frac{\sigma}{\sqrt{n}} = \frac{4.50}{\sqrt{36}} = \$.75$$

(a) The point estimate of the mean price of all such college textbooks is $70.50; that is,

$$\text{Point estimate of } \mu = \bar{x} = \mathbf{\$70.50}$$

The margin of error associated with this point estimate of μ is

$$\text{Margin of error} = \pm 1.96\sigma_{\bar{x}} = \pm 1.96(.75) = \mathbf{\pm\$1.47}$$

The margin of error means that the mean price of all such college textbooks is $70.50, give or take $1.47. Note that the margin of error is simply the maximum error of estimate for a 95% confidence interval.

(b) The confidence level is 90% or .90. First we find the z value for a 90% confidence level. To do so, we divide .90 by 2 to obtain .4500. Then we locate .4500 in the body of the normal distribution table (Table VII of Appendix D). Because .4500 is not in the normal table, we can use the number closest to .4500, which is either .4495 or .4505. If we use .4505 as an approximation for .4500, the value of z for this number is 1.65.[2]

Next, we substitute all the values in the confidence interval formula for μ. The 90% confidence interval for μ is

$$\bar{x} \pm z\sigma_{\bar{x}} = 70.50 \pm 1.65(.75) = 70.50 \pm 1.24$$

$$= (70.50 - 1.24) \text{ to } (70.50 + 1.24) = \mathbf{\$69.26 \text{ to } \$71.74}$$

Thus, we are 90% confident that the mean price of all such college textbooks is between $69.26 and $71.74. Note that we cannot say for sure whether the interval $69.26 to $71.74 contains the true population mean or not. Since μ is a constant, we cannot say that the probability is .90 that this interval contains μ because either it contains μ or it does not. Consequently, the probability is either 1.0 or 0 that this interval contains μ. All we can say is that we are 90% confident that the mean price of all such college textbooks is between $69.26 and $71.74. ∎

How do we interpret a 90% confidence level? In terms of Example 8–1, if we take all possible samples of 36 such college textbooks each and construct a 90% confidence interval

---

[2]Note that there is no apparent reason for choosing .4505 and not choosing .4495. If we choose .4495, the z value will be 1.64. An alternative is to use the average of 1.64 and 1.65, 1.645, which we will not do in this text.

for μ around each sample mean, we can expect that 90% of these intervals will include μ and 10% will not. In Figure 8.4 we show means $\bar{x}_1$, $\bar{x}_2$, and $\bar{x}_3$ of three different samples of the same size drawn from the same population. Also shown in this figure are the 90% confidence intervals constructed around these three sample means. As we observe, the 90% confidence intervals constructed around $\bar{x}_1$ and $\bar{x}_2$ include μ, but the one constructed around $\bar{x}_3$ does not. We can state for a 90% confidence level that if we take many samples of the same size from a population and construct 90% confidence intervals around the means of these samples, then 90% of these confidence intervals will be like the ones around $\bar{x}_1$ and $\bar{x}_2$ in Figure 8.4, which include μ, and 10% will be like the one around $\bar{x}_3$, which does not include μ.

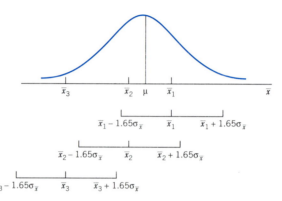

Figure 8.4   Confidence intervals.

In Example 8–1, the value of the population standard deviation, σ, was known. However, more often we do not know the value of σ. In such cases, we estimate the population standard deviation, σ, by the sample standard deviation, s, and estimate the standard deviation of $\bar{x}$, $\sigma_{\bar{x}}$, by $s_{\bar{x}}$. Then we use $s_{\bar{x}}$ for $\sigma_{\bar{x}}$ in the formula for the confidence interval for μ. Owing to the central limit theorem, as long as the sample size is large ($n \geq 30$), even if we do not know σ, we can use the normal distribution.

Example 8–2 illustrates the construction of a confidence interval for μ when σ is not known.

Constructing a confidence interval for μ: σ not known and $n > 30$.

EXAMPLE 8–2   The U.S. Bureau of Labor Statistics regularly collects information on the labor market. According to the bureau, workers employed in manufacturing industries in the United States earned an average of $577 per week in May 1999 (*Bureau of Labor Statistics News,* June 16, 1999). Assume that this mean is based on a random sample of 1600 workers selected from the manufacturing industries and that the standard deviation of weekly earnings for this sample is $80. Make a 99% confidence interval for the mean weekly earnings of all workers employed in manufacturing industries in May 1999.

Solution   From the given information,

$$n = 1600 \qquad \bar{x} = \$577 \qquad s = \$80$$

$$\text{Confidence level} = 99\% \text{ or } .99$$

First we find the standard deviation of $\bar{x}$. Because σ is not known, we will use $s_{\bar{x}}$ as an estimator of $\sigma_{\bar{x}}$. The value of $s_{\bar{x}}$ is

$$s_{\bar{x}} = \frac{s}{\sqrt{n}} = \frac{80}{\sqrt{1600}} = \$2$$

Because the sample size is large ($n > 30$), we will use the normal distribution to determine the confidence interval for μ. To find z for a 99% confidence level, we divide .99 by 2 to obtain .4950. From the normal distribution table, the z value for .4950 is approximately 2.58. Substituting all the values in the formula, we find the 99% confidence interval for μ as follows.

$$\bar{x} \pm z s_{\bar{x}} = 577 \pm 2.58(2) = 577 \pm 5.16 = \textbf{\$571.84 to \$582.16}$$

Thus, we can state with 99% confidence that the average weekly earnings of workers employed in U.S. manufacturing industries in May 1999 were between \$571.84 and \$582.16. ∎

The **width of a confidence interval** depends on the size of the maximum error, $z\sigma_{\bar{x}}$, which depends on the values of z, σ, and n because $\sigma_{\bar{x}} = \sigma/\sqrt{n}$. However, the value of σ is not under the control of the investigator. Hence, the width of a confidence interval depends on

1. The value of z, which depends on the confidence level
2. The sample size n

The confidence level determines the value of z, which in turn determines the size of the maximum error. The value of z increases as the confidence level increases, and it decreases as the confidence level decreases. For example, the value of z is approximately 1.65 for a 90% confidence level, 1.96 for a 95% confidence level, and approximately 2.58 for a 99% confidence level. Hence, the higher the confidence level, the larger the width of the confidence interval, other things remaining the same.

For the same value of σ, an increase in the sample size decreases the value of $\sigma_{\bar{x}}$, which in turn decreases the size of the maximum error when the confidence level remains unchanged. Therefore, an increase in the sample size decreases the width of the confidence interval.

Thus, if we want to decrease the width of a confidence interval, we have two choices:

1. Lower the confidence level
2. Increase the sample size

Lowering the confidence level is not a good choice, however, because a lower confidence level may give less reliable results. Therefore, we should always prefer to increase the sample size if we want to decrease the width of a confidence interval. Next we illustrate, using Example 8–2, how either a decrease in the confidence level or an increase in the sample size decreases the width of the confidence interval.

1. CONFIDENCE LEVEL AND THE WIDTH OF THE CONFIDENCE INTERVAL

Reconsider Example 8–2. Suppose all the information given in that example remains the same. First, let us decrease the confidence level to 95%. From the normal distribution table, $z = 1.96$ for a 95% confidence level. Then, using $z = 1.96$ in the confidence interval for Example 8–2, we obtain

$$\bar{x} \pm z s_{\bar{x}} = 577 \pm 1.96(2) = 577 \pm 3.92 = \textbf{\$573.08 to \$580.92}$$

Comparing this confidence interval to the one obtained in Example 8–2, we observe that the width of the confidence interval for a 95% confidence level is smaller than the one for a 99% confidence level.

2. SAMPLE SIZE AND THE WIDTH OF THE CONFIDENCE INTERVAL

Consider Example 8–2 again. Now suppose the information given in that example is based on a sample size of 2500. Further assume that all other information given in that example, including the confidence level, remains the same. First, we calculate the standard deviation of the sample mean using $n = 2500$:

$$s_{\bar{x}} = \frac{s}{\sqrt{n}} = \frac{80}{\sqrt{2500}} = \$1.60$$

Then, the 99% confidence interval for μ is

$$\bar{x} \pm z s_{\bar{x}} = 577 \pm 2.58(1.60) = 577 \pm 4.13 = \textbf{\$572.87 to \$581.13}$$

Comparing this confidence interval to the one obtained in Example 8–2, we observe that the width of the 99% confidence interval for $n = 2500$ is smaller than the 99% confidence interval for $n = 1600$.

CASE STUDY 8-1   COST OF RUNNING A HOUSEHOLD

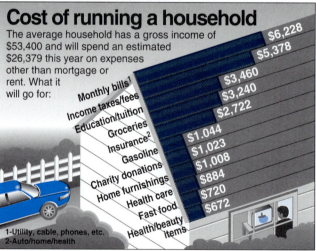

USA SNAPSHOTS®

A look at statistics that shape your finances

Cost of running a household

The average household has a gross income of $53,400 and will spend an estimated $26,379 this year on expenses other than mortgage or rent. What it will go for:

Monthly bills[1]	$6,228
Income taxes/fees	$5,378
Education/tuition	$3,460
Groceries	$3,240
Insurance[2]	$2,722
Gasoline	$1,044
Charity donations	$1,023
Home furnishings	$1,008
Health care	$884
Fast food	$720
Health/beauty items	$672

1-Utility, cable, phones, etc.
2-Auto/home/health

Source: American Express Everyday Spending Index   By Cindy Hall and Gary Visgaitis, USA TODAY

The above chart shows the average annual expenditures for several categories of household expenses. For example, according to the data in the chart, the total of monthly bills for utilities, cable, phones, etc., averages $6228 per household per year. The total expenditure on income taxes/fees averages $5378 per household per year. Note that the average (estimated) expenditures shown in the chart are based on a sample of households. If we know the sample sizes and the standard deviations for these different types of expenses, we can then find confidence intervals for the mean expenditures by all households for these categories of expenses, as shown in the table below.

Category of Expense	Mean Expenditure for Sample	Confidence Interval
Monthly bills	$6228	$6228 \pm zs_{\bar{x}}$
Income taxes/fees	5378	$5378 \pm zs_{\bar{x}}$
Education/tuition	3460	$3460 \pm zs_{\bar{x}}$
Groceries	3240	$3240 \pm zs_{\bar{x}}$
Insurance	2722	$2722 \pm zs_{\bar{x}}$
Gasoline	1044	$1044 \pm zs_{\bar{x}}$
Charity donations	1023	$1023 \pm zs_{\bar{x}}$
Home furnishings	1008	$1008 \pm zs_{\bar{x}}$
Health care	884	$884 \pm zs_{\bar{x}}$
Fast food	720	$720 \pm zs_{\bar{x}}$
Health/beauty items	672	$672 \pm zs_{\bar{x}}$

In the confidence intervals given in the table, we can substitute the value of z for the given confidence level and the value of $s_{\bar{x}}$, which is calculated as $s/\sqrt{n}$. For example, suppose we want to find a 98% confidence interval for the mean annual expenditure on monthly bills for all households. Assume that the average for this category given in the chart is based on a sample of 900 households and that the sample standard deviation for the expenditures on monthly bills is $750. Then a 98% confidence interval for this population mean is calculated as follows:

$$s_{\bar{x}} = \frac{s}{\sqrt{n}} = \frac{750}{\sqrt{900}} = 25.00$$

$$\bar{x} \pm zs_{\bar{x}} = 6228 \pm 2.33(25.00) = 6228 \pm 58.25 = \$6169.75 \text{ to } \$6286.25$$

We can find confidence intervals for the population means for other categories in the same way. Note that the sample means given in the table are the point estimates of the corresponding population means.

Source: The chart is reproduced with permission from *USA TODAY,* June 3, 1999. Copyright © 1999, *USA TODAY.*

EXERCISES

■ **Concepts and Procedures**

8.1 Briefly explain the meaning of an estimator and an estimate.

8.2 Explain the meaning of a point estimate and an interval estimate.

8.3 What is the point estimator of the population mean, μ? How would you calculate the margin of error for a point estimate of μ?

8.4 Explain the various alternatives for decreasing the width of a confidence interval. Which is the best alternative?

8.5 Briefly explain how the width of a confidence interval decreases with an increase in the sample size. Give an example.

8.6 Briefly explain how the width of a confidence interval decreases with a decrease in the confidence level. Give an example.

8.7 Briefly explain the difference between a confidence level and a confidence interval.

8.8 What is the maximum error of estimate for μ for a large sample? How is it calculated?

8.9 How will you interpret a 99% confidence interval for μ for a large sample? Explain.

8.10 Find z for each of the following confidence levels.

a. 90%   **b.** 95%   **c.** 96%   **d.** 97%   **e.** 98%   **f.** 99%

8.11 For a data set obtained from a sample, $n = 64$, $\bar{x} = 24.5$, and $s = 3.1$.

a. What is the point estimate of μ?
b. What is the margin of error associated with the point estimate of μ?
c. Make a 99% confidence interval for μ.
d. What is the maximum error of estimate for part c?

8.12 For a data set obtained from a sample, $n = 81$, $\bar{x} = 48.25$, and $s = 4.8$.

a. What is the point estimate of μ?
b. What is the margin of error associated with the point estimate of μ?
c. Make a 95% confidence interval for μ.
d. What is the maximum error of estimate for part c?

8.13 The standard deviation for a population is $\sigma = 15.3$. A sample of 36 observations selected from this population gave a mean equal to 74.8.

a. Make a 90% confidence interval for μ.
b. Construct a 95% confidence interval for μ.
c. Determine a 99% confidence interval for μ.
d. Does the width of the confidence intervals constructed in parts a through c increase as the confidence level increases? Explain your answer.

8.14 The standard deviation for a population is $\sigma = 14.8$. A sample of 100 observations selected from this population gave a mean equal to 143.72.

a. Make a 99% confidence interval for μ.
b. Construct a 95% confidence interval for μ.
c. Determine a 90% confidence interval for μ.
d. Does the width of the confidence intervals constructed in parts a through c decrease as the confidence level decreases? Explain your answer.

 8.15 The standard deviation for a population is $\sigma = 6.30$. A random sample selected from this population gave a mean equal to 81.90.

a. Make a 99% confidence interval for μ assuming $n = 36$.
b. Construct a 99% confidence interval for μ assuming $n = 81$.
c. Determine a 99% confidence interval for μ assuming $n = 100$.
d. Does the width of the confidence intervals constructed in parts a through c decrease as the sample size increases? Explain.

8.16 The standard deviation for a population is $\sigma = 7.14$. A random sample selected from this population gave a mean equal to 48.52.

a. Make a 95% confidence interval for μ assuming $n = 196$.
b. Construct a 95% confidence interval for μ assuming $n = 100$.
c. Determine a 95% confidence interval for μ assuming $n = 49$.
d. Does the width of the confidence intervals constructed in parts a through c increase as the sample size decreases? Explain.

8.17 a. A sample of 100 observations taken from a population produced a sample mean equal to 55.32 and a standard deviation equal to 8.4. Make a 90% confidence interval for μ.

b. Another sample of 100 observations taken from the same population produced a sample mean equal to 57.40 and a standard deviation equal to 7.5. Make a 90% confidence interval for μ.

c. A third sample of 100 observations taken from the same population produced a sample mean equal to 56.25 and a standard deviation equal to 7.9. Make a 90% confidence interval for μ.

d. The true population mean for this population is 55.80. Which of the confidence intervals constructed in parts a through c cover this population mean and which do not?

8.18 a. A sample of 400 observations taken from a population produced a sample mean equal to 92.45 and a standard deviation equal to 12.20. Make a 97% confidence interval for μ.

b. Another sample of 400 observations taken from the same population produced a sample mean equal to 91.75 and a standard deviation equal to 14.50. Make a 97% confidence interval for μ.

c. A third sample of 400 observations taken from the same population produced a sample mean equal to 89.63 and a standard deviation equal to 13.40. Make a 97% confidence interval for μ.

d. The true population mean for this population is 90.65. Which of the confidence intervals constructed in parts a through c cover this population mean and which do not?

8.19 For a population, the value of the standard deviation is 2.65. A sample of 35 observations taken from this population produced the following data.

42	51	42	31	28	36	49
29	46	37	32	27	33	41
47	41	28	46	34	39	48
26	35	37	38	46	48	39
29	31	44	41	37	38	46

a. What is the point estimate of μ?
b. What is the margin of error associated with the point estimate of μ?
c. Make a 98% confidence interval for μ.
d. What is the maximum error of estimate for part c?

8.20 For a population, the value of the standard deviation is 4.96. A sample of 32 observations taken from this population produced the following data.

74	85	72	73	86	81	77	60
83	78	79	88	76	73	84	78
81	72	82	81	79	83	88	86
78	83	87	82	80	84	76	74

a. What is the point estimate of μ?
b. What is the margin of error associated with the point estimate of μ?
c. Make a 99% confidence interval for μ.
d. What is the maximum error of estimate for part c?

■ Applications

8.21 According to the IRS, the average deduction for medical expenses taken by taxpayers with adjusted gross incomes between $20,001 and $25,000 is $4160 (*USA TODAY,* March 5, 1999). Assume that this mean is based on a random sample of 400 such taxpayers and that the sample standard deviation is $1500. Construct a 99% confidence interval for the mean medical deduction taken by all taxpayers with adjusted gross incomes between $20,001 and $25,000.

8.22 According to the U.S. Commerce Department, the average price of single-family new homes in the United States was $196,900 in September 1999 (*The Wall Street Journal,* November 1, 1999). Suppose that this mean price is based on a random sample of 900 such homes and that the standard deviation for this sample is $31,000. Construct a 99% confidence interval for the corresponding population mean.

8.23 In 1999, Keynote Systems measured the performance of (electronic) commerce sites on the Web and found that the average time for such a site to come up on a personal computer was 21.36 seconds

(*USA TODAY,* December 8, 1999). Suppose that this mean time was based on a random sample of 36 contacts with such sites and that the sample standard deviation was 4.2 seconds. Make a 95% confidence interval for the corresponding population mean.

8.24 According to the Centers for Disease Control and Prevention, the average length of stay for patients in nonfederal short-stay hospitals in the United States was 5.2 days (*Advance Data,* August 31, 1998). Assume that this average was based on a random sample of 800 such hospital stays and that the sample standard deviation was 1.7 days. Construct a 95% confidence interval for the mean length of all such hospital stays.

8.25 According to a 1999 poll of 1014 adults by the National Sleep Foundation, adults sleep an average of (about) 7 hours per night during the workweek (*USA TODAY,* March 25, 1999). Assume that these 1014 adults make up a random sample of all U.S. adults and that the standard deviation for this sample is .75 hour.

a. What is the point estimate of the corresponding population mean? What is the margin of error for this estimate?

b. Make a 98% confidence interval for the corresponding population mean.

8.26 According to a study by Gallup for Pfizer Animal Health, households that own one dog spend an average of $144 per year on veterinary care (*USA TODAY,* July 20, 1998). Assume that this average is based on a random sample of 800 such dog owners and that the sample standard deviation is $45.

a. What is the point estimate of the mean annual expenditure on veterinary care for all such dog owners? What is the margin of error for this estimate?

b. Make a 90% confidence interval for the mean annual expenditure on veterinary care for all such dog owners.

8.27 According to data from Jupiter Communications, the average household income of online shoppers in the United States was $59,000 in 1998 (*Time,* July 20, 1998). Assume that this mean is based on a random sample of 500 online shoppers and that the sample standard deviation is $7000.

a. Make a 97% confidence interval for the mean household income of all online shoppers in the United States.

b. Explain why we need to make the confidence interval. Why can we not say that the mean household income of all online shoppers in the United States is $59,000?

8.28 Computer Action Company sells computers and computer parts by mail. The company assures its customers that products are mailed as soon as possible after an order is placed with the company. A sample of 50 recent orders showed that the mean time taken to mail products for these orders was 70 hours and the standard deviation was 16 hours.

a. Construct a 95% confidence interval for the mean time taken to mail products for all orders received at the office of this company.

b. Explain why we need to make the confidence interval. Why can we not say that the mean time taken to mail products for all orders received at the office of this company is 70 hours?

 8.29 Lazurus Steel Corporation produces iron rods that are supposed to be 36 inches long. The machine that makes these rods does not produce each rod exactly 36 inches long. The lengths of the rods vary slightly. It is known that when the machine is working properly, the mean length of the rods made on this machine is 36 inches. The standard deviation of the lengths of all rods produced on this machine is always equal to .10 inch. The quality control department takes a sample of 40 such rods every week, calculates the mean length of these rods, and makes a 99% confidence interval for the population mean. If either the upper limit of this confidence interval is greater than 36.05 inches or the lower limit of this confidence interval is less than 35.95 inches, the machine is stopped and adjusted. A recent such sample of 40 rods produced a mean length of 36.02 inches. Based on this sample, will you conclude that the machine needs an adjustment?

8.30 At Farmer's Dairy, a machine is set to fill 32-ounce milk cartons. However, this machine does not put exactly 32 ounces of milk into each carton; the amount varies slightly from carton to carton. It is known that when the machine is working properly, the mean net weight of these cartons is 32 ounces. The standard deviation of the amount of milk in all such cartons is always equal to .15 ounce.

The quality control department takes a sample of 35 such cartons every week, calculates the mean net weight of these cartons, and makes a 99% confidence interval for the population mean. If either the upper limit of this confidence interval is greater than 32.15 ounces or the lower limit of this confidence interval is less than 31.85 ounces, the machine is stopped and adjusted. A recent sample of 35 such cartons produced a mean net weight of 31.94 ounces. Based on this sample, will you conclude that the machine needs an adjustment?

8.31 A consumer agency that proposes that lawyers' rates are too high wanted to estimate the mean hourly rate for all lawyers in New York City. A sample of 70 lawyers taken from New York City showed that the mean hourly rate charged by them is $310 and the standard deviation of hourly charges is $75.

a. Construct a 99% confidence interval for the mean hourly charges for all lawyers in New York City.
b. Suppose the confidence interval obtained in part a is too wide. How can the width of this interval be reduced? Discuss all possible alternatives. Which alternative is the best?

8.32 A bank manager wants to know the mean amount of mortgage paid per month by homeowners in an area. A random sample of 40 homeowners selected from this area showed that they pay an average of $1575 per month for their mortgage with a standard deviation of $215.

a. Find a 97% confidence interval for the mean amount of mortgage paid per month by all homeowners in this area.
b. Suppose the confidence interval obtained in part a is too wide. How can the width of this interval be reduced? Discuss all possible alternatives. Which alternative is the best?

*8.33 You are interested in estimating the mean commuting time from home to school for all commuter students at your school. Briefly explain the procedure you will follow to conduct this study. Collect the required data from a sample of 30 or more such students and then estimate the population mean at a 99% confidence level.

*8.34 You are interested in estimating the mean age of cars owned by all people in the United States. Briefly explain the procedure you will follow to conduct this study. Collect the required data on a sample of 30 or more cars and then estimate the population mean at a 95% confidence level.

8.4 INTERVAL ESTIMATION OF A POPULATION MEAN: SMALL SAMPLES

Recall from Section 8.3 that for large samples ($n \geq 30$), whether or not σ is known, the normal distribution is used to estimate the population mean, μ. We use the normal distribution in such cases because, according to the central limit theorem, the sampling distribution of $\bar{x}$ is approximately normal for large samples irrespective of the shape of the population distribution.

Many times we can select only small samples, however. This may be due either to the nature of the experiment or to the cost involved in taking a sample. For example, to test a new drug on patients, research may have to be based on a small sample either because there are not many patients available or willing to participate or because it is too expensive to include enough patients in the research to have a large sample.

If the sample size is small, the normal distribution can still be used to construct a confidence interval for μ if (1) the population from which the sample is drawn is normally distributed and (2) the value of σ is known. But more often we do not know σ and, consequently, we have to use the sample standard deviation, s, as an estimator of σ. In such cases, the normal distribution cannot be used to make confidence intervals about μ. When (1) the population from which the sample is selected is (approximately) normally distributed, (2) the sample size is small (that is, $n < 30$), and (3) the population standard deviation, σ, is not known, the normal distribution is replaced by the *t distribution* to construct confidence intervals about μ. The *t* distribution is described in the next subsection.

> **CONDITIONS UNDER WHICH THE *t* DISTRIBUTION IS USED TO MAKE A CONFIDENCE INTERVAL ABOUT μ**
>
> The *t distribution* is used to make a confidence interval about μ if
>
> 1. The population from which the sample is drawn is (approximately) normally distributed
> 2. The sample size is small (that is, $n < 30$)
> 3. The population standard deviation, σ, is not known

8.4.1 THE *t* DISTRIBUTION

The *t* **distribution** was developed by W. S. Gosset in 1908 and published under the pseudonym *Student.* As a result, the *t* distribution is also called *Student's t distribution.* The *t* distribution is similar to the normal distribution in some respects. Like the normal distribution curve, the *t* distribution curve is symmetric (bell-shaped) about the mean and it never meets the horizontal axis. The total area under a *t* distribution curve is 1.0 or 100%. However, the *t* distribution curve is flatter than the standard normal distribution curve. In other words, the *t* distribution curve has a lower height and a wider spread (or, we can say, larger standard deviation) than the standard normal distribution. However, as the sample size increases, the *t* distribution approaches the standard normal distribution. The units of a *t* distribution are denoted by *t*.

The shape of a particular *t* distribution curve depends on the number of **degrees of freedom (*df*)**. For the purpose of Chapters 8 and 9, the number of degrees of freedom for a *t* distribution is equal to the sample size minus one, that is,

$$df = n - 1$$

The number of degrees of freedom is the only parameter of the *t* distribution. There is a different *t* distribution for each number of degrees of freedom. Like the standard normal distribution, the mean of the *t* distribution is 0. But unlike the standard normal distribution, whose standard deviation is 1, the standard deviation of a *t* distribution is $\sqrt{df/(df - 2)}$, which is always greater than 1. Thus, the standard deviation of a *t* distribution is larger than the standard deviation of the standard normal distribution.

> **THE *t* DISTRIBUTION** The *t distribution* is a specific type of bell-shaped distribution with a lower height and a wider spread than the standard normal distribution. As the sample size becomes larger, the *t* distribution approaches the standard normal distribution. The *t* distribution has only one parameter, called the degrees of freedom (*df*). The mean of the *t* distribution is equal to 0 and its standard deviation is $\sqrt{df/(df - 2)}$.

Figure 8.5 shows the standard normal distribution and the *t* distribution for 9 degrees of freedom. The standard deviation of the standard normal distribution is 1.0, and the standard deviation of the *t* distribution is $\sqrt{df/(df - 2)} = \sqrt{9/(9 - 2)} = 1.134$.

As stated earlier, the number of degrees of freedom for a *t* distribution for the purpose of this chapter is $n - 1$. *The number of degrees of freedom is defined as the number of observations that can be chosen freely.* As an example, suppose we know that the mean of 4

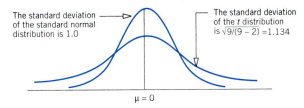

The standard deviation of the standard normal distribution is 1.0

The standard deviation of the t distribution is $\sqrt{9/(9-2)} = 1.134$

$\mu = 0$

Figure 8.5 The t distribution for $df = 9$ and the standard normal distribution.

values is 20. Consequently, the sum of these 4 values is 20(4) = 80. Now, how many values out of 4 can we choose freely so that the sum of these 4 values is 80? The answer is that we can freely choose 4 − 1 = 3 values. Suppose we choose 27, 8, and 19 as the 3 values. Given these 3 values and the information that the mean of the 4 values is 20, the fourth value is 80 − 27 − 8 − 19 = 26. Thus, once we have chosen 3 values, the fourth value is automatically determined. Consequently, the number of degrees of freedom for this example is

$$df = n - 1 = 4 - 1 = 3$$

We subtract 1 from n because we lose 1 degree of freedom to calculate the mean.

Table VIII of Appendix D lists the values of t for the given number of degrees of freedom and areas in the right tail of a t distribution. Because the t distribution is symmetric, these are also the values of $-t$ for the same number of degrees of freedom and the same areas in the left tail of the t distribution. Example 8–3 describes how to read Table VIII of Appendix D.

Reading the t distribution table.

EXAMPLE 8–3 Find the value of t for 16 degrees of freedom and .05 area in the right tail of a t distribution curve.

Solution In Table VIII of Appendix D, we locate 16 in the column of degrees of freedom (labeled *df*) and .05 in the row of *Area in the right tail under the t distribution curve* at the top of the table. The entry at the intersection of the row of 16 and the column of .05, which is 1.746, gives the required value of t. The relevant portion of Table VIII of Appendix D is shown here as Table 8.1. The value of t read from the t distribution table is shown in Figure 8.6 on page 350.

Table 8.1 Determining t for 16 *df* and .05 Area in the Right Tail

Area in the right tail

	Area in the Right Tail Under the t Distribution Curve				
df	.10	.05 ←	.025	...	.001
1	3.078	6.314	12.706	...	318.309
2	1.886	2.920	4.303	...	22.327
3	1.638	2.353	3.182	...	10.215
⋮	...	...	...	...	...
df → **16**	1.337	**1.746** ←	2.120	...	3.686
⋮	...	...	...	...	...
75	1.293	1.665	1.992	...	3.202
∞	1.282	1.645	1.960	...	3.090

The required value of t for 16 *df* and .05 area in the right tail

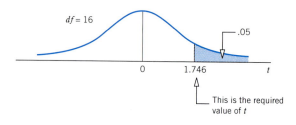

Figure 8.6 The value of t for 16 df and .05 area in the right tail. ■

Because of the symmetric shape of the t distribution curve, the value of t for 16 degrees of freedom and .05 area in the left tail is -1.746. Figure 8.7 illustrates this case.

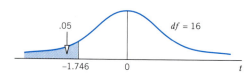

Figure 8.7 The value of t for 16 df and .05 area in the left tail.

8.4.2 CONFIDENCE INTERVAL FOR μ USING THE t DISTRIBUTION

To reiterate, when the following three conditions hold true, we use the t distribution to construct a confidence interval for the population mean, μ.

1. The population from which the sample is drawn is (approximately) normally distributed.
2. The sample size is small (that is, $n < 30$).
3. The population standard deviation, σ, is not known.

CONFIDENCE INTERVAL FOR μ FOR SMALL SAMPLES
The $(1 - \alpha)100\%$ *confidence interval for μ* is

$$\bar{x} \pm ts_{\bar{x}} \quad \text{where} \quad s_{\bar{x}} = \frac{s}{\sqrt{n}}$$

The value of t is obtained from the t distribution table for $n - 1$ degrees of freedom and the given confidence level.

Examples 8–4 and 8–5 describe the procedure of constructing a confidence interval for μ using the t distribution.

Constructing a 95% confidence interval for μ using the t distribution.

EXAMPLE 8–4 Dr. Moore wanted to estimate the mean cholesterol level for all adult men living in Hartford. He took a sample of 25 adult men from Hartford and found that the mean cholesterol level for this sample is 186 with a standard deviation of 12. Assume that the cholesterol levels for all adult men in Hartford are (approximately) normally distributed. Construct a 95% confidence interval for the population mean μ.

Solution From the given information,

$$n = 25 \qquad \bar{x} = 186 \qquad s = 12$$

Confidence level $= 95\%$ or $.95$

The value of $s_{\bar{x}}$ is

$$s_{\bar{x}} = \frac{s}{\sqrt{n}} = \frac{12}{\sqrt{25}} = 2.40$$

To find the value of t, we need to know the degrees of freedom and the area under the t distribution curve in each tail.

$$\text{Degrees of freedom} = n - 1 = 25 - 1 = 24$$

To find the area in each tail, we divide the confidence level by 2 and subtract the number obtained from .5. Thus,

$$\text{Area in each tail} = .5 - (.95/2) = .5 - .4750 = .025$$

From the t distribution table, Table VIII of Appendix D, the value of t for $df = 24$ and .025 area in the right tail is 2.064. The value of t is shown in Figure 8.8.

Figure 8.8 The value of t.

When we substitute all values in the formula for the confidence interval for μ, the 95% confidence interval is

$$\bar{x} \pm ts_{\bar{x}} = 186 \pm 2.064(2.40) = 186 \pm 4.95 = \textbf{181.05 to 190.95}$$

Thus, we can state with 95% confidence that the mean cholesterol level for all adult men living in Hartford lies between 181.05 and 190.95.

Note that $\bar{x} = 186$ is a point estimate of μ in this example. ■

Constructing a 99% confidence interval for μ using the t distribution.

EXAMPLE 8–5   According to an IRS report, the average total itemized deductions taken by individual tax return filers who used this option was $16,615 for 1997 (*The Wall Street Journal,* July 7, 1999). Suppose that this mean is based on a random sample of 25 such individual tax returns and that the standard deviation for the itemized deductions for this sample is $2000. Further assume that the itemized deductions for all such individual tax returns have an approximately normal distribution. Determine a 99% confidence interval for the corresponding population mean.

Solution   From the given information,

$$n = 25 \qquad \bar{x} = \$16,615 \qquad s = \$2000$$

$$\text{Confidence level} = 99\% \text{ or } .99$$

First we calculate the standard deviation of $\bar{x}$, the number of degrees of freedom, and the area in each tail of the t distribution.

$$s_{\bar{x}} = \frac{s}{\sqrt{n}} = \frac{2000}{\sqrt{25}} = \$400$$

$$df = n - 1 = 25 - 1 = 24$$

$$\text{Area in each tail} = .5 - (.99/2) = .5 - .4950 = .005$$

From the t distribution table, $t = 2.797$ for 24 degrees of freedom and .005 area in the right tail. The 99% confidence interval for μ is

$$\bar{x} \pm ts_{\bar{x}} = 16,615 \pm 2.797(400)$$

$$= 16,615 \pm 1118.80 = \textbf{\$15,496.20 to \$17,733.80}$$

Thus, we can state with 99% confidence that based on this sample the mean itemized deductions for individual tax returns for those who used this option was between \$15,496.20 and \$17,733.80 in 1997. ■

Again, we can decrease the width of a confidence interval for μ either by lowering the confidence level or by increasing the sample size, as was done in Section 8.3. However, increasing the sample size is the better alternative.

CASE STUDY 8-2 CARDIAC DEMANDS OF HEAVY SNOW SHOVELING

During the winter months, newspapers often report cases of sudden deaths occurring during snow shoveling. Dr. Barry A. Franklin et al. conducted a controlled experiment to measure the effects of snow shoveling on men. After screening for health problems, the doctors included 10 men, with a mean age of 32.4 years, in the experiment. Each of these men completed exercise tests on a treadmill and on a device known as an arm-crank ergometer. Then, each of these men participated in the snow removal tests. Each man cleared two 15-meter (49.2-feet) long tracts of heavy, wet snow ranging from 3 to 5 inches in depth, using a snow shovel in one phase and an electric snow thrower in the other. The order of these two phases was determined randomly for each man. Each man shoveled snow for 10 minutes at a time, with 10- to 15-minute rest periods. During each 10-minute period of work, heart rates were recorded at 2-minute intervals. Four variables were measured for each man during the experiment: heart rate, systolic blood pressure, oxygen consumption, and the subject's own rating of his level of exertion.

The following table compares the men's reaction to snow removal with shovels and with electric snow throwers. The means and standard deviations are based on the maximum value for each variable for each man during the experiment, and the confidence intervals are calculated using these values of the mean and the standard deviation. It is assumed that the population of each variable has a normal distribution.

Here, heart rate is measured in beats per minute, systolic blood pressure is in millimeters of mercury, oxygen consumption is in metabolic equivalents, and rating of perceived exertion is on a numerical scale ranging from 6 to 20.

Variable	Snow Shoveling			Automated Snow Removal		
	$\bar{x}$	s	90% Confidence Interval	$\bar{x}$	s	90% Confidence Interval
Heart rate	175	15	$175 \pm 1.833(4.7434)$	124	18	$124 \pm 1.833(5.6921)$
Systolic blood pressure	198	17	$198 \pm 1.833(5.3759)$	161	14	$161 \pm 1.833(4.4272)$
Oxygen consumption	5.7	.8	$5.7 \pm 1.833(.2530)$	2.4	.7	$2.4 \pm 1.833(.2214)$
Own rating of exertion	16.7	1.7	$16.7 \pm 1.833(.5376)$	9.9	1.0	$9.9 \pm 1.833(.3162)$

The confidence intervals given in the table are constructed by using the method studied in Section 8.4.2 of this chapter. For example, for the variable on heart rate while shoveling snow manually,

$$\bar{x} = 175, \quad s = 15, \quad \text{and} \quad n = 10$$

Because $n < 30$, we will use the t distribution to make the confidence interval, assuming that the corresponding population has a normal distribution. For the 90% confidence level and $df = 9$, the t value from the t distribution table is 1.833. The standard deviation of $\bar{x}$ is

$$s_{\bar{x}} = \frac{s}{\sqrt{n}} = \frac{15}{\sqrt{10}} = 4.7434$$

Thus, the 90% confidence interval for the corresponding population mean is

$$\bar{x} \pm ts_{\bar{x}} = 175 \pm 1.833(4.7434) = 175 \pm 8.69 = 166.31 \text{ to } 183.69$$

Based on this result, we can state with 90% confidence that the mean heart rate for healthy men in this age group while shoveling snow manually is between 166.31 and 183.69 beats per minute.

Source: Barry A. Franklin et al., "Cardiac Demands of Heavy Snow Shoveling," *The Journal of the American Medical Association* 273[11], March 15, 1995.

EXERCISES

■ **Concepts and Procedures**

8.35 Briefly explain the similarities and the differences between the standard normal distribution and the t distribution.

8.36 What are the parameters of a normal distribution and a t distribution? Explain.

8.37 Briefly explain the meaning of the degrees of freedom for a t distribution. Give one example.

8.38 What assumptions must hold true to use the t distribution to make a confidence interval for μ?

8.39 Find the value of t for the t distribution for each of the following.

a. Area in the right tail $= .05$ and $df = 12$ **b.** Area in the left tail $= .025$ and $n = 26$
c. Area in the left tail $= .001$ and $df = 19$ **d.** Area in the right tail $= .005$ and $n = 24$

8.40 **a.** Find the value of t for a t distribution with a sample size of 21 and area in the left tail equal to .10.

b. Find the value of t for a t distribution with a sample size of 14 and area in the right tail equal to .025.

c. Find the value of t for a t distribution with 16 degrees of freedom and .001 area in the right tail.

d. Find the value of t for a t distribution with 27 degrees of freedom and .005 area in the left tail.

8.41 For each of the following, find the area in the appropriate tail of the t distribution.

a. $t = 2.467$ and $df = 28$ **b.** $t = -2.878$ and $df = 18$
c. $t = -2.145$ and $n = 15$ **d.** $t = 2.508$ and $n = 23$

8.42 For each of the following, find the area in the appropriate tail of the t distribution.

a. $t = -1.717$ and $df = 22$ **b.** $t = 2.797$ and $n = 25$
c. $t = 1.397$ and $n = 9$ **d.** $t = -2.567$ and $df = 17$

8.43 Find the value of t from the t distribution table for each of the following.

a. Confidence level $= 99\%$ and $df = 13$
b. Confidence level $= 95\%$ and $n = 26$
c. Confidence level $= 90\%$ and $df = 16$

8.44 a. Find the value of t from the t distribution table for a sample size of 22 and a confidence level of 95%.

b. Find the value of t from the t distribution table for 10 degrees of freedom and a 90% confidence level.

c. Find the value of t from the t distribution table for a sample size of 24 and a confidence level of 99%.

8.45 A sample of 12 observations taken from a normally distributed population produced the following data.

13	15	9	11	8	19
17	9	10	14	16	12

a. What is the point estimate of μ?
b. Make a 99% confidence interval for μ.
c. What is the maximum error of estimate for part b?

8.46 A sample of 10 observations taken from a normally distributed population produced the following data.

44	52	31	48	46	39	47	36	41	57

a. What is the point estimate of μ?
b. Make a 95% confidence interval for μ.
c. What is the maximum error of estimate for part b?

8.47 Suppose, for a sample selected from a normally distributed population, $\bar{x} = 68.50$ and $s = 8.9$.

a. Construct a 95% confidence interval for μ assuming $n = 16$.
b. Construct a 90% confidence interval for μ assuming $n = 16$. Is the width of the 90% confidence interval smaller than the width of the 95% confidence interval calculated in part a? If yes, explain why.
c. Find a 95% confidence interval for μ assuming $n = 25$. Is the width of the 95% confidence interval for μ with $n = 25$ smaller than the width of the 95% confidence interval for μ with $n = 16$ calculated in part a? If so, why? Explain.

8.48 Suppose, for a sample selected from a normally distributed population, $\bar{x} = 25.5$ and $s = 4.9$.

a. Construct a 95% confidence interval for μ assuming $n = 27$.
b. Construct a 99% confidence interval for μ assuming $n = 27$. Is the width of the 99% confidence interval larger than the width of the 95% confidence interval calculated in part a? If yes, explain why.
c. Find a 95% confidence interval for μ assuming $n = 20$. Is the width of the 95% confidence interval for μ with $n = 20$ larger than the width of the 95% confidence interval for μ with $n = 27$ calculated in part a? If so, why? Explain.

■ **Applications**

8.49 A random sample of 15 statistics students showed that the mean time taken to solve a computer assignment is 21 minutes with a standard deviation of 3 minutes. Construct a 99% confidence interval for the mean time taken by all statistics students to solve this computer assignment. Assume that the time taken to solve this computer assignment by all students follows a normal distribution.

8.50 A random sample of 20 acres gave a mean yield of wheat equal to 41.2 bushels per acre with a standard deviation of 3 bushels. Assuming that the yield of wheat per acre is normally distributed, construct a 90% confidence interval for the population mean μ.

8.51 Liposuction, a surgical procedure for removing fat, has become increasingly popular in the past few years. According to an estimate, the average fee charged by surgeons for such a procedure is $1700 (*Kiplinger's Personal Finance Magazine,* April 1999). Suppose that a random sample of 28 liposuction procedures yielded a mean fee of $1700 with a sample standard deviation of $280. Assuming that such fees are normally distributed, find a 95% confidence interval for the mean fee for all liposuction procedures.

8.52 A company wants to estimate the mean net weight of its Top Taste cereal boxes. A sample of 16 such boxes produced the mean net weight of 31.98 ounces with a standard deviation of .26 ounce. Make a 95% confidence interval for the mean net weight of all Top Taste cereal boxes. Assume that the net weights of all such cereal boxes have a normal distribution.

8.53 The high cost of health care is a matter of major concern for a large number of families. A random sample of 25 families selected from an area showed that they spend an average of $143 per month on health care with a standard deviation of $28. Make a 98% confidence interval for the mean health care expenditure per month incurred by all families in this area. Assume that the monthly health care expenditures of all families in this area have a normal distribution.

8.54 A random sample of 27 auto claims filed with an insurance company produced a mean amount for these claims equal to $2450 with a standard deviation of $460. Construct a 99% confidence interval for the corresponding population mean. Assume that the amounts of all such claims filed with this company have a normal distribution.

8.55 A random sample of 16 mid-sized cars, which were tested for fuel consumption, gave a mean of 26.4 miles per gallon with a standard deviation of 2.3 miles per gallon.

a. Assuming that the miles per gallon given by all mid-sized cars have a normal distribution, find a 99% confidence interval for the population mean, μ.
b. Suppose the confidence interval obtained in part a is too wide. How can the width of this interval be reduced? Describe all possible alternatives. Which alternative is the best and why?

8.56 The mean time taken to design a home plan by 20 architects was found to be 23 hours with a standard deviation of 3.75 hours.

a. Assume that the time taken by all architects to design this home plan is normally distributed. Construct a 98% confidence interval for the population mean μ.
b. Suppose the confidence interval obtained in part a is too wide. How can the width of this interval be reduced? Describe all possible alternatives. Which alternative is the best and why?

8.57 The following data give the speeds (in miles per hour), as measured by radar, of 10 cars traveling on Interstate Highway I-15.

       76   72   80   68   76   74   71   78   82   65

Assuming that the speeds of all cars traveling on this highway have a normal distribution, construct a 90% confidence interval for the mean speed of all cars traveling on this highway.

8.58 A sample of eight adults was taken, and these adults were asked about the time they spend per week on leisure activities. Their responses (in hours) are as follows.

       45   12   29   16   20   14   18   26

Assuming that the times spent on leisure activities by all adults are normally distributed, make a 95% confidence interval for the mean time spent per week on leisure activities by all adults.

*8.59 You are working for a supermarket. The manager has asked you to estimate the mean time taken by a cashier to serve customers at this supermarket. Unfortunately, he has asked you to take a small sample. Briefly explain how you will conduct this study. Collect data on the time taken by any supermarket cashier to serve 10 customers. Then estimate the population mean. Choose your own confidence level.

*8.60 You are working for a bank. The bank manager wants to know the mean waiting time for all customers who visit this bank. She has asked you to estimate this mean by taking a small sample. Briefly explain how you will conduct this study. Collect data on the waiting times for 15 customers who visit a bank. Then estimate the population mean. Choose your own confidence level.

8.5 INTERVAL ESTIMATION OF A POPULATION PROPORTION: LARGE SAMPLES

Often we want to estimate the population proportion or percentage. (Recall that a percentage is obtained by multiplying the proportion by 100.) For example, the production manager of a company may want to estimate the proportion of defective items produced on a machine. A bank manager may want to find the percentage of customers who are satisfied with the service provided by the bank.

Again, if we can conduct a census each time we want to find the value of a population proportion, there is no need to learn the procedures discussed in this section. However, we usually derive our results from sample surveys. Hence, to take into account the variability in the results obtained from different sample surveys, we need to know the procedures for estimating a population proportion.

Recall from Chapter 7 that the population proportion is denoted by p and the sample proportion is denoted by $\hat{p}$. This section explains how to estimate the population proportion, p, using the sample proportion, $\hat{p}$. The sample proportion, $\hat{p}$, is a sample statistic, and it possesses a sampling distribution. From Chapter 7, we know that for large samples:

1. The sampling distribution of the sample proportion, $\hat{p}$, is (approximately) normal.
2. The mean, $\mu_{\hat{p}}$, of the sampling distribution of $\hat{p}$ is equal to the population proportion, p.
3. The standard deviation, $\sigma_{\hat{p}}$, of the sampling distribution of the sample proportion, $\hat{p}$, is $\sqrt{pq/n}$, where $q = 1 - p$.

☞ *Remember* In the case of a proportion, a sample is considered to be large if np and nq are both greater than 5. If p and q are not known, then $n\hat{p}$ and $n\hat{q}$ should each be greater than 5 for the sample to be large.

When estimating the value of a population proportion, we do not know the values of p and q. Consequently, we cannot compute $\sigma_{\hat{p}}$. Therefore, in the estimation of a population proportion, we use the value of $s_{\hat{p}}$ as an estimate of $\sigma_{\hat{p}}$. The value of $s_{\hat{p}}$ is calculated using the following formula.

> **ESTIMATOR OF THE STANDARD DEVIATION OF $\hat{p}$** The value of $s_{\hat{p}}$, which gives a point estimate of $\sigma_{\hat{p}}$, is calculated as
>
> $$s_{\hat{p}} = \sqrt{\frac{\hat{p}\hat{q}}{n}}$$

The sample proportion, $\hat{p}$, is the point estimator of the corresponding population proportion, p. As in the case of the mean, whenever we use point estimation for the population proportion, we calculate the **margin of error** associated with that point estimation. For the estimation of the population proportion, the margin of error is calculated as follows.

$$\text{Margin of error} = \pm 1.96 s_{\hat{p}}$$

That is, we find the standard deviation of the sample proportion and multiply it by 1.96. The $(1 - \alpha)100\%$ confidence interval for p is constructed using the following formula.

> **CONFIDENCE INTERVAL FOR THE POPULATION PROPORTION, p** The $(1 - \alpha)100\%$ *confidence interval for the population proportion, p, is*
>
> $$\hat{p} \pm z s_{\hat{p}}$$
>
> The value of z used here is obtained from the standard normal distribution table for the given confidence level, and $s_{\hat{p}} = \sqrt{\hat{p}\,\hat{q}/n}$.

Examples 8–6 and 8–7 illustrate the procedure for constructing a confidence interval for p.

Finding the point estimate and 99% confidence interval for p: large sample.

EXAMPLE 8-6 In a nationwide telephone poll of 1001 U.S. adults conducted for the Freedom Forum by the Center for Survey Research and Analysis at the University of Connecticut, 53% of those polled said that they think the press has too much freedom in the United States (*USA TODAY,* July 2, 1999).

(a) What is the point estimate of the population proportion? What is the margin of error of this estimate?

(b) Find, with a 99% confidence level, what percentage of all adults would hold this opinion.

Solution Let p be the proportion of all adults who hold the opinion that the press has too much freedom in the United States and let $\hat{p}$ be the corresponding sample proportion. From the given information,

$$n = 1001, \quad \hat{p} = .53, \quad \text{and} \quad \hat{q} = 1 - \hat{p} = 1 - .53 = .47$$

First, we calculate the value of the standard deviation of the sample proportion as follows:

$$s_{\hat{p}} = \sqrt{\frac{\hat{p}\hat{q}}{n}} = \sqrt{\frac{(.53)(.47)}{1001}} = .01577502$$

Note that $n\hat{p}$ and $n\hat{q}$ are both greater than 5. (Readers should check this condition.) Consequently, the sampling distribution of $\hat{p}$ is approximately normal, and we will use the normal distribution to calculate the margin of error for the point estimate of p and to make a confidence interval about p.

(a) The point estimate of the proportion of all adults who hold the opinion that the press has too much freedom in the United States is equal to .53; that is,

$$\text{Point estimate of } p = \hat{p} = \mathbf{.53}$$

The margin of error associated with this point estimate of p is

$$\text{Margin of error} = \pm 1.96 s_{\hat{p}} = \pm 1.96(.01577502) = \pm\mathbf{.031} \text{ or } \pm\mathbf{3.1\%}$$

The margin of error states that the proportion of all adults who hold the opinion that the press has too much freedom in the United States is .53 (or 53%), give or take .031 (or 3.1%). As in the case of the mean, the margin of error is simply the maximum error of estimate for a 95% confidence interval for the population proportion.

(b) The confidence level is 99%, or .99. To find z, we divide the confidence level by 2, look for this number in the body of the normal distribution table, and record the corresponding value of z. We have

$$.99/2 = .4950$$

From the normal distribution table, the z value for .4950 is approximately 2.58. Substituting all the values in the confidence interval formula for p, we obtain

$$\hat{p} \pm zs_{\hat{p}} = .53 \pm 2.58(.01577502) = .53 \pm .041$$

$$= \textbf{.489 to .571} \quad \text{or} \quad \textbf{48.9\% to 57.1\%}$$

Thus, we can state with 99% confidence that .489 to .571, or 48.9% to 57.1%, of all adults hold the opinion that the press has too much freedom in the United States. ∎

Constructing a 95% confidence interval for p: large sample.

EXAMPLE 8–7   In a Gallup poll of 1523 U.S. adults, 63% approved of casino gambling as a way for their states to raise money (*USA TODAY,* June 17, 1999). Construct a 95% confidence interval for the proportion of all U.S. adults who hold this opinion.

Solution   Let p be the proportion of all U.S. adults who approve of casino gambling as a way for their states to raise money, and let $\hat{p}$ be the corresponding sample proportion. From the given information,

$$n = 1523 \qquad \hat{p} = .63 \qquad \hat{q} = 1 - \hat{p} = 1 - .63 = .37$$

$$\text{Confidence level} = 95\% \text{ or } .95$$

The standard deviation of the sample proportion is

$$s_{\hat{p}} = \sqrt{\frac{\hat{p}\hat{q}}{n}} = \sqrt{\frac{(.63)(.37)}{1523}} = .01237147$$

From the normal distribution table, the value of z for .95/2 = .4750 is 1.96. The 95% confidence interval for p is

$$\hat{p} \pm zs_{\hat{p}} = .63 \pm 1.96(.01237147) = .63 \pm .024$$

$$= \textbf{.606 to .654} \quad \text{or} \quad \textbf{60.6\% to 65.4\%}$$

Thus, we can state with 95% confidence that the proportion of all U.S. adults who approve of casino gambling as a way for their states to raise money is between .606 and .654. The confidence interval can be converted into a percentage interval as 60.6% to 65.4%. ∎

Again, we can decrease the width of a confidence interval for p either by lowering the confidence level or by increasing the sample size. However, lowering the confidence level is not a good choice because it simply decreases the likelihood that the confidence interval contains p. Hence, to decrease the width of a confidence interval for p, we should always increase the sample size.

EXERCISES

■ **Concepts and Procedures**

8.61   What assumption(s) must hold true to use the normal distribution to make a confidence interval for the population proportion, p?

8.62   What is the point estimator of the population proportion, p? How is the margin of error for a point estimate of p calculated?

8.63   Check if the sample size is large enough to use the normal distribution to make a confidence interval for p for each of the following cases.

a. $n = 50$ and $\hat{p} = .25$ **b.** $n = 160$ and $\hat{p} = .03$
c. $n = 400$ and $\hat{p} = .65$ **d.** $n = 75$ and $\hat{p} = .06$

8.64 Check if the sample size is large enough to use the normal distribution to make a confidence interval for p for each of the following cases.

a. $n = 120$ and $\hat{p} = .04$ **b.** $n = 60$ and $\hat{p} = .08$
c. $n = 40$ and $\hat{p} = .50$ **d.** $n = 900$ and $\hat{p} = .15$

8.65 a. A sample of 400 observations taken from a population produced a sample proportion of .63. Make a 95% confidence interval for p.

b. Another sample of 400 observations taken from the same population produced a sample proportion of .59. Make a 95% confidence interval for p.

c. A third sample of 400 observations taken from the same population produced a sample proportion of .67. Make a 95% confidence interval for p.

d. The true population proportion for this population is .65. Which of the confidence intervals constructed in parts a through c cover this population proportion and which do not?

8.66 a. A sample of 900 observations taken from a population produced a sample proportion of .32. Make a 90% confidence interval for p.

b. Another sample of 900 observations taken from the same population produced a sample proportion of .36. Make a 90% confidence interval for p.

c. A third sample of 900 observations taken from the same population produced a sample proportion of .30. Make a 90% confidence interval for p.

d. The true population proportion for this population is .34. Which of the confidence intervals constructed in parts a through c cover this population proportion and which do not?

8.67 A sample of 500 observations selected from a population produced a sample proportion equal to .68.

a. Make a 90% confidence interval for p.
b. Construct a 95% confidence interval for p.
c. Make a 99% confidence interval for p.
d. Does the width of the confidence intervals constructed in parts a through c increase as the confidence level increases? If yes, explain why.

8.68 A sample of 200 observations selected from a population gave a sample proportion equal to .27.

a. Make a 99% confidence interval for p.
b. Construct a 97% confidence interval for p.
c. Make a 90% confidence interval for p.
d. Does the width of the confidence intervals constructed in parts a through c decrease as the confidence level decreases? If yes, explain why.

8.69 A sample selected from a population gave a sample proportion equal to .73.

a. Make a 99% confidence interval for p assuming $n = 100$.
b. Construct a 99% confidence interval for p assuming $n = 600$.
c. Make a 99% confidence interval for p assuming $n = 1500$.
d. Does the width of the confidence intervals constructed in parts a through c decrease as the sample size increases? If yes, explain why.

8.70 A sample selected from a population gave a sample proportion equal to .31.

a. Make a 95% confidence interval for p assuming $n = 1200$.
b. Construct a 95% confidence interval for p assuming $n = 500$.
c. Make a 95% confidence interval for p assuming $n = 80$.
d. Does the width of the confidence intervals constructed in parts a through c increase as the sample size decreases? If yes, explain why.

■ **Applications**

8.71 In a telephone poll of 1024 adult Americans taken by Yankelovich Partners, Inc., for *Time*/CNN in January 1999, 44% of adults favored using tax money to buy undeveloped land to keep it free from development (*Time,* March 22, 1999). Assume that these 1024 adults make up a random sample of all American adults.

a. What is the point estimate of the corresponding population proportion? What is the margin of error associated with this point estimate?
b. Construct a 99% confidence interval for the proportion of all American adults who feel that tax money should be used in this way.

8.72 According to data from World Research, 21% of shoppers who never buy online said that they do not do so due to "fear of hackers" (*Time,* July 20, 1998). Suppose that this percentage is based on a random sample of 200 shoppers who never buy online.

a. What is the point estimate of the corresponding population proportion? What is the margin of error associated with this point estimate?
b. Find a 95% confidence interval for the proportion of all such shoppers who never shop online due to fear of hackers.

8.73 In a survey conducted by Roper Starch Worldwide, 41% of Americans polled agree with the statement, "The car I drive says a lot about who I am" (*American Demographics,* October 1998). Suppose that this percentage is based on a random sample of 450 Americans.

a. What is the point estimate of the corresponding population proportion? What is the margin of error associated with this point estimate?
b. Find a 95% confidence interval for the corresponding population proportion.

8.74 It is said that happy and healthy workers are efficient and productive. A company that manufactures exercising machines wanted to know the percentage of large companies that provide on-site health club facilities. A sample of 240 such companies showed that 96 of them provide such facilities on site.

a. What is the point estimate of the percentage of all such companies that provide such facilities on site? What is the margin of error associated with this point estimate?
b. Construct a 97% confidence interval for the percentage of all such companies that provide such facilities on site.

8.75 A mail-order company promises its customers that the products ordered will be mailed within 72 hours after an order is placed. The quality control department at the company checks from time to time to see if this promise is kept. Recently the quality control department took a sample of 50 orders and found that 35 of them were mailed within 72 hours of the placement of the orders.

a. Construct a 98% confidence interval for the percentage of all orders that are mailed within 72 hours of their placement.
b. Suppose the confidence interval obtained in part a is too wide. How can the width of this interval be reduced? Discuss all possible alternatives. Which alternative is the best?

8.76 One of the major problems faced by department stores is a high percentage of returns. The manager of a department store wanted to estimate the percentage of all sales that result in returns. A sample of 40 sales showed that 10 of them had products returned within the time allowed for returns.

a. Make a 99% confidence interval for the percentage of all sales that result in returns.
b. Suppose the confidence interval obtained in part a is too wide. How can the width of this interval be reduced? Discuss all possible alternatives. Which alternative is the best?

8.77 In a survey conducted by Yankelovich Partners for Lutheran Brotherhood, adults were asked what they would first spend money on if they suddenly became wealthy. The most common response was "house," given by 31% of the adults surveyed (*USA TODAY,* March 13, 1999). Suppose that this percentage is based on a random sample of 400 adults.

a. Determine a 99% confidence interval for the percentage of all adults who would first spend suddenly acquired wealth on a house.
b. Explain why we need to make the confidence interval. Why can we not simply say that 31% of adults would first spend suddenly acquired wealth on a house?

8.78 In a 1998 poll of 1151 adults, (about) 44% of respondents were in favor of allowing students and parents to choose private schools for children to attend at public expense (*Phi Delta Kappa*/Gallup Poll of the Public's Attitudes Toward Public Schools, *Education Week,* September 9, 1998).

a. Make a 98% confidence interval for the proportion of all U.S. adults who favor such a choice.
b. Explain why we need to make the confidence interval. Why can we not simply say that 44% of U.S. adults favor such a choice?

8.79 A researcher wanted to know the percentage of judges who are in favor of the death penalty. He took a random sample of 15 judges and asked them whether or not they favor the death penalty. The responses of these judges are given here.

Yes	No	Yes	Yes	No	No	No	Yes
Yes	No	Yes	Yes	Yes	No	Yes	

a. What is the point estimate of the population proportion? What is the margin of error associated with this point estimate?
b. Make a 95% confidence interval for the percentage of all judges who are in favor of the death penalty.

8.80 The manager of a supermarket wanted to know the percentage of shoppers who prefer to buy name brand products. A random sample of 20 shoppers who shopped at this store were asked this question. The following are the responses.

No	No	No	Yes	Yes	No	No	Yes	No	Yes
No	No	Yes	No	No	No	No	Yes	No	Yes

a. What is the point estimate of the population proportion? What is the margin of error associated with this point estimate?
b. Construct a 99% confidence interval for the percentage of all shoppers who prefer to buy name brand products.

***8.81** You want to estimate the proportion of students at your college who hold off-campus (part-time or full-time) jobs. Briefly explain how you will make such an estimate. Collect data from 40 students at your college on whether or not they hold off-campus jobs. Then, calculate the proportion of students in this sample who hold off-campus jobs. Using this information, estimate the population proportion. Select your own confidence level.

***8.82** You want to estimate the percentage of people who are satisfied with the services provided by their banks. Briefly explain how you will make such an estimate. Collect data from 30 people on whether or not they are satisfied with the services provided by their banks. Then, calculate the percentage of people in this sample who are satisfied. Using this information, estimate the population percentage. Select your own confidence level.

8.6 DETERMINING THE SAMPLE SIZE FOR THE ESTIMATION OF MEAN

One reason we usually conduct a sample survey and not a census is that almost always we have limited resources at our disposal. In light of this, if a smaller sample can serve our purpose, then we will be wasting our resources by taking a larger sample. For instance, suppose we want to estimate the mean life of a certain auto battery. If a sample of 40 batteries can give us the confidence interval we are looking for, then we will be wasting money and time if we take a sample of a much larger size—say, 500 batteries. In such cases, if we know the confidence level and the width of the confidence interval that we want, then we can find the (approximate) size of the sample that will produce the required result.

In Section 8.3, we learned that $E = z\sigma_{\bar{x}}$ is called the maximum error of estimate for μ. As we know, the standard deviation of the sample mean is equal to $\sigma/\sqrt{n}$. Therefore, we can write the maximum error of estimate for μ as

$$E = z \cdot \frac{\sigma}{\sqrt{n}}$$

Suppose we predetermine the size of the maximum error, *E,* and want to find the size of the sample that will yield this maximum error. From the expression on the previous page, the following formula is obtained that determines the required sample size *n.*

DETERMINING THE SAMPLE SIZE FOR THE ESTIMATION OF μ   Given the confidence level and the standard deviation of the population, the sample size that will produce a predetermined maximum error E of the confidence interval *estimate of μ* is

$$n = \frac{z^2\sigma^2}{E^2}$$

If we do not know σ, we can take a preliminary sample (of any arbitrarily determined size) and find the sample standard deviation, *s.* Then we can use *s* for σ in the formula. However, note that using *s* for σ may give a sample size that eventually may produce an error much larger (or smaller) than the predetermined maximum error. This will depend on how close *s* and σ are.

Example 8–8 illustrates how we determine the sample size that will produce the maximum error of estimate for μ within a certain limit.

Determining the sample size for the estimation of μ.

EXAMPLE 8–8   Suppose the U.S. Bureau of the Census wants to estimate the mean family size of all U.S. families at a 99% confidence level. It is known that the standard deviation σ for the sizes of all families in the United States is .6. How large a sample should the bureau select if it wants its estimate to be within .01 of the population mean?

Solution   The Bureau of the Census wants the 99% confidence interval for the mean family size to be

$$\bar{x} \pm .01$$

Hence, the maximum size of the error of estimate is to be .01; that is,

$$E = .01$$

The value of *z* for a 99% confidence level is 2.58. The value of σ is given to be .6. Therefore, substituting all values in the formula and simplifying give

$$n = \frac{z^2\sigma^2}{E^2} = \frac{(2.58)^2(.6)^2}{(.01)^2} = \frac{(6.6564)(.36)}{(.0001)} = 23,963.04 \approx \mathbf{23,964}$$

Thus, the required sample size is 23,964. If the Bureau of the Census takes a sample of 23,964 families, computes the mean family size for this sample, and then constructs a 99% confidence interval around this sample mean, the maximum error of the estimate will be approximately .01. Note that we have rounded the final answer for the sample size to the next higher integer. This is always the case when determining the sample size.   ■

EXERCISES

■ **Concepts and Procedures**

8.83   For data on a variable for a population, $\sigma = 12.5$.

a.   How large a sample should be selected so that the maximum error of estimate for a 99% confidence interval for μ is 2.50?

b. How large a sample should be selected so that the maximum error of estimate for a 96% confidence interval for μ is 3.20?

8.84 For data on a variable for a population, $\sigma = 14.50$.

a. What should the sample size be for a 98% confidence interval for μ to have a maximum error of estimate equal to 5.50?

b. What should the sample size be for a 95% confidence interval for μ to have a maximum error of estimate equal to 4.25?

8.85 Determine the sample size for the estimate of μ for the following.

a. $E = 2.3$, $\sigma = 15.40$,   confidence level = 99%
b. $E = 4.1$, $\sigma = 23.45$,   confidence level = 95%
c. $E = 25.9$, $\sigma = 122.25$,   confidence level = 90%

8.86 Determine the sample size for the estimate of μ for the following.

a. $E = .17$, $\sigma = .90$,   confidence level = 99%
b. $E = 1.45$, $\sigma = 5.82$,   confidence level = 95%
c. $E = 5.65$, $\sigma = 18.20$,   confidence level = 90%

■ **Applications**

8.87 A researcher wants to determine a 95% confidence interval for the mean number of hours that high school students spend doing homework per week. She knows that the standard deviation for hours spent per week by all high school students doing homework is 9. How large a sample should the researcher select so that the estimate will be within 1.5 hours of the population mean?

8.88 A company that produces detergents wants to estimate the mean amount of detergent in 64-ounce jugs at a 99% confidence level. The company knows that the standard deviation of amounts of detergent in such jugs is .20 ounce. How large a sample should the company take so that the estimate is within .04 ounce of the population mean?

8.89 A department store manager wants to estimate at a 90% confidence level the mean amount spent by all customers at this store. From an earlier study, the manager knows that the standard deviation of amounts spent by customers at this store is $31. What sample size should he choose so that the estimate is within $3 of the population mean?

8.90 A U.S. government agency wants to estimate at a 95% confidence level the mean speed for all cars traveling on Interstate Highway I-95. From a previous study, the agency knows that the standard deviation of speeds of cars traveling on this highway is 4.2 miles per hour. What sample size should the agency choose for the estimate to be within 1.5 miles per hour of the population mean?

8.7 DETERMINING THE SAMPLE SIZE FOR THE ESTIMATION OF PROPORTION

Just as we did with the mean, we can also determine the sample size for estimating the population proportion, p. This sample size will yield an error of estimate that may not be larger than a predetermined maximum error. By knowing the sample size that can give us the required results, we can save our scarce resources by not taking an unnecessarily large sample. From Section 8.5, the maximum error, E, of the interval estimation of the population proportion is

$$E = z\sigma_{\hat{p}} = z \times \sqrt{\frac{pq}{n}}$$

By manipulating this expression algebraically, we obtain the formula given on the next page to find the required sample size given E, p, q, and z.

> **DETERMINING THE SAMPLE SIZE FOR THE ESTIMATION OF *p***   Given the confidence level and the values of *p* and *q*, the sample size that will produce a predetermined maximum error *E* of the confidence interval *estimate of p* is
>
> $$n = \frac{z^2 pq}{E^2}$$

We can observe from this formula that to find *n,* we need to know the values of *p* and *q*. However, the values of *p* and *q* are not known to us. In such a situation, we can choose one of the following alternatives.

1. We make the *most conservative estimate* of the sample size *n* by using *p* = .50 and *q* = .50. For a given *E,* these values of *p* and *q* will give us the largest sample size in comparison to any other pair of values for *p* and *q* because the product of *p* = .50 and *q* = .50 is greater than the product of any other pair of values for *p* and *q*.

2. We take a *preliminary sample* (of arbitrarily determined size) and calculate $\hat{p}$ and $\hat{q}$ for this sample. Then, we use these values of $\hat{p}$ and $\hat{q}$ to find *n*.

Examples 8–9 and 8–10 illustrate how to determine the sample size that will produce the error of estimation for the population proportion within a predetermined maximum value. Example 8–9 gives the most conservative estimate of *n,* and Example 8–10 uses the results from a preliminary sample to determine the required sample size.

Determining the most conservative estimate of n for the estimation of p.

EXAMPLE 8–9   Lombard Electronics Company has just installed a new machine that makes a part that is used in clocks. The company wants to estimate the proportion of these parts produced by this machine that are defective. The company manager wants this estimate to be within .02 of the population proportion for a 95% confidence level. What is the most conservative estimate of the sample size that will limit the maximum error to within .02 of the population proportion?

Solution   The company manager wants the 95% confidence interval to be

$$\hat{p} \pm .02$$

Therefore, $E = .02$

The value of *z* for a 95% confidence level is 1.96. For the most conservative estimate of the sample size, we will use *p* = .50 and *q* = .50. Hence, the required sample size is

$$n = \frac{z^2 pq}{E^2} = \frac{(1.96)^2(.50)(.50)}{(.02)^2} = \mathbf{2401}$$

Thus, if the company takes a sample of 2401 parts, there is a 95% chance that the estimate of *p* will be within .02 of the population proportion.   ■

Determining n for the estimation of p using preliminary sample results.

EXAMPLE 8–10   Consider Example 8–9 again. Suppose a preliminary sample of 200 parts produced by this machine showed that 7% of them are defective. How large a sample should the company select so that the 95% confidence interval for *p* is within .02 of the population proportion?

Solution   Again, the company wants the 95% confidence interval for *p* to be

$$\hat{p} \pm .02$$

Hence, $$E = .02$$

The value of z for a 95% confidence level is 1.96. From the preliminary sample,

$$\hat{p} = .07 \quad \text{and} \quad \hat{q} = 1 - .07 = .93$$

Using these values of $\hat{p}$ and $\hat{q}$ as estimates of p and q, we obtain

$$n = \frac{z^2 \hat{p} \hat{q}}{E^2} = \frac{(1.96)^2(.07)(.93)}{(.02)^2} = \frac{(3.8416)(.07)(.93)}{.0004} = 625.22 \approx \mathbf{626}$$

Thus, if the company takes a sample of 626 items, there is a 95% chance that the estimate of p will be within .02 of the population proportion. However, we should note that this sample size will produce the maximum error within .02 point only if $\hat{p}$ is .07 or less for the new sample. If $\hat{p}$ for the new sample happens to be much higher than .07, the maximum error will not be within .02. Therefore, to avoid such a situation, we may be more conservative and take a much larger sample than 626 items. ∎

EXERCISES

■ Concepts and Procedures

8.91 a. How large a sample should be selected so that the maximum error of estimate for a 99% confidence interval for p is .035 when the value of the sample proportion obtained from a preliminary sample is .29?
b. Find the most conservative sample size that will produce the maximum error for a 99% confidence interval for p equal to .035.

8.92 a. How large a sample should be selected so that the maximum error of estimate for a 98% confidence interval for p is .045 when the value of the sample proportion obtained from a preliminary sample is .53?
b. Find the most conservative sample size that will produce the maximum error for a 98% confidence interval for p equal to .045.

8.93 Determine the most conservative sample size for the estimation of the population proportion for the following.

a. $E = .03$,   confidence level = 99%
b. $E = .04$,   confidence level = 95%
c. $E = .01$,   confidence level = 90%

8.94 Determine the sample size for the estimation of the population proportion for the following, where $\hat{p}$ is the sample proportion based on a preliminary sample.

a. $E = .03$, $\hat{p} = .32$,   confidence level = 99%
b. $E = .04$, $\hat{p} = .78$,   confidence level = 95%
c. $E = .02$, $\hat{p} = .64$,   confidence level = 90%

■ Applications

8.95 Tony's Pizza guarantees all pizza deliveries within 30 minutes of the placement of orders. An agency wants to estimate the proportion of all pizzas delivered within 30 minutes by Tony's. What is the most conservative estimate of the sample size that would limit the maximum error to within .02 of the population proportion for a 99% confidence interval?

8.96 Refer to Exercise 8.95. Assume that a preliminary study has shown that 93% of all Tony's pizzas are delivered within 30 minutes. How large should the sample size be so that the 99% confidence interval for the population proportion has a maximum error of .02?

8.97 A consumer agency wants to estimate the proportion of all drivers who wear seat belts while driving. Assume that a preliminary study has shown that 76% of drivers wear seat belts while driving.

How large should the sample size be so that the 99% confidence interval for the population proportion has a maximum error of .03?

8.98 Refer to Exercise 8.97. What is the most conservative estimate of the sample size that would limit the maximum error to within .03 of the population proportion for a 99% confidence interval?

GLOSSARY

Confidence interval An interval constructed around the value of a sample statistic to estimate the corresponding population parameter.

Confidence level Confidence level, denoted by $(1 - \alpha)100\%$, states how much confidence we have that a confidence interval contains the true population parameter.

Degrees of freedom (*df*) The number of observations that can be chosen freely. For the estimation of μ using the t distribution, the degrees of freedom are $n - 1$.

Estimate The value of a sample statistic that is used to find the corresponding population parameter.

Estimation A procedure by which a numerical value or values are assigned to a population parameter based on the information collected from a sample.

Estimator The sample statistic that is used to estimate a population parameter.

Interval estimate An interval constructed around the point estimate that is likely to contain the corresponding population parameter. Each interval estimate has a confidence level.

Maximum error of estimate The quantity that is subtracted from and added to the value of a sample statistic to obtain a confidence interval for the corresponding population parameter.

Point estimate The value of a sample statistic assigned to the corresponding population parameter.

_t_ distribution A continuous distribution with a specific type of bell-shaped curve with its mean equal to 0 and standard deviation equal to $\sqrt{df/(df - 2)}$.

KEY FORMULAS

1. **Margin of error associated with the point estimation of μ**

$$\pm 1.96\sigma_{\bar{x}} \quad \text{or} \quad \pm 1.96 s_{\bar{x}}$$

where $\qquad \sigma_{\bar{x}} = \sigma/\sqrt{n} \quad \text{and} \quad s_{\bar{x}} = s/\sqrt{n}$

2. **The $(1 - \alpha)100\%$ confidence interval for μ for a large sample ($n \geq 30$)**

$$\bar{x} \pm z\sigma_{\bar{x}} \quad \text{if } \sigma \text{ is known}$$

$$\bar{x} \pm z s_{\bar{x}} \quad \text{if } \sigma \text{ is not known}$$

where $\qquad \sigma_{\bar{x}} = \sigma/\sqrt{n} \quad \text{and} \quad s_{\bar{x}} = s/\sqrt{n}$

3. **The $(1 - \alpha)100\%$ confidence interval for μ for a small sample ($n < 30$) when the population is (approximately) normally distributed and σ is not known**

$$\bar{x} \pm t s_{\bar{x}} \quad \text{where} \quad s_{\bar{x}} = s/\sqrt{n}$$

4. **Margin of error associated with the point estimation of p**

$$\pm 1.96 s_{\hat{p}} \quad \text{where} \quad s_{\hat{p}} = \sqrt{\frac{\hat{p}\hat{q}}{n}}$$

5. **The $(1 - \alpha)100\%$ confidence interval for p for a large sample**

$$\hat{p} \pm zs_{\hat{p}} \quad \text{where} \quad s_{\hat{p}} = \sqrt{\frac{\hat{p}\hat{q}}{n}}$$

6. **Maximum error, E, of the estimate for μ**

$$E = z\sigma_{\bar{x}} \quad \text{or} \quad zs_{\bar{x}}$$

7. **Required sample size for a predetermined maximum error for estimating μ**

$$n = \frac{z^2\sigma^2}{E^2}$$

8. **Maximum error, E, of the estimate for the population proportion**

$$E = zs_{\hat{p}} \quad \text{where} \quad s_{\hat{p}} = \sqrt{\frac{\hat{p}\hat{q}}{n}}$$

9. **Required sample size for a predetermined maximum error for estimating p**

$$n = \frac{z^2pq}{E^2}$$

Use $p = .50$ and $q = .50$ for the most conservative estimate, and use the values of $\hat{p}$ and $\hat{q}$ if the estimate is to be based on a preliminary sample.

SUPPLEMENTARY EXERCISES

8.99 A company opened a new movie theater. Before setting the price of a movie ticket, the company wants to find the average price charged by other movie theaters. A random sample of 100 movie theaters taken by the company showed that the mean price of a movie ticket is $7.25 with a standard deviation of $.80.

a. What is the point estimate of the mean price of movie tickets for all theaters? What is the margin of error associated with this estimate?

b. Make a 95% confidence interval for the population mean, μ.

8.100 A bank manager wants to know the mean amount owed on credit card accounts that become delinquent. A random sample of 100 delinquent credit card accounts taken by the manager produced a mean amount owed on these accounts equal to $2640 with a standard deviation of $578.

a. What is the point estimate of the mean amount owed on all delinquent credit card accounts at this bank? What is the margin of error associated with this estimate?

b. Construct a 97% confidence interval for the mean amount owed on all delinquent credit card accounts for this bank.

8.101 York Steel Corporation produces iron rings that are supplied to other companies. These rings are supposed to have a diameter of 24 inches. The machine that makes these rings does not produce each ring with a diameter of exactly 24 inches. The diameter of each of the rings varies slightly. It is known that when the machine is working properly, the rings made on this machine have a mean diameter of 24 inches. The standard deviation of the diameters of all rings produced on this machine is always equal to .06 inch. The quality control department takes a sample of 36 such rings every week, calculates the mean of the diameters for these rings, and makes a 99% confidence interval for the population mean. If either the lower limit of this confidence interval is less than 23.975 inches or the upper limit of this confidence interval is greater than 24.025 inches, the machine is stopped and adjusted. A recent such sample of 36 rings produced a mean diameter of 24.015 inches. Based on this sample, can you conclude that the machine needs an adjustment? Explain.

8.102 Yunan Corporation produces bolts that are supplied to other companies. These bolts are supposed to be 4 inches long. The machine that makes these bolts does not produce each bolt exactly 4 inches long. It is known that when the machine is working properly, the mean length of the bolts made on this machine is 4 inches. The standard deviation of the lengths of all bolts produced on this machine is always equal to .04 inch. The quality control department takes a sample of 50 such bolts every week, calculates the mean length of these bolts, and makes a 98% confidence interval for the population mean. If either the upper limit of this confidence interval is greater than 4.02 inches or the lower limit of this confidence interval is less than 3.98 inches, the machine is stopped and adjusted. A recent such sample of 50 bolts produced a mean length of 3.99 inches. Based on this sample, will you conclude that the machine needs an adjustment?

8.103 A magazine wanted to estimate the mean daily cost of a room in all major hotels in the United States. The research department at the magazine took a sample of 33 hotels and obtained the following data on the daily charges per room.

$124	157	105	185	86	210	130	120	195	230	160
145	177	180	205	153	240	210	75	115	265	189
205	143	185	112	254	142	179	190	224	90	125

a. What is the point estimate of the mean daily cost per room for all hotels in the United States? What is the margin of error associated with this estimate?
b. Construct a 98% confidence interval for the mean daily cost per room for all hotels in the United States.

8.104 A travel magazine wanted to estimate the mean amount of leisure time per week enjoyed by adults. The research department at the magazine took a sample of 36 adults and obtained the following data on the weekly leisure time (in hours).

15	12	18	23	11	21	16	13	9	19	26	14
7	18	11	15	23	26	10	8	17	21	12	7
19	21	11	13	21	16	14	9	15	12	10	14

a. What is the point estimate of the mean leisure time per week enjoyed by all adults? What is the margin of error associated with this estimate?
b. Construct a 99% confidence interval for the mean leisure time per week enjoyed by all adults.

8.105 A random sample of 25 life insurance policyholders showed that the average premium they pay on their life insurance policies is $460 per year with a standard deviation of $62. Assuming that the life insurance policy premiums for all life insurance policyholders have a normal distribution, make a 99% confidence interval for the population mean, μ.

8.106 A drug that provides relief from headaches was tried on 18 randomly selected patients. The experiment showed that the mean time to get relief from headache for these patients after taking this drug was 24 minutes with a standard deviation of 4.5 minutes. Assuming that the time taken to get relief from a headache after taking this drug is (approximately) normally distributed, determine a 95% confidence interval for the mean relief time for this drug for all patients.

8.107 A survey of 20 randomly selected adult men showed that the mean time they spend per week watching sports on television is 9.75 hours with a standard deviation of 2.2 hours. Assuming that the time spent per week watching sports on television by all adult men is (approximately) normally distributed, construct a 90% confidence interval for the population mean, μ.

8.108 A random sample of 20 female members of health clubs in Los Angeles showed that they spend, on average, 4.5 hours per week doing physical exercise with a standard deviation of .75 hour. Assume that the times spent doing physical exercise by all female members of health clubs in Los Angeles are (approximately) normally distributed. Find a 98% confidence interval for the population mean.

8.109 A computer company that recently developed a new software product wanted to estimate the mean time taken to learn how to use this software by people who are somewhat familiar with computers. A random sample of 12 such persons was selected. The following data give the times taken (in hours) by these persons to learn how to use this software.

1.75	2.25	2.40	1.90	1.50	2.75
2.15	2.25	1.80	2.20	3.25	2.60

Construct a 95% confidence interval for the population mean. Assume that the times taken by all persons who are somewhat familiar with computers to learn how to use this software are approximately normally distributed.

8.110 A company that produces 8-ounce low-fat yogurt cups wanted to estimate the mean number of calories for such cups. A random sample of 10 such cups produced the following numbers of calories.

147	159	153	146	144	148	163	153	143	158

Construct a 99% confidence interval for the population mean. Assume that the numbers of calories for such cups of yogurt produced by this company have an approximately normal distribution.

8.111 An insurance company selected a sample of 50 auto claims filed with it and investigated those claims carefully. The company found that 12% of those claims were fraudulent.

a. What is the point estimate of the percentage of all auto claims filed with this company that are fraudulent? What is the margin of error associated with this estimate?

b. Make a 99% confidence interval for the percentage of all auto claims filed with this company that are fraudulent.

8.112 An auto company wanted to know the percentage of people who prefer to own safer cars (that is, cars that possess more safety features) even if they have to pay a few thousand dollars more. A random sample of 500 persons showed that 44% of them will not mind paying a few thousand dollars more to have safer cars.

a. What is the point estimate of the percentage of all people who will not mind paying a few thousand dollars more to have safer cars? What is the margin of error associated with this estimate?

b. Construct a 90% confidence interval for the percentage of all people who will not mind paying a few thousand dollars more to have safer cars.

8.113 A sample of 20 managers was taken, and they were asked whether or not they usually take work home. The responses of these managers are given below, where *yes* indicates they usually take work home and *no* means they do not.

Yes	Yes	No	No	No	Yes	No	No	No	No
Yes	Yes	No	Yes	Yes	No	No	No	No	Yes

Make a 99% confidence interval for the percentage of all managers who take work home.

8.114 A sample of 16 salespersons was taken and they were asked whether or not they have ever been fired from their jobs. The following are the responses.

No	No	Yes	No	Yes	No	Yes	Yes
No	No	No	No	Yes	No	Yes	No

Construct a 97% confidence interval for the percentage of all salespersons who have ever been fired from their jobs.

8.115 A researcher wants to determine a 99% confidence interval for the mean number of hours that adults spend per week doing community service. How large a sample should the researcher select so that the estimate is within 1.2 hours of the population mean? Assume that the standard deviation for time spent per week doing community service by all adults is 3 hours.

8.116 An economist wants to find a 90% confidence interval for the mean sale price of houses in a state. How large a sample should he or she select so that the estimate is within $3500 of the population mean? Assume that the standard deviation for the sale prices of all houses in this state is $31,500.

8.117 A telephone company wants to estimate the proportion of all households that own telephone answering machines. What is the most conservative estimate of the sample size that would limit the maximum error to be within .03 of the population proportion for a 95% confidence interval?

8.118 Refer to Exercise 8.117. Assume that a preliminary sample has shown that 67% of the households in the sample own telephone answering machines. How large should the sample size be so that the 95% confidence interval for the population proportion has a maximum error of .03?

■ **Advanced Exercises**

8.119 Let μ be the mean hourly wage of carpenters in the state of Idaho. A random sample of carpenters from Idaho yielded a 95% confidence interval for μ of $11.45 to $14.35. Assume that this sample included more than 30 carpenters.

a. Find the value of $\bar{x}$ for this sample.
b. Find the 99% confidence interval for μ based on this same sample.

8.120 The director of Channel 66, an independent television station, claims that this channel devotes an average of 12 minutes or less per hour to commercials. Students in a communications course at a local college are asked to estimate the mean time that Channel 66 actually devotes per hour to commercials. Since the amount of commercial time seems to vary from hour to hour, each student is asked to watch Channel 66 for a preassigned 20-minute time interval and use a stopwatch to time the commercials. Thirty students participated in this study, which showed that a mean of 4.68 minutes of commercials were aired on this channel per 20-minute interval with a standard deviation of 1.30 minutes.

a. Using this information, find the 90% confidence interval for μ, the mean time devoted to commercials per hour on Channel 66.
b. A student named Waldo was absent when the project was assigned. The instructor has asked him to make his own estimate of μ. Waldo does not have much time to watch television, and, hence, he decides to turn on Channel 66 at randomly chosen times and record whether or not a commercial is on at that time. He does this 30 times and observes a commercial on 7 of those occasions. Use this information to find a point estimate of μ.
c. Compare Waldo's point estimate of μ to the one based on the information collected by the other 30 students.
d. Use Waldo's data to find a 90% confidence interval for μ. (*Hint:* First find a 90% confidence interval for p, the proportion of Channel 66's broadcast time that is devoted to commercials. Then convert this interval to a confidence interval for μ.)
e. Compare the maximum error of Waldo's estimate of μ to that of the class.
f. To achieve the same accuracy as that of the class's estimate, how many times would Waldo have to turn on Channel 66? Do you think this is feasible?

8.121 According to data from the Marist Institute for Public Opinion, 31% of adults rated the public schools excellent or good at "doing their overall job," 24% rated them excellent or good at "preparing students for work," and 33% rated them excellent or good at "preparing students for college" (*USA TODAY,* September 10, 1998). Assume that these percentages are based on a random sample of 1200 American adults. Using each of these three percentages, find a 95% confidence interval for the corresponding true percentage of all American adults who will hold that opinion. Write a one-page report to present these results to a group of college students who have not taken statistics. Your report should answer such questions as: (1) What is a confidence interval? (2) Why is a range of values more informative than a single percentage? (3) What does 95% mean in this context? (4) What assumptions, if any, are you making when you construct the confidence interval?

8.122 A group of veterinarians wants to test a new canine vaccine for Lyme disease. (Lyme disease is transmitted by the bite of an infected tick.) In an area that has a high incidence of Lyme disease, 100 dogs are randomly selected (with their owners' permission) to receive the vaccine. Over a 12-month period, these dogs are periodically examined by veterinarians for symptoms of Lyme disease. At the end of 12 months, 10 of these 100 dogs are diagnosed with the disease. During the same 12-month period, 18% of the unvaccinated dogs in the area have been found to have Lyme disease. Let p be the proportion of all potential vaccinated dogs who would contract Lyme disease in this area.

a. Find a 95% confidence interval for p.

b. Does 18% lie within your confidence interval of part a? Does this suggest the vaccine might or might not be effective to some degree?

c. Write a brief critique of this experiment, pointing out anything that may have distorted the results or conclusions.

8.123 When one is attempting to determine the required sample size for estimating a population mean and the information on the population standard deviation is not available, it may be feasible to take a small preliminary sample and use the sample standard deviation to estimate the required sample size, n. Suppose that we want to estimate μ, the mean commuting distance for students at a community college, to within 1 mile with a confidence level of 95%. A random sample of 20 students yields a standard deviation of 4.1 miles. Use this value of the sample standard deviation, s, to estimate the required sample size, n.

8.124 A gas station attendant would like to estimate p, the proportion of all households that own more than two vehicles. To obtain an estimate, the attendant decides to ask the next 200 gasoline customers how many vehicles their households own. To obtain an estimate of p, the attendant counts the number of customers who say there are more than two vehicles in their households and then divides this number by 200. How would you critique this estimation procedure? Is there anything wrong with this procedure that would result in sampling and/or nonsampling errors? If so, can you suggest a procedure that would reduce this error?

8.125 A couple considering the purchase of a new home would like to estimate the average number of cars that go past the location per day. The couple guesses that the number of cars passing this location per day has a standard deviation of 170.

a. On how many randomly selected days should the number of cars passing the location be observed so that the couple can be 99% certain the estimate will be within 100 cars of the true average?

b. Suppose the couple finds out that the standard deviation of the number of cars passing the location per day is not 170 but is actually 272. If they have already taken a sample of the size computed in part a, what confidence does the couple have that their point estimate is within 100 cars of the true average?

c. If the couple has already taken a sample of the size computed in part a and later finds out that the standard deviation of the number of cars passing the location per day is actually 130, they can be 99% confident their point estimate is within how much of the true average?

8.126 The U.S. Senate just passed a bill by a vote of 55–45 (with all 100 senators voting). A student who took an elementary statistics course last semester says, "We can use these data to make a confidence interval about p. We have $n = 100$ and $\hat{p} = 55/100 = .55$." Hence, according to him, a 95% confidence interval for p is

$$\hat{p} \pm z\sigma_{\hat{p}} = .55 \pm 1.96 \sqrt{\frac{(.55)(.45)}{100}} = .55 \pm .098 = .452 \text{ to } .648$$

Does this make sense? If not, what is wrong with the student's reasoning?

SELF-REVIEW TEST

1. Complete the following sentences using the terms *population parameter* and *sample statistic*.

a. Estimation means assigning values to a _____ based on the value of a _____.

b. An estimator is the _____ used to estimate a _____.

c. The value of a _____ is called the point estimate of the corresponding _____.

2. A 95% confidence interval for μ can be interpreted to mean that if we take 100 samples of the same size and construct 100 such confidence intervals for μ, then

    a. 95 of them will not include μ **b.** 95 will include μ **c.** 95 will include $\bar{x}$

3. The confidence level is denoted by

    a. $(1 - \alpha)100\%$ **b.** $100\alpha\%$ **c.** α

4. The maximum error of the estimate for μ is

    a. $z\sigma_{\bar{x}}$ (or $zs_{\bar{x}}$)   **b.** $\sigma/\sqrt{n}$ (or $s/\sqrt{n}$)   **c.** $\sigma_{\bar{x}}$ (or $s_{\bar{x}}$)

5. Which of the following assumptions is not required to use the t distribution to make a confidence interval for μ?

    a. The population from which the sample is taken is (approximately) normally distributed.

    b. $n < 30$

    c. The population standard deviation, σ, is not known.

    d. The sample size is at least 10.

6. The parameter(s) of the t distribution is (are)

    a. n **b.** degrees of freedom   **c.** μ and degrees of freedom

7. A sample of 50 packages mailed from a specific post office showed a mean mailing charge of $2.85 with a standard deviation of $.72.

    a. What is the point estimate of the population mean? What is the margin of error associated with this estimate?

    b. Construct a 99% confidence interval for the mean mailing charge for all packages mailed from this post office.

8. A sample of 25 malpractice lawsuits filed against doctors showed that the mean compensation awarded to the plaintiffs was $310,425 with a standard deviation of $74,820. Find a 95% confidence interval for the mean compensation awarded to plaintiffs of all such lawsuits. Assume that the compensations awarded to plaintiffs of all such lawsuits are normally distributed.

9. According to a survey by Maritz Marketing Research, Inc., 42% of men reported playing the lottery at least once a month (*American Demographics,* January 1998). Suppose that this percentage is based on a random sample of 600 U.S. men.

    a. What is the point estimate of the corresponding population proportion? What is the margin of error associated with this estimate?

    b. Construct a 95% confidence interval for the proportion of all U.S. men who play the lottery at least once a month.

10. A statistician is interested in estimating at a 95% confidence level the mean number of houses sold per month by all real estate agents in a large city. From an earlier study, it is known that the standard deviation of the number of houses sold per month by all real estate agents in this city is 2.2. How large a sample should be taken so that the estimate is within .65 of the population mean?

11. A company wants to estimate the proportion of all workers who hold more than one job. What is the most conservative estimate of the sample size that would limit the maximum error to be within .025 of the population proportion for a 99% confidence interval?

12. Refer to Problem 11. Assume that a preliminary study has shown that 12% of adults hold more than one job. How large a sample should be taken in this case so that the maximum error is within .025 of the population proportion for a 99% confidence interval?

13. Dr. Garcia estimated the mean stress score before a statistics test for a random sample of 25 students. She found the mean and standard deviation for this sample to be 7.1 (on a scale of 1 to 10) and 1.2, respectively. She used a 97% confidence level. However, she thinks that the confidence interval is too wide. How can she reduce the width of the confidence interval? Describe all possible alternatives. Which alternative do you think is best and why?

*14. You want to estimate the mean number of hours that students at your college work per week. Briefly explain how you will conduct this study using a small sample. Take a sample of 12 students from your college who hold a job. Collect data on the number of hours that these students spent working last week. Then estimate the population mean. Choose your own confidence level. What assumptions will you make to estimate this population mean?

*15. You want to estimate the proportion of people who are happy with their current jobs. Briefly explain how you will conduct this study. Take a sample of 35 persons and collect data on whether or not they are happy with their current jobs. Then estimate the population proportion. Choose your own confidence level.

MINI-PROJECTS

Mini-Project 8–1

Deborah A. Stiles of the School of Education at Webster University and her coauthors reviewed several studies on the occupational goals of adolescent boys. Even though fewer than one in 10,000 American boys ever reach this goal, the most popular occupational choice for these boys in the United States is "professional athlete." In one of the studies, 61 U.S. students (26 boys and 35 girls, aged 12 to 18 years) participated to seek insight into why so many boys aspire to professional athletics. The table, adapted from the paper published by the authors, summarizes the reasons these students offered. Note that each of the 61 students typically gave several reasons or comments. Both boys and girls were asked to answer the question, Why might boys wish to become professional athletes?

Reason/Comment	Number of Comments	Percentage of Total
Well-known and admired	94	35
Money	56	21
Masculine role	29	11
Like sports	27	10
Easy life	24	9
Education not necessary	19	7
Unrealistic fantasy	14	5
Other	5	2

Source: Deborah A. Stiles, Judith L. Gibbons, Daniel L. Sebben, and Deane C. Wiley, "Why Adolescent Boys Dream of Becoming Professional Athletes," *Psychological Reports,* 1999, 84, pp. 1075–1085.

These 61 students, who were attending schools near Boston and near St. Louis, were recruited from two after-school groups and three high school classes, so they were not a random sample of all U.S. students in their age group. If, however, these students had constituted a random sample of all U.S. students aged 12 to 18, explain how you could use the data in the table to construct confidence intervals for the true percentages of total comments for each of the reasons given.

Mini-Project 8–2

Consider the data set on the heights of NBA players that accompanies this text.

a. Take a random sample of 15 players and find a 95% confidence interval for μ. Assume the heights of these players are normally distributed.
b. Repeat part a for samples of size 31 and 45, respectively.
c. Compare the widths of your three confidence intervals.
d. Now calculate the mean, μ, of the heights of all players. Do all of your confidence intervals contain this μ? If not, which ones do not contain μ?

For instructions on using MINITAB for this chapter, please see Section B.7 in Appendix B.

See Appendix C for *Excel Adventure 8* on sampling from a population.

COMPUTER ASSIGNMENTS

CA8.1 The following data give the annual incomes (in thousands of dollars) before taxes for a sample of 36 randomly selected families from a city.

21.6	13.0	25.6	27.9	50.0	18.1
10.1	21.5	10.0	72.8	58.2	15.4
7.2	27.0	32.2	45.0	95.0	27.8
92.8	8.4	45.3	76.0	28.6	9.3
30.6	19.0	25.5	27.5	9.7	15.1
6.3	44.5	24.0	13.0	61.7	16.0

Using MINITAB or any other statistical software, construct a 99% confidence interval for μ assuming that the population standard deviation is $13.75 thousand.

CA8.2 The following data give the checking account balances on a certain day for a randomly selected sample of 30 households.

500	100	650	1917	2200	500	180	3000	1500	1300
319	1500	1102	405	124	1000	134	2000	150	800
200	750	300	2300	40	1200	500	900	20	160

Using MINITAB or any other statistical software, construct a 97% confidence interval for μ assuming that the population standard deviation is unknown.

CA8.3 Refer to Data Set I (that accompanies this text) on the prices of various products in different cities across the country. Using the data on monthly telephone charges given in column C7, make a 98% confidence interval for the population mean μ.

CA8.4 Refer to the Manchester Road Race data set (that accompanies this text) for all participants. Using MINITAB or any other statistical software, take a sample of 100 observations from this data set.

a. Using the sample data, make a 95% confidence interval for the mean time taken to complete this race by all participants.

b. The mean time taken to run this race by all participants is 48.78 minutes. Does the confidence interval made in part a include this population mean?

CA8.5 Repeat Computer Assignment CA8.4 for a sample of 25 observations. Assume that the distribution of time taken to run this race by all participants is approximately normal.

CA8.6 The following data give the prices (in thousands of dollars) of 16 recently sold houses in an area.

141	163	127	104	197	203	113	179
256	228	183	119	133	199	271	191

Using MINITAB or any other statistical software, construct a 99% confidence interval for the mean price of all houses in this area. Assume that the distribution of prices of all houses in the given area is normal.

CA8.7 A researcher wanted to estimate the mean contributions made to charitable causes by major companies. A random sample of 18 companies produced the following data on contributions (in millions of dollars) made by them.

1.8	.6	1.2	.3	2.6	1.9	3.4	2.6	.2
2.4	1.4	2.5	3.1	.9	1.2	2.0	.8	1.1

Using MINITAB or any other statistical software, make a 98% confidence interval for the mean contributions made to charitable causes by all major companies. Assume that the contributions made to charitable causes by all major companies have a normal distribution.

CA8.8   A mail-order company promises its customers that their orders will be processed and mailed within 72 hours after an order is placed. The quality control department at the company checks from time to time to see if this promise is kept. Recently the quality control department took a sample of 200 orders and found that 176 of them were processed and mailed within 72 hours of the placement of the orders. Using MINITAB or any other statistical software, make a 98% confidence interval for the corresponding population proportion.

CA8.9   One of the major problems faced by department stores is a high percentage of returns. The manager of a department store wanted to estimate the percentage of all sales that result in returns. A sample of 500 sales showed that 95 of them had products returned within the time allowed for returns. Using MINITAB or any other statistical software, make a 99% confidence interval for the corresponding population proportion.

CA8.10   One of the major problems faced by auto insurance companies is the filing of fraudulent claims. An insurance company carefully investigated 1000 auto claims filed with it and found 108 of them to be fraudulent. Using MINITAB or any other statistical software, make a 96% confidence interval for the corresponding population proportion.

9

HYPOTHESIS TESTS ABOUT THE MEAN AND PROPORTION

9.1   HYPOTHESIS TESTS: AN INTRODUCTION

9.2   HYPOTHESIS TESTS ABOUT A POPULATION MEAN: LARGE SAMPLES

9.3   HYPOTHESIS TESTS USING THE *p*-VALUE APPROACH

9.4   HYPOTHESIS TESTS ABOUT A POPULATION MEAN: SMALL SAMPLES

9.5   HYPOTHESIS TESTS ABOUT A POPULATION PROPORTION: LARGE SAMPLES

CASE STUDY 9–1   OLDER WORKERS MOST CONTENT

GLOSSARY

KEY FORMULAS

SUPPLEMENTARY EXERCISES

SELF-REVIEW TEST

MINI-PROJECTS

COMPUTER ASSIGNMENTS

This chapter introduces the second topic in inferential statistics: tests of hypotheses. In a test of hypothesis, we test a certain given theory or belief about a population parameter. We may want to find out, using some sample information, whether or not a given claim (or statement) about a population parameter is true. This chapter discusses how to make such tests of hypotheses about the population mean, μ, and the population proportion, p.

As an example, a soft-drink company may claim that, on average, its cans contain 12 ounces of soda. A government agency may want to test whether or not such cans contain, on average, 12 ounces of soda. As another example, according to the U.S. Bureau of the Census, 16.3% of the population in the United States lacked health insurance in 1998. An economist may want to check if this percentage is still true for this year. In the first of these two examples we are to test a hypothesis about the population mean, μ, and in the second example we are to test a hypothesis about the population proportion, p.

9.1 HYPOTHESIS TESTS: AN INTRODUCTION

Why do we need to perform a test of hypothesis? Reconsider the example about soft-drink cans. Suppose we take a sample of 100 cans of the soft drink under investigation. We then find out that the mean amount of soda in these 100 cans is 11.89 ounces. Based on this result, can we state that, on average, all such cans contain less than 12 ounces of soda and that the company is lying to the public? Not until we perform a test of hypothesis can we make such an accusation. The reason is that the mean, $\bar{x} = 11.89$ ounces, is obtained from a sample. The difference between 12 ounces (the required average amount for the population) and 11.89 ounces (the observed average amount for the sample) may have occurred only because of the sampling error. Another sample of 100 cans may give us a mean of 12.04 ounces. Therefore, we make a test of hypothesis to find out how large the difference between 12 ounces and 11.89 ounces is and to investigate whether or not this difference has occurred as a result of chance alone. Now, if 11.89 ounces is the mean for all cans and not for just 100 cans, then we do not need to make a test of hypothesis. Instead, we can immediately state that the mean amount of soda in all such cans is less than 12 ounces. We perform a test of hypothesis only when we are making a decision about a population parameter based on the value of a sample statistic.

9.1.1 TWO HYPOTHESES

Consider a nonstatistical example of a person who has been indicted for committing a crime and is being tried in a court. Based on the available evidence, the judge or jury will make one of two possible decisions:

1. The person is not guilty.
2. The person is guilty.

At the outset of the trial, the person is presumed not guilty. The prosecutor's efforts are to prove that the person has committed the crime and, hence, is guilty.

In statistics, *the person is not guilty* is called the **null hypothesis** and *the person is guilty* is called the **alternative hypothesis**. The null hypothesis is denoted by H_0 and the alternative hypothesis is denoted by H_1. In the beginning of the trial it is assumed that the person is not guilty. The null hypothesis is usually the hypothesis that is assumed to be true to begin with. The two hypotheses for the court case are written as follows (notice the colon after H_0 and H_1):

Null hypothesis: H_0: The person is not guilty

Alternative hypothesis: H_1: The person is guilty

In a statistics example, the null hypothesis states that a given claim (or statement) about a population parameter is true. Reconsider the example of the soft-drink company's claim that, on average, its cans contain 12 ounces of soda. In reality, this claim may or may not be true. However, we will initially assume that the company's claim is true (that is, the company is not guilty of cheating and lying). To test the claim of the soft-drink company, the null hypothesis will be that the company's claim is true. Let μ be the mean amount of soda in all cans. The company's claim will be true if $\mu = 12$ ounces. Thus, the null hypothesis will be written as

H_0: $\mu = 12$ ounces   (The company's claim is true)

In this example, the null hypothesis can also be written as $\mu \geq 12$ ounces because the claim of the company will still be true if the cans contain, on average, more than 12 ounces of soda. The company will be accused of cheating the public only if the cans contain, on average, less than 12 ounces of soda. However, it will not affect the test whether we use an $=$ or a $\geq$ sign in the null hypothesis as long as the alternative hypothesis has a $<$ sign. Remember that in the null hypothesis (and in the alternative hypothesis also) we use the population parameter (such as μ or p), and not the sample statistic (such as $\bar{x}$ or $\hat{p}$).

> **NULL HYPOTHESIS** A *null hypothesis* is a claim (or statement) about a population parameter that is assumed to be true until it is declared false.

The alternative hypothesis in our statistics example will be that the company's claim is false and its soft-drink cans contain, on average, less than 12 ounces of soda—that is, $\mu < 12$ ounces. The alternative hypothesis will be written as

$$H_1: \mu < 12 \text{ ounces} \qquad \text{(The company's claim is false)}$$

> **ALTERNATIVE HYPOTHESIS** An *alternative hypothesis* is a claim about a population parameter that will be true if the null hypothesis is false.

Let us return to the example of the court trial. The trial begins with the assumption that the null hypothesis is true—that is, the person is not guilty. The prosecutor assembles all the possible evidence and presents it in the court to prove that the null hypothesis is false and the alternative hypothesis is true (that is, the person is guilty). In the case of our statistics example, the information obtained from a sample will be used as evidence to decide whether or not the claim of the company is true. In the court case, the decision made by the judge (or jury) depends on the amount of evidence presented by the prosecutor. At the end of the trial, the judge (or jury) will consider whether or not the evidence presented by the prosecutor is sufficient to declare the person guilty. The amount of evidence that will be considered to be sufficient to declare the person guilty depends on the discretion of the judge (or jury).

9.1.2 REJECTION AND NONREJECTION REGIONS

In Figure 9.1, which represents the court case, the point marked 0 indicates that there is no evidence against the person being tried. The farther we move toward the right on the horizontal axis, the more convincing the evidence is that the person has committed the crime.

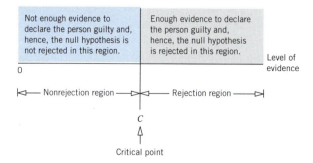

Figure 9.1 Nonrejection and rejection regions for the court case.

We have arbitrarily marked a point C on the horizontal axis. Let us assume that a judge (or jury) considers any amount of evidence to the right of point C to be sufficient and any amount of evidence to the left of C to be insufficient to declare the person guilty. Point C is called the **critical value** or **critical point** in statistics. If the amount of evidence presented by the prosecutor falls in the area to the left of point C, the verdict will reflect that there is not enough evidence to declare the person guilty. Consequently, the accused person will be declared *not guilty*. In statistics, this decision is stated as *do not reject* H_0. It is equivalent to saying that there is not enough evidence to declare the null hypothesis false. The area to the left of point C is called the *nonrejection region;* that is, this is the region where the null hypothesis is not rejected. However, if the amount of evidence falls in the area to the right of point C, the verdict will be that there is sufficient evidence to declare the person guilty. In statistics, this decision is stated as *reject* H_0 or *the null hypothesis is false.* Rejecting H_0 is equivalent to saying that *the alternative hypothesis is true.* The area to the right of point C is called the *rejection region;* that is, this is the region where the null hypothesis is rejected.

9.1.3 TWO TYPES OF ERRORS

We all know that a court's verdict is not always correct. If a person is declared guilty at the end of a trial, there are two possibilities.

1. The person has *not* committed the crime but is declared guilty (because of what may be false evidence).
2. The person *has* committed the crime and is rightfully declared guilty.

In the first case, the court has made an error by punishing an innocent person. In statistics, this kind of error is called a **Type I** or an **α *(alpha)* error**. In the second case, because the guilty person has been punished, the court has made the correct decision. The second row in the shaded portion of Table 9.1 shows these two cases. The two columns of Table 9.1, corresponding to *the person is not guilty* and *the person is guilty,* give the two actual situations. Which one of these is true is known only to the person being tried. The two rows in this table, corresponding to *the person is not guilty* and *the person is guilty,* show the two possible court decisions.

Table 9.1

		Actual Situation	
		The Person Is Not Guilty	**The Person Is Guilty**
Court's decision	The person is not guilty	Correct decision	Type II or β error
	The person is guilty	Type I or α error	Correct decision

In our statistics example, a Type I error will occur when H_0 is actually true (that is, the cans do contain, on average, 12 ounces of soda), but it just happens that we draw a sample with a mean that is much less than 12 ounces and we wrongfully reject the null hypothesis, H_0. The value of α, called the **significance level** of the test, represents the probability of making a Type I error. In other words, α is the probability of rejecting the null hypothesis, H_0, when in fact it is true.

TYPE I ERROR A *Type I error* occurs when a true null hypothesis is rejected. The value of α represents the probability of committing this type of error; that is,

$$\alpha = P(H_0 \text{ is rejected} \mid H_0 \text{ is true})$$

The value of α represents the *significance level* of the test.

The size of the rejection region in a statistics problem of a test of hypothesis depends on the value assigned to α. In a test of hypothesis, we usually assign a value to α before making the test. Although any value can be assigned to α, the commonly used values of α are .01, .025, .05, and .10. Usually the value assigned to α does not exceed .10 (or 10%).

Now, suppose that in the court trial case the person is declared not guilty at the end of the trial. Such a verdict does not indicate that the person has indeed *not* committed the crime. It is possible that the person is guilty but there is not enough evidence to prove the guilt. Consequently, in this situation there are again two possibilities.

1. The person has *not* committed the crime and is declared not guilty.
2. The person *has* committed the crime but, *because of the lack of enough evidence*, is declared not guilty.

In the first case, the court's decision is correct. But in the second case, the court has committed an error by setting a guilty person free. In statistics, this type of error is called a **Type II** or a **β** (the Greek letter *beta*) **error**. These two cases are shown in the first row of the shaded portion of Table 9.1.

In our statistics example, a Type II error will occur when the null hypothesis H_0 is actually false (that is, the soda contained in all cans, on average, is less than 12 ounces), but it happens by chance that we draw a sample with a mean that is close to or greater than 12 ounces and we wrongfully conclude *do not reject H_0*. The value of β represents the probability of making a Type II error. It represents the probability that H_0 is not rejected when actually H_0 is false. The value of $1 - \beta$ is called the **power of the test**. It represents the probability of not making a Type II error.

TYPE II ERROR A *Type II error* occurs when a false null hypothesis is not rejected. The value of β represents the probability of committing a Type II error; that is,

$$\beta = P(H_0 \text{ is not rejected} \mid H_0 \text{ is false})$$

The value of $1 - \beta$ is called the *power of the test*. It represents the probability of not making a Type II error.

The two types of errors that occur in tests of hypotheses depend on each other. We cannot lower the values of α and β simultaneously for a test of hypothesis for a fixed sample size. Lowering the value of α will raise the value of β, and lowering the value of β will raise the value of α. However, we can decrease both α and β simultaneously by increasing the sample size. The explanation of how α and β are related and the computation of β are not within the scope of this text.

Table 9.2, which is similar to Table 9.1, is written for the statistics problem of a test of hypothesis. In Table 9.2 *the person is not guilty* is replaced by *H_0 is true, the person is guilty* by *H_0 is false,* and the *court's decision* by *decision.*

Table 9.2

		Actual Situation	
		H_0 is true	H_0 is false
Decision	Do not reject H_0	Correct decision	Type II or β error
	Reject H_0	Type I or α error	Correct decision

9.1.4 TAILS OF A TEST

The statistical hypothesis-testing procedure is similar to the trial of a person in court but with two major differences. The first major difference is that in a statistical test of hypothesis, the partition of the total region into rejection and nonrejection regions is not arbitrary. Instead, it depends on the value assigned to α (Type I error). As mentioned earlier, α is also called the significance level of the test.

The second major difference relates to the rejection region. In the court case, the rejection region is on the right side of the critical point, as shown in Figure 9.1. However, in statistics, the rejection region for a hypothesis-testing problem can be on both sides, with the nonrejection region in the middle, or it can be on the left side or on the right side of the nonrejection region. These possibilities are explained in the next three parts of this section. A test with two rejection regions is called a **two-tailed test**, and a test with one rejection region is called a **one-tailed test**. The one-tailed test is called a **left-tailed test** if the rejection region is in the left tail of the distribution curve, and it is called a **right-tailed test** if the rejection region is in the right tail of the distribution curve.

> **TAILS OF THE TEST** A *two-tailed test* has rejection regions in both tails, a *left-tailed test* has the rejection region in the left tail, and a *right-tailed test* has the rejection region in the right tail of the distribution curve.

A Two-Tailed Test

According to the U.S. Bureau of the Census, the mean family size in the United States was 3.18 in 1998. A researcher wants to check whether or not this mean has changed since 1998. The key word here is *changed*. The mean family size has changed if it has either increased or decreased during the period since 1998. This is an example of a two-tailed test. Let μ be the current mean family size for all families. The two possible decisions are

1. The mean family size has not changed—that is, $\mu = 3.18$.
2. The mean family size has changed—that is, $\mu \neq 3.18$.

We write the null and alternative hypotheses for this test as

$$H_0: \mu = 3.18 \quad \text{(The mean family size has not changed)}$$

$$H_1: \mu \neq 3.18 \quad \text{(The mean family size has changed)}$$

Whether a test is two-tailed or one-tailed is determined by the sign in the alternative hypothesis. If the alternative hypothesis has a *not equal to* ($\neq$) sign, as in this example, it is a two-tailed test. As shown in Figure 9.2, a two-tailed test has two rejection regions, one in

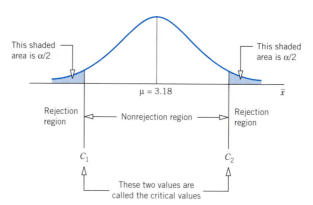

Figure 9.2   A two-tailed test.

each tail of the distribution curve. Figure 9.2 shows the sampling distribution of $\bar{x}$ for a large sample. Assuming H_0 is true, $\bar{x}$ has a normal distribution with its mean equal to 3.18 (the value of μ in H_0). In Figure 9.2, the area of each of the two rejection regions is $\alpha/2$ and the total area of both rejection regions is α (the significance level). As shown in this figure, a two-tailed test of hypothesis has two critical values that separate the two rejection regions from the nonrejection region. We will reject H_0 if the value of $\bar{x}$ obtained from the sample falls in either of the two rejection regions. We will not reject H_0 if the value of $\bar{x}$ lies in the nonrejection region. By rejecting H_0, we are saying that the difference between the value of μ stated in H_0 and the value of $\bar{x}$ obtained from the sample is too large to have occurred because of the sampling error alone. Consequently, this difference is real. By not rejecting H_0, we are saying that the difference between the value of μ stated in H_0 and the value of $\bar{x}$ obtained from the sample is small and it may have occurred because of the sampling error alone.

A Left-Tailed Test

Reconsider the example of the mean amount of soda in all soft-drink cans produced by a company. The company claims that these cans, on average, contain 12 ounces of soda. However, if these cans contain less than the claimed amount of soda, then the company can be accused of cheating. Suppose a consumer agency wants to test whether the mean amount of soda per can is less than 12 ounces. Note that the key phrase this time is *less than*, which indicates a left-tailed test. Let μ be the mean amount of soda in all cans. The two possible decisions are

1.   The mean amount of soda in all cans is not less than 12 ounces—that is, $\mu = 12$ ounces.
2.   The mean amount of soda in all cans is less than 12 ounces—that is, $\mu < 12$ ounces.

The null and alternative hypotheses for this test are written as

$$H_0: \mu = 12 \text{ ounces} \quad \text{(The mean is not less than 12 ounces)}$$

$$H_1: \mu < 12 \text{ ounces} \quad \text{(The mean is less than 12 ounces)}$$

In this case, we can also write the null hypothesis as $H_0: \mu \geq 12$. This will not affect the result of the test as long as the sign in H_1 is *less than* ($<$).

When the alternative hypothesis has a *less than* ($<$) sign, as in this case, the test is always left-tailed. In a left-tailed test, the rejection region is in the left tail of the distribution

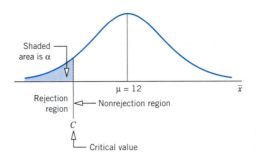

Figure 9.3 A left-tailed test.

curve, as shown in Figure 9.3, and the area of this rejection region is equal to α (the significance level). We can observe from this figure that there is only one critical value in a left-tailed test.

Assuming H_0 is true, $\bar{x}$ has a normal distribution for a large sample with its mean equal to 12 ounces (the value of μ in H_0). We will reject H_0 if the value of $\bar{x}$ obtained from the sample falls in the rejection region; we will not reject H_0 otherwise.

A Right-Tailed Test

To illustrate the third case, according to a 1999 study by the American Federation of Teachers, the mean starting salary of school teachers in the United States was \$25,735 during 1997–98. Suppose we want to test whether the current mean starting salary of all school teachers in the United States is higher than \$25,735. The key phrase in this case is *higher than*, which indicates a right-tailed test. Let μ be the current mean starting salary of school teachers in the United States. The two possible decisions this time are

1. The current mean starting salary of all school teachers in the United States is not higher than \$25,735—that is, $\mu = \$25,735$.

2. The current mean starting salary of all school teachers in the United States is higher than \$25,735—that is, $\mu > \$25,735$.

We write the null and alternative hypotheses for this test as

H_0: $\mu = \$25,735$ (The current mean starting salary is not higher than \$25,735)

H_1: $\mu > \$25,735$ (The current mean starting salary is higher than \$25,735)

In this case, we can also write the null hypothesis as H_0: $\mu \leq \$25,735$, which states that the current mean starting salary of all school teachers in the United States is either equal to or less than \$25,735. Again, the result of the test will not be affected whether we use an *equal to* (=) or a *less than or equal to* ($\leq$) sign in H_0 as long as the alternative hypothesis has a *greater than* (>) sign.

When the alternative hypothesis has a *greater than* (>) sign, the test is always right-tailed. As shown in Figure 9.4 on page 384, in a right-tailed test, the rejection region is in the right tail of the distribution curve. The area of this rejection region is equal to α, the significance level. Like a left-tailed test, a right-tailed test has only one critical value.

Again, assuming H_0 is true, $\bar{x}$ has a normal distribution for a large sample with its mean equal to \$25,735 (the value of μ in H_0). We will reject H_0 if the value of $\bar{x}$ obtained from the sample falls in the rejection region. Otherwise, we will not reject H_0.

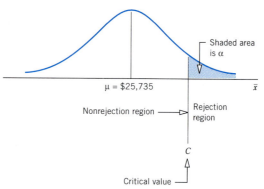

Figure 9.4 A right-tailed test.

Table 9.3 summarizes the foregoing discussion about the relationship between the signs in H_0 and H_1 and the tails of a test.

Table 9.3

	Two-Tailed Test	Left-Tailed Test	Right-Tailed Test
Sign in the null hypothesis H_0	$=$	$=$ or $\geq$	$=$ or $\leq$
Sign in the alternative hypothesis H_1	$\neq$	$<$	$>$
Rejection region	In both tails	In the left tail	In the right tail

Note that the null hypothesis always has an *equal to* ($=$) or a *greater than or equal to* ($\geq$) or a *less than or equal to* ($\leq$) sign, and the alternative hypothesis always has a *not equal to* ($\neq$) or a *less than* ($<$) or a *greater than* ($>$) sign.

A test of hypothesis involves five steps, which are listed next.

STEPS TO PERFORM A TEST OF HYPOTHESIS A statistical test of hypothesis procedure has the following five steps.

1. State the null and alternative hypotheses.
2. Select the distribution to use.
3. Determine the rejection and nonrejection regions.
4. Calculate the value of the test statistic.
5. Make a decision.

With the help of examples, these steps will be described in the next section.

EXERCISES

■ **Concepts and Procedures**

9.1 Briefly explain the meaning of each of the following terms.

a. Null hypothesis   **b.** Alternative hypothesis
c. Critical point(s)   **d.** Significance level
e. Nonrejection region   **f.** Rejection region
g. Tails of a test   **h.** Two types of errors

9.2 What are the four possible outcomes for a test of hypothesis? Show these outcomes by writing a table. Briefly describe the Type I and Type II errors.

9.3 Explain how the tails of a test depend on the sign in the alternative hypothesis. Describe the signs in the null and alternative hypotheses for a two-tailed, a left-tailed, and a right-tailed test, respectively.

9.4 Explain which of the following is a two-tailed test, a left-tailed test, or a right-tailed test.

a. $H_0: \mu = 45,$ $H_1: \mu > 45$
b. $H_0: \mu = 23,$ $H_1: \mu \neq 23$
c. $H_0: \mu \geq 75,$ $H_1: \mu < 75$

Show the rejection and nonrejection regions for each of these cases by drawing a sampling distribution curve for the sample mean, assuming that the sample size is large in each case.

9.5 Explain which of the following is a two-tailed test, a left-tailed test, or a right-tailed test.

a. $H_0: \mu = 12,$ $H_1: \mu < 12$
b. $H_0: \mu \leq 85,$ $H_1: \mu > 85$
c. $H_0: \mu = 33,$ $H_1: \mu \neq 33$

Show the rejection and nonrejection regions for each of these cases by drawing a sampling distribution curve for the sample mean, assuming that the sample size is large in each case.

9.6 Which of the two hypotheses (null and alternative) is initially assumed to be true in a test of hypothesis?

9.7 Consider $H_0: \mu = 20$ versus $H_1: \mu < 20$.

a. What type of error are you making if the null hypothesis is actually false and you fail to reject it?
b. What type of error are you making if the null hypothesis is actually true and you reject it?

9.8 Consider $H_0: \mu = 55$ versus $H_1: \mu \neq 55$.

a. What type of error are you making if the null hypothesis is actually false and you fail to reject it?
b. What type of error are you making if the null hypothesis is actually true and you reject it?

■ **Applications**

9.9 Write the null and alternative hypotheses for each of the following examples. Determine if each is a case of a two-tailed, a left-tailed, or a right-tailed test.

a. To test whether or not the mean price of houses in Connecticut is greater than $143,000
b. To test if the mean number of hours spent working per week by college students who hold jobs is different from 15 hours
c. To test whether the mean life of a particular brand of auto batteries is less than 45 months
d. To test if the mean amount of time spent doing homework by all fourth-graders is different from 5 hours a week
e. To test if the mean age of all college students is different from 24 years

9.10 Write the null and alternative hypotheses for each of the following examples. Determine if each is a case of a two-tailed, a left-tailed, or a right-tailed test.

a. To test if the mean amount of time spent per week watching sports on television by all adult men is different from 9.5 hours
b. To test if the mean amount of money spent by all customers at a supermarket is less than $85

   c. To test whether the mean starting salary of college graduates is higher than \$29,000 per year

   d. To test if the mean GPA of all students at a university is lower than 2.9

   e. To test if the mean cholesterol level of all adult men in the United States is higher than 175

9.2 HYPOTHESIS TESTS ABOUT A POPULATION MEAN: LARGE SAMPLES

From the central limit theorem discussed in Chapter 7, the sampling distribution of $\bar{x}$ is approximately normal for large samples ($n \geq 30$). Consequently, whether or not σ is known, the normal distribution is used to test hypotheses about the population mean when a sample size is large.

> **TEST STATISTIC** In tests of hypotheses about μ for large samples, the random variable
>
> $$z = \frac{\bar{x} - \mu}{\sigma_{\bar{x}}} \quad \text{or} \quad \frac{\bar{x} - \mu}{s_{\bar{x}}}$$
>
> where $\quad \sigma_{\bar{x}} = \sigma/\sqrt{n} \quad$ and $\quad s_{\bar{x}} = s/\sqrt{n}$
>
> is called the *test statistic*. The test statistic can be defined as a rule or criterion that is used to make the decision whether or not to reject the null hypothesis.

At the end of Section 9.1, it was mentioned that a test of hypothesis procedure involves the following five steps.

1. State the null and alternative hypotheses.
2. Select the distribution to use.
3. Determine the rejection and nonrejection regions.
4. Calculate the value of the test statistic.
5. Make a decision.

Examples 9–1 through 9–3 illustrate the use of these five steps to perform tests of hypotheses about the population mean μ. Example 9–1 is concerned with a two-tailed test and Examples 9–2 and 9–3 describe one-tailed tests.

Conducting a two-tailed test of hypothesis about μ for a large sample.

EXAMPLE 9–1 The TIV Telephone Company provides long-distance telephone service in an area. According to the company's records, the average length of all long-distance calls placed through this company in 1999 was 12.44 minutes. The company's management wanted to check if the mean length of the current long-distance calls is different from 12.44 minutes. A sample of 150 such calls placed through this company produced a mean length of 13.71 minutes with a standard deviation of 2.65 minutes. Using the 5% significance level, can you conclude that the mean length of all current long-distance calls is different from 12.44 minutes?

Solution Let μ be the mean length of all current long-distance calls placed through this company and $\bar{x}$ be the corresponding mean for the sample. From the given information,

$$n = 150, \quad \bar{x} = 13.71 \text{ minutes}, \quad \text{and} \quad s = 2.65 \text{ minutes}$$

We are to test whether or not the mean length of all current long-distance calls is different from 12.44 minutes. The significance level α is .05; that is, the probability of rejecting the null hypothesis when it actually is true should not exceed .05. This is the probability of making a Type I error. We perform the test of hypothesis using the five steps.

Step 1. *State the null and alternative hypotheses.*

Notice that we are testing to find whether or not the mean length of all current long-distance calls is different from 12.44 minutes. We write the null and alternative hypotheses as follows.

H_0: $\mu = 12.44$ (The mean length of all current long-distance calls is 12.44 minutes)

H_1: $\mu \neq 12.44$ (The mean length of all current long-distance calls is different from 12.44 minutes)

Step 2. *Select the distribution to use.*

Because the sample size is large ($n > 30$), the sampling distribution of $\bar{x}$ is (approximately) normal. Consequently, we use the normal distribution to make the test.

Step 3. *Determine the rejection and nonrejection regions.*

The significance level is .05. The $\neq$ sign in the alternative hypothesis indicates that the test is two-tailed with two rejection regions, one in each tail of the normal distribution curve of $\bar{x}$. Because the total area of both rejection regions is .05 (the significance level), the area of the rejection region in each tail is .025; that is,

Area in each tail $= \alpha/2 = .05/2 = .025$

These areas are shown in Figure 9.5. Two critical points in this figure separate the two rejection regions from the nonrejection region. Next, we find the z values for the two critical points using the area of the rejection region. To find the z values for these critical points, we first find the area between the mean and one of the critical points. We obtain this area by subtracting .025 (the area in each tail) from .5, which gives .4750. Next we look for .4750 in the standard normal distribution table, Table VII of Appendix D. The value of z for .4750 is 1.96. Hence, the z values of the two critical points, as shown in Figure 9.5, are -1.96 and 1.96.

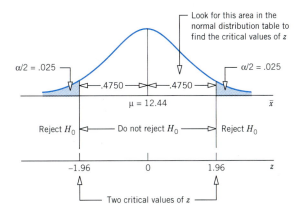

Figure 9.5

Step 4. *Calculate the value of the test statistic.*

The decision to reject or not to reject the null hypothesis will depend on whether the evidence from the sample falls in the rejection or the nonrejection region. If the value of $\bar{x}$ falls in either of the two rejection regions, we reject H_0. Otherwise, we do not reject H_0. The value of $\bar{x}$ obtained from the sample is called the *observed value of* $\bar{x}$. To locate the position of $\bar{x} = 13.71$ on the sampling distribution curve of $\bar{x}$ in Figure 9.5, we first calculate the z value for $\bar{x} = 13.71$. This is called the *value of the test statistic*. Then, we compare the value of the test statistic with the two critical values of z, -1.96 and 1.96, shown in Figure 9.5. If the value of the test statistic is between -1.96 and 1.96, we do not reject H_0. If the value of the test statistic is either greater than 1.96 or less than -1.96, we reject H_0.

CALCULATING THE VALUE OF THE TEST STATISTIC For a large sample, *the value of the test statistic z for $\bar{x}$ for a test of hypothesis about μ is computed as follows:*

$$z = \frac{\bar{x} - \mu}{\sigma_{\bar{x}}} \quad \text{if } \sigma \text{ is known}$$

$$z = \frac{\bar{x} - \mu}{s_{\bar{x}}} \quad \text{if } \sigma \text{ is not known}$$

where $\quad \sigma_{\bar{x}} = \sigma/\sqrt{n} \quad$ and $\quad s_{\bar{x}} = s/\sqrt{n}$

The value of z calculated for $\bar{x}$ using the formula is also called the **observed value of z**.

The value of $\bar{x}$ from the sample is 13.71. Because σ is not known, we calculate the z value using $s_{\bar{x}}$ as follows:

$$s_{\bar{x}} = \frac{s}{\sqrt{n}} = \frac{2.65}{\sqrt{150}} = .21637159$$

From H_0

$$z = \frac{\bar{x} - \mu}{s_{\bar{x}}} = \frac{13.71 - 12.44}{.21637159} = 5.87$$

The value of μ in the calculation of the z value is substituted from the null hypothesis. The value of $z = 5.87$ calculated for $\bar{x}$ is called the *computed value of the test statistic z*. This is the value of z that corresponds to the value of $\bar{x}$ observed from the sample. It is also called the *observed value of z*.

Step 5. *Make a decision.*

In the final step we make a decision based on the location of the value of the test statistic z computed for $\bar{x}$ in Step 4. This value of $z = 5.87$ is greater than the critical value of $z = 1.96$, and it falls in the rejection region in the right tail in Figure 9.5. Hence, we reject H_0 and conclude that based on the sample information, it appears that the mean length of all such calls is not equal to 12.44 minutes.

By rejecting the null hypothesis, we are stating that the difference between the sample mean, $\bar{x} = 13.71$ minutes, and the hypothesized value of the population mean, $\mu = 12.44$ minutes, is too large and may not have occurred because of chance or sampling error alone. This difference seems to be real and, hence, the mean length of all such calls is different from 12.44 minutes. Note that the rejection of the null hypothesis does not necessarily indicate that the mean length of all such calls is definitely different from 12.44 minutes. It simply indicates that there is strong evidence (from the sample) that the mean length of such calls is not equal to 12.44 minutes. There is a possibility that the mean length of all such calls is equal to 12.44 minutes but, by the luck of the draw, we selected a sample with a mean that is too far from the hypothesized mean of 12.44 minutes. If so, we have wrongfully rejected the null hypothesis H_0. This is a Type I error and its probability is .05 in this example. ∎

Conducting a right-tailed test of hypothesis about μ for a large sample.

EXAMPLE 9-2 According to an estimate, the average sale price of homes in the neighborhood with a high concentration of top-ranked professionals and executives in Stamford, Connecticut, was $520,234 in December 1999 (*The Wall Street Journal*, December 10, 1999). A random sample of 50 homes from this neighborhood that were recently sold gave a mean sale price of $565,750 with a standard deviation of $75,210. Using 1% significance level, can you conclude that the current mean sale price of homes in this neighborhood is higher than $520,234?

Solution Let μ be the current mean sale price of all homes in this neighborhood of Stamford, Connecticut, and let $\bar{x}$ be the corresponding mean for the sample. From the given information,

$$n = 50, \quad \bar{x} = \$565,750, \quad \text{and} \quad s = \$75,210$$

The significance level is $\alpha = .01$.

Step 1. *State the null and alternative hypotheses.*

We are to test whether the current mean sale price of homes in the said neighborhood of Stamford, Connecticut, is higher than $520,234. The null and alternative hypotheses are

$$H_0: \mu = \$520,234 \qquad \text{(The current mean is $520,234)}$$

$$H_1: \mu > \$520,234 \qquad \text{(The current mean is greater than $520,234)}$$

Step 2. *Select the distribution to use.*

Because the sample size is large ($n > 30$), the sampling distribution of $\bar{x}$ is (approximately) normal. Consequently, we use the normal distribution to make the test.

Step 3. *Determine the rejection and nonrejection regions.*

The significance level is .01. The $>$ sign in the alternative hypothesis indicates that the test is right-tailed with its rejection region in the right tail of the sampling distribution curve of $\bar{x}$. Because there is only one rejection region, its area is $\alpha = .01$. As shown in Figure 9.6, the critical value of z, obtained from Table VII of Appendix D for .4900, is approximately 2.33.

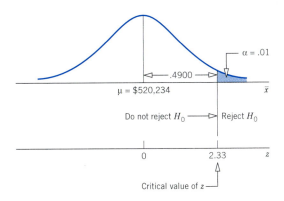

Figure 9.6

Step 4. *Calculate the value of the test statistic.*

The value of the test statistic z for $\bar{x} = \$565,750$ is calculated as follows:

$$s_{\bar{x}} = \frac{s}{\sqrt{n}} = \frac{75,210}{\sqrt{50}} = 10,636.3002$$

$$z = \frac{\bar{x} - \mu}{s_{\bar{x}}} = \frac{565,750 - 520,234}{10,636.3002} = 4.28$$

From H_0

Step 5. *Make a decision.*

The value of the test statistic $z = 4.28$ is greater than the critical value of $z = 2.33$ and it falls in the rejection region. Consequently, we reject H_0. Therefore, we can state that the sample mean $\bar{x} = \$565{,}750$ is too far from the hypothesized population mean $\mu = \$520{,}234$. The difference between the two may not be attributed to chance or sampling error alone. Therefore, the current mean sale price of homes in the said neighborhood of Stamford, Connecticut, is higher than \$520,234. ∎

Conducting a left-tailed test of hypothesis about μ for a large sample.

EXAMPLE 9-3 Because couples are deciding to have fewer children, the family size in the United States has decreased over the past few decades. According to the U.S. Bureau of the Census, the mean family size was 3.18 in 1998. A researcher wanted to check if the current mean family size is less than 3.18. A sample of 900 families taken this year by this researcher produced a mean family size of 3.16 with a standard deviation of .70. Using the .025 significance level, can we conclude that the mean family size has decreased since 1998?

Solution Let μ be the current mean size of all families and $\bar{x}$ the mean family size for the sample. From the given information,

$$n = 900, \quad \bar{x} = 3.16, \quad \text{and} \quad s = .70$$

The mean family size for 1998 is given to be 3.18. The significance level α is .025.

Step 1. *State the null and alternative hypotheses.*

Notice that we are testing for a decrease in the mean family size. The null and alternative hypotheses are written as follows.

$$H_0: \mu = 3.18 \qquad \text{(The mean family size has not decreased)}$$

$$H_1: \mu < 3.18 \qquad \text{(The mean family size has decreased)}$$

Step 2. *Select the distribution to use.*

Because the sample size is large ($n > 30$), the sampling distribution of $\bar{x}$ is (approximately) normal. Consequently, we use the normal distribution to make the test.

Step 3. *Determine the rejection and nonrejection regions.*

The significance level is .025. The $<$ sign in the alternative hypothesis indicates that the test is left-tailed with the rejection region in the left tail of the sampling distribution curve of $\bar{x}$. The critical value of z, obtained from the normal table for .4750, is -1.96, as shown in Figure 9.7.

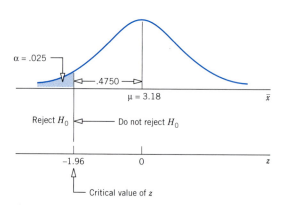

Figure 9.7

Step 4. *Calculate the value of the test statistic.*

The value of the test statistic z for $\bar{x} = 3.16$ is calculated as follows:

$$s_{\bar{x}} = \frac{s}{\sqrt{n}} = \frac{.70}{\sqrt{900}} = .02333333$$

From H_0

$$z = \frac{\bar{x} - \mu}{s_{\bar{x}}} = \frac{3.16 - 3.18}{.02333333} = -.86$$

Step 5. *Make a decision.*

The value of the test statistic $z = -.86$ is greater than the critical value of $z = -1.96$, and it falls in the nonrejection region. As a result, we fail to reject H_0. Consequently, we can state that based on the sample information, it appears that the mean family size has not decreased since 1998. Note that we are not concluding that the mean family size has definitely not decreased. By not rejecting the null hypothesis, we are saying that the information obtained from the sample is not strong enough to reject the null hypothesis and to conclude that the family size has decreased since 1998. ∎

In studies published in various journals, authors usually use the terms *significantly different* and *not significantly different* when deriving conclusions based on hypothesis tests. These terms are short versions of the terms *statistically significantly different* and *statistically not significantly different*. The expression *significantly different* means that the difference between the observed value of the sample mean $\bar{x}$ and the hypothesized value of the population mean μ is so large that it probably did not occur because of the sampling error alone. Consequently, the null hypothesis is rejected. In other words, the difference between $\bar{x}$ and μ is statistically significant. Thus, the statement *significantly different* is equivalent to saying that the *null hypothesis is rejected*. In Example 9–2, we can state as a conclusion that the observed value of $\bar{x} = \$565,750$ is significantly different from the hypothesized value of $\mu = \$520,234$. That is, the current mean sale price of neighborhood homes is significantly different from \$520,234.

On the other hand, the statement *not significantly different* means that the difference between the observed value of the sample mean $\bar{x}$ and the hypothesized value of the population mean μ is so small that it may have occurred just because of chance. Consequently, the null hypothesis is not rejected. Thus, the expression *not significantly different* is equivalent to saying that we *fail to reject the null hypothesis*. In Example 9–3, we can state as a conclusion that the observed value of $\bar{x} = 3.16$ is not significantly different from the hypothesized value of $\mu = 3.18$. In other words, the current mean family size does not seem to be significantly different from 3.18.

EXERCISES

■ **Concepts and Procedures**

9.11   What are the five steps of a test of hypothesis? Explain briefly.

9.12   What does the level of significance represent in a test of hypothesis? Explain.

9.13   By rejecting the null hypothesis in a test of hypothesis example, are you stating that the alternative hypothesis is true?

9.14   What is the difference between the critical value of z and the observed value of z?

9.15 For each of the following examples of tests of hypotheses about μ, show the rejection and non-rejection regions on the sampling distribution of the sample mean.

a. A two-tailed test with $\alpha = .05$ and $n = 40$
b. A left-tailed test with $\alpha = .01$ and $n = 67$
c. A right-tailed test with $\alpha = .02$ and $n = 55$

9.16 For each of the following examples of tests of hypotheses about μ, show the rejection and non-rejection regions on the sampling distribution of the sample mean.

a. A two-tailed test with $\alpha = .01$ and $n = 100$
b. A left-tailed test with $\alpha = .005$ and $n = 60$
c. A right-tailed test with $\alpha = .025$ and $n = 36$

9.17 Consider the following null and alternative hypotheses:

$$H_0: \mu = 25 \quad \text{versus} \quad H_1: \mu \neq 25$$

Suppose you perform this test at $\alpha = .05$ and reject the null hypothesis. Would you state that the difference between the hypothesized value of the population mean and the observed value of the sample mean is "statistically significant" or would you state that this difference is "statistically not significant"? Explain.

9.18 Consider the following null and alternative hypotheses:

$$H_0: \mu = 60 \quad \text{versus} \quad H_1: \mu > 60$$

Suppose you perform this test at $\alpha = .01$ and fail to reject the null hypothesis. Would you state that the difference between the hypothesized value of the population mean and the observed value of the sample mean is "statistically significant" or would you state that this difference is "statistically not significant"? Explain.

9.19 For each of the following significance levels, what is the probability of making a Type I error?

a. $\alpha = .025$ **b.** $\alpha = .05$ **c.** $\alpha = .01$

9.20 For each of the following significance levels, what is the probability of making a Type I error?

a. $\alpha = .10$ **b.** $\alpha = .02$ **c.** $\alpha = .005$

9.21 A random sample of 120 observations produced a sample mean of 32 and a standard deviation of 6. Find the critical and observed values of z for each of the following tests of hypotheses using $\alpha = .05$.

a. $H_0: \mu = 28$ versus $H_1: \mu > 28$
b. $H_0: \mu = 28$ versus $H_1: \mu \neq 28$

9.22 A random sample of 90 observations produced a sample mean of 15 and a standard deviation of 4. Find the critical and observed values of z for each of the following tests of hypotheses using $\alpha = .01$.

a. $H_0: \mu = 20$ versus $H_1: \mu < 20$
b. $H_0: \mu = 20$ versus $H_1: \mu \neq 20$

9.23 Consider the null hypothesis $H_0: \mu = 50$. Suppose a random sample of 120 observations is taken to perform this test. Using $\alpha = .05$, show the rejection and nonrejection regions on the sampling distribution curve of the sample mean and find the critical value(s) of z when the alternative hypothesis is

a. $H_1: \mu < 50$ **b.** $H_1: \mu \neq 50$ **c.** $H_1: \mu > 50$

9.24 Consider the null hypothesis $H_0: \mu = 35$. Suppose a random sample of 70 observations is taken to perform this test. Using $\alpha = .01$, show the rejection and nonrejection regions on the sampling distribution curve of the sample mean and find the critical value(s) of z for a

a. left-tailed test   **b.** two-tailed test   **c.** right-tailed test

9.25 Consider $H_0: \mu = 100$ versus $H_1: \mu \neq 100$.

a. A random sample of 64 observations produced a sample mean of 98 and a standard deviation of 12. Using $\alpha = .01$, would you reject the null hypothesis?

b. Another random sample of 64 observations taken from the same population produced a sample mean of 104 and a standard deviation of 10. Using $\alpha = .01$, would you reject the null hypothesis?

Comment on the results of parts a and b.

9.26 Consider $H_0: \mu = 45$ versus $H_1: \mu < 45$.

a. A random sample of 100 observations produced a sample mean of 43 and a standard deviation of 5. Using $\alpha = .025$, would you reject the null hypothesis?

b. Another random sample of 100 observations taken from the same population produced a sample mean of 43.8 and a standard deviation of 7. Using $\alpha = .025$, would you reject the null hypothesis?

Comment on the results of parts a and b.

9.27 Make the following tests of hypotheses.

a. $H_0: \mu = 25,$ $H_1: \mu \neq 25,$ $n = 81,$ $\bar{x} = 28.5,$ $s = 3,$ $\alpha = .01$
b. $H_0: \mu = 12,$ $H_1: \mu < 12,$ $n = 45,$ $\bar{x} = 11.25,$ $\sigma = 4.5,$ $\alpha = .05$
c. $H_0: \mu = 40,$ $H_1: \mu > 40,$ $n = 100,$ $\bar{x} = 47,$ $s = 7,$ $\alpha = .10$

9.28 Make the following test of hypotheses.

a. $H_0: \mu = 80,$ $H_1: \mu \neq 80,$ $n = 33,$ $\bar{x} = 76.5,$ $\sigma = 15,$ $\alpha = .10$
b. $H_0: \mu = 32,$ $H_1: \mu < 32,$ $n = 75,$ $\bar{x} = 26.5,$ $s = 7.4,$ $\alpha = .01$
c. $H_0: \mu = 55,$ $H_1: \mu > 55,$ $n = 40,$ $\bar{x} = 60.5,$ $s = 4,$ $\alpha = .05$

■ **Applications**

9.29 According to data from the U.S. Center for Health Statistics, Americans aged 18–24 made an annual average of 3.9 visits to physicians (*Statistical Abstract of the United States*, 1998). Last year a random sample of 350 Americans in this age group showed a mean of 3.7 visits per person with a standard deviation of 1.6 visits. Test at the 1% significance level whether the mean number of visits to physicians last year by Americans in this age group differed from 3.9.

9.30 According to data from the U.S. Health Care Financing Administration, the average annual expenditure on health care is $3645 per person in the United States (*Statistical Abstract of the United States*, 1998). A random sample of 200 persons showed that they spent an average of $3950 on health care last year with a standard deviation of $1450. Test at the 2% significance level whether last year's mean annual health expenditure per person for all people in the United States was greater than $3645.

9.31 According to 1999 data from Bruskin-Goldring for Goodyear/Gemini Automotive Care, the average age of the primary vehicles (which include cars, vans, pickups, and SUVs) owned by car owners in the United States was 5.6 years (*USA TODAY*, November 17, 1999). Assume that this result holds true for all primary vehicles in the United States in 1999. A recent random sample of 250 car owners from the United States showed that the average age of their primary vehicles is 6 years with a standard deviation of .75 year. Using the 2.5% significance level, can you conclude that the mean age of such cars has changed since the 1999 report?

9.32 A 1998 news article stated that the consumption of artificial sweeteners in the United States was equivalent to about 24 pounds of sugar per person per year (*Newsweek*, July 13, 1998). Suppose that a random sample of 150 Americans taken this year showed an annual average consumption of artificial sweeteners equivalent to 27 pounds of sugar with a standard deviation of 9.0 pounds. Does the sample support the alternative hypothesis that the current mean annual consumption of artificial sweeteners is more than 24 pounds of sugar? Use $\alpha = .01$. Explain your conclusion.

9.33 According to a 1997 survey of public participation in the arts, Americans who used personal computers at home during their free time averaged 5.2 hours per week on their computers (*American Demographics*, August 1998). A recent random sample of 50 Americans who use computers at home during their free time showed that the mean time these individuals spent on computers is 6.6 hours with a standard deviation of 2.3 hours.

a. At the 5% level of significance, can you conclude that the mean time spent on their computers during their free time at home by all such Americans currently exceeds 5.2 hours per week?

b. What is the Type I error in this case? Explain. What is the probability of making this error?

9.34 According to data from the College Board, the average cost per student for books and supplies for the 1997–1998 academic year was $634 at public four-year colleges in the United States (*The Chronicle of Higher Education*, August 1998). A recent random sample of 340 students at such colleges yielded a mean cost for books and supplies of $683 with a standard deviation of $254.

a. Testing at the 1% significance level, can you conclude that the mean of such costs currently differs from $634?

b. What is the Type I error in this case? Explain. What is the probability of making this error?

9.35 A study conducted a few years ago claims that adult men spend an average of 11 hours a week watching sports on television. A recent sample of 100 adult men showed that the mean time they spend per week watching sports on television is 9 hours with a standard deviation of 2.2 hours.

a. Test at the 1% significance level whether currently all adult men spend less than 11 hours per week watching sports on television.

b. What will your decision be in part a if the probability of making a Type I error is zero? Explain.

9.36 A restaurant franchise company has a policy of opening new restaurants only in those areas that have a mean household income of at least $35,000 per year. The company is currently considering an area in which to open a new restaurant. The company's research department took a sample of 150 households from this area and found that the mean income of these households is $33,400 per year with a standard deviation of $5400.

a. Using the 1% significance level, would you conclude that the company should not open a restaurant in this area?

b. What will your decision be in part a if the probability of making a Type I error is zero? Explain.

9.37 The manufacturer of a certain brand of auto batteries claims that the mean life of these batteries is 45 months. A consumer protection agency that wants to check this claim took a random sample of 36 such batteries and found that the mean life for this sample is 43.75 months with a standard deviation of 4 months.

a. Using the 2.5% significance level, would you conclude that the mean life of these batteries is less than 45 months?

b. Make the test of part a using a 5% significance level. Is your decision different from the one in part a? Comment on the results of parts a and b.

9.38 A study claims that all adults spend an average of 8 hours or more on chores during a weekend. A researcher wanted to check if this claim is true. A random sample of 200 adults taken by this researcher showed that these adults spend an average of 7.68 hours on chores during a weekend with a standard deviation of 2.1 hours.

a. Using the 1% significance level, can you conclude that the claim that all adults spend an average of 8 hours or more on chores during a weekend is false?

b. Make the test of part a using a 2.5% significance level. Is your decision different from the one in part a? Comment on the results of parts a and b.

9.39 Lazurus Steel Corporation produces iron rods that are supposed to be 36 inches long. The machine that makes these rods does not produce each rod exactly 36 inches long. The lengths of the rods vary slightly. It is known that when the machine is working properly, the mean length of the rods is 36 inches. The standard deviation of the lengths of all rods produced on this machine is always equal to .05 inch. The quality control department at the company takes a sample of 40 such rods each week, calculates the mean length of these rods, and tests the null hypothesis $\mu = 36$ inches against the alternative hypothesis $\mu \neq 36$ inches using a 1% significance level. If the null hypothesis is rejected, the machine is stopped and adjusted. A recent sample of 40 such rods produced a mean length of 36.015 inches. Based on this sample, would you conclude that the machine needs an adjustment?

9.40 At Farmer's Dairy, a machine is set to fill 32-ounce milk cartons. However, this machine does not put exactly 32 ounces of milk into each carton; the amount varies slightly from carton to carton.

It is known that when the machine is working properly, the mean net weight of these cartons is 32 ounces. The standard deviation of the milk in all such cartons is always equal to .15 ounce. The quality control inspector at this dairy takes a sample of 35 such cartons each week, calculates the mean net weight of these cartons, and tests the null hypothesis $\mu = 32$ ounces against the alternative hypothesis $\mu \neq 32$ ounces using a 2% significance level. If the null hypothesis is rejected, the machine is stopped and adjusted. A recent sample of 35 such cartons produced a mean net weight of 31.90 ounces. Based on this sample, would you conclude that the machine needs to be adjusted?

9.41 A company claims that the mean net weight of the contents of its All Taste cereal boxes is at least 18 ounces. Suppose you want to test whether or not the claim of the company is true. Explain briefly how you would conduct this test using a large sample.

9.42 A researcher claims that college students spend an average of 45 minutes per week on community service. You want to test if the mean time spent per week on community service by college students is different from 45 minutes. Explain briefly how you would conduct this test using a large sample.

9.3 HYPOTHESIS TESTS USING THE *p*-VALUE APPROACH

In the discussion of tests of hypotheses in Section 9.2, the value of the significance level α was selected before the test was performed. Sometimes we may prefer not to predetermine α. Instead, we may want to find a value such that a given null hypothesis will be rejected for any α greater than this value and it will not be rejected for any α less than this value. The **probability-value approach**, more commonly called the *p-value approach*, gives such a value. In this approach, we calculate the **p-value** for the test, which is defined as the smallest level of significance at which the given null hypothesis is rejected.

> **p-VALUE** The *p-value* is the smallest significance level at which the null hypothesis is rejected.

Using the *p*-value approach, we reject the null hypothesis if

$$p\text{-value} < \alpha$$

and we do not reject the null hypothesis if

$$p\text{-value} \geq \alpha$$

For a one-tailed test, the *p*-value is given by the area in the tail of the sampling distribution curve beyond the observed value of the sample statistic. Figure 9.8 shows the *p*-value for a right-tailed test about μ.

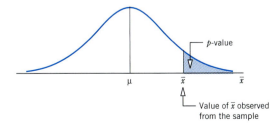

Figure 9.8 The *p*-value for a right-tailed test.

For a two-tailed test, the *p*-value is twice the area in the tail of the sampling distribution curve beyond the observed value of the sample statistic. Figure 9.9 shows the *p*-value for a two-tailed test. Each of the areas in the two tails gives one-half the *p*-value.

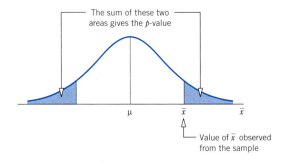

Value of $\bar{x}$ observed from the sample

Figure 9.9 The *p*-value for a two-tailed test.

Examples 9–4 and 9–5 illustrate the calculation and use of the *p*-value.

Calculating the p-value for a one-tailed test of hypothesis.

EXAMPLE 9-4 The management of Priority Health Club claims that its members lose an average of 10 pounds or more within the first month after joining the club. A consumer agency that wanted to check this claim took a random sample of 36 members of this health club and found that they lost an average of 9.2 pounds within the first month of membership with a standard deviation of 2.4 pounds. Find the *p*-value for this test.

Solution Let μ be the mean weight lost during the first month of membership by all members of this health club, and let $\bar{x}$ be the corresponding mean for the sample. From the given information,

$$n = 36, \quad \bar{x} = 9.2 \text{ pounds}, \quad \text{and} \quad s = 2.4 \text{ pounds}$$

The claim of the club is that its members lose, on average, 10 pounds or more within the first month of membership. To calculate the *p*-value, we apply the following three steps.

Step 1. *State the null and alternative hypotheses.*

$$H_0: \mu \geq 10 \quad \text{(The mean weight lost is 10 pounds or more)}$$

$$H_1: \mu < 10 \quad \text{(The mean weight lost is less than 10 pounds)}$$

Step 2. *Select the distribution to use.*

Because the sample size is large, we use the normal distribution to make the test and to calculate the *p*-value.

Step 3. *Calculate the p-value.*

The $<$ sign in the alternative hypothesis indicates that the test is left-tailed. The *p*-value is given by the area to the left of $\bar{x} = 9.2$ under the sampling distribution curve of $\bar{x}$, as shown in Figure 9.10. To find this area, we first find the *z* value for $\bar{x} = 9.2$ as follows:

$$s_{\bar{x}} = \frac{s}{\sqrt{n}} = \frac{2.4}{\sqrt{36}} = .40$$

$$z = \frac{\bar{x} - \mu}{s_{\bar{x}}} = \frac{9.2 - 10}{.40} = -2.00$$

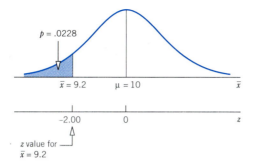

p = .0228

$\bar{x} = 9.2$ $\mu = 10$ $\bar{x}$

−2.00   0   *z*

z value for
$\bar{x} = 9.2$

Figure 9.10 The *p*-value
for a left-tailed test.

The area to the left of $\bar{x} = 9.2$ under the sampling distribution of $\bar{x}$ is equal to the area under the standard normal curve to the left of $z = -2.00$. From the normal distribution table, the area between the mean and $z = -2.00$ is .4772. Hence, the area to the left of $z = -2.00$ is $.5 - .4772 = .0228$. Consequently,

$$p\text{-value} = \textbf{.0228}$$

Thus, based on the *p*-value of .0228 we can state that for any α (significance level) greater than .0228 we will reject the null hypothesis stated in Step 1 and for any α less than .0228 we will not reject the null hypothesis. Suppose we make the test for this example at $\alpha = .01$. Because $\alpha = .01$ is less than the *p*-value of .0228, we will not reject the null hypothesis. Now, suppose we make the test at $\alpha = .05$. This time, because $\alpha = .05$ is greater than the *p*-value of .0228, we will reject the null hypothesis. ∎

The reader should make the test of hypothesis for Example 9–4 at $\alpha = .01$ and at $\alpha = .05$ by using the five steps learned in Section 9.2. The null hypothesis will not be rejected at $\alpha = .01$ (as .01 is less than $p = .0228$), and the null hypothesis will be rejected at $\alpha = .05$ (as .05 is greater than $p = .0228$).

Calculating the p-value for a two-tailed test of hypothesis.

EXAMPLE 9–5 At Canon Food Corporation, it used to take an average of 50 minutes for new workers to learn a food processing job. Recently the company installed a new food processing machine. The supervisor at the company wants to find if the mean time taken by new workers to learn the food processing procedure on this new machine is different from 50 minutes. A sample of 40 workers showed that it took, on average, 47 minutes for them to learn the food processing procedure on the new machine with a standard deviation of 7 minutes. Find the *p*-value for the test that the mean learning time for the food processing procedure on the new machine is different from 50 minutes.

Solution   Let μ be the mean time (in minutes) taken to learn the food processing procedure on the new machine by all workers, and let $\bar{x}$ be the corresponding sample mean. From the given information,

$$n = 40, \quad \bar{x} = 47 \text{ minutes}, \quad \text{and} \quad s = 7 \text{ minutes}$$

To calculate the *p*-value, we apply the following three steps.

Step 1.   *State the null and alternative hypotheses.*

$$H_0: \mu = 50 \text{ minutes}$$

$$H_1: \mu \neq 50 \text{ minutes}$$

Note that the null hypothesis states that the mean time for learning the food processing procedure on the new machine is 50 minutes, and the alternative hypothesis states that this time is different from 50 minutes.

Step 2.   *Select the distribution to use.*

Because the sample size is large, we use the normal distribution to make the test and to calculate the *p*-value.

Step 3.   *Calculate the p-value.*

The $\neq$ sign in the alternative hypothesis indicates that the test is two-tailed. The *p*-value is equal to twice the area in the tail of the sampling distribution curve of $\bar{x}$ to the left of $\bar{x} = 47$, as shown in Figure 9.11. To find this area, we first find the *z* value for $\bar{x} = 47$ as follows:

$$s_{\bar{x}} = \frac{s}{\sqrt{n}} = \frac{7}{\sqrt{40}} = 1.10679718 \text{ minutes}$$

$$z = \frac{\bar{x} - \mu}{s_{\bar{x}}} = \frac{47 - 50}{1.10679718} = -2.71$$

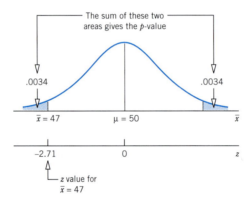

Figure 9.11   The *p*-value for a two-tailed test.

The area to the left of $\bar{x} = 47$ is equal to the area under the standard normal curve to the left of $z = -2.71$. From the normal distribution table, the area between the mean and $z = -2.71$ is .4966. Hence, the area to the left of $z = -2.71$ is

$$.5 - .4966 = .0034$$

Consequently, the *p*-value is

$$p\text{-value} = 2(.0034) = \textbf{.0068}$$

Thus, based on the *p*-value of .0068, we conclude that for any α (significance level) greater than .0068 we will reject the null hypothesis and for any α less than .0068 we will not reject the null hypothesis.   ■

EXERCISES

■ **Concepts and Procedures**

9.43   Briefly explain the procedure used to calculate the *p*-value for a two-tailed and for a one-tailed test, respectively.

9.44 Find the *p*-value for each of the following hypothesis tests.

a. $H_0: \mu = 23$, $\quad H_1: \mu \neq 23$, $\quad n = 50$, $\quad \bar{x} = 21.25$, $\quad s = 5$
b. $H_0: \mu = 15$, $\quad H_1: \mu < 15$, $\quad n = 80$, $\quad \bar{x} = 13.25$, $\quad s = 5.5$
c. $H_0: \mu = 38$, $\quad H_1: \mu > 38$, $\quad n = 35$, $\quad \bar{x} = 40.25$, $\quad s = 7.2$

9.45 Find the *p*-value for each of the following hypothesis tests.

a. $H_0: \mu = 46$, $\quad H_1: \mu \neq 46$, $\quad n = 40$, $\quad \bar{x} = 49.60$, $\quad s = 9.7$
b. $H_0: \mu = 26$, $\quad H_1: \mu < 26$, $\quad n = 33$, $\quad \bar{x} = 24.30$, $\quad s = 4.3$
c. $H_0: \mu = 18$, $\quad H_1: \mu > 18$, $\quad n = 55$, $\quad \bar{x} = 20.50$, $\quad s = 7.8$

9.46 Consider $H_0: \mu = 29$ versus $H_1: \mu \neq 29$. A random sample of 60 observations taken from this population produced a sample mean of 31.4 and a standard deviation of 8.

a. Calculate the *p*-value.
b. Considering the *p*-value of part a, would you reject the null hypothesis if the test were made at the significance level of .05?
c. Considering the *p*-value of part a, would you reject the null hypothesis if the test were made at the significance level of .01?

9.47 Consider $H_0: \mu = 72$ versus $H_1: \mu > 72$. A random sample of 36 observations taken from this population produced a sample mean of 74.07 and a standard deviation of 6.

a. Calculate the *p*-value.
b. Considering the *p*-value of part a, would you reject the null hypothesis if the test were made at the significance level of .01?
c. Considering the *p*-value of part a, would you reject the null hypothesis if the test were made at the significance level of .025?

■ **Applications**

9.48 According to a forecast made in 1999, the average daily rate for a hotel room in the United States in the year 2000 would be $85 (Smith Travel Research, *USA TODAY*, June 28, 1999). Suppose that a random sample of 81 hotel rooms in the year 2000 found a mean daily rate of $88 with a standard deviation of $12. Find the *p*-value for the test of hypothesis with the alternative hypothesis that the actual mean rate in the year 2000 differs from the forecast.

9.49 According to the Energy Information Administration of the U.S. Department of Energy, the average price of unleaded regular gasoline in the United States was 108.2 cents per gallon for the period January–June 1998 (*World Almanac and Book of Facts*, 1999). Suppose that a recent nationwide random sample of 200 gas stations found a mean price for unleaded regular gasoline of 106.5 cents per gallon with a standard deviation of 18.0 cents per gallon. Find the *p*-value for the hypothesis test with the alternative hypothesis that the mean price for unleaded gasoline currently differs from 108.2 cents per gallon.

9.50 The manufacturer of a certain brand of auto batteries claims that the mean life of these batteries is 45 months. A consumer protection agency that wants to check this claim took a random sample of 36 such batteries and found that the mean life for this sample is 43.75 months with a standard deviation of 4.5 months. Find the *p*-value for the test of hypothesis with the alternative hypothesis that the mean life of these batteries is less than 45 months.

9.51 A study claims that all adults spend an average of 14 hours or more on chores during a weekend. A researcher wanted to check if this claim is true. A random sample of 200 adults taken by this researcher showed that these adults spend an average of 13.75 hours on chores during a weekend with a standard deviation of 3.0 hours. Find the *p*-value for the hypothesis test with the alternative hypothesis that all adults spend less than 14 hours on chores during a weekend.

9.52 Data from the Consumer Expenditure Survey conducted by the U.S. Bureau of Labor Statistics showed that U.S. households spend an average of $3028 per year on car loan or lease payments (*Money Magazine,* July 1999). Suppose that a recent random sample of 300 households yielded a mean annual expenditure for car loan or lease payments of $3145 with a standard deviation of $920.

a. Find the *p*-value for the test of hypothesis with the alternative hypothesis that the current mean annual household expenditure on car loan or lease payments exceeds $3028.

b. If $\alpha = .01$, based on the p-value calculated in part a, would you reject the null hypothesis? Explain.
c. If $\alpha = .025$, based on the p-value calculated in part a, would you reject the null hypothesis? Explain.

9.53 A telephone company claims that the mean duration of all long-distance phone calls made by its residential customers is 10 minutes. A random sample of 100 long-distance calls made by its residential customers taken from the records of this company showed that the mean duration of calls for this sample is 9.25 minutes with a standard deviation of 3.75 minutes.

a. Find the p-value for the test that the mean duration of all long-distance calls made by residential customers is less than 10 minutes.
b. If $\alpha = .02$, based on the p-value calculated in part a, would you reject the null hypothesis? Explain.
c. If $\alpha = .05$, based on the p-value calculated in part a, would you reject the null hypothesis? Explain.

9.54 Lazurus Steel Corporation produces iron rods that are supposed to be 36 inches long. The machine that makes these rods does not produce each rod exactly 36 inches long. The lengths of the rods vary slightly. It is known that when the machine is working properly, the mean length of the rods is 36 inches. The standard deviation of the lengths of all rods produced on this machine is always equal to .05 inch. The quality control department at the company takes a sample of 40 such rods every week, calculates the mean length of these rods, and tests the null hypothesis $\mu = 36$ inches against the alternative hypothesis $\mu \neq 36$ inches. If the null hypothesis is rejected, the machine is stopped and adjusted. A recent such sample of 40 rods produced a mean length of 36.015 inches.

a. Calculate the p-value for this test of hypothesis.
b. Based on the p-value calculated in part a, will the quality control inspector decide to stop the machine and adjust it if he chooses the maximum probability of a Type I error to be .02? What if the maximum probability of a Type I error is .10?

9.55 At Farmer's Dairy, a machine is set to fill 32-ounce milk cartons. However, this machine does not put exactly 32 ounces of milk into each carton; the amount varies slightly from carton to carton. It is known that when the machine is working properly, the mean net weight of these cartons is 32 ounces. The standard deviation of the milk in all such cartons is always equal to .15 ounce. The quality control inspector at this company takes a sample of 35 such cartons every week, calculates the mean net weight of these cartons, and tests the null hypothesis $\mu = 32$ ounces against the alternative hypothesis $\mu \neq 32$ ounces. If the null hypothesis is rejected, the machine is stopped and adjusted. A recent sample of 35 such cartons produced a mean net weight of 31.90 ounces.

a. Calculate the p-value for this test of hypothesis.
b. Based on the p-value calculated in part a, will the quality control inspector decide to stop the machine and readjust it if she chooses the maximum probability of a Type I error to be .01? What if the maximum probability of a Type I error is .05?

9.4   HYPOTHESIS TESTS ABOUT A POPULATION MEAN: SMALL SAMPLES

Many times the size of a sample that is used to make a test of hypothesis about μ is small—that is, $n < 30$. This may be the case because we have limited resources and cannot afford to take a large sample or because of the nature of the experiment itself. For example, to test a new model of a car for fuel efficiency (miles per gallon), the company may prefer to use a small sample. All cars included in such a test must be sold as used cars. In the case of a small sample, if the population from which the sample is drawn is (approximately) normally distributed and the population standard deviation σ is known, we can still use the normal distribution to make a test of hypothesis about μ. However, if the population is (approximately) normally distributed, the population standard deviation σ is not known, and the sample size is small ($n < 30$), then the normal distribution is replaced by the t distribution to make a test of hypothesis about μ. In such a case the random variable

$$t = \frac{\bar{x} - \mu}{s_{\bar{x}}} \qquad \text{where } s_{\bar{x}} = \frac{s}{\sqrt{n}}$$

has a t distribution. The t is called the **test statistic** to make a hypothesis test about a population mean for small samples.

CONDITIONS UNDER WHICH THE t DISTRIBUTION IS USED TO MAKE TESTS OF HYPOTHESIS ABOUT μ The t distribution is used to conduct a *test of hypothesis about μ* if

1. The sample size is small ($n < 30$).
2. The population from which the sample is drawn is (approximately) normally distributed.
3. The population standard deviation σ is unknown.

The procedure that is used to make hypothesis tests about μ in the case of small samples is similar to the one for large samples. We perform the same five steps with the only difference being the use of the t distribution in place of the normal distribution.

TEST STATISTIC The value of the *test statistic t* for the sample mean $\bar{x}$ is computed as

$$t = \frac{\bar{x} - \mu}{s_{\bar{x}}} \qquad \text{where } s_{\bar{x}} = \frac{s}{\sqrt{n}}$$

The value of t calculated for $\bar{x}$ by using the above formula is also called the **observed value** of t.

Examples 9–6, 9–7, and 9–8 describe the procedure of testing hypotheses about the population mean using the t distribution.

Conducting a two-tailed test of hypothesis about μ: $n < 30$.

EXAMPLE 9–6 A psychologist claims that the mean age at which children start walking is 12.5 months. Carol wanted to check if this claim is true. She took a random sample of 18 children and found that the mean age at which these children started walking was 12.9 months with a standard deviation of .80 month. Using the 1% significance level, can you conclude that the mean age at which all children start walking is different from 12.5 months? Assume that the ages at which all children start walking have an approximately normal distribution.

Solution Let μ be the mean age at which all children start walking and $\bar{x}$ the corresponding mean for the sample. Then, from the given information,

$$n = 18, \quad \bar{x} = 12.9 \text{ months}, \quad s = .80 \text{ month}, \quad \text{and} \quad \alpha = .01$$

Step 1. *State the null and alternative hypotheses.*

We are to test if the mean age at which all children start walking is different from 12.5 months. The null and alternative hypotheses are

$$H_0: \mu = 12.5 \qquad \text{(The mean walking age is 12.5 months)}$$

$$H_1: \mu \neq 12.5 \qquad \text{(The mean walking age is different from 12.5 months)}$$

Step 2. *Select the distribution to use.*

The sample size is small and the population is approximately normally distributed. However, we do not know the population standard deviation σ. Hence, we use the t distribution to make the test.

Step 3.   *Determine the rejection and nonrejection regions.*

The significance level is .01. The $\neq$ sign in the alternative hypothesis indicates that the test is two-tailed and the rejection region lies in both tails. The area of the rejection region in each tail of the *t* distribution curve is

$$\text{Area in each tail} = \alpha/2 = .01/2 = .005$$

$$df = n - 1 = 18 - 1 = 17$$

From the *t* distribution table, the critical values of *t* for 17 degrees of freedom and .005 area in each tail of the *t* distribution curve are -2.898 and 2.898. These values are shown in Figure 9.12.

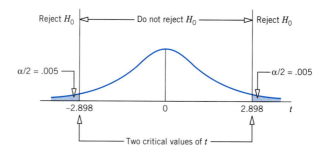

Reject H_0 ◁─────── Do not reject H_0 ──────▷ Reject H_0

$\alpha/2 = .005$ $\alpha/2 = .005$

-2.898 0 2.898 t

Two critical values of t

Figure 9.12

Step 4.   *Calculate the value of the test statistic.*

We calculate the value of the test statistic *t* for $\bar{x} = 12.9$ as follows:

$$s_{\bar{x}} = \frac{s}{\sqrt{n}} = \frac{.80}{\sqrt{18}} = .18856181$$

From H_0

$$t = \frac{\bar{x} - \mu}{s_{\bar{x}}} = \frac{12.9 - 12.5}{.18856181} = 2.121$$

Step 5.   **Make a decision.**

The value of the test statistic $t = 2.121$ falls between the two critical points, -2.898 and 2.898, which is the nonrejection region. Consequently, we fail to reject H_0. As a result, we can state that the difference between the hypothesized population mean and the sample mean is so small that it may have occurred because of sampling error. The mean age at which children start walking is not different from 12.5 months. ■

USING THE *p*-VALUE APPROACH IN EXAMPLE 9–6

Note that using the procedure discussed in Section 9.3, we can use the *p*-value approach to make a decision in all problems relating to hypothesis testing in this and succeeding chapters. For instance, we can find the *p*-value for $\bar{x} = 12.9$ in Example 9–6 and compare it with the given significance level to make a decision. As shown in Step 4 of Example 9–6, the *t* value for $\bar{x} = 12.9$ is 2.121. From the *t* distribution table, for $df = 17$ and $t = 2.121$, the *p*-value for a two-tailed test is approximately .05. (Note that this *p*-value for $df = 17$ and $t = 2.110$ is $2 \times .025 = .05$. However, for the same degrees of freedom but $t = 2.121$, this *p*-value will be slightly less than .05.) Since $\alpha = .01$ is less than the *p*-value of .05, we fail to reject the null hypothesis.

The same procedure can be used to obtain the *p*-value for the test of hypothesis problems in Section 9.5 and in succeeding chapters. To do so, we find the value of the test statistic for the given value of the sample statistic obtained from the sample and then obtain the *p*-value from the corresponding probability distribution table for that value of the test statistic. Finally, we compare that *p*-value with the significance level and make a decision.

All the statistical software packages, including MINITAB, give the *p*-value in the solution to a test of hypothesis problem. Thus, if you are using a statistical software package to solve a test of hypothesis problem, you can compare the *p*-value given in the computer solution to the significance level and make a decision.

Conducting a left-tailed test of hypothesis about μ: n < 30.

EXAMPLE 9-7 Grand Auto Corporation produces auto batteries. The company claims that its top-of-the-line Never Die batteries are good, on average, for at least 65 months. A consumer protection agency tested 15 such batteries to check this claim. It found the mean life of these 15 batteries to be 63 months with a standard deviation of 2 months. At the 5% significance level, can you conclude that the claim of the company is true? Assume that the life of such a battery has an approximately normal distribution.

Solution Let μ be the mean life of all Never Die batteries and $\bar{x}$ the corresponding mean for the sample. Then, from the given information,

$$n = 15, \quad \bar{x} = 63 \text{ months}, \quad \text{and} \quad s = 2 \text{ months}$$

The significance level is $\alpha = .05$. The company's claim is that the mean life of these batteries is at least 65 months.

Step 1. *State the null and alternative hypotheses.*

We are to test whether or not the mean life of Never Die batteries is at least 65 months. The null and alternative hypotheses are as follows:

$$H_0: \mu \geq 65 \qquad \text{(The mean life is at least 65 months)}$$

$$H_1: \mu < 65 \qquad \text{(The mean life is less than 65 months)}$$

Step 2. *Select the distribution to use.*

The sample size is small ($n < 30$) and the life of a battery is approximately normally distributed. However, the population standard deviation is not known. Hence, we use the *t* distribution to make the test.

Step 3. *Determine the rejection and nonrejection regions.*

The significance level is .05. The < sign in the alternative hypothesis indicates that the test is left-tailed with the rejection region in the left tail of the *t* distribution curve. To find the critical value of *t*, we need to know the area in the left tail and the degrees of freedom.

$$\text{Area in the left tail} = \alpha = .05$$

$$df = n - 1 = 15 - 1 = 14$$

From the *t* distribution table, the critical value of *t* for 14 degrees of freedom and an area of .05 in the left tail is −1.761. This value is shown in Figure 9.13.

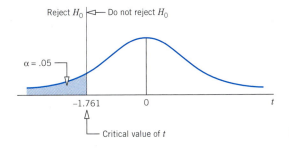

Figure 9.13

Step 4. *Calculate the value of the test statistic.*

The value of the test statistic t for $\bar{x} = 63$ is calculated as follows:

$$s_{\bar{x}} = \frac{s}{\sqrt{n}} = \frac{2}{\sqrt{15}} = .51639778$$

$$t = \frac{\bar{x} - \mu}{s_{\bar{x}}} = \frac{63 - 65}{.51639778} = -3.873$$

From H_0

Step 5. *Make a decision.*

The value of the test statistic $t = -3.873$ is less than the critical value of $t = -1.761$, and it falls in the rejection region. Therefore, we reject H_0 and conclude that the sample mean is too small compared to 65 (company's claimed value of μ) and the difference between the two may not be attributed to chance alone. We can conclude that the mean life of the company's Never Die batteries is less than 65 months. ■

Conducting a right-tailed test of hypothesis about μ: $n < 30$.

EXAMPLE 9–8   The management at Massachusetts Savings Bank is always concerned about the quality of service provided to its customers. With the old computer system, a teller at this bank could serve, on average, 22 customers per hour. The management noticed that with this service rate, the waiting time for customers was too long. Recently the management of the bank installed a new computer system in the bank, expecting that it would increase the service rate and consequently make the customers happier by reducing the waiting time. To check if the new computer system is more efficient than the old system, the management of the bank took a random sample of 18 hours and found that during these hours the mean number of customers served by tellers was 28 per hour with a standard deviation of 2.5. Testing at the 1% significance level, would you conclude that the new computer system is more efficient than the old computer system? Assume that the number of customers served per hour by a teller on this computer system has an approximately normal distribution.

Solution   Let μ be the mean number of customers served per hour by a teller using the new system, and let $\bar{x}$ be the corresponding mean for the sample. Then, from the given information,

$$n = 18 \text{ hours}, \quad \bar{x} = 28 \text{ customers}, \quad s = 2.5 \text{ customers}, \quad \text{and} \quad \alpha = .01$$

Step 1. *State the null and alternative hypotheses.*

We are to test whether or not the new computer system is more efficient than the old system. The new computer system will be more efficient than the old system if the mean number of customers served per hour by using the new computer system is significantly more than 22; otherwise, it will not be more efficient. The null and alternative hypotheses are

$$H_0: \mu = 22 \quad \text{(The new computer system is not more efficient)}$$
$$H_1: \mu > 22 \quad \text{(The new computer system is more efficient)}$$

Step 2. *Select the distribution to use.*

The sample size is small and the population is approximately normally distributed. However, we do not know the population standard deviation σ. Hence, we use the t distribution to make the test.

Step 3. *Determine the rejection and nonrejection regions.*

The significance level is .01. The $>$ sign in the alternative hypothesis indicates that the test is right-tailed and the rejection region lies in the right tail of the t distribution curve.

$$\text{Area in the right tail} = \alpha = .01$$

$$df = n - 1 = 18 - 1 = 17$$

From the t distribution table, the critical value of t for 17 degrees of freedom and .01 area in the right tail is 2.567. This value is shown in Figure 9.14.

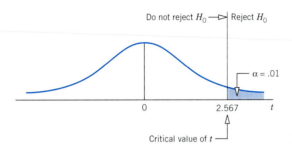

Figure 9.14

Step 4. *Calculate the value of the test statistic.*

The value of the test statistic t for $\bar{x} = 28$ is calculated as follows:

$$s_{\bar{x}} = \frac{s}{\sqrt{n}} = \frac{2.5}{\sqrt{18}} = .58925565$$

$$t = \frac{\bar{x} - \mu}{s_{\bar{x}}} = \frac{28 - 22}{.58925565} = 10.182 \qquad \text{From } H_0$$

Step 5. *Make a decision.*

The value of the test statistic $t = 10.182$ is greater than the critical value of $t = 2.567$, and it falls in the rejection region. Consequently, we reject H_0. As a result, we conclude that the value of the sample mean is too large compared to the hypothesized value of the population mean, and the difference between the two may not be attributed to chance alone. The mean number of customers served per hour using the new computer system is more than 22. The new computer system is more efficient than the old computer system. ∎

EXERCISES

■ Concepts and Procedures

9.56 Briefly explain the conditions that must hold true to use the t distribution to make a test of hypothesis about the population mean.

9.57 For each of the following examples of tests of hypotheses about μ, show the rejection and non-rejection regions on the t distribution curve.

a. A two-tailed test with $\alpha = .02$ and $n = 20$
b. A left-tailed test with $\alpha = .01$ and $n = 16$
c. A right-tailed test with $\alpha = .05$ and $n = 18$

9.58   For each of the following examples of tests of hypotheses about μ, show the rejection and non-rejection regions on the t distribution curve.

a.   A two-tailed test with $\alpha = .01$ and $n = 15$
b.   A left-tailed test with $\alpha = .005$ and $n = 25$
c.   A right-tailed test with $\alpha = .025$ and $n = 22$

9.59   A random sample of 25 observations taken from a population that is normally distributed produced a sample mean of 58.5 and a standard deviation of 7.5. Find the critical and observed values of t for each of the following tests of hypotheses using $\alpha = .01$.

a. H_0: $\mu = 55$ versus H_1: $\mu > 55$
b. H_0: $\mu = 55$ versus H_1: $\mu \neq 55$

9.60   A random sample of 16 observations taken from a population that is normally distributed produced a sample mean of 42.4 and a standard deviation of 8. Find the critical and observed values of t for each of the following tests of hypotheses using $\alpha = .05$.

a. H_0: $\mu = 46$ versus H_1: $\mu < 46$
b. H_0: $\mu = 46$ versus H_1: $\mu \neq 46$

9.61   Consider the null hypothesis H_0: $\mu = 70$ about the mean of a population that is normally distributed. Suppose a random sample of 20 observations is taken from this population to make this test. Using $\alpha = .01$, show the rejection and nonrejection regions and find the critical value(s) of t for a

a.   left-tailed test   **b.**   two-tailed test   **c.**   right-tailed test

9.62   Consider the null hypothesis H_0: $\mu = 35$ about the mean of a population that is normally distributed. Suppose a random sample of 22 observations is taken from this population to make this test. Using $\alpha = .05$, show the rejection and nonrejection regions and find the critical value(s) of t for a

a.   left-tailed test   **b.**   two-tailed test   **c.**   right-tailed test

9.63   Consider H_0: $\mu = 80$ versus H_1: $\mu \neq 80$ for a population that is normally distributed.

a.   A random sample of 25 observations taken from this population produced a sample mean of 77 and a standard deviation of 8. Using $\alpha = .01$, would you reject the null hypothesis?
b.   Another random sample of 25 observations taken from the same population produced a sample mean of 86 and a standard deviation of 6. Using $\alpha = .01$, would you reject the null hypothesis?

Comment on the results of parts a and b.

9.64   Consider H_0: $\mu = 40$ versus H_1: $\mu > 40$ for a population that is normally distributed.

a.   A random sample of 16 observations taken from this population produced a sample mean of 45 and a standard deviation of 5. Using $\alpha = .025$, would you reject the null hypothesis?
b.   Another random sample of 16 observations taken from the same population produced a sample mean of 41.9 and a standard deviation of 7. Using $\alpha = .025$, would you reject the null hypothesis?

Comment on the results of parts a and b.

9.65   Assuming that the respective populations are normally distributed, perform the following hypothesis tests.

a. H_0: $\mu = 24$, H_1: $\mu \neq 24$, $n = 25$, $\bar{x} = 28.5$, $s = 4.9$, $\alpha = .01$
b. H_0: $\mu = 30$, H_1: $\mu < 30$, $n = 16$, $\bar{x} = 27.5$, $s = 6.6$, $\alpha = .025$
c. H_0: $\mu = 18$, H_1: $\mu > 18$, $n = 20$, $\bar{x} = 22.5$, $s = 8$, $\alpha = .10$

9.66   Assuming that the respective populations are normally distributed, perform the following hypothesis tests.

a. H_0: $\mu = 60$, H_1: $\mu \neq 60$, $n = 14$, $\bar{x} = 57$, $s = 9$, $\alpha = .05$
b. H_0: $\mu = 35$, H_1: $\mu < 35$, $n = 24$, $\bar{x} = 28$, $s = 5.4$, $\alpha = .005$
c. H_0: $\mu = 47$, H_1: $\mu > 47$, $n = 18$, $\bar{x} = 50$, $s = 6$, $\alpha = .001$

■ Applications

9.67   According to a basketball coach, the mean height of all female college basketball players is 69.5 inches. A random sample of 25 such players produced a mean height of 70.25 inches with a standard deviation of 2.1 inches. Assuming that the heights of all female college basketball players are normally distributed, test at the 1% significance level whether their mean height is different from 69.5 inches.

9.68 According to the College Board, in 1999 the average SAT score in mathematics for students from Rhode Island was 499. Suppose that a recent random sample of 25 mathematics SAT scores from Rhode Island had a mean of 504 with a standard deviation of 105. Using the 5% significance level, can you conclude that the current mean SAT score in mathematics for students from Rhode Island exceeds 499? Assume that such SAT scores for all students from Rhode Island are normally distributed.

9.69 The president of a university claims that the mean time spent partying by all students at this university is not more than 7 hours per week. A random sample of 20 students taken from this university showed that they spent an average of 10.50 hours partying the previous week with a standard deviation of 2.3 hours. Assuming that the time spent partying by all students at this university is approximately normally distributed, test at the 2.5% significance level whether the president's claim is true. Explain your conclusion in words.

9.70 The mean balance of all checking accounts at a bank on December 31, 1999, was $850. A random sample of 25 checking accounts taken recently from this bank gave a mean balance of $780 with a standard deviation of $230. Assume that the balances of all checking accounts at this bank are normally distributed. Using the 1% significance level, can you conclude that the mean balance of such accounts has decreased during this period? Explain your conclusion in words.

9.71 A soft-drink manufacturer claims that its 12-ounce cans do not contain, on average, more than 30 calories. A random sample of 16 cans of this soft drink, which were checked for calories, contained a mean of 32 calories with a standard deviation of 3 calories. Assume that the number of calories in 12-ounce soda cans is normally distributed. Does the sample information support the alternative hypothesis that the manufacturer's claim is false? Use a significance level of 5%. Explain your conclusion in words.

9.72 According to data from the National Public Transportation Survey, commuters who use public transportation in metropolitan areas with populations of 3 million or more average 43 minutes of travel time (*USA TODAY*, April 14, 1999). A recent random sample of 20 such commuters showed a mean travel time of 52 minutes with a standard deviation of 14 minutes. Using the 2% significance level, can you conclude that the mean travel time for all such commuters currently differs from 43 minutes? Assume that the travel times for all such commuters are normally distributed.

9.73 A paint manufacturing company claims that the mean drying time for its paints is not longer than 45 minutes. A random sample of 20 gallons of paints selected from the production line of this company showed that the mean drying time for this sample is 49.50 minutes with a standard deviation of 3 minutes. Assume that the drying times for these paints have a normal distribution.

a. Using the 1% significance level, would you conclude that the company's claim is true?
b. What is the Type I error in this exercise? Explain in words. What is the probability of making such an error?

9.74 A 1999 poll of 1014 adults by the National Sleep Foundation reported that these adults have an average of (about) 7 hours sleep per night during the workweek (*USA TODAY,* March 25, 1999). A recent random sample of 18 adults showed an average of 6.75 hours of sleep per night during the workweek with a standard deviation of .60 hour. Assume that such sleep times for all adults are normally distributed.

a. Using the 2.5% significance level, can you conclude that the mean sleep duration for all adults during the workweek is currently less than 7 hours?
b. What is the Type I error in this hypothesis test? Explain. What is the probability of making such an error?

9.75 A business school claims that students who complete a three-month typing course can type, on average, at least 1200 words an hour. A random sample of 25 students who completed this course typed, on average, 1125 words an hour with a standard deviation of 85 words. Assume that the typing speeds for all students who complete this course have an approximately normal distribution.

a. Suppose the probability of making a Type I error is selected to be zero. Can you conclude that the claim of the business school is true? Answer without performing the five steps of a test of hypothesis.
b. Using the 5% significance level, can you conclude that the claim of the business school is true?

9.76 According to data from the House Ways and Means Committee, the average time required for taxpayers to prepare IRS Schedule A (itemized deductions) was (about) 4.5 hours (*USA TODAY*, April 20, 1999). A taxpayers' organization claims that the time required to prepare this form averages more than 4.5 hours. Suppose that a random sample of 26 taxpayers shows a mean preparation time of 4.75 hours with a standard deviation of .75 hour. Assume that all such preparation times are normally distributed.

a. Suppose that the probability of the Type I error is selected to be zero. Can you conclude that the claim of the taxpayers' organization is true? Answer without performing the five steps of a test of hypothesis.

b. Using the 5% level of significance, can you conclude that the claim of the taxpayers' organization is true?

9.77 A past study claims that adults in America spend an average of 18 hours a week on leisure activities. A researcher wanted to test this claim. She took a sample of 10 adults and asked them about the time they spend per week on leisure activities. Their responses (in hours) are as follows.

$$14 \quad 25 \quad 22 \quad 38 \quad 16 \quad 26 \quad 19 \quad 23 \quad 41 \quad 33$$

Assume that the times spent on leisure activities by all adults are normally distributed. Using the 5% significance level, can you conclude that the claim of the earlier study is true? (*Hint*: First calculate the sample mean and the sample standard deviation for these data using the formulas learned in Sections 3.1.1 and 3.2.2 of Chapter 3. Then make the test of hypothesis about μ.)

9.78 The past records of a supermarket show that its customers spend an average of $65 per visit at this store. Recently the management of the store initiated a promotional campaign according to which each customer receives points based on the total money spent at the store and these points can be used to buy products at the store. The management expects that as a result of this campaign, the customers should be encouraged to spend more money at the store. To check whether this is true, the manager of the store took a sample of 12 customers who visited the store. The following data give the money (in dollars) spent by these customers at this supermarket during their visits.

$$88 \quad 69 \quad 141 \quad 28 \quad 106 \quad 45 \quad 32 \quad 51 \quad 78 \quad 54 \quad 110 \quad 83$$

Assume that the money spent by all customers at this supermarket has a normal distribution. Using the 1% significance level, can you conclude that the mean amount of money spent by all customers at this supermarket after the campaign was started is more than $65? (*Hint*: First calculate the sample mean and the sample standard deviation for these data using the formulas learned in Sections 3.1.1 and 3.2.2 of Chapter 3. Then make the test of hypothesis about μ.)

*9.79 The manager of a service station claims that the mean amount spent on gas by its customers is $10.90. You want to test if the mean amount spent on gas at this station is different from $10.90. Briefly explain how you would conduct this test by taking a small sample.

*9.80 A tool manufacturing company claims that its top-of-the-line machine that is used to manufacture bolts produces an average of 88 or more bolts per hour. A company that is interested in buying this machine wants to check this claim. Suppose you are asked to conduct this test. Briefly explain how you would do so by taking a small sample.

9.5 HYPOTHESIS TESTS ABOUT A POPULATION PROPORTION: LARGE SAMPLES

Often we want to conduct a test of hypothesis about a population proportion. For example, 33% of the students listed in *Who's Who Among American High School Students* said that drugs and alcohol are the most serious problems facing their high schools. A sociologist may want to check if this percentage still holds. As another example, a mail-order company claims that 90% of all orders it receives are shipped within 72 hours. The company's management may want to determine from time to time whether or not this claim is true.

This section presents the procedure to perform tests of hypotheses about the population proportion, p, for large samples. The procedure to make such tests is similar in many respects to the one for the population mean, μ. The procedure includes the same five steps. Again, the test can be two-tailed or one-tailed. We know from Chapter 7 that when the sample size is large, the sample proportion, $\hat{p}$, is approximately normally distributed with its mean equal to p and standard deviation equal to $\sqrt{pq/n}$. Hence, we use the normal distribution to perform a test of hypothesis about the population proportion, p, for a large sample. As was mentioned in Chapters 7 and 8, in the case of a proportion, the sample size is considered to be large when np and nq are both greater than 5.

TEST STATISTIC   The value of the *test statistic z* for the sample proportion, $\hat{p}$, is computed as

$$z = \frac{\hat{p} - p}{\sigma_{\hat{p}}} \qquad \text{where } \sigma_{\hat{p}} = \sqrt{\frac{pq}{n}}$$

The value of p used in this formula is the one used in the null hypothesis. The value of q is equal to $1 - p$.

The value of z calculated for $\hat{p}$ using the above formula is also called the **observed value of z**.

Examples 9–9, 9–10, and 9–11 describe the procedure to make tests of hypotheses about the population proportion, p.

Conducting a two-tailed test of hypothesis about p: large sample.

EXAMPLE 9-9   According to Hewitt Associates, 36% of the companies surveyed in 1999 paid holiday bonuses to their employees (*The Wall Street Journal*, December 22, 1999). Assume that this result holds true for all U.S. companies in 1999. A recent random sample of 400 companies showed that 33% of them pay holiday bonuses to their employees. Using the 1% significance level, can you conclude that the current percentage of companies that pay holiday bonuses to their employees is different from that for 1999?

Solution   Let p be the proportion of all U.S. companies that currently pay holiday bonuses to their employees, and let $\hat{p}$ be the corresponding sample proportion. Then, from the given information,

$$n = 400, \quad \hat{p} = .33, \quad \text{and} \quad \alpha = .01$$

In 1999, 36% of the companies paid holiday bonuses to their employees. Hence,

$$p = .36 \quad \text{and} \quad q = 1 - p = 1 - .36 = .64$$

Step 1.   *State the null and alternative hypotheses.*

The percentage of all U.S. companies that currently pay holiday bonuses to their employees is not different from that for 1999 if $p = .36$, and the current percentage is different from 1999 if $p \neq .36$. The null and alternative hypotheses are as follows:

H_0: $p = .36$ (The current percentage is not different from 1999)

H_1: $p \neq .36$ (The current percentage is different from 1999)

Step 2. *Select the distribution to use.*

The values of *np* and *nq* are

$$np = 400(.36) = 144 \quad \text{and} \quad nq = 400(.64) = 256$$

Because both *np* and *nq* are greater than 5, the sample size is large. Consequently, we use the normal distribution to make the hypothesis test about *p*.

Step 3. *Determine the rejection and nonrejection regions.*

The ≠ sign in the alternative hypothesis indicates that the test is two-tailed. The significance level is .01. Therefore, the total area of the two rejection regions is .01 and the rejection region in each tail of the sampling distribution of $\hat{p}$ is $\alpha/2 = .01/2 = .005$. The critical values of *z*, obtained from the standard normal distribution table, are −2.58 and 2.58, as shown in Figure 9.15.

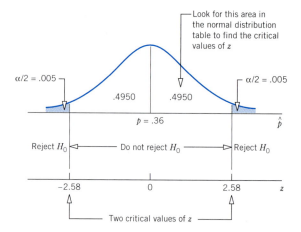

Figure 9.15

Step 4. *Calculate the value of the test statistic.*

The value of the test statistic *z* for $\hat{p} = .33$ is calculated as follows:

$$\sigma_{\hat{p}} = \sqrt{\frac{pq}{n}} = \sqrt{\frac{(.36)(.64)}{400}} = .024$$

$$z = \frac{\hat{p} - p}{\sigma_{\hat{p}}} = \frac{.33 - .36}{.024} = -1.25$$

From H_0

Step 5. *Make a decision.*

The value of the test statistic $z = -1.25$ for $\hat{p}$ lies in the nonrejection region. Consequently, we fail to reject H_0. Therefore, we can state that the sample proportion is not too far from the hypothesized value of the population proportion and the difference between the two can be attributed to chance. We conclude that the percentage of U.S. companies that currently pay holiday bonuses to their employees is not different from that for 1999. ∎

Conducting a right-tailed test of hypothesis about p: large sample.

EXAMPLE 9–10 When working properly, a machine that is used to make chips for calculators does not produce more than 4% defective chips. Whenever the machine produces more than 4% defective chips, it needs an adjustment. To check if the machine is working properly, the quality control department at the company often takes samples of chips and inspects them to determine if they are good or defective. One such random sample of 200 chips taken recently from the production line contained 14 defective chips. Test at the 5% significance level whether or not the machine needs an adjustment.

Solution Let p be the proportion of defective chips in all chips produced by this machine, and let $\hat{p}$ be the corresponding sample proportion. Then, from the given information,

$$n = 200, \quad \hat{p} = 14/200 = .07, \quad \text{and} \quad \alpha = .05$$

When the machine is working properly it does not produce more than 4% defective chips. Consequently, assuming that the machine is working properly,

$$p = .04 \quad \text{and} \quad q = 1 - p = 1 - .04 = .96$$

Step 1. *State the null and alternative hypotheses.*

The machine will not need an adjustment if the percentage of defective chips is 4% or less, and it will need an adjustment if this percentage is greater than 4%. Hence, the null and alternative hypotheses are

$$H_0: p \le .04 \quad \text{(The machine does not need an adjustment)}$$

$$H_1: p > .04 \quad \text{(The machine needs an adjustment)}$$

Step 2. *Select the distribution to use.*

The values of np and nq are

$$np = 200(.04) = 8 > 5 \quad \text{and} \quad nq = 200(.96) = 192 > 5$$

Because the sample size is large, we use the normal distribution to make the hypothesis test about p.

Step 3. *Determine the rejection and nonrejection regions.*

The significance level is .05. The $>$ sign in the alternative hypothesis indicates that the test is right-tailed and the rejection region lies in the right tail of the sampling distribution of $\hat{p}$ with its area equal to .05. As shown in Figure 9.16, the critical value of z, obtained from the normal distribution table for .4500, is approximately 1.65.

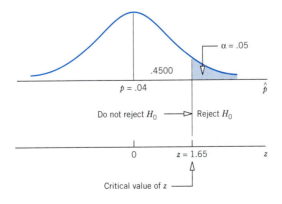

Figure 9.16

Step 4.   *Calculate the value of the test statistic.*

The value of the test statistic z for $\hat{p} = .07$ is calculated as follows:

$$\sigma_{\hat{p}} = \sqrt{\frac{pq}{n}} = \sqrt{\frac{(.04)(.96)}{200}} = .01385641$$

From H_0

$$z = \frac{\hat{p} - p}{\sigma_{\hat{p}}} = \frac{.07 - .04}{.01385641} = 2.17$$

Step 5.   *Make a decision.*

Because the value of the test statistic $z = 2.17$ is greater than the critical value of $z = 1.65$ and it falls in the rejection region, we reject H_0. We conclude that the sample proportion is too far from the hypothesized value of the population proportion and the difference between the two cannot be attributed to chance alone. Therefore, based on the sample information, we conclude that the machine needs an adjustment.   ■

Conducting a left-tailed test of hypothesis about p: large sample.

EXAMPLE 9–11   Direct Mailing Company sells computers and computer parts by mail. The company claims that at least 90% of all orders are mailed within 72 hours after they are received. The quality control department at the company often takes samples to check if this claim is valid. A recently taken sample of 150 orders showed that 129 of them were mailed within 72 hours. Do you think the company's claim is true? Use a 2.5% significance level.

Solution   Let p be the proportion of all orders that are mailed by the company within 72 hours and $\hat{p}$ the corresponding sample proportion. Then, from the given information,

$$n = 150, \quad \hat{p} = 129/150 = .86, \quad \text{and} \quad \alpha = .025$$

The company claims that at least 90% of all orders are mailed within 72 hours. Assuming that this claim is true, the values of p and q are

$$p = .90 \quad \text{and} \quad q = 1 - p = 1 - .90 = .10$$

Step 1.   *State the null and alternative hypotheses.*

The null and alternative hypotheses are

$$H_0: p \geq .90 \quad \text{(The company's claim is true)}$$

$$H_1: p < .90 \quad \text{(The company's claim is false)}$$

Step 2.   *Select the distribution to use.*

We first check whether both np and nq are greater than 5.

$$np = 150(.90) = 135 > 5 \quad \text{and} \quad nq = 150(.10) = 15 > 5$$

Consequently, the sample size is large. Therefore, we use the normal distribution to make the hypothesis test about p.

Step 3.   *Determine the rejection and nonrejection regions.*

The significance level is .025. The $<$ sign in the alternative hypothesis indicates that the test is left-tailed and the rejection region lies in the left tail of the sampling distribution of $\hat{p}$ with its area equal to .025. As shown in Figure 9.17, the critical value of z, obtained from the normal distribution table for .4750, is -1.96.

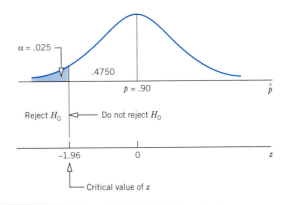

Figure 9.17

Step 4. *Calculate the value of the test statistic.*

The value of the test statistic z for $\hat{p} = .86$ is calculated as follows:

$$\sigma_{\hat{p}} = \sqrt{\frac{pq}{n}} = \sqrt{\frac{(.90)(.10)}{150}} = .02449490$$

From H_0

$$z = \frac{\hat{p} - p}{\sigma_{\hat{p}}} = \frac{.86 - .90}{.02449490} = -1.63$$

Step 5. *Make a decision.*

The value of the test statistic $z = -1.63$ is greater than the critical value of $z = -1.96$, and it falls in the nonrejection region. Therefore, we fail to reject H_0. We can state that the difference between the sample proportion and the hypothesized value of the population proportion is small and this difference may have occurred owing to chance alone. Therefore, the proportion of all orders that are mailed within 72 hours is at least 90%, and the company's claim is true. ∎

We can also use the *p*-value approach to make tests of hypotheses about the population proportion *p*. The procedure to calculate the *p*-value for the sample proportion is similar to the one applied to the sample mean in Section 9.3.

CASE STUDY 9-1   OLDER WORKERS MOST CONTENT

The chart on page 414 shows the percentage of employees in different age groups who are satisfied with their current employer. As we can observe from the chart, only 58% of the employees under age 35 are satisfied with their current employer. The corresponding percentage is 70% for the employees in the age group 35–54, and it jumps to 93% for employees 55 and older. Note that these percentages are based on a sample survey. Suppose that the survey included 1000 employees under 35 years of age. Suppose the overall percentage of

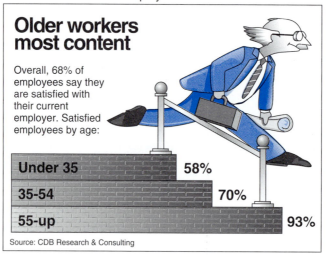

USA SNAPSHOTS®

A look at statistics that shape your finances

Older workers most content

Overall, 68% of employees say they are satisfied with their current employer. Satisfied employees by age:

Under 35	58%
35-54	70%
55-up	93%

Source: CDB Research & Consulting

By Cindy Hall and Web Bryant, USA TODAY

68% is true for the population of all employees. We can conduct a test of hypothesis to find out whether the percentage of employees under 35 years of age who are satisfied with their current employer is less than 68%. Suppose we make this test using a 1% significance level. Below we perform this test of hypothesis.

$$H_0: p = .68$$

$$H_1: p < .68$$

Here, $n = 1000$, $\hat{p} = .58$, and $\alpha = .01$. The test is left-tailed. Using the normal distribution to perform the test, the critical value of z is -2.33. Thus, we will reject the null hypothesis if the observed value of z is -2.33 or smaller. We find the observed value of z as follows.

$$\sigma_{\hat{p}} = \sqrt{\frac{pq}{n}} = \sqrt{\frac{(.68)(.32)}{1000}} = .01475127$$

$$z = \frac{\hat{p} - p}{\sigma_{\hat{p}}} = \frac{.58 - .68}{.01475127} = -6.78$$

The value of the test statistic $z = -6.78$ for $\hat{p}$ lies in the rejection region. Consequently, we reject H_0 and conclude that the percentage of employees under 35 years of age who are satisfied with their current employer is less than 68%.

Similarly, we can test hypotheses about the percentages of employees in the other two age groups who are satisfied with their current employers.

Source: The chart is reproduced with permission from *USA TODAY*, July 8, 1998. Copyright © 1998, *USA TODAY*.

EXERCISES

■ Concepts and Procedures

9.81 Explain when a sample is large enough to use the normal distribution to make a test of hypothesis about the population proportion.

9.82 In each of the following cases, do you think the sample size is large enough to use the normal distribution to make a test of hypothesis about the population proportion? Explain why or why not.

a. $n = 40$ and $p = .11$ **b.** $n = 100$ and $p = .73$
c. $n = 80$ and $p = .05$ **d.** $n = 50$ and $p = .14$

9.83 In each of the following cases, do you think the sample size is large enough to use the normal distribution to make a test of hypothesis about the population proportion? Explain why or why not.

a. $n = 30$ and $p = .65$ **b.** $n = 70$ and $p = .05$
c. $n = 60$ and $p = .06$ **d.** $n = 900$ and $p = .17$

9.84 For each of the following examples of tests of hypotheses about the population proportion, show the rejection and nonrejection regions on the graph of the sampling distribution of the sample proportion.

a. A two-tailed test with $\alpha = .10$
b. A left-tailed test with $\alpha = .01$
c. A right-tailed test with $\alpha = .05$

9.85 For each of the following examples of tests of hypotheses about the population proportion, show the rejection and nonrejection regions on the graph of the sampling distribution of the sample proportion.

a. A two-tailed test with $\alpha = .05$
b. A left-tailed test with $\alpha = .02$
c. A right-tailed test with $\alpha = .025$

9.86 A random sample of 500 observations produced a sample proportion equal to .38. Find the critical and observed values of z for each of the following tests of hypotheses using $\alpha = .05$.

a. $H_0: p = .30$ versus $H_1: p > .30$
b. $H_0: p = .30$ versus $H_1: p \neq .30$

9.87 A random sample of 200 observations produced a sample proportion equal to .60. Find the critical and observed values of z for each of the following tests of hypotheses using $\alpha = .01$.

a. $H_0: p = .63$ versus $H_1: p < .63$
b. $H_0: p = .63$ versus $H_1: p \neq .63$

9.88 Consider the null hypothesis $H_0: p = .65$. Suppose a random sample of 1000 observations is taken to make this test about the population proportion. Using $\alpha = .05$, show the rejection and nonrejection regions and find the critical value(s) of z for a

a. left-tailed test   **b.** two-tailed test   **c.** right-tailed test

9.89 Consider the null hypothesis $H_0: p = .25$. Suppose a random sample of 400 observations is taken to make this test about the population proportion. Using $\alpha = .01$, show the rejection and nonrejection regions and find the critical value(s) of z for a

a. left-tailed test   **b.** two-tailed test   **c.** right-tailed test

9.90 Consider $H_0: p = .70$ versus $H_1: p \neq .70$.

a. A random sample of 600 observations produced a sample proportion equal to .68. Using $\alpha = .01$, would you reject the null hypothesis?
b. Another random sample of 600 observations taken from the same population produced a sample proportion equal to .76. Using $\alpha = .01$, would you reject the null hypothesis?

Comment on the results of parts a and b.

9.91   Consider H_0: $p = .45$ versus H_1: $p < .45$.

a.   A random sample of 400 observations produced a sample proportion equal to .42. Using $\alpha = .025$, would you reject the null hypothesis?

b.   Another random sample of 400 observations taken from the same population produced a sample proportion of .39. Using $\alpha = .025$, would you reject the null hypothesis?

Comment on the results of parts a and b.

9.92   Make the following hypothesis tests about p.

a.	H_0: $p = .45$,	H_1: $p \neq .45$,	$n = 100$,	$\hat{p} = .49$,	$\alpha = .10$
b.	H_0: $p = .72$,	H_1: $p < .72$,	$n = 700$,	$\hat{p} = .64$,	$\alpha = .05$
c.	H_0: $p = .30$,	H_1: $p > .30$,	$n = 200$,	$\hat{p} = .33$,	$\alpha = .01$

9.93   Make the following hypothesis tests about p.

a.	H_0: $p = .57$,	H_1: $p \neq .57$,	$n = 800$,	$\hat{p} = .50$,	$\alpha = .05$
b.	H_0: $p = .26$,	H_1: $p < .26$,	$n = 400$,	$\hat{p} = .23$,	$\alpha = .01$
c.	H_0: $p = .84$,	H_1: $p > .84$,	$n = 250$,	$\hat{p} = .85$,	$\alpha = .025$

■ Applications

9.94   Managers often tolerate poor performance by employees because the termination process is so complicated. In a survey of federal government managers who had not taken action against unsatisfactory employees, 66% stated that they failed to act because of the long process required (*USA TODAY*, July 29, 1998). Suppose that 80 such managers are randomly selected and 48 of them failed to act because of the long process required. Can you conclude that the current percentage of such managers who fail to act because of the long process involved is less than 66%? Use $\alpha = .025$.

9.95   Many traffic accidents are blamed on motorists who are distracted by cell phone use while driving. In a survey conducted by Bruskin/Goldring for Exxon, 29% of adults said that they make phone calls *sometimes* or *frequently* while driving alone (*USA TODAY*, May 20, 1999). Suppose that a recently taken random sample of 500 adults showed that 165 of them make phone calls *sometimes* or *frequently* while driving alone. At the 5% level of significance, can you conclude that the current percentage of adults who make such phone calls exceeds 29%?

9.96   According to a 1998 survey of 1200 adults by the Travel Industry Association of America, 46% made travel plans through the Internet, but only 9% actually used the Internet to make reservations (*USA TODAY*, April 26, 1999). A recent random sample of 400 travelers found that 58 of them made reservations on the Internet. Test at the 1% significance level whether the current percentage of all travelers who make reservations on the Internet exceeds 9%.

9.97   According to a telephone poll of 1049 adult Americans conducted for *Time*/CNN by Yankelovich Partners, Inc., 73% would forgive someone who told lies about them (*Time*, April 5, 1999). In a recent random sample of 1000 adult Americans, 700 indicated that they would forgive someone who lied about them. At the 5% level of significance, can you conclude that the proportion of all American adults who feel this way is less than 73%?

9.98   In a telephone poll of 428 dog owners conducted by Yankelovich Partners, Inc., for *Time*/CNN, 87% of the respondents said that they think dogs have good and bad moods as humans do (*Time*, February 1, 1999). In a recent random sample of 400 dog owners, 336 held this view.

a.   Test at the 2% level of significance whether the current percentage of all dog owners who hold this view is different from 87%.

b.   What is the Type I error in this case? What is the probability of making this error?

9.99   A poll of high school students released by the nonprofit Josephson Institute for Ethics found that 47% of high school students admitted stealing from a store in the past year (*USA TODAY*, October 19, 1998). Assume that this percentage was true for all such students at the time that poll was conducted. A recent survey of a random sample of 1100 high school students found that 550 of them admitted stealing from a store in the past year.

a.   Test at the 5% significance level whether the percentage of all high school students who would admit to such thefts has increased since 1998.

b.   What is the Type I error in this case? What is the probability of making this error?

9.100 A food company is planning to market a new type of frozen yogurt. However, before marketing this yogurt, the company wants to find what percentage of the people like it. The company's management has decided that it will market this yogurt only if at least 35% of the people like it. The company's research department selected a random sample of 400 persons and asked them to taste this yogurt. Of these 400 persons, 112 said they liked it.

a. Testing at the 2.5% significance level, can you conclude that the company should market this yogurt?
b. What will your decision be in part a if the probability of making a Type I error is zero? Explain.

9.101 A mail-order company claims that at least 60% of all orders are mailed within 48 hours. From time to time the quality control department at the company checks if this promise is fulfilled. Recently the quality control department at this company took a sample of 400 orders and found that 208 of them were mailed within 48 hours of the placement of the orders.

a. Testing at the 1% significance level, can you conclude that the company's claim is true?
b. What will your decision be in part a if the probability of making a Type I error is zero? Explain.

9.102 Brooklyn Corporation manufactures computer diskettes. The machine that is used to make these diskettes is known to produce not more than 5% defective diskettes. The quality control inspector selects a sample of 200 diskettes each week and inspects them for being good or defective. Using the sample proportion, the quality control inspector tests the null hypothesis $p \leq .05$ against the alternative hypothesis $p > .05$, where p is the proportion of diskettes that are defective. She always uses a 2.5% significance level. If the null hypothesis is rejected, the production process is stopped to make any necessary adjustments. A recent such sample of 200 diskettes contained 17 defective diskettes.

a. Using the 2.5% significance level, would you conclude that the production process should be stopped to make necessary adjustments?
b. Perform the test of part a using a 1% significance level. Is your decision different from the one in part a?

Comment on the results of parts a and b.

9.103 Shulman Steel Corporation makes bearings that are supplied to other companies. One of the machines makes bearings that are supposed to have a diameter of 4 inches. The bearings that have a diameter of either more or less than 4 inches are considered defective and are discarded. When working properly, the machine does not produce more than 7% of bearings that are defective. The quality control inspector selects a sample of 200 bearings each week and inspects them for the size of their diameters. Using the sample proportion, the quality control inspector tests the null hypothesis $p \leq .07$ against the alternative hypothesis $p > .07$, where p is the proportion of bearings that are defective. He always uses a 2% significance level. If the null hypothesis is rejected, the machine is stopped to make any necessary adjustments. One such sample of 200 bearings taken recently contained 22 defective bearings.

a. Using the 2% significance level, will you conclude that the machine should be stopped to make necessary adjustments?
b. Perform the test of part a using a 1% significance level. Is your decision different from the one in part a?

Comment on the results of parts a and b.

*9.104 Two years ago, 75% of the customers of a bank said that they were satisfied with the services provided by the bank. The manager of the bank wants to know if this percentage of satisfied customers has changed since then. She assigns this responsibility to you. Briefly explain how you would conduct such a test.

*9.105 A study claims that 65% of students at all colleges and universities hold off-campus (part-time or full-time) jobs. You want to check if the percentage of students at your school who hold off-campus jobs is different from 65%. Briefly explain how you would conduct such a test. Collect data from 40 students at your school on whether or not they hold off-campus jobs. Then, calculate the proportion of students in this sample who hold off-campus jobs. Using this information, test the hypothesis. Select your own significance level.

GLOSSARY

α The significance level of a test of hypothesis that denotes the probability of rejecting a null hypothesis when it actually is true. (The probability of committing a Type I error.)

Alternative hypothesis   A claim about a population parameter that will be true if the null hypothesis is false.

β The probability of not rejecting a null hypothesis when it actually is false. (The probability of committing a Type II error.)

Critical value or **critical point**   One or two values that divide the whole region under the sampling distribution of a sample statistic into rejection and nonrejection regions.

Left-tailed test   A test in which the rejection region lies in the left tail of the distribution curve.

Null hypothesis   A claim about a population parameter that is assumed to be true until proven otherwise.

Observed value of z or t   The value of z or t calculated for a sample statistic such as the sample mean or the sample proportion.

One-tailed test   A test in which there is only one rejection region, either in the left tail or in the right tail of the distribution curve.

p-value   The smallest significance level at which a null hypothesis can be rejected.

Right-tailed test   A test in which the rejection region lies in the right tail of the distribution curve.

Significance level   The value of α that gives the probability of committing a Type I error.

Test statistic   The value of z or t calculated for a sample statistic such as the sample mean or the sample proportion.

Two-tailed test   A test in which there are two rejection regions, one in each tail of the distribution curve.

Type I error   An error that occurs when a true null hypothesis is rejected.

Type II error   An error that occurs when a false null hypothesis is not rejected.

KEY FORMULAS

1. **Value of the test statistic z for $\bar{x}$ in a test of hypothesis about μ for a large sample**

$$z = \frac{\bar{x} - \mu}{\sigma_{\bar{x}}} \quad \text{if } \sigma \text{ is known, where } \sigma_{\bar{x}} = \frac{\sigma}{\sqrt{n}}$$

or

$$z = \frac{\bar{x} - \mu}{s_{\bar{x}}} \quad \text{if } \sigma \text{ is not known, where } s_{\bar{x}} = \frac{s}{\sqrt{n}}$$

2. **Value of the test statistic t for $\bar{x}$ in a test of hypothesis about μ for a small sample**

$$t = \frac{\bar{x} - \mu}{s_{\bar{x}}} \quad \text{where} \quad s_{\bar{x}} = \frac{s}{\sqrt{n}}$$

3. **Value of the test statistic z for $\hat{p}$ in a test of hypothesis about p for a large sample**

$$z = \frac{\hat{p} - p}{\sigma_{\hat{p}}} \quad \text{where} \quad \sigma_{\hat{p}} = \sqrt{\frac{pq}{n}}$$

SUPPLEMENTARY EXERCISES

9.106 Consider the following null and alternative hypotheses:

$$H_0: \mu = 120 \quad \text{versus} \quad H_1: \mu > 120$$

A random sample of 81 observations taken from this population produced a sample mean of 123.5 and a sample standard deviation of 15.

a. If this test is made at the 2.5% significance level, would you reject the null hypothesis?
b. What is the probability of making a Type I error in part a?
c. Calculate the p-value for the test. Based on this p-value, would you reject the null hypothesis if $\alpha = .01$? What if $\alpha = .05$?

9.107 Consider the following null and alternative hypotheses:

$$H_0: \mu = 40 \quad \text{versus} \quad H_1: \mu \neq 40$$

A random sample of 64 observations taken from this population produced a sample mean of 38.4 and a sample standard deviation of 6.

a. If this test is made at the 2% significance level, would you reject the null hypothesis?
b. What is the probability of making a Type I error in part a?
c. Calculate the p-value for the test. Based on this p-value, would you reject the null hypothesis if $\alpha = .01$? What if $\alpha = .05$?

9.108 Consider the following null and alternative hypotheses:

$$H_0: p = .82 \quad \text{versus} \quad H_1: p \neq .82$$

A random sample of 600 observations taken from this population produced a sample proportion of .86.

a. If this test is made at the 2% significance level, would you reject the null hypothesis?
b. What is the probability of making a Type I error in part a?
c. Calculate the p-value for the test. Based on this p-value, would you reject the null hypothesis if $\alpha = .025$? What if $\alpha = .005$?

9.109 Consider the following null and alternative hypotheses:

$$H_0: p = .44 \quad \text{versus} \quad H_1: p < .44$$

A random sample of 450 observations taken from this population produced a sample proportion of .39.

a. If this test is made at the 2% significance level, would you reject the null hypothesis?
b. What is the probability of making a Type I error in part a?
c. Calculate the p-value for the test. Based on this p-value, would you reject the null hypothesis if $\alpha = .01$? What if $\alpha = .025$?

9.110 A manufacturer of fluorescent lightbulbs claims that the mean life of these bulbs is at least 2500 hours. A consumer agency wanted to check whether or not this claim is true. The agency took a random sample of 36 such bulbs and tested them. The mean life for the sample was found to be 2457 hours with a standard deviation of 180 hours.

a. Do you think the sample information supports the company's claim? Use $\alpha = .025$.
b. What is the Type I error in this case? Explain. What is the probability of making this error?
c. Will your conclusion of part a change if the probability of making a Type I error is zero?

9.111 According to data from the U.S. Bureau of Justice Statistics, the average length of prison sentences for violent crimes in the United States was 90.7 months in 1997 (*Statistical Abstract of the United States,* 1998). A recent sample of 55 such sentences yielded a mean of 101 months with a standard deviation of 30 months.

a. Does the sample information support the alternative hypothesis that the current mean sentence for such crimes exceeds 90.7 months? Use $\alpha = .02$.

b. What is the Type I error in this case? Explain. What is the probability of making this error?

c. Would your conclusion for part a change if the probability of making a Type I error were zero?

9.112 According to an estimate, the average hourly wage for summer employment for high school and college students was $6.69 in 1999 (*USA TODAY*, June 25, 1999). Suppose that a random sample of 350 such students yielded a mean hourly wage for the past summer employment of $6.90 with a standard deviation of $1.88.

a. Using $\alpha = .05$, can you conclude that the mean hourly wage for the past summer employment for all such students is higher than $6.69?

b. Using $\alpha = .01$, can you conclude that the mean hourly wage for the past summer employment for all such students is higher than $6.69?

Comment on the results of parts a and b.

9.113 According to *Family PC Magazine*, the average computer usage by readers of this magazine is 16 to 18 hours per week (*USA TODAY*, April 14, 1999). Suppose a recent random sample of 80 such readers showed an average weekly computer usage of 20 hours with a standard deviation of 9 hours.

a. Using $\alpha = .05$, does the sample information support the alternative hypothesis that current mean weekly computer usage by such readers exceeds 18 hours?

b. Using $\alpha = .01$, does the sample information support the alternative hypothesis that current mean weekly computer usage by such readers exceeds 18 hours?

Comment on the results of parts a and b.

9.114 Customers often complain about long waiting times at restaurants before the food is served. A restaurant claims that it serves food to its customers, on average, within 15 minutes after the order is placed. A local newspaper journalist wanted to check if the restaurant's claim is true. A sample of 36 customers showed that the mean time taken to serve food to them was 15.75 minutes with a standard deviation of 2.4 minutes. Using the sample mean, the journalist says that the restaurant's claim is false. Do you think the journalist's conclusion is fair to the restaurant? Use the 1% significance level to answer this question.

9.115 The customers at a bank complained about long lines and the time they had to spend waiting for service. It is known that the customers at this bank had to wait 8 minutes, on average, before being served. The management made some changes to reduce the waiting time for its customers. A sample of 32 customers taken after these changes were made produced a mean waiting time of 7.5 minutes with a standard deviation of 2.1 minutes. Using this sample mean, the bank manager displayed a huge banner inside the bank mentioning that the mean waiting time for customers has been reduced by new changes. Do you think the bank manager's claim is justifiable? Use the 2.5% significance level to answer this question.

9.116 According to data from the College Board, the average cost per student for books and supplies in the 1997–1998 academic year was $634 at public four-year colleges in the United States (*The Chronicle of Higher Education*, August 1998). A recent random sample of 250 students at such colleges yielded a mean cost for books and supplies of $660 with a standard deviation of $170. Find the *p*-value for the test with the alternative hypothesis that the current mean for these costs differs from $634.

9.117 The mean consumption of water per household in a city was 1245 cubic feet per month. Due to a water shortage because of a drought, the city council campaigned for water use conservation by households. A few months after the campaign was started, the mean consumption of water for a sample of 100 households was found to be 1175 cubic feet per month with a standard deviation of 250 cubic feet. Find the *p*-value for the hypothesis test that the mean consumption of water per household has decreased due to the campaign by the city council.

9.118 Data furnished by American Express Everyday Spending Index indicated a mean annual household expenditure of $1044 on gasoline (*USA TODAY*, June 3, 1999). A random sample of 28 U.S. households produced a mean annual household expenditure of $921 on gasoline with a standard deviation of $343 for last year. Using the 1% significance level, test whether the mean gasoline expenditure for all U.S. households was less than $1044 last year. Assume that the gasoline expenditures for all U.S. households have a normal distribution.

9.119 The administrative office of a hospital claims that the mean waiting time for patients to get treatment in its emergency ward is 25 minutes. A random sample of 16 patients who received treatment in the emergency ward of this hospital produced a mean waiting time of 27.5 minutes with a standard deviation of 4.8 minutes. Using the 1% significance level, test whether the mean waiting time at the emergency ward is different from 25 minutes. Assume that the waiting times for all patients at this emergency ward have a normal distribution.

9.120 According to data from Runzheimer International, the average cost of suburban day care (for a 3-year-old in a for-profit center for five days a week) in the Minneapolis area was $572 per month (*USA TODAY*, August 18, 1998). A recent random sample of 25 parents in the Minneapolis area with 3-year-old children found a mean day-care cost of $634 per month with a standard deviation of $136. Assume that the monthly costs for all such day care are normally distributed.

a. Using $\alpha = .025$, can you conclude that the mean of all such costs currently exceeds $572?
b. Suppose the probability of making a Type I error is zero. Can you make a decision for the test of part a without going through the five steps of hypothesis testing? If so, what is your decision? Explain.

9.121 An earlier study claims that U.S. adults spend an average of 114 minutes with their families per day. A recently taken sample of 25 adults showed that they spend an average of 109 minutes per day with their families. The sample standard deviation is 11 minutes. Assume that the times spent by adults with their families have an approximately normal distribution.

a. Using the 1% significance level, test whether the mean time spent currently by all adults with their families is less than 114 minutes a day.
b. Suppose the probability of making a Type I error is zero. Can you make a decision for the test of part a without going through the five steps of hypothesis testing? If yes, what is your decision? Explain.

9.122 A computer company that recently introduced a new software product claims that the mean time it takes to learn how to use this software is not more than 2 hours for people who are somewhat familiar with computers. A random sample of 12 such persons was selected. The following data give the times taken (in hours) by these persons to learn how to use this software.

1.75	2.25	2.40	1.90	1.50	2.75
2.15	2.25	1.80	2.20	3.25	2.60

Test at the 1% significance level whether the company's claim is true. Assume that the times taken by all persons who are somewhat familiar with computers to learn how to use this software are approximately normally distributed.

9.123 A company claims that its 8-ounce low-fat yogurt cups contain, on average, at most 150 calories per cup. A consumer agency wanted to check whether or not this claim is true. A random sample of 10 such cups produced the following data on calories.

147	159	153	146	144	161	163	153	143	158

Test at the 2.5% significance level whether the company's claim is true. Assume that the number of calories for such cups of yogurt produced by this company has an approximately normal distribution.

9.124 According to a survey conducted for *Time* magazine, 62% of the respondents said that the government should regulate gene therapy to cure or prevent disease (*Time*, January 11, 1999). In a recent random sample of 200 people, 58% held this view.

a. Test at the 2.5% significance level whether the percentage of all people who favor such government regulation is currently less than 62%.
b. How do you explain the Type I error in this case? What is the probability of making this error in part a?

9.125 According to a *USA TODAY*/Gallup telephone poll, 49% of adults expected that a worldwide collapse of the economy will occur by 2025 (*USA TODAY*, October 13, 1998). Assume that this result was true for the population of all adult Americans at the time of the poll. In a recent random sample of 800 adult Americans, 43% expected such a collapse by 2025.

a. Test at the 1% level of significance whether the current percentage of adult Americans who expect such a collapse has changed since the earlier poll.

b. How do you explain the Type I error in this case? What is the probability of making this error in part a?

9.126 More and more people are abandoning national brand products and buying store brand products to save money. The president of a company that produces national brand coffee claims that 40% of the people prefer to buy national brand coffee. A random sample of 700 people who buy coffee showed that 259 of them buy national brand coffee. Using $\alpha = .01$, can you conclude that the percentage of people who buy national brand coffee is different from 40%?

9.127 According to DATAQUEST, Inc., 26% of adults in the United States had wireless phones in 1999 (*Business Week*, May 3, 1999). A recent random sample of 200 adult Americans showed that 72 of them have wireless phones. At the 2% level of significance, can you conclude that the current percentage of U.S. adults who have wireless phones exceeds 26%?

9.128 Mong Corporation makes auto batteries. The company claims that 80% of its LL70 batteries are good for 70 months or longer. A consumer agency wanted to check if this claim is true. The agency took a random sample of 40 such batteries and found that 75% of them were good for 70 months or longer.

a. Using the 1% significance level, can you conclude that the company's claim is false?

b. What will your decision be in part a if the probability of making a Type I error is zero? Explain.

9.129 Dartmouth Distribution Warehouse makes deliveries of a large number of products to its customers. To keep its customers happy and satisfied, the company's policy is to deliver on time at least 90% of all the orders it receives from its customers. The quality control inspector at the company quite often takes samples of orders delivered and checks if this policy is maintained. A recent such sample of 90 orders taken by this inspector showed that 75 of them were delivered on time.

a. Using the 2% significance level, can you conclude that the company's policy is maintained?

b. What will your decision be in part a if the probability of making a Type I error is zero? Explain.

■ **Advanced Exercises**

9.130 Refer to Exercise 9.125. Find the *p*-value for the test of hypothesis mentioned in that exercise. Using this *p*-value, would you reject the null hypothesis at $\alpha = .05$? What if $\alpha = .01$?

9.131 Refer to Exercise 9.129. Find the *p*-value for the test of hypothesis mentioned in that exercise. Using this *p*-value, would you reject the null hypothesis at $\alpha = .05$? What if $\alpha = .01$?

9.132 Professor Hansen believes that some people have the ability to predict in advance the outcome of a spin of a roulette wheel. He takes 100 student volunteers to a casino. The roulette wheel has 38 numbers, each of which is equally likely to occur. Of these 38 numbers, 18 are red, 18 are black, and 2 are green. Each student is to place a series of five bets, choosing either a red or a black number before each spin of the wheel. Thus, a student who bets on red has an 18/38 chance of winning that bet. The same is true of betting on black.

a. Assuming random guessing, what is the probability that a particular student will win all five of his or her bets?

b. Suppose for each student we formulate the hypothesis test

H_0: The student is guessing
H_1: The student has some predictive ability

Suppose we reject H_0 only if the student wins all five bets. What is the significance level?

c. Suppose that two of the 100 students win all five of their bets. Professor Hansen says, "For these two students we can reject H_0 and conclude that we have found two students with some ability to predict." What do you make of Professor Hansen's conclusion?

9.133 Acme Bicycle Company makes derailleurs for mountain bikes. Usually no more than 4% of these parts are defective, but occasionally the machines that make them get out of adjustment and the rate of defectives exceeds 4%. To guard against this, the chief quality control inspector takes a random sample of 130 derailleurs each week and checks each one for defects. If too many of these parts are defective, the machines are shut down and adjusted. To decide how many parts must be defective in order to shut down the machines, the company's statistician has set up the hypothesis test

$$H_0: p \leq .04 \quad \text{versus} \quad H_1: p > .04$$

where p is the proportion of defectives among all derailleurs being made currently. Rejection of H_0 would call for shutting down the machines. For the inspector's convenience, the statistician would like the rejection region to have the form, "Reject H_0 if the number of defective parts is C or more." Find the value of C that will make the significance level (approximately) .05.

9.134 Alpha Airlines claims that only 15% of its flights arrive more than 10 minutes late. Let p be the proportion of all of Alpha's flights that arrive more than 10 minutes late. Consider the hypothesis test

$$H_0: p \leq .15 \quad \text{versus} \quad H_1: p > .15$$

Suppose we take a random sample of 50 flights by Alpha Airlines and agree to reject H_0 if 9 or more of them arrive late. Find the significance level for this test.

9.135 The standard therapy that is used to treat a disorder cures 60% of all patients in an average of 140 visits. A health care provider considers supporting a new therapy regime for the disorder if it is effective in reducing the number of visits while retaining the cure rate of the standard therapy. A study of 200 patients with the disorder who were treated by the new therapy regime reveals that 108 of them were cured in an average of 132 visits with a standard deviation of 38 visits. What decision should be made using a .01 level of significance?

9.136 The print on the packages of 100-watt General Electric soft-white lightbulbs states that these lightbulbs have an average life of 750 hours. Assume that the standard deviation of the lengths of lives of these lightbulbs is 50 hours. A skeptical consumer does not think these lightbulbs last as long as the manufacturer claims and she decides to test 64 randomly selected lightbulbs. She has set up the decision rule that if the average life of these 64 lightbulbs is less than 735 hours, then she will conclude that GE has printed too high an average length of life on the packages and will write them a letter to that effect. Approximately what significance level is the consumer using? Approximately what significance level is she using if she decides that GE has printed too high an average length of life on the packages if the average life of the 64 lightbulbs is less than 700 hours? Interpret the values you get.

9.137 Thirty percent of all people who are inoculated with the current vaccine that is used to prevent a disease contract the disease within a year. The developer of a new vaccine that is intended to prevent this disease wishes to test for significant evidence that the new vaccine is more effective.

a. Determine the appropriate null and alternative hypotheses.

b. The developer decides to study 100 randomly selected people by inoculating them with the new vaccine. If 84 or more of them do not contract the disease within a year, the developer will conclude that the new vaccine is superior to the old one. What significance level is the developer using for the test?

c. Suppose 20 people inoculated with the new vaccine are studied and the new vaccine is concluded to be better than the old one if fewer than 3 people contract the disease within a year. What is the significance level of the test?

9.138 The Parks and Recreation Department has determined that the thickness of the ice on a town pond must average 5 inches to be safe for ice skating. Consider the decision of whether or not to allow ice skating as a hypothesis-testing problem.

a. What is the meaning of the unknown parameter that is being tested?

b. Set up the appropriate null and alternative hypotheses for this situation.

c. What would it mean to commit a Type I error for your test in part b? What about a Type II error?
d. Would you want the significance level for your test in part b to be large or small? Suggest a significance level and explain what it means in this problem.

9.139 Since 1984, all automobiles have been manufactured with a middle taillight. You have been hired to answer the question, Is the middle taillight effective in reducing the number of rear-end collisions? You have available to you any information you could possibly want about all rear-end collisions involving cars built before 1984. How would you conduct an experiment to answer the question? In your answer, include things like: (a) the precise meaning of the unknown parameter you are testing; (b) H_0 and H_1; (c) a detailed explanation of what sample data you would collect to draw a conclusion; and (d) any assumptions you would make, particularly about the characteristics of cars built before 1984 versus those built since 1984.

9.140 Before a championship football game, the referee is given a special commemorative coin to toss to decide which team will kick the ball first. Two minutes before game time, he receives an anonymous tip that the captain of one of the teams may have substituted a biased coin that has a 70% chance of showing heads each time it is tossed. The referee has time to toss the coin 10 times to test it. He decides that if it shows 8 or more heads in 10 tosses, he will reject this coin and replace it with another coin. Let p be the probability that this coin shows heads when it is tossed once.

a. Formulate the relevant null and alternative hypotheses (in terms of p) for the referee's test.
b. Using the referee's decision rule, find α for this test.

SELF-REVIEW TEST

1. A test of hypothesis is always about

   a. a population parameter   **b.** a sample statistic   **c.** a test statistic

2. A Type I error is committed when

   a. a null hypothesis is not rejected when it is actually false
   b. a null hypothesis is rejected when it is actually true
   c. an alternative hypothesis is rejected when it is actually true

3. A Type II error is committed when

   a. a null hypothesis is not rejected when it is actually false
   b. a null hypothesis is rejected when it is actually true
   c. an alternative hypothesis is rejected when it is actually true

4. A critical value is the value

   a. calculated from sample data
   b. determined from a table (e.g., the normal distribution table or other such tables)
   c. neither a nor b

5. The computed value of a test statistic is the value

   a. calculated for a sample statistic
   b. determined from a table (e.g., the normal distribution table or other such tables)
   c. neither a nor b

6. The observed value of a test statistic is the value

   a. calculated for a sample statistic
   b. determined from a table (e.g., the normal distribution table or other such tables)
   c. neither a nor b

7. The significance level, denoted by α, is
   a. the probability of committing a Type I error
   b. the probability of committing a Type II error
   c. neither a nor b

8. The value of β gives the
   a. probability of committing a Type I error
   b. probability of committing a Type II error
   c. power of the test

9. The value of $1 - \beta$ gives the
   a. probability of committing a Type I error
   b. probability of committing a Type II error
   c. power of the test

10. A two-tailed test is a test with
    a. two rejection regions   b. two nonrejection regions   c. two test statistics

11. A one-tailed test
    a. has one rejection region   b. has one nonrejection region   c. both a and b

12. The smallest level of significance at which a null hypothesis is rejected is called
    a. α b. p-value   c. β

13. Which of the following is not required to apply the t distribution to make a test of hypothesis about μ?
    a. $n < 30$ b. Population is normally distributed
    c. σ is unknown   d. β is known

14. The sign in the alternative hypothesis in a two-tailed test is always
    a. $<$ b. $>$ c. $\neq$

15. The sign in the alternative hypothesis in a left-tailed test is always
    a. $<$ b. $>$ c. $\neq$

16. The sign in the alternative hypothesis in a right-tailed test is always
    a. $<$ b. $>$ c. $\neq$

17. A bank loan officer claims that the mean monthly mortgage payment made by all homeowners in a certain city is $1365. A housing magazine wanted to test this claim. A random sample of 100 homeowners taken by this magazine produced the mean monthly mortgage of $1505 with a standard deviation of $278.
    a. Testing at the 1% significance level, would you conclude that the mean monthly mortgage payment made by all homeowners in this city is different from $1365?
    b. What is the Type I error in part a? What is the probability of making this error?
    c. What will your decision be in part a if the probability of making a Type I error is zero? Explain.

18. An editor of a New York publishing company claims that the mean time it takes to write a textbook is at least 31 months. A sample of 16 textbook authors found that the mean time taken by them to write a textbook was 25 months with a standard deviation of 7.2 months.
    a. Using the 2.5% significance level, would you conclude that the editor's claim is true? Assume that the time taken to write a textbook is normally distributed for all textbook authors.
    b. What is the Type I error in part a? What is the probability of making this error?
    c. What will your decision be in part a if the probability of making a Type I error is .001?

19. In a survey, 66% of the respondents agreed that police should be allowed to collect DNA information from suspected criminals (*Time*, January 11, 1999). In a recently taken random sample of 500 Americans, 305 held this view.

   a. Using a 5% significance level, can you conclude that the current percentage of Americans who hold this view differs from 66%?

   b. What is the Type I error in part a? What is the probability of making this error?

   c. What would your decision be in part a if the probability of making a Type I error were zero? Explain.

20. According to a survey, the starting salary for an entry-level position in investment banking was $37,120 in 1998 (*Newsweek*, February 1, 1999). Assume that $37,120 was the 1998 mean starting salary for all such positions. A recently taken random sample of 100 such positions found a mean starting salary of $38,050 with a standard deviation of $4420.

   a. Find the *p*-value for the test with the alternative hypothesis that the mean starting salary for such positions exceeds $37,120.

   b. Using the *p*-value calculated in part a, would you reject the null hypothesis at a significance level of 1%? What if $\alpha = .05$?

*21. Refer to Problem 19.

   a. Find the *p*-value for the test of hypothesis mentioned in part a of that problem.

   b. Using this *p*-value, will you reject the null hypothesis if $\alpha = .05$? What if $\alpha = .01$?

MINI-PROJECTS

Mini-Project 9–1

The mean height of players who were on the rosters of National Basketball Association teams at the beginning of the 1996–1997 season was 79.22 inches. Let μ denote the mean height of NBA players at the beginning of the 1999–2000 season.

a. Take a random sample of 15 players from the NBA data that accompany this text. Test H_0: $\mu = 79.22$ inches against H_1: $\mu \neq 79.22$ inches using $\alpha = .05$.

b. Repeat part a for samples of 31 and 45 players, respectively.

c. Did any of the three tests in parts a and b lead to the conclusion that the mean height of NBA players in 1999–2000 is different from that in 1996–1997?

Mini-Project 9–2

A thumbtack that is tossed on a desk can land in one of the two ways shown in the illustration.

Heads   Tails

Brad and Dan cannot agree on the likelihood of obtaining a head or a tail. Brad argues that obtaining a tail is more likely than obtaining a head because of the shape of the tack. If the tack had no point at all, it would resemble a coin that has the same probability of coming up heads or tails when tossed. But, the longer the point, the less likely it is that the tack will stand up on its head when tossed. Dan believes that as the tack lands tails, the point causes the tack to jump around and come to rest in the heads position. Brad and Dan need you to settle their dispute. Do you think the tack is equally likely to land heads or tails? To investigate this question, find an ordinary thumbtack and toss it a large number of times (say, 100 times).

a. What is the meaning, in words, of the unknown parameter in this problem?

b. Set up the null and alternative hypotheses and compute the *p*-value based on your results from tossing the tack.

c. How would you answer the original question now? If you decide the tack is not fair, do you side with Brad or Dan?

d. What would you estimate the value of the parameter in part a to be? Find a 90% confidence interval for this parameter.

e. After doing this experiment, do you think 100 tosses are enough to infer the nature of your tack? Using your result as a preliminary estimate, determine how many tosses would be necessary to be 95% certain of having 4% accuracy; that is, the maximum error of estimate is ±4%. Have you observed enough tosses?

For instructions on using MINITAB for this chapter, please see Section B.8 of Appendix B.

See Appendix C for *Excel Adventure 9* on hypothesis testing.

COMPUTER ASSIGNMENTS

CA9.1   According to an earlier study, the mean amount spent on clothes by American women is $575 per year. A researcher wanted to check if this result still holds true. A random sample of 39 women taken recently by this researcher produced the following data on the amounts they spent on clothes last year.

671	584	328	498	827	921	425	204	382	539
1070	854	669	328	537	849	930	1234	695	738
341	189	867	923	721	125	298	473	876	932
573	931	460	1430	391	887	958	674	782	

Using MINITAB or any other statistical software, test at the 1% significance level whether the mean expenditure on clothes for American women for last year is different from $575. Assume that the population standard deviation is $132.

CA9.2   The mean weight of all babies born at a hospital last year was 7.6 pounds. A random sample of 35 babies born at this hospital this year produced the following data.

8.2	9.1	6.9	5.8	6.4	10.3	12.1	9.1	5.9	7.3
11.2	8.3	6.5	7.1	8.0	9.2	5.7	9.5	8.3	6.3
4.9	7.6	10.1	9.2	8.4	7.5	7.2	8.3	7.2	9.7
6.0	8.1	6.1	8.3	6.7					

Using MINITAB or any other statistical software, test at the 2.5% significance level whether the mean weight of babies born at this hospital this year is more than 7.6 pounds.

CA9.3   The president of a large university claims that the mean time spent partying by all students at the university is not more than 7 hours per week. The following data give the times spent partying during the previous week by a random sample of 16 students taken from this university.

12	9	5	15	11	13	10	6
4	11	6	9	13	6	16	8

Using MINITAB or any other statistical software, test at the 1% significance level whether the president's claim is true. Assume that the times spent partying by all students at this university have an approximately normal distribution.

CA9.4   According to a basketball coach, the mean height of all male college basketball players is 74 inches. A random sample of 25 such players produced the following data on their heights.

68	76	74	83	77	76	69	67	71	74	79	85	69
78	75	78	68	72	83	79	82	76	69	70	81	

Using MINITAB or any other statistical software, test at the 2% significance level whether the mean height of all male college basketball players is different from 74 inches. Assume that the heights of all male college basketball players are (approximately) normally distributed.

CA9.5 A past study claims that adults in America spend an average of 18 hours a week on leisure activities. A researcher wanted to test this claim. She took a sample of 10 adults and asked them about the time they spend per week on leisure activities. Their responses (in hours) follow.

14   25   22   38   16   26   19   23   41   33

Assume that the times spent on leisure activities by all adults are normally distributed. Using the 5% significance level, can you conclude that the claim of the earlier study is true? Use MINITAB or any other statistical software to answer this question.

CA9.6 In a *Time* magazine poll, 55% of the respondents stated that smokers should pay higher insurance rates than nonsmokers (*Time*, January 11, 1999). Assume that this result was true for the population of all American adults at the time of the poll. A researcher wants to know whether this result holds true for the current population of Americans. A random sample of 700 American adults taken recently by this researcher showed that 399 of them hold this view. Using MINITAB or any other statistical software and a 2.5% significance level, can you conclude that the current percentage of American adults who think that smokers should pay more for insurance than nonsmokers is different from 55%?

CA9.7 A mail-order company claims that at least 60% of all orders it receives are mailed within 48 hours. From time to time the quality control department at the company checks if this promise is kept. Recently, the quality control department at this company took a sample of 400 orders and found that 224 of them were mailed within 48 hours of the placement of the orders. Using MINITAB or any other statistical software, test at the 1% significance level whether or not the company's claim is true.

10 ESTIMATION AND HYPOTHESIS TESTING: TWO POPULATIONS

10.1 INFERENCES ABOUT THE DIFFERENCE BETWEEN TWO POPULATION MEANS FOR LARGE AND INDEPENDENT SAMPLES

10.2 INFERENCES ABOUT THE DIFFERENCE BETWEEN TWO POPULATION MEANS FOR SMALL AND INDEPENDENT SAMPLES: EQUAL STANDARD DEVIATIONS

10.3 INFERENCES ABOUT THE DIFFERENCE BETWEEN TWO POPULATION MEANS FOR SMALL AND INDEPENDENT SAMPLES: UNEQUAL STANDARD DEVIATIONS

10.4 INFERENCES ABOUT THE DIFFERENCE BETWEEN TWO POPULATION MEANS FOR PAIRED SAMPLES

10.5 INFERENCES ABOUT THE DIFFERENCE BETWEEN TWO POPULATION PROPORTIONS FOR LARGE AND INDEPENDENT SAMPLES

CASE STUDY 10–1 WORRIED ABOUT PAYING THE BILLS

GLOSSARY

KEY FORMULAS

SUPPLEMENTARY EXERCISES

SELF-REVIEW TEST

MINI-PROJECTS

COMPUTER ASSIGNMENTS

Chapters 8 and 9 discussed the estimation and hypothesis-testing procedures for μ and p involving a single population. This chapter extends the discussion of estimation and hypothesis-testing procedures to the difference between two population means and the difference between two population proportions. For example, we may want to make a confidence interval for the difference between mean prices of houses in California and in New York. Or we may want to test the hypothesis that the mean price of houses in California is different from that in New York. As another example, we may want to make a confidence interval for the difference between the proportions of all male and female adults who abstain from drinking. Or we may want to test the hypothesis that the proportion of all adult men who abstain from drinking is different from the proportion of all adult women who abstain from drinking. Constructing confidence intervals and testing hypotheses about population parameters are referred to as *making inferences*.

10.1   INFERENCES ABOUT THE DIFFERENCE BETWEEN TWO POPULATION MEANS FOR LARGE AND INDEPENDENT SAMPLES

Let μ_1 be the mean of the first population and μ_2 be the mean of the second population. Suppose we want to make a confidence interval and test a hypothesis about the difference between these two population means, that is, $\mu_1 - \mu_2$. Let $\bar{x}_1$ be the mean of a sample taken from the first population and $\bar{x}_2$ be the mean of a sample taken from the second population. Then, $\bar{x}_1 - \bar{x}_2$ is the sample statistic that is used to make an interval estimate and to test a hypothesis about $\mu_1 - \mu_2$. This section discusses how to make confidence intervals and test hypotheses about $\mu_1 - \mu_2$ when the two samples are large and independent. As discussed in earlier chapters, in the case of μ, a sample is considered to be large if it contains 30 or more observations. The concept of independent and dependent samples is explained next.

10.1.1   INDEPENDENT VERSUS DEPENDENT SAMPLES

Two samples are **independent** if they are drawn from two different populations and the elements of one sample have no relationship to the elements of the second sample. If the elements of the two samples are somehow related, then the samples are said to be **dependent**. Thus, in two independent samples, the selection of one sample has no effect on the selection of the second sample.

> **INDEPENDENT VERSUS DEPENDENT SAMPLES**   Two samples drawn from two populations are *independent* if the selection of one sample from one population does not affect the selection of the second sample from the second population. Otherwise, the samples are *dependent*.

Examples 10–1 and 10–2 illustrate independent and dependent samples, respectively.

Illustrating two independent samples.

EXAMPLE 10-1   Suppose we want to estimate the difference between the mean salaries of all male and all female executives. To do so, we draw two samples, one from the population of male executives and another from the population of female executives. These two samples are *independent* because they are drawn from two different populations, and the samples have no effect on each other.   ∎

Illustrating two dependent samples.

EXAMPLE 10-2   Suppose we want to estimate the difference between the mean weights of all participants before and after a weight loss program. To accomplish this, suppose we take a sample of 40 participants and measure their weights before and after the completion of this program. Note that these two samples include the same 40 participants. This is an example of two *dependent* samples. Such samples are also called *paired* or *matched* samples.   ∎

This section and Sections 10.2, 10.3, and 10.5 discuss how to make confidence intervals and test hypotheses about the difference between two population parameters when samples are independent. Section 10.4 discusses how to make confidence intervals and test hypotheses about the difference between two population means when samples are dependent.

10.1.2 MEAN, STANDARD DEVIATION, AND SAMPLING DISTRIBUTION OF $\bar{x}_1 - \bar{x}_2$

Suppose we draw two (independent) large samples from two different populations that are referred to as population 1 and population 2. Let

$\mu_1 =$ the mean of population 1

$\mu_2 =$ the mean of population 2

$\sigma_1 =$ the standard deviation of population 1

$\sigma_2 =$ the standard deviation of population 2

$n_1 =$ the size of the sample drawn from population 1 ($n_1 \geq 30$)

$n_2 =$ the size of the sample drawn from population 2 ($n_2 \geq 30$)

$\bar{x}_1 =$ the mean of the sample drawn from population 1

$\bar{x}_2 =$ the mean of the sample drawn from population 2

Then, from the central limit theorem, $\bar{x}_1$ is approximately normally distributed with mean μ_1 and standard deviation $\sigma_1/\sqrt{n_1}$, and $\bar{x}_2$ is approximately normally distributed with mean μ_2 and standard deviation $\sigma_2/\sqrt{n_2}$.

Using these results, we can make the following statements about the mean, the standard deviation, and the shape of the sampling distribution of $\bar{x}_1 - \bar{x}_2$. Figure 10.1 shows the sampling distribution of $\bar{x}_1 - \bar{x}_2$.

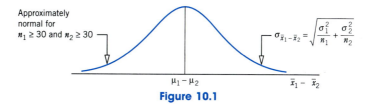

Figure 10.1

1. The mean of $\bar{x}_1 - \bar{x}_2$, denoted by $\mu_{\bar{x}_1 - \bar{x}_2}$, is

$$\mu_{\bar{x}_1 - \bar{x}_2} = \mu_1 - \mu_2$$

2. The standard deviation of $\bar{x}_1 - \bar{x}_2$, denoted by $\sigma_{\bar{x}_1 - \bar{x}_2}$, is[1]

$$\sigma_{\bar{x}_1 - \bar{x}_2} = \sqrt{\frac{\sigma_1^2}{n_1} + \frac{\sigma_2^2}{n_2}}$$

3. Regardless of the shapes of the two populations, the shape of the sampling distribution of $\bar{x}_1 - \bar{x}_2$ is approximately normal. This is so because the difference between two normally distributed random variables is also normally distributed. Note again that for this to hold true, both samples must be large.

---

[1]The formula for the standard deviation of $\bar{x}_1 - \bar{x}_2$ can also be written as

$$\sigma_{\bar{x}_1 - \bar{x}_2} = \sqrt{\sigma_{\bar{x}_1}^2 + \sigma_{\bar{x}_2}^2}$$

where $\sigma_{\bar{x}_1} = \sigma_1/\sqrt{n_1}$ and $\sigma_{\bar{x}_2} = \sigma_2/\sqrt{n_2}$.

SAMPLING DISTRIBUTION, MEAN, AND STANDARD DEVIATION OF $\bar{x}_1 - \bar{x}_2$ For two large and independent samples selected from two different populations, the *sampling distribution* of $\bar{x}_1 - \bar{x}_2$ is (approximately) normal with its *mean* and *standard deviation* as follows:

$$\mu_{\bar{x}_1 - \bar{x}_2} = \mu_1 - \mu_2 \quad \text{and} \quad \sigma_{\bar{x}_1 - \bar{x}_2} = \sqrt{\frac{\sigma_1^2}{n_1} + \frac{\sigma_2^2}{n_2}}$$

However, we usually do not know the standard deviations σ_1 and σ_2 of the two populations. In such cases, we replace $\sigma_{\bar{x}_1 - \bar{x}_2}$ by its point estimator $s_{\bar{x}_1 - \bar{x}_2}$, which is calculated as follows.

ESTIMATE OF THE STANDARD DEVIATION OF $\bar{x}_1 - \bar{x}_2$ The value of $s_{\bar{x}_1 - \bar{x}_2}$, which gives an *estimate* of $\sigma_{\bar{x}_1 - \bar{x}_2}$, is calculated as

$$s_{\bar{x}_1 - \bar{x}_2} = \sqrt{\frac{s_1^2}{n_1} + \frac{s_2^2}{n_2}}$$

where s_1 and s_2 are the standard deviations of the two samples selected from the two populations.

Thus, when both samples are large, the sampling distribution of $\bar{x}_1 - \bar{x}_2$ is approximately normal. Consequently, in such cases, we use the normal distribution to make a confidence interval and to test a hypothesis about $\mu_1 - \mu_2$.

10.1.3   INTERVAL ESTIMATION OF $\mu_1 - \mu_2$

By constructing a confidence interval for $\mu_1 - \mu_2$, we find the difference between the means of two populations. For example, we may want to find the difference between the mean heights of male and female adults. The difference between the two sample means, $\bar{x}_1 - \bar{x}_2$, is the point estimator of the difference between the two population means, $\mu_1 - \mu_2$. Again, in this section we assume that the two samples are large and independent. When these assumptions hold true, we use the normal distribution to make a confidence interval for the difference between the two population means. The following formula gives the interval estimation for $\mu_1 - \mu_2$.

CONFIDENCE INTERVAL FOR $\mu_1 - \mu_2$
   The $(1 - \alpha)100\%$ *confidence interval for $\mu_1 - \mu_2$* is

$$(\bar{x}_1 - \bar{x}_2) \pm z\sigma_{\bar{x}_1 - \bar{x}_2} \quad \text{if } \sigma_1 \text{ and } \sigma_2 \text{ are known}$$

$$(\bar{x}_1 - \bar{x}_2) \pm zs_{\bar{x}_1 - \bar{x}_2} \quad \text{if } \sigma_1 \text{ and } \sigma_2 \text{ are not known}$$

The value of z is obtained from the normal distribution table for the given confidence level. The values of $\sigma_{\bar{x}_1 - \bar{x}_2}$ and $s_{\bar{x}_1 - \bar{x}_2}$ are calculated as explained earlier.

Examples 10–3 and 10–4 illustrate the procedure to construct a confidence interval for $\mu_1 - \mu_2$ for large samples. In Example 10–3 the population standard deviations are known, and in Example 10–4 they are not known.

Constructing a confidence interval for $\mu_1 - \mu_2$: σ_1 and σ_2 known.

EXAMPLE 10-3   According to the American Medical Association, the average annual earnings of radiologists in the United States are $230,000 and those of surgeons are $225,000 (*Fortune*, September 27, 1999). Suppose that these means are based on random samples of 300 radiologists and 400 surgeons and that the population standard deviations of the annual earnings of radiologists and surgeons are $28,000 and $32,000, respectively.

(a)   What is the point estimate of $\mu_1 - \mu_2$? What is the margin of error?

(b)   Construct a 97% confidence interval for the difference between the mean annual earnings of radiologists and surgeons.

Solution   Let us refer to all radiologists as population 1 and all surgeons as population 2. The respective samples, then, are samples 1 and 2. Let μ_1 and μ_2 be the mean annual earnings of populations 1 and 2, and let $\bar{x}_1$ and $\bar{x}_2$ be the means of the respective samples. From the given information,

$$\text{For radiologists:} \quad n_1 = 300, \quad \bar{x}_1 = \$230{,}000, \quad \sigma_1 = \$28{,}000$$

$$\text{For surgeons:} \quad n_2 = 400, \quad \bar{x}_2 = \$225{,}000, \quad \sigma_2 = \$32{,}000$$

(a)   The point estimate of $\mu_1 - \mu_2$ is given by the value of $\bar{x}_1 - \bar{x}_2$. Thus,

$$\text{Point estimate of } \mu_1 - \mu_2 = \$230{,}000 - \$225{,}000 = \textbf{\$5000}$$

To find the margin of error, first we calculate $\sigma_{\bar{x}_1 - \bar{x}_2}$ as follows:

$$\sigma_{\bar{x}_1 - \bar{x}_2} = \sqrt{\frac{\sigma_1^2}{n_1} + \frac{\sigma_2^2}{n_2}} = \sqrt{\frac{(28{,}000)^2}{300} + \frac{(32{,}000)^2}{400}} = \$2274.496281$$

Then,

$$\text{Margin of error} = \pm 1.96\sigma_{\bar{x}_1 - \bar{x}_2} = \pm 1.96(2274.496281) = \pm\textbf{\$4458.01}$$

(b)   The confidence level is $1 - \alpha = .97$. In part a we calculated the standard deviation of $\bar{x}_1 - \bar{x}_2$ as

$$\sigma_{\bar{x}_1 - \bar{x}_2} = \$2274.496281$$

Next, we find the z value for the 97% confidence level. From the normal distribution table, this value of z is 2.17. Finally, substituting all the values in the confidence interval formula, we obtain the 97% confidence interval for $\mu_1 - \mu_2$:

$$(\bar{x}_1 - \bar{x}_2) \pm z\sigma_{\bar{x}_1 - \bar{x}_2} = (230{,}000 - 225{,}000) \pm 2.17(2274.496281)$$

$$= 5000 \pm 4935.66 = \textbf{\$64.34 to \$9935.66}$$

Thus, with 97% confidence we can state that the difference between the mean annual earnings of radiologists and surgeons is $64.34 to $9935.66.   ∎

Constructing a confidence interval for $\mu_1 - \mu_2$: σ_1 and σ_2 not known.

EXAMPLE 10-4   According to the most recent estimates by the U.S. Department of Labor, the average retirement age in the United States is 62.7 years for women and 62.2 years for men. Assume that these means are based on samples of 800 women and 1000 men and that the sample standard deviations for the two samples are 4.5 years and 3 years, respectively.

Find a 99% confidence interval for the difference between the corresponding population means.

Solution   Let μ_1 be the mean retirement age for women and μ_2 the mean retirement age for men. Let $\bar{x}_1$ and $\bar{x}_2$ be the means of the respective samples. From the given information,

$$\text{For women:} \quad n_1 = 800, \quad \bar{x}_1 = 62.7 \text{ years}, \quad s_1 = 4.5 \text{ years}$$

$$\text{For men:} \quad n_2 = 1000, \quad \bar{x}_2 = 62.2 \text{ years}, \quad s_2 = 3 \text{ years}$$

The confidence level is $1 - \alpha = .99$.

Because σ_1 and σ_2 are not known, we use $s_{\bar{x}_1 - \bar{x}_2}$ in the confidence interval formula. The value of $s_{\bar{x}_1 - \bar{x}_2}$ is

$$s_{\bar{x}_1 - \bar{x}_2} = \sqrt{\frac{s_1^2}{n_1} + \frac{s_2^2}{n_2}} = \sqrt{\frac{(4.5)^2}{800} + \frac{(3)^2}{1000}} = .18523634$$

From the normal distribution table, the z value for a 99% confidence level is (approximately) 2.58. The 99% confidence interval for $\mu_1 - \mu_2$ is

$$(\bar{x}_1 - \bar{x}_2) \pm z s_{\bar{x}_1 - \bar{x}_2} = (62.7 - 62.2) \pm 2.58(.18523634)$$

$$= .5 \pm .48 = \textbf{.02 to .98 year}$$

Thus, with 99% confidence we can state that the difference between the two population means is .02 to .98 year. ∎

10.1.4   HYPOTHESIS TESTING ABOUT $\mu_1 - \mu_2$

It is often necessary to compare the means of two populations. For example, we may want to know if the mean price of houses in Chicago is the same as that in Los Angeles. Similarly, we may be interested in knowing if, on average, American children spend fewer hours in school than Japanese children do. In both these cases, we will perform a test of hypothesis about $\mu_1 - \mu_2$. The alternative hypothesis in a test of hypothesis may be that the means of the two populations are different, or that the mean of the first population is greater than the mean of the second population, or that the mean of the first population is less than the mean of the second population. These three situations are described next.

1. Testing an alternative hypothesis that the means of two populations are different is equivalent to $\mu_1 \neq \mu_2$, which is the same as $\mu_1 - \mu_2 \neq 0$.
2. Testing an alternative hypothesis that the mean of the first population is greater than the mean of the second population is equivalent to $\mu_1 > \mu_2$, which is the same as $\mu_1 - \mu_2 > 0$.
3. Testing an alternative hypothesis that the mean of the first population is less than the mean of the second population is equivalent to $\mu_1 < \mu_2$, which is the same as $\mu_1 - \mu_2 < 0$.

The procedure followed to perform a test of hypothesis about the difference between two population means is similar to the one used to test hypotheses about single population parameters in Chapter 9. The procedure involves the same five steps that were used in Chapter 9 to test hypotheses about μ and p. Because we are dealing with large (and independent) samples in this section, we will use the normal distribution to conduct a test of hypothesis about $\mu_1 - \mu_2$.

> **TEST STATISTIC z FOR $\bar{x}_1 - \bar{x}_2$** The value of the *test statistic z for $\bar{x}_1 - \bar{x}_2$* is computed as
>
> $$z = \frac{(\bar{x}_1 - \bar{x}_2) - (\mu_1 - \mu_2)}{\sigma_{\bar{x}_1 - \bar{x}_2}}$$
>
> The value of $\mu_1 - \mu_2$ is substituted from H_0. If the values of σ_1 and σ_2 are not known, we replace $\sigma_{\bar{x}_1 - \bar{x}_2}$ by $s_{\bar{x}_1 - \bar{x}_2}$ in the formula.

Making a two-tailed test of hypothesis about $\mu_1 - \mu_2$: large samples.

EXAMPLE 10–5 Refer to Example 10–3 about the mean annual earnings of radiologists and surgeons. Test at the 1% significance level whether the mean annual earnings of radiologists and surgeons are different.

Solution From the information given in Example 10–3,

> For radiologists: $n_1 = 300$, $\bar{x}_1 = \$230{,}000$, $\sigma_1 = \$28{,}000$
>
> For surgeons: $n_2 = 400$, $\bar{x}_2 = \$225{,}000$, $\sigma_2 = \$32{,}000$

Let μ_1 and μ_2 be the mean annual earnings of radiologists and surgeons, respectively.

Step 1. *State the null and alternative hypotheses.*

We are to test whether the two population means are different. The two possibilities are:

(i) The mean annual earnings of radiologists and surgeons are not different. In other words, $\mu_1 = \mu_2$, which can be written as $\mu_1 - \mu_2 = 0$.

(ii) The mean annual earnings of radiologists and surgeons are different. That is, $\mu_1 \neq \mu_2$, which can be written as $\mu_1 - \mu_2 \neq 0$.

From these two possibilities, the null and alternative hypotheses are

> H_0: $\mu_1 - \mu_2 = 0$ (The two population means are not different)
>
> H_1: $\mu_1 - \mu_2 \neq 0$ (The two population means are different)

Step 2. *Select the distribution to use.*

Because $n_1 > 30$ and $n_2 > 30$, both samples are large. Therefore, the sampling distribution of $\bar{x}_1 - \bar{x}_2$ is approximately normal, and we use the normal distribution to perform the hypothesis test.

Step 3. *Determine the rejection and nonrejection regions.*

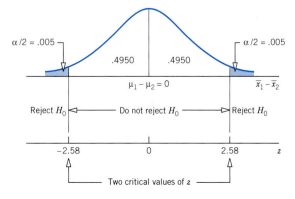

Figure 10.2

The significance level is given to be .01. The $\neq$ sign in the alternative hypothesis indicates that the test is two-tailed. The area in each tail of the normal distribution curve is $\alpha/2 = .01/2 = .005$. The critical values of z for .005 area in each tail of the normal distribution curve are (approximately) 2.58 and -2.58. These values are shown in Figure 10.2.

Step 4. *Calculate the value of the test statistic.*

The value of the test statistic z for $\bar{x}_1 - \bar{x}_2$ is computed as follows:

$$\sigma_{\bar{x}_1-\bar{x}_2} = \sqrt{\frac{\sigma_1^2}{n_1} + \frac{\sigma_2^2}{n_2}} = \sqrt{\frac{(28,000)^2}{300} + \frac{(32,000)^2}{400}} = \$2274.496281$$

$$z = \frac{(\bar{x}_1 - \bar{x}_2) - (\mu_1 - \mu_2)}{\sigma_{\bar{x}_1-\bar{x}_2}} = \frac{(230,000 - 225,000) - \overset{\text{From } H_0}{0}}{2274.496281} = 2.20$$

Step 5. *Make a decision.*

Because the value of the test statistic $z = 2.20$ falls in the nonrejection region, we fail to reject the null hypothesis H_0. Therefore, we conclude that the mean annual earnings of radiologists and surgeons are not different. Note that we cannot be sure that the two population means are not different. All we can say is that the evidence from the two samples indicates that the corresponding population means are not different. ∎

Making a right-tailed test of hypothesis about $\mu_1 - \mu_2$: large samples.

EXAMPLE 10–6 Refer to Example 10–4 about the mean retirement ages for women and men. Test at the 2.5% significance level whether the mean retirement age for all women is higher than the mean retirement age for all men.

Solution From the information given in Example 10–4,

For women: $n_1 = 800,$ $\bar{x}_1 = 62.7$ years, $s_1 = 4.5$ years

For men: $n_2 = 1000,$ $\bar{x}_2 = 62.2$ years, $s_2 = 3$ years

Let μ_1 and μ_2 be the mean retirement ages for women and men, respectively.

Step 1. *State the null and alternative hypotheses.*

The two possibilities are:

(i)   The mean retirement age for all women is not higher than that for all men, which can be written as $\mu_1 = \mu_2$ or $\mu_1 - \mu_2 = 0$.

(ii) The mean retirement age for all women is higher than that for all men, which can be written as $\mu_1 > \mu_2$ or $\mu_1 - \mu_2 > 0$.

The null and alternative hypotheses are

$$H_0: \mu_1 - \mu_2 = 0 \qquad (\mu_1 \text{ is equal to } \mu_2)$$

$$H_1: \mu_1 - \mu_2 > 0 \qquad (\mu_1 \text{ is greater than } \mu_2)$$

Step 2. *Select the distribution to use.*

Because $n_1 > 30$ and $n_2 > 30$, both samples are large. Therefore, the sampling distribution of $\bar{x}_1 - \bar{x}_2$ is approximately normal, and we use the normal distribution to perform the hypothesis test.

Step 3. *Determine the rejection and nonrejection regions.*

The significance level is given to be .025. The $>$ sign in the alternative hypothesis indicates that the test is right-tailed. Consequently, the critical value of z is 1.96, as shown in Figure 10.3.

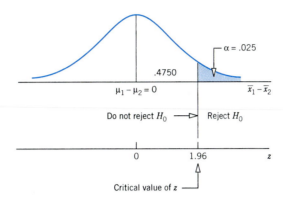

.4750

$\mu_1 - \mu_2 = 0$

$\bar{x}_1 - \bar{x}_2$

$\alpha = .025$

Do not reject H_0 ⟶ ▷ Reject H_0

0   1.96 z

Critical value of z

Figure 10.3

Step 4. *Calculate the value of the test statistic.*

The value of the test statistic z for $\bar{x}_1 - \bar{x}_2$ is computed as follows:

$$s_{\bar{x}_1 - \bar{x}_2} = \sqrt{\frac{s_1^2}{n_1} + \frac{s_2^2}{n_2}} = \sqrt{\frac{(4.5)^2}{800} + \frac{(3)^2}{1000}} = .18523634$$

From H_0

$$z = \frac{(\bar{x}_1 - \bar{x}_2) - (\mu_1 - \mu_2)}{s_{\bar{x}_1 - \bar{x}_2}} = \frac{(62.7 - 62.2) - 0}{.18523634} = 2.70$$

Step 5. *Make a decision.*

Because the value of the test statistic $z = 2.70$ falls in the rejection region, we reject the null hypothesis H_0. Therefore, we conclude that the mean retirement age for women is higher than the mean retirement age for men. ■

EXERCISES

■ **Concepts and Procedures**

10.1 Briefly explain the meaning of independent and dependent samples. Give one example of each.

10.2 Describe the sampling distribution of $\bar{x}_1 - \bar{x}_2$ for large and independent samples. What are the mean and standard deviation of this sampling distribution?

10.3 The following information is obtained from two independent samples selected from two populations.

$$n_1 = 240 \quad \bar{x}_1 = 5.56 \quad s_1 = 1.65$$
$$n_2 = 270 \quad \bar{x}_2 = 4.80 \quad s_2 = 1.58$$

a. What is the point estimate of $\mu_1 - \mu_2$?
b. Construct a 99% confidence interval for $\mu_1 - \mu_2$.

10.4   The following information is obtained from two independent samples selected from two populations.

$$n_1 = 300 \qquad \bar{x}_1 = 22.0 \qquad s_1 = 4.9$$
$$n_2 = 250 \qquad \bar{x}_2 = 27.6 \qquad s_2 = 4.5$$

a.   What is the point estimate of $\mu_1 - \mu_2$?
b.   Construct a 95% confidence interval for $\mu_1 - \mu_2$.

10.5   Refer to the information given in Exercise 10.3. Test at the 5% significance level if the two population means are different.

10.6   Refer to the information given in Exercise 10.4. Test at the 1% significance level if the two population means are different.

10.7   Refer to the information given in Exercise 10.3. Test at the 1% significance level if μ_1 is greater than μ_2.

10.8   Refer to the information given in Exercise 10.4. Test at the 5% significance level if μ_1 is less than μ_2.

■ Applications

10.9   According to the U.S. Bureau of the Census, the average annual salaries of men and women over 25 years of age who hold full-time year-round jobs but do not have a high school diploma are $24,604 and $17,129, respectively (*USA TODAY,* September 11, 1998). Suppose that these mean salaries are based on random samples of 200 men and 210 women, and further assume that the sample standard deviation is $3350 for the annual salaries of such men and $3270 for the annual salaries of such women.

a.   Let μ_1 and μ_2 be the population means of annual salaries for such men and women, respectively. Find the point estimate of $\mu_1 - \mu_2$ and its margin of error.
b.   Find a 99% confidence interval for $\mu_1 - \mu_2$.
c.   Using the 1% level of significance, can you conclude that the mean annual salaries for all such men and for all such women are different?

10.10   According to the U.S. Bureau of the Census, the average annual salaries of men and women over 25 years of age who hold full-time year-round jobs and have a bachelor's degree are $53,102 and $37,304, respectively (*USA TODAY,* September 11, 1998). Suppose that these means are based on random samples of 108 men and 102 women, and further assume that the sample standard deviation is $5040 for the annual salaries of such men and $4910 for the annual salaries of such women.

a.   Let μ_1 and μ_2 be the population means of annual salaries for such men and women, respectively. Find the point estimate of $\mu_1 - \mu_2$ and its margin of error.
b.   Find a 97% confidence interval for $\mu_1 - \mu_2$.
c.   Test at the 2% significance level whether the mean annual salary for all such men is different from the mean annual salary for all such women.

10.11   According to data published in *Newsweek* magazine, in 1998 the average starting salary for history majors was $26,820 and the starting salary for psychology majors was $25,689 (*Newsweek,* February 1, 1999). Suppose that these mean starting salaries are based on random samples of 105 history majors and 111 psychology majors, and further assume that the standard deviations for the starting salaries of these majors were $3850 and $3710, respectively, in 1998.

a.   Let μ_1 and μ_2 be the 1998 mean starting salaries for all history and all psychology majors, respectively. What are the point estimate of $\mu_1 - \mu_2$ and its margin of error?
b.   Find a 90% confidence interval for $\mu_1 - \mu_2$.
c.   Test at the 1% significance level whether the 1998 mean starting salary for all history majors exceeded that for all psychology majors.

10.12   According to the U.S. Department of Labor, the average hourly earnings of workers in the manufacturing sector were $13.40 in July 1998, and the corresponding average for workers in wholesale trade was $13.98 (*Bureau of Labor Statistics News,* August 18, 1998). Suppose that these means are based on random samples of 1100 workers selected from the manufacturing sector and 1200 workers selected from wholesale trade. Assume that the sample standard deviations for the hourly earnings of workers in the manufacturing sector and in wholesale trade were $1.10 and $1.06, respectively, in July 1998.

a. Let μ_1 and μ_2 be the population means of hourly earnings for workers in the manufacturing sector and in wholesale trade, respectively. What are the point estimate of $\mu_1 - \mu_2$ and its margin of error?
b. Construct a 97% confidence interval for $\mu_1 - \mu_2$.
c. Using the 2% significance level, can you conclude that the July 1998 mean hourly earnings of all workers in the manufacturing sector were less than those of all workers in wholesale trade?

10.13 A business consultant wanted to investigate if providing day-care facilities on premises by companies reduces the absentee rate of working mothers with 6-year-old or younger children. She took a sample of 45 such mothers from companies that provide day-care facilities on premises. These mothers missed an average of 6.4 days from work last year with a standard deviation of 1.20 days. Another sample of 50 such mothers taken from companies that do not provide day-care facilities on premises showed that these mothers missed an average of 9.3 days last year with a standard deviation of 1.85 days.

a. Construct a 98% confidence interval for the difference between the two population means.
b. Using the 2.5% significance level, can you conclude that the mean number of days missed per year by mothers working for companies that provide day-care facilities on premises is less than the mean number of days missed per year by mothers working for companies that do not provide day-care facilities on premises?
c. What are the Type I error and its probability for the test of hypothesis in part b? Explain.

10.14 According to the Centers for Disease Control and Prevention/National Center for Health Statistics, the average length of stay in short-stay hospitals was 5.8 days for men and 4.9 days for women (*Advance Data,* August 31, 1998). Suppose that these means were based on random samples of 110 such stays for men and 124 such stays for women. Assume that the population standard deviations were 2.1 days for men and 2.0 days for women.

a. Construct a 99% confidence interval for the difference between the two population means.
b. Using the 1% significance level, can you conclude that the mean stay in such hospitals for all men is longer than that for all women?
c. What are the Type I error and its probability for the hypothesis test in part b? Explain.

10.15 Psychologists Claudia Mueller and Carol Dweck studied the motivational effects of praising students for work well done. In six studies involving 412 fifth-graders, these researchers found that telling students they were smart when they performed a task well tended to lower their motivation for future assignments. When students were told that their success was due to their efforts, however, they were more likely to work hard in the future (*The Hartford Courant,* August 22, 1998). Suppose that 100 fifth-graders who passed a test are randomly divided into two groups of 50 each. Students in the first group are told that they did well because they are intelligent, whereas students in the second group are told that their success was the result of hard work. Subsequently, these 100 students are given another test. The first group's mean score is 72 with a standard deviation of 12 and the second group's mean score is 80 with a standard deviation of 10. Assume that these 100 students are representative of all fifth-graders.

a. Construct a 95% confidence interval for the difference in the mean scores for the populations of the two groups.
b. Using the 2.5% level of significance, can you conclude that the mean score on such a test for all fifth-graders who are praised for their intelligence is lower than that of all fifth-graders who are praised for their efforts?
c. What would your decision in part b be if the probability of making a Type I error were zero? Explain.

10.16 According to IRS data, taxpayers with adjusted annual gross incomes of $60,000 to $75,000 averaged $6486 in deductions for medical expenses, whereas those with adjusted annual gross incomes of $75,001 to $100,000 averaged $6291 in such deductions (*USA TODAY,* March 5, 1999). Suppose that these means were obtained from random samples of 1000 taxpayers in each income group. Further assume that the sample standard deviations for such deductions were $1600 for the taxpayers in the first income group and $1550 for the taxpayers in the second income group.

a. Construct a 99% confidence interval for the difference between the two population means.
b. Using the 1% significance level, can you conclude that the mean deductions for medical expenses are different for the two income groups?
c. What would your decision in part b be if the probability of making a Type I error were zero? Explain.

10.17 The management at the New Century Bank claims that the mean waiting time for all customers at its branches is less than that at the Public Bank, which is its main competitor. A business consulting firm took a sample of 200 customers from the New Century Bank and found that they waited an average of 4.5 minutes with a standard deviation of 1.2 minutes before being served. Another sample of 300 customers taken from the Public Bank showed that these customers waited an average of 4.75 minutes with a standard deviation of 1.5 minutes before being served.

a. Make a 97% confidence interval for the difference between the two population means.
b. Test at the 2.5% significance level whether the claim of the management of the New Century Bank is true.
*c. Calculate the *p*-value for the test of part b. Based on this *p*-value, would you reject the null hypothesis if $\alpha = .01$? What if $\alpha = .05$?

10.18 Maine Mountain Dairy claims that its 8-ounce low-fat yogurt cups contain, on average, fewer calories than the 8-ounce low-fat yogurt cups produced by a competitor. A consumer agency wanted to check this claim. A sample of 50 such yogurt cups produced by this company showed that they contained an average of 141 calories per cup with a standard deviation of 5.5 calories. A sample of 40 such yogurt cups produced by its competitor showed that they contained an average of 144 calories per cup with a standard deviation of 6.4 calories.

a. Make a 98% confidence interval for the difference between the mean number of calories in the 8-ounce low-fat yogurt cups produced by the two companies.
b. Test at the 1% significance level whether Maine Mountain Dairy's claim is true.
*c. Calculate the *p*-value for the test of part b. Based on this *p*-value, would you reject the null hypothesis if $\alpha = .005$? What if $\alpha = .025$?

10.2   INFERENCES ABOUT THE DIFFERENCE BETWEEN TWO POPULATION MEANS FOR SMALL AND INDEPENDENT SAMPLES: EQUAL STANDARD DEVIATIONS

Many times, due to either budget constraints or the nature of the populations, it may not be possible to take large samples to make inferences about the difference between two population means. This section discusses how to make a confidence interval and test a hypothesis about the difference between two population means when the samples are small ($n_1 < 30$ and $n_2 < 30$) and independent. Our main assumption in this case is that the two populations from which the two samples are drawn are (approximately) normally distributed. If this assumption is true and we know the population standard deviations, we can still use the normal distribution to make inferences about $\mu_1 - \mu_2$ when samples are small and independent. However, we usually do not know the population standard deviations σ_1 and σ_2. In such cases, we replace the normal distribution by the *t* distribution to make inferences about $\mu_1 - \mu_2$ for small and independent samples. We will make one more assumption in this section that the standard deviations of the two populations are equal. In other words, we assume that although σ_1 and σ_2 are unknown, they are equal. The case when σ_1 and σ_2 are not equal will be discussed in Section 10.3.

> **WHEN TO USE THE t DISTRIBUTION TO MAKE INFERENCES ABOUT $\mu_1 - \mu_2$**   The t distribution is used to make inferences about $\mu_1 - \mu_2$ when the following assumptions hold true:
>
> 1. The two populations from which the two samples are drawn are (approximately) normally distributed.
> 2. The samples are small ($n_1 < 30$ and $n_2 < 30$) and independent.
> 3. The standard deviations σ_1 and σ_2 of the two populations are unknown but they are assumed to be equal; that is, $\sigma_1 = \sigma_2$.

When the standard deviations of the two populations are equal, we can use σ for both σ_1 and σ_2. Since σ is unknown, we replace it by its point estimator s_p, which is called the **pooled sample standard deviation** (hence, the subscript p). The value of s_p is computed by using the information from the two samples as follows.

> **POOLED STANDARD DEVIATION FOR TWO SAMPLES**   The *pooled standard deviation for two samples* is computed as
>
> $$s_p = \sqrt{\frac{(n_1 - 1)s_1^2 + (n_2 - 1)s_2^2}{n_1 + n_2 - 2}}$$
>
> where n_1 and n_2 are the sizes of the two samples and s_1^2 and s_2^2 are the variances of the two samples.

In this formula, $n_1 - 1$ are the degrees of freedom for sample 1, $n_2 - 1$ are the degrees of freedom for sample 2, and $n_1 + n_2 - 2$ are the *degrees of freedom for the two samples taken together.*

When s_p is used as an estimator of σ, the standard deviation $\sigma_{\bar{x}_1 - \bar{x}_2}$ of $\bar{x}_1 - \bar{x}_2$ is estimated by $s_{\bar{x}_1 - \bar{x}_2}$. The value of $s_{\bar{x}_1 - \bar{x}_2}$ is calculated by using the following formula.

> **ESTIMATOR OF THE STANDARD DEVIATION OF $\bar{x}_1 - \bar{x}_2$**   The *estimator of the standard deviation of $\bar{x}_1 - \bar{x}_2$* is
>
> $$s_{\bar{x}_1 - \bar{x}_2} = s_p \sqrt{\frac{1}{n_1} + \frac{1}{n_2}}$$

Now we are ready to discuss the procedures that are used to make confidence intervals and test hypotheses about $\mu_1 - \mu_2$ for small and independent samples selected from two populations with unknown but equal standard deviations.

10.2.1   INTERVAL ESTIMATION OF $\mu_1 - \mu_2$

As was mentioned earlier, the difference between the two sample means, $\bar{x}_1 - \bar{x}_2$, is the point estimator of the difference between the two population means, $\mu_1 - \mu_2$. The following formula gives the confidence interval for $\mu_1 - \mu_2$ when the t distribution is used.

CONFIDENCE INTERVAL FOR $\mu_1 - \mu_2$

The $(1 - \alpha)100\%$ *confidence interval for* $\mu_1 - \mu_2$ is

$$(\bar{x}_1 - \bar{x}_2) \pm t s_{\bar{x}_1 - \bar{x}_2}$$

where the value of t is obtained from the t distribution table for the given confidence level and $n_1 + n_2 - 2$ degrees of freedom, and $s_{\bar{x}_1 - \bar{x}_2}$ is calculated as explained on page 441.

Example 10–7 describes the procedure to make a confidence interval for $\mu_1 - \mu_2$ using the t distribution.

Constructing a confidence interval for $\mu_1 - \mu_2$: small and independent samples and $\sigma_1 = \sigma_2$.

EXAMPLE 10–7   A consumer agency wanted to estimate the difference in the mean amounts of caffeine in two brands of coffee. The agency took a sample of 15 one-pound jars of Brand I coffee that showed the mean amount of caffeine in these jars to be 80 milligrams per jar with a standard deviation of 5 milligrams. Another sample of 12 one-pound jars of Brand II coffee gave a mean amount of caffeine equal to 77 milligrams per jar with a standard deviation of 6 milligrams. Construct a 95% confidence interval for the difference between the mean amounts of caffeine in one-pound jars of these two brands of coffee. Assume that the two populations are normally distributed and that the standard deviations of the two populations are equal.

Solution   Let μ_1 and μ_2 be the mean amounts of caffeine per jar in all one-pound jars of Brands I and II, respectively, and let $\bar{x}_1$ and $\bar{x}_2$ be the means of the two respective samples. From the given information,

$$n_1 = 15 \qquad \bar{x}_1 = 80 \text{ milligrams} \qquad s_1 = 5 \text{ milligrams}$$
$$n_2 = 12 \qquad \bar{x}_2 = 77 \text{ milligrams} \qquad s_2 = 6 \text{ milligrams}$$

The confidence level is $1 - \alpha = .95$.

First we calculate the standard deviation of $\bar{x}_1 - \bar{x}_2$ as follows:

$$s_p = \sqrt{\frac{(n_1 - 1)s_1^2 + (n_2 - 1)s_2^2}{n_1 + n_2 - 2}} = \sqrt{\frac{(15 - 1)(5)^2 + (12 - 1)(6)^2}{15 + 12 - 2}} = 5.46260011$$

$$s_{\bar{x}_1 - \bar{x}_2} = s_p \sqrt{\frac{1}{n_1} + \frac{1}{n_2}} = (5.46260011)\sqrt{\frac{1}{15} + \frac{1}{12}} = 2.11565593$$

Next, to find the t value from the t distribution table, we need to know the area in each tail of the t distribution curve and the degrees of freedom.

$$\text{Area in each tail} = \alpha/2 = .5 - (.95/2) = .025$$

$$\text{Degrees of freedom} = n_1 + n_2 - 2 = 15 + 12 - 2 = 25$$

The t value for $df = 25$ and .025 area in the right tail of the t distribution curve is 2.060. The 95% confidence interval for $\mu_1 - \mu_2$ is

$$(\bar{x}_1 - \bar{x}_2) \pm t s_{\bar{x}_1 - \bar{x}_2} = (80 - 77) \pm 2.060(2.11565593)$$

$$= 3 \pm 4.36 = \textbf{-1.36 to 7.36 milligrams}$$

Thus, with 95% confidence we can state that based on these two sample results, the difference in the mean amounts of caffeine in one-pound jars of these two brands of coffee lies between -1.36 and 7.36 milligrams. Because the lower limit of the interval is negative, it is possible that the mean amount of caffeine is greater in the second brand than in the first brand of coffee.

Note that the value of $\bar{x}_1 - \bar{x}_2$, which is $80 - 77 = 3$, gives the point estimate of $\mu_1 - \mu_2$. ∎

10.2.2 HYPOTHESIS TESTING ABOUT $\mu_1 - \mu_2$

When the three assumptions mentioned in Section 10.2 are satisfied, the t distribution is applied to make a hypothesis test about the difference between two population means. The test statistic in this case is t, which is calculated as follows.

> **TEST STATISTIC t FOR $\bar{x}_1 - \bar{x}_2$** The value of the *test statistic t for $\bar{x}_1 - \bar{x}_2$* is computed as
>
> $$t = \frac{(\bar{x}_1 - \bar{x}_2) - (\mu_1 - \mu_2)}{s_{\bar{x}_1 - \bar{x}_2}}$$
>
> The value of $\mu_1 - \mu_2$ in this formula is substituted from the null hypothesis and $s_{\bar{x}_1 - \bar{x}_2}$ is calculated as explained on page 441.

Examples 10–8 and 10–9 illustrate how a test of hypothesis about the difference between two population means for small and independent samples that are selected from two populations with equal standard deviations is conducted using the t distribution.

Making a two-tailed test of hypothesis about $\mu_1 - \mu_2$: small and independent samples and $\sigma_1 = \sigma_2$.

EXAMPLE 10–8 A sample of 14 cans of Brand I diet soda gave the mean number of calories of 23 per can with a standard deviation of 3 calories. Another sample of 16 cans of Brand II diet soda gave the mean number of calories of 25 per can with a standard deviation of 4 calories. At the 1% significance level, can you conclude that the mean numbers of calories per can are different for these two brands of diet soda? Assume that the calories per can of diet soda are normally distributed for each of the two brands and that the standard deviations for the two populations are equal.

Solution Let μ_1 and μ_2 be the mean number of calories per can for diet soda of Brand I and Brand II, respectively, and let $\bar{x}_1$ and $\bar{x}_2$ be the means of the respective samples. From the given information,

$$n_1 = 14 \quad \bar{x}_1 = 23 \quad s_1 = 3$$
$$n_2 = 16 \quad \bar{x}_2 = 25 \quad s_2 = 4$$

The significance level is $\alpha = .01$.

Step 1. *State the null and alternative hypotheses.*

We are to test for the difference in the mean numbers of calories per can for the two brands. The null and alternative hypotheses are

$$H_0: \mu_1 - \mu_2 = 0 \quad \text{(The mean numbers of calories are not different)}$$

$$H_1: \mu_1 - \mu_2 \neq 0 \quad \text{(The mean numbers of calories are different)}$$

Step 2. *Select the distribution to use.*

The two populations are normally distributed, the samples are small and independent, and the standard deviations of the two populations are unknown but equal. Consequently, we will use the *t* distribution.

Step 3. *Determine the rejection and nonrejection regions.*

The ≠ sign in the alternative hypothesis indicates that the test is two-tailed. The significance level is .01. Hence,

$$\text{Area in each tail} = \alpha/2 = .01/2 = .005$$

$$\text{Degrees of freedom} = n_1 + n_2 - 2 = 14 + 16 - 2 = 28$$

The critical values of *t* for *df* = 28 and .005 area in each tail of the *t* distribution curve are −2.763 and 2.763, as shown in Figure 10.4.

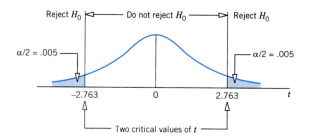

Figure 10.4

Step 4. *Calculate the value of the test statistic.*

The value of the test statistic *t* for $\bar{x}_1 - \bar{x}_2$ is computed as follows:

$$s_p = \sqrt{\frac{(n_1 - 1)s_1^2 + (n_2 - 1)s_2^2}{n_1 + n_2 - 2}} = \sqrt{\frac{(14 - 1)(3)^2 + (16 - 1)(4)^2}{14 + 16 - 2}} = 3.57071421$$

$$s_{\bar{x}_1 - \bar{x}_2} = s_p \sqrt{\frac{1}{n_1} + \frac{1}{n_2}} = (3.57071421) \sqrt{\frac{1}{14} + \frac{1}{16}} = 1.30674760$$

$$t = \frac{(\bar{x}_1 - \bar{x}_2) - (\mu_1 - \mu_2)}{s_{\bar{x}_1 - \bar{x}_2}} = \frac{(23 - 25) - 0}{1.30674760} = -1.531$$

From H_0

Step 5. *Make a decision.*

Because the value of the test statistic *t* = −1.531 for $\bar{x}_1 - \bar{x}_2$ falls in the nonrejection region, we fail to reject the null hypothesis. Consequently we conclude that there is no difference in the mean numbers of calories per can for the two brands of diet soda. The difference in $\bar{x}_1$ and $\bar{x}_2$ observed for the two samples may have occurred due to sampling error only. ∎

Making a right-tailed test of hypothesis about $\mu_1 - \mu_2$: small and independent samples and $\sigma_1 = \sigma_2$.

EXAMPLE 10-9   A sample of 15 children from New York State showed that the mean time they spend watching television is 28.5 hours per week with a standard deviation of 4 hours. Another sample of 16 children from California showed that the mean time spent by them watching television is 23.25 hours per week with a standard deviation of 5 hours. Using a 2.5% significance level, can you conclude that the mean time spent watching television by children in New York State is greater than that for children in California? Assume that the times spent watching television by children have a normal distribution for both populations and that the standard deviations for the two populations are equal.

Solution Let the children from New York State be referred to as population 1 and those from California as population 2. Let μ_1 and μ_2 be the mean time spent watching television by children in populations 1 and 2, respectively, and let $\bar{x}_1$ and $\bar{x}_2$ be the mean time spent watching television by children in the respective samples. From the given information,

$$n_1 = 15 \qquad \bar{x}_1 = 28.5 \text{ hours} \qquad s_1 = 4 \text{ hours}$$
$$n_2 = 16 \qquad \bar{x}_2 = 23.25 \text{ hours} \qquad s_2 = 5 \text{ hours}$$

The significance level is $\alpha = .025$.

Step 1. *State the null and alternative hypotheses.*

The two possible decisions are:

(i) The mean time spent watching television by children in New York State is not greater than that for children in California. This can be written as $\mu_1 = \mu_2$ or $\mu_1 - \mu_2 = 0$.

(ii) The mean time spent watching television by children in New York State is greater than that for children in California. This can be written as $\mu_1 > \mu_2$ or $\mu_1 - \mu_2 > 0$.

Hence, the null and alternative hypotheses are

$$H_0: \mu_1 - \mu_2 = 0$$
$$H_1: \mu_1 - \mu_2 > 0$$

Note that the null hypothesis can also be written as $\mu_1 - \mu_2 \le 0$.

Step 2. *Select the distribution to use.*

The two populations are normally distributed, the samples are small and independent, and the standard deviations of the two populations are unknown but equal. Consequently, we use the t distribution to make the test.

Step 3. *Determine the rejection and nonrejection regions.*

The $>$ sign in the alternative hypothesis indicates that the test is right-tailed. The significance level is .025.

$$\text{Area in the right tail of the } t \text{ distribution} = \alpha = .025$$

$$\text{Degrees of freedom} = n_1 + n_2 - 2 = 15 + 16 - 2 = 29$$

From the t distribution table, the critical value of t for $df = 29$ and .025 area in the right tail of the t distribution is 2.045. This value is shown in Figure 10.5.

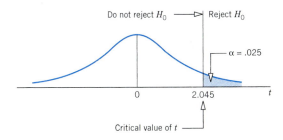

Figure 10.5

Step 4. *Calculate the value of the test statistic.*

The value of the test statistic t for $\bar{x}_1 - \bar{x}_2$ is computed as follows:

$$s_p = \sqrt{\frac{(n_1 - 1)s_1^2 + (n_2 - 1)s_2^2}{n_1 + n_2 - 2}} = \sqrt{\frac{(15 - 1)(4)^2 + (16 - 1)(5)^2}{15 + 16 - 2}} = 4.54479619$$

$$s_{\bar{x}_1 - \bar{x}_2} = s_p\sqrt{\frac{1}{n_1} + \frac{1}{n_2}} = (4.54479619)\sqrt{\frac{1}{15} + \frac{1}{16}} = 1.63338904$$

From H_0

$$t = \frac{(\bar{x}_1 - \bar{x}_2) - (\mu_1 - \mu_2)}{s_{\bar{x}_1 - \bar{x}_2}} = \frac{(28.5 - 23.25) - 0}{1.63338904} = 3.214$$

Step 5. *Make a decision.*

Because the value of the test statistic $t = 3.214$ for $\bar{x}_1 - \bar{x}_2$ falls in the rejection region, we reject the null hypothesis H_0. Hence, we conclude that children in New York State spend, on average, more time watching TV than children in California. ∎

EXERCISES

■ **Concepts and Procedures**

10.19 Explain what conditions must hold true to use the t distribution to make a confidence interval and to test a hypothesis about $\mu_1 - \mu_2$ for two independent samples selected from two populations with unknown but equal standard deviations.

10.20 The following information was obtained from two independent samples selected from two normally distributed populations with unknown but equal standard deviations.

$$n_1 = 25 \qquad \bar{x}_1 = 12.50 \qquad s_1 = 3.75$$
$$n_2 = 20 \qquad \bar{x}_2 = 14.60 \qquad s_2 = 3.10$$

a. What is the point estimate of $\mu_1 - \mu_2$?
b. Construct a 95% confidence interval for $\mu_1 - \mu_2$.

10.21 The following information was obtained from two independent samples selected from two normally distributed populations with unknown but equal standard deviations.

$$n_1 = 18 \qquad \bar{x}_1 = 33.75 \qquad s_1 = 5.25$$
$$n_2 = 20 \qquad \bar{x}_2 = 28.50 \qquad s_2 = 4.55$$

a. What is the point estimate of $\mu_1 - \mu_2$?
b. Construct a 99% confidence interval for $\mu_1 - \mu_2$.

10.22 Refer to the information given in Exercise 10.20. Test at the 5% significance level if the two population means are different.

10.23 Refer to the information given in Exercise 10.21. Test at the 1% significance level if the two population means are different.

10.24 Refer to the information given in Exercise 10.20. Test at the 1% significance level if μ_1 is less than μ_2.

10.25 Refer to the information given in Exercise 10.21. Test at the 5% significance level if μ_1 is greater than μ_2.

10.26 The following information was obtained from two independent samples selected from two normally distributed populations with unknown but equal standard deviations.

Sample 1:	27	39	25	33	21	35	30	26	25	31	35	30	28
Sample 2:	24	28	23	25	24	22	29	26	29	28	19	29	

a. Let μ_1 be the mean of population 1 and μ_2 be the mean of population 2. What is the point estimate of $\mu_1 - \mu_2$?
b. Construct a 98% confidence interval for $\mu_1 - \mu_2$.
c. Test at the 1% significance level if μ_1 is greater than μ_2.

10.27 The following information was obtained from two independent samples selected from two normally distributed populations with unknown but equal standard deviations.

Sample 1:	13	14	9	12	8	10	5	10	9	12	16	
Sample 2:	16	18	11	19	14	17	13	16	17	18	22	12

a. Let μ_1 be the mean of population 1 and μ_2 be the mean of population 2. What is the point estimate of $\mu_1 - \mu_2$?
b. Construct a 99% confidence interval for $\mu_1 - \mu_2$.
c. Test at the 2.5% significance level if μ_1 is lower than μ_2.

■ **Applications**

10.28 The management of a supermarket wanted to investigate whether the male customers spend less money, on average, than the female customers. A sample of 25 male customers who shopped at this supermarket showed that they spent an average of $80 with a standard deviation of $17.50. Another sample of 20 female customers who shopped at the same supermarket showed that they spent an average of $96 with a standard deviation of $14.40. Assume that the amounts spent at this supermarket by all male and all female customers are normally distributed with equal but unknown population standard deviations.

a. Construct a 99% confidence interval for the difference between the mean amounts spent by all male and all female customers at this supermarket.
b. Using the 5% significance level, can you conclude that the mean amount spent by all male customers at this supermarket is less than that of all female customers?

10.29 A manufacturing company is interested in buying one of two different kinds of machines. The company tested the two machines for production purposes. The first machine was run for 8 hours. It produced an average of 126 items per hour with a standard deviation of 9 items. The second machine was run for 10 hours. It produced an average of 117 items per hour with a standard deviation of 6 items. Assume that the production per hour for each machine is (approximately) normally distributed. Further assume that the standard deviations of the hourly productions of the two populations are equal.

a. Make a 95% confidence interval for the difference between the two population means.
b. Using the 2.5% significance level, can you conclude that the mean number of items produced per hour by the first machine is greater than that of the second machine?

10.30 An insurance company wants to know if the average speed at which men drive cars is greater than that of women drivers. The company took a random sample of 27 cars driven by men on a highway and found the mean speed to be 72 miles per hour with a standard deviation of 2.2 miles per hour. Another sample of 18 cars driven by women on the same highway gave a mean speed of 68 miles per hour with a standard deviation of 2.5 miles per hour. Assume that the speeds at which all men and all women drive cars on this highway are both normally distributed with the same population standard deviation.

a. Construct a 98% confidence interval for the difference between the mean speeds of cars driven by all men and all women on this highway.
b. Test at the 1% significance level whether the mean speed of cars driven by all men drivers on this highway is greater than that of cars driven by all women drivers.

10.31 According to the U.S. Census Bureau, the mean annual salary for men over 25 years old with a doctoral degree is $86,436, whereas that for women of the same age and with the same degree is $62,169 (*USA TODAY,* September 11, 1998). Suppose that these means are based on random samples of 27 men and 25 women and that the sample standard deviations are $5210 for men and $5080 for women. Further assume that the annual salaries of all such men and women have normal distributions with equal but unknown standard deviations.

a. Construct a 90% confidence interval for the difference between the mean annual salaries for all such men and all such women.

b. Test at the 5% significance level whether the mean annual salaries for all such men and all such women are different.

10.32 Quadro Corporation has two supermarket stores in a city. The company's quality control department wanted to check if the customers are equally satisfied with the service provided at these two stores. A sample of 25 customers selected from Supermarket I produced a mean satisfaction index of 7.6 (on a scale of 1 to 10, 1 being the lowest and 10 being the highest) with a standard deviation of .75. Another sample of 28 customers selected from Supermarket II produced a mean satisfaction index of 8.1 with a standard deviation of .59. Assume that the customer satisfaction index for each supermarket has a normal distribution with the same population standard deviation.

a. Construct a 98% confidence interval for the difference between the mean satisfaction indexes for all customers for the two supermarkets.

b. Test at the 1% significance level whether the mean satisfaction indexes for all customers for the two supermarkets are different.

10.33 A company claims that its medicine, Brand A, provides faster relief from pain than another company's medicine, Brand B. A researcher tested both brands of medicine on two groups of randomly selected patients. The results of the test are given in the following table. The mean and standard deviation of relief times are in minutes.

Brand	Sample Size	Mean of Relief Times	Standard Deviation of Relief Times
A	25	44	11
B	22	49	9

a. Construct a 99% confidence interval for the difference between the mean relief times for the two brands of medicine.

b. Test at the 1% significance level whether the mean relief time for Brand A is less than that for Brand B.

Assume that the two populations are normally distributed with equal standard deviations.

10.34 A study of time spent on housework found that employed married men spent an average of 8.8 hours per week on housework, whereas employed unmarried men averaged 6.9 hours per week on housework (America's Use of Time Project, University of Maryland, *American Demographics,* November 1998). Suppose that these means were based on random samples of 21 employed married men and 23 employed unmarried men with sample standard deviations of 2.2 hours and 1.9 hours, respectively. Assume that the two populations from which these samples were selected are normally distributed with equal but unknown standard deviations.

a. Construct a 90% confidence interval for the difference between the population means for these two groups of men.

b. Test at the 5% significance level whether the mean time spent on housework per week is greater for the population of employed married men than for the population of employed unmarried men.

10.35 According to data from the College Board, the average cost for room and board at public four-year residential colleges for 1997–1998 was $4361 and the corresponding average cost for private colleges was $5549 (*The Chronicle of Higher Education Almanac,* August 28, 1998). Suppose that these means were obtained from random samples of 15 public and 14 private four-year residential colleges. Further assume that the sample standard deviations of such costs were $1200 for public colleges and

$1540 for private colleges. Assume that the 1997–1998 room and board costs for both groups of colleges are normally distributed with equal but unknown standard deviations.

a. Construct a 95% confidence interval for the difference between the two population means.
b. Using the 1% significance level, can you conclude that the 1997–1998 mean cost of room and board at all public four-year residential colleges was lower than the 1997–1998 mean of such costs at all private four-year residential colleges?

10.3 INFERENCES ABOUT THE DIFFERENCE BETWEEN TWO POPULATION MEANS FOR SMALL AND INDEPENDENT SAMPLES: UNEQUAL STANDARD DEVIATIONS

Section 10.2 explained how to make inferences about the difference between two population means using the t distribution when the standard deviations of the two populations are unknown but equal and certain other assumptions hold true. Now, what if all other assumptions of Section 10.2 hold true but the population standard deviations are not only unknown but also unequal? In this case, the procedures used to make confidence intervals and to test hypotheses about $\mu_1 - \mu_2$ remain similar to the ones we learned in Sections 10.2.1 and 10.2.2 except for two differences. When the population standard deviations are unknown and not equal, the degrees of freedom are no longer given by $n_1 + n_2 - 2$ and the standard deviation of $\bar{x}_1 - \bar{x}_2$ is not calculated using the pooled standard deviation s_p.

DEGREES OF FREEDOM If

1. the two populations from which the samples are drawn are (approximately) normally distributed

2. the two samples are small (that is, $n_1 < 30$ and $n_2 < 30$) and independent

3. the two population standard deviations are unknown and unequal

then the t distribution is used to make inferences about $\mu_1 - \mu_2$ and the *degrees of freedom* for the t distribution are given by

$$df = \frac{\left(\dfrac{s_1^2}{n_1} + \dfrac{s_2^2}{n_2}\right)^2}{\dfrac{\left(\dfrac{s_1^2}{n_1}\right)^2}{n_1 - 1} + \dfrac{\left(\dfrac{s_2^2}{n_2}\right)^2}{n_2 - 1}}$$

The number given by this formula is always rounded down for *df*.

Because the standard deviations of the two populations are not known, we use $s_{\bar{x}_1 - \bar{x}_2}$ as a point estimator of $\sigma_{\bar{x}_1 - \bar{x}_2}$. The following formula is used to calculate the standard deviation $s_{\bar{x}_1 - \bar{x}_2}$ of $\bar{x}_1 - \bar{x}_2$.

ESTIMATE OF THE STANDARD DEVIATION OF $\bar{x}_1 - \bar{x}_2$ The value of $s_{\bar{x}_1 - \bar{x}_2}$ is calculated as

$$s_{\bar{x}_1 - \bar{x}_2} = \sqrt{\frac{s_1^2}{n_1} + \frac{s_2^2}{n_2}}$$

10.3.1   INTERVAL ESTIMATION OF $\mu_1 - \mu_2$

Again, the difference between the two sample means, $\bar{x}_1 - \bar{x}_2$, is the point estimator of the difference between the two population means, $\mu_1 - \mu_2$. The following formula gives the confidence interval for $\mu_1 - \mu_2$ when the t distribution is used and the population standard deviations are unknown and presumed to be unequal.

> **CONFIDENCE INTERVAL FOR $\mu_1 - \mu_2$** The $(1 - \alpha)100\%$ *confidence interval for* $\mu_1 - \mu_2$ is
>
> $$(\bar{x}_1 - \bar{x}_2) \pm ts_{\bar{x}_1 - \bar{x}_2}$$
>
> where the value of t is obtained from the t distribution table for a given confidence level and the degrees of freedom given by the formula mentioned on page 449, and $s_{\bar{x}_1 - \bar{x}_2}$ is calculated as explained on the same page.

Example 10–10 describes how to construct a confidence interval for $\mu_1 - \mu_2$ when the standard deviations of the two populations are unknown and unequal.

Constructing a confidence interval for $\mu_1 - \mu_2$: small and independent samples and $\sigma_1 \neq \sigma_2$.

EXAMPLE 10-10   According to Example 10–7 of Section 10.2.1, a sample of 15 one-pound jars of coffee of Brand I showed that the mean amount of caffeine in these jars is 80 milligrams per jar with a standard deviation of 5 milligrams. Another sample of 12 one-pound coffee jars of Brand II gave a mean amount of caffeine equal to 77 milligrams per jar with a standard deviation of 6 milligrams. Construct a 95% confidence interval for the difference between the mean amounts of caffeine in one-pound coffee jars of these two brands. Assume that the two populations are normally distributed and that the standard deviations of the two populations are not equal.

Solution   Let μ_1 and μ_2 be the mean amount of caffeine per jar in all one-pound jars of Brands I and II, respectively, and let $\bar{x}_1$ and $\bar{x}_2$ be the means of the two respective samples.

From the given information,

$$n_1 = 15 \quad \bar{x}_1 = 80 \text{ milligrams} \quad s_1 = 5 \text{ milligrams}$$
$$n_2 = 12 \quad \bar{x}_2 = 77 \text{ milligrams} \quad s_2 = 6 \text{ milligrams}$$

The confidence level is $1 - \alpha = .95$.

First, we calculate the standard deviation of $\bar{x}_1 - \bar{x}_2$ as follows:

$$s_{\bar{x}_1 - \bar{x}_2} = \sqrt{\frac{s_1^2}{n_1} + \frac{s_2^2}{n_2}} = \sqrt{\frac{(5)^2}{15} + \frac{(6)^2}{12}} = 2.16024690$$

Next, to find the t value from the t distribution table, we need to know the area in each tail of the t distribution curve and the degrees of freedom.

$$\text{Area in each tail} = \alpha/2 = .5 - (.95/2) = .025$$

$$df = \frac{\left(\frac{s_1^2}{n_1} + \frac{s_2^2}{n_2}\right)^2}{\frac{\left(\frac{s_1^2}{n_1}\right)^2}{n_1 - 1} + \frac{\left(\frac{s_2^2}{n_2}\right)^2}{n_2 - 1}} = \frac{\left(\frac{(5)^2}{15} + \frac{(6)^2}{12}\right)^2}{\frac{\left(\frac{(5)^2}{15}\right)^2}{15 - 1} + \frac{\left(\frac{(6)^2}{12}\right)^2}{12 - 1}} = 21.42 \approx 21$$

Note that the degrees of freedom are always rounded down as in this calculation. From the

t distribution table, the t value for $df = 21$ and .025 area in the right tail of the t distribution curve is 2.080. The 95% confidence interval for $\mu_1 - \mu_2$ is

$$(\bar{x}_1 - \bar{x}_2) \pm ts_{\bar{x}_1 - \bar{x}_2} = (80 - 77) \pm 2.080(2.16024690)$$

$$= 3 \pm 4.49 = -1.49 \text{ to } 7.49$$

Thus, with 95% confidence we can state that based on these two sample results, the difference in the mean amounts of caffeine in one-pound jars of these two brands of coffee is between -1.49 and 7.49 milligrams. ∎

Comparing this confidence interval with the one obtained in Example 10–7, we observe that the two confidence intervals are very close. From this we can conclude that even if the standard deviations of the two populations are not equal and we use the procedure of Section 10.2.1 to make a confidence interval for $\mu_1 - \mu_2$, the margin of error will be small as long as the difference between the two standard deviations is not too large.

10.3.2 HYPOTHESIS TESTING ABOUT $\mu_1 - \mu_2$

When the standard deviations of the two populations are unknown and unequal, with the other conditons of Section 10.2.2 holding true, we use the t distribution to make a test of hypothesis about $\mu_1 - \mu_2$. This procedure differs from the one in Section 10.2.2 only in the calculation of degrees of freedom for the t distribution and the standard deviation of $\bar{x}_1 - \bar{x}_2$. The df and the standard deviation of $\bar{x}_1 - \bar{x}_2$ in this case are given by the formulas on page 449.

> **TEST STATISTIC t FOR $\bar{x}_1 - \bar{x}_2$** The value of the *test statistic t for* $\bar{x}_1 - \bar{x}_2$ is computed as
>
> $$t = \frac{(\bar{x}_1 - \bar{x}_2) - (\mu_1 - \mu_2)}{s_{\bar{x}_1 - \bar{x}_2}}$$
>
> The value of $\mu_1 - \mu_2$ in this formula is substituted from the null hypothesis and $s_{\bar{x}_1 - \bar{x}_2}$ is calculated as explained earlier.

Example 10–11 illustrates the procedure used to conduct a test of hypothesis about $\mu_1 - \mu_2$ when the standard deviations of the two populations are unknown and unequal.

Making a two-tailed test of hypothesis about $\mu_1 - \mu_2$: small and independent samples and $\sigma_1 \neq \sigma_2$.

EXAMPLE 10–11 According to Example 10–8 of Section 10.2.2, a sample of 14 cans of Brand I diet soda gave the mean number of calories per can as 23 with a standard deviation of 3 calories. Another sample of 16 cans of Brand II diet soda gave the mean number of calories of 25 per can with a standard deviation of 4 calories. Test at the 1% significance level whether the mean numbers of calories per can of diet soda are different for these two brands. Assume that the calories per can of diet soda are normally distributed for each of these two brands and that the standard deviations for the two populations are not equal.

Solution Let μ_1 and μ_2 be the mean number of calories for all cans of diet soda of Brand I and Brand II, respectively, and let $\bar{x}_1$ and $\bar{x}_2$ be the means of the respective samples. From the given information,

$$n_1 = 14 \quad \bar{x}_1 = 23 \quad s_1 = 3$$
$$n_2 = 16 \quad \bar{x}_2 = 25 \quad s_2 = 4$$

The significance level is $\alpha = .01$.

Step 1. *State the null and alternative hypotheses.*

We are to test for the difference in the mean numbers of calories per can for the two brands. The null and alternative hypotheses are

$$H_0: \mu_1 - \mu_2 = 0 \qquad \text{(The mean numbers of calories are not different)}$$

$$H_1: \mu_1 - \mu_2 \neq 0 \qquad \text{(The mean numbers of calories are different)}$$

Step 2. *Select the distribution to use.*

The two populations are normally distributed, the samples are small and independent, and the standard deviations of the two populations are unknown and unequal. Consequently, we use the *t* distribution to make the test.

Step 3. *Determine the rejection and nonrejection regions.*

The $\neq$ sign in the alternative hypothesis indicates that the test is two-tailed. The significance level is .01. Hence,

$$\text{Area in each tail} = \alpha/2 = .01/2 = .005$$

The degrees of freedom are calculated as follows:

$$df = \frac{\left(\dfrac{s_1^2}{n_1} + \dfrac{s_2^2}{n_2}\right)^2}{\dfrac{\left(\dfrac{s_1^2}{n_1}\right)^2}{n_1 - 1} + \dfrac{\left(\dfrac{s_2^2}{n_2}\right)^2}{n_2 - 1}} = \frac{\left(\dfrac{(3)^2}{14} + \dfrac{(4)^2}{16}\right)^2}{\dfrac{\left(\dfrac{(3)^2}{14}\right)^2}{14 - 1} + \dfrac{\left(\dfrac{(4)^2}{16}\right)^2}{16 - 1}} = 27.41 \approx 27$$

From the *t* distribution table, the critical values of *t* for *df* = 27 and .005 area in each tail of the *t* distribution curve are −2.771 and 2.771. These values are shown in Figure 10.6.

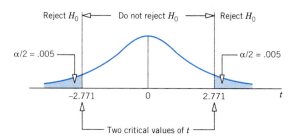

Figure 10.6

Step 4. *Calculate the value of the test statistic.*

The value of the test statistic *t* for $\bar{x}_1 - \bar{x}_2$ is computed as follows:

$$s_{\bar{x}_1 - \bar{x}_2} = \sqrt{\frac{s_1^2}{n_1} + \frac{s_2^2}{n_2}} = \sqrt{\frac{(3)^2}{14} + \frac{(4)^2}{16}} = 1.28173989$$

$$t = \frac{(\bar{x}_1 - \bar{x}_2) - (\mu_1 - \mu_2)}{s_{\bar{x}_1 - \bar{x}_2}} = \frac{(23 - 25) - 0}{1.28173989} = -1.560$$

From H_0

Step 5. *Make a decision.*

Because the value of the test statistic *t* = −1.560 for $\bar{x}_1 - \bar{x}_2$ falls in the nonrejection region, we fail to reject the null hypothesis. Hence, there is no difference in the mean numbers of calories per can for the two brands of diet soda. The difference in $\bar{x}_1$ and $\bar{x}_2$ observed for the two samples may have occurred due to sampling error only. ∎

 Remember The degrees of freedom for the procedures to make a confidence interval and to test a hypothesis about $\mu_1 - \mu_2$ learned in Sections 10.3.1 and 10.3.2 are always rounded down.

EXERCISES

■ Concepts and Procedures

10.36 Assuming that the two populations are normally distributed with unequal and unknown population standard deviations, construct a 95% confidence interval for $\mu_1 - \mu_2$ for the following.

$$n_1 = 24 \qquad \bar{x}_1 = 20.50 \qquad s_1 = 3.90$$
$$n_2 = 16 \qquad \bar{x}_2 = 22.60 \qquad s_2 = 5.15$$

 10.37 Assuming that the two populations are normally distributed with unequal and unknown population standard deviations, construct a 99% confidence interval for $\mu_1 - \mu_2$ for the following.

$$n_1 = 15 \qquad \bar{x}_1 = 52.61 \qquad s_1 = 3.55$$
$$n_2 = 19 \qquad \bar{x}_2 = 43.75 \qquad s_2 = 5.40$$

10.38 Refer to Exercise 10.36. Test at the 5% significance level if the two population means are different.

 10.39 Refer to Exercise 10.37. Test at the 1% significance level if the two population means are different.

10.40 Refer to Exercise 10.36. Test at the 1% significance level if μ_1 is less than μ_2.

10.41 Refer to Exercise 10.37. Test at the 5% significance level if μ_1 is greater than μ_2.

■ Applications

10.42 According to the information given in Exercise 10.28, a sample of 25 male customers who shopped at a supermarket showed that they spent an average of $80 with a standard deviation of $17.50. Another sample of 20 female customers who shopped at the same supermarket showed that they spent an average of $96 with a standard deviation of $14.40. Assume that the amounts spent at this supermarket by all male and all female customers are normally distributed with unequal and unknown population standard deviations.

a. Construct a 99% confidence interval for the difference between the mean amounts spent by all male and all female customers at this supermarket.
b. Using the 5% significance level, can you conclude that the mean amount spent by all male customers at this supermarket is less than that of all female customers?

10.43 According to Exercise 10.29, a manufacturing company is interested in buying one of two different kinds of machines. The company tested the two machines for production purposes. The first machine was run for 8 hours. It produced an average of 126 items per hour with a standard deviation of 9 items. The second machine was run for 10 hours. It produced an average of 117 items per hour with a standard deviation of 6 items. Assume that the production per hour for each machine is (approximately) normally distributed. Further assume that the standard deviations of the hourly productions of the two populations are unequal.

a. Make a 95% confidence interval for the difference between the two population means.
b. Using the 2.5% significance level, can you conclude that the mean number of items produced per hour by the first machine is greater than that of the second machine?

10.44　According to Exercise 10.30, an insurance company wants to know if the average speed at which men drive cars is higher than that of women drivers. The company took a random sample of 27 cars driven by men on a highway and found the mean speed to be 72 miles per hour with a standard deviation of 2.2 miles. Another sample of 18 cars driven by women on the same highway gave a mean speed of 68 miles per hour with a standard deviation of 2.5 miles. Assume that the speeds at which all men and all women drive cars on this highway are both normally distributed with unequal population standard deviations.

a.　Construct a 98% confidence interval for the difference between the mean speeds of cars driven by all men and all women on this highway.

b.　Test at the 1% significance level whether the mean speed of cars driven by all men drivers on this highway is higher than that of cars driven by all women drivers.

10.45　Refer to Exercise 10.31. According to the U.S. Census Bureau, the mean annual salary for men over 25 years old with a doctoral degree is $86,436, whereas that for women of the same age and with the same degree is $62,169 (*USA TODAY,* September 11, 1998). Suppose that these means are based on random samples of 27 men and 25 women and that the sample standard deviations are $5210 for men and $5080 for women. Further assume that the annual salaries of all such men and women have normal distributions with unequal and unknown standard deviations.

a.　Construct a 90% confidence interval for the difference between the mean annual salaries for all such men and all such women.

b.　Test at the 5% significance level whether the mean annual salaries for all such men and all such women are different.

10.46　As mentioned in Exercise 10.32, Quadro Corporation has two supermarket stores in a city. The company's quality control department wanted to check if the customers are equally satisfied with the service provided at these two stores. A sample of 25 customers selected from Supermarket I produced a mean satisfaction index of 7.6 (on a scale of 1 to 10, 1 being the lowest and 10 being the highest) with a standard deviation of .75. Another sample of 28 customers selected from Supermarket II produced a mean satisfaction index of 8.1 with a standard deviation of .59. Assume that the customer satisfaction index for each supermarket has a normal distribution with a different population standard deviation.

a.　Construct a 98% confidence interval for the difference between the mean satisfaction indexes for all customers for the two supermarkets.

b.　Test at the 1% significance level whether the mean satisfaction indexes for all customers for the two supermarkets are different.

10.47　As mentioned in Exercise 10.33, a company claims that its medicine, Brand A, provides faster relief from pain than another company's medicine, Brand B. A researcher tested both brands of medicine on two groups of randomly selected patients. The results of the test are given in the following table. The mean and standard deviation of relief times are in minutes.

Brand	Sample Size	Mean of Relief Times	Standard Deviation of Relief Times
A	25	44	11
B	22	49	9

a.　Construct a 99% confidence interval for the difference between the mean relief times for the two brands of medicine.

b.　Test at the 1% significance level whether the mean relief time for Brand A is less than that for Brand B.

Assume that the two populations are normally distributed with unequal standard deviations.

10.48　Refer to Exercise 10.34. A study of time spent on housework found that employed married men spent an average of 8.8 hours per week on housework, whereas employed unmarried men averaged 6.9 hours per week on housework (America's Use of Time Project, University of Maryland, *American De-*

mographics, November 1998). Suppose that these means were based on random samples of 21 employed married men and 23 employed unmarried men with sample standard deviations of 2.2 hours and 1.9 hours, respectively. Assume that the two populations from which these samples were selected are normally distributed with unequal and unknown standard deviations.

a. Construct a 90% confidence interval for the difference between the population means for these two groups of men.

b. Test at the 5% significance level whether the mean time spent on housework per week is greater for the population of employed married men than that for the population of employed unmarried men.

10.49 Refer to Exercise 10.35. According to data from the College Board, the average cost for room and board at public four-year residential colleges for 1997–1998 was $4361 and the corresponding average cost for private colleges was $5549 (*The Chronicle of Higher Education Almanac,* August 28, 1998). Suppose that these means were obtained from random samples of 15 public and 14 private four-year residential colleges. Further assume that the sample standard deviations of such costs were $1200 for public colleges and $1540 for private colleges. Assume that the 1997–1998 room and board costs for both groups of colleges are normally distributed with unequal and unknown standard deviations.

a. Construct a 95% confidence interval for the difference between the two population means.

b. Using the 1% significance level, can you conclude that the 1997–1998 mean cost of room and board at all public four-year residential colleges was lower than the 1997–1998 mean of such costs at all private four-year residential colleges?

10.4 INFERENCES ABOUT THE DIFFERENCE BETWEEN TWO POPULATION MEANS FOR PAIRED SAMPLES

Sections 10.1, 10.2, and 10.3 were concerned with estimation and hypothesis testing about the difference between two population means when the two samples were drawn independently from two different populations. This section describes estimation and hypothesis-testing procedures for the difference between two population means when the samples are dependent.

In a case of two dependent samples, two data values—one for each sample—are collected from the same source (or element) and, hence, these are also called **paired** or **matched samples.** For example, we may want to make inferences about the mean weight loss for members of a health club after they have gone through an exercise program for a certain period of time. To do so, suppose we select a sample of 15 members of this health club and record their weights before and after the program. In this example, both sets of data are collected from the same 15 persons, once before and once after the program. Thus, although there are two samples, they contain the same 15 persons. This is an example of paired (or dependent or matched) samples. The procedures to make confidence intervals and test hypotheses in the case of paired samples are different from the ones for independent samples discussed in earlier sections of this chapter.

> **PAIRED OR MATCHED SAMPLES** Two samples are said to be *paired* or *matched samples* when for each data value collected from one sample there is a corresponding data value collected from the second sample, and both these data values are collected from the same source.

As another example of paired samples, suppose an agronomist wants to measure the effect of a new brand of fertilizer on the yield of potatoes. To do so, he selects 10 pieces of land and divides each piece into two portions. Then he randomly assigns one of the two portions from each piece of land to grow potatoes without using fertilizer (or using some other brand of fertilizer). The second portion from each piece of land is used to grow potatoes with the new brand of fertilizer. Thus, he will have 10 pairs of data values. Then, using the procedure to be discussed in this section, he will make inferences about the difference in the mean yields of potatoes with and without the new fertilizer.

The question arises, why does the agronomist not choose 10 pieces of land on which to grow potatoes without using the new brand of fertilizer and another 10 pieces of land to grow potatoes by using the new brand of fertilizer? If he does so, the effect of the fertilizer might be confused with the effects due to soil differences at different locations. Thus, he will not be able to isolate the effect of the new brand of fertilizer on the yield of potatoes. Consequently, the results will not be reliable. By choosing 10 pieces of land and then dividing each of them into two portions, the researcher decreases the possibility that the difference in the productivities of different pieces of land affects the results.

In paired samples, the difference between the two data values for each element of the two samples is denoted by d. This value of d is called the **paired difference**. We then treat all the values of d as one sample and make inferences applying procedures similar to the ones used for one-sample cases in Chapters 8 and 9. Note that because each source (or element) gives a pair of values (one for each of the two data sets), each sample contains the same number of values. That is, both samples are the same size. Therefore, we denote the (common) **sample size** by n, which gives the number of paired difference values denoted by d. The **degrees of freedom** for the paired samples are $n - 1$. Let

μ_d = the mean of the paired differences for the population

σ_d = the standard deviation of the paired differences for the population

$\bar{d}$ = the mean of the paired differences for the sample

s_d = the standard deviation of the paired differences for the sample

n = the number of paired difference values

> **MEAN AND STANDARD DEVIATION OF THE PAIRED DIFFERENCES FOR SAMPLES** The values of $\bar{d}$ and s_d are calculated as[2]
>
> $$\bar{d} = \frac{\Sigma d}{n}$$
>
> $$s_d = \sqrt{\frac{\Sigma d^2 - \frac{(\Sigma d)^2}{n}}{n - 1}}$$

In paired samples, instead of using $\bar{x}_1 - \bar{x}_2$ as the sample statistic to make inferences about $\mu_1 - \mu_2$, we use the sample statistic $\bar{d}$ to make inferences about μ_d. Actually the value of $\bar{d}$ is always equal to $\bar{x}_1 - \bar{x}_2$, and the value of μ_d is always equal to $\mu_1 - \mu_2$.

[2]The basic formula to calculate s_d is

$$s_d = \sqrt{\frac{\Sigma(d - \bar{d})^2}{n - 1}}$$

However, we will not use this formula to make calculations in this chapter.

> **SAMPLING DISTRIBUTION, MEAN, AND STANDARD DEVIATION OF $\bar{d}$** If the number of paired values is large ($n \geq 30$), then because of the central limit theorem, the *sampling distribution* of $\bar{d}$ is approximately normal with its *mean* and *standard deviation* given as
>
> $$\mu_{\bar{d}} = \mu_d \quad \text{and} \quad \sigma_{\bar{d}} = \frac{\sigma_d}{\sqrt{n}}$$

In cases when $n \geq 30$, the normal distribution can be used to make inferences about μ_d. However, in cases of paired samples, the sample sizes are usually small and σ_d is unknown. Then, assuming that the paired differences for the population are (approximately) normally distributed, the normal distribution is replaced by the t distribution to make inferences about μ_d. When σ_d is not known, the standard deviation of $\bar{d}$ is estimated by $s_{\bar{d}} = s_d/\sqrt{n}$.

> **ESTIMATE OF THE STANDARD DEVIATION OF PAIRED DIFFERENCES** If
>
> 1. n is less than 30
> 2. σ_d is not known
> 3. the population of paired differences is (approximately) normally distributed
>
> then the t distribution is used to make inferences about μ_d. The standard deviation $\sigma_{\bar{d}}$ of $\bar{d}$ is estimated by $s_{\bar{d}}$, which is calculated as
>
> $$s_{\bar{d}} = \frac{s_d}{\sqrt{n}}$$

Sections 10.4.1 and 10.4.2 describe the procedures used to make a confidence interval and test a hypothesis about μ_d when σ_d is unknown and n is small. The inferences are made using the t distribution. However, if n is large, even if σ_d is unknown, the normal distribution can be used to make inferences about μ_d.

10.4.1 INTERVAL ESTIMATION OF μ_d

The mean $\bar{d}$ of paired differences for paired samples is the point estimator of μ_d. The following formula is used to construct a confidence interval for μ_d in the case of (approximately) normally distributed populations.

> **CONFIDENCE INTERVAL FOR μ_d** The $(1 - \alpha)100\%$ *confidence interval for μ_d* is
>
> $$\bar{d} \pm ts_{\bar{d}}$$
>
> where the value of t is obtained from the t distribution table for the given confidence level and $n - 1$ degrees of freedom, and $s_{\bar{d}}$ is calculated as explained earlier.

Example 10–12 illustrates the procedure to construct a confidence interval for μ_d.

*Constructing a
confidence interval for
μ_d: paired samples.*

EXAMPLE 10-12   A researcher wanted to find the effect of a special diet on systolic blood pressure. She selected a sample of seven adults and put them on this dietary plan for three months. The following table gives the systolic blood pressures of these seven adults before and after the completion of this plan.

Before	210	180	195	220	231	199	224
After	193	186	186	223	220	183	233

Let μ_d be the mean reduction in the systolic blood pressures due to this special dietary plan for the population of all adults. Construct a 95% confidence interval for μ_d. Assume that the population of paired differences is (approximately) normally distributed.

Solution   Because the information obtained is from paired samples, we will make the confidence interval for the paired difference mean μ_d of the population using the paired difference mean $\bar{d}$ of the sample. Let d be the difference in the systolic blood pressure of an adult before and after this special dietary plan. Then, d is obtained by subtracting the systolic blood pressure after the plan from the systolic blood pressure before the plan. The third column of Table 10.1 lists the values of d for the seven adults. The fourth column of the table records the values of d^2, which are obtained by squaring each of the d values.

Table 10.1

		Difference	
Before	**After**	**d**	**d^2**
210	193	17	289
180	186	−6	36
195	186	9	81
220	223	−3	9
231	220	11	121
199	183	16	256
224	233	−9	81
		$\Sigma d = 35$	$\Sigma d^2 = 873$

The values of $\bar{d}$ and s_d are calculated as follows:

$$\bar{d} = \frac{\Sigma d}{n} = \frac{35}{7} = 5.00$$

$$s_d = \sqrt{\frac{\Sigma d^2 - \frac{(\Sigma d)^2}{n}}{n-1}} = \sqrt{\frac{873 - \frac{(35)^2}{7}}{7-1}} = 10.78579312$$

Hence, the standard deviation of $\bar{d}$ is

$$s_{\bar{d}} = \frac{s_d}{\sqrt{n}} = \frac{10.78579312}{\sqrt{7}} = 4.07664661$$

For the 95% confidence interval, the area in each tail of the t distribution curve is

$$\text{Area in each tail} = \alpha/2 = .5 - (.95/2) = .025$$

The degrees of freedom are

$$df = n - 1 = 7 - 1 = 6$$

From the t distribution table, the t value for $df = 6$ and .025 area in the right tail of the t distribution curve is 2.447. Therefore, the 95% confidence interval for μ_d is

$$\bar{d} \pm ts_{\bar{d}} = 5.00 \pm 2.447(4.07664661) = 5.00 \pm 9.98 = \mathbf{-4.98 \text{ to } 14.98}$$

Thus, we can state with 95% confidence that the mean difference between systolic blood pressures before and after the given dietary plan for all adult participants is between -4.98 and 14.98. ∎

10.4.2   HYPOTHESIS TESTING ABOUT μ_d

A hypothesis about μ_d is tested by using the sample statistic $\bar{d}$. If n is 30 or larger, we can use the normal distribution to test a hypothesis about μ_d. However, if n is less than 30, we replace the normal distribution by the t distribution. To use the t distribution, we assume that the population of all paired differences is (approximately) normally distributed and that the population standard deviation, σ_d, of paired differences is not known. This section illustrates the case of the t distribution only. The following formula is used to calculate the value of the test statistic t when testing a hypothesis about μ_d.

TEST STATISTIC t FOR $\bar{d}$ The value of the *test statistic t for $\bar{d}$* is computed as follows:

$$t = \frac{\bar{d} - \mu_d}{s_{\bar{d}}}$$

The critical value of t is found from the t distribution table for the given significance level and $n - 1$ degrees of freedom.

Examples 10–13 and 10–14 illustrate the hypothesis-testing procedure for μ_d.

Conducting a left-tailed test of hypothesis about μ_d: paired samples.

EXAMPLE 10–13   A company wanted to know if attending a course on "how to be a successful salesperson" can increase the average sales of its employees. The company sent six of its salespersons to attend this course. The following table gives the one-week sales of these salespersons before and after they attended this course.

Before	12	18	25	9	14	16
After	18	24	24	14	19	20

Using the 1% significance level, can you conclude that the mean weekly sales for all salespersons increase as a result of attending this course? Assume that the population of paired differences has a normal distribution.

Solution   Because the data are for paired samples, we test a hypothesis about the paired differences mean μ_d of the population using the paired differences mean $\bar{d}$ of the sample.

Let

$$d = (\text{Weekly sales before the course}) - (\text{Weekly sales after the course})$$

In Table 10.2, we calculate d for each of the six salespersons by subtracting the sales after the course from the sales before the course. The fourth column of the table lists the values of d^2.

Table 10.2

Before	After	Difference d	d^2
12	18	−6	36
18	24	−6	36
25	24	1	1
9	14	−5	25
14	19	−5	25
16	20	−4	16
		$\Sigma d = -25$	$\Sigma d^2 = 139$

The values of $\bar{d}$ and s_d are calculated as follows:

$$\bar{d} = \frac{\Sigma d}{n} = \frac{-25}{6} = -4.17$$

$$s_d = \sqrt{\frac{\Sigma d^2 - \frac{(\Sigma d)^2}{n}}{n-1}} = \sqrt{\frac{139 - \frac{(-25)^2}{6}}{6-1}} = 2.63944439$$

The standard deviation of $\bar{d}$ is

$$s_{\bar{d}} = \frac{s_d}{\sqrt{n}} = \frac{2.63944439}{\sqrt{6}} = 1.07754866$$

Step 1. *State the null and alternative hypotheses.*

We are to test if the mean weekly sales for all salespersons increase as a result of taking the course. Let μ_1 be the mean weekly sales for all salespersons before the course and μ_2 the mean weekly sales for all salespersons after the course. Then $\mu_d = \mu_1 - \mu_2$. The mean weekly sales for all salespersons will increase due to attending the course if μ_1 is less than μ_2, which can be written as $\mu_1 - \mu_2 < 0$ or $\mu_d < 0$. Consequently, the null and alternative hypotheses are

$H_0: \mu_d = 0$ ($\mu_1 - \mu_2 = 0$ or the mean weekly sales do not increase)

$H_1: \mu_d < 0$ ($\mu_1 - \mu_2 < 0$ or the mean weekly sales do increase)

Note that we can also write the null hypothesis as $\mu_d \geq 0$.

Step 2. *Select the distribution to use.*

The sample size is small ($n < 30$), the population of paired differences is normal, and σ_d is unknown. Therefore, we use the t distribution to conduct the test.

Step 3. *Determine the rejection and nonrejection regions.*

The $<$ sign in the alternative hypothesis indicates that the test is left-tailed. The significance level is .01. Hence,

Area in left tail $= \alpha = .01$

Degrees of freedom $= n - 1 = 6 - 1 = 5$

The critical value of t for $df = 5$ and .01 area in the left tail of the t distribution curve is -3.365. This value is shown in Figure 10.7.

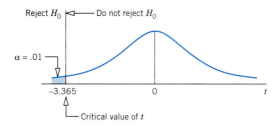

Figure 10.7

Step 4. *Calculate the value of the test statistic.*

The value of the test statistic t for $\bar{d}$ is computed as follows:

$$t = \frac{\bar{d} - \mu_d}{s_{\bar{d}}} = \frac{-4.17 - 0}{1.07754866} = -3.870$$

From H_0

Step 5. *Make a decision.*

Because the value of the test statistic $t = -3.870$ for $\bar{d}$ falls in the rejection region, we reject the null hypothesis. Consequently, we conclude that the mean weekly sales for all salespersons increase as a result of this course. ■

Making a two-tailed test of hypothesis about μ_d: paired samples.

EXAMPLE 10–14   Refer to Example 10–12. The table that gives the blood pressures of seven adults before and after the completion of a special dietary plan is reproduced here.

Before	210	180	195	220	231	199	224
After	193	186	186	223	220	183	233

Let μ_d be the mean of the differences between the systolic blood pressures before and after completing this special dietary plan for the population of all adults. Using the 5% significance level, can we conclude that the mean of the paired differences μ_d is different from zero? Assume that the population of paired differences is (approximately) normally distributed.

Solution   Table 10.3 gives d and d^2 for each of the seven adults.

Table 10.3

Before	After	Difference d	d^2
210	193	17	289
180	186	-6	36
195	186	9	81
220	223	-3	9
231	220	11	121
199	183	16	256
224	233	-9	81
		$\Sigma d = 35$	$\Sigma d^2 = 873$

The values of $\bar{d}$ and s_d are calculated as follows:

$$\bar{d} = \frac{\Sigma d}{n} = \frac{35}{7} = 5.00$$

$$s_d = \sqrt{\frac{\Sigma d^2 - \frac{(\Sigma d)^2}{n}}{n - 1}} = \sqrt{\frac{873 - \frac{(35)^2}{7}}{7 - 1}} = 10.78579312$$

Hence, the standard deviation of $\bar{d}$ is

$$s_{\bar{d}} = \frac{s_d}{\sqrt{n}} = \frac{10.78579312}{\sqrt{7}} = 4.07664661$$

Step 1.   *State the null and alternative hypotheses.*

$H_0: \mu_d = 0$ (The mean of the paired differences is not different from zero)

$H_1: \mu_d \neq 0$ (The mean of the paired differences is different from zero)

Step 2.   *Select the distribution to use.*

Because the sample size is small, the population of paired differences is (approximately) normal, and σ_d is not known, we use the t distribution to make the test.

Step 3.   *Determine the rejection and nonrejection regions.*

The $\neq$ sign in the alternative hypothesis indicates that the test is two-tailed. The significance level is .05.

Area in each tail of the curve $= \alpha/2 = .05/2 = .025$

Degrees of freedom $= n - 1 = 7 - 1 = 6$

The two critical values of t for $df = 6$ and .025 area in each tail of the t distribution curve are -2.447 and 2.447. These values are shown in Figure 10.8.

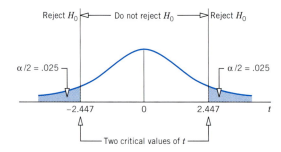

Figure 10.8

Step 4.   *Calculate the value of the test statistic.*

The value of the test statistic t for $\bar{d}$ is computed as follows:

$$t = \frac{\bar{d} - \mu_d}{s_{\bar{d}}} = \frac{\overbrace{5.00 - 0}^{\text{From } H_0}}{4.07664661} = 1.226$$

Step 5.   *Make a decision.*

Because the value of the test statistic $t = 1.226$ for $\bar{d}$ falls in the nonrejection region, we fail to reject the null hypothesis. Hence, we conclude that the mean of the population paired differences is not different from zero. In other words, we can state that the mean of the differences between the systolic blood pressures before and after completing this special dietary plan for the population of all adults is not different from zero. ∎

EXERCISES

■ Concepts and Procedures

10.50 Explain when you would use the paired samples procedure to make confidence intervals and test hypotheses.

10.51 Find the following confidence intervals for μ_d assuming that the populations of paired differences are normally distributed.

a. $n = 11$, $\bar{d} = 25.4$, $s_d = 13.5$,   confidence level = 99%
b. $n = 23$, $\bar{d} = 13.2$, $s_d = 4.8$,   confidence level = 95%
c. $n = 18$, $\bar{d} = 34.6$, $s_d = 11.7$,   confidence level = 90%

10.52 Find the following confidence intervals for μ_d assuming that the populations of paired differences are normally distributed.

a. $n = 12$, $\bar{d} = 17.5$, $s_d = 6.3$,   confidence level = 99%
b. $n = 27$, $\bar{d} = 55.9$, $s_d = 14.7$,   confidence level = 95%
c. $n = 16$, $\bar{d} = 29.3$, $s_d = 8.3$,   confidence level = 90%

10.53 Perform the following tests of hypotheses assuming that the populations of paired differences are normally distributed.

a. H_0: $\mu_d = 0$, H_1: $\mu_d \neq 0$, $n = 9$, $\bar{d} = 6.7$, $s_d = 2.5$, $\alpha = .10$
b. H_0: $\mu_d = 0$, H_1: $\mu_d > 0$, $n = 22$, $\bar{d} = 14.8$, $s_d = 6.4$, $\alpha = .05$
c. H_0: $\mu_d = 0$, H_1: $\mu_d < 0$, $n = 17$, $\bar{d} = -9.3$, $s_d = 4.8$, $\alpha = .01$

10.54 Conduct the following tests of hypotheses assuming that the populations of paired differences are normally distributed.

a. H_0: $\mu_d = 0$, H_1: $\mu_d \neq 0$, $n = 26$, $\bar{d} = 9.6$, $s_d = 3.9$, $\alpha = .05$
b. H_0: $\mu_d = 0$, H_1: $\mu_d > 0$, $n = 15$, $\bar{d} = 8.8$, $s_d = 4.7$, $\alpha = .01$
c. H_0: $\mu_d = 0$, H_1: $\mu_d < 0$, $n = 20$, $\bar{d} = -7.4$, $s_d = 2.3$, $\alpha = .10$

■ Applications

10.55 A company sent seven of its employees to attend a course in building self-confidence. These employees were evaluated for their self-confidence before and after attending this course. The following table gives the scores (on a scale of 1 to 15, 1 being the lowest and 15 being the highest score) of these employees before and after they attended the course.

Before	8	5	4	9	6	9	5
After	10	8	5	11	6	7	9

a. Construct a 95% confidence interval for the mean μ_d of the population paired differences, where a paired difference is equal to the score of an employee before attending the course minus the score of the same employee after attending the course.
b. Test at the 1% significance level whether attending this course increases the mean score of employees.

Assume that the population of paired differences has a normal distribution.

10.56 Many students suffer from math anxiety. A professor who teaches statistics offered her students a 2-hour lecture on math anxiety and ways to overcome it. The following table gives the test scores in statistics of seven students before and after they attended this lecture.

Before	56	69	48	74	65	71	60
After	62	73	44	85	71	70	73

a. Construct a 99% confidence interval for the mean μ_d of the population paired differences, where a paired difference is equal to the score before attending this lecture minus the score after attending this lecture.

b. Test at the 2.5% significance level whether attending this lecture increases the average score in statistics.

Assume that the population of paired differences is (approximately) normally distributed.

10.57 A private agency claims that the crash course it offers significantly increases the writing speed of secretaries. The following table gives the scores of eight secretaries before and after they attended this course.

Before	81	75	89	91	65	70	90	64
After	97	72	93	110	78	69	115	72

a. Make a 90% confidence interval for the mean μ_d of the population paired differences, where a paired difference is equal to the score before attending the course minus the score after attending the course.

b. Using the 5% significance level, can you conclude that attending this course increases the writing speed of secretaries?

Assume that the population of paired differences is (approximately) normally distributed.

10.58 A company claims that its 12-week special exercise program significantly reduces weight. A random sample of six persons was selected, and these persons were put on this exercise program for 12 weeks. The following table gives the weights (in pounds) of those six persons before and after the program.

Before	180	195	177	221	208	199
After	183	187	161	204	197	189

a. Make a 95% confidence interval for the mean μ_d of the population paired differences, where a paired difference is equal to the weight before joining this exercise program minus the weight at the end of the 12-week program.

b. Using the 1% significance level, can you conclude that the mean weight loss for all persons due to this special exercise program is greater than zero?

Assume that the population of all paired differences is (approximately) normally distributed.

10.59 The manufacturer of a gasoline additive claims that the use of this additive increases gasoline mileage. A random sample of six cars was selected and these cars were driven for one week without the gasoline additive and then for one week with the gasoline additive. The following table gives the miles per gallon for these cars without and with the gasoline additive.

Without	24.6	28.3	18.9	23.7	15.4	29.5
With	26.3	31.7	18.2	25.3	18.3	30.9

a. Construct a 99% confidence interval for the mean μ_d of the population paired differences, where a paired difference is equal to the miles per gallon without the gasoline additive minus the miles per gallon with the gasoline additive.

b. Using the 2.5% significance level, can you conclude that the use of the gasoline additive increases the gasoline mileage?

Assume that the population of paired differences is (approximately) normally distributed.

 10.60 A company is considering installing new machines to assemble its products. The company is considering two types of machines, but it will buy only one type. The company selected eight assembly workers and asked them to use these two types of machines to assemble products. The following table gives the time taken (in minutes) to assemble one unit of the product on each type of machine for each of these eight workers.

Machine I	23	26	19	24	27	22	20	18
Machine II	21	24	23	25	24	28	24	23

a. Construct a 98% confidence interval for the mean μ_d of the population paired differences, where a paired difference is equal to the time taken to assemble a unit of the product on Machine I minus the time taken to assemble a unit of the product on Machine II.
b. Test at the 5% significance level whether the mean times taken to assemble a unit of the product are different for the two types of machines.

Assume that the population of paired differences is (approximately) normally distributed.

10.5 INFERENCES ABOUT THE DIFFERENCE BETWEEN TWO POPULATION PROPORTIONS FOR LARGE AND INDEPENDENT SAMPLES

Quite often we need to construct a confidence interval and test a hypothesis about the difference between two population proportions. For instance, we may want to estimate the difference between the proportions of defective items produced on two different machines. If p_1 and p_2 are the proportions of defective items produced on the first and second machine, respectively, then we are to make a confidence interval for $p_1 - p_2$. Or we may want to test the hypothesis that the proportion of defective items produced on Machine I is different from the proportion of defective items produced on Machine II. In this case, we are to test the null hypothesis $p_1 - p_2 = 0$ against the alternative hypothesis $p_1 - p_2 \neq 0$.

This section discusses how to make a confidence interval and test a hypothesis about $p_1 - p_2$ for two large and independent samples. The sample statistic that is used to make inferences about $p_1 - p_2$ is $\hat{p}_1 - \hat{p}_2$, where $\hat{p}_1$ and $\hat{p}_2$ are the proportions for two large and independent samples. As discussed in Section 7.6 of Chapter 7, we determine a sample proportion by dividing the number of elements in the sample that possess a given attribute by the sample size. Thus,

$$\hat{p}_1 = x_1/n_1 \quad \text{and} \quad \hat{p}_2 = x_2/n_2$$

where x_1 and x_2 are the number of elements that possess a given characteristic in the two samples and n_1 and n_2 are the sizes of the two samples, respectively.

10.5.1 MEAN, STANDARD DEVIATION, AND SAMPLING DISTRIBUTION OF $\hat{p}_1 - \hat{p}_2$

As discussed in Chapter 7, for a large sample the sample proportion $\hat{p}$ is (approximately) normally distributed with mean p and standard deviation $\sqrt{pq/n}$. Hence, for two large and independent samples of sizes n_1 and n_2, respectively, their sample proportions $\hat{p}_1$ and $\hat{p}_2$ are (approximately) normally distributed with means p_1 and p_2 and standard deviations $\sqrt{p_1 q_1/n_1}$

and $\sqrt{p_2 q_2 / n_2}$, respectively. Using these results, we can make the following statements about the shape of the sampling distribution of $\hat{p}_1 - \hat{p}_2$ and its mean and standard deviation.

MEAN, STANDARD DEVIATION, AND SAMPLING DISTRIBUTION OF $\hat{p}_1 - \hat{p}_2$ For two large and independent samples, the *sampling distribution* of $\hat{p}_1 - \hat{p}_2$ is (approximately) normal with its *mean* and *standard deviation* given as

$$\mu_{\hat{p}_1 - \hat{p}_2} = p_1 - p_2$$

and

$$\sigma_{\hat{p}_1 - \hat{p}_2} = \sqrt{\frac{p_1 q_1}{n_1} + \frac{p_2 q_2}{n_2}}$$

respectively, where $q_1 = 1 - p_1$ and $q_2 = 1 - p_2$.

Thus, to construct a confidence interval and test a hypothesis about $p_1 - p_2$ for large and independent samples, we use the normal distribution. As was indicated in Chapter 7, in the case of proportions, the sample is large if np and nq are both greater than 5. In the case of two samples, both sample sizes are large if $n_1 p_1$, $n_1 q_1$, $n_2 p_2$, and $n_2 q_2$ are all greater than 5.

10.5.2 INTERVAL ESTIMATION OF $p_1 - p_2$

The difference between two sample proportions $\hat{p}_1 - \hat{p}_2$ is the point estimator for the difference between two population proportions $p_1 - p_2$. Because we do not know p_1 and p_2 when we are making a confidence interval for $p_1 - p_2$, we cannot calculate the value of $\sigma_{\hat{p}_1 - \hat{p}_2}$. Therefore, we use $s_{\hat{p}_1 - \hat{p}_2}$ as the point estimator of $\sigma_{\hat{p}_1 - \hat{p}_2}$ in the interval estimation. We construct the confidence interval for $p_1 - p_2$ using the following formula.

CONFIDENCE INTERVAL FOR $p_1 - p_2$
The $(1 - \alpha)100\%$ *confidence interval for $p_1 - p_2$* is

$$(\hat{p}_1 - \hat{p}_2) \pm z s_{\hat{p}_1 - \hat{p}_2}$$

where the value of z is read from the normal distribution table for the given confidence level, and $s_{\hat{p}_1 - \hat{p}_2}$ is calculated as

$$s_{\hat{p}_1 - \hat{p}_2} = \sqrt{\frac{\hat{p}_1 \hat{q}_1}{n_1} + \frac{\hat{p}_2 \hat{q}_2}{n_2}}$$

Example 10–15 describes the procedure used to make a confidence interval for the difference between two population proportions for large samples.

Constructing a confidence interval for $p_1 - p_2$: large and independent samples.

EXAMPLE 10–15   A researcher wanted to estimate the difference between the percentages of users of two toothpastes who will never switch to another toothpaste. In a sample of 500 users of Toothpaste A taken by this researcher, 100 said that they will never switch to another toothpaste. In another sample of 400 users of Toothpaste B taken by the same researcher, 68 said that they will never switch to another toothpaste.

(a)   Let p_1 and p_2 be the proportions of all users of Toothpastes A and B, respectively, who will never switch to another toothpaste. What is the point estimate of $p_1 - p_2$?

(b)   Construct a 97% confidence interval for the difference between the proportions of all users of the two toothpastes who will never switch.

Solution   Let p_1 and p_2 be the proportions of all users of Toothpastes A and B, respectively, who will never switch to another toothpaste and let $\hat{p}_1$ and $\hat{p}_2$ be the respective sample proportions. Let x_1 and x_2 be the number of users of Toothpastes A and B, respectively, in the two samples who said that they will never switch to another toothpaste. From the given information,

$$\text{Toothpaste A:} \quad n_1 = 500 \quad \text{and} \quad x_1 = 100$$

$$\text{Toothpaste B:} \quad n_2 = 400 \quad \text{and} \quad x_2 = 68$$

The two sample proportions are calculated as follows:

$$\hat{p}_1 = x_1/n_1 = 100/500 = .20$$

$$\hat{p}_2 = x_2/n_2 = 68/400 = .17$$

Then,

$$\hat{q}_1 = 1 - .20 = .80 \quad \text{and} \quad \hat{q}_2 = 1 - .17 = .83$$

(a)   The point estimate of $p_1 - p_2$ is as follows:

$$\text{Point estimate of } p_1 - p_2 = \hat{p}_1 - \hat{p}_2 = .20 - .17 = \mathbf{.03}$$

(b)   The values of $n_1\hat{p}_1$, $n_1\hat{q}_1$, $n_2\hat{p}_2$, and $n_2\hat{q}_2$ are

$$n_1\hat{p}_1 = 500(.20) = 100 \quad n_1\hat{q}_1 = 500(.80) = 400$$

$$n_2\hat{p}_2 = 400(.17) = 68 \quad n_2\hat{q}_2 = 400(.83) = 332$$

Since each of these values is greater than 5, both sample sizes are large. Consequently we use the normal distribution to make a confidence interval for $p_1 - p_2$. The standard deviation of $\hat{p}_1 - \hat{p}_2$ is

$$s_{\hat{p}_1-\hat{p}_2} = \sqrt{\frac{\hat{p}_1\hat{q}_1}{n_1} + \frac{\hat{p}_2\hat{q}_2}{n_2}} = \sqrt{\frac{(.20)(.80)}{500} + \frac{(.17)(.83)}{400}} = .02593742$$

The z value for a 97% confidence level, obtained from the normal distribution table for $.97/2 = .4850$, is 2.17. The 97% confidence interval for $p_1 - p_2$ is

$$(\hat{p}_1 - \hat{p}_2) \pm zs_{\hat{p}_1-\hat{p}_2} = (.20 - .17) \pm 2.17(.02593742)$$

$$= .03 \pm .056 = \mathbf{-.026 \text{ to } .086}$$

Thus, with 97% confidence we can state that the difference between the two population proportions is between $-.026$ and $.086$.   ∎

10.5.3   HYPOTHESIS TESTING ABOUT $p_1 - p_2$

In this section we learn how to test a hypothesis about $p_1 - p_2$ for two large and independent samples. The procedure involves the same five steps that we have used previously. Once again, we calculate the standard deviation of $\hat{p}_1 - \hat{p}_2$ as

$$\sigma_{\hat{p}_1-\hat{p}_2} = \sqrt{\frac{p_1q_1}{n_1} + \frac{p_2q_2}{n_2}}$$

When a test of hypothesis about $p_1 - p_2$ is performed, usually the null hypothesis is $p_1 = p_2$ and the values of p_1 and p_2 are not known. Assuming that the null hypothesis is true and $p_1 = p_2$, a common value of p_1 and p_2, denoted by $\bar{p}$, is calculated by using one of the following formulas:

$$\bar{p} = \frac{x_1 + x_2}{n_1 + n_2} \quad \text{or} \quad \frac{n_1\hat{p}_1 + n_2\hat{p}_2}{n_1 + n_2}$$

Which of these formulas is used depends on whether the values of x_1 and x_2 or the values of $\hat{p}_1$ and $\hat{p}_2$ are known. Note that x_1 and x_2 are the number of elements in each of the two samples that possess a certain characteristic. This value of $\bar{p}$ is called the **pooled sample proportion**. Using the value of the pooled sample proportion, we compute an estimate of the standard deviation of $\hat{p}_1 - \hat{p}_2$ as follows:

$$s_{\hat{p}_1 - \hat{p}_2} = \sqrt{\bar{p}\,\bar{q}\left(\frac{1}{n_1} + \frac{1}{n_2}\right)}$$

where $\bar{q} = 1 - \bar{p}$.

TEST STATISTIC z FOR $\hat{p}_1 - \hat{p}_2$ The value of the *test statistic z for $\hat{p}_1 - \hat{p}_2$* is calculated as

$$z = \frac{(\hat{p}_1 - \hat{p}_2) - (p_1 - p_2)}{s_{\hat{p}_1 - \hat{p}_2}}$$

The value of $p_1 - p_2$ is substituted from H_0, which usually is zero.

Examples 10–16 and 10–17 illustrate the procedure to test hypotheses about the difference between two population proportions for large samples.

Making a right-tailed test of hypothesis about $p_1 - p_2$: large and independent samples.

EXAMPLE 10–16   Reconsider Example 10–15 about the percentages of users of two toothpastes who will never switch to another toothpaste. At the 1% significance level, can we conclude that the proportion of users of Toothpaste A who will never switch to another toothpaste is higher than the proportion of users of Toothpaste B who will never switch to another toothpaste?

Solution   Let p_1 and p_2 be the proportions of all users of Toothpastes A and B, respectively, who will never switch to another toothpaste and let $\hat{p}_1$ and $\hat{p}_2$ be the corresponding sample proportions. Let x_1 and x_2 be the number of users of Toothpastes A and B, respectively, in the two samples who said that they will never switch to another toothpaste. From the given information,

$$\text{Toothpaste A:} \quad n_1 = 500 \quad \text{and} \quad x_1 = 100$$

$$\text{Toothpaste B:} \quad n_2 = 400 \quad \text{and} \quad x_2 = 68$$

The significance level is $\alpha = .01$. The two sample proportions are calculated as follows:

$$\hat{p}_1 = x_1/n_1 = 100/500 = .20$$

$$\hat{p}_2 = x_2/n_2 = 68/400 = .17$$

Step 1.   *State the null and alternative hypotheses.*

We are to test if the proportion of users of Toothpaste A who will never switch to another toothpaste is higher than the proportion of users of Toothpaste B who will never switch to another toothpaste. In other words, we are to test whether p_1 is greater than p_2. This can be written as $p_1 - p_2 > 0$. Thus, the two hypotheses are

$$H_0: p_1 - p_2 = 0 \qquad (p_1 \text{ is not greater than } p_2)$$

$$H_1: p_1 - p_2 > 0 \qquad (p_1 \text{ is greater than } p_2)$$

Step 2. *Select the distribution to use.*

As shown in Example 10–15, $n_1\hat{p}_1$, $n_1\hat{q}_1$, $n_2\hat{p}_2$, and $n_2\hat{q}_2$ are all greater than 5. Consequently both samples are large, and we apply the normal distribution to make the test.

Step 3. *Determine the rejection and nonrejection regions.*

The $>$ sign in the alternative hypothesis indicates that the test is right-tailed. From the normal distribution table, for a .01 significance level, the critical value of z is 2.33. This is shown in Figure 10.9.

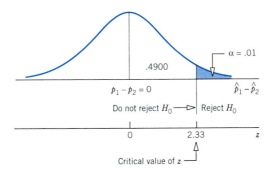

Figure 10.9

Step 4. *Calculate the value of the test statistic.*

The pooled sample proportion is

$$\bar{p} = \frac{x_1 + x_2}{n_1 + n_2} = \frac{100 + 68}{500 + 400} = .187$$

$$\bar{q} = 1 - \bar{p} = 1 - .187 = .813$$

The estimate of the standard deviation of $\hat{p}_1 - \hat{p}_2$ is

$$s_{\hat{p}_1 - \hat{p}_2} = \sqrt{\bar{p}\,\bar{q}\left(\frac{1}{n_1} + \frac{1}{n_2}\right)} = \sqrt{(.187)\,(.813)\left(\frac{1}{500} + \frac{1}{400}\right)} = .02615606$$

The value of the test statistic z for $\hat{p}_1 - \hat{p}_2$ is

$$z = \frac{(\hat{p}_1 - \hat{p}_2) - (p_1 - p_2)}{s_{\hat{p}_1 - \hat{p}_2}} = \frac{(.20 - .17) - 0}{.02615606} = 1.15 \qquad \text{From } H_0$$

Step 5. *Make a decision.*

Since the value of the test statistic $z = 1.15$ for $\hat{p}_1 - \hat{p}_2$ falls in the nonrejection region, we fail to reject the null hypothesis. Therefore, we conclude that the proportion of users of Toothpaste A who will never switch to another toothpaste is not greater than the proportion of users of Toothpaste B who will never switch to another toothpaste. ■

Conducting a two-tailed
test of hypothesis about
$p_1 - p_2$: *large and*
independent samples.

EXAMPLE 10-17 According to a poll conducted for *Men's Health* magazine/CNN by Opinion Research, 67% of men and 77% of women said that a good diet is very important to good health (*USA TODAY,* January 25, 1999). Suppose that this study is based on samples of 900 men and 1200 women. Test whether the percentages of all men and all women who hold this view are different. Use the 1% significance level.

Solution Let p_1 and p_2 be the proportions of all men and all women, respectively, who hold the view that a good diet is very important to good health. Let $\hat{p}_1$ and $\hat{p}_2$ be the corresponding sample proportions. From the given information,

$$\text{For men:} \quad n_1 = 900 \quad \text{and} \quad \hat{p}_1 = .67$$

$$\text{For women:} \quad n_2 = 1200 \quad \text{and} \quad \hat{p}_2 = .77$$

The significance level is $\alpha = .01$.

Step 1. *State the null and alternative hypotheses.*

The null and alternative hypotheses are

$$H_0: p_1 - p_2 = 0 \quad \text{(The two population proportions are not different)}$$

$$H_1: p_1 - p_2 \neq 0 \quad \text{(The two population proportions are different)}$$

Step 2. *Select the distribution to use.*

Because the samples are large and independent, we apply the normal distribution to make the test. (The reader should check that $n_1\hat{p}_1$, $n_1\hat{q}_1$, $n_2\hat{p}_2$, and $n_2\hat{q}_2$ are all greater than 5.)

Step 3. *Determine the rejection and nonrejection regions.*

The $\neq$ sign in the alternative hypothesis indicates that the test is two-tailed. For a 1% significance level, the critical values of z are -2.58 and 2.58. These values are shown in Figure 10.10.

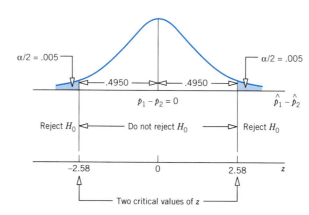

Figure 10.10

Step 4. *Calculate the value of the test statistic.*

The pooled sample proportion is

$$\bar{p} = \frac{n_1\hat{p}_1 + n_2\hat{p}_2}{n_1 + n_2} = \frac{900(.67) + 1200(.77)}{900 + 1200} = .727$$

$$\bar{q} = 1 - \bar{p} = 1 - .727 = .273$$

The estimate of the standard deviation of $\hat{p}_1 - \hat{p}_2$ is

$$s_{\hat{p}_1 - \hat{p}_2} = \sqrt{\bar{p}\,\bar{q}\left(\frac{1}{n_1} + \frac{1}{n_2}\right)} = \sqrt{(.727)(.273)\left(\frac{1}{900} + \frac{1}{1200}\right)} = .01964474$$

The value of the test statistic z for $\hat{p}_1 - \hat{p}_2$ is

$$z = \frac{(\hat{p}_1 - \hat{p}_2) - (p_1 - p_2)}{s_{\hat{p}_1 - \hat{p}_2}} = \frac{(.67 - .77) - 0}{.01964474} = -5.09$$

From H_0

Step 5. *Make a decision.*

The value of the test statistic $z = -5.09$ for $\hat{p}_1 - \hat{p}_2$ falls in the rejection region. Consequently, we reject the null hypothesis. As a result, we conclude that the percentages of all men and all women who hold the view that good diet is very important to good health are different.

■

CASE STUDY 10-1   WORRIED ABOUT PAYING THE BILLS

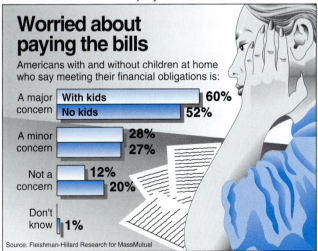

USA SNAPSHOTS®

A look at statistics that shape your finances

Worried about paying the bills

Americans with and without children at home who say meeting their financial obligations is:

| A major concern | With kids | 60% |
| | No kids | 52% |

A minor concern — 28% / 27%

Not a concern — 12% / 20%

Don't know — 1%

Source: Fleishman-Hillard Research for MassMutual

By Anne R. Carey and Gary Visgaitis, USA TODAY

The above chart shows the percentages of households with and without children with different levels of concern over meeting their financial obligations. For example, 60% of the households with children say that meeting their financial obligations is a major concern, whereas 52% of the households without children feel this way. If we know the sample sizes for the two types of households, we can construct a confidence interval and conduct a test of hypothesis for the difference in the two percentages for each category.

Suppose that this survey is based on a random sample of 1000 households, of which 495 have children and 505 do not. For a particular category of concern, let p_1 be the proportion of households with children who share that level of concern and p_2 be the proportion of households without children who feel that way. Let $\hat{p}_1$ and $\hat{p}_2$ be the corresponding sample proportions. Then for the first category (that meeting their financial obligations is a major concern):

$$n_1 = 495 \qquad \hat{p}_1 = .60$$

$$n_2 = 505 \qquad \hat{p}_2 = .52$$

Below we find a confidence interval and test a hypothesis for the difference $p_1 - p_2$ for this category.

1. Confidence interval for $p_1 - p_2$

Suppose we want to construct a 97% confidence interval for $p_1 - p_2$. The z value for the 97% confidence level is 2.17. The standard deviation of $\hat{p}_1 - \hat{p}_2$ is

$$s_{\hat{p}_1 - \hat{p}_2} = \sqrt{\frac{\hat{p}_1 \hat{q}_1}{n_1} + \frac{\hat{p}_2 \hat{q}_2}{n_2}} = \sqrt{\frac{(.60)(.40)}{495} + \frac{(.52)(.48)}{505}} = .03129067$$

Hence, a 97% confidence interval for $p_1 - p_2$ is

$$(\hat{p}_1 - \hat{p}_2) \pm z s_{\hat{p}_1 - \hat{p}_2} = (.60 - .52) \pm 2.17(.03129067) = .08 \pm .068$$

$$= .012 \text{ to } .148 \quad \text{or} \quad 1.2\% \text{ to } 14.8\%$$

Thus, we can say with 97% confidence that the difference in the proportions of all households with and without children who say that meeting their financial obligations is a major concern is in the interval .012 to .148 or 1.2% to 14.8%.

2. Test of hypothesis about $p_1 - p_2$

Suppose we want to test, at the 1% significance level, whether p_1 is greater than p_2. Then,

$$H_0: p_1 = p_2 \text{ or } p_1 - p_2 = 0$$

$$H_1: p_1 > p_2 \text{ or } p_1 - p_2 > 0$$

Note that the test is right-tailed. For $\alpha = .01$, we will reject the null hypothesis if the observed value of z is 2.33 or larger. The pooled sample proportion is

$$\bar{p} = \frac{n_1 \hat{p}_1 + n_2 \hat{p}_2}{n_1 + n_2} = \frac{495(.60) + 505(.52)}{495 + 505} = .5596$$

and

$$\bar{q} = 1 - \bar{p} = 1 - .5596 = .4404$$

The estimate of the standard deviation of $\hat{p}_1 - \hat{p}_2$ is

$$s_{\hat{p}_1 - \hat{p}_2} = \sqrt{\bar{p}\bar{q}\left(\frac{1}{n_1} + \frac{1}{n_2}\right)} = \sqrt{(.5596)(.4404)\left(\frac{1}{495} + \frac{1}{505}\right)} = .03139888$$

The value of the test statistic z for $\hat{p}_1 - \hat{p}_2$ is

$$z = \frac{(\hat{p}_1 - \hat{p}_2) - (p_1 - p_2)}{s_{\hat{p}_1 - \hat{p}_2}} = \frac{(.60 - .52) - 0}{.03139888} = 2.55$$

Since the observed value of $z = 2.55$ is greater than the critical value of 2.33, we reject the null hypothesis. As a result, we conclude that p_1 is greater than p_2.

Source: The chart reproduced with permission from *USA TODAY*, May 3, 1999. Copyright © 1999, *USA TODAY.*

EXERCISES

■ **Concepts and Procedures**

10.61   What is the shape of the sampling distribution of $\hat{p}_1 - \hat{p}_2$ for two large samples? What are the mean and standard deviation of this sampling distribution?

10.62   When are the samples considered large enough for the sampling distribution of the difference between two sample proportions to be (approximately) normal?

10.63   Construct a 99% confidence interval for $p_1 - p_2$ for the following.

$$n_1 = 300 \qquad \hat{p}_1 = .55 \qquad n_2 = 200 \qquad \hat{p}_2 = .62$$

10.64   Construct a 95% confidence interval for $p_1 - p_2$ for the following.

$$n_1 = 100 \qquad \hat{p}_1 = .81 \qquad n_2 = 150 \qquad \hat{p}_2 = .77$$

10.65   Refer to the information given in Exercise 10.63. Test at the 1% significance level if the two population proportions are different.

10.66   Refer to the information given in Exercise 10.64. Test at the 5% significance level if $p_1 - p_2$ is different from zero.

10.67   Refer to the information given in Exercise 10.63. Test at the 1% significance level if p_1 is less than p_2.

10.68   Refer to the information given in Exercise 10.64. Test at the 2.5% significance level if p_1 is greater than p_2.

10.69   A sample of 500 observations taken from the first population gave $x_1 = 305$. Another sample of 600 observations taken from the second population gave $x_2 = 348$.

a.   Find the point estimate of $p_1 - p_2$.
b.   Make a 97% confidence interval for $p_1 - p_2$.
c.   Show the rejection and nonrejection regions on the sampling distribution of $\hat{p}_1 - \hat{p}_2$ for $H_0: p_1 = p_2$ versus $H_1: p_1 > p_2$. Use a significance level of 2.5%.
d.   Find the value of the test statistic z for the test of part c.
e.   Will you reject the null hypothesis mentioned in part c at a significance level of 2.5%?

10.70   A sample of 1000 observations taken from the first population gave $x_1 = 290$. Another sample of 1200 observations taken from the second population gave $x_2 = 396$.

a.   Find the point estimate of $p_1 - p_2$.
b.   Make a 98% confidence interval for $p_1 - p_2$.
c.   Show the rejection and nonrejection regions on the sampling distribution of $\hat{p}_1 - \hat{p}_2$ for $H_0: p_1 = p_2$ versus $H_1: p_1 < p_2$. Use a significance level of 1%.
d.   Find the value of the test statistic z for the test of part c.
e.   Will you reject the null hypothesis mentioned in part c at a significance level of 1%?

■ **Applications**

10.71　In a 1998 survey, teachers and members of the public were asked, How much do you think computers have helped improve student learning? Fifty-eight percent of the members of the public and 31% of the teachers stated that computers have helped a "great amount" (MCI National Poll on the Internet in Education, 1998, *Education Week,* October 1, 1998). Assume that these estimates were based on random samples of 200 members of the public and 192 teachers.

a.　Determine a 99% confidence interval for the difference between the two population proportions.
b.　At the 1% significance level, can you conclude that the proportion of all members of the public who hold this view is greater than the proportion of all teachers who hold this view?

10.72　A company has two restaurants in two different areas of New York City. The company wants to estimate the percentages of patrons who think that the food and service at each of these restaurants are excellent. A sample of 200 patrons taken from the restaurant in Area A showed that 118 of them think that the food and service are excellent at this restaurant. Another sample of 250 patrons taken from the restaurant in Area B showed that 160 of them think that the food and service are excellent at this restaurant.

a.　Construct a 97% confidence interval for the difference between the two population proportions.
b.　Testing at the 2.5% significance level, can you conclude that the proportion of patrons at the restaurant in Area A who think that the food and service are excellent is lower than the corresponding proportion at the restaurant in Area B?

10.73　According to two polls conducted for AARP (American Association of Retired Persons), 14% of grown children say that they provide some housekeeping help for their elderly parents, whereas only 9% of parents agree with this statement (*USA TODAY,* January 28, 1999). These estimates are based on samples of 1689 grown children and 896 parents for the two polls.

a.　Construct a 98% confidence interval for the difference between the two population proportions.
b.　Test at the 1% level of significance whether the two population proportions are different.

10.74　In a survey conducted for *Men's Health* magazine/CNN by Opinion Research, 61% of men and 68% of women rated exercise as "very important" to good health (*USA TODAY,* January 25, 1999). Assume that the percentages were based on random samples of 250 men and 250 women.

a.　Construct a 95% confidence interval for the difference between the proportions of all men and all women who rate exercise as "very important" to good health.
b.　Testing at the 2.5% significance level, can you conclude that the proportion of all men who consider exercise "very important" to good health is lower than the proportion of all women who hold this view?

10.75　The management of a supermarket wanted to investigate if the percentages of men and women who prefer to buy national brand products over the store brand products are different. A sample of 600 men shoppers at the company's supermarkets showed that 246 of them prefer to buy national brand products over the store brand products. Another sample of 700 women shoppers at the company's supermarkets showed that 266 of them prefer to buy national brand products over the store brand products.

a.　What is the point estimate of the difference between the two population proportions?
b.　Construct a 95% confidence interval for the difference between the proportions of all men and all women shoppers at these supermarkets who prefer to buy national brand products over the store brand products.
c.　Testing at the 5% significance level, can you conclude that the proportions of all men and all women shoppers at these supermarkets who prefer to buy national brand products over the store brand products are different?

10.76　The lottery commissioner's office in a state wanted to find if the percentages of men and women who play the lottery often are different. A sample of 500 men taken by the commissioner's office showed that 160 of them play the lottery often. Another sample of 300 women showed that 66 of them play the lottery often.

a.　What is the point estimate of the difference between the two population proportions?
b.　Construct a 99% confidence interval for the difference between the proportions of all men and all women who play the lottery often.

c. Testing at the 1% significance level, can you conclude that the proportions of all men and all women who play the lottery often are different?

10.77 A mail-order company has two warehouses, one on the West Coast and the second on the East Coast. The company's policy is to mail all orders placed with it within 72 hours. The company's quality control department checks quite often whether or not this policy is maintained at the two warehouses. A recently taken sample of 400 orders placed with the warehouse on the West Coast showed that 364 of them were mailed within 72 hours. Another sample of 300 orders placed with the warehouse on the East Coast showed that 279 of them were mailed within 72 hours.

a. Construct a 97% confidence interval for the difference between the proportions of all orders placed at the two warehouses that are mailed within 72 hours.

b. Using the 2.5% significance level, can you conclude that the proportion of all orders placed at the warehouse on the West Coast that are mailed within 72 hours is lower than the corresponding proportion for the warehouse on the East Coast?

*c. Find the *p*-value for the test mentioned in part b.

10.78 A company that has many department stores in the southern states wanted to find the percentage of sales at two such stores for which at least one of the items was returned. A sample of 800 sales randomly selected from Store A showed that for 240 of them at least one item was returned. Another sample of 900 sales randomly selected from Store B showed that for 279 of them at least one item was returned.

a. Construct a 98% confidence interval for the difference between the proportions of all sales at the two stores for which at least one item is returned.

b. Using the 1% significance level, can you conclude that the proportions of all sales at the two stores for which at least one item is returned are different?

*c. Find the *p*-value for the test mentioned in part b.

GLOSSARY

d The difference between two matched values in two samples collected from the same source. It is called the paired difference.

$\bar{d}$ The mean of the paired differences for a sample.

Independent samples Two samples drawn from two populations such that the selection of one does not affect the selection of the other.

Paired or **matched samples** Two samples drawn in such a way that they include the same elements and two data values are obtained from each element, one for each sample. Also called **dependent samples**.

μ_d The mean of the paired differences for the population.

s_d The standard deviation of the paired differences for a sample.

σ_d The standard deviation of the paired differences for the population.

KEY FORMULAS

1. **Mean of the sampling distribution of** $\bar{x}_1 - \bar{x}_2$

$$\mu_{\bar{x}_1 - \bar{x}_2} = \mu_1 - \mu_2$$

2. **Standard deviation of** $\bar{x}_1 - \bar{x}_2$ **for two large and independent samples**

$$\sigma_{\bar{x}_1 - \bar{x}_2} = \sqrt{\frac{\sigma_1^2}{n_1} + \frac{\sigma_2^2}{n_2}}$$

3. The $(1 - \alpha)100\%$ confidence interval for $\mu_1 - \mu_2$ for two large and independent samples

$$(\bar{x}_1 - \bar{x}_2) \pm z\sigma_{\bar{x}_1 - \bar{x}_2}$$

If σ_1 and σ_2 are not known, then $\sigma_{\bar{x}_1 - \bar{x}_2}$ is replaced by its point estimator $s_{\bar{x}_1 - \bar{x}_2}$, which is calculated as

$$s_{\bar{x}_1 - \bar{x}_2} = \sqrt{\frac{s_1^2}{n_1} + \frac{s_2^2}{n_2}}$$

4. Value of the test statistic z for $\bar{x}_1 - \bar{x}_2$ for two large and independent samples

$$z = \frac{(\bar{x}_1 - \bar{x}_2) - (\mu_1 - \mu_2)}{\sigma_{\bar{x}_1 - \bar{x}_2}}$$

If σ_1 and σ_2 are not known, then $\sigma_{\bar{x}_1 - \bar{x}_2}$ is replaced by $s_{\bar{x}_1 - \bar{x}_2}$.

5. Pooled standard deviation for two small and independent samples taken from two populations with equal standard deviations

$$s_p = \sqrt{\frac{(n_1 - 1)s_1^2 + (n_2 - 1)s_2^2}{n_1 + n_2 - 2}}$$

6. Estimator of the standard deviation of $\bar{x}_1 - \bar{x}_2$ for two small and independent samples taken from two populations with equal standard deviations

$$s_{\bar{x}_1 - \bar{x}_2} = s_p \sqrt{\frac{1}{n_1} + \frac{1}{n_2}}$$

7. The $(1 - \alpha)100\%$ confidence interval for $\mu_1 - \mu_2$ for two small and independent samples taken from two normally distributed populations with equal standard deviations

$$(\bar{x}_1 - \bar{x}_2) \pm ts_{\bar{x}_1 - \bar{x}_2}$$

8. Value of the test statistic t for $\bar{x}_1 - \bar{x}_2$ for two small and independent samples taken from two normally distributed populations with equal standard deviations

$$t = \frac{(\bar{x}_1 - \bar{x}_2) - (\mu_1 - \mu_2)}{s_{\bar{x}_1 - \bar{x}_2}}$$

9. Degrees of freedom to make inferences about $\mu_1 - \mu_2$ for two small and independent samples taken from two normally distributed populations with unequal standard deviations

$$df = \frac{\left(\dfrac{s_1^2}{n_1} + \dfrac{s_2^2}{n_2}\right)^2}{\dfrac{\left(\dfrac{s_1^2}{n_1}\right)^2}{n_1 - 1} + \dfrac{\left(\dfrac{s_2^2}{n_2}\right)^2}{n_2 - 1}}$$

10. Estimate of the standard deviation of $\bar{x}_1 - \bar{x}_2$ for two small and independent samples taken from two normally distributed populations with unequal standard deviations

$$s_{\bar{x}_1 - \bar{x}_2} = \sqrt{\frac{s_1^2}{n_1} + \frac{s_2^2}{n_2}}$$

11. The $(1 - \alpha)100\%$ confidence interval for $\mu_1 - \mu_2$ for two small and independent samples taken from two normally distributed populations with unequal standard deviations

$$(\bar{x}_1 - \bar{x}_2) \pm ts_{\bar{x}_1 - \bar{x}_2}$$

12. Value of the test statistic t for $\bar{x}_1 - \bar{x}_2$ for two small and independent samples taken from two normally distributed populations with unequal standard deviations

$$t = \frac{(\bar{x}_1 - \bar{x}_2) - (\mu_1 - \mu_2)}{s_{\bar{x}_1 - \bar{x}_2}}$$

13. **Sample mean for paired differences**

$$\bar{d} = \frac{\Sigma d}{n}$$

14. **Sample standard deviation for paired differences**

$$s_d = \sqrt{\frac{\Sigma d^2 - \frac{(\Sigma d)^2}{n}}{n - 1}}$$

15. **Mean and standard deviation of the sampling distribution of $\bar{d}$**

$$\mu_{\bar{d}} = \mu_d \quad \text{and} \quad s_{\bar{d}} = \frac{s_d}{\sqrt{n}}$$

16. **The $(1 - \alpha)100\%$ confidence interval for μ_d**

$$\bar{d} \pm ts_{\bar{d}}$$

17. **Value of the test statistic t for $\bar{d}$**

$$t = \frac{\bar{d} - \mu_d}{s_{\bar{d}}}$$

18. **Mean of the sampling distribution of $\hat{p}_1 - \hat{p}_2$**

$$\mu_{\hat{p}_1 - \hat{p}_2} = p_1 - p_2$$

19. **Estimate of the standard deviation of $\hat{p}_1 - \hat{p}_2$ for two large and independent samples**

$$s_{\hat{p}_1 - \hat{p}_2} = \sqrt{\frac{\hat{p}_1 \hat{q}_1}{n_1} + \frac{\hat{p}_2 \hat{q}_2}{n_2}}$$

20. **The $(1 - \alpha)100\%$ confidence interval for $p_1 - p_2$ for two large and independent samples**

$$(\hat{p}_1 - \hat{p}_2) \pm zs_{\hat{p}_1 - \hat{p}_2}$$

21. **Pooled sample proportion for two samples**

$$\bar{p} = \frac{x_1 + x_2}{n_1 + n_2} \quad \text{or} \quad \frac{n_1 \hat{p}_1 + n_2 \hat{p}_2}{n_1 + n_2}$$

22. **Estimator of the standard deviation of $\hat{p}_1 - \hat{p}_2$ using the pooled sample proportion**

$$s_{\hat{p}_1 - \hat{p}_2} = \sqrt{\bar{p}\bar{q}\left(\frac{1}{n_1} + \frac{1}{n_2}\right)}$$

23. **Value of the test statistic z for $\hat{p}_1 - \hat{p}_2$ for large and independent samples**

$$z = \frac{(\hat{p}_1 - \hat{p}_2) - (p_1 - p_2)}{s_{\hat{p}_1 - \hat{p}_2}}$$

SUPPLEMENTARY EXERCISES

10.79 According to the College and University Personnel Association, in 1997–1998 the average salary of economics professors was $60,021 at four-year public institutions of higher education and $61,515 at similar private institutions (*The Chronicle of Higher Education Almanac,* August 28, 1998). Suppose that these results are based on random samples of 400 economics professors from public institutions and 360 economics professors from private institutions. Further assume that the two sample standard deviations are $9400 and $11,000, respectively.

a.   Construct a 99% confidence interval for the difference between the 1997–1998 mean salaries of economics professors at four-year public and private institutions of higher education.
b.   Using the 1% significance level, can you conclude that the 1997–1998 mean salary of economics professors at four-year public institutions of higher education was lower than this mean at four-year private institutions?
c.   What would your decision in part b be if the probability of making a Type I error were zero? Explain.

10.80   According to the Centers for Disease Control and Prevention/National Center for Health Statistics, the average length of stay for cases of appendicitis in short-stay hospitals in the United States was 3.5 days for patients under 15 years of age and 3.0 days for patients aged 15–44 (*Advance Data,* August 31, 1998). Suppose that these averages are based on random samples of 160 patients under 15 years of age and 200 patients aged 15–44. Further assume that the standard deviations for the two samples are 1.1 days and 1.0 day, respectively.

a.   Construct a 95% confidence interval for the difference between the two population means.
b.   Test at the 5% significance level whether the mean length of stay for cases of appendicitis for all patients under 15 years of age is greater than that for all patients in the 15–44 age group.
c.   What would your decision in part b be if the probability of making a Type I error were zero? Explain.

10.81   A consulting agency was asked by a large insurance company to investigate if business majors were better salespersons than those with other majors. A sample of 40 salespersons with a business degree showed that they sold an average of 11 insurance policies per week with a standard deviation of 1.80 policies. Another sample of 45 salespersons with a degree other than business showed that they sold an average of 9 insurance policies per week with a standard deviation of 1.35 policies.

a.   Construct a 99% confidence interval for the difference between the two population means.
b.   Using the 1% significance level, can you conclude that persons with a business degree are better salespersons than those who have a degree in another area?

10.82   According to an estimate, the average earnings of female workers who are not union members are $388 per week and those of female workers who are union members are $505 per week. Suppose that these average earnings are calculated based on random samples of 1500 female workers who are not union members and 2000 female workers who are union members. Further assume that the standard deviations for these two samples are $30 and $35, respectively.

a.   Construct a 95% confidence interval for the difference between the two population means.
b.   Test at the 2.5% significance level whether the mean weekly earnings of female workers who are not union members are less than those of female workers who are union members.

10.83   A researcher wants to test if the mean GPAs (grade point averages) of all male and all female college students who actively participate in sports are different. She took a random sample of 28 male students and 24 female students who are actively involved in sports. She found the mean GPAs of the two groups to be 2.62 and 2.74, respectively, with the corresponding standard deviations equal to .43 and .38.

a.   Test at the 5% significance level whether the mean GPAs of the two populations are different.
b.   Construct a 90% confidence interval for the difference between the two population means.

Assume that the GPAs of all male and all female student athletes both have a normal distribution with the same standard deviation.

10.84   According to data from Runzheimer International, the average monthly cost of day care for a three-year-old in a for-profit center, 5 days per week, 8 hours a day, was $606 in New York and $622 in Boston (*USA TODAY,* August 18, 1998). Suppose that these means were obtained from random samples of 23 day-care centers in New York and 27 day-care centers in Boston. Further assume that the two sample standard deviations are $110 and $105, respectively, and that the two populations are normally distributed with equal but unknown standard deviations.

a.   Construct a 90% confidence interval for the difference in the mean monthly day-care costs for the two cities.

b. Test at the 5% significance level whether the mean monthly day-care costs are different for the two cities.

10.85 An agency wanted to estimate the difference between the auto insurance premiums paid by drivers insured with two different insurance companies. A random sample of 25 drivers insured with insurance company A showed that they paid an average monthly insurance premium of $97 with a standard deviation of $14. Another random sample of 20 drivers insured with insurance company B showed that these drivers paid an average monthly insurance premium of $89 with a standard deviation of $12. Assume that the insurance premiums paid by all drivers insured with companies A and B are both normally distributed with equal standard deviations.

a. Construct a 99% confidence interval for the difference between the two population means.
b. Test at the 1% significance level whether the mean monthly insurance premium paid by drivers insured with company A is higher than that of drivers insured with company B.

10.86 A random sample of 28 children selected from families with only one child gave a mean tolerance level of 2.8 (on a scale of 1 to 8) with a standard deviation of .62. Another random sample of 25 children selected from families with more than one child gave a mean tolerance level of 4.0 with a standard deviation of .47.

a. Construct a 99% confidence interval for the difference between the two population means.
b. Test at the 5% significance level whether the mean tolerance level for children from families with only one child is lower than that for children from families with more than one child.

Assume that the tolerance levels for all children in both groups have normal distributions with the same standard deviation.

10.87 Repeat Exercise 10.83, but now assume that the GPAs of all male and all female student athletes are both normally distributed with unequal standard deviations.

10.88 Repeat Exercise 10.84, but now assume that the two populations have normal distributions with unequal standard deviations.

10.89 Repeat Exercise 10.85, but now assume that the insurance premiums paid by all drivers insured with companies A and B both are normally distributed with unequal standard deviations.

10.90 Repeat Exercise 10.86, but now assume that the tolerance levels for all children in both groups are normally distributed with unequal standard deviations.

10.91 Which day of the week is most productive in the workplace? In a 1998 survey of company executives by Accountemps, over half of those executives said that more work was accomplished on Tuesday than on any other day of the week (*The Willimantic Chronicle*, October 13, 1998). A health insurance company executive wanted to compare the productivity of claim processors on Mondays to that on Tuesdays. He took a random sample of eight such workers and recorded the numbers of claims each of them processed on four Mondays and on four Tuesdays. The table below shows the total number of claims processed on these four Mondays and four Tuesdays by each worker.

Worker	A	B	C	D	E	F	G	H
Mondays	47	41	51	48	44	50	47	48
Tuesdays	53	49	44	53	49	53	55	49

a. Construct a 95% confidence interval for the mean μ_d of population paired differences, where a paired difference is defined as the number of claims processed on Mondays minus the number of claims processed on Tuesdays.
b. Test at the 2.5% significance level whether the average productivity of all such workers is lower on Mondays than on Tuesdays.

10.92 A random sample of nine students was selected to test for the effectiveness of a special course designed to improve memory. The following table gives the results of a memory test given to these students before and after this course.

Before	43	57	48	65	81	49	38	69	58
After	49	56	55	77	89	57	36	64	69

a. Construct a 95% confidence interval for the mean μ_d of the population paired differences, where a paired difference is defined as the difference between the memory test scores of a student before and after attending this course.

b. Test at the 1% significance level whether this course makes any statistically significant improvement in the memory of all students.

Assume that the population of the paired differences has a normal distribution.

10.93 According to a survey of 1000 men and 900 women, 21% of men and 28% of women read for fun almost every day.

a. Construct a 95% confidence interval for the difference between the proportions of all men and all women who read for fun every day.

b. Test at the 2% significance level whether the proportions of all men and all women who read for fun every day are different.

10.94 The National Television Violence Study, released in 1998, analyzed 2750 different programs (*Education Week*, April 22, 1998). According to this study, 53% of music videos and 67% of children's programs contained at least one violent act. Suppose that these percentages were based on random samples of 380 music videos and 440 children's programs.

a. Construct a 99% confidence interval for the difference between the two population proportions.

b. At the 2.5% significance level, can you conclude that the percentage of music videos with violent acts is lower than the percentage of children's programs with violent acts?

10.95 In a 1998 poll, 1151 adults were asked whether teacher unionization had helped, hurt, or made no difference in the quality of public education in the United States (1998 Phi Delta Kappa/Gallup Poll of the Public's Attitudes Toward the Public Schools, *Education Week*, October 7, 1998). Twenty-eight percent of public school parents and 23% of nonpublic school parents said that unionization had helped. Suppose that the sample of 1151 adults included 410 public school parents and 245 nonpublic school parents (the remaining adults had no children in school).

a. Find a 95% confidence interval for the difference between the two population proportions.

b. At the 2.5% significance level, can you conclude that the percentage of all public school parents who feel that unionization has helped is different from the percentage of all nonpublic school parents who hold this view?

10.96 According to data from Teenage Research Unlimited, 21% of U.S. teenage boys described Levi's as a "cool" brand in 1994; however, only 7% held this opinion in 1998 (*Fortune*, April 12, 1999). Suppose that these percentages are based on random samples of 300 U.S. teenage boys in 1994 and 290 in 1998.

a. Construct a 98% confidence interval for the difference between the two population proportions.

b. Using the 1% significance level, can you conclude that the proportion of all U.S. teenage boys who considered Levi's a "cool" brand in 1998 was less than the corresponding proportion in 1994?

■ **Advanced Exercises**

10.97 Manufacturers of two competing automobile models, Gofer and Diplomat, each claim to have the lowest mean fuel consumption. Let μ_1 be the mean fuel consumption in miles per gallon (mpg) for the Gofer and μ_2 the mean fuel consumption in mpg for the Diplomat. The two manufacturers have agreed to a test in which several cars of each model will be driven on a 100-mile test run. Then the fuel consumption, in mpg, will be calculated for each test run. The average of the mpg for all 100-mile test runs for each model gives the corresponding mean. Assume that for each model the gas mileages for the test runs are normally distributed with $\sigma = 2$ mpg. Note that each car is driven for one and only one 100-mile test run.

a. How many cars (i.e., sample size) for each model are required to estimate $\mu_1 - \mu_2$ with a 90% confidence level and with a maximum error of estimate of 1.5 mpg? Use the same number of cars (i.e., sample size) for each model.

b. If μ_1 is actually 33 mpg and μ_2 is actually 30 mpg, what is the probability that five cars for each model would yield $\bar{x}_1 \geq \bar{x}_2$?

10.98 Maria and Ellen both specialize in throwing the javelin. Maria throws the javelin a mean distance of 200 feet with a standard deviation of 10 feet, whereas Ellen throws the javelin a mean distance of 210 feet with a standard deviation of 12 feet. Assume that the distances each of these athletes throws the javelin are normally distributed with these means and standard deviations. If Maria and Ellen each throw the javelin once, what is the probability that Maria's throw is longer than Ellen's?

10.99 A new type of sleeping pill is tested against an older standard pill. Two thousand insomniacs are randomly divided into two equal groups. The first group is given the old standard pill, while the second group receives the new pill. The time required to fall asleep after the pill is administered is recorded for each person. The results of the experiment are given in the following table, where $\bar{x}$ and s represent the mean and standard deviation, respectively, for the times required to fall asleep for people in each group after the pill is taken.

	Group 1 (Old Pill)	Group 2 (New Pill)
n	1000	1000
$\bar{x}$	15.4 minutes	15.0 minutes
s	3.5 minutes	3.0 minutes

Consider the test of hypothesis $H_0: \mu_1 - \mu_2 = 0$ versus $H_1: \mu_1 - \mu_2 > 0$, where μ_1 and μ_2 are the mean times required for all potential users to fall asleep using the old pill and the new pill, respectively.

a. Find the p-value for this test.
b. Does your answer to part a indicate that the result is statistically significant? Use $\alpha = .025$.
c. Find the 95% confidence interval for $\mu_1 - \mu_2$.
d. Does your answer to part c imply that this result is of great *practical* significance?

10.100 Gamma Corporation is considering the installation of governors on cars driven by its sales staff. These devices would limit the car speeds to a preset level, which is expected to improve fuel economy. The company is planning to test several cars for fuel consumption without governors for one week. Then governors would be installed in the same cars, and fuel consumption will be monitored for another week. Gamma Corporation wants to estimate the mean difference in fuel consumption with a maximum error of estimate of 2 mpg with a 90% confidence level. Assume that the differences in fuel consumption are normally distributed and that previous studies suggest that an estimate of $s_d = 3$ mpg is reasonable. How many cars should be tested? (Note that the critical value of t will depend on n, so it will be necessary to use trial and error.)

10.101 Refer to Exercise 10.100. Suppose Gamma Corporation decides to test governors on seven cars. However, the management is afraid that the speed limit imposed by the governors will reduce the number of contacts the salespersons can make each day. Thus, both the fuel consumption and the number of contacts made are recorded for each car/salesperson for each week of the testing period, both before and after the installation of governors.

Salesperson	Number of Contacts Before	Number of Contacts After	Fuel Consumption (mpg) Before	Fuel Consumption (mpg) After
A	50	49	25	26
B	63	60	21	24
C	42	47	27	26
D	55	51	23	25
E	44	50	19	24
F	65	60	18	22
G	66	58	20	23

Suppose that as a statistical analyst with the company, you are directed to prepare a brief report that includes statistical analysis and interpretation of the data. Management will use your report to help decide whether or not to install governors on all salespersons' cars. Use the 90% confidence intervals and .05 significance levels for any hypothesis tests to make suggestions. Assume that the differences in fuel consumption and the differences in the number of contacts are both normally distributed.

10.102   Two competing airlines, Alpha and Beta, fly a route between Des Moines, Iowa, and Wichita, Kansas. Each airline claims to have a lower percentage of flights that arrive late. Let p_1 be the proportion of Alpha's flights that arrive late and p_2 the proportion of Beta's flights that arrive late.

a.   You are asked to observe a random sample of arrivals for each airline to estimate $p_1 - p_2$ with a 90% confidence level and a maximum error of estimate of .05. How many arrivals for each airline would you have to observe? (Assume that you will observe the same number of arrivals, n, for each airline. To be sure of taking a large enough sample, use $p_1 = p_2 = .50$ in your calculations for n.)

b.   Suppose that p_1 is actually .30 and p_2 is actually .23. What is the probability that a sample of 100 flights for each airline (200 in all) would yield $\hat{p}_1 \geq \hat{p}_2$?

10.103   Refer to Exercise 10.59, in which a random sample of six cars was selected to test a gasoline additive. The six cars were driven for one week without the gasoline additive and then for one week with the additive. The data reproduced here from that exercise show miles per gallon without and with the additive.

Without	24.6	28.3	18.9	23.7	15.4	29.5
With	26.3	31.7	18.2	25.3	18.3	30.9

Suppose that instead of the study with six cars, a random sample of 12 cars is selected and these cars are divided randomly into two groups of six cars each. The cars in the first group are driven for one week without the additive, and the cars in the second group are driven for one week with the additive. Suppose that the top row of the table lists the gas mileages for the six cars without the additive, and the bottom row gives the gas mileages for the cars with the additive. Assume that the distributions of the gas mileages with or without the additive are (approximately) normal with equal standard deviations.

a.   Would a paired sample test as described in Section 10.4 be appropriate in this case? Why or why not? Explain.

b.   If the paired sample test is inappropriate here, carry out a suitable test whether the mean gas mileage is lower without the additive. Use $\alpha = .025$.

c.   Compare your conclusion in part b with the result of the hypothesis test in Exercise 10.59.

10.104   Does the use of cellular telephones increase the risk of brain tumors? Suppose that a manufacturer of cellular telephones hires you to answer this question because of concern about public liability suits. How would you conduct an experiment to address this question? Be specific. Explain how you would observe, how many observations you would take, and how you would analyze the data once you collect them. What are your null and alternative hypotheses? Would you want to use a higher or a lower significance level for the test? Explain.

10.105   We wish to estimate the difference between the mean scores on a standardized test of students taught by Instructors A and B. The scores of all students taught by Instructor A have a normal distribution with a standard deviation of 15, and the scores of all students taught by Instructor B have a normal distribution with a standard deviation of 10. To estimate the difference between the two means, you decide that the same number of students from each instructor's class should be observed.

a.   Assuming that the sample size is the same for each instructor's class, how large a sample should be taken from each class to estimate the difference between the mean scores of the two populations to within 5 points with 90% confidence?

b.   Suppose that samples of the size computed in part a will be selected in order to test for the difference between the two population mean scores using a .05 level of significance. How large does the difference between the two sample means have to be for us to conclude that the two population means are different?

10.106 The weekly weight losses of all dieters on Diet I have a normal distribution with a mean of 1.3 pounds and a standard deviation of .4 pound. The weekly weight losses of all dieters on Diet II have a normal distribution with a mean of 1.5 pounds and a standard deviation of .7 pound. A random sample of 25 dieters on Diet I and another sample of 36 dieters on Diet II are observed.

a. What is the probability that the difference between the two sample means, $\bar{x}_1 - \bar{x}_2$, will be less than .15 pound?

b. What is the probability that the average weight loss $\bar{x}_1$ for dieters on Diet I will be greater than the average weight loss $\bar{x}_2$ for dieters on Diet II?

c. If the average weight loss of the 25 dieters using Diet I is computed to be 2.0 pounds, what is the probability that the difference between the two sample means, $\bar{x}_1 - \bar{x}_2$, will be less than .15 pound?

10.107 Sixty-five percent of all male voters and 40% of all female voters favor a particular candidate. A sample of 100 male voters and another sample of 100 female voters will be polled. What is the probability that at least 10 more male voters than female voters will favor this candidate?

SELF-REVIEW TEST

1. To test the hypothesis that the mean blood pressure of university professors is lower than that of company executives, which of the following would you use?

    a. A left-tailed test   **b.** A two-tailed test   **c.** A right-tailed test

2. Briefly explain the meaning of independent and dependent samples. Give one example of each of these cases.

3. A company psychologist wanted to test if company executives have job-related stress scores higher than those of university professors. He took a sample of 40 executives and 50 professors and tested them for job-related stress. The sample of 40 executives gave a mean stress score of 7.6 with a standard deviation of .8. The sample of 50 professors produced a mean stress score of 5.4 with a standard deviation of 1.3.

    a. Construct a 99% confidence interval for the difference between the mean stress scores of all executives and all professors.

    b. Test at the 2.5% significance level whether the mean stress score of all executives is higher than that of all professors.

4. A sample of 20 alcoholic fathers showed that they spend an average of 2.3 hours per week playing with their children with a standard deviation of .54 hour. A sample of 25 nonalcoholic fathers gave a mean of 4.6 hours per week with a standard deviation of .8 hour.

    a. Construct a 95% confidence interval for the difference between the mean times spent per week playing with their children by all alcoholic and all nonalcoholic fathers.

    b. Test at the 1% significance level whether the mean time spent per week playing with their children by all alcoholic fathers is less than that of nonalcoholic fathers.

Assume that the times spent per week playing with their children by all alcoholic and all nonalcoholic fathers both are normally distributed with equal but unknown standard deviations.

5. Repeat Problem 4 assuming that the times spent per week playing with their children by all alcoholic and all nonalcoholic fathers both are normally distributed with unequal and unknown standard deviations.

6. The table on page 484 gives the number of items made in one hour by seven randomly selected workers on two different machines.

Worker	1	2	3	4	5	6	7
Machine I	15	18	14	20	16	18	21
Machine II	16	20	13	23	19	18	20

a. Construct a 99% confidence interval for the mean μ_d of the population paired differences, where a paired difference is equal to the number of items made by an employee in one hour on Machine I minus the number of items made by the same employee in one hour on Machine II.

b. Test at the 5% significance level whether the mean μ_d of the population paired differences is different from zero.

Assume that the population of paired differences is (approximately) normally distributed.

7. A sample of 500 male registered voters showed that 57% of them voted in the last presidential election. Another sample of 400 female registered voters showed that 55% of them voted in the same election.

a. Construct a 97% confidence interval for the difference between the proportion of all male and all female registered voters who voted in the last presidential election.

b. Test at the 1% significance level whether the proportion of all male voters who voted in the last presidential election is different from that of all female voters.

MINI-PROJECTS

Mini-Project 10–1

Medical science is continually seeking safe and effective drugs for the prevention of influenza. Zanamivir, a promising antiviral drug, was tested in a double-blind, randomized, placebo-controlled clinical trial for 28 days during the 1997–1998 flu season (Arnold S. Monto et al., "Zanamivir in the Prevention of Influenza Among Healthy Adults: A Randomized Controlled Trial," *JAMA* 282 [1], July 7, 1999). The researchers recruited 1107 healthy adults for the experiment and divided them randomly into two groups: a treatment group of 553 participants who received zanamivir and a control group of 554 who were given a placebo. During the experiment, 11 participants in the treatment group and 34 participants in the control group contracted influenza. Explain how to construct a 95% confidence interval for the difference between the relevant population proportions. Also describe an appropriate hypothesis test, using the given data, to evaluate the effectiveness of zanamivir in preventing flu.

Find a similar article in a journal of medicine, psychology, or another field that lends itself to confidence intervals and hypothesis tests for differences in two means or proportions. First explain how to make the confidence intervals and hypothesis tests; then do so using the data given in that article.

Mini-Project 10–2

A researcher conjectures that cities in the more populous states of the United States tend to have higher costs for hospital rooms. Using "CITY DATA" that accompany this text, select a random sample of 10 cities from the six most populous states (California, Texas, New York, Florida, Pennsylvania, and Illinois). Then take a random sample of 10 cities from the remaining states in the data set. For each of the 20 cities, record the average daily cost of a semiprivate hospital room. Assume that such costs are approximately normally distributed for all cities in each of the two groups of states. Further assume that the cities you selected make random samples of all cities for the two groups of states.

a. Construct a 95% confidence interval for the difference in the means of such hospital costs for all cities in the two groups of states.

b. At the 5% level of significance, can you conclude that the average daily cost of a semiprivate hospital room for all cities in the six most populous states is higher than that of such a room for all cities in the remaining states?

For instructions on using MINITAB for this chapter, please see Section B.9 of Appendix B.

See Appendix C for *Excel Adventure 10* on inferences on two population means.

COMPUTER ASSIGNMENTS

CA10.1   A random sample of 13 male college students who hold jobs gave the following data on their GPAs.

3.12	2.84	2.43	2.15	3.92	2.45	2.73
3.06	2.36	1.93	2.81	3.27	1.83	

Another random sample of 16 female college students who also hold jobs gave the following data on their GPAs.

2.76	3.84	2.24	2.81	1.79	3.89	2.96	3.77
2.36	2.81	3.29	2.08	3.11	1.69	2.84	3.02

a.   Using MINITAB or any other statistical software, construct a 99% confidence interval for the difference between the mean GPAs of all male and all female college students who hold jobs.

b.   Using MINITAB or any other statistical software, test at the 5% significance level whether the mean GPAs of all male and all female college students who hold jobs are different.

Assume that the GPAs of all such male and female college students are normally distributed with equal but unknown population standard deviations.

CA10.2   A company recently opened two supermarkets in two different areas. The management wants to know if the mean sales per day for these two supermarkets are different. A sample of 10 days for Supermarket A produced the following data on daily sales (in thousand dollars).

47.56	57.66	51.23	58.29	43.71
49.33	52.35	50.13	47.45	53.86

A sample of 12 days for Supermarket B produced the following data on daily sales (in thousand dollars).

56.34	63.55	61.64	63.75	54.78	58.19
55.40	59.44	62.33	67.82	56.65	67.90

Assume that the daily sales of the two supermarkets are both normally distributed with equal but unknown standard deviations.

a.   Using MINITAB or any other statistical software, construct a 99% confidence interval for the difference between the mean daily sales for these two supermarkets.

b.   Using MINITAB or any other statistical software, test at the 1% significance level whether the mean daily sales for these two supermarkets are different.

CA10.3   Refer to Computer Assignment CA10.1. Now do that assignment assuming the GPAs of all such male and female college students are normally distributed with unequal and unknown population standard deviations.

CA10.4   Refer to Computer Assignment CA10.2. Now do that assignment assuming the daily sales of the two supermarkets are both normally distributed with unequal and unknown standard deviations.

CA10.5   The manufacturer of a gasoline additive claims that the use of this additive increases gasoline mileage. A random sample of six cars was selected. These cars were driven for one week without the gasoline additive and then for one week with the gasoline additive. The table at the top of page 486 gives the miles per gallon for these cars without and with the gasoline additive.

Without	24.6	28.3	18.9	23.7	15.4	29.5
With	26.3	31.7	18.2	25.3	18.3	30.9

a. Using MINITAB or any other statistical software, construct a 99% confidence interval for the mean μ_d of the population paired differences.

b. Using MINITAB or any other statistical software, test at the 1% significance level whether the use of the gasoline additive increases the gasoline mileage.

Assume that the population of paired differences is (approximately) normally distributed.

CA10.6 A company is considering installing new machines to assemble its products. The company is considering two types of machines, but it will buy only one type. The company selected eight assembly workers and asked them to use these two types of machines to assemble products. The following table gives the time taken (in minutes) to assemble one unit of the product on each type of machine for each of these eight workers.

Machine I	23	26	19	24	27	22	20	18
Machine II	21	24	23	25	24	28	24	23

a. Using MINITAB or any other statistical software, construct a 98% confidence interval for the mean μ_d of the population paired differences, where a paired difference is equal to the time taken to assemble a unit of the product on Machine I minus the time taken to assemble a unit of the product on Machine II by the same worker.

b. Using MINITAB or any other statistical software, test at the 5% significance level whether the mean time taken to assemble a unit of the product is different for the two types of machines.

Assume that the population of paired differences is (approximately) normally distributed.

CA10.7 A company has two restaurants in two different areas of New York City. The company wants to estimate the percentages of patrons who think that the food and service at each of these restaurants are excellent. A sample of 200 patrons taken from the restaurant in Area A showed that 118 of them think that the food and service are excellent at this restaurant. Another sample of 250 patrons selected from the restaurant in Area B showed that 160 of them think that the food and service are excellent at this restaurant.

a. Construct a 97% confidence interval for the difference between the two population proportions.

b. Testing at the 2.5% significance level, can you conclude that the proportion of patrons at the restaurant in Area A who think that the food and service are excellent is lower than the corresponding proportion at the restaurant in Area B?

CA10.8 The management of a supermarket wanted to investigate whether the percentages of all men and women who prefer to buy national-brand products over the store-brand products are different. A sample of 600 men shoppers at the company's supermarkets showed that 246 of them prefer to buy national-brand products over the store-brand products. Another sample of 700 women shoppers at the company's supermarkets showed that 266 of them prefer to buy national-brand products over the store-brand products.

a. Construct a 99% confidence interval for the difference between the proportions of all men and all women shoppers at these supermarkets who prefer to buy national-brand products over the store-brand products.

b. Testing at the 2% significance level, can you conclude that the proportions of all men and all women shoppers at these supermarkets who prefer to buy national-brand products over the store-brand products are different?

11

CHI-SQUARE TESTS

11.1 THE CHI-SQUARE DISTRIBUTION

11.2 A GOODNESS-OF-FIT TEST

CASE STUDY 11-1 OFFICE ORDER

11.3 CONTINGENCY TABLES

11.4 A TEST OF INDEPENDENCE OR HOMOGENEITY

11.5 INFERENCES ABOUT THE POPULATION VARIANCE

GLOSSARY

KEY FORMULAS

SUPPLEMENTARY EXERCISES

SELF-REVIEW TEST

MINI-PROJECTS

COMPUTER ASSIGNMENTS

The tests of hypotheses about the mean, the difference between two means, the proportion, and the difference between two proportions were discussed in Chapters 9 and 10. The tests about proportions dealt with countable or categorical data. In the case of a proportion and the difference between two proportions, the tests concerned experiments with only two categories. Recall from Chapter 5 that such experiments are called binomial experiments.

This chapter describes three types of tests:

1. Tests of hypotheses for experiments with more than two categories, called goodness-of-fit tests

2. Tests of hypotheses about contingency tables, called independence and homogeneity tests

3. Tests of hypotheses about the variance and standard deviation of a single population

All of these tests are performed by using the **chi-square distribution**, which is sometimes written as χ^2 _distribution_ and is read as "chi-square distribution." The symbol χ is the Greek letter **chi,** pronounced "ki." The values of a chi-square distribution are denoted by the symbol χ^2 (read as "chi-square"), just as the values of the standard normal distribution and the t distribution are denoted by z and t, respectively. Section 11.1 describes the chi-square distribution.

11.1   THE CHI-SQUARE DISTRIBUTION

Like the t distribution, the chi-square distribution has only one parameter called the degrees of freedom (df). The shape of a specific chi-square distribution depends on the number of degrees of freedom.[1] (The degrees of freedom for a chi-square distribution are calculated by using different formulas for different tests. This will be explained when we discuss those tests.) The random variable χ^2 assumes nonnegative values only. Hence, a chi-square distribution curve starts at the origin (zero point) and lies entirely to the right of the vertical axis. Figure 11.1 shows three chi-square distribution curves. They are for 2, 7, and 12 degrees of freedom.

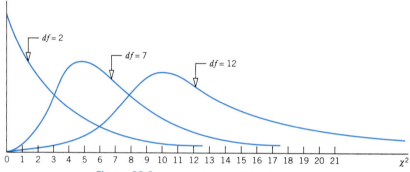

Figure 11.1   Three chi-square distribution curves.

As we can see from Figure 11.1, the shape of a chi-square distribution curve is skewed for very small degrees of freedom, and it changes drastically as the degrees of freedom increase. Eventually, for large degrees of freedom, the chi-square distribution curve looks like a normal distribution curve. The peak (or mode) of a chi-square distribution curve with 1 or 2 degrees of freedom occurs at zero and for a curve with 3 or more degrees of freedom at $df - 2$. For instance, the peak of the chi-square distribution curve with $df = 2$ in Figure 11.1 occurs at zero. The peak for the curve with $df = 7$ occurs at $7 - 2 = 5$. Finally, the peak for the curve with $df = 12$ occurs at $12 - 2 = 10$. Like all other continuous distribution curves, the total area under a chi-square distribution curve is 1.0.

> **THE CHI-SQUARE DISTRIBUTION**   The *chi-square distribution* has only one parameter called the degrees of freedom. The shape of a chi-square distribution curve is skewed to the right for small df and becomes symmetric for large df. The entire chi-square distribution curve lies to the right of the vertical axis. The chi-square distribution assumes nonnegative values only, and these are denoted by the symbol χ^2 (read as "chi-square").

If we know the degrees of freedom and the area in the right tail of a chi-square distribution curve, we can find the value of χ^2 from Table IX of Appendix D. Examples 11–1 and 11–2 show how to read that table.

---

[1]The mean of a chi-square distribution is equal to its df, and the standard deviation is equal to $\sqrt{2\ df}$.

Reading the chi-square distribution table: area in the right tail known.

EXAMPLE 11–1   Find the value of χ^2 for 7 degrees of freedom and an area of .10 in the right tail of the chi-square distribution curve.

Solution   To find the required value of χ^2, we locate 7 in the column for df and .100 in the top row in Table IX of Appendix D. The required χ^2 value is given by the entry at the intersection of the row for 7 and the column for .100. This value is 12.017. The relevant portion of Table IX is presented as Table 11.1.

Table 11.1 χ^2 for $df = 7$ and .10 Area in the Right Tail

df	Area in the Right Tail Under the Chi-square Distribution Curve				
	.995	...	.100	...	.005
1	0.000	...	2.706	...	7.879
2	0.010	...	4.605	...	10.597
.	...	...	...	...	...
.	...	...	...	...	...
7	0.989	...	12.017 ←	...	20.278
.	...	...	...	...	...
.	...	...	...	...	...
100	67.328	...	118.498	...	140.169

Required value of χ^2

As shown in Figure 11.2, the χ^2 value for $df = 7$ and an area of .10 in the right tail of the chi-square distribution curve is **12.017**.

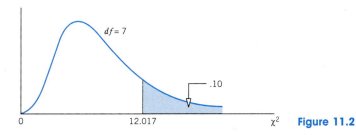

$df = 7$

.10

0   12.017 χ^2 **Figure 11.2**

Reading the chi-square distribution table: area in the left tail known.

EXAMPLE 11–2   Find the value of χ^2 for 12 degrees of freedom and an area of .05 in the left tail of the chi-square distribution curve.

Solution   We can read Table IX of Appendix D only when an area in the right tail of the chi-square distribution curve is known. When the given area is in the left tail, as in this example, the first step is to find the area in the right tail of the chi-square distribution curve as follows.

Area in the right tail $= 1 -$ Area in the left tail

Therefore, for our example,

Area in the right tail $= 1 - .05 = .95$

Next, we locate 12 in the column for df and .950 in the top row in Table IX of Appendix D. The required value of χ^2, given by the entry at the intersection of the row for 12 and the column for .950, is 5.226. The relevant portion of Table IX is presented as Table 11.2 on page 490.

Table 11.2 χ^2 for df = 12 and .95 Area in the Right Tail

df	Area in the Right Tail Under the Chi-square Distribution Curve				
	.995	**...**	**.950**	**...**	**.005**
1	0.000	...	0.004	...	7.879
2	0.010	...	0.103	...	10.597
.	...	...	...	...	...
.	...	...	...	...	...
.	...	...	...	...	...
12	3.074	...	**5.226** ←	...	28.300
.	...	...	...	...	...
.	...	...	...	...	...
100	67.328	...	77.929	...	140.169

Required value of χ^2

As shown in Figure 11.3, the χ^2 value for df = 12 and .05 area in the left tail is **5.226**.

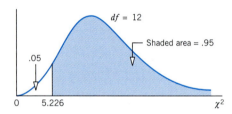

Figure 11.3

EXERCISES

■ **Concepts and Procedures**

11.1   Describe the chi-square distribution. What is the parameter (parameters) of such a distribution?

11.2   Find the value of χ^2 for 12 degrees of freedom and an area of .025 in the right tail of the chi-square distribution curve.

11.3   Find the value of χ^2 for 28 degrees of freedom and an area of .05 in the right tail of the chi-square distribution curve.

11.4   Determine the value of χ^2 for 14 degrees of freedom and an area of .10 in the left tail of the chi-square distribution curve.

11.5   Determine the value of χ^2 for 23 degrees of freedom and an area of .990 in the left tail of the chi-square distribution curve.

11.6   Find the value of χ^2 for 4 degrees of freedom and

a.   .005 area in the right tail of the chi-square distribution curve
b.   .05 area in the left tail of the chi-square distribution curve

11.7   Determine the value of χ^2 for 13 degrees of freedom and

a.   .025 area in the left tail of the chi-square distribution curve
b.   .995 area in the right tail of the chi-square distribution curve

11.2   A GOODNESS-OF-FIT TEST

This section explains how to make tests of hypotheses about experiments with more than two possible outcomes (or categories). Such experiments, called **multinomial experiments**, possess four characteristics. Note that a binomial experiment is a special case of a multinomial experiment.

> **A MULTINOMIAL EXPERIMENT** An experiment with the following characteristics is called a *multinomial experiment*.
>
> 1. It consists of n identical trials (repetitions).
> 2. Each trial results in one of k possible outcomes (or categories), where $k > 2$.
> 3. The trials are independent.
> 4. The probabilities of the various outcomes remain constant for each trial.

An experiment of many rolls of a die is an example of a multinomial experiment. It consists of many identical rolls (trials); each roll (trial) results in one of the six possible outcomes; each roll is independent of the other rolls; and the probabilities of the six outcomes remain constant for each roll.

As a second example of a multinomial experiment, suppose we select a random sample of people and ask them whether or not the quality of American cars is better than that of Japanese cars. The response of a person can be *yes, no,* or *does not know*. Each person included in the sample can be considered as one trial (repetition) of the experiment. There will be as many trials for this experiment as the number of persons selected. Each person can belong to any of the three categories—*yes, no,* or *does not know.* The response of each selected person is independent of the responses of other persons. Given that the population is large, the probabilities of a person belonging to the three categories remain the same for each trial. Consequently, this is an example of a multinomial experiment.

The frequencies obtained from the actual performance of an experiment are called the **observed frequencies**. In a **goodness-of-fit test**, we test the null hypothesis that the observed frequencies for an experiment follow a certain pattern or theoretical distribution. The test is called a goodness-of-fit test because the hypothesis tested is how *good* the observed frequencies *fit* a given pattern.

For our first example involving the experiment of many rolls of a die, we may test the null hypothesis that the given die is fair. The die will be fair if the observed frequency for each outcome is close to one-sixth of the total number of rolls.

For our second example involving opinions of people on the quality of American cars, suppose such a survey was conducted in 1999 and in that survey 41% of the people said *yes,* 48% said *no,* and 11% said *do not know.* We want to test if these percentages still hold true. Suppose we take a random sample of 1000 adults and observe that 536 of them think that the quality of American cars is better than that of Japanese cars, 362 say it is worse, and 102 have no opinion. The frequencies 536, 362, and 102 are the observed frequencies. These frequencies are obtained by actually performing the survey. Now, assuming that the 1999 percentages are still true (which will be our null hypothesis), in a sample of 1000 adults we will expect 410 to say *yes,* 480 to say *no,* and 110 to say *do not know.* These frequencies are obtained by multiplying the sample size (1000) by the 1999 proportions. These frequencies are called the **expected frequencies**. Then, we will make a decision to reject or not to reject the null hypothesis based on how large the difference between the observed frequencies and the expected frequencies is. To perform this test, we will use the chi-square distribution. Note that in this case, we are testing the null hypothesis that all three percentages (or proportions) are unchanged. However, if we want to make a test for only one of the three proportions, we use the procedure learned in Section 9.5 of Chapter 9. For example, if we are testing the hypothesis that the percentage of people who think the quality of American cars is better than that of the Japanese cars is different from 41%, then we will test the null hypothesis H_0: $p = .41$ against the alternative hypothesis H_1: $p \neq .41$. This test will be conducted using the procedure discussed in Section 9.5 of Chapter 9.

As mentioned earlier, the frequencies obtained from the performance of an experiment are called the observed frequencies. They are denoted by O. To make a goodness-of-fit test, we calculate the expected frequencies for all categories of the experiment. The expected frequency for a category, denoted by E, is given by the product of n and p, where n is the total number of trials and p is the probability for that category.

> **OBSERVED AND EXPECTED FREQUENCIES** The frequencies obtained from the performance of an experiment are called the *observed frequencies* and are denoted by O. The *expected frequencies*, denoted by E, are the frequencies that we expect to obtain if the null hypothesis is true. The expected frequency for a category is obtained as
>
> $$E = np$$
>
> where n is the sample size and p is the probability that an element belongs to that category if the null hypothesis is true.

> **DEGREES OF FREEDOM FOR A GOODNESS-OF-FIT TEST** In a goodness-of-fit test, the *degrees of freedom* are
>
> $$df = k - 1$$
>
> where k denotes the number of possible outcomes (or categories) for the experiment.

The procedure to make a goodness-of-fit test involves the same five steps that we used in the preceding chapters. *The chi-square goodness-of-fit test is always a right-tailed test.*

> **TEST STATISTIC FOR A GOODNESS-OF-FIT TEST** The *test statistic for a goodness-of-fit test* is χ^2 and its value is calculated as
>
> $$\chi^2 = \sum \frac{(O - E)^2}{E}$$
>
> where O = observed frequency for a category
>
> E = expected frequency for a category = np
>
> Remember that a chi-square goodness-of-fit test is always a right-tailed test.

Whether or not the null hypothesis is rejected depends on how much the observed and expected frequencies differ from each other. To find how large the difference between the observed frequencies and the expected frequencies is, we do not look at just $\Sigma(O - E)$ because some of the $O - E$ values will be positive and others will be negative. The net result of the sum of these differences will always be zero. Therefore, we square each of the $O - E$ values to obtain $(O - E)^2$ and then we weight them according to the reciprocals of their expected frequencies. The sum of the resulting numbers gives the computed value of the test statistic χ^2.

To make a goodness-of-fit test, the sample size should be large enough so that the expected frequency for each category is at least 5. If there is a category with an expected frequency of less than 5, either increase the sample size or combine two or more categories to make each expected frequency at least 5.

Examples 11–3 and 11–4 describe the procedure for performing goodness-of-fit tests using the chi-square distribution.

Conducting a goodness-of-fit test: equal proportions for all categories.

EXAMPLE 11–3 The following table lists the age distribution for a sample of 100 persons arrested for drunk driving.

Age	16–25	26–35	36–45	46–55	56 and older
Arrests	32	25	19	16	8

Using the 1% significance level, can we reject the null hypothesis that the proportion of people arrested for drunk driving is the same for all age groups?

Solution To make this test, we proceed as follows.

Step 1. *State the null and alternative hypotheses.*

Because five categories are listed in the table, the proportion of drunk drivers will be the same for all age groups if each group contains one-fifth of the total drunk drivers. The null and alternative hypotheses are as follows.

H_0: The proportion of people arrested for drunk driving is the same for all age groups

H_1: The proportion of people arrested for drunk driving is not the same for all age groups

If the proportion of people arrested for drunk driving is the same for all age groups, then the probability for any randomly selected drunk driver to belong to any of the five age groups listed in the table will be 1/5 or .20. Let p_1, p_2, p_3, p_4, and p_5 be the probabilities that any randomly selected drunk driver will belong to each of the five age groups, respectively. Then, the null hypothesis can also be written as

$$H_0: p_1 = p_2 = p_3 = p_4 = p_5 = .20$$

and the alternative hypothesis can be stated as

$$H_1: \text{At least two of the five probabilities are not equal to .20}$$

Step 2. *Select the distribution to use.*

Because there are five categories (i.e., five age groups listed in the table), this is a multinomial experiment. Consequently, we use the chi-square distribution to make this test.

Step 3. *Determine the rejection and nonrejection regions.*

The significance level is given to be .01, and the goodness-of-fit test is always right-tailed. Therefore, the area in the right tail of the chi-square distribution curve is

$$\text{Area in the right tail} = \alpha = .01$$

The degrees of freedom are calculated as follows.

$$k = \text{number of categories} = 5$$

$$df = k - 1 = 5 - 1 = 4$$

From the chi-square distribution table (Table IX of Appendix D), the critical value of χ^2 for $df = 4$ and .01 area in the right tail of the chi-square distribution curve is 13.277, as shown in Figure 11.4.

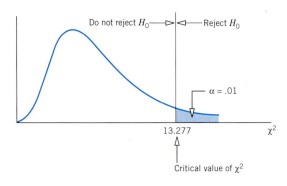

Figure 11.4

Step 4. *Calculate the value of the test statistic.*

All the required calculations to find the value of the test statistic χ^2 are shown in Table 11.3. The calculations made in Table 11.3 are explained next.

Table 11.3

Category (age)	Observed Frequency O	p	Expected Frequency $E = np$	$(O - E)$	$(O - E)^2$	$\dfrac{(O - E)^2}{E}$
16–25	32	.20	$100(.20) = 20$	12	144	7.200
26–35	25	.20	$100(.20) = 20$	5	25	1.250
36–45	19	.20	$100(.20) = 20$	−1	1	.050
46–55	16	.20	$100(.20) = 20$	−4	16	.800
56 and older	8	.20	$100(.20) = 20$	−12	144	7.200
	$n = 100$					Sum = 16.500

1. The first two columns in Table 11.3 list the five categories (age groups) and the observed frequencies for the sample of 100 drunk drivers, respectively. The third column contains the probabilities for the five categories assuming that the null hypothesis is true.

2. The fourth column contains the expected frequencies. These frequencies are obtained by multiplying the sample size ($n = 100$) by the probabilities listed in the third column. If the null hypothesis is true (i.e., the drunk drivers are equally distributed over all categories), then we will expect 20 out of 100 drunk drivers to belong to each category. Consequently, each category in the fourth column has the same expected frequency.

3. The fifth column lists the differences between the observed and expected frequencies, that is, $O - E$. These values are squared and recorded in the sixth column.

4. Finally, we divide the squared differences (that appear in the sixth column) by the corresponding expected frequencies (listed in the fourth column) and write the resulting numbers in the seventh column.

5. The sum of the seventh column gives the value of the test statistic χ^2. Thus,

$$\chi^2 = \sum \frac{(O - E)^2}{E} = 16.500$$

Step 5. *Make a decision.*

The value of the test statistic $\chi^2 = 16.500$ is greater than the critical value of $\chi^2 = 13.277$ and it falls in the rejection region. Hence, we reject the null hypothesis and state that the proportion of drunk drivers is not the same for all age groups listed in the given table. In other words, we conclude that a higher percentage of drunk drivers belong to one or more of the age groups. However, based on this test we cannot say which age groups contain higher percentages of drunk drivers. ∎

Conducting a goodness-of-fit test: testing if results of a survey fit a given distribution.

EXAMPLE 11–4   In a 1995 National Public Transportation survey conducted in 1995 on the modes of transportation used to commute to work, 79.6% of the respondents said that they drive alone, 11.1% car pool, 5.1% use public transit, and 4.2% depend on other modes of transportation (*USA TODAY*, April 14, 1999). Assume that these percentages hold true for the 1995 population of all commuting workers. Recently 1000 randomly selected workers were asked what mode of transportation they use to commute to work. The following table lists the results of this survey.

Mode of transportation	Drive alone	Car pool	Public transit	Other
Number of workers	812	102	57	29

Test at the 2.5% significance level whether the current pattern of use of transportation modes is different from that for 1995.

Solution   We perform the following five steps for this test of hypothesis.

Step 1. *State the null and alternative hypotheses.*

The null and alternative hypotheses are

H_0: The current percentage distribution of the use of transportation modes is the same as that for 1995

H_1: The current percentage distribution of the use of transportation modes is different from that for 1995

Step 2. *Select the distribution to use.*

Because this experiment has four categories (*drive alone, car pool, public transit,* and *other*), it is a multinomial experiment. Consequently, we use the chi-square distribution to make the test.

Step 3. *Determine the rejection and nonrejection regions.*

The significance level is .025. Because a goodness-of-fit test is right-tailed, the area in the right tail of the chi-square distribution curve is

$$\text{Area in the right tail} = \alpha = .025$$

The degrees of freedom are calculated as follows:

$$k = \text{number of categories} = 4$$

$$df = k - 1 = 4 - 1 = 3$$

From the chi-square distribution table (Table IX of Appendix D), the critical value of χ^2 for $df = 3$ and .025 area in the right tail of the chi-square distribution curve is 9.348. This value is shown in Figure 11.5 on page 496.

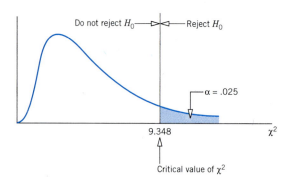

Figure 11.5

Step 4. *Calculate the value of the test statistic.*

All the required calculations to find the value of the test statistic χ^2 are shown in Table 11.4. Note that the given percentages for 1995 have been converted into probabilities and recorded in the third column of Table 11.4. The value of the test statistic χ^2 is given by the sum of the last column. Thus,

$$\chi^2 = \sum \frac{(O - E)^2}{E} = 5.782$$

Table 11.4

Category	Observed Frequency O	p	Expected Frequency $E = np$	$(O - E)$	$(O - E)^2$	$\dfrac{(O - E)^2}{E}$
Drive alone	812	.796	1000(.796) = 796	16	256	.322
Car pool	102	.111	1000(.111) = 111	−9	81	.730
Public transit	57	.051	1000(.051) = 51	6	36	.706
Other	29	.042	1000(.042) = 42	−13	169	4.024
	$n = 1000$					Sum = 5.782

Step 5. *Make a decision.*

The value of the test statistic $\chi^2 = 5.782$ is less than the critical value of $\chi^2 = 9.348$ and it falls in the nonrejection region. Hence, we fail to reject the null hypothesis and state that the current percentage distribution of the use of transportation modes is not different from that for 1995. The difference between the observed frequencies and the expected frequencies seems to have occurred only because of sampling error. ■

CASE STUDY 11-1 OFFICE ORDER

The bar graph shows how office workers described their organizational habits in a 1999 survey. Note that the percentages shown in the graph add up to 97%. Consequently, 3% of the workers must have selected categories that are not listed in the graph. For convenience we will

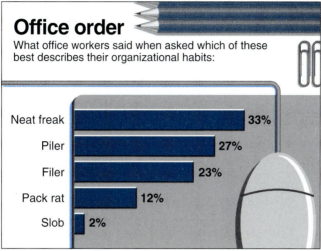

USA SNAPSHOTS®

A look at statistics that shape your finances

Office order

What office workers said when asked which of these best describes their organizational habits:

Neat freak **33%**
Piler **27%**
Filer **23%**
Pack rat **12%**
Slob **2%**

Source: Opinion Research for Steelcase   By Anne R. Carey and Genevieve Lynn, USA TODAY

assign these 3% of the workers to a new class called "other." Assume that the percentages given in the table are true for all workers' organizational habits for the year 1999. Suppose that we want to find out whether today's office workers describe themselves the same way. Suppose that in a recent random sample of 400 office workers, 144 describe themselves as *Neat freaks,* 98 as *Pilers,* 101 as *Filers,* 47 as *Pack rats,* 4 as *Slobs,* and 6 as *Other.* To test whether today's office workers describe themselves differently from those in 1999, we will use the chi-square goodness-of-fit test. The null and alternative hypotheses are as follows:

H_0: The current percentage distribution is the same as that for 1999

H_1: The current percentage distribution is different from that for 1999

Suppose we use the 5% significance level to perform this test. Then for $df = 6 - 1 = 5$ and .05 area in the right tail, the observed value of χ^2 from Table IX in Appendix D is 11.070. Thus, we will reject the null hypothesis if the observed value of χ^2 is 11.070 or greater. The following table shows all the calculations required to find the observed value of χ^2.

Category	Observed Frequency O	p	Expected Frequency $E = np$	$(O - E)$	$(O - E)^2$	$\dfrac{(O - E)^2}{E}$
Neat freak	144	.33	$400(.33) = 132$	12	144	1.091
Piler	98	.27	$400(.27) = 108$	−10	100	.926
Filer	101	.23	$400(.23) = 92$	9	81	.880
Pack rat	47	.12	$400(.12) = 48$	−1	1	.021
Slob	4	.02	$400(.02) = 8$	−4	16	2.000
Other	6	.03	$400(.03) = 12$	−6	36	3.000
	$n = 400$					Sum = 7.918

Because the observed value of $\chi^2 = 7.918$ is less than the critical value of $\chi^2 = 11.070$, we fail to reject the null hypothesis. Thus we can conclude that office workers' descriptions of their organizational habits have not changed since 1999.

Source: The chart is reproduced with permission from *USA TODAY,* May 24, 1999. Copyright © 1999, *USA TODAY.*

EXERCISES

■ **Concepts and Procedures**

11.8   Describe the four characteristics of a multinomial experiment.

11.9   What is a goodness-of-fit test and when is it applied? Explain.

11.10   Explain the difference between the observed and expected frequencies for a goodness-of-fit test.

11.11   How is the expected frequency of a category calculated for a goodness-of-fit test? What are the degrees of freedom for such a test?

11.12   To make a goodness-of-fit test, what should be the minimum expected frequency for each category? What are the alternatives if this condition is not satisfied?

11.13   The following table lists the frequency distribution for 60 rolls of a die.

Outcome	1-spot	2-spot	3-spot	4-spot	5-spot	6-spot
Frequency	7	12	8	15	11	7

Test at the 5% significance level whether the null hypothesis that the given die is fair is true.

■ **Applications**

11.14   A 1998 survey asked workers aged 18–29 to state the factor that was most important to them in a job or career (Princeton Survey Research for Domino's Pizza, *USA TODAY,* June 3, 1998). Twenty-six percent of these workers said that the most important factor was advancement opportunity, 26% mentioned benefits, 18% indicated money, 11% said that being their own boss was most important, 10% preferred flexible hours, 5% indicated number of hours, and 4% mentioned none or don't know. A recent random sample of 250 such workers were asked the same question. Their responses are summarized in the table.

Factor	Advancement	Benefits	Money	Own boss	Flexible hours	Number of hours	None
Frequency	70	60	52	18	32	11	7

Test at the 5% significance level whether the current distribution of priorities of all such workers differs from the distribution of priorities in the 1998 survey.

11.15   In a poll of *Frequent Flyer* readers, business travelers were asked how often they check their bags when flying (*USA TODAY,* March 25, 1998). The following table gives the responses and the percentage of respondents that belong to each of the responses.

Response	Always	Often	Seldom	Never
Percentage	14	28	49	9

In a recent random sample of 200 business travelers, respondents were asked the same question. Of them, 16 said always, 50 mentioned often, 110 indicated seldom, and 24 said never. At the 5% significance level, can you conclude that the current distribution of responses differs from the distribution in the earlier survey?

11.16 On which day of the week is the most work accomplished in the workplace? In a 1998 Accountemps survey of company executives, 17% indicated Monday, 51% said Tuesday, 15% said Wednesday, 5% said Thursday, 1% said Friday, and 11% did not know (*The Willimantic Chronicle,* October 13, 1998). Assume that this was the true distribution of opinions of all U.S. company executives in 1998. A recent random sample of 160 U.S. company executives yielded the responses given in the table.

Opinion	Monday	Tuesday	Wednesday	Thursday	Friday	Don't know
Frequency	32	75	24	10	1	18

Using the 1% significance level, can you conclude that the distribution of opinions of all U.S. company executives concerning the most productive day of the week has changed since 1998? Note that the expected frequency for Friday is less than 5. Combine "Friday" and "Don't know" into a single category before proceeding with your analysis.

11.17 Home Mail Corporation sells products by mail. The company's management wants to find out if the number of orders received at the company's office on each of the five days of the week is the same. The company took a sample of 400 orders received during a four-week period. The following table lists the frequency distribution for these orders by the day of the week.

Day of the week	Mon	Tue	Wed	Thu	Fri
Number of orders received	92	71	65	83	89

Test at the 5% significance level whether the null hypothesis that the orders are evenly distributed over all days of the week is true.

11.18 The following table lists the frequency distribution for a sample of 50 absences of college students from classes according to the day of occurrence.

Day of the week	Mon	Tue	Wed	Thu	Fri
Number of absences	14	7	4	9	16

Test at the 5% significance level whether the null hypothesis that the absences are equally distributed over all days of the week is true.

11.19 The following table lists the frequency distribution of cars sold at an auto dealership during the past 12 months.

Month	Jan	Feb	Mar	Apr	May	Jun	Jul	Aug	Sep	Oct	Nov	Dec
Cars sold	23	17	15	10	14	12	13	15	23	26	27	29

Using the 10% significance level, will you reject the null hypothesis that the number of cars sold at this dealership is the same for each month?

11.20 Of all students enrolled at a large undergraduate university, 19% are seniors, 23% are juniors, 27% are sophomores, and 31% are freshmen. A sample of 200 students taken from this university by the student senate to conduct a survey includes 50 seniors, 46 juniors, 55 sophomores, and 49 freshmen. Using the 10% significance level, test the null hypothesis that this sample is a random sample. (*Hint:* This sample will be a random sample if it includes approximately 19% seniors, 23% juniors, 27% sophomores, and 31% freshmen.)

11.21 Chance Corporation produces beauty products. Two years ago the quality control department at the company conducted a survey of users of one of the company's products. The survey revealed that

53% of the users said the product was excellent, 31% said it was satisfactory, 7% said it was unsatisfactory, and 9% had no opinion. Assume that these percentages were true for the population of all users of this product at that time. After this survey was conducted, the company redesigned this product. A recent survey of 800 users of the redesigned product conducted by the quality control department at the company showed that 495 of the users think the product is excellent, 255 think it is satisfactory, 35 think it is unsatisfactory, and 15 have no opinion. Do you think the percentage distribution of the opinions of users of the redesigned product is different from the percentage distribution of users of this product before it was redesigned? Use $\alpha = .025$.

11.22 Henderson Corporation makes metal sheets among other products. When the process that is used to make metal sheets works properly, 92% of the metal sheets contain no defects, 5% have one defect each, and 3% have two or more defects each. The quality control inspectors at the company take samples of metal sheets quite often and check them for defects. If the distribution of defects for a sample is significantly different from the above mentioned percentage distribution, the process is stopped and adjusted. A recent sample of 300 sheets produced the frequency distribution of defects listed in the following table.

Number of defects	None	One	Two or More
Number of metal sheets	262	24	14

Does the evidence from this sample suggest that the process needs an adjustment? Use $\alpha = .01$.

11.3 CONTINGENCY TABLES

We often may have information on more than one variable for each element. Such information can be summarized and presented using a two-way classification table, which is also called a *contingency table* or *cross-tabulation*. Table 11.5 is an example of a contingency table, which contains data collected by the U.S. Department of Education. It gives information on the gender of all students enrolled in 1999 at institutions of higher education and on whether these students were full-time or part-time. Note that the numbers in the table give enrollments in thousands. Table 11.5 has two rows (one for males and the second for females) and two columns (one for full-time and the second for part-time students). Hence, it is also called a 2 × 2 (read as "two by two") contingency table.

Table 11.5 Total 1999 Enrollment (in thousands) at Institutions of Higher Education

	Full-time	**Part-time**
Male	3768	2615
Female	4658	3717

Students who are male and enrolled part-time

Source: U.S. Department of Education, National Center for Education Statistics.

A contingency table can be of any size. For example, it can be 2 × 3, 3 × 2, 3 × 3, or 4 × 2. Note that in these notations, the first digit refers to the number of rows in the table, and the second digit refers to the number of columns. For example, a 3 × 2 table will contain three rows and two columns. In general, an $R \times C$ table contains R rows and C columns.

Each of the four boxes that contain numbers in Table 11.5 is called a *cell*. The number of cells in a contingency table is obtained by multiplying the number of rows by the number of columns. Thus, Table 11.5 contains 2 × 2 = 4 cells. The subjects that belong to a cell of a contingency table possess two characteristics. For example, 2615 thousand students listed in the second cell of the first row in Table 11.5 are *male* and *part-time*. The numbers written inside the cells are usually called the *joint frequencies*. For example, 2615 thousand students belong to the joint category of *male* and *part-time*. Hence, it is referred to as the joint frequency of this category.

11.4 A TEST OF INDEPENDENCE OR HOMOGENEITY

This section is concerned with tests of independence and homogeneity, which are performed using contingency tables. Except for a few modifications, the procedure used to make such tests is almost the same as the one applied in Section 11.2 for a goodness-of-fit test.

11.4.1 A TEST OF INDEPENDENCE

In a **test of independence** for a contingency table, we test the null hypothesis that the two attributes (characteristics) of the elements of a given population are not related (that is, they are independent) against the alternative hypothesis that the two characteristics are related (that is, they are dependent). For example, we may want to test if the affiliation of people with the Democratic and Republican parties is independent of their income levels. We perform such a test by using the chi-square distribution. As another example, we may want to test if there is an association between being a man or a woman and having a preference for watching sports or soap operas on television.

> **DEGREES OF FREEDOM FOR A TEST OF INDEPENDENCE** A test of independence involves a test of the null hypothesis that two attributes of a population are not related. The *degrees of freedom for a test of independence* are
>
> $$df = (R - 1)(C - 1)$$
>
> where R and C are the number of rows and the number of columns, respectively, in the given contingency table.

The value of the test statistic χ^2 in a test of independence is obtained using the same formula as in the goodness-of-fit test described in Section 11.2.

> **TEST STATISTIC FOR A TEST OF INDEPENDENCE** The value of the *test statistic χ^2 for a test of independence* is calculated as
>
> $$\chi^2 = \sum \frac{(O - E)^2}{E}$$
>
> where O and E are the observed and expected frequencies, respectively, for a cell.

The null hypothesis in a test of independence is always that the two attributes are not related. The alternative hypothesis is that the two attributes are related.

The frequencies obtained from the performance of an experiment for a contingency table are called the **observed frequencies**. The procedure to calculate the **expected frequencies** for a contingency table for a test of independence is different from the one for a goodness-of-fit test. Example 11–5 describes this procedure.

EXAMPLE 11–5   Violence and lack of discipline have become major problems in schools in the United States. A random sample of 300 adults was selected, and they were asked if they favor giving more freedom to schoolteachers to punish students for violence and lack of discipline. The two-way classification of the responses of these adults is presented in the following table.

	In Favor (F)	Against (A)	No Opinion (N)
Men (M)	93	70	12
Women (W)	87	32	6

Calculate the expected frequencies for this table assuming that the two attributes, gender and opinions on the issue, are independent.

Solution   The preceding table is reproduced in Table 11.6. Note that Table 11.6 includes the row and column totals.

Table 11.6

	In Favor (F)	Against (A)	No Opinion (N)	Row Totals
Men (M)	93	70	12	175
Women (W)	87	32	6	125
Column Totals	180	102	18	300

The numbers 93, 70, 12, 87, 32, and 6 listed inside the six cells of Table 11.6 are called the *observed frequencies* of the respective cells.

As mentioned earlier, the null hypothesis in a test of independence is that the two attributes (or classifications) are independent. In an independence test of hypothesis, first we assume that the null hypothesis is true and that the two attributes are independent. Assuming that the null hypothesis is true and that gender and opinions are not related in this example, the expected frequency for the cell corresponding to *Men* and *In Favor* is calculated as shown next. From Table 11.6,

$$P(\text{a person is a } Man) = P(M) = 175/300$$

$$P(\text{a person is } In\ Favor) = P(F) = 180/300$$

Because we are assuming that *M* and *F* are independent (by assuming that the null hypothesis is true), from the formula learned in Chapter 4, the joint probability of these two events is

$$P(M \text{ and } F) = P(M) \times P(F) = (175/300) \times (180/300)$$

Then, assuming that *M* and *F* are independent, the number of persons expected to be *Men* and *In Favor* in a sample of 300 is

$$E \text{ for } Men \text{ and } In\ Favor = 300 \times P(M \text{ and } F)$$

$$= 300 \times \frac{175}{300} \times \frac{180}{300} = \frac{175 \times 180}{300}$$

$$= \frac{(\text{Row total})(\text{Column total})}{\text{Sample size}}$$

Thus, the rule for obtaining the expected frequency for a cell is to divide the product of the corresponding row and column totals by the sample size.

EXPECTED FREQUENCIES FOR A TEST OF INDEPENDENCE The expected frequency E for a cell is calculated as

$$E = \frac{\text{(Row total)(Column total)}}{\text{Sample size}}$$

Using this rule, we calculate the expected frequencies of the six cells of Table 11.6 as follows.

E for *Men* and *In Favor* cell = (175)(180)/300 = **105.00**

E for *Men* and *Against* cell = (175)(102)/300 = **59.50**

E for *Men* and *No Opinion* cell = (175)(18)/300 = **10.50**

E for *Women* and *In Favor* cell = (125)(180)/300 = **75.00**

E for *Women* and *Against* cell = (125)(102)/300 = **42.50**

E for *Women* and *No Opinion* cell = (125)(18)/300 = **7.50**

The expected frequencies are usually written in parentheses below the observed frequencies within the corresponding cells, as shown in Table 11.7.

Table 11.7

	In Favor (F)	Against (A)	No Opinion (N)	Row Totals
Men (M)	93 (105.00)	70 (59.50)	12 (10.50)	175
Women (W)	87 (75.00)	32 (42.50)	6 (7.50)	125
Column Totals	180	102	18	300 ■

Like a goodness-of-fit test, *a test of independence is always right-tailed.* To apply a chi-square test of independence, *the sample size should be large enough so that the expected frequency for each cell is at least 5.* If the expected frequency for a cell is not at least 5, we either increase the sample size or combine some categories. Examples 11–6 and 11–7 describe the procedure to make tests of independence using the chi-square distribution.

Making a test of independence: 2 × 3 table.

EXAMPLE 11–6 Reconsider the two-way classification table given in Example 11–5. In that example, a random sample of 300 adults was selected, and they were asked if they favor giving more freedom to schoolteachers to punish students for violence and lack of discipline. Based on the results of the survey, a two-way classification table was prepared and presented in Example 11–5. Does the sample provide sufficient information to conclude that the two attributes, gender and opinions of adults, are dependent? Use a 1% significance level.

Solution   The test involves the following five steps.

Step 1.   *State the null and alternative hypotheses.*

As mentioned earlier, the null hypothesis must be that the two attributes are independent. Consequently, the alternative hypothesis is that these attributes are dependent.

H_0: Gender and opinions of adults are independent

H_1: Gender and opinions of adults are dependent

Step 2.   *Select the distribution to use.*

We use the chi-square distribution to make a test of independence for a contingency table.

Step 3.   *Determine the rejection and nonrejection regions.*

The significance level is 1%. Because a test of independence is always right-tailed, the area of the rejection region is .01, and it falls in the right tail of the chi-square distribution curve. The contingency table contains two rows (*Men* and *Women*) and three columns (*In Favor, Against,* and *No Opinion*). Note that we do not count the row and column of totals. The degrees of freedom are

$$df = (R - 1)(C - 1) = (2 - 1)(3 - 1) = 2$$

From Table IX of Appendix D, the critical value of χ^2 for $df = 2$ and $\alpha = .01$ is 9.210. This value is shown in Figure 11.6.

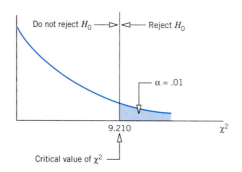

Figure 11.6

Step 4.   *Calculate the value of the test statistic.*

Table 11.7, with the observed and expected frequencies constructed in Example 11–5, is reproduced as Table 11.8.

Table 11.8

	In Favor (F)	Against (A)	No Opinion (N)	Row Totals
Men (M)	93 (105.00)	70 (59.50)	12 (10.50)	175
Women (W)	87 (75.00)	32 (42.50)	6 (7.50)	125
Column Totals	180	102	18	300

To compute the value of the test statistic χ^2, we take the difference between each pair of observed and expected frequencies listed in Table 11.8, square those differences, and then divide each of the squared differences by the respective expected frequencies. The sum of the resulting numbers gives the value of the test statistic χ^2. All these calculations are made as follows.

$$\chi^2 = \sum \frac{(O - E)^2}{E}$$

$$= \frac{(93 - 105.00)^2}{105.00} + \frac{(70 - 59.50)^2}{59.50} + \frac{(12 - 10.50)^2}{10.50}$$

$$+ \frac{(87 - 75.00)^2}{75.00} + \frac{(32 - 42.50)^2}{42.50} + \frac{(6 - 7.50)^2}{7.50}$$

$$= 1.371 + 1.853 + .214 + 1.920 + 2.594 + .300 = 8.252$$

Step 5. *Make a decision.*

The value of the test statistic $\chi^2 = 8.252$ is less than the critical value of $\chi^2 = 9.210$, and it falls in the nonrejection region. Hence, we fail to reject the null hypothesis and state that there is not enough evidence from the sample to conclude that the two characteristics, *gender* and *opinions of adults,* are dependent for this issue. ∎

Making a test of independence: 2 × 3 table.

EXAMPLE 11-7 The U.S. Bureau of Labor Statistics compiles data on the employment status of high school dropouts aged 18–24. The following table classifies such individuals by gender and employment status as of October 1997. Note that the numbers in the table are in thousands. In the table, *Not in Labor Force* means that these individuals are neither working nor looking for a job.

	Employed	**Unemployed**	**Not in Labor Force**
Men	311	52	497
Women	440	49	506

Source: U.S. Department of Labor News, May 1, 1998.

Suppose that this table was based on a random sample of high school dropouts. At the 5% level of significance, can you conclude that employment status and gender are related for all such dropouts?

Solution We perform the following five steps to make this test of hypothesis.

Step 1. *State the null and alternative hypotheses.*

The null and alternative hypotheses are

H_0: Gender and employment status are not related

H_1: Gender and employment status are related

Step 2. *Select the distribution to use.*

Because we are performing a test of independence, we use the chi-square distribution to make the test.

Step 3. *Determine the rejection and nonrejection regions.*

With a significance level of 5%, the area of the rejection region is .05, and it falls in the right tail of the chi-square distribution curve. The contingency table contains two rows (*Men* and *Women*) and three columns (*Employed, Unemployed,* and *Not in Labor Force*). The degrees of freedom are

$$df = (R - 1)(C - 1) = (2 - 1)(3 - 1) = 2$$

From Table IX of Appendix D, the critical value of χ^2 for $df = 2$ and $\alpha = .05$ is 5.991. This value is shown in Figure 11.7.

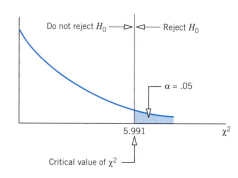

Figure 11.7

Step 4. *Calculate the value of the test statistic.*

The expected frequencies for the various cells are calculated as follows, and they are listed in parentheses in Table 11.9.

E for *Men* and *Employed* cell = (860)(751)/1855 = 348.17

E for *Men* and *Unemployed* cell = (860)(101)/1855 = 46.82

E for *Men* and *Not in Labor Force* cell = (860)(1003)/1855 = 465.00

E for *Women* and *Employed* cell = (995)(751)/1855 = 402.83

E for *Women* and *Unemployed* cell = (995)(101)/1855 = 54.18

E for *Women* and *Not in Labor Force* cell = (995)(1003)/1855 = 538.00

Table 11.9

	Employed (*E*)	Unemployed (*U*)	Not in Labor Force (*N*)	Row Totals
Men (*M*)	311 (348.17)	52 (46.82)	497 (465.00)	860
Women (*W*)	440 (402.83)	49 (54.18)	506 (538.00)	995
Column Totals	751	101	1003	1855

The value of the test statistic χ^2 is calculated as follows:

$$\chi^2 = \sum \frac{(O - E)^2}{E}$$

$$= \frac{(311 - 348.17)^2}{348.17} + \frac{(52 - 46.82)^2}{46.82} + \frac{(497 - 465.00)^2}{465.00}$$

$$+ \frac{(440 - 402.83)^2}{402.83} + \frac{(49 - 54.18)^2}{54.18} + \frac{(506 - 538.00)^2}{538.00}$$

$$= 3.968 + .573 + 2.202 + 3.430 + .495 + 1.903 = 12.571$$

Step 5. *Make a decision.*

The value of the test statistic $\chi^2 = 12.571$ is greater than the critical value of $\chi^2 = 5.991$ and it falls in the rejection region. Hence, we reject the null hypothesis and state that there is strong evidence from the sample to conclude that the two characteristics, *gender* and *employment status,* are related for all high school dropouts. ∎

11.4.2 A TEST OF HOMOGENEITY

In a **test of homogeneity**, we test if two (or more) populations are homogeneous (similar) with regard to the distribution of a certain characteristic. For example, we might be interested in testing the null hypothesis that the proportions of households that belong to different income groups are the same in California and Wisconsin. Or we may want to test whether or not the preferences of people in Florida, Arizona, and Vermont arc similar with regard to Coke, Pepsi, and 7-Up.

> **A TEST OF HOMOGENEITY** A *test of homogeneity* involves testing the null hypothesis that the proportions of elements with certain characteristics in two or more different populations are the same against the alternative hypothesis that these proportions are not the same.

Let us consider the example of testing the null hypothesis that the proportions of households in California and Wisconsin who belong to various income groups are the same. (Note that in a test of homogeneity, the null hypothesis is always that the proportions of elements with certain characteristics are the same in two or more populations. The alternative hypothesis is that these proportions are not the same.) Suppose we define three income strata: high income group (with an income of more than $60,000), medium income group (with an income of $30,000 to $60,000), and low income group (with an income of less than $30,000). Furthermore, assume that we take one sample of 250 households from California and another sample of 150 households from Wisconsin, collect the information on the incomes of these households, and prepare the contingency Table 11.10.

Table 11.10

	California	Wisconsin	Row Totals
High income	70	34	104
Medium income	80	40	120
Low income	100	76	176
Column Totals	250	150	400

Note that in this example the column totals are fixed. That is, we decided in advance to take samples of 250 households from California and 150 from Wisconsin. However, the row totals (of 104, 120, and 176) are determined randomly by the outcomes of the two samples. If we compare this example to the one about violence and lack of discipline in schools in the previous section, we will notice that neither the column nor the row totals were fixed in that example. Instead, the researcher took just one sample of 300 adults, collected the information on gender and opinions, and prepared the contingency table. Thus, in that example, the row and column totals were all determined randomly. Thus, when both the row and column totals are determined randomly, we make a test of independence. However, when either the column totals or the row totals are fixed, we make a test of homogeneity. In the case of income groups in California and Wisconsin, we will make a test of homogeneity to test for the similarity of income groups in the two states.

The procedure to make a test of homogeneity is similar to the procedure used to make a test of independence discussed earlier. Like a test of independence, a test of homogeneity is right-tailed. Example 11–8 illustrates the procedure to make a homogeneity test.

Making a test of homogeneity.

EXAMPLE 11–8   Consider the data on income distribution for households in California and Wisconsin given in Table 11.10. Using the 2.5% significance level, test the null hypothesis that the distribution of households with regard to income levels is similar (homogeneous) for the two states.

Solution   We perform the following five steps to make this test of hypothesis.

Step 1.   *State the null and alternative hypotheses.*

The two hypotheses are[2]

H_0: The proportions of households that belong to different income groups are the same in both states

H_1: The proportions of households that belong to different income groups are not the same in both states

Step 2.   *Select the distribution to use.*

We use the chi-square distribution to make a homogeneity test.

Step 3.   *Determine the rejection and nonrejection regions.*

The significance level is 2.5%. Because the homogeneity test is right-tailed, the area of the rejection region is .025, and it lies in the right tail of the chi-square distribution curve. The contingency table for income groups in California and Wisconsin contains three rows and two columns. Hence, the degrees of freedom are

$$df = (R - 1)(C - 1) = (3 - 1)(2 - 1) = 2$$

From Table IX of Appendix D, the value of χ^2 for $df = 2$ and .025 area in the right tail of the chi-square distribution curve is 7.378. This value is shown in Figure 11.8.

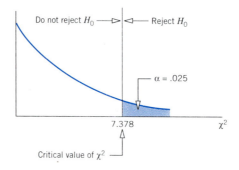

Figure 11.8

---

[2]Let p_{HC}, p_{MC}, and p_{LC} be the proportions of households in California who belong to high, middle, and low income groups, respectively. Let p_{HW}, p_{MW}, and p_{LW} be the corresponding proportions for Wisconsin. Then, we can also write the null hypothesis as

$$H_0: p_{HC} = p_{HW}, p_{MC} = p_{MW}, \text{ and } p_{LC} = p_{LW}$$

and the alternative hypothesis as

$$H_1: \text{At least two of the equalities mentioned in } H_0 \text{ are not true.}$$

Step 4. *Calculate the value of the test statistic.*

To compute the value of the test statistic χ^2, we need to calculate the expected frequencies first. Table 11.11 lists the observed as well as the expected frequencies. The numbers in parentheses in this table are the expected frequencies, which are calculated using the formula

$$E = \frac{(\text{Row total})(\text{Column total})}{\text{Total of both samples}}$$

Thus, for instance,

$$E \text{ for } \textit{High income} \text{ and } \textit{California} \text{ cell} = \frac{(104)(250)}{400} = 65$$

The remaining expected frequencies are calculated in the same way. Note that the expected frequencies in a test of homogeneity are calculated in the same way as in a test of independence. The value of the test statistic χ^2 is computed as follows.

$$\chi^2 = \sum \frac{(O - E)^2}{E}$$

$$= \frac{(70 - 65)^2}{65} + \frac{(34 - 39)^2}{39} + \frac{(80 - 75)^2}{75} + \frac{(40 - 45)^2}{45}$$

$$+ \frac{(100 - 110)^2}{110} + \frac{(76 - 66)^2}{66}$$

$$= .385 + .641 + .333 + .556 + .909 + 1.515 = \mathbf{4.339}$$

Table 11.11

	California	Wisconsin	Row Totals
High income	70 (65)	34 (39)	104
Medium income	80 (75)	40 (45)	120
Low income	100 (110)	76 (66)	176
Column Totals	250	150	400

Step 5. *Make a decision.*

The value of the test statistic $\chi^2 = 4.339$ is less than the critical value of $\chi^2 = 7.378$, and it falls in the nonrejection region. Hence, we fail to reject the null hypothesis and state that the distribution of households with regard to income appears to be similar (homogeneous) in California and Wisconsin. ■

EXERCISES

■ **Concepts and Procedures**

11.23 Describe in your own words a test of independence and a test of homogeneity. Give one example of each.

11.24 Explain how the expected frequencies for cells of a contingency table are calculated in a test of independence or homogeneity. How do you find the degrees of freedom for such tests?

11.25 To make a test of independence or homogeneity, what should be the minimum expected frequency for each cell? What are the alternatives if this condition is not satisfied?

11.26 Consider the following contingency table that is based on a sample survey.

	Column 1	Column 2	Column 3
Row 1	137	64	105
Row 2	98	71	65
Row 3	115	81	115

a. Write the null and alternative hypotheses for a test of independence for this table.
b. Calculate the expected frequencies for all cells assuming that the null hypothesis is true.
c. For $\alpha = .01$, find the critical value of χ^2. Show the rejection and nonrejection regions on the chi-square distribution curve.
d. Find the value of the test statistic χ^2.
e. Using $\alpha = .01$, would you reject the null hypothesis?

11.27 Consider the following contingency table that records the results obtained for four samples of fixed sizes selected from four populations.

	Sample Selected from			
	Population 1	**Population 2**	**Population 3**	**Population 4**
Row 1	24	81	60	121
Row 2	46	64	91	72
Row 3	20	37	105	93

a. Write the null and alternative hypotheses for a test of homogeneity for this table.
b. Calculate the expected frequencies for all cells assuming that the null hypothesis is true.
c. For $\alpha = .025$, find the critical value of χ^2. Show the rejection and nonrejection regions on the chi-square distribution curve.
d. Find the value of the test statistic χ^2.
e. Using $\alpha = .025$, would you reject the null hypothesis?

■ **Applications**

11.28 During the recession in the early 1990s, many families faced hard times financially. Some studies observed that more and more people stopped buying national brand products and started buying less expensive store brand products instead. Data produced by a recent sample of 700 adults on whether they usually buy store brand or name brand products are recorded in the following table.

	More Often Buy	
	National Brand	**Store Brand**
Men	170	145
Women	182	203

Using the 1% significance level, can you reject the null hypothesis that the two attributes, gender and buying national or store brand products, are independent?

11.29 One hundred auto drivers who were stopped by police for some violation were also checked to see if they were wearing their seat belts. The following table records the results of this survey.

	Wearing Seat Belt	Not Wearing Seat Belt
Men	34	21
Women	32	13

Test at the 2.5% significance level whether being a man or a woman and wearing or not wearing a seat belt are related.

11.30 A sample of 15,000 college students produced the classification of these students by gender and full-time or part-time status as shown in the following table.

	Full-time	Part-time
Male	3859	2628
Female	4784	3729

Test at the 1% significance level whether full-time or part-time status and gender are related.

11.31 In a 1999 poll conducted for *Newsweek* magazine by Princeton Survey Research Associates, women were asked whether they would have some form of cosmetic surgery if cost were not an issue (*Newsweek* Special Issue, Spring/Summer 1999). The data given in the magazine were presented separately for 18–29-year-olds and 30–39-year-olds. Assuming that the survey was based on random samples of 500 women in each age group, the percentages given in the magazine would produce the numbers shown in the table below.

		Yes	No	Do Not Know
	18–29	335	155	10
Age				
	30–39	290	210	0

At the 5% level of significance, can you conclude that the desire for cosmetic surgery is related to age?

11.32 The following table gives the two-way classification of 400 randomly selected persons based on their status as a smoker or a nonsmoker and on the number of visits they made to their physicians last year.

	Visits to the Physician		
	0–1	2–4	≥ 5
Smoker	25	60	75
Nonsmoker	110	90	40

Test at the 5% significance level whether smoking and visits to the physician are related for all persons.

11.33 In 1998, a poll of 1024 adult Americans was conducted by Yankelovich Partners for *Time*/CNN to examine attitudes about health insurance. Of them, 848 had health insurance. These 848 adults were asked whether it had become more of a hassle to deal with their health care plans over the past 5 years.

If the pollsters had selected 424 adults with managed care and 424 with traditional plans, the percentages given in the magazine article would have yielded the following table.

	Managed Care	**Traditional**
More hassle	178	106
Less hassle	72	51
No change or not sure	174	267

At the 1% significance level, can you conclude that difficulty in dealing with a health care plan is dependent on whether the plan is managed care or traditional?

11.34 National Electronics Company buys parts from two subsidiaries. The quality control department at this company wanted to check if the distribution of good and defective parts is the same for the supplies of parts received from both subsidiaries. The quality control inspector selected a sample of 300 parts received from Subsidiary A and a sample of 400 parts received from Subsidiary B. These parts were checked for being good or defective. The following table records the results of this investigation.

	Subsidiary A	**Subsidiary B**
Good	284	381
Defective	16	19

Using the 5% significance level, test the null hypothesis that the distributions of good and defective parts are the same for both subsidiaries.

11.35 Two drugs were administered to two groups of randomly assigned patients to cure the same disease. One group of 60 patients and another group of 40 patients were selected. The following table gives information about the number of patients who were cured and not cured by each of the two drugs.

	Cured	**Not Cured**
Drug I	44	16
Drug II	18	22

Test at the 1% significance level whether or not the two drugs are similar in curing and not curing the patients.

11.36 A company introduced a new product in the market a few months ago. The management wants to determine the reaction of customers in different regions to this product. The research department at the company selected four different samples of 400 users of this product from four regions—East, South, Midwest, and West. The users of the product were asked whether or not they like the product. The responses of these people are recorded in the following table.

	East	**South**	**Midwest**	**West**
Like	274	206	291	254
Do not like	126	194	109	146

Based on the evidence from these samples, can you conclude that the distributions of opinions of users of this product are not homogeneous for all four regions with regard to liking and not liking this product? Use $\alpha = .01$.

11.37 Carpal tunnel syndrome (CTS) is a painful and debilitating disorder of the wrist that often occurs in workers whose jobs require repetitive flexing of the wrist. A study was conducted to assess the effectiveness of a yoga-based treatment regimen. Patients were randomly assigned to the yoga group or to a control group, in which case they received conventional treatments. A measure known as Tinel sign is often used to evaluate the condition of a CTS patient. Patients were tested in the beginning of the experiment and again after eight weeks of treatment. In some patients only one wrist was affected, whereas in others both wrists were involved. The following table gives the number of wrists in each category. The table summarizes changes in Tinel sign for all affected wrists.

Tinel Sign	Yoga Group	Control Group
Improved	7	3
Same	22	21
Worsened	4	6

Source: Marian S. Garfinkel et al., "Yoga-Based Intervention for Carpal Tunnel Syndrome," *Journal of the American Medical Association,* November 11, 1998.

Using the 5% level of significance, test the null hypothesis that the response measured by the Tinel sign is the same for both modes of treatment. Note that two expected frequencies will be slightly less than 5. Assume that this does not seriously affect the validity of the test.

11.38 The following table gives the distributions of grades for three professors for a few randomly selected classes that each of them taught during the past two years.

		Professor		
		Miller	**Smith**	**Moore**
	A	18	36	20
	B	25	44	15
Grade	C	85	73	82
	D&F	17	12	8

Using the 2.5% significance level, test the null hypothesis that the grade distributions are homogeneous for these three professors.

11.39 Two random samples, one of 95 blue-collar workers and a second of 50 white-collar workers, were taken from a large company. These workers were asked about their views on a certain company issue. The following table gives the results of the survey.

	Opinion		
	Favor	**Oppose**	**Uncertain**
Blue-collar workers	44	39	12
White-collar workers	21	26	3

Using the 2.5% significance level, test the null hypothesis that the distributions of opinions are homogeneous for the two groups of workers.

11.5 INFERENCES ABOUT THE POPULATION VARIANCE

Earlier chapters explained how to make inferences (confidence intervals and hypothesis tests) about the population mean and population proportion. However, we may often need to control the variance (or standard deviation). Consequently, there may be a need to estimate and to test a hypothesis about the population variance σ^2. Section 11.5.1 describes how to make a confidence interval for the population variance (or standard deviation). Section 11.5.2 explains how to test a hypothesis about the population variance.

As an example, suppose a machine is set up to fill packages of cookies so that the net weight of cookies per package is 32 ounces. Note that the machine will not put exactly 32 ounces of cookies into each package. Some of the packages will contain less and some will contain more than 32 ounces. However, if the variance (and, hence, the standard deviation) is too large, some of the packages will contain quite a bit less than 32 ounces of cookies and some others will contain quite a bit more than 32 ounces. The manufacturer will not want a large variation in the amounts of cookies put into different packages. To keep this variation within some specified acceptable limit, the machine will be adjusted from time to time. Before the manager decides to adjust the machine at any time, he must estimate the variance or test a hypothesis or do both to find out if the variance exceeds the maximum acceptable value.

Like every sample statistic, the sample variance is a random variable, and it possesses a sampling distribution. If all the possible samples of a given size are taken from a population and their variances are calculated, the probability distribution of these variances is called the *sampling distribution of the sample variance.*

SAMPLING DISTRIBUTION OF $(n - 1)s^2/\sigma^2$ If the population from which the sample is selected is (approximately) normally distributed, then

$$\frac{(n - 1)\, s^2}{\sigma^2}$$

has a chi-square distribution with $n - 1$ degrees of freedom.

Thus, the chi-square distribution is used to construct a confidence interval and test a hypothesis about the population variance σ^2.

11.5.1 ESTIMATION OF THE POPULATION VARIANCE

The value of the sample variance s^2 is a point estimate of the population variance σ^2. The $(1 - \alpha)100\%$ confidence interval for σ^2 is given by the following formula.

CONFIDENCE INTERVAL FOR THE POPULATION VARIANCE σ^2 Assuming that the population from which the sample is selected is (approximately) normally distributed, the $(1 - \alpha)100\%$ *confidence interval for the population variance σ^2* is

$$\frac{(n - 1)\, s^2}{\chi^2_{\alpha/2}} \quad \text{to} \quad \frac{(n - 1)\, s^2}{\chi^2_{1-\alpha/2}}$$

where $\chi^2_{\alpha/2}$ and $\chi^2_{1-\alpha/2}$ are obtained from the chi-square distribution table for $\alpha/2$ and $1 - \alpha/2$ areas in the right tail of the chi-square distribution curve, respectively, and for $n - 1$ degrees of freedom.

The confidence interval for the population standard deviation can be obtained by simply taking the positive square root of the two limits of the confidence interval for the population variance.

The procedure for making a confidence interval for σ^2 involves the following three steps.

1. Take a sample of size n and compute s^2 using the formula learned in Chapter 3. However, if n and s^2 are given, then perform only steps 2 and 3.

2. Calculate $\alpha/2$ and $1 - \alpha/2$. Find two values of χ^2 from the chi-square distribution table (Table IX of Appendix D): one for $\alpha/2$ area in the right tail of the chi-square distribution curve and $df = n - 1$, and the second for $1 - \alpha/2$ area in the right tail and $df = n - 1$.

3. Substitute all the values in the formula for the confidence interval for σ^2 and simplify.

Example 11–9 illustrates the estimation of the population variance and population standard deviation.

Constructing confidence intervals for σ^2 and σ.

EXAMPLE 11-9 One type of cookie manufactured by Haddad Food Company is Cocoa Cookies. The machine that fills packages of these cookies is set up in such a way that the average net weight of these packages is 32 ounces with a variance of .015 square ounce. From time to time the quality control inspector at the company selects a sample of a few such packages, calculates the variance of the net weights of these packages, and constructs a 95% confidence interval for the population variance. If either both or one of the two limits of this confidence interval is not in the interval .008 to .030, the machine is stopped and adjusted. A recently taken random sample of 25 packages from the production line gave a sample variance of .029 square ounce. Based on this sample information, do you think the machine needs an adjustment? Assume that the net weights of cookies in all packages are normally distributed.

Solution The following three steps are performed to estimate the population variance and to make a decision.

Step 1. From the given information,

$$n = 25 \quad \text{and} \quad s^2 = .029$$

Step 2. The confidence level is $1 - \alpha = .95$. Hence, $\alpha = 1 - .95 = .05$.

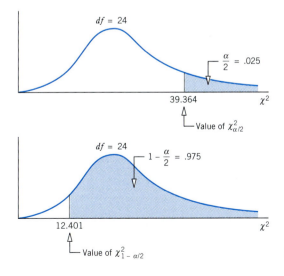

Figure 11.9

We obtain:

$$\alpha/2 = .05/2 = .025$$

$$1 - \alpha/2 = 1 - .025 = .975$$

From Table IX of Appendix D, with $df = n - 1 = 25 - 1 = 24$,

χ^2 for 24 df and .025 area in the right tail $= 39.364$

χ^2 for 24 df and .975 area in the right tail $= 12.401$

These values are shown in Figure 11.9.

Step 3.   The 95% confidence interval for σ^2 is

$$\frac{(n-1)\,s^2}{\chi^2_{\alpha/2}} \quad \text{to} \quad \frac{(n-1)\,s^2}{\chi^2_{1-\alpha/2}}$$

$$\frac{(25-1)\,(.029)}{39.364} \quad \text{to} \quad \frac{(25-1)\,(.029)}{12.401}$$

.0177   to   .0561

Thus, with 95% confidence, we can state that the variance for all packages of Cocoa Cookies lies between .0177 and .0561 square ounce. Note that the lower limit (.0177) of this confidence interval is between .008 and .030, but the upper limit (.0561) is larger than .030 and falls outside the interval .008 to .030. Since the upper limit is larger than .030, we can state that the machine needs to be stopped and adjusted.

We can obtain the confidence interval for the population standard deviation σ by taking the positive square root of the two limits of the above confidence interval for the population variance. Thus, a 95% confidence interval for the population standard deviation is

$$\sqrt{.0177} \text{ to } \sqrt{.0561} \quad \text{or} \quad \textbf{.133 to .237}$$

Hence, the standard deviation of all packages of Cocoa Cookies is between .133 and .237 ounce at a 95% confidence level. ∎

11.5.2   HYPOTHESIS TESTS ABOUT THE POPULATION VARIANCE

A test of hypothesis about the population variance can be one-tailed or two-tailed. To make a test of hypothesis about σ^2, we perform the same five steps we have used earlier in hypothesis-testing examples. The procedure to test a hypothesis about σ^2 discussed in this section is applied only when the population from which a sample is selected is (approximately) normally distributed.

> **TEST STATISTIC FOR A TEST OF HYPOTHESIS ABOUT σ^2**   The value of the *test statistic* χ^2 is calculated as
>
> $$\chi^2 = \frac{(n-1)\,s^2}{\sigma^2}$$
>
> where s^2 is the sample variance, σ^2 is the hypothesized value of the population variance, and $n - 1$ represents the degrees of freedom. The population from which the sample is selected is assumed to be (approximately) normally distributed.

Examples 11–10 and 11–11 illustrate the procedure to make tests of hypothesis about σ^2.

Making a right-tailed test of hypothesis about σ^2.

EXAMPLE 11–10   One type of cookie manufactured by Haddad Food Company is Cocoa Cookies. The machine that fills packages of these cookies is set up in such a way that the average net weight of these packages is 32 ounces with a variance of .015 square ounce. From time to time the quality control inspector at the company selects a sample of a few such packages, calculates the variance of the net weights of these packages, and makes a test of hypothesis about the population variance. She always uses $\alpha = .01$. The acceptable value of the population variance is .015 square ounce or less. If the conclusion from the test of hypothesis is that the population variance is not within the acceptable limit, the machine is stopped and adjusted. A recently taken random sample of 25 packages from the production line gave a sample variance of .029 square ounce. Based on this sample information, do you think the machine needs an adjustment? Assume that the net weights of cookies in all packages are normally distributed.

Solution   From the given information,

$$n = 25, \quad \alpha = .01, \quad \text{and} \quad s^2 = .029$$

The population variance should not exceed .015 square ounce.

Step 1.   *State the null and alternative hypotheses.*

We are to test whether or not the population variance is within the acceptable limit. The population variance is within the acceptable limit if it is less than or equal to .015; otherwise, it is not. Thus, the two hypotheses are

$H_0: \sigma^2 \leq .015$ (The population variance is within the acceptable limit)

$H_1: \sigma^2 > .015$ (The population variance exceeds the acceptable limit)

Step 2.   *Select the distribution to use.*

We use the chi-square distribution to test a hypothesis about σ^2.

Step 3.   *Determine the rejection and nonrejection regions.*

The significance level is 1% and, because of the $>$ sign in H_1, the test is right-tailed. The rejection region lies in the right tail of the chi-square distribution curve with its area equal to .01. The degrees of freedom for a chi-square test about σ^2 are $n - 1$; that is,

$$df = n - 1 = 25 - 1 = 24$$

From Table IX of Appendix D, the critical value of χ^2 for 24 degrees of freedom and .01 area in the right tail is 42.980. This value is shown in Figure 11.10.

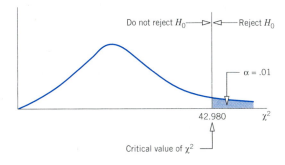

Figure 11.10

Step 4. *Calculate the value of the test statistic.*

The value of the test statistic χ^2 for the sample variance is calculated as follows:

$$\chi^2 = \frac{(n-1)\,s^2}{\sigma^2} = \frac{(25-1)\,(.029)}{.015} = 46.400$$

$\uparrow$ From H_0

Step 5. *Make a decision.*

The value of the test statistic $\chi^2 = 46.400$ is greater than the critical value of $\chi^2 = 42.980$, and it falls in the rejection region. Consequently, we reject H_0 and conclude that the population variance is not within the acceptable limit. The machine should be stopped and adjusted. ∎

Conducting a two-tailed test of hypothesis about σ^2.

EXAMPLE 11-11 The variance of scores on a standardized mathematics test for all high school seniors was 150 in 1999. A sample of scores for 20 high school seniors who took this test this year gave a variance of 170. Test at the 5% significance level if the variance of current scores of all high school seniors on this test is different from 150. Assume that the scores of all high school seniors on this test are (approximately) normally distributed.

Solution From the given information,

$$n = 20, \quad \alpha = .05, \quad \text{and} \quad s^2 = 170$$

The population variance was 150 in 1999.

Step 1. *State the null and alternative hypotheses.*

The null and alternative hypotheses are

H_0: $\sigma^2 = 150$ (The population variance is not different from 150)

H_1: $\sigma^2 \neq 150$ (The population variance is different from 150)

Step 2. *Select the distribution to use.*

We use the chi-square distribution to test a hypothesis about σ^2.

Step 3. *Determine the rejection and nonrejection regions.*

The significance level is 5%. The $\neq$ sign in H_1 indicates that the test is two-tailed. The rejection region lies in both tails of the chi-square distribution curve with its total area equal to .05. Consequently, the area in each tail of the distribution curve is .025. The values of $\alpha/2$ and $1 - \alpha/2$ are

$$\frac{\alpha}{2} = \frac{.05}{2} = .025 \quad \text{and} \quad 1 - \frac{\alpha}{2} = 1 - .025 = .975$$

The degrees of freedom are

$$df = n - 1 = 20 - 1 = 19$$

From Table IX of Appendix D, the critical values of χ^2 for 19 degrees of freedom and for $\alpha/2$ and $1 - \alpha/2$ areas in the right tail are

$$\chi^2 \text{ for 19 } df \text{ and .025 area in the right tail} = 32.852$$

$$\chi^2 \text{ for 19 } df \text{ and .975 area in the right tail} = 8.907$$

These two values are shown in Figure 11.11.

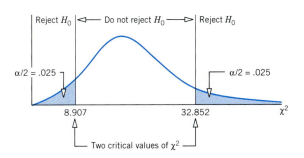

Figure 11.11

Step 4. *Calculate the value of the test statistic.*

The value of the test statistic χ^2 for the sample variance is calculated as follows:

$$\chi^2 = \frac{(n-1)\,s^2}{\sigma^2} = \frac{(20-1)\,(170)}{150} = 21.533$$

From H_0

Step 5. *Make a decision.*

The value of the test statistic $\chi^2 = 21.533$ is between the two critical values of χ^2, 8.907 and 32.852, and it falls in the nonrejection region. Consequently, we fail to reject H_0 and conclude that the population variance of the current scores of high school seniors on this standardized mathematics test does not appear to be different from 150. ∎

Note that we can make a test of hypothesis about the population standard deviation σ using the same procedure as that for the population variance σ^2. To make a test of hypothesis about σ, the only change will be mentioning the values of σ in H_0 and H_1. The rest of the procedure remains the same as in case of σ^2.

EXERCISES

■ **Concepts and Procedures**

11.40 A sample of certain observations selected from a normally distributed population produced a sample variance of 46. Construct a 95% confidence interval for σ^2 for each of the following cases and comment on what happens to the confidence interval of σ^2 when the sample size increases.

a. $n = 12$ **b.** $n = 16$ **c.** $n = 25$

11.41 A sample of 25 observations selected from a normally distributed population produced a sample variance of 35. Construct a confidence interval for σ^2 for each of the following confidence levels and comment on what happens to the confidence interval of σ^2 when the confidence level decreases.

a. $1 - \alpha = .99$ **b.** $1 - \alpha = .95$ **c.** $1 - \alpha = .90$

11.42 A sample of 22 observations selected from a normally distributed population produced a sample variance of 18.

a. Write the null and alternative hypotheses to test whether the population variance is different from 14.
b. Using $\alpha = .05$, find the critical values of χ^2. Show the rejection and nonrejection regions on a chi-square distribution curve.
c. Find the value of the test statistic χ^2.
d. Using the 5% significance level, will you reject the null hypothesis stated in part a?

11.43 A sample of 16 observations selected from a normally distributed population produced a sample variance of 1.10.

a. Write the null and alternative hypotheses to test whether the population variance is greater than .80.
b. Using $\alpha = .01$, find the critical value of χ^2. Show the rejection and nonrejection regions on a chi-square distribution curve.
c. Find the value of the test statistic χ^2.
d. Using the 1% significance level, will you reject the null hypothesis stated in part a?

11.44 A sample of 25 observations selected from a normally distributed population produced a sample variance of .70.

a. Write the null and alternative hypotheses to test whether the population variance is less than 1.25.
b. Using $\alpha = .025$, find the critical value of χ^2. Show the rejection and nonrejection regions on a chi-square distribution curve.
c. Find the value of the test statistic χ^2.
d. Using the 2.5% significance level, will you reject the null hypothesis stated in part a?

11.45 A sample of 18 observations selected from a normally distributed population produced a sample variance of 4.6.

a. Write the null and alternative hypotheses to test whether the population variance is different from 2.2.
b. Using $\alpha = .05$, find the critical values of χ^2. Show the rejection and nonrejection regions on a chi-square distribution curve.
c. Find the value of the test statistic χ^2.
d. Using the 5% significance level, will you reject the null hypothesis stated in part a?

■ **Applications**

11.46 The management of a soft-drink company does not want the variance of the amounts of soda in 12-ounce cans to be more than .01 square ounce. (Recall from Chapter 3 that the variance is always in square units.) The company manager takes a sample of certain cans and estimates the population variance quite often. A random sample of twenty 12-ounce cans taken from the production line of this company showed that the variance for this sample was .014 square ounce.

a. Construct the 99% confidence intervals for the population variance and standard deviation. Assume that the amounts of soda in all 12-ounce cans have a normal distribution.
b. Test at the 2.5% significance level whether the variance of the amounts of soda in all 12-ounce cans for this company is greater than .01 square ounce.

11.47 The 2-inch-long bolts manufactured by a company must have a variance of .003 square inch or less for acceptance by a buyer. A random sample of 26 such bolts gave a variance of .0061 square inch.

a. Test at the 1% significance level whether the variance of all such bolts is greater than .003 square inch. Assume that the lengths of all 2-inch-long bolts manufactured by this company are (approximately) normally distributed.
b. Make the 98% confidence intervals for the population variance and standard deviation.

11.48 An auto manufacturing company wants to estimate the variance of miles per gallon for its auto model AST727. A random sample of 22 cars of this model showed that the variance of miles per gallon for these cars is .62.

a. Construct the 95% confidence intervals for the population variance and standard deviation. Assume that the miles per gallon for all such cars are (approximately) normally distributed.
b. Test at the 1% significance level whether the sample result indicates that the population variance is different from .30.

11.49 The manufacturer of a certain brand of lightbulbs claims that the variance of the lives of these bulbs is 4200 square hours. A consumer agency took a random sample of 25 such bulbs and tested them. The variance of the lives of these bulbs was found to be 5200 square hours. Assume that the lives of all such bulbs are (approximately) normally distributed.

a. Make the 99% confidence intervals for the variance and standard deviation of the lives of all such bulbs.
b. Test at the 5% significance level whether the variance of such bulbs is different from 4200 square hours.

GLOSSARY

Chi-square distribution   A distribution, with degrees of freedom as the only parameter, that is skewed to the right for small df and looks like a normal curve for large df.

Expected frequencies   The frequencies for different categories of a multinomial experiment or for different cells of a contingency table that are expected to occur when a given null hypothesis is true.

Goodness-of-fit test   A test of the null hypothesis that the observed frequencies for an experiment follow a certain pattern or theoretical distribution.

Multinomial experiment   An experiment with n trials for which (1) the trials are identical, (2) there are more than two possible outcomes per trial, (3) the trials are independent, and (4) the probabilities of the various outcomes remain constant for each trial.

Observed frequencies   The frequencies actually obtained from the performance of an experiment.

Test of homogeneity   A test of the null hypothesis that the proportions of elements that belong to different groups in two (or more) populations are similar.

Test of independence   A test of the null hypothesis that two attributes of a population are not related.

KEY FORMULAS

1.   **Expected frequency for a category for a goodness-of-fit test**

$$E = np$$

where n is the sample size and p is the probability that an element belongs to this category.

2.   **Value of the test statistic χ^2 for a goodness-of-fit test and a test of independence or homogeneity**

$$\chi^2 = \sum \frac{(O - E)^2}{E}$$

where O and E are the observed and expected frequencies, respectively, for a category or cell.

3.   **Degrees of freedom for a goodness-of-fit test**

$$df = k - 1$$

where k denotes the number of categories for the experiment.

4.   **Degrees of freedom for a test of independence or homogeneity**

$$df = (R - 1)(C - 1)$$

where R and C are, respectively, the number of rows and columns for the contingency table.

5.   **Expected frequency for a cell for an independence or homogeneity test**

$$E = \frac{\text{(Row total) (Column total)}}{\text{Sample size}}$$

6.   **The $(1 - \alpha)100\%$ confidence interval for population variance σ^2**

$$\frac{(n - 1) s^2}{\chi^2_{\alpha/2}} \quad \text{to} \quad \frac{(n - 1) s^2}{\chi^2_{1-\alpha/2}}$$

where $\chi^2_{\alpha/2}$ and $\chi^2_{1-\alpha/2}$ are obtained from the chi-square table for $\alpha/2$ and $1 - \alpha/2$ areas in the right tail of the chi-square distribution curve and $n - 1$ degrees of freedom.

7. **Value of the test statistic χ^2 in a hypothesis test about σ^2**

$$\chi^2 = \frac{(n-1)\,s^2}{\sigma^2}$$

where s^2 is the sample variance, σ^2 is the hypothesized value of the population variance, and $n-1$ are the degrees of freedom.

SUPPLEMENTARY EXERCISES

11.50 In a survey by Robert Half International, Inc., human resource executives were asked to state the most likely reason for a good employee to leave his or her current position (*USA TODAY,* November 5, 1998). The table records the percentage of respondents who chose each reason.

Reason	Percentage
Limited advancement potential	41
Lack of recognition	25
Low salary/benefits	15
Unhappy with management	10
Bored with job	5
Don't know	4

A recent random sample of 280 such executives yielded the following frequencies for the reasons given in the table, listed in that order: 118, 85, 33, 32, 10, and 2. Test at the 1% significance level whether the current distribution of such executives' opinions on why workers leave jobs differs from that found in the 1998 survey.

11.51 One of the products produced by Branco Food Company is All-Bran Cereal, which competes with three other brands of similar all-bran cereals. The company's research office wants to investigate if the percentage of people who consume all-bran cereal is the same for each of these four brands. Let us denote the four brands of cereal by A, B, C, and D. A sample of 1000 persons who consume all-bran cereal was taken, and they were asked which brand they most often consume. Of the respondents, 212 said they usually consume Brand A, 284 consume Brand B, 254 consume Brand C, and 250 consume Brand D. Does the sample provide enough evidence to reject the null hypothesis that the percentage of people who consume all-bran cereal is the same for all four brands? Use $\alpha = .05$.

11.52 In a U.S. Census Bureau survey, registered voters who did not vote in the 1996 elections were asked to give the main reason for their failure to vote (*USA TODAY,* November 2, 1998). The reasons and the percentage of nonvoters who gave each reason are summarized in the table.

Reason	Percentage
No time off/busy	22
No interest/didn't care	17
Illness/family emergency	15
Disliked candidates	13
Traveling	11
No transportation	4
Forgot	4
Lines too long	1
Other/don't know	13

A random sample of 500 registered voters who failed to vote in a recent election produced the following frequencies for the reasons given in the table, listed in that order: 98, 77, 82, 76, 64, 27, 20,

10, and 46. Does the sample information provide sufficient evidence to reject the null hypothesis that the current distribution of reasons given for not voting is the same as in 1996? Use the 2.5% level of significance.

11.53 In a 1999 survey, adults were asked how often they usually have an eye examination. The following table lists the percentages of adults with different responses.

Category	Percentage
More than once a year	10
Once a year	38
Once every 2 years	25
Once every 3 years	6
Once every 4 years	3
Once every 5 years	12
Never	6

Source: AMD Alliance International, *USA TODAY,* December 8, 1999.

Recently a random sample of 100 adults were asked the same question. The percentages of responses for the categories given in the table, listed in that order, are as follows: 7, 40, 36, 4, 2, 7, and 4. At the 5% level of significance, can you conclude that the present frequency distribution for eye examinations of adults differs from that found in the 1999 survey? Note that the expected value for the *Once every 4 years* category will be less than 5. Combine this category with *Once every 3 years* to create a single category.

11.54 The following table shows the number of persons in a random sample of 210 listed according to the day of the week on which they prefer to do their grocery shopping.

Day	Mon	Tue	Wed	Thu	Fri	Sat	Sun
Number of persons	9	17	12	26	36	69	41

Using the 2.5% significance level, test the null hypothesis that the proportion of persons who prefer to do their grocery shopping on a particular day is the same for all days of the week.

11.55 A randomly selected sample of 100 persons who suffer from allergies were asked during what season they suffer the most. The results of the survey are recorded in the following table.

Season	Fall	Winter	Spring	Summer
Persons allergic	18	13	31	38

Using the 1% significance level, test the null hypothesis that the proportions of all allergic persons are the same over the four seasons.

11.56 The president of a bank selected a sample of 200 loan applications to check if the approval or rejection of an application depends on which one of the two loan officers, Thurow or Webber, handles that application. The information obtained from the sample is summarized in the following table.

	Approved	Rejected
Thurow	54	41
Webber	69	36

Test at the 2.5% significance level whether the approval or rejection of a loan application depends on which loan officer handles the application.

11.57 The U.S. Bureau of Labor Statistics compiles data on the current employment status of "displaced workers" who have lost or left a job for reasons given in the table below. The table gives the number of workers (in thousands) who lost jobs but are now employed, unemployed, or not in the labor force.

Reason for Job Loss	Employed	Unemployed	Not in Labor Force
Plant or company closed down or moved	2532	340	429
Insufficient work	1892	379	258
Position or shift abolished	1635	212	276

Source: U.S. Department of Labor News, August 19, 1998.

Suppose that these data were obtained from a random sample of displaced workers. Using the 1% significance level, can you conclude that the current employment status and reason for job loss are related?

11.58 A random sample of 100 jurors was selected and asked whether or not each of them had ever been a victim of crime. The jurors were also asked whether they are strict, fair, or lenient regarding punishment for crime. The following table gives the results of the survey.

	Strict	Fair	Lenient
Have been a victim	20	8	3
Have never been a victim	22	33	14

Test at the 5% significance level whether the two attributes for all jurors are dependent.

11.59 A recession and bad economic conditions forced many people to hold more than one job to make ends meet. A sample of 500 persons who held more than one job produced the following two-way table.

	Single	Married	Other
Male	72	209	39
Female	33	102	45

Test at the 10% significance level whether gender and marital status are related for all people who hold more than one job.

11.60 The following table gives the classification, by gender and living arrangements, of noninstitutionalized U.S. patients who were under hospice care.

Living Arrangements	Male	Female
With family members	19,200	17,500
Alone	1300	5100
With nonfamily members only	1500	1700

Source: Centers for Disease Control and Prevention Advance Data, August 28, 1998.

Suppose that these data were obtained from random samples of male and female noninstitutionalized patients. Using the 2.5% significance level, test to determine whether the distributions of living arrangements for males and females are different.

11.61 A random sample of 100 persons was selected from each of four regions in the United States. These people were asked whether or not they support a certain farm subsidy program. The results of the survey are summarized in the following table.

	Favor	Oppose	Uncertain
Northeast	56	33	11
Midwest	73	23	4
South	67	28	5
West	59	35	6

Using the 1% significance level, test the null hypothesis that the percentages of people with different opinions are similar for all four regions.

11.62 Construct the 98% confidence intervals for the population variance and standard deviation for the following data assuming that the respective populations are (approximately) normally distributed.

a. $n = 21$, $s^2 = 9.2$ **b.** $n = 17$, $s^2 = 1.7$

11.63 Construct the 95% confidence intervals for the population variance and standard deviation for the following data assuming that the respective populations are (approximately) normally distributed.

a. $n = 10$, $s^2 = 7.2$ **b.** $n = 18$, $s^2 = 14.8$

11.64 Refer to Exercise 11.62a. Test at the 5% significance level if the population variance is different from 6.5.

11.65 Refer to Exercise 11.62b. Test at the 2.5% significance level if the population variance is greater than 1.1.

11.66 Refer to Exercise 11.63a. Test at the 1% significance level if the population variance is greater than 4.2.

11.67 Refer to Exercise 11.63b. Test at the 5% significance level if the population variance is different from 10.4.

11.68 Usually people do not like waiting in line a long time for service. A bank manager does not want the variance of the waiting times for her customers to be greater than 4.0 square minutes. A random sample of 25 customers taken from this bank gave the variance of the waiting times equal to 8.3 square minutes.

a. Test at the 1% significance level whether the variance of the waiting times for all customers at this bank is greater than 4.0 square minutes. Assume that the waiting times for all customers are normally distributed.

b. Construct a 99% confidence interval for the population variance.

11.69 The variance of the SAT scores for all students who took that test this year is 5000. The variance of the SAT scores for a random sample of 20 students from one school is equal to 3175.

a. Test at the 2.5% significance level whether the variance of the SAT scores for students from this school is lower than 5000. Assume that the SAT scores for all students at this school are (approximately) normally distributed.

b. Construct the 98% confidence intervals for the variance and the standard deviation of SAT scores for all students at this school.

11.70 A company manufactures ball bearings that are supplied to other companies. The machine that is used to manufacture these ball bearings produces them with a variance of diameters of .025 square millimeter or less. The quality control officer takes a sample of such ball bearings quite often and checks, using confidence intervals and tests of hypotheses, whether or not the variance of these bearings is within .025 square millimeter. If it is not, the machine is stopped and adjusted. A recently taken random sample of 23 ball bearings gave a variance of the diameters equal to .034 square millimeter.

a. Using the 5% significance level, can you conclude that the machine needs an adjustment? Assume that the diameters of all ball bearings have a normal distribution.

b. Construct a 95% confidence interval for the population variance.

11.71 A random sample of 25 students taken from a university gave the variance of their GPAs equal to .19.

a. Construct the 99% confidence intervals for the population variance and standard deviation. Assume that the GPAs of all students are (approximately) normally distributed.

b. The variance of GPAs of all students at this university was .13 two years ago. Test at the 1% significance level whether the variance of GPAs now is different from .13.

11.72 A sample of seven students selected from a small college produced the following data on their heights (in inches).

$$65 \qquad 74 \qquad 70 \qquad 62 \qquad 60 \qquad 67 \qquad 66$$

a. Using the formula from Chapter 3, find the sample variance s^2 for these data.

b. Make the 98% confidence intervals for the population variance and standard deviation. Assume that the population from which this sample is selected is normally distributed.

c. Test at the 1% significance level whether the population variance is greater than 15 square inches.

11.73 The following are the prices of the same brand of camcorder found at eight stores in Los Angeles.

$$\$755 \qquad 815 \qquad 789 \qquad 799 \qquad 732 \qquad 835 \qquad 799 \qquad 769$$

a. Using the formula from Chapter 3, find the sample variance s^2 for these data.

b. Make the 95% confidence intervals for the population variance and standard deviation. Assume that the prices of this camcorder at all stores in Los Angeles follow a normal distribution.

c. Test at the 5% significance level whether the population variance is different from 500 square dollars.

■ **Advanced Exercises**

11.74 A poll conducted in 1998 by CBD Research and Consulting was concerned with employees' job satisfaction (*USA TODAY,* July 8, 1998). Assuming that this poll included 200 employees in each of the three age groups listed in the table, the percentages given in the newspaper would yield the number of employees who are satisfied with their current employer as reported in the table.

Age	Satisfied with Current Employer
Under 35	116
35 to 54	140
55 and up	186

Using the 1% significance level, test the null hypothesis that all three age groups have the same degree of satisfaction with their current employers. Assume that all employees reported that either they are satisfied or not satisfied with their current employer, and no employee said *no opinion.*

11.75 A chemical manufacturing company wants to locate a hazardous waste disposal site near a city of 50,000 residents and has offered substantial financial inducements to the city. Two hundred adults (110 women and 90 men) who are residents of this city are chosen at random. Sixty percent of those polled oppose the site, 32% are in favor, and 8% are undecided. Of those who oppose the site, 65% are women; of those in favor, 62.5% are men. Using the 5% level of significance, can you conclude that opinions on the disposal site are dependent on gender?

11.76 A student who needs to pass an elementary statistics course wonders whether it will make a difference if she takes the course with one instructor rather than another. Observing the final grades given by each instructor in a recent elementary statistics course, she finds that Instructor A gave 48 passing grades in a class of 52 students and Instructor B gave 44 passing grades in a class of 54 students.

a. Compute the value of the standard normal test statistic z of Section 10.5.3 for the data and use it to find the *p*-value when testing for a difference between the proportions of passing grades given by these instructors.

b. Construct a 2 × 2 contingency table for these data. Compute the value of the χ^2 test statistic for the test of homogeneity and use it to find the *p*-value.

c. How do the test statistics in parts a and b compare? How do the *p*-values for the tests in parts a and b compare? Do you think this is a coincidence, or do you think this will always happen?

11.77 Each of five boxes contains a large (but unknown) number of red and green marbles. You have been asked to find if the proportions of red and green marbles are the same for each of the five boxes. You sample 50 times, with replacement, from each of the five boxes and observe 20, 14, 23, 30, and 18 red marbles, respectively. Can you conclude that the five boxes have the same proportions of red and green marbles? Use a .05 level of significance.

SELF-REVIEW TEST

1. The random variable χ^2 assumes only

   a. positive   **b.** nonnegative   **c.** nonpositive values

2. The parameter(s) of the chi-square distribution is (are)

   a. degrees of freedom   **b.** *df* and *n*   **c.** χ^2

3. Which of the following is *not* a characteristic of a multinomial experiment?

   a. It consists of *n* identical trials.
   b. There are *k* possible outcomes for each trial and $k > 2$.
   c. The trials are random.
   d. The trials are independent.
   e. The probabilities of outcomes remain constant for each trial.

4. The observed frequencies for a goodness-of-fit test are

   a. the frequencies obtained from the performance of an experiment
   b. the frequencies given by the product of *n* and *p*
   c. the frequencies obtained by adding the results of a and b

5. The expected frequencies for a goodness-of-fit test are

   a. the frequencies obtained from the performance of an experiment
   b. the frequencies given by the product of *n* and *p*
   c. the frequencies obtained by adding the results of a and b

6. The degrees of freedom for a goodness-of-fit test are

   a. $n - 1$ **b.** $k - 1$ **c.** $n + k - 1$

7. The chi-square goodness-of-fit test is always

   a. two-tailed   **b.** left-tailed   **c.** right-tailed

8. To apply a goodness-of-fit test, the expected frequency of each category must be at least

   a. 10   **b.** 5   **c.** 8

9. The degrees of freedom for a test of independence are

   a. $(R - 1)(C - 1)$ **b.** $n - 2$ **c.** $(n - 1)(k - 1)$

10. In a 1998 *USA TODAY*/CNN/Gallup poll of American adults, 53% of the respondents thought that the quality of life for an average American would be better in 2025 than in 1998, 42% thought that it would be worse, and 5% thought that it would be the same or had no opinion (*USA TODAY*, October 13, 1998). Assume that these results were true for the entire population of American adults at the time of the poll. In a recent sample of 1000 American adults, 510 thought that life would be better in 2025, 455 thought that it would be worse, and 35 thought that it would be the same or had no opinion. At the 1% significance level, can you conclude that the current distribution of opinions on this matter differs from that for 1998?

11. The following table gives the two-way classification of 1000 persons who have been married at least once. They are classified by educational level and marital status.

	Educational Level			
	Less Than High School	High School Degree	Some College	College Degree
Divorced	173	158	95	53
Never divorced	162	126	110	123

Test at the 1% significance level whether educational level and ever being divorced are dependent.

12. A researcher wanted to investigate if people who belong to different income groups are homogeneous with regard to playing lotteries. She took a sample of 600 people from the low income group, another sample of 500 people from the middle income group, and a third sample of 400 people from the high income group. All these people were asked whether they play the lottery often, sometimes, or never. The results of the survey are summarized in the following table.

	Income Group		
	Low	Middle	High
Play often	174	163	90
Play sometimes	286	217	120
Never play	140	120	190

Using the 5% significance level, can you reject the null hypothesis that the percentages of people who play the lottery often, sometimes, and never are the same for each income group?

13. A cough syrup drug manufacturer requires that the variance for a chemical contained in the bottles of this drug should not exceed .03 square ounce. A sample of 25 such bottles gave the variance for this chemical as .06 square ounce.

a. Construct the 99% confidence intervals for the population variance and the population standard deviation. Assume that the amounts of this chemical in all such bottles are (approximately) normally distributed.

b. Test at the 1% significance level whether the variance of this chemical in all such bottles exceeds .03 square ounce.

MINI-PROJECTS

Mini-Project 11–1

In recent years drivers have become careless about signaling their turns. To study this problem, go to a busy intersection and observe at least 75 vehicles that make left turns. Divide these vehicles into three or four classes. For example, you might use cars, trucks, and others, where "others" include minivans and sport-utility vehicles, as classes. For each left turn made by a vehicle, record the type of vehicle and whether or not the driver used the left turn signal before making this turn. It would be better to avoid intersections that have designated left-turn lanes or green arrows for left turns because drivers in these situations often assume that their intent to turn left is obvious. Carry out an appropriate test at the 1% level of significance to determine if signaling behavior and vehicle type are dependent.

Mini-Project 11–2

Select two foreign currencies that presently have about the same exchange rate against the U.S. dollar. For example, in the middle of December 1999 the Deutschmark and the New Zealand dollar were each equivalent to about 51 U.S. cents.

a. Using the financial pages of a newspaper, record the exchange rates of these two currencies against the U.S. dollar each business day for four weeks.

b. Calculate the sample standard deviation for the exchange rates of each currency. Then construct a 90% confidence interval for each population standard deviation. What assumptions must you make to construct these confidence intervals?

c. Which currency shows the greatest variation against the U.S. dollar? Do the two confidence intervals overlap?

For instructions on using MINITAB for this chapter, please see Section B.10 in Appendix B.

See Appendix C for *Excel Adventure 11* on chi-square applications.

COMPUTER ASSIGNMENTS

CA11.1 A 1998 survey asked workers aged 18–29 to state the factor that was most important to them in a job or career (Princeton Survey Research for Domino's Pizza, *USA TODAY,* June 3, 1998). Twenty-six percent of these workers said that the most important factor was advancement opportunity, 26% mentioned benefits, 18% indicated money, 11% said that being their own boss was most important, 10% preferred flexible hours, 5% indicated number of hours, and 4% mentioned none or don't know. A recent random sample of 250 such workers were asked the same question. Their responses are summarized in the table below.

Factor	Advancement	Benefits	Money	Own Boss	Flexible Hours	Number of Hours	None
Frequency	70	60	52	18	32	11	7

Using MINITAB or any other statistical software, test at the 5% significance level whether the current distribution of priorities of all such workers differs from the distribution of priorities in the 1998 survey.

CA11.2 A sample of 4000 persons aged 18 and older produced the following two-way classification table.

	Men	**Women**
Single	531	357
Married	1375	1179
Widowed	55	195
Divorced	139	169

Using MINITAB or any other statistical software, test at the 10% significance level whether gender and marital status are dependent for all persons aged 18 and older.

CA11.3 Two samples, one of 3000 students from urban high schools and another of 2000 students from rural high schools, were taken. These students were asked if they have ever smoked. The following table lists the summary of the results.

	Urban	**Rural**
Have never smoked	1448	1228
Have smoked	1552	772

Using the 5% significance level, test the null hypothesis that the proportions of urban and rural students who have smoked and who have never smoked are homogeneous. Use MINITAB or any other statistical software.

12

ANALYSIS OF VARIANCE

12.1 THE *F* DISTRIBUTION

12.2 ONE-WAY ANALYSIS OF VARIANCE

GLOSSARY

KEY FORMULAS

SUPPLEMENTARY EXERCISES

SELF-REVIEW TEST

MINI-PROJECTS

COMPUTER ASSIGNMENTS

Chapter 10 described the procedures that are used to test hypotheses about the difference between two population means using the normal and *t* distributions. Also described in that chapter were the hypothesis-testing procedures for the difference between two population proportions using the normal distribution. Then, Chapter 11 explained the procedures used to test hypotheses about the equality of more than two population proportions using the chi-square distribution.

This chapter explains how to test the null hypothesis that the means of more than two populations are equal. For example, suppose that teachers at a school have devised three different methods to teach arithmetic. They want to find out if these three methods produce different mean scores. Let μ_1, μ_2, and μ_3 be the mean scores of all students who will be taught by Methods I, II, and III, respectively. To test whether or not the three teaching methods produce different means, we test the null hypothesis

$$H_0\text{: } \mu_1 = \mu_2 = \mu_3 \qquad \text{(All three population means are equal)}$$

against the alternative hypothesis

$$H_1\text{: Not all three population means are equal}$$

We use the analysis of variance procedure to perform such a test of hypothesis.

Note that the analysis of variance procedure can be used to compare two population means. However, the procedures learned in Chapter 10 are more efficient for performing tests of hypotheses about the difference between two population means, and the analysis of variance procedure, to be discussed in this chapter, is used to compare three or more population means.

An *analysis of variance* test is performed using the *F* distribution. First, the *F* distribution is described in Section 12.1 of this chapter. Then, Section 12.2 discusses the application of the one-way analysis of variance procedure to perform tests of hypotheses.

12.1   THE *F* DISTRIBUTION

Like the *t* and chi-square distributions, the shape of a particular ***F* distribution**[1] curve depends on the number of degrees of freedom. However, the *F* distribution has *two* numbers of degrees of freedom: *degrees of freedom for the numerator* and *degrees of freedom for the denominator*. These two numbers representing two types of degrees of freedom are the *parameters of the F distribution*. Each combination of degrees of freedom for the numerator and for the denominator gives a different *F* distribution curve. The units of an *F* distribution are denoted by *F*, which assumes only nonnegative values. Like the normal, *t*, and chi-square distributions, the *F* distribution is a continuous distribution. The shape of an *F* distribution curve is skewed to the right, but the skewness decreases as the number of degrees of freedom increases.

THE *F* DISTRIBUTION

1. The *F distribution* is continuous and skewed to the right.
2. The *F* distribution has two numbers of degrees of freedom: *df* for the numerator and *df* for the denominator.
3. The units of an *F* distribution, denoted by *F*, are nonnegative.

For an *F* distribution, degrees of freedom for the numerator and degrees of freedom for the denominator are usually written as follows:

$$df = (8, 14)$$

First number denotes the   Second number denotes the
df for the numerator   *df* for the denominator

Figure 12.1 gives three *F* distribution curves for three sets of degrees of freedom for the numerator and for the denominator. In the figure, the first number gives the degrees of freedom associated with the numerator and the second number gives the degrees of freedom associated with the denominator. We can observe from this figure that as the degrees of freedom increase, the peak of the curve moves to the right; that is, the skewness decreases.

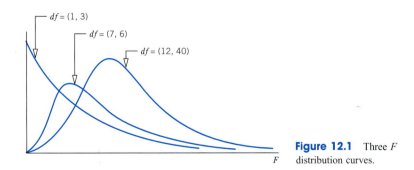

df = (1, 3)
df = (7, 6)
df = (12, 40)

F

Figure 12.1   Three *F* distribution curves.

Table X in Appendix D lists the values of *F* for the *F* distribution. To read Table X, we need to know three quantities: the degrees of freedom for the numerator, the degrees of free-

---

[1]The *F* distribution is named after Sir Ronald Fisher.

dom for the denominator, and an area in the right tail of an F distribution curve. Note that the F distribution table (Table X) is read only for an area in the right tail of the F distribution curve. Also note that Table X has four parts. These four parts give the F values for areas of .01, .025, .05, and .10, respectively, in the right tail of the F distribution curve. Example 12–1 illustrates how to read Table X.

Reading the F distribution table.

EXAMPLE 12-1   Find the F value for 8 degrees of freedom for the numerator, 14 degrees of freedom for the denominator, and .05 area in the right tail of the F distribution curve.

Solution   To find the required value of F, we consult the portion of Table X of Appendix D that corresponds to .05 area in the right tail of the F distribution curve. The relevant portion of that table is shown here as Table 12.1. To find the required F value, we locate 8 in the row for degrees of freedom for the numerator (at the top of Table X) and 14 in the column for degrees of freedom for the denominator (the first column on the left side in Table X). The entry where the column for 8 and the row for 14 intersect gives the required F value. This value of F is **2.70**, as shown in Table 12.1 and Figure 12.2. The F value taken from this table for a test of hypothesis is called the critical value of F.

Table 12.1

		Degrees of Freedom for the Numerator				
		1	**2**	...	**8**	...	**100**
	1	161.5	199.5	...	238.9	...	253.0
Degrees of Freedom for the Denominator	2	18.51	19.00	...	19.37	...	19.49
	.	...	...	...	...	...	...
	14	4.60	3.74	...	**2.70**	...	2.19
	.	...	...	...	...	...	...
	100	3.94	3.09	...	2.03	...	1.39

The F value for 8 df for the numerator, 14 df for the denominator, and .05 area in the right tail

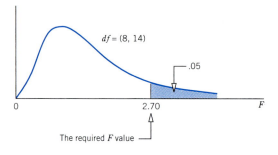

df = (8, 14)

.05

0   2.70 F

The required F value

Figure 12.2   The critical value of F for 8 df for the numerator, 14 df for the denominator, and .05 area in the right tail. ∎

EXERCISES

■ **Concepts and Procedures**

12.1   Describe the main characteristics of an F distribution.

12.2   Find the critical value of F for the following.
a. $df = (6, 12)$ and area in the right tail $= .05$
b. $df = (4, 18)$ and area in the right tail $= .05$
c. $df = (12, 6)$ and area in the right tail $= .05$

12.3 Find the critical value of F for the following.

a. $df = (6, 12)$ and area in the right tail $= .025$
b. $df = (4, 18)$ and area in the right tail $= .025$
c. $df = (12, 6)$ and area in the right tail $= .025$

12.4 Determine the critical value of F for the following.

a. $df = (7, 11)$ and area in the right tail $= .01$
b. $df = (5, 12)$ and area in the right tail $= .01$
c. $df = (15, 6)$ and area in the right tail $= .01$

12.5 Determine the critical value of F for the following.

a. $df = (4, 14)$ and area in the right tail $= .10$
b. $df = (9, 11)$ and area in the right tail $= .10$
c. $df = (11, 5)$ and area in the right tail $= .10$

12.6 Find the critical value of F for an F distribution with $df = (3, 12)$ and

a. area in the right tail $= .05$
b. area in the right tail $= .10$

12.7 Find the critical value of F for an F distribution with $df = (11, 5)$ and

a. area in the right tail $= .01$
b. area in the right tail $= .025$

12.8 Find the critical value of F for an F distribution with .025 area in the right tail and

a. $df = (4, 11)$ **b.** $df = (15, 3)$

12.9 Find the critical value of F for an F distribution with .01 area in the right tail and

a. $df = (10, 10)$ **b.** $df = (9, 25)$

12.2 ONE-WAY ANALYSIS OF VARIANCE

As mentioned in the beginning of this chapter, the analysis of variance procedure is used to test the null hypothesis that the means of three or more populations are the same against the alternative hypothesis that not all population means are the same. The analysis of variance procedure can be used to compare two population means. However, the procedures learned in Chapter 10 are more efficient for performing tests of hypotheses about the difference between two population means, and the analysis of variance procedure is used to compare three or more population means.

Reconsider the example of teachers at a school who have devised three different methods to teach arithmetic. They want to find out if these three methods produce different mean scores. Let μ_1, μ_2, and μ_3 be the mean scores of all students who are taught by Methods I, II, and III, respectively. To test if the three teaching methods produce different means, we test the null hypothesis

$$H_0: \mu_1 = \mu_2 = \mu_3 \quad \text{(All three population means are equal)}$$

against the alternative hypothesis

$$H_1: \text{Not all three population means are equal}$$

One method to test such a hypothesis is to test the three hypotheses $H_0: \mu_1 = \mu_2$, $H_0: \mu_1 = \mu_3$, and $H_0: \mu_2 = \mu_3$ separately using the procedure discussed in Chapter 10. Besides being time consuming, such a procedure has other disadvantages. First, if we reject even one of these three hypotheses, then we must reject the null hypothesis $H_0: \mu_1 = \mu_2 = \mu_3$. Second, combining the Type I error probabilities for the three tests (one for each test) will give a very large Type I error probability for the test H_0: $\mu_1 = \mu_2 = \mu_3$. Hence, we should prefer a procedure that can test the equality of three means in one test. The **ANOVA**, short for **analysis of variance**, provides such a procedure. It is used to compare three or more population means in a single test.

> **ANOVA** *ANOVA* is a procedure used to test the null hypothesis that the means of three or more populations are equal.

This section discusses the **one-way ANOVA** procedure to make tests comparing the means of several populations. By using a one-way ANOVA test, we analyze only one factor or variable. For instance, in the example of testing for the equality of mean arithmetic scores of students taught by each of the three different methods, we are considering only one factor, which is the effect of different teaching methods on the scores of students. Sometimes we may analyze the effects of two factors. For example, if different teachers teach arithmetic using these three methods, we can analyze the effects of teachers and teaching methods on the scores of students. This is done by using a two-way ANOVA. The procedure under discussion in this chapter is called the analysis of variance because the test is based on the analysis of variation in the data obtained from different samples. The application of one-way ANOVA requires that the following assumptions hold true.

> **ASSUMPTIONS OF ONE-WAY ANOVA** The following assumptions must hold true to use *one-way ANOVA*.
>
> 1. The populations from which the samples are drawn are (approximately) normally distributed.
> 2. The populations from which the samples are drawn have the same variance (or standard deviation).
> 3. The samples drawn from different populations are random and independent.

For instance, in the example about three methods of teaching arithmetic, we first assume that the scores of all students taught by each method are (approximately) normally distributed. Second, the means of the distributions of scores for the three teaching methods may or may not be the same, but all three distributions have the same variance σ^2. Third, when we take samples to make an ANOVA test, these samples are drawn independently and randomly from three different populations.

The ANOVA test is applied by calculating two estimates of the variance, σ^2, of population distributions: the **variance between samples** and the **variance within samples**. The variance between samples is also called the **mean square between samples** or **MSB**. The variance within samples is also called the **mean square within samples** or **MSW**.

The variance between samples, MSB, gives an estimate of σ^2 based on the variation among the means of samples taken from different populations. For the example of three teaching methods, MSB will be based on the values of the mean scores of three samples of students taught by three different methods. If the means of all populations under consideration are equal, the means of the respective samples will still be different but the variation among them is expected to be small and, consequently, the value of MSB is expected to be small.

However, if the means of populations under consideration are not all equal, the variation among the means of respective samples is expected to be large and, consequently, the value of MSB is expected to be large.

The variance within samples, MSW, gives an estimate of σ^2 based on the variation within the data of different samples. For the example of three teaching methods, MSW will be based on the scores of individual students included in the three samples taken from three populations. The concept of MSW is similar to the concept of the pooled standard deviation, s_p, for two samples discussed in Section 10.2 of Chapter 10.

The one-way ANOVA test is always right-tailed with the rejection region in the right tail of the F distribution curve. The hypothesis-testing procedure using ANOVA involves the same five steps that were used in earlier chapters. The next subsection explains how to calculate the value of the test statistic F for an ANOVA test.

12.2.1 CALCULATING THE VALUE OF THE TEST STATISTIC

The value of the test statistic F for a test of hypothesis using ANOVA is given by the ratio of two variances, the variance between samples (MSB) and the variance within samples (MSW).

> **TEST STATISTIC F FOR A ONE-WAY ANOVA TEST** The value of the *test statistic F* for an ANOVA test is calculated as
>
> $$F = \frac{\text{Variance between samples}}{\text{Variance within samples}} \quad \text{or} \quad \frac{\text{MSB}}{\text{MSW}}$$
>
> The calculation of MSB and MSW is explained in Example 12–2.

Example 12–2 describes the calculation of MSB, MSW, and the value of the test statistic F. Since the basic formulas are laborious to use, they are not presented here. We have used only the short-cut formulas to make calculations in this chapter.

Calculating the value of the test statistic F.

EXAMPLE 12–2 Fifteen fourth-grade students were randomly assigned to three groups in order to experiment with three different methods of teaching arithmetic. At the end of the semester, the same test was given to all 15 students. The table gives the scores of students in the three groups.

Method I	Method II	Method III
48	55	84
73	85	68
51	70	95
65	69	74
87	90	67

Calculate the value of the test statistic F. Assume that all the required assumptions mentioned in Section 12.2 hold true.

Solution In ANOVA terminology, the three methods used to teach arithmetic are called **treatments**. The table contains data on the scores of fourth-graders included in the three samples. Each sample of students is taught by a different method. Let

x = the score of a student

k = the number of different samples (or treatments)

n_i = the size of sample i

T_i = the sum of the values in sample i

n = the number of values in all samples = $n_1 + n_2 + n_3 + \cdots$

Σx = the sum of the values in all samples = $T_1 + T_2 + T_3 + \cdots$

Σx^2 = the sum of the squares of the values in all samples

To calculate MSB and MSW, we first compute the **between-samples sum of squares** denoted by **SSB** and the **within-samples sum of squares** denoted by **SSW**. The sum of SSB and SSW is called the **total sum of squares** and it is denoted by **SST**; that is,

$$SST = SSB + SSW$$

The values of SSB and SSW are calculated using the following formulas.

BETWEEN- AND WITHIN-SAMPLES SUMS OF SQUARES The *between-samples sum of squares*, denoted by SSB, is calculated as

$$SSB = \left(\frac{T_1^2}{n_1} + \frac{T_2^2}{n_2} + \frac{T_3^2}{n_3} + \cdots \right) - \frac{(\Sigma x)^2}{n}$$

The *within-samples sum of squares*, denoted by SSW, is calculated as

$$SSW = \Sigma x^2 - \left(\frac{T_1^2}{n_1} + \frac{T_2^2}{n_2} + \frac{T_3^2}{n_3} + \cdots \right)$$

Table 12.2 lists the scores of 15 students who were taught arithmetic by each of the three different methods, the values of T_1, T_2, and T_3, and the values of n_1, n_2, and n_3.

Table 12.2

Method I	Method II	Method III
48	55	84
73	85	68
51	70	95
65	69	74
87	90	67
$T_1 = 324$	$T_2 = 369$	$T_3 = 388$
$n_1 = 5$	$n_2 = 5$	$n_3 = 5$

In Table 12.2, T_1 is obtained by adding the five scores of the first sample. Thus, $T_1 = 48 + 73 + 51 + 65 + 87 = 324$. Similarly, the sums of the values in the second and third samples give $T_2 = 369$ and $T_3 = 388$, respectively. Because there are five observations in each sample, $n_1 = n_2 = n_3 = 5$. The values of Σx and n are

$$\Sigma x = T_1 + T_2 + T_3 = 324 + 369 + 388 = 1081$$

$$n = n_1 + n_2 + n_3 = 5 + 5 + 5 = 15$$

To calculate Σx^2, we square all the scores included in all three samples and then add them. Thus,

$$\Sigma x^2 = (48)^2 + (73)^2 + (51)^2 + (65)^2 + (87)^2 + (55)^2 + (85)^2 + (70)^2$$
$$+ (69)^2 + (90)^2 + (84)^2 + (68)^2 + (95)^2 + (74)^2 + (67)^2$$
$$= 80,709$$

Substituting all the values in the formulas for SSB and SSW, we obtain the following values of SSB and SSW:

$$SSB = \left(\frac{(324)^2}{5} + \frac{(369)^2}{5} + \frac{(388)^2}{5} \right) - \frac{(1081)^2}{15} = 432.1333$$

$$SSW = 80,709 - \left(\frac{(324)^2}{5} + \frac{(369)^2}{5} + \frac{(388)^2}{5} \right) = 2372.8000$$

The value of SST is obtained by adding the values of SSB and SSW. Thus,

$$SST = 432.1333 + 2372.8000 = 2804.9333$$

The variance between samples (MSB) and the variance within samples (MSW) are calculated using the following formulas.

> **CALCULATING THE VALUES OF MSB AND MSW**   *MSB* and *MSW* are calculated as
>
> $$MSB = \frac{SSB}{k - 1} \quad \text{and} \quad MSW = \frac{SSW}{n - k}$$
>
> where $k - 1$ and $n - k$ are, respectively, the *df* for the numerator and the *df* for the denominator for the F distribution.

Consequently, the variance between samples is

$$MSB = \frac{SSB}{k - 1} = \frac{432.1333}{3 - 1} = 216.0667$$

The variance within samples is

$$MSW = \frac{SSW}{n - k} = \frac{2372.8000}{15 - 3} = 197.7333$$

The value of the test statistic F is given by the ratio of MSB and MSW. Therefore,

$$F = \frac{MSB}{MSW} = \frac{216.0667}{197.7333} = \mathbf{1.09}$$

For convenience, all these calculations are often recorded in a table called the *ANOVA table*. Table 12.3 on page 538 gives the general form of an ANOVA table.

Table 12.3   ANOVA Table

Source of Variation	Degrees of Freedom	Sum of Squares	Mean Square	Value of the Test Statistic
Between	$k - 1$	SSB	MSB	$F = \dfrac{\text{MSB}}{\text{MSW}}$
Within	$n - k$	SSW	MSW	
Total	$n - 1$	SST		

Substituting the values of the various quantities into Table 12.3, we write the ANOVA table for our example as Table 12.4.

Table 12.4   ANOVA Table for Example 12–2

Source of Variation	Degrees of Freedom	Sum of Squares	Mean Square	Value of the Test Statistic
Between	2	432.1333	216.0667	$F = \dfrac{216.0667}{197.7333} = 1.09$
Within	12	2372.8000	197.7333	
Total	14	2804.9333		

12.2.2   ONE-WAY ANOVA TEST

Now suppose we want to test the null hypothesis that the mean scores are equal for all three groups of fourth-graders taught by three different methods of Example 12–2 against the alternative hypothesis that the mean scores of all three groups are not equal. Note that in a one-way ANOVA test, the null hypothesis is that the means for all populations are equal. The alternative hypothesis is that not all population means are equal. In other words, the alternative hypothesis states that at least one of the population means is different from the others. Example 12–3 demonstrates how we use the one-way ANOVA procedure to make such a test.

Performing a one-way ANOVA test: all samples the same size.

EXAMPLE 12–3   Reconsider Example 12–2 about the scores of 15 fourth-grade students who were randomly assigned to three groups in order to experiment with three different methods of teaching arithmetic. At the 1% significance level, can we reject the null hypothesis that the mean arithmetic score of all fourth-grade students taught by each of these three methods is the same? Assume that all the assumptions required to apply the one-way ANOVA procedure hold true.

Solution   To make a test about the equality of the means of three populations, we follow our standard procedure with five steps.

Step 1.   *State the null and alternative hypotheses.*

Let μ_1, μ_2, and μ_3 be the mean arithmetic scores of all fourth-grade students who are taught, respectively, by Methods I, II, and III. The null and alternative hypotheses are

$$H_0\text{: } \mu_1 = \mu_2 = \mu_3 \quad \text{(The mean scores of the three groups are equal)}$$

$$H_1\text{: Not all three means are equal}$$

Note that the alternative hypothesis states that at least one population mean is different from the other two.

Step 2. *Select the distribution to use.*

Because we are comparing the means for three normally distributed populations, we use the F distribution to make this test.

Step 3. *Determine the rejection and nonrejection regions.*

The significance level is .01. Because a one-way ANOVA test is always right-tailed, the area in the right tail of the F distribution curve is .01, which is the rejection region in Figure 12.3.

Next we need to know the degrees of freedom for the numerator and the denominator. In our example, the students were assigned to three different methods. As mentioned earlier, these methods are called treatments. The number of treatments is denoted by k. The total number of observations in all samples taken together is denoted by n. Then the number of degrees of freedom for the numerator is equal to $k - 1$ and the number of degrees of freedom for the denominator is equal to $n - k$. In our example, there are 3 treatments (methods of teaching) and 15 total observations (total number of students) in all three samples. Thus,

$$\text{Degrees of freedom for the numerator} = k - 1 = 3 - 1 = 2$$

$$\text{Degrees of freedom for the denominator} = n - k = 15 - 3 = 12$$

From Table X of Appendix D, we find the critical value of F for 2 df for the numerator, 12 df for the denominator, and .01 area in the right tail of the F distribution curve. This value is shown in Figure 12.3. The required value of F is 6.93.

Thus, we will fail to reject H_0 if the calculated value of the test statistic F is less than 6.93 and we will reject H_0 if it is greater than 6.93.

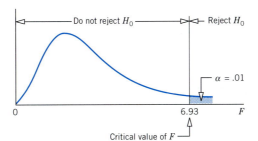

Figure 12.3 Critical value of F for $df = (2, 12)$ and $\alpha = .01$.

Step 4. *Calculate the value of the test statistic.*

We computed the value of the test statistic F for these data in Example 12–2. This value is

$$F = 1.09$$

Step 5. *Make a decision.*

Because the value of the test statistic $F = 1.09$ is less than the critical value of $F = 6.93$, it falls in the nonrejection region. Hence, we fail to reject the null hypothesis and conclude that the means of the three populations are equal. In other words, the three different methods of teaching arithmetic do not seem to affect the mean scores of students. The difference in the three mean scores in the case of our three samples occurred only because of sampling error. ■

In Example 12–3, the sample sizes were the same for all treatments. Example 12–4 describes a case in which the sample sizes are not the same for all treatments.

Performing a one-way ANOVA test: all samples not the same size.

EXAMPLE 12–4 From time to time, unknown to its employees, the research department at Post Bank observes various employees for their work productivity. Recently this department wanted to check whether the four tellers at a branch of this bank serve, on average, the same number of customers per hour. The research manager observed each of the four tellers for a certain number of hours. The following table gives the number of customers served by the four tellers during each of the observed hours.

Teller A	Teller B	Teller C	Teller D
19	14	11	24
21	16	14	19
26	14	21	21
24	13	13	26
18	17	16	20
	13	18	

At the 5% significance level, test the null hypothesis that the mean number of customers served per hour by each of these four tellers is the same. Assume that all the assumptions required to apply the one-way ANOVA procedure hold true.

Solution To make a test about the equality of means of four populations, we follow our standard procedure with five steps.

Step 1. *State the null and alternative hypotheses.*

Let μ_1, μ_2, μ_3, and μ_4 be the mean number of customers served per hour by tellers A, B, C, and D, respectively. The null and alternative hypotheses are

H_0: $\mu_1 = \mu_2 = \mu_3 = \mu_4$ (The mean number of customers served per hour by each of the four tellers is the same)

H_1: Not all four population means are equal

Step 2. *Select the distribution to use.*

Because we are testing for the equality of four means for four normally distributed populations, we use the F distribution to make the test.

Step 3. *Determine the rejection and nonrejection regions.*

The significance level is .05, which means the area in the right tail of the F distribution curve is .05. In this example, there are four treatments (tellers) and 22 total observations in all four samples. Thus,

$$\text{Degrees of freedom for the numerator} = k - 1 = 4 - 1 = 3$$

$$\text{Degrees of freedom for the denominator} = n - k = 22 - 4 = 18$$

The critical value of F from Table X for 3 *df* for the numerator, 18 *df* for the denominator, and .05 area in the right tail of the F distribution curve is 3.16. This value is shown in Figure 12.4.

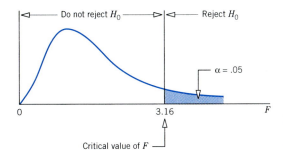

Figure 12.4 Critical value of F for $df = (3, 18)$ and $\alpha = .05$.

Step 4. *Calculate the value of the test statistic.*

First we calculate SSB and SSW. Table 12.5 lists the numbers of customers served by the four tellers during the selected hours, the values of T_1, T_2, T_3, and T_4, and the values of n_1, n_2, n_3, and n_4.

Table 12.5

Teller A	Teller B	Teller C	Teller D
19	14	11	24
21	16	14	19
26	14	21	21
24	13	13	26
18	17	16	20
	13	18	
$T_1 = 108$	$T_2 = 87$	$T_3 = 93$	$T_4 = 110$
$n_1 = 5$	$n_2 = 6$	$n_3 = 6$	$n_4 = 5$

The values of Σx and n are

$$\Sigma x = T_1 + T_2 + T_3 + T_4 = 108 + 87 + 93 + 110 = 398$$

$$n = n_1 + n_2 + n_3 + n_4 = 5 + 6 + 6 + 5 = 22$$

The value of Σx^2 is calculated as follows:

$$\Sigma x^2 = (19)^2 + (21)^2 + (26)^2 + (24)^2 + (18)^2 + (14)^2 + (16)^2 + (14)^2$$
$$+ (13)^2 + (17)^2 + (13)^2 + (11)^2 + (14)^2 + (21)^2 + (13)^2$$
$$+ (16)^2 + (18)^2 + (24)^2 + (19)^2 + (21)^2 + (26)^2 + (20)^2$$
$$= 7614$$

Substituting all the values in the formulas for SSB and SSW, we obtain the following values of SSB and SSW:

$$SSB = \left(\frac{T_1^2}{n_1} + \frac{T_2^2}{n_2} + \frac{T_3^2}{n_3} + \frac{T_4^2}{n_4} \right) - \frac{(\Sigma x)^2}{n}$$

$$= \left(\frac{(108)^2}{5} + \frac{(87)^2}{6} + \frac{(93)^2}{6} + \frac{(110)^2}{5} \right) - \frac{(398)^2}{22} = 255.6182$$

$$SSW = \Sigma x^2 - \left(\frac{T_1^2}{n_1} + \frac{T_2^2}{n_2} + \frac{T_3^2}{n_3} + \frac{T_4^2}{n_4} \right)$$

$$= 7614 - \left(\frac{(108)^2}{5} + \frac{(87)^2}{6} + \frac{(93)^2}{6} + \frac{(110)^2}{5} \right) = 158.2000$$

Hence, the variance between samples MSB and the variance within samples MSW are

$$MSB = \frac{SSB}{k-1} = \frac{255.6182}{4-1} = 85.2061$$

$$MSW = \frac{SSW}{n-k} = \frac{158.2000}{22-4} = 8.7889$$

The value of the test statistic F is given by the ratio of MSB and MSW, which is

$$F = \frac{MSB}{MSW} = \frac{85.2061}{8.7889} = 9.69$$

Writing the values of the various quantities in the ANOVA table, we obtain Table 12.6.

Table 12.6 ANOVA Table for Example 12–4

Source of Variation	Degrees of Freedom	Sum of Squares	Mean Square	Value of the Test Statistic
Between	3	255.6182	85.2061	$F = \dfrac{85.2061}{8.7889} = 9.69$
Within	18	158.2000	8.7889	
Total	21	413.8182		

Step 5. *Make a decision.*

Because the value of the test statistic $F = 9.69$ is greater than the critical value of $F = 3.16$, it falls in the rejection region. Consequently, we reject the null hypothesis and conclude that the mean number of customers served per hour by each of the four tellers is not the same. In other words, at least one of the four means is different from the other three. ∎

EXERCISES

■ **Concepts and Procedures**

12.10 Briefly explain when a one-way ANOVA procedure is used to make a test of hypothesis.

12.11 Describe the assumptions that must hold true to apply the one-way analysis of variance procedure to test hypotheses.

12.12 Consider the following data obtained for two samples selected at random from two populations that are independent and normally distributed with equal variances.

Sample I	Sample II
32	27
26	35
31	33
20	40
27	38
34	31

a. Calculate the means and standard deviations for these samples using the formulas from Chapter 3.
b. Using the procedure learned in Section 10.2 of Chapter 10, test at the 1% significance level whether the means of the populations from which these samples are drawn are equal.

c. Using the one-way ANOVA procedure, test at the 1% significance level whether the means of the populations from which these samples are drawn are equal.

d. Are the conclusions reached in parts b and c the same?

12.13 Consider the following data obtained for two samples selected at random from two populations that are independent and normally distributed with equal variances.

Sample I	Sample II
14	11
21	8
11	12
9	18
13	15
20	7
17	6

a. Calculate the means and standard deviations for these samples using the formulas from Chapter 3.

b. Using the procedure learned in Section 10.2 of Chapter 10, test at the 5% significance level whether the means of the populations from which these samples are drawn are equal.

c. Using the one-way ANOVA procedure, test at the 5% significance level whether the means of the populations from which these samples are drawn are equal.

d. Are the conclusions reached in parts b and c the same?

12.14 The following ANOVA table, based on information obtained for three samples selected from three independent populations that are normally distributed with equal variances, has a few missing values.

Source of Variation	Degrees of Freedom	Sum of Squares	Mean Square	Value of the Test Statistic
Between	2		19.2813	
Within		89.3677		$F =$ ——— $=$
Total	12			

a. Find the missing values and complete the ANOVA table.

b. Using $\alpha = .01$, what is your conclusion for the test with the null hypothesis that the means of the three populations are all equal against the alternative hypothesis that the means of the three populations are not all equal?

12.15 The following ANOVA table, based on information obtained for four samples selected from four independent populations that are normally distributed with equal variances, has a few missing values.

Source of Variation	Degrees of Freedom	Sum of Squares	Mean Square	Value of the Test Statistic
Between				
Within	15		9.2154	$F =$ ——— $= 4.07$
Total	18			

a. Find the missing values and complete the ANOVA table.

b. Using $\alpha = .05$, what is your conclusion for the test with the null hypothesis that the means of the four populations are all equal against the alternative hypothesis that the means of the four populations are not all equal?

■ **Applications**

For the following exercises assume that all the assumptions required to apply the one-way ANOVA procedure hold true.

12.16 A dietitian wanted to test three different diets to find out if the mean weight loss for each of these diets is the same. She randomly selected 21 overweight persons, randomly divided them into three groups, and put each group on one of the three diets. The following table records the weights (in pounds) lost by these persons after being on these diets for two months.

Diet I	Diet II	Diet III
15	11	9
8	16	17
17	9	11
7	16	8
26	24	15
12	20	6
8	19	14

a. We are to test if the mean weights lost by all persons on each of the three diets are the same. Write the null and alternative hypotheses.
b. Show the rejection and nonrejection regions on the F distribution curve for $\alpha = .025$.
c. Calculate SSB, SSW, and SST.
d. What are the degrees of freedom for the numerator and the denominator?
e. Calculate the between-samples and within-samples variances.
f. What is the critical value of F for $\alpha = .025$?
g. What is the calculated value of the test statistic F?
h. Write the ANOVA table for this exercise.
i. Will you reject the null hypothesis stated in part a at a significance level of 2.5%?

12.17 The following table gives the numbers of classes missed during one semester by 25 randomly selected college students drawn from three different age groups.

Below 25	25 to 30	31 and Above
19	9	5
13	6	8
25	11	2
10	14	3
19	5	10
4	9	9
15	3	
10	11	
16	18	
9		

a. We are to test if the mean numbers of classes missed during the semester by all students in each of these three age groups are the same. Write the null and alternative hypotheses.
b. What are the degrees of freedom for the numerator and the denominator?
c. Calculate SSB, SSW, and SST.
d. Show the rejection and nonrejection regions on the F distribution curve for $\alpha = .01$.
e. Calculate the between-samples and within-samples variances.
f. What is the critical value of F for $\alpha = .01$?
g. What is the calculated value of the test statistic F?

h. Write the ANOVA table for this exercise.

i. Will you reject the null hypothesis stated in part a at a significance level of 1%?

12.18 A consumer agency wanted to investigate if four insurance companies differed with regard to the premiums they charge for auto insurance. The agency randomly selected a few auto drivers who were insured by each of these four companies and had similar driving records, autos, and insurance policies. The following table gives the premiums paid per month by these drivers insured with these four insurance companies.

Company A	Company B	Company C	Company D
75	59	65	76
83	75	70	60
68	100	97	52
52		90	58
		73	

Using the 1% significance level, test the null hypothesis that the mean auto insurance premiums paid per month by all drivers insured by each of these four companies are the same.

12.19 Three new brands of fertilizer that a farmer can use to grow crops came on the market recently. Before deciding which brand he should use permanently for all crops, a farmer decided to experiment for one season. To do so, he randomly assigned each fertilizer to eight 1-acre tracts of land that he used to grow wheat. The following table gives the production of wheat (in bushels) for each acre for the three brands of fertilizer.

Fertilizer I	Fertilizer II	Fertilizer III
72	58	61
69	40	55
75	53	63
57	47	70
61	45	55
68	52	64
73	45	57
65	57	63

At the 5% significance level, can you conclude that the mean yields of wheat for each of these three brands of fertilizer are the same?

12.20 A consumer agency wanted to find out if the mean time it takes for each of three brands of medicines to provide relief from a headache is the same. The first drug was administered to six randomly selected patients, the second to four randomly selected patients, and the third to five randomly selected patients. The following table gives the time (in minutes) taken by each patient to get relief from a headache after taking the medicine.

Drug I	Drug II	Drug III
25	15	44
38	21	39
42	19	54
65	25	58
47		73
52		

At the 2.5% significance level, will you conclude that the mean times taken to provide relief from a headache are the same for all three drugs?

 12.21   A large company buys thousands of lightbulbs every year. The company is currently considering four brands of lightbulbs to choose from. Before the company decides which lightbulbs to buy, it wants to investigate if the mean lifetimes of the four types of lightbulbs are the same. The company's research department randomly selected a few bulbs of each type and tested them. The following table lists the number of hours (in thousands) that each of the bulbs in each brand lasted before being burned out.

Brand I	Brand II	Brand III	Brand IV
23	19	23	26
24	23	27	24
19	18	25	21
26	24	26	29
22	20	23	28
23	22	21	24
25	19	27	28

At the 2.5% significance level, test the null hypothesis that the mean lifetimes of bulbs for each of these four brands are the same.

GLOSSARY

Analysis of variance (ANOVA)   A statistical technique used to test whether the means of three or more populations are equal.

F distribution   A continuous distribution that has two parameters: df for the numerator and df for the denominator.

Mean square between samples or **MSB**   A measure of the variation among the means of samples taken from different populations.

Mean square within samples or **MSW**   A measure of the variation within the data of all samples taken from different populations.

One-way ANOVA   The analysis of variance technique that analyzes one variable only.

SSB   The sum of squares between samples. Also called the sum of squares of the factor or treatment.

SST   The total sum of squares given by the sum of SSB and SSW.

SSW   The sum of squares within samples. Also called the sum of squares of errors.

KEY FORMULAS

Let

$$k = \text{the number of different samples (or treatments)}$$

$$n_i = \text{the size of sample } i$$

$$T_i = \text{the sum of the values in sample } i$$

$$n = \text{the number of values in all samples} = n_1 + n_2 + n_3 + \cdots$$

$$\Sigma x = \text{the sum of the values in all samples} = T_1 + T_2 + T_3 + \cdots$$

$$\Sigma x^2 = \text{the sum of the squares of values in all samples}$$

1. **Degrees of freedom for the F distribution**

$$\text{Degrees of freedom for the numerator} = k - 1$$

$$\text{Degrees of freedom for the denominator} = n - k$$

2. **Between-samples sum of squares**

$$\text{SSB} = \left(\frac{T_1^2}{n_1} + \frac{T_2^2}{n_2} + \frac{T_3^2}{n_3} + \cdots \right) - \frac{(\Sigma x)^2}{n}$$

3. **Within-samples sum of squares**

$$\text{SSW} = \Sigma x^2 - \left(\frac{T_1^2}{n_1} + \frac{T_2^2}{n_2} + \frac{T_3^2}{n_3} + \cdots \right)$$

4. **Total sum of squares**

$$\text{SST} = \text{SSB} + \text{SSW}$$

5. **Variance between samples**

$$\text{MSB} = \frac{\text{SSB}}{k - 1}$$

6. **Variance within samples**

$$\text{MSW} = \frac{\text{SSW}}{n - k}$$

7. **Value of the test statistic F**

$$F = \frac{\text{Variance between samples}}{\text{Variance within samples}} \quad \text{or} \quad \frac{\text{MSB}}{\text{MSW}}$$

SUPPLEMENTARY EXERCISES

For the following exercises, assume that all the assumptions required to apply the one-way ANOVA procedure hold true.

12.22 The following table lists the numbers of violent crimes reported to police on randomly selected days for this year. The data are taken from three large cities of about the same size.

City A	City B	City C
5	2	8
9	4	12
12	1	10
3	13	3
9	7	9
7	6	14
13		

Using the 5% significance level, test the null hypothesis that the mean numbers of violent crimes reported per day are the same for each of these three cities.

12.23   A consumer agency wants to check if the mean lives of four brands of auto batteries, which sell for nearly the same price, are the same. The agency randomly selected a few batteries of each brand and tested them. The following table gives the lives of these batteries in thousands of hours.

Brand A	Brand B	Brand C	Brand D
74	53	57	56
78	51	71	51
51	47	81	49
56	59	77	43
65		68	

a.   At the 5% significance level, will you reject the null hypothesis that the mean lifetimes of each of these four brands of batteries are the same?
b.   What is the Type I error in this case and what is the probability of committing such an error? Explain.

12.24   A researcher wanted to investigate if recent college graduates with different majors who got jobs in San Francisco are commanding the same average salary. She selected random samples of recent graduates in four areas—engineering, business, mathematics, and sociology. The following table gives the starting salaries (in thousands of dollars) for these samples.

Engineering	Business	Mathematics	Sociology
39.2	33.3	23.3	24.6
36.5	29.8	31.4	19.2
44.3	35.4	30.3	23.9
39.4	47.5	29.6	
42.1	34.2		

a.   At the 1% significance level, test the null hypothesis that the mean starting salaries of all recent college graduates with these four majors who got jobs in San Francisco are equal.
b.   What is the Type I error in this case and what is the probability of committing such an error? Explain.

12.25   A farmer wants to test three brands of weight-gain diets for chickens to determine if the mean weight gains for each of these brands are the same. He selected 15 chickens and randomly put each of them on one of these three brands of diet. The following table lists the weights (in pounds) gained by these chickens after a period of one month.

Brand A	Brand B	Brand C
.8	.6	1.2
1.3	1.3	.8
1.7	.6	.7
.9	.4	1.5
.6	.7	.9

a.   At the 1% significance level, can you conclude that the mean weight gains by all chickens are the same for each of these three diets?

b. If you did not reject the null hypothesis in part a, explain the Type II error that you may have made in this case. Note that you cannot calculate the probability of committing a Type II error without additional information.

12.26 The following table lists the prices of certain randomly selected college textbooks in statistics, psychology, economics, and business.

Statistics	Psychology	Economics	Business
67	88	75	84
81	75	64	75
69	80	80	62
62	86	72	70
71		82	82

a. Using the 5% significance level, test the null hypothesis that the mean prices of college textbooks in statistics, psychology, economics, and business are all equal.

b. If you did not reject the null hypothesis in part a, explain the Type II error that you may have made in this case. Note that you cannot calculate the probability of committing a Type II error without additional information.

12.27 A consumer agency that wanted to compare drying times for paints made by three companies tested a few samples of paints from each of these three companies. The following table records the drying times (in minutes) for these samples of paints.

Company A	Company B	Company C
42	57	45
53	63	45
43	61	51
47	54	58
40	49	44
51	60	41
56		47

a. Using the 5% significance level, test the null hypothesis that the mean drying times for paints of these three companies are equal.

b. What will your decision be if the probability of making a Type I error is zero? Explain.

12.28 According to a survey about public participation in the arts, among adults who use a computer at home for hobbies or enjoyment, those with less than a high school education average 6.2 hours a week on such activities, high school graduates average 5.0 hours, and college graduates average 4.6 hours (*American Demographics,* August 1998). Suppose that these means are based on independent random samples of 20 such computer users with less than a high school education, 17 users who are high school graduates, and 16 who are college graduates. Assume that for these 53 computer users, $\Sigma x^2 = 1792$. Assume that the times spent using computers are normally distributed with equal variances for each of the three groups of users. Using the 5% level of significance, can you conclude that the mean times spent using computers for such activities are the same for all three populations?

■ Advanced Exercises

12.29 A national public transportation survey collected information for metropolitan areas of five different sizes on commuting times for people who use public transportation (*USA TODAY,* April 14, 1999). The table on page 550 shows the mean travel time (in minutes) and the population size for each of these metropolitan areas. Suppose that these means were based on random samples of the sizes shown in the third column of the table.

Population	Mean Commuting Time	Sample Size
Less than 250,000	44	14
250,000 to 499,999	37	12
500,000 to 999,999	42	15
1 million to less than 3 million	35	14
3 million or more	43	11

Assume that the commuting times for each size of metropolitan area are normally distributed and that the standard deviations are the same for the cities of all five sizes. Assume that $\Sigma x^2 = 110{,}130$. At the 1% significance level, can you conclude that the mean travel times are the same for each size of metropolitan area?

12.30 Suppose that you are a newspaper reporter whose editor has asked you to compare the hourly wages of carpenters, plumbers, electricians, and masons in your city. Since many of these workers are not union members, the wages vary considerably among individuals in the same trade.

a. What data should you gather and how would you collect them? What statistics would you present in your article and how would you calculate them? Assume that your newspaper is not intended for technical readers.

b. Suppose that you must submit your findings to a technical journal that requires statistical analysis of your data. If you want to determine whether or not the mean hourly wages are the same for all four trades, briefly describe how you would analyze the data. Assume that hourly wages in each trade are normally distributed and that the four variances are equal.

12.31 The editor of an automotive magazine has asked you to compare the mean gas mileages in city driving of three makes of compact cars. The editor has made available to you one car of each of the three makes, three drivers, and a budget sufficient to buy gas and pay the drivers for approximately 500 miles of city driving for each car.

a. Explain how you would conduct an experiment and gather the data for a magazine article comparing the gas mileage.

b. Suppose you wish to test the null hypothesis that the mean gas mileages in city driving are the same for all three makes. Outline the procedure for using your data to conduct this test. Assume that the assumptions for applying analysis of variance are satisfied.

12.32 Do rock music CDs and country music CDs give the consumer the same amount of music listening time? A sample of 12 randomly selected single rock music CDs and a sample of 14 randomly selected single country music CDs have the following total lengths (in minutes).

Rock Music	Country Music
43.0	45.3
44.3	40.2
63.8	42.8
32.8	33.0
54.2	33.5
51.3	37.7
64.8	36.8
36.1	34.6
33.9	33.4
51.7	36.5
36.5	43.3
59.7	31.7
	44.0
	42.7

Assume that the two populations are normally distributed with equal standard deviations.

a. Compute the value of the test statistic t for testing the null hypothesis that the mean lengths of the rock and country music single CDs are the same against the alternative hypothesis that these mean lengths are not the same. Use the value of this t statistic to compute the (approximate) p-value.

b. Compute the value of the (one-way ANOVA) test statistic F for performing the test of equality of the mean lengths of the rock and country music single CDs and use it to find the (approximate) p-value.

c. How do the test statistics in parts a and b compare? How do the p-values computed in parts a and b compare? Do you think that this is a coincidence, or will this always happen?

12.33 A dietitian wanted to determine whether the mean weight losses for each of three diet plans are the same. She took a random sample of 12 people who wanted to lose weight. Then she randomly assigned four of these people to Diet A, four to Diet B, and four to Diet C. The following table gives the MINITAB output for a one-way ANOVA test based on the mean weight lost by people on each diet plan at the end of a three-week period. Find the missing entries.

```
ANALYSIS OF VARIANCE

SOURCE      DF      SS       MS          F        P
FACTOR      ___     ___      _____       .65      .545
ERROR       ___     ___      56.92
TOTAL       ___     ___


LEVEL       N       MEAN     STDEV
C1          ___     12.25    9.323
C2          ___     11.75    5.560
C3          ___      6.75    7.274


POOLED  STDEV  =  _____
```

SELF-REVIEW TEST

1. The F distribution is

   a. continuous   **b.** discrete   **c.** neither

2. The F distribution is always

   a. symmetric   **b.** skewed to the right   **c.** skewed to the left

3. The units of the F distribution, denoted by F, are always

   a. nonpositive   **b.** positive   **c.** nonnegative

4. The one-way ANOVA test analyzes only one

   a. variable   **b.** population   **c.** sample

5. The one-way ANOVA test is always

   a. right-tailed   **b.** left-tailed   **c.** two-tailed

6. For a one-way ANOVA with k treatments and n observations in all samples taken together, the degrees of freedom for the numerator are

   a. $k - 1$ **b.** $n - k$ **c.** $n - 1$

7. For a one-way ANOVA with k treatments and n observations in all samples taken together, the degrees of freedom for the denominator are

    a. $k - 1$ **b.** $n - k$ **c.** $n - 1$

8. The ANOVA test can be applied to compare

    a. three or more population means
    b. more than four population means only
    c. more than three population means only

9. Briefly describe the assumptions that must hold true to apply the one-way ANOVA procedure as mentioned in this chapter.

10. The following table gives the hourly wages of computer programmers for samples taken from three cities.

New York	Boston	Los Angeles
$19.45	$23.50	$31.75
28.80	18.60	11.40
36.45	29.75	29.40
33.10	30.00	33.30
31.50	35.40	24.60
39.30	26.40	19.50
35.50		28.30

    a. Using the 1% significance level, test the null hypothesis that the mean hourly wages for all computer programmers in each of these three cities are the same.

    b. Is it a Type I error or a Type II error that may have been committed in part a? Explain.

MINI-PROJECTS

Mini-Project 12–1

Are some days of the week busier than others on the New York Stock Exchange? Record the number of shares traded on the NYSE each day for a period of six weeks (round the number of shares to the nearest million). You will have five samples—first for shares traded on six Mondays, second for shares traded on six Tuesdays, and so forth. Assume that these days make up random samples for the respective populations. Further assume that each of the five populations from which these five samples are taken follows a normal distribution with the same variance. Test if the mean numbers of shares traded are the same for all five populations. Use a 1% significance level.

Mini-Project 12–2

In January 2000, the U.S. national debt was approximately $5.76 trillion (*The Hartford Courant,* January 14, 2000). Take a random sample of 10 or more students from each of three different majors and ask these students to estimate the size of the national debt. Assume that the assumptions of one-way ANOVA given in Section 12.2 are satisfied.

a. At the 5% level of significance, can you conclude that the perceived mean values of the national debt are the same for all three majors?

b. Find the current value of the national debt. Overall, did the students in your samples tend to overestimate its value, or did they tend to underestimate it?

Mini-Project 12–3

Pick at least 30 students at random and divide them randomly into three groups (A, B, and C) of approximately equal size. Take the students one by one, ring a bell, and 17 seconds later ring another bell. Then ask the students to estimate the elapsed time between the first and second rings. For group A, tell each student before the experiment starts that people tend to underestimate the elapsed time. Tell each student in group B that people tend to overestimate the time. Do not make any such statement to the students in group C. Record the estimates for all students, and then conduct an appropriate hypothesis test to see if there is a difference in the mean estimates of elapsed time for the populations represented by these groups. Use the 5% level of significance and assume that the three populations of elapsed time are normally distributed with equal standard deviation.

For instructions on using MINITAB for this chapter, please see Section B.11 of Appendix B.

See Appendix C for *Excel Adventure 12* on analysis of variance.

COMPUTER ASSIGNMENTS

CA12.1 Refer to Exercise 12.18. Solve that exercise using MINITAB or any other statistical software.

CA12.2 Refer to Exercise 12.26. Solve that exercise using MINITAB or any other statistical software.

13

SIMPLE LINEAR REGRESSION

13.1 SIMPLE LINEAR REGRESSION MODEL

13.2 SIMPLE LINEAR REGRESSION ANALYSIS

CASE STUDY 13-1 REGRESSION OF HEIGHTS AND WEIGHTS OF NBA PLAYERS

13.3 STANDARD DEVIATION OF RANDOM ERRORS

13.4 COEFFICIENT OF DETERMINATION

13.5 INFERENCES ABOUT *B*

13.6 LINEAR CORRELATION

13.7 REGRESSION ANALYSIS: A COMPLETE EXAMPLE

13.8 USING THE REGRESSION MODEL

13.9 CAUTIONS IN USING REGRESSION

GLOSSARY

KEY FORMULAS

SUPPLEMENTARY EXERCISES

SELF-REVIEW TEST

MINI-PROJECTS

COMPUTER ASSIGNMENTS

This chapter considers the relationship between two variables in two ways: (1) by using regression analysis and (2) by computing the correlation coefficient. By using the regression model, we can evaluate the magnitude of change in one variable due to a certain change in another variable. For example, an economist can estimate the amount of change in food expenditure due to a certain change in the income of a household by using the regression model. A sociologist may want to estimate the increase in the crime rate due to a particular increase in the unemployment rate. Besides answering these questions, a regression model also helps to predict the value of one variable for a given value of another variable. For example, by using the regression line, we can predict the (approximate) food expenditure of a household with a given income.

The correlation coefficient, on the other hand, simply tells us how strongly two variables are related. It does not provide any information about the size of the change in one variable as a result of a certain change in the other variable. For example, the correlation coefficient tells us how strongly income and food expenditure or crime rate and unemployment rate are related.

13.1 SIMPLE LINEAR REGRESSION MODEL

Only simple linear regression will be discussed in this chapter.[1] In the next two subsections the meaning of the words *simple* and *linear* as used in *simple linear regression* is explained.

13.1.1 SIMPLE REGRESSION

Let us return to the example of an economist investigating the relationship between food expenditure and income. What factors or variables does a household consider when deciding how much money it should spend on food every week or every month? Certainly, income of the household is one factor. However, many other variables also affect food expenditure. For instance, the assets owned by the household, the size of the household, the preferences and tastes of household members, and any special dietary needs of household members are some of the variables that influence a household's decision about food expenditure. These variables are called **independent** or **explanatory variables** because they all vary independently, and they explain the variation in food expenditures among different households. In other words, these variables explain why different households spend different amounts of money on food. Food expenditure is called the **dependent variable** because it depends on the independent variables. Studying the effect of two or more independent variables on a dependent variable using regression analysis is called **multiple regression**. However, if we choose only one (usually the most important) independent variable and study the effect of that single variable on a dependent variable, it is called a **simple regression**. Thus, a simple regression includes only two variables: one independent and one dependent. Note that whether it is a simple or a multiple regression analysis, it always includes one and only one dependent variable. It is the number of independent variables that changes in simple and multiple regressions.

> **SIMPLE REGRESSION** A regression model is a mathematical equation that describes the relationship between two or more variables. A *simple regression* model includes only two variables: one independent and one dependent. The dependent variable is the one being explained, and the independent variable is the one used to explain the variation in the dependent variable.

13.1.2 LINEAR REGRESSION

The relationship between two variables in a regression analysis is expressed by a mathematical equation called a **regression equation** or **model**. A regression equation, when plotted, may assume one of many possible shapes, including a straight line. A regression equation that gives a straight-line relationship between two variables is called a **linear regression model**; otherwise, the model is called a **nonlinear regression model**. In this chapter, only linear regression models are studied.

> **LINEAR REGRESSION** A (simple) regression model that gives a straight-line relationship between two variables is called a *linear regression* model.

---

[1]The term *regression* was first used by Sir Francis Galton (1822–1911), who studied the relationship between the heights of children and the heights of their parents.

The two diagrams in Figure 13.1 show a linear and a nonlinear relationship between the dependent variable food expenditure and the independent variable income. A linear relationship between income and food expenditure, shown in Figure 13.1a, indicates that as income increases, the food expenditure always increases at a constant rate. A nonlinear relationship between income and food expenditure, as depicted in Figure 13.1b, shows that as income increases, the food expenditure increases, although, after a point, the rate of increase in food expenditure is lower for every subsequent increase in income.

Figure 13.1 Relationship between food expenditure and income. (a) Linear relationship. (b) Nonlinear relationship.

The **equation of a linear relationship** between two variables x and y is written as

$$y = a + bx$$

Each set of values of a and b gives a different straight line. For instance, when $a = 50$ and $b = 5$, this equation becomes

$$y = 50 + 5x$$

To plot a straight line, we need to know two points that lie on that line. We can find two points on a line by assigning any two values to x and then calculating the corresponding values of y. For the equation $y = 50 + 5x$,

1. When $x = 0$, then $y = 50 + 5(0) = 50$.
2. When $x = 10$, then $y = 50 + 5(10) = 100$.

These two points are plotted in Figure 13.2. By joining these two points, we obtain the line representing the equation $y = 50 + 5x$.

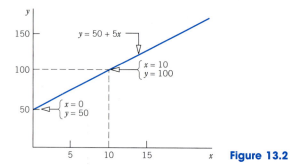

Figure 13.2

Note that in Figure 13.2 the line intersects the y (vertical) axis at 50. Consequently, 50 is called the **y-intercept**. The y-intercept is given by the constant term in the equation. It is the value of y when x is zero.

In the equation $y = 50 + 5x$, 5 is called the **coefficient of x** or the **slope** of the line. It gives the amount of change in y due to a change of one unit in x. For example,

$$\text{If } x = 10, \text{ then } y = 50 + 5(10) = 100.$$

$$\text{If } x = 11, \text{ then } y = 50 + 5(11) = 105.$$

Hence, as x increases by 1 unit (from 10 to 11), y increases by 5 units (from 100 to 105). This is true for any value of x. Such changes in x and y are shown in Figure 13.3.

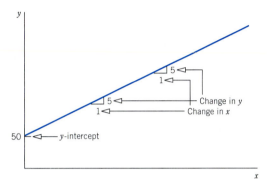

Figure 13.3

In general, when an equation is written in the form

$$y = a + bx$$

a gives the y-intercept and b represents the slope of the line. In other words, a represents the point where the line intersects the y-axis and b gives the amount of change in y due to a change of one unit in x. Note that b is also called the coefficient of x.

13.2 SIMPLE LINEAR REGRESSION ANALYSIS

In a regression model, the independent variable is usually denoted by x and the dependent variable is usually denoted by y. The x variable, with its coefficient, is written on the right side of the $=$ sign, whereas the y variable is written on the left side of the $=$ sign. The y-intercept and the slope, which we earlier denoted by a and b, respectively, can be represented by any of the many commonly used symbols. Let us denote the y-intercept (which is also called the *constant term*) by A, and the slope (or the coefficient of the x variable) by B. Then, our simple linear regression model is written as

Constant term or y-intercept —┐   ┌— Slope

$$y = A + Bx \tag{1}$$

Dependent variable   Independent variable

In model (1), A gives the value of y for $x = 0$, and B gives the change in y due to a change of one unit in x.

Model (1) is called a **deterministic model**. It gives an **exact relationship** between x and y. This model simply states that y is determined exactly by x and for a given value of x there is one and only one (unique) value of y.

However, in many cases the relationship between variables is not exact. For instance, if y is food expenditure and x is income, then model (1) would state that food expenditure is determined by income only and that all households with the same income spend the same amount on food. But as mentioned earlier, food expenditure is determined by many variables, only one of which is included in model (1). In reality, different households with the same income spend different amounts of money on food because of the differences in the sizes of the household, the assets they own, and their preferences and tastes. Hence, to take these variables into consideration and to make our model complete, we add another term to the right side of model (1). This term is called the **random error term**. It is denoted by ϵ (Greek letter *epsilon*). The complete regression model is written as

$$y = A + Bx + \epsilon \tag{2}$$

$$\uparrow$$

Random error term

The regression model (2) is called a **probabilistic model** (or a **statistical relationship**).

> **EQUATION OF A REGRESSION MODEL** In the *regression model* $y = A + Bx + \epsilon$, A is called the y-intercept or constant term, B is the slope, and ϵ is the random error term. The dependent and independent variables are y and x, respectively.

The random error term ϵ is included in the model to represent the following two phenomena.

1. *Missing or omitted variables.* As mentioned earlier, food expenditure is affected by many variables other than income. The random error term ϵ is included to capture the effect of all those missing or omitted variables that have not been included in the model.

2. *Random variation.* Human behavior is unpredictable. For example, a household may have many parties during one month and spend more than usual on food during that month. The same household may spend less than usual during another month because it spent quite a bit of money to buy furniture. The variation in food expenditure for such reasons may be called random variation.

In model (2), A and B are the **population parameters**. The regression line obtained for model (2) by using the population data is called the **population regression line**. The values of A and B in the population regression line are called the **true values of the y-intercept and slope**.

However, population data are difficult to obtain. As a result, we almost always use sample data to estimate model (2). The values of the y-intercept and slope calculated from sample data on x and y are called the **estimated values of A and B** and are denoted by a and b. Using a and b, we write the estimated regression model as

$$\hat{y} = a + bx \tag{3}$$

where $\hat{y}$ (read as *y hat*) is the **estimated** or **predicted value of y** for a given value of x. Equation (3) is called the **estimated regression model**; it gives the **regression of y on x.**

> **ESTIMATES OF A AND B** In the model $\hat{y} = a + bx$, a and b, which are calculated using sample data, are called the *estimates of A and B.*

13.2.1 SCATTER DIAGRAM

Suppose we take a sample of seven households and collect information on their incomes and food expenditures for the past month. The information obtained (in hundreds of dollars) is given in Table 13.1.

Table 13.1 Incomes and Food Expenditures of Seven Households

Income (hundreds)	Food Expenditure (hundreds of dollars)
$35	9
49	15
21	7
39	11
15	5
28	8
25	9

In Table 13.1, we have a pair of observations for each of the seven households. Each pair consists of one observation on income and a second on food expenditure. For example, the first household's income for the past month was $3500 and its food expenditure was $900. By plotting all seven pairs of values, we obtain a **scatter diagram** or **scattergram**. Figure 13.4 gives the scatter diagram for the data of Table 13.1. Each dot in this diagram represents one household. A scatter diagram is helpful in detecting a relationship between two variables. For example, by looking at the scatter diagram of Figure 13.4, we can observe that there exists a strong linear relationship between food expenditure and income. If a straight line is drawn through the points, the points will be scattered closely around the line.

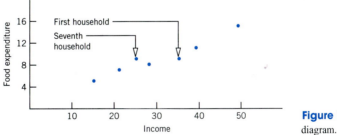

Figure 13.4 Scatter diagram.

> **SCATTER DIAGRAM** A plot of paired observations is called a *scatter diagram*.

As shown in Figure 13.5 on page 560, a large number of straight lines can be drawn through the scatter diagram of Figure 13.4. Each of these lines will give different values for *a* and *b* of model (3).

In regression analysis, we try to find a line that best fits the points in the scatter diagram. Such a line provides the best possible description of the relationship between the dependent and independent variables. The **least squares method**, discussed in the next section, gives such a line. The line obtained by using the least squares method is called the **least squares regression line**.

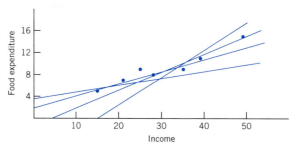

Figure 13.5

13.2.2 LEAST SQUARES LINE

The value of y obtained for a member from the survey is called the **observed or actual value of y**. As mentioned earlier in Section 13.2, the value of y, denoted by $\hat{y}$, obtained for a given x by using the regression line is called the **predicted value of y**. The random error ϵ denotes the difference between the actual value of y and the predicted value of y for population data. For example, for a given household, ϵ is the difference between what this household actually spent on food during the past month and what is predicted using the population regression line. The ϵ is also called the *residual* because it measures the surplus (positive or negative) of actual food expenditure over what is predicted by using the regression model. If we estimate model (2) by using sample data, the difference between the actual y and the predicted y based on this estimation cannot be denoted by ϵ. *The random error for the sample regression model is denoted by e.* Thus, e is an estimator of ϵ. If we estimate model (2) using sample data, then the value of e is given by

$$e = \text{Actual food expenditure} - \text{Predicted food expenditure} = y - \hat{y}$$

In Figure 13.6, e is the vertical distance between the actual position of a household and the point on the regression line. Note that in such a diagram, we always measure the dependent variable on the vertical axis and the independent variable on the horizontal axis.

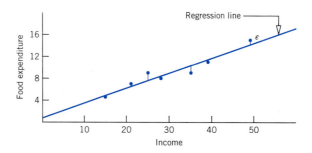

Figure 13.6

The value of an error is positive if the point that gives the actual food expenditure is above the regression line and negative if it is below the regression line. *The sum of these errors is always zero.* In other words, the sum of the actual food expenditures for seven households included in the sample will be the same as the sum of the food expenditures predicted from the regression model. Thus,

$$\Sigma e = \Sigma(y - \hat{y}) = 0$$

Hence, to find the line that best fits the scatter of points, we cannot minimize the sum of errors. Instead, we minimize the **error sum of squares**, denoted by **SSE**, which is obtained by adding the squares of errors. Thus,

$$\text{SSE} = \Sigma e^2 = \Sigma(y - \hat{y})^2$$

The least squares method gives the values of a and b for model (3) such that the sum of squared errors (SSE) is minimum.

ERROR SUM OF SQUARES (SSE) The *error sum of squares*, denoted by SSE, is

$$\text{SSE} = \Sigma e^2 = \Sigma(y - \hat{y})^2$$

The values of a and b that give the minimum SSE are called the *least squares estimates* of A and B, and the regression line obtained with these estimates is called the *least squares line*.

The least squares values of a and b are computed using the following formulas.

THE LEAST SQUARES LINE For the least squares regression line $\hat{y} = a + bx$

$$b = \frac{\text{SS}_{xy}}{\text{SS}_{xx}} \quad \text{and} \quad a = \bar{y} - b\bar{x}$$

where
$$\text{SS}_{xy} = \Sigma xy - \frac{(\Sigma x)(\Sigma y)}{n} \quad \text{and} \quad \text{SS}_{xx} = \Sigma x^2 - \frac{(\Sigma x)^2}{n}$$

and SS stands for "sum of squares."[2] The least squares regression line $\hat{y} = a + bx$ is also called the regression of y on x.

The preceding formulas are for estimating a sample regression line. Suppose we have access to a population data set. We can find the population regression line by using the same formulas with a little adaptation. If we have access to population data, we replace a by A, b by B, and n by N in these formulas, and use the values of Σx, Σy, Σxy, and Σx^2 calculated for population data to make the required computations. The population regression line is written as

$$\mu_{y|x} = A + Bx$$

where $\mu_{y|x}$ is read as "the mean value of y for a given x." When plotted on a graph, the points on this population regression line give the average values of y for the corresponding values of x. These average values of y are denoted by $\mu_{y|x}$.

Example 13–1 illustrates how to estimate a regression line for sample data.

---

[2]The values of SS_{xy} and SS_{xx} can also be obtained by using the following basic formulas:

$$\text{SS}_{xy} = \Sigma(x - \bar{x})(y - \bar{y}) \quad \text{and} \quad \text{SS}_{xx} = \Sigma(x - \bar{x})^2$$

However, these formulas take longer to make calculations.

Estimating the least squares regression line.

EXAMPLE 13–1   Find the least squares regression line for the data on incomes and food expenditures on the seven households given in Table 13.1. Use income as an independent variable and food expenditure as a dependent variable.

Solution   We are to find the values of a and b for the regression model $\hat{y} = a + bx$. Table 13.2 shows the calculations required for the computation of a and b. We denote the independent variable (income) by x and the dependent variable (food expenditure) by y.

Table 13.2

Income	Food Expenditure		
x	y	xy	x^2
35	9	315	1225
49	15	735	2401
21	7	147	441
39	11	429	1521
15	5	75	225
28	8	224	784
25	9	225	625
$\Sigma x = 212$	$\Sigma y = 64$	$\Sigma xy = 2150$	$\Sigma x^2 = 7222$

The following steps are performed to compute a and b.

Step 1.   *Compute Σx, Σy, $\bar{x}$, and $\bar{y}$.*

$$\Sigma x = 212 \qquad \Sigma y = 64$$

$$\bar{x} = \Sigma x/n = 212/7 = 30.2857$$

$$\bar{y} = \Sigma y/n = 64/7 = 9.1429$$

Step 2.   *Compute Σxy and Σx^2.*

To calculate Σxy, we multiply the corresponding values of x and y. Then, we sum all the products. The products of x and y are recorded in the third column of Table 13.2. To compute Σx^2, we square each of the x values and then add them. The squared values of x are listed in the fourth column of Table 13.2. From these calculations,

$$\Sigma xy = 2150 \quad \text{and} \quad \Sigma x^2 = 7222$$

Step 3.   *Compute SS_{xy} and SS_{xx}.*

$$SS_{xy} = \Sigma xy - \frac{(\Sigma x)(\Sigma y)}{n} = 2150 - \frac{(212)(64)}{7} = 211.7143$$

$$SS_{xx} = \Sigma x^2 - \frac{(\Sigma x)^2}{n} = 7222 - \frac{(212)^2}{7} = 801.4286$$

Step 4.   *Compute a and b.*

$$b = \frac{SS_{xy}}{SS_{xx}} = \frac{211.7143}{801.4286} = .2642$$

$$a = \bar{y} - b\bar{x} = 9.1429 - (.2642)(30.2857) = 1.1414$$

Thus, our estimated regression model $\hat{y} = a + bx$ is

$$\hat{y} = 1.1414 + .2642x$$

This regression line is called the least squares regression line. It gives the *regression of food expenditure on income.*

Note that we have rounded all calculations to four decimal places. We can round the values of a and b in the regression equation to two decimal places, but it is not done here because we will use this regression equation for prediction and estimation purposes later on. ■

Using this estimated regression model, we can find the predicted value of y for any specific value of x. For instance, suppose we randomly select a household whose monthly income is $3500 so that $x = 35$ (recall that x denotes income in hundreds of dollars). The predicted value of food expenditure for this household is

$$\hat{y} = 1.1414 + (.2642)(35) = \$10.3884 \text{ hundred} = \$1038.84$$

In other words, based on our regression line, we predict that a household with a monthly income of $3500 is expected to spend $1038.84 per month on food. This value of $\hat{y}$ can also be interpreted as a point estimator of the mean value of y for $x = 35$. Thus, we can state that, on average, all households with a monthly income of $3500 spend about $1038.84 per month on food.

In our data on seven households, there is one household whose income is $3500. The actual food expenditure for that household is $900 (see Table 13.1). The difference between the actual and predicted values gives the error of prediction. Thus, the error of prediction for this household, which is shown in Figure 13.7, is

$$e = y - \hat{y} = 9.00 - 10.3884 = -\$1.3884 \text{ hundred} = -\$138.84$$

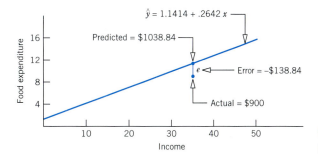

Figure 13.7 Error of prediction.

Therefore, the error of prediction is $-\$138.84$. The negative error indicates that the predicted value of y is greater than the actual value of y. Thus, if we use the regression model, this household's food expenditure is overestimated by $138.84.

13.2.3 INTERPRETATION OF *a* AND *b*

How do we interpret $a = 1.1414$ and $b = .2642$ obtained in Example 13–1 for the regression of food expenditure on income? A brief explanation of the y-intercept and the slope of a regression line was given in Section 13.1.2. The next two parts of this subsection explain the meaning of a and b in more detail.

Interpretation of *a*

Consider a household with zero income. Using the estimated regression line obtained in Example 13–1, we get the predicted value of y for $x = 0$:

$$\hat{y} = 1.1414 + .2642(0) = \$1.1414 \text{ hundred} = \$114.14$$

Thus, we can state that a household with no income is expected to spend $114.14 per month on food. Alternatively, we can also state that the point estimate of the average monthly food expenditure for all households with zero income is $114.14. Note that here we have used $\hat{y}$ as a point estimate of $\mu_{y|x}$. Thus, $a = 1.1414$ gives the predicted or mean value of y for $x = 0$ based on the regression model estimated for the sample data.

However, we should be very careful when making this interpretation of a. In our sample of seven households, the incomes vary from a minimum of $1500 to a maximum of $4900. (Note that in Table 13.1, the minimum value of x is 15 and the maximum value is 49.) Hence, our regression line is valid only for the values of x between 15 and 49. If we predict y for a value of x outside this range, the prediction usually will not hold true. Thus, since $x = 0$ is outside the range of household incomes that we have in the sample data, the prediction that a household with zero income spends $114.14 per month on food does not carry much credibility. The same is true if we try to predict y for an income greater than $4900, which is the maximum value of x in Table 13.1.

Interpretation of *b*

The value of b in a regression model gives the change in y (dependent variable) due to a change of one unit in x (independent variable). For example, by using the regression equation obtained in Example 13–1, we see:

$$\text{When } x = 30, \hat{y} = 1.1414 + .2642(30) = 9.0674$$

$$\text{When } x = 31, \hat{y} = 1.1414 + .2642(31) = 9.3316$$

Hence, when x increased by one unit, from 30 to 31, $\hat{y}$ increased by $9.3316 - 9.0674 = .2642$, which is the value of b. Because our unit of measurement is hundreds of dollars, we can state that, on average, a $100 increase in income will result in a $26.42 increase in food expenditure. We can also state that, on average, a $1 increase in income of a household will increase the food expenditure by $.2642. Note the phrase "on average" in these statements. The regression line is seen as a measure of the mean value of y for a given value of x. If one household's income is increased by $100, that household's food expenditure may or may not increase by $26.42. However, if the incomes of all households are increased by $100 each, the average increase in their food expenditures will be very close to $26.42.

Note that when b is positive, an increase in x will lead to an increase in y and a decrease in x will lead to a decrease in y. In other words, when b is positive, the movements in x and y are in the same direction. Such a relationship between x and y is called a **positive linear relationship**. The regression line in this case slopes upward from left to right. On the other hand, if the value of b is negative, an increase in x will lead to a decrease in y and a decrease in x will cause an increase in y. The changes in x and y in this case are in opposite directions. Such a relationship between x and y is called a **negative linear relationship**. The regression line in this case slopes downward from left to right. The two diagrams in Figure 13.8 show these two cases.

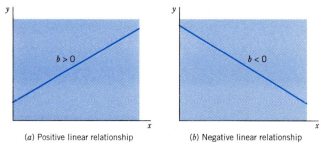

(a) Positive linear relationship   (b) Negative linear relationship

Figure 13.8   Positive and negative linear relationships between x and y.

 ***Remember*** For a regression model, b is computed as $b = SS_{xy}/SS_{xx}$. The value of SS_{xx} is always positive and that of SS_{xy} can be positive or negative. Hence, the sign of b depends on the sign of SS_{xy}. If SS_{xy} is positive (as in our example on the incomes and food expenditures of seven households), then b will be positive, and if SS_{xy} is negative, then b will be negative.

Case Study 13–1 illustrates the difference between the population regression line and a sample regression line.

CASE STUDY 13-1   REGRESSION OF HEIGHTS AND WEIGHTS OF NBA PLAYERS

Data Set III that accompanies this text lists the heights and weights of all National Basketball Association players who were on the rosters of all NBA teams at the beginning of the 1999–2000 season. These data comprise the population of NBA players for that point in time. We postulate the following simple linear regression model for these data:

$$y = A + Bx + \epsilon$$

where y is the weight (in pounds) and x is the height (in inches) of an NBA player.

Using the population data, we obtain the following regression line:

$$\mu_{y|x} = -285 + 6.41\, x$$

This equation gives the population regression line because it is obtained by using the population data. (Note that in the population regression line we write $\mu_{y|x}$ instead of $\hat{y}$.) Thus, the true values of A and B are

$$A = -285 \quad \text{and} \quad B = 6.41$$

The value of B indicates that for every one-inch increase in the height of an NBA player, weight increases on average by 6.41 pounds. However, $A = -285$ does not make any sense. It states that the weight of a player with zero height is -285 pounds. (Recall from Section 13.2.3 that we cannot apply the regression equation to predict y for values of x outside the range of data used to find the regression line.) Figure 13.9, constructed using MINITAB, gives the scatter diagram for the heights and weights of all NBA players.

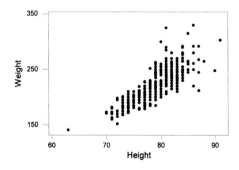

Figure 13.9 Scatter diagram for the data on heights and weights of all NBA players.

Next, we selected a random sample of 30 players and estimated the regression model for this sample. The estimated regression line for the sample is

$$\hat{y} = -351 + 7.22x$$

The values of a and b are $a = -351$ and $b = 7.22$. These values of a and b give the estimates of A and B based on sample data. The scatter diagram for the sample observations on heights and weights is given in Figure 13.10.

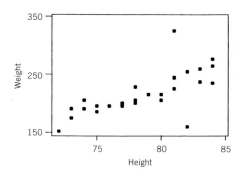

Figure 13.10 Scatter diagram for the data on heights and weights of 30 NBA players.

As we can observe from Figures 13.9 and 13.10, the scatter diagrams for population and sample data both show a (positive) linear relationship between the heights and weights of NBA players.

Source: National Basketball Association.

13.2.4 ASSUMPTIONS OF THE REGRESSION MODEL

Like any other theory, the linear regression analysis is also based on certain assumptions. Consider the population regression model

$$y = A + Bx + \epsilon \tag{4}$$

Four assumptions are made about this model. These assumptions are explained next with reference to the example on incomes and food expenditures of households. Note that these assumptions are made about the population regression model and not about the sample regression model.

Assumption 1: The random error term ϵ has a mean equal to zero for each x. In other words, among all households with the same income, some spend more than the predicted food expenditure (and, hence, have positive errors) and others spend less than the predicted food expenditure (and, consequently, have negative errors). This assumption simply states that the sum of the positive errors is equal to the sum of the negative errors, so that the mean of errors for all households with the same income is zero. Thus, when the mean value of ϵ is zero, the mean value of y for a given x is equal to $A + Bx$ and it is written as

$$\mu_{y|x} = A + Bx$$

As mentioned earlier in this chapter, $\mu_{y|x}$ is read as "the mean value of y for a given value of x." When we find the values of A and B for model (4) using the population data, the points on the regression line give the average values of y, denoted by $\mu_{y|x}$, for the corresponding values of x.

Assumption 2: The errors associated with different observations are independent. According to this assumption, the errors for any two households in our example are independent. In other words, all households decide independently how much to spend on food.

Assumption 3: For any given x, the distribution of errors is normal. The corollary of this assumption is that the food expenditures for all households with the same income are normally distributed.

Assumption 4: The distribution of population errors for each x has the same (constant) standard deviation, which is denoted by σ_ϵ. This assumption indicates that the spread of points around the regression line is similar for all x values.

Figure 13.11 illustrates the meanings of the first, third, and fourth assumptions for households with incomes of \$2000 and \$3500 per month. The same assumptions hold true for any other income level. In the population of all households, there will be many households with a monthly income of \$2000. Using the population regression line, if we calculate the errors for all these households and prepare the distribution of these errors, it will look like the distribution given in Figure 13.11a. Its standard deviation will be σ_ϵ. Similarly, Figure 13.11b gives the distribution of errors for all those households in the population whose monthly income is \$3500. Its standard deviation is also σ_ϵ. Both these distributions are identical. Note that the mean of both of these distributions is $E(\epsilon) = 0$.

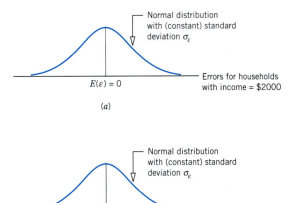

Figure 13.11 (a) Errors for households with an income of \$2000 per month. (b) Errors for households with an income of \$3500 per month.

Figure 13.12 on page 568 shows how these distributions look when they are plotted on the same diagram with the population regression line. The points on the vertical line through $x = 20$ give the food expenditures for various households in the population, each of which has the same monthly income of \$2000. The same is true about the vertical line through $x = 35$ or any other vertical line for some other value of x.

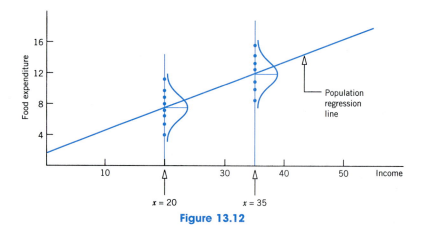

Figure 13.12

13.2.5 A NOTE ON THE USE OF SIMPLE LINEAR REGRESSION

We should apply linear regression with caution. When we use simple linear regression, we assume that the relationship between two variables is described by a straight line. In the real world, the relationship between variables may not be linear. Hence, before we use a simple linear regression, it is better to construct a scatter diagram and look at the plot of the data points. We should estimate a linear regression model only if the scatter diagram indicates such a relationship. The scatter diagrams of Figure 13.13 give two examples where the relationship between x and y is not linear. Consequently, fitting linear regression in such cases would be wrong.

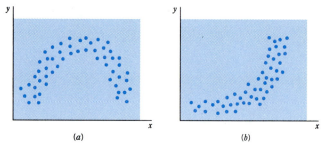

Figure 13.13   Nonlinear relationship between x and y.

EXERCISES

■ **Concepts and Procedures**

13.1 Explain the meaning of the words *simple* and *linear* as used in *simple linear regression*.

13.2 Explain the meaning of independent and dependent variables for a regression model.

13.3 Explain the difference between exact and nonexact relationships between two variables.

13.4 Explain the difference between linear and nonlinear relationships between two variables.

13.5 Explain the difference between a simple and a multiple regression model.

13.6 Briefly explain the difference between a deterministic and a probabilistic regression model.

13.7 Why is the random error term included in a regression model?

13.8 Explain the least squares method and least squares regression line. Why are they called by these names?

13.9 Explain the meaning and concept of SSE. You may use a graph for illustration purposes.

13.10 Explain the difference between y and $\hat{y}$.

13.11 Two variables x and y have a positive linear relationship. Explain what happens to the value of y when x increases.

13.12 Two variables x and y have a negative linear relationship. Explain what happens to the value of y when x increases.

13.13 Explain the following.

a. Population regression line
b. Sample regression line
c. True values of A and B
d. Estimated values of A and B that are denoted by a and b, respectively

13.14 Briefly explain the assumptions of the population regression model.

13.15 Plot the following straight lines. Give the values of the y-intercept and slope for each of these lines and interpret them. Indicate whether each of the lines gives a positive or a negative relationship between x and y.

a. $y = 100 + 5x$ **b.** $y = 400 - 4x$

13.16 Plot the following straight lines. Give the values of the y-intercept and slope for each of these lines and interpret them. Indicate whether each of the lines gives a positive or a negative relationship between x and y.

a. $y = -60 + 8x$ **b.** $y = 300 - 6x$

13.17 A population data set produced the following information.

$$N = 250, \quad \Sigma x = 9880, \quad \Sigma y = 1456, \quad \Sigma xy = 85,080, \quad \Sigma x^2 = 485,870$$

Find the population regression line.

13.18 A population data set produced the following information.

$$N = 460, \quad \Sigma x = 3920, \quad \Sigma y = 2650, \quad \Sigma xy = 26,570, \quad \Sigma x^2 = 48,530$$

Find the population regression line.

13.19 The following information is obtained from a sample data set.

$$n = 10, \quad \Sigma x = 100, \quad \Sigma y = 220, \quad \Sigma xy = 3680, \quad \Sigma x^2 = 1140$$

Find the estimated regression line.

13.20 The following information is obtained from a sample data set.

$$n = 12, \quad \Sigma x = 66, \quad \Sigma y = 588, \quad \Sigma xy = 2244, \quad \Sigma x^2 = 396$$

Find the estimated regression line.

■ **Applications**

13.21 A car rental company charges $30 a day and 15 cents per mile for renting a car. Let y be the total rental charges (in dollars) for a car for one day and x be the miles driven. The equation for the relationship between x and y is

$$y = 30 + .15x$$

a. How much will a person pay who rents a car for one day and drives it 100 miles?
b. Suppose each of 20 persons rents a car from this agency for one day and drives it 100 miles. Will each of them pay the same amount for renting a car for a day or is each person expected to pay a different amount? Explain.
c. Is the relationship between x and y exact or nonexact?

13.22 Ben is an electrician who makes house calls for electrical repairs. He charges $30 to go to a house plus $25 per hour. Let y be the total amount (in dollars) paid by a household that uses Ben's services and x the number of hours Ben spends doing repairs in that household's home. The equation for the relationship between x and y is

$$y = 30 + 25x$$

a. Ben spent 6 hours doing repairs in Kristine's home. How much will he be paid?
b. Suppose seven persons called Ben for repairs during a week. Surprisingly, each of these jobs took 6 hours. Will each of these homeowners pay the same amount for repairs or do you expect each to pay a different amount? Explain.
c. Is the relationship between x and y exact or nonexact?

13.23 A researcher took a sample of 25 electronics companies and found the following relationship between x and y where x is the amount of money (in millions of dollars) spent on advertising by a company in 1999 and y represents the total gross sales (in millions of dollars) of that company for 1999.

$$\hat{y} = 3.6 + 11.75x$$

a. An electronics company spent $2 million on advertising in 1999. What are its expected gross sales for 1999?
b. Suppose four electronics companies spent $2 million each on advertising in 1999. Do you expect these four companies to have the same actual gross sales for 1999? Explain.
c. Is the relationship between x and y exact or nonexact?

13.24 A researcher took a sample of 10 years and found the following relationship between x and y, where x is the number of major natural calamities (such as tornadoes, hurricanes, earthquakes, floods, etc.) that occurred during a year and y represents the average annual total profits (in millions of dollars) of all insurance companies in the United States.

$$\hat{y} = 342.6 - 2.10x$$

a. A randomly selected year had 24 major calamities. What are the expected average profits of U.S. insurance companies for that year?
b. Suppose the number of major calamities was the same for each of three years. Do you expect the average profits for all U.S. insurance companies to be the same for each of these three years? Explain.
c. Is the relationship between x and y exact or nonexact?

13.25 An auto manufacturing company wanted to investigate how the price of one of its car models depreciates with age. The research department at the company took a sample of eight cars of this model and collected the following information on the ages (in years) and prices (in hundreds of dollars) of these cars.

Age	8	3	6	9	2	5	6	3
Price	18	94	50	21	145	42	36	99

a. Construct a scatter diagram for these data. Does the scatter diagram exhibit a linear relationship between ages and prices of cars?
b. Find the regression line with price as a dependent variable and age as an independent variable.
c. Give a brief interpretation of the values of a and b calculated in part b.
d. Plot the regression line on the scatter diagram of part a and show the errors by drawing vertical lines between scatter points and the regression line.
e. Predict the price of a 7-year-old car of this model.
f. Estimate the price of an 18-year-old car of this model. Comment on this finding.

13.26 Seven students were tested for stress before a mathematics test. The following table gives the stress scores (on a scale of 1 to 10) of these students and their scores on the math test.

Stress score	6.5	4.0	2.5	7.2	8.1	3.4	5.5
Test score	81	96	93	68	63	84	71

a. Construct a scatter diagram for these data. Does the scatter diagram exhibit a linear relationship between stress scores and test scores?
b. Find the regression of test scores on stress scores.
c. Give a brief interpretation of the values of a and b calculated in part b.
d. Plot the regression line on the scatter diagram of part a and show the errors by drawing vertical lines between scatter points and the regression line.
e. Predict the test score of a student with a 7.5 stress score before a math test.
f. Estimate the test score of a student with a stress score of 9.5 before a math test. Comment on this finding.

13.27 An insurance company wants to know how the amount of life insurance depends on the income of persons. The research department at the company collected information on six persons. The following table lists the annual incomes (in thousands of dollars) and amounts (in thousands of dollars) of life insurance policies for these six persons.

Annual income	52	58	31	43	65	24
Life insurance	250	300	100	150	500	75

a. Construct a scatter diagram for these data. Does the scatter diagram show a linear relationship between annual incomes and amounts of life insurance policies?
b. Find the regression line $\hat{y} = a + bx$ with annual income as an independent variable and amount of life insurance policy as a dependent variable.
c. Give a brief interpretation of the values of a and b calculated in part b.
d. Plot the regression line on the scatter diagram of part a and show the errors by drawing vertical lines between the scatter points and the regression line.
e. What is the estimated value of life insurance for a person with an annual income of $45,000?
f. One of the persons in our sample has an annual income of $58,000 and $300,000 of life insurance. What is the predicted value of life insurance for this person? Find the error for this observation.

13.28 A consumer welfare agency wants to investigate the relationship between the sizes of houses and the rents paid by tenants in a small city. The agency collected the following information on the sizes (in hundreds of square feet) of six houses and the monthly rents (in dollars) paid by tenants.

Size of house	21	13	19	27	34	23
Monthly rent	700	580	720	1000	1250	800

a. Construct a scatter diagram for these data. Does the scatter diagram show a linear relationship between sizes of houses and monthly rents?
b. Find the regression line $\hat{y} = a + bx$ with the size of a house as an independent variable and monthly rent as a dependent variable.
c. Give a brief interpretation of the values of a and b calculated in part b.
d. Plot the regression line on the scatter diagram of part a and show the errors by drawing vertical lines between the scatter points and the regression line.
e. Predict the monthly rent for a house with 2500 square feet.
f. One of the houses in our sample is 2700 square feet and its rent is $1000. What is the predicted rent for this house? Find the error of estimation for this observation.

13.29   The following table gives the total 1999 payroll (rounded to the nearest million dollars) and the percentage of games won during the 1999 season by each of the National League baseball teams.

Team	Total Payroll (millions)	Percentage of Games Won
Arizona Diamondbacks	$70	61.7
Atlanta Braves	75	63.6
Chicago Cubs	55	41.4
Cincinnati Reds	42	58.9
Colorado Rockies	54	44.4
Florida Marlins	15	39.5
Houston Astros	56	59.9
Los Angeles Dodgers	71	47.5
Milwaukee Brewers	43	46.0
Montreal Expos	16	42.0
New York Mets	71	59.5
Philadelphia Phillies	31	47.5
Pittsburgh Pirates	24	48.4
St. Louis Cardinals	46	46.6
San Diego Padres	46	45.7
San Francisco Giants	46	53.1

a.   Find the least squares regression line with total payroll as an independent variable and percentage of games won as a dependent variable.
b.   Is the regression line obtained in part a the population regression line? Why or why not? Do the values of the y-intercept and the slope of the regression line give A and B or a and b?
c.   Give a brief interpretation of the values of the y-intercept and the slope.
d.   Predict the percentage of games won for a team with a total payroll of $35 million.

13.30   The following table gives the total 1999 payroll (rounded to the nearest million dollars) and the percentage of games won during the 1999 season by each of the American League baseball teams.

Team	Total Payroll (millions)	Percentage of Games Won
Anaheim Angels	$50	43.2
Baltimore Orioles	71	48.1
Boston Red Sox	72	58.0
Chicago White Sox	25	46.6
Cleveland Indians	74	59.9
Detroit Tigers	35	42.9
Kansas City Royals	17	39.8
Minnesota Twins	16	39.4
New York Yankees	88	60.5
Oakland A's	24	53.7
Seattle Mariners	44	48.8
Tampa Bay Devil Rays	38	42.6
Texas Rangers	81	58.6
Toronto Blue Jays	48	51.9

a.   Find the least squares regression line with total payroll as an independent variable and percentage of games won as a dependent variable.
b.   Is the regression line obtained in part a the population regression line? Why or why not? Do the values of the y-intercept and the slope in the regression line give A and B or a and b?
c.   Give a brief interpretation of the values of the y-intercept and the slope.
d.   Predict the percentage of games won for a team with a total payroll of $38 million.

13.3 STANDARD DEVIATION OF RANDOM ERRORS

When we consider income and food expenditures, all households with the same income are expected to spend different amounts on food. Consequently, the random error ϵ will assume different values for these households. The standard deviation σ_ϵ measures the spread of these errors around the population regression line. The **standard deviation of errors** tells us how widely the errors and, hence, the values of y are spread for a given x. In Figure 13.12, which is reproduced as Figure 13.14, the points on the vertical line through $x = 20$ give the monthly food expenditures for all households with a monthly income of \$2000. The distance of each dot from the point on the regression line gives the value of the corresponding error. The standard deviation of errors σ_ϵ measures the spread of such points around the population regression line. The same is true for $x = 35$ or any other value of x.

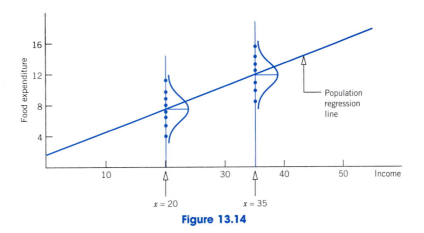

Figure 13.14

Note that σ_ϵ denotes the standard deviation of errors for the population. However, usually σ_ϵ is unknown. In such cases, it is estimated by s_e, which is the standard deviation of errors for the sample data. The following is the basic formula to calculate s_e:

$$s_e = \sqrt{\frac{\text{SSE}}{n-2}} \qquad \text{where SSE} = \Sigma(y - \hat{y})^2$$

In the above formula, $n - 2$ represents the **degrees of freedom** for the regression model. The reason $df = n - 2$ is that we lose one degree of freedom to calculate $\bar{x}$ and one for $\bar{y}$.

> **DEGREES OF FREEDOM FOR A SIMPLE LINEAR REGRESSION MODEL** The *degrees of freedom* for a simple linear regression model are
>
> $$df = n - 2$$

For computational purposes, it is more convenient to use the following formula to calculate the standard deviation of errors s_e.

STANDARD DEVIATION OF ERRORS The *standard deviation of errors* is calculated as[3]

$$s_e = \sqrt{\frac{SS_{yy} - b\, SS_{xy}}{n - 2}}$$

where

$$SS_{yy} = \Sigma y^2 - \frac{(\Sigma y)^2}{n}$$

The calculation of SS_{xy} was discussed earlier in this chapter.[4]

Like the value of SS_{xx}, the value of SS_{yy} is always positive.

Example 13–2 illustrates the calculation of the standard deviation of errors for the data of Table 13.1.

Calculating the standard deviation of errors.

EXAMPLE 13–2 Compute the standard deviation of errors s_e for the data on monthly incomes and food expenditures of the seven households given in Table 13.1.

Solution To compute s_e, we need to know the values of SS_{yy}, SS_{xy}, and b. Earlier, in Example 13–1, we computed SS_{xy} and b. These values are

$$SS_{xy} = 211.7143 \quad \text{and} \quad b = .2642$$

To compute SS_{yy}, we calculate Σy^2 as shown in Table 13.3.

Table 13.3

Income	Food Expenditure	
x	y	y^2
35	9	81
49	15	225
21	7	49
39	11	121
15	5	25
28	8	64
25	9	81
$\Sigma x = 212$	$\Sigma y = 64$	$\Sigma y^2 = 646$

The value of SS_{yy} is

$$SS_{yy} = \Sigma y^2 - \frac{(\Sigma y)^2}{n} = 646 - \frac{(64)^2}{7} = 60.8571$$

---

[3]If we have access to population data, the value of σ_ϵ is calculated using the formula

$$\sigma_\epsilon = \sqrt{\frac{SS_{yy} - B\, SS_{xy}}{N}}$$

[4]The basic formula to calculate SS_{yy} is $\Sigma(y - \bar{y})^2$.

Hence, the standard deviation of errors is

$$s_e = \sqrt{\frac{SS_{yy} - b\,SS_{xy}}{n-2}} = \sqrt{\frac{60.8571 - .2642(211.7143)}{7-2}} = .9922$$ ∎

13.4 COEFFICIENT OF DETERMINATION

We may ask the question, How good is the regression model? In other words, How well does the independent variable explain the dependent variable in the regression model? The *coefficient of determination* is one concept that answers this question.

For a moment, assume that we possess information only on the food expenditures of households and not on their incomes. Hence, in this case, we cannot use the regression line to predict the food expenditure for any household. As we did in earlier chapters, in the absence of a regression model, we use $\bar{y}$ to estimate or predict every household's food expenditure. Consequently, the error of prediction for each household is now given by $y - \bar{y}$, which is the difference between the actual food expenditure of a household and the mean food expenditure. If we calculate such errors for all households and then square and add them, the resulting sum is called the **total sum of squares** and is denoted by **SST**. Actually SST is the same as SS_{yy} and is defined as

$$SST = SS_{yy} = \Sigma(y - \bar{y})^2$$

However, for computational purposes, SST is calculated using the following formula.

> **TOTAL SUM OF SQUARES (SST)** The *total sum of squares*, denoted by SST, is
>
> $$SST = \Sigma y^2 - \frac{(\Sigma y)^2}{n}$$
>
> Note that this is the same formula that we used to calculate SS_{yy}.

The value of SS_{yy}, which is 60.8571, was calculated in Example 13–2. Consequently, the value of SST is

$$SST = 60.8571$$

From Example 13–1, $\bar{y} = 9.1429$. Figure 13.15 shows the error for each of the seven households in our sample using the scatter diagram of Figure 13.4.

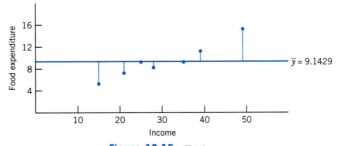

Figure 13.15 Total errors.

Now suppose we use the simple linear regression model to predict the food expenditure of each of the seven households in our sample. In this case, we predict each household's food expenditure by using the regression line we estimated earlier in Example 13–1, which is

$$\hat{y} = 1.1414 + .2642x$$

The predicted food expenditures, denoted by $\hat{y}$, for the seven households are listed in Table 13.4. Also given are the errors and error squares.

Table 13.4

x	y	$\hat{y} = 1.1414 + .2642x$	$e = y - \hat{y}$	$e^2 = (y - \hat{y})^2$
35	9	10.3884	−1.3884	1.9277
49	15	14.0872	.9128	.8332
21	7	6.6896	.3104	.0963
39	11	11.4452	−.4452	.1982
15	5	5.1044	−.1044	.0109
28	8	8.5390	−.5390	.2905
25	9	7.7464	1.2536	1.5715

$$\Sigma e^2 = \Sigma(y - \hat{y})^2 = 4.9283$$

We calculate the values of $\hat{y}$ (given in the third column of Table 13.4) by substituting the values of x in the estimated regression model. For example, the value of x for the first household is 35. Substituting this value of x in the regression equation, we obtain

$$\hat{y} = 1.1414 + .2642(35) = 10.3884$$

Similarly we find the other values of $\hat{y}$. The error sum of squares SSE is given by the sum of the fifth column in Table 13.4. Thus,

$$\text{SSE} = \Sigma(y - \hat{y})^2 = 4.9283$$

The errors of prediction for the regression model for the seven households are shown in Figure 13.16.

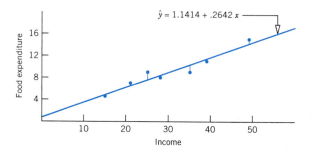

Figure 13.16 Errors of prediction when regression model is used.

Thus, from the foregoing calculations,

$$\text{SST} = 60.8571 \quad \text{and} \quad \text{SSE} = 4.9283$$

These values indicate that the sum of squared errors decreased from 60.8571 to 4.9283 when we used $\hat{y}$ in place of $\bar{y}$ to predict food expenditures. This reduction in squared errors is called the **regression sum of squares** and is denoted by **SSR**. Thus,

$$\text{SSR} = \text{SST} - \text{SSE} = 60.8571 - 4.9283 = 55.9288$$

The value of SSR can also be computed by using the formula

$$SSR = \Sigma(\hat{y} - \bar{y})^2$$

REGRESSION SUM OF SQUARES (SSR) The *regression sum of squares*, denoted by SSR, is

$$SSR = SST - SSE$$

Thus, SSR is the portion of SST that is explained by the use of the regression model, and SSE is the portion of SST that is not explained by the use of the regression model. The sum of SSR and SSE is always equal to SST. Thus,

$$SST = SSR + SSE$$

The ratio of SSR to SST gives the **coefficient of determination**. The coefficient of determination calculated for population data is denoted by ρ^2 (ρ is the Greek letter *rho*) and the one calculated for sample data is denoted by r^2. The coefficient of determination gives the proportion of SST that is explained by the use of the regression model. The value of the coefficient of determination always lies in the range zero to 1. The coefficient of determination can be calculated by using the formula

$$r^2 = \frac{SSR}{SST} \quad or \quad \frac{SST - SSE}{SST}$$

However, for computational purposes, the formula given next is more efficient to use to calculate the coefficient of determination.

COEFFICIENT OF DETERMINATION The *coefficient of determination*, denoted by r^2, represents the proportion of SST that is explained by the use of the regression model. The computational formula for r^2 is[5]

$$r^2 = \frac{b\,SS_{xy}}{SS_{yy}}$$

and

$$0 \le r^2 \le 1$$

Example 13–3 illustrates the calculation of the coefficient of determination for a sample data set.

---

[5]If we have access to population data, the value of ρ^2 is calculated using the formula

$$\rho^2 = \frac{B\,SS_{xy}}{SS_{yy}}$$

The values of SS_{xy} and SS_{yy} used here are calculated for the population data set.

EXAMPLE 13-3   For the data of Table 13.1 on monthly incomes and food expenditures of seven households, calculate the coefficient of determination.

Solution   From earlier calculations made in Examples 13–1 and 13–2,

$$b = .2642, \quad SS_{xy} = 211.7143, \quad \text{and} \quad SS_{yy} = 60.8571$$

Hence,

$$r^2 = \frac{b \, SS_{xy}}{SS_{yy}} = \frac{(.2642)(211.7143)}{60.8571} = .92$$

Thus, we can state that SST is reduced by approximately 92% (from 60.8571 to 4.9283) when we use $\hat{y}$, instead of $\bar{y}$, to predict the food expenditures of households. Note that r^2 is usually rounded to two decimal places.   ■

The total sum of squares SST is a measure of the total variation in food expenditures, the regression sum of squares SSR is the portion of total variation explained by the regression model (or by income), and the error sum of squares SSE is the portion of total variation not explained by the regression model. Hence, for Example 13–3 we can state that 92% of the total variation in food expenditures of households occurs because of the variation in their incomes, and the remaining 8% is due to randomness and other variables.

Usually, the higher the value of r^2, the better the regression model. This is so because if r^2 is larger, a greater portion of the total errors is explained by the included independent variable and a smaller portion of errors is attributed to other variables and randomness.

EXERCISES

■ **Concepts and Procedures**

13.31   What are the degrees of freedom for a simple linear regression model?

13.32   Explain the meaning of coefficient of determination.

13.33   Explain the meaning of SST and SSR. You may use graphs for illustration purposes.

13.34   A population data set produced the following information.

$$N = 250, \quad \Sigma x = 9880, \quad \Sigma y = 1456, \quad \Sigma xy = 85{,}080,$$

$$\Sigma x^2 = 485{,}870, \quad \text{and} \quad \Sigma y^2 = 135{,}675$$

Find the values of σ_ϵ and ρ^2.

13.35   A population data set produced the following information.

$$N = 460, \quad \Sigma x = 3920, \quad \Sigma y = 2650, \quad \Sigma xy = 26{,}570,$$

$$\Sigma x^2 = 48{,}530, \quad \text{and} \quad \Sigma y^2 = 39{,}347$$

Find the values of σ_ϵ and ρ^2.

13.36   The following information is obtained from a sample data set.

$$n = 10, \quad \Sigma x = 100, \quad \Sigma y = 220, \quad \Sigma xy = 3680,$$

$$\Sigma x^2 = 1140, \quad \text{and} \quad \Sigma y^2 = 25{,}272$$

Find the values of s_e and r^2.

13.37 The following information is obtained from a sample data set.

$$n = 12, \quad \Sigma x = 66, \quad \Sigma y = 588, \quad \Sigma xy = 2244,$$

$$\Sigma x^2 = 396, \quad \text{and} \quad \Sigma y^2 = 58,734$$

Find the values of s_e and r^2.

■ **Applications**

13.38 The following table gives information on the monthly incomes (in hundreds of dollars) and monthly telephone bills (in dollars) for a random sample of 10 households.

Income	20	54	46	42	36	23	61	26	49	64
Phone bill	45	165	175	90	95	35	210	50	110	190

Find the following.

a. SS_{xx}, SS_{yy}, and SS_{xy} **b.** Standard deviation of errors
c. SST, SSE, and SSR **d.** Coefficient of determination

13.39 The following table gives information on the average saturated fat (in grams) consumed per day and the cholesterol level (in milligrams per hundred milliliters) for eight men.

Fat consumption	55	68	50	34	43	58	77	36
Cholesterol level	180	215	195	165	170	204	235	150

Compute the following.

a. SS_{xx}, SS_{yy}, and SS_{xy} **b.** Standard deviation of errors
c. SST, SSE, and SSR **d.** Coefficient of determination

13.40 Refer to Exercise 13.25. The following table, which gives the ages (in years) and prices (in hundreds of dollars) of eight cars of a specific model, is reproduced from that exercise.

Age	8	3	6	9	2	5	6	3
Price	18	94	50	21	145	42	36	99

a. Calculate the standard deviation of errors.
b. Compute the coefficient of determination and give a brief interpretation of it.

13.41 The following data on the stress scores before a math test and the math test scores for seven students are reproduced from Exercise 13.26.

Stress score	6.5	4.0	2.5	7.2	8.1	3.4	5.5
Test score	81	96	93	68	63	84	71

a. Determine the standard deviation of errors.
b. Find the coefficient of determination and give a brief interpretation of it.

13.42 The following data set on annual incomes (in thousands of dollars) and amounts (in thousands of dollars) of life insurance policies for six persons is reproduced from Exercise 13.27.

Annual income	52	58	31	43	65	24
Life insurance	250	300	100	150	500	75

a. Find the standard deviation of errors.

b. Compute the coefficient of determination. What percentage of the variation in life insurance amounts is explained by the annual incomes? What percentage of this variation is not explained?

13.43 The following table, reproduced from Exercise 13.28, lists the sizes of six houses (in hundreds of square feet) and the monthly rents (in dollars) paid by tenants for those houses.

Size of house	21	13	19	27	34	23
Monthly rent	700	580	720	1000	1250	800

a. Compute the standard deviation of errors.

b. Calculate the coefficient of determination. What percentage of the variation in monthly rents is explained by the sizes of the houses? What percentage of this variation is not explained?

13.44 Refer to data given in Exercise 13.29 on the total 1999 payroll and the percentage of games won during the 1999 season by each of the National League baseball teams.

a. Find the standard deviation of errors, σ_ϵ. (Note that this data set belongs to a population.)

b. Compute the coefficient of determination, ρ^2.

13.45 Refer to data given in Exercise 13.30 on the total 1999 payroll and the percentage of games won during the 1999 season by each of the American League baseball teams.

a. Find the standard deviation of errors, σ_ϵ. (Note that this data set belongs to a population.)

b. Compute the coefficient of determination, ρ^2.

13.5   INFERENCES ABOUT *B*

This section is concerned with estimation and tests of hypotheses about the population regression slope B. We can also make confidence intervals and test hypotheses about the y-intercept A of the population regression line. However, making inferences about A is beyond the scope of this text.

13.5.1   SAMPLING DISTRIBUTION OF *b*

One of the main purposes for determining a regression line is to find the true value of the slope B of the population regression line. However, in almost all cases, the regression line is estimated using sample data. Then, based on the sample regression line, inferences are made about the population regression line. The slope b of a sample regression line is a point estimator of the slope B of the population regression line. The different sample regression lines estimated for different samples taken from the same population will give different values of b. If only one sample is taken and the regression line for that sample is estimated, the value of b will depend on which elements are included in the sample. Thus, b is a random variable and it possesses a probability distribution that is more commonly called its sampling distribution. The shape of the sampling distribution of b, its mean, and standard deviation are given next.

> **MEAN, STANDARD DEVIATION, AND SAMPLING DISTRIBUTION OF *b*** Because of the assumption of normally distributed random errors, the sampling distribution of b is normal. The mean and standard deviation of b, denoted by μ_b and σ_b, respectively, are
>
> $$\mu_b = B \quad \text{and} \quad \sigma_b = \frac{\sigma_\epsilon}{\sqrt{SS_{xx}}}$$

However, usually the standard deviation of population errors σ_ϵ is not known. Hence, the sample standard deviation of errors s_e is used to estimate σ_ϵ. In such a case, when σ_ϵ is unknown, the standard deviation of b is estimated by s_b, which is calculated as

$$s_b = \frac{s_e}{\sqrt{SS_{xx}}}$$

If σ_ϵ is not known and the sample size is large ($n \geq 30$), the normal distribution can be used to make inferences about B. However, if σ_ϵ is not known and the sample size is small ($n < 30$), the normal distribution is replaced by the t distribution to make inferences about B.

13.5.2 ESTIMATION OF *B*

The value of b obtained from the sample regression line is a point estimate of the slope B of the population regression line. As mentioned in Section 13.5.1, if σ_ϵ is not known and the sample size is small, the t distribution is used to make a confidence interval for B.

> **CONFIDENCE INTERVAL FOR *B*** The $(1 - \alpha)$ 100% *confidence interval for B* is given by
>
> $$b \pm ts_b$$
>
> where $$s_b = \frac{s_e}{\sqrt{SS_{xx}}}$$
>
> and the value of t is obtained from the t distribution table for $\alpha/2$ area in the right tail of the t distribution and $n - 2$ degrees of freedom.

Example 13–4 describes the procedure for making a confidence interval for B.

Constructing a confidence interval for B.

EXAMPLE 13–4 Construct a 95% confidence interval for B for the data on incomes and food expenditures of seven households given in Table 13.1.

Solution From the given information and earlier calculations in Examples 13–1 and 13–2,

$$n = 7, \quad b = .2642, \quad SS_{xx} = 801.4286, \quad \text{and} \quad s_e = .9922$$

The confidence level is 95%.

$$s_b = \frac{s_e}{\sqrt{SS_{xx}}} = \frac{.9922}{\sqrt{801.4286}} = .0350$$

$$df = n - 2 = 7 - 2 = 5$$

$$\alpha/2 = .5 - (.95/2) = .025$$

From the t distribution table, the value of t for 5 df and .025 area in the right tail of the t distribution curve is 2.571. The 95% confidence interval for B is

$$b \pm ts_b = .2642 \pm 2.571(.0350) = .2642 \pm .0900 = \textbf{.17 to .35}$$

Thus, we are 95% confident that the slope B of the population regression line is between .17 and .35.　■

13.5.3   HYPOTHESIS TESTING ABOUT B

Testing a hypothesis about B when the null hypothesis is $B = 0$ (that is, the slope of the regression line is zero) is equivalent to testing that x does not determine y and that the regression line is of no use in predicting y for a given x. However, we should remember that we are testing for a linear relationship between x and y. It is possible that x may determine y nonlinearly. Hence, a nonlinear relationship may exist between x and y.

To test the hypothesis that x does not determine y linearly, we will test the null hypothesis that the slope of the regression line is zero; that is, $B = 0$. The alternative hypothesis can be: (1) x determines y; that is, $B \neq 0$; (2) x determines y positively; that is, $B > 0$; or (3) x determines y negatively; that is, $B < 0$.

The procedure used to make a hypothesis test about B is similar to the one used in earlier chapters. It involves the same five steps.

TEST STATISTIC FOR b   The value of the *test statistic t for b* is calculated as

$$t = \frac{b - B}{s_b}$$

The value of B is substituted from the null hypothesis.

Example 13–5 illustrates the procedure for testing a hypothesis about B.

Conducting a test of hypothesis about B.

EXAMPLE 13–5   Test at the 1% significance level whether the slope of the regression line for the example on incomes and food expenditures of seven households is positive.

Solution   From the given information and earlier calculations in Examples 13–1 and 13–4,

$$n = 7, \quad b = .2642, \quad \text{and} \quad s_b = .0350$$

Step 1.   *State the null and alternative hypotheses.*

We are to test whether or not the slope B of the population regression line is positive. Hence, the two hypotheses are

$$H_0: B = 0 \qquad \text{(The slope is zero)}$$

$$H_1: B > 0 \qquad \text{(The slope is positive)}$$

Note that we can also write the null hypothesis as $H_0: B \leq 0$, which states that the slope is either zero or negative.

Step 2.   *Select the distribution to use.*

The sample size is small ($n < 30$) and σ_ϵ is not known. Hence, we will use the t distribution to make the test about B.

Step 3.   *Determine the rejection and nonrejection regions.*

The significance level is .01. The $>$ sign in the alternative hypothesis indicates that the test is right-tailed. Therefore,

$$\text{Area in the right tail of the } t \text{ distribution} = \alpha = .01$$

$$df = n - 2 = 7 - 2 = 5$$

From the t distribution table, the critical value of t for 5 df and .01 area in the right tail of the t distribution is 3.365, as shown in Figure 13.17.

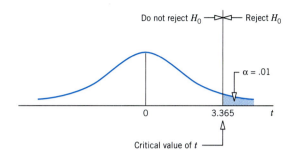

Do not reject H_0 —▷◁— Reject H_0

$\alpha = .01$

0   3.365 t

Critical value of t

Figure 13.17

Step 4. *Calculate the value of the test statistic.*

The value of the test statistic t for b is calculated as follows:

$$t = \frac{b - B}{s_b} = \frac{.2642 - \overset{\text{From } H_0}{0}}{.0350} = 7.549$$

Step 5. *Make a decision.*

The value of the test statistic $t = 7.549$ is greater than the critical value of $t = 3.365$, and it falls in the rejection region. Hence, we reject the null hypothesis and conclude that x (income) determines y (food expenditure) positively. That is, food expenditure increases with an increase in income and it decreases with a decrease in income. ■

Note that the null hypothesis does not always have to be $B = 0$. We may test the null hypothesis that B is equal to a certain value. See Exercises 13.47 to 13.50, 13.57, and 13.58 for such cases.

EXERCISES

■ **Concepts and Procedures**

13.46 Describe the mean, standard deviation, and shape of the sampling distribution of the slope b of the simple linear regression model.

13.47 The following information is obtained for a sample of 16 observations taken from a population.

$$SS_{xx} = 340.700, \quad s_e = 1.951, \quad \text{and} \quad \hat{y} = 12.45 + 6.32x$$

a. Make a 99% confidence interval for B.
b. Using a significance level of .025, can you conclude that B is positive?
c. Using a significance level of .01, can you conclude that B is different from zero?
d. Using a significance level of .02, test whether B is different from 4.50. (*Hint:* The null hypothesis here will be H_0: $B = 4.50$, and the alternative hypothesis will be H_1: $B \neq 4.50$. Notice that the value of $B = 4.50$ will be used to calculate the value of the test statistic t.)

13.48 The following information is obtained for a sample of 25 observations taken from a population.

$$SS_{xx} = 274.600, \quad s_e = .932, \quad \text{and} \quad \hat{y} = 280.56 - 3.77x$$

a. Make a 95% confidence interval for B.
b. Using a significance level of .01, test whether B is negative.

c. Testing at the 5% significance level, can you conclude that B is different from zero?
d. Test if B is different from -5.20. Use $\alpha = .01$.

13.49 The following information is obtained for a sample of 100 observations taken from a population. (Note that because $n > 30$, we can use the normal distribution to make a confidence interval and test a hypothesis about B.)

$$SS_{xx} = 524.884 \quad s_e = 1.464, \quad \text{and} \quad \hat{y} = 5.48 + 2.50x$$

a. Make a 98% confidence interval for B.
b. Test at the 2% significance level whether B is positive.
c. Can you conclude that B is different from zero? Use $\alpha = .01$.
d. Using a significance level of .01, test whether B is greater than 1.75.

13.50 The following information is obtained for a sample of 80 observations taken from a population.

$$SS_{xx} = 380.592, \quad s_e = .961, \quad \text{and} \quad \hat{y} = 160.24 - 2.70x$$

a. Make a 97% confidence interval for B.
b. Test at the 1% significance level whether B is negative.
c. Can you conclude that B is different from zero? Use $\alpha = .01$.
d. Using a significance level of .02, test whether B is less than -1.25.

■ **Applications**

13.51 Refer to Exercise 13.25. The data on ages (in years) and prices (in hundreds of dollars) for eight cars of a specific model are reproduced from that exercise.

Age	8	3	6	9	2	5	6	3
Price	18	94	50	21	145	42	36	99

a. Construct a 95% confidence interval for B. You can use results obtained in Exercises 13.25 and 13.40 here.
b. Test at the 5% significance level whether B is negative.

13.52 The data given in the table below are the midterm scores in a course for a sample of 10 students and the scores of student evaluations of the instructor. (In the instructor evaluation scores, 1 is the lowest and 4 is the highest score.)

Instructor score	3	2	3	1	2	4	3	4	4	2
Midterm score	90	75	97	64	47	99	75	88	93	81

a. Find the regression of instructor scores on midterm scores.
b. Construct a 99% confidence interval for B.
c. Test at the 1% significance level whether B is positive.

13.53 The following data give the experience (in years) and monthly salaries (in hundreds of dollars) of nine randomly selected secretaries.

Experience	14	3	5	6	4	9	18	5	16
Monthly salary	42	24	33	31	29	39	47	30	43

a. Find the least squares regression line with experience as an independent variable and monthly salary as a dependent variable.
b. Construct a 98% confidence interval for B.
c. Test at the 2.5% significance level whether B is greater than zero.

13.54 The following data on the stress scores before a math test and the math test scores for seven students are reproduced from Exercise 13.26.

Stress score	6.5	4.0	2.5	7.2	8.1	3.4	5.5
Test score	81	96	93	68	63	84	71

a. Make a 95% confidence interval for B. You can use the calculations made in Exercises 13.26 and 13.41 here.
b. Test at the 2.5% significance level whether B is negative.

13.55 The following data set on annual incomes (in thousands of dollars) and amounts (in thousands of dollars) of life insurance policies for six persons is reproduced from Exercise 13.27.

Annual income	52	58	31	43	65	24
Life insurance	250	300	100	150	500	75

a. Construct a 99% confidence interval for B. You can use the calculations made in Exercises 13.27 and 13.42 here.
b. Test at the 1% significance level whether B is different from zero.

13.56 The data on the sizes of six houses (in hundreds of square feet) and the monthly rents (in dollars) paid by tenants for those houses are reproduced from Exercise 13.28.

Size of house	21	13	19	27	34	23
Monthly rent	700	580	720	1000	1250	800

a. Construct a 98% confidence interval for B. You can use the calculations made in Exercises 13.28 and 13.43 here.
b. Testing at the 5% significance level, can you conclude that B is different from zero?

13.57 Refer to Exercise 13.38. The following table gives information on the monthly incomes (in hundreds of dollars) and monthly telephone bills (in dollars) for a random sample of 10 households.

Income	20	54	46	42	36	23	61	26	49	64
Phone bill	45	165	175	90	95	35	210	50	110	190

a. Find the least squares regression line with monthly income as an independent variable and monthly telephone bill as a dependent variable.
b. Make a 95% confidence interval for B. You can use the results obtained in Exercise 13.38 here.
c. An earlier study claims that for the relationship between income and amount of phone bills, $B = 1.50$. Test at the 1% significance level whether B is greater than 1.50. (*Hint:* Here the null hypothesis will be H_0: $B = 1.50$ and the alternative hypothesis will be H_1: $B > 1.50$. Notice that the value of $B = 1.50$ will be used to calculate the value of the test statistic t.)

13.58 The following table, reproduced from Exercise 13.39, gives information on the average saturated fat (in grams) consumed per day and the cholesterol level (in milligrams per hundred milliliters) of eight men.

Fat consumption	55	68	50	34	43	58	77	36
Cholesterol level	180	215	195	165	170	204	235	150

a. Find the regression line $\hat{y} = a + bx$, where x is the fat consumption and y is the cholesterol level. You can use the results obtained in Exercise 13.39 here.
b. Construct a 90% confidence interval for B.
c. An earlier study claims that B is 1.75. Test at the 5% significance level whether B is different from 1.75.

13.6   LINEAR CORRELATION

This section describes the meaning and calculation of the linear correlation coefficient and the procedure to conduct a test of hypothesis about it.

13.6.1   LINEAR CORRELATION COEFFICIENT

Another measure of the relationship between two variables is the correlation coefficient. This section describes the simple linear correlation, for short **linear correlation**, which measures the strength of the linear association between two variables. In other words, the linear correlation coefficient measures how closely the points in a scatter diagram are spread around the regression line. The correlation coefficient calculated for the population data is denoted by ρ (Greek letter *rho*) and the one calculated for sample data is denoted by r. (Note that the square of the correlation coefficient is equal to the coefficient of determination.)

> **VALUE OF THE CORRELATION COEFFICIENT**   The *value of the correlation coefficient* always lies in the range -1 to 1; that is,
>
> $$-1 \leq \rho \leq 1 \quad \text{and} \quad -1 \leq r \leq 1$$

Although we can explain the linear correlation using the population correlation coefficient ρ, we will do so using the sample correlation coefficient r.

If $r = 1$, it is said to be a *perfect positive linear correlation*. In such a case, all points in the scatter diagram lie on a straight line that slopes upward from left to right, as shown in Figure 13.18a. If $r = -1$, the correlation is said to be a *perfect negative linear correlation*. In this case, all points in the scatter diagram fall on a straight line that slopes downward from left to right, as shown in Figure 13.18b. If the points are scattered all over the diagram, as shown in Figure 13.18c, then there is *no linear correlation* between the two variables and consequently r is close to 0.

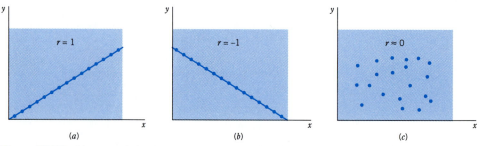

Figure 13.18   Linear correlation between two variables. (*a*) Perfect positive linear correlation, $r = 1$. (*b*) Perfect negative linear correlation, $r = -1$. (*c*) No linear correlation, $r \approx 0$.

We do not usually encounter an example with perfect positive or perfect negative correlation. What we observe in real-world problems is either a positive linear correlation with $0 < r < 1$ (that is, the correlation coefficient is greater than zero but less than 1) or a negative linear correlation with $-1 < r < 0$ (that is, the correlation coefficient is greater than -1 but less than zero).

If the correlation between two variables is positive and close to 1, we say that the variables have a *strong positive linear correlation*. If the correlation between two variables is positive but close to zero, then the variables have a *weak positive linear correlation*. In contrast, if the correlation between two variables is negative and close to -1, then the variables are said to have a *strong negative linear correlation*. If the correlation between variables is negative but close to zero, there exists a *weak negative linear correlation* between the variables. Graphically, a strong correlation indicates that the points in the scatter diagram are very close to the regression line, and a weak correlation indicates that the points in the scatter diagram are widely spread around the regression line. These four cases are shown in Figure 13.19a–d.

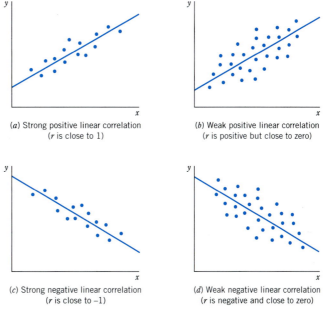

(a) Strong positive linear correlation
(r is close to 1)

(b) Weak positive linear correlation
(r is positive but close to zero)

(c) Strong negative linear correlation
(r is close to –1)

(d) Weak negative linear correlation
(r is negative and close to zero)

Figure 13.19 Linear correlation between variables.

The linear correlation coefficient is calculated by using the following formula. (This correlation coefficient is also called the *Pearson product moment correlation coefficient*.)

> **LINEAR CORRELATION COEFFICIENT** The *simple linear correlation*, denoted by r, measures the strength of the linear relationship between two variables for a sample and is calculated as[6]
>
> $$r = \frac{SS_{xy}}{\sqrt{SS_{xx}\, SS_{yy}}}$$

[6]If we have access to population data, the value of ρ is calculated using the formula

$$\rho = \frac{SS_{xy}}{\sqrt{SS_{xx}\, SS_{yy}}}$$

Here the values of SS_{xy}, SS_{xx}, and SS_{yy} are calculated using the population data.

Because both SS_{xx} and SS_{yy} are always positive, the sign of the correlation coefficient r depends on the sign of SS_{xy}. If SS_{xy} is positive, then r will be positive, and if SS_{xy} is negative, then r will be negative. Another important observation to remember is that *r and b, calculated for the same sample, will always have the same sign. That is, both r and b are either positive or negative.* This is so because both r and b provide information about the relationship between x and y. Likewise, the corresponding population parameters ρ and B will always have the same sign.

Example 13–6 illustrates the calculation of the linear correlation coefficient r.

Calculating the linear correlation coefficient.

EXAMPLE 13–6   Calculate the correlation coefficient for the example on incomes and food expenditures of seven households.

Solution   From earlier calculations made in Examples 13–1 and 13–2,

$$SS_{xy} = 211.7143, \quad SS_{xx} = 801.4286, \quad \text{and} \quad SS_{yy} = 60.8571$$

Substituting these values in the formula for r, we obtain

$$r = \frac{SS_{xy}}{\sqrt{SS_{xx}SS_{yy}}} = \frac{211.7143}{\sqrt{(801.4286)(60.8571)}} = .96$$

Thus, the linear correlation coefficient is .96. The correlation coefficient is usually rounded to two decimal places. ∎

The linear correlation coefficient simply tells us how strongly the two variables are (linearly) related. The correlation coefficient of .96 for incomes and food expenditures of seven households indicates that income and food expenditure are very strongly and positively correlated. This correlation coefficient does not, however, provide us with any more information.

The square of the correlation coefficient gives the coefficient of determination, which was explained in Section 13.4. Thus, $(.96)^2$ is .92, which is the value of r^2 calculated in Example 13–3.

Sometimes the calculated value of r may indicate that the two variables are very strongly linearly correlated but in reality they are not. For example, if we calculate the correlation coefficient between the price of Coke and the size of families in the United States using data for the past 30 years, we will find a strong negative linear correlation. Over time, the price of Coke has increased and the size of families has decreased. This finding does not mean that family size and the price of Coke are related. As a result, before we calculate the correlation coefficient, we must seek help from a theory or from common sense to postulate whether or not the two variables have a causal relationship.

Another point to note is that in a simple regression model, one of the two variables is categorized as an independent variable and the other is classified as a dependent variable. However, no such distinction is made between the two variables when the correlation coefficient is calculated.

13.6.2   HYPOTHESIS TESTING ABOUT THE LINEAR CORRELATION COEFFICIENT

This section describes how to perform a test of hypothesis about the population correlation coefficient ρ using the sample correlation coefficient r. We can use the t distribution to make this test. However, to use the t distribution, both variables should be normally distributed.

Usually (although not always), the null hypothesis is that the linear correlation coefficient between the two variables is zero, that is $\rho = 0$. The alternative hypothesis can be one of the following: (1) the linear correlation coefficient between the two variables is less than zero, that is $\rho < 0$; (2) the linear correlation coefficient between the two variables is greater than zero, that is $\rho > 0$; or (3) the linear correlation coefficient between the two variables is not equal to zero, that is $\rho \neq 0$.

TEST STATISTIC FOR r If both variables are normally distributed and the null hypothesis is H_0: $\rho = 0$, then the value of the test statistic t is calculated as

$$t = r\sqrt{\frac{n-2}{1-r^2}}$$

Here $n - 2$ are the degrees of freedom.

Example 13–7 describes the procedure to perform a test of hypothesis about the linear correlation coefficient.

Performing a test of hypothesis about the correlation coefficient.

EXAMPLE 13-7 Using the 1% level of significance and the data from Example 13–1, test whether the linear correlation coefficient between incomes and food expenditures is positive. Assume that the populations of both variables are normally distributed.

Solution From Examples 13–1 and 13–6:

$$n = 7 \quad \text{and} \quad r = .96$$

Below we use the five steps to perform this test of hypothesis.

Step 1. *State the null and alternative hypotheses.*

We are to test whether the linear correlation coefficient between incomes and food expenditures is positive. Hence, the null and alternative hypotheses are:

$$H_0: \rho = 0 \quad \text{(The linear correlation coefficient is zero)}$$

$$H_1: \rho > 0 \quad \text{(The linear correlation coefficient is positive)}$$

Step 2. *Select the distribution to use.*

The population distributions for both variables are normally distributed. Hence, we can use the t distribution to perform this test about the linear correlation coefficient.

Step 3. *Determine the rejection and nonrejection regions.*

The significance level is 1%. From the alternative hypothesis we know that the test is right-tailed. Hence,

$$\text{Area in the right tail of the } t \text{ distribution} = .01$$

$$df = n - 2 = 7 - 2 = 5$$

From the t distribution table, the critical value of t is 3.365. The rejection and nonrejection regions for this test are shown in Figure 13.20 on page 590.

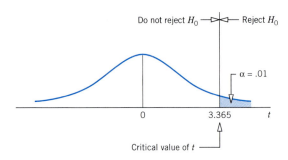

Figure 13.20

Step 4. *Calculate the value of the test statistic.*

The value of the test statistic t for r is calculated as follows:

$$t = r\sqrt{\frac{n-2}{1-r^2}} = .96\sqrt{\frac{7-2}{1-(.96)^2}} = 7.667$$

Step 5. *Make a decision.*

The value of the test statistic $t = 7.667$ is greater than the critical value of $t = 3.365$ and it falls in the rejection region. Hence, we reject the null hypothesis and conclude that there is a positive linear relationship between incomes and food expenditures. ■

EXERCISES

■ **Concepts and Procedures**

13.59   What does a linear correlation coefficient tell about the relationship between two variables? Within what range can a correlation coefficient assume a value?

13.60   What is the difference between ρ and r? Explain.

13.61   Explain each of the following concepts. You may use graphs to illustrate each concept.

a.   Perfect positive linear correlation
b.   Perfect negative linear correlation
c.   Strong positive linear correlation
d.   Strong negative linear correlation
e.   Weak positive linear correlation
f.   Weak negative linear correlation
g.   No linear correlation

13.62   Can the values of B and ρ calculated for the same population data have different signs? Explain.

13.63   For a sample data set, the linear correlation coefficient r has a positive value. Which of the following is true about the slope b of the regression line estimated for the same sample data?

a.   The value of b will be positive.
b.   The value of b will be negative.
c.   The value of b can be positive or negative.

13.64   For a sample data set, the slope b of the regression line has a positive value. Which of the following is true about the linear correlation coefficient r calculated for the same sample data?

a.   The value of r will be positive.
b.   The value of r will be negative.
c.   The value of r can be positive or negative.

13.65   For a sample data set on two variables, the value of the linear correlation coefficient is zero. Does this mean that these variables are not related? Explain.

13.66 Will you expect a positive, zero, or negative linear correlation between the two variables for each of the following examples?

a. Grade of a student and hours spent studying
b. Incomes and entertainment expenditures of households
c. Ages of women and makeup expenses per month
d. Price of a computer and consumption of Coke
e. Price and consumption of wine

13.67 Will you expect a positive, zero, or negative linear correlation between the two variables for each of the following examples?

a. SAT scores and GPAs of students
b. Stress level and blood pressure of individuals
c. Amount of fertilizer used and yield of corn per acre
d. Ages and prices of houses
e. Heights of husbands and incomes of their wives

13.68 A population data set produced the following information.

$$N = 250, \quad \Sigma x = 9880, \quad \Sigma y = 1456, \quad \Sigma xy = 85,080,$$

$$\Sigma x^2 = 485,870, \quad \text{and} \quad \Sigma y^2 = 135,675$$

Find the linear correlation coefficient ρ.

13.69 A population data set produced the following information.

$$N = 460, \quad \Sigma x = 3920, \quad \Sigma y = 2650, \quad \Sigma xy = 26,570,$$

$$\Sigma x^2 = 48,530, \quad \text{and} \quad \Sigma y^2 = 39,347$$

Find the linear correlation coefficient ρ.

13.70 A sample data set produced the following information.

$$n = 10, \quad \Sigma x = 100, \quad \Sigma y = 220, \quad \Sigma xy = 3680,$$

$$\Sigma x^2 = 1140, \quad \text{and} \quad \Sigma y^2 = 25,272$$

a. Calculate the linear correlation coefficient r.
b. Using the 2% significance level, can you conclude that ρ is different from zero?

13.71 A sample data set produced the following information.

$$n = 12, \quad \Sigma x = 66, \quad \Sigma y = 588, \quad \Sigma xy = 2244,$$

$$\Sigma x^2 = 396, \quad \text{and} \quad \Sigma y^2 = 58,734$$

a. Calculate the linear correlation coefficient r.
b. Using the 1% significance level, can you conclude that ρ is negative?

■ **Applications**

13.72 Refer to Exercise 13.25. The data on ages (in years) and prices (in hundreds of dollars) for eight cars of a specific model are reproduced from that exercise.

Age	8	3	6	9	2	5	6	3
Price	18	94	50	21	145	42	36	99

a. Do you expect the ages and prices of cars to be positively or negatively related? Explain.
b. Calculate the linear correlation coefficient.
c. Test at the 2.5% significance level whether ρ is negative.

13.73 The following table, reproduced from Exercise 13.53, gives the experience (in years) and monthly salaries (in hundreds of dollars) of nine randomly selected secretaries.

Experience	14	3	5	6	4	9	18	5	16
Monthly salary	42	24	33	31	29	39	47	30	43

a. Do you expect the experience and monthly salaries to be positively or negatively related? Explain.
b. Compute the linear correlation coefficient.
c. Test at the 5% significance level whether ρ is positive.

13.74 The following table lists the midterm and final exam scores for seven students in a statistics class.

Midterm score	79	95	81	66	87	94	59
Final exam score	85	97	78	76	94	84	67

a. Do you expect the midterm and final exam scores to be positively or negatively related?
b. Plot a scatter diagram. By looking at the scatter diagram, do you expect the correlation coefficient between these two variables to be close to zero, 1, or −1?
c. Find the correlation coefficient. Is the value of r consistent with what you expected in parts a and b?
d. Using the 1% significance level, test whether the linear correlation coefficient is positive.

13.75 The following data give the ages of husbands and wives for six couples.

Husband's age	43	57	28	19	35	39
Wife's age	37	51	32	20	33	38

a. Do you expect the ages of husbands and wives to be positively or negatively related?
b. Plot a scatter diagram. By looking at the scatter diagram, do you expect the correlation coefficient between these two variables to be close to zero, 1, or −1?
c. Find the correlation coefficient. Is the value of r consistent with what you expected in parts a and b?
d. Using the 5% significance level, test whether the correlation coefficient is different from zero.

13.76 The following data on stress scores before a math test and math test scores for seven students are reproduced from Exercise 13.26.

Stress score	6.5	4.0	2.5	7.2	8.1	3.4	5.5
Test score	81	96	93	68	63	84	71

a. Find the correlation coefficient. Is the sign of the correlation coefficient the same as that of b calculated in Exercise 13.26?
b. Test at the 2.5% significance level whether the linear correlation coefficient is negative. Is your decision the same as in the test of B in Exercise 13.54?

13.77 The following table, reproduced from Exercise 13.58, gives information on average saturated fat (in grams) consumed per day and cholesterol level (in milligrams per hundred milliliters) of eight men.

Fat consumption	55	68	50	34	43	58	77	36
Cholesterol level	180	215	195	165	170	204	235	150

a. Find the correlation coefficient. Is the sign of the correlation coefficient the same as that of b calculated in Exercise 13.58?
b. Test at the 1% significance level whether ρ is different from zero.

13.78 Refer to data given in Exercise 13.29 on the total 1999 payroll and the percentage of games won during the 1999 season by each of the National League baseball teams. Compute the linear correlation coefficient, ρ.

13.79 Refer to data given in Exercise 13.30 on the total 1999 payroll and the percentage of games won during the 1999 season by each of the American League baseball teams. Compute the linear correlation coefficient, ρ.

13.7 REGRESSION ANALYSIS: A COMPLETE EXAMPLE

This section works out an example that includes all the topics we have discussed so far in this chapter.

A complete example of regression analysis.

EXAMPLE 13–8   A random sample of eight auto drivers insured with a company and having similar auto insurance policies was selected. The following table lists their driving experience (in years) and monthly auto insurance premium.

Driving Experience (years)	Monthly Auto Insurance Premium
5	$64
2	87
12	50
9	71
15	44
6	56
25	42
16	60

(a)   Does the insurance premium depend on the driving experience or does the driving experience depend on the insurance premium? Do you expect a positive or a negative relationship between these two variables?

(b)   Compute SS_{xx}, SS_{yy}, and SS_{xy}.

(c)   Find the least squares regression line by choosing appropriate dependent and independent variables based on your answer in part a.

(d)   Interpret the meaning of the values of a and b calculated in part c.

(e)   Plot the scatter diagram and the regression line.

(f)   Calculate r and r^2 and explain what they mean.

(g)   Predict the monthly auto insurance premium for a driver with 10 years of driving experience.

(h)   Compute the standard deviation of errors.

(i)   Construct a 90% confidence interval for B.

(j)   Test at the 5% significance level whether B is negative.

(k)   Using $\alpha = .05$, test whether ρ is different from zero.

Solution

(a)   Based on theory and intuition, we expect the insurance premium to depend on driving experience. Consequently, the insurance premium is a dependent variable and driving experience is an independent variable in the regression model. A new driver is considered a high risk by the insurance companies, and he or she has to pay a higher premium for auto insurance. On average, the insurance premium is expected to decrease with an increase in the years of driving experience. Therefore, we expect a negative relationship between these two variables. In other words, both the population correlation coefficient ρ and the population regression slope B are expected to be negative.

(b) Table 13.5 shows the calculation of Σx, Σy, Σxy, Σx^2, and Σy^2.

Table 13.5

Experience x	Premium y	xy	x^2	y^2
5	64	320	25	4096
2	87	174	4	7569
12	50	600	144	2500
9	71	639	81	5041
15	44	660	225	1936
6	56	336	36	3136
25	42	1050	625	1764
16	60	960	256	3600
$\Sigma x = 90$	$\Sigma y = 474$	$\Sigma xy = 4739$	$\Sigma x^2 = 1396$	$\Sigma y^2 = 29{,}642$

The values of $\bar{x}$ and $\bar{y}$ are

$$\bar{x} = \Sigma x/n = 90/8 = 11.25$$

$$\bar{y} = \Sigma y/n = 474/8 = 59.25$$

The values of SS_{xy}, SS_{xx}, and SS_{yy} are computed as follows:

$$SS_{xy} = \Sigma xy - \frac{(\Sigma x)(\Sigma y)}{n} = 4739 - \frac{(90)(474)}{8} = -593.5000$$

$$SS_{xx} = \Sigma x^2 - \frac{(\Sigma x)^2}{n} = 1396 - \frac{(90)^2}{8} = 383.5000$$

$$SS_{yy} = \Sigma y^2 - \frac{(\Sigma y)^2}{n} = 29{,}642 - \frac{(474)^2}{8} = 1557.5000$$

(c) To find the regression line, we calculate a and b as follows:

$$b = \frac{SS_{xy}}{SS_{xx}} = \frac{-593.5000}{383.5000} = -1.5476$$

$$a = \bar{y} - b\bar{x} = 59.25 - (-1.5476)(11.25) = 76.6605$$

Thus, our estimated regression line $\hat{y} = a + bx$ is

$$\hat{y} = 76.6605 - 1.5476x$$

(d) The value of $a = 76.6605$ gives the value of $\hat{y}$ for $x = 0$; that is, it gives the monthly auto insurance premium for a driver with no driving experience. However, as mentioned earlier in this chapter, we should not attach much importance to this statement because the sample contains drivers with only two or more years of experience. The value of b gives the change in $\hat{y}$ due to a change of one unit in x. Thus, $b = -1.5476$ indicates that, on average, for every extra year of driving experience, the monthly auto insurance premium decreases by \$1.55. Note that when b is negative, y decreases as x increases.

(e) Figure 13.21 shows the scatter diagram and the regression line for the data on eight auto drivers. Note that the regression line slopes downward from left to right. This result is consistent with the negative relationship we anticipated between driving experience and insurance premium.

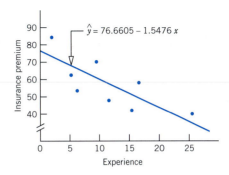

Figure 13.21

(f) The values of r and r^2 are computed as follows:

$$r = \frac{SS_{xy}}{\sqrt{SS_{xx}\,SS_{yy}}} = \frac{-593.5000}{\sqrt{(383.5000)(1557.5000)}} = -.77$$

$$r^2 = \frac{b\,SS_{xy}}{SS_{yy}} = \frac{(-1.5476)(-593.5000)}{1557.5000} = .59$$

The value of $r = -.77$ indicates that the driving experience and the monthly auto insurance premium are negatively related. The (linear) relationship is strong but not very strong. The value of $r^2 = .59$ states that 59% of the total variation in insurance premiums is explained by years of driving experience and 41% is not. The low value of r^2 indicates that there may be many other important variables that contribute to the determination of auto insurance premiums. For example, the premium is expected to depend on the driving record of a driver and the type and age of the car.

(g) Using the estimated regression line, we find the predicted value of y for $x = 10$ is

$$\hat{y} = 76.6605 - 1.5476x = 76.6605 - 1.5476(10) = \textbf{\$61.18}$$

Thus, we expect the monthly auto insurance premium of a driver with 10 years of driving experience to be \$61.18.

(h) The standard deviation of errors is

$$s_e = \sqrt{\frac{SS_{yy} - b\,SS_{xy}}{n-2}} = \sqrt{\frac{1557.5000 - (-1.5476)(-593.5000)}{8-2}} = \textbf{10.3199}$$

(i) To construct a 90% confidence interval for B, first we calculate the standard deviation of b:

$$s_b = \frac{s_e}{\sqrt{SS_{xx}}} = \frac{10.3199}{\sqrt{383.5000}} = .5270$$

For a 90% confidence level, the area in each tail of the t distribution is

$$\alpha/2 = .5 - (.90/2) = .05$$

The degrees of freedom are

$$df = n - 2 = 8 - 2 = 6$$

From the t distribution table, the t value for .05 area in the right tail of the t distribution and 6 df is 1.943. The 90% confidence interval for B is

$$b \pm ts_b = -1.5476 \pm 1.943(.5270)$$

$$= -1.5476 \pm 1.0240 = \mathbf{-2.57 \text{ to } -.52}$$

Thus, we can state with 90% confidence that B lies in the interval -2.57 to $-.52$. That is, on average, the monthly auto insurance premium of a driver decreases by an amount between \$.52 and \$2.57 for every extra year of driving experience.

(j)   We perform the following five steps to test the hypothesis about B.

Step 1.   *State the null and alternative hypotheses.*

The null and alternative hypotheses are written as follows:

$$H_0: B = 0 \qquad (B \text{ is not negative})$$

$$H_1: B < 0 \qquad (B \text{ is negative})$$

Note that the null hypothesis can also be written as $H_0: B \geq 0$.

Step 2.   *Select the distribution to use.*

Because the sample size is small ($n < 30$) and σ_ϵ is not known, we use the t distribution to make the hypothesis test.

Step 3.   *Determine the rejection and nonrejection regions.*

The significance level is .05. The $<$ sign in the alternative hypothesis indicates that it is a left-tailed test.

$$\text{Area in the left tail of the } t \text{ distribution} = \alpha = .05$$

$$df = n - 2 = 8 - 2 = 6$$

From the t distribution table, the critical value of t for .05 area in the left tail of the t distribution and 6 df is -1.943, as shown in Figure 13.22.

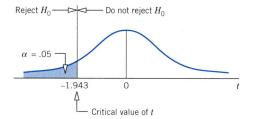

Figure 13.22

Step 4.   *Calculate the value of the test statistic.*

The value of the test statistic t for b is calculated as follows:

$$t = \frac{b - B}{s_b} = \frac{-1.5476 - \overset{\text{From } H_0}{0}}{.5270} = -2.937$$

Step 5. *Make a decision.*

The value of the test statistic $t = -2.937$ falls in the rejection region. Hence, we reject the null hypothesis and conclude that B is negative. That is, the monthly auto insurance premium decreases with an increase in years of driving experience.

(k) We perform the following five steps to test the hypothesis about the linear correlation coefficient ρ.

Step 1. *State the null and alternative hypotheses.*

The null and alternative hypotheses are:

$$H_0: \rho = 0 \quad \text{(The linear correlation coefficient is zero)}$$

$$H_1: \rho \neq 0 \quad \text{(The linear correlation coefficient is different from zero)}$$

Step 2. *Select the distribution to use.*

We will use the t distribution to perform this test about the linear correlation coefficient.

Step 3. *Determine the rejection and nonrejection regions.*

The significance level is 5%. From the alternative hypothesis we know that the test is two-tailed. Hence,

$$\text{Area in each tail of the } t \text{ distribution} = .05/2 = .025$$

$$df = n - 2 = 8 - 2 = 6$$

From the t distribution table, the critical values of t are -2.447 and 2.447. The rejection and nonrejection regions for this test are shown in Figure 13.23.

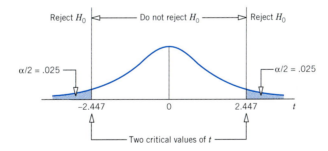

Figure 13.23

Step 4. *Calculate the value of the test statistic.*

The value of the test statistic t for r is calculated as follows:

$$t = r\sqrt{\frac{n-2}{1-r^2}} = (-.77)\sqrt{\frac{8-2}{1-(-.77)^2}} = -2.956$$

Step 5. *Make a decision.*

The value of the test statistic $t = -2.956$ falls in the rejection region. Hence, we reject the null hypothesis and conclude that the linear correlation coefficient between driving experience and auto insurance premium is different from zero. ■

EXERCISES

■ **Applications**

13.80 Eight students, randomly selected from a large class, were asked to keep a record of the hours they spent studying before the midterm examination. The following table gives the numbers of hours these eight students studied before the midterm and their scores on the midterm.

Hours studied	15	7	12	8	18	6	9	11
Midterm score	97	78	87	92	93	57	74	67

a. Do the midterm scores depend on hours studied or do hours studied depend on the midterm scores? Do you expect a positive or a negative relationship between these two variables?

b. Taking hours studied as an independent variable and midterm scores as a dependent variable, compute SS_{xx}, SS_{yy}, and SS_{xy}.

c. Find the least squares regression line.

d. Interpret the meaning of the values of a and b calculated in part c.

e. Plot the scatter diagram and the regression line.

f. Calculate r and r^2 and briefly explain what they mean.

g. Predict the midterm score of a student who studied for 14 hours.

h. Compute the standard deviation of errors.

i. Construct a 99% confidence interval for B.

j. Test at the 1% significance level whether B is positive.

k. Using $\alpha = .01$, test whether ρ is positive.

13.81 The following table gives information on ages and cholesterol levels for a random sample of 10 men.

Age	58	69	43	39	63	52	47	31	74	36
Cholesterol level	189	235	193	177	154	191	213	165	198	181

a. Taking age as an independent variable and cholesterol level as a dependent variable, compute SS_{xx}, SS_{yy}, and SS_{xy}.

b. Find the regression of cholesterol level on age.

c. Briefly explain the meaning of the values of a and b calculated in part b.

d. Calculate r and r^2 and explain what they mean.

e. Plot the scatter diagram and the regression line.

f. Predict the cholesterol level of a 60-year-old man.

g. Compute the standard deviation of errors.

h. Construct a 95% confidence interval for B.

i. Test at the 5% significance level if B is positive.

j. Using $\alpha = .025$, can you conclude that the linear correlation coefficient is positive?

13.82 A farmer wanted to find the relationship between the amount of fertilizer used and the yield of corn. He selected seven acres of his land on which he used different amounts of fertilizer to grow corn. The following table gives the amount (in pounds) of fertilizer used and the yield (in bushels) of corn for each of the seven acres.

Fertilizer Used	Yield of Corn
120	142
80	112
100	132
70	96
88	119
75	104
110	136

a. With the amount of fertilizer used as an independent variable and yield of corn as a dependent variable, compute SS_{xx}, SS_{yy}, and SS_{xy}.
b. Find the least squares regression line.
c. Interpret the meaning of the values of a and b calculated in part b.
d. Calculate r and r^2 and explain what they mean.
e. Compute the standard deviation of errors.
f. Predict the yield of corn per acre for $x = 105$.
g. Construct a 98% confidence interval for B.
h. Test at the 5% significance level whether B is different from zero.
i. Using $\alpha = .05$, can you conclude that ρ is different from zero?

13.83 The following table gives information on the incomes (in thousands of dollars) and charitable contributions (in hundreds of dollars) for the past year for a random sample of 10 households.

Income (thousands)	Charitable Contributions (hundreds)
$36	$10
26	4
82	29
52	23
32	3
71	28
30	8
42	16
60	18
20	1

a. With income as an independent variable and charitable contributions as a dependent variable, compute SS_{xx}, SS_{yy}, and SS_{xy}.
b. Find the regression of charitable contributions on income.
c. Briefly explain the meaning of the values of a and b.
d. Calculate r and r^2 and briefly explain what they mean.
e. Compute the standard deviation of errors.
f. Construct a 99% confidence interval for B.
g. Test at the 1% significance level whether B is positive.
h. Using the 1% significance level, can you conclude that the linear correlation coefficient is different from zero?

13.84 The following data give information on the lowest cost ticket price (in dollars) and the average attendance (rounded to the nearest thousand) for the past year for six football teams.

Ticket price	28.50	16.50	24.00	35.50	39.50	26.00
Attendance	56	65	71	69	55	42

a. Taking ticket price as an independent variable and attendance as a dependent variable, compute SS_{xx}, SS_{yy}, and SS_{xy}.
b. Find the least squares regression line.
c. Briefly explain the meaning of the values of a and b calculated in part b.
d. Calculate r and r^2 and briefly explain what they mean.
e. Compute the standard deviation of errors.
f. Construct a 90% confidence interval for B.
g. Test at the 2.5% significance level whether B is negative.
h. Using the 2.5% significance level, test whether ρ is negative.

13.85 The following table gives information on GPAs (grade point averages) and starting salaries (rounded to the nearest thousand dollars) of seven recent college graduates.

GPA	2.90	3.81	3.20	2.42	3.94	2.05	2.25
Starting salary	28	38	28	25	40	21	27

a. With GPA as an independent variable and starting salary as a dependent variable, compute SS_{xx}, SS_{yy}, and SS_{xy}.
b. Find the least squares regression line.
c. Interpret the meaning of the values of a and b calculated in part b.
d. Calculate r and r^2 and briefly explain what they mean.
e. Compute the standard deviation of errors.
f. Construct a 95% confidence interval for B.
g. Test at the 1% significance level whether B is different from zero.
h. Test at the 1% significance level whether ρ is positive.

13.8 USING THE REGRESSION MODEL

Let us return to the example on incomes and food expenditures to discuss two major uses of a regression model:

1. Estimating the mean value of y for a given value of x. For instance, we can use our food expenditure regression model to estimate the mean food expenditure of all households with a specific income (say, $3500 per month).

2. Predicting a particular value of y for a given value of x. For instance, we can determine the expected food expenditure of a randomly selected household with a particular monthly income (say, $3500) using our food expenditure regression model.

13.8.1 USING THE REGRESSION MODEL FOR ESTIMATING THE MEAN VALUE OF y

Our population regression model is

$$y = A + Bx + \epsilon$$

As mentioned earlier in this chapter, the mean value of y for a given x is denoted by $\mu_{y|x}$, read as "the mean value of y for a given value of x." Because of the assumption that the mean value of ϵ is zero, the mean value of y is given by

$$\mu_{y|x} = A + Bx$$

Our objective is to estimate this mean value. The value of $\hat{y}$, obtained from the sample regression line by substituting the value of x, is the *point estimate of* $\mu_{y|x}$ for that x.

For our example on incomes and food expenditures, the estimated sample regression line is (see Example 13–1)

$$\hat{y} = 1.1414 + .2642x$$

Suppose we want to estimate the mean food expenditure for all households with a monthly income of $3500. We will denote this population mean by $\mu_{y|x=35}$ or $\mu_{y|35}$. Note that we have written $x = 35$ and not $x = 3500$ in $\mu_{y|35}$ because the units of measurement for the data used to estimate the above regression line in Example 13–1 were hundreds of dollars. Using the regression line, we find that the point estimate of $\mu_{y|35}$ is

$$\hat{y} = 1.1414 + .2642(35) = \$10.3884 \text{ hundred}$$

Thus, based on the sample regression line, the point estimate for the mean food expenditure $\mu_{y|35}$ for all households with a monthly income of $3500 is $1038.84 per month.

However, suppose we take a second sample of seven households from the same population and estimate the regression line for this sample. The point estimate of $\mu_{y|35}$ obtained from the regression line for the second sample is expected to be different. All possible samples of the same size taken from the same population will give different regression lines as shown in Figure 13.24, and, consequently, a different point estimate of $\mu_{y|x}$. Therefore, a confidence interval constructed for $\mu_{y|x}$ based on one sample will give a more reliable estimate of $\mu_{y|x}$ than will a point estimate.

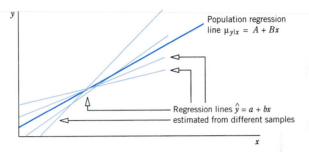

Figure 13.24 Population and sample regression lines.

To construct a confidence interval for $\mu_{y|x}$, we must know the mean, the standard deviation, and the shape of the sampling distribution of its point estimator $\hat{y}$.

The point estimator $\hat{y}$ of $\mu_{y|x}$ is normally distributed with a mean of $A + Bx$ and a standard deviation of

$$\sigma_{\hat{y}_m} = \sigma_\epsilon \sqrt{\frac{1}{n} + \frac{(x_0 - \bar{x})^2}{SS_{xx}}}$$

where $\sigma_{\hat{y}_m}$ is the standard deviation of $\hat{y}$ when it is used to estimate $\mu_{y|x}$, x_0 is the value of x for which we are estimating $\mu_{y|x}$, and σ_ϵ is the population standard deviation of ϵ.

However, usually σ_ϵ is not known. Rather, it is estimated by the standard deviation of sample errors s_e. In this case, we replace σ_ϵ by s_e and $\sigma_{\hat{y}_m}$ by $s_{\hat{y}_m}$ in the foregoing expression. To make a confidence interval for $\mu_{y|x}$, we use the normal distribution when the sample size is large ($n \geq 30$) and the t distribution when the sample size is small ($n < 30$).

CONFIDENCE INTERVAL FOR $\mu_{y|x}$

The $(1 - \alpha)100\%$ *confidence interval for $\mu_{y|x}$* for $x = x_0$ is

$$\hat{y} \pm t s_{\hat{y}_m}$$

where the value of t is obtained from the t distribution table for $\alpha/2$ area in the right tail of the t distribution curve and $df = n - 2$. The value of $s_{\hat{y}_m}$ is calculated as follows:

$$s_{\hat{y}_m} = s_e \sqrt{\frac{1}{n} + \frac{(x_0 - \bar{x})^2}{SS_{xx}}}$$

Example 13–9 illustrates how to make a confidence interval for the mean value of y, $\mu_{y|x}$.

EXAMPLE 13-9   Refer to Example 13–1 on incomes and food expenditures. Find a 99% confidence interval for the mean food expenditure for all households with a monthly income of $3500.

Solution   Using the regression line estimated in Example 13–1, we find the point estimate of the mean food expenditure for $x = 35$ is

$$\hat{y} = 1.1414 + .2642(35) = \$10.3884 \text{ hundred}$$

The confidence level is 99%. Hence, the area in each tail of the t distribution is

$$\alpha/2 = .5 - (.99/2) = .005$$

The degrees of freedom are

$$df = n - 2 = 7 - 2 = 5$$

From the t distribution table, the t value for .005 area in the right tail of the t distribution and 5 df is 4.032. From calculations in Examples 13–1 and 13–2, we know that

$$s_e = .9922, \quad \bar{x} = 30.2857, \quad \text{and} \quad SS_{xx} = 801.4286$$

The standard deviation of $\hat{y}$ as an estimate of $\mu_{y|x}$ for $x = 35$ is calculated as follows:

$$s_{\hat{y}_m} = s_e \sqrt{\frac{1}{n} + \frac{(x_0 - \bar{x})^2}{SS_{xx}}} = (.9922) \sqrt{\frac{1}{7} + \frac{(35 - 30.2857)^2}{801.4286}} = .4098$$

Hence, the 99% confidence interval for $\mu_{y|35}$ is

$$\hat{y} \pm ts_{\hat{y}_m} = 10.3884 \pm 4.032(.4098)$$

$$= 10.3884 \pm 1.6523 = \textbf{8.7361 to 12.0407}$$

Thus, with 99% confidence we can state that the mean food expenditure for all households with a monthly income of $3500 is between $873.61 and $1204.07.   ■

13.8.2   USING THE REGRESSION MODEL FOR PREDICTING A PARTICULAR VALUE OF *y*

The second major use of a regression model is to predict a particular value of y for a given value of x—say, x_0. For example, we may want to predict the food expenditure of a randomly selected household with a monthly income of $3500. In this case, we are not interested in the mean food expenditure of all households with a monthly income of $3500 but in the food expenditure of one particular household with a monthly income of $3500. This predicted value of y is denoted by y_p. Again, to predict a single value of y for $x = x_0$ from the estimated sample regression line, we use the value of $\hat{y}$ as *a point estimate of* y_p. Using the estimated regression line, we find that $\hat{y}$ for $x = 35$ is

$$\hat{y} = 1.1414 + .2642(35) = \$10.3884 \text{ hundred}$$

Thus, based on our regression line, the point estimate for the food expenditure of a given household with a monthly income of $3500 is $1038.84 per month. Note that $\hat{y} = 1038.84$ is the point estimate for the mean food expenditure for all households with $x = 35$ as well as for the predicted value of food expenditure of one household with $x = 35$.

Different regression lines estimated by using different samples of seven households each taken from the same population will give different values of the point estimator for the predicted value of y for $x = 35$. Hence, a confidence interval constructed for y_p based on one sample will give a more reliable estimate of y_p than will a point estimate. The confidence interval constructed for y_p is more commonly called a **prediction interval**.

The procedure to construct a prediction interval for y_p is similar to that for constructing a confidence interval for $\mu_{y|x}$ except that the standard deviation of $\hat{y}$ is larger when we predict a single value of y than when we estimate $\mu_{y|x}$.

The point estimator $\hat{y}$ of y_p is normally distributed with a mean of $A + Bx$ and a standard deviation of

$$\sigma_{\hat{y}_p} = \sigma_\epsilon \sqrt{1 + \frac{1}{n} + \frac{(x_0 - \bar{x})^2}{SS_{xx}}}$$

where $\sigma_{\hat{y}_p}$ is the standard deviation of the predicted value of y, x_0 is the value of x for which we are predicting y, and σ_ϵ is the population standard deviation of ϵ.

However, usually σ_ϵ is not known. In this case, we replace σ_ϵ by s_e and $\sigma_{\hat{y}_p}$ by $s_{\hat{y}_p}$ in the foregoing expression. To make a prediction interval for y_p, we use the normal distribution when the sample size is large ($n \geq 30$) and the t distribution when the sample size is small ($n < 30$).

PREDICTION INTERVAL FOR y_p The $(1 - \alpha)100\%$ *prediction interval* for the predicted value of y, denoted by y_p, for $x = x_0$ is

$$\hat{y} \pm t s_{\hat{y}_p}$$

where the value of t is obtained from the t distribution table for $\alpha/2$ area in the right tail of the t distribution curve and $df = n - 2$. The value of $s_{\hat{y}_p}$ is calculated as follows:

$$s_{\hat{y}_p} = s_e \sqrt{1 + \frac{1}{n} + \frac{(x_0 - \bar{x})^2}{SS_{xx}}}$$

Example 13–10 illustrates the procedure to make a prediction interval for a particular value of y.

Making a prediction interval for a particular value of y.

EXAMPLE 13–10 Refer to Example 13–1 on incomes and food expenditures. Find a 99% prediction interval for the predicted food expenditure for a randomly selected household with a monthly income of $3500.

Solution Using the regression line estimated in Example 13–1, we find the point estimate of the predicted food expenditure for $x = 35$:

$$\hat{y} = 1.1414 + .2642(35) = \$10.3884 \text{ hundred}$$

The area in each tail of the t distribution for a 99% confidence level is

$$\alpha/2 = .5 - (.99/2) = .005$$

The degrees of freedom are

$$df = n - 2 = 7 - 2 = 5$$

From the t distribution table, the t value for .005 area in the right tail of the t distribution curve and 5 df is 4.032. From calculations in Examples 13–1 and 13–2,

$$s_e = .9922, \quad \bar{x} = 30.2857, \quad \text{and} \quad SS_{xx} = 801.4286$$

The standard deviation of $\hat{y}$ as an estimator of y_p for $x = 35$ is calculated as follows:

$$s_{\hat{y}_p} = s_e \sqrt{1 + \frac{1}{n} + \frac{(x_0 - \bar{x})^2}{SS_{xx}}}$$

$$= (.9922) \sqrt{1 + \frac{1}{7} + \frac{(35 - 30.2857)^2}{801.4286}} = 1.0735$$

Hence, the 99% prediction interval for y_p for $x = 35$ is

$$\hat{y} \pm t s_{\hat{y}_p} = 10.3884 \pm 4.032(1.0735)$$

$$= 10.3884 \pm 4.3284 = \textbf{6.0600 to 14.7168}$$

Thus, with 99% confidence we can state that the predicted food expenditure of a household with a monthly income of $3500 is between $606.00 and $1471.68. ∎

As we can observe, this interval is much wider than the one for the mean value of y for $x = 35$ calculated in Example 13–9, which was $873.61 to $1204.07. This is always true. The prediction interval for predicting a single value of y is always larger than the confidence interval for estimating the mean value of y for a certain value of x.

13.9 CAUTIONS IN USING REGRESSION

When carefully applied, regression is a very helpful technique for making predictions and estimations about one variable for a certain value of another variable. However, we need to be cautious while using the regression analysis, for it can give us misleading results and predictions. The following are the two most important points to remember when using regression.

Extrapolation

The regression line estimated for the sample data is true only for the range of x values observed in the sample. For example, the values of x in our example on incomes and food expenditures vary from a minimum of 15 to a maximum of 49. Hence, our estimated regression line is applicable only for values of x between 15 and 49; that is, we should use this regression line to estimate the mean food expenditure or to predict the food expenditure of a single household only for income levels between $1500 and $4900. If we estimate or predict y for a value of x either less than 15 or greater than 49, it is called *extrapolation*. This does not mean that we should never use the regression line for extrapolation. Instead, we should interpret such predictions cautiously and not attach much value to them.

Similarly, if the data used for the regression estimation are time-series data, the predicted values of y for periods outside the time interval used for the estimation of the regression line should be interpreted very cautiously. When using the estimated regression line for extrapolation, we are assuming that the same linear relationship between the two variables holds true for values of x outside the given range. It is possible that the relationship between the two variables may not be linear outside that range. Nonetheless, even if it is linear, adding a few more observations at either end will probably give a new estimation of the regression line.

Causality

The regression line does not prove causality between two variables; that is, it does not predict that a change in y is *caused* by a change in x. The information about causality is based on theory or common sense. A regression line describes only whether or not a significant quantitative relationship between x and y exists. Significant relationship means that we reject the null hypothesis H_0: $B = 0$ at a given significance level. The estimated regression line gives the change in y due to a change of one unit in x. Note that it does not indicate that the reason y has changed is that x has changed. In our example on incomes and food expenditures, it is economic theory and common sense, not the regression line, that tell us that food expenditure depends on income. The regression analysis simply helps to determine whether or not this dependence is significant.

EXERCISES

■ Concepts and Procedures

13.86 Briefly explain the difference between estimating the mean value of y and predicting a particular value of y using a regression model.

13.87 Construct a 99% confidence interval for the mean value of y and a 99% prediction interval for the predicted value of y for the following.

a. $\hat{y} = 3.25 + .80x$ for $x = 15$ given $s_e = .954$, $\bar{x} = 18.52$, $SS_{xx} = 144.65$, and $n = 10$
b. $\hat{y} = -27 + 7.67x$ for $x = 12$ given $s_e = 2.46$, $\bar{x} = 13.43$, $SS_{xx} = 369.77$, and $n = 10$

13.88 Construct a 95% confidence interval for the mean value of y and a 95% prediction interval for the predicted value of y for the following.

a. $\hat{y} = 13.40 + 2.58x$ for $x = 8$ given $s_e = 1.29$, $\bar{x} = 11.30$, $SS_{xx} = 210.45$, and $n = 12$
b. $\hat{y} = -8.6 + 3.72x$ for $x = 24$ given $s_e = 1.89$, $\bar{x} = 19.70$, $SS_{xx} = 315.40$, and $n = 10$

■ Applications

13.89 Refer to Exercise 13.53. Construct a 90% confidence interval for the mean monthly salary of all secretaries with 10 years of experience. Construct a 90% prediction interval for the monthly salary of a randomly selected secretary with 10 years of experience.

13.90 Refer to the data on hours studied and the midterm scores of eight students given in Exercise 13.80. Construct a 99% confidence interval for $\mu_{y|x}$ for $x = 13$ and a 99% prediction interval for y_p for $x = 13$.

13.91 Refer to Exercise 13.82. Construct a 99% confidence interval for the mean yield of corn per acre for all acres on which 90 pounds of fertilizer are used. Determine a 99% prediction interval for the yield of corn for a randomly selected acre on which 90 pounds of fertilizer are used.

13.92 Using the data on ages and cholesterol levels of 10 men given in Exercise 13.81, find a 95% confidence interval for the mean cholesterol level for all 53-year-old men. Make a 95% prediction interval for the cholesterol level for a randomly selected 53-year-old man.

13.93 Refer to Exercise 13.83. Construct a 95% confidence interval for the mean charitable contributions made by all households with an income of \$64,000. Make a 95% prediction interval for the charitable contributions made by a randomly selected household with an income of \$64,000.

13.94 Refer to Exercise 13.85. Construct a 98% confidence interval for the mean starting salary of recent college graduates with a GPA of 3.15. Construct a 98% prediction interval for the starting salary of a randomly selected recent college graduate with a GPA of 3.15.

GLOSSARY

Coefficient of determination A measure that gives the proportion (or percentage) of the total variation in a dependent variable that is explained by a given independent variable.

Degrees of freedom for a simple linear regression model Sample size minus 2; that is, $n - 2$.

Dependent variable The variable to be predicted or explained.

Deterministic model   A model in which the independent variable determines the dependent variable exactly. Such a model gives an **exact relationship** between two variables.

Estimated or **predicted value of y**   The value of the dependent variable, denoted by $\hat{y}$, that is calculated for a given value of x using the estimated regression model.

Independent or **explanatory variable**   The variable included in a model to explain the variation in the dependent variable.

Least squares estimates of A and B   The values of a and b that are calculated by using the sample data.

Least squares method   The method used to fit a regression line through a scatter diagram such that the error sum of squares is minimum.

Least squares regression line   A regression line obtained by using the least squares method.

Linear correlation coefficient   A measure of the strength of the linear relationship between two variables.

Linear regression model   A regression model that gives a straight-line relationship between two variables.

Multiple regression model   A regression model that contains two or more independent variables.

Negative relationship between two variables   The value of the slope in the regression line and the correlation coefficient between two variables are both negative.

Nonlinear (simple) regression model   A regression model that does not give a straight-line relationship between two variables.

Population parameters for a simple regression model   The values of A and B for the regression model $y = A + Bx + \epsilon$ that are obtained by using population data.

Positive relationship between two variables   The value of the slope in the regression line and the correlation coefficient between two variables are both positive.

Prediction interval   The confidence interval for a particular value of y for a given value of x.

Probabilistic or **statistical model**   A model in which the independent variable does not determine the dependent variable exactly.

Random error term (ϵ)   The difference between the actual and predicted values of y.

Scatter diagram or **scattergram**   A plot of the paired observations of x and y.

Simple linear regression   A regression model with one dependent and one independent variable that assumes a straight-line relationship.

Slope   The coefficient of x in a regression model that gives the change in y for a change of one unit in x.

SSE (error sum of squares)   The sum of the squared differences between the actual and predicted values of y. It is the portion of the SST that is not explained by the regression model.

SSR (regression sum of squares)   The portion of the SST that is explained by the regression model.

SST (total sum of squares)   The sum of the squared differences between actual y values and $\bar{y}$.

Standard deviation of errors   A measure of spread for the random errors.

y-intercept   The point at which the regression line intersects the vertical axis on which the dependent variable is marked. It is the value of y when x is zero.

KEY FORMULAS

1. **Simple linear regression model**

$$y = A + Bx + \epsilon$$

2. **Estimated simple linear regression model**

$$\hat{y} = a + bx$$

3. **Least squares estimates of A and B**

$$b = \frac{SS_{xy}}{SS_{xx}} \quad \text{and} \quad a = \bar{y} - b\bar{x}$$

where

$$SS_{xy} = \Sigma xy - \frac{(\Sigma x)(\Sigma y)}{n}$$

$$SS_{xx} = \Sigma x^2 - \frac{(\Sigma x)^2}{n}$$

4. **Standard deviation of the sample errors**

$$s_e = \sqrt{\frac{SS_{yy} - b\,SS_{xy}}{n-2}}$$

where

$$SS_{yy} = \Sigma y^2 - \frac{(\Sigma y)^2}{n}$$

5. **Error sum of squares**

$$SSE = \Sigma e^2 = \Sigma(y - \hat{y})^2$$

6. **Total sum of squares**

$$SST = \Sigma y^2 - \frac{(\Sigma y)^2}{n}$$

7. **Regression sum of squares**

$$SSR = SST - SSE$$

8. **Coefficient of determination**

$$r^2 = \frac{b\,SS_{xy}}{SS_{yy}}$$

9. **Standard deviation of b**

$$s_b = \frac{s_e}{\sqrt{SS_{xx}}}$$

10. **The $(1 - \alpha)100\%$ confidence interval for B**

$$b \pm t s_b$$

11. **Value of the test statistic t for b**

$$t = \frac{b - B}{s_b}$$

12. **Linear correlation coefficient**

$$r = \frac{SS_{xy}}{\sqrt{SS_{xx}\,SS_{yy}}}$$

13. **Value of the test statistic t for r**

$$t = r\sqrt{\frac{n-2}{1-r^2}}$$

14. The $(1 - \alpha)100\%$ confidence interval for $\mu_{y|x}$

$$\hat{y} \pm ts_{\hat{y}_m}$$

where

$$s_{\hat{y}_m} = s_e \sqrt{\frac{1}{n} + \frac{(x_0 - \bar{x})^2}{SS_{xx}}}$$

15. The $(1 - \alpha)100\%$ prediction interval for y_p

$$\hat{y} \pm ts_{\hat{y}_p}$$

where

$$s_{\hat{y}_p} = s_e \sqrt{1 + \frac{1}{n} + \frac{(x_0 - \bar{x})^2}{SS_{xx}}}$$

SUPPLEMENTARY EXERCISES

13.95 The following data give information on the ages (in years) and the number of breakdowns during the past month for a sample of seven machines at a large company.

Age	12	7	2	8	13	9	4
Number of breakdowns	10	5	1	4	12	7	2

a. Taking age as an independent variable and number of breakdowns as a dependent variable, what is your hypothesis about the sign of B in the regression line? (In other words, do you expect B to be positive or negative?)
b. Find the least squares regression line. Is the sign of b the same as you hypothesized for B in part a?
c. Give a brief interpretation of the values of a and b calculated in part b.
d. Compute r and r^2 and explain what they mean.
e. Compute the standard deviation of errors.
f. Construct a 99% confidence interval for B.
g. Test at the 2.5% significance level whether B is positive.
h. At the 2.5% significance level, can you conclude that ρ is positive? Is your conclusion the same as in part g?

13.96 The following table gives information on the number of hours that eight bank loan officers slept the previous night and the number of loan applications that they processed the next day.

Number of hours slept	8	5	7	6	4	8	6	5
Number of applications processed	16	11	18	13	8	17	10	8

a. Taking the number of hours slept as an independent variable and the number of applications processed as a dependent variable, do you expect B to be positive or negative in the regression model $y = A + Bx + \epsilon$?
b. Find the least squares regression line. Is the sign of b the same as you hypothesized for B in part a?
c. Compute r and r^2 and explain what they mean.
d. Compute the standard deviation of errors.
e. Construct a 90% confidence interval for B.
f. Test at the 5% significance level whether B is positive.
g. Test at the 5% significance level whether ρ is positive. Is your conclusion the same as in part f?

13.97 The management of a supermarket wants to find if there is a relationship between the number of times a specific product is promoted on the intercom system in the store and the number of units of that product sold. To experiment, the management selected a product and promoted it on the intercom system for seven days. The following table gives the number of times this product was promoted each day and the number of units sold.

Number of Promotions Per Day	Number of Units Sold Per Day (hundreds)
15	11
22	22
42	30
30	26
18	17
12	15
38	23

a. With the number of promotions as an independent variable and the number of units sold as a dependent variable, what do you expect the sign of B in the regression line $y = A + Bx + \epsilon$ will be?

b. Find the least squares regression line $\hat{y} = a + bx$. Is the sign of b the same as you hypothesized for B in part a?

c. Give a brief interpretation of the values of a and b calculated in part b.

d. Compute r and r^2 and explain what they mean.

e. Predict the number of units of this product sold on a day with 35 promotions.

f. Compute the standard deviation of errors.

g. Construct a 98% confidence interval for B.

h. Testing at the 1% significance level, can you conclude that B is positive?

i. Using $\alpha = .02$, can you conclude that the correlation coefficient is different from zero?

 13.98 The following table gives information on the temperature in a city and the volume of ice cream (in pounds) sold at an ice cream parlor for a random sample of eight days during the summer of 1999.

Temperature	93	86	77	89	98	102	87	79
Ice cream sold	208	175	123	198	232	277	158	117

a. Find the least squares regression line $\hat{y} = a + bx$. Take temperature as an independent variable and volume of ice cream sold as a dependent variable.

b. Give a brief interpretation of the values of a and b.

c. Compute r and r^2 and explain what they mean.

d. Predict the amount of ice cream sold on a day with a temperature of 95°.

e. Compute the standard deviation of errors.

f. Construct a 99% confidence interval for B.

g. Testing at the 1% significance level, can you conclude that B is different from zero?

h. Using $\alpha = .01$, can you conclude that the correlation coefficient is different from zero?

13.99 The following table gives the number of U.S. households (in millions) with cable TV for the years 1987–1997.

Year	Number of Households with Cable TV (millions)
1987	45.0
1988	48.6
1989	52.6
1990	54.9
1991	55.8
1992	57.2
1993	58.8
1994	60.5
1995	63.0
1996	64.7
1997	65.9

Source: Nielsen Media Research, *The World Almanac and Book of Facts*, 1999.

a. Assign a value of 0 to 1987, 1 to 1988, 2 to 1989, and so on. Call this new variable *time*. Make a new table with the variables *Time* and *Number of Households with Cable TV*.

b. With time as an independent variable and number of households as the dependent variable, compute SS_{xx}, SS_{yy}, and SS_{xy}.

c. Construct a scatter diagram for these data. Does the scatter diagram exhibit a linear positive relationship between time and number of households with cable TV?

d. Find the least squares regression line $\hat{y} = a + bx$.

e. Give a brief interpretation of the values of a and b calculated in part d.

f. Compute the correlation coefficient r.

g. Predict the number of households with cable TV for the year 2002. Comment on this prediction.

13.100 The following table gives the public debt (in billions of dollars) of the United States for the years 1987–1997.

Year	Public Debt (billions)
1987	$2350.3
1988	2602.3
1989	2857.4
1990	3233.3
1991	3665.3
1992	4064.6
1993	4411.5
1994	4692.8
1995	4974.0
1996	5224.8
1997	5413.1

Source: Bureau of the Public Debt, U.S. Department of the Treasury.

a. Assign a value of 0 to 1987, 1 to 1988, 2 to 1989, and so on. Call this new variable *time*. Make a new table with the variables *Time* and *Public Debt*.

b. With time as an independent variable and public debt as the dependent variable, compute SS_{xx}, SS_{yy}, and SS_{xy}.

c. Construct a scatter diagram for these data. Does the scatter diagram exhibit a linear positive relationship between time and public debt?

d. Find the least squares regression line $\hat{y} = a + bx$.

e. Give a brief interpretation of the values of a and b calculated in part d.

f. Compute the correlation coefficient r.

g. Predict the amount of public debt for $x = 15$. Comment on this prediction.

13.101 Because of rapid changes in technology, personal computers (PCs) tend to become obsolete within a few years. The table below gives the (approximate) average estimated useful lives of PCs for the years 1992 through 2005. The average lives given in the table are estimated from a graph given in the newspaper article.

Year	Average Useful Life (years)	Year	Average Useful Life (years)
1992	4.5	1999	3.1
1993	4.2	2000	2.8
1994	4.1	2001	2.6
1995	3.8	2002	2.4
1996	3.6	2003	2.2
1997	3.4	2004	2.1
1998	3.2	2005	2.0

Source: Stanford Resources Inc., *USA TODAY*, June 22, 1999.

a. Assign a value of 0 to 1992, 1 to 1993, 2 to 1994, and so on. Call this new variable *time*. Write a new table with the variables *Time* and *Average Useful Life*.

b. With time as an independent variable and average useful life as the dependent variable, compute SS_{xx}, SS_{yy}, and SS_{xy}.

c. Construct a scatter diagram for these data. Does the scatter diagram exhibit a linear negative relationship between the variables time and average useful life?

d. Find the least squares regression line $\hat{y} = a + bx$.

e. Give a brief interpretation of the values of a and b calculated in part d.

f. Compute the correlation coefficient r.

g. Predict the average useful life of a PC for the year 2010. Comment on this prediction.

13.102 Refer to the data on ages and numbers of breakdowns for seven machines given in Exercise 13.95. Construct a 99% confidence interval for the mean number of breakdowns per month for all machines with an age of 8 years. Find a 99% prediction interval for the number of breakdowns per month for a randomly selected machine with an age of 8 years.

13.103 Refer to the data on the number of hours slept and the number of loan applications processed by bank loan officers given in Exercise 13.96. Determine a 95% confidence interval for the mean number of applications processed by all bank loan officers who slept for 6 hours. Make a 95% prediction interval for the number of applications processed by a bank loan officer who slept for 6 hours.

13.104 Refer to the data given in Exercise 13.97 on the number of times a specific product is promoted on the intercom system in a supermarket and the number of units of that product sold. Make a 90% confidence interval for the mean number of units of that product sold on days with 35 promotions. Construct a 90% prediction interval for the number of units of that product sold on a randomly selected day with 35 promotions.

13.105 Refer to the data given in Exercise 13.98 on temperatures and the volumes of ice cream sold at an ice cream parlor for a sample of eight days. Construct a 98% confidence interval for the mean volume of ice cream sold at this parlor on all days with a temperature of 95°. Determine a 98% prediction interval for the volume of ice cream sold at this parlor on a randomly selected day with a temperature of 95°.

■ **Advanced Exercises**

13.106 Consider the data given in the following table.

x	10	20	30	40	50	60
y	12	15	19	21	25	30

a. Find the least squares regression line and the linear correlation coefficient r.

b. Suppose that each value of y given in the table is increased by 5 and the x values remain unchanged. Would you expect r to increase, decrease, or remain the same? How do you expect the least squares regression line to change?

c. Increase each value of y given in the table by 5 and find the new least squares regression line and the correlation coefficient r. Do these results agree with your expectation in part b?

13.107 Suppose that you work part-time at a bowling alley that is open daily from noon to midnight. Although business is usually slow from noon to 6:00 P.M., the owner has noticed that it is better on hotter days during the summer, perhaps because the premises are comfortably air-conditioned. The owner shows you some data that she gathered last summer. This data set includes the maximum temperature and the number of lines bowled between noon and 6:00 P.M. for each of 20 days. (The maximum temperatures ranged from 77° to 95° Fahrenheit during this period.) The owner would like to know if she can estimate tomorrow's business from noon to 6:00 P.M. by looking at tomorrow's weather forecast. She asks you to analyze the data. Let x be the maximum temperature for a day and y the number of lines bowled between noon and 6:00 P.M. on that day. The computer output based on the data for 20 days provided the following results:

$$\hat{y} = -432 + 7.7x \qquad s_e = 28.17 \qquad SS_{xx} = 607 \qquad \bar{x} = 87.5$$

Assume that the weather forecasts are reasonably accurate.

a. Does the maximum temperature seem to be a useful predictor of bowling activity between noon and 6:00 P.M.? Use an appropriate statistical procedure based on the information given. Use $\alpha = .05$.

b. The owner wants to know how many lines of bowling she can expect, on average, for days with a maximum temperature of 90°. Answer using a 95% confidence level.

c. The owner has seen tomorrow's weather forecast, which predicts a high of 90°. About how many lines of bowling can she expect? Answer using a 95% confidence level.

d. Give a brief commonsense explanation to the owner for the difference in the interval estimates of parts b and c.

e. The owner asks you how many lines of bowling she could expect if the high temperature were 100°. Give a point estimate, together with an appropriate warning to the owner.

13.108   An economist wanted to study the relationship between the incomes of 30-year-old persons employed full-time and the incomes of their fathers. Let x be the mean of the three highest annual incomes for a 30-year-old person employed full-time and y the mean of the three highest annual incomes for the person's father. Suppose that after collecting data for a random sample of 300 such 30-year-olds and their fathers, the economist finds a linear correlation coefficient of .60. A friend of yours, who has read about this research, asks you several questions, such as: Does the positive value of the correlation coefficient suggest that the 30-year-olds tend to earn more than their fathers? Does the correlation coefficient reveal anything at all about the difference between the incomes of 30-year-olds and their fathers? If not, what other information would we need from this study? What does the correlation coefficient tell us about the relationship between the two variables in this example? Write a short note to your friend answering these questions.

13.109   For the past 25 years Burton Hodge has been keeping track of how many times he mows his lawn and the average size of the ears of corn in his garden. Hearing about the Pearson correlation coefficient from a statistician buddy of his, Burton decides to substantiate his suspicion that the more often he mows his lawn, the bigger the ears of corn are. He does so by computing the correlation coefficient. Lo and behold, Burton finds a .93 coefficient of correlation! Elated, he calls his friend the statistician to thank him and announce that next year he will have prize winning ears of corn because he plans to mow his lawn every day. Do you think Burton's logic is correct? If not, how would you explain to Burton the mistake he is making in his presumption (without eroding his new opinion of statistics)? Suggest what Burton could do next year to make the ears of corn large, and relate this to the Pearson correlation coefficient.

13.110   It seems reasonable that the more hours per week a full-time college student works at a job, the less time he or she will have to study and, consequently, the lower his or her GPA would be.

a. Assuming a linear relationship, suggest specifically what the equation relating x and y would be, where x is the average number of hours a student works per week and y represents a student's GPA. Try several values of x and see if your equation gives reasonable values of y.

b. Using the following observations taken from 10 randomly selected students, compute the regression equation and compare it to yours of part a.

Average number of hours worked	20	28	10	35	5	14	0	40	8	23
GPA	2.8	2.5	3.1	2.1	3.4	3.3	2.8	2.5	3.6	1.8

13.111   You are hired by the local telephone company to study the monthly telephone bills at residences of college students serviced by a single phone line. The phone company wants you to relate the number of students living at the address and the size of the phone bill.

a. Before taking any observations, suggest a specific relationship between these two variables. Try some values to see if that relationship seems reasonable.

b. A random sample of 15 residences of students revealed the following monthly phone bills (in dollars) and numbers of students living at those addresses. Compute the regression equation based on these data and compare it to your answer in part a.

Phone Bill	Number of Students
$67.40	2
171.12	5
53.67	1
101.07	2
84.26	3
152.71	3
123.22	5
168.21	4
82.55	3
126.98	4
41.85	1
239.01	6
75.18	2
114.38	4
82.27	3

13.112 Was Leo Durocher right when he said, "Nice guys finish last"? Cornell University and the "head hunters" Ray & Berndtson gave a personality test to 3600 U.S. and European executives (*Business Week*, July 27, 1998). One aspect of this test was a measure of how "agreeable" each executive was, measured on a scale of 1 to 60. U.S. executives' average score was 44 points. It was calculated that every five points scored above the average were associated with a loss in annual salary of $16,836 for U.S. executives. Suppose that these conclusions were based on a random sample of U.S. executives and that their mean annual salary was $200,000. Let x be an executive's "agreeableness" score on such a test and let y be the executive's annual salary.

a. Write a regression equation that is consistent with the information above.
b. Over what range of values of x would your equation be valid?

13.113 The owner of a bowling establishment is interested in the relationship between the price she charges for a game of bowling and the number of games bowled per day. She collects data on the number of games bowled per day at 15 different prices. Fill in the missing entries in the following MINITAB output that was obtained for these data. In this output, X represents the price of a game and Y is the number of games bowled per day.

```
THE REGRESSION EQUATION IS

Y = _____ _____ X

PREDICTOR      COEF     STDEV   T-RATIO      P
CONSTANT      691.02    21.70   _____      .000
X            -141.30    _____  -16.83      .000

S = _____     R-SQ = _____     R-SQ(ADJ)=95.3%

ANALYSIS OF VARIANCE

SOURCE        DF       SS       MS        F        P
REGRESSION   _____   148484.0  _____   283.13    .000
ERROR        _____   _____     _____
TOTAL        _____   _____
```

SELF-REVIEW TEST

1. A simple regression is a regression model that contains
   a. only one independent variable
   b. only one dependent variable
   c. more than one independent variable
   d. both a and b

2. The relationship between independent and dependent variables represented by the (simple) linear regression is that of
   a. a straight line   b. a curve   c. both a and b

3. A deterministic regression model is a model that
   a. contains the random error term
   b. does not contain the random error term
   c. gives a nonlinear relationship

4. A probabilistic regression model is a model that
   a. contains the random error term
   b. does not contain the random error term
   c. shows an exact relationship

5. The least squares regression line minimizes the sum of
   a. errors   b. squared errors   c. predictions

6. The degrees of freedom for a simple regression model are
   a. $n - 1$ b. $n - 2$ c. $n - 5$

7. Indicate if the following statement is true or false.

   The coefficient of determination gives the proportion of total squared errors (SST) that is explained by the use of the regression model.

8. Indicate if the following statement is true or false.

   The linear correlation coefficient measures the strength of the linear association between two variables.

9. The value of the coefficient of determination is always in the range
   a. 0 to 1   b. -1 to 1   c. -1 to 0

10. The value of the correlation coefficient is always in the range
    a. 0 to 1   b. -1 to 1   c. -1 to 0

11. Explain why the random error term ϵ is added to the regression model.

12. Explain the difference between A and a and between B and b for a regression model.

13. Briefly explain the assumptions of a regression model.

14. Briefly explain the difference between the population regression line and a sample regression line.

15. The following table gives the experience (in years) and the number of computers sold during the previous three months by seven salespersons.

Experience	4	12	9	6	10	16	7
Computers sold	19	42	28	33	40	37	23

a. Do you think experience depends on the number of computers sold or the number of computers sold depends on experience?

b. With experience as an independent variable and the number of computers sold as a dependent variable, what is your hypothesis about the sign of B in the regression model?

c. Construct a scatter diagram for these data. Does the scatter diagram exhibit a linear relationship between the two variables?

d. Find the least squares regression line. Is the sign of b the same as the one you hypothesized for B in part b?

e. Give a brief interpretation of the values of the y-intercept and slope calculated in part d.

f. Compute r and r^2 and explain what they mean.

g. Predict the number of computers sold during the past three months by a salesperson with 11 years of experience.

h. Compute the standard deviation of errors.

i. Construct a 99% confidence interval for B.

j. Testing at the 1% significance level, can you conclude that B is positive?

k. Construct a 95% confidence interval for the mean number of computers sold in a three-month period by all salespersons with 8 years of experience.

l. Make a 95% prediction interval for the number of computers sold in a three-month period by a randomly selected salesperson with 8 years of experience.

m. Using the 1% significance level, can you conclude that the linear correlation coefficient is positive?

MINI-PROJECTS

Mini-Project 13–1

Do the more prosperous states in the United States tend to have a higher per capita energy expenditure, or do other factors such as industry, transportation patterns, and heating and cooling requirements overshadow such a relationship?

a. Let x be a state's per capita income and y be that state's per capita energy expenditure. Take a random sample of 10 of the 50 states. Using Data on States that accompany this text, calculate the linear correlation coefficient between x and y.

b. Does your value of r indicate a positive relationship between x and y?

c. Find the least squares regression line.

d. Using the 5% level of significance, can you conclude that states with higher per capita income tend to have higher per capita energy expenditures? Use the slope of the regression line to make this test.

e. Using the 5% level of significance, can you conclude that the linear correlation coefficient is positive?

Mini-Project 13–2

Two friends are arguing about the relationship between the prices of soft drinks and wine in U.S. cities. Justin thinks that the prices of any two types of beverages (a soft drink and wine) should be positively related. Ivan disagrees, arguing that the prices of alcoholic beverages in a city depend primarily on state and local taxes.

a. Take a random sample of 15 U.S. cities from CITY DATA that accompany this text. Let x be the price of a 2-liter bottle of Coca-Cola and y the price of a 1.5-liter bottle of Gallo Chablis Blanc wine. Calculate the linear correlation coefficient between x and y.

b. Does your value of r suggest a positive linear relationship between x and y?

c. Find the least squares regression line.

d. Using $\alpha = .01$, would you support Justin's view? Use the slope of the regression line to make this test.

e. Using the 1% level of significance, can you conclude that the linear correlation coefficient is positive?

For instructions on using MINITAB for this chapter, please see Section B.12 in Appendix B.

See Appendix C for *Excel Adventure 13* on regression analysis.

COMPUTER ASSIGNMENTS

CA13.1 Professor Hamid Zangenehzadeh studied the relationship between student evaluations of a teacher and the expected grades of students in a course. The following table lists the (average of the) ratings of teachers by students and (the average of the) students' expected grades for 39 faculty members of the three departments of the School of Management at Widener University as reported in this study. (*Note*: The ratings of teachers reported in this table are what the author called the *unadjusted ratings of teachers*.)

Ratings of Teachers	Students' Expected Grades	Ratings of Teachers	Students' Expected Grades
3.833	3.500	2.739	3.000
3.769	3.769	2.543	2.829
3.642	3.214	2.286	3.143
3.625	3.250	2.278	2.833
3.529	3.529	2.133	2.800
3.500	3.300	2.103	2.620
3.500	3.500	2.053	2.368
3.409	3.864	2.043	2.696
3.380	3.048	1.944	2.944
3.333	3.200	1.923	2.846
3.294	3.059	1.800	2.800
3.267	3.000	1.800	3.000
3.263	3.368	1.692	2.769
3.120	3.440	1.692	3.462
3.045	2.909	1.688	3.125
3.000	3.500	1.667	4.000
3.000	3.500	1.625	2.375
2.923	2.538	1.333	3.555
2.826	3.086	.521	2.652
2.778	3.111		

Source: Hamid Zangenehzadeh, "Grade Inflation: A Way Out." *Journal of Economic Education*, summer 1988, 217–226. Reprinted with permission of the Helen Dwight Reid Educational Foundation. Published by Heldref Publications, Washington, D.C. Copyright © 1988.

Use MINITAB or any other statistical software to do the following.

a. Construct a scatter diagram for these data.

b. Find the correlation between the two variables.

c. Find the regression line with ratings of teachers as a dependent variable and students' expected grades as an independent variable.

d. Make a 95% confidence interval for B.
e. Test at the 1% significance level whether B is positive.
(*Note*: Because the sample size is large, $n = 39$, the normal distribution can be applied to construct a confidence interval and to make a test of hypothesis about B in parts d and e if we so desire.)

CA13.2 Refer to Data Set III on the heights and weights of NBA players. Select a random sample of 25 players from that population. Use MINITAB or any other statistical software to do the following.

a. Construct a scatter diagram for these data.
b. Find the correlation between these two variables.
c. Find the regression line with weight as a dependent variable and height as an independent variable.
d. Make a 90% confidence interval for B.
e. Test at the 5% significance level whether B is positive.
f. Make a 95% confidence interval for the mean weight of all NBA players who are 78 inches tall. Construct a 95% prediction interval for the weight of a randomly selected NBA player with a height of 78 inches.

CA13.3 Refer to the data on the ages and the numbers of breakdowns for a sample of seven machines given in Exercise 13.95. Use MINITAB or any other statistical software to answer the following questions.

a. Construct a scatter diagram for these data.
b. Find the least squares regression line with age as an independent variable and the number of breakdowns as a dependent variable.
c. Compute the correlation coefficient.
d. Construct a 99% confidence interval for B.
e. Test at the 2.5% significance level whether B is positive.

14 | NONPARAMETRIC METHODS

14.1 THE SIGN TEST

14.2 THE WILCOXON SIGNED-RANK TEST FOR TWO DEPENDENT SAMPLES

14.3 THE WILCOXON RANK SUM TEST FOR TWO INDEPENDENT SAMPLES

14.4 THE KRUSKAL–WALLIS TEST

14.5 THE SPEARMAN RHO RANK CORRELATION COEFFICIENT TEST

14.6 THE RUNS TEST FOR RANDOMNESS

GLOSSARY

KEY FORMULAS

SUPPLEMENTARY EXERCISES

SELF-REVIEW TEST

MINI-PROJECTS

COMPUTER ASSIGNMENTS

The hypothesis tests discussed so far in this text are called **parametric tests**. In those tests, we used the normal, t, chi-square, and F distributions to make tests about population parameters such as means, proportions, variances, and standard deviations. In doing so, we made some assumptions, such as the assumption that the population from which the sample was drawn was normally distributed. This chapter discusses a few **nonparametric tests**. These tests do not require the same kinds of assumptions, and hence, they are also called **distribution-free tests**.

Nonparametric tests have several advantages over parametric tests: They are easier to use and understand; they can be applied to situations in which parametric tests cannot be used; and they do not require that the population being sampled is normally distributed. However, a major problem with nonparametric tests is that they are less efficient than parametric tests. The sample size must be larger for a nonparametric test to have the same probability of committing the two types of errors.

Although there are a large number of nonparametric tests that can be applied to conduct tests of hypothesis, this chapter discusses only six of them: the sign test, the Wilcoxon signed-rank test, the Wilcoxon rank sum test, the Kruskal–Wallis test, the Spearman rho rank correlation coefficient test, and the runs test for randomness.

14.1 THE SIGN TEST

The **sign test** is one of the easiest tests to apply to test hypotheses. It uses only plus and minus signs. The sign test can be used to perform the following types of tests:

1. To determine the preference for one product or item over another, or to determine whether one outcome occurs more often than another outcome in categorical data. For example, we may test whether or not people prefer one kind of soft drink over another kind.

2. To conduct a test for the median of a single population. For example, we may use this procedure to test whether the median rent paid by all tenants in a city is different from $1250.

3. To perform a test for the median of paired differences using data from two dependent samples. For example, we may use this procedure to test whether the median score on a standardized test increases after a preparatory course is taken.

> **SIGN TEST** The *sign test* is used to make hypothesis tests about preferences, a single median, and the median of paired differences for two dependent populations. We use only plus and minus signs to perform the tests.

In the following subsections we discuss these tests for small and large samples.

14.1.1 TESTS ABOUT CATEGORICAL DATA

Data that are divided into different categories for identification purposes are called **categorical data**. For example, people's opinions about a certain issue—in favor, against, or no opinion—produce categorical data. This subsection discusses how to perform tests about such data using the sign test procedure. We discuss two situations in which such tests can be performed: the small-sample case and the large-sample case.

The Small-Sample Case

When we apply the sign test for categorical data, if the *sample size is 25 or less* (i.e., $n \leq 25$), we consider it a small sample. Table XI: Critical Values for the Sign Test, which appears in Appendix D, is based on the binomial probability distribution. This table gives the critical values of the test statistic for the sign test when $n \leq 25$ using the binomial probability distribution.

The sign test can be used to test whether or not customers prefer one brand of a product over another brand of the same type of product. For example, we can test whether customers have a higher preference for Coke or for Pepsi. This procedure can also be used to test whether people prefer one of two alternatives over the other. For example, given a choice, do people prefer to live in New York City or in Los Angeles?

Performing sign test with categorical data: small sample.

EXAMPLE 14–1 The Top Taste Water Company produces and distributes Top Taste bottled water. The company wants to determine whether customers have a higher preference for its bottled water than for its main competitor, Spring Hill bottled water. The Top Taste Water Company hired a statistician to conduct this study. The statistician selected a random sample of 10 people and asked each of them to taste one sample of each of the two brands of water. The customers did not know the brand of each water sample. Also, the order in which each person tasted the two brands of water was determined randomly. Each person was asked to indicate which of the two samples of water he or she preferred. The following table shows the preferences of these 10 individuals.

Person	Brand Preferred
1	Spring Hill
2	Top Taste
3	Top Taste
4	Neither
5	Top Taste
6	Spring Hill
7	Spring Hill
8	Top Taste
9	Top Taste
10	Top Taste

Based on these results, can the statistician conclude that people prefer one brand of bottled water over the other? Use the significance level of 5%.

Solution We use the same five steps to perform this test of hypothesis that we used in earlier chapters.

Step 1. *State the null and alternative hypotheses.*

If we assume that people do not prefer either brand of water over the other, we would expect about 50% of the people (who show a preference) to indicate a preference for Top Taste water and the other 50% to indicate a preference for Spring Hill water. Let p be the proportion of all people who prefer Top Taste bottled water. The two hypotheses are as follows:

H_0: $p = .50$ (People do not prefer either of the two brands of water)

H_1: $p \neq .50$ (People prefer one brand of water over the other)

The null hypothesis states that 50% of the people prefer Top Taste water over Spring Hill water (and, hence, the other 50% prefer Spring Hill water). Note that we do not consider people who have no preference, and we drop them from the sample. If we fail to reject H_0, we will conclude that people do not prefer one brand over the other. However, if we reject H_0, we will conclude that the percentage of people who prefer Top Taste water over Spring Hill water is different from 50%. Thus, the conclusion will be that people prefer one brand over the other.

Step 2. *Select the distribution to use.*

We use the binomial probability distribution to make the test. Note that here there is only one sample and each member of the sample is asked to indicate a preference if he or she has one. We drop the members who do not indicate a preference and then compare the preferences of the remaining members. Also note that there are three outcomes for each person: (1) prefers Top Taste water, (2) prefers Spring Hill water, or (3) has no preference. We are to compare the two outcomes with preferences and determine whether more people belong to one of these two outcomes than to the other. All such tests of preferences are conducted by using the binomial probability distribution. If we assume that H_0 is true, then the number of people who indicate a preference for Top Taste bottled water (the number of successes) follows the binomial distribution, with $p = .50$.

Step 3. *Determine the rejection and nonrejection regions.*

Note that 10 people were selected to taste the two brands of water and indicate their preferences. However, one of these individuals stated no preference. Hence, only 9 of the 10 people have indicated a preference for one or the other of the two brands of bottled water. To conduct the test, the person who has shown no preference is dropped from the sample. Thus, the *true sample size* is 9; that is, $n = 9$.

The significance level for the test is .05. Let X be the number of people in the sample of 9 who prefer Top Taste bottled water. Here X is called the *test statistic.* To establish a decision rule, we find the critical values of X from Table XI for $n = 9$. From that table, for $n = 9$ and $\alpha = .05$ for a two-tailed test, the critical values of X are 1 and 8. Note that in a two-tailed test we read both the lower and the upper critical values, which are shown in Figure 14.1. Thus, we will reject the null hypothesis if either fewer than 2 or more than 7 people in 9 indicate a preference for Top Taste bottled water.

0 or 1	2 to 7	8 or 9	
Rejection region	Nonrejection region	Rejection region	X

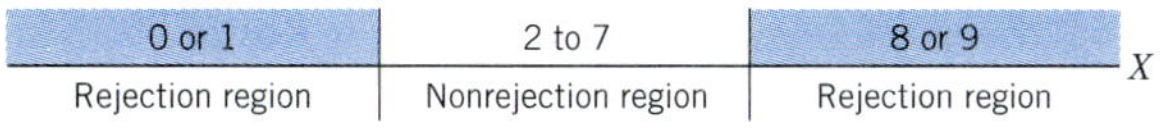

Figure 14.1

> **CRITICAL VALUE OF X** In a sign test for a small sample, the *critical value of X* is obtained from Table XI. If the test is two-tailed, we read both the lower and the upper critical values from that table. However, we read only the lower critical value if the test is left-tailed, and only the upper critical value if the test is right-tailed. Also note that which column we use to obtain this critical value depends on the given significance level and on whether the test is two-tailed or one-tailed.

Step 4. *Calculate the value of the test statistic.*

To record the results of the experiment, we mark a plus sign for each person who prefers Top Taste bottled water, a minus sign for each person who prefers Spring Hill bottled water, and a zero for the one who indicates no preference. This listing is shown in Table 14.1.

Table 14.1

Person	Brand Preferred	Sign
1	Spring Hill	−
2	Top Taste	+
3	Top Taste	+
4	Neither	0
5	Top Taste	+
6	Spring Hill	−
7	Spring Hill	−
8	Top Taste	+
9	Top Taste	+
10	Top Taste	+

Now, we count the number of plus signs (the sign that belongs to Top Taste bottled water because p in H_0 refers to Top Taste water). There are 6 plus signs, indicating that 6 of 9 people in the sample stated a preference for Top Taste bottled water. Note that the sample size is 9, not 10, because we drop the person with zero sign. Thus,

$$\text{Observed value of } X = 6$$

> **OBSERVED VALUE OF X** The *observed value of X* is given by the number of signs that belong to the category whose proportion we are testing for.

Step 5. *Make a decision.*

Since the observed value of $X = 6$ falls in the nonrejection region (see Figure 14.1), we fail to reject H_0. Hence, we conclude that our sample does not indicate that people prefer either of these two brands of bottled water over the other.

Note that it does not matter which outcome p refers to. If we assume that p is the proportion of people who prefer Spring Hill water, then X will denote the number of people in a sample of n who prefer Spring Hill water. The observed value of X this time would be 3, which is the number of minus signs in Table 14.1. From Figure 14.1, $X = 3$ also falls in the nonrejection region. Hence, we again fail to reject the null hypothesis. ∎

The Large-Sample Case

If we are testing a hypothesis about preference for categorical data and $n > 25$, we can use the normal probability distribution as an approximation to the binomial probability distribution.

> **THE LARGE-SAMPLE CASE** If $n > 25$, the normal distribution can be used as an approximation to the binomial probability distribution to perform a test of hypothesis about the preference for categorical data. The *observed value* of the test statistic z, in this case, is calculated as
>
> $$z = \frac{(X \pm .5) - \mu}{\sigma}$$
>
> where X is the number of units in the sample that belong to the outcome referring to p. We either add .5 to X or subtract .5 from X to correct for continuity (see Section 6.7 of Chapter 6). We will add .5 to X if the value of X is less than or equal to $n/2$, and we will subtract .5 from X if the value of X is greater than $n/2$. The values of the mean and standard deviation are calculated as
>
> $$\mu = np \quad \text{and} \quad \sigma = \sqrt{npq}$$

Example 14–2 illustrates the procedure for the large-sample case.

Performing sign test with categorical data: large sample.

EXAMPLE 14–2 A developer is interested in building a shopping mall adjacent to a residential area. Before granting or denying permission to build such a mall, the town council took a random sample of 75 adults from adjacent areas and asked them whether they favor or oppose construction of this mall. Of these 75 adults, 40 opposed construction of the mall,

30 favored it, and 5 had no opinion. Can you conclude that the number of adults in this area who oppose construction of the mall is higher than the number who favor it? Use $\alpha = .01$.

Solution Again, each adult in the sample has to pick one of three choices: oppose, favor, or have no opinion. And we are to compare two outcomes—oppose and favor—to find out whether more adults belong to the outcome indicated by *oppose*. We can use the sign test here. To do so, we drop the subjects who have no opinion—that is, the adults who belong to the outcome that is not being compared. In our example, 5 adults have no opinion. Thus, we drop these adults from our sample and use the sample size of $n = 70$ for the purposes of this test. Let p be the proportion of adults who oppose construction of the mall and q be the proportion who favor it. We apply the five steps to make this test.

Step 1. *State the null and alternative hypotheses.*

$H_0: p = q = .50$ (The two proportions are equal)

$H_1: p > q$ (The proportion of adults who oppose the mall is greater than the proportion who favor it)

The null hypothesis here states that the proportion of adults who oppose and the proportion who favor construction of the mall are both .50, which means that $p = .50$ and $q = .50$. The alternative hypothesis is that $p > q$, which means that more adults oppose construction of the mall than favor it. Note that H_1 states that $p > .50$ and $q < .50$. In other words, of those who have an opinion, more than 50% oppose and less than 50% favor construction of the shopping mall.

Step 2. *Select the distribution to use.*

As explained earlier, we will use the sign test to perform this test. Although 75 adults were asked their opinion, only 70 of them offered it and 5 did not. Hence, our sample size is 70; that is, $n = 70$. Since it is a large sample ($n > 25$), we can use the normal approximation to perform the test.

Step 3. *Determine the rejection and nonrejection regions.*

Because H_1 states that $p > .50$, our test is right-tailed. Also, $\alpha = .01$. From Table VII (the standard normal distribution table) in Appendix D, the z value for $.5 - .01 = .4900$ is approximately 2.33. Thus, the decision rule is that we will not reject H_0 if $z < 2.33$ and we will reject H_0 if $z \geq 2.33$. Thus, the nonrejection region lies to the left of $z = 2.33$ and the rejection region to the right of $z = 2.33$, as shown in Figure 14.2.

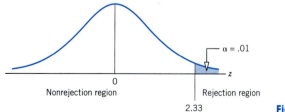

$\alpha = .01$

0
Nonrejection region

Rejection region

2.33   **Figure 14.2**

Step 4. *Calculate the value of the test statistic.*

Assuming that the null hypothesis is true, we expect (about) half of the adults in the population to oppose construction of the mall and the other half to favor it. Thus, we expect

$p = .50$ and $q = .50$. Note that we do not count the people who have no opinion. The mean and standard deviation of the binomial distribution are

$$\mu = np = 70(.50) = 35$$

$$\sigma = \sqrt{npq} = \sqrt{70(.50)(.50)} = \sqrt{17.5} = 4.18330013$$

In this example, p refers to the proportion of adults who oppose construction of the mall. Hence, X refers to the number in 70 who oppose the mall. Thus,

$$X = 40 \quad \text{and} \quad \frac{n}{2} = \frac{70}{2} = 35$$

Because X is greater than $n/2$, the observed value of the test statistic z is

$$z = \frac{(X - .5) - \mu}{\sigma} = \frac{(40 - .5) - 35}{4.18330013} = 1.08$$

Step 5. *Make a decision.*

Since the observed value of $z = 1.08$ is less than the critical value of $z = 2.33$, it falls in the nonrejection region. Hence, we do not reject H_0. Consequently, we conclude that the number of adults who oppose construction of the mall is not higher than the number who favor its construction. ∎

14.1.2 TESTS ABOUT THE MEDIAN OF A SINGLE POPULATION

The sign test can be used to test a hypothesis about a population median. Recall from Chapter 3 that the median is the value that divides a ranked data set into two equal parts. For example, if the median age of students in a class is 24, half of the students are younger than 24 and half are older. This section discusses how to make a test of hypothesis about the median of a population.

The Small-Sample Case

If $n \le 25$, we use the binomial probability distribution to test a hypothesis about the median of a population. The procedure used to conduct such a test is very similar to the one explained in Example 14–1.

Performing sign test about a population median: small sample.

EXAMPLE 14–3 A real estate agent claims that the median price of homes in a small town is $137,000. A sample of 10 houses selected by a statistician produced the following data on their prices.

Home	1	2	3	4	5	6	7	8	9	10
Price ($)	147,500	123,600	139,000	168,200	129,450	132,400	156,400	188,210	198,425	215,300

Using the 5% significance level, can you conclude that the median price of homes in this town is different from $137,000?

Solution Using the given data, we prepare Table 14.2, which contains a sign row. In this row, we assign a positive sign to each price that is above the claimed median price of $137,000 and a negative sign to each price that is below the claimed median price.

Table 14.2

Home	1	2	3	4	5	6	7	8	9	10
Sign	+	−	+	+	−	−	+	+	+	+

In Table 14.2, there are seven plus signs, indicating that the prices of seven houses are higher than the claimed median price of $137,000, and there are three minus signs, showing that the prices of three homes are lower than the claimed median price. Note that if one or more values in a data set are equal to the median, then each of them is assigned a zero value and dropped from the sample. Next, we perform the following five steps to perform the test of hypothesis.

Step 1. *State the null and alternative hypotheses.*

$$H_0: \text{Median price} = \$137,000$$

$$H_1: \text{Median price} \neq \$137,000$$

Step 2. *Select the distribution to use.*

For a test of the median of a population, we employ the sign test procedure by using the binomial probability distribution if $n \leq 25$. Since in our example $n = 10$, which is less than 25, we use the binomial probability distribution to conduct the test.

Step 3. *Determine the rejection and nonrejection regions.*

In our example, $n = 10$ and $\alpha = .05$. The test is two-tailed. Let X be the test statistic that represents the number of plus signs in Table 14.2. From Table XI, the (lower and upper) critical values of X are 1 and 9. Using these critical values, Figure 14.3 shows the rejection and nonrejection regions. Thus, we will reject the null hypothesis if the observed value of X is either 1 or 0, or 9 or 10. Note that because X represents the number of plus signs in the sample, its lowest possible value is 0 and its highest possible value is 10.

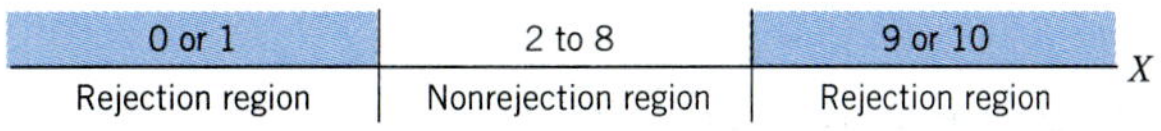

0 or 1	2 to 8	9 or 10	
Rejection region	Nonrejection region	Rejection region	X

Figure 14.3

Step 4. *Calculate the value of the test statistic.*

The observed value of X is given by the number of plus signs in Table 14.2. Thus,

$$\text{Observed value of } X = 7$$

> **OBSERVED VALUE OF X** When using the sign test to perform a test about a median, we can use either the number of positive signs or the number of negative signs as the observed value of X if the test is two-tailed. However, the *observed value of X* is equal to the larger of these two numbers (the number of positive and negative signs) if the test is right-tailed, and equal to the smaller of these two numbers if the test is left-tailed.

Step 5. *Make a decision.*

The observed value of $X = 7$ falls in the nonrejection region in Figure 14.3. Hence, we do not reject H_0 and conclude that the median price of homes in this town is not different from $137,000. ∎

The Large-Sample Case

For a test of the median of a single population, we can use the normal approximation to the binomial probability distribution when $n > 25$. The observed value of z, in this case, is calculated as in a test of hypothesis about the preference for categorical data (see Example 14–2). Example 14–4 explains the procedure for such a test.

Performing sign test about a population median: large sample.

EXAMPLE 14–4 A long-distance phone company believes that the median phone bill (for long distance calls) is at least $70 for all the families in New Haven, Connecticut. A random sample of 90 families selected from New Haven showed that the phone bills of 51 of them were less than $70 and those of 38 of them were more than $70, while one family had a phone bill of exactly $70. Using the 1% significance level, can you conclude that the company's claim is true?

Solution We use the usual five steps to test this hypothesis.

Step 1. *State the null and alternative hypotheses.*

The company's claim is that the median phone bill is at least $70. Hence, the two hypotheses are as follows:

$$H_0: \text{Median} \geq \$70$$

$$H_1: \text{Median} < \$70$$

Step 2. *Select the distribution to use.*

This is a test about the median and $n > 25$. Hence, to conduct this test, we can use the normal distribution as an approximation to the binomial probability distribution.

Step 3. *Determine the rejection and nonrejection regions.*

The test is left-tailed and $\alpha = .01$. From Table VII (the standard normal distribution table), the z value for $.5 - .01 = .4900$ is -2.33. Note that the z value is negative because it is a left-tailed test. Thus, we will reject H_0 if the observed value of z is -2.33 or lower, and we will not reject H_0 if the observed value of z is greater than -2.33. Figure 14.4 shows the rejection and nonrejection regions.

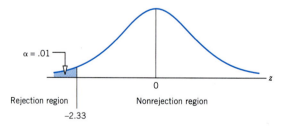

Figure 14.4

Step 4. *Calculate the value of the test statistic.*

In our example, 51 phone bills out of 90 are below the hypothesized median, 38 are above it, and one is exactly equal to the median. When we perform this test, we drop the value or values that are exactly equal to the median. Thus, after dropping one value that is equal to the median, our sample size is $51 + 38 = 89$; that is, $n = 89$. Let a phone bill below the median be represented by a minus sign and one above the median by a plus sign. Then, in these 89 bills there are 51 minus signs (for values less than the median) and 38 plus signs (for values greater than the median). If the given claim is true, we would expect (about) half plus signs and half minus signs. Let p be the proportion of plus signs in 89. Then we would expect $p = .50$ if H_0 is true. Hence, the mean and the standard deviation of the binomial distribution are calculated as follows:

$$n = 89 \qquad p = .50 \qquad q = 1 - p = .50$$

$$\mu = np = 89(.50) = 44.50$$

$$\sigma = \sqrt{npq} = \sqrt{89(.50)(.50)} = 4.71699057$$

In our example, 51 phone bills are below the median and 38 are above the median. Because it is a left-tailed test, $X = 38$, which is the smaller of the two numbers (51 and 38). Consequently, the z value is calculated as follows. Note that we have added .5 to X because the value of X is less than $n/2$, which is $89/2 = 44.5$.

$$z = \frac{(X + .5) - \mu}{\sigma} = \frac{(38 + .5) - 44.50}{4.71699057} = -1.27$$

Step 5. *Make a decision.*

Since $z = -1.27$ is greater than the critical value of $z = -2.33$, we do not reject H_0. Hence, we conclude that the company's claim that the median phone bill is at least \$70 seems to be true. ∎

☞ OBSERVATION

Note that in Example 14–4 there were 51 minus signs and 38 plus signs. We assigned the smaller of these two numbers to X so that $X = 38$. We did so in order to obtain a negative value of the observed z because the test is left-tailed and the critical value of z is negative. If we assigned 51 as the value of X, we would obtain $z = +1.27$ as the observed value of z, which does not make sense. Let X_1 be the number of plus signs and X_2 the number of minus signs in a test about the median. Then we can establish the following rules to calculate the observed value of X.

1. If the test is two-tailed, it does not matter which of the two values, X_1 and X_2, is assigned to X to calculate the observed value of z.
2. If the test is left-tailed, X should be assigned a value equal to the smaller of the values of X_1 and X_2.
3. If the test is right-tailed, X should be assigned a value equal to the larger of the values of X_1 and X_2.

14.1.3 TESTS ABOUT THE MEDIAN DIFFERENCE BETWEEN PAIRED DATA

We can use the sign test to perform a test of hypothesis about the difference between the medians of two dependent populations using the data obtained from paired samples. We learned in Section 10.4 of Chapter 10 that two samples are paired samples when, for each data value collected from one sample, there is a corresponding data value collected from the second sample, and both data values are collected from the same source. In this section we discuss the small-sample and the large-sample cases to conduct such tests.

The Small-Sample Case

If $n \leq 25$, we use the binomial probability distribution to perform a test about the difference between the medians of paired data. In such a case, Table XI is used to find the critical values of the test statistic. Example 14–5 illustrates this procedure.

Performing sign test about the median of paired differences: small samples.

EXAMPLE 14–5 A researcher wanted to find the effects of a special diet on systolic blood pressure in adults. She selected a sample of 12 adults and put them on this dietary plan for three months. The following table gives the systolic blood pressure of each adult before and after the completion of the plan.

Before	210	185	215	198	187	225	234	217	212	191	226	238
After	196	192	204	193	181	233	208	211	190	186	218	236

Using the 2.5% significance level, can we conclude that the dietary plan reduces the median systolic blood pressure of adults?

Solution We find the sign of the difference between the two blood pressure readings of each adult by subtracting the blood pressure after completion of the dietary plan from the blood pressure before the plan. A plus sign indicates that the plan reduced that person's blood pressure and a minus sign means that it increased blood pressure. Table 14.3 gives the signs of the differences.

Table 14.3

Before	210	185	215	198	187	225	234	217	212	191	226	238
After	196	192	204	193	181	233	208	211	190	186	218	236
Sign of difference (before − after)	+	−	+	+	+	−	+	+	+	+	+	+

Next we perform the five steps for testing the hypothesis.

Step 1. *State the null and alternative hypotheses.*

Let M denote the difference in median blood pressure readings before and after the dietary plan. The null and alternative hypotheses are as follows:

$H_0: M = 0$ (The dietary plan does not reduce the median blood pressure)

$H_1: M > 0$ (The dietary plan reduces the median blood pressure)

The alternative hypothesis is that the dietary plan reduces the median blood pressure, which means that the median systolic blood pressure of all adults after the completion of the dietary plan is lower than the median systolic blood pressure before the completion of the dietary plan. In this case, the median of the paired differences will be greater than zero.

Step 2. *Select the distribution to use.*

The sample size is small (that is, $n = 12 < 25$), and we do not know the shape of the distribution of the population of paired differences. Hence, we use the sign test with the binomial probability distribution.

Step 3. *Determine the rejection and nonrejection regions.*

Because the test is right-tailed, $n = 12$, and $\alpha = .025$, the critical value of X from Table XI is 10. Note that we use the upper critical value of X in Table XI because, as indicated by the sign in H_1, the test is right-tailed. Thus, we will reject the null hypothesis if the observed value of X is greater than or equal to 10, and we will not reject H_0 otherwise. The rejection and nonrejection regions are shown in Figure 14.5.

0 to 9	10 to 12	
Nonrejection region	Rejection region	X

Figure 14.5

Step 4. *Calculate the value of the test statistic.*

In our sample data, the blood pressure of 10 adults decreases and that of 2 adults increases after the dietary plan. Note that there are 10 plus signs and 2 minus signs in Table 14.3. Whenever the test is right-tailed, the observed value of X is equal to the larger of these two numbers. Thus, for our example,

$$\text{Observed value of } X = 10$$

Step 5. *Make a decision.*

Since the observed value of $X = 10$ falls in the rejection region, we reject H_0. Hence, we conclude that the dietary plan reduces the median blood pressure of adults. ■

The Large-Sample Case

In Example 14–5, we used Table XI to find the critical value of the test statistic X. However, Table XI goes up to only $n = 25$. If $n > 25$, we can use the normal distribution as an approximation of the binomial distribution to conduct a test about the difference between the medians of paired data. The following example illustrates such a case.

Performing sign test about the median of paired differences: large samples.

EXAMPLE 14–6 Many students suffer from math anxiety. A statistics professor offered her students a 2-hour lecture on math anxiety and ways to overcome it. A total of 42 students attended this lecture. The students were given similar statistics tests before and after the lecture. Thirty-three of the 42 students scored higher on the test after the lecture, 7 scored lower after the lecture, and 2 scored the same on both tests. Using the 1% significance level, can you conclude that the median score of students increases as a result of attending this lecture? Assume that these 42 students constitute a random sample of all students who suffer from math anxiety.

Solution Let M be the median of the paired differences between scores of students before and after the test, where a paired difference is obtained by subtracting the score after the lecture from the score before the lecture. In other words,

$$\text{Paired difference} = \text{Score before} - \text{Score after}$$

Thus, a positive paired difference means that the score before the lecture is higher than the score after the lecture for that student, and a negative paired difference indicates that the score before the lecture is lower than the score after the lecture for that student. Thus, there are 33 minus signs, 7 plus signs, and 2 zeros.

Step 1. *State the null and alternative hypotheses.*

$$H_0: M = 0 \quad \text{(The lecture does not increase the median score)}$$

$$H_1: M < 0 \quad \text{(The lecture increases the median score)}$$

The alternative hypothesis is that the lecture increases the median score, which means that the median score after the lecture is higher than the median score before the lecture. In this case, the median of the paired differences will be less than zero.

Step 2. *Select the distribution to use.*

Here, $n = 40$. Note that to find the sample size, we drop the students whose score did not change. Because $n > 25$, we can use the normal distribution to test this hypothesis about the median of paired differences.

Step 3. *Determine the rejection and nonrejection regions.*

The test is left-tailed and $\alpha = .01$. From Table VII, the critical value of z for $.5 - .01 = .4900$ is -2.33. Thus, the decision rule is that we will reject the null hypothesis if the observed value of z is -2.33 or smaller, and we will not reject H_0 otherwise. The rejection and nonrejection regions are shown in Figure 14.6.

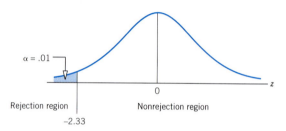

$\alpha = .01$

Rejection region

Nonrejection region

-2.33

0

z

Figure 14.6

Step 4. *Calculate the value of the test statistic.*

If the null hypothesis is true (that is, the lecture does not increase the median score), then we would expect (about) half of the students to score higher and the other half to score lower after the lecture than before. Thus, we would expect (about) half plus signs and half minus signs in the population. In other words, if p is the proportion of plus signs, we would expect $p = .50$ when H_0 is true. Hence, the mean and standard deviation of the binomial distribution are

$$\mu = np = 40(.50) = 20$$

$$\sigma = \sqrt{npq} = \sqrt{40(.50)(.50)} = 3.16227766$$

In our example, 33 of the students scored higher after the lecture and 7 scored lower after the lecture. Thus, there are 33 minus signs and 7 plus signs. We assign the smaller of these two values to X when the test is left-tailed. Here $X = 7$, and the observed value of z is calculated as follows:

$$z = \frac{(X + .5) - \mu}{\sigma} = \frac{(7 + .5) - 20}{3.16227766} = -3.95$$

Step 5. *Make a decision.*

Since the observed value of $z = -3.95$ is less than the critical value of $z = -2.33$, it falls in the rejection region. Consequently, we reject H_0 and conclude that attending the math anxiety lecture increases the median test score. ∎

☞ *Remember* Again, remember that if the test is left-tailed, X is assigned the value equal to the smaller number of plus or minus signs. On the other hand, if the test is right-tailed, X is assigned the value equal to the larger number of plus or minus signs.

EXERCISES

■ **Concepts and Procedures**

14.1 Briefly explain the meaning of categorical data and give two examples.

14.2 When we use the sign test for categorical data, how large a sample size is required to permit the use of the normal distribution for determining the rejection region?

14.3 When we use the sign test for the median of a single population, how small must the sample size be to require the use of Table XI ?

14.4 When we use the sign test for the difference between the medians of two dependent populations, how large must n be for the large-sample case?

14.5 Determine the rejection region for each of the following sign tests for categorical data.

a. $H_0: p = .50,$ $H_1: p > .50,$ $n = 15,$ $\alpha = .05$
b. $H_0: p = .50,$ $H_1: p \neq .50,$ $n = 20,$ $\alpha = .01$
c. $H_0: p = .50,$ $H_1: p < .50,$ $n = 30,$ $\alpha = .05$

14.6 In each case below, n is the sample size, p is the proportion of the population possessing a certain characteristic, and X is the number of items in the sample that possess that characteristic. In each case, perform the appropriate sign test using $\alpha = .05$.

a. $n = 14,$ $X = 10,$ $H_0: p = .50,$ $H_1: p > .50$
b. $n = 10,$ $X = 1,$ $H_0: p = .50,$ $H_1: p \neq .50$
c. $n = 30,$ $X = 12,$ $H_0: p = .50,$ $H_1: p < .50$
d. $n = 27,$ $X = 20,$ $H_0: p = .50,$ $H_1: p > .50$

14.7 In each case below, n is the sample size and X is the appropriate number of plus or minus signs as defined in Section 14.1.2. In each case, perform the appropriate sign test using $\alpha = .05$.

a. $n = 10,$ $X = 8,$ $H_0:$ Median $= 28,$ $H_1:$ Median > 28
b. $n = 11,$ $X = 1,$ $H_0:$ Median $= 100,$ $H_1:$ Median < 100
c. $n = 26,$ $X = 3,$ $H_0:$ Median $= 180,$ $H_1:$ Median $\neq 180$
d. $n = 30,$ $X = 6,$ $H_0:$ Median $= 55,$ $H_1:$ Median < 55

14.8 In each case below, M is the difference between two population medians, n is the sample size, and X is the appropriate number of plus or minus signs as defined at the end of Section 14.1.3. In each case, perform the appropriate sign test using $\alpha = .01$.

a. $n = 20,$ $X = 6,$ $H_0: M = 0,$ $H_1: M < 0$
b. $n = 8,$ $X = 8,$ $H_0: M = 0,$ $H_1: M > 0$
c. $n = 29,$ $X = 4,$ $H_0: M = 0,$ $H_1: M \neq 0$

■ **Applications**

14.9 A company has developed two new video games, A and B. The company takes a random sample of 10 children and allows each child to play each game for 2 hours. The order in which each child plays the games is determined randomly. The preferences of the 10 children are shown here.

A A B A A A B A A A

At the 5% significance level, can you conclude that children prefer either of these two games over the other?

14.10 A consumer organization wanted to compare two rival brands of infant car seats, Brand A and Brand B. Fifteen families, each with a child under 12 months of age, were selected at random. Each family tested each of the two brands of car seats for one week. The order in which each family tried the two brands was decided by a coin toss. At the end of two weeks, each family indicated which brand it preferred. The preferences of these 15 families are listed here. The 0 indicates that one family had no preference.

A A A B A A B A A A A 0 A B A

At the 5% level of significance, can you conclude that families prefer Brand A over Brand B?

14.11   Twenty randomly chosen loyal drinkers of JW's Beer are tested to see if they can distinguish between JW's and its chief rival. Each of the 20 drinkers is given two unmarked cups, one containing JW's and the other containing the rival brand. Thirteen of the drinkers correctly indicate which cup contains JW's, but the other seven are incorrect. At the 2.5% level of significance, can you conclude that drinkers of JW's Beer are more likely to correctly identify it than not?

14.12   Three weeks before an election for state senator, a poll of 200 randomly selected voters shows that 95 voters favor the Republican candidate, 85 favor the Democratic candidate, and the remaining 20 have no opinion. Using the sign test, can you conclude that voters prefer one candidate over the other? Use $\alpha = .01$.

14.13   One hundred randomly chosen adult residents of North Dakota are asked whether they would prefer to live in another state or to stay in North Dakota. Of these 100 adults, 55 indicate that they would like to move to another state, 41 would prefer to stay, and 4 have no preference. At the 2.5% level of significance, can you conclude that less than half of all adult residents of North Dakota would prefer to stay?

14.14   Three hundred randomly chosen doctors were asked, Which is the most important single factor in weight control: diet or exercise? Of these 300 doctors, 162 felt that diet was more important, 117 favored exercise, and 21 thought that diet and exercise were equally important. At the 1% level of significance, can you conclude that for all doctors, the number who favor diet exceeds the number who favor exercise?

14.15   In a 1999 poll of Americans' attitudes on health insurance, about half (51%) of adults surveyed said that treatment for common mental health conditions, such as depression and eating disorders, is absolutely essential under the basic health care coverage (*Newsweek*, November 8, 1999). Suppose that in a recent random sample of 500 adults, 270 indicated that such coverage is absolutely essential. Using the sign test, can you conclude that more than half of all adults consider such coverage absolutely essential? Use $\alpha = .01$.

14.16   A past study claims that adults in the United States spend a median of 18 hours a week on leisure activities. A researcher took a sample of 10 adults and asked them how many hours they spend per week on leisure activities. She obtained the following data:

$$14 \quad 25 \quad 22 \quad 38 \quad 16 \quad 26 \quad 19 \quad 23 \quad 41 \quad 33$$

Using $\alpha = .05$, can you conclude that the median amount of time spent per week on leisure activities by all adults is more than 18 hours?

14.17   The manager of a soft-drink bottling plant wants to see if the median amount of soda in 12-ounce bottles differs from 12 ounces. Ten filled bottles are selected at random from the bottling machine, and the amount of soda in each is carefully measured. The results (in ounces) follow:

$$12.10 \quad 11.95 \quad 12.00 \quad 12.01 \quad 12.02 \quad 12.05 \quad 12.02 \quad 12.03 \quad 12.04 \quad 12.06$$

Using the 5% level of significance, can you conclude that the median amount of soda in all such bottles differs from 12 ounces?

14.18   On March 8, 1999, ESPN Sports reported that the median salary of players in the National Hockey League (NHL) was $600,000. Suppose that a recent random sample of 9 NHL players yielded the following salaries (in thousands of dollars):

$$800 \quad 1280 \quad 460 \quad 3500 \quad 700 \quad 580 \quad 910 \quad 2200 \quad 1250$$

Using the 5% level of significance, can you conclude that the median salary of all NHL players now exceeds $600,000?

14.19   A city police department claims that its median response time to 911 calls in the inner city is 4 minutes or less. Shown below is a random sample of 28 response times (in minutes) to 911 calls in the inner city.

$$6 \quad 5 \quad 7 \quad 12 \quad 2 \quad 1.5 \quad 3.5 \quad 4 \quad 10 \quad 11 \quad 4.5 \quad 6 \quad 5 \quad 8.5$$
$$7 \quad 15 \quad 9 \quad 8 \quad 3 \quad 10 \quad 8 \quad 4.5 \quad 9 \quad 4 \quad 6 \quad 3 \quad 6 \quad 7.5$$

Using $\alpha = .01$, can you conclude that the median response time to all 911 calls in the inner city is longer than 4 minutes?

14.20 According to data from the U.S. Bureau of the Census, the median age of men marrying for the first time was 26.8 years in 1997 (*The World Almanac and Book of Facts*, 1999). The following data give the ages of a recent random sample of 30 men marrying for the first time:

28	25	31	42	32	17	21	36	29	24
19	27	35	45	31	50	23	40	26	33
26	22	33	37	32	28	24	21	41	19

Using the 2% level of significance, can you conclude that the current median age of men marrying for the first time differs from 26.8 years?

14.21 The following numbers are the times served (in months) by 35 prison inmates who were released recently.

37	6	20	5	25	30	24	10	12	20
24	8	26	15	13	22	72	80	96	33
84	86	70	40	92	36	28	90	36	32
72	45	38	18	9					

Using $\alpha = .01$, test the null hypothesis that the median time served by all such prisoners is 42 months against the alternative hypothesis that the median time served is less than 42 months.

14.22 Twelve sixth-grade boys who are underweight are put on a special diet for one month. Each boy is weighed before and after the one-month dietary regime. The weights (in pounds) of these boys are recorded here.

Before	65	63	71	60	66	72	78	74	58	59	77	65
After	70	68	75	60	69	70	81	81	66	56	79	71

Can you conclude that this diet increases the median weight of all such boys? Use the 2.5% level of significance. Assume that these 12 boys constitute a random sample of all underweight sixth-grade boys.

14.23 Refer to Exercise 10.55 of Chapter 10. The following table shows the self-confidence test scores of seven employees before and after they attended a course on building self-confidence.

Before	8	5	4	9	6	9	5
After	10	8	5	11	6	7	9

At the 5% level of significance, can you conclude that attending this course increases the median self-confidence test score of all employees?

14.24 The manager at a large factory suspects that the night-shift workers use more hours of sick leave than the day-shift workers. The following table gives the number of hours of sick leave used by workers in each shift for a period of 12 working days. Note that both shifts have the same number of workers.

Day shift	20	32	12	24	16	0	22	8	10	38	16	12
Night shift	16	56	0	28	36	24	40	29	30	26	32	20

Using the 5% level of significance, can you conclude that the median number of hours of sick leave used by workers is lower for the day shift than for the night shift?

14.25   A real estate brokerage agency wants to compare the work of two appraisers, A and B. Each appraiser is given the same list of 27 houses and asked to assess the value of each house. The following table gives the results of the 27 assessments for each appraiser. (The numbers are the appraised values of homes in thousands of dollars.)

House	Appraiser A	Appraiser B	House	Appraiser A	Appraiser B
1	98	100	15	200	215
2	80	85	16	199.5	210
3	77	75	17	179	190
4	101	102	18	165	159.5
5	105	108	19	133	139.5
6	110	116	20	139	151
7	125	120.5	21	149.5	153
8	130	137	22	177	190
9	108	116	23	225	217.5
10	99	98.5	24	239	253
11	143	149	25	181	179
12	165	175	26	173	189
13	150	159	27	199	182
14	145	142.5			

Using the 2% level of significance, can you conclude that the median assessed values for all such houses given by the two appraisers are the same?

14.26   A researcher suspects that two medical laboratories, A and B, tend to give different results when determining the cholesterol content of blood samples. The researcher obtains blood samples from 30 randomly selected adults and divides each sample into two parts. One part of each blood sample is sent to Lab A, the other part to Lab B. Each lab determines the cholesterol content of each of its 30 samples and reports the results to the researcher. The following table gives the cholesterol levels (in milligrams per hundred milliliters) reported by the two labs.

Sample	Lab A	Lab B	Sample	Lab A	Lab B
1	135	137	16	214	202
2	202	195	17	255	242
3	239	250	18	233	217
4	210	202	19	246	231
5	180	185	20	292	262
6	195	195	21	229	212
7	188	177	22	170	172
8	200	204	23	261	243
9	320	300	24	310	281
10	290	269	25	302	277
11	285	271	26	283	264
12	210	216	27	221	199
13	185	176	28	208	211
14	194	184	29	344	321
15	181	182	30	170	164

Can you conclude that the median cholesterol level for all such adults as determined by Lab A is higher than that determined by Lab B? Use a significance level of 1%.

14.27   A dairy agency wants to test a hormone that may increase cows' milk production. Some members of the group fear that the hormone could actually decrease production, so a "matched pairs" test is arranged. Thirty randomly selected cows are given the hormone, and their milk production is recorded for four weeks. Each of these 30 cows is matched with another cow of similar size, age, and prior record of milk production. This second group of 30 cows do not receive the hormone. The milk production of these cows is recorded for the same time period. In 19 of these 30 pairs, the cow taking the hormone produced more milk; in 9 of the pairs, the cow taking the hormone produced less; and in two of the pairs, there was no difference. Using the 5% level of significance, can you conclude that the hormone changes the median milk production of such cows?

14.2 THE WILCOXON SIGNED-RANK TEST FOR TWO DEPENDENT SAMPLES

The **Wilcoxon signed-rank test** for two dependent (paired) samples is used to test whether or not the two populations from which these samples are drawn are identical. We can also test the alternative hypothesis that one population distribution lies to the left or to the right of the other. Actually, the null hypothesis in this test states that the medians of the two population distributions are equal. The alternative hypothesis states that the medians of the two populations are not equal, or that the median of the first population is less than that of the second population, or that the median of the first population is greater than that of the second population. This test is an alternative to the paired-samples test discussed in Section 10.4.2 of Chapter 10. In that section, we assumed that the paired differences have a normal distribution. Here, in the Wilcoxon signed-rank test, we do not make that assumption. In this test, we rank the absolute differences between the pairs of data values collected from two samples and then assign them the sign based on which of the paired data values is larger. Then we compare the sums of the ranks with plus and minus signs and make a decision.

The Small-Sample Case

If the *sample size is 15 or smaller*, we find the critical value of the test statistic, denoted by T, from Table XII in Appendix D, which gives the critical values of T for the Wilcoxon signed-rank test. We also calculate the observed value of the test statistic differently in this test. However, when $n > 15$, we can use the normal distribution to perform the test. Example 14–7 describes the small-sample case for the Wilcoxon signed-rank test.

Performing the Wilcoxon signed-rank test for two dependent populations: small samples.

EXAMPLE 14–7 A private agency claims that the crash course it offers significantly increases the writing speed of secretaries. The following table gives the writing speeds of eight secretaries before and after they attended this course.

Before	84	75	88	91	65	71	90	75
After	97	72	93	110	78	69	115	75

Using the 2.5% significance level, can you conclude that attending this course increases the writing speed of secretaries? Use the Wilcoxon signed-rank test.

Solution We use the five steps to make the hypothesis test.

Step 1. *State the null and alternative hypotheses.*

> H_0: The crash course does not increase the writing speed of secretaries
>
> H_1: The crash course does increase the writing speed of secretaries

Note that the alternative hypothesis states that the population distribution of writing speeds of secretaries moves to the right after they attend the crash course. In other words, the center of the population distribution of writing speeds after the crash course is greater than the center of the population distribution of writing speeds before the crash course. If we measure the centers of the two populations by their respective medians, with M_A the median of the population distribution after the course and M_B the median of the population distribution before the course, we can rewrite the two hypotheses as follows:

$$H_0: M_A = M_B$$

$$H_1: M_A > M_B$$

Step 2. *Select the distribution to use.*

We are making a test for paired samples, and the distribution of paired differences is unknown. Since $n < 15$, we use the Wilcoxon signed-rank test procedure for the small-sample case.

Step 3. *Determine the rejection and nonrejection regions.*

As mentioned earlier, we denote the test statistic in this case by T. The critical value of T is found from Table XII, which lists the critical values of T for the Wilcoxon signed-rank test for small samples ($n \leq 15$). Our test is right-tailed because the alternative hypothesis is that the "after" distribution lies to the right of the "before" distribution. Also, $\alpha = .025$ and $n = 7$. Note that for one pair of data, both values are the same, 75. We drop such cases when determining the sample size for the test. From Table XII, the critical value of T is 2. Thus, our decision rule will be: Reject H_0 if the observed value of T is less than or equal to the critical value of T, which is 2. Note that in the Wilcoxon signed-rank test, the null hypothesis is rejected if the observed value of T is less than or equal to the critical value of T. This rule is true for a two-tailed, a right-tailed, or a left-tailed test. The observed value of T is calculated differently, depending on whether the test is two-tailed or one-tailed. This is explained in the next step. Figure 14.7 shows the rejection and nonrejection regions.

0, 1, or 2	3 or higher	
Rejection region	Nonrejection region	T

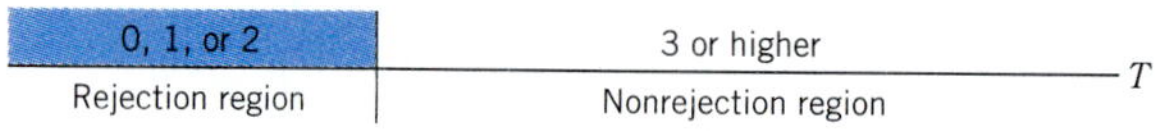

Figure 14.7

> **DECISION RULE** For the Wilcoxon signed-rank test for small samples ($n \leq 15$), the critical value of T is obtained from Table XII. Note that in the Wilcoxon signed-rank test, the decision rule is to reject the null hypothesis if the observed value of T is less than or equal to the critical value of T. This rule is true for a two-tailed, a right-tailed, or a left-tailed test.

Step 4. *Calculate the value of the test statistic.*

The observed value of the test statistic, T, is calculated as follows. The given data on writing speeds before and after the course are reproduced in the first two columns of Table 14.4.

Table 14.4

Before	After	Differences (Before − After)	Absolute Differences	Ranks of Differences	Signed Ranks
84	97	−13	13	4.5	−4.5
75	72	+3	3	2	+2
88	93	−5	5	3	−3
91	110	−19	19	6	−6
65	78	−13	13	4.5	−4.5
71	69	+2	2	1	+1
90	115	−25	25	7	−7
75	75	0	0	—	—

1. We obtain the differences column by subtracting each data value after the course from the corresponding data value before the course. Thus,

   Difference = Writing speed before the course − Writing speed after the course

   These differences are listed in the third column of Table 14.4.

2. In the fourth column, we write the absolute values of differences. In other words, the numbers in the fourth column of the table are the same as those in the third column, but without plus and minus signs.

3. Next, we rank the absolute differences listed in the fourth column from lowest to highest. These ranks are listed in the fifth column. Note that the difference of zero is not ranked and is dropped from the sample. Among the remaining absolute differences, the smallest difference is 2, which is assigned a rank of 1. The next smallest absolute difference is 3, which is assigned a rank of 2. Next, the absolute difference of 5 is given a rank of 3. Then, two absolute differences have the same value, which is 13. We assign the average of the next two ranks, $(4 + 5)/2 = 4.5$, to these two values. Thus, as a rule, whenever some of the absolute differences have the same value, they are all assigned the average of their ranks.

4. In the last column of Table 14.4, we write the ranks of the fifth column with the signs of the corresponding paired differences. For example, the first difference of -13 has a minus sign in the third column. Consequently, we assign a minus sign to its rank of 4.5 in the sixth column. The second difference of 3 has a positive sign. Hence, its rank of 2 is assigned a positive sign.

5. Next, we add all the positive ranks and we add the absolute values of the negative ranks separately. Thus, we obtain:

   Sum of the positive ranks $= 2 + 1 = 3$

   Sum of the absolute values of the negative ranks $= 4.5 + 3 + 6 + 4.5 + 7 = 25$

The observed value of the test statistic is determined as shown in the next box.

OBSERVED VALUE OF THE TEST STATISTIC T

I. If the test is two-tailed with the alternative hypothesis that the two distributions are not the same, then the observed value of T is given by the smaller of the two sums, the sum of the positive ranks and the sum of the absolute values of the negative ranks. We will reject H_0 if the observed value of T is less than or equal to the critical value of T.

II. If the test is right-tailed with the alternative hypothesis that the distribution of *after* values is to the right of the distribution of *before* values, then the observed value of T is given by the sum of the values of the positive ranks. We will reject H_0 if the observed value of T is less than or equal to the critical value of T.

III. If the test is left-tailed with the alternative hypothesis that the distribution of *after* values is to the left of the distribution of *before* values, then the observed value of T is given by the sum of the absolute values of the negative ranks. We will reject H_0 if the observed value of T is less than or equal to the critical value of T.

Remember, for the above to be true, the paired difference is defined as the *before* value minus the *after* value. In other words, the differences are obtained by subtracting the *after* values from the *before* values.

Our example is a right-tailed test. Hence,

$$\text{Observed value of } T = \text{sum of the positive ranks} = 3$$

Step 5.   *Make a decision.*

Whether the test is two-tailed, left-tailed, or right-tailed, we will reject the null hypothesis if:

$$\text{Observed value of } T \leq \text{Critical value of } T$$

where the observed value of T is calculated as explained in Step 4. In this example, the observed value of T is 3 and the critical value of T is 2. Since the observed value of T is greater than the critical value of T, we do not reject H_0. Hence, we conclude that the crash course does not seem to increase the writing speed of secretaries. ∎

The Large-Sample Case

If $n > 15$, we can use the normal distribution to make a test of hypothesis about the paired differences. Example 14–8 illustrates the procedure for making such a test.

Performing the Wilcoxon signed-rank test for paired populations: large samples.

EXAMPLE 14–8   The manufacturer of a gasoline additive claims that the use of its additive increases gasoline mileage. A random sample of 25 cars was selected, and these cars were driven for one week without the gasoline additive and then for one week with the additive. Then the miles per gallon (mpg) were estimated for these cars without and with the additive. Next, the paired differences were calculated for these 25 cars, where a paired difference is defined as

$$\text{Paired difference} = \text{mpg without additive} - \text{mpg with additive}$$

The differences were positive for 4 cars, negative for 19 cars, and zero for 2 cars. First, the absolute values of the paired differences were ranked, and then these ranks were assigned the signs of the corresponding paired differences. The sum of the ranks of the positive paired differences was 58, and the sum of the absolute values of the ranks of the negative paired differences was 218. Can you conclude that the use of the additive increases gasoline mileage? Use the 1% significance level.

Solution   We perform the five steps to conduct this test of hypothesis.

Step 1.   *State the null and alternative hypotheses.*

We are to test whether or not the gasoline additive increases gasoline mileage. This will be true if the distribution of gasoline mileages with the additive lies to the right of the distribution of gasoline mileage without the additive. The median mileage with the additive will be higher than the median mileage without the additive. Let M_A and M_B be the median mileage after (with) and before (without) the gasoline additive. Then the null and the alternative hypotheses can be written as follows:

$$H_0: M_A = M_B$$
$$H_1: M_A > M_B$$

Step 2.   *Select the distribution to use.*

We are given information about the sums of positive and negative ranks. The sample size is greater than 15. We use the Wilcoxon signed-rank test procedure with the normal distribution approximation.

Step 3. *Determine the rejection and nonrejection regions.*

We are using the normal distribution as an approximation to make this test. Hence, we will find the critical value of *z* from Table VII in Appendix D. The test is right-tailed. The significance level is .01, which gives the area between the mean and the critical point as .5 − .01 = .4900. Therefore, the critical value of *z* is 2.33. The rejection and nonrejection regions are shown in Figure 14.8.

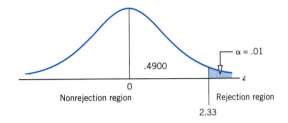

Figure 14.8

Step 4. *Calculate the value of the test statistic.*

Because the sample size is larger than 15, the test statistic *T* follows a normal distribution.

OBSERVED VALUE OF *z* In a Wilcoxon signed-rank test for two dependent samples, when the sample size is large ($n > 15$), the observed value of *z* for the test statistic *T* is calculated as

$$z = \frac{T - \mu_T}{\sigma_T}$$

where

$$\mu_T = \frac{n(n + 1)}{4} \quad \text{and} \quad \sigma_T = \sqrt{\frac{n(n + 1)(2n + 1)}{24}}$$

The value of *T* that is used to calculate the value of *z* is determined based on the alternative hypothesis, as explained next.

1. If the test is two-tailed with the alternative hypothesis that the two distributions are not the same, then the value of *T* may be equal to either of the two sums, the sum of the positive ranks or the sum of the absolute values of the negative ranks. We will reject H_0 if the observed value of *z* falls in either of the rejection regions.

2. If the test is right-tailed with the alternative hypothesis that the distribution of *after* values is to the right of the distribution of *before* values, then the value of *T* is equal to the sum of the absolute values of the negative ranks. We will reject H_0 if the observed value of *z* is greater than or equal to the critical value of *z*.

3. If the test is left-tailed with the alternative hypothesis that the distribution of *after* values is to the left of the distribution of *before* values, then the value of *T* is equal to the sum of the absolute values of the negative ranks. We will reject H_0 if the observed value of *z* is less than or equal to the critical value of *z*.

Remember, for the above to be true, the paired difference is defined as the *before* value minus the *after* value. In other words, the differences are obtained by subtracting the *after* values from the *before* values.

Using the given information, we calculate the values of μ_T and σ_T and the observed value of z as follows. Note that here $n = 23$ because two of the paired differences are zero.

$$\mu_T = \frac{n(n + 1)}{4} = \frac{23(23 + 1)}{4} = 138$$

$$\sigma_T = \sqrt{\frac{n(n + 1)(2n + 1)}{24}} = \sqrt{\frac{23(23 + 1)(46 + 1)}{24}} = 32.87856445$$

$$z = \frac{T - \mu_T}{\sigma_T} = \frac{218 - 138}{32.87856445} = 2.43$$

Step 5. *Make a decision.*

The observed value of $z = 2.43$ falls in the rejection region. Hence, we reject the null hypothesis and conclude that the gasoline additive increases mileage. ∎

EXERCISES

■ **Concepts and Procedures**

14.28 When would you use the Wilcoxon signed-rank test procedure instead of the paired-samples test of Chapter 10?

14.29 Explain how the null hypothesis is usually stated in the Wilcoxon signed-rank test.

14.30 How are ranks assigned to two or more absolute differences that have the same value in the Wilcoxon signed-rank test?

14.31 Determine the rejection region for the Wilcoxon signed-rank test for each of the following. Indicate whether the rejection region is based on T or z.

a. $n = 10$, $H_0: M_A = M_B$, $H_1: M_A > M_B$, $\alpha = .05$
b. $n = 12$, $H_0: M_A = M_B$, $H_1: M_A \neq M_B$, $\alpha = .01$
c. $n = 20$, $H_0: M_A = M_B$, $H_1: M_A < M_B$, $\alpha = .025$
d. $n = 30$, $H_0: M_A = M_B$, $H_1: M_A > M_B$, $\alpha = .01$

14.32 In each case, perform the Wilcoxon signed-rank test.

a. $n = 8$, $T = 5$,   left-tailed test using $\alpha = .05$
b. $n = 15$, $T = 20$,   right-tailed test using $\alpha = .01$
c. $n = 25$, $T = 51$,   two-tailed test using $\alpha = .02$
d. $n = 36$, $T = 238$,   left-tailed test using $\alpha = .01$

■ **Applications**

14.33 Refer to Exercise 10.101 of Chapter 10, which deals with Gamma Corporation's installation of governors on its salespersons' cars to regulate their speeds. The following table gives the number of contacts made by each of seven randomly selected sales representatives during the week before governors were installed and the number of contacts made during the week after installation.

Salesperson	A	B	C	D	E	F	G
Before	50	63	42	55	44	65	66
After	49	60	47	51	50	60	58

a. Using the Wilcoxon signed-rank test at the 5% level of significance, can you conclude that the use of governors tends to reduce the number of contacts made per week by Gamma Corporation's sales representatives?

b. Compare your conclusions of part a with the result of the hypothesis test that was performed (using the t distribution) in Exercise 10.101.

14.34 Refer to Exercise 10.101 of Chapter 10. The following table gives the gas mileage (in miles per gallon) for each of seven randomly selected sales representatives' cars during the week before governors were installed and the gas mileage in the week after installation.

Salesperson	A	B	C	D	E	F	G
Before	25	21	27	23	19	18	20
After	26	24	26	25	24	22	23

a. Using the Wilcoxon signed-rank test at the 5% level of significance, can you conclude that the use of governors tends to increase the median gas mileage for Gamma Corporation's sales representatives?
b. Compare your conclusion of part a with the result of the hypothesis test that was performed (using the *t* distribution) in Exercise 10.101.

14.35 Refer to Exercise 14.23. The following table shows the self-confidence test scores of seven employees before and after they attended a course designed to build self-confidence.

Before	8	5	4	9	6	9	5
After	10	8	5	11	6	7	9

a. Using the Wilcoxon signed-rank test at the 5% level of significance, can you conclude that attending this course increases the median self-confidence test score of employees?
b. Compare your conclusion of part a with the result of Exercise 14.23.

14.36 Refer to Exercise 14.25, which gives the values assigned to 27 houses by two appraisers, A and B.

a. Using the Wilcoxon signed-rank test at the 2% level of significance, can you conclude that the median assessed values for all such houses are the same for both appraisers?
b. Compare your conclusion of part a with the result of the sign test performed in Exercise 14.25.

14.37 Twenty randomly selected adults who describe themselves as "couch potatoes" were given a six-week course in physical fitness. Before starting the course, each adult took a 2-mile hike on the same trail. The time required to complete the hike was recorded for each adult. After finishing the course, they all took the same hike again and their times were recorded again. The following table lists the times recorded (in minutes) before and after the course for each of the 20 adults.

Before	After	Before	After	Before	After
41	37	64	55	100	78
91	71	37	31	48	40
35	30	54	57	50	48
58	64.5	70	59	94	102
45	44	40	33	42.5	40
48.5	44	78	70.5	75	63
84	78	66	56		

Does the fitness course appear to reduce the time required to complete the 2-mile hike? Use the Wilcoxon signed-rank test at the 2.5% level of significance.

14.38 Many adults in the United States have accumulated excessive balances on their credit cards. Ninety such adults were randomly chosen to participate in a group therapy program designed to reduce this debt. Each adult's total credit card balance was recorded twice: before the program began and three months after the program ended. The paired difference was then calculated for each adult by subtracting the balance after the program ended from the balance before the program began. These paired differences were positive for 49 adults and negative for 41 adults. The sum of the ranks of the positive paired differences was 2507, and the sum of the absolute values of the ranks of the negative paired differences was 1588. Can you conclude that this group therapy program reduces credit card debt? Use the 5% level of significance.

14.3 THE WILCOXON RANK SUM TEST FOR TWO INDEPENDENT SAMPLES

In Chapter 10 we discussed tests of hypotheses about the difference between the means of two independent populations using the normal and t distributions. In this section we compare two independent populations using the results obtained from samples drawn from these populations. In a **Wilcoxon rank sum test**, we assume that the two populations have identical shapes but differ only in location, which is measured by the median. Note that identical shapes do not mean that they have to have a normal distribution. To apply this test, we must be able to rank the given data. Note that the Wilcoxon rank sum test is almost identical to the **Mann–Whitney test**.

In the hypothesis tests discussed in this section, the null hypothesis is usually that the two population distributions are identical. The alternative hypothesis can be that the two population distributions are not identical or that one distribution is to the right of the other or that one distribution is to the left of the other. Assuming that the null hypothesis is true and that the two populations are identical, we rank all the (combined) data values of the two samples as if they were drawn from the same population. Any tied data values are assigned the ranks in the same manner as in the preceding section. Then we sum the ranks for the data values of each sample separately. If the two populations are identical, the ranks should be spread randomly (and evenly) between the two samples. In this case, the sums of the ranks for the two samples should be almost equal, given that the sizes of the two samples are almost the same. However, if one of the two samples contains mostly lower ranks and the other contains mostly higher ranks, then the sums of the ranks for the two samples will be quite different. The larger the difference in the sums of the ranks of the two populations, the more convincing is the evidence that the two populations are not identical and that the null hypothesis is not true.

In this section, we discuss the Wilcoxon rank sum test first for small samples and then for large samples.

The Small-Sample Case

If *the sizes of both samples are 10 or less*, we use the Wilcoxon rank sum test for small samples. Example 14–9 illustrates how the test is performed. To make this test, the population that corresponds to the smaller sample is labeled population 1 and the one that corresponds to the larger sample is called population 2. The respective samples are sample 1 and sample 2. If the sizes of the two samples are equal, either of the two populations can be labeled population 1.

Performing the Wilcoxon rank sum test for two independent populations: small samples.

EXAMPLE 14–9 A researcher wants to determine whether the distributions of daily crimes in two cities are identical. The following data give the numbers of violent crimes on eight randomly selected days for City A and on nine days for City B.

City A	12	21	16	8	26	13	19	23	
City B	18	25	14	16	23	19	28	20	31

Using the 5% significance level, can you conclude that the distributions of daily crimes in the two cities are different?

Solution We apply the following five steps to perform the hypothesis test.

Step 1. *State the null and alternative hypotheses.*

We are to test if the two populations are identical or different. Hence, the two hypotheses are as follows:

H_0: The population distributions of daily crimes in the two cities are identical

H_1: The population distributions of daily crimes in the two cities are different

Step 2. *Select the distribution to use.*

Let the distribution of daily crimes in City A be called population 1 (note that it corresponds to the smaller sample) and that in City B be called population 2. The respective samples are called sample 1 and sample 2. Because $n_1 < 10$ and $n_2 < 10$, we use the Wilcoxon rank sum test for small samples.

Step 3. *Determine the rejection and nonrejection regions.*

The test statistic in Wilcoxon's rank sum test is denoted by T. The critical value or values of T are obtained from Table XIII in Appendix D. In this table, T_L gives the lower critical value and T_U gives the upper critical value. If the test is two-tailed, we use both T_L and T_U. For a left-tailed test we use T_L only, and for a right-tailed test we use T_U only.

In our example, the test is two-tailed. Also, $\alpha = .05$, $n_1 = 8$, and $n_2 = 9$. Hence, from Table XIII, the values of T_L and T_U are 51 and 93, respectively. We will reject the null hypothesis if the observed value of T is either less than or equal to T_L or greater than or equal to T_U. The rejection and nonrejection regions are shown in Figure 14.9. Thus, the decision rule is that we will reject H_0 if either the observed value of $T \leq 51$ or the observed value of $T \geq 93$.

51 or lower	52 to 92	93 or higher	
Rejection region	Nonrejection region	Rejection region	**Figure 14.9**

Step 4. *Calculate the value of the test statistic.*

To find the observed value of T, first we rank all the data values of both samples as if they belonged to the same population. Then we find the sum of the ranks for each sample separately. The observed value of the test statistic T is given by the sum of the ranks for the smaller sample. If the sizes of the samples are the same, we can use either of the rank sums as the observed value of T.

In Table 14.5, we rank all the values of both samples and find the sum of the ranks for each sample. Note that 8 is the smallest data value in both samples. Hence, it is assigned a rank of 1. The next smallest value in both samples is 12, which is assigned a rank of 2. The remaining values are assigned ranks in the same way.

Table 14.5

City A		City B	
Crimes	**Rank**	**Crimes**	**Rank**
12	2	18	7
21	11	25	14
16	5.5	14	4
8	1	16	5.5
26	15	23	12.5
13	3	19	8.5
19	8.5	28	16
23	12.5	20	10
		31	17
	Sum = 58.5		Sum = 94.5

Since $n_1 = 8$ and $n_2 = 9$, the sample size for City A is smaller. Hence, the observed value of T is given by the sum of the ranks for city A. Thus,

$$\text{Observed value of } T = 58.5$$

Step 5.   *Make a decision.*

Comparing the observed value of T with T_L and T_U (obtained from Table XIII in Step 3), we see that the observed value of $T = 58.5$ is between $T_L = 51$ and $T_U = 93$. Hence, we do not reject H_0, and we conclude that the two population distributions seem to be identical.   ∎

In the next box, the Wilcoxon rank sum test procedure for small samples is described for two-tailed, right-tailed, and left-tailed tests.

WILCOXON RANK SUM TEST FOR SMALL INDEPENDENT SAMPLES

1. *A two-tailed test:* The null hypothesis is that the two population distributions are identical, and the alternative hypothesis is that the two population distributions are different. The critical values of T, T_L and T_U, for this test are obtained from Table XIII for the given significance level and sample sizes. The observed value of T is given by the sum of the ranks for the smaller sample. The null hypothesis is rejected if $T \leq T_L$ or $T \geq T_U$. Otherwise, the null hypothesis is not rejected.

   Note that if the two sample sizes are equal, the observed value of T is given by the sum of the ranks for either sample.

2. *A right-tailed test:* The null hypothesis is that the two population distributions are identical, and the alternative hypothesis is that the distribution of population 1 (the population that corresponds to the smaller sample) lies to the right of the distribution of population 2. The critical value of T is given by T_U in Table XIII for the given α for a one-tailed test and the given sample sizes. The observed value of T is given by the sum of the ranks for the smaller sample. The null hypothesis is rejected if $T \geq T_U$. Otherwise, the null hypothesis is not rejected.

   Note that if the two sample sizes are equal, the observed value of T is given by the sum of the ranks for sample 1.

3. *A left-tailed test:* The null hypothesis is that the two population distributions are identical, and the alternative hypothesis is that the distribution of population 1 (the population that corresponds to the smaller sample) lies to the left of the distribution of population 2. The critical value of T in this case is given by T_L in Table XIII for the given α for a one-tailed test and the given sample sizes. The observed value of T is given by the sum of the ranks for the smaller sample. The null hypothesis is rejected if $T \leq T_L$. Otherwise the null hypothesis is not rejected.

   Note that if the two sample sizes are equal, the observed value of T is given by the sum of the ranks for sample 1.

The Large-Sample Case

If *either n_1 or n_2 or both n_1 and n_2 are greater than 10,* we use the normal distribution as an approximation to the Wilcoxon rank sum test for two independent samples.

> **OBSERVED VALUE OF z** In the case of a large sample, the observed value of z is calculated as
>
> $$z = \frac{T - \mu_T}{\sigma_T}$$
>
> Here, the sampling distribution of the test statistic T is approximately normal with mean μ_T and standard deviation σ_T. The values of μ_T and σ_T are calculated as
>
> $$\mu_T = \frac{n_1(n_1 + n_2 + 1)}{2} \quad \text{and} \quad \sigma_T = \sqrt{\frac{n_1 n_2(n_1 + n_2 + 1)}{12}}$$
>
> Note that in these calculations sample 1 refers to the smaller sample and sample 2 to the larger sample. However, if the two samples are of the same size, we can label either one sample 1. The value of T used in the calculation of z is given by the sum of the ranks for sample 1.

The critical value or values of z are obtained from Table VII in Appendix D for the given significance level. We will reject the null hypothesis if the observed value of z is in the rejection region. Otherwise, we will not reject H_0. Example 14–10 illustrates the procedure for performing such a test.

Performing the Wilcoxon rank sum test for two independent populations: large samples.

EXAMPLE 14–10   A researcher wanted to find out whether job-related stress is lower for college and university professors than for physicians. She took random samples of 14 professors and 11 physicians and tested them for job-related stress. The following data give the stress levels for professors and physicians on a scale of 1 to 20, where 1 is the lowest level of stress and 20 is the highest.

Professors	5	9	4	12	6	15	2	8	10	4	6	11	8	3
Physicians	10	18	12	5	13	18	14	9	6	16	11			

Using the 1% significance level, can you conclude that the job-related stress level for professors is lower than that for physicians?

Solution   Because the smaller sample should be labeled sample 1, the sample of 11 physicians will be called sample 1 and that of 14 professors will be called sample 2. The respective populations are populations 1 and 2. Thus, $n_1 = 11$ and $n_2 = 14$. We perform the five steps of the hypothesis test.

Step 1.   *State the null and alternative hypotheses.*

We are to test whether or not professors have lower job-related stress than physicians. Because physicians are labeled population 1 and professors population 2, professors will have a lower stress level if the distribution of population 1 is to the right of the distribution of population 2. Thus, we can state the two hypotheses as follows.

H_0: The two population distributions are identical

H_1: The distribution of population 1 is to the right of the distribution of population 2

Step 2.   *Select the distribution to use.*

Because $n_1 > 10$ and $n_2 > 10$, we use the normal distribution to make this test as the test statistic T follows an approximately normal distribution.

Step 3.   *Determine the rejection and nonrejection regions.*

The test is right-tailed and $\alpha = .01$. The area between the mean and the critical point under the normal distribution curve is $.5 - .01 = .4900$. From Table VII in Appendix D, the critical value of z for .4900 is 2.33. The rejection and nonrejection regions are shown in Figure 14.10. Thus, we will reject H_0 if the observed value of z is 2.33 or greater. Otherwise, we will not reject H_0.

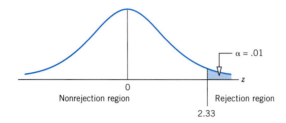

Figure 14.10

Step 4.   *Calculate the value of the test statistic.*

Table 14.6 shows the rankings of all the data values for the two samples and the sums of these ranks for each sample separately.

Table 14.6

Physicians		Professors	
Stress Level	**Rank**	**Stress Level**	**Rank**
10	14.5	5	5.5
18	24.5	9	12.5
12	18.5	4	3.5
5	5.5	12	18.5
13	20	6	8
18	24.5	15	22
14	21	2	1
9	12.5	8	10.5
6	8	10	14.5
16	23	4	3.5
11	16.5	6	8
		11	16.5
		8	10.5
		3	2
	Sum = 188.5		Sum = 136.5

Hence, we calculate the value of the test statistic as follows:

$$\mu_T = \frac{n_1(n_1 + n_2 + 1)}{2} = \frac{11(11 + 14 + 1)}{2} = 143$$

$$\sigma_T = \sqrt{\frac{n_1 n_2(n_1 + n_2 + 1)}{12}} = \sqrt{\frac{11(14)(11 + 14 + 1)}{12}} = 18.26654501$$

$$z = \frac{T - \mu_T}{\sigma_T} = \frac{188.5 - 143}{18.26654501} = 2.49$$

Thus, the observed value of z is 2.49. Note that in the calculation of z, we used the value of T that belongs to sample 1, which should always be the case.

Step 5. *Make a decision.*

Because the observed value of $z = 2.49$ is greater than the critical value of $z = 2.33$, it falls in the rejection region. Hence, we reject H_0 and conclude that the distribution of population 1 is to the right of the distribution of population 2. Thus, the job-related stress level of physicians is higher than that of professors. This can also be stated as the job-related stress level of professors is lower than that of physicians. ■

In the next box, the Wilcoxon rank sum test procedure for large samples is described for two-tailed, right-tailed, and left-tailed tests.

WILCOXON RANK SUM TEST FOR LARGE INDEPENDENT SAMPLES When $n_1 > 10$ *or* $n_2 > 10$ (or both samples are greater than 10), the distribution of T (the sum of the ranks of the smaller of the two samples) is approximately normal with mean and standard deviation as follows:

$$\mu_T = \frac{n_1(n_1 + n_2 + 1)}{2} \quad \text{and} \quad \sigma_T = \sqrt{\frac{n_1 n_2(n_1 + n_2 + 1)}{12}}$$

For two-tailed, right-tailed, and left-tailed tests, first calculate T, μ_T, σ_T, and the value of the test statistic, $z = (T - \mu_T)/\sigma_T$. If $n_1 = n_2$, T can be calculated from either sample 1 or sample 2.

1. *A two-tailed test:* The null hypothesis is that the two population distributions are identical, and the alternative hypothesis is that the two population distributions are different. At significance level α, the critical values of z are obtained from Table VII in Appendix D. The null hypothesis is rejected if the observed value of z falls in the rejection region.

2. *A right-tailed test:* The null hypothesis is that the two population distributions are identical, and the alternative hypothesis is that the distribution of population 1 (the population with the smaller sample size) lies to the right of the distribution of population 2. At significance level α, the critical value of z is obtained from Table VII in Appendix D. The null hypothesis is rejected if the observed value of z falls in the rejection region.

3. *A left-tailed test:* The null hypothesis is that the two population distributions are identical, and the alternative hypothesis is that the distribution of population 1 (the population with the smaller sample size) lies to the left of the distribution of population 2. At significance level α, the critical value of z is found from Table VII of Appendix D. The null hypothesis is rejected if the observed value of z falls in the rejection region.

EXERCISES

■ **Concepts and Procedures**

14.39 Explain what determines whether to use the Wilcoxon signed-rank test or the Wilcoxon rank sum test.

14.40 Find the rejection region for the Wilcoxon rank sum test in each of the following cases.

a. $n_1 = 7$, $n_2 = 8$,   right-tailed test using $\alpha = .05$
b. $n_1 = 10$, $n_2 = 10$,   two-tailed test using $\alpha = .10$

c. $n_1 = 18,$ $\qquad n_2 = 20,$ $\qquad$ left-tailed test using $\alpha = .05$
d. $n_1 = 25,$ $\qquad n_2 = 25,$ $\qquad$ two-tailed test using $\alpha = .01$

14.41 In each of the following cases, perform the Wilcoxon rank sum test.

a. $n_1 = 6,$ $\qquad n_2 = 7,$ $\qquad T = 22,$ $\qquad$ two-tailed test with $\alpha = .05$
b. $n_1 = 10,$ $\qquad n_2 = 12,$ $\qquad T = 137,$ $\qquad$ right-tailed test with $\alpha = .025$
c. $n_1 = 9,$ $\qquad n_2 = 11,$ $\qquad T = 68,$ $\qquad$ left-tailed test with $\alpha = .05$
d. $n_1 = 22,$ $\qquad n_2 = 23,$ $\qquad T = 638,$ $\qquad$ two-tailed test with $\alpha = .01$

■ **Applications**

14.42 A consumer agency wants to compare the caffeine content of two brands of coffee. Eight jars of each brand are analyzed, and the amount of caffeine found in each jar is recorded as shown in the table.

Brand I	82	77	85	73	84	79	81	82
Brand II	75	80	76	81	72	74	73	78

Using $\alpha = .10$, can you conclude that the two brands have different median caffeine contents per jar?

14.43 In a winter Olympics trial for women's speed skating, seven skaters use a new type of skate, while eight others use the traditional type. Each skater is timed (in seconds) in the 500-meter event. The results are given in the following table.

New skates	40.5	40.3	39.5	39.7	40.0	39.9	41.5	
Traditional skates	41.0	40.8	40.9	39.8	40.6	40.7	41.1	40.5

Assuming that these 15 skaters make up a random sample of all Olympic-class 500-meter female speed skaters, can you conclude that the new skates tend to produce faster times in this event? Use the 5% level of significance.

14.44 A new employee of a downtown insurance firm is given the choice to work either from 8:00 A.M. to 5:00 P.M. or from 8:30 A.M. to 5:30 P.M. The main factor considered by this employee in making a decision is the total commute time. He decides to spend four weeks comparing the two schedules. The first day he starts work at 8:00, the next day at 8:30, and he continues alternating his work schedule this way for a total of 20 workdays. The following table shows his total (morning plus evening) commute time (in minutes) for each day.

8:00–5:00	61	57	53	80	60	59	55	58	87	61
8:30–5:30	50	65	59	78	52	63	64	57	66	56

At the 5% level of significance, can you conclude that the median commute times are different for the two work schedules?

14.45 A factory's management is concerned about the number of defective parts produced by its machinists. Management suspects that production may be improved by giving machinists frequent breaks to reduce fatigue. Twenty-four randomly chosen machinists are randomly divided into two groups (A and B) of 12 each. During the next week all 24 machinists work to manufacture similar parts. The workers in Group A get a 5-minute break every hour, whereas the workers in Group B stay on the usual schedule. The number of good parts produced by each machinist during the week is recorded in the following table.

Group A	157	139	188	143	172	144	191	128	177	160	175	162
Group B	160	118	150	165	158	159	127	133	170	164	152	142

At the 1% level of significance, can you conclude that the median number of good parts produced by machinists who take a 5-minute break every hour is higher than the median number of good parts produced by the machinists who do not take a break?

14.46 Two brands of tires are tested to compare their durability. Eleven Brand X tires and 12 Brand Y tires are tested on a machine that simulates road conditions. The mileages (in thousands of miles for each tire) are shown in the following table.

Brand X	51	55	53	49	50.5	57	54.5	48.5	51.5	52	53.5	
Brand Y	48	47	54	55.5	50	51	46	49.5	52.5	51	49	45

Using the 5% level of significance, can you conclude that the median mileage for Brand X tires is greater than the median mileage for Brand Y tires?

14.47 During the first two weeks of July, Lakeside Summer Camp hosts 10- to 12-year-old children. Between 2:00 and 5:00 P.M. each day, campers are free to choose among outdoor activities (such as boating or volleyball) and indoor pastimes (such as video games or television). A child may devote any, all, or none of this 3-hour period to outdoor activities each day. For five days of this two-week session, 25 randomly selected girls and 23 randomly selected boys were monitored. All 48 children were in good health, and all five days were suitable for outdoor activities. The total amounts of time spent on outdoor activities during these hours for the five days were recorded for each child. All 48 of these times were ranked. The sum of the ranks for the 25 girls was 674; the sum of the ranks for the boys was 502. Using $\alpha = .05$, can you conclude that the median times spent on outdoor activities differed for boys and girls?

14.4 THE KRUSKAL-WALLIS TEST

In Chapter 12 we used the one-way analysis of variance (ANOVA) procedure to test whether or not the means of three or more populations are all equal. To apply the ANOVA procedure using the F distribution, we assumed that the populations from which the samples were drawn were normally distributed with equal variance, σ^2. However, if the populations being sampled are not normally distributed, then we cannot apply the ANOVA procedure of Chapter 12. In such cases, we can use the **Kruskal–Wallis test**, also called the Kruskal–Wallis H test. This is a nonparametric test because to use it we do not make any assumptions about the distributions of the populations being sampled. The only assumption we make is that all populations under consideration have identical shapes but differ only in location, which is measured by the median. Note that identical shapes do not mean that they have to have a normal distribution.

In a Kruskal–Wallis test, the null hypothesis is that the population distributions under consideration are all identical. The alternative hypothesis is that at least one of the population distributions differs and that, therefore, not all of the population distributions are identical. Note that we use the Kruskal–Wallis test to compare three or more populations. Also note that to apply the Kruskal–Wallis test, the size of each sample must be at least five.

> **KRUSKAL-WALLIS TEST** To perform the Kruskal–Wallis test, we use the chi-square distribution that was discussed in Chapter 11. The test statistic in this test is denoted by H, which follows (approximately) the chi-square distribution. The critical value of H is obtained from Table IX in Appendix D for the given level of significance and $df = k - 1$, where k is the number of populations under consideration. *Note that the Kruskal–Wallis test is always right-tailed.*

To find the observed value of the test statistic H, we first rank the combined data from all samples in the same way as in a Wilcoxon rank sum test. The tied data values are handled in the same way as in a Wilcoxon test. Then the observed value of H is calculated as explained in the next box.

OBSERVED VALUE OF THE TEST STATISTIC H   The observed value of the test statistic H is calculated using the following formula:

$$H = \frac{12}{n(n+1)} \left(\frac{R_1^2}{n_1} + \frac{R_2^2}{n_2} + \cdots + \frac{R_k^2}{n_k} \right) - 3(n+1)$$

where

$R_1 = $ sum of the ranks for sample 1

$R_2 = $ sum of the ranks for sample 2
.
.
.

$R_k = $ sum of the ranks for sample k

$n_1 = $ sample size for sample 1

$n_2 = $ sample size for sample 2
.
.
.

$n_k = $ sample size for sample k

$n = n_1 + n_2 + \cdots + n_k$

$k = $ number of samples

The test statistic H measures the extent to which the k samples differ with regard to the ranks assigned to their data values. Basically, H is a measure of the variance of ranks (or of the variance of the means of ranks) for different samples. If all k samples have exactly the same mean of ranks, H will have a value of zero. The value of H becomes larger as the difference between the means of ranks for different samples increases. Thus, a larger observed value of H indicates that the distributions of the given populations do not seem to be identical.

Example 14–11 illustrates the procedure for applying the Kruskal–Wallis test.

Performing the Kruskal–Wallis test.

EXAMPLE 14-11   A researcher wanted to find out whether the population distributions of salaries of computer programmers are identical in three cities, Boston, San Francisco, and Atlanta. Three different samples—one from each city—produced the following data on the annual salaries (in thousands of dollars) of computer programmers.

Boston	San Francisco	Atlanta
43	54	57
39	33	68
62	58	60
73	38	44
51	43	39
46	55	28
	34	49
		57

Using the 2.5% significance level, can you conclude that the population distributions of salaries for computer programmers in these three cities are all identical?

Solution   We apply the five steps to perform this hypothesis test.

Step 1.   *State the null and alternative hypotheses.*

H_0: The population distributions of salaries of computer programmers in the three cities are all identical

H_1: The population distributions of salaries of computer programmers in the three cities are not all identical

Note that the alternative hypothesis states that the population distribution of at least one city is different from those of the other two cities.

Step 2.   *Select the distribution to use.*

The shapes of the population distributions are unknown. We are comparing three populations. Hence, we apply the Kruskal–Wallis procedure to perform this test, and we use the chi-square distribution.

Step 3.   *Determine the rejection and nonrejection regions.*

In this example,

$$\alpha = .025 \quad \text{and} \quad df = k - 1 = 3 - 1 = 2$$

Hence, from Table IX in Appendix D, the critical value of χ^2 is 7.378, as shown in Figure 14.11.

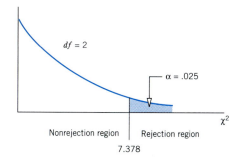

Figure 14.11

Step 4.   *Calculate the value of the test statistic.*

To calculate the observed value of the test statistic H, we first rank the combined data for all three samples and find the sum of ranks for each sample separately. This is done in Table 14.7.

Table 14.7

Boston		San Francisco		Atlanta	
Salary	**Rank**	**Salary**	**Rank**	**Salary**	**Rank**
43	7.5	54	13	57	15.5
39	5.5	33	2	68	20
62	19	58	17	60	18
73	21	38	4	44	9
51	12	43	7.5	39	5.5
46	10	55	14	28	1
		34	3	49	11
				57	15.5
$n_1 = 6$	$R_1 = 75$	$n_2 = 7$	$R_2 = 60.5$	$n_3 = 8$	$R_3 = 95.5$

We have

$$n = n_1 + n_2 + n_3 = 6 + 7 + 8 = 21$$

and

$$
\begin{aligned}
H &= \frac{12}{n(n+1)} \left(\frac{R_1^2}{n_1} + \frac{R_2^2}{n_2} + \cdots + \frac{R_k^2}{n_k} \right) - 3(n+1) \\
&= \frac{12}{21(21+1)} \left(\frac{(75)^2}{6} + \frac{(60.5)^2}{7} + \frac{(95.5)^2}{8} \right) - 3(21+1) \\
&= 1.543
\end{aligned}
$$

Step 5. *Make a decision.*

Since the observed value of $H = 1.543$ is less than the critical value of $H = 7.378$ and it falls in the nonrejection region, we do not reject the null hypothesis. Consequently, we conclude that the population distributions of salaries of computer programmers in the three cities seem to be all identical. ∎

EXERCISES

■ **Concepts and Procedures**

14.48 Briefly explain when the Kruskal–Wallis test is used to make a test of hypothesis.

14.49 What assumption that is required for the ANOVA procedure of Chapter 12 is not necessary for the Kruskal–Wallis test?

14.50 Describe the form of the null and alternative hypotheses for a Kruskal–Wallis test.

14.51 Here, n_i is the size of the ith sample and R_i is the sum of ranks for the ith sample. For each of the following cases, perform the Kruskal–Wallis test using the 5% level of significance.

a. $n_1 = 9,$ $n_2 = 8,$ $n_3 = 5,$ $R_1 = 81,$ $R_2 = 102,$ $R_3 = 70$
b. $n_1 = n_2 = n_3 = n_4 = 5,$ $R_1 = 27,$ $R_2 = 30,$ $R_3 = 83,$ $R_4 = 70$
c. $n_1 = 6,$ $n_2 = 10,$ $n_3 = 6,$ $R_1 = 93,$ $R_2 = 70,$ $R_3 = 90$
d. $n_1 = 8,$ $n_2 = 9,$ $n_3 = 8,$ $n_4 = 10,$ $n_5 = 9,$ $R_1 = 210,$
   $R_2 = 195,$ $R_3 = 178,$ $R_4 = 212,$ $R_5 = 195$

14.52 The following table gives the ranked data for three samples. Perform the Kruskal–Wallis test using the 1% level of significance.

Sample I	Sample II	Sample III
3	14	2
1	11	4.5
10	16	13
7	15	4.5
9	12	8
6		

■ **Applications**

14.53 Refer to Examples 12–2 and 12–3 of Chapter 12. Fifteen randomly selected fourth-grade students were randomly assigned to three groups, and each group was taught arithmetic by a different method. At the end of the semester, all 15 students took the same arithmetic test. Their test scores are given in the following table.

Method I	Method II	Method III
48	55	84
73	85	68
51	70	95
65	69	74
87	90	67

a. At the 1% level of significance, can you reject the null hypothesis that the median arithmetic test scores of all fourth-grade students taught by each of these three methods are the same?

b. Compare your answer to part a with the result of the hypothesis test in Example 12–3.

14.54 A consumer agency investigated the premiums charged by four auto insurance companies. The agency randomly selected five drivers insured by each company who had similar driving records, autos, and insurance coverage. The following table gives the monthly premiums paid by the 20 drivers.

Company A	Company B	Company C	Company D
$65	$48	$57	$62
73	69	61	53
54	88	89	45
43	75	77	51
70	72	69	44

Can you reject the null hypothesis that the distributions of auto insurance premiums paid per month by all such drivers are the same for all four companies? Use $\alpha = .05$.

14.55 Refer to Problem 10 of the Self-Review Test in Chapter 12. The following table gives the hourly wages (in dollars) of computer programmers for random samples taken in three cities.

New York	Boston	Los Angeles
19.45	23.50	31.75
28.80	18.60	11.40
36.45	29.75	29.40
33.10	30.00	33.30
31.50	35.40	24.60
39.30	26.40	19.50
35.50		28.30

a. Test the null hypothesis that the median hourly wage for all computer programmers in each of these three cities is the same. Use the 1% level of significance.

b. Compare your conclusion of part a with the answer to part a of Problem 10 of the Self-Review Test in Chapter 12.

14.56 Refer to Exercise 12.27 of Chapter 12. A consumer agency tested a few samples of paints from each of three companies to compare their drying times. The following table lists the drying times (in minutes) for these samples.

Company A	Company B	Company C
42	57	45
53	63	45
43	61	51
47	54	58
40	49	44
51	60	41
56		47

a. Using the 5% significance level, test the null hypothesis that the population distributions of drying times for paints from these three companies are identical.

b. Compare your conclusion of part a with that of part a of Exercise 12.27 of Chapter 12.

c. What would your decision be if the probability of making a Type I error was zero? Explain.

14.57 A factory operates three shifts a day, five days per week, each with the same number of workers and approximately the same level of production. The following table gives the number of defective parts produced during each shift over a period of five days.

First Shift	Second Shift	Third Shift
23	25	33
36	35	44
32	41	50
40	38	52
45	50	60

At the 5% level of significance, can you conclude that the median number of defective parts is the same for all three shifts?

14.58 A consumer group wanted to compare the service time at three fast-food restaurants, Al's, Eduardo's, and Patel's. Every Tuesday and Wednesday for four weeks, three staff members of the group were randomly assigned to these three restaurants. Each staff member went to his or her assigned restaurant and ordered a hamburger, fries, and a Coke and then recorded the time that elapsed from entering the restaurant until receiving the food. The service times (in minutes) for these eight days for the three restaurants are listed.

Al's	Eduardo's	Patel's
7.0	3.3	1.1
8.3	11.0	2.4
6.9	5.7	1.8
1.3	8.1	3.0
6.7	6.6	4.1
7.1	13.0	12.0
5.5	2.3	1.5
6.6	5.9	3.1

Assume that these service times make up random samples of all service times at the respective restaurants. At the 10% level of significance, can you conclude that there is any difference in the median service times at these three restaurants?

14.5 THE SPEARMAN RHO RANK CORRELATION COEFFICIENT TEST

In Chapter 13 we discussed the linear correlation coefficient between two variables x and y. We also learned how to make a test of hypothesis about the population correlation coefficient ρ using the information from a sample. In that chapter we used the t distribution to perform this test about ρ. However, to use the procedure of Chapter 13 and to use the t distribution to make this test of hypothesis about ρ require that both variables x and y are normally distributed.

The **Spearman rho rank correlation coefficient** (Spearman's rho) is a nonparametric analog of the linear correlation coefficient of Chapter 13. It helps us decide what type of relationship, if any, exists between data from populations with unknown distributions. The Spearman rho rank correlation coefficient is denoted by r_s for sample data and by ρ_s for population data. This correlation coefficient is simply the linear correlation coefficient between the ranks of the data on variables x and y. To make a test of hypothesis about the Spearman rho rank correlation coefficient, we do not need to make any assumptions about the populations of x and y variables.

> **SPEARMAN RHO RANK CORRELATION COEFFICIENT** The Spearman rho rank correlation coefficient is denoted by r_s for sample data and by ρ_s for population data. This correlation coefficient is simply the linear correlation coefficient between the ranks of the data. To calculate the value of r_s, we rank the data for each variable, x and y, separately and denote those ranks by u and v, respectively. Then we take the difference between each pair of ranks and denote it by d. Thus,
>
> $$\text{Difference between each pair of ranks} = d = u - v$$
>
> Next, we square each difference d and add these squared differences to find Σd^2. Finally, we calculate the value of r_s using the formula:
>
> $$r_s = 1 - \frac{6\Sigma d^2}{n(n^2 - 1)}$$
>
> In a test of hypothesis about the Spearman rho rank correlation coefficient ρ_s, the test statistic is r_s and its observed value is calculated by using the above formula.

Example 14–12 shows how to calculate the Spearman rho rank correlation coefficient ρ_s and how to perform a test of hypothesis about ρ_s.

Performing the Spearman rho rank correlation coefficient test.

EXAMPLE 14–12 Suppose we want to investigate the relationship between the per capita income (in thousands of dollars) and the infant mortality rate (in percent) for different states. The following table gives data on these two variables for a random sample of eight states.

Per capita income (x)	29.85	19.0	19.18	31.78	25.22	16.68	23.98	26.33
Infant mortality (y)	8.3	10.1	10.3	7.1	9.9	11.5	8.7	9.8

Based on these data, can you conclude that there is no significant (linear) correlation between the per capita incomes and the infant mortality rates for all states? Use $\alpha = .05$.

Solution We perform the five steps to test the null hypothesis that there is no correlation between the two variables against the alternative hypothesis that there is a significant correlation.

Step 1. *State the null and alternative hypotheses.*

The null and alternative hypotheses are as follows:

H_0: There is no correlation between per capita incomes and infant mortality rates in all states

H_1: There is a correlation between per capita incomes and infant mortality rates in all states

If we denote the Spearman correlation coefficient by ρ_s, the null hypothesis and the alternative hypothesis can be written as

$$H_0: \rho_s = 0$$

$$H_1: \rho_s \neq 0$$

Note that this is a two-tailed test.

Step 2. *Select the distribution to use.*

Because the sample is taken from a small population and the variables do not follow a normal distribution, we use the Spearman rho rank correlation coefficient test procedure to make this test.

Step 3. *Determine the rejection and nonrejection regions.*

The test statistic that is used to make this test is r_s, and its critical values are given in Table XIV in Appendix D. Note that, for this example,

$$n = 8 \quad \text{and} \quad \alpha = .05$$

To read the critical value of r_s from Table XIV in Appendix D, we locate 8 in the column labeled n and .05 in the top row of the table for a two-tailed test. The critical values of r_s are $\pm.738$, or $+.738$ and $-.738$. Thus, we will reject the null hypothesis if the observed value of r_s is either $-.738$ or less, or $+.738$ or greater. The rejection and nonrejection regions for this example are shown in Figure 14.12.

Rejection region   Nonrejection region   Rejection region r_s

$-.738$ $+.738$ **Figure 14.12**

CRITICAL VALUE OF r_s The critical value of r_s is obtained from Table XIV in Appendix D for the given sample size and significance level. If the test is two-tailed, we use two critical values, one negative and one positive. However, we use only the negative value of r_s if the test is left-tailed, and only the positive value of r_s if the test is right-tailed.

Step 4. *Calculate the value of the test statistic.*

In the Spearman rho rank correlation coefficient test, the test statistic is denoted by r_s, which is simply the linear correlation coefficient between the ranks of the data. As explained in the beginning of this section, to calculate the observed value of r_s, we use the following formula:

$$r_s = 1 - \frac{6\Sigma d^2}{n(n^2 - 1)}$$

where $d = u - v$, and u and v are the ranks of variables x and y, respectively.

Table 14.8 shows the ranks for x and y, which are denoted by u and v, respectively. The table also lists the values of d, d^2, and Σd^2. Note that if two or more values are equal, we use the average of their ranks for all of them. Hence, the observed value of r_s is

$$r_s = 1 - \frac{6(160)}{8(64 - 1)} = 1 - \frac{960}{504} = -.905$$

Table 14.8

u	7	2	3	8	5	1	4	6	
v	2	6	7	1	5	8	3	4	
d	5	−4	−4	7	0	−7	1	2	
d^2	25	16	16	49	0	49	1	4	$\Sigma d^2 = 160$

Note that Spearman's rho rank correlation coefficient has the same properties as the linear correlation coefficient (discussed in Chapter 13). Thus, $-1 \leq r_s \leq 1$ or $-1 \leq \rho_s \leq 1$, depending on whether sample or population data are used to calculate the Spearman rho rank correlation coefficient. If $\rho_s = 0$, there is no relationship between the x and y data. If $0 < \rho_s \leq 1$, on average a larger value of x is associated with a larger value of y. Similarly, if $-1 \leq \rho_s < 0$, on average a larger value of x is associated with a smaller value of y.

Step 5. *Make a decision.*

Since $r_s = -.905$ is less than $-.738$ and it falls in the rejection region, we reject H_0 and conclude that there is a correlation between the per capita incomes and the infant mortality rates in all states. Because the value of r_s from the sample is negative, we can also state that as per capita income increases, infant mortality tends to decrease. ∎

DECISION RULE FOR THE SPEARMAN RHO RANK CORRELATION COEFFICIENT   The null hypothesis is always H_0: $\rho_s = 0$. The observed value of the test statistic is always the value of r_s computed from the sample data. Let α denote the significance level, and $-c$ and $+c$ are the critical values for the Spearman rho rank correlation coefficient test obtained from Table XIV.

1.   For a two-tailed test, the alternative hypothesis is H_1: $\rho_s \neq 0$. If $\pm c$ are the critical values corresponding to sample size n and two-tailed α, we reject H_0 if either $r_s \leq -c$ or $r_s \geq +c$; that is, reject H_0 if r_s is "too small" or "too large."
2.   For a right-tailed test, the alternative hypothesis is H_1: $\rho_s > 0$. If $+c$ is the critical value corresponding to sample size n and one-sided α, we reject H_0 if $r_s \geq +c$; that is, reject H_0 if r_s is "too large."
3.   For a left-tailed test, the alternative hypothesis is H_1: $\rho_s < 0$. If $-c$ is the critical value corresponding to sample size n and one-sided α, we reject H_0 if $r_s \leq -c$; that is, reject H_0 if r_s is "too small."

EXERCISES

■ **Concepts and Procedures**

14.59   What assumptions that are required for hypothesis tests about the linear correlation coefficient ρ in Chapter 13 are not required for testing a hypothesis about the Spearman rho rank correlation coefficient?

14.60   Two sets of paired data on two variables, x and y, have been ranked. In each case, the ranks for x and y are denoted by u and v, respectively, and are shown in the tables at the top of page 658. Calculate the Spearman rho rank correlation coefficient for each case.

a.

u	2	1	3	4	6	5	7	8
v	8	6	7	4	5	2	1	3

b.

u	1	2	3	4	5	6	7
v	4	2	1	5	3	7	6

14.61   Calculate the Spearman rho rank correlation coefficient for each of the following data sets.

a.

x	5	10	15	20	25	30
y	17	15	12	14	10	9

b.

x	27	15	32	21	16	40	8
y	95	81	102	88	75	120	62

14.62   Perform the indicated hypothesis test in each of the following cases.

a. $n = 9$, $r_s = .575$, $H_0: \rho_s = 0$, $H_1: \rho_s > 0$, $\alpha = .025$
b. $n = 15$, $r_s = -.575$, $H_0: \rho_s = 0$, $H_1: \rho_s < 0$, $\alpha = .005$
c. $n = 20$, $r_s = .554$, $H_0: \rho_s = 0$, $H_1: \rho_s \neq 0$, $\alpha = .01$
d. $n = 20$, $r_s = .554$, $H_0: \rho_s = 0$, $H_1: \rho_s > 0$, $\alpha = .01$

■ **Applications**

14.63   The following data are a random sample of the heights (in inches) and weights (in pounds) of 10 NBA players selected at random.

Height	84	76	79	79	84	74	83	81	83	75
Weight	240	208	205	215	265	182	225	220	250	190

a.   Based on the reasonable assumption that as height increases, weight tends to increase, do you expect the value of r_s to be positive or negative? Why?
b.   Compute the value of r_s. Does it agree with your expectation of its value in part a?

14.64   Let ρ_s be the Spearman rho rank correlation coefficient between heights (in inches) and weights (in pounds) for the entire population of NBA players listed in Data Set II. Using the value of r_s computed from the sample data in Exercise 14.63, test the null hypothesis $H_0: \rho_s = 0$ against the alternative hypothesis $H_1: \rho_s > 0$ at the significance level of $\alpha = .01$.

14.65   In Example 13–1 of Chapter 13, we estimated the regression line for the data given in Table 13.2 on food expenditures (in hundred dollars) and incomes (in hundred dollars). Those data are reproduced here.

Income (x)	35	49	21	39	15	28	25
Food expenditure (y)	9	15	7	11	5	8	9

The estimated regression line in Example 13–1 had a slope of .2642.

a.   Do you expect the Spearman rho rank correlation coefficient for these data to be positive or negative? Why?
b.   Compute r_s for these data. Did it come out as expected?

14.66 In Example 13–7 of Chapter 13, the null hypothesis H_0: $\rho = 0$ was tested against the alternative hypothesis H_1: $\rho > 0$, where ρ is the linear correlation coefficient for the population. At a significance level of $\alpha = .01$, H_0: $\rho = 0$ was rejected there. Hence, it seemed that in fact $\rho > 0$.

a. If ρ_s is the Spearman rho rank correlation coefficient for the entire population for food expenditure and income data, and you test H_0: $\rho_s = 0$ against H_1: $\rho_s > 0$ at the significance level of $\alpha = .01$, based on the results of Example 13–7, do you expect to reject or accept H_0? Why?

b. Perform the hypothesis test stated in part a.

14.67 The following table shows the combined (math and verbal) SAT scores (denoted by x) and college grade point averages (denoted by y) upon completion of a bachelor's degree for nine randomly chosen recent college graduates who had taken the SAT test.

x	1105	990	1040	1215	1405	975	1300	1010	1080
y	3.33	2.62	3.05	3.60	3.85	2.43	3.90	2.40	2.95

a. Find the Spearman rho rank correlation coefficient for this data set.

b. Test H_0: $\rho_s = 0$ against H_1: $\rho_s > 0$ using the 5% level of significance.

c. Does your test result indicate a positive relationship between the variables x and y?

14.68 A day-care center operator is concerned about the aggressive behavior of young boys left in her care. She feels that prolonged watching of television tends to promote aggressive behavior. She selects seven boys at random and ranks them according to the level of aggressiveness in their behavior, with a rank of 7 indicating most aggressive and a rank of 1 denoting least aggressive. Then she asks each boy's parent(s) to estimate the average number of hours per week the boy spends watching television. The following table shows the data collected on the aggressiveness ranks of these boys and the number of hours spent watching TV per week.

Aggressiveness rank	1	2	3	4	5	6	7
Weekly TV hours	15	21	28	8	24	32	20

a. Calculate r_s for these data.

b. At the 5% level of significance, can you conclude that there is a positive relationship between aggressiveness and hours spent watching television?

14.6 THE RUNS TEST FOR RANDOMNESS

The **runs test for randomness** tests the null hypothesis that a sequence of events has occurred randomly against the alternative hypothesis that this sequence of events has not occurred randomly.

The Small-Sample Case

As an example of a runs test, in apartment complexes families with children are often assigned to units close to one another to lessen the impact of noise on childless tenants. We would like to determine whether a landlord has randomly assigned units to tenants irrespective of whether they have children, or has tried to cluster tenants with children. Here what we really mean by *random* is independence in the sense that if, say, a childless tenant lives in unit 3, this provides no additional information about whether the tenants in units 2 and 4 are childless. With nonrandomness, we would have some additional information.

Suppose that a part of the apartment complex consists of a single building with 10 adjacent units numbered 1 to 10. We specify the family status of the 10 tenants via a string of

10 letters, using C for "has children" and D for "no children." One possible string is D D C D C C D D C C, which would mean that the tenants in units 3, 5, 6, 9, and 10 have children and those in 1, 2, 4, 7, and 8 do not. (Note that in our example the numbers of tenants with and without children are equal, but they do not have to be equal.)

Two extreme arrangements that may not be random are C C C C C D D D D D and C D C D C D C D C D. In the first arrangement, all the tenants with children are adjacent, as are those without children. In the second case, tenants with and without children alternate perfectly. Note that in these two examples exactly half of the tenants have children and half do not.

A characteristic of a string of the letters C and D that helps determine the randomness of the string is called a **run**. A run is a sequence of the same symbol (in this case the letter C or D) appearing one or more times. The arrangement C C C C C D D D D D has two runs; the arrangement C D C D C D C D C D has 10 runs. Intuitively, we know that in the string with two runs the arrangement is not random because there are too few runs, whereas in the string with 10 runs the arrangement is not random due to too many runs.

If the number of tenants with children, the number without children, and their arrangement in the 10 units are all random, then the number of runs in the arrangement, which we denote by R, will also be random. Thus, R is a statistic with its own sampling distribution. Table XV in Appendix D gives the critical values of R for a significance level of 5%—that is, for $\alpha = .05$. There are two parameters associated with the distribution of R, n_1 and n_2. Here n_1 is the number of times the first symbol (in our example the letter C) appears in the string, and n_2 is the number of times the second symbol (in our example the letter D) appears. Table XV provides critical values of R for values of n_1 and n_2 up to 15. If either $n_1 > 15$ or $n_2 > 15$, we can apply the normal approximation (discussed later in the section on the large-sample case) to perform the test. For each pair of n_1 and n_2, there are two critical values: a smaller value (denoted by c_1) and a larger value (denoted by c_2).

RUN A *run* is a sequence of one or more consecutive occurrences of the same outcome in a sequence of occurrences in which there are only two outcomes. The number of runs in a sequence is denoted by R. The value of R obtained for a sequence of outcomes for a sample gives the observed value of the test statistic for the runs test for randomness.

Suppose we formally set up the following hypotheses:

H_0: Tenants with and without children are randomly mixed among the 10 units

H_1: These tenants are not randomly mixed

We will reject H_0 if either of the following occurs:

$$R \leq c_1 \text{ (too few runs)} \quad \text{or} \quad R \geq c_2 \text{ (too many runs)}$$

Let's apply these rules to the hypothetical strings listed earlier to determine whether or not to reject H_0 at a significance level of $\alpha = .05$.

1. Let the string of letters be C C C C C D D D D D. Here $n_1 = 5$, $n_2 = 5$, and $R = 2$. From Table XV, $c_1 = 2$ and $c_2 = 10$. Since $R \leq c_1$, we reject H_0 on the basis that there are too few runs.

2. Let the string of letters be C D C D C D C D C D. Here $n_1 = 5$, $n_2 = 5$, and $R = 10$. From Table XV, $c_1 = 2$ and $c_2 = 10$. Since $R \geq c_2$, we reject H_0 on the basis that there are too many runs.

3. Let the string of letters be D D C D C C D D C C. Here $n_1 = 5$, $n_2 = 5$, and $R = 6$. From Table XV, $c_1 = 2$ and $c_2 = 10$. Since the value of $R = 6$ is between $c_1 = 2$ and $c_2 = 10$, we do not reject H_0.

Example 14–13 illustrates the application of the runs test for randomness.

Performing the runs test for randomness: small sample.

EXAMPLE 14-13 A college admissions office is interested in knowing whether applications for admission arrive randomly with respect to gender. The genders of 25 consecutively arriving applications were found to arrive in the following order (here M denotes a male applicant and F a female applicant).

M F M M F F F M F M M M F F F F M M M F F M F M M

Can you conclude that the applications for admission arrive randomly with respect to gender? Use $\alpha = .05$.

Solution   We perform the following five steps in this hypothesis test.

Step 1.   *State the null and alternative hypotheses.*

> H_0: Applications arrive in a random order with respect to gender
>
> H_1: Applications do not arrive in a random order with respect to gender

Step 2.   *Select the distribution to use.*

Let n_1 and n_2 be the number of male and female applicants, respectively. Then

$$n_1 = 13 \quad \text{and} \quad n_2 = 12$$

Since both n_1 and n_2 are less than 15, we use the runs test to check for randomness.

Step 3.   *Determine the rejection and nonrejection regions.*

For $n_1 = 13$, $n_2 = 12$, and $\alpha = .05$, the critical values from Table XV are $c_1 = 8$ and $c_2 = 19$. Thus, we will not reject the null hypothesis if the observed value of R is in the interval 9 to 18. We will reject the null hypothesis if the observed value of R is either 8 or lower or 19 or higher. The rejection and nonrejection regions are shown in Figure 14.13.

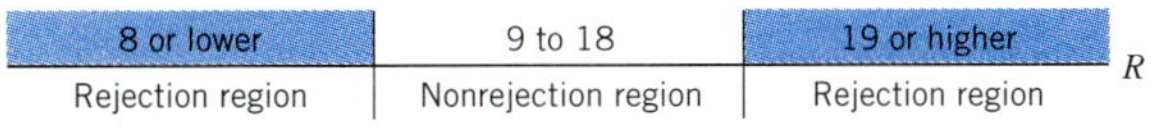

Figure 14.13

Step 4.   *Calculate the value of the test statistic.*

As the given data show, the 25 applications included in the sample were received in the following order with respect to gender:

M F M M F F F M F M M M F F F F M M M F F M F M M

Because this string of the letters M and F has 13 runs,

Observed value of $R = 13$

Step 5.   *Make a decision.*

Since $R = 13$ is between 9 and 18, we do not reject H_0. Hence, we conclude that the applications for admission arrive in a random order with respect to gender.   ∎

The Large-Sample Case

If *either* $n_1 > 15$ *or* $n_2 > 15$, the sample is considered large for the purpose of applying the runs test for randomness and we use the normal distribution to perform the test.

> **OBSERVED VALUE OF z** For large values of n_1 and n_2, the distribution of R (the number of runs in the sample) is approximately normal with its mean and standard deviation given as
>
> $$\mu_R = \frac{2n_1n_2}{n_1 + n_2} + 1 \quad \text{and} \quad \sigma_R = \sqrt{\frac{2n_1n_2(2n_1n_2 - n_1 - n_2)}{(n_1 + n_2)^2(n_1 + n_2 - 1)}}$$
>
> The observed value of z for R is calculated using the formula
>
> $$z = \frac{R - \mu_R}{\sigma_R}$$

In this case, rather than using Table XV to find the critical values of R, we use the standard normal distribution table (Table VII in Appendix D) to find the critical values of z for the given significance level. Then, we make a decision to reject or not to reject the null hypothesis based on whether the observed value of z falls in the rejection or the nonrejection region. Example 14–14 describes the application of this procedure.

Performing the runs test for randomness: large sample.

EXAMPLE 14–14 Refer to Example 14–13. Suppose that the admissions officer examines 50 consecutive applications and observes that $n_1 = 22$, $n_2 = 28$, and $R = 20$, where n_1 is the number of male applicants, n_2 the number of female applicants, and R the number of runs. Can we conclude that the applications for admission arrive randomly with respect to gender? Use $\alpha = .01$.

Solution We perform the following five steps to make this test.

Step 1. *State the null and alternative hypotheses.*

H_0: Applications arrive in a random order with respect to gender

H_1: Applications do not arrive in a random order with respect to gender

Step 2. *Select the distribution to use.*

Here, $n_1 = 22$ and $n_2 = 28$. Since n_1 and n_2 are both greater than 15, we use the normal distribution to make the runs test. Note that only one of n_1 and n_2 has to be greater than 15 to apply the normal distribution.

Step 3. *Determine the rejection and nonrejection regions.*

The significance level is .01 and the test is two-tailed. From Table VII in Appendix D, the critical values of z are -2.58 and 2.58. The rejection and nonrejection regions are shown in Figure 14.14.

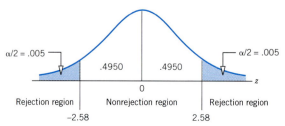

Figure 14.14

Step 4. *Calculate the value of the test statistic.*

To find the observed value of z, we first find the mean and standard deviation of R as follows:

$$\mu_R = \frac{2n_1 n_2}{n_1 + n_2} + 1 = \frac{(2)(22)(28)}{22 + 28} + 1 = 25.64$$

$$\sigma_R = \sqrt{\frac{2n_1 n_2 (2n_1 n_2 - n_1 - n_2)}{(n_1 + n_2)^2 (n_1 + n_2 - 1)}} = \sqrt{\frac{2(22)(28)(2 \cdot 22 \cdot 28 - 22 - 28)}{(22 + 28)^2 (22 + 28 - 1)}} = 3.44783162$$

The observed value of the test statistic z is

$$z = \frac{R - \mu_R}{\sigma_R} = \frac{20 - 25.64}{3.44783162} = -1.64$$

Note that the value of R is given to be 20.

Step 5. *Make a decision.*

Since $z = -1.64$ is between -2.58 and 2.58, we do not reject H_0, and we conclude that the applications for admission arrive in a random order with respect to gender. ∎

EXERCISES

■ **Concepts and Procedures**

14.69   Briefly explain the term *run* as used in a runs test.

14.70   What is the usual form of the null hypothesis in a runs test for randomness?

14.71   Under what conditions may we use the normal distribution to perform a runs test?

14.72   Using the randomness test, indicate whether the null hypothesis should be rejected in each of the following cases.

a. $n_1 = 10,$ $n_2 = 12,$ $R = 17,$ $\alpha = .05$
b. $n_1 = 20,$ $n_2 = 23,$ $R = 35,$ $\alpha = .01$
c. $n_1 = 15,$ $n_2 = 17,$ $R = 7,$ $\alpha = .05$
d. $n_1 = 14,$ $n_2 = 13,$ $R = 21,$ $\alpha = .05$

14.73   In Example 14–13, if we use the symbol 0 for male and 1 for female instead of M and F, would this affect the test in any way? Why or why not?

14.74   For each of the following sequences of observations, determine the values of n_1, n_2, and R.

a. X X Y X Y Y X Y X Y X X X Y Y
b. F M F F F F M M F F F F F F
c. + + + − − − − − − − + + − + − + − + + + + + +
d. 1 1 0 0 0 0 1 1 0 0 1 1 1 1

■ **Applications**

14.75   A psychic claims that she can cause a nonrandom sequence of heads (H) and tails (T) to appear when a coin is tossed a number of times. A fair coin was tossed 20 times in her presence, and the following sequence of heads and tails was obtained.

H H T H T H T T H T H H H T T H H H T T H

Can you conclude that the psychic's claim is true? Use $\alpha = .05$.

14.76   At a small soda factory, the amount of soda put into each 12-ounce bottle by the bottling machine varies slightly for each filling. The plant manager suspects that the machine has a random pattern of overfilling and underfilling the bottles. The following are the results of filling 18 bottles, where O denotes 12 ounces or more of soda in a bottle and U denotes less than 12 ounces of soda.

$$U\ U\ U\ O\ O\ O\ O\ U\ U\ O\ O\ O\ U\ O\ U\ U\ U\ U$$

Using the runs test at the 5% significance level, can you conclude that there is a nonrandom pattern of overfilling and underfilling such bottles?

14.77 An experimental planting of a new variety of pear trees consists of a single row of 20 trees. Several of these trees were affected by an unknown disease. The order of diseased and normal trees is shown below, where D denotes a diseased and N denotes a normal tree.

$$N\ N\ N\ D\ D\ D\ N\ N\ N\ N\ N\ D\ D\ D\ D\ N\ N\ N$$

If the sequence of diseased and normal trees falls into a nonrandom pattern with clusters of diseased trees, that would suggest that the disease may be contagious. Perform the runs test at the 5% significance level to determine if there is evidence of a nonrandom pattern in the sequence.

14.78 Table I of Appendix D is a table of random numbers. The first 25 numbers in the first column are examined and are noted as being even (E) or odd (O). The results are as follows.

$$E\ E\ O\ E\ E\ E\ O\ O\ E\ E\ E\ O\ E\ E\ E\ O\ E\ E\ E\ O\ O\ O\ E\ E\ O$$

At the 5% significance level, test the null hypothesis that the odd (and even) numbers are randomly distributed in the population against the alternative hypothesis that the odd (and even) numbers are not randomly distributed in the population. You may use the normal approximation.

14.79 Do baseball players' hits come in streaks? Seventy-five consecutive at-bats of a baseball player were recorded to determine whether the hits were randomly distributed or nonrandomly distributed for him, thus possibly indicating the presence of "hitting streaks." The observation of these 75 at-bats produced the following data.

$$n_1 = \text{number of hits} = 22 \qquad n_2 = \text{number of nonhits} = 53 \qquad R = \text{number of runs} = 37$$

Can you conclude that hits occur randomly for this player? Use $\alpha = .01$.

14.80 A researcher wanted to determine whether the stock market moves up and down randomly. He recorded the movement of the Dow Jones Industrial Average for 40 consecutive business days. The observed data showed that the market moved up 16 times, moved down 24 times, and had 11 runs during these 40 days. Using a significance level of 5%, do you think that the movements in the stock market are random?

14.81 Many state lotteries offer a "daily number" game, in which a three-digit number is randomly drawn by the state every day. Suppose that a certain state generates its daily number by a computer program, and the state lottery commissioner suspects that the process is flawed. Specifically, she feels that if today's lucky number is higher than 500, tomorrow's number is more likely to exceed 500; and if today's number is below 500, tomorrow's number is more likely to be under 500. Suppose that a sequence of the state's lucky numbers for a period of consecutive days is analyzed for runs above 500 and runs below 500. If the commissioner is right, there should be fewer runs than expected by chance. Analysis of a sequence for 50 consecutive days produced the following information.

$$n_1 = 27 \text{ numbers above } 500 \qquad n_2 = 23 \text{ numbers below } 500 \qquad R = \text{number of runs} = 11$$

Using the 2.5% level of significance, can you conclude that the sequence of all this state's daily numbers for this game is nonrandom?

GLOSSARY

Categorical data   Data divided into different categories for identification purposes only.

Distribution-free test   A hypothesis test in which no assumptions are made about the specific population distribution from which the sample is selected.

H   The test statistic used in the Kruskal–Wallis test.

Kruskal–Wallis test   A distribution-free method used to test the hypothesis that three or more populations have identical distributions.

Nonparametric test   A hypothesis test in which the sample data are not assumed to come from a specific type of population distribution, such as the normal distribution.

ρ_s The value of the Spearman rho rank correlation coefficient between the ranks of the values of two variables for population data.

r_s The value of the Spearman rho rank correlation coefficient between the ranks of the values of two variables for sample data.

R   The number of runs in a runs test used to test for randomness.

Run   A sequence of one or more consecutive occurrences of the same outcome in a sequence of occurrences in which there are only two possible outcomes.

Runs test for randomness   A test that is used to test the null hypothesis that a sequence of events has occurred randomly.

Sign test   A nonparametric test that is used to test a population proportion (with categorical data), a population median (with numerical data), or the difference in population medians for two dependent and paired sets of numerical data.

Spearman rho rank correlation coefficient   The linear correlation coefficient between the ranks of paired data for two samples or populations.

T   The test statistic used in the Wilcoxon signed-rank test and Wilcoxon rank sum test.

T_U, T_L The upper and lower critical values for the Wilcoxon rank sum test obtained from the table.

Wilcoxon rank sum test   A nonparametric test that is used to test whether two independent samples come from identically distributed populations by analyzing the ranks of the pooled sample data.

Wilcoxon signed-rank test   A nonparametric test that is used to test whether two paired and dependent samples come from identically distributed populations by analyzing the ranks of the paired differences of the samples.

KEY FORMULAS

1. **Value of the test statistic z in the sign test about the population proportion for a large sample**

$$z = \frac{(X \pm .5) - \mu}{\sigma} \quad \text{where} \quad \mu = np \text{ and } \sigma = \sqrt{npq}$$

Use $(X + .5)$ if $X \leq n/2$ and use $(X - .5)$ if $X > n/2$.

2. **Value of the test statistic z in the sign test about the median for a large sample**

$$z = \frac{(X + .5) - \mu}{\sigma} \quad \text{where} \quad \mu = np, p = .5, \text{ and } \sigma = \sqrt{npq}$$

Let X_1 be the number of plus signs and X_2 the number of minus signs in a test about the median. Then, if the test is two-tailed, it does not matter which of the two values, X_1 and X_2, is assigned to X to calculate the observed value of z. If the test is left-tailed, X should be assigned the smaller of the values of X_1 and X_2. If the test is right-tailed, X should be assigned the larger of the values of X_1 and X_2.

3. **Value of the test statistic z in the Wilcoxon signed-rank test for a large sample**

$$z = \frac{T - \mu_T}{\sigma_T} \quad \text{where} \quad \mu_T = \frac{n(n + 1)}{4} \quad \text{and} \quad \sigma_T = \sqrt{\frac{n(n + 1)(2n + 1)}{24}}$$

4. **Value of the test statistic z in the Wilcoxon rank sum test for large independent samples**

$$z = \frac{T - \mu_T}{\sigma_T} \quad \text{where} \quad \mu_T = \frac{n_1(n_1 + n_2 + 1)}{2} \quad \text{and} \quad \sigma_T = \sqrt{\frac{n_1 n_2(n_1 + n_2 + 1)}{12}}$$

5. **Value of the test statistic H in the Kruskal–Wallis test**

$$H = \frac{12}{n(n + 1)} \left[\frac{R_1^2}{n_1} + \frac{R_2^2}{n_2} + \cdots + \frac{R_k^2}{n_k} \right] - 3(n + 1)$$

where n_i = size of ith sample, $n = n_1 + n_2 + \cdots + n_k$, k = number of samples, and R_i = sum of ranks for ith sample

6. **Value of the test statistic r_s for the Spearman rho rank correlation coefficient test**

$$r_s = 1 - \frac{6\Sigma d^2}{n(n^2 - 1)}$$

where d_i = difference in ranks of x_i and y_i

7. **Value of the test statistic z in the runs test for randomness for a large sample**

$$z = \frac{R - \mu_R}{\sigma_R}$$

where
$$R = \text{number of runs in the sequence}$$

$$\mu_R = \frac{2n_1 n_2}{n_1 + n_2} + 1$$

$$\sigma_R = \sqrt{\frac{2n_1 n_2(2n_1 n_2 - n_1 - n_2)}{(n_1 + n_2)^2(n_1 + n_2 - 1)}}$$

SUPPLEMENTARY EXERCISES

14.82 Fifteen cola drinkers are given two paper cups, one containing Brand A cola and the other containing Brand B cola. Each person tries both drinks and then indicates which one he or she prefers. The drinks are offered in random order (some people are given Brand A first, and others get Brand B first). Ten of the people prefer Brand A, while five prefer Brand B. Using the sign test at the .05 significance level, can you conclude that among all cola drinkers there is a preference for Brand A?

14.83 Twenty-four randomly selected people are given samples of two brands of low-fat ice cream. Seventeen of them prefer Brand B, while 7 prefer Brand A. Using the sign test at the .05 significance level, can you conclude that among all people there is a preference for Brand B?

14.84 A random sample of 400 adults are asked whether they prefer watching evening news on Network A or Network B. Of these adults, 225 favor Network A, 160 prefer Network B, and 15 have no preference. At the 2% level of significance, can you conclude that among all adults there is a preference for either network over the other?

14.85 A random sample of 200 customers of a large bank are asked whether they prefer using an automatic teller machine (ATM) or seeing a human teller for deposits and withdrawals. Of these customers, 122 said they prefer an ATM, 66 prefer a teller, and 12 have no opinion. At the 1% level of significance, can you conclude that more than half of all customers of this bank prefer an ATM?

14.86 Suppose that a polling agency is conducting a telephone survey. When prospective participants answer the phone, they are told that the survey will take just 5 minutes of their time. Ten randomly selected calls are monitored. The lengths of time (in minutes) required for the survey in these 10 cases are shown here.

| 7.1 | 6.3 | 4.9 | 5.0 | 5.7 | 9.0 | 8.2 | 5.9 | 6.5 | 7.7 |

Using the sign test at the 5% level of significance, can you conclude that the median time for the survey exceeds 5 minutes?

14.87 Suppose that in the town of River City, the median household electric bill in January 1999 was $73.57. A random sample of 20 River City households showed that for 14 of them January 2000 electric bills exceeded $73.57 and for 6 it was less than $73.57. At the 5% level of significance, can you conclude that the median household electric bill for River City households in January 2000 exceeded $73.57?

14.88 A state motor vehicle department requires auto owners to bring their autos to state emission centers periodically for testing. State officials claim that the median waiting time between the hours of 8:00 A.M. and 11:00 A.M. on weekdays at a particular site is 25 minutes. In a check of 30 randomly selected motorists during this time period at this site, 9 motorists waited for less than 25 minutes, 2 waited exactly 25 minutes, and 19 waited longer than 25 minutes.

a. Using the sign test at the 5% significance level, can you conclude that the median waiting time at this site during these hours exceeds 25 minutes?
b. Perform the test of part a at the 2.5% level of significance.
c. Comment on the results of parts a and b.

14.89 The following data give the amounts (in dollars) spent on textbooks by 35 college students during the 1998–1999 academic year.

475	418	180	110	155	288	110	175	250
295	420	610	380	98	230	415	357	357
409	611	455	318	395	612	468	610	380
450	280	490	490	626	350	188	388	

Using $\alpha = .05$, can you conclude that the median expenditure on textbooks by all such students in 1998–1999 was different from $250?

14.90 Botanists at an agricultural research station have designed an experiment to compare the yield of a genetically engineered variety of wheat (A) with the yield of a conventional variety (B). Twenty test plots are split into equal halves. Variety A is grown on one half of each plot, while Variety B is grown on the other half. For 15 of these plots, the yield of Variety A exceeds that of Variety B. For the remaining 5 plots, Variety B has the higher yield. At the 5% level of significance, can you conclude that the median yield of Variety B for all acres of land is less than that of Variety A?

14.91 Refer to Exercise 14.34 and to Exercise 10.101 of Chapter 10, which concern Gamma Corporation's installation of governors on its salespersons' cars to regulate their speeds. The following table gives the gas mileage (in miles per gallon) for each of seven sales representatives' cars during the week before governors were installed and the gas mileage in the week after installation.

Salesperson	A	B	C	D	E	F	G
Before	25	21	27	23	19	18	20
After	26	24	26	25	24	22	23

a. Using the sign test at the 5% level of significance, can you conclude that the use of governors tends to increase the median gas mileage for Gamma Corporation's sales representatives' cars?

b. Compare your conclusion of part a with the result of the Wilcoxon signed-rank test that was performed in part a of Exercise 14.34 and with the result of the corresponding hypothesis test (using the t distribution) of Exercise 10.101.

c. If there is a difference in the three conclusions, how can you account for it?

14.92 A reporter for a travel magazine wanted to compare the effectiveness of two large travel agencies (X and Y) in finding the lowest airfares to given destinations. She randomly chose 32 destinations from the many offered by both agencies. She and her assistants requested the lowest available fare for each destination from each agency. For 18 of these destinations Agency X quoted a fare lower than that of Agency Y, for 8 destinations Agency Y found a lower fare, and in 6 cases the fares were the same. At the 2% level of significance, can you conclude that there is any difference in the median fares quoted by Agency X and Agency Y for all destinations they both offer?

14.93 Thirty-five patients with high blood pressure are given medication to lower their blood pressure. For all 35 patients, their blood pressure is measured before they begin the medication and again after they have finished taking medication for 30 days. For 25 patients the blood pressure was lower after finishing the medication, in 7 cases it was higher, and for 3 patients there was no change. Assume that these 35 patients make up a random sample of all people suffering from high blood pressure. At the 2.5% level of significance, can you conclude that the median blood pressure in all such patients is lower after the medication than before?

14.94 The following table shows the one-week sales of six salespersons before and after they attend a course on "how to be a successful salesperson."

Before	12	18	25	9	14	16
After	18	24	24	14	19	20

a. Using the Wilcoxon signed-rank test at the 5% significance level, can you conclude that the weekly sales for all salespersons tend to increase as a result of attending this course?

b. Perform the test of part a using the sign test at the 5% significance level.

c. Compare your conclusions from parts a and b.

14.95 An official at a figure skating competition thinks that two of the judges tend to score skaters differently. Shown next are the two judges' scores for eight skaters.

Skater	1	2	3	4	5	6	7	8
Judge A	5.8	5.7	5.6	5.9	5.8	5.9	5.8	5.6
Judge B	5.4	5.5	5.7	5.4	5.6	5.3	5.4	5.6

Using the Wilcoxon signed-rank test at the 5% level of significance, can you conclude that either judge tends to give higher median scores than the other?

14.96 Refer to Exercise 14.26. Consider the data given in that exercise on the cholesterol levels (in milligrams per hundred milliliters) for 30 randomly selected adults as determined by two laboratories, A and B.

a. Using the Wilcoxon signed-rank test at the 1% level of significance, can you conclude that the median cholesterol level for all such adults as determined by Lab A is higher than that determined by Lab B?

b. Compare your conclusion in part a to that of Exercise 14.26.

14.97 A consumer agency conducts a fuel economy test on two new subcompact cars, the Mouse (M) and the Road Runner (R). Each of 18 randomly selected drivers takes both cars on an 80-mile test run. For each driver, the gas mileage (in miles per gallon) is recorded for both cars; then the gas mileage for the R car is subtracted from the gas mileage for the M car. Thus, a minus difference indicates a higher gas mileage for the R car. One of the 18 drivers obtains exactly the same gas mileage for both cars. For the other drivers, the differences are ranked. The sum of the positive ranks is 31, and the sum of the absolute values of the negative ranks is 122. Can you conclude that the R car gets better gas mileage than the M car? Use the 2.5% level of significance.

14.98 Each of the two supermarkets, Al's and Bart's, in River City claims to offer lower-cost shopping. Fifty people who normally do the grocery shopping for their families are chosen at random. Each shopper makes up a list for a week's supply of groceries. Then these items are priced and the total cost is computed for each store. The paired differences are then calculated for each of the 50 shoppers, where a paired difference is defined as the cost of a cart of groceries at Al's minus the cost of the same cart of groceries at Bart's. These paired differences were positive for 21 shoppers and negative for 29 shoppers. The sum of ranks of the positive paired differences was 527 and the sum of the absolute values of the ranks of the negative paired differences was 748. Using the 1% level of significance, can you conclude that either store is less expensive than the other?

14.99 A consumer advocate is comparing the prices of eggs at supermarkets in the suburbs with the prices of eggs at supermarkets in the cities. The following data give the prices (in dollars) of a dozen large eggs in 13 supermarkets, 6 of which are in cities and 7 are in suburbs.

City	1.49	1.29	1.35	1.58	1.33	1.47	
Suburb	.99	1.09	1.39	1.28	1.16	1.44	1.05

Using the .05 level of significance and the Wilcoxon rank sum test, can you conclude that egg prices tend to be higher in the city?

14.100 Two drugs for headache were tested to see how long the patients must wait for relief. Six persons suffering from headache were given Brand A, while seven were given Brand B. The following data give the times (in minutes) required for relief from headache for these persons.

Brand A	4.5	7.7	3.8	9.9	8.8	5.0	
Brand B	7.0	8.1	7.7	12.1	8.5	8.1	20.0

Using $\alpha = .05$ and the Wilcoxon rank sum test, can you conclude that relief times tend to be longer for Brand B?

14.101 A researcher obtains a random sample of 24 students taking elementary statistics at a large university and divides them randomly into two groups. Group A receives instruction to use Software A to do a statistics assignment, while Group B is taught to use Software B to do the same statistics assignment. The time (in minutes) taken by each student to complete this assignment is given in the table.

Group A	123	101	112	85	87	133	129	114	150	110	180	115
Group B	65	115	95	100	94	72	60	110	99	102	88	97

a. Using the 5% level of significance and the Wilcoxon rank sum test, can you conclude that the median time required for all students taking elementary statistics at this university to complete this assignment is longer for Software A than for Software B?
b. Would a paired-samples sign test be appropriate here? Why or why not?

14.102 Refer to Exercise 14.101. The scores on the homework assignment for the 24 students are given in the table.

Group A	48	38	45	31	42	25	40	43	50	30	33	46
Group B	37	21	40	27	49	44	36	41	20	39	18	40

Using the 10% significance level and the Wilcoxon rank sum test, can you conclude that there is a difference in the median scores for all students using Software A and all students using Software B?

14.103 A factory manager compared the work of three machinists who produce the same type of precision parts. The following table gives the number of defective parts produced by each machinist on five randomly chosen days.

Machinist		
A	**B**	**C**
10	17	15
12	7	9
8	11	10
14	13	18
16	10	16

At the 5% level of significance, can you reject the null hypothesis that the distribution of defective parts produced per day is the same for all three machinists?

14.104 An academic employment service compared the starting salaries of May 2000 graduates in three major fields. Random samples were taken of 8 engineering majors, 10 business majors, and 7 mathematics majors. The starting salaries of all 25 graduates were determined and then ranked, yielding the following rank sums:

Engineering: 137   Business: 126   Mathematics: 62

At the 5% level of significance, can you reject the null hypothesis that the median salary is the same for graduates in all three major fields?

14.105 A sports magazine conducted a test of three brands (A, B, and C) of golf balls by having a professional golfer drive six balls of each brand. The lengths of the drives (in yards) from this test are listed in the table.

Brand A	**Brand B**	**Brand C**
275	245	267
266	256	283
301	261	259
281	270	250
288	259	263
277	262	256

At the 5% level of significance, can you reject the null hypothesis that the median distance of drives by this golfer is the same for all three brands of golf balls?

14.106 A students' group at a state university wanted to compare textbook costs for students majoring in economics, history, and psychology. The group obtained data from random samples of 10 economics majors, 9 history majors, and 11 psychology majors, all in the second semester of their junior year. The total textbook costs of the 30 students were recorded and ranked. The rank sums for economics and history majors were 134 and 157, respectively.

a. Find the rank sum for psychology majors. [*Hint:* The sum of n integers from 1 through n is given by $n(n + 1)/2$.]

b. At the 2.5% significance level, can you reject the null hypothesis that the median textbook costs are the same for students in all three majors who are in the second semester of the junior year?

14.107 The following table shows the average mathematics SAT score and the percentage of high school graduates who took the SAT in 1998 for a random sample of 10 states.

State	Math SAT Score	Percentage of Graduates Taking SAT
Louisiana	558	8
Kansas	585	9
Washington	526	53
North Dakota	599	5

(*continued*)

State	Math SAT Score	Percentage of Graduates Taking SAT
New Jersey	508	79
Wisconsin	594	7
Florida	501	52
Maine	501	68
Pennsylvania	495	71
New York	503	76

a. For all 50 states, would you expect ρ_s to be positive, negative, or near zero? Why?

b. Calculate r_s for the sample of 10 states and indicate whether its value is consistent with your answer to part a.

c. Using the value of r_s calculated in part b, test H_0: $\rho_s = 0$ against H_1: $\rho_s \neq 0$ using the 5% level of significance.

14.108 The Spearman rho rank correlation coefficient may be used in cases where data for one or both variables are given in the form of ranks. Suppose that a film critic views 10 randomly chosen new movies and ranks them, with rank 10 being assigned to the film that he thinks will have the highest box office receipts, rank 9 to the next most profitable, and so forth. Three months after each film is released, its total box office receipts (in millions of dollars) are tabulated. The following table shows the ranking and receipts for each of these 10 films.

Rank	7	3	10	1	4	5	2	6	8	9
Receipts	40	5	66	2	3	10	28	15	30	17

a. Calculate r_s for the sample of 10 films.

b. Using the 5% level of significance, test H_0: $\rho_s = 0$ against H_1: $\rho_s > 0$.

c. Based on your conclusion in part b, is there sufficient evidence that this critic can predict the box office performance of a film?

14.109 Refer to Example 13–8, which contained data on monthly auto insurance premiums and years of driving experience. Those data are reproduced here.

Driving Experience (years)	Monthly Auto Insurance Premium
5	$64
2	87
12	50
9	71
15	44
6	56
25	42
16	60

The estimated regression line was found to be $\hat{y} = 76.6605 - 1.5476x$ in Example 13–8 and the simple linear correlation coefficient for the sample data was $-.77$. The true regression slope B was found to be significantly less than zero at the significance level of 5%.

a. Based on this information, if ρ_s is the Spearman rho rank correlation coefficient for the entire population from which this sample was taken, what would you expect from a test of H_0: $\rho_s = 0$ against H_1: $\rho_s < 0$ at the significance level of 5%?

b. Perform the hypothesis test mentioned in part a.

14.110   A researcher wonders if men still tend to stand aside and let women board elevators ahead of them. She observes 10 men and 10 women boarding the same elevator. The order in which they boarded is given here.

W W W M M W W M W W W M M M M W W M M M

Using the 5% level of significance, can you conclude that the order of boarding is nonrandom with respect to gender?

14.111   A machinist is making precision cutting tools. Because of the exacting specifications for these tools, about 20% of them fail to pass inspection and are judged defective. The shop supervisor feels that the machinist tends to produce defective tools in clusters, perhaps due to fatigue or distraction. If this is true, then a sequence of tools produced by this machinist will tend to have fewer runs of defective and good tools than expected by chance. The supervisor chooses a day at random and observes the following sequence of 18 tools, where G denotes a tool that passed inspection and D indicates a defective tool.

G G G G G D D G G G G D G G G D D G

Do you think that there is evidence of nonrandomness in this sequence? Use the 5% level of significance.

14.112   Some states require periodic testing of cars to monitor the emission of pollutants. A state official suspects that the inspection process at a particular station is faulty, that a car's test result may be affected by the tests of preceding cars. Analysis of the sequence of test results for a random sample chosen on a day yields the following information:

$$n_1 = \text{number of cars that passed the test} = 157$$

$$n_2 = \text{number of cars that failed the test} = 143$$

$$R = \text{number of runs} = 41$$

Using the 1% level of significance, can you conclude that the test results for this emissions test station are not random?

14.113   The following data give the sequence of wins and losses in 30 consecutive games for a baseball team during a season.

L W L W L W L W L L W W W L L W L L L L L W L L L L L W L W L L

Can you conclude that the wins are randomly distributed for this baseball team? Use the 2.0% significance level.

■ **Advanced Exercises**

14.114   A medical researcher wants to study the effects of a low-calorie diet on the longevity of laboratory mice. She randomly divides 20 mice into two groups. Group A gets a standard diet, while Group B receives a diet that contains all the necessary nutrients but provides only 70% as many calories as Group A's diet. The experiment is conducted for 36 months and the length of life (in days) of each mouse is recorded. The data obtained on the lives of these mice are shown in the following table. In these data, the asterisk (*) indicates that this mouse was still alive at the end of the 36-month experiment.

Group A	900	907	751	833	920	787	850	877	848	901
Group B	1037	905	1023	988	1078	1011	*	1063	898	1033

a.   Using the 2.5% level of significance and the Wilcoxon rank sum test, does the low-calorie diet appear to lengthen the longevity of laboratory mice? Should the mouse that was still alive at the end of the experiment be eliminated from your analysis or is there a way to include it?
b.   Would a Wilcoxon signed-rank test be appropriate in this example? Why or why not?

14.115 The editor of an automotive magazine has asked you to compare the median gas mileages for driving in the city for three models of compact cars. The editor has made available to you one car of each of the three models, three drivers, and a budget sufficient to buy gas and pay the drivers for approximately 500 miles of city driving for each car.

a. Explain how you would conduct an experiment and gather the data for a magazine article comparing gas mileages.

b. Suppose you wish to test the null hypothesis that the median gas mileages for driving in the city are the same for all three models of cars. Outline the procedure for using your data to conduct this test. Do not assume that the gas mileages for all cars of each model are normally distributed.

14.116 Refer to Exercise 10.101 in Chapter 10. Suppose Gamma Corporation decides to test the governors on seven cars. However, management is afraid that the speed limit imposed by the governors will reduce the number of contacts the salespersons can make each day. Thus, both the fuel consumption and the number of contacts made are recorded for each car/salesperson for each week of the testing period, both before and after the installation of governors.

Salesperson	Number of Contacts		Fuel Consumption (mpg)	
	Before	**After**	**Before**	**After**
A	50	49	25	26
B	63	60	21	24
C	42	47	27	26
D	55	51	23	25
E	44	50	19	24
F	65	60	18	22
G	66	58	20	23

Suppose that you are directed to prepare a brief report that includes statistical analysis and interpretation of the data. Management will use your report to help decide whether or not to install governors on all salespersons' cars. Use 5% significance levels for any hypothesis tests you perform to make suggestions. In contrast to Exercise 10.101, do not assume that the numbers of contacts, fuel consumption, or differences are normally distributed.

14.117 Suppose that you are a newspaper reporter and your editor has asked you to compare the hourly wages of carpenters, plumbers, electricians, and masons in your city. Since many of these workers are not union members, the wages may vary considerably among individuals in the same trade.

a. What data should you gather to make this statistical analysis and how would you collect them? What sample statistics would you present in your article and how would you calculate them? Assume that your newspaper is not intended for technical readers.

b. Suppose that you must submit your findings to a technical journal that does require statistical analysis of your data. If you want to determine whether or not the median hourly wages are the same for all four trades, briefly describe how you would analyze the data. Do not assume that the hourly wages for these populations are normally distributed.

14.118 Consider the data in the following table.

x	10	20	30	40	50	60
y	12	15	19	21	25	30

a. Suppose that each value of y in the table is increased by 5 but the x values remain unchanged. What effect will this have on the rank of each value of y? Do you expect the value of r_s to increase, decrease, or remain the same? Explain why.

b. Now, first calculate the value of r_s for the data in the table, and then increase each value of y by 5 and recalculate the value of r_s. Does the value of r_s increase, decrease, or remain the same? Does this result agree with your expectation in part a?

14.119   The English department at a college has hired a new instructor to teach the composition course to first-year students. The department head is concerned that the new instructor's grading practices might not be consistent with those of the professor (let us call him Professor A) who taught this course previously. She randomly selects 10 essays written by students for this class and makes two copies of each essay. She asks Professor A and this instructor (working independently) to assign a numerical grade to each of the 10 essays. The results are shown in the following table.

Essay	1	2	3	4	5	6	7	8	9	10
Professor A	75	62	90	48	67	82	94	76	78	84
Instructor	80	50	85	55	63	78	89	81	75	83

a.   Suppose the department head wants to determine whether the instructor tends to grade higher or lower than Professor A. Which of the statistical tests discussed in this chapter could she use? Note that more than one test may be appropriate.

b.   Using an appropriate test from your answer in part a, can you conclude that the instructor tends to grade higher or lower than Professor A? Use $\alpha = .05$.

c.   Suppose the department head wants to determine whether the instructor is consistent with Professor A in the sense that they tend to agree on which paper is best, which is second best, and so forth. Which test from this chapter would be appropriate to use? State the relevant null and alternative hypotheses.

d.   Using the test you chose in part c, can you conclude that Professor A and the instructor are consistent in their grading? Use the 5% level of significance.

14.120   Three doctors are employed at a large clinic. The manager at the clinic wants to know whether these three doctors spend the same amount of time per patient. The manager randomly chooses 10 routine appointments of patients with each of the three doctors and times them. Thus, the data set consists of 10 observations on the time spent with patients by each doctor.

a.   To test the null hypothesis that the mean or median times are equal for all three doctors against the alternative hypothesis that they are not all equal, which tests from Chapters 12 and 14 are appropriate?

b.   For each test that you indicated in part a, specify whether the test is about the means or the medians.

c.   What assumptions are required for the test from Chapter 12?

14.121   An educational researcher is studying the relationship between high school grade point averages (GPAs) and SAT scores. She obtains GPAs and SAT scores for a random sample of 25 students and wants to test the null hypothesis that there is no correlation between GPAs and SAT scores against the alternative hypothesis that these variables are positively correlated.

a.   If she wants to base her test on the linear correlation coefficient of Chapter 13, what assumptions are required about the two variables (GPAs and SAT scores)?

b.   If the assumptions required in part a are not satisfied, what other test(s) might she use?

14.122   To test the effectiveness of a new six-week body-building course, 12 tenth-grade boys are randomly selected. Each boy is tested before and after the course to see how much weight he can lift.

a.   To test whether or not the mean or median weight lifted by all such boys tends to be greater after the course than before, which tests from Chapters 10 and 14 might be used?

b.   For each test that you indicated in part a, specify whether it involves the mean or median.

c.   If the paired differences in weights lifted before and after the test are not normally distributed, which of the tests indicated in part a could be used?

14.123   Suppose in a sample we have 10 A's and 15 B's. What is the maximum number of runs possible in a sequence of these 25 letters?

SELF-REVIEW TEST

1. Nonparametric tests

   a. are more efficient than the corresponding parametric tests

   b. do not require that the population being sampled has a normal distribution

   c. generally require more assumptions about the population than parametric tests do

2. For small samples ($n \leq 25$), the critical value(s) for the sign test are based on the

   a. binomial distribution   **b.** normal distribution   **c.** t distribution

3. Which of the following tests may be used to test hypotheses about one median?

   a. The sign test   **b.** The Kruskal–Wallis test   **c.** The Wilcoxon rank sum test

4. The Wilcoxon signed-rank test may be used to test

   a. for a difference between the medians of two independent samples

   b. for a preference for one product over another

   c. hypotheses involving paired samples

5. When we use the Wilcoxon signed-rank test,

   a. all observations are ranked

   b. the difference for each pair is calculated and then all the differences are ranked according to their absolute values

   c. only the signs of the differences are used to calculate the value of the test statistic

6. In order to perform a Wilcoxon rank sum test, one must calculate the

   a. standard deviation of each sample

   b. range of the data

   c. rank of each observation

7. Which of the following tests may be used with paired samples? Circle all that apply.

   a. Sign test

   b. Wilcoxon signed-rank test

   c. Wilcoxon rank sum test

   d. Spearman rho rank correlation coefficient test

8. The Spearman rho rank correlation coefficient is calculated as the

   a. simple linear correlation coefficient between the two sets of observations

   b. simple linear correlation coefficient between the ranks of the two sets of observations

   c. square of the simple linear correlation coefficient between two sets of observations

9. In order to test a hypothesis about the Spearman rho rank correlation coefficient

   a. both sets of data must come from normally distributed populations

   b. one set of data must come from a normally distributed population

   c. either set of data can have any distribution

10. The Spearman rho rank correlation coefficient is positive when

    a. there is no relationship between the two sets of observations

    b. the values in one set of observations increase as the values of the corresponding observations in the other set decrease

    c. the values in one set of observations increase as the values of the corresponding observations in the other set increase

11. For the runs test for randomness, which of the following statements are true?

    a. We notice which of the two possible outcomes has occurred at each stage in a list of consecutive outcomes.

    b. A run is one or more consecutive occurrences of either one of the two possible outcomes.

    c. We are testing the hypothesis that one of the two possible outcomes occurred significantly more frequently than the other.

12. In the runs test for randomness, we reject the null hypothesis

    a. only when there is a very large number of runs

    b. only when there is a very small number of runs

    c. if there is either a very large or a very small number of runs

13. In the runs test for randomness, the distribution of R (the total number of runs) is approximately normal when

    a. R is greater than 10

    b. at least one of the two possible outcomes occurs more than 15 times

    c. each of the two possible outcomes occurs more than 15 times

14. A large pool of prospective jurors is made up of an equal number of men and women. A 12-person jury selected from this pool consists of 2 women and 10 men. At the 5% level of significance, can we reject the null hypothesis that the selection process is unbiased in terms of gender?

15. In January 1999 a telephone poll of 428 dog owners was conducted by Yankelovich Partners for *Time*/CNN (*Time*, February 1, 1999). These owners were asked, Does your dog get in a bad mood when you leave the home? Fifty-two percent of the owners who had an opinion answered *yes*. Suppose the answers to the question were distributed as follows: 207 of the 428 owners said *yes,* 191 said *no,* and 30 were not sure. Assume that these 428 dog owners are a random sample of all dog owners in the United States. Using the 2.5% level of significance, can you conclude that more than half of all dog owners in the United States who have an opinion on this question would answer *yes*?

16. The past records of a supermarket show that its customers spent a median of $65 per visit. After a promotional campaign designed to increase spending, the store took a sample of 12 customers and recorded the amounts (in dollars) they spent. The amounts are listed here.

$$88 \quad 69 \quad 141 \quad 28 \quad 106 \quad 45 \quad 32 \quad 51 \quad 78 \quad 54 \quad 110 \quad 83$$

Using $\alpha = .05$, can you conclude that the median amount spent by all customers at this store after the campaign exceeds $65?

17. According to the U.S. Commerce Department, in September 1999 the median price of new homes in the United States was $160,000 (*The Wall Street Journal*, November 11, 1999). In a random sample of 544 recently sold new homes, 302 were sold for more than $160,000, 241 were sold for less than $160,000, and one was sold for exactly $160,000. At the 1% level of significance, can you conclude that the median price of new homes in the United States has increased since September 1999?

18. The following table gives the cholesterol levels for seven adults before and after they completed a special dietary plan.

Before	210	180	195	220	231	199	224
After	193	186	186	223	220	183	233

    a. Using the sign test at the 5% significance level, can you conclude that the median cholesterol level before the diet is the same as the level after the diet?

    b. Using the Wilcoxon signed-rank test at the 5% significance level, can you conclude that the median cholesterol level before the diet is the same as the level after the diet?

    c. Compare your conclusions for parts a and b.

19. An archeologist wants to compare two methods (I and II) of radioactive dating of artifacts. He submits a random sample of 33 artifacts that are suitable for radioactive dating. Each one of these artifacts is dated by both methods. The paired differences are then calculated for each of the 33 artifacts, where a paired difference is defined as the age of an artifact dated by Method I minus the age of the same artifact dated by Method II. These paired differences were positive for 11 of the artifacts, negative for 20, and zero for 2 artifacts. Using the sign test at the 2% level of significance, can you conclude that the median estimated ages of such artifacts differ for the two methods?

20. A professor at a large university suspects that the grades of engineering majors tend to be lower in the spring semester than in the fall semester. He randomly selects 10 sophomore electrical engineering majors and records their grade point averages (GPAs) for the fall and the spring semesters. The data obtained are shown in the table.

Student	1	2	3	4	5	6	7	8	9	10
Fall GPA	3.20	3.56	3.05	3.78	4.00	2.85	3.33	2.67	3.00	3.67
Spring GPA	3.15	3.40	2.88	3.67	4.00	3.00	3.30	3.05	2.95	3.50

Using the Wilcoxon signed-rank test at the 5% level of significance, can you conclude that the median GPA of all sophomore electrical engineering majors at this university tends to be lower in the spring semester than in the fall semester?

21. A random sample of 30 students was selected to test the effectiveness of a course designed to improve memory. Each student was given a memory test before and after taking the course. Each student's score after taking the course was subtracted from his or her score before the course; then the 30 differences were ranked. Thus, a negative rank denotes an improved score after taking the course. The sum of the positive ranks was 102; the sum of the absolute values of the negative ranks was 276. Three students scored exactly the same on both tests. Using the 2.5% level of significance, can you conclude that the course tends to improve scores on memory tests?

22. A commuter has two alternative routes (Route 1 and Route 2) to drive to work. Picking days at random, she drives to work using each route for eight days and records the time (in minutes) taken to commute from home to work on each day. These times are shown in the following table.

Route 1	45	43	38	56	41	43	46	44
Route 2	38	40	39	42	50	37	46	36

Using the Wilcoxon rank sum test at the 5% level of significance, can you reject the null hypothesis that the median commuting time is the same for both routes?

23. An accounting firm has hired two temporary employees, A and B, to prepare individual federal income tax returns during the tax season. Clients who have relatively simple tax situations are randomly assigned to either A or B. The firm randomly selected 18 income tax returns prepared by each of these two employees and recorded the times taken to prepare these tax returns. After these times taken to prepare 36 tax returns were ranked, the sum of the ranks for A was found to be 298 and the sum of the ranks for B was equal to 368. Using the Wilcoxon rank sum test at the 2.5% level of significance, can you conclude that there is a difference in the median times taken to prepare such income tax returns by A and B?

24. The following table lists the numbers of cases of telemarketing fraud reported to law-enforcement officials during several randomly chosen weeks in 1999 for three large cities of approximately equal populations.

City A	City B	City C
53	29	75
46	35	49
59	44	62
33	31	68
60	50	52
	48	

    a. At the 2.5% level of significance, can you reject the null hypothesis that the distributions of the numbers of such reported cases are identical for all three cities?

    b. Can you reject the null hypothesis of part a at the 1% level of significance?

    c. Comment on the results of parts a and b.

25. The following is a list of home runs (denoted by x) and runs batted in (denoted by y) for 10 players selected at random from the American League on June 20, 1999 (*New York Times,* June 20, 1999).

Player	1	2	3	4	5	6	7	8	9	10
x	3	15	10	12	7	2	6	5	7	19
y	22	49	42	39	27	29	45	26	30	57

    a. As home runs increase, runs batted in tend to increase, so do you expect the value of the Spearman rho rank correlation coefficient to be positive or negative?

    b. Compute r_s for the data.

    c. Suppose ρ_s is the value of the Spearman rho rank correlation coefficient for all players in the American League. Using the 2.5% significance level, test the null hypothesis H_0: $\rho_s = 0$ against the alternative hypothesis H_1: $\rho_s > 0$.

26. Ramon fishes in a lake where the minimum size for bass to be kept is 12 inches long; all smaller bass must be returned to the water. He thinks that most of the "keepers" (bass 12 inches or longer) are caught early in the morning. If he is right, there should be a few long runs of keepers caught in the early morning followed by a few long runs of smaller bass caught later on during the day. Thus, if the fish are recorded in sequence, there should be fewer runs than expected by chance. Last Saturday Ramon fished from 6:00 A.M. to 11:00 A.M. and caught 14 bass in the following order, where K denotes a keeper and S denotes a bass shorter than 12 inches.

K K K K S K K S K S S S S S

Using the 5% significance level, does this sequence support Ramon's theory?

27. As of June 23, 1999, the won–lost record of the St. Louis Cardinals for the 1999 season was 34 wins and 37 losses, for a total of 71 games. In these 71 consecutive games, there were 41 runs (in the statistical sense). Using the runs test for randomness, can we assert that the 34 wins and 37 losses are randomly spread out among the 71 games? Use a significance level of 5%.

MINI-PROJECTS

Mini-Project 14–1

For a period of 30 business days, record the daily price of crude oil and the price of a stock that you think might be affected by the oil price (for example, an oil company or an alternative energy company).

a. Compute the Spearman rho rank correlation coefficient for these data.

b. Can you conclude that there is a relationship between the two sets of prices? Use $\alpha = .05$.

Mini-Project 14–2

For a period of 30 business days, record whether the Dow Jones Industrial Average moves up or down. Use your data to perform an appropriate test at the 1% level of significance to see if the sequence of upward and downward movements of the Dow appears to be random over this period. (As an alternative, you might use the NASDAQ index, the price of an individual stock, or the price of gold.)

Mini-Project 14–3

In January 2000, the U.S. national debt was approximately $5.76 trillion (*The Hartford Courant*, January 14, 2000). Take random samples of 10 or more students from each of three different majors and ask each student to estimate the size of the national debt.

a. At the 5% level of significance and using the Kruskal–Wallis test, can you conclude that the median perceived values of the national debt are the same for all three majors?

b. Find the current value of the national debt. On the whole, did the students in your sample tend to overestimate its value, or did they tend to underestimate?

For instructions on using MINITAB for this chapter, please see Section B.13 of Appendix B.

See Appendix C for *Excel Adventure 14* on the Spearman rank correlation coefficient.

COMPUTER ASSIGNMENTS

CA14.1 Fifteen coffee drinkers are selected at random and asked to test and state their preference for Brand X coffee, Brand Y coffee, or neither (N). The results are as follows:

X   X   Y   X   N   Y   Y   X   Y   X   X   Y   Y   Y   X

Let p be the proportion of coffee drinkers in the population who prefer Brand X. Using MINITAB or any other statistical software and the sign test, perform the test H_0: $p = .50$ against H_1: $p > .50$. Use a significance level of 2.5%.

CA14.2 Twelve sixth-graders were selected at random and asked how many hours per week they spend watching television. The data obtained are shown here.

23   30   22.5   28   29   24.5   25   32   31   26   27   21

Using MINITAB or any other statistical software and the sign test, can you conclude that the median number of hours spent per week watching television by all sixth-graders is less than 28? Use a significance level of 5%.

CA14.3 The manufacturer of an engine oil additive, Hyper-Slick, claims that this product reduces the engine friction and, consequently, increases the miles per gallon (mpg). To test this claim, 10 cars are driven on a fixed 300-mile course without the oil additive, and each car's mpg is calculated and recorded. Then the engine oil additive is added to each car and the process is repeated. The data obtained are shown in the table.

Car	MPG without Additive	MPG with Additive
1	20.00	19.90
2	23.60	27.85
3	29.40	28.70
4	25.70	28.20
5	35.80	37.30
6	32.20	31.30
7	26.30	26.10
8	31.80	36.80
9	29.00	32.75
10	24.70	29.20

Using MINITAB or any other statistical software and the sign test, can you conclude that the manufacturer's claim is true? Use a significance level of 5%.

CA14.4 Do Computer Assignment CA14.3 using the Wilcoxon signed-rank test and a significance level of 5%. Compare your conclusion with that of Computer Assignment CA14.3 and comment.

CA14.5 Refer to Exercise 14.43. In a winter Olympics trial for women's speed skating, seven skaters use a new type of skate, while eight others use the traditional type. Each skater is timed (in seconds) in the 500-meter event. The results are given in the table.

New skates	40.5	40.3	39.5	39.7	40.0	39.9	41.5	
Traditional skates	41.0	40.8	40.9	39.8	40.6	40.7	41.1	40.5

Assume that these 15 skaters make up a random sample of all Olympic-class 500-meter female speed skaters. Using MINITAB or any other statistical software and the Wilcoxon rank sum test (Mann–Whitney test), can you conclude that the new skates tend to produce faster times in this event? Use the 5% significance level.

CA14.6 Refer to Exercise 14.46. Two brands of tires are tested to compare their durability. Eleven Brand X tires and 12 Brand Y tires are tested on a machine that simulates road conditions. The mileages (in thousands of miles for each tire) are shown in the following table.

Brand X	51	55	53	49	50.5	57	54.5	48.5	51.5	52	53.5	
Brand Y	48	47	54	55.5	50	51	46	49.5	52.5	51	49	45

Using MINITAB or any other statistical software and the Wilcoxon rank sum (Mann–Whitney) test, can you conclude that the median mileage for Brand X tires is greater than the median mileage for Brand Y tires? Use the 5% level of significance.

CA14.7 Three brands of 60-watt lightbulbs—Brand A, Brand B, and a generic brand—are tested for their lives. The following table shows the lives (in hours) of these bulbs.

Brand A	Brand B	Generic
975	1001	899
1050	1099	789
890	915	824
933	959	1011
962	986	907
925	957	923
1007	987	937
855	881	865
	1025	1024

Using MINITAB or any other statistical software and the Kruskal–Wallis test with a significance level of .05, can you conclude that the distributions of the lives of lightbulbs are the same for all three brands?

CA14.8 Refer to Exercise 14.54. A consumer agency investigated the premiums charged by four auto insurance companies. The agency randomly selected five drivers insured by each company who had similar driving records, autos, and insurance coverage. The following table gives the monthly premiums paid by these 20 drivers.

Company A	Company B	Company C	Company D
$65	$48	$57	$62
73	69	61	53
54	88	89	45
43	75	77	51
70	72	69	44

Using MINITAB or any other statistical software and the Kruskal–Wallis test at the 5% significance level, can you reject the null hypothesis that the distributions of auto insurance premiums paid per month by all such drivers are the same for all four companies?

CA14.9   Refer to Exercise 14.75. A fair coin is tossed 20 times in the presence of a psychic who claims that she can cause a nonrandom sequence of heads (H) and tails (T) to appear. The following sequence of heads and tails is obtained in these 20 tosses.

H H T H T H T T H T H H T T H H H T T H

Using MINITAB or any other statistical software and the runs test, can you conclude that the psychic's claim is true? Use a significance level of 5%.

CA14.10   At a small soda factory, the amount of soda put into each 12-ounce bottle by the bottling machine varies slightly for each filling. The plant manager suspects that the machine has a random pattern of overfilling and underfilling the bottles. The following are the results of filling 18 bottles, where O denotes 12 ounces or more of soda in a bottle and U denotes less than 12 ounces of soda.

U U U O O O O U U O O O U O U U U U

Using MINITAB or any other statistical software and the runs test at the 5% significance level, can you conclude that there is a nonrandom pattern of overfilling and underfilling such bottles?

A

SAMPLE SURVEYS, SAMPLING TECHNIQUES, AND DESIGN OF EXPERIMENTS

(The following content is not included in this text but is available for download on the Web site at www.wiley.com/college/mann. A printed copy is available in the Instructor Solutions Manual and the Student Solutions Manual.)

A.1   SOURCES OF DATA

CASE STUDY A–1   IS IT A SIMPLE QUESTION?

A.2   SAMPLE SURVEYS AND SAMPLING TECHNIQUES

A.3   DESIGN OF EXPERIMENTS

CASE STUDY A–2   SCIENTISTS PROBE MYSTERY OF THE PLACEBO EFFECT

EXERCISES

GLOSSARY

COMPUTER ASSIGNMENTS

B USING MINITAB

B.1 AN INTRODUCTION (Chapter 1)

B.2 ORGANIZING DATA (Chapter 2)

B.3 NUMERICAL DESCRIPTIVE MEASURES (Chapter 3)

B.4 THE BINOMIAL AND POISSON PROBABILITY DISTRIBUTIONS (Chapter 5)

B.5 THE NORMAL PROBABILITY DISTRIBUTION (Chapter 6)

B.6 SAMPLING DISTRIBUTIONS (Chapter 7)

B.7 ESTIMATION OF THE MEAN AND PROPORTION (Chapter 8)

B.8 HYPOTHESIS TESTS ABOUT THE MEAN AND PROPORTION (Chapter 9)

B.9 ESTIMATION AND HYPOTHESIS TESTING: TWO POPULATIONS (Chapter 10)

B.10 CHI-SQUARE TESTS (Chapter 11)

B.11 ANALYSIS OF VARIANCE (Chapter 12)

B.12 SIMPLE LINEAR REGRESSION (Chapter 13)

B.13 NONPARAMETRIC METHODS (Chapter 14)

B.14 GENERATING RANDOM NUMBERS AND SELECTING A SAMPLE (Appendix A)

B.1 AN INTRODUCTION

In recent years the use of computers has significantly increased in almost every aspect of life. Such usage of computers has reduced the computation time for quantitative analysis to a negligible amount.

In the real world, when doing a statistical analysis, we usually deal with hundreds or thousands of observations. For this reason, it is either very time consuming or almost impossible to make all the required calculations manually. The computer is of invaluable assistance in such situations. Consequently, learning to use a statistical software package has become an important part of learning statistics.

A large number of statistical software packages have been developed in recent years. Most of these software packages are user friendly. Four of the major statistical software packages are BMDP, MINITAB, SAS, and SPSS.[1] Besides these four packages, a large number of software packages have been developed for personal computers. This appendix provides brief instructions on how to use MINITAB to solve statistical problems.

When using MINITAB on a personal computer, you have the option of using the **pull-down menus**, the **command language**, or both together. The pull-down menus appear at the top of the screen. The command language refers to typing the MINITAB commands on the screen instead of using the pull-down menus.

The MINITAB sections in this text are based on **MINITAB Release 13 for Windows**, **Windows 95/98**, and **Windows NT**. Most of the MINITAB applications described in the text will also work on Macintosh computers.

In the following sections, we first describe pull-down menus and then explain the command language.

USING PULL-DOWN MENUS

Assuming you know how to use Windows and you are in the Windows environment, double-click the blue MINITAB icon to load MINITAB and to enter the MINITAB environment. Once you have entered the MINITAB environment, you will see the heading **MINITAB—Untitled** at the upper left side of the screen. Below **MINITAB—Untitled** you will see the following row of pull-down menus:

File   Edit   Manip   Calc   Stat   Graph   Editor   Window   Help

By clicking any of these menus, you can obtain the options included in that menu. For example, if you click the **File** menu, the following options will appear on the screen. You can select any of these options by clicking it.

New	Ctrl + N
Open Project	Ctrl + O
Save Project	Ctrl + S
Save Project As	
Project Description	
Open Worksheet	
Query Database (ODBC)	
Save Current Worksheet	
Save Current Worksheet As	
Close Worksheet	
Open Graph	
Save Window As	
Other Files	▶
Print Worksheet	Ctrl + P
Print Setup	
Exit	

[1]BMDP is a registered trademark of BMDP Statistical Software, Inc. MINITAB is a registered trademark of Minitab, Inc. SAS is a registered trademark of SAS Institute, Inc. SPSS is a registered trademark of SPSS, Inc.

The computer screen below the row of pull-down menus is divided into two parts—**Session** and **Data** windows. The Session window is used to type the MINITAB commands if you are using the command language. Also, the MINITAB output will appear in this part of the screen. You can move the Session window up and down by clicking the bar to the right of it.

Entering Data

You will usually enter data in the Data window. The Data window contains columns denoted by C1, C2, . . . , and so on. In MINITAB, data are always entered in columns. For example, if you have data on heights of people, you will enter these data in column C1 or any other column. Similarly, if you have information on weights of people, you can enter these data in column C2. You can use any column to enter data on a variable. You can enter data on a qualitative variable (also called an alpha variable), such as names, in a column in the Data window. You enter only one name or data value in a cell of a column. In the cell immediately below C1 or C2 (etc.) you can enter the name of the variable such as Height or Weight (etc.). Note that this cell is not numbered. You can move the cursor to any cell by clicking that cell. You can move up or down in the Data window by clicking the bar that appears on the right side.

To enter new data into the MINITAB worksheet, click the cell in the row labeled 1 and the column in which you want to enter data and then start entering data. Once you have finished entering data, you can use the pull-down menus to do any statistical analysis. Remember, *do not use commas when entering data*. For example, to enter 45,763, you must type **45763** and not **45,763**. Illustration B1–1 shows how to enter data in the MINITAB Data window.

ILLUSTRATION B1–1   Suppose you want to enter the following data on the names and scores of eight students into the MINITAB Data window in columns C1 and C2.

Name	Score	Name	Score
Susan	91	Allison	81
Mark	85	Deborah	73
Robert	62	Scott	90
Peter	96	Maureen	88

To enter these data in columns C1 and C2 of the Data window, first click the unnumbered cell below C1 and type **Name** in it. Then move the cursor to the cell below it either by hitting the Enter/Return key or by pressing the downward arrow (⬇) key. Next, type the first name, **Susan**. Continue this process until all names have been entered. Then click in the cell below C2 and type **Score**. Note that this is the unnumbered cell in column C2. Next, move the cursor to the next cell down and type the first score, which is **91**. Continue this process until all scores have been entered. After you have entered all eight names and scores, the Data window will look as follows:

	C1-T	C2	C3	C4	C5	C6	C7
	Name	Score					
1	Susan	91					
2	Mark	85					
3	Robert	62					
4	Peter	96					
5	Allison	81					
6	Deborah	73					
7	Scott	90					
8	Maureen	88					
9							

If you make any mistakes entering information, move the cursor to the desired cell by clicking that cell and then enter the correct information. The old information will be replaced by the new information.

Saving a MINITAB Data File

After you have entered data into the columns of the Data window, you can save them as a MINITAB data file for future use. Suppose you want to save the file containing the data on the names and scores of the eight students in Illustration B1–1 as the SCORES file. To do so, perform the following steps.

Step 1. Click the **File** pull-down menu at the top of the screen.

Step 2. Click **Save Current Worksheet As** from the selections available in the **File** menu.

Step 3. A dialog box entitled **Save Worksheet As** will appear on the screen. Type the file name in the box next to **File Name**. Note that this file name will usually have the extension ".MTW," which indicates that it is a MINITAB worksheet file. For example, for the data entered in Illustration B1–1, type **SCORES.MTW** in this box. Next, decide where to save the worksheet by clicking the down arrow to the right of **Save in**. Among the choices that appear will be $3\frac{1}{2}$ **Floppy [A:]** and **[C:]**, along with any other drives you can save to. Click the drive you want to save to. Any directories on this drive will appear below the **Save in** box. If you wish to save in one of these directories, click it to highlight it and then click **Open**. The directory you highlighted will appear next to the **Save in** box. Make sure the box next to the **Save as type** has the word MINITAB, indicating that it is a MINITAB data file.

Step 4. After you have made all the selections in the dialog box of Step 3, click the **Save** button. The file will be saved as a MINITAB worksheet. The file containing the data of Illustration B1–1 will be saved as SCORES.MTW.

Retrieving a Previously Saved MINITAB Data File

You can retrieve a previously saved MINITAB data file by performing the following steps.

Step 1. Click the **File** pull-down menu at the top of the screen.

Step 2. Click **Open Worksheet** from the selections available in the **File** menu.

Step 3. A dialog box entitled **Open Worksheet** will appear on the screen. To select the drive and directory containing the desired file, click the down arrow next to the **Look in** box. A list of drives and directories will appear. Click the drive you want. A list of directories on that drive will appear. Click the directory containing the MINITAB file to be retrieved and then click **Open**. A list of files in that directory will appear. Click the MINITAB file to be retrieved. For example, to retrieve the SCORES.MTW file, click **Scores**.

Step 4. After you have made all the selections in the dialog box of Step 3, click **Open**. The SCORES.MTW file will be opened, and the data you entered earlier will appear in columns C1 and C2.

Saving a MINITAB Output File

A MINITAB output file contains answers to all data analysis procedures you perform. You can save this output file for future use as a *text* file. Suppose you performed some statistical analysis on the data of Illustration B1–1 and you want to save this output file as SCORES.TXT. To do so, perform the following steps.

Step 1. Click the **File** pull-down menu at the top of the screen.

Step 2. Click **Save Session Window As** from the **File** menu.

Step 3. A dialog box entitled **Save As** will appear on the screen. Type the file name in the box next to **File Name**. Note that this file name will usually have the extension ".TXT," which indicates that it is a MINITAB output file. For example, to save the entire Session window plus all output from the data of Illustration B1–1, type **SCORES.TXT** in this box. Next, decide where to save the output file by clicking the down arrow to the right of **Save in**. Among the choices that appear will be $3\frac{1}{2}$ **Floppy [A:]** and **[C:]**, along with any other drives you can save to. Click the drive you want to save to. Any directories on this drive will appear below the **Save in** box. If you wish to save in one of these directories, click to highlight it and then click **Open**. The directory you highlighted will appear next to the **Save in** box. Make sure the box next to **Save as type** has the words **Minitab Session [*.TXT]**, indicating that it is a MINITAB Session window output file.

Step 4.　After you have made all the selections in the dialog box of Step 3, click the **Save** button. The Session window plus all output will be saved as a text file under the name SCORES.TXT.

You can open a text (output) file in any word processor, such as Microsoft Word, WordPerfect, and so on, and also either insert it in another document or print it from the word processor.

If you constructed a graph using MINITAB, you can save it using the four steps just described, with some minor changes: In Step 2, click **Save Graph As**. In Step 3, the file name entered should have the extension ".MGF," and the box next to **Save as type** should have the words **Minitab Graph [*MGF]**, indicating that it is a MINITAB graph file.

You can copy a graph from the MINITAB window and paste it in any other document in a word processor. When you make a graph for a data set, the graph appears on the screen in what we can call the *Graph window*. To copy a graph, click anywhere inside the Graph window, then click the **Edit** pull-down menu, and finally click **Copy Graph** from the options available in the **Edit** pull-down menu. Then you can exit MINITAB and paste the graph in any other document—for example, a Microsoft Word or WordPerfect document.

To close this Graph window after you have either copied the graph or printed it, click the top of the left-hand corner of the Graph window. Then click the **Close** option.

You can copy any part of the MINITAB output into another file outside MINITAB. To do so, highlight the required portion of the MINITAB output, click the **Edit** pull-down menu, and click the **Copy** option. Then exit MINITAB and paste this text in any document you want.

Printing a MINITAB Output File

You can print three types of MINITAB files: the Data window (also called the worksheet), which is usually at the bottom of the screen; the Session window, which is usually at the top of the screen; and any graph you may have created, which, if it is showing, will cover parts of both the Data and Session windows. The procedure for printing any of these three files is the same.

Step 1.　Make sure the window or graph you want to print is in what could be called the *active foreground*. This means that the window or graph you want to print is showing on the screen and has a blue bar at the top. If the window or graph you want to print is already showing, make it active by clicking anywhere in it. If it is not showing, minimize the Data window by clicking its **Minimize** icon. Then click the icon for the graph or window you want brought into the active foreground.

Step 2.　Now that the graph or window you want to print is in the active foreground, click the **File** pull-down menu at the top left of the screen. If the Data window is active, click the menu choice **Print Worksheet**. If the Session window is active, click the **Print Session Window**. If a graph is active, click **Print Graph**. Note that only one of these three choices will appear on the menu, depending on what is in the active foreground.

Step 3.　If you are printing the Data window (worksheet), a dialog box entitled **Data Window Print Options** will appear. You can select or deselect a number of self-explanatory options by clicking them. You can also enter this dialog box by clicking **OK**. This dialog box appears only when you are printing the Data window.

Step 4.　After you have completed Step 2 for printing a Graph or Session window, or Step 3 for printing the Data Window, a dialog box entitled **Print** will appear. By clicking the various options, you can specify any or all of the pages to print, the number of copies, the paper size, paper orientation, and so on. When you have made your selections, click **OK** and the graph or window (as the case may be) will be printed.

You can also save a MINITAB output file as described earlier, open it in a word processor, and then print it from there.

Selecting a Sample Using MINITAB

MINITAB can be used to select a sample from a population. Suppose the data on names and scores of eight students entered in Illustration B1–1 belong to a population of eight students in a small class. You want to select a sample of three observations from this population and store the sample data in columns C3 and C4. Note that the population data are in columns C1 and C2. You can select this sample by performing the following steps in MINITAB for Windows.

Step 1. Click the **Calc** pull-down menu at the top of the screen.

Step 2. Click **Random Data** from the selections available in the **Calc** menu.

Step 3. Click **Sample From Columns** from the selections available in the **Random Data** menu.

Step 4. You will see a dialog box entitled **Sample From Columns** appear on the screen. Type **3** in the box next to **Sample** and **C1–C2** in the box below **Samples**. This tells MINITAB to select a sample of three rows (observations) from the data entered in columns C1 and C2. Next, type **C3–C4** in the box below **Store samples in**. If you want the sample to be selected with replacement, then check the box next to **Sample with replacement**. Otherwise, make sure this box is not checked.

Step 5. Click the **OK** button at the bottom of this dialog box. A sample of three observations will appear in columns C3 and C4 of the Data window.

Exiting MINITAB

Before you exit MINITAB, make sure you have saved the output and data files if you want to use them at a later date. To exit MINITAB, perform the following steps.

Step 1. Click the **File** pull-down menu at the top of the screen.

Step 2. Click **Exit** from the selections available in the **File** menu.

Step 3. MINITAB will ask you one or more questions. Answer *yes* or *no* as you like. After you answer these questions, the MINITAB session will end.

USING COMMAND LANGUAGE

The alternative to using the pull-down menus is to use the **command language**. Here, you type MINITAB commands in the Session window and obtain MINITAB output in response to a single command or a set of commands.

When you enter the MINITAB environment, the cursor is in the Data window. To be able to use the command language, perform the following steps.

Step 1. Click anywhere inside the Session window. The cursor will move from the Data window to the Session window.

Step 2. Click the **Editor** pull-down menu at the top of the screen.

Step 3. Click the **Enable Commands** option from the selections available in the **Editor** menu. Note that a check mark (✓) next to **Enable Commands** indicates that you are able to use the command language.

As a result of these steps, the following **MINITAB prompt** will appear on the screen of your computer.

MTB >

At this point, the computer is ready to receive MINITAB commands. The MINITAB commands are entered next to the MINITAB prompt, and the Enter/Return key is pressed after you complete each command. The MINITAB commands can be entered in uppercase letters, lowercase letters, or any combination of the two.

To disable the command language, first click the **Editor** pull-down menu, and then click the **Enable Commands** option from the selections available in the **Editor** menu. The check mark next to **Enable Commands** will be removed. Note that you can use the pull-down menus and the command language at the same time after you enable the command language.

You can enter data in the Session window using the **SET** or **READ** MINITAB command. However, it is easier to enter data in the spreadsheet in the Data window. You can enter the data of Illustration B1–1 in columns C1 and C2 in the Data window. Either of the following MINITAB commands typed in the Session window next to the MINITAB prompt will enable you to see these data on the screen:

```
MTB > PRINT C1-C2
MTB > PRINT 'NAME' 'SCORE'
```

Note that in Illustration B1–1, the two variables are called "Name" and "Score." Also note that when you use the assigned name for a column, such as **Name** or **Score** in that example, this name must always be enclosed within single quotes in any MINITAB command.

You can save a data or an output file, retrieve a previously saved file, or print a file using the command language. But it is easier to do so using the pull-down menus. The following *information* command helps you know what is in a retrieved file:

```
MTB > INFO
```

You can use the command language to select a sample as follows. Suppose the data entered on the names and scores of eight students in Illustration B1–1 belong to a population of eight students in a small class. You want to select a sample of three observations from this population and store the sample data in columns C3 and C4. The following MINITAB command will select a sample of three observations from columns C1 and C2 and put the sample data in columns C3 and C4:

```
MTB > SAMPLE 3 C1-C2 C3-C4
```

To print the sample data on the screen, enter the following MINITAB command:

```
MTB > PRINT C3-C4
```

The following are some additional MINITAB commands and their explanations:

`MTB > HELP HELP`	← Seeks help about MINITAB commands
`MTB > HELP COMMANDS`	← Provides help about MINITAB commands
`MTB > HELP OVERVIEW`	← Gives an overview of MINITAB
`MTB > COPY C1 C2`	← Copies all data values from column C1 to column C2
`MTB > ERASE C2`	← Deletes all data values entered in column C2
`MTB > DELETE 2 C1`	← Deletes the second value entered in column C1
`MTB > DELETE 2 C1-C2`	← Deletes the data values entered in the second row of columns C1 and C2
`MTB > SORT C1 C3`	← Sorts the data of column C1 in increasing order and puts them in column C3
`MTB > LET C4 = C1*C2`	← Multiplies the corresponding values of columns C1 and C2 and puts the new data in column C4
`MTB > Let C5 = C1**2`	← Squares each value entered in column C1 and puts the new data in column C5

The MINITAB **HELP** command followed by a specific command provides information about that command. For example, the command **HELP DELETE** can be used to find information about the **DELETE** command and its usage.

B.2 ORGANIZING DATA

This section illustrates how to use MINITAB to obtain a bar graph, a pie chart, a histogram, and a stem-and-leaf display. We will illustrate the required steps using examples.

BAR GRAPH AND PIE CHART

Illustration B2–1 shows how to make a bar graph and a pie chart for the frequency distribution of a qualitative data set. The procedure is the same to construct a bar graph and a pie chart for quantitative data.

ILLUSTRATION B2-1 Refer to the frequency distribution of student majors recorded in Table 2.4 of Example 2–1. Using MINITAB, prepare a bar graph and a pie chart for that frequency distribution.

Solution If you are using the pull-down menus, perform the following steps to construct a *bar graph* for the frequency distribution of Table 2.4.

Step 1. Enter the data containing categories and frequencies (from the first and third columns of Table 2.4) into columns C1 and C2, respectively, of the Data window.

Step 2. Click the **Graph** pull-down menu at the top of the screen.

Step 3. Click **Chart** from the selections available in the **Graph** menu.

Step 4. You will see a dialog box entitled **Chart** appear on the screen. Type **C2** in the box below **Y** and **C1** in the box below **X**. You can change some of the features of the bar graph by clicking any of the buttons marked **Edit Attributes**, **Annotation**, **Frame**, **Regions**, and **Options** and making changes in the boxes that appear on the screen in response to each of these choices. By clicking the **Options** button and choosing one of the three options—**Default**, **Increasing Y**, **Decreasing Y**—you can change the order in which the categories are listed in the bar graph. If you choose **Default**, MINITAB will list the categories in alphabetical order in the bar graph.

Step 5. Click the **OK** button at the bottom of the dialog box. The bar graph will appear on the screen.

If you are using the command language, enter the given data on categories and frequencies (from the first and third columns of Table 2.4) into columns C1 and C2, respectively. Then, type the following MINITAB command to obtain the bar graph:

```
MTB > CHART C2*C1
```

Whether you use the pull-down menus or the command language, you will obtain the bar graph given in Figure B2.1 for the data of Table 2.4.

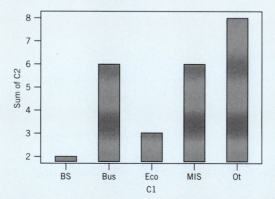

Figure B2.1 Bar graph for the frequency distribution of Table 2.4.

To obtain a *pie chart* for the frequency distribution of Table 2.4 of Example 2–1, perform the following steps if you are using the pull-down menus.

Step 1.   Enter the data containing categories and frequencies (from the first and third columns of Table 2.4) into columns C1 and C2, respectively, of the Data window.

Step 2.   Click the **Graph** pull-down menu at the top of the screen.

Step 3.   Click **Pie chart** from the selections available in the **Graph** menu.

Step 4.   You will see a dialog box entitled **Pie Chart** appear on the screen. Click inside the circle next to the **Chart table**. Then, type **C1** in the box next to **Categories in** and **C2** in the box next to **Frequencies in**. You can choose one of the four options listed under **Order of categories** in this dialog box. If you choose the **Worksheet** option (which we used to construct the pie chart given in Figure B2.2), MINITAB will list the categories in the pie chart in the same order you used in column C1. You can also explode one or more of the categories (to signify their importance) on the pie chart by listing the numbers of those categories in the box next to **Explode slice number(s)**. Here "number" refers to the number at which a category appears in C1. Next, you can type a title for your pie chart in the box next to **Title**. You can change some of the other options by clicking the **Options** button.

Step 5.   Click the **OK** button at the bottom of this dialog box. The pie chart will appear on the screen.

If you are using the command language, enter the given data on categories and frequencies (from the first and third columns of Table 2.4) into columns C1 and C2. Then, type the following MINITAB commands to obtain the pie chart:

```
MTB  > %PIE C1;
SUBC > COUNTS C2.
```

Here, the first command tells MINITAB to make a pie chart for the categories listed in column C1. The subcommand tells it to use the frequencies listed in column C2.

The pie chart given in Figure B2.2 for the data of Table 2.4 is obtained by using the pull-down menus with the **Worksheet** option. In the pie chart, MINITAB lists frequencies and percentages within parentheses next to the categories. The pie chart obtained by using the command language may list the categories in a different order.

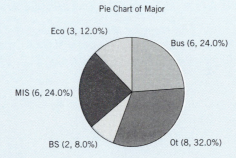

Pie Chart of Major

Figure B2.2 Pie chart for the frequency distribution of Table 2.4.

HISTOGRAM

Illustration B2–2 shows how to construct a histogram for a quantitative data set.

ILLUSTRATION B2–2   Refer to the data on the 1999 payrolls of all major league baseball teams given in Example 2–3. Construct a histogram for that data set.

Solution   Suppose you want to use the same classes for the histogram you used in Example 2–3, which are 15–30, 31–46, 47–62, 63–78, and 79–94. The midpoints of these classes are 22.5, 38.5, 54.5, 70.5, and 86.5, respectively. The width of each class is 16 units. If you are using the pull-down menus, perform the following steps to obtain a histogram.

Step 1. Enter the data on the payrolls of baseball teams given in Example 2–3 into column C1 of the Data window.

Step 2. Click the **Graph** pull-down menu at the top of the screen.

Step 3. Click **Histogram** from the selections available in the **Graph** menu.

Step 4. A dialog box entitled **Histogram** will appear on the screen. Type **C1** in the first box below **X**.

Step 5. Click the **Options** button that appears at the bottom of this dialog box. A second dialog box entitled **Histogram Options** will appear on the screen.

Step 6. Select the type of histogram by selecting the appropriate option from the selections below **Type of Histogram**. Next, click inside the circle next to **Midpoint** below **Type of Intervals**, and then click inside the circle next to **Midpoint/cutpoint positions** below **Definition of Intervals**. Next, type **22.5:86.5/16** in the box next to **Midpoint/cutpoint positions**. Note that, here, 22.5 is the midpoint of the first class, 86.5 is the midpoint of the last class, and 16 is the width of each class.

Step 7. Click the **OK** button at the bottom of the second dialog box. Then click the **OK** button at the bottom of the first dialog box. The histogram will appear on the screen. Note that before clicking this **OK** button, you can make other changes to the histogram by clicking the button **Edit Attributes**, **Annotation**, **Frame**, or **Regions** and making any necessary changes.

If you are using the command language, enter the given data on payrolls into column C1 of the Data window. Then type the following MINITAB commands in the Session window to obtain the histogram:

```
MTB  > HISTOGRAM C1;
SUBC > MIDPOINTS 22.5:86.5/16.
```

Observe these MINITAB commands carefully for semicolons and periods. Note that a semicolon at the end of a MINITAB command instructs MINITAB that a subcommand (SUBC) is to follow with some additional information. A semicolon at the end of a subcommand indicates that another subcommand is to follow with some more information. A period at the end of a subcommand instructs MINITAB that all MINITAB commands and subcommands have been entered.

Whether you use the pull-down menus or the command language, you will obtain the histogram given in Figure B2.3 for the data on the payrolls of baseball teams from Example 2–3.

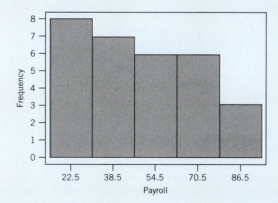

Figure B2.3 Histogram for 1999 payrolls of major league baseball teams. ■

In the MINITAB commands, we selected our own classes. If you want MINITAB to choose classes and widths, do not enter the values of the midpoints and width of classes in Step 6. In the command language, do not type a semicolon after the first command and, thus, skip the subcommand.

STEM-AND-LEAF DISPLAY

Illustration B2–3 shows how to construct a stem-and-leaf display for a data set.

ILLUSTRATION B2–3   Refer to the data on test scores for 30 students given in Example 2–8. Prepare a stem-and-leaf display for that data set.

Solution   If you are using the pull-down menus, perform the following steps to obtain a stem-and-leaf display.

Step 1.   Enter the data on scores given in Example 2–8 into column C1 of the Data window.

Step 2.   Click the **Graph** pull-down menu at the top of the screen.

Step 3.   Click **Stem-and-Leaf** from the selections available in the **Graph** menu.

Step 4.   A dialog box entitled **Stem-and-Leaf** will appear on the screen. Type **C1** in the box below **Variables** and **10** in the box next to **increment**.

Step 5.   Click the **OK** button at the bottom of this dialog box. The stem-and-leaf display will appear on the screen.

If you are using the command language, enter the given data on scores into column C1. Then, type the following MINITAB commands to obtain the stem-and-leaf display:

```
MTB  > STEM-AND-LEAF C1;
SUBC > INCREMENT 10.
```

In Step 4 of the pull-down menu and in the MINITAB subcommand, the INCREMENT 10 indicates the distance between any two consecutive stems. As a consequence of this command, the stems will be 5 (for numbers in the 50s), 6 (for numbers in the 60s), and so forth.

Whether you use the pull-down menus or the command language, you will obtain the output given in Figure B2.4 for the data on scores from Example 2–8.

```
Stem-and-Leaf Display: C1

Stem-and-leaf of C1      N = 30
Leaf Unit = 1.0

     3      5 027
     8      6 14589
    (9)     7 112256799
    13      8 0134677
     6      9 223568
```

Figure B2.4   Stem-and-leaf display for Illustration B2–3.

In the MINITAB printout of Figure B2.4, N = 30 is the number of observations in the data. Leaf Unit = 1.0 means that the decimal point is after one leaf digit in the printout. Thus, the first number is 50, the second is 52, and so on. If the leaf unit is .10, the numbers in the stem-and-leaf display will be 5.0, 5.2, 5.7, and so on. (See Computer Assignment CA2.8 as an example of this case.) On the other hand, if the leaf unit is 10, then the numbers will be 500, 520, 570, and so on.

The numbers in the first column of the stem-and-leaf display of Figure B2.4 are called depths, which give the cumulative frequencies from above and below. The depth, which appears in parentheses (9 in this MINITAB output), gives the number of leaves in the row that contains the median value. The depths before this row give the total number of leaves in the corresponding row and the row or rows before it. Thus, the first depth (which is 3) gives the number of leaves belonging to stem

5. The second number (which is 8) gives the cumulative number of leaves belonging to the first two stems. After the stem that contains the median data value, the depths are cumulative from the bottom of the stem-and-leaf display. For example, the depth of 13 for the stem of 8 indicates the total number of leaves belonging to stems 8 and 9. The last depth, which is 6, indicates the number of leaves for stem 9. ∎

B.3 NUMERICAL DESCRIPTIVE MEASURES

This section illustrates how to use MINITAB to calculate some of the summary measures for a data set and to prepare a box-and-whisker plot. We will illustrate the required steps for these data analyses using examples.

CALCULATING SUMMARY MEASURES

Illustration B3–1 shows how to calculate the mean, median, range, standard deviation, and some of the other summary measures.

ILLUSTRATION B3–1   Refer to the data on the 1998 total pay of the 10 top-paid chief executives in the United States given in Example 3–5. Using MINITAB, find the mean, median, range, and standard deviation for those data.

Solution   You can use either the **Stat** or **Calc** pull-down menu to find the mean, median, range, and standard deviation for any data set. The **Stat** menu gives the values of many summary measures at the same time, whereas the **Calc** menu helps to find the value of one summary measure at a time. If you are using the pull-down menu, perform the following steps to find these summary measures using the **Stat** pull-down menu.

Step 1.   Enter the data on total pay into column C1 of the Data window.

Step 2.   Click the **Stat** pull-down menu at the top of the screen.

Step 3.   Click **Basic Statistics** from the selections available in the **Stat** menu.

Step 4.   Click **Display Descriptive Statistics** from the selections available in the **Basic Statistics** menu.

Step 5.   You will see a dialog box entitled **Display Descriptive Statistics** appear on the screen. Type **C1** in the box below **Variables**. You can also construct graphs for the entered data by clicking the **Graphs** option in this box and making the necessary choices.

Step 6.   Click the **OK** button at the bottom of this dialog box. The MINITAB output will appear on the screen.

If you are using the command language, enter the given data into column C1. Then, type the following MINITAB command to obtain the various summary measures:

```
MTB > DESCRIBE C1
```

In response to the above, you will obtain the MINITAB output shown in Figure B3.1.

Most of the entries in the MINITAB output of Figure B3.1 are self-explanatory. For example, from this output, the mean is 155.8, the median is 100.1, the standard deviation is 156.2, the minimum value in the data set is 57.3, the maximum value in the data set is 575.6, the first quartile (Q1) is 65.0, and the third quartile (Q3) is 175.8. The fourth entry, labeled TrMean (called the trimmed mean and briefly explained in Exercise 3.33), is the mean of the data when approximately 5% of the values at each end of the ranked data are dropped. In this example, it is calculated by dropping the smallest and the largest values of the data set. This sixth entry, SE Mean, gives the standard error (also called the standard deviation) of the sample mean, which is discussed in Chapter 7.

```
Descriptive Statistics: C1

Variable    N      Mean     Median    TrMean    StDev    SE Mean
C1          10     155.8    100.1     115.6     156.2    49.4

Variable    Minimum    Maximum    Q1      Q3
C1          57.3       575.6      65.0    175.8
```

Figure B3.1   MINITAB output for Illustration B3–1.

As mentioned earlier, you can also use the **Calc** pull-down menu to find the various summary measures for a data set. Note that you will find the value of one summary measure at a time if you use the **Calc** menu. If you are using the pull-down menus, perform the following steps to find these summary measures using the **Calc** pull-down menu.

Step 1.   Enter the given data into column C1 of the Data window.

Step 2.   Click the **Calc** pull-down menu at the top of the screen.

Step 3.   Click **Column Statistics** from the selections available in the **Calc** menu.

Step 4.   You will see a dialog box entitled **Column Statistics** appear on the screen. Click next to the summary measure you want to find. For example, click inside the circle next to **Mean** to find the value of the mean. Note that you can choose only one of the options given there.

Step 5.   Type **C1** in the box next to **Input variable**.

Step 6.   If you want to store the answer permanently as a constant, type **K1** in the box next to **Store result in**. Remember, if you have used K1 earlier to store another answer, then type **K2** in this box, and so on. However, if you do not want to store the answer, skip this step.

Step 7.   Click the **OK** button at the bottom of this dialog box. The MINITAB output will appear on the screen.

If you are using the command language, enter the given data into column C1. Then, type the following MINITAB command to obtain the mean:

$$MTB > MEAN \ C1$$

In response to the above, you will obtain the MINITAB output shown in Figure B3.2.

```
MEAN of C1

Mean of C1 = 155.75
```

Figure B3.2   Mean for the 1998 total pay of 10 top-paid chief executives.

Similarly, you can calculate the median, range, or standard deviation by clicking next to the required summary measure in Step 4 above. In the case of command language, you type the command **MEDIAN C1**, **RANGE C1**, or **STANDARD DEVIATION C1** next to the MTB prompt to find the required measure. ■

BOX-AND-WHISKER PLOT

Illustration B3–2 shows how to prepare a box-and-whisker plot using MINITAB.

ILLUSTRATION B3-2   Refer to the data on incomes (in thousands of dollars) for a sample of 12 households given in Example 3–24. Using MINITAB, construct a box-and-whisker plot for those data.

Solution   To obtain a box-and-whisker plot for the data on incomes given in Example 3–24, perform the following steps if you are using the pull-down menus.

Step 1.   Enter the data on incomes into column C1 of the Data window.

Step 2.   Click the **Graph** pull-down menu at the top of the screen.

Step 3.   Click **Boxplot** from the selections available in the **Graph** menu.

Step 4.   You will see a dialog box entitled **Boxplot** appear on the screen. Type **C1** in the box below **Y**. You can change some of the features of the box-and-whisker plot by clicking any of the buttons marked **Edit Attributes**, **Annotation**, **Frame**, **Regions**, and **Options** and making changes in the boxes that appear on the screen in response to each of these selections.

Step 5.   Click the **OK** button at the bottom of this dialog box. The boxplot will appear on the screen.

If you are using the command language, enter the given data on incomes into column C1. Then, type the following MINITAB command to obtain the box-and-whisker plot:

```
MTB > BOXPLOT C1
```

In response to the above steps, you will obtain the box-and-whisker plot given in Figure B3.3, for the data on incomes. The asterisk in this graph indicates an outlier. Note that the box-and-whisker plot here appears vertically.

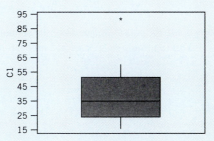

Figure B3.3   Box-and-whisker plot for incomes.

B.4   THE BINOMIAL AND POISSON PROBABILITY DISTRIBUTIONS

Using MINITAB, we can find the probability of a single outcome for the binomial and Poisson probability distributions or we can list the probabilities for all outcomes of a binomial or a Poisson experiment.

THE BINOMIAL PROBABILITY DISTRIBUTION

Illustration B4–1 shows how to find the probability of a single outcome for the binomial probability distribution, Illustration B4–2 explains how to find the cumulative probability for the binomial probability distribution, and Illustration B4–3 describes the procedure for obtaining the probability distribution of x for a binomial distribution.

ILLUSTRATION B4-1   Reconsider Example 5–18. Five percent of all VCRs manufactured by an electronics company are defective. Find the probability that if three VCRs are randomly selected from the production line of this company, exactly one of them will be defective.

Solution   From the given information,

$$n = \text{total number of VCRs in the sample} = 3$$

$$x = \text{defective VCRs in the sample} = 1$$

$$p = P(\text{a VCR is defective}) = .05$$

If you are using the pull-down menus, perform the following steps to find the probability of $x = 1$ when $n = 3$ and $p = .05$.

Step 1.   Click the **Calc** pull-down menu at the top of the screen.

Step 2.   Click **Probability Distributions** from the selections available in the **Calc** menu.

Step 3.   Click **Binomial** from the selections available in the **Probability Distributions** menu.

Step 4.   You will see a dialog box entitled **Binomial Distribution** appear on the screen. Click inside the circle next to **Probability**. Type **3** (the value of n) in the box next to **Number of trials** and **.05** (the value of p) in the box next to **Probability of success**. Then click inside the circle next to **Input constant** and type **1** (the value of x) in the box next to it.

Step 5.   Click the **OK** button at the bottom of this dialog box. The probability of $x = 1$ will appear on the screen.

If you are using the command language, type the following MINITAB commands to obtain the probability of $x = 1$. Note that in the following commands, 1 is the value of x, 3 is the value of n, and .05 is the value of p.

```
MTB  >  PDF 1;
SUBC >  BINOMIAL 3 .05.
```

Whether you use the pull-down menus or the command language, you will obtain the output given in Figure B4.1. From this output, the required probability is .1354.

```
Probability Density Function

Binomial with n = 3 and p = 0.0500000

        x         P( X = x)
     1.00            0.1354
```

Figure B4.1   Binomial probability of $x = 1$ for $n = 3$ and $p = .05$.   ■

ILLUSTRATION B4–2   Refer to Illustration B4–1. Find $P(x \leq 2)$.

Solution   Here, you want to find the cumulative probability $P(x \leq 2)$ for the binomial distribution problem of Illustration B4–1. In other words, you need to find the probability that at most two VCRs are defective in a sample of three when the probability of success is .05. To find this probability using the pull-down menus, click inside the circle next to **Cumulative probability** instead of **Probability** in Step 4 of the steps described in Illustration B4–1. Then, type **3** (the value of n) in the box next to **Number of trials** and **.05** (the value of p) in the box next to **Probability of success**. Then click inside the circle next to **Input constant** and type **2** (the value of x) in the box next to it. The remaining steps are the same as in Illustration B4–1.

If you are using the command language, type the following commands:

```
MTB  >  CDF 2;
SUBC >  BINOMIAL 3 .05.
```

Either of these procedures will produce the output shown in Figure B4.2, which gives

$$P(x \leq 2) = .9999$$

```
Cumulative Distribution Function

Binomial with n = 3 and p = 0.0500000

        x        P( X <= x)
     2.00           0.9999
```

Figure B4.2   Cumulative binomial probability of $P(x \leq 2)$ for $n = 3$ and $p = .05$. ■

ILLUSTRATION B4–3   Reconsider Illustration B4–1. Let x denote the number of defective VCRs in a random sample of three. Using MINITAB, list the probability distribution of x.

Solution   As we know, $n = 3$ and $p = .05$

If you are using the pull-down menus, perform the following steps to obtain the probability distribution of x for $n = 3$ and $p = .05$.

Step 1.   Enter the values 0, 1, 2, and 3 in column C1 of the Data window. Note that these are the possible values that x can assume in this example.

Step 2.   Click the **Calc** pull-down menu at the top of the screen.

Step 3.   Click **Probability Distributions** from the selections available in the **Calc** menu.

Step 4.   Click **Binomial** from the selections available in the **Probability Distributions** menu.

Step 5.   You will see a dialog box entitled **Binomial Distribution** appear on the screen. Click inside the circle next to **Probability**. Type **3** (the value of n) in the box next to **Number of trials** and **.05** (the value of p) in the box next to **Probability of success**. Then click inside the circle next to **Input column** and type **C1** inside the box next to it.

Step 6.   Click the **OK** button at the bottom of this dialog box. The probability distribution of x will appear on the screen.

If you are using the command language, type the following MINITAB commands to obtain the probability distribution of x:

```
MTB  > PDF;
SUBC > BINOMIAL 3  .05.
```

Whether you use the pull-down menus or the command language, you will obtain the output given in Figure B4.3. In this figure, the column labeled x lists the possible values of x, and the column labeled $P(X = x)$ gives their corresponding probabilities.

```
Probability Density Function

Binomial with n = 3 and p = 0.0500000

         x        P( X = x)
      0.00          0.8574
      1.00          0.1354
      2.00          0.0071
      3.00          0.0001
```

Figure B4.3   Binomial probability distribution of x. ■

Using the MINITAB output of Figure B4.3, you can write the probability of any value of x for this example. For instance,

$$P(1 \leq x \leq 3) = P(1) + P(2) + P(3) = .1354 + .0071 + .0001 = .1426$$

$$P(\text{at least 2 defective VCRs}) = P(2) + P(3) = .0071 + .0001 = .0072$$

$$P(\text{at most 1 defective VCR}) = P(0) + P(1) = .8574 + .1354 = .9928$$

THE POISSON PROBABILITY DISTRIBUTION

Illustration B4–4 shows how to find the probability of a single outcome for the Poisson probability distribution, Illustration B4–5 explains how to find the cumulative probability for the Poisson probability distribution, and Illustration B4–6 describes the procedure for obtaining the probability distribution of x for a Poisson probability distribution.

ILLUSTRATION B4-4   A washing machine in a laundromat breaks down an average of 1.2 times per month. Find the probability that during a given month it will have exactly three breakdowns.

Solution   From the given information,

$$\lambda = \text{mean number of breakdowns per month} = 1.2$$

$$x = \text{number of breakdowns during a given month} = 3$$

If you are using the pull-down menus, perform the following steps to find the probability of $x = 3$ when $\lambda = 1.2$.

Step 1.   Click the **Calc** pull-down menu at the top of the screen.

Step 2.   Click **Probability Distributions** from the selections available in the **Calc** menu.

Step 3.   Click **Poisson** from the selections available in the **Probability Distributions** menu.

Step 4.   You will see a dialog box entitled **Poisson Distribution** appear on the screen. Click inside the circle next to **Probability**. Type **1.2** (the value of λ) in the box next to **Mean**. Then click inside the circle next to **Input constant** and type **3** (the value of x) in the box next to it.

Step 5.   Click the **OK** button at the bottom of this dialog box. The probability of $x = 3$ will appear on the screen.

If you are using the command language, type the following MINITAB commands to obtain the probability of $x = 3$. Note that in these commands, 3 is the value of x and 1.2 is the value of the mean.

```
MTB  > PDF 3;
SUBC > POISSON 1.2.
```

Whether you use the pull-down menus or the command language, you will obtain the output given in Figure B4.4. From this output, the required probability is .0867. Note that in the MINITAB output, the statement Poisson with mu = 1.20000 refers to $\lambda = 1.2$.

```
Probability Density Function

Poisson with mu = 1.20000

        x        P( X = x)
     3.00           0.0867
```

Figure B4.4   Poisson probability of $x = 3$ when $\lambda = 1.2$.

ILLUSTRATION B4-5   Refer to Illustration B4–4. Find $P(x \leq 2)$.

Solution   Here, you want to find the cumulative probability $P(x \leq 2)$ for the Poisson distribution problem in Illustration B4–4. In other words, you need to find the probability that there will be at most two breakdowns during a given month when $\lambda = 1.2$. To find this probability using the pull-down menus, click inside the circle next to **Cumulative probability** instead of **Probability** in Step 4 of the steps described in Illustration B4–4. Then, type **1.2** (the value of λ) in the box next to **Mean**. Click inside the circle next to **Input constant** and type **2** (the value of x) in the box next to it. The remaining steps are the same as in Illustration B4–4.

If you are using the command language, use the following commands to find this probability:

```
MTB  > CDF 2;
SUBC > POISSON 1.2.
```

Either of these procedures will produce the MINITAB output shown in Figure B4.5, which gives $P(x \leq 2) = .8795$.

```
Cumulative Distribution Function

Poisson with mu = 1.20000

        x       P(X <= x)
     2.00         0.8795
```

Figure B4.5   Cumulative Poisson probability $P(x \leq 2)$ when $\lambda = 1.2$.   ∎

ILLUSTRATION B4-6   Reconsider Illustration B4–4. Let x denote the number of breakdowns observed during a given month. Using MINITAB, list the probability distribution of x.

Solution   From the given information, $\lambda = 1.2$.

If you are using the pull-down menus, perform the following steps to obtain the probability distribution of x for $\lambda = 1.2$.

Step 1.   Enter the values 0, 1, 2, . . . in column C1 of the Data window. Note that these are the possible values that x can assume in this example. Because it is difficult to know the largest value that x can assume in a Poisson probability distribution, you may enter as many values as you want.

Step 2.   Click the **Calc** pull-down menu at the top of the screen.

Step 3.   Click **Probability Distributions** from the selections available in the **Calc** menu.

Step 4.   Click **Poisson** from the selections available in the **Probability Distributions** menu.

Step 5.   You will see a dialog box entitled **Poisson Distribution** appear on the screen. Click inside the circle next to **Probability**. Type **1.2** (the value of λ) in the box next to **Mean**. Then click inside the circle next to **Input column** and type **C1** in the box next to it.

Step 6.   Click the **OK** button at the bottom of this dialog box. The probability distribution of x will appear on the same screen.

If you are using the command language, type the following MINITAB commands to obtain the probability distribution of x:

```
MTB  > PDF;
SUBC > POISSON 1.2.
```

Whether you use the pull-down menus or the command language, you will obtain the output given in Figure B4.6. In this figure, the column labeled x lists the possible values of x, and the column labeled $P(X = x)$ gives their corresponding probabilities.

```
Probability Density Function

Poisson with mu = 1.20000

          x        P(X = x)
       0.00          0.3012
       1.00          0.3614
       2.00          0.2169
       3.00          0.0867
       4.00          0.0260
       5.00          0.0062
       6.00          0.0012
       7.00          0.0002
       8.00          0.0000
```

Figure B4.6   Poisson probability distribution of x when $\lambda = 1.2$.

The problem with using the pull-down menus in Illustration B4–6 is that you may not know how many values of x should be entered in column C1. In this example, if you enter only up to $x = 6$ in column C1, then the probability distribution will list the probabilities only up to $x = 6$. Hence, in a case where you need to list the probability distribution of x, it is easier to use the command language.

Using the output given in Figure B4.6, you can determine the probability of any value of x for this Poisson probability distribution problem. For example, the probability that there will be fewer than three breakdowns during a given month is

$$P(x < 3) = P(0) + P(1) + P(2) = .3012 + .3614 + .2169 = .8795$$

Similarly, the probability that at least four breakdowns will occur during a given month is

$$P(x \geq 4) = P(4) + P(5) + P(6) + P(7) = .0260 + .0062 + .0012 + .0002 = .0336$$

B.5   THE NORMAL PROBABILITY DISTRIBUTION

This section explains the procedure for finding an area under the normal distribution curve using MINITAB. Illustration B5–1 describes the steps involved in finding such areas.

ILLUSTRATION B5–1   According to the records of an electric company serving the Boston area, the mean electric consumption during winter for all households is 1650 kilowatt-hours per month. Assume that the monthly electric consumption during winter by all households in this area has a normal distribution with a mean of 1650 kilowatt-hours and a standard deviation of 320 kilowatt-hours. Using MINITAB, find the probability that the monthly electric consumption during winter of a randomly selected household from this area is less than 1800 kilowatt-hours.

Solution   The probability that the monthly electric consumption during winter of a randomly selected household from this area is less than 1800 kilowatt-hours will be given by the area under the normal distribution curve to the left of $x = 1800$. The normal distribution of x has $\mu = 1650$ and $\sigma = 320$.

If you are using the pull-down menus, perform the following steps to find the probability $P(x < 1800)$ when $\mu = 1650$ and $\sigma = 320$.

Step 1.   Click the **Calc** pull-down menu at the top of the screen.

Step 2.   Click **Probability Distributions** from the selections available in the **Calc** menu.

Step 3. Click **Normal** from the selections available in the **Probability Distributions** menu.

Step 4. You will see a dialog box entitled **Normal Distribution** appear on the screen. Click inside the circle next to **Cumulative probability**. Then, type **1650** (the value of μ) in the box next to **Mean** and **320** (the value of σ) in the box next to **Standard deviation**. Click inside the circle next to **Input constant** and type **1800** (the value of x) in the box next to it.

Step 5. Click the **OK** button at the bottom of this dialog box. The required probability will appear on the screen.

If you are using the command language, type the following MINITAB commands to obtain the probability $P(x < 1800)$. Note that in the first MINITAB command you enter the value of x, and in the second command you enter the values of the mean and the standard deviation in that order.

```
MTB  > CDF 1800;
SUBC > NORMAL 1650 320.
```

Whether you use the pull-down menus or the command language, you will obtain the output given in Figure B5.1. From this output, the required probability is .6804. The number 1.80E + 03 below x in Figure B5.1 refers to 1800. Here, +03 means that the decimal in 1.80 should be moved three digits to the right, which makes it 1800.

```
Cumulative Distribution Function

Normal with mean = 1650.00 and standard deviation = 320.000

           x         P(X <= x)
    1.80E + 03         0.6804
```

Figure B5.1 Normal distribution probability $P(x < 1800)$ when $\mu = 1650$ and $\sigma = 320$.

Note that when you use the **Cumulative probability** option in the above commands, MINITAB gives the whole area under the normal curve to the left of the given x value. For example, in Step 4 of the pull-down menus procedure just presented, you clicked **Cumulative probability**. Also, in the command language you used **CDF** in the first MINITAB command. Both of these commands helped us to find the cumulative probability to the left of $x = 1800$. ∎

If you want to find $P(x > a)$ for a normal distribution, you first find $P(x \leq a)$ and then subtract this probability from 1.0, which is the total area under the normal curve. For example, to find the probability that the monthly electric consumption during winter of a randomly selected household from this area is greater than 1800 kilowatt-hours, you first find $P(x \leq 1800)$ and then subtract this probability from 1.0. Thus,

$$P(x > 1800) = 1 - P(x \leq 1800) = 1 - .6804 = .3196$$

To find the probability that x is between two numbers such as $P(a \leq x \leq b)$, you first find $P(x \leq a)$ and $P(x \leq b)$. Then you subtract the smaller probability from the larger probability. For example, suppose you need to find the probability that the monthly electric consumption during winter of a randomly selected household from this area is between 1500 and 1900 kilowatt-hours. Using the procedures just described, you first find $P(x \leq 1500)$ and $P(x \leq 1900)$ and then take the difference between these two probabilities. Using the MINITAB procedures explained earlier, you will find

$$P(x \leq 1500) = .3196 \quad \text{and} \quad P(x \leq 1900) = .7827$$

Hence,

$$P(1500 \leq x \leq 1900) = P(x \leq 1900) - P(x \leq 1500) = .7827 - .3196 = .4631$$

The procedure for finding an area under the standard normal curve is exactly the same. In this case you enter the value of z instead of the value of x and enter the values of the mean and standard deviation as 0 and 1, respectively, in the pull-down menus and the command language steps.

B.6   SAMPLING DISTRIBUTIONS

This MINITAB section describes how to construct the sampling distribution of the sample mean by simulating the sampling procedure. Illustration B6–1 describes the procedure to do so. Note that the pull-down menus do not contain this procedure. It can be done only by using the command language as shown in the following illustration.

ILLUSTRATION B6–1   Consider the experiment of rolling a die. Obtain 1000 samples, each containing the results of 40 rolls of the die. Then find the sample mean for each of these 1000 samples and prepare a sampling distribution of $\bar{x}$ by using these 1000 sample means.

Solution   Each roll of the die has six possible outcomes: 1 spot, 2 spots, 3 spots, 4 spots, 5 spots, and 6 spots. Rolling a die can be simulated by randomly selecting a number from 1, 2, 3, 4, 5, and 6. The MINITAB commands given below will produce 1000 samples, with each sample containing the results of 40 rolls. Note that a roll of the die is simulated by randomly selecting a number from 1 through 6. The set of values in the first row of the 40 columns gives the first sample, the set of values in the second row of the 40 columns produces the second sample, and so on. Thus, you obtain 1000 samples, each containing the results of 40 rolls of the die.

```
MTB  >  RANDOM 1000 C1-C40;
SUBC >  INTEGER 1 6.
```

To calculate the sample means for the 1000 samples you obtained, use the **RMEANS** command, as shown below. The **RMEANS** command calculates the means of rows. The following command will calculate the means of 1000 samples given in 1000 rows of the 40 columns and will put these means in column C41:

```
MTB >  RMEANS C1-C40 C41
```

Now, by using the following MINITAB commands, you can calculate the mean and standard deviation and construct the histogram of the sample means given in column C41.

```
MTB >  MEAN C41
MTB >  STDEV C41
MTB >  HISTOGRAM C41
```

By combining all the above commands, the mean, the standard deviation, and the histogram of the sample means of 1000 samples (each sample containing the results of 40 rolls of a die) are obtained. One such MINITAB output is shown in Figure B6.1, as are all the MINITAB commands mentioned in this section.

The frequency histogram of Figure B6.1 can be used to construct the sampling distribution of the sample mean $\bar{x}$. The shape of the sampling distribution of $\bar{x}$ for these 1000 samples will look exactly like the histogram of Figure B6.1. The mean and standard deviation of this sampling distribution of $\bar{x}$ are 3.4857 and .26677, respectively; that is,

$$\mu_{\bar{x}} = 3.4857 \quad \text{and} \quad \sigma_{\bar{x}} = .26677$$

Note that each time you repeat this experiment, you will obtain different values of $\mu_{\bar{x}}$ and $\sigma_{\bar{x}}$ and a different histogram.

```
MTB > random 1000 c1-c40;
SUBC> integer 1 6.
MTB > rmeans c1-c40 c41
MTB > mean c41
```

Mean of C41

```
Mean of C41 = 3.4857

MTB > stdev c41
```

Standard Deviation of C41

```
Standard deviation of C41 = 0.26677

MTB > histogram c41
```

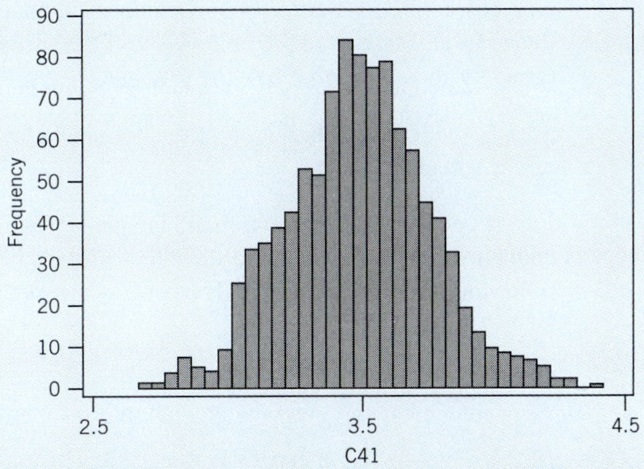

Figure B6.1 Mean, standard deviation, and histogram of $\bar{x}$.

B.7 ESTIMATION OF THE MEAN AND PROPORTION

This section describes how to use MINITAB to make confidence intervals for the population mean for large and small samples, and for the population proportion for large samples. Remember that when the sample size is large ($n \geq 30$), whether σ is known or not, the normal distribution can be used to make a confidence interval for the population mean. However, if the sample size is small ($n < 30$), the population standard deviation is not known, and the population has a normal distribution, then the t distribution is used to make a confidence interval for the population mean.

INTERVAL ESTIMATION OF A POPULATION MEAN: LARGE SAMPLES

Illustration B7–1 shows how to make a confidence interval for the population mean for a large sample.

ILLUSTRATION B7–1   The following data give the heights (in inches) of 36 randomly selected adults:

65	68	71	66	69	60	62	69	74	68	70	63
62	72	75	69	65	62	71	68	63	60	66	68
70	68	63	67	71	61	66	61	69	72	71	67

Assuming that the population standard deviation is not known, use MINITAB to construct a 95% confidence interval for the mean height of all adults.

Solution   If you are using the pull-down menus, perform the following steps to find the 95% confidence interval for the mean height of all adults.

Step 1.   Enter the data on heights of 36 adults in column C1 of the Data window.

Step 2.   Calculate the standard deviation of the sample data using the MINITAB procedure explained in Section B.3. You will obtain $s = 4.0356$.

Step 3.   Click the **Stat** pull-down menu at the top of the screen.

Step 4.   Click **Basic Statistics** from the selections available in the **Stat** menu.

Step 5.   Click **1-Sample Z** from the selections available in the **Basic Statistics** menu.

Step 6.   A dialog box entitled **1-Sample Z (Test and Confidence Interval)** will appear on the screen. Type **C1** in the box below **Variables**. Enter **4.0356** (the value of *s* obtained in Step 2) in the box next to **Sigma**. Note that you are using the value of *s* for σ here. Next, click the **Options** button. A new dialog box entitled **1-Sample Z – Options** will appear on the screen. Type **95** in the box next to **Confidence level**. Make sure the box next to **Alternative** in this dialog box shows **not equal**. If it does not, change it to *not equal*. Click the **OK** button at the bottom of this second box.

Step 7.   Click the **OK** button at the bottom of the first dialog box. The MINITAB output will appear on the screen.

If you are using the command language, first enter the given data on the heights of 36 adults into column C1, find the standard deviation of the sample data (which will be 4.0356), and then type the following MINITAB commands to obtain the required confidence interval for μ:

```
MTB  > ONEZ C1;
SUBC > SIGMA 4.0356;
SUBC > CONFIDENCE 95.
```

Here, ONEZ instructs MINITAB to use the normal distribution, C1 instructs MINITAB to construct a confidence interval for μ using the value of $\bar{x}$ calculated for the data entered in column C1, SIGMA 4.0356 represents the value of the standard deviation, and CONFIDENCE 95 indicates the confidence level in percent.

Whether you use the pull-down menus or the command language, you will obtain the output given in Figure B7.1.

```
One-Sample Z: C1

The assumed sigma = 4.0356

Variable     N      Mean    StDev   SE Mean        95.0% CI
C1          36    67.000    4.036     0.673   ( 65.682, 68.318)
```

Figure B7.1   Confidence interval for μ for a large sample.

In the MINITAB solution of Figure B7.1, The assumed sigma = 4.0356 is the value of *s* used for σ, 36 gives the number of values in the data entered in column C1, 67.000 is the sample mean ($\bar{x}$) of that data set, 4.036 is the standard deviation (*s*) of the sample data, 0.673 is the standard error (or standard deviation) of the sample mean calculated as $s/\sqrt{n} = 4.036/\sqrt{36} = .673$, 65.682 is the lower limit of the 95% confidence interval for μ, and 68.318 is the upper limit of that confidence interval.

Thus, the 95% confidence interval for the mean height (in inches) of all adults based on the heights of adults included in this sample is

$$65.682 \text{ to } 68.318$$

■

If you know the population standard deviation when estimating the population mean for a large sample, skip Step 2 and use the given value of σ in Step 6 of the pull-down menus procedure explained above. In the command language as well, use the given values of σ instead of using the value of s.

INTERVAL ESTIMATION OF A POPULATION MEAN: SMALL SAMPLES

You know from the discussion in Chapter 8 that if the sample size is small, the population is normally distributed, and the population standard deviation is not known, then the t distribution is used to construct a confidence interval for μ. Illustration B7–2 shows how to find the confidence interval for the population mean in such a case.

ILLUSTRATION B7–2   The following data give the yearly earnings (in thousands of dollars) of 12 randomly selected households from a small town:

36.75	52.43	18.82	28.45	39.50	22.65
14.30	46.75	24.48	31.70	17.25	40.27

Using MINITAB, construct a 99% confidence interval for the mean yearly earnings of all households in this town. Assume that the distribution of yearly earnings of all households in this town is approximately normal.

Solution   Because all three conditions of the t distribution are satisfied, the t distribution will be used to make the confidence interval for μ. If you are using the pull-down menus, perform the following steps to find the 99% confidence interval for the mean yearly earnings of all households in this town using the t distribution.

Step 1.   Enter the given data on yearly earnings of 12 households in column C1 of the Data window.

Step 2.   Click the **Stat** pull-down menu at the top of the screen.

Step 3.   Click **Basic Statistics** from the selections available in the **Stat** menu.

Step 4.   Click **1-Sample t** from the selections available in the **Basic Statistics** menu.

Step 5.   A dialog box entitled **1-Sample t (Test and Confidence Interval)** will appear on the screen. Type **C1** in the box below **Variables**. Next, click the **Options** button. A new dialog box entitled **1-Sample t – Options** will appear on the screen. Type **99** in the box next to **Confidence level**. Make sure the box next to **Alternative** in this dialog box shows **not equal**. Click the **OK** button at the bottom of this second box.

Step 6.   Click the **OK** button at the bottom of the first dialog box. The MINITAB output will appear on the screen.

If you are using the command language, first enter the given data on the yearly earnings of 12 households into column C1 and then type the following MINITAB commands to obtain the required confidence interval for μ:

```
MTB  > ONET C1;
SUBC > CONFIDENCE 99.
```

Here, ONET instructs MINITAB to use the t distribution with one sample, CONFIDENCE 99 indicates the confidence level as a percentage, and C1 instructs MINITAB to construct a confidence interval for μ using the value of $\bar{x}$ calculated for the data entered in column C1.

Whether you use the pull-down menus or the command language, you will obtain the output given in Figure B7.2.

```
One-Sample T: C1

Variable     N     Mean    StDev    SE Mean      99.0% CI
C1          12    31.11    12.19       3.52    ( 20.18,  42.05)
```

Figure B7.2   Confidence interval for μ for a small sample.

Thus, the 99% confidence interval for the yearly earnings of all households in this town based on the earnings of 12 households included in this sample is

$$20.18 \text{ to } 42.05$$

Because the given data are in thousands of dollars, the confidence interval can be written as

$$\$20{,}180 \text{ to } \$42{,}050 \qquad \blacksquare$$

INTERVAL ESTIMATION OF A POPULATION PROPORTION: LARGE SAMPLES

This section describes how to use MINITAB to make confidence intervals for the population proportion for large samples. Remember that in case of the proportion, the sample is large if $np > 5$ and $nq > 5$. In such a situation, you use the normal distribution to make a confidence interval for the population proportion.

Illustration B7–3 shows how to make a confidence interval for the population proportion for a large sample.

ILLUSTRATION B7–3   Shopping on the Internet is becoming more and more popular. In a random sample of 2000 U.S. adults, 780 said that they have shopped on the Internet. Using MINITAB, construct a 98% confidence interval for the proportion of all U.S. adults who have shopped on the Internet.

Solution   To use MINITAB to make a confidence interval for the population proportion, you need to know the total number of trials, the number of successes, and the confidence level. If you do not know the number of successes but you know the number of trials and the sample proportion, you can find the number of successes by multiplying the number of trials by the sample proportion. In this illustration, you do know the number of successes. From the given information,

$$n = \text{number of trials} = 2000$$

$$x = \text{number of successes} = 780$$

$$\text{Confidence level} = 98\%$$

If you are using the pull-down menus, perform the following steps to make the 98% confidence interval for the proportion of all U.S. adults who have shopped on the Internet.

Step 1.   Click the **Stat** pull-down menu at the top of the screen.

Step 2.   Click **Basic Statistics** from the selections available in the **Stat** menu.

Step 3.   Click **1 Proportion** from the selections available in the **Basic Statistics** menu.

Step 4.   A dialog box entitled **1 Proportion (Test and Confidence Interval)** will appear on the screen. Click inside the circle next to **Summarized data**. Enter **2000** (the value of n) in the box next to **Number of trials** and **780** (the value of x) in the box next to **Number of successes**.

Step 5. Click the **Options** button that appears toward the bottom of this dialog box. A second dialog box entitled **1 Proportion – Options** will appear on the screen. Type **98** in the box next to **Confidence level**. Make sure the box next to **Alternative** shows **not equal**. Check the box next to **Use test and interval based on normal distribution**. Click the **OK** button at the bottom of this second dialog box.

Step 6. Click the **OK** button at the bottom of the first dialog box. The MINITAB output will appear on the screen.

If you are using the command language, type the following MINITAB commands to obtain the required confidence interval for the population proportion. Here PONE indicates that you are making a confidence interval for one proportion, 2000 represents the number of trials, 780 is the number of successes, CONFIDENCE 98 gives the confidence level, and USEZ tells MINITAB to use the normal distribution.

```
MTB  > PONE 2000 780;
SUBC > CONFIDENCE 98;
SUBC > USEZ.
```

Whether you use the pull-down menus or the command language, you will obtain the output given in Figure B7.3. Note that MINITAB gives you some additional output such as the statement "Test of $p = 0.5$ vs p not $= 0.5$," the value of z and the P-Value even though you did not ask for this information. The meaning of these entries is discussed in Chapter 9.

```
Test and CI for One Proportion

Test of p = 0.5 vs p not = 0.5

Sample    X     N  Sample p         98.0% CI          Z-Value  P-Value
1       780  2000  0.390000  (0.364628, 0.415372)      -9.84    0.000
```

Figure B7.3 Confidence interval for *p* for a large sample.

From Figure B7.3, the sample proportion is .39, and the 98% confidence interval for the proportion of all U.S. adults who have shopped on the Internet is

0.364628 to 0.415372 or .36 to .42 or 36% to 42% ■

B.8 HYPOTHESIS TESTS ABOUT THE MEAN AND PROPORTION

This section describes how to use MINITAB to test hypotheses about the population mean for large and small samples and about the population proportion for large samples. Remember that when the sample size is large ($n \geq 30$), whether σ is known or not, the normal distribution is used to make a hypothesis test about a population mean. However, if the sample size is small ($n < 30$), the population standard deviation is not known, and the population has a normal distribution, then the t distribution is used to make a hypothesis test about a population mean.

HYPOTHESIS TESTS ABOUT A POPULATION MEAN: LARGE SAMPLES

Illustration B8–1 shows how to test a hypothesis about the population mean for a large sample.

ILLUSTRATION B8-1   An earlier study showed that the mean time spent by all college students on community service is 50 minutes a week. The following data give the time spent on community service by each of 34 college students based on a recently taken random sample.

34	56	74	23	12	89	87	56	48	13	9	85
76	56	17	28	66	38	46	81	29	33	41	78
11	57	17	22	91	54	19	35	65	47		

Using MINITAB, test at the 5% significance level whether the mean time spent per week doing community service by all college students is less than 50 minutes.

Solution   You are to test whether or not the mean time spent doing community service by all college students is less than 50 minutes. Hence, the null and alternative hypotheses are

$$H_0: \mu = 50$$

$$H_1: \mu < 50$$

The significance level is .05 and the test is left-tailed.

If you are using the pull-down menu, perform the following steps to make the test of hypothesis about the population mean for this example.

Step 1.   Enter the given data on community service in column C1 of the Data window.

Step 2.   Calculate the standard deviation of the sample data using the MINITAB procedure explained in Section B.3. You will obtain $s = 25.659$.

Step 3.   Click the **Stat** pull-down menu at the top of the screen.

Step 4.   Click **Basic Statistics** from the selections available in the **Stat** menu.

Step 5.   Click **1-Sample Z** from the selections available in the **Basic Statistics** menu.

Step 6.   A dialog box entitled **1-Sample Z (Test and Confidence Interval)** will appear on the screen. Type **C1** in the box below **Variables**. Enter **25.659** (the value of *s* obtained in Step 2) in the box next to **Sigma**. Note that you are using the value of *s* for σ here. Then enter **50** (the value of μ in the null hypothesis) in the box next to **Test mean**. Next, click the **Options** button. A new dialog box entitled **1-Sample Z – Options** will appear on the screen. Select the option **less than** in the box next to **Alternative**. Click the **OK** button at the bottom of this second box.

Step 7.   Click the **OK** button at the bottom of the first dialog box. The MINITAB output will appear on the screen.

If you are using the command language, first enter the given data on community service into column C1, find the standard deviation of the sample data (which will be 25.659), and then type the following MINITAB commands to make the required test of hypothesis about μ:

```
MTB  > ONEZ C1;
SUBC > SIGMA 25.659;
SUBC > TEST 50;
SUBC > ALTERNATIVE -1.
```

Here, in the first MINITAB command, ONEZ instructs MINITAB to use the normal distribution for one population; C1 instructs MINITAB to make the test about μ using the value of $\bar{x}$ calculated for the data entered in column C1; 25.659 represents the value of the sample standard deviation, which is used as a value of σ; TEST 50 indicates the value of μ in the null hypothesis; ALTERNATIVE -1 tells MINITAB that it is a left-tailed test.

Note that the value of ALTERNATIVE used in the subcommand will be -1, 0, or 1 depending on whether the test is *left-tailed*, *two-tailed*, or *right-tailed*, respectively. In other words, the last subcommand will be one of the following:

ALTERNATIVE -1 if the test is left-tailed

ALTERNATIVE 0   if the test is two-tailed

ALTERNATIVE 1   if the test is right-tailed

Whether you use the pull-down menus or the command language, you will obtain the output given in Figure B8.1.

```
One-Sample Z: C1

Test of mu = 50 vs mu < 50
The assumed sigma = 25.659

Variable      N      Mean     StDev     SE Mean
C1            34     46.85    25.66        4.40

Variable     95.0% Upper Bound         Z        P
C1                        54.09     -0.72    0.237
```

Figure B8.1   MINITAB output for a test of hypothesis about μ for a large sample.

In the MINITAB solution given in Figure B8.1, Test of mu = 50 vs mu < 50 indicates that the null hypothesis is $\mu = 50$ and the alternative hypothesis is $\mu < 50$. In the row of C1 in the MINITAB output, 34 is the sample size, 46.85 is the mean of the sample data, 25.66 is the standard deviation of the sample data, 4.40 is the standard deviation (or standard error) of the sample mean, -0.72 is the observed value of the test statistic z, and 0.237 is the p-value discussed in Section 9.3.

To make a decision, find the critical value of z from the normal distribution table for a left-tailed test with $\alpha = .05$. This value (for .4500 area between the mean and z) is approximately -1.65. Because the observed value of the test statistic, $z = -.72$, is greater than the critical value of $z = -1.65$ and it falls in the nonrejection region, the null hypothesis is not rejected. (The reader is advised to draw a graph that shows the rejection and nonrejection regions.) Based on this sample information, it can be concluded that the mean time spent doing community service by college students is not less than 50 minutes a week.

You can also use the p-value, printed in the MINITAB solution, to make the decision. As you know from Section 9.3, you will reject H_0 if the value of α is greater than the p-value, and you will fail to reject H_0 if the value of α is less than the p-value. In the MINITAB solution in Figure B8.1, the p-value is .24. Because $\alpha = .05$ is less than the p-value of .24, you fail to reject H_0. ■

In Illustration B8–1, the population standard deviation was unknown. However, if the population standard deviation is known, you will use that value of σ in the test of hypothesis, and you do not need to calculate the sample standard deviation before making the test.

HYPOTHESIS TESTS ABOUT A POPULATION MEAN: SMALL SAMPLES

We know from the discussion in Chapter 9 that we apply the t distribution to make a hypothesis test about the population mean when

1. the sample size is small—that is, $n < 30$
2. the population is (approximately) normally distributed
3. the population standard deviation, σ, is not known

Illustration B8–2 shows how to test a hypothesis about the population mean for a small sample.

ILLUSTRATION B8–2   A psychologist claims that the mean age at which children start to walk is 12.5 months. The following data give the ages (in months) at which 18 randomly selected children started to walk.

15	11	13	14	15	12	15	10	16
17	14	16	13	15	15	14	11	13

Using MINITAB, test at the 1% significance level if the mean age at which children start to walk is different from 12.5 months. Assume that the ages at which all children start to walk are normally distributed.

Solution   You are to test if the mean age at which children start to walk is different from 12.5 months. The null and alternative hypotheses are

$$H_0: \mu = 12.5$$

$$H_1: \mu \neq 12.5$$

The significance level is .01 and the test is two-tailed.

If you are using the pull-down menus, perform the following steps to make the test of hypothesis about the population mean for this example.

Step 1.   Enter the given data on 18 children into column C1 of the Data window.

Step 2.   Click the **Stat** pull-down menu at the top of the screen.

Step 3.   Click **Basic Statistics** from the selections available in the **Stat** menu.

Step 4.   Click **1-Sample t** from the selections available in the **Basic Statistics** menu.

Step 5.   A dialog box entitled **1-Sample t (Test and Confidence Interval)** will appear on the screen. Type **C1** in the box below **Variables**. Then enter **12.5** (the value of μ in the null hypothesis) in the box next to **Test mean**. Next, click the **Options** button. A new dialog box entitled **1-Sample t – Options** will appear on the screen. Select the option **not equal** in the box next to **Alternative**. Click the **OK** button at the bottom of this second box.

Step 6.   Click the **OK** button at the bottom of the first dialog box. The MINITAB output will appear on the screen.

If you are using the command language, first enter the given data on 18 children into column C1 and then type the following MINITAB commands to make the required test of hypothesis about μ:

```
MTB  > ONET C1;
SUBC > TEST 12.5;
SUBC > ALTERNATIVE 0.
```

In these MINITAB commands ONET C1 instructs MINITAB to use the t distribution for a single population, using data in column C1, TEST 12.5 indicates the value of μ in the null hypothesis, and ALTERNATIVE 0 tells MINITAB that it is a two-tailed test.

Whether you use the pull-down menus or the command language, you will obtain the output given in Figure B8.2.

```
One-Sample T: C1

Test of mu = 12.5 vs mu not = 12.5

Variable    N      Mean    StDev   SE Mean
C1         18    13.833    1.917     0.452

Variable         95.0% CI          T        P
C1          ( 12.880, 14.787)    2.95    0.009
```

Figure B8.2   MINITAB output for a test of hypothesis about μ for a small sample.

In the MINITAB solution given in Figure B8.2, Test of mu = 12.5 vs mu not = 12.5 indicates that the null hypothesis is $\mu = 12.5$ and the alternative hypothesis is $\mu \neq 12.5$. In the row of C1 in MINITAB output, 18 is the sample size, 13.833 is the mean of the sample data, 1.917 is the standard deviation of the sample data, 0.452 is the standard deviation (or standard error) of the sample mean, 2.95 is the observed value of the test statistic t, and 0.009 is the p-value discussed in Section 9.3, 12.880 to 14.787 is the 95% confidence interval for μ.

To make a decision, find the critical values of t from the t distribution table for a two-tailed test with $\alpha/2 = .005$ and $df = 18 - 1 = 17$. These values of t are -2.898 and 2.898. Because the value of the test statistic, $t = 2.95$, is greater than 2.898, it falls in the rejection region. (The reader is advised to draw a graph showing the rejection and nonrejection regions.) Consequently, the null hypothesis is rejected. Thus, based on this sample information, it is concluded that the mean age at which all children start to walk is different from 12.5 months.

You can reach the same conclusion by using the p-value. Because $\alpha = .01$ is greater than the p-value of .0090, you reject the null hypothesis. ∎

HYPOTHESIS TESTS ABOUT A POPULATION PROPORTION: LARGE SAMPLES

This section describes how to use MINITAB to make a test of hypothesis about the population proportion for large samples. Remember that in case of the proportion, the sample is large if $np > 5$ and $nq > 5$. In such a situation, you use the normal distribution to make a test of hypothesis about the population proportion. Illustration B8–3 shows how to make a test of hypothesis about the population proportion for a large sample.

ILLUSTRATION B8–3   Refer to Example 9–10 of Chapter 9. When working properly, a machine that is used to make chips for calculators does not produce more than 4% defective chips. Whenever the machine produces more than 4% defective chips, it needs an adjustment. To check if the machine is working properly, the quality control department at the company often takes samples of chips and inspects them to determine if they are good or defective. One such random sample of 200 chips taken recently from the production line contained 14 defective chips. Using MINITAB, test at the 5% significance level whether or not the machine needs an adjustment. Based on this sample information, make a 99% confidence interval for the population proportion.

Solution   To use the MINITAB pull-down menus to make a test of hypothesis about the population proportion, you need to know the number of trials, the number of successes, and the significance level. If you do not know the number of successes but you know the number of trials and the sample proportion, you can find the number of successes by multiplying the number of trials by the sample proportion. In this illustration, you know the number of successes. Let n be the total number of chips in the sample and x be the number of defective chips. Then, from the given information,

$$n = \text{number of trials} = 200 \qquad x = \text{number of successes} = 14$$

For the test of hypothesis, $\alpha = .05$. The null and alternative hypotheses are

$$H_0: p \leq .04 \qquad \text{(The machine does not need an adjustment)}$$

$$H_1: p > .04 \qquad \text{(The machine needs an adjustment)}$$

You can perform the following steps using the pull-down menus to make the test of hypothesis about the population proportion in this illustration.

Step 1.   Click the **Stat** pull-down menu at the top of the screen.

Step 2.   Click **Basic Statistics** from the selections available in the **Stat** menu.

Step 3.   Click **1 Proportion** from the selections available in the **Basic Statistics** menu.

Step 4.   A dialog box entitled **1 Proportion (Test and Confidence Interval)** will appear on the screen. Click inside the circle next to **Summarized data**. Enter **200** (the value of *n*) in the box next to **Number of trials** and **14** (the value of *x*) in the box next to **Number of successes**.

Step 5.   Click the **Options** button that appears toward the bottom of this dialog box. A second dialog box entitled **1 Proportion – Options** will appear on the screen. Type **.04** (the value of *p* in the null hypothesis) in the box next to **Test proportion**. Select the **greater than** option in the box next to **Alternative**. Check the box next to **Use test and interval based on normal distribution**. Click the **OK** button at the bottom of this second dialog box.

Step 6.   Click the **OK** button at the bottom of the first dialog box. The MINITAB output will appear on the screen.

If you are using the command language, type the following MINITAB commands to perform this test of hypothesis about the population proportion. Here PONE indicates that you are making a test of hypothesis about the population proportion of a single population, 200 is the number of trials, 14 represents the number of successes, TEST .04 is the value of *p* in the null hypothesis, ALTERNATIVE 1 indicates that it is a right-tailed test, and USEZ tells MINITAB to use the normal distribution.

```
MTB  > PONE 200 14;
SUBC > TEST .04;
SUBC > ALTERNATIVE 1;
SUBC > USEZ.
```

In response to the above steps or MINITAB commands, you will obtain the output given in Figure B8.3.

```
Test and CI for One Proportion

Test of p = 0.04 vs p > 0.04

Sample     X     N   Sample p   95% Lower Bound   Z-Value    P-Value
1         14   200   0.070000          0.040324      2.17      0.015
```

Figure B8.3   MINITAB output for a test of hypothesis for *p* for a large sample.

From Figure B8.3, $\hat{p}$ (the sample proportion) is .07, the observed value of *z* is 2.17, and the *p*-value for the test is .015. Since $\alpha = .05$ and the test is right-tailed, the critical value of *z* is 1.65. Hence, the null hypothesis is rejected. Also, $\alpha = .05$ is greater than $p = .015$, which again indicates to reject the null hypothesis.   ■

B.9   ESTIMATION AND HYPOTHESIS TESTING: TWO POPULATIONS

INFERENCES ABOUT THE DIFFERENCE BETWEEN TWO POPULATION MEANS FOR LARGE AND INDEPENDENT SAMPLES

MINITAB does not contain a procedure that can be used to make a confidence interval and a test of hypothesis about the difference between two population means for large and independent samples using the normal distribution. The simplest way to make such inferences with MINITAB is to use the *t* distribution irrespective of the sample sizes. This procedure is explained next.

INFERENCES ABOUT THE DIFFERENCE BETWEEN TWO POPULATION MEANS FOR SMALL AND INDEPENDENT SAMPLES: EQUAL POPULATION STANDARD DEVIATIONS

Illustration B9–1 shows how to use MINITAB to make a confidence interval and test a hypothesis for the difference between two population means for small and independent samples when the two population standard deviations are equal and the two populations are normally distributed. Note that you can also use this procedure to make a confidence interval and test a hypothesis for the difference between two population means for large and independent samples. In such a case, you do not have to assume that the two populations are normally distributed and that the two population standard deviations are equal.

ILLUSTRATION B9–1   An insurance company wanted to find the difference between the mean speeds of cars driven by all men drivers and by all women drivers on a highway. A random sample of 16 men who were driving on this highway produced the following data on the speeds of their cars at the time of the survey:

75	72	70	77	71	63	79	74
68	67	69	81	70	65	73	69

Another random sample of 14 women who were driving on the same highway produced the following data on the speeds of their cars at the time of the survey:

71	65	68	73	75	64	67
70	67	77	69	66	68	69

(a) Construct a 99% confidence interval for the difference between the mean speeds of cars driven by all men drivers and by all women drivers on this highway.

(b) Test at the 1% significance level whether the mean speed of cars driven by all men drivers on this highway is higher than that of cars driven by all women drivers. Assume that the speeds at which all men and all women drive cars on this highway are normally distributed with equal but unknown population standard deviations.

Solution   Let μ_1 and μ_2 be the mean speeds of cars driven by all men and all women, respectively, on this highway, and let $\bar{x}_1$ and $\bar{x}_2$ be the means of the respective samples.

(a) First we explain the MINITAB procedure to obtain a confidence interval for $\mu_1 - \mu_2$. If you are using the MINITAB pull-down menus, perform the following steps to obtain a 99% confidence interval for $\mu_1 - \mu_2$.

Step 1.   Enter the given data into columns C1 and C2 of the data window. Suppose C1 contains data on men drivers and C2 on women drivers.

Step 2.   Click the **Stat** pull-down menu at the top of the screen.

Step 3.   Click **Basic Statistics** from the selections available in the **Stat** menu.

Step 4.   Click **2-Sample t** from the selections available in the **Basic Statistics** menu.

Step 5.   A dialog box entitled **2-Sample t (Test and Confidence Interval)** will appear on the screen. Click inside the circle next to **Samples in different columns**. Enter **C1** in the box next to **First** and **C2** in the box next to **Second**. Then check the box next to **Assume equal variances**. Next, click the **Options** button. A new dialog box entitled **2-Sample t – Options** will appear on the screen. Type **99** in the box next to **Confidence level**. Make sure the box next to **Alternative** shows **not equal**. Click the **OK** button at the bottom of this second box.

Step 6.   Click the **OK** button at the bottom of the first dialog box. The MINITAB output will appear on the screen.

If you are using the command language, first enter the given data on speeds into columns C1 and C2 and then type the following MINITAB commands to obtain the required confidence interval. In these commands, TWOSAMPLE 99 C1 C2 indicates that we are to make a confidence interval for the difference between the two population means using the data entered in columns C1 and C2, and POOLED means that we are assuming equal variances (or standard deviations).

```
MTB  > TWOSAMPLE 99 C1 C2;
SUBC > POOLED.
```

In response to the above mentioned MINITAB pull-down menu steps or the command language, you will obtain the MINITAB output shown in Figure B9.1.

```
Two-Sample T-Test and CI: C1, C2

Two-sample T for C1 vs C2

       N     Mean     StDev    SE Mean
C1    16    71.44     4.91       1.2
C2    14    69.21     3.72       1.0


Difference = mu C1 - mu C2
Estimate for difference: 2.22
99% CI for difference: (-2.23, 6.67)
T-Test of difference = 0 (vs not = ): T-Value = 1.38  P-Value = 0.178  DF = 28
Both use Pooled StDev = 4.40
```

Figure B9.1 Confidence interval for $\mu_1 - \mu_2$.

From the MINITAB output given in Figure B9.1, the 99% confidence interval for $\mu_1 - \mu_2$ is

$$-2.23 \text{ to } 6.67$$

(b)   To perform a hypothesis test at the 1% significance level to test whether or not μ_1 is greater than μ_2, we use the following procedure. Here, the null and alternative hypotheses are

$$H_0: \mu_1 - \mu_2 = 0$$

$$H_1: \mu_1 - \mu_2 > 0$$

The significance level is .01 and the test is right-tailed.

If you are using the pull-down menus, perform the following steps to perform the required test of hypothesis.

Step 1.   Enter the given data into columns C1 and C2 of the Data window. Suppose C1 contains data on men drivers and C2 on women drivers.

Step 2.   Click the **Stat** pull-down menu at the top of the screen.

Step 3.   Click **Basic Statistics** from the selections available in the **Stat** menu.

Step 4.   Click **2-Sample t** from the selections available in the **Basic Statistics** menu.

Step 5.   A dialog box entitled **2-Sample t (Test and Confidence Interval)** will appear on the screen. Click inside the circle next to **Samples in different columns**. Enter **C1** in the box next to **First** and **C2** in the box next to **Second**. Then check the box next to **Assume equal variances**. Next, click the **Options** button. A new dialog box entitled **2-Sample t – Options** will appear on the screen. Enter **0** in the box next to **Test mean**. This indicates that the null hypothesis is $\mu_1 - \mu_2 = 0$. Next, select the

greater than option in the box next to **Alternative**, indicating that the test is right-tailed. Click the **OK** button at the bottom of this second box.

Step 6. Click the **OK** button at the bottom of the first dialog box. The MINITAB output will appear on the screen.

If you are using the command language, first enter the given data on speeds into columns C1 and C2 and then type the following MINITAB commands to make the required test of hypothesis about $\mu_1 - \mu_2$:

```
MTB  > TWOSAMPLE C1 C2;
SUBC > POOLED;
SUBC > TEST 0;
SUBC > ALTERNATIVE 1.
```

Here, in the first command, TWOSAMPLE C1 C2 instructs MINITAB to perform a test of hypothesis about the difference between the two population means using the data from two samples entered in columns C1 and C2. The subcommand POOLED indicates that MINITAB should use the pooled standard deviation procedure (see Section 10.2) assuming that the standard deviations of the two populations are equal. The subcommand TEST 0 indicates that the null hypothesis is $\mu_1 - \mu_2 = 0$. The subcommand ALTERNATIVE 1 tells MINITAB that the test is right-tailed. Remember that ALTERNATIVE will be equal to -1, 0, or 1, depending on whether the alternative hypothesis is $\mu_1 - \mu_2 < 0$, $\mu_1 - \mu_2 \neq 0$, or $\mu_1 - \mu_2 > 0$, respectively.

In response to the above steps or commands, you will obtain the output given in Figure B9.2.

```
Two Sample T-Test and CI: C1, C2

Two sample T for C1 vs C2

        N     Mean    StDev    SE Mean
C1      16    71.44   4.91        1.2
C2      14    69.21   3.72        1.0


Difference = mu c1 - mu c2
Estimate for difference: 2.22
95% lower bound for difference: -0.52
T-Test of difference = 0 (vs >): T-Value = 1.38  P-Value = 0.089  DF = 28
Both use Pooled StDev = 4.40
```

Figure B9.2   Test of hypothesis about $\mu_1 - \mu_2$.

From Figure B9.2,

The row of C1 gives $n_1 = 16$, $\bar{x}_1 = 71.44$, $s_1 = 4.91$,   and $s_{\bar{x}_1} = 1.2$

The row of C2 gives $n_2 = 14$, $\bar{x}_2 = 69.21$, $s_2 = 3.72$,   and $s_{\bar{x}_2} = 1.0$

The pooled standard deviation is $s_p = 4.40$

T-Test of difference mu C1 = mu C2 (vs >) indicates that it is a test about H_0: $\mu_1 - \mu_2 = 0$ against H_1: $\mu_1 - \mu_2 > 0$ using the t distribution. Also, from Figure B9.2,

The value of the test statistic t for $\bar{x}_1 - \bar{x}_2$ is 1.38

The p-value is .089

Degrees of freedom = $df = 28$

The test is right-tailed. The significance level is given to be 1%.

From the t distribution table, the critical value of t for $df = 28$ and .01 area in the right tail is 2.467. Because the value of the test statistic, $t = 1.38$, is less than the critical value of $t = 2.467$, it falls in the nonrejection region. (The reader should draw a graph showing the rejection and non-rejection regions.) Consequently, the null hypothesis is not rejected.

You can reach the same conclusion using the p-value printed in the MINITAB solution. From the MINITAB solution, the p-value is .089. Since the value of $\alpha = .01$ is less than the p-value of .089, you fail to reject the null hypothesis. ■

INFERENCES ABOUT THE DIFFERENCE BETWEEN TWO POPULATION MEANS FOR SMALL AND INDEPENDENT SAMPLES: UNEQUAL POPULATION STANDARD DEVIATIONS

The procedure to make a confidence interval and to test a hypothesis about $\mu_1 - \mu_2$ for small and independent samples when the standard deviations of the two populations are unknown and unequal is similar to the one used earlier in this section for the case of equal population standard deviations with these exceptions: **Do not click inside the box next to Assume equal variances** in Step 5 of the pull-down menus procedure, and **do not include the subcommand POOLED** in the command language of the previous section. We will not show an illustration of this procedure here.

INFERENCES ABOUT THE DIFFERENCE BETWEEN TWO POPULATION MEANS FOR PAIRED SAMPLES

Illustration B9–2 explains how to make a confidence interval and test a hypothesis about μ_d.

ILLUSTRATION B9–2   Recall Example 10–12 of Chapter 10. A researcher wanted to find the effect of a special diet on systolic blood pressure. She selected a sample of seven adults and put them on this dietary program for three months. The following table gives the systolic blood pressures of these seven adults before and after the completion of the program.

Before	210	180	195	220	231	199	224
After	193	186	186	223	220	183	233

(a) Construct a 99% confidence interval for μ_d, where μ_d is the mean reduction in the systolic blood pressure due to this special dietary program for the population of all adults.

(b) Test at the 5% significance level if the mean of the population paired differences, μ_d, is different from zero. Assume that the population of paired differences has a normal distribution.

Solution   Let d be the reduction in the systolic blood pressure of a person due to this special dietary program. Then

$$d = \text{Systolic blood pressure before} - \text{Systolic blood pressure after}$$

Thus, the differences in the two blood pressures will be obtained by subtracting C2 from C1.

(a) First we explain the procedure to make a confidence interval for μ_d. If you are using the pull-down menus, perform the following steps to obtain the 99% confidence interval for μ_d.

Step 1.   Enter the given data into columns C1 and C2 of the data window. Note that C1 will contain the blood pressure before, and C2 will contain the blood pressure after the completion of the program.

Step 2.   Click the **Stat** pull-down menu at the top of the screen.

Step 3.   Click **Basic Statistics** from the selections available in the **Stat** menu.

Step 4.   Click **Paired t** from the selections available in the **Basic statistics** menu.

Step 5.   A dialog box entitled **Paired t (Test and Confidence Interval)** will appear on the screen. Enter **C1** in the box next to **First sample** and **C2** in the box next to **Second sample**. Next, click the **Options** button. A new dialog box entitled **Paired t – Options** will appear on the screen. Type **99** in the box next to **Confidence level**. Make sure that the box next to **Alternative** shows **not equal**. Click the **OK** button at the bottom of this second box.

Step 6. Click the **OK** button at the bottom of the first box. The MINITAB output will appear on the screen.

If you are using the command language, use the following MINITAB commands. In these MINITAB commands, PAIRED C1 C2 indicates that we are to use the paired differences for the data entered in columns C1 and C2, and CONFIDENCE 99 tells MINITAB to make a 99% confidence interval for the mean of the population paired difference using the mean of these sample paired differences.

```
MTB  > PAIRED C1 C2;
SUBC > CONFIDENCE 99.
```

In response to these pull-down menu steps or the MINITAB commands, you will obtain the output shown in Figure B9.3. Note that MINITAB will obtain the paired differences by subtracting the values of column C2 from the values in column C1.

```
Paired T-Test and CI: C1, C2

Paired T for C1 - C2

              N      Mean     StDev    SE Mean
C1            7     208.43    18.10      6.84
C2            7     203.43    21.08      7.97
Difference    7       5.00    10.79      4.08


99% CI for mean difference: (-10.11, 20.11)
T-Test of mean difference = 0 (vs not = 0): T-Value = 1.23 P-Value = 0.266
```

Figure B9.3   MINITAB output for the confidence interval for μ_d for a small sample.

From the MINITAB output in Figure B9.3, the 99% confidence interval for the mean paired difference μ_d is

$$-10.11 \text{ to } 20.11$$

(b)   To perform the given test of hypothesis about μ_d, the null and alternative hypotheses are

$$H_0: \mu_d = 0$$

$$H_1: \mu_d \neq 0$$

If you are using the pull-down menus, perform the following steps to make this test of hypothesis.

Step 1. Enter the given data into columns C1 and C2 of the data window. Note that C1 will contain the blood pressure before, and C2 will contain the blood pressure after the completion of the program.

Step 2. Click the **Stat** pull-down menu at the top of the screen.

Step 3. Click **Basic Statistics** from the selections available in the **Stat** menu.

Step 4. Click **Paired t** from the selections available in the **Basic statistics** menu.

Step 5. A dialog box entitled **Paired t (Test and Confidence Interval)** will appear on the screen. Enter **C1** in the box next to **First sample** and **C2** in the box next to **Second sample**. Next, click the **Options** button. A new dialog box entitled **Paired t – Options** will appear on the screen. Enter **0** in the box next to **Test mean**. This indicates that the null hypothesis is $\mu_d = 0$. Next, select the **not equal** option in the box next to **Alternative** indicating that the test is two-tailed. Click the **OK** button at the bottom of this second box.

Step 6.   Click the **OK** button at the bottom of the first box. The MINITAB output will appear on the screen.

If you are using the command language, type the following MINITAB command.

$$\text{MTB} > \text{PAIRED C1 C2}$$

In response to the above steps or MINITAB command, you will obtain the MINITAB output given in Figure B9.4.

```
Paired T-Test and CI: C1, C2

Paired T for c1 - c2

            N      Mean    StDev    SE Mean
C1          7    208.43    18.10       6.84
C2          7    203.43    21.08       7.97
Difference  7      5.00    10.79       4.08

95% CI for mean difference: (-4.98, 14.98)
T-Test of mean difference = 0 (vs not = 0): T-Value = 1.23 P-Value = 0.266
```

Figure B9.4   MINITAB output for a test of hypothesis about μ_d for a small sample.

In Figure B9.4, T-Test of mean difference = 0 in the last row indicates that the null hypothesis is $\mu_d = 0$, and (vs not = 0) means that the alternative hypothesis is $\mu_d \neq 0$. Also, from the row of Difference in the output,

$$n = 7, \quad \bar{d} = 5.00, \quad s_d = 10.79, \quad \text{and} \quad s_{\bar{d}} = 4.08$$

From the last row in Figure B9.4, the observed value of the test statistic is $t = 1.23$ and the p-value is .266.

To make a decision, find the critical values of t from the t distribution table for a two-tailed test with $\alpha/2 = .025$ and $df = 7 - 1 = 6$. These values of t are -2.447 and 2.447. Because the value of the test statistic, $t = 1.23$, is between the two critical values, it falls in the nonrejection region. (The reader is advised to draw a graph showing the rejection and nonrejection regions.) Consequently, the null hypothesis is not rejected. Thus, based on this sample information, it is concluded that the mean of the paired differences, μ_d, is equal to zero. In other words, it is concluded that the mean systolic blood pressure for all adults does not decrease due to this special dietary program.

You can reach the same conclusion by using the p-value. Because $\alpha = .05$ is less than the p value of .266, the null hypothesis is not rejected. ∎

INFERENCES ABOUT THE DIFFERENCE BETWEEN TWO POPULATION PROPORTIONS FOR LARGE AND INDEPENDENT SAMPLES

In this section we will learn how to use MINITAB to make a confidence interval and test a hypothesis about the difference between two population proportions $p_1 - p_2$ for large and independent samples. Illustration B9–3 explains these procedures.

ILLUSTRATION B9-3   Refer to Example 10–15 of Chapter 10. In a sample of 500 users of Toothpaste A taken by a researcher, 100 said that they would never switch to another toothpaste. In another sample of 400 users of Toothpaste B taken by the same researcher, 68 said that they would never switch to another toothpaste. Let p_1 and p_2 be the proportions of all users of Toothpastes A and B, respectively, who will never switch to another toothpaste.

(a)   Obtain a 97% confidence interval for $p_1 - p_2$.

(b)   Using the 1% significance level, can you conclude that the proportion of users of Toothpaste A who will never switch to another toothpaste is higher than the proportion of users of Toothpaste B who will never switch to another toothpaste?

Solution   To use the MINITAB pull-down menus to make a confidence interval and to perform a test of hypothesis about the difference between two population proportions, we need to know the total number of trials and the number of successes for both samples. If we do not know the number of successes but we know the number of trials and the sample proportion for each sample, we can find the number of successes by multiplying the number of trials by the respective sample proportions. In this illustration we do know the number of successes for both samples. From the given information:

$$\text{For Toothpaste A:}\quad \text{Number of trials} = 500 \quad \text{and} \quad \text{Number of successes} = 100$$

$$\text{For Toothpaste B:}\quad \text{Number of trials} = 400 \quad \text{and} \quad \text{Number of successes} = 68$$

(a)   First we explain the MINITAB procedure to obtain a confidence interval for $p_1 - p_2$. If you are using the MINITAB pull-down menus, perform the following steps to obtain a 97% confidence interval for $p_1 - p_2$.

Step 1.   Click the **Stat** pull-down menu at the top of the screen.

Step 2.   Click **Basic Statistics** from the selections available in the **Stat** menu.

Step 3.   Click **2 Proportions** from the selections available in the **Basic Statistics** menu.

Step 4.   A dialog box entitled **2 Proportions (Test and Confidence Interval)** will appear on the screen. Click inside the circle next to **Summarized data**. Enter the values of the **number of trials** and the **number of successes** for the two samples in the respective boxes. For this illustration, you will enter the numbers **500**, **100**, **400**, and **68** in the four boxes.

Step 5.   Click the **Options** button that appears towards the bottom of this dialog box. A second dialog box entitled **2 Proportions – Options** will appear on the screen. Type **97** (the value of the confidence level) in the box next to **Confidence level**. Make sure the box next to **Alternative** shows **not equal**. Click the **OK** button at the bottom of this second dialog box.

Step 6.   Click the **OK** button at the bottom of the first dialog box. The MINITAB output will appear on the screen.

If you are using the command language, type the following MINITAB commands to obtain the required confidence interval for $p_1 - p_2$. In these commands, PTWO indicates that we are to use the procedure for the difference between two proportions, the numbers 500 100 400 68 represent the number of trials and the number of successes for the two samples, CONFIDENCE 97 indicates that the confidence level is 97%.

```
MTB  >  PTWO 500 100 400 68;
SUBC >  CONFIDENCE 97.
```

In response to the above steps or commands, you will obtain the output given in Figure B9.5.

```
Test and CI for Two Proportions

Sample     X      N     Sample p
1         100    500    0.200000
2          68    400    0.170000


Estimate for p(1) - p(2): 0.03
97% CI for p(1) - p(2): (-0.0262866, 0.0862866)
Test for p(1) - p(2) = 0 (vs not = 0): Z = 1.16  P-Value = 0.247
```

Figure B9.5   Confidence interval for $p_1 - p_2$.

From the MINITAB output given in Figure B9.5, the two sample proportions are .20 and .17, respectively, and the 97% confidence interval for $p_1 - p_2$ is

$$-.026 \text{ to } .086 \quad \text{or} \quad -2.6\% \text{ to } 8.6\%$$

(b) Now we explain the MINITAB procedure to perform a test of hypothesis for this illustration. The null and alternative hypotheses are

$$H_0: p_1 - p_2 = 0$$

$$H_1: p_1 - p_2 > 0$$

If you are using the MINITAB pull-down menus, perform the following steps to make this test of hypothesis.

Step 1. Click the **Stat** pull-down menu at the top of the screen.

Step 2. Click **Basic Statistics** from the selections available in the **Stat** menu.

Step 3. Click **2 Proportions** from the selections available in the **Basic Statistics** menu.

Step 4. A dialog box entitled **2 Proportions (Test and Confidence Interval)** will appear on the screen. Click inside the circle next to **Summarized data**. Enter the values of the **number of trials** and the **number of successes** for the two samples in the respective boxes. For this illustration, you will enter the numbers **500**, **100**, **400**, and **68** in the four boxes.

Step 5. Click the **Options** button that appears toward the bottom of this dialog box. A second dialog box entitled **2 Proportions – Options** will appear on the screen. Type **0** (the value of $p_1 - p_2$ in the null hypothesis) in the box next to **Test difference**. Select the **greater than** option in the box next to **Alternative**. Check the box next to **Use pooled estimate of p for test**. Click the **OK** button at the bottom of this second dialog box.

Step 6. Click the **OK** button at the bottom of the first dialog box. The MINITAB output will appear on the screen.

If you are using the command language, type the following MINITAB commands to perform the test of hypothesis about $p_1 - p_2$. In these commands, PTWO indicates that we are to use the procedure for the difference between two proportions, the numbers 500 100 400 68 represent the number of trials and the number of successes for the two samples, ALTERNATIVE 1 indicates that the test is right-tailed, and POOLED means that the test is to be made using the pooled sample proportion.

```
MTB  > PTWO 500 100 400 68;
SUBC > ALTERNATIVE 1;
SUBC > POOLED.
```

In response to the above steps or commands, you will obtain the output given in Figure B9.6.

```
Test and CI for Two Proportions

Sample    X      N     Sample p
1        100    500    0.200000
2         68    400    0.170000


Estimate for p(1) - p(2): 0.03
95% lower bound for p(1) - p(2): -0.0126633
Test for p(1) - p(2) = 0  (vs > 0): Z = 1.15  P-Value = 0.126
```

Figure B9.6 Hypothesis test about $p_1 - p_2$.

From Figure B9.6, the observed value of z is 1.15, and the p-value for the test is .126. Since $\alpha = .01$ and the test is right-tailed, the critical value of z is 2.33. Hence, we fail to reject the null hypothesis. Also, $\alpha = .01$ is smaller than $p = .126$, which again indicates that we fail to reject the null hypothesis. ∎

B.10 CHI-SQUARE TESTS

In this section we first discuss how MINITAB is used to make a goodness-of-fit test and then how to make a test of independence.

A GOODNESS-OF-FIT TEST

MINITAB does not have a direct command to make a goodness-of-fit test. However, by combining a few commands, you can make such a test. Illustration B10–1 describes the procedure to perform a goodness-of-fit test using MINITAB. Note that in Illustration B10–1, you use the command language to make the goodness-of-fit test because the pull-down menus do not have a procedure to perform this test.

ILLUSTRATION B10–1   Refer to Example 11–4. In a National Public Transportation survey conducted in 1995 on the modes of transportation used to commute to work, 79.6% of the respondents said that they drive alone, 11.1% car pool, 5.1% use public transit, and 4.2% depend on other modes of transportation (*USA TODAY*, April 14, 1999). Assume that these percentages hold true for the 1995 population of all commuting workers. Recently 1000 randomly selected workers were asked what mode of transportation they use to commute to work. The following table lists the results of this survey.

Mode of transportation	Drive alone	Car pool	Public transit	Other
Number of workers	812	102	57	29

Test at the 2.5% significance level whether the current pattern of use of transportation modes is different from that for 1995.

Solution   The null and alternative hypotheses are

H_0: The current percentage distribution of the use of transportation modes is the same as that for 1995

H_1: The current percentage distribution of the use of transportation modes is different from that for 1995

To make a goodness-of-fit test for this illustration, perform the following steps.

Step 1.   Enter the data on observed frequencies and probabilities into columns C1 and C2, respectively, of the data window.

Step 2.   Find the sum of column C1 (the column of observed frequencies) and put it in K1. This gives the sample size.

Step 3.   Create column C3 by multiplying K1 by the probabilities listed in column C2. Column C3 lists the expected frequencies.

Step 4.   Create column C4 using the formula $(O - E)^2/E$. In MINITAB, this formula is written as **LET C4 = (C1 − C3)**2/C3**.

Step 5.   Find the sum of the values listed in column C4, which gives the value of the test statistic χ^2.

All these steps except step 1 are shown in Figure B10.1, which presents the MINITAB input and output display. Note that it is easier to perform steps 2–5 by using the command language.

```
MTB > sum c1 k1  ←——  ⎰This command calculates the sum of the values
                      ⎱entered in column C1 and puts it in K1.

Column Sum

    Sum of C1   =   1000.0  ←—— This is the sample size.

MTB > let c3 = k1 * c2  ←——  ⎰This command creates column C3 of expected
                             ⎱frequencies by multiplying K1 and column C2 values.

MTB > let c4 = (c1 - c3)**2/c3  ←——  ⎰This command calculates
                                     ⎱$\frac{(O - E)^2}{E}$ for each category.

MTB > print c1 - c4

Data Display

Row     C1        C2       C3        C4
 1      812      0.796     796     0.32161
 2      102      0.111     111     0.72973   ⎰Compare this table
 3       57      0.051      51     0.70588   ⎨with Table 11.4 of
 4       29      0.042      42     4.02381   ⎱Example 11–4.

MTB > sum c4  ←——  ⎰This command prints the sum of column C4,
                   ⎱which is the value of the test statistic $\chi^2$.

Column Sum

    Sum of C4 = 5.7810  ←—— The value of the test statistic $\chi^2$
```

Figure B10.1   MINITAB input and output for goodness-of-fit test.

From the MINITAB solution given in Figure B10.1, the value of the test statistic χ^2 is 5.7810. For this illustration, $\alpha = .025$ and $df = 4 - 1 = 3$. From the chi-square distribution table, the critical value of χ^2 for $\alpha = .025$ and $df = 3$ is 9.348. (See Figure 11.5 of Example 11–4.) The value of the test statistic, $\chi^2 = 5.7810$, is less than the critical value of $\chi^2 = 9.348$, and it falls in the nonrejection region. Consequently, you fail to reject the null hypothesis and state that the current percentage distribution of the use of transportation modes is not different from that in 1995. The difference between the observed frequencies and the expected frequencies seems to have occurred only because of sampling error. ■

A TEST OF INDEPENDENCE OR HOMOGENEITY

Illustration B10–2 describes the procedure used to make a chi-square test of independence. The procedure for a chi-square test of homogeneity is similar.

ILLUSTRATION B10-2   Refer to Examples 11–5 and 11–6 of Chapter 11. Violence and lack of discipline have become major problems in schools in the United States. A random sample of 300 adults were asked if they favor giving more freedom to schoolteachers to punish students for violence and lack of discipline. The two-way classification of the responses of these adults is presented in the following table.

	In Favor (F)	Against (A)	No Opinion (N)
Men (M)	93	70	12
Women (W)	87	32	6

Does the sample provide sufficient evidence to conclude that the two attributes, gender and opinion, are dependent? Use $\alpha = .01$.

Solution   The null and alternative hypotheses are

$$H_0: \text{Gender and opinion are independent}$$

$$H_1: \text{Gender and opinion are dependent}$$

If you are using the pull-down menus, perform the following steps to obtain the solution.

Step 1.   Enter the data given in the three columns of the contingency table into columns C1, C2, and C3 of the Data window.

Step 2.   Click the **Stat** pull-down menu at the top of the screen.

Step 3.   Click **Tables** from the selections available in the **Stat** menu.

Step 4.   Click **Chi-Square Test** from the options available in the **Tables** menu.

Step 5.   You will see a dialog box entitled **Chi-Square Test** appear on the screen. Type **C1–C3** in the box below **Columns containing the table**.

Step 6.   Click the **OK** button at the bottom of this dialog box. The MINITAB output will appear on the screen.

If you are using the command language, enter the given data in columns C1, C2, and C3. Then, type the following MINITAB command to obtain the solution:

```
MTB > CHISQUARE C1-C3
```

Whether you use the pull-down menus or the command language, you will obtain the MINITAB output given in Figure B10.2 for the data of Illustration B10–2.

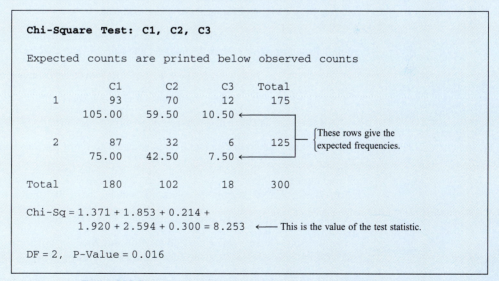

Figure B10.2   MINITAB output for the test of independence of Illustration B10–2.

From the MINITAB output presented in Figure B10.2, the value of the test statistic is $\chi^2 = 8.253$. The critical value of χ^2 from Table IX of Appendix D for $\alpha = .01$ and $df = (2 - 1)(3 - 1) = 2$ is 9.210. Because the value of the test statistic, $\chi^2 = 8.253$, is less than the critical value of $\chi^2 = 9.210$ and it falls in the nonrejection region, you fail to reject the null hypothesis (see Figure 11.6 of Example 11–6). Consequently, you conclude that the two characteristics, *gender* and *opinion*, are independent.

You can also use the *p*-value to make this decision. From Figure B10.2, the *p*-value is .016, which is greater than $\alpha = .01$. Hence, you fail to reject the null hypothesis and conclude that the two characteristics, *gender* and *opinion*, are independent.   ■

B.11   ANALYSIS OF VARIANCE

Illustration B11–1 describes the use of MINITAB to perform a test of hypothesis using the one-way analysis of variance procedure.

ILLUSTRATION B11-1   According to Example 12–4 of Chapter 12, the research department at Post Bank wants to know if the mean number of customers served per hour by each of the four tellers at a branch of this bank is the same. The research manager observed each of the four tellers for a certain number of hours. The following table gives the number of customers served by each of the four tellers during each of the observed hours.

Teller A	Teller B	Teller C	Teller D
19	14	11	24
21	16	14	19
26	14	21	21
24	13	13	26
18	17	16	20
	13	18	

Using MINITAB, test the null hypothesis that the mean number of customers served per hour by each of these four tellers is the same. Use $\alpha = .05$ and assume that all the assumptions required to apply the one-way analysis of variance procedure hold true.

Solution   Let μ_1, μ_2, μ_3, and μ_4 be the mean number of customers served per hour by each of the four tellers, respectively. Then the null and alternative hypotheses are

$$H_0: \mu_1 = \mu_2 = \mu_3 = \mu_4 \quad \text{(All four population means are equal)}$$

H_1: All four population means are not equal

If you are using the pull-down menus, perform the following steps to obtain the ANOVA solution.

Step 1.   Enter the given data into four columns C1 to C4 of the Data window.

Step 2.   Click the **Stat** pull-down menu at the top of the screen.

Step 3.   Click **ANOVA** from the selections available in the **Stat** menu.

Step 4.   Click **One-way (Unstacked)** from the options available in the **ANOVA** menu.

Step 5.   You will see a dialog box entitled **One-way Analysis of Variance** appear on the screen. Type **C1–C4** in the box below **Responses (in separate columns)**.

Step 6.   Click the **OK** button at the bottom of this dialog box. The MINITAB output will appear on the screen.

If you are using the command language, first enter the given data into four columns C1 to C4. Then, type the following MINITAB command to obtain the solution:

```
MTB > AOVONEWAY C1-C4
```

In the MINITAB command AOVONEWAY, AOV stands for analysis of variance and ONEWAY stands for one-way.

Whether you use the pull-down menus or the command language, you will obtain the MINITAB output given in Figure B11.1 for the data of Illustration B11–1.

```
One-way ANOVA: C1, C2, C3, C4

Analysis of Variance
Source    DF        SS        MS        F        P      ← Compare this table
Factor     3    255.62     85.21     9.69    0.000 }      to the ANOVA Table
Error     18    158.20      8.79                  }      12.6 of Example 12–4.
Total     21    413.82


                                  Individual 95% CIs For Mean
                                  Based on Pooled StDev
Level    N      Mean    StDev    ------+---------+---------+---------+
C1       5    21.600    3.362                            (-------*-------)
C2       6    14.500    1.643    (------*-------)
C3       6    15.500    3.619      (------*-------)
C4       5    22.000    2.915                          (------*-------)
                                  ------+---------+---------+---------+
Pooled StDev =   2.965           14.0      17.5      21.0      24.5
```

Figure B11.1   MINITAB output for Illustration B11–1.

Compare the analysis of variance table in the MINITAB solution of Figure B11.1 with Table 12.6 of Example 12–4. In Table 12.6, the two sources of variation were called the variations between and within samples in the column labeled *Source of Variation*. In the MINITAB solution of Figure B11.1, these two sources are called the *Factor* and *Error*, respectively. Also, MINITAB prints the *p*-value for the test.

The MINITAB solution also gives the following information.

1.   The mean and standard deviation for data contained in each of the columns C1, C2, C3, and C4. Thus, for example, the mean and standard deviation for the data of column C1 (for teller A) are 21.6 and 3.362, respectively.

2.   The 95% confidence interval for the mean of the population corresponding to each of the four samples—that is, the 95% confidence interval for the mean number of customers served per hour by each of the four tellers.

3.   The pooled standard deviation, which is 2.965. This pooled standard deviation is nothing but the square root of what we called MSW in Chapter 12. The MSW calculated in Example 12–4 was 8.7889. The square root of 8.7889 is 2.965, which is printed as the pooled standard deviation in the MINITAB solution.

From the MINITAB printout of Figure B11.1, the value of the test statistic F is 9.69. The critical value of F from the F distribution table for $\alpha = .05$, *df* for the numerator = 3, and *df* for the denominator = 18 is 3.16 (see Figure 12.4 of Example 12–4). The value of the test statistic, $F = 9.69$, is greater than the critical value of $F = 3.16$ and it falls in the rejection region. Consequently, reject the null hypothesis and conclude that the mean number of customers served per hour by each of the four tellers is not the same.

You can reach the same conclusion by considering the *p*-value. The *p*-value from the MINITAB solution is 0.000. Because this *p*-value is less than $\alpha = .05$, reject the null hypothesis. ■

B.12   SIMPLE LINEAR REGRESSION

In this section we describe the use of MINITAB for doing regression analysis. We use Illustration B12–1 to show how the pull-down menus and the command language can be used to obtain a computer solution.

ILLUSTRATION B12-1   Refer to the data of Example 13–8 on driving experience (in years) and monthly auto insurance premiums (in dollars) for eight auto drivers. That data set is reproduced here.

Driving Experience (years)	Monthly Auto Insurance Premium
5	$64
2	87
12	50
9	71
15	44
6	56
25	42
16	60

Use MINITAB to do the following.

(a)   Construct a scatter diagram for these data.

(b)   Find the correlation between these two variables.

(c)   Find the regression line with experience as an independent variable and premium as a dependent variable.

(d)   Make a 90% confidence interval for B.

(e)   Test at the 5% significance level whether B is negative.

(f)   Make a 95% confidence interval for the mean monthly auto insurance premium for all drivers with 10 years of driving experience. Construct a 95% prediction interval for the monthly auto insurance premium for a randomly selected driver with 10 years of driving experience.

(g)   Test at the 5% significance level whether the correlation coefficient is different from zero.

Solution

(a)   Note that in this example experience is the independent variable and premium is the dependent variable. If you are using the pull-down menus, perform the following steps to obtain the scatter diagram.

Step 1.   Enter the given data into columns C1 and C2 of the Data window. You can name these columns EXPER and PREMIUM. Note that the data entered in column C1 represent the x variable and those entered in column C2 represent the y variable.

Step 2.   Click the **Graph** pull-down menu at the top of the screen.

Step 3.   Click **Plot** from the selections available in the **Graph** menu.

Step 4.   You will see a dialog box entitled **Plot** appear on the screen. Type **C2** or **Premium** in the box below **Y** and **C1** or **Exper** in the box below **X**.

Step 5.   Click the **OK** button at the bottom of this dialog box. The scatter diagram will appear on the screen.

If you are using the command language, first enter the given data into columns C1 and C2. Then, type the following MINITAB command to obtain the scatter diagram:

```
MTB > PLOT C2 * C1
```

If you named the two columns EXPER and PREMIUM, then you can type the following command:

```
MTB > PLOT 'PREMIUM' * 'EXPER'
```

Whether you use the pull-down menus or the command language, you will obtain the scatter diagram given in Figure B12.1.

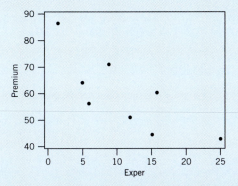

Figure B12.1 Scatter diagram for Illustration B12–1.

To do the remaining parts of this illustration, you have already entered the given data in columns C1 and C2. Also, you have named these columns EXPER and PREMIUM. If you did not, then replace EXPER by C1 and PREMIUM by C2 in the MINITAB commands in sections b through g below.

(b) If you are using the pull-down menus, perform the following steps to obtain the correlation coefficient.

Step 1. Click the **Stat** pull-down menu at the top of the screen.

Step 2. Click **Basic Statistics** from the selections available in the **Stat** menu.

Step 3. Click **Correlation** from the options available in the **Basic Statistics** menu.

Step 4. You will see a dialog box entitled **Correlation** appear on the screen. Type **C1 C2** or **EXPER PREMIUM** in the box below **variables**.

Step 5. Click the **OK** button at the bottom of this dialog box. The MINITAB output will appear on the screen.

If you are using the command language, type the following MINITAB command to obtain the correlation coefficient:

```
MTB > CORRELATION 'EXPER' 'PREMIUM'
```

Whether you use the pull-down menus or command language, you will obtain the MINITAB output shown in Figure B12.2. Note that MINITAB gives the *p*-value for the test of correlation coefficient also. This will be the case if the box next to **Display p-values** in the **Correlation** dialog box is checked. Hence, the linear correlation coefficient between experience and premium is

$$r = -.768$$

```
Correlations: exper, premium

Pearson Correlation of EXPER and PREMIUM = -0.768
P-Value = 0.026
```

Figure B12.2 Correlation coefficient.

(c) If you are using the pull-down menus, perform the following steps to obtain the regression line.

Step 1. Click the **Stat** pull-down menu at the top of the screen.

Step 2. Click **Regression** from the selections available in the **Stat** menu.

Step 3. Click **Regression** again from the options available in the **Regression** menu.

Step 4. You will see a dialog box entitled **Regression** appear on the screen. Type **C2** or **PRE-MIUM** in the box next to **Response** and **C1** or **EXPER** in the box next to **Predictors**.

Step 5. Click the **OK** button at the bottom of this dialog box. The MINITAB output will appear on the screen.

If you are using the command language, type the following MINITAB command to obtain the regression line:

```
MTB > REGRESS 'PREMIUM' 1 'EXPER'
```

In this MINITAB command, 1 indicates that there is only one independent variable.

Whether you use the pull-down menus or the command language, you will obtain the MINITAB output given in Figure B12.3.

```
Regression Analysis: premium versus exper

The regression equation is
premium = 76.7 - 1.55 exper

Predictor        Coef     SE Coef         T         P
Constant       76.660       6.961     11.01     0.000
exper          -1.5476      0.5270     -2.94     0.026

S = 10.32     R-Sq = 59.0%     R-sq(adj) = 52.1%

Analysis of Variance

Source              DF          SS         MS        F        P
Regression           1       918.5      918.5     8.62    0.026
Residual Error       6       639.0      106.5
Total                7      1557.5
```

Figure B12.3 MINITAB regression output.

The MINITAB output given in Figure B12.3 has four main parts. The first part gives the regression equation, which is

$$\text{premium} = 76.7 - 1.55 \ \text{exper}$$

Thus,

$$a = 76.7 \quad \text{and} \quad b = -1.55$$

The second part of the MINITAB output gives the values of a and b, their standard deviations, the values of the test statistic t for the hypothesis tests about A and B, and the p-values for these two tests. Figure B12.4 explains this part of the output.

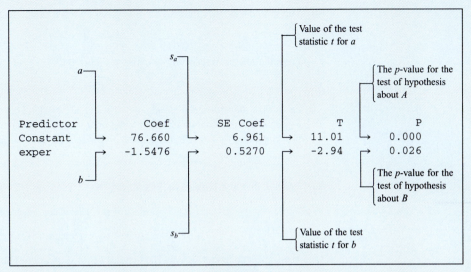

Figure B12.4

Note that we did not discuss the standard deviation s_a of a and the test of hypothesis about a (and, hence, the value of the test statistic t for a and the p-value for the test of hypothesis about a) in Chapter 13. Therefore, in the row for Constant in Figure B12.4, all the values except the coefficient value are irrelevant for this chapter. If you compare the various values in this figure with the ones calculated in Example 13–8, you will notice a slight difference due to rounding.

The third part of the MINITAB output in Figure B12.3 gives the standard deviation of errors s_e, the coefficient of determination r^2, and the adjusted r^2 (the value of r^2 adjusted for the degrees of freedom). However, the concept of adjusted r^2 was not discussed in Chapter 13. The explanation of this part of the output appears in Figure B12.5.

```
  S = 10.32          R-Sq = 59.0%            R-Sq(adj) = 52.1%
     ↑                   ↑                        ↑
    s_e                 r²                  r² adjusted for df
```

Figure B12.5

The fourth part of the MINITAB output gives the analysis of variance table. Chapter 13 did not discuss such a table except for the SS column. However, this table is similar to the analysis of variance table discussed in Chapter 12. The column of SS gives the values of SSR, SSE, and SST as shown in Figure B12.6.

```
Source                SS
Regression          918.5  ←── SSR
Residual  Error     639.0  ←── SSE
Total              1557.5  ←── SST
```

Figure B12.6

(d)   The 90% confidence interval for B is given by the formula

$$b \pm ts_b$$

From the row corresponding to exper in Figure B12.4,

$$b = -1.5476 \quad \text{and} \quad s_b = 0.5270$$

For a 90% confidence interval, $\alpha = .10$ and $\alpha/2 = .05$. The degrees of freedom are $n - 2 = 8 - 2 = 6$. From the t distribution table, the t value for .05 area in the right tail of the t distribution curve and 6 df is 1.943. The 90% confidence interval for B is

$$b \pm ts_b = -1.5476 \pm 1.943(.5270) = -1.5476 \pm 1.0240 = -2.57 \text{ to } -.52$$

(e)   The null and alternative hypotheses are

$$H_0: B = 0 \quad (B \text{ is not negative})$$

$$H_1: B < 0 \quad (B \text{ is negative})$$

The significance level is .05. The degrees of freedom are 6. The test is left-tailed. From the t distribution table, the critical value of t for $\alpha = .05$ and $df = 6$ is -1.943.

We again use the information given in the row corresponding to exper in Figure B12.4. From that row of information,

$$\text{Value of the test statistic } t \text{ for } b = -2.94$$

Because the value of the test statistic, $t = -2.94$, is less than the critical value of $t = -1.943$, it falls in the rejection region, which is in the left tail of the t distribution curve (see Figure 13.22 of Example 13–8). Consequently, reject the null hypothesis and state that the sample information supports the alternative hypothesis that B is negative.

(f)   To make a 95% confidence interval for $\mu_{y|x}$ and a 95% prediction interval for y_p for EXPER = 10, you cannot use the pull-down menus. In this case use the command language as follows:

```
MTB  > REGRESS 'PREMIUM' 1 'EXPER';
SUBC > PREDICT 10.
```

Note that when you use the above command and subcommand, MINITAB will give all the output listed in Figure B12.3 and the output given in Figure B12.7. We have omitted the output of Figure B12.3 from Figure B12.7.

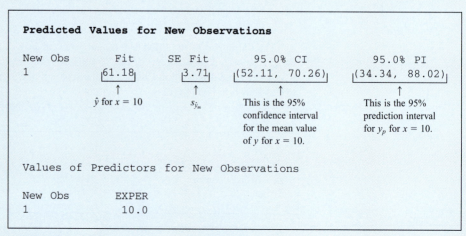

Predicted Values for New Observations

New Obs	Fit	SE Fit	95.0% CI	95.0% PI
1	61.18	3.71	(52.11, 70.26)	(34.34, 88.02)

$\hat{y}$ for $x = 10$ $s_{\hat{y}_m}$ This is the 95% confidence interval for the mean value of y for $x = 10$.   This is the 95% prediction interval for y_p for $x = 10$.

Values of Predictors for New Observations

New Obs	EXPER
1	10.0

Figure B12.7

Thus, from Figure B12.7, the value of $\hat{y}$ for $x = 10$ is 61.18. This can be considered as the point estimate of the mean value of y as well as the point estimate of the predicted value of y for $x = 10$. The interval 52.11 to 70.26 in the printout gives a 95% confidence interval for the mean value $\mu_{y|10}$ of y for $x = 10$ as was discussed in Section 13.8.1. The interval 34.34 to 88.02 gives a 95% prediction interval for y_p for $x = 10$ as was discussed in Section 13.8.2.

(g) Minitab does not contain a procedure to make a test of hypothesis about the correlation coefficient. However, we can use the p-value obtained from MINITAB in part b to perform the test about the correlation coefficient. Here the null and alternative hypotheses are

$$H_0: \rho = 0 \quad \text{(The linear correlation coefficient is zero)}$$

$$H_1: \rho \neq 0 \quad \text{(The linear correlation coefficient is not zero)}$$

The significance level is .05. In part b of this illustration, when we calculated the correlation coefficient, we checked the *Display p-values* option in the *Correlation* dialog box. That option gave us a p-value of .026. This p-value provided by MINITAB for the correlation coefficient is always for a two-tailed test. If your test is one-tailed, you must divide this p-value by 2 to compare it to the significance level to make a decision. However, in our illustration, the test is two-tailed. Since the p-value of .026 is smaller than the significance level of .05, we reject the null hypothesis. Hence, we conclude that the correlation coefficient between driving experience and auto insurance premium is different from zero. ∎

B.13 NONPARAMETRIC METHODS

This section describes how to use MINITAB to test hypotheses using nonparametric methods. The current version of MINITAB does not contain a procedure to test hypotheses about the Spearman rho rank correlation coefficient. Thus, this section contains MINITAB procedures for all nonparametric methods discussed in Chapter 14 except the Spearman rho rank correlation coefficient.

HYPOTHESIS TESTS ABOUT THE MEDIAN USING THE SIGN TEST PROCEDURE

MINITAB does not contain a procedure to test a hypothesis about whether people prefer one brand of product over another brand. Hence, we start with the procedure to perform a test about the population median. Illustration B13–1 shows how to use MINITAB to perform a hypothesis test about the *population median of a single population* using the sign test procedure.

ILLUSTRATION B13–1 Refer to Example 14–3 of Chapter 14. A real estate agent claims that the median price of homes in a small town is $137,000. A sample of 10 houses selected by a statistician produced the following data on their prices.

Home	1	2	3	4	5	6	7	8	9	10
Price ($)	147,500	123,600	139,000	168,200	129,450	132,400	156,400	188,210	198,425	215,300

Using the 5% significance level, can you conclude that the median price of homes in this town is different from $137,000?

Solution In Example 14–3 the null and alternative hypotheses were stated as follows:

$$H_0: \text{Median price} = \$137,000$$

$$H_1: \text{Median price} \neq \$137,000$$

If you are using the pull-down menus, perform the following steps to test the above hypothesis.

Step 1.   Enter the data on the prices of 10 homes into column C1 in the MINITAB Data window.

Step 2.   Click the **Stat** pull-down menu from the top of the screen.

Step 3.   Click **Nonparametrics** from the selections available in the **Stat** menu.

Step 4.   Click **1-Sample Sign** from the selections available in the **Nonparametrics** menu.

Step 5.   A dialog box entitled **1-Sample Sign** will appear on the screen. Type **C1** in the box below **Variables**. Click inside the circle next to **Test median**. Enter **137000** in the box next to **Test median**. Note that this is the value of the median in the null hypothesis. Next, select the **not equal** option in the box next to **Alternative**.

Step 6.   Click the **OK** button at the bottom of this dialog box. The MINITAB output will appear on the screen.

If you are using the command language, enter the data into column C1 as described in Step 1 above and then type the following MINITAB commands:

```
MTB > STEST 137000 C1;
SUBC > ALTERNATIVE 0.
```

Here, the first command tells MINITAB to use the sign test, with the value of the median equal to $137,000 in the null hypothesis, and to use the data of C1 to make the test. The subcommand states that the alternative hypothesis is "not equal to."

Whether you use the pull-down menus or the command language, you will obtain the output given in Figure B13.1.

```
Sign Test for Median: C1

Sign test of median = 137000 versus not = 137000

       N    Below    Equal    Above        P    Median
C1    10        3        0        7   0.3438    151950
```

Figure B13.1   MINITAB output for the sign test about a population median.

Since the *p*-value in the MINITAB output in Figure B13.1 is .3438, which is greater than the given significance level of .05, you fail to reject the null hypothesis H_0. Based on the information given in the MINITAB output, three of the prices in the sample are less than $137,000, seven are greater than $137,000, and the median price of homes in the sample is $151,950. Having three prices less than $137,000 is equivalent to having three minus signs, and having seven prices greater than $137,000 is equivalent to having seven positive signs. Thus, the observed value of X is 7, which is the number of plus signs. Since $X = 7$ falls in the nonrejection region for $\alpha = .05$, shown in Figure 14.3 of Chapter 14, you fail to reject the null hypothesis. ■

MINITAB can be used to perform the sign test to test a hypothesis about the *median difference between paired data*. Illustration B13–2 shows how to use this procedure.

ILLUSTRATION B13–2   Refer to Example 14–5 of Chapter 14. A researcher wanted to find the effects of a special diet on systolic blood pressure in adults. She selected a sample of 12 adults and put them on this dietary plan for three months. The following table gives the systolic blood pressure of each adult before and after the completion of the plan:

Before	210	185	215	198	187	225	234	217	212	191	226	238
After	196	192	204	193	181	233	208	211	190	186	218	236

Using the 2.5% significance level, can you conclude that the dietary plan reduces the median systolic blood pressure of adults?

Solution   Let M denote the median of the paired differences in the blood pressures before and after the dietary plan. Note that here the paired difference is defined as "Before $-$ After." In Example 14–5, the null and alternative hypotheses were stated as follows:

$H_0: M = 0$ (The dietary plan does not reduce the median blood pressure)

$H_1: M > 0$ (The dietary plan reduces the median blood pressure)

If you are using the pull-down menus, perform the following steps to test the above hypothesis.

Step 1.   Enter the *Before* data into column C1 and the *After* data into column C2.

Step 2.   Compute the paired differences "Before $-$ After" and enter them into column C3. To do so, first select the **Calc** pull-down menu from the top of the screen. Then click **Calculator** from the selections available in the **Calc** menu. A dialog box entitled **Calculator** will appear on the screen. Enter **C3** in the box next to **Store result in variable**. Enter **C1–C2** in the box below **Expression**. Then, click the **OK** button that appears at the bottom of the dialog box. The paired differences C1 $-$ C2 will now appear in column C3. Now you can test the hypothesis using the data of column C3.

Step 3.   Click the **Stat** pull-down menu from the top of the screen.

Step 4.   Click **Nonparametrics** from the selections available in the **Stat** menu.

Step 5.   Click **1-Sample Sign** from the selections available in the **Nonparametrics** menu.

Step 6.   A dialog box entitled **1-Sample Sign** will appear on the screen. Type **C3** in the box below **Variables**. Click inside the circle next to **Test median**. Enter **0** in the box next to **Test median**. Note that this is the value of the median in the null hypothesis. Next, select the **greater than** option in the box next to **Alternative**.

Step 7.   Click the **OK** button at the bottom of this dialog box. The MINITAB output will appear on the screen.

If you are using the command language, enter the *Before* and *After* data into columns C1 and C2, respectively, of the MINITAB Data window and then enter the following commands:

```
MTB  > LET C3=C1-C2
MTB  > STEST 0 C3;
SUBC > ALTERNATIVE 1.
```

Whether you use the pull-down menus or the command language, you will obtain the MINITAB output given in Figure B13.2.

```
Sign Test for Median: C3

Sign test of median = 0.00000 versus > 0.00000

        N    Below   Equal   Above        P   Median
C3     12        2       0      10   0.0193    6.000
```

Figure B13.2   MINITAB output for the sign test about the median of paired differences.

From the MINITAB output in Figure B13.2, the *p*-value is .0193, so you reject H_0 at any significance level greater than .0193. Thus, reject the null hypothesis at the 2.5% significance level and conclude that the median of the paired differences is greater than zero, which indicates that the systolic blood pressure decreases due to this diet. Also, *Above = 10* in the MINITAB solution indicates that there are 10 plus signs, which gives the observed value of *X*. Since $X = 10$ falls in the rejection region for $\alpha = .025$ that is shown in Figure 14.5 of Chapter 14, reject the null hypothesis. ■

THE WILCOXON SIGNED-RANK TEST FOR TWO DEPENDENT SAMPLES

As was the case with the sign test, when performing the Wilcoxon signed-rank test, MINITAB always uses the small-sample procedure to make the test when you choose this procedure. Illustration B13–3 shows how to use MINITAB to perform a hypothesis test about the *median of the paired differences for two dependent samples* using the Wilcoxon signed-rank test procedure.

ILLUSTRATION B13–3 Refer to Example 14–7 of Chapter 14. A private agency claims that the crash course it offers significantly increases the writing speed of secretaries. The following table gives the writing speeds of eight secretaries before and after they attended this course:

Before	84	75	88	91	65	71	90	75
After	97	72	93	110	78	69	115	75

Using the 2.5% significance level, can you conclude that attending this course increases the writing speed of secretaries? Use the Wilcoxon signed-rank test.

Solution Let *M* denote the median of the paired differences in the writing speeds of the secretaries before and after they attend the course. Note that here the paired difference is defined as "Before − After." The null and alternative hypotheses are stated as follows:

> $H_0: M = 0$ (Attending this course does not increase the median writing speed)
>
> $H_1: M < 0$ (Attending this course does increase the median writing speed)

If you are using the pull-down menus, perform the following steps to test the hypothesis.

Step 1. Enter the *Before* data into column C1 and the *After* data into column C2.

Step 2. Compute the paired differences "Before − After" and enter them into column C3. To do so, first select the **Calc** pull-down menu from the top of the screen. Then click **Calculator** from the selections available in the **Calc** menu. A dialog box entitled **Calculator** will appear on the screen. Enter **C3** in the box next to **Store result in variable**. Enter **C1 − C2** in the box below **Expression**. Then, click the **OK** button that appears at the bottom of this dialog box. The paired differences C1 − C2 will now appear in column C3. Now you can test the hypothesis using the data of column C3.

Step 3. Click the **Stat** pull-down menu from the top of the screen.

Step 4. Click **Nonparametrics** from the selections available in the **Stat** menu.

Step 5. Click **1-Sample Wilcoxon** from the selections available in the **Nonparametrics** menu.

Step 6. A dialog box entitled **1-Sample Wilcoxon** will appear on the screen. Type **C3** in the box below **Variables**. Click inside the circle next to **Test median**. Enter **0** in the box next to **Test median**. Note that this is the value of the median in the null hypothesis. Next, select the **less than** option in the box next to **Alternative**.

Step 7. Click the **OK** button at the bottom of this dialog box. The MINITAB output will appear on the screen.

If you are using the command language, enter the *Before* and *After* data into columns C1 and C2, respectively, of the MINITAB Data window and then enter the following commands:

```
MTB  > LET C3=C1-C2
MTB  > WTEST 0 C3;
SUBC > ALTERNATIVE -1.
```

Whether you use the pull-down menus or the command language, you will obtain the MINITAB output given in Figure B13.3.

Wilcoxon Signed Rank Test: C3

Test of median = 0.00000 versus median < 0.00000

	N	N for Test	Wilcoxon Statistic	P	Estimated Median
C3	8	7	3.0	0.038	-8.750

Figure B13.3 MINITAB output for the Wilcoxon signed-rank test about the median of paired differences.

Since, from the MINITAB output, the *p*-value is .038, you will reject H_0 at any significance level greater than .038. The significance level in our example is .025, which is less than the *p*-value. Thus, you fail to reject the null hypothesis at the 2.5% significance level and conclude that the mean of the paired differences is not less than zero, which indicates that attending this course does not seem to increase the median writing speed.

Also, from the MINITAB output in Figure B13.3, the observed value of the test statistic is 3.0. This value is the same as in Example 14–7 and it falls in the nonrejection region shown in Figure 14.7. Consequently, you fail to reject the null hypothesis. ■

THE WILCOXON RANK SUM TEST FOR TWO INDEPENDENT SAMPLES

In the statistics literature, this test is also referred to as the **Mann–Whitney test** and that is how MINITAB refers to it. In the Mann–Whitney test, the population medians are usually denoted by the Greek letter η (pronounced *eta*), and MINITAB refers to the medians of the two populations as ETA1 (η_1) and ETA2 (η_2). MINITAB provides a confidence interval for the difference between the two medians ($\eta_1 - \eta_2$) as well as the test of hypothesis. Illustration B13–4 shows how to use MINITAB to perform such a test.

ILLUSTRATION B13–4 Refer to Example 14–9. A researcher wants to determine whether the distributions of daily crimes in two cities are identical. The following data give the numbers of violent crimes for eight randomly selected days for City A and for nine randomly selected days for City B.

City A	12	21	16	8	26	13	19	23	
City B	18	25	14	16	23	19	28	20	31

Using the 5% significance level, can you conclude that the distributions of daily crimes in the two cities are different?

Solution Since you are to test whether the two distributions are different, this is equivalent to the following null and alternative hypotheses:

H_0: Median daily crimes are equal in both cities

H_1: Median daily crimes are not equal in both cities

Note that H_0 can also be written as $\eta_1 = \eta_2$ and H_1 as $\eta_1 \neq \eta_2$, where η_1 is the median of the distribution of crimes in City A and η_2 is the median of the distribution of crimes in City B.

If you are using the pull-down menus, perform the following steps to test the hypothesis.

Step 1.   Enter the data for City A into column C1. Enter the data for City B into column C2.

Step 2.   Click the **Stat** pull-down menu from the top of the screen.

Step 3.   Click **Nonparametrics** from the selections available in the **Stat** menu.

Step 4.   Click **Mann–Whitney** from the selections available in the **Nonparametrics** menu.

Step 5.   A dialog box entitled **Mann–Whitney** will appear on the screen. Enter **C1** in the box labeled **First Sample** and **C2** in the box labeled **Second Sample**. Note that MINITAB asks for the confidence level and not the significance level. Since the significance level is .05 and it is a two-tailed test, enter **1 − .05 = .95** or **95** in the box labeled **Confidence level**. Select **not equal** in the box labeled **Alternative**.

Step 6.   Click the **OK** button at the bottom of the screen. The MINITAB output will appear on the screen.

If you are using the command language, enter the data for City A into column C1 and the data for City B into column C2, respectively, of the MINITAB data window and then enter the following commands:

```
MTB  > MANN-WHITNEY 95 C1 C2;
SUBC > ALTERNATIVE 0.
```

Whether you use the pull-down menus or the command language, you will obtain the MINITAB output given in Figure B13.4.

```
Mann-Whitney Test and CI: C1, C2

C1      N = 8    Median = 17.50
C2      N = 9    Median = 20.00

Point estimate for ETA1 - ETA2 is -4.00
95.1 Percent CI for ETA1 - ETA2 is (-11.00, 3.00)
W = 58.5
Test of ETA1 = ETA2 vs ETA1 not = ETA2 is significant at 0.2110
The test is significant at 0.2101 (adjusted for ties)
Cannot reject at alpha = 0.05
```

Figure B13.4   MINITAB output for the Mann–Whitney test.

Note that, as mentioned earlier, in the MINITAB output, ETA1 is the median for City A and ETA2 is the median for City B. The p-value for the test is .211, which is greater than the significance level of .05. Hence, you fail to reject H_0 at $\alpha = .05$ and conclude that these two cities seem to have identical distributions for daily crimes.

Also, from the MINITAB output in Figure B13.4, the observed value of the test statistic is $W = 58.5$. Note that the notation used in the text was T for this test statistic. This value is the same as the observed value of T in Example 14–9 and it falls in the nonrejection region shown in Figure 14.9. Consequently, you fail to reject the null hypothesis. ∎

THE KRUSKAL–WALLIS TEST

The Kruskal–Wallis test is used to test the null hypothesis that all k populations are identical, where $k \geq 3$. MINITAB tests the hypothesis that all k populations have the same median, which is equivalent to testing the null hypothesis that all k populations have identical distributions. Note that it is assumed that all k populations have the same shape and they may differ only in the locations of their centers. MINITAB calculates and prints out the value of the Kruskal–Wallis test statistic H and the p-value for the test.

ILLUSTRATION B13-5   Refer to Example 14–11 of Chapter 14. A researcher wanted to find out whether the population distributions of salaries of computer programmers are identical in three cities, Boston, San Francisco, and Atlanta. Three different samples—one from each city—produced the following data on the annual salaries (in thousands of dollars) of computer programmers.

Boston	San Francisco	Atlanta
43	54	57
39	33	68
62	58	60
73	38	44
51	43	39
46	55	28
	34	49
		57

Using the 2.5% significance level, can you conclude that the population distributions of salaries for computer programmers in these three cities are all identical?

Solution   The null and alternative hypotheses for the test are as follows:

H_0: The population distributions of salaries of computer programmers in the three cities are all identical

H_1: The population distributions of salaries of computer programmers in the three cities are not all identical

If you are using the pull-down menus, perform the following steps to test the hypothesis.

Step 1.   Enter the salaries for Boston into rows 1 through 6 of column C1, for San Francisco into rows 7 through 13 of Column C1, and for Atlanta into rows 14 through 21 of column C1. Note that the salaries for all three cities are entered into column C1—first for Boston, then for San Francisco, and next for Atlanta.

Step 2.   Enter **1** (for Boston) into each of the rows 1 through 6 of column C2, **2** (for San Francisco) into each of the rows 7 through 13 of column C2, and **3** (for Atlanta) into each of the rows 14 through 21 of column C2. These numbers entered into column C2 tell MINITAB that the first six numbers (salaries) entered into column C1 belong to sample 1 (Boston), the next seven numbers (salaries) belong to sample 2 (San Francisco), and the last eight numbers (salaries) belong to sample 3 (Atlanta).

Step 3.   Click the **Stat** pull-down menu from the top of the screen.

Step 4.   Click **Nonparametrics** from the selections available in the **Stat** menu.

Step 5.   Click **Kruskal–Wallis** from the selections available in the **Nonparametrics** menu.

Step 6.   A dialog box entitled **Kruskal–Wallis** will appear on the screen. Enter **C1** in the box next to **Response** and **C2** in the box next to **Factor**.

Step 7. Click the **OK** button at the bottom of the screen. The MINITAB output will appear on the screen.

If you are using the command language, first enter data into columns C1 and C2 as described in Step 1 above. Then enter the following MINITAB command:

```
MTB > KRUSKAL-WALLIS C1 C2
```

Whether you use the pull-down menus or the command language, you will obtain the MINITAB output given in Figure B13.5.

```
Kruskal-Wallis Test: C1 versus C2

Kruskal-Wallis Test on C1

C2           N     Median    Ave Rank        z
1            6     48.50         12.5     0.70
2            7     43.00          8.6    -1.23
3            8     53.00         11.9     0.54
Overall     21                   11.0

H = 1.54     DF = 2    P = 0.462
H = 1.55     DF = 2    P = 0.462  (adjusted for ties)
```

Figure B13.5 MINITAB output for the Kruskal–Wallis test.

From the MINITAB output, the value of the test statistic for the Kruskal–Wallis test is $H = 1.54$. Comparing it with the critical value of 7.378 shown in Figure 14.11, you fail to reject the null hypothesis. Also, from the MINITAB output, the *p*-value is .462, which is greater than the significance level of .025. Hence, you fail to reject the null hypothesis. ∎

THE RUNS TEST FOR RANDOMNESS

To use MINITAB to perform a runs test, first assign 0 and 1 to the two outcomes because MINITAB does not accept outcome strings that consist of letters. Illustration B13–6 shows how to use MINITAB to perform a runs test.

ILLUSTRATION B13–6   Refer to Example 14–13 of Chapter 14. A college admissions officer is interested in knowing whether applications for admission arrive randomly with respect to gender. The genders of 25 consecutively arriving applications were found to arrive in the following order, where M denotes a male applicant and F a female applicant:

M F M M F F F M F M M M F F F F M M M F F M F M M

Can you conclude that the applications for admission arrive randomly with respect to gender? Use $\alpha = .05$.

Solution   The null and alternative hypotheses for this test are

H_0: Applications arrive in a random order with respect to gender

H_1: Applications do not arrive in a random order with respect to gender

Assign a value of 0 to each M and a value of 1 to each F in the data set. Then, you can write the given string of letters as follows:

0 1 0 0 1 1 1 0 1 0 0 0 1 1 1 1 0 0 0 1 1 0 1 0 0

If you are using the pull-down menus, perform the following steps for the runs test.

Step 1. Enter the data containing strings of 0 and 1 into column C1.

Step 2. Click the **Stat** pull-down menu from the top of the screen.

Step 3. Click **Nonparametrics** from the selections available in the **Stat** menu.

Step 4. Click **Runs Test** from the selections available in the **Nonparametrics** menu.

Step 5. A dialog box entitled **Runs Test** will appear on the screen. Enter **C1** in the box below **Variable**.

Step 6. Click the **OK** button at the bottom of the screen. The MINITAB output will appear on the screen.

If you are using the command language, first enter data containing strings of 0 and 1 into column C1 as described in Step 1 above. Then enter the following MINITAB command:

MTB > RUNS C1

Whether you use the pull-down menus or the command language, you will obtain the MINITAB output given in Figure B13.6.

```
Runs Test: C1
C1
K = 0.4800
The observed number of runs = 13
The expected number of runs = 13.4800
12 Observations above K    13 below
        The test is significant at 0.8443
        Cannot reject at alpha = 0.05
```

Figure B13.6 MINITAB output for a runs test.

In the MINITAB output, K refers to the mean value of column C1, which is .48, but it has no practical meaning in this test. The 12 Observations above K refers to the 12 values of 1, or 12 female applicants. The 13 below refers to the 13 values of 0, or 13 male applicants. The observed value of the test statistic is $R = 13$, which is the same as in Example 14–13. As you can observe, this observed value falls in the nonrejection region in Figure 14.13. Consequently, you fail to reject the null hypothesis.

Also, from the MINITAB solution, the *p*-value is .8443, which is greater than the significance level of 5%. Hence, again you fail to reject the null hypothesis at the significance level of .05. ∎

B.14 GENERATING RANDOM NUMBERS AND SELECTING A SAMPLE

This section describes how to generate random numbers and select a sample from a population using MINITAB.

GENERATING RANDOM NUMBERS

Illustration B14–1 describes the procedure to generate random numbers using MINITAB.

ILLUSTRATION B14-1   Using MINITAB, generate eight three-digit random numbers.

Solution   Note that 100 is the smallest three-digit number and 999 is the largest three-digit number. If you are using the pull-down menus, perform the following steps to generate eight three-digit random numbers.

Step 1.   Click the **Calc** pull-down menu at the top of the screen.

Step 2.   Click **Random Data** from the selections available in the **Calc** menu.

Step 3.   Click **Integer** from the selections available in the **Random Data** menu.

Step 4.   You will see a dialog box entitled **Integer Distribution** appear on the screen. Type **8** in the box next to **Generate**. This tells MINITAB to generate eight rows of data. Next, enter **C1** in the box below **Store in column(s)**. Since the smallest and the largest values of three-digit numbers are 100 and 999, respectively, enter **100** in the box next to **Minimum value** and **999** in the box next to **Maximum value**.

Step 5.   Click the **OK** button at the bottom of this dialog box. Eight three-digit random numbers will appear in column C1 of the Data window.

If you are using the command language, type the following MINITAB commands to obtain eight three-digit random numbers:

```
MTB  > RANDOM 8 C1;
SUBC > INTEGERS 100 999.
MTB  > PRINT C1
```

Each time you use the above set of MINITAB commands or the pull-down menus steps, you will obtain a different set of eight three-digit random numbers. One such MINITAB output is given in Figure B14.1.

```
Data Display

C1
      781    819    183    589    469    853    198    401
```

Figure B14.1   MINITAB output for eight three-digit random numbers.   ■

Suppose you want to generate two-digit random numbers. Because 10 is the smallest two-digit number and 99 is the largest two-digit number, replace 100 by 10 and 999 by 99 in Step 4 of the pull-down menus procedure and in the command language subcommand. Similarly, to generate four-digit random numbers, replace 100 by 1000 and 999 by 9999.

SELECTING A SAMPLE

MINITAB can be used to select a sample from a population. Suppose you have some population data entered in column C1 and want to select a sample of 16 observations from this population and store the sample data in column C2. If you are using the pull-down menus, perform the following steps to select a sample.

Step 1.   Click the **Calc** pull-down menu at the top of the screen.

Step 2.   Click **Random Data** from the selections available in the **Calc** menu.

Step 3.   Click **Sample From Columns** from the selections available in the **Random Data** menu.

Step 4. You will see a dialog box entitled **Sample From Columns** appear on the screen. Type **16** in the box next to **Sample** and **C1** in the box below **Sample rows from column(s)**. This tells MINITAB to select a sample of 16 rows (observations) from the data entered in column C1. Then type **C2** in the box below **Store samples in**. If you want the sample to be selected with replacement, then check the box next to **Sample with replacement**. Otherwise, make sure this box is not checked.

Step 5. Click the **OK** button at the bottom of this dialog box. A sample of 16 observations will appear in column C2 of the Data window.

If you are using the command language, type the following MINITAB commands to obtain a sample of 16 observations from the population data of column C1 and store the sample data in column C2:

```
MTB > SAMPLE 16 C1 C2
MTB > PRINT C2
```

Each time you use the above MINITAB commands or the pull-down menus procedure, you will obtain a different sample of 16 observations.

Now suppose the population data contain a large number of observations on three variables and these data are entered in columns C1 to C3. Further suppose that you want to select a sample of 12 observations, without replacement, from these data and store the sample data in columns C4 to C6.

If you are using the pull-down menus, Step 4 in the above procedure will be changed as follows: Type **12** in the box next to **Sample** and **C1–C3** in the box below it. Then type **C4–C6** in the box below **Store samples in**. Make sure the box next to **Sample with replacement** is not checked. This will give you a sample of 12 observations; store those data in columns C4–C6.

If you are using the command language, type the following MINITAB commands to obtain a sample of 12 observations from the population data of columns C1–C3 and store the sample data in columns C4–C6:

```
MTB > SAMPLE 12 C1-C3 C4-C6
MTB > PRINT C4-C6
```

Section B.1 of this appendix also shows how we can select a sample from a population using MINITAB.

C USING EXCEL

C.1 AN INTRODUCTION

C.2 ORGANIZING DATA

C.3 NUMERICAL DESCRIPTIVE MEASURES

C.4 PROBABILITY

C.5 DISCRETE PROBABILITY DISTRIBUTIONS

C.6 THE NORMAL DISTRIBUTION

C.7 SAMPLING FROM A POPULATION

C.8 CONFIDENCE INTERVAL

C.9 HYPOTHESIS TESTING

C.10 INFERENCES ON TWO POPULATION MEANS

C.11 CHI-SQUARE APPLICATIONS

C.12 ANALYSIS OF VARIANCE

C.13 REGRESSION ANALYSIS

C.14 THE SPEARMAN RANK CORRELATION COEFFICIENT

C.1   USING EXCEL IN CHAPTER 1: AN INTRODUCTION

Another useful software program to help you manipulate, organize, and display data is the spreadsheet program Excel by Microsoft. In this chapter, we use the menus and commands in Excel to solve this adventure. The objective is to enter Excel's mathematical functions such as count, max, min, and average directly into a cell.

EXCEL ADVENTURE 1:   **HOW MUCH CAFFEINE IS IN A 16-OZ CUP OF COFFEE?**

A random sample of 56 coffee houses were tested for the amount of caffeine in a 16-oz cup of coffee. (See Figure 1.) Use this sample and Excel's formulas to summarize the data to reveal:

(a)   The number of data points

(b)   The maximum amount of caffeine in a 16-oz cup of coffee

(c)   The minimum amount of caffeine in a 16-oz cup of coffee

(d)   The mean amount of caffeine in a 16-oz cup of coffee in all the coffee houses

adv1-01

	A	B	C	D	E	F	G	H
1	Adventure 1-01 Caffeine in mg in a 16 oz cup of coffee in local coffee houses							
2								
3	549	550	547	548	559	543	543	
4	543	554	553	544	556	544	556	
5	556	551	549	545	552	556	550	
6	541	559	553	548	557	541	550	
7	543	547	554	557	546	550	551	
8	554	550	550	544	559	550	547	
9	551	542	542	553	541	551	543	
10	558	547	545	546	551	550	545	
11								

Figure 1   Data for Adventure 1.

The data fall in the range from A3 to G10, indicated in Excel as **A3:G10**.

The command to find the number of data points is **=count(A3:G10)**.

The command to find the largest data value is **=max(A3:G10)**.

The command to find the smallest data value is **=min(A3:G10)**.

The command to find the mean or average is **=average(A3:G10)**.

Enter the appropriate formulas in the cells shown below. Compare your answers with those in Figure 2.

adv1-01

	A	B
12	n=	56
13	max=	559
14	min=	542
15	average=	551.5357
16		

Figure 2   Formulas for Adv1-01.

Print the results and save the file as **ADV01.XLS** on your disk.

C.2   USING EXCEL IN CHAPTER 2: ORGANIZING DATA

Excel contains a Data Analysis ToolPak to help you organize raw data into a frequency table. In this chapter, you will learn how to use the Data Analysis ToolPak. Within Excel, click the **Tools** menu; if the **Data Analysis** option does not appear as an option, click **Add-In** . . . , check the box before the option **Analysis-ToolPak**, and click **OK**. If you still cannot access the add-in from the tools menu, run the Setup program from your Excel CD and install the **Analysis ToolPak**.

EXCEL ADVENTURE 2:

HOW LONG DO HONEYBEES LIVE?

Sensors are planted on a group of bees and their lifespan is monitored during one summer. Their longevity in days is recorded in the spreadsheet (**Adv02**) in Figure 1.

	A	B	C	D	E	F	G
1							
2	Adv02	Life Expectancy(in days) of a Honey Bee during the summer months					
3							
4	39	36	31	30	35		
5	41	39	37	41	30		
6	35	36	29	35	30		
7	35	38	34	39	38		
8	35	35	31	32	30		
9	39	35	31	35	31		
10	35	35	30	39	35		
11	33	33	33	30	30		
12	29	37	36	40	39		
13	32	35	39	36	35		
14	37	35	37	36	40		
15	35	38	37	35	35		
16	39	41	32	35	39		
17	30	41	37	39	33		
18							

Figure 1   Data on honeybee longevity.

(a)   Create formulas to find the number of data points, the maximum, and the minimum of the data set. Compare with Figure 2.

	G	H
5	n =	70
6	max=	41
7	min=	29
8		

Figure 2.   Formulas for count, maximum, and minimum.

(b)   Enter data and formulas to compute the number of classes and the width of a class.

In cell **I10**, enter the number of classes. Let's enter the number 5.   ENTER

In cell **I11**, enter a formula that computes the estimated class width. (See Figure 3.) This formula should be

$$w = \frac{max - min}{no.\ of\ classes}$$

	F	G	H	I
10	Number of classes ==>			5
11	Estimated class width=>			2.4
12				
13	Enter class width ====>			
14				

Figure 3 Computing class width.

You should have obtained the value 2.4, which is an estimated value for the class width.

Use the estimated value of the class width as a gauge to choose the value you will actually use in the table. Increase the number 2.4 to **2.5** to ensure that every value will fall into a class. Enter in cell **I13** the value **2.5**. ENTER

(c) Establish bin numbers. Bin numbers in Excel are placed in a column (see Figure 4). The values must be in ascending order and represent the upper limit/boundary of each class.

	G	H	I
13	class width ====>		2.5
14			
15			
16	31.0		
17	33.5		
18	36.0		
19	38.5		
20	41.0		

Figure 4 Bin numbers are upper class limits.

The minimum value in the data is 29, so the beginning lower limit of the first class must include 29. Choose any value of 29 or less. There is no fast rule to start the lower limit of the first class. Let's start the lower class limit at 28.5; then the lower class limit of the second class is 31. This is also the upper class limit of the first class.

Enter in cell **G16** the upper limit of the first class, which is **31**. Create bin numbers so that the class width is added to each of the previous class limits. The last number should be 41 in cell **G20**.

Note that the bin numbers are upper class limits. The bin range starts in cell **G16** and ends in cell **G20**.

(d) Invoke the Tools Analysis command to create a frequency table.

Step 1. Issue the command by clicking on the Data Analysis option under the Tools menu.

Tools → Data Analysis

See Figure 5.

Step 2. Choose the option Histogram and click OK. See Figure 6.

Step 3. Determine the input range (see Figure 7). You must now inform Excel where the raw data are located. This is called the **Input Range** in the Histogram box. There are two ways to select the input range. The first method is to type the range. The range of cells is **A4** to **E17** or **A4:E17**. Excel automatically will place '$' to make them absolute. The second method is to use the mouse and click the spreadsheet icon located beside the Input Range; then with the mouse highlight the range A4:E17.

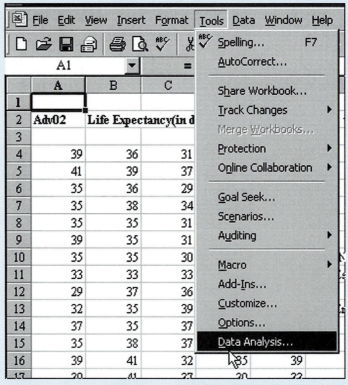

Figure 5   Data Analysis add-in.

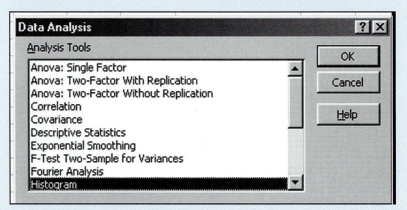

Figure 6   Histogram tool.

Step 4.   Set up the bin range (see Figure 8). The bin range is the column where you entered the upper class limits. Place the mouse pointer in the white rectangular region and click the spread-sheet icon. Next, Excel displays a Histogram dialog box. Type the range of the bin numbers or select with the mouse, the range is **G16:G20**, and press ⎣ENTER⎦.

Step 5.   Placement of the results. There are three output options: (a) in the current worksheet, (b) in a worksheet as part of this workbook, or (c) in a new workbook.

Let's place the results in this workbook, but in a new worksheet. This is called the New Work-sheet Ply: Click on the button under the ⎣Output Options⎦ and select ⎣New Worksheet Ply⎦.

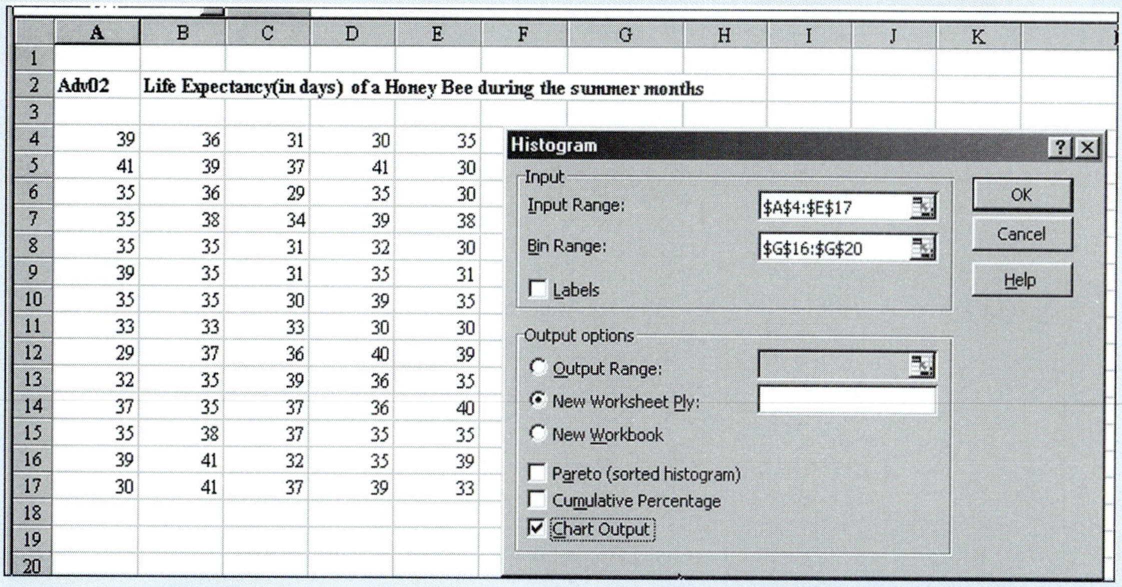

	A	B	C	D	E	F	G	H	I	J	K
1											
2	Adv02	Life Expectancy(in days) of a Honey Bee during the summer months									
3											
4	39	36	31	30	35						
5	41	39	37	41	30						
6	35	36	29	35	30						
7	35	38	34	39	38						
8	35	35	31	32	30						
9	39	35	31	35	31						
10	35	35	30	39	35						
11	33	33	33	30	30						
12	29	37	36	40	39						
13	32	35	39	36	35						
14	37	35	37	36	40						
15	35	38	37	35	35						
16	39	41	32	35	39						
17	30	41	37	39	33						
18											
19											
20											

Histogram [?][X]

Input
Input Range: $A$4:$E$17
Bin Range: $G$16:$G$20
□ Labels

Output options
○ Output Range:
● New Worksheet Ply:
○ New Workbook

□ Pareto (sorted histogram)
□ Cumulative Percentage
☑ Chart Output

OK
Cancel
Help

Figure 7   Input range (raw data).

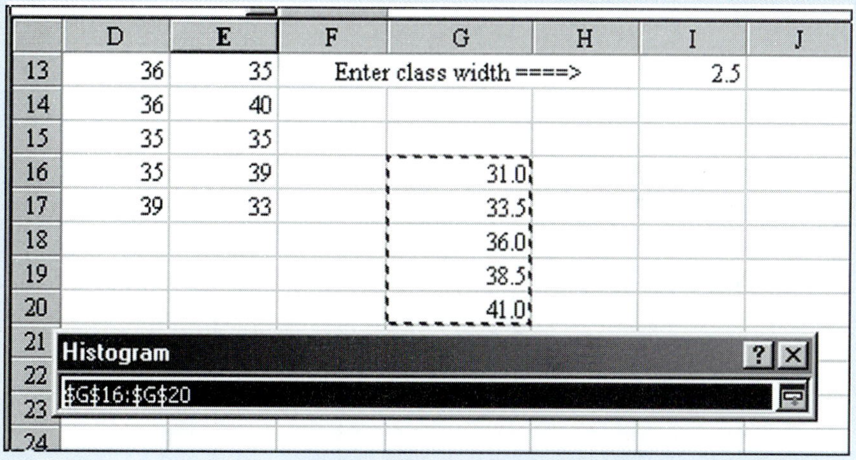

	D	E	F	G	H	I	J
13	36	35	Enter class width ====>			2.5	
14	36	40					
15	35	35					
16	35	39		31.0			
17	39	33		33.5			
18				36.0			
19				38.5			
20				41.0			
21	Histogram					[?][X]	
22							
23	$G$16:$G$20						
24							

Figure 8   Bin number range 1.

Step 6.   Draw a histogram. Click the histogram option on the Data Analysis screen as shown in Figure 9 on page C6. Click this option and select OK.

Frequency Table and Histogram

Excel places the new frequency table and histogram on a new spreadsheet. The two spreadsheets together form a workbook, which is saved like a worksheet with the extension **.xls**. Excel automatically formats the chart with a legend, background color, and axis labels. To delete or change any of these options, just click the object and modify. To delete the legend, click the legend box and press the delete key. To eliminate the darkened plot area, click the region and select the area to automatic.

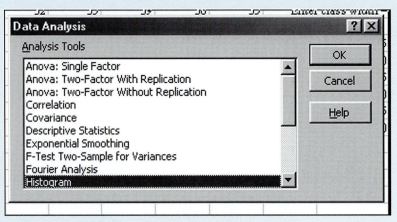

Figure 9   Worksheet ply and Histogram option.

The title, axis labels, and plot area were automatically formatted to the text shown in Figure 10. Also the legend box was deleted.

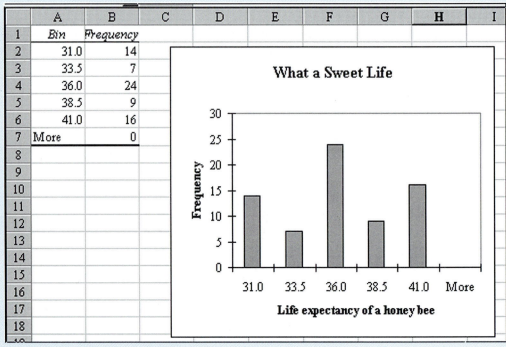

Figure 10   Frequency table and histogram 1.

The life expectancy of the honeybee is not very long. So be nice to this creature whose average lifespan is a little more than a month in the summer and even less in the winter.

Print the results and save on your disk as **ADV2.XLS**.

C.3 USING EXCEL IN CHAPTER 3: NUMERICAL DESCRIPTIVE MEASURES

EXCEL ADVENTURE 3: **INVESTING IN STOCK**

TOMATO

Your grandparents are proud that you are getting a college degree and they know how smart you are. They want to invest in a stock that can grow the most, and they are asking for your advice. Based on the last 13 months of stock data for Campbell Soup and Pfizer Drugs (Figure 1), compute the following measurements using Excel's Data Analysis command:

(a) The range for the price of the stock

(b) The median stock prices of both companies

(c) The mean stock prices of both companies

(d) The variance of the stock prices of both companies

(e) The standard deviation of the stock prices of the two companies

(f) Which company has the least fluctuations? Justify your answer.

(g) Create a line chart for each company.

Figure 2 Pfizer makes Bengay products.

	A	B	C	D	E	F	G
1	ADV3 Comparision between Campbell and Pfizer Inc						
2							
3	Campbell Soup Stock Data				Pfizer Inc Stock Data		
4	Date	Stock Price			Date	Stock Price	
5	Jan-99	45.88			Jan-99	42.52	
6	Feb-99	39.28			Feb-99	43.69	
7	Mar-99	39.77			Mar-99	45.95	
8	Apr-99	40.05			Apr-99	38.1	
9	May-99	43.49			May-99	35.5	
10	Jun-99	45.58			Jun-99	36.16	
11	Jul-99	43.47			Jul-99	33.72	
12	Aug-99	43.66			Aug-99	37.66	
13	Sep-99	38.66			Sep-99	35.79	
14	Oct-99	44.72			Oct-99	39.6	
15	Nov-99	44.35			Nov-99	36.63	
16	Dec-99	38.44			Dec-99	32.44	
17	Jan-00	35.88			Jan-00	34.5	

Figure 1 Stock data for Campbell Soup and Pfizer.

To analyze the stock price of Campbell Soup, perform the following steps:

Step 1. Issue the $\boxed{\text{Tools}}$ → $\boxed{\text{Data Analysis}}$ commands. See Figure 3 on page C8.

Step 2. From the list, select the option $\boxed{\text{Descriptive Statistics}}$. See Figure 4.

Step 3. The Input Range is the location of stock price for Campbell Soup Company. The range is **B5:B17**. Click the spreadsheet icon on the far right beside the $\boxed{\text{Input Range:}}$ and select the range on the spreadsheet. See Figure 5.

Figure 3   Tools and Data Analysis commands.

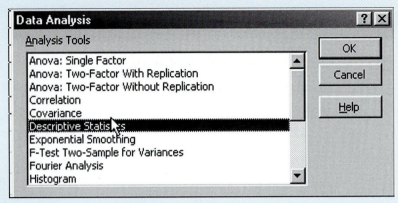

Figure 4   Descriptive Statistics option.

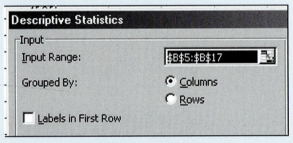

Figure 5   Input Range.

Step 4.   The results of Descriptive Statistics can be placed in a new worksheet, in a new workbook, or on the spreadsheet containing the data. Let's place the results below the Campbell Soup data, starting in cell A21. First, click the button beside ⌞Output Range:⌟. Second, type the cell location where you want the top left corner of the data to reside. This is in cell **A21**. See Figure 6.

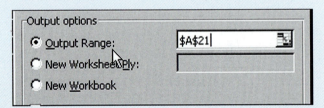

Figure 6   Top left corner of output 1.

Step 5.   For the options of Descriptive Statistics output, click the Summary statistics box, the Kth Largest, and the Kth Smallest. Click ⌞OK⌟. See Figure 7.

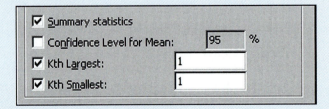

Figure 7   Option to display.

Compare your results with those in Figure 8. There is no mode because no data are repeated.

	A	B	C
20	**Campbell Soup Descriptive Statistic**		
21	*Column1*		
22			
23	Mean	41.78692	
24	Standard Error	0.894411	
25	Median	43.47	
26	Mode	#N/A	
27	Standard Deviation	3.224845	
28	Sample Variance	10.39962	
29	Kurtosis	-1.20258	
30	Skewness	-0.35789	
31	Range	10	
32	Minimum	35.88	
33	Maximum	45.88	
34	Sum	543.23	
35	Count	13	
36	Largest(1)	45.88	
37	Smallest(1)	35.88	

Figure 8 Results from Tools–Statistical Analysis in Excel.

Repeat the five steps with the Pfizer stock data and answer questions a–g on the spreadsheet. Print the spreadsheet and graphs. Save the workbook as **ADV3.XLS**.

C.4 USING EXCEL IN CHAPTER 4: PROBABILITY

EXCEL ADVENTURE 4:

DATABASE OF BIRTHS AT GENERAL HOSPITAL IN 2000

It is easier to answer some probability questions if the data can be summarized in a table. That normally takes a bit of time. Excel can accurately and quickly summarize data in a table or database format like that shown in Figure 1 on page C10 through the use of the Pivot Table command.

Use the Pivot Table command in Excel to summarize the number of babies that are of a certain gender and eye color. First mentally draw a table and decide what features you want for the rows and what features you want for the columns. A table is formed where the rows represent the gender and the columns represent the eye color. Within the data table are the counts of those babies that are female or male and have a certain eye color.

Gender/Eye	Brown	Green	Black	Blue
Male				
Female				

Next, issue from Excel the `Data` → `Pivot Table and PivotChart Report` command as shown in Figure 2. Excel will prompt you from the Pivot Table Wizard with three important questions.

	A	B	C	D	E	F
1	Adv4-Database of Births at General Hospital in 2000					
2						
3	Baby	Gender	Eye Color	Weight	Days in Hospital	Length(in)
4	1	M	Brown	light	3	18.05
5	2	F	Brown	light	5	12.98
6	3	F	Black	medium	6	19.44
7	4	F	Green	medium	7	19.6
8	5	M	Black	medium	5	20.4
9	6	F	Black	heavy	6	16.43
10	7	F	Blue	heavy	4	15.89
11	8	F	Blue	light	5	17.83
12	9	M	Blue	light	6	16.07
13	10	M	Brown	medium	7	21.52
14	11	M	Brown	medium	7	17.63
15	12	F	Brown	medium	5	21.25
16	13	M	Blue	heavy	5	10.73
17	14	F	Brown	medium	4	12.81
18	15	M	Blue	medium	4	14.9
19	16	F	Green	heavy	6	19.26
20	17	M	Green	medium	2	19.89
21	18	M	Black	medium	5	21.01
22	19	F	Black	medium	3	20.21
23	20	M	Green	heavy	2	14.16
24	21	F	Brown	medium	3	12.62
25	22	F	Black	medium	5	12.73
26	23	F	Black	light	3	20.46
27	24	F	Brown	medium	5	21.22
28	25	M	Brown	light	6	20.37
29	26	M	Green	light	3	11.07
30	27	M	Brown	medium	3	13.52

Figure 1   Database of births at General Hospital.

Step 1 of 3.   Where are the data? Click the button **Microsoft Excel list** to select that the data are in the current worksheet. Click the button **PivotTable** and click ⌈ Next ⌉. See Figure 3.

Step 2 of 3.   What is the range in the current database? Click the spreadsheet icon and highlight the range of the database. This should include the column headings and also the row headings. The range is A3:F30. Click ⌈ Next ⌉. See Figure 4 on page C12.

Step 3 of 3.   Where do you want to put the pivot table? Place the table on the existing worksheet. Click the button beside **Existing worksheet**. Let's place the table so that the top left-hand corner is in cell A32. Enter A32 in the Existing worksheet box or select the cell A32 using the spreadsheet icon.

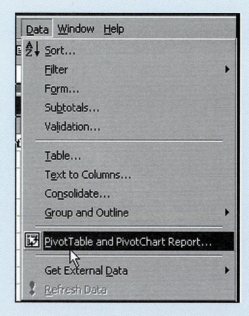

Figure 2 Data Pivot Table.

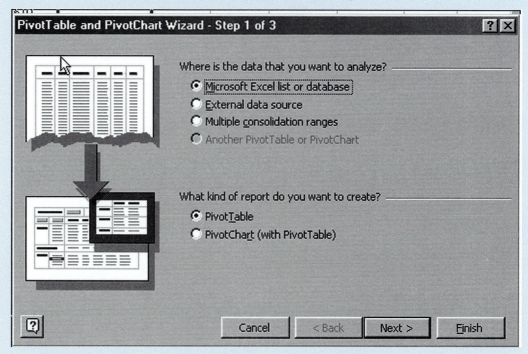

Figure 3 Step 1 of 3 of Pivot Table command.

The final step is very tricky, so be careful. You haven't told Excel what attributes are along the rows (gender) or what attributes are along the column (eye color) yet. To do so, you must click the Layout button in the **Step 3 of 3 Pivot Table** prompt. See Figure 5.

Drag the attributes Gender and Eye Color to the Row and Column region of the layout screen. Place the mouse pointer over attribute, hold left mouse button down, and drag to desired region. See Figure 6.

Figure 4   Database range.

Figure 5   Output of PivotTable.

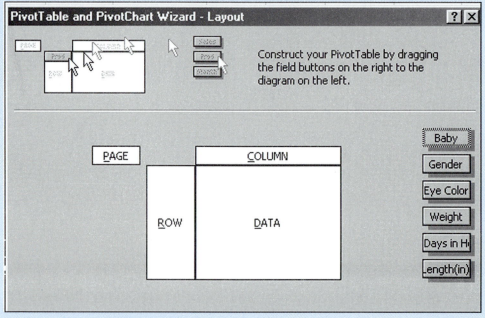

Figure 6   Drag attributes to Row, Column, Data region of screen.

The result should look like Figure 7.

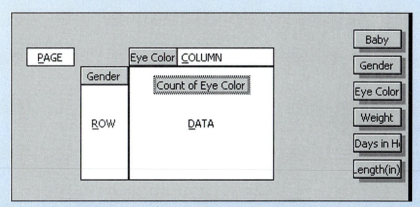

Figure 7   Field buttons dragged to Row, Column, and Data ranges.

Click the OK and Finish buttons. The results are shown in Figure 8.

	A	B	C	D	E	F
31						
32	Count of Eye Color	Eye Color ▼				
33	Gender ▼	Black	Blue	Brown	Green	Grand Total
34	F	5	2	5	2	14
35	M	2	3	5	3	13
36	Grand Total	7	5	10	5	27
37						
38						

Figure 8   Results of pivot table.

Use the pivot table to answer the following probability questions about a baby selected from those born at General Hospital in the year 2000.

(a)   Find the probability that a baby is female and has blue eyes.

$$P(\text{female and blue-eyed}) = \frac{2}{27}$$

(b)   Find the probability that a baby is female given that it has blue eyes.

$$P(\text{female} \mid \text{blue-eyed}) = \frac{2}{5}$$

(c)   Find the probability that a baby is female or has blue eyes.

$$P(\text{female or blue-eyed}) = \frac{14}{27} + \frac{5}{27} - \frac{2}{27} = \frac{17}{27}$$

Use the pivot table to create a summary table where the rows are eye color and the columns denote weight. Your results should look like Figure 9 on page C14.

	A	B	C	D	E
38	Count of Weight	Weight ▼			
39	Eye Color ▼	heavy	light	medium	Grand Total
40	Black	1	1	5	7
41	Blue	2	2	1	5
42	Brown		3	7	10
43	Green	2	1	2	5
44	Grand Total	5	7	15	27
45					

Figure 9   Pivot table of eye color and weight.

(d)   Find the probability that a baby is blue-eyed and heavy.

(e)   Find the probability that a baby is female given that it is heavy.

(f)   Find the probability that a baby is female or heavy.

(g)   Find the probability that a baby is heavy given that it is a female.

(h)   Find the probability that a baby is light-weight and has green eyes.

Write your results on the disk and print out worksheets. Save as **ADV4.XLS**.

C.5   USING EXCEL IN CHAPTER 5: DISCRETE PROBABILITY DISTRIBUTIONS

EXCEL ADVENTURE 5:   **APPLICATION TO QUEUING THEORY**

A new microcomputer processor can process six computer instructions in 1 nanosecond. When six or fewer instructions are sent to the processor, each instruction is processed immediately. When there are more than six requests, instructions must wait in queue.

Theoretically, the number of instructions, x, sent to the processor follows a Poisson distribution. In a particular network environment, the average number of arrivals is 5.26 instructions each nanosecond.

(a)   Create a probability distribution table. Compute the mean and standard deviation of the distribution.

(b)   Find the probability that there is a wait, if the processor can handle at most six requests.

(c)   Make a bar chart of the distribution.

Solution   Let x = number of instructions sent to the processor. Given $\lambda = 5.26$ instructions per nanosecond, find $P(\text{wait}) = P(x \geq 7)$.

In Excel, to find $P(x = a)$, where a is a whole number, use the command

$$\boxed{\text{Insert}} \rightarrow \boxed{\text{Function}} \rightarrow \boxed{\text{Statistical}} \rightarrow \boxed{\text{Poisson}}$$

as shown in Figure 1, or you can enter the command **POISSON(x, λ,0)**, where the values of **x** and λ are numerical values or cell locations where numerical values reside.

If the value of 1 is used in place of 0, then the function returns the cumulative distribution $P(x \leq a)$.

Let's set up a spreadsheet displaying the Poisson probability distribution with $\lambda = 5.26$ and $x = 0, 1, 2, 3, 4, \ldots$.

Step 1.   Enter column headings along row 5, and titles in rows 1 and 3 as shown in Figure 2. Enter the values of x from 0 to 16 along column A.

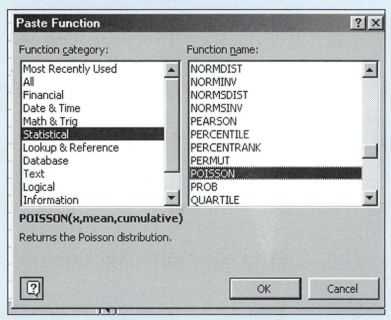

Figure 1 Entering the function POISSON.

	A	B	C	D
1	**Adv5- Poisson Distribution using Excel**			
2				
3	Let the mean =	$\lambda=$	5.26	
4				
5	x	P(x)	xp(x)	x*x*p
6	0			
7	1			
8	2			
9	3			
10	4			
11	5			
12	6			
13	7			
14	8			
15	9			
16	10			
17	11			
18	12			
19	13			
20	14			
21	15			
22	16			
23				

Figure 2 Poisson distribution with mean = 5.26.

Step 2.   Enter formulas along column B. In cell B6, enter the formula

$$= \textbf{poisson(A6, \$C\$3,0)} \boxed{\textsf{ENTER}}$$

or use the Insert Function command. Dollar signs (\$) are placed around C and 3 so that when this formula is copied down, the cell address C3 will not change. This is called absolute cell referencing. The cell address A6 is called relative cell addressing because when cell 6 is copied down, it will change from A6 to A7 to A8, etc. Copy cell B6 down to cell B22.

Step 3.   Compute the necessary formulas in the table and find the column sums.

Step 4.   Use the formula for the mean of a probability distribution, $\mu = \Sigma xp$. Use the command **=sum(beginning cell:ending cell)** to find the sum of values from cells B6 to B22.
    The standard deviation of a probability distribution is $\sqrt{\Sigma x^2 p(x) - \mu^2}$. Use the Excel command **=sqrt()** to find the square root of an expression.
    Compare your results with Figure 3.

	A	B	C	D	
4					
5	x	P(x)	xp(x)	x*x*p	
6	0	0.0051953	0	0	
7	1	0.0273273	0.027327	0.027327	
8	2	0.0718708	0.143742	0.287483	
9	3	0.1260135	0.37804	1.134121	
10	4	0.1657077	0.662831	2.651324	
11	5	0.1743245	0.871623	4.358113	
12	6	0.1528245	0.916947	5.501682	
13	7	0.1148367	0.803857	5.626998	
14	8	0.0755051	0.604041	4.832328	
15	9	0.0441286	0.397157	3.574413	
16	10	0.0232116	0.232116	2.321162	
17	11	0.0110994	0.122093	1.343024	
18	12	0.0048652	0.058383	0.700592	
19	13	0.0019685	0.025591	0.332684	
20	14	0.0007396	0.010355	0.144964	
21	15	0.0002594	0.00389	0.058355	
22	16	8.526E-05	0.001364	0.021827	
23		0.999963	5.259357	32.9164	
24					

Figure 3   Poisson probability distribution table.

(a)   Find the mean and standard deviation of the Poisson distribution; enter the results on the spreadsheet. Do they agree with the formulas in Figure 3?

(b)   Since the processor can handle at most six requests in a nanosecond, what is the probability that there is a wait? Type your answer in the spreadsheet.

(c)   Create a bar chart of the distribution. To do so in Excel, perform the following steps.
    Select the frequency or probabilities. These are values along columns B6 to B22. Click the command $\boxed{\textsf{Insert}} \rightarrow \boxed{\textsf{Chart}}$. The chart wizard will assist you through the four steps of inserting a chart.

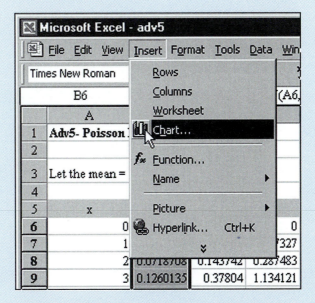

Figure 4 Insert a column chart.

Step 1 of 4—Chart Type. Excel displays a variety of different types of charts. Select the **Column** chart to represent our bar chart. And click the type of column chart desired.

To display the frequencies or probabilities, click the first chart among the various column charts shown in the first row in Figure 5 and click Next .

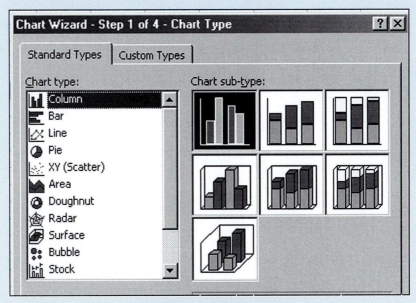

Figure 5 Insert a column chart.

Step 2 of 4—Chart Source Data. If you highlighted the range of probabilities, it should appear in the Data range area. If you did not, just enter the range B6:B22 or select the area in the spreadsheet with the mouse and click Next .

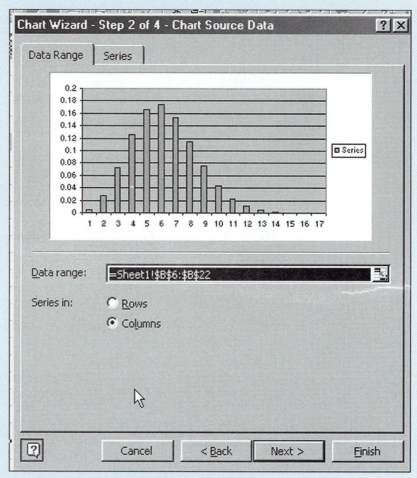

Figure 6   Source data.

Step 3 of 4—Chart Options.   Enter the titles for the main chart, the *x*-axis, and the *y*-axis. Select the Legend option, and deselect the display option. Click Next . See Figure 7.

Step 4 of 4—Insert chart as an object in the existing spreadsheet.   Click the option to insert the object into the current worksheet. Click Finish .

***Final Step.* Insert labels along the *x*-axis.**   We're almost there. Excel inserts along the *x*-axis a sequence of natural numbers, 1, 2, 3, ... To display labels or other values, click the frame of the chart, and select Chart → Source Data as shown in Figure 8.

Next, enter the range where the data values appear in the worksheet beside the Category (X) axis labels as shown in Figure 9 and click OK .

Print out the results and save on your disk as **ADV5.XLS**.

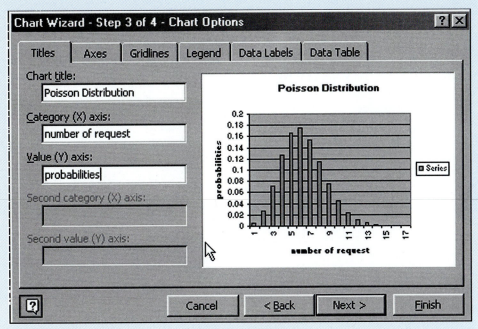

Figure 7   Chart Titles.

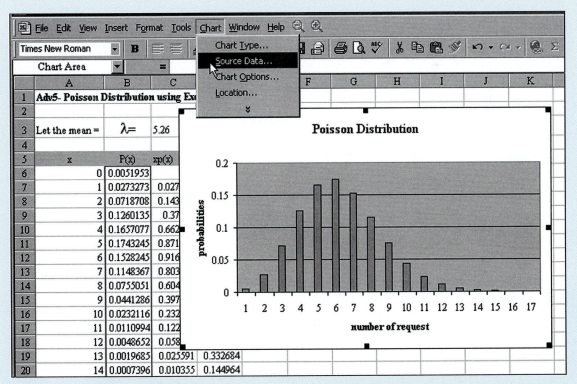

Figure 8   Modifying Source Data.

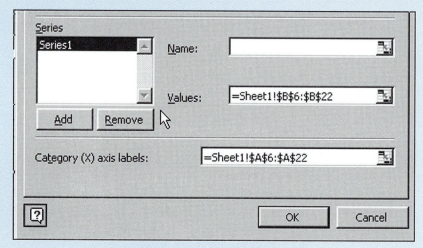

Figure 9   Location of *x*-axis labels.

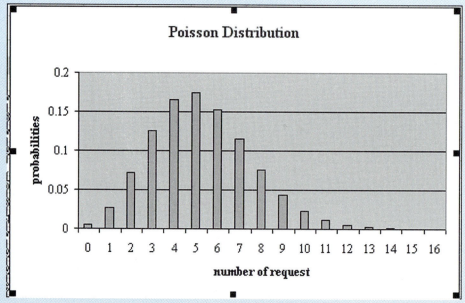

Figure 10   Poisson probability distribution chart.

C.6   USING EXCEL IN CHAPTER 6: THE NORMAL DISTRIBUTION

In this section you will learn how to use Excel's normal probability functions to compute probabilities and to find *z* values.

EXCEL ADVENTURE 6:   **HOW MUCH DOES IT TAKE?**

Dow and Jones were the best of friends in high school. Dow decided to go to college, while Jones went to work as a used-car salesman. Jones wanted to party every night, but Dow had to study in order to do well in college. Consequently, they drifted apart. When Dow graduated from college, she applied

Figure 1 Dow studying for an exam in college.

for a job at a company that paid all its workers an average of $56,200 a year, with a standard deviation of $12,000. Assume that the wages at this company are normally distributed. The company offered Dow $72,000. Use Excel to find the following:

(a) What percent of the company's workers make less than $72,000?

(b) Jones is now assistant manager of the used-car lot, making an annual salary of $45,000. What percent of Dow's company's workers make less than the salary of Jones?

(c) What percent of Dow's company's workers make a salary between $60,000 and $75,000?

(d) What is the cutoff salary for the top 3% of the workers at Dow's company?

Solution

(a) To find the percent of the company's workforce that makes less than $72,000, you want to find $P(x < \$72{,}000)$, where x is a random variable denoting the salary of a worker at the company. Excel has a built-in function

$$\textbf{NORMDIST(x-value, mean, standard deviation, 1)}$$

that denotes the area under the normal curve to the left of x.

Create the worksheet shown in Figure 2. In cell D7, enter the formula

$$\textbf{=normdist(72000,c3,c4,1)} \boxed{\text{ENTER}}$$

Did you get the solution that **90.6%** of the workforce in the company makes less than $72,000 a year?

	A	B	C	D	E	F	G
1	Adv6 Distribution of Salaries at the Company						
2							
3		μ=	56,200				
4		σ=	12,000				
5							
6		a) What % of the firm works for less than your salary of $72,000					
7			P(x<72,000)=				

Figure 2 Using NORMDIST(x, μ, σ, 1).

(b) Jones now has an annual salary of $45,000. You want to compute $P(x < \$45{,}000)$. Enter a formula and compare your answer with Figure 3.

	A	B	C	D	E	F	G	H
9		b) What % of the firm works for less than your friend, whose salary is $45,000)						
10			P(x<45,000)=	0.1753239				

Figure 3 $P(x < \$45{,}000)$.

(c) What percent make a salary between $60,000 and $75,000? *Hint:* Use the normdist function twice. Compare your result with that shown in Figure 4.

	A	B	C	D	E	F	G
12		c) What % of the firm works for a salary between $60,000 and $75,000?					
13			P(60,000<x<75,000)=	0.317152			

Figure 4 Use of normdist.

(d) Here you want to find an x value so that the area to the right is 3%. Excel has a function that will find the x value if you give the area to the left of the x value. The formula is

$$=\text{NORMINV(area to the left, } \mu, \sigma)$$

Example 1. **=norminv(.8413,0,1)** is 1. The area to the left of $x = 1$, for a normal distribution with mean 0 and standard deviation of 1, is about 84.13%.

Example 2. To find a so that $P(x < a) = 20\%$ with mean $56,200 and standard deviation $12,000, you would enter the formula

$$=\text{norminv}(.20, 56200,12000) \boxed{\text{ENTER}}$$

Excel will return a value of $46,101. So, 20% of the workforce makes a salary of less than $46,101.

Use the **norminv** function to complete Figure 5. *Hint*: To find a so that $P(x > a) = .03$, change the problem so it is in the form $P(x < a) = ?$.

	A	B	C	D	E	F
15		d) What is the cutoff salary for the top 3% of the workforce?				
16		Find c so P(x>a)=3%=>				
17						

Figure 5   Solve using norminv function.

Dow takes the job and lives happily ever after. It is not too late for Jones, who starts college next semester as a freshman. Dow promises to wait until Jones finishes college to get married. Will Dow become Dow Jones?

Print out the worksheet and save it as **ADV6.XLS** on your disk.

Figure 6   Jones decides to go to college!

C.7   USING EXCEL IN CHAPTER 7: SAMPLING FROM A POPULATION

EXCEL ADVENTURE 7:   **SIMULATION OF SAMPLING FROM A POPULATION**

Suppose there are 10 balls in a box and each ball is numbered sequentially from 0 to 9. Two balls are selected from the box at random without replacement and the average of the two numbers is computed. Demonstrate the central limit theorem using Excel.

Step 1.   Set up the worksheet shown in Figure 1 and compute the population measurements.

	A	B	C	D	E	F	G
1	Adv7 Simulation of choosing from population a set of 2 numbers without replacement						
2	Population =====>	0	1	2	3	4	
3		5	6	7	8	9	
4							
5	N=no. in pop =>						
6	No. of Samples =>						
7	Population mean=>						
8	Population s.d .=>						
9							
10							

Figure 1   Enter formulas in cells B5 to B8.

In cell B5, to have Excel count the number of values in the range B2 to F3, enter the formula

$$=\textbf{count(B2:F3)} \quad \boxed{\text{ENTER}}$$

In cell B7, enter the formula to find the number of samples of size 2 that can be formed from 10 digits. The formula is the combination of 2 things taken from 10 or $\binom{10}{2}$. The Excel formula for $\binom{n}{r}$ is *combin(n,r)*. So enter the formula

$$=\textbf{combin(10,2)} \quad \boxed{\text{ENTER}}$$

In cell B8, enter the formula to find the mean of the population, μ:

$$=\textbf{average(b2:f3)} \quad \boxed{\text{ENTER}}$$

In cell B9, enter the formula to find the standard deviation of the population, σ:

$$=\textbf{stdevp(b2:f3)} \quad \boxed{\text{ENTER}}$$

Step 2. Generate the 45 samples by systematically entering all possible samples of size 2. *Hint:* Use the $\boxed{\text{Edit}} \rightarrow \boxed{\text{Fill}} \rightarrow \boxed{\text{Down}}$ and the $\boxed{\text{Edit}} \rightarrow \boxed{\text{Copy}}$ and $\boxed{\text{Edit}} \rightarrow \boxed{\text{Paste}}$ commands. Also enter the formulas to compute the sample means. See Figure 2. Compare your results with those shown in Figure 2. There should be 45 samples, with the sample means computed in column E.

	A	B	C	D	E
11	Sample	1st no.	2nd no.		sample mean
12	1	0	1		0.5
13	2	0	2		1
14	3	0	3		1.5
15	4	0	4		2
16	5	0	5		2.5
17	6	0	6		3
18	7	0	7		3.5
19	8	0	8		4
20	9	0	9		4.5
21	10	1	2		1.5
22	11	1	3		2
23	12	1	4		2.5
24	13	1	5		3
25	14	1	6		3.5
26	15	1	7		4
27	16	1	8		4.5
28	17	1	9		5
29	18	2	3		2.5
30	19	2	4		3
31	20	2	5		3.5
32	21	2	6		4

Figure 2   Generate all possible samples of size 2.

Step 3. The central limit theorem states that if samples of size 2 are taken from a population, the mean of all possible sample means is the same as the mean of the population.

About $\mu_{\bar{x}}$:   The mean of the population is $\mu = 4.5$. Enter the formula for the mean of all sample means $\mu_{\bar{x}}$ in cell **F5**. The formula should compute the average of all the samples from cell E12 to E56. See Figure 3.

	A	B	C	D	E
1	Adv7 Simulation of choosing from population a set of 2 numbers wi'				
2	Population =====>	0	1	2	
3		5	6	7	
4					
5	N=no. in pop =>	10		mu xbar===>	
6	No. of Samples =>	45		sigma xbar=>	
7	Population mean=>	4.5		sigma xbar=>	
8	Population s.d .=>	2.872281		with correction factor	

Figure 3 Formula for mean of sample means.

Did you get a value of 4.5? The central limit theorem states that $\mu_{\bar{x}} = \mu$. So $\mu = 4.5$ and $\mu_{\bar{x}} = 4.5$.

About $\sigma_{\bar{x}}$:   The second part of the central limit theorem states that $\sigma_{\bar{x}} = \sigma/\sqrt{n}$. Since σ was computed in cell B8 and n = size of sample = 2, we have

$$\sigma_{\bar{x}} = \frac{2.872281}{\sqrt{2}} = 2.031009601$$

Compute the standard deviation of all the sample means from E12 to E56. Place this formula in cell **F6**:

$$\text{=STDEVP(E12:E56)} \boxed{\text{ENTER}}$$

There are actually two formulas in Excel for the standard deviation. **STDEVP** is for the standard deviation of the population, whereas **STDEV** is for the standard deviation of a sample.

Did you get a value of 1.914854? This is not the same as the theoretical value of 2.031009601!

The reason for this discrepancy is that the sampling is done without replacement, where the sample is more than 5% of the population. A sample size $n = 2$ is more than 5% of the population size $N = 10$. So, as the text suggests, to compute $\sigma_{\bar{x}}$, use the formula involving the finite population correction factor:

$$\sigma_{\bar{x}} = \frac{\sigma}{\sqrt{n}} \sqrt{\frac{N - n}{N - 1}}$$

Enter this formula in cell **F7**. Compare your results to Figure 4.

	A	B	C	D	E	F	G
1	Adv7 Simulation of choosing from population a set of 2 numbers without replacement						
2	Population =====>	0	1	2	3	4	
3		5	6	7	8	9	
4							
5	N=no. in pop =>	10		mu xbar===>		4.5	
6	No. of Samples =>	45		sigma xbar=>		1.914854	
7	Population mean=>	4.5		sigma xbar=>		1.914854	
8	Population s.d .=>	2.872281		with correction factor			

Figure 4 Standard deviation of sample means.

Wow! The theoretical formula really works. The standard deviation of the sample means can be found without really drawing every possible sample, computing all possible sample means, and then finally computing the standard deviation of all the sample means.

Print out your results and save as **ADV7.XLS** on your disk.

C.8   USING EXCEL IN CHAPTER 8: CONFIDENCE INTERVAL

EXCEL ADVENTURE 8:

WHAT IS THE AVERAGE WEIGHT, μ, OF BASS AT LAKE MEAD?

Lake Mead is famous for its large wide-mouth bass, which is advertised throughout the world as being an average of 15 pounds. There is a conjecture that a chemical spill killed a great number of bass.

Biologists are sent to the lake to take a sample by electrocuting an area of the lake and then measuring the bass that are temporarily immobilized. Use the sample in Figure 1 to find a 98% confidence interval for the average weight of all bass at the lake.

	A	B	C	D	E	F	G	H	I
1	Adv8 What is the averge weight of bass at Lake Mead?								
2	..								
3	Data in pounds								
4		9.89	13.36	8.35	7.96	11.66	16.69	14.74	14.28
5		7.78	14.47	15.37	7.82	8.52	7.72	7.89	8.81
6		14.34	11.06	9.55	7.67	11.79	10.91	8.2	16.67
7		11.45	16.2	16.05	14.38	8.89	9.38	15.69	8.77
8									
9	Desired Confidence interval =======>			98%					

Figure 1   Weight in pounds of a sample of bass.

Step 1.   Enter formulas to compute the sample measurements as shown in Figure 2.

	A	B	C	D	E
10					
11	**Formulas**				
12		**Sample Mean ===>**			
13		**Sample s.d. ====>**			
14		**Sample size ====>**			
15					

Figure 2   Sample statistics.

(a)   Compute the sample mean by entering in cell D12 the formula:

$$\text{=average(b4:i7)} \boxed{\text{ENTER}}$$

(b)   Compute the sample standard deviation by entering formula in cell D13.

$$\text{=stdev(b4:i7)} \boxed{\text{ENTER}}$$

Notice that the formula for the standard deviation of a sample is stdev, whereas the formula for the standard deviation of a population is stdevp.

(c)   Enter a formula in cell D14 to count the number of values in the sample in cell D14.

Step 2.   The size of the sample n is 32, which is a large sample. Use the normal distribution to find the confidence interval for the population mean. See Figure 3 on page C26.

	A	B	C	D	E
8					
9	Desired Confidence interval ========>				98%
10					
11	Formulas				
12		Sample Mean ===>		11.44719	
13		Sample s.d. ====>		3.214917	
14		Sample size ====>		32	
15					
16	Computed statistical values				
17		z-value ========>			
18					
19		standard error =>			
20		of the mean			

Figure 3   Compute z and standard error of mean.

(a)   Compute the z value so that the confidence level is 98%. Instead of looking up the z value in your text, you can issue a command in Excel to find the z value. Use the Excel function

=NORMSINV(area under normal curve to the left of z+)

This function returns the $z+$ value when the argument is an area (see Figure 4).

=normsinv(.5)   returns a $z+$ value of 0

=normsinv(.5 + .98/2)   returns a $z+$ value of 2.326342

In cell **D17**, enter the formula to compute the $z+$ value, where the area to the left of the $z+$ value is $.5 + 1/2$ of the confidence level:

=NORMSINV(0.5+0.5*E9) `ENTER`

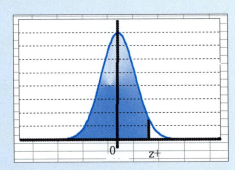

Figure 4   Function normsinv.

(b)   Compute the standard error of the mean, $\sigma_{\bar{x}} = \sigma/\sqrt{n}$, in cell **D19**. Since the sample size $n > 30$, an approximation for σ is s, the value in cell **D13**. Place the cell pointer in cell D19, and enter the formula:

=D13/SQRT(D14) `ENTER`

Step 3.   Enter the formulas for the upper and lower limits. The formula is

$$\bar{x} - z\,\frac{s}{\sqrt{n}} < \mu < \bar{x} + z\,\frac{s}{\sqrt{n}}$$

In cell D22, enter the lower limit $\bar{x} - z\,\dfrac{s}{\sqrt{n}}$.

In cell D23, enter the upper limit $\bar{x} + z\,\dfrac{s}{\sqrt{n}}$.

Given the 32 data values, the 98% confidence interval for the average weight of bass in Lake Mead is between 10.125 and 12.7693 pounds. (See Figure 5.) Thus, the interval is far lower than the historical average of 15 pounds.

	A	B	C	D	E
8					
9	Desired Confidence interval =======>				98%
10					
11	Formulas				
12		Sample Mean ===>		11.44719	
13		Sample s.d. ====>		3.214917	
14		Sample size ====>		32	
15					
16	Computed statistical values				
17		z-value ========>		2.326342	
18					
19		standard error =>		0.568322	
20		of the mean			
21					
22		Lower Limit ===>		10.12508	
23		Upper Limit ===>		12.7693	
24					

Figure 5 Confidence interval for population mean.

Print out the results and save the file as **ADV8.XLS** on your disk.

C.9 USING EXCEL IN CHAPTER 9: HYPOTHESIS TESTING

EXCEL ADVENTURE 9:   LIVING ALONE

In the last census, it was found that 25% of all households were made up of one person. With the high cost of rent and the expense of owning a home, has the percentage of people living alone decreased? Test this theory at the 5% level.

Step 1.   Excel to summarize the data in Figure 1.

	A	B	C	D	E	F	G	H	I	J	K
1	Adv9- Test if the percentage of people living alone has decreased.										
2		Answer to question if you are living alone									
3	Data==	N	N	Y	N	N	N	N	N	N	Y
4		N	N	N	Y	Y	N	N	N	Y	N
5		N	N	N	N	N	N	N	N	N	N
6		N	N	N	Y	N	Y	N	N	N	N
7		Y	N	N	Y	N	N	N	N	N	N
8		N	N	N	N	N	N	N	N	N	N

Figure 1 Do you live alone?

It is painstaking to summarize the data visually. Leave it to Excel to do the grunt work.

(a) In cell C11, enter the formula **=counta(B3:k8)** ENTER. This formula counts all cells that contain alphabetic or numerical data.

(b) To count up all the Y's, use the formula **=countif(range, condition)**. In cell C12, enter the formula **=countif(B3:K8, "Y")** ENTER.

(c) Count the number of N's in the data by entering another **countif** formula.

(d) In cell F12, enter the formula for the sample proportion, p', denoting the percentage of Y's to the total. This value should be 22%.

Also enter the titles to make the worksheet readable. See Figure 2.

	A	B	C	D	E	F	G	H	I
10	Formulas to summarize data								
11		n=	60		Let p' = proportion in sample that answered Yes				
12		number of Y	9		p'=	15%			
13		number of N	51		p=	25%			
14									

Figure 2 Summarize data with countif.

Step 2. Set up the hypotheses H_0 and H_1. See Figure 3.

	A	B	C	D	E
15	Setup hypotheses				
16		H0: p = .25			
17		H1: p < .25			
18		α=	0.05	level of significance	
19					
20		p' =	15%		
21		p=	25%		

Figure 3 Null and alternative hypotheses.

Also compute p' (sample proportion) and p (population proportion).

Step 3. Compute the standard error, $\sigma_{p'} = \sqrt{\dfrac{p(1-p)}{n}}$.

(a) Enter this formula in cell C24. Recall that the command to find square roots is **sqrt**.

(b) Compute the test statistic, z^*.

$$z^* = \frac{p' - p}{\sqrt{\dfrac{p(1-p)}{n}}}$$

Enter the formula in cell C27.

(c) Compute the p-value using the function **=NORMSDIST(z-value)**. The z value is the value found in cell C27. See Figure 4.

	A	B	C
23	Compute the z* test statistic		
24		s.d. of p'=	0.055901699
25			
26			
27		z*=	-1.788854382
28		p-value is	0.036819084
30			

Figure 4 Test statistic, z^*.

Step 4. State the conclusion.

Since the *p*-value is less than α, we can reject H_0. The conclusion is that there is a decrease in the percentage of people living alone.

Print out the results and save the worksheet on your disk as **ADV9.XLS**.

C.10 USING EXCEL IN CHAPTER 10: INFERENCES ON TWO POPULATION MEANS

EXCEL ADVENTURE 10: **DO JOGGERS HAVE BETTER MEMORIES?**

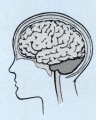

A prominent doctor claims that joggers have better memories than non-joggers. Perhaps jogging is all you need to do to remember all the statistical formulas you have learned this semester. To test the doctor's claim, a blind study is conducted in which 55 students are randomly selected and divided into two groups: those who jog and those who don't. Twenty-five don't jog, and 30 do. Each subject is shown 10 digits for 30 seconds and then has to recite the sequence back. A score of 1 means only the first digit was recalled accurately. The higher the number, the better is the memory for that subject. Test at the 5% significance level the claim that joggers have better memories.

To solve this problem in Excel, one mode of attack is to use the **Tools → Data Analysis** command.

Step 1. Prepare the data. The values for each sample must be in one column. Arrange Sample 1 for the non-joggers along column B: cells B3 to B27. Arrange Sample 2 for the joggers in column D: cells D3 to D32. See Figure 1 on page C30.

Step 2. Reflect on the type of inferences about two population means that you will use in this adventure. There are three types.

Type 1. Two population means for small and independent samples with equal population standard deviations

Type 2. Two population means for small and independent samples with unequal population standard deviations

Type 3. Two population means using paired samples

Because we don't know whether the population standard deviations are the same or different, let's solve this problem both ways.

Step 3. Use the $\boxed{\text{Tools}}$ → $\boxed{\text{Data Analysis}}$ command. See Figure 2 on page C31.

Case 1. Assume that the variances between the two populations are unequal. Here you will use the *t*-test on unequal variances covered in Section 10.3.

After selecting $\boxed{\text{Data Analysis}}$, a menu appears for you to select the type of test. Click the scroll bar until the choice

$$\boxed{\text{t-test: Two-Sample Assuming Unequal Variances}}$$

appears. See Figure 3. Click $\boxed{\text{OK}}$.

	A	B	C	D	E
1	Adventure 10. Do joggers have a better memory than non-joggers?				
2	Data	Sample A: Non-joggers memory of no. of digits		Sample B: joggers memory of no. of digits	
3		7		10	
4		5		8	
5		8		8	
6		6		8	
7		7		10	
8		5		8	
9		5		9	
10		6		10	
11		5		8	
12		5		8	
13		7		9	
14		6		8	
15		7		10	
16		8		10	
17		7		8	
18		7		10	
19		8		10	
20		6		8	
21		7		8	
22		8		9	
23		5		8	
24		5		9	
25		6		10	
26		7		9	
27		6		8	
28				8	
29				9	
30				9	
31				10	
32				9	

Figure 1   Digits memorized.

Excel prompts you to fill in four boxes with questions (see Figure 4 on page C32):

Input Question Box   Enter the range for variable 1, the data containing the number of digits memorized by the non-joggers, B3:B27. Enter the range for variable 2, the data containing the digits for the joggers, D3:D32.

Hypothesized Mean Difference Box   Since the null hypothesis is H_0: $\mu_{nj} - \mu_j = 0$, the assumed mean of the non-joggers is the same as the mean of the joggers. Enter **0** in this box.

Alpha (α *level*) Box   This is the level of significance, $\alpha = .05$.

Figure 2 Tools → Data Analysis.

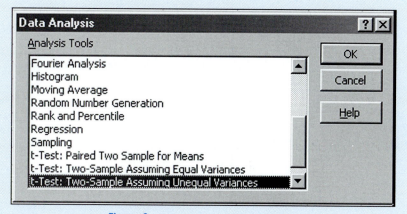

Figure 3 *t*-test: unequal population variances.

Output Range Box   This is the location where you want the output to occur. Let's place the results on the worksheet. Click the circle beside **Output Range** and click the top left corner where the output data will reside. Click ⎡OK⎤. Verify your results with Figure 5 on page C32.

Now, let's work case 2, where the variances of the populations are assumed to be equal, before using the results on this screen.

Case 2.   Assume that the variances between the two populations are equal. Here you will use the *t*-test on equal variances covered in Section 10.2.

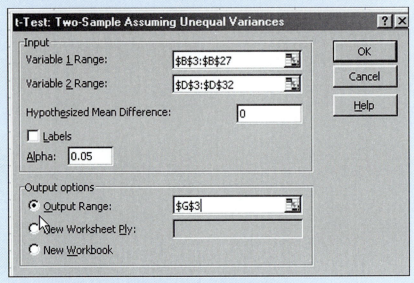

Figure 4   Fill in location of data.

	G	H	I
3	t-Test: Two-Sample Assuming Unequal Variances		
4			
5		Variable 1	Variable 2
6	Mean	6.36	8.866666667
7	Variance	1.156666667	0.740229885
8	Observations	25	30
9	Hypothesized Mean Difference	0	
10	df	46	
11	t Stat	-9.411263748	
12	P(T<=t) one-tail	1.35175E-12	
13	t Critical one-tail	1.678658919	
14	P(T<=t) two-tail	2.7035E-12	
15	t Critical two-tail	2.012893674	

Figure 5   Results of t-test for two samples with unequal population variances.

Issue the command ⎡Tools⎤ → ⎡Data Analysis⎤ again. Select from the Data Analysis box, the **t-test: Two-Sample Assuming Equal Variance** and click ⎡OK⎤. See Figure 6.

Fill in the question box concerning the location of the data and level of significance. Place the output on the current worksheet where the top left corner is cell **G18**. (See Figure 7.) Click ⎡OK⎤.

Step 4.   Set up the hypotheses and formalize the analysis (see Figure 8 on page C34).

H_0: $\mu_{nj} - \mu_j = 0$ (Memory averages the same among non-joggers and joggers)

H_1: $\mu_{nj} - \mu_j < 0$ (Memory average among non-joggers is less than memory average among joggers)

$\alpha = .05$ level of significance

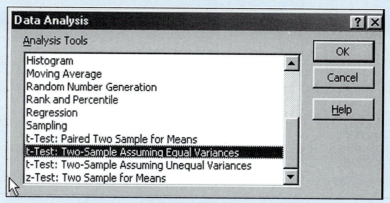

Figure 6   *t*-test: Assume population variances are equal.

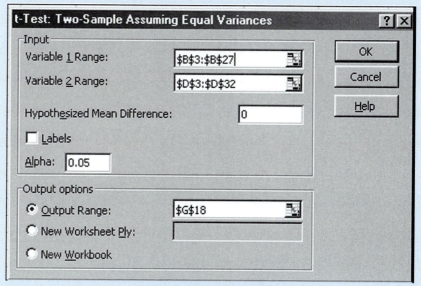

Figure 7   Equal variance question box.

Case 1.   Variances are unequal. If the population variances are unequal, then $t_c = -1.678658919$, with $df = 46$. The test statistic is $t^* = -9.411$ with a p-value of 1.35×10^{-12}. This is highly significant, leading to the conclusion that joggers do have better memories.

Case 2.   Variances are equal. If population variances are equal, then $t_c = -1.674115993$, with $df = 53$. The pooled variance $= .928805031$. The test statistic is $t^* = -9.60469$ with a p-value of 1.69079×10^{-13}. This is highly significant, leading to the conclusion that joggers do have better memories.

Write your results on the worksheet and print them out. Also save the file as **ADV10.XLS** on your disk and go on a short jog.

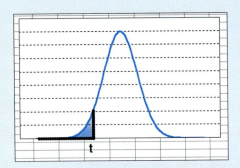

	G	H	I
3	t-Test: Two-Sample Assuming Unequal Variances		
4			
5		*Variable 1*	*Variable 2*
6	Mean	6.36	8.866666667
7	Variance	1.156666667	0.740229885
8	Observations	25	30
9	Hypothesized Mean Difference	0	
10	df	46	
11	t Stat	-9.411263748	
12	P(T<=t) one-tail	1.35175E-12	
13	t Critical one-tail	1.678658919	
14	P(T<=t) two-tail	2.7035E-12	
15	t Critical two-tail	2.012893674	
16			
17			
18	t-Test: Two-Sample Assuming Equal Variances		
19			
20		*Variable 1*	*Variable 2*
21	Mean	6.36	8.866666667
22	Variance	1.156666667	0.740229885
23	Observations	25	30
24	Pooled Variance	0.928805031	
25	Hypothesized Mean Difference	0	
26	df	53	
27	t Stat	-9.604695329	
28	P(T<=t) one-tail	1.69079E-13	
29	t Critical one-tail	1.674115993	
30	P(T<=t) two-tail	3.38158E-13	
31	t Critical two-tail	2.005745046	

Figure 8   Results from Tools-Data analysis.

C.11   USING EXCEL IN CHAPTER 11: CHI-SQUARE APPLICATIONS

EXCEL ADVENTURE 11:

YOU ARE WHAT YOU EAT!

As a waiter in a gourmet restaurant can attest, patrons are either nice or nasty. A customer who is an expert psychologist claims that the amount of meat one eats determines the mood of an individual. Test at the 2.5% level whether there is a relationship between personality and the amount of meat eaten during a week using the database in Figure 1.

This adventure is an application of the chi-square test of independence. Let's state the null and alternative hypotheses:

H_0: Personality and meat consumption are independent

H_1: There is some relationship between personality and meat consumption

	A	B	C	D
1	Adventure 11 You are what you eat			
2			ounces	
3	subject	sex	meat	personality
4	Joe	M	10	Nice
5	Anne	F	12	Nasty
6	Bill	M	16	Nice
7	Ken	M	20	Nasty
8	Jenny	F	20	Nice
9	Carol	F	32	Nasty
10	Tom	M	5	Nasty
11	Alice	F	12	Nice
12	Ted	M	25	Nasty
13	Jerry	M	18	Nasty
14	Ben	M	12	Nice
15	Cindy	F	15	Nasty
16	Pete	M	5	Nice
17	Sylvia	F	20	Nasty
18	Helen	F	8	Nice
19	Sabrina	F	5	Nice
20	Jeff	M	11	Nice
21	Lonnie	F	13	Nasty
22	Cary	M	18	Nasty
23	Debbie	F	8	Nice
24	Cloe	F	0	Nasty
25	Anna	F	0	Nice
26	Dexter	M	13	Nice
27	Paul	M	25	Nice
28	Mary	F	10	Nice
29	David	M	30	Nice
30	June	F	0	Nice
31	Betty	F	19	Nasty
32	Theresa	F	10	Nice
33	George	M	25	Nasty
34	John	M	2	Nasty
35	Dan	M	5	Nasty
36	Harry	M	20	Nice
37	Roberta	F	25	Nice
38	Heffy	M	30	Nasty
39				

Figure 1 Personality and amount of meat eaten during the week.

Step 1. Before a chi-square test can be used, you must arrange the data into a two-way classification. Create a table where the rows denote the personality of the patron and the columns denote the amount of meat (in ounces) eaten during a week. To do this quickly and accurately, use the command $\boxed{\text{Data}} \rightarrow \boxed{\text{Pivot Table}}$. (See Figure 2 on page C36.) Select option **Microsoft Excel list or database**. Select the report to be a **PivotTable**. Click $\boxed{\text{Next>}}$.

Step 2. Select the range of the database. See Figure 3. Be sure to include the labels (field names) in the first row. Click the spreadsheet icon, and then select range A3:D38. Click $\boxed{\text{Next>}}$.

In the PivotTable Wizard, click the option to place results on **Existing worksheet** where the top left corner is F4. Next click the $\boxed{\text{Layout}}$ box to set up the two-way classification table. See Figure 4.

In the Layout screen, drag the field buttons $\boxed{\text{personality}}$ to the Row and $\boxed{\text{meat}}$ to the Column. Drag the **personality** field button into the Data region. See Figure 5 on page C37. Click $\boxed{\text{OK}}$ and click $\boxed{\text{Finish}}$.

Step 3. Group the columns in the contingency table. (See Figure 6.) The column fields are numerical and include ounces of meat from 0 to 30. There are 31 columns—far too many! You can group the columns together where Excel will sum the frequencies in the columns that are grouped together.

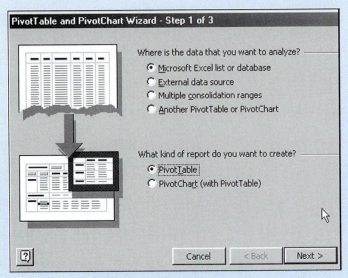

Figure 2   Select options in step 1 of 3.

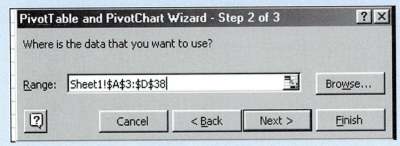

Figure 3   Database range.

Figure 4   PivotTable Wizard.

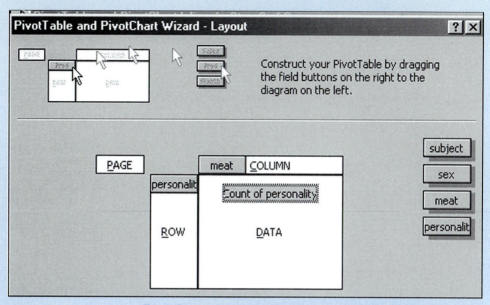

Figure 5   Drag field names to rows and columns.

F	G	H	I	J	K	L	M	N	O	P
Count of personality	meat ▼									
personality ▼	0	2	5	8	10	11	12	13	15	16
Nasty	1	1	2				1	1	1	
Nice	2		2	2	3	1	2	1		1
Grand Total	3	1	4	2	3	1	3	2	1	1

Figure 6   Table created with PivotTable.

Let's group the columns from 0 to 15 as group 1, and group columns 16 to 32 as group 2. First, highlight the columns you want to group—say, G5 to O5. Second, select

Data → Group and Outline → Group (See Figure 7 on page C38.)

Excel groups the highlighted columns denoted by 0 ounces to 15 ounces as group 1. Repeat the command to group the columns denoted by 16 to 32 ounces of meat using the above method. You should end up with two groups:

Group 1: 0 to 15

Group 2: 16 to 32

See Figure 8.

Step 4.   There are still 17 columns. A contingency table sums the frequencies in each group. Excel has a neat way of doing just that. (See Figure 9 on page C39.)

Highlight the cells in group 1—say, G6:O6. Click the command

Data → Group and Outline → Hide Detail

Repeat this again to group the frequencies within group 2. Notice that Excel actually sums the columns in each group and produces a nice two-way classification table. See Figure 10.

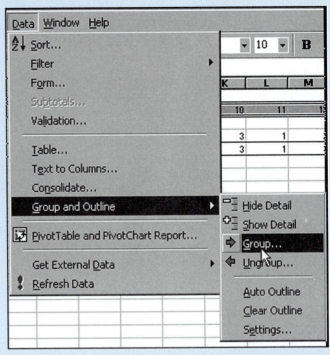

Figure 7   Grouping the columns.

	F	G	H	I	J	K	L	M	N	O	P	Q	R	S	T	U	V
4	Count of personality	meat2 ▼	meat ▼														
5		Group1									Group2						
6	personality ▼	0	2	5	8	10	11	12	13	15	16	18	19	20	25	30	32
7	Nasty	1	1	2			1	1	1			2	1	2	2	1	1
8	Nice	2		2	2	3	1	2	1		1			2	2	1	
9	Grand Total	3	1	4	2	3	1	3	2	1	1	2	1	4	4	2	1

Figure 8   Frequencies shown in Group 1 and Group 2.

Step 5.   Create within the Excel worksheet observed and expected frequencies. Excel does not generate the expected values; you need to apply the techniques from Chapter 11 to find the expected values using the table of observed values. In cell H12, the formula for expected value is **=(I7/I9)*H9** ⎡ENTER⎤. Enter the formulas now and compare your answers to those in Figure 11.

Step 6.   Compute $\chi^2 = \Sigma \dfrac{(O - E)^2}{E}$.

Next, create a table with headings for observed (O), expected (E), and $(O - E)^2/E$. The sum of the range **I18:I21** is entered in cell **I23**. The computed chi-square value from the sample is $\chi^2 =$ **2.158717105**. (See Figure 12 on page C40.)

Step 7.   Set up the null and alternative hypotheses and finish the test. There are three functions in Excel that deal with chi-square:

=CHIDIST(chi-square value, deg of freedom) returns the area under the chi-square distribution to the right of the chi-square value.

=CHIINV(area to the right, deg of freedom) returns the chi-square value given the area on the right tail, and degrees of freedom.

=CHITEST(range for the observed values, range for the expected values) returns the area to the right under the chi-square probability distribution. This is the *p*-value.

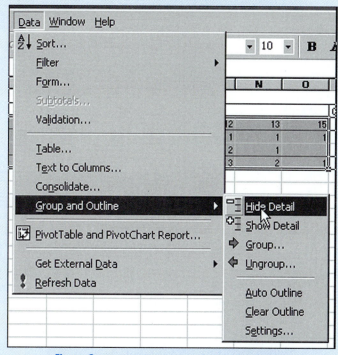

Figure 9   Have Excel sum the frequencies in each group.

	F	G	H	I
4	Count of personality	meat2 ▼	meat ▼	
5		Group1	Group2	Grand Total
6	personality ▼			
7	Nasty	7	9	16
8	Nice	13	6	19
9	Grand Total	20	15	35
10				

Figure 10   Two-way classification table.

	F	G	H
10			
11		Group1	Group2
12	Nasty	9.14285714	6.85714286
13	Nice	10.8571429	8.14285714

Figure 11   Formulas to find expected values.

Use the chi-square functions to complete the test. Compare your results with those in Figure 13 on page C40.

Since the χ^2 computed value of 2.158 is less than the χ^2 critical value of 5.02, the test statistic does not fall in the critical region (right tail). We can't conclude that personality and meat consumption are related. It's OK to order that 32-oz steak on your next trip to the restaurant.

Print out your results and save them as **ADV11.XLS** on your disk.

	G	H	I
17	observed	expected	(o-e)^2/e
18	7	9.14285714	0.502232143
19	9	6.85714286	0.669642857
20	13	10.8571429	0.422932331
21	6	8.14285714	0.563909774
22			
23		chisq=	2.158717105
24			

Figure 12   Compare expected and observed frequencies.

	F	G	H	I	J	K
25	Ho: Personality independent of the amount of meat eaten					
26	H1: There is some relationship between type of personality and amount of meat eaten					
27						
28	df=	1				
29	alpha=	0.025				
30						
31	Computed chisq =	2.15871711				
32						
33	Critical chisq =	5.02390259				
34						
35	Conclusion ==>	Computed chisquare value does not fall in the area of rejection.				
36		Fail to reject Ho.				

Figure 13   Test of independence.

C.12   USING EXCEL IN CHAPTER 12: ANALYSIS OF VARIANCE

EXCEL ADVENTURE 12:   **GERM WARFARE**

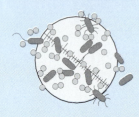

Germs are everywhere. These creepy creatures are on surfaces of ATM machines, on pens and countertops, on armchairs, and on people's hands.

An environmental microbiologist examined four major sources of contamination in banks and counted the number of germs per square millimeter: ATM machines, telephones, tellers' countertops, and doorknobs. Random samples were taken at various locations and tested. The number of germs for each sample is recorded as shown in Figure 1. Test at the 1% level of significance whether the number of germs per square millimeter is different among the sources of germ contamination.

Set up the population parameters and the hypotheses. Let

μ_A = average number of germs found on all **ATM** machines

μ_T = average number of germs found on all **telephones**

μ_C = average number of germs found on all **countertops**

μ_D = average number of germs found on all **doorknobs**

$H_0: \mu_A = \mu_T = \mu_C = \mu_D$

H_1: At least one mean is different

$\alpha = .01$

Excel has a quick and accurate routine for comparing the means of each factor by using ANOVA.

	A	B	C	D	E
1	Adventure 12. Germ Warfare				
2					
3	Number of germs per square millimeter on surfaces				
4					
5	ATM	Telephones	Counter-tops	Door-knobs	
6	526	521	541	589	
7	577	560	541	573	
8	569	515	585	576	
9	547	540	589	588	
10	521	522	569	565	
11	510	513	536	570	
12	532	510	576	545	
13	553	557	540	601	
14	554	550	587	605	
15	568	512	524	543	
16	561	557	545	579	
17	553	534	549	588	
18	574	503	528	583	
19	500	491	592	557	
20	521	514	573	573	
21	500	511	560	613	
22	551	494	526	590	
23	525	555	535	543	
24	564	564	552	577	
25	516	504	558	613	
26	522	534	556	601	
27	516	545	541	540	
28	534	547	584	606	
29	509	526	587	605	
30	521	499	542	565	
31	577	515	580	614	
32	556	512		575	
33	538	525		568	
34	548	522		563	
35	507	508			
36	513				
37	535				
38	561				
39	563				
40	575				

Figure 1 Germ data.

Step 1.   Enter the command

Tools → Data Analysis

From the Data Analysis menu, choose Anova: Single Factor and click OK . (See Figure 2.)

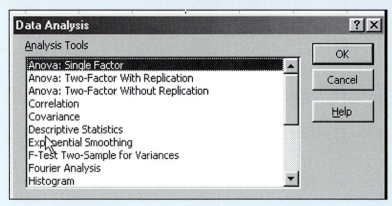

Figure 2   Choose ANOVA from Data Analysis box.

Step 2.   In the **Anova: Single Factor** menu, enter the ranges for the following:

Input Range (the range where the values exist in each sample): In our problem, the range is from A6 to D40 even though the sample sizes are not all equal.

The level of significance:

$$\alpha = .01$$

Output range: Indicate the top left corner of where you want the data to appear—say, **F6**. See Figure 3. Click OK .

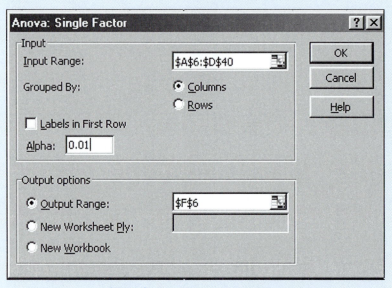

Figure 3   Input Range and Output Range.

Step 3.   Excel performs the computations to determine the sample size, mean, and variance for each sample. These measurements are used and the *F* value is found as shown in Figure 4.

	F	G	H	I	J	K	L
6	Anova: Single Factor						
7							
8	SUMMARY						
9	*Groups*	*Count*	*Sum*	*Average*	*Variance*		
10	Column 1	35	18897	539.9143	580.6689		
11	Column 2	30	15760	525.3333	443.1954		
12	Column 3	26	14496	557.5385	487.4585		
13	Column 4	29	16808	579.5862	488.5369		
14							
15							
16	ANOVA						
17	*Source of Variation*	*SS*	*df*	*MS*	*F*	*P-value*	*F crit*
18	Between Groups	48663.419	3	16221.14	32.1865	3.27089E-15	3.955051
19	Within Groups	58460.906	116	503.9733			
20							
21	Total	107124.33	119				
22							

Figure 4 ANOVA table generated by Excel.

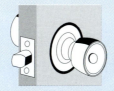

Excel computes MS(factor) = MSB = 16,221.14 and MS(error) = MSW = 503.9733. The table also shows the computed value $F* = 32.1865$ with a p-value of $3.27 \times 10^{-15} \approx 0$. The table also displays the critical F value of 3.955.

Since $F* > 3.955$, we conclude that the means are not all equal; we can reject H_0. Note that the samples seem to show that there is a higher concentration of germs on doorknobs. So the next time you visit your friendly neighborhood bank, carry some disinfectants.

Print out the results and save them as **ADV12.XLS** on your disk.

C.13   USING EXCEL IN CHAPTER 13: REGRESSION ANALYSIS

EXCEL ADVENTURE 13:   **IT'S GETTING CROWDED IN HERE!**

With information from the U.S. Census Bureau, the populations of the United States and five regions for the last 10 years are shown in Figure 1 on page C44.

The U.S. population grew 9%, California's population grew 11%, Florida's population grew 16%, and Arizona's population grew a whopping 30% between 1990 and 1999. But two regions' populations actually decreased: the District of Columbia and North Dakota.

Using the data for North Dakota for the last 10 years, do the following:

(a)   Find the equation of the regression line.

(b)   Generate estimated y values for the line of best fit.

(c)   Plot the regression line on the scatter diagram.

Excel can perform regression analysis on a time series arranged in column format. The x data are located in the range A5:A14; the y data for North Dakota are located in the range G5:G14. Issue the command Tools → Data Analysis .

	A	B	C	D	E	F	G
1	Adv13 Its Getting Crowded Here!!						
2							
3	Population from 1990 to 1999						
4	Date	United States	California	Arizona	Florida	District of Columbia	North Dakota
5	1990	2,494.64	299.50	36.79	130.18	6.04	6.37
6	1991	2,521.53	304.14	37.62	132.89	5.93	6.34
7	1992	2,550.30	308.76	38.67	135.05	5.84	6.35
8	1993	2,577.83	311.47	39.93	137.14	5.76	6.37
9	1994	2,603.27	313.17	41.48	139.62	5.65	6.40
10	1995	2,628.03	314.94	43.07	141.85	5.51	6.42
11	1996	2,652.29	317.81	44.32	144.27	5.38	6.43
12	1997	2,677.84	322.18	45.52	146.83	5.29	6.41
13	1998	2,702.48	326.83	46.67	149.08	5.21	6.38
14	1999	2,726.91	331.45	47.78	151.11	5.19	6.34

Figure 1   Populations in units of 100,000.

Step 1.   After the Data Analysis dialog box appears, select the option **Regression** and click [OK]. See Figure 2.

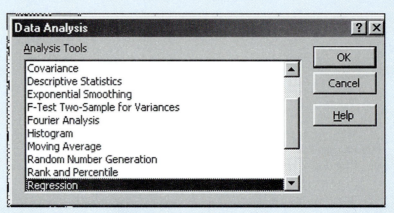

Figure 2   Select the Regression option.

Step 2.   In the Regression dialog box, enter the ranges for the input and output regions. The Input range contains two series: the x values and the y values. The y values (population in North Dakota) are located in the range **G5:G14**. The x values are the years, located in **A5:A14**.

The Output region can be one of three choices: within the spreadsheet, as a new spreadsheet within the workbook, or in a new spreadsheet. Let's select the first choice and show the results within the existing spreadsheet with the top left corner at cell **I5**. Enter the three responses as shown in Figure 3 and click [OK].

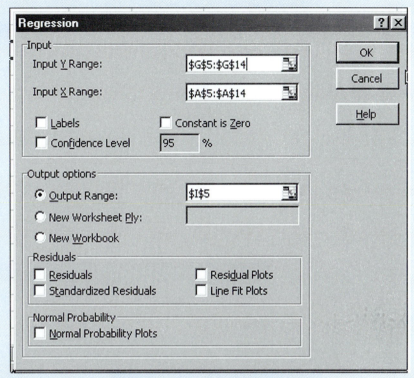

Figure 3   Enter the ranges for the three regions.

Step 3.   Excel displays the summary of the regression analysis shown in Figure 4.

	I	J	K	L	M	N	O	P	Q
5	SUMMARY OUTPUT								
6									
7	*Regression Statistics*								
8	Multiple R	0.223423							
9	R Square	0.049918							
10	Adjusted F	-0.06884							
11	Standard E	0.032472							
12	Observatio	10							
13									
14	ANOVA								
15		*df*	*SS*	*MS*	*F*	*ignificance F*			
16	Regressior	1	0.000443	0.000443	0.420325	0.534937			
17	Residual	8	0.008436	0.001054					
18	Total	9	0.008879						
19									
20		*Coefficients*	*andard Err*	*t Stat*	*P-value*	*Lower 95%*	*Upper 95%*	*.ower 95.0%*	*lpper 95.0%*
21	Intercept	1.757918	7.130523	0.246534	0.811477	-14.6851	18.20094	-14.6851	18.20094
22	X Variable	0.002318	0.003575	0.648325	0.534937	-0.00593	0.010562	-0.00593	0.010562

Figure 4   Results from the regression analysis.

(a)   Now find the equation of the regression line. The slope of the regression line is the coefficient of the x variable. Excel displays the slope in cell J22. The slope is .002318, positive but very close to 0. The slope describes the change in population per year; there is an increase of 231 (.002318 times 100,000) people per year.

  The y-intercept is located in cell J21. This value is 1.757918, so the equation of the regression line is $y = .002318x + 1.757918$.

  Make a sketch of the regression equation on the scatter diagram with the following commands. Highlight the y values by selecting **G5:G14**. First, make a scatter diagram.

$$\boxed{\text{Insert}} \rightarrow \boxed{\text{Chart}} \rightarrow \boxed{\text{XY}}$$

Step 1 of 4:   Chart Type screen.

Click the chart displaying the scatter points (this is the first chart). Click $\boxed{\text{Next}}$.

Step 2 of 4:   Chart Source Data screen. Click $\boxed{\text{Series}}$. Input the x values in the Source Data dialog box (see Figure 5). This box is on the right side just above the y values. Click $\boxed{\text{Next}}$.

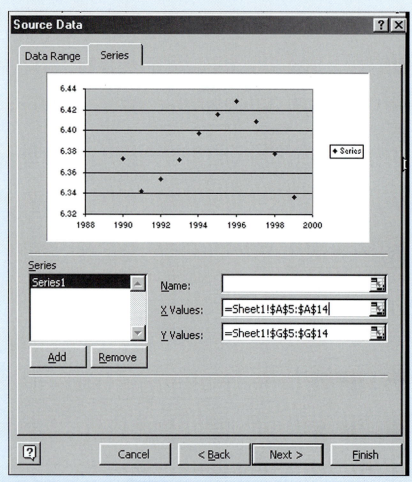

Figure 5   Enter x and y value ranges.

Step 3 of 4:   The Chart Option screen.

Enter a main title for the sketch: **Population in North Dakota**.

Enter a title for the *x*-axis: **years**.

Enter a title for the *y*-axis: **in 100,000**. Click Next .

Step 4 of 4: Chart location.

Place the chart within the current worksheet. Select the option As Object in . Click Finish . Excel will place the chart on the existing worksheet. You can size the chart and drag it anywhere in the worksheet. Let's drag and place the chart as shown in Figure 6.

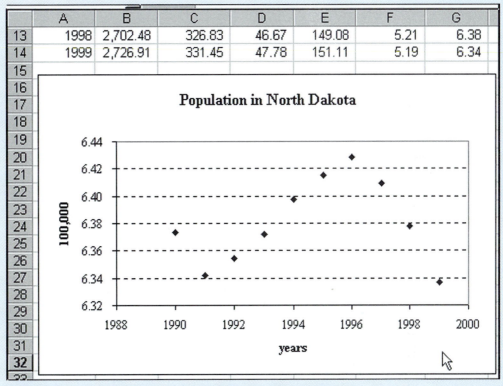

Figure 6 Scatter diagram.

(b) Use the regression line to generate *y* values on the line. First, enter the label "**Estimated y-value**" in cell **H4**. Second, enter the formula representing the equation of the line from the summary output: $y = .002318x + 1.757918$. Enter the formula in cell **H5**:

$$=.002318*A5+1.757918 \quad \boxed{\text{ENTER}}$$

Third, copy this formula down from cell H5 to cell H14. See Figure 7 on page C48.

(c) Graph the regression line on the scatter diagram. First, click a point on the scatter plot. Second, click Chart → Add Data. . . . (See Figure 8.) Third, Excel displays the **Add Data** dialog box. Enter the range **H5:H14**. Fourth, to connect the second series with a straight line, double-click the series. Excel displays the **Format Data Point dialog box**, where you will turn off the marker for the point by clicking the button **None**. Also click the Line option to Custom and select the Color and Style as shown in Figure 9.

The result is the scatter diagram and regression line in Figure 10 on page C49. Print out the results and save them as **ADV13.XLS**.

	G	H
4	North Dakota	Estimated y-value
5	6.37	6.370375818
6	6.34	6.372693636
7	6.35	6.375011455
8	6.37	6.377329273
9	6.40	6.379647091
10	6.42	6.381964909
11	6.43	6.384282727
12	6.41	6.386600545
13	6.38	6.388918364
14	6.34	6.391236182
15		

Figure 7   Estimated *y* values.

Figure 8   Add the regression points.

Figure 9   Formatting the regression line to pass through points.

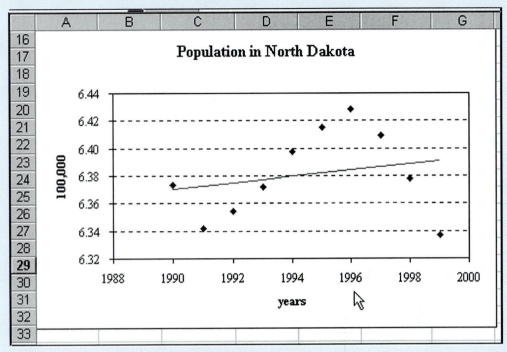

Figure 10 Regression line drawn on scatter diagram.

C.14 USING EXCEL IN CHAPTER 14: THE SPEARMAN RANK CORRELATION COEFFICIENT

EXCEL ADVENTURE 14: **IS THERE REALLY A COLD WAR?**

Skaters are given a score from 0 to 6 in competitions. A score of 0 indicates the skater did not skate well, whereas a score of 6 means the skater's performance was perfect and faultless. Suppose that at the trials, most of the scores were fairly close to the U.S. judge's, but the Russian judge's score was different. Use Spearman's rank correlation coefficient to test at the 5% significance level whether the Russian judge scored the skaters differently than the U.S. judge did.

Step 1. Set up the null and alternative hypotheses:

H_0: There is no correlation between the scores of the two judges

H_1: There is a correlation between the scores of the two judges

$\alpha = .05$

Step 2. Rank the scores for each judge (u for the Russian judge and v for the U.S. judge) in columns D and E. (See Figure 1 on page C50.)

Create the ranking for the Russian judge in column D. Use the function =RANK(cell of particular score, range of all scores). The result is the ranking. In cell D5, enter the formula.

=RANK($B5,$B$5:$B$27) [ENTER]

Copy the formula in cell D5 down to cell D27.

Repeat the procedure to rank the scores of the U.S. judge in column E.

	A	B	C	D
1	ADV14 Judging the Olympics at Salt Lake			
2				
3		Russian	U.S.	
4	Contestant	Judge	Judge	
5	Tara	6.0	6.0	
6	Kristie	5.6	5.4	
7	Michelle	5.5	4.9	
8	Nancy	5.4	4.4	
9	Nicole	5.0	4.3	
10	Lu	5.3	4.4	
11	Mairia	5.6	5.2	
12	Irina	5.9	5.7	
13	Vanessa	5.0	4.8	
14	Elena	5.1	4.2	
15	Tatyana	5.2	4.4	
16	Surya	4.4	3.6	
17	Yulia	4.5	4.1	
18	Joanne	2.9	2.4	
19	Julia	4.2	3.9	
20	Yulia	4.1	4.0	
21	Lenka	4.9	4.9	
22	Anna	5.6	5.0	
23	Laetitia	5.6	4.9	
24	Alisa	4.2	3.8	
25	Marta	4.5	3.6	
26	Mojca	4.3	3.7	
27	Shirene	4.2	3.5	
28				

Figure 1   Scores from two judges.

Step 3.   Form the columns for formulas $d = (u - v)$ and d^2 in columns F and G.

Step 4.   Compute the Spearman r_s correlation coefficient:

$$r_s = 1 - \frac{6\Sigma d^2}{n(n^2 - 1)}$$

by first finding Σd^2 in cell G28. Did you get the sum equal to 188? Enter the formula for r_s in cell C31. Did you get $r_s = .907$?

	D	E	F	G
3				
4	u	v	d	d^2
5	1	1	0	0
6	3	3	0	0
7	7	6	1	1
8	8	10	-2	4
9	12	13	-1	1
10	9	10	-1	1
11	3	4	-1	1
12	2	2	0	0
13	12	9	3	9
14	11	14	-3	9
15	10	10	0	0
16	17	20	-3	9
17	15	15	0	0
18	23	23	0	0
19	19	17	2	4
20	22	16	6	36
21	14	6	8	64
22	3	5	-2	4
23	3	6	-3	9
24	19	18	1	1
25	15	20	-5	25
26	18	19	-1	1
27	19	22	-3	9
28				

Figure 2 Rankings of scores of two judges.

SALT LAKE 2002

Step 5. Find the critical value of the Spearman rho rank correlation coefficient in Table XIV. The critical value of r_s for a two-tailed test when $n = 23$ and $\alpha = .05$ is $\pm.418$.

Step 6. Form the conclusion. We have $r_s = .907$. So in spite of the difference in the absolute number scores, the rankings are in line with each other. We reject H_0 and conclude that the Russian and U.S. judges' rankings are in agreement.

Print out the results and save them as **ADV14.XLS** on your disk.

D STATISTICAL TABLES

TABLE I RANDOM NUMBERS

TABLE II FACTORIALS

TABLE III VALUES OF $\binom{n}{x}$ (COMBINATION)

TABLE IV TABLE OF BINOMIAL PROBABILITIES

TABLE V VALUES OF $e^{-\lambda}$

TABLE VI TABLE OF POISSON PROBABILITIES

TABLE VII STANDARD NORMAL DISTRIBUTION TABLE

TABLE VIII THE t DISTRIBUTION TABLE

TABLE IX CHI-SQUARE DISTRIBUTION TABLE

TABLE X THE F DISTRIBUTION TABLE

TABLE XI CRITICAL VALUES OF X FOR THE SIGN TEST

TABLE XII CRITICAL VALUES OF T FOR THE WILCOXON SIGNED-RANK TEST

TABLE XIII CRITICAL VALUES OF T FOR THE WILCOXON RANK SUM TEST

TABLE XIV CRITICAL VALUES FOR THE SPEARMAN RHO RANK CORRELATION COEFFICIENT TEST

TABLE XV CRITICAL VALUES FOR A TWO-TAILED RUNS TEST WITH $\alpha = .05$

TABLE I   RANDOM NUMBERS

D1

TABLE I   RANDOM NUMBERS

57728	16308	27337	53884	60742	61693	39887	81779	36354
63962	45765	75060	46767	28844	32354	91463	25057	91907
51041	22252	38447	71567	95103	11124	34960	35710	91098
84048	53578	67379	42605	59122	39415	82869	86971	64817
17736	34458	67227	97041	77846	20338	52372	34645	56563
82238	83763	45464	18493	98489	72138	38942	97661	95788
28853	61793	44664	69427	68144	71949	57192	25592	49835
22251	73098	68108	87626	76724	56495	87357	83065	95316
66236	46591	69225	29867	60815	51931	40507	52568	47097
50006	91666	86406	92778	51232	38761	21861	98596	42673
68328	12840	61206	64298	27378	61452	13349	27223	79637
83039	25015	95983	82835	67268	23355	44647	25542	10536
53158	82329	81756	81429	54366	97530	51447	11324	49939
46802	61720	97508	73339	29277	17964	35421	39880	38180
25162	78468	44303	14425	42587	37212	58866	39008	91938
65957	15171	22417	95571	90679	54774	43979	71017	49647
10876	36062	91375	90128	14906	81447	49158	14703	89517
35354	66633	62311	58185	67310	95474	21878	89101	38299
70822	69983	23726	97422	46713	20340	42807	10859	26897
64299	12987	60370	70165	43306	14417	79261	53891	72816
74007	61658	86698	31571	75098	11676	35867	39764	47504
70909	68300	55074	42093	55745	80364	18488	47981	18702
67898	98830	97705	10723	82370	45586	19013	60915	84961
59386	25440	92441	14265	26123	85453	57326	72790	55243
71469	49833	95737	84195	78444	32104	89917	88361	35344
34064	12993	23818	28197	33755	96438	84223	10400	36797
86492	25367	65712	81581	89579	31759	56108	24476	47696
86914	87565	20344	39027	98338	95171	75562	54283	35342
88418	58064	13624	32978	90704	56218	84064	69990	45354
87948	83451	96217	40534	40775	74376	43157	74856	13950
13049	85293	32747	17728	50495	34617	73707	33976	86177
86544	52703	74990	98288	61833	48803	75258	83382	79099
77295	70694	97326	35430	53881	94007	70471	66815	73042
54637	32831	59063	72353	87365	15322	33156	40331	93942
50938	12004	18585	23896	62559	44470	27701	66780	56157
80999	49724	76745	25232	74291	74184	91055	58903	18172
71303	36255	77310	95847	30282	77207	34439	47763	99697
79264	16901	55814	89734	30255	87209	31629	19328	42532
30235	69368	38685	32790	58980	42159	88577	18427	73504
59110	69783	93713	29151	34933	95745	72271	38684	15426
28094	19560	27259	82736	49700	37876	52322	69562	75837
40341	20666	26662	16422	76351	70520	36890	86559	89160
30117	68850	28319	44992	68110	47007	22243	72813	60934
62287	44957	47690	79484	69449	27981	34770	34228	81686
96976	77830	61746	67846	15584	28070	79200	12663	63273
82584	34789	33494	55533	25040	84187	14479	26286	10665
35728	87881	70271	13115	35745	99145	92717	74357	16716

TABLE I   (*Continued*)

88458	63625	59577	92037	99012	40836	58817	30757	37934
49789	20873	53858	91356	11387	75208	33643	88210	42440
49131	34078	45396	56884	81416	46292	36012	30806	65220
96256	82566	34796	88012	43066	35786	93715	15550	16690
43742	97487	68089	69887	23737	71136	21108	85204	60726
34527	87490	81183	95864	59430	19473	57978	39853	47877
58906	37390	88924	80917	58840	29907	99098	33761	50335
30438	12056	12104	61012	44674	49815	85298	94129	59542
79149	98261	48599	54336	71894	82889	51219	70291	60922
48703	25290	13835	35695	15440	52533	82849	25504	81623
96050	74505	18706	10572	66015	53509	48115	87578	86099
98859	48791	15048	73300	48045	33559	98939	98003	39453
32758	55597	11686	18385	31103	87621	39659	81413	68625
17238	73653	15557	79374	60965	75564	15872	34611	86497
79748	21687	94964	43348	26957	27085	81760	29099	23553
89199	75213	37815	99891	60990	37062	80331	54009	26812
87491	62544	51229	13028	81370	16309	28493	98555	24278
87338	17647	40018	48386	49992	44304	23330	38730	21601
48635	73063	37450	65403	65134	83119	16341	95766	83949
32197	94930	50586	88559	48025	97023	15372	18847	97168
83421	68819	69623	45088	54839	70855	86714	38202	98163
66167	84791	40631	33428	78200	41145	57816	86795	31646
26555	70521	69140	93495	38179	43253	78172	44239	60701
47786	37539	17452	88719	24423	59201	24979	51019	35458
47775	42564	15665	92454	98345	87963	81142	34356	41518
49414	83761	74309	82620	53677	34575	81871	76615	27653
75918	39825	60958	96584	26872	15379	84080	40371	35019
13440	85096	85668	24896	65261	83757	68388	29797	48376
39614	53926	97122	85279	15622	29329	59579	60250	73895
40067	48944	98882	39023	31677	41118	52818	29586	43848
65350	11148	63012	59418	54688	83692	95840	91627	84057
98902	62170	49281	29406	63143	43722	35838	98979	67024
10529	81048	29639	50740	93253	77339	80328	88580	30970
99123	48497	35247	33488	63781	19388	98534	27479	44269
65147	42913	50654	64220	13950	74293	53489	39014	86040
86886	15231	43834	88205	87159	30789	10959	81631	15575
17264	57846	52347	96649	69212	28053	62290	93328	98520
36110	87509	95913	66687	67149	81500	44107	27546	94868
58288	91109	66433	75388	80441	95720	64891	63049	68237
67834	18606	88840	39705	17329	15690	90382	35725	21362
67746	23016	87357	89427	98266	39452	58011	86665	70716
32196	36633	63350	73154	47699	15479	63905	81186	67181
33646	35175	41141	75793	58908	80681	88974	84611	46634
66973	98812	21094	45209	52503	51038	96306	75653	32482
74819	41419	91296	31736	99727	68791	93588	99566	98413
76495	37282	43051	17275	30370	76105	55926	98910	84767
39350	85262	59225	47343	63449	47004	94970	77067	16857
29657	66820	47420	37404	80296	94070	54249	60378	54670

TABLE I RANDOM NUMBERS

D3

TABLE I (*Continued*)

98059	31868	86468	80389	66521	23304	99582	48791	74154
78131	47852	62735	79575	48757	25712	22468	66035	29237
36071	71312	33098	38558	57088	26162	32752	24827	95562
23742	21969	82378	28923	19944	91024	63237	36022	76979
48418	36759	92342	93571	86923	26627	46138	86343	21083
74678	64188	39402	65189	30854	65086	43052	54042	79127
45693	86700	39667	53646	11663	86785	84727	83728	21758
68539	65113	68955	71627	38626	57160	63171	41707	51634
56807	61373	21941	71481	88523	72157	92088	41244	75735
29320	38387	89881	59789	50099	64811	31131	74334	65674
63399	73318	61578	28141	21655	65378	56261	69795	67096
33316	23627	55609	48463	92502	64287	99853	54497	85985
15403	81891	20190	72235	85636	16745	99483	43583	89137
76057	62447	71848	60035	76280	38017	26998	82690	16512
93885	29489	26222	77121	83244	74614	62527	36019	31265
66312	17182	96913	10736	52184	57082	58901	87749	34684
88960	52088	92432	18463	25562	20674	48988	41829	98681
38903	23457	87215	47089	38395	21929	94929	59489	79066
88281	90912	22965	16428	32289	99354	87068	55884	69518
98627	47123	32667	69196	70158	19828	42793	54593	53682
13183	37010	53184	53434	53631	96983	21201	10236	90134
69177	18284	72840	68433	88300	85396	10298	15680	13859
79501	74784	50483	37213	67077	59481	43976	35404	90683
52734	93419	47519	85203	27665	97179	47002	41258	39219
70256	58003	11565	67432	56505	18468	10293	46490	98191
47603	62142	37636	43374	19773	10538	27243	20800	50383
41507	43884	18253	81908	22803	84840	38968	70176	59393
56464	84865	65387	97484	95349	55548	54214	86814	20654
54523	35676	93542	20744	23942	22935	19794	53413	12979
86965	67669	42284	68532	24766	41411	97597	34998	70248
80707	35351	29958	12270	76227	31529	26105	94145	30469
57351	44045	67826	18191	75712	86420	36234	93377	20205
56860	89252	53362	82306	24114	91538	49114	67506	73489
87844	42168	83234	59134	17403	20418	65647	14702	59080
47157	27318	46686	59507	31598	16152	41184	95641	58835
63530	51286	46562	43739	51259	39836	72962	96998	89257
74595	35004	61728	28879	60412	81320	99003	20824	47086
26649	55512	58180	87954	91885	22660	31132	15752	11807
63596	44068	12648	91827	35448	59307	64466	68502	36292
66621	89136	46721	43322	78706	60249	90841	79917	18000
98128	99125	86432	87068	88376	65121	64402	55931	45748
85968	99264	64582	85694	29027	62883	53615	26692	73490
93011	71694	78514	63842	33754	84577	78698	38667	54673
36994	29619	36095	44782	85794	28498	25870	83655	97905
58857	32343	61392	65331	66939	51145	77060	85743	85278
97430	82854	28720	52153	37246	87152	95563	51769	79320
64642	69774	67582	95955	91433	95515	35211	39734	82631
56789	90056	28697	88922	39250	66008	55324	39129	63408

TABLE I   (*Continued*)

78707	36317	69939	63529	88044	66897	16846	67664	99997
20752	71605	38186	18221	79499	14660	86115	80339	34321
20794	82021	37432	97568	85812	97016	15655	40601	39475
84832	45347	60186	66673	62148	37683	79034	46572	69243
39960	63046	99657	28301	19953	84261	30215	52274	91374
51835	19676	40685	45677	57150	73208	59526	76240	24209
88213	83367	21935	72494	87548	43000	72275	81974	54718
39746	38989	28721	38803	89668	57496	97127	59364	83335
95915	51291	14163	40972	33163	85169	66522	72010	53429
43937	78760	47672	69700	87058	19072	89435	13390	72315
28633	29330	20463	89033	16968	62815	65802	53006	70674
27415	32278	61924	61670	38880	13911	85037	93738	94913
10157	74513	43054	44601	35689	54559	91660	60035	83733
18041	40798	39274	72760	83644	48960	52193	95674	22516
22679	12792	60046	80515	12962	57351	36431	52277	50567
81468	99534	30455	17430	92600	85813	90223	15335	97102
50636	87932	25489	29395	87683	84579	10396	38276	33729
60635	13409	81824	77150	51472	65915	62520	33839	52209
68336	29892	94343	37822	55260	97321	20488	50172	45199
79273	96036	89979	78654	38959	36250	91126	90337	91381
71942	89335	75664	75278	40445	12818	24033	11809	44129
87842	60665	73523	55824	61257	22080	74425	54851	84786

TABLE II   FACTORIALS

D5

TABLE II   FACTORIALS

n	$n!$
0	1
1	1
2	2
3	6
4	24
5	120
6	720
7	5,040
8	40,320
9	362,880
10	3,628,800
11	39,916,800
12	479,001,600
13	6,227,020,800
14	87,178,291,200
15	1,307,674,368,000
16	20,922,789,888,000
17	355,687,428,096,000
18	6,402,373,705,728,000
19	121,645,100,408,832,000
20	2,432,902,008,176,640,000
21	51,090,942,171,709,440,000
22	1,124,000,727,777,607,680,000
23	25,852,016,738,884,976,640,000
24	620,448,401,733,239,439,360,000
25	15,511,210,043,330,985,984,000,000

TABLE III VALUES OF $\binom{n}{x}$ (COMBINATION)

n	0	1	2	3	4	5	6	7	8	9	10	11	12	13	14	15	16	17	18	19	20
1	1	1																			
2	1	2	1																		
3	1	3	3	1																	
4	1	4	6	4	1																
5	1	5	10	10	5	1															
6	1	6	15	20	15	6	1														
7	1	7	21	35	35	21	7	1													
8	1	8	28	56	70	56	28	8	1												
9	1	9	36	84	126	126	84	36	9	1											
10	1	10	45	120	210	252	210	120	45	10	1										
11	1	11	55	165	330	462	462	330	165	55	11	1									
12	1	12	66	220	495	792	924	792	495	220	66	12	1								
13	1	13	78	286	715	1,287	1,716	1,716	1,287	715	286	78	13	1							
14	1	14	91	364	1,001	2,002	3,003	3,432	3,003	2,002	1,001	364	91	14	1						
15	1	15	105	455	1,365	3,003	5,005	6,435	6,435	5,005	3,003	1,365	455	105	15	1					
16	1	16	120	560	1,820	4,368	8,008	11,440	12,870	11,440	8,008	4,368	1,820	560	120	16	1				
17	1	17	136	680	2,380	6,188	12,376	19,448	24,310	24,310	19,448	12,376	6,188	2,380	680	136	17	1			
18	1	18	153	816	3,060	8,568	18,564	31,824	43,758	48,620	43,758	31,824	18,564	8,568	3,060	816	153	18	1		
19	1	19	171	969	3,876	11,628	27,132	50,388	75,582	92,378	92,378	75,582	50,388	27,132	11,628	3,876	969	171	19	1	
20	1	20	190	1,140	4,845	15,504	38,760	77,520	125,970	167,960	184,756	167,960	125,970	77,520	38,760	15,504	4,845	1,140	190	20	1
21	1	21	210	1,330	5,985	20,349	54,264	116,280	203,490	293,930	352,716	352,716	293,930	203,490	116,280	54,264	20,349	5,985	1,330	210	21
22	1	22	231	1,540	7,315	26,334	74,613	170,544	319,770	497,420	646,646	705,432	646,646	497,420	319,770	170,544	74,613	26,334	7,315	1,540	231
23	1	23	253	1,771	8,855	33,649	100,947	245,157	490,314	817,190	1,144,066	1,352,078	1,352,078	1,144,066	817,190	490,314	245,157	100,947	33,649	8,855	1,771
24	1	24	276	2,024	10,626	42,504	134,596	346,104	735,471	1,307,504	1,961,256	2,496,144	2,704,156	2,496,144	1,961,256	1,307,504	735,471	346,104	134,596	42,504	10,626
25	1	25	300	2,300	12,650	53,130	177,100	480,700	1,081,575	2,042,975	3,268,760	4,457,400	5,200,300	5,200,300	4,457,400	3,268,760	2,042,975	1,081,575	480,700	177,100	53,130

TABLE IV   TABLE OF BINOMIAL PROBABILITIES

D7

TABLE IV   TABLE OF BINOMIAL PROBABILITIES

							p					
n	*x*	.05	.10	.20	.30	.40	.50	.60	.70	.80	.90	.95
1	0	.9500	.9000	.8000	.7000	.6000	.5000	.4000	.3000	.2000	.1000	.0500
	1	.0500	.1000	.2000	.3000	.4000	.5000	.6000	.7000	.8000	.9000	.9500
2	0	.9025	.8100	.6400	.4900	.3600	.2500	.1600	.0900	.0400	.0100	.0025
	1	.0950	.1800	.3200	.4200	.4800	.5000	.4800	.4200	.3200	.1800	.0950
	2	.0025	.0100	.0400	.0900	.1600	.2500	.3600	.4900	.6400	.8100	.9025
3	0	.8574	.7290	.5120	.3430	.2160	.1250	.0640	.0270	.0080	.0010	.0001
	1	.1354	.2430	.3840	.4410	.4320	.3750	.2880	.1890	.0960	.0270	.0071
	2	.0071	.0270	.0960	.1890	.2880	.3750	.4320	.4410	.3840	.2430	.1354
	3	.0001	.0010	.0080	.0270	.0640	.1250	.2160	.3430	.5120	.7290	.8574
4	0	.8145	.6561	.4096	.2401	.1296	.0625	.0256	.0081	.0016	.0001	.0000
	1	.1715	.2916	.4096	.4116	.3456	.2500	.1536	.0756	.0256	.0036	.0005
	2	.0135	.0486	.1536	.2646	.3456	.3750	.3456	.2646	.1536	.0486	.0135
	3	.0005	.0036	.0256	.0756	.1536	.2500	.3456	.4116	.4096	.2916	.1715
	4	.0000	.0001	.0016	.0081	.0256	.0625	.1296	.2401	.4096	.6561	.8145
5	0	.7738	.5905	.3277	.1681	.0778	.0312	.0102	.0024	.0003	.0000	.0000
	1	.2036	.3280	.4096	.3602	.2592	.1562	.0768	.0284	.0064	.0005	.0000
	2	.0214	.0729	.2048	.3087	.3456	.3125	.2304	.1323	.0512	.0081	.0011
	3	.0011	.0081	.0512	.1323	.2304	.3125	.3456	.3087	.2048	.0729	.0214
	4	.0000	.0004	.0064	.0283	.0768	.1562	.2592	.3601	.4096	.3281	.2036
	5	.0000	.0000	.0003	.0024	.0102	.0312	.0778	.1681	.3277	.5905	.7738
6	0	.7351	.5314	.2621	.1176	.0467	.0156	.0041	.0007	.0001	.0000	.0000
	1	.2321	.3543	.3932	.3025	.1866	.0937	.0369	.0102	.0015	.0001	.0000
	2	.0305	.0984	.2458	.3241	.3110	.2344	.1382	.0595	.0154	.0012	.0001
	3	.0021	.0146	.0819	.1852	.2765	.3125	.2765	.1852	.0819	.0146	.0021
	4	.0001	.0012	.0154	.0595	.1382	.2344	.3110	.3241	.2458	.0984	.0305
	5	.0000	.0001	.0015	.0102	.0369	.0937	.1866	.3025	.3932	.3543	.2321
	6	.0000	.0000	.0001	.0007	.0041	.0156	.0467	.1176	.2621	.5314	.7351
7	0	.6983	.4783	.2097	.0824	.0280	.0078	.0016	.0002	.0000	.0000	.0000
	1	.2573	.3720	.3670	.2471	.1306	.0547	.0172	.0036	.0004	.0000	.0000
	2	.0406	.1240	.2753	.3177	.2613	.1641	.0774	.0250	.0043	.0002	.0000
	3	.0036	.0230	.1147	.2269	.2903	.2734	.1935	.0972	.0287	.0026	.0002
	4	.0002	.0026	.0287	.0972	.1935	.2734	.2903	.2269	.1147	.0230	.0036
	5	.0000	.0002	.0043	.0250	.0774	.1641	.2613	.3177	.2753	.1240	.0406
	6	.0000	.0000	.0004	.0036	.0172	.0547	.1306	.2471	.3670	.3720	.2573
	7	.0000	.0000	.0000	.0002	.0016	.0078	.0280	.0824	.2097	.4783	.6983
8	0	.6634	.4305	.1678	.0576	.0168	.0039	.0007	.0001	.0000	.0000	.0000
	1	.2793	.3826	.3355	.1977	.0896	.0312	.0079	.0012	.0001	.0000	.0000
	2	.0515	.1488	.2936	.2965	.2090	.1094	.0413	.0100	.0011	.0000	.0000
	3	.0054	.0331	.1468	.2541	.2787	.2187	.1239	.0467	.0092	.0004	.0000
	4	.0004	.0046	.0459	.1361	.2322	.2734	.2322	.1361	.0459	.0046	.0004

TABLE IV   (*Continued*)

n	x	.05	.10	.20	.30	.40	.50	.60	.70	.80	.90	.95
	5	.0000	.0004	.0092	.0467	.1239	.2187	.2787	.2541	.1468	.0331	.0054
	6	.0000	.0000	.0011	.0100	.0413	.1094	.2090	.2965	.2936	.1488	.0515
	7	.0000	.0000	.0001	.0012	.0079	.0312	.0896	.1977	.3355	.3826	.2793
	8	.0000	.0000	.0000	.0001	.0007	.0039	.0168	.0576	.1678	.4305	.6634
9	0	.6302	.3874	.1342	.0404	.0101	.0020	.0003	.0000	.0000	.0000	.0000
	1	.2985	.3874	.3020	.1556	.0605	.0176	.0035	.0004	.0000	.0000	.0000
	2	.0629	.1722	.3020	.2668	.1612	.0703	.0212	.0039	.0003	.0000	.0000
	3	.0077	.0446	.1762	.2668	.2508	.1641	.0743	.0210	.0028	.0001	.0000
	4	.0006	.0074	.0661	.1715	.2508	.2461	.1672	.0735	.0165	.0008	.0000
	5	.0000	.0008	.0165	.0735	.1672	.2461	.2508	.1715	.0661	.0074	.0006
	6	.0000	.0001	.0028	.0210	.0743	.1641	.2508	.2668	.1762	.0446	.0077
	7	.0000	.0000	.0003	.0039	.0212	.0703	.1612	.2668	.3020	.1722	.0629
	8	.0000	.0000	.0000	.0004	.0035	.0176	.0605	.1556	.3020	.3874	.2985
	9	.0000	.0000	.0000	.0000	.0003	.0020	.0101	.0404	.1342	.3874	.6302
10	0	.5987	.3487	.1074	.0282	.0060	.0010	.0001	.0000	.0000	.0000	.0000
	1	.3151	.3874	.2684	.1211	.0403	.0098	.0016	.0001	.0000	.0000	.0000
	2	.0746	.1937	.3020	.2335	.1209	.0439	.0106	.0014	.0001	.0000	.0000
	3	.0105	.0574	.2013	.2668	.2150	.1172	.0425	.0090	.0008	.0000	.0000
	4	.0010	.0112	.0881	.2001	.2508	.2051	.1115	.0368	.0055	.0001	.0000
	5	.0001	.0015	.0264	.1029	.2007	.2461	.2007	.1029	.0264	.0015	.0001
	6	.0000	.0001	.0055	.0368	.1115	.2051	.2508	.2001	.0881	.0112	.0010
	7	.0000	.0000	.0008	.0090	.0425	.1172	.2150	.2668	.2013	.0574	.0105
	8	.0000	.0000	.0001	.0014	.0106	.0439	.1209	.2335	.3020	.1937	.0746
	9	.0000	.0000	.0000	.0001	.0016	.0098	.0403	.1211	.2684	.3874	.3151
	10	.0000	.0000	.0000	.0000	.0001	.0010	.0060	.0282	.1074	.3487	.5987
11	0	.5688	.3138	.0859	.0198	.0036	.0005	.0000	.0000	.0000	.0000	.0000
	1	.3293	.3835	.2362	.0932	.0266	.0054	.0007	.0000	.0000	.0000	.0000
	2	.0867	.2131	.2953	.1998	.0887	.0269	.0052	.0005	.0000	.0000	.0000
	3	.0137	.0710	.2215	.2568	.1774	.0806	.0234	.0037	.0002	.0000	.0000
	4	.0014	.0158	.1107	.2201	.2365	.1611	.0701	.0173	.0017	.0000	.0000
	5	.0001	.0025	.0388	.1321	.2207	.2256	.1471	.0566	.0097	.0003	.0000
	6	.0000	.0003	.0097	.0566	.1471	.2256	.2207	.1321	.0388	.0025	.0001
	7	.0000	.0000	.0017	.0173	.0701	.1611	.2365	.2201	.1107	.0158	.0014
	8	.0000	.0000	.0002	.0037	.0234	.0806	.1774	.2568	.2215	.0710	.0137
	9	.0000	.0000	.0000	.0005	.0052	.0269	.0887	.1998	.2953	.2131	.0867
	10	.0000	.0000	.0000	.0000	.0007	.0054	.0266	.0932	.2362	.3835	.3293
	11	.0000	.0000	.0000	.0000	.0000	.0005	.0036	.0198	.0859	.3138	.5688
12	0	.5404	.2824	.0687	.0138	.0022	.0002	.0000	.0000	.0000	.0000	.0000
	1	.3413	.3766	.2062	.0712	.0174	.0029	.0003	.0000	.0000	.0000	.0000
	2	.0988	.2301	.2835	.1678	.0639	.0161	.0025	.0002	.0000	.0000	.0000
	3	.0173	.0852	.2362	.2397	.1419	.0537	.0125	.0015	.0001	.0000	.0000
	4	.0021	.0213	.1329	.2311	.2128	.1208	.0420	.0078	.0005	.0000	.0000
	5	.0002	.0038	.0532	.1585	.2270	.1934	.1009	.0291	.0033	.0000	.0000
	6	.0000	.0005	.0155	.0792	.1766	.2256	.1766	.0792	.0155	.0005	.0000
	7	.0000	.0000	.0033	.0291	.1009	.1934	.2270	.1585	.0532	.0038	.0002
	8	.0000	.0000	.0005	.0078	.0420	.1208	.2128	.2311	.1329	.0213	.0021
	9	.0000	.0000	.0001	.0015	.0125	.0537	.1419	.2397	.2362	.0852	.0173

TABLE IV   TABLE OF BINOMIAL PROBABILITIES

D7

TABLE IV   TABLE OF BINOMIAL PROBABILITIES

							p					
n	x	.05	.10	.20	.30	.40	.50	.60	.70	.80	.90	.95
1	0	.9500	.9000	.8000	.7000	.6000	.5000	.4000	.3000	.2000	.1000	.0500
	1	.0500	.1000	.2000	.3000	.4000	.5000	.6000	.7000	.8000	.9000	.9500
2	0	.9025	.8100	.6400	.4900	.3600	.2500	.1600	.0900	.0400	.0100	.0025
	1	.0950	.1800	.3200	.4200	.4800	.5000	.4800	.4200	.3200	.1800	.0950
	2	.0025	.0100	.0400	.0900	.1600	.2500	.3600	.4900	.6400	.8100	.9025
3	0	.8574	.7290	.5120	.3430	.2160	.1250	.0640	.0270	.0080	.0010	.0001
	1	.1354	.2430	.3840	.4410	.4320	.3750	.2880	.1890	.0960	.0270	.0071
	2	.0071	.0270	.0960	.1890	.2880	.3750	.4320	.4410	.3840	.2430	.1354
	3	.0001	.0010	.0080	.0270	.0640	.1250	.2160	.3430	.5120	.7290	.8574
4	0	.8145	.6561	.4096	.2401	.1296	.0625	.0256	.0081	.0016	.0001	.0000
	1	.1715	.2916	.4096	.4116	.3456	.2500	.1536	.0756	.0256	.0036	.0005
	2	.0135	.0486	.1536	.2646	.3456	.3750	.3456	.2646	.1536	.0486	.0135
	3	.0005	.0036	.0256	.0756	.1536	.2500	.3456	.4116	.4096	.2916	.1715
	4	.0000	.0001	.0016	.0081	.0256	.0625	.1296	.2401	.4096	.6561	.8145
5	0	.7738	.5905	.3277	.1681	.0778	.0312	.0102	.0024	.0003	.0000	.0000
	1	.2036	.3280	.4096	.3602	.2592	.1562	.0768	.0284	.0064	.0005	.0000
	2	.0214	.0729	.2048	.3087	.3456	.3125	.2304	.1323	.0512	.0081	.0011
	3	.0011	.0081	.0512	.1323	.2304	.3125	.3456	.3087	.2048	.0729	.0214
	4	.0000	.0004	.0064	.0283	.0768	.1562	.2592	.3601	.4096	.3281	.2036
	5	.0000	.0000	.0003	.0024	.0102	.0312	.0778	.1681	.3277	.5905	.7738
6	0	.7351	.5314	.2621	.1176	.0467	.0156	.0041	.0007	.0001	.0000	.0000
	1	.2321	.3543	.3932	.3025	.1866	.0937	.0369	.0102	.0015	.0001	.0000
	2	.0305	.0984	.2458	.3241	.3110	.2344	.1382	.0595	.0154	.0012	.0001
	3	.0021	.0146	.0819	.1852	.2765	.3125	.2765	.1852	.0819	.0146	.0021
	4	.0001	.0012	.0154	.0595	.1382	.2344	.3110	.3241	.2458	.0984	.0305
	5	.0000	.0001	.0015	.0102	.0369	.0937	.1866	.3025	.3932	.3543	.2321
	6	.0000	.0000	.0001	.0007	.0041	.0156	.0467	.1176	.2621	.5314	.7351
7	0	.6983	.4783	.2097	.0824	.0280	.0078	.0016	.0002	.0000	.0000	.0000
	1	.2573	.3720	.3670	.2471	.1306	.0547	.0172	.0036	.0004	.0000	.0000
	2	.0406	.1240	.2753	.3177	.2613	.1641	.0774	.0250	.0043	.0002	.0000
	3	.0036	.0230	.1147	.2269	.2903	.2734	.1935	.0972	.0287	.0026	.0002
	4	.0002	.0026	.0287	.0972	.1935	.2734	.2903	.2269	.1147	.0230	.0036
	5	.0000	.0002	.0043	.0250	.0774	.1641	.2613	.3177	.2753	.1240	.0406
	6	.0000	.0000	.0004	.0036	.0172	.0547	.1306	.2471	.3670	.3720	.2573
	7	.0000	.0000	.0000	.0002	.0016	.0078	.0280	.0824	.2097	.4783	.6983
8	0	.6634	.4305	.1678	.0576	.0168	.0039	.0007	.0001	.0000	.0000	.0000
	1	.2793	.3826	.3355	.1977	.0896	.0312	.0079	.0012	.0001	.0000	.0000
	2	.0515	.1488	.2936	.2965	.2090	.1094	.0413	.0100	.0011	.0000	.0000
	3	.0054	.0331	.1468	.2541	.2787	.2187	.1239	.0467	.0092	.0004	.0000
	4	.0004	.0046	.0459	.1361	.2322	.2734	.2322	.1361	.0459	.0046	.0004

TABLE IV   (*Continued*)

						p						
n	*x*	**.05**	**.10**	**.20**	**.30**	**.40**	**.50**	**.60**	**.70**	**.80**	**.90**	**.95**
	5	.0000	.0004	.0092	.0467	.1239	.2187	.2787	.2541	.1468	.0331	.0054
	6	.0000	.0000	.0011	.0100	.0413	.1094	.2090	.2965	.2936	.1488	.0515
	7	.0000	.0000	.0001	.0012	.0079	.0312	.0896	.1977	.3355	.3826	.2793
	8	.0000	.0000	.0000	.0001	.0007	.0039	.0168	.0576	.1678	.4305	.6634
9	0	.6302	.3874	.1342	.0404	.0101	.0020	.0003	.0000	.0000	.0000	.0000
	1	.2985	.3874	.3020	.1556	.0605	.0176	.0035	.0004	.0000	.0000	.0000
	2	.0629	.1722	.3020	.2668	.1612	.0703	.0212	.0039	.0003	.0000	.0000
	3	.0077	.0446	.1762	.2668	.2508	.1641	.0743	.0210	.0028	.0001	.0000
	4	.0006	.0074	.0661	.1715	.2508	.2461	.1672	.0735	.0165	.0008	.0000
	5	.0000	.0008	.0165	.0735	.1672	.2461	.2508	.1715	.0661	.0074	.0006
	6	.0000	.0001	.0028	.0210	.0743	.1641	.2508	.2668	.1762	.0446	.0077
	7	.0000	.0000	.0003	.0039	.0212	.0703	.1612	.2668	.3020	.1722	.0629
	8	.0000	.0000	.0000	.0004	.0035	.0176	.0605	.1556	.3020	.3874	.2985
	9	.0000	.0000	.0000	.0000	.0003	.0020	.0101	.0404	.1342	.3874	.6302
10	0	.5987	.3487	.1074	.0282	.0060	.0010	.0001	.0000	.0000	.0000	.0000
	1	.3151	.3874	.2684	.1211	.0403	.0098	.0016	.0001	.0000	.0000	.0000
	2	.0746	.1937	.3020	.2335	.1209	.0439	.0106	.0014	.0001	.0000	.0000
	3	.0105	.0574	.2013	.2668	.2150	.1172	.0425	.0090	.0008	.0000	.0000
	4	.0010	.0112	.0881	.2001	.2508	.2051	.1115	.0368	.0055	.0001	.0000
	5	.0001	.0015	.0264	.1029	.2007	.2461	.2007	.1029	.0264	.0015	.0001
	6	.0000	.0001	.0055	.0368	.1115	.2051	.2508	.2001	.0881	.0112	.0010
	7	.0000	.0000	.0008	.0090	.0425	.1172	.2150	.2668	.2013	.0574	.0105
	8	.0000	.0000	.0001	.0014	.0106	.0439	.1209	.2335	.3020	.1937	.0746
	9	.0000	.0000	.0000	.0001	.0016	.0098	.0403	.1211	.2684	.3874	.3151
	10	.0000	.0000	.0000	.0000	.0001	.0010	.0060	.0282	.1074	.3487	.5987
11	0	.5688	.3138	.0859	.0198	.0036	.0005	.0000	.0000	.0000	.0000	.0000
	1	.3293	.3835	.2362	.0932	.0266	.0054	.0007	.0000	.0000	.0000	.0000
	2	.0867	.2131	.2953	.1998	.0887	.0269	.0052	.0005	.0000	.0000	.0000
	3	.0137	.0710	.2215	.2568	.1774	.0806	.0234	.0037	.0002	.0000	.0000
	4	.0014	.0158	.1107	.2201	.2365	.1611	.0701	.0173	.0017	.0000	.0000
	5	.0001	.0025	.0388	.1321	.2207	.2256	.1471	.0566	.0097	.0003	.0000
	6	.0000	.0003	.0097	.0566	.1471	.2256	.2207	.1321	.0388	.0025	.0001
	7	.0000	.0000	.0017	.0173	.0701	.1611	.2365	.2201	.1107	.0158	.0014
	8	.0000	.0000	.0002	.0037	.0234	.0806	.1774	.2568	.2215	.0710	.0137
	9	.0000	.0000	.0000	.0005	.0052	.0269	.0887	.1998	.2953	.2131	.0867
	10	.0000	.0000	.0000	.0000	.0007	.0054	.0266	.0932	.2362	.3835	.3293
	11	.0000	.0000	.0000	.0000	.0000	.0005	.0036	.0198	.0859	.3138	.5688
12	0	.5404	.2824	.0687	.0138	.0022	.0002	.0000	.0000	.0000	.0000	.0000
	1	.3413	.3766	.2062	.0712	.0174	.0029	.0003	.0000	.0000	.0000	.0000
	2	.0988	.2301	.2835	.1678	.0639	.0161	.0025	.0002	.0000	.0000	.0000
	3	.0173	.0852	.2362	.2397	.1419	.0537	.0125	.0015	.0001	.0000	.0000
	4	.0021	.0213	.1329	.2311	.2128	.1208	.0420	.0078	.0005	.0000	.0000
	5	.0002	.0038	.0532	.1585	.2270	.1934	.1009	.0291	.0033	.0000	.0000
	6	.0000	.0005	.0155	.0792	.1766	.2256	.1766	.0792	.0155	.0005	.0000
	7	.0000	.0000	.0033	.0291	.1009	.1934	.2270	.1585	.0532	.0038	.0002
	8	.0000	.0000	.0005	.0078	.0420	.1208	.2128	.2311	.1329	.0213	.0021
	9	.0000	.0000	.0001	.0015	.0125	.0537	.1419	.2397	.2362	.0852	.0173

TABLE IV   TABLE OF BINOMIAL PROBABILITIES

D7

TABLE IV   TABLE OF BINOMIAL PROBABILITIES

							p					
n	x	.05	.10	.20	.30	.40	.50	.60	.70	.80	.90	.95
1	0	.9500	.9000	.8000	.7000	.6000	.5000	.4000	.3000	.2000	.1000	.0500
	1	.0500	.1000	.2000	.3000	.4000	.5000	.6000	.7000	.8000	.9000	.9500
2	0	.9025	.8100	.6400	.4900	.3600	.2500	.1600	.0900	.0400	.0100	.0025
	1	.0950	.1800	.3200	.4200	.4800	.5000	.4800	.4200	.3200	.1800	.0950
	2	.0025	.0100	.0400	.0900	.1600	.2500	.3600	.4900	.6400	.8100	.9025
3	0	.8574	.7290	.5120	.3430	.2160	.1250	.0640	.0270	.0080	.0010	.0001
	1	.1354	.2430	.3840	.4410	.4320	.3750	.2880	.1890	.0960	.0270	.0071
	2	.0071	.0270	.0960	.1890	.2880	.3750	.4320	.4410	.3840	.2430	.1354
	3	.0001	.0010	.0080	.0270	.0640	.1250	.2160	.3430	.5120	.7290	.8574
4	0	.8145	.6561	.4096	.2401	.1296	.0625	.0256	.0081	.0016	.0001	.0000
	1	.1715	.2916	.4096	.4116	.3456	.2500	.1536	.0756	.0256	.0036	.0005
	2	.0135	.0486	.1536	.2646	.3456	.3750	.3456	.2646	.1536	.0486	.0135
	3	.0005	.0036	.0256	.0756	.1536	.2500	.3456	.4116	.4096	.2916	.1715
	4	.0000	.0001	.0016	.0081	.0256	.0625	.1296	.2401	.4096	.6561	.8145
5	0	.7738	.5905	.3277	.1681	.0778	.0312	.0102	.0024	.0003	.0000	.0000
	1	.2036	.3280	.4096	.3602	.2592	.1562	.0768	.0284	.0064	.0005	.0000
	2	.0214	.0729	.2048	.3087	.3456	.3125	.2304	.1323	.0512	.0081	.0011
	3	.0011	.0081	.0512	.1323	.2304	.3125	.3456	.3087	.2048	.0729	.0214
	4	.0000	.0004	.0064	.0283	.0768	.1562	.2592	.3601	.4096	.3281	.2036
	5	.0000	.0000	.0003	.0024	.0102	.0312	.0778	.1681	.3277	.5905	.7738
6	0	.7351	.5314	.2621	.1176	.0467	.0156	.0041	.0007	.0001	.0000	.0000
	1	.2321	.3543	.3932	.3025	.1866	.0937	.0369	.0102	.0015	.0001	.0000
	2	.0305	.0984	.2458	.3241	.3110	.2344	.1382	.0595	.0154	.0012	.0001
	3	.0021	.0146	.0819	.1852	.2765	.3125	.2765	.1852	.0819	.0146	.0021
	4	.0001	.0012	.0154	.0595	.1382	.2344	.3110	.3241	.2458	.0984	.0305
	5	.0000	.0001	.0015	.0102	.0369	.0937	.1866	.3025	.3932	.3543	.2321
	6	.0000	.0000	.0001	.0007	.0041	.0156	.0467	.1176	.2621	.5314	.7351
7	0	.6983	.4783	.2097	.0824	.0280	.0078	.0016	.0002	.0000	.0000	.0000
	1	.2573	.3720	.3670	.2471	.1306	.0547	.0172	.0036	.0004	.0000	.0000
	2	.0406	.1240	.2753	.3177	.2613	.1641	.0774	.0250	.0043	.0002	.0000
	3	.0036	.0230	.1147	.2269	.2903	.2734	.1935	.0972	.0287	.0026	.0002
	4	.0002	.0026	.0287	.0972	.1935	.2734	.2903	.2269	.1147	.0230	.0036
	5	.0000	.0002	.0043	.0250	.0774	.1641	.2613	.3177	.2753	.1240	.0406
	6	.0000	.0000	.0004	.0036	.0172	.0547	.1306	.2471	.3670	.3720	.2573
	7	.0000	.0000	.0000	.0002	.0016	.0078	.0280	.0824	.2097	.4783	.6983
8	0	.6634	.4305	.1678	.0576	.0168	.0039	.0007	.0001	.0000	.0000	.0000
	1	.2793	.3826	.3355	.1977	.0896	.0312	.0079	.0012	.0001	.0000	.0000
	2	.0515	.1488	.2936	.2965	.2090	.1094	.0413	.0100	.0011	.0000	.0000
	3	.0054	.0331	.1468	.2541	.2787	.2187	.1239	.0467	.0092	.0004	.0000
	4	.0004	.0046	.0459	.1361	.2322	.2734	.2322	.1361	.0459	.0046	.0004

TABLE IV   (*Continued*)

n	x	.05	.10	.20	.30	.40	.50	.60	.70	.80	.90	.95
	5	.0000	.0004	.0092	.0467	.1239	.2187	.2787	.2541	.1468	.0331	.0054
	6	.0000	.0000	.0011	.0100	.0413	.1094	.2090	.2965	.2936	.1488	.0515
	7	.0000	.0000	.0001	.0012	.0079	.0312	.0896	.1977	.3355	.3826	.2793
	8	.0000	.0000	.0000	.0001	.0007	.0039	.0168	.0576	.1678	.4305	.6634
9	0	.6302	.3874	.1342	.0404	.0101	.0020	.0003	.0000	.0000	.0000	.0000
	1	.2985	.3874	.3020	.1556	.0605	.0176	.0035	.0004	.0000	.0000	.0000
	2	.0629	.1722	.3020	.2668	.1612	.0703	.0212	.0039	.0003	.0000	.0000
	3	.0077	.0446	.1762	.2668	.2508	.1641	.0743	.0210	.0028	.0001	.0000
	4	.0006	.0074	.0661	.1715	.2508	.2461	.1672	.0735	.0165	.0008	.0000
	5	.0000	.0008	.0165	.0735	.1672	.2461	.2508	.1715	.0661	.0074	.0006
	6	.0000	.0001	.0028	.0210	.0743	.1641	.2508	.2668	.1762	.0446	.0077
	7	.0000	.0000	.0003	.0039	.0212	.0703	.1612	.2668	.3020	.1722	.0629
	8	.0000	.0000	.0000	.0004	.0035	.0176	.0605	.1556	.3020	.3874	.2985
	9	.0000	.0000	.0000	.0000	.0003	.0020	.0101	.0404	.1342	.3874	.6302
10	0	.5987	.3487	.1074	.0282	.0060	.0010	.0001	.0000	.0000	.0000	.0000
	1	.3151	.3874	.2684	.1211	.0403	.0098	.0016	.0001	.0000	.0000	.0000
	2	.0746	.1937	.3020	.2335	.1209	.0439	.0106	.0014	.0001	.0000	.0000
	3	.0105	.0574	.2013	.2668	.2150	.1172	.0425	.0090	.0008	.0000	.0000
	4	.0010	.0112	.0881	.2001	.2508	.2051	.1115	.0368	.0055	.0001	.0000
	5	.0001	.0015	.0264	.1029	.2007	.2461	.2007	.1029	.0264	.0015	.0001
	6	.0000	.0001	.0055	.0368	.1115	.2051	.2508	.2001	.0881	.0112	.0010
	7	.0000	.0000	.0008	.0090	.0425	.1172	.2150	.2668	.2013	.0574	.0105
	8	.0000	.0000	.0001	.0014	.0106	.0439	.1209	.2335	.3020	.1937	.0746
	9	.0000	.0000	.0000	.0001	.0016	.0098	.0403	.1211	.2684	.3874	.3151
	10	.0000	.0000	.0000	.0000	.0001	.0010	.0060	.0282	.1074	.3487	.5987
11	0	.5688	.3138	.0859	.0198	.0036	.0005	.0000	.0000	.0000	.0000	.0000
	1	.3293	.3835	.2362	.0932	.0266	.0054	.0007	.0000	.0000	.0000	.0000
	2	.0867	.2131	.2953	.1998	.0887	.0269	.0052	.0005	.0000	.0000	.0000
	3	.0137	.0710	.2215	.2568	.1774	.0806	.0234	.0037	.0002	.0000	.0000
	4	.0014	.0158	.1107	.2201	.2365	.1611	.0701	.0173	.0017	.0000	.0000
	5	.0001	.0025	.0388	.1321	.2207	.2256	.1471	.0566	.0097	.0003	.0000
	6	.0000	.0003	.0097	.0566	.1471	.2256	.2207	.1321	.0388	.0025	.0001
	7	.0000	.0000	.0017	.0173	.0701	.1611	.2365	.2201	.1107	.0158	.0014
	8	.0000	.0000	.0002	.0037	.0234	.0806	.1774	.2568	.2215	.0710	.0137
	9	.0000	.0000	.0000	.0005	.0052	.0269	.0887	.1998	.2953	.2131	.0867
	10	.0000	.0000	.0000	.0000	.0007	.0054	.0266	.0932	.2362	.3835	.3293
	11	.0000	.0000	.0000	.0000	.0000	.0005	.0036	.0198	.0859	.3138	.5688
12	0	.5404	.2824	.0687	.0138	.0022	.0002	.0000	.0000	.0000	.0000	.0000
	1	.3413	.3766	.2062	.0712	.0174	.0029	.0003	.0000	.0000	.0000	.0000
	2	.0988	.2301	.2835	.1678	.0639	.0161	.0025	.0002	.0000	.0000	.0000
	3	.0173	.0852	.2362	.2397	.1419	.0537	.0125	.0015	.0001	.0000	.0000
	4	.0021	.0213	.1329	.2311	.2128	.1208	.0420	.0078	.0005	.0000	.0000
	5	.0002	.0038	.0532	.1585	.2270	.1934	.1009	.0291	.0033	.0000	.0000
	6	.0000	.0005	.0155	.0792	.1766	.2256	.1766	.0792	.0155	.0005	.0000
	7	.0000	.0000	.0033	.0291	.1009	.1934	.2270	.1585	.0532	.0038	.0002
	8	.0000	.0000	.0005	.0078	.0420	.1208	.2128	.2311	.1329	.0213	.0021
	9	.0000	.0000	.0001	.0015	.0125	.0537	.1419	.2397	.2362	.0852	.0173

TABLE IV   TABLE OF BINOMIAL PROBABILITIES   **D9**

TABLE IV   (*Continued*)

n	x	.05	.10	.20	.30	.40	.50	.60	.70	.80	.90	.95
	10	.0000	.0000	.0000	.0002	.0025	.0161	.0639	.1678	.2835	.2301	.0988
	11	.0000	.0000	.0000	.0000	.0003	.0029	.0174	.0712	.2062	.3766	.3413
	12	.0000	.0000	.0000	.0000	.0000	.0002	.0022	.0138	.0687	.2824	.5404
13	0	.5133	.2542	.0550	.0097	.0013	.0001	.0000	.0000	.0000	.0000	.0000
	1	.3512	.3672	.1787	.0540	.0113	.0016	.0001	.0000	.0000	.0000	.0000
	2	.1109	.2448	.2680	.1388	.0453	.0095	.0012	.0001	.0000	.0000	.0000
	3	.0214	.0997	.2457	.2181	.1107	.0349	.0065	.0006	.0000	.0000	.0000
	4	.0028	.0277	.1535	.2337	.1845	.0873	.0243	.0034	.0001	.0000	.0000
	5	.0003	.0055	.0691	.1803	.2214	.1571	.0656	.0142	.0011	.0000	.0000
	6	.0000	.0008	.0230	.1030	.1968	.2095	.1312	.0442	.0058	.0001	.0000
	7	.0000	.0001	.0058	.0442	.1312	.2095	.1968	.1030	.0230	.0008	.0000
	8	.0000	.0000	.0011	.0142	.0656	.1571	.2214	.1803	.0691	.0055	.0003
	9	.0000	.0000	.0001	.0034	.0243	.0873	.1845	.2337	.1535	.0277	.0028
	10	.0000	.0000	.0000	.0006	.0065	.0349	.1107	.2181	.2457	.0997	.0214
	11	.0000	.0000	.0000	.0001	.0012	.0095	.0453	.1388	.2680	.2448	.1109
	12	.0000	.0000	.0000	.0000	.0001	.0016	.0113	.0540	.1787	.3672	.3512
	13	.0000	.0000	.0000	.0000	.0000	.0001	.0013	.0097	.0550	.2542	.5133
14	0	.4877	.2288	.0440	.0068	.0008	.0001	.0000	.0000	.0000	.0000	.0000
	1	.3593	.3559	.1539	.0407	.0073	.0009	.0001	.0000	.0000	.0000	.0000
	2	.1229	.2570	.2501	.1134	.0317	.0056	.0005	.0000	.0000	.0000	.0000
	3	.0259	.1142	.2501	.1943	.0845	.0222	.0033	.0002	.0000	.0000	.0000
	4	.0037	.0349	.1720	.2290	.1549	.0611	.0136	.0014	.0000	.0000	.0000
	5	.0004	.0078	.0860	.1963	.2066	.1222	.0408	.0066	.0003	.0000	.0000
	6	.0000	.0013	.0322	.1262	.2066	.1833	.0918	.0232	.0020	.0000	.0000
	7	.0000	.0002	.0092	.0618	.1574	.2095	.1574	.0618	.0092	.0002	.0000
	8	.0000	.0000	.0020	.0232	.0918	.1833	.2066	.1262	.0322	.0013	.0000
	9	.0000	.0000	.0003	.0066	.0408	.1222	.2066	.1963	.0860	.0078	.0004
	10	.0000	.0000	.0000	.0014	.0136	.0611	.1549	.2290	.1720	.0349	.0037
	11	.0000	.0000	.0000	.0002	.0033	.0222	.0845	.1943	.2501	.1142	.0259
	12	.0000	.0000	.0000	.0000	.0005	.0056	.0317	.1134	.2501	.2570	.1229
	13	.0000	.0000	.0000	.0000	.0001	.0009	.0073	.0407	.1539	.3559	.3593
	14	.0000	.0000	.0000	.0000	.0000	.0001	.0008	.0068	.0440	.2288	.4877
15	0	.4633	.2059	.0352	.0047	.0005	.0000	.0000	.0000	.0000	.0000	.0000
	1	.3658	.3432	.1319	.0305	.0047	.0005	.0000	.0000	.0000	.0000	.0000
	2	.1348	.2669	.2309	.0916	.0219	.0032	.0003	.0000	.0000	.0000	.0000
	3	.0307	.1285	.2501	.1700	.0634	.0139	.0016	.0001	.0000	.0000	.0000
	4	.0049	.0428	.1876	.2186	.1268	.0417	.0074	.0006	.0000	.0000	.0000
	5	.0006	.0105	.1032	.2061	.1859	.0916	.0245	.0030	.0001	.0000	.0000
	6	.0000	.0019	.0430	.1472	.2066	.1527	.0612	.0116	.0007	.0000	.0000
	7	.0000	.0003	.0138	.0811	.1771	.1964	.1181	.0348	.0035	.0000	.0000
	8	.0000	.0000	.0035	.0348	.1181	.1964	.1771	.0811	.0138	.0003	.0000
	9	.0000	.0000	.0007	.0116	.0612	.1527	.2066	.1472	.0430	.0019	.0000
	10	.0000	.0000	.0001	.0030	.0245	.0916	.1859	.2061	.1032	.0105	.0006
	11	.0000	.0000	.0000	.0006	.0074	.0417	.1268	.2186	.1876	.0428	.0049
	12	.0000	.0000	.0000	.0001	.0016	.0139	.0634	.1700	.2501	.1285	.0307
	13	.0000	.0000	.0000	.0000	.0003	.0032	.0219	.0916	.2309	.2669	.1348
	14	.0000	.0000	.0000	.0000	.0000	.0005	.0047	.0305	.1319	.3432	.3658
	15	.0000	.0000	.0000	.0000	.0000	.0000	.0005	.0047	.0352	.2059	.4633

TABLE IV   (*Continued*)

							p					
n	x	.05	.10	.20	.30	.40	.50	.60	.70	.80	.90	.95
16	0	.4401	.1853	.0281	.0033	.0003	.0000	.0000	.0000	.0000	.0000	.0000
	1	.3706	.3294	.1126	.0228	.0030	.0002	.0000	.0000	.0000	.0000	.0000
	2	.1463	.2745	.2111	.0732	.0150	.0018	.0001	.0000	.0000	.0000	.0000
	3	.0359	.1423	.2463	.1465	.0468	.0085	.0008	.0000	.0000	.0000	.0000
	4	.0061	.0514	.2001	.2040	.1014	.0278	.0040	.0002	.0000	.0000	.0000
	5	.0008	.0137	.1201	.2099	.1623	.0667	.0142	.0013	.0000	.0000	.0000
	6	.0001	.0028	.0550	.1649	.1983	.1222	.0392	.0056	.0002	.0000	.0000
	7	.0000	.0004	.0197	.1010	.1889	.1746	.0840	.0185	.0012	.0000	.0000
	8	.0000	.0001	.0055	.0487	.1417	.1964	.1417	.0487	.0055	.0001	.0000
	9	.0000	.0000	.0012	.0185	.0840	.1746	.1889	.1010	.0197	.0004	.0000
	10	.0000	.0000	.0002	.0056	.0392	.1222	.1983	.1649	.0550	.0028	.0001
	11	.0000	.0000	.0000	.0013	.0142	.0666	.1623	.2099	.1201	.0137	.0008
	12	.0000	.0000	.0000	.0002	.0040	.0278	.1014	.2040	.2001	.0514	.0061
	13	.0000	.0000	.0000	.0000	.0008	.0085	.0468	.1465	.2463	.1423	.0359
	14	.0000	.0000	.0000	.0000	.0001	.0018	.0150	.0732	.2111	.2745	.1463
	15	.0000	.0000	.0000	.0000	.0000	.0002	.0030	.0228	.1126	.3294	.3706
	16	.0000	.0000	.0000	.0000	.0000	.0000	.0003	.0033	.0281	.1853	.4401
17	0	.4181	.1668	.0225	.0023	.0002	.0000	.0000	.0000	.0000	.0000	.0000
	1	.3741	.3150	.0957	.0169	.0019	.0001	.0000	.0000	.0000	.0000	.0000
	2	.1575	.2800	.1914	.0581	.0102	.0010	.0001	.0000	.0000	.0000	.0000
	3	.0415	.1556	.2393	.1245	.0341	.0052	.0004	.0000	.0000	.0000	.0000
	4	.0076	.0605	.2093	.1868	.0796	.0182	.0021	.0001	.0000	.0000	.0000
	5	.0010	.0175	.1361	.2081	.1379	.0472	.0081	.0006	.0000	.0000	.0000
	6	.0001	.0039	.0680	.1784	.1839	.0944	.0242	.0026	.0001	.0000	.0000
	7	.0000	.0007	.0267	.1201	.1927	.1484	.0571	.0095	.0004	.0000	.0000
	8	.0000	.0001	.0084	.0644	.1606	.1855	.1070	.0276	.0021	.0000	.0000
	9	.0000	.0000	.0021	.0276	.1070	.1855	.1606	.0644	.0084	.0001	.0000
	10	.0000	.0000	.0004	.0095	.0571	.1484	.1927	.1201	.0267	.0007	.0000
	11	.0000	.0000	.0001	.0026	.0242	.0944	.1839	.1784	.0680	.0039	.0001
	12	.0000	.0000	.0000	.0006	.0081	.0472	.1379	.2081	.1361	.0175	.0010
	13	.0000	.0000	.0000	.0001	.0021	.0182	.0796	.1868	.2093	.0605	.0076
	14	.0000	.0000	.0000	.0000	.0004	.0052	.0341	.1245	.2393	.1556	.0415
	15	.0000	.0000	.0000	.0000	.0001	.0010	.0102	.0581	.1914	.2800	.1575
	16	.0000	.0000	.0000	.0000	.0000	.0001	.0019	.0169	.0957	.3150	.3741
	17	.0000	.0000	.0000	.0000	.0000	.0000	.0002	.0023	.0225	.1668	.4181
18	0	.3972	.1501	.0180	.0016	.0001	.0000	.0000	.0000	.0000	.0000	.0000
	1	.3763	.3002	.0811	.0126	.0012	.0001	.0000	.0000	.0000	.0000	.0000
	2	.1683	.2835	.1723	.0458	.0069	.0006	.0000	.0000	.0000	.0000	.0000
	3	.0473	.1680	.2297	.1046	.0246	.0031	.0002	.0000	.0000	.0000	.0000
	4	.0093	.0700	.2153	.1681	.0614	.0117	.0011	.0000	.0000	.0000	.0000
	5	.0014	.0218	.1507	.2017	.1146	.0327	.0045	.0002	.0000	.0000	.0000
	6	.0002	.0052	.0816	.1873	.1655	.0708	.0145	.0012	.0000	.0000	.0000
	7	.0000	.0010	.0350	.1376	.1892	.1214	.0374	.0046	.0001	.0000	.0000
	8	.0000	.0002	.0120	.0811	.1734	.1669	.0771	.0149	.0008	.0000	.0000
	9	.0000	.0000	.0033	.0386	.1284	.1855	.1284	.0386	.0033	.0000	.0000
	10	.0000	.0000	.0008	.0149	.0771	.1669	.1734	.0811	.0120	.0002	.0000
	11	.0000	.0000	.0001	.0046	.0374	.1214	.1892	.1376	.0350	.0010	.0000
	12	.0000	.0000	.0000	.0012	.0145	.0708	.1655	.1873	.0816	.0052	.0002

TABLE IV   TABLE OF BINOMIAL PROBABILITIES

D11

TABLE IV   (*Continued*)

n	x	.05	.10	.20	.30	.40	.50	.60	.70	.80	.90	.95
	13	.0000	.0000	.0000	.0002	.0045	.0327	.1146	.2017	.1507	.0218	.0014
	14	.0000	.0000	.0000	.0000	.0011	.0117	.0614	.1681	.2153	.0700	.0093
	15	.0000	.0000	.0000	.0000	.0002	.0031	.0246	.1046	.2297	.1680	.0473
	16	.0000	.0000	.0000	.0000	.0000	.0006	.0069	.0458	.1723	.2835	.1683
	17	.0000	.0000	.0000	.0000	.0000	.0001	.0012	.0126	.0811	.3002	.3763
	18	.0000	.0000	.0000	.0000	.0000	.0000	.0001	.0016	.0180	.1501	.3972
19	0	.3774	.1351	.0144	.0011	.0001	.0000	.0000	.0000	.0000	.0000	.0000
	1	.3774	.2852	.0685	.0093	.0008	.0000	.0000	.0000	.0000	.0000	.0000
	2	.1787	.2852	.1540	.0358	.0046	.0003	.0000	.0000	.0000	.0000	.0000
	3	.0533	.1796	.2182	.0869	.0175	.0018	.0001	.0000	.0000	.0000	.0000
	4	.0112	.0798	.2182	.1491	.0467	.0074	.0005	.0000	.0000	.0000	.0000
	5	.0018	.0266	.1636	.1916	.0933	.0222	.0024	.0001	.0000	.0000	.0000
	6	.0002	.0069	.0955	.1916	.1451	.0518	.0085	.0005	.0000	.0000	.0000
	7	.0000	.0014	.0443	.1525	.1797	.0961	.0237	.0022	.0000	.0000	.0000
	8	.0000	.0002	.0166	.0981	.1797	.1442	.0532	.0077	.0003	.0000	.0000
	9	.0000	.0000	.0051	.0514	.1464	.1762	.0976	.0220	.0013	.0000	.0000
	10	.0000	.0000	.0013	.0220	.0976	.1762	.1464	.0514	.0051	.0000	.0000
	11	.0000	.0000	.0003	.0077	.0532	.1442	.1797	.0981	.0166	.0002	.0000
	12	.0000	.0000	.0000	.0022	.0237	.0961	.1797	.1525	.0443	.0014	.0000
	13	.0000	.0000	.0000	.0005	.0085	.0518	.1451	.1916	.0955	.0069	.0002
	14	.0000	.0000	.0000	.0001	.0024	.0222	.0933	.1916	.1636	.0266	.0018
	15	.0000	.0000	.0000	.0000	.0005	.0074	.0467	.1491	.2182	.0798	.0112
	16	.0000	.0000	.0000	.0000	.0001	.0018	.0175	.0869	.2182	.1796	.0533
	17	.0000	.0000	.0000	.0000	.0000	.0003	.0046	.0358	.1540	.2852	.1787
	18	.0000	.0000	.0000	.0000	.0000	.0000	.0008	.0093	.0685	.2852	.3774
	19	.0000	.0000	.0000	.0000	.0000	.0000	.0001	.0011	.0144	.1351	.3774
20	0	.3585	.1216	.0115	.0008	.0000	.0000	.0000	.0000	.0000	.0000	.0000
	1	.3774	.2702	.0576	.0068	.0005	.0000	.0000	.0000	.0000	.0000	.0000
	2	.1887	.2852	.1369	.0278	.0031	.0002	.0000	.0000	.0000	.0000	.0000
	3	.0596	.1901	.2054	.0716	.0123	.0011	.0000	.0000	.0000	.0000	.0000
	4	.0133	.0898	.2182	.1304	.0350	.0046	.0003	.0000	.0000	.0000	.0000
	5	.0022	.0319	.1746	.1789	.0746	.0148	.0013	.0000	.0000	.0000	.0000
	6	.0003	.0089	.1091	.1916	.1244	.0370	.0049	.0002	.0000	.0000	.0000
	7	.0000	.0020	.0545	.1643	.1659	.0739	.0146	.0010	.0000	.0000	.0000
	8	.0000	.0004	.0222	.1144	.1797	.1201	.0355	.0039	.0001	.0000	.0000
	9	.0000	.0001	.0074	.0654	.1597	.1602	.0710	.0120	.0005	.0000	.0000
	10	.0000	.0000	.0020	.0308	.1171	.1762	.1171	.0308	.0020	.0000	.0000
	11	.0000	.0000	.0005	.0120	.0710	.1602	.1597	.0654	.0074	.0001	.0000
	12	.0000	.0000	.0001	.0039	.0355	.1201	.1797	.1144	.0222	.0004	.0000
	13	.0000	.0000	.0000	.0010	.0146	.0739	.1659	.1643	.0545	.0020	.0000
	14	.0000	.0000	.0000	.0002	.0049	.0370	.1244	.1916	.1091	.0089	.0003
	15	.0000	.0000	.0000	.0000	.0013	.0148	.0746	.1789	.1746	.0319	.0022
	16	.0000	.0000	.0000	.0000	.0003	.0046	.0350	.1304	.2182	.0898	.0133
	17	.0000	.0000	.0000	.0000	.0000	.0011	.0123	.0716	.2054	.1901	.0596
	18	.0000	.0000	.0000	.0000	.0000	.0002	.0031	.0278	.1369	.2852	.1887
	19	.0000	.0000	.0000	.0000	.0000	.0000	.0005	.0068	.0576	.2702	.3774
	20	.0000	.0000	.0000	.0000	.0000	.0000	.0000	.0008	.0115	.1216	.3585

TABLE IV   (*Continued*)

n	x		.05	.10	.20	.30	.40	.50	.60	.70	.80	.90	.95
								p						
21	0		.3406	.1094	.0092	.0006	.0000	.0000	.0000	.0000	.0000	.0000	.0000
	1		.3764	.2553	.0484	.0050	.0003	.0000	.0000	.0000	.0000	.0000	.0000
	2		.1981	.2837	.1211	.0215	.0020	.0001	.0000	.0000	.0000	.0000	.0000
	3		.0660	.1996	.1917	.0585	.0086	.0006	.0000	.0000	.0000	.0000	.0000
	4		.0156	.0998	.2156	.1128	.0259	.0029	.0001	.0000	.0000	.0000	.0000
	5		.0028	.0377	.1833	.1643	.0588	.0097	.0007	.0000	.0000	.0000	.0000
	6		.0004	.0112	.1222	.1878	.1045	.0259	.0027	.0001	.0000	.0000	.0000
	7		.0000	.0027	.0655	.1725	.1493	.0554	.0087	.0005	.0000	.0000	.0000
	8		.0000	.0005	.0286	.1294	.1742	.0970	.0229	.0019	.0000	.0000	.0000
	9		.0000	.0001	.0103	.0801	.1677	.1402	.0497	.0063	.0002	.0000	.0000
	10		.0000	.0000	.0031	.0412	.1342	.1682	.0895	.0176	.0008	.0000	.0000
	11		.0000	.0000	.0008	.0176	.0895	.1682	.1342	.0412	.0031	.0000	.0000
	12		.0000	.0000	.0002	.0063	.0497	.1402	.1677	.0801	.0103	.0001	.0000
	13		.0000	.0000	.0000	.0019	.0229	.0970	.1742	.1294	.0286	.0005	.0000
	14		.0000	.0000	.0000	.0005	.0087	.0554	.1493	.1725	.0655	.0027	.0000
	15		.0000	.0000	.0000	.0001	.0027	.0259	.1045	.1878	.1222	.0112	.0004
	16		.0000	.0000	.0000	.0000	.0007	.0097	.0588	.1643	.1833	.0377	.0028
	17		.0000	.0000	.0000	.0000	.0001	.0029	.0259	.1128	.2156	.0998	.0156
	18		.0000	.0000	.0000	.0000	.0000	.0006	.0086	.0585	.1917	.1996	.0660
	19		.0000	.0000	.0000	.0000	.0000	.0001	.0020	.0215	.1211	.2837	.1981
	20		.0000	.0000	.0000	.0000	.0000	.0000	.0003	.0050	.0484	.2553	.3764
	21		.0000	.0000	.0000	.0000	.0000	.0000	.0000	.0006	.0092	.1094	.3406
22	0		.3235	.0985	.0074	.0004	.0000	.0000	.0000	.0000	.0000	.0000	.0000
	1		.3746	.2407	.0406	.0037	.0002	.0000	.0000	.0000	.0000	.0000	.0000
	2		.2070	.2808	.1065	.0166	.0014	.0001	.0000	.0000	.0000	.0000	.0000
	3		.0726	.2080	.1775	.0474	.0060	.0004	.0000	.0000	.0000	.0000	.0000
	4		.0182	.1098	.2108	.0965	.0190	.0017	.0001	.0000	.0000	.0000	.0000
	5		.0034	.0439	.1898	.1489	.0456	.0063	.0004	.0000	.0000	.0000	.0000
	6		.0005	.0138	.1344	.1808	.0862	.0178	.0015	.0000	.0000	.0000	.0000
	7		.0001	.0035	.0768	.1771	.1314	.0407	.0051	.0002	.0000	.0000	.0000
	8		.0000	.0007	.0360	.1423	.1642	.0762	.0144	.0009	.0000	.0000	.0000
	9		.0000	.0001	.0140	.0949	.1703	.1186	.0336	.0032	.0001	.0000	.0000
	10		.0000	.0000	.0046	.0529	.1476	.1542	.0656	.0097	.0003	.0000	.0000
	11		.0000	.0000	.0012	.0247	.1073	.1682	.1073	.0247	.0012	.0000	.0000
	12		.0000	.0000	.0003	.0097	.0656	.1542	.1476	.0529	.0046	.0000	.0000
	13		.0000	.0000	.0001	.0032	.0336	.1186	.1703	.0949	.0140	.0001	.0000
	14		.0000	.0000	.0000	.0009	.0144	.0762	.1642	.1423	.0360	.0007	.0000
	15		.0000	.0000	.0000	.0002	.0051	.0407	.1314	.1771	.0768	.0035	.0001
	16		.0000	.0000	.0000	.0000	.0015	.0178	.0862	.1808	.1344	.0138	.0005
	17		.0000	.0000	.0000	.0000	.0004	.0063	.0456	.1489	.1898	.0439	.0034
	18		.0000	.0000	.0000	.0000	.0001	.0017	.0190	.0965	.2108	.1098	.0182
	19		.0000	.0000	.0000	.0000	.0000	.0004	.0060	.0474	.1775	.2080	.0726
	20		.0000	.0000	.0000	.0000	.0000	.0001	.0014	.0166	.1065	.2808	.2070
	21		.0000	.0000	.0000	.0000	.0000	.0000	.0002	.0037	.0406	.2407	.3746
	22		.0000	.0000	.0000	.0000	.0000	.0000	.0000	.0004	.0074	.0985	.3235
23	0		.3074	.0886	.0059	.0003	.0000	.0000	.0000	.0000	.0000	.0000	.0000
	1		.3721	.2265	.0339	.0027	.0001	.0000	.0000	.0000	.0000	.0000	.0000
	2		.2154	.2768	.0933	.0127	.0009	.0000	.0000	.0000	.0000	.0000	.0000
	3		.0794	.2153	.1633	.0382	.0041	.0002	.0000	.0000	.0000	.0000	.0000

TABLE IV   TABLE OF BINOMIAL PROBABILITIES   **D13**

TABLE IV (*Continued*)

n	x	.05	.10	.20	.30	.40	.50	.60	.70	.80	.90	.95
							p					
	4	.0209	.1196	.2042	.0818	.0138	.0011	.0000	.0000	.0000	.0000	.0000
	5	.0042	.0505	.1940	.1332	.0350	.0040	.0002	.0000	.0000	.0000	.0000
	6	.0007	.0168	.1455	.1712	.0700	.0120	.0008	.0000	.0000	.0000	.0000
	7	.0001	.0045	.0883	.1782	.1133	.0292	.0029	.0001	.0000	.0000	.0000
	8	.0000	.0010	.0442	.1527	.1511	.0584	.0088	.0004	.0000	.0000	.0000
	9	.0000	.0002	.0184	.1091	.1679	.0974	.0221	.0016	.0000	.0000	.0000
	10	.0000	.0000	.0064	.0655	.1567	.1364	.0464	.0052	.0001	.0000	.0000
	11	.0000	.0000	.0019	.0332	.1234	.1612	.0823	.0142	.0005	.0000	.0000
	12	.0000	.0000	.0005	.0142	.0823	.1612	.1234	.0332	.0019	.0000	.0000
	13	.0000	.0000	.0001	.0052	.0464	.1364	.1567	.0655	.0064	.0000	.0000
	14	.0000	.0000	.0000	.0016	.0221	.0974	.1679	.1091	.0184	.0002	.0000
	15	.0000	.0000	.0000	.0004	.0088	.0584	.1511	.1527	.0442	.0010	.0000
	16	.0000	.0000	.0000	.0001	.0029	.0292	.1133	.1782	.0883	.0045	.0001
	17	.0000	.0000	.0000	.0000	.0008	.0120	.0700	.1712	.1455	.0168	.0007
	18	.0000	.0000	.0000	.0000	.0002	.0040	.0350	.1332	.1940	.0505	.0042
	19	.0000	.0000	.0000	.0000	.0000	.0011	.0138	.0818	.2042	.1196	.0209
	20	.0000	.0000	.0000	.0000	.0000	.0002	.0041	.0382	.1633	.2153	.0794
	21	.0000	.0000	.0000	.0000	.0000	.0000	.0009	.0127	.0933	.2768	.2154
	22	.0000	.0000	.0000	.0000	.0000	.0000	.0001	.0027	.0339	.2265	.3721
	23	.0000	.0000	.0000	.0000	.0000	.0000	.0000	.0003	.0059	.0886	.3074
24	0	.2920	.0798	.0047	.0002	.0000	.0000	.0000	.0000	.0000	.0000	.0000
	1	.3688	.2127	.0283	.0020	.0001	.0000	.0000	.0000	.0000	.0000	.0000
	2	.2232	.2718	.0815	.0097	.0006	.0000	.0000	.0000	.0000	.0000	.0000
	3	.0862	.2215	.1493	.0305	.0028	.0001	.0000	.0000	.0000	.0000	.0000
	4	.0238	.1292	.1960	.0687	.0099	.0006	.0000	.0000	.0000	.0000	.0000
	5	.0050	.0574	.1960	.1177	.0265	.0025	.0001	.0000	.0000	.0000	.0000
	6	.0008	.0202	.1552	.1598	.0560	.0080	.0004	.0000	.0000	.0000	.0000
	7	.0001	.0058	.0998	.1761	.0960	.0206	.0017	.0000	.0000	.0000	.0000
	8	.0000	.0014	.0530	.1604	.1360	.0438	.0053	.0002	.0000	.0000	.0000
	9	.0000	.0003	.0236	.1222	.1612	.0779	.0141	.0008	.0000	.0000	.0000
	10	.0000	.0000	.0088	.0785	.1612	.1169	.0318	.0026	.0000	.0000	.0000
	11	.0000	.0000	.0028	.0428	.1367	.1488	.0608	.0079	.0002	.0000	.0000
	12	.0000	.0000	.0008	.0199	.0988	.1612	.0988	.0199	.0008	.0000	.0000
	13	.0000	.0000	.0002	.0079	.0608	.1488	.1367	.0428	.0028	.0000	.0000
	14	.0000	.0000	.0000	.0026	.0318	.1169	.1612	.0785	.0088	.0000	.0000
	15	.0000	.0000	.0000	.0008	.0141	.0779	.1612	.1222	.0236	.0003	.0000
	16	.0000	.0000	.0000	.0002	.0053	.0438	.1360	.1604	.0530	.0014	.0000
	17	.0000	.0000	.0000	.0000	.0017	.0206	.0960	.1761	.0998	.0058	.0001
	18	.0000	.0000	.0000	.0000	.0004	.0080	.0560	.1598	.1552	.0202	.0008
	19	.0000	.0000	.0000	.0000	.0001	.0025	.0265	.1177	.1960	.0574	.0050
	20	.0000	.0000	.0000	.0000	.0000	.0006	.0099	.0687	.1960	.1292	.0238
	21	.0000	.0000	.0000	.0000	.0000	.0001	.0028	.0305	.1493	.2215	.0862
	22	.0000	.0000	.0000	.0000	.0000	.0000	.0006	.0097	.0815	.2718	.2232
	23	.0000	.0000	.0000	.0000	.0000	.0000	.0001	.0020	.0283	.2127	.3688
	24	.0000	.0000	.0000	.0000	.0000	.0000	.0000	.0002	.0047	.0798	.2920
25	0	.2774	.0718	.0038	.0001	.0000	.0000	.0000	.0000	.0000	.0000	.0000
	1	.3650	.1994	.0236	.0014	.0000	.0000	.0000	.0000	.0000	.0000	.0000
	2	.2305	.2659	.0708	.0074	.0004	.0000	.0000	.0000	.0000	.0000	.0000
	3	.0930	.2265	.1358	.0243	.0019	.0001	.0000	.0000	.0000	.0000	.0000

TABLE IV   (*Continued*)

							p					
n	*x*	.05	.10	.20	.30	.40	.50	.60	.70	.80	.90	.95
	4	.0269	.1384	.1867	.0572	.0071	.0004	.0000	.0000	.0000	.0000	.0000
	5	.0060	.0646	.1960	.1030	.0199	.0016	.0000	.0000	.0000	.0000	.0000
	6	.0010	.0239	.1633	.1472	.0442	.0053	.0002	.0000	.0000	.0000	.0000
	7	.0001	.0072	.1108	.1712	.0800	.0143	.0009	.0000	.0000	.0000	.0000
	8	.0000	.0018	.0623	.1651	.1200	.0322	.0031	.0001	.0000	.0000	.0000
	9	.0000	.0004	.0294	.1336	.1511	.0609	.0088	.0004	.0000	.0000	.0000
	10	.0000	.0001	.0118	.0916	.1612	.0974	.0212	.0013	.0000	.0000	.0000
	11	.0000	.0000	.0040	.0536	.1465	.1328	.0434	.0042	.0001	.0000	.0000
	12	.0000	.0000	.0012	.0268	.1140	.1550	.0760	.0115	.0003	.0000	.0000
	13	.0000	.0000	.0003	.0115	.0760	.1550	.1140	.0268	.0012	.0000	.0000
	14	.0000	.0000	.0001	.0042	.0434	.1328	.1465	.0536	.0040	.0000	.0000
	15	.0000	.0000	.0000	.0013	.0212	.0974	.1612	.0916	.0118	.0001	.0000
	16	.0000	.0000	.0000	.0004	.0088	.0609	.1511	.1336	.0294	.0004	.0000
	17	.0000	.0000	.0000	.0001	.0031	.0322	.1200	.1651	.0623	.0018	.0000
	18	.0000	.0000	.0000	.0000	.0009	.0143	.0800	.1712	.1108	.0072	.0001
	19	.0000	.0000	.0000	.0000	.0002	.0053	.0442	.1472	.1633	.0239	.0010
	20	.0000	.0000	.0000	.0000	.0000	.0016	.0199	.1030	.1960	.0646	.0060
	21	.0000	.0000	.0000	.0000	.0000	.0004	.0071	.0572	.1867	.1384	.0269
	22	.0000	.0000	.0000	.0000	.0000	.0001	.0019	.0243	.1358	.2265	.0930
	23	.0000	.0000	.0000	.0000	.0000	.0000	.0004	.0074	.0708	.2659	.2305
	24	.0000	.0000	.0000	.0000	.0000	.0000	.0000	.0014	.0236	.1994	.3650
	25	.0000	.0000	.0000	.0000	.0000	.0000	.0000	.0001	.0038	.0718	.2774

TABLE V   VALUES OF $e^{-\lambda}$

D15

TABLE V   VALUES OF $e^{-\lambda}$

λ	$e^{-\lambda}$	λ	$e^{-\lambda}$
0.0	1.00000000	5.5	.00408677
0.1	.90483742	5.6	.00369786
0.2	.81873075	5.7	.00334597
0.3	.74081822	5.8	.00302755
0.4	.67032005	5.9	.00273944
0.5	.60653066	6.0	.00247875
0.6	.54881164	6.1	.00224287
0.7	.49658530	6.2	.00202943
0.8	.44932896	6.3	.00183630
0.9	.40656966	6.4	.00166156
1.0	.36787944	6.5	.00150344
1.1	.33287108	6.6	.00136037
1.2	.30119421	6.7	.00123091
1.3	.27253179	6.8	.00111378
1.4	.24659696	6.9	.00100779
1.5	.22313016	7.0	.00091188
1.6	.20189652	7.1	.00082510
1.7	.18268352	7.2	.00074659
1.8	.16529889	7.3	.00067554
1.9	.14956862	7.4	.00061125
2.0	.13533528	7.5	.00055308
2.1	.12245643	7.6	.00050045
2.2	.11080316	7.7	.00045283
2.3	.10025884	7.8	.00040973
2.4	.09071795	7.9	.00037074
2.5	.08208500	8.0	.00033546
2.6	.07427358	8.1	.00030354
2.7	.06720551	8.2	.00027465
2.8	.06081006	8.3	.00024852
2.9	.05502322	8.4	.00022487
3.0	.04978707	8.5	.00020347
3.1	.04504920	8.6	.00018411
3.2	.04076220	8.7	.00016659
3.3	.03688317	8.8	.00015073
3.4	.03337327	8.9	.00013639
3.5	.03019738	9.0	.00012341
3.6	.02732372	9.1	.00011167
3.7	.02472353	9.2	.00010104
3.8	.02237077	9.3	.00009142
3.9	.02024191	9.4	.00008272
4.0	.01831564	9.5	.00007485
4.1	.01657268	9.6	.00006773
4.2	.01499558	9.7	.00006128
4.3	.01356856	9.8	.00005545
4.4	.01227734	9.9	.00005017
4.5	.01110900	10.0	.00004540
4.6	.01005184	11.0	.00001670
4.7	.00909528	12.0	.00000614
4.8	.00822975	13.0	.00000226
4.9	.00744658	14.0	.00000083
5.0	.00673795	15.0	.00000031
5.1	.00609675	16.0	.00000011
5.2	.00551656	17.0	.00000004
5.3	.00499159	18.0	.000000015
5.4	.00451658	19.0	.000000006
		20.0	.000000002

TABLE VI   TABLE OF POISSON PROBABILITIES

					λ					
x	0.1	0.2	0.3	0.4	0.5	0.6	0.7	0.8	0.9	1.0
0	.9048	.8187	.7408	.6703	.6065	.5488	.4966	.4493	.4066	.3679
1	.0905	.1637	.2222	.2681	.3033	.3293	.3476	.3595	.3659	.3679
2	.0045	.0164	.0333	.0536	.0758	.0988	.1217	.1438	.1647	.1839
3	.0002	.0011	.0033	.0072	.0126	.0198	.0284	.0383	.0494	.0613
4	.0000	.0001	.0003	.0007	.0016	.0030	.0050	.0077	.0111	.0153
5	.0000	.0000	.0000	.0001	.0002	.0004	.0007	.0012	.0020	.0031
6	.0000	.0000	.0000	.0000	.0000	.0000	.0001	.0002	.0003	.0005
7	.0000	.0000	.0000	.0000	.0000	.0000	.0000	.0000	.0000	.0001

					λ					
x	1.1	1.2	1.3	1.4	1.5	1.6	1.7	1.8	1.9	2.0
0	.3329	.3012	.2725	.2466	.2231	.2019	.1827	.1653	.1496	.1353
1	.3662	.3614	.3543	.3452	.3347	.3230	.3106	.2975	.2842	.2707
2	.2014	.2169	.2303	.2417	.2510	.2584	.2640	.2678	.2700	.2707
3	.0738	.0867	.0998	.1128	.1255	.1378	.1496	.1607	.1710	.1804
4	.0203	.0260	.0324	.0395	.0471	.0551	.0636	.0723	.0812	.0902
5	.0045	.0062	.0084	.0111	.0141	.0176	.0216	.0260	.0309	.0361
6	.0008	.0012	.0018	.0026	.0035	.0047	.0061	.0078	.0098	.0120
7	.0001	.0002	.0003	.0005	.0008	.0011	.0015	.0020	.0027	.0034
8	.0000	.0000	.0001	.0001	.0001	.0002	.0003	.0005	.0006	.0009
9	.0000	.0000	.0000	.0000	.0000	.0000	.0001	.0001	.0001	.0002

					λ					
x	2.1	2.2	2.3	2.4	2.5	2.6	2.7	2.8	2.9	3.0
0	.1225	.1108	.1003	.0907	.0821	.0743	.0672	.0608	.0550	.0498
1	.2572	.2438	.2306	.2177	.2052	.1931	.1815	.1703	.1596	.1494
2	.2700	.2681	.2652	.2613	.2565	.2510	.2450	.2384	.2314	.2240
3	.1890	.1966	.2033	.2090	.2138	.2176	.2205	.2225	.2237	.2240
4	.0992	.1082	.1169	.1254	.1336	.1414	.1488	.1557	.1622	.1680
5	.0417	.0476	.0538	.0602	.0668	.0735	.0804	.0872	.0940	.1008
6	.0146	.0174	.0206	.0241	.0278	.0319	.0362	.0407	.0455	.0504
7	.0044	.0055	.0068	.0083	.0099	.0118	.0139	.0163	.0188	.0216
8	.0011	.0015	.0019	.0025	.0031	.0038	.0047	.0057	.0068	.0081
9	.0003	.0004	.0005	.0007	.0009	.0011	.0014	.0018	.0022	.0027
10	.0001	.0001	.0001	.0002	.0002	.0003	.0004	.0005	.0006	.0008
11	.0000	.0000	.0000	.0000	.0000	.0001	.0001	.0001	.0002	.0002
12	.0000	.0000	.0000	.0000	.0000	.0000	.0000	.0000	.0000	.0001

					λ					
x	3.1	3.2	3.3	3.4	3.5	3.6	3.7	3.8	3.9	4.0
0	.0450	.0408	.0369	.0334	.0302	.0273	.0247	.0224	.0202	.0183
1	.1397	.1304	.1217	.1135	.1057	.0984	.0915	.0850	.0789	.0733
2	.2165	.2087	.2008	.1929	.1850	.1771	.1692	.1615	.1539	.1465

TABLE VI   TABLE OF POISSON PROBABILITIES   **D17**

TABLE VI   (*Continued*)

					λ					
x	3.1	3.2	3.3	3.4	3.5	3.6	3.7	3.8	3.9	4.0
3	.2237	.2226	.2209	.2186	.2158	.2125	.2087	.2046	.2001	.1954
4	.1733	.1781	.1823	.1858	.1888	.1912	.1931	.1944	.1951	.1954
5	.1075	.1140	.1203	.1264	.1322	.1377	.1429	.1477	.1522	.1563
6	.0555	.0608	.0662	.0716	.0771	.0826	.0881	.0936	.0989	.1042
7	.0246	.0278	.0312	.0348	.0385	.0425	.0466	.0508	.0551	.0595
8	.0095	.0111	.0129	.0148	.0169	.0191	.0215	.0241	.0269	.0298
9	.0033	.0040	.0047	.0056	.0066	.0076	.0089	.0102	.0116	.0132
10	.0010	.0013	.0016	.0019	.0023	.0028	.0033	.0039	.0045	.0053
11	.0003	.0004	.0005	.0006	.0007	.0009	.0011	.0013	.0016	.0019
12	.0001	.0001	.0001	.0002	.0002	.0003	.0003	.0004	.0005	.0006
13	.0000	.0000	.0000	.0000	.0001	.0001	.0001	.0001	.0002	.0002
14	.0000	.0000	.0000	.0000	.0000	.0000	.0000	.0000	.0000	.0001

					λ					
x	4.1	4.2	4.3	4.4	4.5	4.6	4.7	4.8	4.9	5.0
0	.0166	.0150	.0136	.0123	.0111	.0101	.0091	.0082	.0074	.0067
1	.0679	.0630	.0583	.0540	.0500	.0462	.0427	.0395	.0365	.0337
2	.1393	.1323	.1254	.1188	.1125	.1063	.1005	.0948	.0894	.0842
3	.1904	.1852	.1798	.1743	.1687	.1631	.1574	.1517	.1460	.1404
4	.1951	.1944	.1933	.1917	.1898	.1875	.1849	.1820	.1789	.1755
5	.1600	.1633	.1662	.1687	.1708	.1725	.1738	.1747	.1753	.1755
6	.1093	.1143	.1191	.1237	.1281	.1323	.1362	.1398	.1432	.1462
7	.0640	.0686	.0732	.0778	.0824	.0869	.0914	.0959	.1002	.1044
8	.0328	.0360	.0393	.0428	.0463	.0500	.0537	.0575	.0614	.0653
9	.0150	.0168	.0188	.0209	.0232	.0255	.0281	.0307	.0334	.0363
10	.0061	.0071	.0081	.0092	.0104	.0118	.0132	.0147	.0164	.0181
11	.0023	.0027	.0032	.0037	.0043	.0049	.0056	.0064	.0073	.0082
12	.0008	.0009	.0011	.0014	.0016	.0019	.0022	.0026	.0030	.0034
13	.0002	.0003	.0004	.0005	.0006	.0007	.0008	.0009	.0011	.0013
14	.0001	.0001	.0001	.0001	.0002	.0002	.0003	.0003	.0004	.0005
15	.0000	.0000	.0000	.0000	.0001	.0001	.0001	.0001	.0001	.0002

					λ					
x	5.1	5.2	5.3	5.4	5.5	5.6	5.7	5.8	5.9	6.0
0	.0061	.0055	.0050	.0045	.0041	.0037	.0033	.0030	.0027	.0025
1	.0311	.0287	.0265	.0244	.0225	.0207	.0191	.0176	.0162	.0149
2	.0793	.0746	.0701	.0659	.0618	.0580	.0544	.0509	.0477	.0446
3	.1348	.1293	.1239	.1185	.1133	.1082	.1033	.0985	.0938	.0892
4	.1719	.1681	.1641	.1600	.1558	.1515	.1472	.1428	.1383	.1339
5	.1753	.1748	.1740	.1728	.1714	.1697	.1678	.1656	.1632	.1606
6	.1490	.1515	.1537	.1555	.1571	.1584	.1594	.1601	.1605	.1606
7	.1086	.1125	.1163	.1200	.1234	.1267	.1298	.1326	.1353	.1377

TABLE VI   (*Continued*)

					λ					
x	5.1	5.2	5.3	5.4	5.5	5.6	5.7	5.8	5.9	6.0
8	.0692	.0731	.0771	.0810	.0849	.0887	.0925	.0962	.0998	.1033
9	.0392	.0423	.0454	.0486	.0519	.0552	.0586	.0620	.0654	.0688
10	.0200	.0220	.0241	.0262	.0285	.0309	.0334	.0359	.0386	.0413
11	.0093	.0104	.0116	.0129	.0143	.0157	.0173	.0190	.0207	.0225
12	.0039	.0045	.0051	.0058	.0065	.0073	.0082	.0092	.0102	.0113
13	.0015	.0018	.0021	.0024	.0028	.0032	.0036	.0041	.0046	.0052
14	.0006	.0007	.0008	.0009	.0011	.0013	.0015	.0017	.0019	.0022
15	.0002	.0002	.0003	.0003	.0004	.0005	.0006	.0007	.0008	.0009
16	.0001	.0001	.0001	.0001	.0001	.0002	.0002	.0002	.0003	.0003
17	.0000	.0000	.0000	.0000	.0000	.0001	.0001	.0001	.0001	.0001

					λ					
x	6.1	6.2	6.3	6.4	6.5	6.6	6.7	6.8	6.9	7.0
0	.0022	.0020	.0018	.0017	.0015	.0014	.0012	.0011	.0010	.0009
1	.0137	.0126	.0116	.0106	.0098	.0090	.0082	.0076	.0070	.0064
2	.0417	.0390	.0364	.0340	.0318	.0296	.0276	.0258	.0240	.0223
3	.0848	.0806	.0765	.0726	.0688	.0652	.0617	.0584	.0552	.0521
4	.1294	.1249	.1205	.1162	.1118	.1076	.1034	.0992	.0952	.0912
5	.1579	.1549	.1519	.1487	.1454	.1420	.1385	.1349	.1314	.1277
6	.1605	.1601	.1595	.1586	.1575	.1562	.1546	.1529	.1511	.1490
7	.1399	.1418	.1435	.1450	.1462	.1472	.1480	.1486	.1489	.1490
8	.1066	.1099	.1130	.1160	.1188	.1215	.1240	.1263	.1284	.1304
9	.0723	.0757	.0791	.0825	.0858	.0891	.0923	.0954	.0985	.1014
10	.0441	.0469	.0498	.0528	.0558	.0588	.0618	.0649	.0679	.0710
11	.0244	.0265	.0285	.0307	.0330	.0353	.0377	.0401	.0426	.0452
12	.0124	.0137	.0150	.0164	.0179	.0194	.0210	.0227	.0245	.0263
13	.0058	.0065	.0073	.0081	.0089	.0099	.0108	.0119	.0130	.0142
14	.0025	.0029	.0033	.0037	.0041	.0046	.0052	.0058	.0064	.0071
15	.0010	.0012	.0014	.0016	.0018	.0020	.0023	.0026	.0029	.0033
16	.0004	.0005	.0005	.0006	.0007	.0008	.0010	.0011	.0013	.0014
17	.0001	.0002	.0002	.0002	.0003	.0003	.0004	.0004	.0005	.0006
18	.0000	.0001	.0001	.0001	.0001	.0001	.0001	.0002	.0002	.0002
19	.0000	.0000	.0000	.0000	.0000	.0000	.0001	.0001	.0001	.0001

					λ					
x	7.1	7.2	7.3	7.4	7.5	7.6	7.7	7.8	7.9	8.0
0	.0008	.0007	.0007	.0006	.0006	.0005	.0005	.0004	.0004	.0003
1	.0059	.0054	.0049	.0045	.0041	.0038	.0035	.0032	.0029	.0027
2	.0208	.0194	.0180	.0167	.0156	.0145	.0134	.0125	.0116	.0107
3	.0492	.0464	.0438	.0413	.0389	.0366	.0345	.0324	.0305	.0286
4	.0874	.0836	.0799	.0764	.0729	.0696	.0663	.0632	.0602	.0573
5	.1241	.1204	.1167	.1130	.1094	.1057	.1021	.0986	.0951	.0916
6	.1468	.1445	.1420	.1394	.1367	.1339	.1311	.1282	.1252	.1221

TABLE VI   TABLE OF POISSON PROBABILITIES

D19

TABLE VI (*Continued*)

x	7.1	7.2	7.3	7.4	λ 7.5	7.6	7.7	7.8	7.9	8.0
7	.1489	.1486	.1481	.1474	.1465	.1454	.1442	.1428	.1413	.1396
8	.1321	.1337	.1351	.1363	.1373	.1381	.1388	.1392	.1395	.1396
9	.1042	.1070	.1096	.1121	.1144	.1167	.1187	.1207	.1224	.1241
10	.0740	.0770	.0800	.0829	.0858	.0887	.0914	.0941	.0967	.0993
11	.0478	.0504	.0531	.0558	.0585	.0613	.0640	.0667	.0695	.0722
12	.0283	.0303	.0323	.0344	.0366	.0388	.0411	.0434	.0457	.0481
13	.0154	.0168	.0181	.0196	.0211	.0227	.0243	.0260	.0278	.0296
14	.0078	.0086	.0095	.0104	.0113	.0123	.0134	.0145	.0157	.0169
15	.0037	.0041	.0046	.0051	.0057	.0062	.0069	.0075	.0083	.0090
16	.0016	.0019	.0021	.0024	.0026	.0030	.0033	.0037	.0041	.0045
17	.0007	.0008	.0009	.0010	.0012	.0013	.0015	.0017	.0019	.0021
18	.0003	.0003	.0004	.0004	.0005	.0006	.0006	.0007	.0008	.0009
19	.0001	.0001	.0001	.0002	.0002	.0002	.0003	.0003	.0003	.0004
20	.0000	.0000	.0001	.0001	.0001	.0001	.0001	.0001	.0001	.0002
21	.0000	.0000	.0000	.0000	.0000	.0000	.0000	.0000	.0001	.0001

x	8.1	8.2	8.3	8.4	λ 8.5	8.6	8.7	8.8	8.9	9.0
0	.0003	.0003	.0002	.0002	.0002	.0002	.0002	.0002	.0001	.0001
1	.0025	.0023	.0021	.0019	.0017	.0016	.0014	.0013	.0012	.0011
2	.0100	.0092	.0086	.0079	.0074	.0068	.0063	.0058	.0054	.0050
3	.0269	.0252	.0237	.0222	.0208	.0195	.0183	.0171	.0160	.0150
4	.0544	.0517	.0491	.0466	.0443	.0420	.0398	.0377	.0357	.0337
5	.0882	.0849	.0816	.0784	.0752	.0722	.0692	.0663	.0635	.0607
6	.1191	.1160	.1128	.1097	.1066	.1034	.1003	.0972	.0941	.0911
7	.1378	.1358	.1338	.1317	.1294	.1271	.1247	.1222	.1197	.1171
8	.1395	.1392	.1388	.1382	.1375	.1366	.1356	.1344	.1332	.1318
9	.1255	.1269	.1280	.1290	.1299	.1306	.1311	.1315	.1317	.1318
10	.1017	.1040	.1063	.1084	.1104	.1123	.1140	.1157	.1172	.1186
11	.0749	.0775	.0802	.0828	.0853	.0878	.0902	.0925	.0948	.0970
12	.0505	.0530	.0555	.0579	.0604	.0629	.0654	.0679	.0703	.0728
13	.0315	.0334	.0354	.0374	.0395	.0416	.0438	.0459	.0481	.0504
14	.0182	.0196	.0210	.0225	.0240	.0256	.0272	.0289	.0306	.0324
15	.0098	.0107	.0116	.0126	.0136	.0147	.0158	.0169	.0182	.0194
16	.0050	.0055	.0060	.0066	.0072	.0079	.0086	.0093	.0101	.0109
17	.0024	.0026	.0029	.0033	.0036	.0040	.0044	.0048	.0053	.0058
18	.0011	.0012	.0014	.0015	.0017	.0019	.0021	.0024	.0026	.0029
19	.0005	.0005	.0006	.0007	.0008	.0009	.0010	.0011	.0012	.0014
20	.0002	.0002	.0002	.0003	.0003	.0004	.0004	.0005	.0005	.0006
21	.0001	.0001	.0001	.0001	.0001	.0002	.0002	.0002	.0002	.0003
22	.0000	.0000	.0000	.0000	.0001	.0001	.0001	.0001	.0001	.0001

TABLE VI   (*Continued*)

					λ					
x	9.1	9.2	9.3	9.4	9.5	9.6	9.7	9.8	9.9	10
0	.0001	.0001	.0001	.0001	.0001	.0001	.0001	.0001	.0001	.0000
1	.0010	.0009	.0009	.0008	.0007	.0007	.0006	.0005	.0005	.0005
2	.0046	.0043	.0040	.0037	.0034	.0031	.0029	.0027	.0025	.0023
3	.0140	.0131	.0123	.0115	.0107	.0100	.0093	.0087	.0081	.0076
4	.0319	.0302	.0285	.0269	.0254	.0240	.0226	.0213	.0201	.0189
5	.0581	.0555	.0530	.0506	.0483	.0460	.0439	.0418	.0398	.0378
6	.0881	.0851	.0822	.0793	.0764	.0736	.0709	.0682	.0656	.0631
7	.1145	.1118	.1091	.1064	.1037	.1010	.0982	.0955	.0928	.0901
8	.1302	.1286	.1269	.1251	.1232	.1212	.1191	.1170	.1148	.1126
9	.1317	.1315	.1311	.1306	.1300	.1293	.1284	.1274	.1263	.1251
10	.1198	.1209	.1219	.1228	.1235	.1241	.1245	.1249	.1250	.1251
11	.0991	.1012	.1031	.1049	.1067	.1083	.1098	.1112	.1125	.1137
12	.0752	.0776	.0799	.0822	.0844	.0866	.0888	.0908	.0928	.0948
13	.0526	.0549	.0572	.0594	.0617	.0640	.0662	.0685	.0707	.0729
14	.0342	.0361	.0380	.0399	.0419	.0439	.0459	.0479	.0500	.0521
15	.0208	.0221	.0235	.0250	.0265	.0281	.0297	.0313	.0330	.0347
16	.0118	.0127	.0137	.0147	.0157	.0168	.0180	.0192	.0204	.0217
17	.0063	.0069	.0075	.0081	.0088	.0095	.0103	.0111	.0119	.0128
18	.0032	.0035	.0039	.0042	.0046	.0051	.0055	.0060	.0065	.0071
19	.0015	.0017	.0019	.0021	.0023	.0026	.0028	.0031	.0034	.0037
20	.0007	.0008	.0009	.0010	.0011	.0012	.0014	.0015	.0017	.0019
21	.0003	.0003	.0004	.0004	.0005	.0006	.0006	.0007	.0008	.0009
22	.0001	.0001	.0002	.0002	.0002	.0002	.0003	.0003	.0004	.0004
23	.0000	.0001	.0001	.0001	.0001	.0001	.0001	.0001	.0002	.0002
24	.0000	.0000	.0000	.0000	.0000	.0000	.0000	.0001	.0001	.0001

					λ					
x	11	12	13	14	15	16	17	18	19	20
0	.0000	.0000	.0000	.0000	.0000	.0000	.0000	.0000	.0000	.0000
1	.0002	.0001	.0000	.0000	.0000	.0000	.0000	.0000	.0000	.0000
2	.0010	.0004	.0002	.0001	.0000	.0000	.0000	.0000	.0000	.0000
3	.0037	.0018	.0008	.0004	.0002	.0001	.0000	.0000	.0000	.0000
4	.0102	.0053	.0027	.0013	.0006	.0003	.0001	.0001	.0000	.0000
5	.0224	.0127	.0070	.0037	.0019	.0010	.0005	.0002	.0001	.0001
6	.0411	.0255	.0152	.0087	.0048	.0026	.0014	.0007	.0004	.0002
7	.0646	.0437	.0281	.0174	.0104	.0060	.0034	.0019	.0010	.0005
8	.0888	.0655	.0457	.0304	.0194	.0120	.0072	.0042	.0024	.0013
9	.1085	.0874	.0661	.0473	.0324	.0213	.0135	.0083	.0050	.0029
10	.1194	.1048	.0859	.0663	.0486	.0341	.0230	.0150	.0095	.0058
11	.1194	.1144	.1015	.0844	.0663	.0496	.0355	.0245	.0164	.0106
12	.1094	.1144	.1099	.0984	.0829	.0661	.0504	.0368	.0259	.0176
13	.0926	.1056	.1099	.1060	.0956	.0814	.0658	.0509	.0378	.0271
14	.0728	.0905	.1021	.1060	.1024	.0930	.0800	.0655	.0514	.0387

TABLE VI   TABLE OF POISSON PROBABILITIES

D21

TABLE VI   (*Continued*)

x	λ 11	12	13	14	15	16	17	18	19	20
15	.0534	.0724	.0885	.0989	.1024	.0992	.0906	.0786	.0650	.0516
16	.0367	.0543	.0719	.0866	.0960	.0992	.0963	.0884	.0772	.0646
17	.0237	.0383	.0550	.0713	.0847	.0934	.0963	.0936	.0863	.0760
18	.0145	.0255	.0397	.0554	.0706	.0830	.0909	.0936	.0911	.0844
19	.0084	.0161	.0272	.0409	.0557	.0699	.0814	.0887	.0911	.0888
20	.0046	.0097	.0177	.0286	.0418	.0559	.0692	.0798	.0866	.0888
21	.0024	.0055	.0109	.0191	.0299	.0426	.0560	.0684	.0783	.0846
22	.0012	.0030	.0065	.0121	.0204	.0310	.0433	.0560	.0676	.0769
23	.0006	.0016	.0037	.0074	.0133	.0216	.0320	.0438	.0559	.0669
24	.0003	.0008	.0020	.0043	.0083	.0144	.0226	.0328	.0442	.0557
25	.0001	.0004	.0010	.0024	.0050	.0092	.0154	.0237	.0336	.0446
26	.0000	.0002	.0005	.0013	.0029	.0057	.0101	.0164	.0246	.0343
27	.0000	.0001	.0002	.0007	.0016	.0034	.0063	.0109	.0173	.0254
28	.0000	.0000	.0001	.0003	.0009	.0019	.0038	.0070	.0117	.0181
29	.0000	.0000	.0001	.0002	.0004	.0011	.0023	.0044	.0077	.0125
30	.0000	.0000	.0000	.0001	.0002	.0006	.0013	.0026	.0049	.0083
31	.0000	.0000	.0000	.0000	.0001	.0003	.0007	.0015	.0030	.0054
32	.0000	.0000	.0000	.0000	.0001	.0001	.0004	.0009	.0018	.0034
33	.0000	.0000	.0000	.0000	.0000	.0001	.0002	.0005	.0010	.0020
34	.0000	.0000	.0000	.0000	.0000	.0000	.0001	.0002	.0006	.0012
35	.0000	.0000	.0000	.0000	.0000	.0000	.0000	.0001	.0003	.0007
36	.0000	.0000	.0000	.0000	.0000	.0000	.0000	.0001	.0002	.0004
37	.0000	.0000	.0000	.0000	.0000	.0000	.0000	.0000	.0001	.0002
38	.0000	.0000	.0000	.0000	.0000	.0000	.0000	.0000	.0000	.0001
39	.0000	.0000	.0000	.0000	.0000	.0000	.0000	.0000	.0000	.0001

TABLE VII   STANDARD NORMAL DISTRIBUTION TABLE

The entries in this table give the areas under
the standard normal curve from 0 to z.

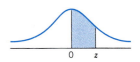

z	.00	.01	.02	.03	.04	.05	.06	.07	.08	.09
0.0	.0000	.0040	.0080	.0120	.0160	.0199	.0239	.0279	.0319	.0359
0.1	.0398	.0438	.0478	.0517	.0557	.0596	.0636	.0675	.0714	.0753
0.2	.0793	.0832	.0871	.0910	.0948	.0987	.1026	.1064	.1103	.1141
0.3	.1179	.1217	.1255	.1293	.1331	.1368	.1406	.1443	.1480	.1517
0.4	.1554	.1591	.1628	.1664	.1700	.1736	.1772	.1808	.1844	.1879
0.5	.1915	.1950	.1985	.2019	.2054	.2088	.2123	.2157	.2190	.2224
0.6	.2257	.2291	.2324	.2357	.2389	.2422	.2454	.2486	.2517	.2549
0.7	.2580	.2611	.2642	.2673	.2704	.2734	.2764	.2794	.2823	.2852
0.8	.2881	.2910	.2939	.2967	.2995	.3023	.3051	.3078	.3106	.3133
0.9	.3159	.3186	.3212	.3238	.3264	.3289	.3315	.3340	.3365	.3389
1.0	.3413	.3438	.3461	.3485	.3508	.3531	.3554	.3577	.3599	.3621
1.1	.3643	.3665	.3686	.3708	.3729	.3749	.3770	.3790	.3810	.3830
1.2	.3849	.3869	.3888	.3907	.3925	.3944	.3962	.3980	.3997	.4015
1.3	.4032	.4049	.4066	.4082	.4099	.4115	.4131	.4147	.4162	.4177
1.4	.4192	.4207	.4222	.4236	.4251	.4265	.4279	.4292	.4306	.4319
1.5	.4332	.4345	.4357	.4370	.4382	.4394	.4406	.4418	.4429	.4441
1.6	.4452	.4463	.4474	.4484	.4495	.4505	.4515	.4525	.4535	.4545
1.7	.4554	.4564	.4573	.4582	.4591	.4599	.4608	.4616	.4625	.4633
1.8	.4641	.4649	.4656	.4664	.4671	.4678	.4686	.4693	.4699	.4706
1.9	.4713	.4719	.4726	.4732	.4738	.4744	.4750	.4756	.4761	.4767
2.0	.4772	.4778	.4783	.4788	.4793	.4798	.4803	.4808	.4812	.4817
2.1	.4821	.4826	.4830	.4834	.4838	.4842	.4846	.4850	.4854	.4857
2.2	.4861	.4864	.4868	.4871	.4875	.4878	.4881	.4884	.4887	.4890
2.3	.4893	.4896	.4898	.4901	.4904	.4906	.4909	.4911	.4913	.4916
2.4	.4918	.4920	.4922	.4925	.4927	.4929	.4931	.4932	.4934	.4936
2.5	.4938	.4940	.4941	.4943	.4945	.4946	.4948	.4949	.4951	.4952
2.6	.4953	.4955	.4956	.4957	.4959	.4960	.4961	.4962	.4963	.4964
2.7	.4965	.4966	.4967	.4968	.4969	.4970	.4971	.4972	.4973	.4974
2.8	.4974	.4975	.4976	.4977	.4977	.4978	.4979	.4979	.4980	.4981
2.9	.4981	.4982	.4982	.4983	.4984	.4984	.4985	.4985	.4986	.4986
3.0	.4987	.4987	.4987	.4988	.4988	.4989	.4989	.4989	.4990	.4990

2.33

TABLE VIII   THE *t* DISTRIBUTION TABLE

D23

TABLE VIII   THE *t* DISTRIBUTION TABLE

The entries in this table give the critical values
of *t* for the specified number of degrees
of freedom and areas in the right tail.

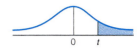

0   *t*

df			Area in the Right Tail under the *t* Distribution Curve			
	.10	**.05**	**.025**	**.01**	**.005**	**.001**
1	3.078	6.314	12.706	31.821	63.657	318.309
2	1.886	2.920	4.303	6.965	9.925	22.327
3	1.638	2.353	3.182	4.541	5.841	10.215
4	1.533	2.132	2.776	3.747	4.604	7.173
5	1.476	2.015	2.571	3.365	4.032	5.893
6	1.440	1.943	2.447	3.143	3.707	5.208
7	1.415	1.895	2.365	2.998	3.499	4.785
8	1.397	1.860	2.306	2.896	3.355	4.501
9	1.383	1.833	2.262	2.821	3.250	4.297
10	1.372	1.812	2.228	2.764	3.169	4.144
11	1.363	1.796	2.201	2.718	3.106	4.025
12	1.356	1.782	2.179	2.681	3.055	3.930
13	1.350	1.771	2.160	2.650	3.012	3.852
14	1.345	1.761	2.145	2.624	2.977	3.787
15	1.341	1.753	2.131	2.602	2.947	3.733
16	1.337	1.746	2.120	2.583	2.921	3.686
17	1.333	1.740	2.110	2.567	2.898	3.646
18	1.330	1.734	2.101	2.552	2.878	3.610
19	1.328	1.729	2.093	2.539	2.861	3.579
20	1.325	1.725	2.086	2.528	2.845	3.552
21	1.323	1.721	2.080	2.518	2.831	3.527
22	1.321	1.717	2.074	2.508	2.819	3.505
23	1.319	1.714	2.069	2.500	2.807	3.485
24	1.318	1.711	2.064	2.492	2.797	3.467
25	1.316	1.708	2.060	2.485	2.787	3.450
26	1.315	1.706	2.056	2.479	2.779	3.435
27	1.314	1.703	2.052	2.473	2.771	3.421
28	1.313	1.701	2.048	2.467	2.763	3.408
29	1.311	1.699	2.045	2.462	2.756	3.396
30	1.310	1.697	2.042	2.457	2.750	3.385
31	1.309	1.696	2.040	2.453	2.744	3.375
32	1.309	1.694	2.037	2.449	2.738	3.365
33	1.308	1.692	2.035	2.445	2.733	3.356
34	1.307	1.691	2.032	2.441	2.728	3.348
35	1.306	1.690	2.030	2.438	2.724	3.340
36	1.306	1.688	2.028	2.434	2.719	3.333
37	1.305	1.687	2.026	2.431	2.715	3.326

TABLE VIII   (*Continued*)

| | Area in the Right Tail under the *t* Distribution Curve | | | | | |
df	.10	.05	.025	.01	.005	.001
38	1.304	1.686	2.024	2.429	2.712	3.319
39	1.304	1.685	2.023	2.426	2.708	3.313
40	1.303	1.684	2.021	2.423	2.704	3.307
41	1.303	1.683	2.020	2.421	2.701	3.301
42	1.302	1.682	2.018	2.418	2.698	3.296
43	1.302	1.681	2.017	2.416	2.695	3.291
44	1.301	1.680	2.015	2.414	2.692	3.286
45	1.301	1.679	2.014	2.412	2.690	3.281
46	1.300	1.679	2.013	2.410	2.687	3.277
47	1.300	1.678	2.012	2.408	2.685	3.273
48	1.299	1.677	2.011	2.407	2.682	3.269
49	1.299	1.677	2.010	2.405	2.680	3.265
50	1.299	1.676	2.009	2.403	2.678	3.261
51	1.298	1.675	2.008	2.402	2.676	3.258
52	1.298	1.675	2.007	2.400	2.674	3.255
53	1.298	1.674	2.006	2.399	2.672	3.251
54	1.297	1.674	2.005	2.397	2.670	3.248
55	1.297	1.673	2.004	2.396	2.668	3.245
56	1.297	1.673	2.003	2.395	2.667	3.242
57	1.297	1.672	2.002	2.394	2.665	3.239
58	1.296	1.672	2.002	2.392	2.663	3.237
59	1.296	1.671	2.001	2.391	2.662	3.234
60	1.296	1.671	2.000	2.390	2.660	3.232
61	1.296	1.670	2.000	2.389	2.659	3.229
62	1.295	1.670	1.999	2.388	2.657	3.227
63	1.295	1.669	1.998	2.387	2.656	3.225
64	1.295	1.669	1.998	2.386	2.655	3.223
65	1.295	1.669	1.997	2.385	2.654	3.220
66	1.295	1.668	1.997	2.384	2.652	3.218
67	1.294	1.668	1.996	2.383	2.651	3.216
68	1.294	1.668	1.995	2.382	2.650	3.214
69	1.294	1.667	1.995	2.382	2.649	3.213
70	1.294	1.667	1.994	2.381	2.648	3.211
71	1.294	1.667	1.994	2.380	2.647	3.209
72	1.293	1.666	1.993	2.379	2.646	3.207
73	1.293	1.666	1.993	2.379	2.645	3.206
74	1.293	1.666	1.993	2.378	2.644	3.204
75	1.293	1.665	1.992	2.377	2.643	3.202
∞	1.282	1.645	1.960	2.326	2.576	3.090

TABLE IX CHI-SQUARE DISTRIBUTION TABLE

D25

TABLE IX CHI-SQUARE DISTRIBUTION TABLE

The entries in this table give the critical values of χ^2 for the specified number of degrees of freedom and areas in the right tail.

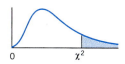

df	Area in the Right Tail under the Chi-square Distribution Curve									
	.995	.990	.975	.950	.900	.100	.050	.025	.010	.005
1	0.000	0.000	0.001	0.004	0.016	2.706	3.841	5.024	6.635	7.879
2	0.010	0.020	0.051	0.103	0.211	4.605	5.991	7.378	9.210	10.597
3	0.072	0.115	0.216	0.352	0.584	6.251	7.815	9.348	11.345	12.838
4	0.207	0.297	0.484	0.711	1.064	7.779	9.488	11.143	13.277	14.860
5	0.412	0.554	0.831	1.145	1.610	9.236	11.070	12.833	15.086	16.750
6	0.676	0.872	1.237	1.635	2.204	10.645	12.592	14.449	16.812	18.548
7	0.989	1.239	1.690	2.167	2.833	12.017	14.067	16.013	18.475	20.278
8	1.344	1.646	2.180	2.733	3.490	13.362	15.507	17.535	20.090	21.955
9	1.735	2.088	2.700	3.325	4.168	14.684	16.919	19.023	21.666	23.589
10	2.156	2.558	3.247	3.940	4.865	15.987	18.307	20.483	23.209	25.188
11	2.603	3.053	3.816	4.575	5.578	17.275	19.675	21.920	24.725	26.757
12	3.074	3.571	4.404	5.226	6.304	18.549	21.026	23.337	26.217	28.300
13	3.565	4.107	5.009	5.892	7.042	19.812	22.362	24.736	27.688	29.819
14	4.075	4.660	5.629	6.571	7.790	21.064	23.685	26.119	29.141	31.319
15	4.601	5.229	6.262	7.261	8.547	22.307	24.996	27.488	30.578	32.801
16	5.142	5.812	6.908	7.962	9.312	23.542	26.296	28.845	32.000	34.267
17	5.697	6.408	7.564	8.672	10.085	24.769	27.587	30.191	33.409	35.718
18	6.265	7.015	8.231	9.390	10.865	25.989	28.869	31.526	34.805	37.156
19	6.844	7.633	8.907	10.117	11.651	27.204	30.144	32.852	36.191	38.582
20	7.434	8.260	9.591	10.851	12.443	28.412	31.410	34.170	37.566	39.997
21	8.034	8.897	10.283	11.591	13.240	29.615	32.671	35.479	38.932	41.401
22	8.643	9.542	10.982	12.338	14.041	30.813	33.924	36.781	40.289	42.796
23	9.260	10.196	11.689	13.091	14.848	32.007	35.172	38.076	41.638	44.181
24	9.886	10.856	12.401	13.848	15.659	33.196	36.415	39.364	42.980	45.559
25	10.520	11.524	13.120	14.611	16.473	34.382	37.652	40.646	44.314	46.928
26	11.160	12.198	13.844	15.379	17.292	35.563	38.885	41.923	45.642	48.290
27	11.808	12.879	14.573	16.151	18.114	36.741	40.113	43.195	46.963	49.645
28	12.461	13.565	15.308	16.928	18.939	37.916	41.337	44.461	48.278	50.993
29	13.121	14.256	16.047	17.708	19.768	39.087	42.557	45.722	49.588	52.336
30	13.787	14.953	16.791	18.493	20.599	40.256	43.773	46.979	50.892	53.672
40	20.707	22.164	24.433	26.509	29.051	51.805	55.758	59.342	63.691	66.766
50	27.991	29.707	32.357	34.764	37.689	63.167	67.505	71.420	76.154	79.490
60	35.534	37.485	40.482	43.188	46.459	74.397	79.082	83.298	88.379	91.952
70	43.275	45.442	48.758	51.739	55.329	85.527	90.531	95.023	100.425	104.215
80	51.172	53.540	57.153	60.391	64.278	96.578	101.879	106.629	112.329	116.321
90	59.196	61.754	65.647	69.126	73.291	107.565	113.145	118.136	124.116	128.299
100	67.328	70.065	74.222	77.929	82.358	118.498	124.342	129.561	135.807	140.169

TABLE X   THE F DISTRIBUTION TABLE

Area in the Right Tail under the F Distribution Curve = .01

Degrees of Freedom for the Denominator	Degrees of Freedom for the Numerator																		
	1	2	3	4	5	6	7	8	9	10	11	12	15	20	25	30	40	50	100
1	4052	5000	5403	5625	5764	5859	5928	5981	6022	6056	6083	6106	6157	6209	6240	6261	6287	6303	6334
2	98.50	99.00	99.17	99.25	99.30	99.33	99.36	99.37	99.39	99.40	99.41	99.42	99.43	99.45	99.46	99.47	99.47	99.48	99.49
3	34.12	30.82	29.46	28.71	28.24	27.91	27.67	27.49	27.35	27.23	27.13	27.05	26.87	26.69	26.58	26.50	26.41	26.35	26.24
4	21.20	18.00	16.69	15.98	15.52	15.21	14.98	14.80	14.66	14.55	14.45	14.37	14.20	14.02	13.91	13.84	13.75	13.69	13.58
5	16.26	13.27	12.06	11.39	10.97	10.67	10.46	10.29	10.16	10.05	9.96	9.89	9.72	9.55	9.45	9.38	9.29	9.24	9.13
6	13.75	10.92	9.78	9.15	8.75	8.47	8.26	8.10	7.98	7.87	7.79	7.72	7.56	7.40	7.30	7.23	7.14	7.09	6.99
7	12.25	9.55	8.45	7.85	7.46	7.19	6.99	6.84	6.72	6.62	6.54	6.47	6.31	6.16	6.06	5.99	5.91	5.86	5.75
8	11.26	8.65	7.59	7.01	6.63	6.37	6.18	6.03	5.91	5.81	5.73	5.67	5.52	5.36	5.26	5.20	5.12	5.07	4.96
9	10.56	8.02	6.99	6.42	6.06	5.80	5.61	5.47	5.35	5.26	5.18	5.11	4.96	4.81	4.71	4.65	4.57	4.52	4.41
10	10.04	7.56	6.55	5.99	5.64	5.39	5.20	5.06	4.94	4.85	4.77	4.71	4.56	4.41	4.31	4.25	4.17	4.12	4.01
11	9.65	7.21	6.22	5.67	5.32	5.07	4.89	4.74	4.63	4.54	4.46	4.40	4.25	4.10	4.01	3.94	3.86	3.81	3.71
12	9.33	6.93	5.95	5.41	5.06	4.82	4.64	4.50	4.39	4.30	4.22	4.16	4.01	3.86	3.76	3.70	3.62	3.57	3.47
13	9.07	6.70	5.74	5.21	4.86	4.62	4.44	4.30	4.19	4.10	4.02	3.96	3.82	3.66	3.57	3.51	3.43	3.38	3.27
14	8.86	6.51	5.56	5.04	4.69	4.46	4.28	4.14	4.03	3.94	3.86	3.80	3.66	3.51	3.41	3.35	3.27	3.22	3.11
15	8.68	6.36	5.42	4.89	4.56	4.32	4.14	4.00	3.89	3.80	3.73	3.67	3.52	3.37	3.28	3.21	3.13	3.08	2.98
16	8.53	6.23	5.29	4.77	4.44	4.20	4.03	3.89	3.78	3.69	3.62	3.55	3.41	3.26	3.16	3.10	3.02	2.97	2.86
17	8.40	6.11	5.18	4.67	4.34	4.10	3.93	3.79	3.68	3.59	3.52	3.46	3.31	3.16	3.07	3.00	2.92	2.87	2.76
18	8.29	6.01	5.09	4.58	4.25	4.01	3.84	3.71	3.60	3.51	3.43	3.37	3.23	3.08	2.98	2.92	2.84	2.78	2.68
19	8.18	5.93	5.01	4.50	4.17	3.94	3.77	3.63	3.52	3.43	3.36	3.30	3.15	3.00	2.91	2.84	2.76	2.71	2.60
20	8.10	5.85	4.94	4.43	4.10	3.87	3.70	3.56	3.46	3.37	3.29	3.23	3.09	2.94	2.84	2.78	2.69	2.64	2.54
21	8.02	5.78	4.87	4.37	4.04	3.81	3.64	3.51	3.40	3.31	3.24	3.17	3.03	2.88	2.79	2.72	2.64	2.58	2.48
22	7.95	5.72	4.82	4.31	3.99	3.76	3.59	3.45	3.35	3.26	3.18	3.12	2.98	2.83	2.73	2.67	2.58	2.53	2.42
23	7.88	5.66	4.76	4.26	3.94	3.71	3.54	3.41	3.30	3.21	3.14	3.07	2.93	2.78	2.69	2.62	2.54	2.48	2.37
24	7.82	5.61	4.72	4.22	3.90	3.67	3.50	3.36	3.26	3.17	3.09	3.03	2.89	2.74	2.64	2.58	2.49	2.44	2.33
25	7.77	5.57	4.68	4.18	3.85	3.63	3.46	3.32	3.22	3.13	3.06	2.99	2.85	2.70	2.60	2.54	2.45	2.40	2.29
30	7.56	5.39	4.51	4.02	3.70	3.47	3.30	3.17	3.07	2.98	2.91	2.84	2.70	2.55	2.45	2.39	2.30	2.25	2.13
40	7.31	5.18	4.31	3.83	3.51	3.29	3.12	2.99	2.89	2.80	2.73	2.66	2.52	2.37	2.27	2.20	2.11	2.06	1.94
50	7.17	5.06	4.20	3.72	3.41	3.19	3.02	2.89	2.78	2.70	2.63	2.56	2.42	2.27	2.17	2.10	2.01	1.95	1.82
100	6.90	4.82	3.98	3.51	3.21	2.99	2.82	2.69	2.59	2.50	2.43	2.37	2.22	2.07	1.97	1.89	1.80	1.74	1.60

Degrees of Freedom for the Denominator

TABLE X   THE *F* DISTRIBUTION TABLE   **D27**

TABLE X (*Continued*)

Area in the Right Tail under the *F* Distribution Curve = .025

Degrees of Freedom for the Numerator

	1	2	3	4	5	6	7	8	9	10	11	12	15	20	25	30	40	50	100
1	647.8	799.5	864.2	899.6	921.8	937.1	948.2	956.7	963.3	968.6	973.0	976.7	984.9	993.1	998.1	1001	1006	1008	1013
2	38.51	39.00	39.17	39.25	39.30	39.33	39.36	39.37	39.39	39.40	39.41	39.41	39.43	39.45	39.46	39.46	39.47	39.48	39.49
3	17.44	16.04	15.44	15.10	14.88	14.73	14.62	14.54	14.47	14.42	14.37	14.34	14.25	14.17	14.12	14.08	14.04	14.01	13.96
4	12.22	10.65	9.98	9.61	9.36	9.20	9.07	8.98	8.90	8.84	8.79	8.75	8.66	8.56	8.50	8.46	8.41	8.38	8.32
5	10.01	8.43	7.76	7.39	7.15	6.98	6.85	6.76	6.68	6.62	6.57	6.52	6.43	6.33	6.27	6.23	6.18	6.14	6.08
6	8.81	7.26	6.60	6.23	5.99	5.82	5.70	5.60	5.52	5.46	5.41	5.37	5.27	5.17	5.11	5.07	5.01	4.98	4.92
7	8.07	6.54	5.89	5.52	5.29	5.12	4.99	4.90	4.82	4.76	4.71	4.67	4.57	4.47	4.40	4.36	4.31	4.28	4.21
8	7.57	6.06	5.42	5.05	4.82	4.65	4.53	4.43	4.36	4.30	4.24	4.20	4.10	4.00	3.94	3.89	3.84	3.81	3.74
9	7.21	5.72	5.08	4.72	4.48	4.32	4.20	4.10	4.03	3.96	3.91	3.87	3.77	3.67	3.60	3.56	3.51	3.47	3.40
10	6.94	5.46	4.83	4.47	4.24	4.07	3.95	3.85	3.78	3.72	3.66	3.62	3.52	3.42	3.35	3.31	3.26	3.22	3.15
11	6.72	5.26	4.63	4.28	4.04	3.88	3.76	3.66	3.59	3.53	3.47	3.43	3.33	3.23	3.16	3.12	3.06	3.03	2.96
12	6.55	5.10	4.47	4.12	3.89	3.73	3.61	3.51	3.44	3.37	3.32	3.28	3.18	3.07	3.01	2.96	2.91	2.87	2.80
13	6.41	4.97	4.35	4.00	3.77	3.60	3.48	3.39	3.31	3.25	3.20	3.15	3.05	2.95	2.88	2.84	2.78	2.74	2.67
14	6.30	4.86	4.24	3.89	3.66	3.50	3.38	3.29	3.21	3.15	3.09	3.05	2.95	2.84	2.78	2.73	2.67	2.64	2.56
15	6.20	4.77	4.15	3.80	3.58	3.41	3.29	3.20	3.12	3.06	3.01	2.96	2.86	2.76	2.69	2.64	2.59	2.55	2.47
16	6.12	4.69	4.08	3.73	3.50	3.34	3.22	3.12	3.05	2.99	2.93	2.89	2.79	2.68	2.61	2.57	2.51	2.47	2.40
17	6.04	4.62	4.01	3.66	3.44	3.28	3.16	3.06	2.98	2.92	2.87	2.82	2.72	2.62	2.55	2.50	2.44	2.41	2.33
18	5.98	4.56	3.95	3.61	3.38	3.22	3.10	3.01	2.93	2.87	2.81	2.77	2.67	2.56	2.49	2.44	2.38	2.35	2.27
19	5.92	4.51	3.90	3.56	3.33	3.17	3.05	2.96	2.88	2.82	2.76	2.72	2.62	2.51	2.44	2.39	2.33	2.30	2.22
20	5.87	4.46	3.86	3.51	3.29	3.13	3.01	2.91	2.84	2.77	2.72	2.68	2.57	2.46	2.40	2.35	2.29	2.25	2.17
21	5.83	4.42	3.82	3.48	3.25	3.09	2.97	2.87	2.80	2.73	2.68	2.64	2.53	2.42	2.36	2.31	2.25	2.21	2.13
22	5.79	4.38	3.78	3.44	3.22	3.05	2.93	2.84	2.76	2.70	2.65	2.60	2.50	2.39	2.32	2.27	2.21	2.17	2.09
23	5.75	4.35	3.75	3.41	3.18	3.02	2.90	2.81	2.73	2.67	2.62	2.57	2.47	2.36	2.29	2.24	2.18	2.14	2.06
24	5.72	4.32	3.72	3.38	3.15	2.99	2.87	2.78	2.70	2.64	2.59	2.54	2.44	2.33	2.26	2.21	2.15	2.11	2.02
25	5.69	4.29	3.69	3.35	3.13	2.97	2.85	2.75	2.68	2.61	2.56	2.51	2.41	2.30	2.23	2.18	2.12	2.08	2.00
30	5.57	4.18	3.59	3.25	3.03	2.87	2.75	2.65	2.57	2.51	2.46	2.41	2.31	2.20	2.12	2.07	2.01	1.97	1.88
40	5.42	4.05	3.46	3.13	2.90	2.74	2.62	2.53	2.45	2.39	2.33	2.29	2.18	2.07	1.99	1.94	1.88	1.83	1.74
50	5.34	3.97	3.39	3.05	2.83	2.67	2.55	2.46	2.38	2.32	2.26	2.22	2.11	1.99	1.92	1.87	1.80	1.75	1.66
100	5.18	3.83	3.25	2.92	2.70	2.54	2.42	2.32	2.24	2.18	2.12	2.08	1.97	1.85	1.77	1.71	1.64	1.59	1.48

Degrees of Freedom for the Denominator

TABLE X (*Continued*)

Area in the Right Tail under the *F* Distribution Curve = .05

	Degrees of Freedom for the Numerator																		
	1	2	3	4	5	6	7	8	9	10	11	12	15	20	25	30	40	50	100
1	161.5	199.5	215.7	224.6	230.2	234.0	236.8	238.9	240.5	241.9	243.0	243.9	246.0	248.0	249.3	250.1	251.1	251.8	253.0
2	18.51	19.00	19.16	19.25	19.30	19.33	19.35	19.37	19.38	19.40	19.40	19.41	19.43	19.45	19.46	19.46	19.47	19.48	19.49
3	10.13	9.55	9.28	9.12	9.01	8.94	8.89	8.85	8.81	8.79	8.76	8.74	8.70	8.66	8.63	8.62	8.59	8.58	8.55
4	7.71	6.94	6.59	6.39	6.26	6.16	6.09	6.04	6.00	5.96	5.94	5.91	5.86	5.80	5.77	5.75	5.72	5.70	5.66
5	6.61	5.79	5.41	5.19	5.05	4.95	4.88	4.82	4.77	4.74	4.70	4.68	4.62	4.56	4.52	4.50	4.46	4.44	4.41
6	5.99	5.14	4.76	4.53	4.39	4.28	4.21	4.15	4.10	4.06	4.03	4.00	3.94	3.87	3.83	3.81	3.77	3.75	3.71
7	5.59	4.74	4.35	4.12	3.97	3.87	3.79	3.73	3.68	3.64	3.60	3.57	3.51	3.44	3.40	3.38	3.34	3.32	3.27
8	5.32	4.46	4.07	3.84	3.69	3.58	3.50	3.44	3.39	3.35	3.31	3.28	3.22	3.15	3.11	3.08	3.04	3.02	2.97
9	5.12	4.26	3.86	3.63	3.48	3.37	3.29	3.23	3.18	3.14	3.10	3.07	3.01	2.94	2.89	2.86	2.83	2.80	2.76
10	4.96	4.10	3.71	3.48	3.33	3.22	3.14	3.07	3.02	2.98	2.94	2.91	2.85	2.77	2.73	2.70	2.66	2.64	2.59
11	4.84	3.98	3.59	3.36	3.20	3.09	3.01	2.95	2.90	2.85	2.82	2.79	2.72	2.65	2.60	2.57	2.53	2.51	2.46
12	4.75	3.89	3.49	3.26	3.11	3.00	2.91	2.85	2.80	2.75	2.72	2.69	2.62	2.54	2.50	2.47	2.43	2.40	2.35
13	4.67	3.81	3.41	3.18	3.03	2.92	2.83	2.77	2.71	2.67	2.63	2.60	2.53	2.46	2.41	2.38	2.34	2.31	2.26
14	4.60	3.74	3.34	3.11	2.96	2.85	2.76	2.70	2.65	2.60	2.57	2.53	2.46	2.39	2.34	2.31	2.27	2.24	2.19
15	4.54	3.68	3.29	3.06	2.90	2.79	2.71	2.64	2.59	2.54	2.51	2.48	2.40	2.33	2.28	2.25	2.20	2.18	2.12
16	4.49	3.63	3.24	3.01	2.85	2.74	2.66	2.59	2.54	2.49	2.46	2.42	2.35	2.28	2.23	2.19	2.15	2.12	2.07
17	4.45	3.59	3.20	2.96	2.81	2.70	2.61	2.55	2.49	2.45	2.41	2.38	2.31	2.23	2.18	2.15	2.10	2.08	2.02
18	4.41	3.55	3.16	2.93	2.77	2.66	2.58	2.51	2.46	2.41	2.37	2.34	2.27	2.19	2.14	2.11	2.06	2.04	1.98
19	4.38	3.52	3.13	2.90	2.74	2.63	2.54	2.48	2.42	2.38	2.34	2.31	2.23	2.16	2.11	2.07	2.03	2.00	1.94
20	4.35	3.49	3.10	2.87	2.71	2.60	2.51	2.45	2.39	2.35	2.31	2.28	2.20	2.12	2.07	2.04	1.99	1.97	1.91
21	4.32	3.47	3.07	2.84	2.68	2.57	2.49	2.42	2.37	2.32	2.28	2.25	2.18	2.10	2.05	2.01	1.96	1.94	1.88
22	4.30	3.44	3.05	2.82	2.66	2.55	2.46	2.40	2.34	2.30	2.26	2.23	2.15	2.07	2.02	1.97	1.94	1.91	1.85
23	4.28	3.42	3.03	2.80	2.64	2.53	2.44	2.37	2.32	2.27	2.24	2.20	2.13	2.05	2.00	1.96	1.91	1.88	1.82
24	4.26	3.40	3.01	2.78	2.62	2.51	2.42	2.36	2.30	2.25	2.22	2.18	2.11	2.03	1.97	1.94	1.89	1.86	1.80
25	4.24	3.39	2.99	2.76	2.60	2.49	2.40	2.34	2.28	2.24	2.20	2.16	2.09	2.01	1.96	1.92	1.87	1.84	1.78
30	4.17	3.32	2.92	2.69	2.53	2.42	2.33	2.27	2.21	2.16	2.13	2.09	2.01	1.93	1.88	1.84	1.79	1.76	1.70
40	4.08	3.23	2.84	2.61	2.45	2.34	2.25	2.18	2.12	2.08	2.04	2.00	1.92	1.84	1.78	1.74	1.69	1.66	1.59
50	4.03	3.18	2.79	2.56	2.40	2.29	2.20	2.13	2.07	2.03	1.99	1.95	1.87	1.78	1.73	1.69	1.63	1.60	1.52
100	3.94	3.09	2.70	2.46	2.31	2.19	2.10	2.03	1.97	1.93	1.89	1.85	1.77	1.68	1.62	1.57	1.52	1.48	1.39

Degrees of Freedom for the Denominator

TABLE X   THE F DISTRIBUTION TABLE

D29

TABLE X (*Continued*)

Area in the Right Tail under the F Distribution Curve = .10

Degrees of Freedom for the Numerator

	1	2	3	4	5	6	7	8	9	10	11	12	15	20	25	30	40	50	100
1	39.86	49.50	53.59	55.83	57.24	58.20	58.91	59.44	59.86	60.19	60.47	60.71	61.22	61.74	62.05	62.26	62.53	62.69	63.01
2	8.53	9.00	9.16	9.24	9.29	9.33	9.35	9.37	9.38	9.39	9.40	9.41	9.42	9.44	9.45	9.46	9.47	9.47	9.48
3	5.54	5.46	5.39	5.34	5.31	5.28	5.27	5.25	5.24	5.23	5.22	5.22	5.20	5.18	5.17	5.17	5.16	5.15	5.14
4	4.54	4.32	4.19	4.11	4.05	4.01	3.98	3.95	3.94	3.92	3.91	3.90	3.87	3.84	3.83	3.82	3.80	3.80	3.78
5	4.06	3.78	3.62	3.52	3.45	3.40	3.37	3.34	3.32	3.30	3.28	3.27	3.24	3.21	3.19	3.17	3.16	3.15	3.13
6	3.78	3.46	3.29	3.18	3.11	3.05	3.01	2.98	2.96	2.94	2.92	2.90	2.87	2.84	2.81	2.80	2.78	2.77	2.75
7	3.59	3.26	3.07	2.96	2.88	2.83	2.78	2.75	2.72	2.70	2.68	2.67	2.63	2.59	2.57	2.56	2.54	2.52	2.50
8	3.46	3.11	2.92	2.81	2.73	2.67	2.62	2.59	2.56	2.54	2.52	2.50	2.46	2.42	2.40	2.38	2.36	2.35	2.32
9	3.36	3.01	2.81	2.69	2.61	2.55	2.51	2.47	2.44	2.42	2.40	2.38	2.34	2.30	2.27	2.25	2.23	2.22	2.19
10	3.29	2.92	2.73	2.61	2.52	2.46	2.41	2.38	2.35	2.32	2.30	2.28	2.24	2.20	2.17	2.16	2.13	2.12	2.09
11	3.23	2.86	2.66	2.54	2.45	2.39	2.34	2.30	2.27	2.25	2.23	2.21	2.17	2.12	2.10	2.08	2.05	2.04	2.01
12	3.18	2.81	2.61	2.48	2.39	2.33	2.28	2.24	2.21	2.19	2.17	2.15	2.10	2.06	2.03	2.01	1.99	1.97	1.94
13	3.14	2.76	2.56	2.43	2.35	2.28	2.23	2.20	2.16	2.14	2.12	2.10	2.05	2.01	1.98	1.96	1.93	1.92	1.88
14	3.10	2.73	2.52	2.39	2.31	2.24	2.19	2.15	2.12	2.10	2.07	2.05	2.01	1.96	1.93	1.91	1.89	1.87	1.83
15	3.07	2.70	2.49	2.36	2.27	2.21	2.16	2.12	2.09	2.06	2.04	2.02	1.97	1.92	1.89	1.87	1.85	1.83	1.79
16	3.05	2.67	2.46	2.33	2.24	2.18	2.13	2.09	2.06	2.03	2.01	1.99	1.94	1.89	1.86	1.84	1.81	1.79	1.76
17	3.03	2.64	2.44	2.31	2.22	2.15	2.10	2.06	2.03	2.00	1.98	1.96	1.91	1.86	1.83	1.81	1.78	1.76	1.73
18	3.01	2.62	2.42	2.29	2.20	2.13	2.08	2.04	2.00	1.98	1.95	1.93	1.89	1.84	1.80	1.78	1.75	1.74	1.70
19	2.99	2.61	2.40	2.27	2.18	2.11	2.06	2.02	1.98	1.96	1.93	1.91	1.86	1.81	1.78	1.76	1.73	1.71	1.67
20	2.97	2.59	2.38	2.25	2.16	2.09	2.04	2.00	1.96	1.94	1.91	1.89	1.84	1.79	1.76	1.74	1.71	1.69	1.65
21	2.96	2.57	2.36	2.23	2.14	2.08	2.02	1.98	1.95	1.92	1.90	1.87	1.83	1.78	1.74	1.72	1.69	1.67	1.63
22	2.95	2.56	2.35	2.22	2.13	2.06	2.01	1.97	1.93	1.90	1.88	1.86	1.81	1.76	1.73	1.70	1.67	1.65	1.61
23	2.94	2.55	2.34	2.21	2.11	2.05	1.99	1.95	1.92	1.89	1.87	1.84	1.80	1.74	1.71	1.69	1.66	1.64	1.59
24	2.93	2.54	2.33	2.19	2.10	2.04	1.98	1.94	1.91	1.88	1.85	1.83	1.78	1.73	1.70	1.67	1.64	1.62	1.58
25	2.92	2.53	2.32	2.18	2.09	2.02	1.97	1.93	1.89	1.87	1.84	1.82	1.77	1.72	1.68	1.66	1.63	1.61	1.56
30	2.88	2.49	2.28	2.14	2.05	1.98	1.93	1.88	1.85	1.82	1.79	1.77	1.72	1.67	1.63	1.61	1.57	1.55	1.51
40	2.84	2.44	2.23	2.09	2.00	1.93	1.87	1.83	1.79	1.76	1.74	1.71	1.66	1.61	1.57	1.54	1.51	1.48	1.43
50	2.81	2.41	2.20	2.06	1.97	1.90	1.84	1.80	1.76	1.73	1.70	1.68	1.63	1.57	1.53	1.50	1.46	1.44	1.39
100	2.76	2.36	2.14	2.00	1.91	1.83	1.78	1.73	1.69	1.66	1.64	1.61	1.56	1.49	1.45	1.42	1.38	1.35	1.29

Degrees of Freedom for the Denominator

TABLE XI   CRITICAL VALUES OF *X* FOR THE SIGN TEST

	One tail $\alpha = .005$ Two tail $\alpha = .01$		One tail $\alpha = .01$ Two tail $\alpha = .02$		One tail $\alpha = .025$ Two tail $\alpha = .05$		One tail $\alpha = .05$ Two tail $\alpha = .10$	
n	Lower critical value	Upper critical value	Lower critical value	Upper critical value	Lower critical value	Upper critical value	Lower critical value	Upper critical value
1	—	—	—	—	—	—	—	—
2	—	—	—	—	—	—	—	—
3	—	—	—	—	—	—	—	—
4	—	—	—	—	—	—	—	—
5	—	—	—	—	—	—	0	5
6	—	—	—	—	0	6	0	6
7	—	—	0	7	0	7	0	7
8	0	8	0	8	0	8	1	7
9	0	9	0	9	1	8	1	8
10	0	10	0	10	1	9	1	9
11	0	11	1	10	1	10	2	9
12	1	11	1	11	2	10	2	10
13	1	12	1	12	2	11	3	10
14	1	13	2	12	2	12	3	11
15	2	13	2	13	3	12	3	12
16	2	14	2	14	3	13	4	12
17	2	15	3	14	4	13	4	13
18	3	15	3	15	4	14	5	13
19	3	16	4	15	4	15	5	14
20	3	17	4	16	5	15	5	15
21	4	17	4	17	5	16	6	15
22	4	18	5	17	5	17	6	16
23	4	19	5	18	6	17	7	16
24	5	19	5	19	6	18	7	17
25	5	20	6	19	7	18	7	18

Source: D. B. Owen, *Handbook of Statistical Tables*. Addison-Wesley Publishing Company, Inc., 1962. Reprinted with permission.

TABLE XII   CRITICAL VALUES OF *T* FOR THE WILCOXON SIGNED-RANK TEST   **D31**

TABLE XII   CRITICAL VALUES OF *T* FOR THE WILCOXON SIGNED-RANK TEST

n	One-tailed $\alpha = .005$ Two-tailed $\alpha = .01$	One-tailed $\alpha = .01$ Two-tailed $\alpha = .02$	One-tailed $\alpha = .025$ Two-tailed $\alpha = .05$	One-tailed $\alpha = .05$ Two-tailed $\alpha = .10$
1	—	—	—	—
2	—	—	—	—
3	—	—	—	—
4	—	—	—	—
5	—	—	—	1
6	—	—	1	2
7	—	0	2	4
8	0	2	4	6
9	2	3	6	8
10	3	5	8	11
11	5	7	11	14
12	7	10	14	17
13	10	13	17	21
14	13	16	21	26
15	16	20	25	30

Source: Some Rapid Approximate Statistical Procedures, 1964, Lederle Laboratories, A Division of American Cyanamid Company, Pearl River, New York. Reprinted with the permission of Lederle Laboratories/Wyeth-Ayerst Laboratories.

TABLE XIII   CRITICAL VALUES OF T FOR THE WILCOXON RANK SUM TEST

a. One-tailed $\alpha = .025$; Two-tailed $\alpha = .05$

n_1 n_2	3		4		5		6		7		8		9		10	
	T_L	T_U	T_L	T_U	T_L	T_U	T_L	T_U	T_L	T_U	T_L	T_U	T_L	T_U	T_L	T_U
3	5	16	6	18	6	21	7	23	7	26	8	28	8	31	9	33
4	6	18	11	25	12	28	12	32	13	35	14	38	15	41	16	44
5	6	21	12	28	18	37	19	41	20	45	21	49	22	53	24	56
6	7	23	12	32	19	41	26	52	28	56	29	61	31	65	32	70
7	7	26	13	35	20	45	28	56	37	68	39	73	41	78	43	83
8	8	28	14	38	21	49	29	61	39	73	49	87	51	93	54	98
9	8	31	15	41	22	53	31	65	41	78	51	93	63	108	66	114
10	9	33	16	44	24	56	32	70	43	83	54	98	66	114	79	131

b. One-tailed $\alpha = .05$; Two-tailed $\alpha = .10$

n_1 n_2	3		4		5		6		7		8		9		10	
	T_L	T_U	T_L	T_U	T_L	T_U	T_L	T_U	T_L	T_U	T_L	T_U	T_L	T_U	T_L	T_U
3	6	15	7	17	7	20	8	22	9	24	9	27	10	29	11	31
4	7	17	12	24	13	27	14	30	15	33	16	36	17	39	18	42
5	7	20	13	27	19	36	20	40	22	43	24	46	25	50	26	54
6	8	22	14	30	20	40	28	50	30	54	32	58	33	63	35	67
7	9	24	15	33	22	43	30	54	39	66	41	71	43	76	46	80
8	9	27	16	36	24	46	32	58	41	71	52	84	54	90	57	95
9	10	29	17	39	25	50	33	63	43	76	54	90	66	105	69	111
10	11	31	18	42	26	54	35	67	46	80	57	95	69	111	83	127

Source: Some Rapid Approximate Statistical Procedures, 1964, Lederle Laboratories, A Division of American Cyanamid Company, Pearl River, New York. Reprinted with the permission of Lederle Laboratories/Wyeth-Ayerst Laboratories.

TABLE XIV   CRITICAL VALUES FOR THE SPEARMAN RHO RANK CORRELATION COEFFICIENT TEST

	One-tailed α			
	.05	.025	.01	.005
	Two-tailed α			
n	.10	.05	.02	.01
5	±.900	—	—	—
6	±.829	±.886	±.943	—
7	±.714	±.786	±.893	±.929
8	±.643	±.738	±.833	±.881
9	±.600	±.700	±.783	±.833
10	±.564	±.648	±.745	±.794
11	±.536	±.618	±.709	±.755
12	±.503	±.587	±.678	±.727
13	±.475	±.566	±.672	±.744
14	±.456	±.544	±.645	±.714
15	±.440	±.524	±.622	±.688
16	±.425	±.506	±.601	±.665
17	±.411	±.490	±.582	±.644
18	±.399	±.475	±.564	±.625
19	±.388	±.462	±.548	±.607
20	±.377	±.450	±.534	±.591
21	±.368	±.438	±.520	±.576
22	±.359	±.428	±.508	±.562
23	±.351	±.418	±.496	±.549
24	±.343	±.409	±.485	±.537
25	±.336	±.400	±.475	±.526
26	±.329	±.392	±.465	±.515
27	±.323	±.384	±.456	±.505
28	±.317	±.377	±.448	±.496
29	±.311	±.370	±.440	±.487
30	±.305	±.364	±.432	±.478

TABLE XV   CRITICAL VALUES FOR A TWO-TAILED RUNS TEST WITH $\alpha = .05$

n_1 \ n_2	5	6	7	8	9	10	11	12	13	14	15
2	—	—	—	—	—	—	—	2 / 6	2 / 6	2 / 6	2 / 6
3	—	2 / 8	2 / 8	2 / 8	2 / 8	2 / 8	2 / 8	2 / 8	2 / 8	2 / 8	3 / 8
4	2 / 9	2 / 9	2 / 10	3 / 10	3 / 10	3 / 10	3 / 10	3 / 10	3 / 10	3 / 10	3 / 10
5	2 / 10	3 / 10	3 / 11	3 / 11	3 / 12	3 / 12	4 / 12	4 / 12	4 / 12	4 / 12	4 / 12
6	3 / 10	3 / 11	3 / 12	3 / 12	4 / 13	4 / 13	4 / 13	4 / 13	5 / 14	5 / 14	5 / 14
7	3 / 11	3 / 12	3 / 13	4 / 13	4 / 14	5 / 14	5 / 14	5 / 14	5 / 15	5 / 15	6 / 15
8	3 / 11	3 / 12	4 / 13	4 / 14	5 / 14	5 / 15	5 / 15	6 / 16	6 / 16	6 / 16	6 / 16
9	3 / 12	4 / 13	4 / 14	5 / 14	5 / 15	5 / 16	6 / 16	6 / 16	6 / 17	7 / 17	7 / 18
10	3 / 12	4 / 13	5 / 14	5 / 15	5 / 16	6 / 16	6 / 17	7 / 17	7 / 18	7 / 18	7 / 18
11	4 / 12	4 / 13	5 / 14	5 / 15	6 / 16	6 / 17	7 / 17	7 / 18	7 / 19	8 / 19	8 / 19
12	4 / 12	4 / 13	5 / 14	6 / 16	6 / 16	7 / 17	7 / 18	7 / 19	8 / 19	8 / 20	8 / 20
13	4 / 12	5 / 14	5 / 15	6 / 16	6 / 17	7 / 18	7 / 19	8 / 19	8 / 20	9 / 20	9 / 21
14	4 / 12	5 / 14	5 / 15	6 / 16	7 / 17	7 / 18	8 / 19	8 / 20	9 / 20	9 / 21	9 / 22
15	4 / 12	5 / 14	6 / 15	6 / 16	7 / 18	7 / 18	8 / 19	8 / 20	9 / 21	9 / 22	10 / 22

Source: Frieda S. Swed and C. Eisenhart, "Tables for Testing Randomness of Grouping in a Sequence of Alternatives," *The Annals of Statistics* 14(1943). Reprinted with permission of the Institute of Mathematical Statistics.

ANSWERS TO SELECTED ODD-NUMBERED EXERCISES AND SELF-REVIEW TESTS

(*Note:* Due to differences in rounding, the answers obtained by readers may differ slightly from the ones given in this Appendix.)

CHAPTER 1

1.7   **a.** population   **b.** sample   **c.** population   **d.** sample   **e.** population

1.11   **a.** score   **b.** five observations   **c.** five elements

1.15   **a.** quantitative   **b.** quantitative   **c.** qualitative   **d.** quantitative   **e.** qualitative

1.17   **a.** discrete   **b.** continuous   **d.** continuous

1.21   **a.** cross-section data   **b.** cross-section data   **c.** time-series data   **d.** time-series data

1.23   **a.** $\Sigma f = 69$ **b.** $\Sigma m^2 = 1363$ **c.** $\Sigma mf = 922$ **d.** $\Sigma m^2 f = 17{,}128$

1.25   **a.** $\Sigma x = 88$ **b.** $\Sigma y = 58$ **c.** $\Sigma xy = 855$ **d.** $\Sigma x^2 = 1590$ **e.** $\Sigma y^2 = 622$

1.27   **a.** $\Sigma x = 13$ **b.** $(\Sigma x)^2 = 169$ **c.** $\Sigma x^2 = 35$ **1.29**   **a.** $\Sigma x = 61$ **b.** $(\Sigma x)^2 = 3721$ **c.** $\Sigma x^2 = 555$

1.33   **a.** sample   **b.** population for the year   **c.** sample   **d.** population

1.35   **a.** sampling without replacement   **b.** sampling with replacement

1.37   **a.** $\Sigma x = 56$ **b.** $(\Sigma x)^2 = 3136$ **c.** $\Sigma x^2 = 586$

1.39   **a.** $\Sigma m = 59$ **b.** $\Sigma f^2 = 2662$ **c.** $\Sigma mf = 1508$ **d.** $\Sigma m^2 f = 24{,}884$ **e.** $\Sigma m^2 = 867$

Self-Review Test

1. b   **2.** c   **3.** **a.** sampling without replacement   **b.** sampling with replacement

4. **a.** qualitative   **b.** quantitative (discrete)   **c.** quantitative (continuous)

6. **a.** $\Sigma x = 106$ **b.** $\Sigma x^2 = 3208$ **c.** $(\Sigma x)^2 = 11{,}236$

7. **a.** $\Sigma m = 45$ **b.** $\Sigma f = 112$ **c.** $\Sigma m^2 = 495$ **d.** $\Sigma mf = 975$ **e.** $\Sigma m^2 f = 9855$ **f.** $\Sigma f^2 = 2994$

CHAPTER 2

2.3   **c.** 26.7%   **d.** 73.4%   **2.5** **c.** 52%   **2.7 c.** 40.0%   **2.15 d.** 62%

2.17   **a.** class limits: $1–$4, $5–$8, $9–$12, $13–$16, $17–$20
     b. class boundaries: $0.5, $4.5, $8.5, $12.5, $16.5, $20.5; width = $4
     c. class midpoints: $2.5, $6.5, $10.5, $14.5, $18.5

2.19   **d.** 30%   **2.29 c.** 12   **2.35 c.** 38%   **e.** about 52%

2.47   218, 245, 256, 329, 367, 383, 397, 404, 427, 433, 471, 523, 537, 551, 563, 581, 592, 622, 636, 647, 655, 678, 689, 810, 841

2.57   **d.** 17.5%   **2.59 c.** 50.0%   **2.61 c.** 56.7%

2.63   **d.** Boundaries of the fourth class are $4200.5 and $5600.5; width = $1400.

2.73   No. The older group may drive more miles per week than the younger group.

2.75   **b.** No, because the attack rates are not given as percentages of a whole population, but as percentages for each country.
     c. No. We would have to know the percentages of males and females in the French workforce.

2.77   **b.** **ii.** The display shows a bimodal distribution due to the presence of both males and females in the sample.

Self-Review Test

2. a. 5   **b.** 5   **c.** 12   **d.** 4.5   **e.** 9   **f.** 100   **g.** .08   **4. c.** 35%   **5. c.** 45.8%
6. about 61%   **8.** 30, 33, 37, 42, 44, 46, 47, 49, 51, 53, 53, 56, 60, 67, 67, 71, 79

CHAPTER 3

3.5 mode   **3.9** mean = 3.00; median = 3.50; no mode   **3.11** mean = $736.83; median = $702.00

3.13 mean = $21.03; median = $21.09; no mode   **3.15** mean = $5.58; median = $5.44

3.17 mean = 3.59; median = 4.65; no mode   **3.19** mean = 6.40 hours; median = 7 hours; mode = 0 and 7 hours

3.21 mean = 29.4; median = 28.5; mode = 23

3.23 **a.** mean = 4409.88; median = 2457
b. outlier = 17,487; when the outlier is dropped: mean = 2541.71; median = 2298; mean changes by a larger amount
c. median

3.25 combined mean = $72.78   **3.27** total = $855   **3.29** age of the sixth person = 48 years

3.31 mean for data set I = 24.60; mean for data set II = 31.60
The mean of the second data set is equal to the mean of the first data set plus 7.

3.33 10% trimmed mean = 38.25 years   **3.39** range = 25; $\sigma^2 = 61.5$; $\sigma = 7.84$

3.41 **a.** $\bar{x} = 105$; deviations from the mean: $-23, 11, -40, 65, -13$. The sum of these deviations is zero.
b. range = 105; $s^2 = 1661.000$; $s = 40.76$

3.43 range = 13; $s^2 = 13.8409$; $s = 3.72$ **3.45** range = 27 pieces; $s^2 = 78.1$; $s = 8.84$ pieces

3.47 range = 20 pounds; $s^2 = 32.9714$; $s = 5.74$ pounds   **3.49** range = 30; $s^2 = 107.4286$; $s = 10.365$

3.51 range = 24.0; $s^2 = 58.5938$; $s = 7.65$ **3.53** $s = 0$

3.55 CV for salaries = 9.02%;   CV for years of schooling = 13.33%; The relative variation in salaries is lower.

3.57 $s = 14.64$ for both data sets   **3.61** $\bar{x} = 9.40$; $s^2 = 37.7114$; $s = 6.14$

3.63 $\mu = 74.40$ hours; $\sigma^2 = 493.4400$; $\sigma = 22.21$ hours

3.65 $\bar{x} = 24.65$ computers; $s^2 = 114.4385$; $s = 10.70$ computers

3.67 $\bar{x} = 36.80$ minutes; $s^2 = 597.7143$; $s = 24.45$ minutes   **3.69** $\bar{x} = 1.74$ sets; $s^2 = .9302$; $s = .96$ set

3.73 at least 75%;   at least 84%;   at least 88.89%   **3.75** 68%;   95%;   99.7%

3.77 **a.** at least 75%   **b.** at least 84%   **c.** at least 88.89%

3.79 **a. i.** at least 75%   **ii.** at least 88.89%   **b.** $765 to $1965

3.81 **a.** 95%   **b.** 68%   **c.** 99.7%   **3.83** **a. i.** 68%   **ii.** 99.7%   **b.** 3 to 11 years old

3.89 **a.** $Q_1 = 69$; $Q_2 = 73$; $Q_3 = 76.5$; $IQR = 7.5$ **b.** $P_{35} = 71$ **c.** 30.77%

3.91 **a.** $Q_1 = 35$; $Q_2 = 40$; $Q_3 = 43$; $IQR = 8$ **b.** $P_{79} = 45$ **c.** 36.67%

3.93 **a.** $Q_1 = 25$; $Q_2 = 28.5$; $Q_3 = 33$; $IQR = 8$ **b.** $P_{65} = 31$ **c.** 33.33%

3.95 **a.** $Q_1 = 69$; $Q_2 = 78$; $Q_3 = 88$; $IQR = 19$ **b.** $P_{93} = 97$ **c.** 57.89%   **3.97** no outlier

3.107 **a.** mean = 40.44; median = 29
b. outliers = -29 and 139; when the outliers are dropped: mean = 36.29; median = 29; mean changes by a larger amount
c. median

3.109 **a.** mean = 33.64; median = 20; modes = 12 and 16   **b.** range = 97; $s^2 = 876.6545$; $s = 29.61$

3.111 $\bar{x} = 5.08$ inches; $s^2 = 6.8506$; $s = 2.62$ inches

3.113 **a. i.** at least 75%   **ii.** at least 88.89%   **b.** 160 to 240 minutes

3.115 **a. i.** 68%   **ii.** 95%   **b.** 140 to 260 minutes

3.117 **a.** $Q_1 = 19$; $Q_2 = 29$; $Q_3 = 59$; $IQR = 40$ **b.** $P_{70} = 45$ **c.** 11%

3.119 The data set is skewed slightly to the right.   **3.121** The minimum score is 169.

3.123 **a.** new mean = 76.4 inches; new median = 78 inches; new range = 13 inches   **b.** new mean = 75.2 inches

3.125 mean = $22.31 per barrel   **3.127** **a.** trimmed mean = 9.5   **b.** 14.3%

3.129 **a.** low-risk: rate for A = .01; rate for B = .02
    b. high-risk: rate for A = .05; rate for B = .08
    c. overall: rate for A = 04; rate for B = .03
    d. A was used in more of the high-risk surgeries, for which both anesthetics had higher death rates.

3.131 **a.** $k = 1.41$ **b.** $k = 2.24$ **3.133** **b.** median

3.135 **b.** For men: mean = 82, median = 79, modes = 75, 79, and 92, $s = 12.08$, $Q_1 = 73.5$, $Q_3 = 89.5$, and $IQR = 16$.
    For women: mean = 97.53, median = 98, modes = 94 and 100, $s = 8.44$, $Q_1 = 94$, $Q_3 = 101$, and $IQR = 7$

3.137 **a.** mean = 30   **b.** mean = 50

3.139 **a.** at least 55.6%   **b.** 1 to 11 inches   **c.** 2.66 to 9.34 inches

Self-Review Test

1. b   **2.** a and d   **3.** c   **4.** c   **5.** b   **6.** b   **7.** a   **8.** a

9. b   **10.** a   **11.** b   **12.** c   **13.** a   **14.** a

15. mean = 10.9; median = 8; mode = 6; range = 26; $s^2 = 65.2111$; $s = 8.08$

19. **b.** $\bar{x} = 19.46$; $s^2 = 44.0400$; $s = 6.64$

20. **a.** **i.** at least 84%   **ii.** at least 88.89%   **b.** 2.9 to 11.7 years

21. **a.** **i.** 68%   **ii.** 99.7%   **b.** 2.9 to 11.7 years

22. $Q_1 = 7$; $Q_2 = 13.5$; $Q_3 = 17.5$; $IQR = 10.5$ **b.** $P_{68} = 16$ **c.** 68.75%

23. Data are skewed slightly to the right.   **24.** combined mean = $466.43

25. GPA of fifth student = 3.17   **26.** 10% trimmed mean = 197.75; trimmed mean is a better measure

27. **a.** mean for data set I = 19.75; mean for data set II = 16.75. The mean of the second data set is equal to the mean of the first data set minus 3.
    b. $s = 11.32$ for both data sets.

CHAPTER 4

4.3 $S = \{AB, AC, BA, BC, CA, CB\}$ **4.5** four possible outcomes; $S = \{RR, RG, GR, GG\}$

4.7 four possible outcomes; $S = \{DD, DG, GD, GG\}$ **4.9** $S = \{HHH, HHT, HTH, HTT, THH, THT, TTH, TTT\}$

4.11 **a.** {RG and GR}; a compound event   **b.** {RG, GR, and RR}; a compound event
    c. {RG, GR, and RR}; a compound event   **d.** {GR}; a simple event

4.13 **a.** {DG, GD, and GG}; a compound event   **b.** {DG and GD}; a compound event
    c. {GD}; a simple event   **d.** {DD, DG, and GD}; a compound event

4.19 $-.55$, 1.56, 5/3, $-2/7$ **4.21** not equally likely events; use relative frequency approach

4.23 subjective probability   **4.25** **a.** .450   **b.** .550   **4.27** .430   **4.29** .580

4.31 **a.** .200   **b.** .800   **4.33** .6667; .3333   **4.35** .325; .675   **4.37** **a.** .376   **b.** .117

4.39 use relative frequency approach   **4.45** 1296

4.47 **a.** no   **b.** no   **c.** $\bar{A} = \{1, 3, 4, 6, 8\}$; $\bar{B} = \{1, 3, 5, 6, 7\}$; $P(\bar{A}) = .625$; $P(\bar{B}) = .625$ **4.49** 20

4.51 960

4.53 **a.** **i.** .750   **ii.** .600   **iii.** .250   **iv.** .600
    b. Events "male" and "female" are mutually exclusive. Events "have shopped" and "male" are not mutually exclusive.
    c. Events "female" and "have shopped" are independent.

4.55 **a.** **i.** .3475   **ii.** .5425   **iii.** .2727   **iv.** .4545
    b. Events "male" and "in favor" are not mutually exclusive. Events "in favor" and "against" are mutually exclusive.
    c. Events "female" and "no opinion" are dependent.

4.57 **a.** **i.** .1725   **ii.** .4685   **iii.** .6337   **iv.** .1550
    b. Events "man" and "commutes for more than one hour" are not mutually exclusive. Events "less than 30 minutes" and "more than one hour" are mutually exclusive.
    c. Events "woman" and "commutes for 30 minutes to one hour" are dependent.

4.59 Events "female" and "pediatrician" are dependent but not mutually exclusive.

4.61 Events "female" and "business major" are dependent but not mutually exclusive.

4.63 $P(A) = .3333$; $P(\overline{A}) = .6667$ **4.65** .35 **4.71** a. .4543 b. .0980

4.73 a. .1520 b. .1824 **4.75** a. .2462 b. .1086 **4.77** .6923 **4.79** .725

4.81 a. i. .1958 ii. .2028 b. .0000 **4.83** a. i. .450 ii. .100

4.85 a. i. .225 ii. .035 b. .0000 **4.87** .2571 **4.89** .2667 **4.91** .0576

4.93 a. .0025 b. .9025 **4.95** .5120 **4.97** .5278 **4.99** .0299

4.105 a. .56 b. .76 **4.107** a. .52 b. .67 **4.109** a. .4825 b. .9161

4.111 a. .850 b. .700 c. 1.0 **4.113** a. .780 b. .550 c. .790 **4.115** .910

4.117 .77 **4.119** .9548 **4.121** .85 **4.123** .9744 **4.125** a. .2571 b. .1429

4.127 a. i. .4360 ii. .4800 iii. .3462 iv. .6809 v. .3400 vi. .6600
b. Events "female" and "prefers watching sports" are dependent but not mutually exclusive.

4.129 a. i. .750 ii. .700 iii. .225 iv. .775
b. Events "student athlete" and "should be paid" are dependent but not mutually exclusive.

4.131 a. .2809 b. .7191 **4.133** .0605 **4.135** .0048 **4.137** a. 17,576,000 b. 2600

4.139 a. 1/80,089,128 = .000000012 b. 1/1,953,393 = .000000511

4.141 a. .5000 b. .3333 c. No; the sixth toss is independent of the first five tosses. Equivalent to part a.

4.143 a. .030 b. .150 **4.145** b. .5687 **4.147** .30

4.149 a. .07 b. .17 c. .23 **4.151** .20

Self-Review Test

1. a **2.** b **3.** c **4.** a **5.** a **6.** b **7.** c **8.** b

9. b **10.** c **11.** b **12.** 36 **13.** a. .3333 b. .6667

14. a. Events "male" and "married" are dependent but not mutually exclusive. b. i. .4000 ii. .6786

15. .780 **16.** .1595 **17.** .4225 **18.** .600 **19.** a. .168 b. .688

20. a. i. .360 ii. .275 iii. .350 iv. .650
b. Events "male" and "yes" are dependent but not mutually exclusive.

CHAPTER 5

5.3 a. discrete random variable b. continuous random variable c. continuous random variable
d. discrete random variable e. discrete random variable f. continuous random variable

5.5 discrete random variable

5.9 a. not a valid probability distribution
b. a valid probability distribution
c. not a valid probability distribution

5.11 a. .17 b. .20 c. .58 d. .42 e. .42 f. .27 g. .68

5.13 b. i. .14 ii. .50 iii. .50 iv. .46

5.15 a.

x	2	3	4	5	6
$P(x)$	.12	.21	.34	.19	.14

b. approximate c. i. .21 ii. .88 iii. .67 iv. .33

5.17

x	0	1	2
$P(x)$	.1024	.4352	.4624

5.19

x	0	1	2
$P(x)$	.360	.480	.160

5.21

x	0	1	2
$P(x)$	.1474	.5052	.3474

5.23 a. $\mu = 1.590$; $\sigma = .960$ b. $\mu = 7.070$; $\sigma = 1.061$

5.25 $\mu = .440$ error; σ .852 error   **5.27** $\mu = 2.94$ camcorders; $s = 1.44$ camcorders

5.29 $\mu = 1.000$ heads; $\sigma = .707$ head   **5.31** $\mu = 1.896$ sets; $\sigma = 1.079$ sets

5.33 $\mu = .100$ car; $\sigma = .308$ car   **5.35** $\mu = \$38,250$; $\sigma = \$18,660$

5.37 $\mu = .500$ person; $\sigma = .584$ person

5.39 $3! = 6$; $(9 - 3)! = 720$; $9! = 362,880$; $(14 - 12)! = 2$;

$\binom{5}{3} = 10$; $\binom{7}{4} = 35$; $\binom{9}{3} = 84$; $\binom{4}{0} = 1$; $\binom{3}{3} = 1$

5.41 120   **5.43** 220   **5.45** 38,760   **5.47** 167,960

5.51 a. not a binomial experiment   b. a binomial experiment   c. a binomial experiment

5.53 a. .2541   b. .1536   c. .3241   **5.55** b. $\mu = 2.100$; $\sigma = 1.212$

5.59 a. 0, 1, 2, 3, 4, 5, 6, 7, 8, 9, 10, 11, 12   b. .2055

5.61 a. .5000   b. .6371   c. .1509   **5.63** a. .0357   b. .0098   c. .1364

5.65 a. .2725   b. .0839

5.67 a. i. .1115   ii. .0060

b. i. .3670   ii. .6330   iii. .7780

5.69 a. $\mu = 5.600$ customers; $\sigma = 1.296$ customers   b. .0467   **5.73** a. .1992   b. .0849

5.75 a. $\mu = .600$; $\sigma = .775$ b. $\mu = 1.800$; $\sigma = 1.342$ **5.77** .0709   **5.79** .0012

5.81 a. .1378   b. i. .5814   ii. .0787   iii. .7833

5.83 a. .0793   b. i. .2793   ii. .1551   iii. .3614

5.85 a. .0203   b. i. .0257   ii. .9743

5.87 a. .4493   c. $\mu = .8$ accident; $\sigma = .894$ accident

5.89 a. .0418   b. i. .0697   ii. .5942   iii. .1249

5.91 $\mu = 2.180$ trips; $\sigma = 1.276$ trips

5.93 a.

x	$\$350$	$-\$99,650$
$P(x)$	.998	.002

b. $\mu = \$150.00$; $\sigma = \$4467.66$

5.95 a. .0532   b. .3837   c. .3677   **5.97** a. $\mu = 1.500$ returns; $\sigma = 1.162$ returns   b. .0105

5.99 a. .0765   b. i. .0278   ii. .1263   iii. .5448   **5.101** a. .0383   **5.103** $p = .3214$

5.105 a. .1840   b. assume that conditions for binomial random variable are satisfied

5.107 a. 36.5   b. no, his expected score would decrease   c. no, her expected score would be unchanged

5.109 a. $-12¢$

b.

Outcome	$2	$5	$10	$20	Flag	Joker
Expected net payoff	$-16¢$	$-16¢$	$-34¢$	$-16¢$	$-18¢$	$-18¢$

c. $1 is best $(-12¢)$; $10 is worst $(-34¢)$

5.111 $E(x) = -8¢$ **5.113** a. .75   b. 1.25   c. .829   **5.115** a. .0680   b. .3380   c. yes

Self-Review Test

2. probability distribution table   **3.** a   **4.** b   **5.** a   **7.** b   **8.** a   **9.** b

10. a   **11.** c   **12.** a   **14.** $\mu = 2.040$ sales; $\sigma = 1.449$ sales

15. a. i. .1678   ii. .8821   iii. .0017   b. $\mu = 8.400$; $\sigma = 1.587$

16. a. i. .0668   ii. .8912   iii. .1085

CHAPTER 6

6.11   .8664   **6.13**   .9876

6.15   **a.**   .4744   **b.**   .4678   **c.**   .1162   **d.**   .0610   **e.**   .9452

6.17   **a.**   .0594   **b.**   .0244   **c.**   .9798   **d.**   .9686

6.19   **a.**   .5 approximately   **b.**   .5 approximately   **c.**   .00 approximately   **d.**   .00 approximately

6.21   **a.**   .9613   **b.**   .4783   **c.**   .4767   **d.**   .0694

6.23   **a.**   .0162   **b.**   .2450   **c.**   .1570   **d.**   .9625

6.25   **a.**   .8365   **b.**   .8762   **c.**   approximately .5   **d.**   approximately .5
      e.   approximately 0   **f.**   approximately 0

6.27   **a.**   1.80   **b.** -2.60 **c.** -1.60 **d.**   2.40   **6.29**   **a.**   .4599   **b.**   .1210   **c.**   .2223

6.31   **a.**   .3336   **b.**   .9564   **c.**   .9564   **d.**   approximately 0

6.33   **a.**   .2178   **b.**   .5997   **6.35**   **a.**   .8212   **b.**   .3085   **c.**   .0401   **d.**   .7486

6.37   **a.**   .0287   **b.**   .2345   **6.39**   **a.**   .3821   **b.**   .4338

6.41   **a.**   12.20%   **b.**   82.93%   **6.43**   **a.**   .2558   **b.**   43.71%

6.45   **a.**   .7357   **b.**   15.64%   **6.47**   **a.**   .2483 or about 25%   **b.**   .1717 or about 17%

6.49   **a.**   29.81%   **b.**   42.86%   **c.**   13.14%   **6.51**   2.64%

6.53   **a.**   2.00   **b.** -2.02 approximately   **c.** $-.37$ approximately   **d.**   1.02 approximately

6.55   **a.**   1.65   **b.** -1.96 **c.** -2.33 **d.**   2.58

6.57   **a.**   208.50   **b.**   241.25   **c.**   178.50   **d.**   145.75   **e.**   158.25   **f.**   251.25

6.59   19 minutes approximately   **6.61**   2060 kilowatt-hours   **6.63**   \$72.02 approximately

6.65 $np > 5$ and $nq > 5$ **6.67**   **a.**   .7688   **b.**   .7697; difference is .0009

6.69   **a.** $\mu = 72$; $\sigma = 5.36656315$ **b.**   .5359   **c.**   .4564

6.71   **a.**   .0764   **b.**   .6793   **c.**   .8413   **d.**   .8238

6.73   .1967

6.75   **a.**   .0339   **b.**   .0084   **c.**   .6421

6.77   **a.**   .0373   **b.**   .0322   **c.**   .1603   **6.79**   **a.**   .7549   **b.**   .2451

6.81   **a.**   .0485   **b.**   4.00%   **c.**   23.95%   **d.**   It is possible, but its probability is close to zero.

6.83   .0124 or 1.24%   **6.85**   **a.**   848 hours   **b.**   792 hours approximately

6.87   **a.**   .0151   **b.**   .0465   **c.**   .8340   **d.**   .2540

6.89   16.23 ounces   **6.91**   .0039   **6.93**   Plant A: .6826, Plant B: .8400

6.95   49.95 minutes or about 8:10 A.M.   **6.97**   **a.**   95.24 approximately   **b.**   41.68%

6.99   **b.**   The probability is .4866 for single-number bets; .3974 for color bets.   **6.101**   **a.**   .0005   **b.**   .7714

Self-Review Test

1. a   **2.** a   **3.** d   **4.** b   **5.** a   **6.** c   **7.** b   **8.** b

9.   **a.**   .1878   **b.**   .9304   **c.**   .0985   **d.**   .7704

10.   **a.** -1.28 approximately   **b.**   .61   **c.**   1.65 approximately   **d.** -1.07 approximately

11.   **a.**   .6582   **b.**   .3520   **c.**   .2843   **d.**   .3169

12.   **a.**   4.74 hours approximately   **b.**   10.89 hours approximately

13.   **a.**   .0035   **b.**   .1994   **c.**   .4247   **d.**   .0154   **e.**   .5337   **f.**   .6179   **g.**   .4168

CHAPTER 7

7.5   **a.** 16.60   **b.** sampling error $= -.27$ **c.** sampling error $= -.27$; nonsampling error $= 1.11$
  d. $\bar{x}_1 = 16.22$; $\bar{x}_2 = 15.67$; $\bar{x}_3 = 17.00$; $\bar{x}_4 = 16.34$; $\bar{x}_5 = 17.45$; $\bar{x}_6 = 16.78$; $\bar{x}_7 = 17.22$; $\bar{x}_8 = 17.67$;
  $\bar{x}_9 = 16.56$; $\bar{x}_{10} = 15.11$

7.7   **b.** $\bar{x}_1 = 28.4$; $\bar{x}_2 = 28.8$; $\bar{x}_3 = 33.8$; $\bar{x}_4 = 34.4$; $\bar{x}_5 = 35.2$; $\bar{x}_6 = 36.4$;   **c.** $\mu = 32.83$

7.13   **a.** $\mu_{\bar{x}} = 60$; $\sigma_{\bar{x}} = 2.357$ **b.** $\mu_{\bar{x}} = 60$; $\sigma_{\bar{x}} = 1.054$

7.15   **a.** $\sigma_{\bar{x}} = 1.400$ **b.** $\sigma_{\bar{x}} = 2.500$ **7.17** **a.** $n = 100$ **b.** $n = 256$

7.19 $\mu_{\bar{x}} = \$3.12$; $\sigma_{\bar{x}} = \$.038$ **7.21** $\mu_{\bar{x}} = \$51,400$; $\sigma_{\bar{x}} = \$1850$ **7.23** $n = 400$

7.25   **a.** $\mu_{\bar{x}} = 80.60$ **b.** $\sigma_{\bar{x}} = 3.302$ **d.** $\sigma_{\bar{x}} = 3.302$

7.33 $\mu_{\bar{x}} = 71$ miles per hour; $\sigma_{\bar{x}} = .716$ mile per hour; the normal distribution

7.35 $\mu_{\bar{x}} = 3.020$; $\sigma_{\bar{x}} = .042$; approximately normal distribution

7.37 $\mu_{\bar{x}} = \$75$; $\sigma_{\bar{x}} = \$2.846$; approximately normal distribution

7.39 $\mu_{\bar{x}} = \$33,702$; $\sigma_{\bar{x}} = \$147.406$; approximately normal distribution   **7.41** 86.64%

7.43   **a.** $z = 2.44$ **b.** $z = -7.25$ **c.** $z = -3.65$ **d.** $z = 5.82$

7.45   **a.** .1940   **b.** .8749   **7.47** **a.** .00 approximately   **b.** .9292

7.49   **a.** .1093   **b.** .0322   **c.** .7776   **7.51** **a.** .3983   **b.** .9398   **c.** .1056

7.53   **a.** .8929   **b.** .0202   **7.55** **a.** .2373   **b.** .9624   **c.** .0418

7.57   **a.** .3085   **b.** .8543   **c.** .8664   **d.** .1056   **7.59** .0124   **7.61** $p = .12$; $\hat{p} = .15$

7.63   7125 subjects in the population; 312 subjects in the sample   **7.65** sampling error $= -.05$

7.71   **a.** $\mu_{\hat{p}} = .21$; $\sigma_{\hat{p}} = .020$ **b.** $\mu_{\hat{p}} = .21$; $\sigma_{\hat{p}} = .015$ **7.73** **a.** $\sigma_{\hat{p}} = .051$ **b.** $\sigma_{\hat{p}} = .071$

7.77   **a.** $p = .667$ **b.** 6   **d.** $-.067, -.067, .133, .133, -.067, -.067$

7.79 $\mu_{\hat{p}} = .66$; $\sigma_{\hat{p}} = .027$; approximately normal distribution

7.81 $\mu_{\hat{p}} = .20$; $\sigma_{\hat{p}} = .057$; approximately normal distribution

7.83   95.44%   **7.85** **a.** $z = -.61$ **b.** $z = 1.83$ **c.** $z = -1.22$ **d.** $z = 1.22$

7.87   **a.** .9734   **b.** .3121

7.89   **a.** .3868   **b.** .6480   **7.91** .2005   **7.93** $\mu_{\bar{x}} = 750$ hours; $\sigma_{\bar{x}} = 11$ hours; the normal distribution

7.95   **a.** .9131   **b.** .1698   **c.** .8262   **d.** .0344

7.97   **a.** .0274   **b.** .8388   **c.** .7108   **d.** .0166

7.99 $\mu_{\hat{p}} = .10$; $\sigma_{\hat{p}} = .034$; approximately normal distribution

7.101 **a.** **i.** .0146   **ii.** .0907   **b.** .9912   **c.** .0146   **7.103** .8764   **7.105** 10 approximately

7.107 **a.** .8023   **b.** 754 approximately   **7.109** .0035

Self-Review Test

1. b   **2.** b   **3.** a   **4.** a   **5.** b   **6.** b   **7.** c   **8.** a   **9.** a

10. a   **11.** c   **12.** a

14. **a.** $\mu_{\bar{x}} = 145$ pounds; $\sigma_{\bar{x}} = 3.600$ pounds; approximately normal distribution
    b. $\mu_{\bar{x}} = 145$ pounds; $\sigma_{\bar{x}} = 1.800$ pounds; approximately normal distribution

15. **a.** $\mu_{\bar{x}} = 47$ minutes; $\sigma_{\bar{x}} = 1.878$ minutes; unknown distribution
    b. $\mu_{\bar{x}} = 47$ minutes; $\sigma_{\bar{x}} = 1.004$ minutes; approximately normal distribution

16. **a.** .1905   **b.** .0233   **c.** .8770   **d.** .8104

17. **a.** **i.** .1203   **ii.** .1335   **iii.** .7486   **b.** .9736   **c.** .0013

18. **a.** $\mu_{\hat{p}} = .08$; $\sigma_{\hat{p}} = .038$; unknown distribution
    b. $\mu_{\hat{p}} = .08$; $\sigma_{\hat{p}} = .019$; approximately normal distribution
    c. $\mu_{\hat{p}} = .08$; $\sigma_{\hat{p}} = .006$; approximately normal distribution

19. **a.** **i.** .0951   **ii.** .9198   **iii.** .9049   **iv.** .1776   **b.** .9708   **c.** .0401   **d.** .0951

CHAPTER 8

8.11 **a.** 24.5   **b.** ±.760   **c.** 23.50 to 25.50   **d.** 1.00

8.13 **a.** 70.59 to 79.01   **b.** 69.80 to 79.80   **c.** 68.22 to 81.38   **d.** yes

8.15 **a.** 79.19 to 84.61   **b.** 80.09 to 83.71   **c.** 80.27 to 83.53   **d.** yes

8.17 **a.** 53.93 to 56.71   **b.** 56.16 to 58.64   **c.** 54.95 to 57.55
d. Confidence intervals of parts a and c cover μ, but the confidence interval of part b does not cover μ.

8.19 **a.** 38.34   **b.** ±.88   **c.** 37.30 to 39.38   **d.** 1.04   **8.21** $3966.50 to $4353.50

8.23 19.99 to 22.73 seconds   **8.25** **a.** 7.0 hours; margin of error = ±.05 hour   **b.** 6.94 to 7.06 hours

8.27 **a.** $58,320.68 to $59,679.32

8.29 35.98 to 36.06 inches; the machine needs to be adjusted

8.31 **a.** $286.87 to $333.13   **8.39** **a.** 1.782   **b.** −2.060   **c.** −3.579   **d.** 2.807

8.41 **a.** $\alpha = .01$; right tail   **b.** $\alpha = .005$; left tail   **c.** $\alpha = .025$; left tail   **d.** $\alpha = .01$; right tail

8.43 **a.** 3.012   **b.** 2.060   **c.** 1.746   **8.45** **a.** 12.75   **b.** 9.59 to 15.91   **c.** 3.16

8.47 **a.** 63.76 to 73.24   **b.** 64.60 to 72.40   **c.** 64.83 to 72.17

8.49 18.69 to 23.31 minutes   **8.51** $1591.42 to $1808.58   **8.53** $129.04 to $156.96

8.55 **a.** 24.71 to 28.09 miles per gallon   **8.57** 71.12 to 77.28 mph

8.63 **a.** yes, sample size is large   **b.** no, sample size is not large   **c.** yes, sample size is large
d. no, sample size is not large

8.65 **a.** .583 to .677   **b.** .542 to .638   **c.** .624 to .716
d. Confidence intervals of parts a and c cover p, but the confidence interval of part b does not.

8.67 **a.** .646 to .714   **b.** .639 to .721   **c.** .626 to .734   **d.** yes

8.69 **a.** .615 to .845   **b.** .683 to .777   **c.** .700 to .760   **d.** yes

8.71 **a.** .44; margin of error = ±.030   **b.** .400 to .480

8.73 **a.** .41; margin of error = ±.045   **b.** .365 to .455

8.75 **a.** 54.9% to 85.1%   **8.77** **a.** 25.0% to 37.0%

8.79 **a.** .60; margin of error = ±.248   **b.** 35.2% to 84.8%   **8.83** **a.** $n = 167$ **b.** $n = 65$

8.85 **a.** $n = 299$ **b.** $n = 126$ **c.** $n = 61$ **8.87** $n = 139$ **8.89** $n = 291$

8.91 **a.** $n = 1119$ **b.** $n = 1359$ **8.93** **a.** $n = 1849$ **b.** $n = 601$ **c.** $n = 6807$

8.95 $n = 4161$ **8.97** $n = 1350$ **8.99** **a.** $7.25; margin of error = ± $.16   **b.** $7.09 to $7.41

8.101 23.989 to 24.041; the machine needs to be adjusted

8.103 **a.** $166.82; margin of error = ± $16.94   **b.** $146.68 to $186.96

8.105 $425.32 to $494.68   **8.107** 8.90 to 10.60 hours   **8.109** 1.92 to 2.54 hours

8.111 **a.** 12%; margin of error = ± 9%   **b.** .1% to 23.9%   **8.113** 11.7% to 68.3%

8.115 $n = 42$ **8.117** $n = 1068$ **8.119** **a.** $\bar{x} = 12.90 **b.** $10.99 to $14.81

8.121 28.4% to 33.6%; 21.6% to 26.4%; 30.3% to 35.7%   **8.123** $n = 65$

8.125 **a.** $n = 20$ days   **b.** 90%   **c.** ±75 cars

Self-Review Test

1. **a.** population parameter; sample statistic   **b.** sample statistic; population parameter
c. sample statistic; population parameter

2. b   **3.** a   **4.** a   **5.** d   **6.** b   **7.** **a.** $2.85; margin of error = ±$.20   **b.** $2.59 to $3.11

8. $279,593.30 to $341,310.70   **9.** **a.** .42; margin of error = ±.039   **b.** .381 to .459

10. $n = 45$ **11.** $n = 2663$ **12.** $n = 1125$

CHAPTER 9

9.5 **a.** a left-tailed test   **b.** a right-tailed test   **c.** a two-tailed test

9.7 **a.** Type II error   **b.** Type I error

9.9 **a.** H_0: $\mu = \$143{,}000$; H_1: $\mu > \$143{,}000$;   a right-tailed test
   b. H_0: $\mu = 15$ hours; H_1: $\mu \neq 15$ hours;   a two-tailed test
   c. H_0: $\mu = 45$ months; H_1: $\mu < 45$ months;   a left-tailed test
   d. H_0: $\mu = 5$ hours; H_1: $\mu \neq 5$ hours;   a two-tailed test
   e. H_0: $\mu = 24$ years; H_1: $\mu \neq 24$ years;   a two-tailed test

9.15 **a.** rejection region is to the left of -1.96 and to the right of 1.96;   nonrejection region is between -1.96 and 1.96
   b. rejection region is to the left of -2.33;   nonrejection region is to the right of -2.33
   c. rejection region is to the right of 2.05;   nonrejection region is to the left of 2.05

9.17 statistically significant   **9.19** **a.** .025   **b.** .05   **c.** .01

9.21 **a.** observed value of z is 7.30; critical value of z is 1.65
   b. observed value of z is 7.30; critical values of z are -1.96 and 1.96

9.23 **a.** reject H_0 if $z < -1.65$ **b.** reject H_0 if $z < -1.96$ or $z > 1.96$ **c.** reject H_0 if $z > 1.65$

9.25 **a.** critical values: $z = -2.58$ and 2.58;   test statistic: $z = -1.33$;   do not reject H_0
   b. critical values: $z = -2.58$ and 2.58;   test statistic: $z = 3.20$;   reject H_0

9.27 **a.** critical values: $z = -2.58$ and 2.58;   test statistic: $z = 10.50$;   reject H_0
   b. critical value: $z = -1.65$;   test statistic: $z = -1.12$;   do not reject H_0
   c. critical value: $z = 1.28$;   test statistic: $z = 10.00$;   reject H_0

9.29 H_0: $\mu = 3.9$ visits; H_1: $\mu \neq 3.9$ visits;   critical values: $z = -2.58$ and 2.58;   test statistic: $z = -2.34$;   do not reject H_0

9.31 H_0: $\mu = 5.6$ years; H_1: $\mu \neq 5.6$ years;   critical values: $z = -2.24$ and 2.24;   test statistic: $z = 8.43$;   reject H_0

9.33 **a.** H_0: $\mu = 5.2$ hours; H_1: $\mu > 5.2$ hours;   critical value: $z = 1.65$;   test statistic: $z = 4.30$;   reject H_0
   b. P(Type I error) $= .05$

9.35 **a.** H_0: $\mu = 11$ hours; H_1: $\mu < 11$ hours;   critical value: $z = -2.33$;   test statistic: $z = -9.09$;   reject H_0
   b. do not reject H_0

9.37 **a.** H_0: $\mu = 45$ months; H_1: $\mu < 45$ months;   critical value: $z = -1.96$;   test statistic: $z = -1.88$;   do not reject H_0
   b. critical value: $z = -1.65$;   test statistic: $z = -1.88$;   reject H_0

9.39 critical values: $z = -2.58$ and 2.58;   test statistic: $z = 1.90$;   do not reject H_0;   the machine does not need adjustment

9.45 **a.** p-value $= .0188$ **b.** p-value $= .0116$ **c.** p-value $= .0087$

9.47 **a.** p-value $= .0192$ **b.** no, do not reject H_0 **c.** yes, reject H_0

9.49 H_0: $\mu = 108.2$¢; H_1: $\mu < 108.2$¢;   test statistic: $z = -1.34$; p-value $= .1802$

9.51 H_0: $\mu \geq 14$ hours; H_1: $\mu < 14$ hours;   test statistic: $z = -1.18$; p-value $= .1190$

9.53 **a.** H_0: $\mu = 10$ minutes; H_1: $\mu < 10$ minutes;   test statistic: $z = -2.00$; p-value $= .0228$
   b. $.0228 > .02$, do not reject H_0 **c.** $.0228 < .05$, reject H_0

9.55 **a.** test statistic: $z = -3.94$; p-value $= .00$ approximately   **b.** yes, adjust for $\alpha = .01$;   yes, adjust for $\alpha = .05$

9.57 **a.** reject H_0 if $t < -2.539$ or $t > 2.539$ **b.** reject H_0 if $t < -2.602$ **c.** reject H_0 if $t > 1.740$

9.59 **a.** critical value: $t = 2.492$;   observed value: $t = 2.333$
   b. critical values: $t = -2.797$ and 2.797;   observed value: $t = 2.333$

9.61 **a.** reject H_0 if $t < -2.539$ **b.** reject H_0 if $t < -2.861$ or $t > 2.861$ **c.** reject H_0 if $t > 2.539$

9.63 **a.** critical values: $t = -2.797$ and 2.797;   test statistic: $t = -1.875$;   do not reject H_0
   b. critical values: $t = -2.797$ and 2.797;   test statistic: $t = 5.000$;   reject H_0

9.65 **a.** critical values: $t = -2.797$ and 2.797;   test statistic: $t = 4.592$;   reject H_0
   b. critical value: $t = -2.131$;   test statistic: $t = -1.515$;   do not reject H_0
   c. critical value: $t = 1.328$;   test statistic: $t = 2.516$;   reject H_0

9.67 H_0: $\mu = 69.5$ inches; H_1: $\mu \neq 69.5$ inches;   critical values: $t = -2.797$ and 2.797; test statistic: $t = 1.786$;   do not reject H_0

9.69 H_0: $\mu \leq 7$ hours; H_1: $\mu > 7$ hours; critical value: $t = 2.093$; test statistic: $t = 6.805$; reject H_0

9.71 H_0: $\mu \leq 30$ calories; H_1: $\mu > 30$ calories; critical value: $t = 1.753$; test statistic: $t = 2.667$; reject H_0

9.73 **a.** H_0: $\mu \leq 45$ minutes; H_1: $\mu > 45$ minutes; critical value: $t = 2.539$; test statistic: $t = 6.708$; reject H_0
  b. P(Type I error) = .01

9.75 **a.** yes, the claim is true
  b. H_0: $\mu \geq 1200$; H_1: $\mu < 1200$; critical value: $t = -1.711$; test statistic: $t = -4.412$; reject H_0

9.77 H_0: $\mu = 18$ hours; H_1: $\mu \neq 18$ hours; critical values: $t = -2.262$ and 2.262; test statistic: $t = 2.692$; reject H_0

9.83 **a.** large enough   **b.** not large enough   **c.** not large enough   **d.** large enough

9.85 **a.** reject H_0 if $z < -1.96$ or $z > 1.96$ **b.** reject H_0 if $z < -2.05$ **c.** reject H_0 if $z > 1.96$

9.87 **a.** critical value: $z = -2.33$; observed value: $z = -.88$
  b. critical values: $z = -2.58$ and 2.58; observed value: $z = -.88$

9.89 **a.** reject H_0 if $z < -2.33$
  b. reject H_0 if $z < -2.58$ or $z > 2.58$
  c. reject H_0 if $z > 2.33$

9.91 **a.** critical value: $z = -1.96$; test statistic: $z = -1.21$; do not reject H_0
  b. critical value: $z = -1.96$; test statistic: $z = -2.41$; reject H_0

9.93 **a.** critical values: $z = -1.96$ and 1.96; test statistic: $z = -4.000$; reject H_0
  b. critical value: $z = -2.33$; test statistic: $z = -1.37$; do not reject H_0
  c. critical value: $z = 1.96$; test statistic: $z = .43$; do not reject H_0

9.95 H_0: $p = .29$; H_1: $p > .29$; critical value: $z = 1.65$; test statistic: $z = 1.97$, reject H_0

9.97 H_0: $p = .73$; H_1: $p < .73$; critical value: $z = -1.65$; test statistic: $z = -2.14$; reject H_0

9.99 **a.** H_0: $p = .47$; H_1: $p > .47$; critical value: $z = 1.65$; test statistic: $z = 1.99$; reject H_0 **b.** P(Type I error) = .05

9.101 **a.** H_0: $p \geq .60$; H_1: $p < .60$; critical value: $z = -2.33$; test statistic: $z = -3.27$; reject H_0 **b.** do not reject H_0

9.103 **a.** critical value: $z = 2.05$; test statistic: $z = 2.22$; reject H_0; adjust machine
  b. critical value: $z = 2.33$; test statistic: $z = 2.22$; do not reject H_0; do not adjust the machine

9.107 **a.** critical values: $z = -2.33$ and 2.33; test statistic: $z = -2.13$; do not reject H_0
  b. P(Type I error) = .02   **c.** p-value = .0332; do not reject H_0 if $\alpha = .01$; reject H_0 if $\alpha = .05$

9.109 **a.** critical value: $z = -2.05$; test statistic: $z = -2.14$; reject H_0 **b.** P(Type I error) = .02
  c. p-value = .0162; do not reject H_0 if $\alpha = .01$; reject H_0 if $\alpha = .025$

9.111 **a.** H_0: $\mu = 90.7$ months; H_1: $\mu > 90.7$ months; critical value: $z = 2.05$; test statistic: $z = 2.55$; reject H_0
  b. P(Type I error) = .02   **c.** yes

9.113 **a.** H_0: $\mu = 18$ hours; H_1: $\mu > 18$ hours; critical value of $z = 1.65$; test statistic: $z = 1.99$; reject H_0
  b. do not reject H_0 if $\alpha = .01$

9.115 H_0: $\mu = 8$ minutes; H_1: $\mu < 8$ minutes; critical value: $z = -1.96$; test statistic: $z = -1.35$; do not reject H_0

9.117 H_0: $\mu = 1245$ cubic feet; H_1: $\mu < 1245$ cubic feet; test statistic: $z = -2.80$; p-value = .0026

9.119 H_0: $\mu = 25$ minutes; H_1: $\mu \neq 25$ minutes; critical values: $t = -2.947$ and 2.947; test statistic: $t = 2.083$; do not reject H_0

9.121 **a.** H_0: $\mu = 114$ minutes; H_1: $\mu < 114$ minutes; critical value: $t = -2.492$; test statistic: $t = -2.273$; do not reject H_0
  b. do not reject H_0

9.123 H_0: $\mu \leq 150$ calories; H_1: $\mu > 150$ calories; critical value: $t = 2.262$; test statistic: $t = 1.157$; do not reject H_0

9.125 **a.** H_0: $p = .49$; H_1: $p \neq .49$; critical values: $z = -2.58$ and 2.58; test statistic: $z = -3.39$; reject H_0
  b. P(Type I error) = .01

9.127 H_0: $p = .26$; H_1: $p > .26$; critical value: $z = 2.05$; test statistic: $z = 3.22$; reject H_0

9.129 **a.** H_0: $p \geq .90$; H_1: $p < .90$; critical value: $z = -2.05$; test statistic: $z = -2.11$; reject H_0 **b.** do not reject H_0

9.131 H_0: $p \geq .90$; H_1: $p < .90$; p-value = .0174; reject H_0 if $\alpha = .05$; do not reject H_0 if $\alpha = .01$

9.133 $C = 9$ approximately

9.135 For cure rate: H_0: $p = .60$; H_1: $p < .60$; critical value: $z = -2.33$; test statistic: $z = -1.73$; do not reject H_0.
  For number of visits: H_0: $\mu = 140$; H_1: $\mu < 140$; critical value: $z = -2.33$; test statistic: $z = -2.98$; reject H_0

9.137 **a.** H_0: $p = .30$; H_1: $p < .30$ **b.** $\alpha = .0016$ **c.** $\alpha = .0354$

Self-Review Test

1. a   2. b   3. a   4. b   5. a   6. a   7. a   8. b
9. c   10. a   11. c   12. b   13. d   14. c   15. a   16. b

17. **a.** H_0: μ = \$1365; H_1: $\mu \neq$ \$1365; critical values: $z = -2.58$ and 2.58; test statistic: $z = 5.04$; reject H_0
    b. P(Type I error) = .01   **c.** do not reject H_0

18. **a.** H_0: $\mu \geq 31$ months; H_1: $\mu < 31$ months; critical value: $t = -2.131$; test statistic: $t = -3.333$; reject H_0
    b. P(Type I error) = .025   **c.** critical value: $t = -3.733$; do not reject H_0

19. **a.** H_0: p = .66; H_1: $p \neq$ .66; critical values: $z = -1.96$ and 1.96; test statistic: $z = -2.36$; reject H_0
    b. P(Type I error) = .05   **c.** do not reject H_0

20. **a.** H_0: μ = \$37,120; H_1: $\mu >$ \$37,120; test statistic: $z = 2.10$; p-value = .0179
    b. do not reject H_0 if α = .01; reject H_0 if α = .05

21. **a.** p-value = .0182   **b.** reject H_0 if α = .05; do not reject H_0 if α = .01

CHAPTER 10

10.3   **a.** .76   **b.** .39 to 1.13

10.5 H_0: $\mu_1 - \mu_2 = 0$; H_1: $\mu_1 - \mu_2 \neq 0$; critical values: $z = -1.96$ and 1.96; test statistic: $z = 5.30$; reject H_0

10.7 H_0: $\mu_1 - \mu_2 = 0$; H_1: $\mu_1 - \mu_2 > 0$; critical value: $z = 2.33$; test statistic: $z = 5.30$; reject H_0

10.9   **a.** \$7475; margin of error = ± \$641.23   **b.** \$6330.94 to \$8319.06
    c. H_0: $\mu_1 - \mu_2 = 0$; H_1: $\mu_1 - \mu_2 \neq 0$; critical values: $z = -2.58$ and 2.58; test statistic: $z = 22.85$; reject H_0

10.11   **a.** \$1131; margin of error = ± \$1009.29   **b.** \$281.34 to \$1980.66
    c. H_0: $\mu_1 - \mu_2 = 0$; H_1: $\mu_1 - \mu_2 > 0$; critical value: $z = 2.33$; test statistic: $z = 2.20$; do not reject H_0

10.13   **a.** -3.64 to -2.16 days
    b. H_0: $\mu_1 - \mu_2 = 0$; H_1: $\mu_1 - \mu_2 < 0$; critical value: $z = -1.96$; test statistic: $z = -9.15$; reject H_0
    c. P(Type I error) = .025

10.15   **a.** -12.33 to -3.67
    b. H_0: $\mu_1 - \mu_2 = 0$; H_1: $\mu_1 - \mu_2 < 0$; critical value: $z = -1.96$; test statistic: $z = -3.62$; reject H_0
    c. do not reject H_0

10.17   **a.** $-.51$ to .01 minute
    b. H_0: $\mu_1 - \mu_2 = 0$; H_1: $\mu_1 - \mu_2 < 0$; critical value: $z = -1.96$; test statistic: $z = -2.06$; reject H_0
    c. p-value = .0197; do not reject H_0 if α = .01; reject H_0 if α = .05

10.21   **a.** 5.25   **b.** .93 to 9.57

10.23 H_0: $\mu_1 - \mu_2 = 0$; H_1: $\mu_1 - \mu_2 \neq 0$; critical values: $t = -2.719$ and 2.719; test statistic: $t = 3.302$; reject H_0

10.25 H_0: $\mu_1 - \mu_2 = 0$; H_1: $\mu_1 - \mu_2 > 0$; critical value: $t = 1.688$; test statistic: $t = 3.302$; reject H_0

10.27   **a.** -5.35 **b.** -9.03 to -1.67
    c. H_0: $\mu_1 - \mu_2 = 0$; H_1: $\mu_1 - \mu_2 < 0$; critical value: $t = -2.080$; test statistic: $t = -4.121$; reject H_0

10.29   **a.** 1.50 to 16.50 items
    b. H_0: $\mu_1 - \mu_2 = 0$; H_1: $\mu_1 - \mu_2 > 0$; critical value: $t = 2.120$; test statistic: $t = 2.543$; reject H_0

10.31   **a.** \$21,872.23 to \$26,661.77
    b. H_0: $\mu_1 - \mu_2 = 0$; H_1: $\mu_1 - \mu_2 \neq 0$; critical values: $t = -2.009$ and 2.009; test statistic: $t = 16.983$; reject H_0

10.33   **a.** -12.95 to 2.95 minutes
    b. H_0: $\mu_1 - \mu_2 = 0$; H_1: $\mu_1 - \mu_2 < 0$; critical value: $t = -2.412$; test statistic: $t = -1.691$; do not reject H_0

10.35   **a.** $-\$2235.93$ to $-\$140.07$
    b. H_0: $\mu_1 - \mu_2 = 0$; H_1: $\mu_1 - \mu_2 < 0$; critical value: $t = -2.473$; test statistic: $t = -2.326$; do not reject H_0

10.37   4.63 to 13.09

10.39 H_0: $\mu_1 - \mu_2 = 0$; H_1: $\mu_1 - \mu_2 \neq 0$; critical values: $t = -2.744$ and 2.744; test statistic: $t = 5.749$; reject H_0

10.41 H_0: $\mu_1 - \mu_2 = 0$; H_1: $\mu_1 - \mu_2 > 0$; critical value: $t = 1.696$; test statistic: $t = 5.749$; reject H_0

10.43 **a.** .85 to 17.15 items
 b. H_0: $\mu_1 - \mu_2 = 0$; H_1: $\mu_1 - \mu_2 > 0$; critical value: $t = 2.201$; test statistic: $t = 2.429$; reject H_0

10.45 **a.** $21,873.18 to $26,660.82
 b. H_0: $\mu_1 - \mu_2 = 0$; H_1: $\mu_1 - \mu_2 \neq 0$; critical values: $t = -2.010$ and 2.010; test statistic: $t = 17.000$; reject H_0

10.47 **a.** −12.86 to 2.86 minutes
 b. H_0: $\mu_1 - \mu_2 = 0$; H_1: $\mu_1 - \mu_2 < 0$; critical value: $t = -2.414$; test statistic: $t = -1.713$; do not reject H_0

10.49 **a.** −$2251.31 to −$124.69
 b. H_0: $\mu_1 - \mu_2 = 0$; H_1: $\mu_1 - \mu_2 < 0$; critical value: $t = -2.492$; test statistic: $t = -2.306$; do not reject H_0

10.51 **a.** 12.50 to 38.30   **b.** 11.12 to 15.28   **c.** 29.80 to 39.40

10.53 **a.** critical values: $t = -1.860$ and 1.860; test statistic: $t = 8.040$; reject H_0
 b. critical value: $t = 1.721$; test statistic: $t = 10.847$; reject H_0
 c. critical value: $t = -2.583$; test statistic: $t = -7.989$; reject H_0

10.55 **a.** −3.27 to .41
 b. H_0: $\mu_d = 0$; H_1: $\mu_d < 0$; critical value: $t = -3.143$; test statistic: $t = -1.903$; do not reject H_0

10.57 **a.** −16.74 to −3.52   **b.** H_0: $\mu_d = 0$; H_1: $\mu_d < 0$; critical value: $t = -1.895$; test statistic: $t = -2.905$; reject H_0

10.59 **a.** −4.07 to .63 miles per gallon
 b. H_0: $\mu_d = 0$; H_1: $\mu_d < 0$; critical value: $t = -2.571$; test statistic: $t = -2.951$; reject H_0

10.63 −.185 to .045

10.65 H_0: $p_1 - p_2 = 0$; H_1: $p_1 - p_2 \neq 0$; critical values: $z = -2.58$ and 2.58; test statistic: $z = -1.5526$; do not reject H_0

10.67 H_0: $p_1 - p_2 = 0$; H_1: $p_1 - p_2 < 0$; critical value: $z = -2.33$; test statistic: $z = -1.55$; do not reject H_0

10.69 **a.** .03
 b. −.034 to .094
 c. rejection region to the right of $z = 1.96$; non-rejection region to the left of $z = 1.96$
 d. test statistic: $z = 1.01$
 e. do not reject H_0

10.71 **a.** .145 to .395
 b. H_0: $p_1 - p_2 = 0$; H_1: $p_1 - p_2 > 0$; critical value: $z = 2.33$; test statistic: $z = 5.37$; reject H_0

10.73 **a.** .020 to .080
 b. H_0: $p_1 - p_2 = 0$; H_1: $p_1 - p_2 \neq 0$; critical values: $z = -2.58$ and 2.58; test statistic: $z = 3.68$; reject H_0

10.75 **a.** .03   **b.** −.023 to .083
 c. H_0: $p_1 - p_2 = 0$; H_1: $p_1 - p_2 \neq 0$; critical values: $z = -1.96$ and 1.96; test statistic: $z = 1.10$; do not reject H_0

10.77 **a.** −.065 to .025
 b. H_0: $p_1 - p_2 = 0$; H_1: $p_1 - p_2 < 0$; critical value: $z = -1.96$; test statistic: $z = -.96$; do not reject H_0
 c. p-value $= .1685$

10.79 **a.** −$3419.54 to $431.54
 b. H_0: $\mu_1 - \mu_2 = 0$; H_1: $\mu_1 - \mu_2 < 0$; critical value: $z = -2.33$; test statistic: $z = -2.00$; do not reject H_0
 c. do not reject H_0

10.81 **a.** 1.10 to 2.90 policies
 b. H_0: $\mu_1 - \mu_2 = 0$; H_1: $\mu_1 - \mu_2 > 0$; critical value: $z = 2.33$; test statistic: $z = 5.74$; reject H_0

10.83 **a.** H_0: $\mu_1 - \mu_2 = 0$; H_1: $\mu_1 - \mu_2 \neq 0$; critical values: $t = -2.009$ and 2.009; test statistic: $t = -1.058$; do not reject H_0
 b. −.31 to .07

10.85 **a.** −$2.63 to $18.63
 b. H_0: $\mu_1 - \mu_2 = 0$; H_1: $\mu_1 - \mu_2 > 0$; critical value: $t = 2.416$; test statistic: $t = 2.027$; do not reject H_0

10.87 **a.** H_0: $\mu_1 - \mu_2 = 0$; H_1: $\mu_1 - \mu_2 \neq 0$; critical values: $t = -2.010$ and 2.010; test statistic: $t = -1.068$; do not reject H_0
 b. −.31 to .07

10.89 **a.** −$2.46 to $18.46
 b. H_0: $\mu_1 - \mu_2 = 0$; H_1: $\mu_1 - \mu_2 > 0$; critical value: $t = 2.418$; test statistic: $t = 2.063$; do not reject H_0

10.91 **a.** −7.72 to .46
 b. H_0: $\mu_d = 0$; H_1: $\mu_d < 0$; critical value: $t = -2.365$; test statistic: $t = -2.097$; do not reject H_0

10.93   **a.** $-.109$ to $-.031$
        b. $H_0: p_1 - p_2 = 0$; $H_1: p_1 - p_2 \neq 0$; critical values: $z = -2.33$ and 2.33; test statistic: $z = -3.55$; reject H_0

10.95   **a.** $-.183$ to $.118$
        b. $H_0: p_1 - p_2 = 0$; $H_1: p_1 - p_2 \neq 0$; critical values: $z = -2.24$ and 2.24; test statistic: $z = 1.41$; do not reject H_0

10.97   **a.**   10   **b.**   .9911   **10.99**   **a.**   .0031   **b.**   yes   **c.**   .11 to .69 minute   **d.**   no

10.101   The following should be included in the report:
        number of contacts: 90% confidence interval for μ_d: -2.38 to 5.24 contacts; $H_0: \mu_d = 0$; $H_1: \mu_d > 0$; critical value: $t = 1.943$; test statistic: $t = .729$; do not reject H_0.
        Gas mileage: 90% confidence interval for μ_d: -3.89 to $-.97$ MPG; $H_0: \mu_d = 0$; $H_1: \mu_d < 0$; critical value: $t = -1.943$; test statistic: $t = -3.234$; reject H_0

10.103   **a.**   no
         b. $H_0: \mu_1 - \mu_2 = 0$; $H_1: \mu_1 - \mu_2 < 0$; critical value: $t = -2.228$; test statistic: $t = -.527$; do not reject H_0
         c.   conclusions differ

10.105   **a.**   36   **b.**   5.89   **10.107**   .9857

Self-Review Test

1.   a

3.   **a.**   1.62 to 2.78
      b. $H_0: \mu_1 - \mu_2 = 0$; $H_1: \mu_1 - \mu_2 > 0$; critical value: $z = 1.96$; test statistic: $z = 9.86$; reject H_0

4.   **a.** -2.72 to -1.88 hours
      b. $H_0: \mu_1 - \mu_2 = 0$; $H_1: \mu_1 - \mu_2 < 0$; critical value: $t = -2.416$; test statistic: $t = -10.997$; reject H_0

5.   **a.** -2.70 to -1.90 hours
      b. $H_0: \mu_1 - \mu_2 = 0$; $H_1: \mu_1 - \mu_2 < 0$; critical value: $t = -2.421$; test statistic: $t = -11.474$; reject H_0

6.   **a.** -3.43 to 1.43 items
      b. $H_0: \mu_d = 0$; $H_1: \mu_d \neq 0$; critical values: $t = -2.447$ and 2.447; test statistic: $t = -1.528$; do not reject H_0

7.   **a.** $-.052$ to $.092$
      b. $H_0: p_1 - p_2 = 0$; $H_1: p_1 - p_2 \neq 0$; critical values: $z = -2.58$ and 2.58; test statistic: $z = .60$; do not reject H_0

CHAPTER 11

11.3 $\chi^2 = 41.337$ **11.5** $\chi^2 = 41.638$ **11.7**   **a.** $\chi^2 = 5.009$ **b.** $\chi^2 = 3.565$

11.13   critical value: $\chi^2 = 11.070$; test statistic: $\chi^2 = 5.200$; do not reject H_0

11.15   critical value: $\chi^2 = 7.815$; test statistic: $\chi^2 = 9.255$; reject H_0

11.17   critical value: $\chi^2 = 9.488$; test statistic: $\chi^2 = 4.389$; do not reject H_0

11.19   critical value: $\chi^2 = 17.275$; test statistic: $\chi^2 = 25.209$; reject H_0

11.21   critical value: $\chi^2 = 9.348$; test statistic: $\chi^2 = 65.087$; reject H_0

11.27   **a.** H_0: the proportion in each row is the same for all four populations;
         H_1: the proportion in each row is not the same for all four populations
      c.   critical value: $\chi^2 = 14.449$ **d.**   test statistic: $\chi^2 = 52.469$ **e.**   reject H_0

11.29   critical value: $\chi^2 = 5.024$; test statistic: $\chi^2 = .953$; do not reject H_0

11.31   critical value: $\chi^2 = 5.991$; test statistic: $\chi^2 = 21.528$; reject H_0

11.33 critical value: $\chi^2 = 9.210$; test statistic: $\chi^2 = 41.452$; reject H_0

11.35 critical value: $\chi^2 = 6.635$; test statistic: $\chi^2 = 8.178$; reject H_0

11.37 critical value: $\chi^2 = 5.991$; test statistic: $\chi^2 = 1.883$; do not reject H_0

11.39 critical value: $\chi^2 = 7.378$; test statistic: $\chi^2 = 2.404$; do not reject H_0

11.41   **a.**   18.4376 to 84.9686   **b.**   21.3393 to 67.7365   **c.**   23.0674 to 60.6586

11.43   **a.** H_0: $\sigma^2 = .80$; H_1: $\sigma^2 > .80$ **b.**   reject H_0 if $\chi^2 > 30.578$
    c.   test statistic: $\chi^2 = 20.625$ **d.**   do not reject H_0

11.45   **a.** H_0: $\sigma^2 \leq 2.2$; H_1: $\sigma^2 \neq 2.2$ **b.**   reject H_0 if $\chi^2 < 7.564$ or $\chi^2 > 30.191$
    c.   test statistic: $\chi^2 = 35.545$ **d.**   reject H_0

11.47   **a.** H_0: $\sigma^2 \leq .003$; H_1: $\sigma^2 > .003$; critical value: $\chi^2 = 44.314$; test statistic: $\chi^2 = 50.833$; reject H_0
    b.   .0034 to .0132; .058 to .115

11.49   **a.**   2739.3051 to 12,623.9126; 52.338 to 112.3562
    b. H_0: $\sigma^2 = 4200$; H_1: $\sigma^2 \neq 4200$; critical values: $\chi^2 = 12.401$ and 39.364; test statistic: $\chi^2 = 29.714$; do not reject H_0

11.51 critical value: $\chi^2 = 7.815$; test statistic: $\chi^2 = 10.464$; reject H_0; p-value = .015

11.53 critical value: $\chi^2 = 11.070$; test statistic: $\chi^2 = 9.595$; do not reject H_0

11.55 critical value: $\chi^2 = 11.345$; test statistic: $\chi^2 = 15.920$; reject H_0

11.57 critical value: $\chi^2 = 13.277$; test statistic: $\chi^2 = 46.255$; reject H_0

11.59 critical value: $\chi^2 = 4.605$; test statistic: $\chi^2 = 13.593$; reject H_0

11.61 critical value: $\chi^2 = 16.812$; test statistic: $\chi^2 = 10.181$; do not reject H_0

11.63   **a.**   3.4064 to 24.0000; 1.846 to 4.899   **b.**   8.3336 to 33.2628; 2.887 to 5.767

11.65 H_0: $\sigma^2 = 1.1$; H_1: $\sigma^2 > 1.1$; critical value: $\chi^2 = 28.845$; test statistic: $\chi^2 = 24.727$; do not reject H_0

11.67 H_0: $\sigma^2 = 10.4$; H_1: $\sigma^2 \neq 10.4$; critical values: $\chi^2 = 7.564$ and 30.191; test statistic: $\chi^2 = 24.192$; do not reject H_0

11.69   **a.** H_0: $\sigma^2 = 5000$; H_1: $\sigma^2 < 5000$; critical value: $\chi^2 = 8.907$; test statistic: $\chi^2 = 12.065$; do not reject H_0
    b.   1668.8509 to 8319.1406; 40.852 to 91.209

11.71   **a.**   .1001 to .4613; .316 to .679
    b. H_0: $\sigma^2 = .13$; H_1: $\sigma^2 \neq .13$; critical values: $\chi^2 = 9.886$ and 45.559; test statistic: $\chi^2 = 35.077$; do not reject H_0

11.73   **a.** $s^2 = 1107.4107$ **b.**   484.0989 to 4586.9082; 22.002 to 67.727
    c. H_0: $\sigma^2 = 500$; H_1: $\sigma^2 \neq 500$; critical values: $\chi^2 = 1.690$ and 16.013; test statistic: $\chi^2 = 15.504$; do not reject H_0

11.75 critical value: $\chi^2 = 5.991$; test statistic: $\chi^2 = 12.931$; reject H_0

11.77 critical value: $\chi^2 = 9.488$; test statistic: $\chi^2 = 11.822$; reject H_0

Self-Review Test

1. b   **2.** a   **3.** c   **4.** a   **5.** b   **6.** b   **7.** c   **8.** b   **9.** a

10.   critical value: $\chi^2 = 9.210$; test statistic: $\chi^2 = 8.172$; do not reject H_0

11.   critical value: $\chi^2 = 11.345$; test statistic: $\chi^2 = 31.188$; reject H_0

12.   critical value: $\chi^2 = 9.488$; test statistic: $\chi^2 = 82.450$; reject H_0

13.   **a.**   .0316 to .1457; .178 to .382
    b. H_0: $\sigma^2 = .03$; H_1: $\sigma^2 > .03$; critical value: $\chi^2 = 42.980$; test statistic: $\chi^2 = 48.000$; reject H_0

CHAPTER 12

12.3   **a.**   3.73   **b.**   3.61   **c.**   5.37   **12.5**   **a.**   2.39   **b.**   2.27   **c.**   3.28

12.7   **a.**   9.96   **b.**   6.57   **12.9**   **a.**   4.85   **b.**   3.22

12.13 **a.** $\bar{x}_1 = 15$; $\bar{x}_2 = 11$; $s_1 = 4.50924975$; $s_2 = 4.39696865$
b. H_0: $\mu_1 = \mu_2$; H_1: $\mu_1 \neq \mu_2$; critical values: $t = -2.179$ and 2.179; test statistic: $t = 1.680$; do not reject H_0
c. critical value: $F = 4.75$; test statistic: $F = 2.82$; do not reject H_0
d. conclusions are the same

12.15 critical value: $F = 3.29$; test statistic: $F = 4.07$; reject H_0

12.17 **a.** H_0: $\mu_1 = \mu_2 = \mu_3$; H_1: all three population means are not equal
b. numerator: $df = 2$; denominator: $df = 22$
c. $SSB = 243.1844$; $SSW = 561.0556$; $SST = 804.2400$
d. reject H_0 if $F > 5.72$
e. $MSB = 121.5922$; $MSW = 25.5025$ **f.** critical value: $F = 5.72$
g. test statistic: $F = 4.77$ **i.** do not reject H_0

12.19 critical value: $F = 3.47$; test statistic: $F = 18.64$; reject H_0

12.21 critical value: $F = 3.72$; test statistic: $F = 5.44$; reject H_0

12.23 **a.** critical value: $F = 3.34$; test statistic: $F = 6.08$; reject H_0 **b.** .05

12.25 critical value: $F = 6.93$; test statistic: $F = 1.24$; do not reject H_0

12.27 **a.** critical value: $F = 3.59$; test statistic: $F = 6.53$; reject H_0 **b.** do not reject H_0

12.29 critical value: F is between 3.51 and 3.72; test statistic: $F = 4.83$; reject H_0

12.33 factor MS $= 36.998$; factor $df = 2$, error $df = 9$, total $df = 11$; factor SS $= 73.996$, error SS $= 512.280$, total SS $= 586.276$; $n_1 = n_2 = n_3 = 4$; POOLED STDEV $= 7.545$

Self-Review Test

1. a   **2.** b   **3.** c   **4.** a   **5.** a   **6.** a   **7.** b   **8.** a
10. **a.** critical value: $F = 6.11$; test statistic: $F = 1.73$; do not reject H_0 **b.** Type II error

CHAPTER 13

13.15 **a.** y-intercept $= 100$; slope $= 5$; positive relationship   **b.** y-intercept $= 400$; slope $= -4$; negative relationship

13.17 $\mu_{y|x} = -5.5815 + .2886x$ **13.19** $\hat{y} = -83.7140 + 10.5714x$

13.21 **a.** $45.00   **b.** the same amount   **c.** exact relationship

13.23 **a.** $27.10 million   **b.** different amounts   **c.** nonexact relationship

13.25 **b.** $\hat{y} = 150.4136 - 16.6264x$ **e.** $3402.88   **f.** $-$14,886.16

13.27 **a.** $\hat{y} = -193.7194 + 9.2942x$ **e.** $224,520   **f.** error $= $45,344

13.29 **a.** $\mu_{y|x} = 37.8949 + .2620x$
b. population regression line because data set includes all 16 National League teams; values of A and B
d. 47.06%

13.35 $\sigma_\epsilon = 7.0756$; $\rho^2 = .04$ **13.37** $s_e = 4.7117$; $r^2 = .99$

13.39 **a.** $SS_{xx} = 1587.8750$; $SS_{yy} = 5591.5000$; $SS_{xy} = 2842.7500$ **b.** $s_\epsilon = 9.1481$
c. SST $= 5591.5000$; SSE $= 502.1652$; SSR $= 5089.3348$ **d.** $r^2 = .91$

13.41 **a.** $s_e = 7.0269$ **b.** $r^2 = .74$ **13.43** **a.** $s_e = 60.8274$ **b.** $r^2 = .95$

13.45 **a.** $\sigma_\epsilon = 4.4479$ **b.** $\rho^2 = .63$

13.47 **a.** 6.01 to 6.63   **b.** H_0: $B = 0$; H_1: $B > 0$; critical value: $t = 2.145$; test statistic: $t = 59.792$; reject H_0
c. H_0: $B = 0$; H_1: $B \neq 0$; critical values: $t = -2.977$ and 2.977; test statistic: $t = 59.792$; reject H_0
d. H_0: $B = 4.50$; H_1: $B \neq 4.50$; critical values: $t = -2.624$ and 2.624; test statistic: $t = 17.219$; reject H_0

13.49 **a.** 2.35 to 2.65   **b.** $H_0: B = 0$; $H_1: B > 0$; critical value: $z = 2.05$; test statistic: $z = 39.12$; reject H_0
    c. $H_0: B = 0$; $H_1: B \neq 0$; critical values: $z = -2.58$ and 2.58; test statistic: $z = 39.12$; reject H_0
    d. $H_0: B = 1.75$; $H_1: B > 1.75$; critical value: $z = 2.33$; test statistic: $z = 11.74$; reject H_0

13.51 **a.** -23.5406 to -9.7122
    b. $H_0: B = 0$; $H_1: B < 0$; critical value: $t = -1.943$; test statistic: $t = -5.884$; reject H_0

13.53 **a.** $\hat{y} = 23.7297 + 1.3054x$ **b.** .88 to 1.73
    c. $H_0: B = 0$; $H_1: B > 0$; critical value: $t = 2.365$; test statistic: $t = 9.193$; reject H_0

13.55 **a.** .86 to 17.73
    b. $H_0: B = 0$; $H_1: B \neq 0$; critical values: $t = -4.604$ and 4.604; test statistic: $t = 5.071$; reject H_0

13.57 **a.** $\hat{y} = -46.4228 + 3.8699x$ **b.** 2.70 to 5.04
    c. $H_0: B = 1.50$; $H_1: B > 1.50$; critical value: $t = 2.896$; test statistic: $t = 4.682$; reject H_0

13.63 a   **13.67** **a.** positive   **b.** positive   **c.** positive   **d.** negative   **e.** zero

13.69 $\rho = .21$

13.71 **a.** $r = -.996$
    b. $H_0: \rho = 0$; $H_1: \rho < 0$; critical value: $t = -2.764$; test statistic: $t = -35.249$; reject H_0

13.73 **a.** positively   **b.** $r = .96$
    c. $H_0: \rho = 0$; $H_1: \rho > 0$; critical value: $t = 1.895$; test statistic: $t = 9.071$; reject H_0

13.75 **a.** positively   **b.** close to 1   **c.** $r = .97$
    d. $H_0: \rho = 0$; $H_1: \rho \neq 0$; critical values: $t = -2.776$ and 2.776; test statistic: $t = 7.980$; reject H_0

13.77 **a.** $r = .95$
    b. $H_0: \rho = 0$; $H_1: \rho \neq 0$; critical values: $t = -3.707$ and 3.707; test statistic:
    $t = 7.452$; reject H_0

13.79 $\rho = .79$

13.81 **a.** $SS_{xx} = 1895.6000$; $SS_{yy} = 4798.4000$; $SS_{xy} = 1231.8000$ **b.** $\hat{y} = 156.3302 + .6498x$ **d.** $r = .41$; $r^2 = .17$
    f. 195.3182   **g.** $s_e = 22.3550$ **h.** $-.53$ to 1.83
    i. $H_0: B = 0$; $H_1: B > 0$; critical value: $t = 1.860$; test statistic: $t = 1.265$; do not reject H_0
    j. $H_0: \rho = 0$; $H_1: \rho > 0$; critical value: $t = 2.306$; test statistic: $t = 1.271$; do not reject H_0

13.83 **a.** $SS_{xx} = 3788.9000$; $SS_{yy} = 964.0000$; $SS_{xy} = 1820.0000$ **b.** $\hat{y} = -7.6660 + .4804x$ **d.** $r = .95$; $r^2 = .91$
    e. $s_e = 3.3480$ **f.** .30 to .66
    g. $H_0: B = 0$; $H_1: B > 0$; critical value: $t = 2.896$; test statistic: $t = 8.831$; reject H_0
    h. $H_0: \rho = 0$; $H_1: \rho \neq 0$; critical values: $t = -3.355$ and 3.355; test statistic: $t = 8.605$; reject H_0

13.85 **a.** $SS_{xx} = 3.3647$; $SS_{yy} = 285.7143$; $SS_{xy} = 29.1957$ **b.** $\hat{y} = 4.0729 + 8.6771x$ **d.** $r = .94$; $r^2 = .89$
    e. $s_e = 2.5448$ **f.** 5.11 to 12.24
    g. $H_0: B = 0$; $H_1: B \neq 0$; critical values: $t = -4.032$ and 4.032; test statistic: $t = 6.255$; reject H_0
    h. $H_0: \rho = 0$; $H_1: \rho > 0$; critical value: $t = 3.365$; test statistic: $t = 6.161$; reject H_0

13.87 **a.** 13.8708 to 16.6292; 11.7648 to 18.7352   **b.** 62.3590 to 67.7210; 56.3623 to 73.7177

13.89 37.3155 to 38.2519; 32.2279 to 41.3395   **13.91** 112.9148 to 124.0182; 102.8425 to 134.0905

13.93 $1967.66 to $2648.26; $1464.24 to $3151.68

13.95 **a.** positive relationship   **b.** $\hat{y} = -1.9175 + .9895x$ **d.** $r = .97$; $r^2 = .94$ **e.** $s_e = 1.0941$
    f. .54 to 1.44   **g.** $H_0: B = 0$; $H_1: B > 0$; critical value: $t = 2.571$; test statistic: $t = 8.811$; reject H_0
    h. $H_0: \rho = 0$; $H_1: \rho > 0$; critical value: $t = 2.571$; test statistic: $t = 8.922$; reject H_0; same conclusion

13.97 **a.** positive   **b.** $\hat{y} = 7.8299 + .5039x$ **d.** $r = .89$; $r^2 = .79$ **e.** 2547   **f.** $s_e = 3.3525$
    g. .11 to .90   **h.** $H_0: B = 0$; $H_1: B > 0$; critical value: $t = 3.365$; test statistic: $t = 4.278$; reject H_0
    i. $H_0: \rho = 0$; $H_1: \rho \neq 0$; critical values: $t = -3.365$ and 3.365; test statistic: $t = 4.365$; reject H_0

13.99 **b.** $SS_{xx} = 110.0000$; $SS_{yy} = 429.8000$; $SS_{xy} = 214.3000$ **d.** $\hat{y} = 47.2590 + 1.9482x$ **f.** $r = .99$
    g. 76.48 million households

13.101 **b.** $SS_{xx} = 227.5000$; $SS_{yy} = 8.8343$; $SS_{xy} = -44.7000$ **c.** yes   **d.** $\hat{y} = 4.3630 - .1965x$ **f.** $r = -.997$
    g. .8260 years

13.103 10.6211 to 14.0175; 7.2440 to 17.3946   **13.105** 207.3197 to 240.1195; 183.2730 to 264.1662

13.107 **a.** yes   **b.** 246.4670 to 275.5330 lines   **c.** 200.0567 to 321.9433 lines   **e.** 338 lines

13.111 **b.** $\hat{y} = 12.9099 + 31.0465x$

Self-Review Test

1. d **2.** a **3.** b **4.** a **5.** b **6.** b **7.** true **8.** true
9. a **10.** b
15. **a.** The number of computers sold depends on experience. **b.** positive **d.** $\hat{y} = 17.2732 + 1.5826x$
 f. $r = .73$; $r^2 = .53$ **g.** 34.68 computers **h.** $s_e = 6.5244$ **i.** -1.09 to 4.25
 j. H_0: $B = 0$; H_1: $B > 0$; critical value: $t = 3.365$; test statistic: $t = 2.387$; do not reject H_0
 k. 23.3013 to 36.5667 **l.** 11.8961 to 47.9719
 m. H_0: $\rho = 0$; H_1: $\rho > 0$; critical value: $t = 3.365$; test statistic: $t = 2.388$; do not reject H_0

CHAPTER 14

14.5 **a.** rejection region is $X \geq 12$ **b.** rejection region is $X \leq 3$ and $X \geq 17$
 c. rejection region is to the left of $z = -1.65$
14.7 **a.** H_0: Median $= 28$; H_1: Median > 28; rejection region: $X \geq 9$; observed value of $X = 8$; do not reject H_0
 b. H_0: Median $= 100$; H_1: Median < 100; rejection region: $X \leq 2$; observed value of $X = 1$; reject H_0
 c. H_0: Median $= 180$; H_1: Median $\neq 180$; critical values: $z = -1.96$ and 1.96; test statistic: $z = -3.73$; reject H_0
 d. H_0: Median $= 55$; H_1: Median < 55; critical value: $z = -1.65$; test statistic: $z = -3.10$; reject H_0
14.9 H_0: $p = .50$; H_1: $p \neq 50$; rejection region: $X \leq 1$ and $X \geq 9$; observed value of $X = 8$; do not reject H_0
14.11 H_0: $p = .50$; H_1: $p > .50$; rejection region: $X \geq 15$; observed value of $X = 13$; do not reject H_0
14.13 H_0: $p = .50$; H_1: $p < .50$; critical value: $z = -1.96$; test statistic: $z = -1.33$; do not reject H_0
14.15 H_0: $p = .50$; H_1: $p > .50$; critical value: $z = 2.33$; test statistic: $z = 1.74$; do not reject H_0
14.17 H_0: Median $= 12$ ounces; H_1: Median $\neq 12$ ounces; rejection region: $X \leq 1$ and $X \geq 8$; observed value of $X = 8$; reject H_0
14.19 H_0: Median $= 4$ minutes; H_1: Median > 4 minutes; critical value: $z = 2.33$; test statistic: $z = 2.94$; reject H_0
14.21 H_0: Median $= 42$ months; H_1: Median < 42 months; critical value: $z = -2.33$; test statistic: $z = -2.37$; reject H_0
14.23 H_0: $M = 0$; H_1: $M < 0$; rejection region: $X = 0$; observed value of $X = 1$; do not reject H_0
14.25 H_0: $M = 0$; H_1: $M \neq 0$; critical values: $z = -2.33$ and 2.33; test statistic: $z = 1.92$; do not reject H_0
14.27 H_0: $M = 0$; H_1: $M \neq 0$; critical values: $z = -1.96$ and 1.96; test statistic: $z = 1.70$; do not reject H_0
14.31 **a.** rejection region: $T \leq 11$ **b.** rejection region: $T \leq 7$
 c. rejection region is to the left of $z = -1.96$ **d.** rejection region is to the right of $z = 2.33$
14.33 **a.** H_0: $M_A = M_B$; H_1: $M_A < M_B$; rejection region: $T \leq 4$; observed value of $T = 10.5$; do not reject H_0
 b. conclusions are the same
14.35 **a.** H_0: $M_A = M_B$; H_1: $M_A > M_B$; rejection region: $T \leq 2$; observed value of $T = 3.0$; do not reject H_0
 b. conclusions are the same
14.37 H_0: $M_A = M_B$; H_1: $M_A < M_B$; critical value: $z = -1.96$; test statistic: $z = -2.89$; reject H_0
14.41 **a.** H_0: population distributions are identical; H_1: population distributions are different; rejection region: $T \leq 28$ and $T \geq 56$; observed value of $T = 22$; reject H_0
 b. H_0: population distributions are identical; H_1: distribution of population 1 lies to right of the distribution of population 2; critical value: $z = 1.96$; test statistic: $z = 1.45$; do not reject H_0
 c. H_0: population distributions are identical; H_1: distribution of population 1 lies to left of distribution of population 2; critical value: $z = -1.65$; test statistic: $z = -2.01$; reject H_0
 d. H_0: population distributions are identical; H_1: population distributions are different; critical values: $z = -2.58$ and 2.58; test statistic: $z = 3.00$; reject H_0
14.43 rejection region: $T \leq 41$; observed value of $T = 40.5$; reject H_0
14.45 critical value: $z = 2.33$; test statistic: $z = 1.41$; do not reject H_0
14.47 critical values: $z = -1.96$ and 1.96; test statistic: $z = -1.27$; do not reject H_0
14.51 **a.** critical value: $\chi^2 = 5.991$; test statistic: $H = 2.372$; do not reject H_0
 b. critical value: $\chi^2 = 7.815$; test statistic: $H = 13.674$; reject H_0
 c. critical value: $\chi^2 = 5.991$; test statistic: $H = 8.822$; reject H_0
 d. critical value: $\chi^2 = 9.488$; test statistic: $H = .863$; do not reject H_0

14.53 **a.** critical value: $\chi^2 = 9.210$; test statistic: $H = 1.860$; do not reject H_0 **b.** conclusions are the same

14.55 **a.** critical value: $\chi^2 = 9.210$; test statistic: $H = 3.498$; do not reject H_0
b. conclusions are the same

14.57 critical value: $\chi^2 = 5.991$; test statistic: $H = 3.815$; do not reject H_0

14.61 **a.** $-.943$ **b.** $.964$ **14.63** **a.** positive **b.** $.948$; yes **14.65** **a.** positive **b.** $.955$; yes

14.67 **a.** $.917$ **b.** rejection region: $r_s \geq .600$; test statistic: $r_s = .917$; reject H_0 **c.** yes **14.73** no

14.75 H_0: the sequence of heads and tails is random; H_1: the sequence of heads and tails is not random; rejection region: $R \leq 6$ and $R \geq 16$; observed value of $R = 13$; do not reject H_0

14.77 rejection region: $R \leq 5$ and $R \geq 15$; observed value of $R = 5$; reject H_0

14.79 critical values: $z = -2.58$ and 2.58; test statistic: $z = 1.38$; do not reject H_0

14.81 critical values: $z = -2.24$ and 2.24; test statistic: $z = -4.27$; reject H_0

14.83 H_0: $p = .50$; H_1: $p < .50$; rejection region: $X \leq 7$; observed value of $X = 7$; reject H_0

14.85 H_0: $p = .50$; H_1: $p > .50$; critical value: $z = 2.33$; test statistic: $z = 4.01$; reject H_0

14.87 H_0: Median $= \$73.57$; H_1: Median $> \$73.57$; rejection region: $X \geq 15$; observed value of $X = 14$; do not reject H_0

14.89 H_0: Median $= \$250$; H_1: Median $\neq \$250$; critical values: $z = -1.96$ and 1.96; test statistic: $z = -2.92$; reject H_0

14.91 **a.** H_0: $M = 0$; H_1: $M < 0$; rejection region: $X = 0$; observed value of $X = 1$; do not reject H_0
b. part a: do not reject H_0; Exercises 14.34 and 10.101: reject H_0

14.93 H_0: $M = 0$; H_1: $M > 0$; critical value: $z = 1.96$; test statistic: $z = 3.01$; reject H_0

14.95 H_0: $M_A = M_B$; H_1: $M_A \neq M_B$; rejection region: $T \leq 2$; observed value of $T = 1$; reject H_0

14.97 H_0: $M_R = M_M$; H_1: $M_R > M_M$; critical value: $z = 1.96$; test statistic: $z = 2.15$; reject H_0

14.99 rejection region: $T \geq 54$; observed value of $T = 57$; reject H_0

14.101 **a.** critical value: $z = 1.65$; test statistic: $z = 2.71$; reject H_0 **b.** no

14.103 critical value: $\chi^2 = 5.991$; test statistic: $H = .645$; do not reject H_0

14.105 critical value: $\chi^2 = 5.991$; test statistic: $H = 8.187$; reject H_0

14.107 **a.** negative **b.** $r_s = -.742$
c. rejection region: $r_s \leq -.648$ and $r_s \geq .648$; test statistic: $r_s = -.742$; reject H_0

14.109 **a.** reject H_0 **b.** rejection region: $r_s \leq -.643$; test statistic: $r_s = -.762$; reject H_0

14.111 rejection region: $R \leq 4$ and $R \geq 12$; observed value of $R = 7$; do not reject H_0

14.113 critical values: $z = -2.33$ and 2.33; test statistic: $z = 1.12$; do not reject H_0

14.119 **a.** sign test for paired data or Wilcoxon signed-rank test
b. H_0: $M_A = M_B$; H_1: $M_A \neq M_B$; rejection region: $T \leq 8$; observed value of $T = 22$; do not reject H_0
c. H_0: $\rho_s = 0$; H_1: $\rho_s > 0$
d. rejection region: $r_s \geq .564$; test statistic: $r_s = .891$; reject H_0

14.123 21

Self-Review Test

1. b **2.** a **3.** a **4.** c **5.** b **6.** c **7.** a, b **8.** b
9. c **10.** c **11.** a, b **12.** c **13.** b

14. H_0: $p = .50$; H_1: $p \neq .50$; rejection region: $X \leq 2$ and $X \geq 10$; observed value of $X = 2$; reject H_0

15. H_0: $p = .50$; H_1: $p > .50$; critical value: $z = 1.96$; test statistic: $z = .75$; do not reject H_0

16. H_0: Median $= \$65$; H_1: Median $> \$65$; rejection region: $X \geq 10$; observed value of $X = 7$; do not reject H_0

17. H_0: Median $= \$160,000$; H_1: Median $> \$160,000$; critical value: $z = 2.33$; test statistic: $z = 2.57$; reject H_0

18. **a.** H_0: $M = 0$; H_1: $M \neq 0$; rejection region: $X = 0$ and $X = 7$; observed value of $X = 3$; do not reject H_0
b. H_0: $M_A = M_B$; H_1: $M_A \neq M_B$; rejection region: $T \leq 2$; observed value of $T = 6.5$; do not reject H_0
c. conclusions are the same

19. H_0: $M = 0$; H_1: $M \neq 0$; critical values: $z = -2.33$ and 2.33; test statistic: $z = 1.44$; do not reject H_0
20. H_0: $M_S = M_F$; H_1: $M_S < M_F$; rejection region: $T \leq 8$; observed value of $T = 14$; do not reject H_0
21. H_0: $M_A = M_B$; H_1: $M_A > M_B$; critical value: $z = 1.96$; test statistic: $z = 2.09$; reject H_0
22. rejection region: $T \leq 49$ and $T \geq 87$; observed value of $T = 82$; do not reject H_0
23. critical values: $z = -2.24$ and 2.24; test statistic: $z = -1.11$; do not reject H_0
24. **a.** critical value: $\chi^2 = 7.378$; test statistic: $H = 7.653$; reject H_0
    b. critical value: $\chi^2 = 9.210$; test statistic: $H = 7.653$; do not reject H_0
25. **a.** positive **b.** $r_s = .773$
    c. rejection region: $r_s \geq .648$; test statistic: $r_s = .773$; reject H_0
26. rejection region: $R \leq 3$ or $R \geq 13$; observed value of $R = 6$; do not reject H_0
27. critical values: $z = -1.96$ and 1.96; test statistic: $z = 1.09$; do not reject H_0

APPENDIX A

A.7 simple random sample **A.9 a.** nonrandom sample **b.** judgment sample **c.** selection error
A.11 a. random sample **b.** simple random sample **c.** no **A.13** voluntary response error
A.15 response error **A.17 a.** observational study **b.** not a double-blind study
A.19 a. designed experiment **b.** double-blind study **A.21** designed experiment **A.23** yes
A.25 b. observational study **c.** not a double-blind study
A.27 a. designed experiment **b.** double-blind study
A.29 a. no **b.** no **c.** convenience sample **A.33 a.** no **b.** nonresponse error and response error
c. above

ANSWERS TO ODD-NUMBERED GRAPHING CALCULATOR EXERCISES

Answers to the graphing calculator exercises have been generated using a graphing calculator, and the following rules regarding rounding have been followed: (1) When the solution requires more than one step, do not round the result for each step. Instead, use the result to the number of places stored in the calculator. (2) When the numbers from statistical tables that appear anywhere in the text are used in computing solutions, use the number exactly as it appears in the table. Do not round the number to fewer places prior to using it in the calculation.

CHAPTER 5

5.27 $\mu = 2.94$ camcorders; $\sigma = 1.44$ camcorders   **5.47** 167,960   **5.65** **a.** .2725   **b.** .0839
5.87 **a.** .4493   **c.** $\mu = .8$ accident; $\sigma = .8944$ accident

CHAPTER 6

6.15 **a.** .4744   **b.** .4678   **c.** .1162   **d.** .0610   **e.** .9452
6.23 **a.** .0162   **b.** .2450   **c.** .1570   **d.** .9625
6.25 **a.** .8365   **b.** .8762   **c.** approximately .5   **d.** approximately .5
    e. approximately 0   **f.** approximately 0
6.31 **a.** .3341   **b.** .9568   **c.** .9568   **d.** approximately 0   **6.33** **a.** .2191   **b.** .6003
6.39 **a.** .3839   **b.** .4308   **6.41** **a.** 12.15%   **b.** 83.02%
6.47 **a.** .2483 or about 25%   **b.** .1717 or about 17%
6.55 **a.** 1.645   **b.** -1.96 **c.** -2.326 **d.** 2.576
6.57 **a.** 208.50   **b.** 241.12   **c.** 178.50   **d.** 145.75   **e.** 158.26   **f.** 251.34
6.61 2060 kilowatt-hours
6.71 **a.** approximation: .0765, actual: .0792   **b.** approximation: .6784, actual: .6769
    c. approximation: .8405, actual: .8404   **d.** approximation: .8226, actual: .8228
6.75 **a.** approximation: .0348, actual: .0347   **b.** approximation: .0085, actual: .0082
    c. approximation: .6409, actual: .6402
6.83 .0124 or 1.24%
6.87 **a.** approximation: .0148, actual: .0149   **b.** approximation: .0465, actual: .0430
    c. approximation: .8345, actual: .8344   **d.** approximation: .2545, actual: .2608
6.93 Plant A: .6827, Plant B: .8400   **6.95** 49.93456 minutes or about 8:10 A.M.

CHAPTER 7

7.15 **a.** $\sigma_{\bar{x}} = 1.3995$ **b.** $\sigma_{\bar{x}} = 2.5$ **7.87** **a.** .9736   **b.** .3117
7.97 **a.** .0276   **b.** .8399   **c.** .7130   **d.** .0166

CHAPTER 8

8.15 **a.** 79.195 to 84.605   **b.** 80.097 to 83.703   **c.** 80.277 to 83.523   **d.** yes
8.29 35.979 to 36.061 inches; the machine needs to be adjusted
8.47 **a.** 63.758 to 73.242   **b.** 64.599 to 72.401   **c.** 64.826 to 72.174   **8.57** 71.1 to 77.3 mph
8.67 **a.** .646 to .714   **b.** .639 to .721   **c.** .626 to .734   **d.** yes   **8.109** 1.928 to 2.539 hours
8.111 **a.** 12%; margin of error $= \pm 9\%$ **b.** .0016 to .238

CHAPTER 9

9.37   **a.** H_0: $\mu = 45$ months; H_1: $\mu < 45$ months; critical value: $z = -1.96$; test statistic: $z = -1.875$; do not reject H_0

       b.   critical value: $z = -1.65$; test statistic: $z = -1.875$; reject H_0

9.49 H_0: $\mu = 108.2$¢; H_1: $\mu < 108.2$¢; test statistic: $z = -1.34$; p-value $= .1817$

9.53   **a.** H_0: $\mu = 10$ minutes; H_1: $\mu < 10$ minutes; test statistic: $z = -2.00$; p-value $= .0213$

       b. $.0213 > .02$, do not reject H_0 **c.** $.0213 < .05$, reject H_0

9.93   **a.**   critical values: $z = -1.96$ and 1.96; test statistic: $z = -3.999$; reject H_0; p-value $= .00006$

       b.   critical value: $z = -2.33$; test statistic: $z = -1.368$; do not reject H_0; p-value $= .0857$

       c.   critical value: $z = 1.96$; test statistic: $z = .43$; do not reject H_0

9.95 H_0: $p = .29$; H_1: $p > .29$; critical value: $z = 1.65$; test statistic: $z = 1.971$, reject H_0; p-value $= .0244$

9.121   **a.** H_0: $\mu = 114$ minutes; H_1: $\mu < 114$ minutes; critical value: $t = -2.492$; test statistic: $t = -2.273$; do not reject H_0; p-value $= .0161$

9.123 H_0: $\mu \leq 150$ calories; H_1: $\mu > 150$ calories; critical value: $t = 2.262$; test statistic: $t = 1.157$; do not reject H_0; p-value $= .1385$

9.127 H_0: $p = .26$; H_1: $p > .26$; critical value: $z = 2.05$; test statistic: $z = 3.224$; reject H_0; p-value $= .00063$

CHAPTER 10

10.11   **a.** 1131; margin of error $= \pm\$1009.29$ **b.** 283.99 to 1978

       c. H_0: $\mu_1 - \mu_2 = 0$; H_1: $\mu_1 - \mu_2 > 0$; critical value: $z = 2.33$; test statistic: $z = 2.1964$; do not reject H_0; p-value $= .0140$

10.17   **a.** $-.51$ to $.01$ minute

       b. H_0: $\mu_1 - \mu_2 = 0$; H_1: $\mu_1 - \mu_2 < 0$; critical value: $z = -1.96$; test statistic: $z = -2.06$; reject H_0

       c.   p-value $= .0197$; do not reject H_0 if $\alpha = .01$; reject H_0 if $\alpha = .05$

10.21   **a.** 5.25 **b.** $.926$ to 9.573

10.23 H_0: $\mu_1 - \mu_2 = 0$; H_1: $\mu_1 - \mu_2 \neq 0$; critical values: $t = -2.719$ and 2.719; test statistic: $t = 3.3025$; reject H_0; p-value $= .0022$

10.27   **a.** -5.35 **b.** -9.03 to -1.67

       c. H_0: $\mu_1 - \mu_2 = 0$; H_1: $\mu_1 - \mu_2 < 0$; critical value: $t = -2.080$; test statistic: $t = -4.121$; reject H_0

10.37 4.6322 to 13.088

10.39 H_0: $\mu_1 - \mu_2 = 0$; H_1: $\mu_1 - \mu_2 \neq 0$; critical values: $t = -2.744$ and 2.744; test statistic: $t = 5.7492$; reject H_0; p-value $= .0000025$

10.47   **a.** -12.85 to 2.8531 minutes

       b. H_0: $\mu_1 - \mu_2 = 0$; H_1: $\mu_1 - \mu_2 < 0$; critical value: $t = -2.414$; test statistic: $t = -1.7128$; do not reject H_0; p-value $= .0468$

10.51   **a.** 12.5 to 38.3 **b.** 11.124 to 15.276 **c.** 29.803 to 39.397

10.53   **a.**   critical values: $t = -1.860$ and 1.860; test statistic: $t = 8.04$; reject H_0; p-value $= .00004$

       b.   critical value: $t = 1.721$; test statistic: $t = 10.8466$; reject H_0; p-value $= .0000$

       c.   critical value: $t = -2.583$; test statistic: $t = -7.989$; reject H_0

10.63 $-.1853$ to $.04528$

10.65 H_0: $p_1 - p_2 = 0$; H_1: $p_1 - p_2 \neq 0$; critical values: $z = -2.58$ and 2.58; test statistic: $z = -1.5526$; do not reject H_0; p-value $= .1205$

10.83   **a.** H_0: $\mu_1 - \mu_2 = 0$; H_1: $\mu_1 - \mu_2 \neq 0$; critical values: $t = -2.009$ and 2.009; test statistic: $t = -1.0579$; do not reject H_0; p-value $= .29517$

       b. $-.3101$ to $.0701$

10.91   **a.** -7.719 to $.469$

       b. H_0: $\mu_d = 0$; H_1: $\mu_d < 0$; critical value: $t = -2.365$; test statistic: $t = -2.0937$; do not reject H_0; p-value $= .0373$

CHAPTER 11

11.19 critical value: $\chi^2 = 17.275$; test statistic: $\chi^2 = 25.2143$; reject H_0; p-value $= .0085$

11.27 **a.** H_0: the proportion in each row is the same for all four populations; H_1: the proportion in each row is not the same for all four populations
     c. critical value: $\chi^2 = 14.449$ **d.** test statistic: $\chi^2 = 52.4543$ **e.** reject H_0; p-value $= .0000000015$

11.33 critical value: $\chi^2 = 9.210$; test statistic: $\chi^2 = 41.4511$; reject H_0; p-value $= .000000001$

11.39 critical value: $\chi^2 = 7.378$; test statistic: $\chi^2 = 1.5874$; do not reject H_0; p-value $= .4522$

11.51 critical value: $\chi^2 = 7.815$; test statistic: $\chi^2 = 10.464$; reject H_0; p-value $= .015$

11.57 critical value: $\chi^2 = 13.277$; test statistic: $\chi^2 = 46.2561$; reject H_0; p-value $= .000000002$

CHAPTER 12

12.13 **a.** $\overline{x}_1 = 15$; $\overline{x}_2 = 11$; $s_1 = 4.50924975$; $s_2 = 4.39696865$
     b. H_0: $\mu_1 = \mu_2$; H_1: $\mu_1 \neq \mu_2$; critical values: $t = -2.179$ and 2.179; test statistic: $t = 1.6803$; do not reject H_0; p-value $= .1187$
     c. critical value: $F = 4.75$; test statistic: $F = 2.8235$; do not reject H_0; p-value $= .1187$
     d. conclusions are the same

12.21 critical value: $F = 3.72$; test statistic: $F = 5.440$; reject H_0; p-value $= .0053$

CHAPTER 13

13.27 **a.** $\hat{y} = -193.7194 + 9.2942x$ **e.** $\$224,520$ **f.** error $= -\$45,344$

13.41 **a.** $s_e = 7.0268$ **b.** $r^2 = .7411$ **13.43** **a.** $s_e = 60.827$ **b.** $r^2 = .9501$

13.55 **a.** .86 to 17.73
     b. H_0: $B = 0$; H_1: $B \neq 0$; critical values: $t = -4.604$ and 4.604; test statistic: $t = 5.071$; reject H_0; p-value $= .0071$

13.77 **a.** $r = .9540$

INDEX

Addition rule of probability, 166
  for mutually exclusive events, 168
Alpha (α), 337
  level of significance, 337
  probability of Type I error, 379–380
Alternative hypothesis, 377–378
Analysis of variance (ANOVA), one-way, 533–535, 538
  alternative hypothesis, 538
  assumptions of, 534
  between-samples sum of squares (SSB), 536
  null hypothesis, 538
  table, 538
  test statistic, 535
  total sum of squares (SST), 536
  variance (or mean square):
    between samples (MSB), 534, 537
    within samples (MSW), 534, 537
  within-samples sum of squares (SSW), 536
Arithmetic mean, 70
Average, 70

Bar chart, 29
Bar graph, 29
Bell-shaped distribution/curve, 100
Bernoulli trials, 209
Beta (β), probability of Type II error, 380
Bimodal distribution, 77
Binomial distribution, 209–211
  mean of, 219
  normal approximation to, 275–276
  probability of failure for, 209
  probability graph, 218
  probability of success for, 209
  standard deviation of, 219
  table, D7
  using the table of, 215
Binomial experiment, 209
  conditions of, 209
Binomial formula, 210–211
Binomial parameters, 211
Binomial random variable, 210
Boundaries, class, 35
Box-and-whisker plot, 109–110

Categorical data, 619
Categorical variable, 13–14
Cell, 142, 500
Census, 6–7, 334
Central limit theorem, 304, 318
Central tendency, measures of, 70
Chance experiment, 185
Chance variable, 185
Chebyshev's theorem, 98
Chi-square distribution, 487–488
  degrees of freedom, 488
  goodness-of-fit test, 490–492
  table of, D25
  test of homogeneity, 507
  test of independence, 501
  using the table of, 489–490
Class, 34
  boundaries, 35
  frequency, 35
  limits, 35
  midpoint/mark, 36
  open-ended, 42
  width/size, 36
Classical probability, 134–135
  concept, 134
  rule, 135
Coefficient of determination, 575–577
Coefficient of linear correlation, 586–587
Coefficient of variation, 90
  population, 90
  sample, 90
Combinations, 204
  table of, 207
Combined mean, 81
Complementary events, 149
Composite events, 130
Compound events, 130–131
Conditional probability, 143, 158
Confidence coefficient, 337
Confidence interval, 336–337
  for the difference between two means
    large and independent samples, 432
    small and independent samples, 442, 450
  for the difference between two proportions
    large and independent samples, 466

Confidence interval (*continued*)
  lower limit of, 336
  maximum error, 338, 363
  for the mean, large sample, 337
  for the mean, small sample, 350
  for the mean of paired differences, 457
  for the mean value of y for a given x, 601
  for the predicted value of y for a given x, 602–603
  for the proportion, large sample, 356–357
  for the slope of regression line, 581
  for the standard deviation, 574
  upper limit of, 336
  for the variance, 514
  width of, 341
Confidence level, 336–337
Consistent estimator, 299
Constant, 10
Contingency table, 142, 500
Continuous probability distribution, 241
Continuous random variable, 186, 241
Continuous variable, 13
Correction factor, finite population, 297, 317
Correction for continuity, 277
Correlation coefficient, 586–587
  population, 586
  sample, 586–587
  Spearman rho rank, 654–655
Counting rule, 141
Critical value, 378
Cross-section data, 15
Cross-tabulation, 500
Cumulative frequency distribution, 52
Cumulative percentage distribution, 53
Cumulative relative frequency distribution, 53

Data, 11
  categorical, 619
  cross-section, 15
  grouped, 35, 40, 70
  qualitative, 14
  quantitative, 12

Data (*continued*)
  ranked, 74
  raw, 26
  sources of, 16
  time-series, 15
  ungrouped, 26, 70
Data set, 3, 11
Decision rule, 636
Degrees of freedom, defined,
  for the chi-square distribution, 488
  for the F distribution, 531
  for simple linear regression, 573
  for the t distribution, 348
Dependent events, 148
Dependent samples, 430
Dependent variable in regression, 555
Descriptive statistics, 3
Determining the sample size:
  for estimating the mean, 362
  for estimating the proportion, 364
Deterministic regression model, 557
Deviation from the mean, 84
Difference between two population means:
  confidence interval for the, 432, 442, 450
  test of hypothesis about the, 434–435,
    443, 451, 453
Difference between two population
    proportions:
  confidence interval for the, 466
  test of hypothesis about the, 467–468
Difference between two sample means:
  mean of the, 431–432
  sampling distribution of the, 431
  standard deviation of the, 431
Difference between two sample proportions:
  mean of the, 465–466
  sampling distribution of the, 465–466
  standard deviation of the, 465–466
Discrete random variable, 185
  expected value of a, 194
  mean of a, 194–195
  probability distribution of a, 187
  standard deviation of a, 198
  variance of a, 198
Discrete variable, 13
Dispersion, measures of, 82–83
Distribution:
  bell-shaped, 100
  binomial, 209–211
  chi-square, 487–488
  cumulative frequency, 52
  cumulative percentage, 53
  cumulative relative frequency, 53
  F, 531–532
  frequency, 27, 35
  normal, 247
  percentage, 28, 39
  Poisson, 222–223
  population, 291
  probability, 187, 241

Distribution (*continued*)
  relative frequency, 28, 39
  sampling, 292
  standard normal, 249–250
  t, 347–348
Distribution-free tests, 618

Element, 3, 10
Elementary event, 130
Empirical rule, 100
Equally likely outcomes or events, 134
Equation of a regression model, 558
Error:
  nonsampling, 294
  random, 558
  sampling, 294
  standard, 297
  sum of squares (SSE), 561
  Type I, 379–380
  Type II, 380
Estimate, 334–335
  interval, 336
  point, 335
Estimated regression line, 558
Estimates of regression coefficients, 558
Estimation, 334
Estimator, 298, 317, 335
  consistent, 299, 317
  unbiased, 298, 317
Event, 130
  complement of an, 149
  composite, 130
  compound, 130–131
  elementary, 130
  impossible, 134
  simple, 130
  sure, 134
Events:
  complementary, 149
  dependent, 148
  equally likely, 134–135
  independent, 147–148
  intersection of, 154
  mutually exclusive, 146
  union of, 165
EXCEL:
  ANOVA, C42
  Chart, C16
  CHIDIST, C38
  CHIINV, C38
  CHITEST, C38
  Data Analysis, C2
  Descriptive Statistics option, C7
  Input Range, C3
  New Worksheet Ply, C4
  Normal distribution, C20
  Normal probabilities, C20
  NORMDIST, C21
  NORMINV, C22

EXCEL (*continued*)
  Output options, C4
  PivotTable/PivotChart Report, C9
  Poisson, C14
  Queuing Theory, C14
  RANK, C49
  Regression, C44
  STDEV, C24
  STDEVP, C24
  t-test, C32
  Tools, C2
Expected frequencies:
  for a goodness-of-fit test, 491–492
  for a test of homogeneity, 501–503
  for a test of independence, 501–503
Expected value of a discrete random
    variable, 194
Experiment, 128
  binomial, 209
  multinomial, 490
  random or chance, 185
  statistical, 128
Extrapolation, 604
Extreme values, 72

F distribution, 531–532
  degrees of freedom for the, 531–532
  table of the, D26
  using the table of, 531–532
Factorial formula, 203
Factorials, 203
  table of, D5
Failure, probability of, 209
Fences, 110–111
  inner, 110–111
  outer, 111
Finite population correction factor, 298, 317
First quartile, 104
Fisher, Sir Ronald, 531
Frequency, 27, 35
  class, 35
  cumulative, 52
  cumulative relative, 53
  expected, 491–492, 501
  histogram, 40
  joint, 501
  observed, 491–492, 501
  polygon, 42
  relative, 28, 39
Frequency density, relative, 241
Frequency distribution:
  curve, 42
  for qualitative data, 27
  for quantitative data, 34–35

Galton, Sir Francis, 555
Goodness-of-fit test, 490–492
  degrees of freedom for a, 491–492

Probability (*continued*)
  marginal, 142
  of mutually exclusive events, 161
  properties of, 134
  relative frequency concept of, 136
  subjective, 138
  of Type I error, 379–380
  of Type II error, 380
  of the union of events, 165
Probability density function, 242
Probability distribution, 241
  binomial, 209–211
  characteristics of a, 188
  of a continuous random variable, 242
  of a discrete random variable, 187
  normal, 247
  Poisson, 222–223
Proportion, population, 314
  confidence interval for the, 356–357
  maximum error of estimate for the,
    361–362
  test of hypothesis about the, 408–409
Proportion, sample, 314
  mean of the sampling distribution, 317
  standard distribution of the, 315
  standard deviation of the, 317
p-value, 395
  for a one-tailed test, 395
  for a two-tailed test, 396
p-value approach to hypothesis testing,
  395

Qualitative data, 14
  frequency distribution for, 27
  percentage distribution for, 28
  relative frequency distribution for, 28
Qualitative variable, 13–14
Quantitative data, 12
  cumulative frequency distribution for,
    34–35
  cumulative percentage distribution for, 39
  cumulative relative frequency distribution
    for, 39
  frequency distribution for, 36–37
  percentage distribution, 39
  relative frequency distribution for, 39
Quantitative variable, 12
Quartiles, 104
  first, 104
  second, 104
  third, 104

Random error, 558
Random experiment, 185
Random numbers, table of, D1
Random sample, 5
  simple, 9
Random variable, 185
  continuous, 186, 241
  discrete, 185–186

Range, 83
Rank, percentile, 107
Ranked data, 74
Raw data, 26
Real class limits, 35
Rectangular histogram, 46
Regression, 555–557
  analysis, 557
  assumptions of the, 566–567
  coefficient of determination, 575–577
  confidence interval for the mean value of
    y for a given *x*, 601
  confidence interval for the slope, 581
  degrees of freedom, 573
  dependent variable, 555
  deterministic model, 557
  error sum of squares (SSE), 561
  independent variable, 555
  least squares line, 559–561
  least squares method, 559–561
  linear, 555–557
  model, 555–557
  multiple, 555
  observed or actual value of *y*, 560
  predicted value of *y*, 560
  prediction interval for *y* for a given *x*,
    603
  probabilistic model, 558
  regression sum of squares (SSR),
    576–577
  scatter diagram, 559
  simple, 555
  standard deviation of errors, 573–574
  test of hypothesis about the slope,
    582–583
  total sum of squares (SST), 575
Regression, simple linear, *see* Regression
Regression line/model, simple linear:
  estimated, 558
  population, 558
  random error of, 558
  slope of the, 557
  true, 558
  y-intercept of the, 556
Regression model, equation of the, 558
Regression sum of squares (SSR), 577
Rejection region, 378
Relative frequency, 28, 39
  cumulative, 53
  histogram, 40
  polygon, 42
  for qualitative data, 28
  for quantitative data, 39
Relative frequency concept of probability,
  136
Relative frequency density, 241
Relative frequency distribution, *see* Relative
  frequency
Representative sample, 8–9
Residual, 560

Right-tailed test, 381, 383
Run, 660
Runs test for randomness, 659
  large-sample case, 662
  observed value, 662
  small-sample case, 659

Sample, 4, 6
  mean, 70, 92
  proportion, 314
  random, 9
  representative, 8–9
  simple random, 9
  standard deviation, 85, 93
  variance, 85, 93
Sample points, 128
Sample size, 456
Sample size, determination of:
  for estimating the mean, 361–362
  for estimating the proportion,
    363–364
Sample space, 128
Sample statistic, 88
Sample survey, 7
Samples:
  dependent, 430
  independent, 430
  matched, 430, 455
  paired, 430, 455
Sampling:
  error, 294
  with replacement, 9
  without replacement, 9
Sampling distribution, 292
  of the difference between two sample
    means for large and independent
    samples, 431–432
  of the difference between two sample
    proportions for large and
    independent samples, 465–466
  of the mean of sample paired differences,
    456
  of the sample mean, 292
  of the sample proportion, 315
  of the slope of a regression line, 580
Scatter diagram, 559
Scattergram, 559
Second quartile, 104
Sign test, 619
  about categorical data, 619
    critical value of *X*, 621
    large-sample case, 622
    observed value of *X*, 622
    small-sample case, 619
  about median of a single population, 624
    large-sample case, 626
    observed value of *X*, 625
    small-sample case, 624

Sign test (*continued*)
    about median difference between
        paired data, 627
        large-sample case, 629
        small-sample case, 627
Significance level, 379
Simple event, 130
Simple linear regression, *see* Regression
Simple random sample, 9
Single-valued classes, 44
Skewed histogram, 45
    to the left, 45
    to the right, 45
Slope of the regression line, 557
    confidence interval for the, 581
    test of hypothesis about the, 581–582
Sources of data, 16
    external, 16
    internal, 16
Spearman rho rank correlation coefficient,
    654–655
    critical value, 656
    decision rule, 657
Standard deviation, 84
    of the binomial distribution, 219
    confidence interval for the, 514
    of the difference of two sample means,
        431–432, 441
    of the difference between sample
        proportions, 465–466
    of a discrete random variable, 198
    of the errors of a simple regression
        model, 573
    for grouped data, 91
    of the mean of paired differences, 456
    of the normal distribution, 248
    of the Poisson distribution, 228–229
    population, 93
    sample, 85, 93
    of the sample mean, 297–299
    of the sample proportion, 317
    of the slope of a regression line, 580
    of the standard normal distribution, 249
    for ungrouped data, 85
    use of, 98
Standard error, 297
Standard normal distribution, 249–250
    table of the, D22
    using the table of, 250–256
Standard score, 249
Standard units, 249
Standardizing a normal distribution,
    257–258
Statistic, 88
Statistically not significant, 391
Statistically significant, 391
Statistics, 2
    descriptive, 3
    inferential, 4

Stem, 56
Stem-and-leaf display, 55–56
    grouped, 57
    ranked, 56
Subjective probability, 138
Success, probability, 209
Sum of squares
    error (SSE), 561
    regression (SSR), 576–577
    total (SST), 575
Summation notation, 17
Sure event, 134
Survey, 6
Symmetric histogram, 45

t distribution, 347–348
    degrees of freedom for the, 348
    table of the, D23
Table of:
    binomial probabilities, D7
    chi-square distribution, D25
    combinations, D6
    exponential, D15
    factorials, D5
    the *F* distribution, D26
    Poisson probabilities, D16
    random numbers, D1
    the standard normal distribution, D22
    the *t* distribution, D23
Tallies, 27
Target population, 6
Test of hypothesis:
    ANOVA, 538
    about the difference between two means:
        large and independent samples,
            434–435
        small and independent samples, 443,
            451–453
    about the difference between two
        proportions:
        large and independent samples,
            467–468
    about the linear correlation coefficient,
        588–589
    goodness-of-fit, 490–492
    homogeneity, 507
    independence, 501
    Kruskal–Wallis test, 649–652
    left-tailed, 381–383
    about the mean, large sample, 386–388
    about the mean of paired differences, 459
    about the mean, small sample, 400–401
    nonrejection region for a, 378–379
    one-tailed, 381
    power of a, 380
    about the proportion, large samples,
        408–409
    p-value approach to make a, 395

Test of hypothesis (*continued*)
    rejection region for a, 378–379
    right-tailed, 381, 383
    runs test for randomness, 659–663
    sign test about categorical data, 619–624
    sign test about median of a single
        population, 624–627
    sign test about median difference
        between paired data, 627–630
    significance level for a, 379
    Spearman's rho rank correlation
        coefficient test, 654–657
    steps of a, 384
    two-tailed, 381
    Type I error for a, 379–380
    Type II error for a, 380
    about the variance, 516
    Wilcoxon rank sum test for two
        independent samples, 642–647
    Wilcoxon signed-rank test for two
        dependent samples, 635–640
Test statistic, tests of hypotheses:
    ANOVA, 535
    difference between two means:
        large and independent samples, 435
        small and independent samples, 443,
            451–453
    difference between two proportions:
        large and independent samples,
            467–468
    goodness-of-fit, 492
    independence or homogeneity, 501
    Kruskal–Wallis test, 650
    linear correlation coefficient, 589
    mean, large samples, 386, 388
    mean, small samples, 400–401
    mean of paired differences, 459
    proportion, large samples, 408–409
    runs test, 662
    sign test, 622, 625, 627
    slope of regression line, 582
    Spearman's rho rank correlation
        coefficient, 655
    variance, 516
    Wilcoxon rank sum test, 644–645
    Wilcoxon signed-rank test, 637, 639
Third quartile, 104
Time-series data, 15
Total sum of squares:
    ANOVA, 535–536
    regression, 575
Treatment, 536
Tree diagram, 128
Trial, 209
Trimmed mean, 82
True population mean, 334
True population proportion, 334
Truncation, 40
Two-tailed test of hypothesis, 381

Type I error, 379–380
    probability of, 379–380
Type II error, 380
    probability of, 380
Typical value, 70

Unbiased estimator, 298
Ungrouped data, 26, 70
Uniform histogram, 46
Unimodal distribution, 77
Union of events, 165
    probability of, 165–166
Upper class boundary, 35
Upper class limit, 35

Variable, 10, 12–14
    categorical, 13–14
    continuous, 13

Variable (*continued*)
    dependent, 555
    discrete, 13
    independent, 555
    qualitative, 13–14
    quantitative, 12
    random, 185
Variance, 84
    between samples, 534, 537
    confidence interval for the, 514
    of a discrete random variable, 198
    for grouped data, 93
    population, 85, 93
    sample, 85, 93
    test of hypothesis about the, 516
    for ungrouped data, 85
    within samples, 534, 537
Variance, analysis of, 533–535, 538

Variation, measures, 82
Venn diagram, 128

Whiskers, 111
Wilcoxon rank sum test, 642
    large-sample case, 644–645, 647
    observed value of z, 645
    small-sample case, 642, 644
Wilcoxon signed-rank test, 635
    decision rule, 636
    large-sample case, 638
    observed value of test statistic T, 637
    observed value of z, 639
    small-sample case, 635

y-intercept in a regression model, 556

z-score, 249–250
z-value, 249–250